IL	interleukin
IMP	inosine-5'-monophosphate
IP_3	inositol-1,4,5-triphosphate
K_m	Michaelis constant
kb	kilobases
kD	kilodalton
LDL	low-density lipoprotein
LHC	light harvesting complex
Man	mannose
NAA	nonessential amino acids
NAD^+	nicotinamide adenine dinucleotide (oxidized form)
NADH	nicotinamide adenine dinucleotide (reduced form)
$NADP^+$	nicotinamide adenine dinucleotide phosphate (oxidized form)
NADPH	nicotinamide adenine dinucleotide phosphate (reduced form)
NDP	nucleoside-5'-diphosphate
NMR	nuclear magnetic resonance
NO	nitric oxide
NTP	nucleoside-5'-triphosphate
P_i	orthophosphate (inorganic phosphate)
PAPS	3'-phosphoadenosine-5'-phosphosulfate
PC	plastocyanin
PDGF	platelet-derived growth factor
PEP	phosphoenolpyruvate
PFK	phosphofructokinase
PIP_2	phosphatidylinositol-4,5-bisphosphate
PP_i	pyrophosphate
PRPP	phosphoribosylpyrophosphate
PS	photosystem
PQ(Q)	plastoquinone (oxidized)
PQH_2 (QH_2)	plastoquinone (reduced)
RER	rough endoplasmic reticulum
RF	releasing factor
RFLP	restriction-frament length polymorphism
RNA	ribonucleic acid
dsRNA	double-stranded RNA
hnRNA	heterogenous nuclear RNA
mRNA	messenger RNA
rRNA	ribosomal RNA
snRNA	small nuclear RNA
ssRNA	single-stranded RNA
tRNA	transfer RNA
snRNP	small ribonucleoprotein particles
RNase	ribonuclease
S	Svedberg unit
SAH	S-adenosylhomocysteine
SAM	S-adenosylmethionine
SDS	sodium dodecyl sulfate
SER	smooth endoplasmic reticulum
SRP	signal recognition particle
T	thymine
THF	tetrahydrofolate
TPP	thiamine pyrophosphate
U	uracil
UDP	uridine-5'-diphosphate
UMP	uridine-5'-monophosphate
UTP	uridine-5'-triphosphate
UQ	ubiquinone (coenzyme Q)(oxidized form)
UQH_2	ubiquinone (reduced form)
VLDL	very low density lipoprotein
XMP	xanthosine-5' monophosphate

IMPORTANT:

HERE IS YOUR REGISTRATION CODE TO ACCESS

YOUR PREMIUM McGRAW-HILL ONLINE RESOURCES.

For key premium online resources you need THIS CODE to gain access. Once the code is entered, you will be able to use the Web resources for the length of your course.

If your course is using **WebCT** or **Blackboard**, you'll be able to use this code to access the McGraw-Hill content within your instructor's online course.

Access is provided if you have purchased a new book. If the registration code is missing from this book, the registration screen on our Website, and within your WebCT or Blackboard course, will tell you how to obtain your new code.

Registering for McGraw-Hill Online Resources

To gain access to your McGraw-Hill web resources simply follow the steps below:

1. USE YOUR WEB BROWSER TO GO TO: **http://www.mhhe.com/mckee/**

2. CLICK ON **FIRST TIME USER**.

3. ENTER THE REGISTRATION CODE* PRINTED ON THE TEAR-OFF BOOKMARK ON THE RIGHT.

4. AFTER YOU HAVE ENTERED YOUR REGISTRATION CODE, CLICK **REGISTER**.

5. FOLLOW THE INSTRUCTIONS TO SET-UP YOUR PERSONAL UserID AND PASSWORD.

6. WRITE YOUR UserID AND PASSWORD DOWN FOR FUTURE REFERENCE.
   KEEP IT IN A SAFE PLACE.

TO GAIN ACCESS to the McGraw-Hill content in your instructor's **WebCT** or **Blackboard** course simply log in to the course with the UserID and Password provided by your instructor. Enter the registration code exactly as it appears in the box to the right when prompted by the system. You will only need to use the code the first time you click on McGraw-Hill content.

Thank you, and welcome to your McGraw-Hill online Resources!

*YOUR REGISTRATION CODE CAN BE USED ONLY ONCE TO ESTABLISH ACCESS. IT IS NOT TRANSFERABLE.

0-07-231592-X MCKEE/MCKEE, BIOCHEMISTRY: THE MOLECULAR BASIS OF LIFE, THIRD EDITION

REGISTRATION CODE

pertaining-10694885

Mc Graw Hill Higher Education

Biochemistry

Biochemistry

THE MOLECULAR BASIS OF LIFE *Third Edition*

Trudy McKee

James R. McKee
University of the Sciences in Philadelphia

Boston   Burr Ridge, IL   Dubuque, IA   Madison, WI   New York   San Francisco   St. Louis
Bangkok   Bogotá   Caracas   Kuala Lumpur   Lisbon   London   Madrid   Mexico City
Milan   Montreal   New Delhi   Santiago   Seoul   Singapore   Sydney   Taipei   Toronto

McGraw-Hill Higher Education

A Division of The **McGraw-Hill** *Companies*

BIOCHEMISTRY: THE MOLECULAR BASIS OF LIFE, THIRD EDITION

Published by McGraw-Hill, a business unit of The McGraw-Hill Companies, Inc., 1221 Avenue of the Americas, New York, NY 10020. Copyright © 2003, 1999, 1996 by The McGraw-Hill Companies, Inc. All rights reserved. No part of this publication may be reproduced or distributed in any form or by any means, or stored in a database or retrieval system, without the prior written consent of The McGraw-Hill Companies, Inc., including, but not limited to, in any network or other electronic storage or transmission, or broadcast for distance learning.

Some ancillaries, including electronic and print components, may not be available to customers outside the United States.

This book is printed on acid-free paper.

International   1 2 3 4 5 6 7 8 9 0 QPV/QPV 0 9 8 7 6 5 4 3 2
Domestic   1 2 3 4 5 6 7 8 9 0 QPV/QPV 0 9 8 7 6 5 4 3 2

ISBN 0–07–231592–X
ISBN 0–07–112248–6 (ISE)

Publisher: *Kent A. Peterson*
Developmental editor: *Spencer J. Cotkin, Ph.D.*
Marketing manager: *Thomas D. Timp*
Senior project manager: *Jill R. Peter*
Production supervisor: *Enboge Chong*
Senior media project manager: *Stacy A. Patch*
Associate media technology producer: *Janna Martin*
Design manager: *Stuart D. Paterson*
Cover/interior designer: *Ellen Pettengell*
Senior photo research coordinator: *John C. Leland*
Photo research: *Alexandra Truitt & Jerry Marshall*/www.pictureresearching.com
Supplement producer: *Brenda A. Ernzen*
Compositor: *Electronic Publishing Services Inc., NYC*
Typeface: *10.5/12 Times Roman*
Printer: *Quebecor World Versailles Inc.*

The cover image shows a structural model of a K+ channel, a type of pore-forming transmembrane protein that allows the rapid movement of specific ions across cell membranes. This image first appeared on the cover of the August, 2000, issue of *Trends in Biochemical Sciences* (vol. 25, no. 8) and was designed by Geraldine Woods.

The credits section for this book begins on page 747 and is considered an extension of the copyright page.

Library of Congress Cataloging-in-Publication Data

McKee, Trudy.
   Biochemistry : the molecular basis of life / Trudy McKee, James R. McKee. — 3rd ed.
      p. cm.
   Includes bibliographical references and index.
   ISBN 0–07–231592–X (acid-free paper)
   1. Biochemistry. I. McKee, James R. (James Robert), 1946–. II. Title.

QD415 .M36   2003
572—dc21   2001044172
                                                                                  CIP

INTERNATIONAL EDITION ISBN 0–07–112248–6
Copyright © 2003. Exclusive rights by The McGraw-Hill Companies, Inc., for manufacture and export. This book cannot be re-exported from the country to which it is sold by McGraw-Hill. The International Edition is not available in North America.

www.mhhe.com

This book is dedicated to the biochemical researchers

whose prodigious efforts have revealed the astonishingly

beautiful and intricate nature of living organisms

Brief Contents

Contents

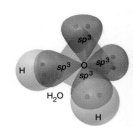

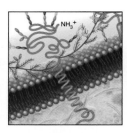

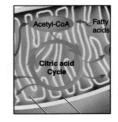

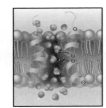

CHAPTER TWELVE

Lipid Metabolism 373

CHAPTER THIRTEEN

Photosynthesis 417

CHAPTER FOURTEEN

Nitrogen Metabolism I: Synthesis 449

CHAPTER FIFTEEN

Nitrogen Metabolism II: Degradation 502

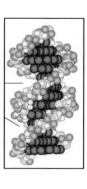

Preface

Biochemistry is the investigation of the molecular basis of life. Throughout the history of this scientific discipline, biochemists have worked to reveal the fundamental chemical and physical principles that underlie living processes. Their success is demonstrated in the enormous impact that the biochemical approach has had on the life sciences. At the beginning of the twenty-first century, the depth and breadth of this influence is astonishing. The progress in our understanding of living organisms, already enormous by the early 1990s, is currently being exceeded, in no small measure, because of the spectacular advances in DNA-based technologies that are based on biochemical research. The access that life scientists now have to the genetic information of entire organisms has resulted in previously unimaginable insights into the inner workings of living organisms and the causes of disease. The challenge in life and physical science; education is how to prepare students for careers in diverse fields in which the pace of knowledge accumulation will only accelerate in the foreseeable future. The most important tool that teachers can offer these students is a coherent understanding of biochemistry. The third edition of *Biochemistry: The Molecular Basis of Life* has been revised and updated to provide a logical and accessible introduction to biochemical principles.

ORGANIZATION AND APPROACH

This textbook is designed for life science students and chemistry and other physical science majors. Few assumptions have been made about the chemistry and biology backgrounds the students have. To ensure that all students are sufficiently prepared for acquiring a meaningful understanding of biochemistry, the first four chapters review the principles of relevant topics, such as organic functional groups, noncovalent bonding, thermodynamics, and cell structure. Several themes are introduced in these early chapters and continued throughout the book. Emphasis is placed on the relationship between molecular architecture and the functional properties of biomolecules, and the dynamic, unceasing, and self-regulating nature of living processes. Students are also provided with overviews of the major physical and chemical techniques that biochemists have used to explore life at the molecular level.

WHAT IS NEW IN THIS EDITION

The rapid pace of discovery in the life sciences has necessitated several notable changes and improvements in this edition. Among these are the following:

- Biochemical Methods Boxes are now integrated within chapters. Materials from Appendix B in the 2nd edition have been updated, rewritten, and inserted at appropriate places within the text. These essays focus on the most important classical and current laboratory techniques (see examples on pages 152, 358, and 630).

- The emerging understanding that the coding of information into the three-dimensional structure of *all* biomolecules is emphasized in this edition (see page 229). This approach introduces students to the most basic and accessible features of biological information theory and makes several topics (e.g., cell signaling mechanisms) more comprehensible.

- A variety of new Special Interest Boxes have been added that introduce students to up-to-date biochemical topics. Examples include "Protein Folding and Human Disease," "The Origin of Life," and "Biological Information and the Sugar Code."

- The art program has been completely reevaluated. More than half of the 700 figures are new or have been substantially modified in order to improve clarity and information content (see examples on pages 24, 40, 109, 311, 359).

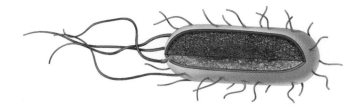

- The design of the book has been completely revised in order to make it more visually interesting, easier to read, and contemporary in appearance.

SUPPLEMENTARY AIDS

The third edition of *Biochemistry: The Molecular Basis of Life* has a variety of ancillary materials, including the following:

- **Digital Content Manager:** This is the primary instructor supplement and offers over 300 text images and nearly 1000 PowerPoint lecture slides prepared by Patrick Flash of Kent State University-Ashtabula Campus. The text images are full color and can be readily incorporated into lecture presentations, exams, or classroom materials. The sets of PowerPoint slides cover all 19 chapters. These slides, which combine art and lecture notes, can be used as is or tailored by instructor preferences.
- **Online Learning Center:** The Online Learning Center offers students a variety of resources and can be found at www.mhhe.com/mckee. At this web site students will be able to take quizzes written by Dan Sullivan of the University of Nebraska-Omaha, and study the structure of many of the most significant biochemical molecules with the interactive Chime plug-in prepared by Todd Carlson of Grand Valley State University.
- **Instructor's Manual/Test Item File:** Written by the authors, this manual is designed to help instructors plan and prepare for classes using *Biochemistry: The Molecular Basis of Life*. For each chapter in the text, this manual provides a chapter outline and an extended lecture outline. Answers for the even-numbered questions appear at the end of each part. The test item file contains approximately thirty-five multiple choice and critical thinking problems per chapter.
- **Student Study Guide/Solutions Manual:** This guide accompanies the text and was written by Patricia Depra of Westfield State College. For each text chapter, a corresponding study guide chapter offers comprehensive reviews, study tips, and additional questions for biochemistry students.
- **Transparencies:** Accompanying this text, 100 transparencies of key illustrations in the text help the instructor coordinate the lecture to the text.
- **Brownstone Diploma computerized classroom management system:** This service includes a database of test questions, reproducible student self-quizzes, and a grade-recording program.

ACKNOWLEDGEMENTS

The authors wish to express their appreciation for the efforts of the individuals who provided detailed reviews of the third edition:

Gul Afshan   *Milwaukee School of Engineering*
Donald R. Babin   *Creighton University*
Oscar P. Chilson   *Washington University*
Danny J. Davis   *University of Arkansas*
Patricia Depra   *Westfield State College*
Robert P. Dixon   *Southern Illinois University-Edwardsville*

Patricia Draves   *University of Central Arkansas*
Nick Flynn   *Angelo State University*
Larry L. Jackson   *Montana State University*
Michael Kalafatis   *Cleveland State University*
Hugh Lawford   *University of Toronto*
Maria O. Longas   *Purdue University-Calumet*
Cran Lucas   *Louisiana State University-Shreveport*
Robin Miskimins   *University of South Dakota*
Tom Rutledge   *Urinus College*
Edward Senkbeil   *Salisbury State University*
Ralph Stephani   *St. Johns University*
Dan M. Sullivan   *University of Nebraska*
John M. Tomich   *Kansas State University*
Shashi Unnithan   *Front Range Community College*
Alexandre G. Volkov   *Oakwood College*
Linette M. Watkins   *Southwest Texas State University*
Lisa Wen Western   *Illinois University*
Kenneth Wunch   *Tulane University*

We would also like to thank those who reviewed the first and second editions of this text:

Richard Saylor   *Shelton State Community College*
Craig R. Johnson   *Carlow College*
Larry D. Martin   *Morningside College*
Arnulfo Mar   *University of Texas at Brownsville*
Terry Helser   *SUNY College at Oneonta*
Edward G. Senkbeil   *Salisbury State University*
Martha McBride   *Norwich University*
Ralph Shaw   *Southeastern Louisiana University*
Clarence Fouche   *Virginia Intermont College*
Jerome Maas   *Oakton Community College*
Justine Walhout   *Rockford College*
William Voige   *James Madison University*
Carol Leslie   *Union University*
Harvey Nikkei   *Grand Valley State University*
Brenda Braaten   *Framingham State College*
Duane LeTourneau   *University of Idaho*
William Sweeney   *Hunter College*
Charles Hosler   *University of Wisconsin*
Mark Annstrong   *Blackbum College*
Treva Pamer   *Jersey City State College*
Bruce Banks   *University of North Carolina*
David Speckhard   *Loras College*
Joyce Miller   *University of Wisconsin-Platteville*
Beulah Woodfin   *University of New Mexico*
Robley J. Light   *Florida State University*
Anthony P. Toste   *Southwest Missouri State University*
Les Wynston   *California State University-Long Beach*
Alfred Winer   *University of Kentucky*
Larry L. Jackson   *Montana State University*
Ivan Kaiser   *University of Wyoming*
Allen T. Phillips   *Pennsylvania State University*
Bruce Morimoto   *Purdue University*
John R. Jefferson   *Luther College*
Ram P. Singhal   *Wichita State University*
Craig Tuerk   *Morehead State University*

Alan Myers   *Iowa State University*
Allan Bieber   *Arizona State University*
Scott Pattison   *Ball State University*
P. Shing Ho   *Oregon State University*
Charles Englund   *Bethany College*
Lawrence K. Duffy   *University of Alaska-Fairbanks*
Paul Kline   *Middle Tennessee State University*
Christine Tachibana   *Pennsylvania State University*

We wish to express our appreciation to Kent Peterson, our sponsoring editor, and Spencer Cotkin, our developmental editor, for their generous support and encouragement in all of our endeavors. We would also like to thank James M. Smith, who provided invaluable guidance during the early months of the revision process. The excellent efforts of the McGraw-Hill production team and the staff of Electronic Publishing Services are gratefully acknowledged. We are especially appreciative of the efforts of Jill Peter, the production manager. We give a very special thank you to Joseph Rabinowitz (Professor Emeritus, University of Pennsylvania), Ann Randolph (Rosemont College), and Diane Stroup (Kent State University), whose consistent diligence on this project has ensured the accuracy of the text. We also thank Michael Kalafatis (Cleveland State University) and Patricia Draves (University of Central Arkansas) for reviewing some of the supplementary material. In addition to their efforts and those of the reviewers, all of the manuscript's narrative and artwork have been reviewed by professional proofreaders. Every word, example, and figure has been independently checked by many individuals.

Finally, we wish to extend our deep appreciation to those individuals who have helped and encouraged us and made this project possible: Nicholas Rosa, Ira and Jean Cantor, and Joseph and Josephine Rabinowitz.

Finally, we thank our son, James Adrian McKee, for his unfailing patience and encouragement.

Trudy McKee

James R. McKee

Guided Tour Through the Biochemistry Learning System

Carbohydrates

OUTLINE

MONOSACCHARIDES
Monosaccharide Stereoisomers
Cyclic Structure of Monosaccharides
Reactions of Monosaccharides
Important Monosaccharides
SPECIAL INTEREST BOX 7.1
ASCORBIC ACID
Monosaccharide Derivatives

**DISACCHARIDES AND
OLIGOSACCHARIDES**

POLYSACCHARIDES
Homopolysaccharides
SPECIAL INTEREST BOX 7.2
LINEN
Heteropolysaccharides

GLYCOCONJUGATES
Proteoglycans
Glycoproteins
SPECIAL INTEREST BOX 7.3
**BIOLOGICAL INFORMATION
AND THE SUGAR CODE**

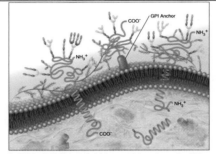

The Cell Surface Significant amounts of carbohydrate are attached to membrane protein and lipids on the external surface of cells. (Colored balls = sugar residues; GPI anchor = a complex lipid molecule that connects many cell surface proteins to the phospholipid bilayer of the plasma membrane)

Carbohydrates are not just an important source of rapid energy production for living cells. They are also structural building blocks of cells and components of numerous metabolic pathways. A broad range of cellular phenomena, such as cell recognition and binding (e.g., by other cells, hormones, and viruses), depend on carbohydrates. Chapter 7 describes the structures and chemistry of typical carbohydrate molecules found in living organisms.

200

CHAPTER OUTLINES AND OVERVIEWS

Each chapter begins with an outline that introduces students to the topics to be presented. This outline also provides the instructor with a quick topic summary that helps to organize lecture material. A one-paragraph overview places the content of the chapter in context and highlights the significance of the material.

DRAMATIC VISUAL PROGRAM

Colorful and informative photographs, illustrations, and tables enhance the learning program. Each chapter begins with an attractive opening photograph or illustration that visually introduces the topics to be discussed.

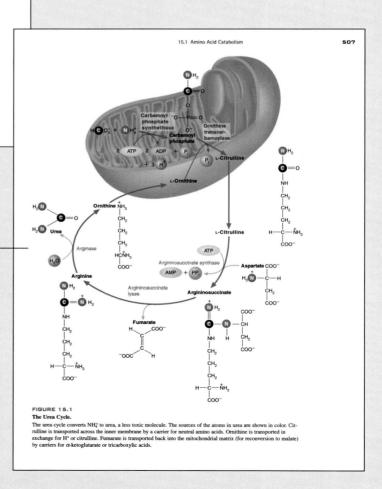

FIGURE 15.1
The Urea Cycle.
The urea cycle converts NH$_4^+$ to urea, a less toxic molecule. The sources of the atoms in urea are shown in color. Citrulline is transported across the inner membrane by a carrier for neutral amino acids. Ornithine is transported in exchange for H$^+$ or citrulline. Fumarate is transported back into the mitochondrial matrix (for reconversion to malate) by carriers for α-ketoglutarate or tricarboxylic acids.

Despite the diversity in the structures of viruses and the types of host cell that are infected, there are several basic steps in the life cycle of all viruses: infection (penetration of the virion or its nucleic acid into the host cell), replication (expression of the viral genome), maturation (assembly of viral components into virions), and release (the emission of new virions from the host cell). Because viruses usually possess only enough genetic information to specify the synthesis of their own components, each type must exploit some of the normal metabolic reactions of its host cell to complete the life cycle. For this reason there are numerous variations on these basic steps. This point can be illustrated by comparing the life cycles of two well-researched viruses: the T4 bacteriophage and the human immunodeficiency virus (HIV).

Bacteriophage T4
The T4 bacteriophage (Figure 17K) is a large virus with an icosahedral head and a long, complex tail similar in structure to T2 (p. 573). The head contains dsDNA, and the tail attaches to the host cell and injects the viral DNA into the host cell.

The life cycle of T4 (Figure 17L) begins with adsorbing the virion to the surface of an *E. coli* cell. Because the bacterial cell wall is rigid, the entire virion cannot penetrate into the cell's interior. Instead, the DNA is injected by flexing and constricting the tail apparatus. Once the DNA has entered the cell, the infective process is complete, and the next phase (replication) begins.

Within 2 minutes after the injection of T4 phage DNA into an *E. coli* cell, synthesis of host DNA, RNA, and protein stops and phage mRNA synthesis begins. Phage mRNA codes for the synthesis of capsid proteins and some of the enzymes required for the replication of the viral genome and the assembly of virion components. In addition, other enzymes are synthesized that weaken the host cell's cell wall, so that new phage can be released for new rounds of infection. Approximately 22 minutes after viral DNA (vDNA) is injected, the host cell, now filled with several hundred new virions, lyses. Upon release, the virions attach to nearby bacteria, thus initiating new infections.

Bacteriophage that initiate this so-called **lytic cycle** are referred to as *virulent* because they destroy their host cells. Many phage, however, do not initially kill their hosts. So-called *temperate* or *lysogenic* phage integrate their genome into that of the host cell. (The term **lysogeny** describes a condition in which the phage genome is integrated into a host chromosome.) The integrated viral genome (called the **prophage**) is copied along with host DNA during cell division for an indefinite time. Occasionally, lysogenic phage can enter a *lytic* phase. Certain external conditions, such as UV or ionizing radiation, activate the prophage, which directs the synthesis of new virions. Sometimes, a lysing bacterial cell releases a few virions that contain some bacterial DNA instead of phage genetic material. When such a virion infects a new host cell, this DNA is introduced into the host genome. This process is referred to as **transduction**.

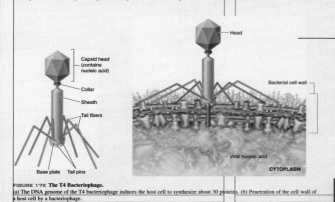

Head
Capsid head (contains nucleic acid)
Collar
Sheath
Tail fibers
Base plate
Tail pins
Bacterial cell wall
Viral nucleic acid
CYTOPLASM

FIGURE 17K The T4 Bacteriophage.
(a) The DNA genome of the T4 bacteriophage induces the host cell to synthesize about 30 proteins. (b) Penetration of the cell wall of a host cell by a bacteriophage.

BOLDFACED KEY WORDS

Key words appear in boldface when they are introduced within the text and are immediately defined by the context. All key words are also defined in the glossary.

SPECIAL INTEREST BOXES

These essays, which appear throughout the text, help students connect biochemical principles with everyday applications, and expand the discussion of many topics.

BIOCHEMICAL METHODS

These all-new boxes occur throughout the text and review the principal research techniques used to investigate living processes. From the information in these boxes students can appreciate the relationship between technology and scientific knowledge. Some of this material will also aid students in answering some in-chapter and end-of-chapter questions.

 Plant Biochemistry

 Medical

 Metabolic Regulation Mechanism

CONCEPT AND APPLICATION ICONS

Throughout the text, students find graphic devices that easily mark several important concepts and applications.

During the past 50 years, our understanding of the functioning of living organisms has undergone a revolution. Much of our current knowledge of biochemical processes is due directly to technological innovations. For example, the development of the electron microscope (EM) as a biological instrument by Keith Porter and his colleagues in the 1940s was responsible for the resolution of the fine structure of organelles. Structures such as the mitochondria and lysosomes were not discovered until this time; under the light microscope they appeared as mere granules. The electron microscope also revealed that the membranes of the Golgi apparatus are often continuous with those of the ER. This discovery is especially significant because of the role both organelles play in protein synthesis. In this box three of the most important cellular techniques used in biochemical research are briefly described: cell fractionation, electron microscopy, and autoradiography.

Cell Fractionation
Cell fractionation techniques (Figure 2D) allow the study of cell organelles in a relatively intact form outside of cells. For example, functioning mitochondria can be used to study cellular energy generation. In these techniques, cells are gently disrupted and separated into several organelle-containing fractions. Cells may be disrupted by several methods, but homogenization is the most commonly used. In this process a cell suspension is placed either in a glass tube fitted with a specially designed glass pestle, or into an electric blender. The resulting homogenate is then separated into several fractions during a procedure called **differential centrifugation**. A refrigerated instrument called the *ultracentrifuge* generates enormous centrifugal forces that separate cell components on the basis of size, surface area, and relative density. (Forces as large as 500,000 times the force of gravity, or 500,000 g, can be generated in unbreakable test tubes placed in the rotor of an ultracentrifuge.) Initially, the homogenate is spun in the ultracentrifuge at low speed (700–1,000 g) for 10 to 20 minutes. The heavier particles, such as the nuclei, form a sediment, or pellet. Lighter particles, such as mitochondria and lysosomes, remain suspended in the *supernatant*, the liquid above the pellet. The supernatant is then transferred to another centrifuge tube and spun at a higher speed (15,000 to 20,000 g) for 10–20 minutes. The resulting pellet contains mitochondria, lysosomes, and peroxisomes. The supernatant, which contains **microsomes** (small closed vesicles formed from ER during homogenization), is transferred to another tube and spun at 100,000 g for 60–120 minutes. Microsomes are deposited in the pellet, and the supernatant contains ribosomes, various cellular membranes, and granules such as glycogen, a carbohydrate polymer. After this latest supernatant is recentrifuged at 200,000 g for 2 to 3 hours, ribosomes and large macromolecules are recovered from the pellet.

Often, the organelle fractions obtained with this technique are not sufficiently pure for research purposes. One method that is often

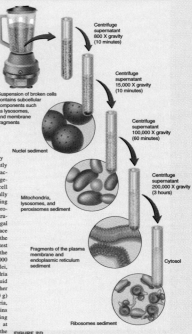

Suspension of broken cells contains subcellular components such as lysosomes, and membrane fragments
Centrifuge supernatant 800 X gravity (10 minutes)
Centrifuge supernatant 15,000 X gravity (10 minutes)
Centrifuge supernatant 100,000 X gravity (60 minutes)
Centrifuge supernatant 200,000 X gravity (3 hours)
Nuclei sediment
Mitochondria, lysosomes, and peroxisomes sediment
Fragments of the plasma membrane and endoplasmic reticulum sediment
Ribosomes sediment
Cytosol

FIGURE 2D
Cell Fractionation.
After the homogenization of cells in a blender, cell components are separated in a series of centrifugations at increasing speeds. As each centrifugation ends, the supernatant is removed, placed into a new centrifuge tube and then subjected to greater centrifugal force. The collected pellet can be resuspended in liquid and then examined by microscopy or biochemical tests.

IN-CHAPTER PROBLEMS AND SOLUTIONS

Because problem solving is most easily learned by studying examples and practicing, problems with solutions are provided wherever appropriate.

QUESTIONS

Numerous in-chapter questions help students integrate newly learned material with timely and interesting related information.

Glyceraldehyde-3-phosphate dehydrogenase, an enzyme in the glycolytic pathway (Chapter 8), is inactivated by alkylation with iodoacetate. Enzymes that use sulfhydryl groups to form covalent bonds with metal cofactors are often irreversibly inhibited by heavy metals (e.g., mercury and lead). The anemia in lead poisoning is caused in part because of lead binding to a sulfhydryl group of ferrochelatase. Ferrochelatase catalyzes the insertion of Fe^{2+} into heme.

Problem 6.4 is concerned with enzyme inhibition.

PROBLEM 6.4

Consider the Lineweaver-Burk plot illustrated in Figure 6.12.

Line A = Normal enzyme-catalyzed reaction
Line B = Compound B added
Line C = Compound C added
Line D = Compound D added

Identify the type of inhibitory action shown by compounds B, C, and D.

Solution
Compound B is a competitive inhibitor because the K_m only has changed. Compound C is a pure noncompetitive inhibitor because the V_{max} only has changed. Compound D is an uncompetitive inhibitor because both K_m and V_{max} have changed.

QUESTION 6.4

Iodoacetamide is an irreversible inhibitor of several enzymes that have a cysteine residue in their active sites. After examining its structure predict the products of the reaction of iodoacetamide with such an enzyme.

ALLOSTERIC ENZYMES Although the Michaelis-Menten model is an invaluable tool, it does not explain the kinetic properties of many enzymes. For example, plots of reaction velocity versus substrate concentration for many enzymes with multiple subunits are often sigmoidal rather than hyperbolic, as predicted by the Michaelis-Menten model (Figure 6.13). Such effects are seen in an important group of enzymes called the **allosteric enzymes**. The substrate-binding curve in Figure 6.13 resembles the oxygen-binding curve of hemoglobin. There are several other similarities between allosteric enzymes and hemoglobin.

FIGURE 6.12
A Lineweaver-Burk Plot.

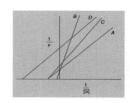

TABLES

A wide variety of information is supplied in tabular form, from numerical data to molecular structures. This information supplements the text and is useful for solving many problems in the text.

KEY CONCEPTS

At the end of sections is a brief summary to help students understand the essential ideas in the section.

TABLE 19.1
The Genetic Code

		U		C		A		G		
	UUU		UCU		UAU	Tyr	UGU	Cys	U
U	UUC	Phe	UCC	Ser	UAC		UGC		C
	UUA	Leu	UCA		UAA	STOP	UGA	STOP	A
	UUG		UCG		UAG		UGG	Trp	G
	CUU		CCU		CAU	His	CGU		U
C	CUC	Leu	CCC	Pro	CAC		CGC	Arg	C
	CUA		CCA		CAA	Gln	CGA		A
	CUG		CCG		CAG		CGG		G
	AUU		ACU		AAU	Asn	AGU	Ser	U
A	AUC	Ile	ACC	Thr	AAC		AGC		C
	AUA		ACA		AAA	Lys	AGA	Arg	A
	AUG	Met	ACG		AAG		AGG		G
	GUU		GCU		GAU	Asp	GGU		U
G	GUC	Val	GCC	Ala	GAC		GGC	Gly	C
	GUA		GCA		GAA	Glu	GGA		A
	GUG		GCG		GAG		GGG		G

First position (5' end) / Second Position / Third position (3' end)

KEY CONCEPTS 19.2

The genetic code is a mechanism by which ribosomes translate nucleotide base sequences into the primary sequence of polypeptides.

3. **Nonoverlapping and without punctuation.** The mRNA coding sequence is "read" by a ribosome starting from the initiating codon (AUG) as a continuous sequence taken three bases at a time until a stop codon is reached. A set of contiguous triplet codons in an mRNA is called a **reading frame**. The term **open reading frame** describes a series of triplet base sequences in mRNA that do not contain a stop codon.

4. **Universal.** With a few minor exceptions the genetic code is universal. In other words, examinations of the translation process in the species that have been investigated have revealed that the coding signals for amino acids are always the same.

QUESTION 19.1

The term *translation* refers to which of the following?

a. DNA $\longrightarrow$ RNA
b. RNA $\longrightarrow$ DNA
c. proteins $\longrightarrow$ RNA
d. RNA $\longrightarrow$ proteins

QUESTION 19.2

Explain the following terms:

a. codon
b. degenerate code
c. reading frame
d. open reading frame
e. universal code

SUMMARY

1. Biochemistry may be defined as the study of the molecular basis of life. Biochemists have contributed to the following insights into life: (1) life is complex and dynamic, (2) life is organized and self-sustaining, (3) life is cellular, (4) life is information-based, and (5) life adapts and evolves.

2. The molecular evidence concerning the evolutionary relationships of living species is sufficiently compelling that many life scientists now classify all living organisms into three domains: the bacteria, the archaea, and the eukarya.

3. All living things are composed of either prokaryotic cells or eukaryotic cells. Prokaryotes, which include bacteria and the archaea, lack a membrane-bound cellular organelle called a nucleus. The eukaryotes consist of all the remaining species. These cells contain a nucleus and complex structures that are not observed in prokaryotes.

4. Many eukaryotes are multicellular. Multicellular organisms have several advantages over unicellular ones. These include (1) the provision of a relatively stable environment for most of the organism's cells, (2) the capacity for greater complexity in an organism's form and function, and (3) the ability to exploit environmental resources more effectively than individual single-celled organisms can.

5. Animal and plant cells contain thousands of molecules. Water constitutes 50% to 90% of a cell's content by weight, and ions such as Na^+, K^+, and Ca^{2+} may account for another 1%. Almost all the other kinds of biomolecules are organic. The properties of the element carbon are responsible for the vast variety of organic molecules. The chemical properties of organic molecules are determined by specific arrangements of atoms called functional groups. Different families of organic molecules result when hydrogen atoms on hydrocarbon molecules are replaced by functional groups. Most biomolecules contain more than one functional group.

6. Many of the biomolecules found in cells are relatively small, with molecular weights of less than 1,000 D. Cells contain four families of small molecules: amino acids, sugars, fatty acids, and nucleotides. Members of each group serve several functions: (1) they are used in the synthesis of larger molecules, (2) some small molecules have special biological functions, and (3) many small molecules are components in complex reaction pathways.

7. All life processes consist of chemical reactions catalyzed by enzymes. The reactions of a living cell, which are known collectively as metabolism, result in highly coordinated and purposeful activity. Among the most common reaction types encountered in biochemical processes are (1) nucleophilic substitution, (2) elimination, (3) addition, (4) isomerization, and (5) oxidation-reduction.

8. Living cells are inherently unstable. Only a constant flow of energy prevents them from becoming disorganized. One of the means by which cells obtain energy is oxidation of biomolecules or certain minerals.

9. Metabolism is the sum total of all the enzyme-catalyzed reactions in a living organism. Many of these reactions are organized into pathways. There are two major types of biochemical pathways: anabolic and catabolic.

10. The complex structure of cells requires a high degree of internal order. This is accomplished by four primary means: (1) synthesis of biomolecules, (2) transport of ions and molecules across cell membranes, (3) production of movement, and (4) removal of metabolic waste products and other toxic substances.

11. In genetic information flow the shape and chemical properties of the bases in the nucleotide residues of DNA direct the assembly of polypeptides from amino acids. There are two phases of gene expression: transcription and translation. In transcription, RNA polymerases use the capacity of nucleotide bases to form base pairs to copy the base sequence of genes to synthesize RNA molecules. During translation, ribosomes use the base sequence information in mRNA to construct polypeptides.

SUGGESTED READINGS

Goodsell, D. S., *The Machinery of Life*, Springer-Verlag, New York, 1993.

Lewis, R., *Life*, 3rd ed., WCB/McGraw-Hill, Dubuque, Iowa, 1998.

Mader, S., *Biology*, 6th ed., WCB/McGraw-Hill, Dubuque, Iowa, 1998.

Raven, P. H., and Johnson, G. B., *Biology*, 5th ed., McGraw-Hill, Dubuque, Iowa, 1999.

Tudge, C., *The Variety of Life: A Survey and a Celebration of All the Creatures that Have Ever Lived*, Oxford University Press, New York, 2000.

KEY WORDS

active transport, 22	biomolecule, 2	energy, 20	gene, 3
addition reaction, 19	bioremediation, 6	enzyme, 2	gene expression, 25
amino acid, 10	catabolic pathway, 21	eukarya, 5	heterotroph, 20
anabolic pathway, 21	chemoautotroph, 20	eukaryotic cell, 4	homeostasis, 2
anticodon, 25	chemoheterotroph, 20	extremophile, 6	hydration, 19
archaea, 5	codon, 25	extremozyme, 6	hydrocarbon, 8
autotroph, 20	electrophile, 17	fatty acid, 13	hydrolysis, 17
bacteria, 5	elimination, 18	functional group, 9	hydrophilic, 10

CHAPTER SUMMARIES

At the end of each chapter is a summary designed to help students more easily identify important concepts and help them review for quizzes and tests.

SUGGESTED READINGS

At the end of each chapter are suggested references for further study of topics in the text or timely related topics.

KEY WORDS

Words and terms introduced in the chapter are listed along with the page number of first appearance.

REVIEW QUESTIONS

1. Define the following terms:
   a. photosystem
   b. reaction center
   c. light reaction
   d. dark reaction
   e. chloroplast
   f. photorespiration

2. What was the most significant contribution of early photosynthetic organisms to the Earth's environment?

3. List the three primary photosynthetic pigments and describe the role each plays in photosynthesis.

4. List five ways in which chloroplasts resemble mitochondria.

5. Excited molecules can return to the ground state by several means. Describe each briefly. Which of these processes are important in photosynthesis? Describe how they function in a living organism.

6. What is the final electron acceptor in photosynthesis?

7. What reactions occur during the light reactions of photosynthesis?

8. What reactions occur during the light-independent reactions of photosynthesis?

9. Why is the oxygen-evolving system referred to as a clock?

10. If the rate of photosynthesis versus the incident wavelength of light is plotted, an action spectrum is obtained. How can the action spectrum provide information about the nature of the light-absorbing pigments involved in photosynthesis?

11. Using the action spectrum for photosynthesis on p.445, determine what wavelengths of light appear to be optimal for photosynthesis.

12. What is the Emerson enhancement effect? How was it used to demonstrate the existence of two different photosystems? (*Hint:* Refer to Biochemical Methods 13.1.)

13. List the types of metals that are components of the photosynthesis mechanism. What functions do they serve?

14. Explain the following observation. When a photosynthetic system is exposed to a brief flash of light, no oxygen is evolved. Only after several bursts of light is oxygen evolved.

15. What is the Z scheme of photosynthesis? How are the products of this reaction used to fix carbon dioxide?

16. Where does carbon dioxide fixation take place in the cell?

17. The chloroplast has a highly organized structure. How does this structure help make photosynthesis possible?

THOUGHT QUESTIONS

1. Without carbon dioxide, chlorophyll fluoresces. How does carbon dioxide prevent this fluorescence?

2. The statement has been made that the more extensively conjugated a chromophore is, the less energy a photon needs to excite it. What is conjugation and how does it contribute to this phenomenon?

3. Increasing the intensity of the incident light but not its energy increases the rate of photosynthesis. Why is this so?

4. Both oxidative phosphorylation and photophosphorylation trap energy in high-energy bonds. How are these processes different? How are they the same?

5. In C3 plants, high concentrations of oxygen inhibit photosynthesis. Why is this so?

6. Generally, increasing the concentration of carbon dioxide increases the rate of photosynthesis. What conditions could prevent this effect?

7. It has been suggested that chloroplasts, like mitochondria, evolved from living organisms. What features of the chloroplast suggest that this is true?

8. Explain why photorespiration is repressed by high concentrations of carbon dioxide.

9. Why does exposing C3 plants to high temperatures raise the carbon dioxide compensation point?

10. Certain herbicides act by promoting photorespiration. These herbicides are lethal to C3 plants but do not affect C4 plants. Why is this so?

11. Corn, a grain of major economic importance, is a C4 plant, and many weeds in temperate climates are C3 plants. Therefore the herbicides described in Question 10 are widely used. What effect is likely if these materials are not degraded before they wash into the ocean?

REVIEW QUESTIONS

The first questions to appear at the end of each chapter are basic questions that help students test their understanding of the material.

THOUGHT QUESTIONS

More advanced questions that require students to think more deeply about biochemistry also appear at the end of each chapter.

Glossary

abiogenesis the mechanism by which inanimate material was transformed on the early Earth into the first primitive living organisms

absorption spectrum a graph of a sample's absorption of electromagnetic radiation

acetal the family of organic compounds with the general formula $RCH(OR_,)_2$; formed from the reaction of a hemiacetal with an alcohol

acid a molecule that can donate hydrogen ions

acidosis a condition in which the pH of the blood is below 7.35 for a prolonged time

aldaric acid the product formed when the aldehyde and CH_2OH groups of a monosaccharide are oxidized

alditol a sugar alcohol; the product when the aldehyde or ketone group of a monosaccharide is reduce

aldol addition a reaction between two aldehyde molecules (or two ketone molecules) in which a bond is formed between the α-carbon of one and the carbonyl carbon of the other

aldol cleavage the reverse of an aldol condensation

aldol condensation an aldol addition involv-

amino acid pool the amino acid molecules that are immediately available in an organism for use in metabolic processes

amino acid residue an amino acid that has been incorporated into a polypeptide molecule

ammonia intoxication elevated concentration of ammonia in the body that causes lethargy, tremors, slurred speech, protein-induced vomiting, and death

amphibolic pathway a metabolic pathway that functions in both anabolism and catabolism

END-OF-BOOK GLOSSARY OF KEY WORDS

All key words in boldface in the text are defined in the glossary at the end of the textbook.

Biochemistry: An Introduction

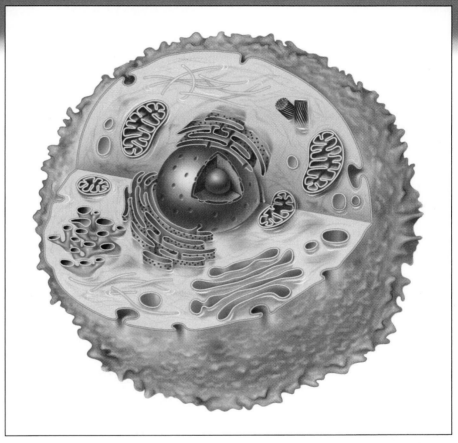

The Living Cell. Living organisms consist of one or more cells. The capacity of living cells to perform functions as energy generation, growth, and reproduction is made possible by their complex structures.

Life has proven to be far more complex than the human imagination could have conceived. The structure of an individual cell is a case in point. Cells are not the bags of protoplasm that scientists envisioned over a century ago; rather, they are structurally complex and dynamic. The story of how our current knowledge about living processes has been acquired, with its myriad plot twists, rivals any piece of detective fiction. The scientists who work to understand the physical reality of the natural world are often amazed at how sophisticated even the simplest organisms are. This chapter and the ones that follow focus on the basic life-sustaining mechanisms that have been discovered thus far.

What is life? The answer to this deceptively simple question has been elusive despite the work of life scientists over several centuries. Much of the difficulty in delineating the precise nature of living organisms lies in the overwhelming diversity of the living world and the apparent overlap in several properties of living and nonliving matter. Consequently, life has been viewed as an intangible property that defies simple explanation. It is usually described in operational terms, for example, as movement, reproduction, adaptation, and responsiveness to external stimuli. Since the end of the nineteenth century, however, the science of biochemistry (the investigation of the molecular basis of life) has provided new insights. Biochemists have investigated living organisms with a unique experimental approach based on the concepts of biology, chemistry, physics, and mathematics, as well as increasingly more sophisticated technologies. Their work has revealed that despite the rich diversity of living organisms, from the blue whale to the smallest of microorganisms, all obey the same chemical and physical laws that rule the universe. All are composed of the same types of molecules and their methods for sustaining biological processes are similar. Among the most important insights gained from the work of biochemists are the following:

1. **Life is complex and dynamic**. All organisms are primarily composed of organic (carbon-based) molecules that have intricate, three-dimensional shapes. Living processes, such as growth and development, involve thousands of chemical reactions in which vast quantities and varieties of vibrating and rotating molecules interact, collide, and rearrange into new molecules.

2. **Life is organized and self-sustaining**. Living organisms are hierarchically organized systems; that is, each level is based on the one below (Figure 1.1). The molecules that make up living organisms, referred to as **biomolecules**, are composed of atoms, which in turn are formed from subatomic particles. Certain biomolecules become linked to form polymers called **macromolecules**. Examples include nucleic acids, proteins, and polysaccharides, which are formed from nucleotides, amino acids, and sugars, respectively. Various combinations of biomolecules and macromolecules form a myriad of larger and more complex supramolecular structures that together make up cells. In multicellular organisms other levels of organization include tissues, organs, and organ systems. At each level of organization the whole is greater than the sum of the parts. In other words, new properties emerge at each level that cannot be predicted from the analysis of component parts. For example, hemoglobin, the protein that transports molecular oxygen in vertebrate blood, is composed of carbon, hydrogen, oxygen, nitrogen, and iron. The heme component of hemoglobin that is responsible for oxygen transport is not oxidized by the oxygen as it would be in the absence of the protein component. The properties of the heme ring system and its protection from oxidation by the protein that surrounds it are examples of *emergent properties*. The organization and ordered functioning of living organisms requires the continuous acquisition of both energy and matter, and the removal of waste molecules. These tasks are accomplished by hundreds of biochemical reactions that are catalyzed by **enzymes**. The sum total of all of the reactions in a living organism is referred to as **metabolism**. The capacity of living organisms to regulate metabolic processes despite variability in their internal and external environments is called **homeostasis**.

3. **Life is cellular**. Cells differ widely in structure and function, but each is surrounded by a membrane that controls the transport of some chemical substances into and out of the cell. The membrane also mediates the response of the cell to components of the extracellular environment. If a cell is divided into its component parts, it will cease to function in a life-sustaining way. Cells arise only from the division of existing cells.

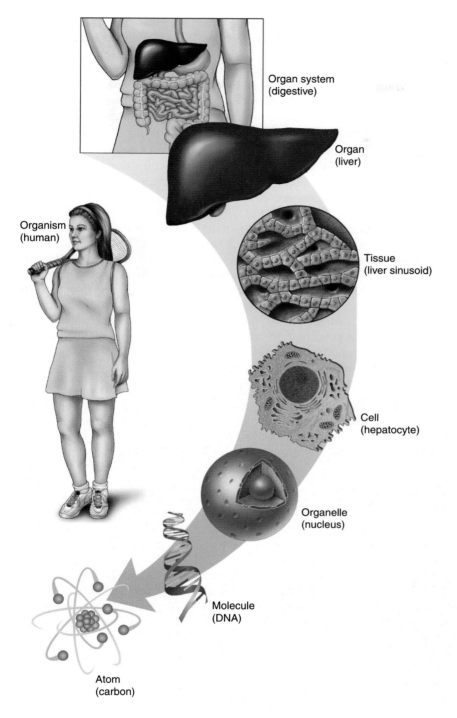

Organ system
(digestive)

Organ
(liver)

Tissue
(liver sinusoid)

Organism
(human)

Cell
(hepatocyte)

Organelle
(nucleus)

Molecule
(DNA)

Atom
(carbon)

FIGURE 1.1

Hierarchical Organization of a Multicellular Organism: The Human Being

Multicellular organisms have several levels of organization: Organ systems, organs, tissues, cells, organelles, molecules, and atoms. The digestive system and one of its component organs (the liver) are shown. The liver is a multifunctional organ that has several digestive functions. For example, it produces bile, which facilitates fat digestion, and it processes and distributes the food molecules absorbed in the small intestine to other parts of the body. DNA, one molecule found in cells, contains the genetic information that controls cell function.

4. **Life is information-based**. Organization requires information. Living organisms can be considered to be information-processing systems because maintenance of their structural integrity and metabolic processes requires the timely management of a vast array of interacting molecules within cells and between cells and generations of future cells. Biological information is in the form of coded messages that are inherent in the unique three-dimensional structure of biomolecules. Genetic information, which is stored in the linear sequences of nucleotides in the nucleic acid deoxyribonucleic acid (DNA), called **genes**, in turn specifies the linear sequence of amino acids in proteins, and how and when those proteins are synthesized. Proteins

perform their function by interacting with other molecules. The unique three-dimensional structure of each type of protein allows it to bind to and interact with a specific type of molecule with a precise complementary shape. Information is transferred during the binding process. For example, the binding of insulin, a protein released by the pancreas of vertebrates, to insulin receptor molecules on the surface of certain cells is a signal that initiates the uptake of the nutrient molecule glucose. The transport of amino acids is insulin sensitive as well.

5. **Life adapts and evolves.** All life on earth has a common origin, with new forms arising from other forms. When an individual organism in a population reproduces itself, stress-induced DNA modifications and errors that occur when DNA molecules are copied can result in **mutations** or sequence changes. Most mutations are silent; that is, they are either repaired by the cell or have no effect on the functioning of the organism. Some, however, are deleterious, serving to limit the reproductive success of the offspring. On rare occasions mutations may contribute to an increased ability of the organism to survive, to adapt to new circumstances, and to reproduce. The principal driving force in this process is the capacity to exploit energy sources. Individuals that possess traits that allow them to better exploit a specific energy source within their habitat may have a competitive advantage when resources are limited. Over many generations the interplay of environmental change and genetic variation can lead to the accumulation of favorable traits, and eventually to increasingly different forms of life.

The life sciences are currently undergoing a revolution created by the application of biochemical techniques to genetics. The technologies developed by this relatively new science—molecular biology—have provided previously unimaginable insight into the inner workings of living organisms. It is now routine to identify specific genes and trace their effects in living organisms. This new approach has already yielded avalanches of information that have transformed such diverse disciplines as agriculture, archaeology, botany, developmental biology, ecology, forensics, medicine, pharmacology, and nutrition. The capacity of future life scientists to deal with the sheer volume of new and ever increasing knowledge begins with mastery of the basic principles of biochemistry. This chapter presents an overview of biochemistry and an introduction to the fundamental concepts of this scientific discipline. After a brief review of the diversity of life, we begin our examination of biochemistry with an introduction to the structure and function of the major biomolecules. This is followed by discussion of the most important biochemical processes. The chapter concludes with an overview of genetic information processing and a brief introduction to the basic concepts of modern experimental biochemistry. Throughout this chapter and for the remainder of the book, we strive to illustrate that the basic processes of living organisms, the structure of biochemical compounds, the thousands of biochemical reactions within organisms, and genetic inheritance are intimately related. Understanding of any one of these topics is inextricably linked to an understanding of the other topics

KEY CONCEPTS 1.1

All living organisms obey the same chemical and physical laws. Life is complex, dynamic, organized, and self-sustaining. Life is cellular and information-based. Life adapts and evolves.

1.1 THE LIVING WORLD

Estimates of the number of living species currently range from several million to tens of millions. All are composed of either **prokaryotic** or **eukaryotic cells**. Most organisms are prokaryotes; that is, their cells lack a nucleus (*pro* = "before," *karyon* = "nucleus" or "kernal"). The eukaryotes (*eu* = "true") are composed of relatively large and exceedingly complex cells that possess a nucleus, a membrane-bound compartment that contains the genetic material.

The prokaryotes are not only the oldest forms of life on earth, but from about 3.8 billion years ago until about 1.8 billion years ago, they were the only forms of life. Until the 1980s, the prokaryotes were believed to consist only of bacteria. Analysis of the nucleotide sequences of ribonucleic acid (RNA), a type of nucleic acid involved in the synthesis of proteins, has revealed that there are two quite distinct groups of prokaryotes: the bacteria and the archaea. Their external appearances are similar, but the differences in the molecular properties of the bacteria and archaea are more pronounced than the differences of either of them from the eukaryotes. The single-celled prokaryotes are the smallest living organisms. Nevertheless, their combined biomass exceeds that of the larger eukaryotic organisms (animals, plants, fungi, and single-celled protists) by at least tenfold. Prokaryotes occupy virtually every niche on earth. In addition to air, soil, and water, various prokaryotic species live on the skin and in the digestive tracts of animals, within hot springs, and to a depth of several kilometers inside the crust of the earth.

The molecular evidence concerning the evolutionary relationships of living species is sufficiently compelling that many life scientists now classify all living organisms into three domains: the **bacteria,** the **archaea**, and the **eukarya** (Figure 1.2). Each domain is briefly discussed. This is followed by an introduction to the viruses, genetic entities that straddle the border between the living and the nonliving.

FIGURE 1.2

The Domains of Life on Earth

Molecular evidence now indicates that all life forms investigated so far can be classified into three domains.

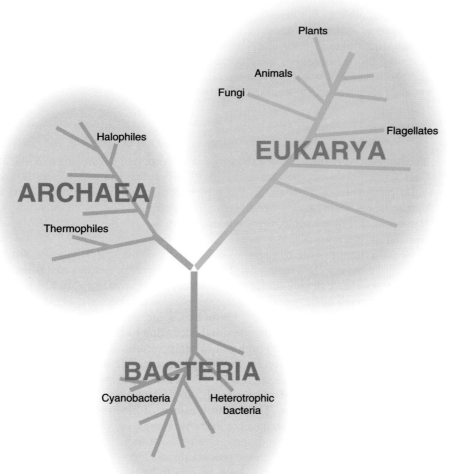

Bacteria

Bacteria are so diverse in their habitats and nutritional capacities that only general statements can be made about them. As a group, bacteria are especially known for their biochemical diversity. Various species possess traits that allow them to exploit virtually every conceivable energy source, nutrient, and habitat. For example, some bacteria species can use light energy to convert CO_2 into organic molecules. Others use energy extracted from inorganic or organic molecules.

Some bacterial species have a well-deserved reputation for causing disease (e.g., cholera, tuberculosis, syphilis, and tetanus). The vast majority, however, play vital roles in sustaining life on earth. The activity of many types of bacteria is required in the cycling of nutrients containing carbon, nitrogen, and sulfur. For example, the bacterium *Rhizobium* converts inert molecular nitrogen (N_2) into ammonia (NH_3) that can then be assimilated by other organisms such as leguminous plants. One of the most important roles of bacteria is in decomposition, a process that releases nutrients from dead organisms so they become available for the living.

Many bacterial species are of great practical interest to humans. Foods such as yogurt, cheese, sourdough bread, and sauerkraut are manufactured with the aid of certain bacteria. Other types of bacteria such as the actinomycetes produce the antibiotics used to cure bacterial infections. Bacteria have especially been valuable in biochemical research. Because of their rapid growth rates and the relative ease of culturing, certain species (especially *Escherichia coli*) have proven to be invaluable in the investigation of most basic biochemical processes. The information acquired in research studies of pathogenic microorganisms has been used in medicine to both alleviate and prevent much human suffering. More recently, biotechnologists have taken advantage of bacterial growth rates and metabolic flexibility by inserting genes that code for hormones, vaccines, and other products of use to humans into bacterial cells.

Archaea

The archaea were only recognized as a distinct group of organisms in 1977 when Carl Woese analyzed specific nucleic acid molecules. Comparison of the molecular properties of archeans to those of bacteria and eukaryotes has revealed that archeans are in many ways closer to the eukaryotes than to the outwardly similar bacteria. For example, the archean system for synthesizing protein is more like that of eukaryotes.

A prominent feature of many of the archaea is their capacity to occupy and even thrive in very challenging habitats. Often referred to as **extremophiles**, some archean species can live under circumstances that would easily kill most life forms. Although other types of organisms (e.g., certain bacteria, algae, and fungi) can live in extreme conditions, the archeans include the most extremophilic species. Extremophiles can be classified according to the types of exotic conditions in which they live: very high or low temperatures, high salt concentrations, or high pressure. In addition to providing considerable insight into the history of life on earth (see Special Interest Box 2.1), investigations of the extremophilic archeans have also allowed unique insights into adaptations of biomolecular structure to extreme conditions. The research efforts of biochemists and biotechnologists have been concentrated on the **extremozymes**, enzymes (protein catalysts) that work under noxious conditions. Examples of industrial applications of this work include enzymes used in food processing and laundry detergents. Along with many bacteria species, archeans have proven to be useful in **bioremediation**, a process in which microorganisms are used to degrade or remove pollutants from toxic waste sites and oil spills.

Eukarya

The third domain of living organisms, the eukarya, is composed of all the remaining species on earth (Figure 1.3). Although the presence or absence of a nucleus is the most notable difference between prokaryotes and eukaryotes, there are other significant distinctions:

1. **Size**. Eukaryotic cells are substantially larger than prokaryotic cells. The diameter of animal cells, for example, varies between 10 and 30 μm. Such values are approximately 10 times higher than those for prokaryotes. Size disparity between the two cell types is more obvious, however, when volume is considered. For example, the volume of a typical eukaryotic cells such as a liver cell (hepatocyte) is between 6,000 and 10,000 μm^3. The volume of *E. coli* cells is several hundred times smaller.

2. **Complexity**. The structural complexity of eukaryotes is remarkable. In addition to a well-formed nucleus, a number of other subcellular structures called **organelles** are present. Each organelle is specialized to perform specific tasks. The compartmentalization afforded by organelles permits the concentration of reactant and product molecules at sites where they can be efficiently used. This plus other factors makes intricate regulatory mechanisms possible. Consequently, the cells of multicellular eukaryotes are able to respond quickly and effectively to the intercellular communications that are required for growth and development.

3. **Multicellularity**. True multicellularity is found only in the eukarya. Although the highly complex single-celled protists make up the largest biomass of the eukarya, all of the remaining categories are multicellular. Although some bacteria exhibit the habit of colonial living, especially on solid media, the cooperativity and specialization of multicellularity is rarely achieved. Multicellular organisms are not just collections of cells: they are highly ordered living systems that together form a coherent entity. The structural complexity of eukaryotic cells provides the capacity for the intricate mechanisms of intercellular communication required in these organisms.

FIGURE 1.3

Biodiversity

The earth is populated with many millions of species that inhabit every niche on the planet. Most species, however, are too small to be seen with the unaided human eye. Only representatives of the large multicellular eukaryotes, the most recognizable to humans, are illustrated in this figure.

Viruses

Viruses are not alive, yet they can disrupt the processes within living organisms by means of biochemical reactions. Composed of DNA or RNA wrapped in protein or a membrane, viruses are infectious agents that reproduce themselves by inserting their genetic information into susceptible host cells. Viral genes may remain dormant in the cell's DNA for long periods or may immediately begin directing the production of massive quantities of nucleic acid and protein components that are assembled into new virus particles. Once an overt viral infection has begun, the cell is, in effect, hijacked. Viral genes subvert the cell's machinery for their own use. As a result of viral infection, cells may be damaged or killed.

Viruses are intracellular parasites known to infect nearly all types of organisms. Each type of virus, however, usually infects only one or a few related species. Viral infections often result in disease. Notable human infectious diseases caused by viruses include acquired immune deficiency syndrome (AIDS), polio, rabies, several types of hepatitis, and the common cold. Plant and animal diseases caused by viruses cause immense damage to agriculture. Viruses, however, do more than cause disease. They can also be agents of genetic change. Occasionally, when new virioids (viral particles) are produced they inadvertently incorporate some genetic material from the host cell. Subsequently, this information is transmitted to a new host cell. In rare circumstances, it is transferred to an unrelated species. This process, called **transduction**, is one source of the genetic variation that drives evolutionary change.

Biochemists have used viruses as tools in their investigation of numerous living processes. For example, investigations of bacteriophages (viruses that infect bacteria) have provided inestimable insight into basic genetic mechanisms. Animal viruses have been used in investigations of the details of genetic information processing expression and other aspects of cell metabolism. Currently, the capacity of viruses to infect certain human cells is being used as a mechanism in gene therapy protocols.

KEY CONCEPTS 1.2

Living organisms have been classified into three domains: bacteria, archaea, and eukarya. Viruses are intracellular parasites that can only reproduce themselves by inserting their genetic information into a living cell.

1.2 BIOMOLECULES

Living organisms are composed of thousands of different kinds of inorganic and organic molecules. Water, an inorganic molecule, may constitute 50% to 95% of a cell's content by weight, and ions such as sodium (Na^+), potassium (K^+), magnesium (Mg^{2+}), and calcium (Ca^{2+}) may account for another 1%. Almost all the other kinds of molecules in living organisms are organic. Organic molecules are principally composed of six elements: carbon, hydrogen, oxygen, nitrogen, phosphorus, and sulfur, and contain trace amounts of certain metallic and other nonmetallic elements. It is noteworthy that the atoms of each of the most common elements found in living organisms can readily form stable covalent bonds. In covalent bonds, atoms are held together by a sharing of electrons that completes the outer orbital shells of each atom.

The remarkable structural complexity and diversity of organic molecules are made possible by the capacity of carbon atoms to form four strong, single covalent bonds either to other carbon atoms or to atoms of other elements. Organic molecules with many carbon atoms can form complicated shapes such as long, straight structures or branched chains and rings.

Functional Groups of Organic Biomolecules

Most biomolecules can be considered to be derived from the simplest type of organic molecules, called the **hydrocarbons**. Hydrocarbons (Figure 1.4) are

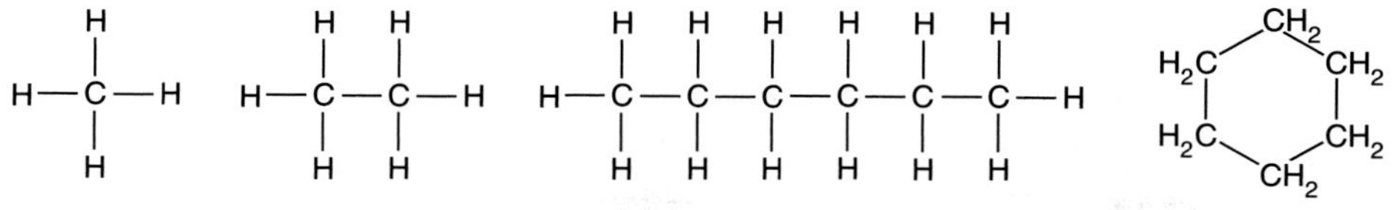

FIGURE 1.4

Structural Formulas of Several Hydrocarbons

carbon- and hydrogen-containing molecules that are **hydrophobic**, meaning they are insoluble in water. All other organic molecules are formed by attaching other atoms or groups of atoms to the carbon backbone of a hydrocarbon. The chemical properties of these derivative molecules are determined by the specific arrangement of atoms called **functional groups** (Table 1.1). For example, alcohols result when hydrogen atoms are replaced by hydroxyl groups (—OH). Thus methane (CH_4), a component of natural gas, can be converted into methanol (CH_3OH), a toxic liquid that is used as a solvent in many industrial processes.

Most biomolecules contain more than one functional group. For example, many simple sugar molecules have several hydroxyl groups and an aldehyde group. Amino acids, the building block molecules of proteins, have both an amino group and a carboxyl group. The distinct chemical properties of each functional group contribute to the behavior of any molecule that contains it.

TABLE 1.1

Important Functional Groups in Biomolecules

Family Name	Group Structure	Group Name	Significance
Alcohol	R—OH	Hydroxyl	Polar (and therefore water-soluble), forms hydrogen bonds
Aldehyde	R—C(=O)—H	Carbonyl	Polar, found in some sugars
Ketone	R—C(=O)—R'	Carbonyl	Polar, found in some sugars
Acids	R—C(=O)—OH	Carboxyl	Weakly acidic, bears a negative charge when it donates a proton
Amines	R—NH₂	Amino	Weakly basic, bears a positive charge when it accepts a proton
Amides	R—C(=O)—NH₂	Amido	Polar but does not bear a charge
Thiols	R—SH	Thiol	Easily oxidized; can form -S-S- (disulfide) bonds readily
Esters	R—C(=O)—O—R	Ester	Found in certain lipid molecules
Double bond	RCH=CHR	Alkene	Important structural component of many biomolecules; e.g., found in lipid molecules

Major Classes of Biomolecules

Many of the organic compounds found in cells are relatively small, with molecular weights of less than 1,000 daltons (D). (One dalton, 1 atomic mass unit is equal to $\frac{1}{12}$, the mass of one atom of ^{12}C.) Cells contain four families of small molecules: amino acids, sugars, fatty acids, and nucleotides (Table 1.2). Members of each group serve several functions. First, they are used in the synthesis of larger molecules, many of which are polymers. For example, proteins, certain carbohydrates, and nucleic acids are polymers composed of amino acids, sugars, and nucleotides, respectively. Fatty acids are components of several types of lipid (water-insoluble) molecules.

Second, some molecules have special biological functions. For example, the nucleotide adenosine triphosphate (ATP) serves as a cellular reservoir of chemical energy. Finally, many small organic molecules are involved in complex reaction pathways. Examples of each class are described in the next four sections.

AMINO ACIDS AND PROTEINS There are hundreds of naturally occurring amino acids, each of which contains an amino group and a carboxyl group. Amino acids are classified α, β, or γ according to the location of the amino group in reference to the carboxyl group. In α-amino acids, the most common type, the amino group is attached to the carbon atom (the α-carbon) immediately adjacent to the carboxyl group (Figure 1.5). In β- and γ-amino acids, the amino group is attached to the second and third carbons, respectively, from the carboxyl group. Also attached to the α-carbon is another group, referred to as the side chain or R group. The chemical properties of each amino acid are determined largely by the properties of its side chain. For example, some side chains are hydrophobic, whereas others are **hydrophilic** (i.e., they dissolve easily in water). The general formula for α-amino acids is

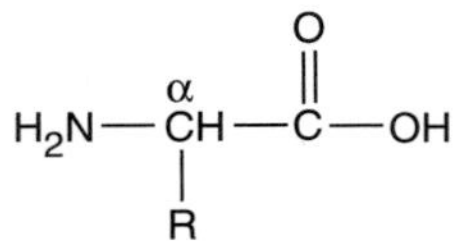

There are 20 standard α-amino acids that occur in proteins. Some standard amino acids have unique functions in living organisms. For example, glycine and glutamic acid function as neurotransmitters in animals. (**Neurotransmit-**

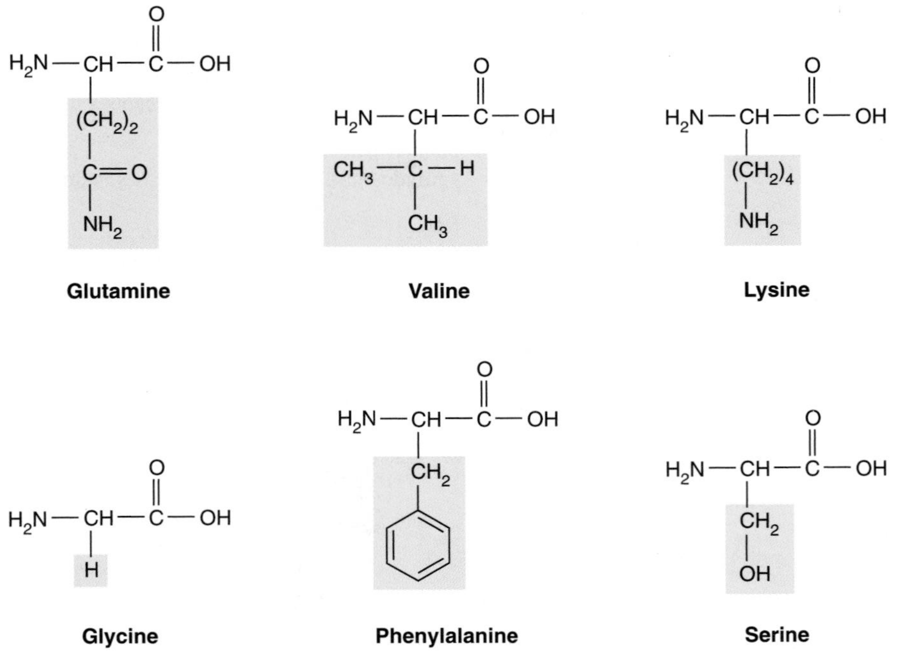

FIGURE 1.5

Structural Formulas for Several α-Amino Acids

R groups are highlighted. An R group in amino acid structures can be a hydrogen atom (e.g., in glycine), a hydrocarbon group (e.g., the isopropyl group in valine), or a hydrocarbon derivative (e.g., the hydroxymethyl group in serine).

TABLE 1.2

Major Classes of Biomolecules

Small Molecule	Polymer	General Functions
Amino acids	Proteins	Catalysts and structural elements
Sugars	Carbohydrates	Energy sources, and structural elements
Fatty acids	N.A.	Energy sources and structural elements of complex lipid molecules
Nucleotides	DNA	Genetic information
	RNA	Protein synthesis

ters are signal molecules released by nerve cells.) Proteins also contain nonstandard amino acids that are modified versions of the standard amino acids. The structure and function of protein molecules are often altered by conversion of certain amino acid residues to phosphorylated, hydroxylated, or other types of derivatives. (The term "residue" refers to a small biomolecule that is incorporated in a macromolecule, e.g., amino acid residues in a protein.) For example, many of the residues of proline are hydroxylated in collagen, the connective tissue protein. Many naturally occurring amino acids are not α-amino acids. Prominent examples include β-alanine, a precursor of the vitamin pantothenic acid, and γ-aminobutyric acid (GABA), a neurotransmitter found in the brain (Figure 1.6).

Amino acid molecules are used primarily in the synthesis of long, complex polymers known as **polypeptides**. Short polypeptides, up to a length of about 50 amino acids, are called **peptides**. Longer polypeptides are often referred to as **proteins**. Polypeptides play a variety of roles in living organisms. Examples of molecules composed of polypeptides include transport proteins, structural proteins, and the enzymes.

The individual amino acids are connected in peptides (Figure 1.7) and polypeptides by the **peptide bond**, an amide linkage that forms in a type of nucleophilic substitution reaction (p. 17) involving a carbonyl group, which occurs between the amino group of one amino acid and the carboxyl group of another. The partial double-bond character of the peptide bond results in a planar arrangement of the HN—CO atoms that provides polypeptide molecules with significant stability under physiological conditions. The α-carbon provides flexibility to the polypeptide chain because of free rotation of the bonds to this carbon.

The final three-dimensional structure, and therefore biological function, of polypeptides results largely from interactions among the R groups (Figure 1.8).

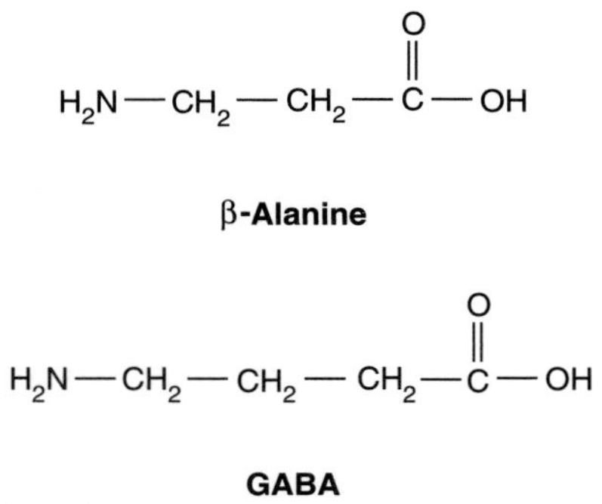

$$H_2N-CH_2-CH_2-\overset{\overset{\textstyle O}{\|}}{C}-OH$$

β-Alanine

$$H_2N-CH_2-CH_2-CH_2-\overset{\overset{\textstyle O}{\|}}{C}-OH$$

GABA

FIGURE 1.6

Selected Examples of Naturally Occurring Amino Acids that Are Not α-Amino Acids: β-Alanine and γ-Aminobutyric Acid (GABA)

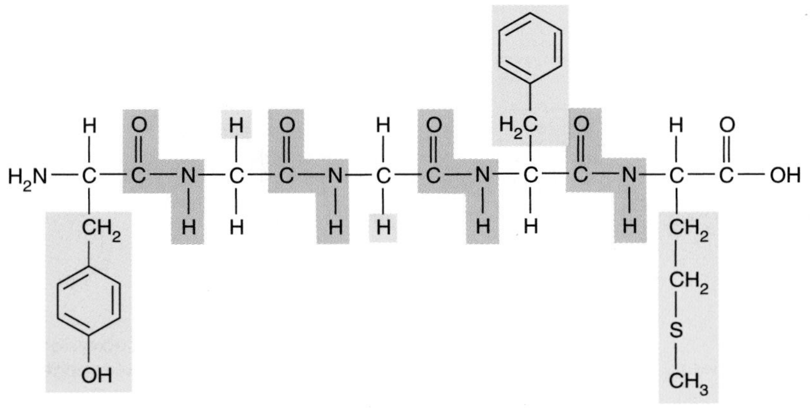

FIGURE 1.7

Structure of Met-Enkephalin, a Pentapeptide

Met-enkephalin is one of a class of molecules that have opiate-like activity. Found in the brain, met-enkephalin inhibits pain perception. (The peptide bonds are colored. The R groups are highlighted.)

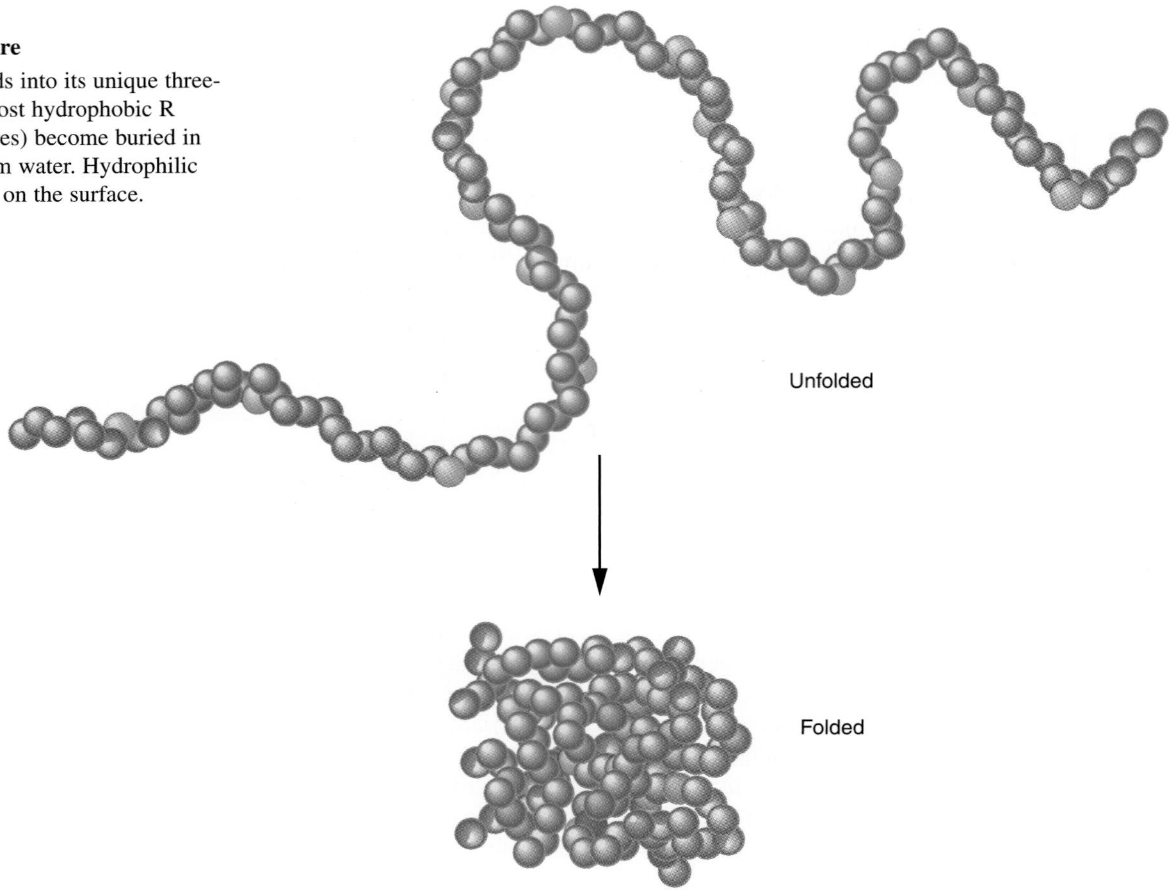

FIGURE 1.8

Polypeptide Structure

As a polypeptide folds into its unique three-dimensional form, most hydrophobic R groups (yellow-spheres) become buried in the interior away from water. Hydrophilic groups usually occur on the surface.

Unfolded

Folded

SUGARS AND CARBOHYDRATES Sugars contain both alcohol and carbonyl functional groups. They are described in terms of both carbon number and the type of carbonyl group they contain. Sugars that possess an aldehyde group are called aldoses and those that possess a ketone group are called ketoses. For example, the six-carbon sugar glucose (an important energy source in most living organisms) is an aldohexose (Figure 1.9). Fructose (fruit sugar) is a ketohexose.

Sugars are the basic units of carbohydrates, the most abundant organic molecules found in nature. Carbohydrates range from the simple sugars, or **monosaccharides**, such as glucose and fructose, to the **polysaccharides**, polymers that contain thou-

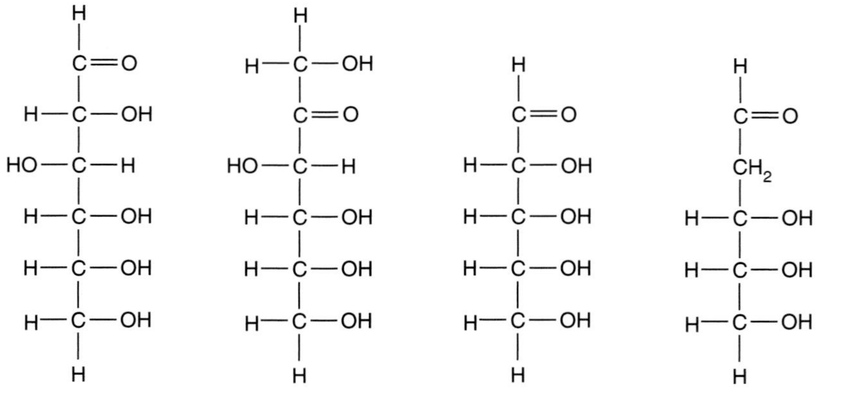

FIGURE 1.9

Examples of Several Biologically Important Monosaccharides

Glucose and fructose are important sources of energy in plants and animals. Ribose and deoxyribose are components of nucleic acids. These monosaccharides occur as ring structures in nature.

Glucose
(an aldohexose)

Fructose
(a ketohexose)

Ribose
(an aldopentose)

2-Deoxyribose
(an aldopentose)

sands of sugar units. Examples of the latter include starch and cellulose in plants and glycogen in animals. Carbohydrates serve a variety of functions in living organisms. Certain sugars are important energy sources. Glucose is the principal carbohydrate energy source in animals and plants. Sucrose is used by many plants as an efficient means of transporting energy throughout their tissues. Some carbohydrates serve as structural materials. Cellulose is the major structural component of wood and certain plant fibers. Chitin, another type of polysaccharide, is found in the protective outer coverings of insects and crustaceans.

Some biomolecules contain carbohydrate components. Nucleotides, the building block molecules of the nucleic acids, contain either of the sugars ribose or deoxyribose. Certain proteins and lipids also contain carbohydrate. Glycoproteins and glycolipids occur on the external surface of cell membranes in multicellular organisms, where they play critical roles in the interactions between cells.

FATTY ACIDS Fatty acids are monocarboxylic acids that usually contain an even number of carbon atoms. In some organisms they serve as energy sources. Fatty acids are represented by the chemical formula R—COOH, in which R is an alkyl group that contains carbon and hydrogen atoms. There are two types of fatty acids: **saturated** fatty acids, which contain no carbon-carbon double bonds, and **unsaturated** fatty acids, which have one or more double bonds (Figure 1.10). Under physiological conditions the carboxyl group of fatty acids exists in the ionized state, R—COO$^-$. For example, the 16-carbon saturated fatty acid called palmitic acid usually exists as palmitate, $CH_3(CH_2)_{14}COO^-$. Although the charged carboxyl group has an affinity for water, the long nonpolar hydrocarbon chains render most fatty acids insoluble in water.

Fatty acids occur as independent (free) molecules in only trace amounts in living organisms. Most often they are components of several types of **lipid** molecules (Figure 1.11). Lipids are a diverse group of substances that are soluble

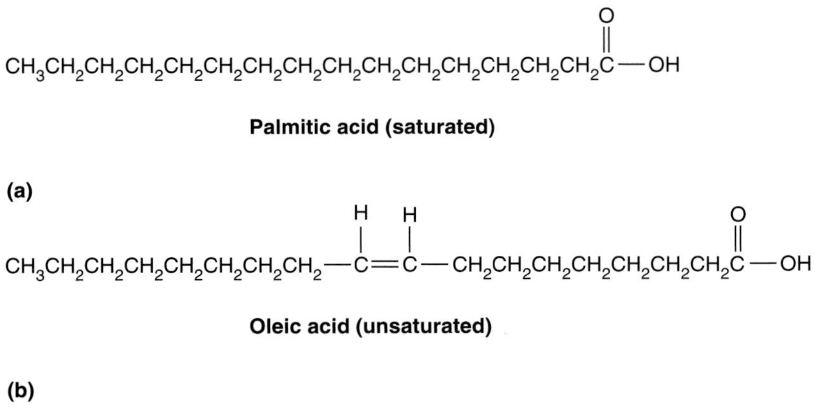

FIGURE 1.10

Fatty Acid Structure

(a) A saturated fatty acid; (b) an unsaturated fatty acid.

Palmitic acid (saturated)

(a)

Oleic acid (unsaturated)

(b)

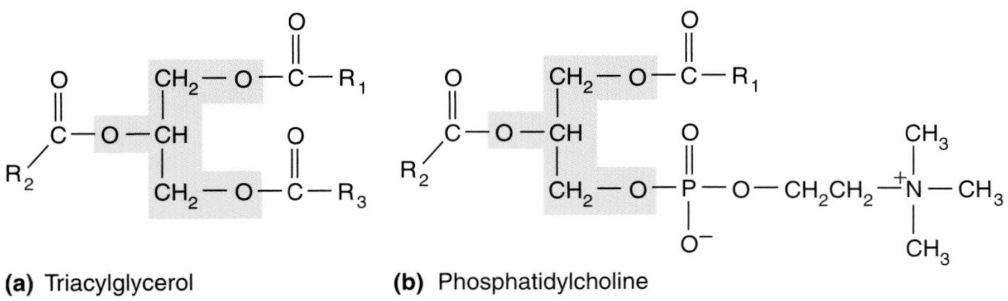

FIGURE 1.11

Lipid Molecules that Contain Fatty Acids

(a) Triacylglycerol; (b) phosphatidylcholine, a type of phosphoglyceride.

(a) Triacylglycerol   **(b)** Phosphatidylcholine

in organic solvents such as chloroform or acetone, but are not soluble in water. For example, triacylglycerols (fats and oils) are esters containing glycerol (a three-carbon alcohol with three hydroxyl groups) and three fatty acids. Certain lipid molecules that resemble triacylglycerols, called phosphoglycerides, contain two fatty acids. In these molecules the third hydroxyl group of glycerol is coupled with phosphate, which is in turn attached to small polar compounds such as choline. Phosphoglycerides are an important structural component of cell membranes.

NUCLEOTIDES AND NUCLEIC ACIDS Each nucleotide contains three components: a five-carbon sugar (either ribose or deoxyribose), a nitrogenous base, and one or more phosphate groups (Figure 1.12). The bases in nucleotides are heterocyclic aromatic rings with a variety of substituents. There are two classes of base: the bicyclic purines and the monocyclic pyrimidines (Figure 1.13).

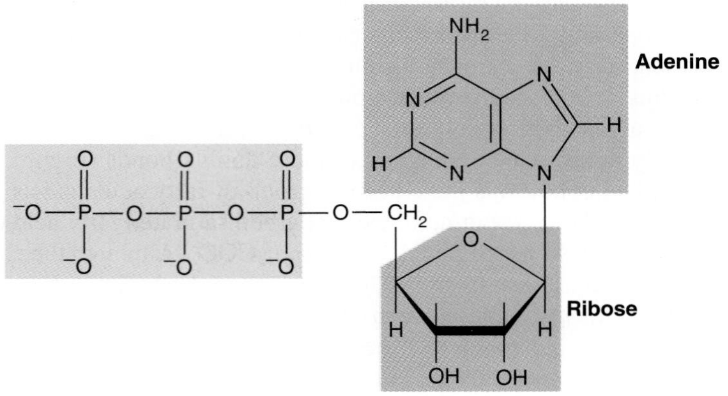

FIGURE 1.12

Nucleotide Structure

Each nucleotide contains a nitrogenous base (in this case, adenine), a pentose sugar (ribose), and one or more phosphates. This nucleotide is adenosine triphosphate.

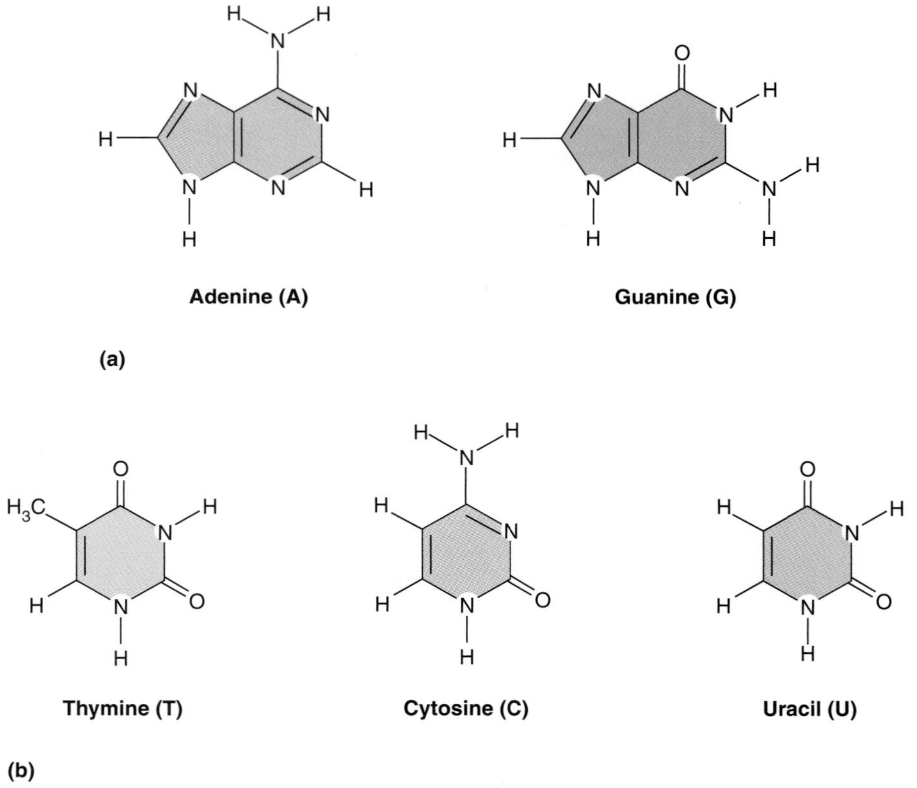

FIGURE 1.13

The Nitrogenous Bases

(a) The purines and (b) the pyrimidines.

Adenine (A)   Guanine (G)

(a)

Thymine (T)   Cytosine (C)   Uracil (U)

(b)

Nucleotides participate in a wide variety of biosynthetic and energy-generating reactions. For example, a substantial proportion of the energy obtained from food molecules is used to form the high-energy phosphate bonds of adenosine triphosphate (ATP). The most important role of nucleotides, however, is their role as the building block molecules of the nucleic acids. In a nucleic acid molecule, large numbers of nucleotides (from hundreds to millions) are linked by phosphodiester linkages to form long polynucleotide chains or strands. There are two types of nucleic acid:

1. **DNA**. DNA is the repository of genetic information. Its structure consists of two polynucleotide strands wound around each other to form a right-handed double helix (Figure 1.14). In addition to the pentose sugar deoxyribose and phosphate, DNA contains four types of base: the **purines** adenine and guanine and the **pyrimidines** thymine and cytosine. The double helix forms because of complementary pairing between the bases made possible by the formation of hydrogen bonds. (A hydrogen bond is a force of attraction between hydrogen and electronegative atoms such as oxygen or nitrogen.) Adenine pairs with thymine and guanine pairs with cytosine. Each gene is composed of a specific and unique linear sequence of bases. Although most genes code for the linear sequence of amino acids in proteins, DNA is not directly involved in protein synthesis. Instead another type of nucleic acid is used to convert DNA's coded instructions into polypeptide products.

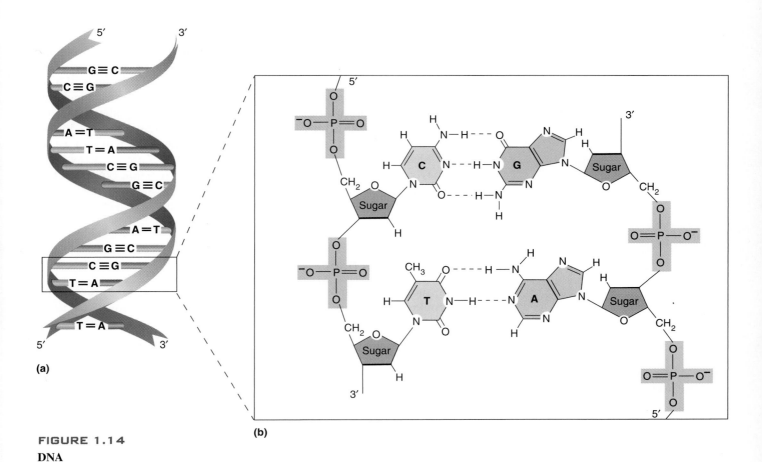

(a)

(b)

FIGURE 1.14

DNA

(a) A diagrammatic view of DNA. The sugar-phosphate backbones of the double helix are represented by colored ribbons. The bases attached to the sugar deoxyribose are on the inside of the helix. (b) An enlarged view of two base pairs. Note that the two DNA strands run in opposite directions defined by the 5′ and 3′ groups of deoxyribose. The bases on opposite strands form pairs because of hydrogen bonds. Cytosine always pairs with guanine; thymine always pairs with adenine.

2. **RNA.** RNA is a polynucleotide that differs from DNA in that it contains the sugar ribose instead of deoxyribose, and the base uracil instead of thymine. In RNA, as in DNA, the nucleotides are linked by phosphodiester linkages. In contrast to the double helix of DNA, RNA is single stranded. RNA molecules fold into complex three-dimensional structures created by local regions of complementary base pairing. During a complex process the DNA double helix partially unwinds and RNA molecules are synthesized using one DNA strand as a template. There are three major types of RNA: messenger RNA (mRNA), ribosomal RNA (rRNA), and transfer RNA (tRNA). Each unique sequence or molecule of mRNA possesses the information that codes directly for the amino acid sequence in a specific polypeptide. Ribosomes, the large, complex, supramolecular structures composed of rRNA and protein molecules, convert the mRNA base sequence into the amino acid sequence of a polypeptide. Transfer RNA molecules function as adapters during protein synthesis. Each type of tRNA molecule bonds to a specific amino acid. Each polypeptide is manufactured as the base sequence information in mRNA is translated by a ribosome as base pairing occurs between the mRNA and tRNA molecules. As the amino acids are brought into close proximity, peptide bonds are formed.

KEY CONCEPTS 1.3

Most molecules in living organisms are organic. The chemical properties or organic molecules are determined by specific arrangements of atoms called functional groups. Cells contain four families of small molecules: amino acids, sugars, fatty acids, and nucleotides.

1.3 BIOCHEMICAL PROCESSES

All of the characteristics of living organisms—their complex organization and their capacity to grow and reproduce—are the result of coordinated and purposeful biochemical processes. Metabolism, the sum total of these processes, is made possible by the flow of energy and nutrients and by thousands of biochemical reactions, each catalyzed by a specific enzyme. The primary functions of metabolism are (1) acquisition and utilization of energy, (2) synthesis of molecules needed for cell structure and functioning (i.e., proteins, carbohydrates, lipids, and nucleic acids), (3) growth and development, and (4) removal of waste products. Metabolic processes require significant amounts of useful energy. This section begins with a review of the primary chemical reaction types and the essential features of energy-generating strategies observed in living organisms. Metabolic processes and the means by which living organisms maintain ordered systems are then briefly outlined.

Biochemical Reactions

At first glance the thousands of reactions that occur in cells appear overwhelmingly complex. However, several characteristics of metabolism allow us to vastly simplify this picture:

1. Although the number of reactions is very large, the number of reaction types is relatively small.

2. Biochemical reactions have simple organic reaction mechanisms (i.e., an enzyme usually only does one thing).

3. Reactions of central importance in biochemistry (i.e., those used in energy production and the synthesis and degradation of major cell components) are relatively few.

Among the most common reaction types encountered in biochemical processes are the following: (1) nucleophilic substitution, (2) elimination, (3) addition, (4) isomerization, and (5) oxidation-reduction. Each will be briefly described.

NUCLEOPHILIC SUBSTITUTION REACTIONS In **nucleophilic substitution** reactions, as the name suggests, one atom or group is substituted for another:

$$A: \quad + \quad B\text{---}X \quad \longrightarrow \quad A\text{---}B \quad + \quad X:$$

In the general reaction shown above, the attacking species (A) is called a **nucleophile** ("nucleus-lover"). Nucleophiles are anions (negatively charged atoms or groups) or neutral species possessing non-bonding electron pairs. **Electrophiles** ("electron-lover") are deficient in electron density and are therefore easily attacked by a nucleophile. As the new bond forms between A and B, the old one between B and X breaks. The outgoing nucleophile (in this case, X) is called a **leaving group**.

The reaction of glucose with ATP provides an important example of nucleophilic substitution (Figure 1.15). In this reaction, which is the first step in the utilization of glucose as an energy source, the hydroxyl oxygen on carbon 6 of the sugar molecule is the nucleophile and phosphorus is the electrophile. Adenosine diphosphate is the leaving group.

Hydrolysis reactions are a kind of nucleophilic substitution reaction in which the oxygen of a water molecule serves as the nucleophile. The electrophile is usually the carbonyl group of an ester, amide, or anhydride.

$$R\text{---}\overset{\displaystyle O}{\underset{\displaystyle \|}{C}}\text{---}O\text{---}R' \; + \; H_2O \; \longrightarrow \; R\text{---}\overset{\displaystyle O}{\underset{\displaystyle \|}{C}}\text{---}OH \; + \; R'OH$$

The digestion of many food molecules involves hydrolysis. For example, proteins are degraded in the stomach in an acid-catalyzed reaction. Another important example is breaking the phosphate bonds of ATP (Figure 1.16). The energy obtained during this reaction is used to drive many cellular processes.

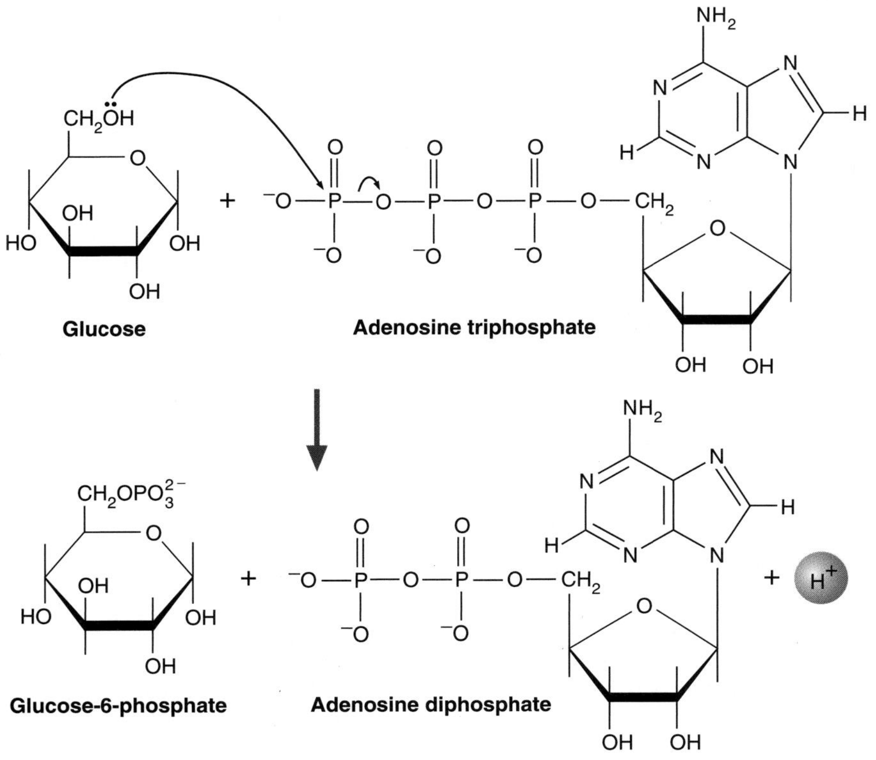

Glucose

Adenosine triphosphate

Glucose-6-phosphate

Adenosine diphosphate

FIGURE 1.15

Example of Nucleophilic Substitution

In the reaction of glucose with ATP, the hydroxyl oxygen of glucose is the nucleophile. The phosphorus atom (the electrophile) is polarized by the oxygens bonded to it so that it bears a partial positive charge. As the reaction occurs the unshared pair of electrons on the CH_2OH of the sugar attacks the phosphorus, resulting in the expulsion of ADP, the leaving group.

FIGURE 1.16

A Hydrolysis Reaction

The hydrolysis of ATP is used to drive an astonishing diversity of energy-requiring biochemical reactions.

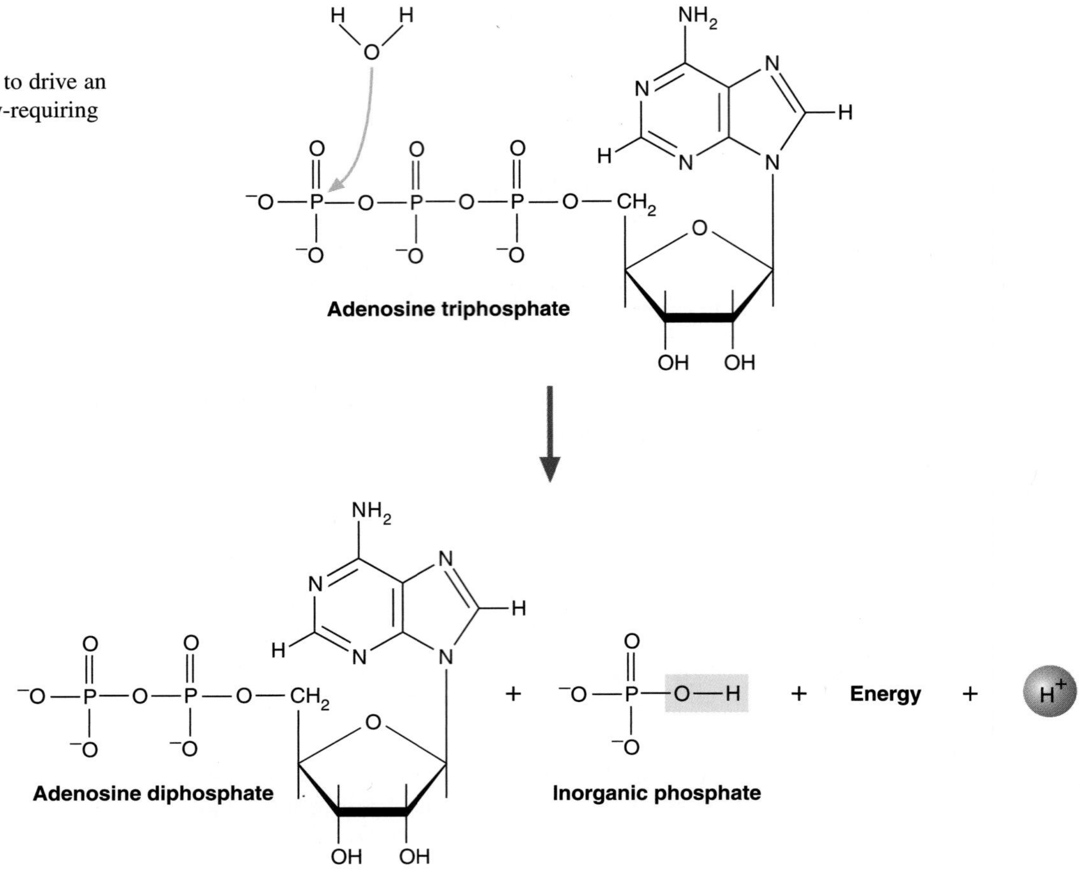

Adenosine triphosphate

Adenosine diphosphate

Inorganic phosphate

ELIMINATION REACTIONS In **elimination** reactions a double bond is formed when atoms in a molecule are removed.

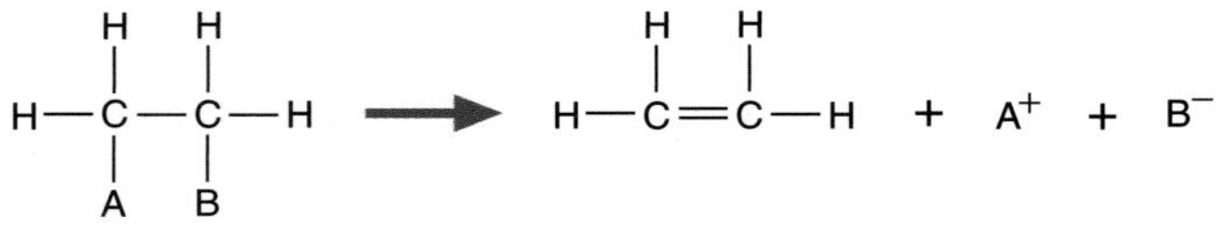

The removal of H_2O from biomolecules containing alcohol functional groups is a commonly encountered reaction. A prominent example of this reaction is the dehydration of 2-phosphoglycerate, an important step in carbohydrate metabolism (Figure 1.17). Other products of elimination reactions include ammonia (NH_3), amines (RNH_2), and alcohols (ROH).

FIGURE 1.17

An Elimination Reaction

When 2-phosphoglycerate is dehydrated, a double bond is formed.

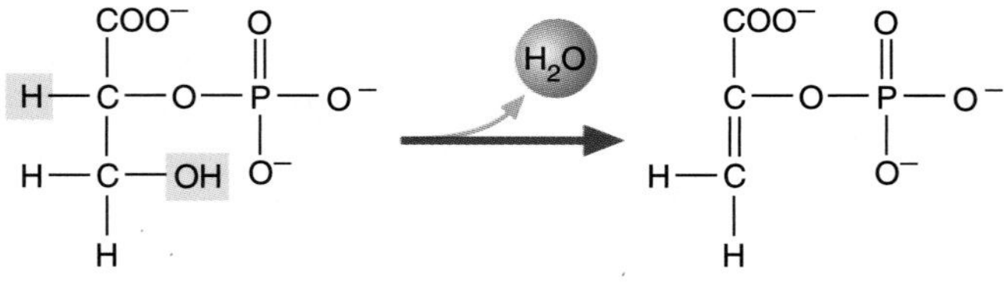

2-Phosphoglycerate

Phosphoenolpyruvate

ADDITION REACTIONS In **addition reactions** two molecules combine to form a single product.

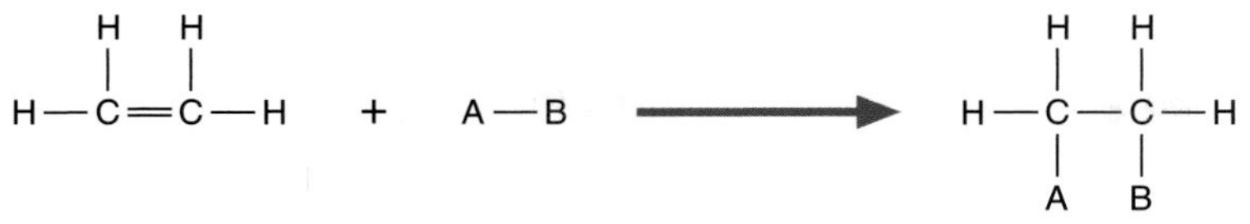

Hydration is one of the most common addition reactions. When water is added to an alkene an alcohol results. The hydration of the metabolic intermediate fumarate to form malate is a typical example (Figure 1.18).

ISOMERIZATION REACTIONS **Isomerization** reactions involve the intramolecular shift of atoms or groups. One of the most common biochemical isomerizations is the interconversion between aldose and ketose sugars (Figure 1.19).

OXIDATION-REDUCTION REACTIONS **Oxidation-reduction reactions** (also called redox reactions) occur when there is a transfer of electrons from a donor (called the **reducing agent**) to an electron acceptor (called the **oxidizing agent**). When reducing agents donate their electrons, they become **oxidized**. As oxidizing agents accept electrons, they become **reduced**. The two processes always occur simultaneously.

It is not always easy to determine whether biomolecules have gained or lost electrons. However, there are two simple rules that may be used to ascertain whether a molecule has been oxidized or reduced:

1. Oxidation has occurred if a molecule gains oxygen or loses hydrogen:

Ethyl alcohol   **Acetic acid**

2. Reduction has occurred if a molecule loses oxygen or gains hydrogen:

Acetic acid   **Ethyl alcohol**

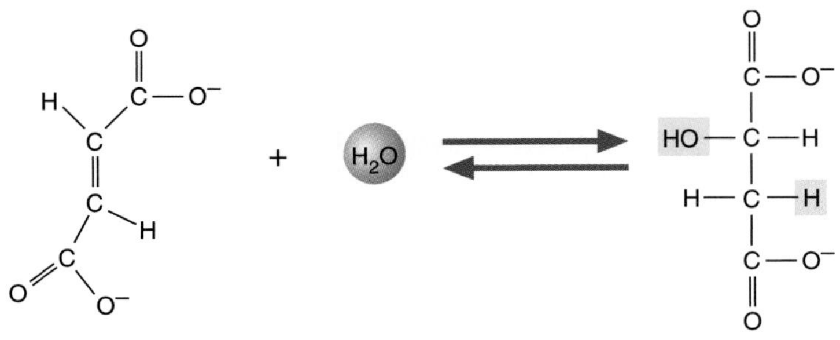

Fumarate   **Malate**

FIGURE 1.18

An Addition Reaction

When water is added to a molecule that contains a double bond, such as fumarate, an alcohol results.

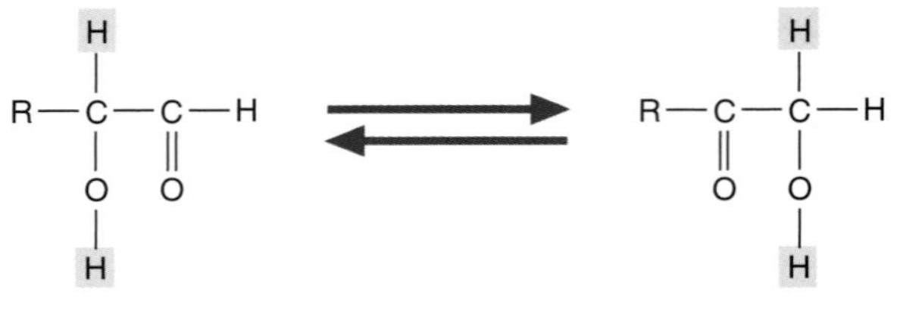

FIGURE 1.19

An Isomerization Reaction

The reversible interconversion of aldose and ketose isomers is a commonly observed biochemical reaction type.

Aldose   **Ketose**

KEY CONCEPTS 1.4

The most common reaction types encountered in biochemical processes are nucleophilic substitution, elimination, addition, isomerization, and oxidation-reduction.

In biological redox reactions, electrons are transferred to electron acceptors such as the nucleotide NAD^+/NADH (nicotinamide adenine dinucleotide in its oxidized/reduced form).

Energy

Energy is defined as the capacity to do work, that is, to move matter. In contrast to human-made machines, which generate and use energy under harsh conditions such as high temperature, pressures, and electrical currents, the relatively fragile molecular machines within living organisms must use more subtle mechanisms. Cells generate most of their energy by using redox reactions in which electrons are transferred from an oxidizable molecule to an electron-deficient molecule. In these reactions electrons are often removed or added as hydrogen atoms (H•) or hydride ions (H$^-$). The more reduced a molecule is, that is, the more hydrogen atoms it possesses, the more energy it contains. For example, fatty acids contain proportionally more hydrogen atoms than sugars do and therefore yield more energy upon oxidation than do sugar molecules. When fatty acids and sugars are oxidized, their hydrogen atoms are removed by the redox coenzymes FAD (flavin adenine dinucleotide) or NAD^+, respectively. (Coenzymes are small molecules that function in association with enzymes by serving as carriers of small molecular groups, or in this case, electrons.) The reduced products of this process ($FADH_2$ or NADH, respectively) can then transfer the electrons to another electron acceptor.

Whenever an electron is transferred, energy is lost. Cells have complex mechanisms for exploiting this phenomenon so that some of the released energy can be captured for cellular work. The most prominent feature of energy generation in most cells is the electron transport pathway, a series of linked membrane-embedded electron carrier molecules. During a regulated process, energy is released as electrons are transferred from one electron carrier molecule to another. During several of these redox reactions, the energy released is sufficient to drive the synthesis of ATP, the energy carrier molecule that directly supplies the energy used to maintain highly organized cellular structures and functions.

Despite their many similarities, groups of living organisms differ in the precise strategies they use to acquire energy from their environment. **Autotrophs** are organisms that transform the energy of the sun (the **photoautotrophs**) or various chemicals (the **chemoautotrophs**) into chemical bond energy. The **heterotrophs** obtain energy by degrading preformed food molecules obtained by consuming other organisms. **Chemoheterotrophs** use preformed food molecules as their sole source of energy. Some prokaryotes and a small number of plants; (e.g., the pitcher plant that digests captured insects) are **photoheterotrophs**, that is, they use both light and organic biomolecules as energy sources.

The ultimate source of the energy used by most life forms on earth is the sun. Photosynthetic organisms such as plants, certain prokaryotes, and algae capture light energy and use it to transform carbon dioxide (CO_2) into sugar and other biomolecules. Among the prokaryotes, there are chemotrophic species that do not obtain energy from the sun. Located in exotic locations such as hot springs, or deep within rock strata, they instead derive the energy required to incorporate CO_2 into organic biomolecules by oxidizing inorganic substances such as hydrogen sulfide (H_2S), nitrite (NO_2^-), or hydrogen gas (H_2). The biomass produced in both types of process is, in turn, consumed by heterotrophic organisms that use it as sources of energy and structural materials. At each step, as molecular bonds are rearranged, some energy is lost as heat. Before this happens, cells use captured energy to maintain their complex structures and activities. The metabolic pathways by which energy is generated and used by living organisms are briefly outlined in the Metabolism section. Descriptions of the basic mechanisms by which cellular order is maintained make up the Biological Order section.

Metabolism

Metabolism, the sum of all the enzyme-catalyzed reactions in a living organism, is a dynamic, coordinated activity. Many of these reactions are organized into pathways. Each biochemical pathway consists of several reactions that occur sequentially; that is, the product of one reaction is reactant for the one that follows. There are two major types of biochemical pathways: anabolic and catabolic. In **anabolic** or biosynthetic pathways, large complex molecules are synthesized from smaller precursors. Building block molecules (e.g., amino acids, sugars, and fatty acids) produced or acquired from the diet are incorporated into larger, more complex molecules. Because biosynthesis increases order and complexity, anabolic pathways require an input of energy. Examples of anabolic processes include the synthesis of polysaccharides and proteins from sugars and amino acids, respectively. During **catabolic pathways** large complex molecules are degraded into smaller, simpler products. Some catabolic pathways release energy. A fraction of this energy is captured and used to drive anabolic reactions.

The relationship between anabolic and catabolic processes is illustrated in Figure 1.20. As nutrient molecules are degraded, energy and reducing power are conserved in ATP and NADH molecules, respectively. Biosynthetic processes use metabolites of catabolism, synthesized ATP and NADPH (reduced nicotinamide adenine dinucleotide phosphate, a source of reducing power, i.e., high-energy electrons), to create complex structure and function.

Biological Order

The coherent unity that is observed in all living organisms involves the functional integration of millions of molecules. In other words, life is highly organized complexity. Despite the rich diversity of living processes that contribute to generating and maintaining biological order, most can be classified into the following categories: (1) synthesis of biomolecules, (2) transport of ions and molecules across cell membranes, (3) production of force and movement, and (4) removal of metabolic waste products and other toxic substances. Each will be discussed briefly.

SYNTHESIS OF BIOMOLECULES Cellular components are synthesized in a vast array of chemical reactions. Many of these reactions are integrated into

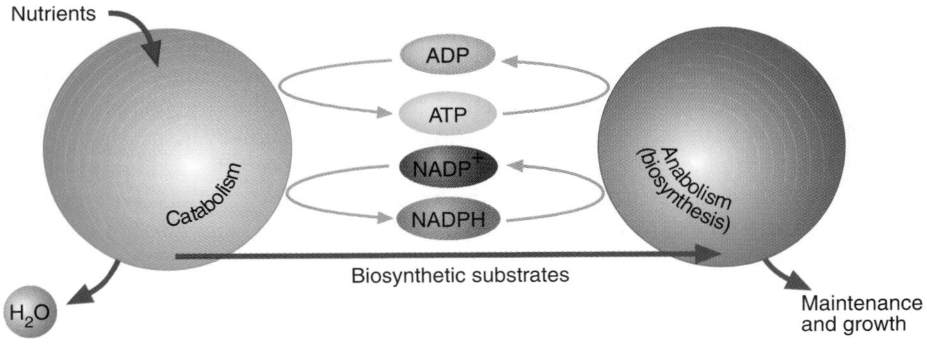

KEY CONCEPTS 1.5

In living organisms energy, the capacity to move matter, is usually generated by redox reactions.

KEY CONCEPTS 1.6

Metabolism is the sum of all reactions in a living organism. Most biochemical reactions can be classified as anabolic (biosynthetic) or catabolic (degradative).

FIGURE 1.20

Anabolism and Catabolism

In organisms that use oxygen to generate energy, catabolic pathways convert nutrients to small-molecule starting materials. The energy (ATP) and reducing power (NADPH) that drive biosynthetic reactions are generated during catabolic processes as certain nutrient molecules are converted to waste products such as carbon dioxide and water.

carefully regulated pathways that involve numerous steps. For example, the nucleotide adenosine monophosphate is synthesized in a 12-step pathway. It should be noted that a large number of biosynthetic reactions require energy, which is supplied directly or indirectly by the simultaneous breaking of the phosphoanhydride bonds of ATP molecules.

The molecules formed in biosynthetic reactions perform several functions. They can be assembled into supramolecular structures (e.g., the proteins and lipids that constitute membranes), serve as informational molecules (e.g., DNA and RNA), or catalyze chemical reactions (i.e., the enzymes).

TRANSPORT ACROSS MEMBRANES Cell membranes regulate the passage of ions and molecules from one compartment to another. For example, the plasma membrane (the cell's outer membrane) is a selective barrier. It is responsible for the transport of certain substances such as nutrients from a relatively disorganized environment into the more orderly cellular interior. Similarly, ions and molecules are transported into and out of organelles during biochemical processes. For example, fatty acids are transported into an organelle known as the mitochondrion so that they may be broken down to generate energy.

Much of the cell's transport work is accomplished by membrane-bound protein molecules. When substances are transported against a gradient (i.e., from an area of low concentration to an area of high concentration), energy is required. This process is referred to as **active transport**. For example, the Na^+-K^+ pump, a protein complex in the plasma membrane, is responsible for maintaining an ion gradient across the cell membrane. This ion gradient supplies the energy needed for many active transport processes and the resting membrane potential for excitable cells (nerve and muscle cells). The Na^+-K^+ pump uses at least one-third of available energy to pump Na^+ out of and K^+ into the cell. For every molecule of ATP hydrolyzed, three Na^+ ions are pumped out of the cell and two K^+ ions are pumped into the cell.

CELL MOVEMENT Organized movement is one of the most obvious characteristics of living organisms. The intricate and coordinated activities required to sustain life require the movement of cell components. Examples include cell division and organelle movement. Both of these processes depend to a large extent on the structure and function of a complex network of protein filaments known as the *cytoskeleton.*

The forms of cellular motion profoundly influence the ability of all organisms to grow, reproduce, and compete for limited resources. As examples, consider the movement of protists as they search for food in a pond or the migration of human white blood cells as they pursue foreign cells during an infection. More subtle examples include the movement of specific enzymes along a DNA molecule during the chromosome replication that precedes cell division and the secretion of insulin by certain pancreatic cells.

WASTE REMOVAL All living cells produce waste products. For example, animal cells ultimately convert food molecules, such as sugars and amino acids, into CO_2, H_2O, and NH_3. These molecules, if not disposed of properly, can be toxic. Some substances are readily removed. In animals, for example, CO_2 diffuses out of cells and (after a brief and reversible conversion to bicarbonate by red blood cells) is quickly exhaled through the respiratory system. Excess H_2O is excreted through the kidneys. Other molecules, however, are sufficiently toxic that elaborate mechanisms have been evolved to provide for their disposal. The urea cycle (described in Chapter 15), used in many animals to dispose of NH_3, converts this extremely harmful substance into urea, a less toxic molecule.

Living cells also contain a wide variety of complex organic molecules that must be disposed of. Plant cells solve this problem by transporting such molecules into a vacuole, where they are either broken down or stored. Animals, however, must use disposal mechanisms that depend on water solubility (e.g., the formation of urine by the kidney). Hydrophobic substances such as steroid hormones, which cannot be broken down into simpler molecules, are converted during a series of reactions into water-soluble derivatives. This mechanism is also used to solubilize some organic molecules such as drugs and environmental contaminants.

1.4 OVERVIEW OF GENETIC INFORMATION PROCESSING

The coding and replication of genetic information are some of the most important topics in biochemistry, and their study involves many of the concepts introduced in this chapter: life, biomolecules, chemical reactions, energetics. In this final section of the first chapter we present a brief overview of the genetic aspects of biochemistry (also known as molecular biology). The material in this overview will help place the material in Chapters 2 through 16 in perspective. Genetic information flow is the subject of the final three chapters of the book.

The means of reproducing genetic information is summarized in the "central dogma" of molecular biology (Figure 1.21). According to this principle, the information encoded in the chemical sequences of genes is used to direct the assembly of amino acids in polypeptides. An essential feature that allows the flow of genetic information is the capacity of the bases in the nucleotides to enter into specific base pairings. As indicated in the schematic diagram in Figure 1.22, there are two phases in the transfer of genetic information: transcription and translation. During **transcription**, enzymes called the RNA polymerases, along with other proteins, copy or *transcribe* the coded instructions in genes into the base sequences of RNA molecules. RNA molecules are synthesized by the formation of phosphodiester bonds between adjacent base-paired ribonucleotides. The incorporation of ribonucleotides into RNA can be summarized as follows:

$$\text{Ribonucleotide triphosphate} + \text{RNA}_n \longrightarrow \text{RNA}_{n+1} + \text{PP}_i$$

in which RNA_{n+1} is RNA that has grown from RNA_n by one unit of ribonucleotide, and PP_i is pyrophosphate ($\text{HP}_2\text{O}_7^{3-}$). This reaction is catalyzed by the enzyme RNA

FIGURE 1.21

The Central Dogma

Under most circumstances genetic information flows from DNA (a) to RNA (b) to protein (c). DNA stores information that directs the synthesis of itself and the RNA molecules involved in protein synthesis. An important exception to the central dogma is the capacity of a small group of viruses, called the retroviruses, to synthesize DNA using an RNA template.

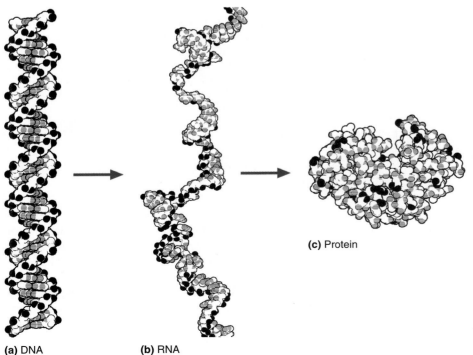

(a) DNA   **(b)** RNA   **(c)** Protein

FIGURE 1.22

An Overview of Genetic Information Flow

The genetic information in DNA is converted into the linear sequence of amino acids in polypeptides in a two-phase process. During *transcription*, RNA molecules are synthesized from a DNA strand through complementary base pairing between the bases in DNA and the bases in free ribonucleoside triphosphate molecules. During the second phase, called *translation*, mRNA molecules bind to ribosomes that are composed of rRNA and ribosomal proteins. Transfer RNA–aminoacyl complexes position their amino acid cargo in the catalytic site within the ribosome in a process that involves complementary base pairing between the mRNA codons and tRNA anticodons. Once the amino acids are correctly positioned within the catalytic site, a peptide bond is formed. After the mRNA molecule moves relative to the ribosome, a new codon enters the ribosome's catalytic site and base pairs with the appropriate anticodon on another aminoacyl-tRNA complex. After a stop codon in the mRNA enters the catalytic site, the newly formed polypeptide is released from the ribosome.

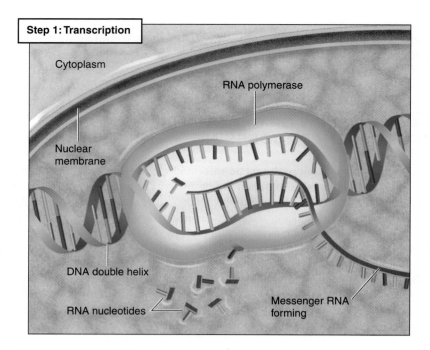

Step 1: Transcription

Cytoplasm
RNA polymerase
Nuclear membrane
DNA double helix
RNA nucleotides
Messenger RNA forming

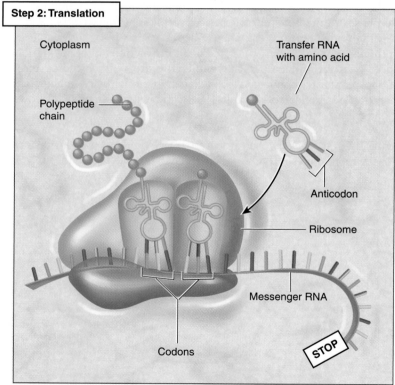

Step 2: Translation

Cytoplasm
Transfer RNA with amino acid
Polypeptide chain
Anticodon
Ribosome
Messenger RNA
Codons
STOP

polymerase. The energy that drives this reaction is generated by the cleavage of the phosphoanhydride bonds of the nucleotides and the subsequent hydrolysis of pyrophosphate. With few exceptions, the sole purpose of RNA molecules is the synthesis of proteins.

During **translation**, messenger RNA (mRNA) binds to ribosomes, where its nucleotide base code is decoded into the amino acid sequence of polypeptides. Ribosomes, which are located free in the cytoplasm or on the rough endoplasmic reticulum (an organelle located within cells), consist of ribosomal RNA (rRNA) and several proteins.

These are exciting times for biochemists! The knowledge now available concerning the inner workings of living organisms is providing life scientists with unprecedented opportunities to solve problems that have long afflicted humans. Our ever-growing knowledge about living processes is based on modern concepts made possible by the research efforts of biochemists throughout the twentieth century. This remarkably effective enterprise is largely the result of intellectual and technological breakthroughs based on a philosophical premise referred to as **reductionism**. According to reductionists, a complex phenomenon, such as life, can be eventually understood by analyzing its simplest components. By developing and using techniques that exploit the chemical and physical properties of component biomolecules, biochemists have provided a coherent view of the organizing principles of living processes. Examples of the properties of biomolecules that have proven to be useful include size, net electrical charge, solubility, capacity to move in an electric field, and ability to absorb or deflect electromagnetic radiation.

As research in the life sciences has progressed, it has become increasingly obvious that biochemistry has become an interdisciplinary science. For example, many of the techniques used by biochemists, such as X-ray diffraction, electron microscopy, and radioisotope labeling, used in investigations of the structure of macromolecules, cellular structure, and metabolic pathways, respectively, were first developed by physicists. In addition, knowledge developed in other related disciplines, such as cell biology and genetics, has become so interwoven with biochemistry that the intellectual barriers within the life sciences themselves have gradually been eroded. The recent explosion of biological information created by recombinant DNA techniques developed by molecular biologists has also created new and unanticipated problems. Molecular biology, the science that uses biochemical techniques to investigate the molecular aspects of genetics, has generated massive amounts of information about the structures of genes and proteins. The challenge that now confronts life scientists is to develop the means by which they use this information to identify and define solvable problems that confront humans, such as improving the nutrition of an ever-increasing population and developing more effective treatments for disease. To aid them in this work, many life scientists are now using the services of computer scientists and mathematicians. Also in response to these issues, many university and federal agencies have formed multidisciplinary teams of scientists. In this relatively new approach, research groups that tackle previously intractable problems consist of such diverse specialists as biochemists, biophysicists, and other life scientists along with engineers and computer scientists. The goal of these efforts is to analyze huge amounts of gene and protein structure information to deepen our understanding of living organisms.

The basic principles of biochemical research are an important part of an education in biochemistry because they provide a deeper understanding of how scientific knowledge is generated. Throughout this book, an introduction to the basic features of the most valuable biochemical techniques is provided in boxes called *Biochemical Methods*. Some of these techniques have great historical interest, whereas others are among the most current methods used by life scientists. The focus of each discussion is on how useful insights into living processes are gained by exploiting the physical and chemical properties of biomolecules.

Translation begins as each ribosome binds an mRNA molecule and proceeds to convert its base sequence into a polymer of amino acids linked by peptide bonds. Each amino acid is specified by a code word, called a **codon**, that consists of three sequential bases. The actual transfer of information occurs when each mRNA codon interacts and forms complementary base pairs with a three-base sequence in a transfer RNA (tRNA) molecule called an **anticodon**.

When codon-anticodon base pairing occurs the amino acid attached to the tRNA is correctly positioned within the ribosome for peptide bond formation. As each peptide bond is formed, the newly incorporated amino acid is released from its tRNA and the mRNA moves relative to the ribosome so that a new codon enters the catalytic site. The latter process is called **translocation**. Translation continues one codon at a time until a special base sequence, called a termination or stop codon, is reached. The polypeptide is then released from the ribosome, and folds into its biologically active conformation. Depending on the type of polypeptide, it may then bind to other folded polypeptides to form larger complexes.

Gene expression is a term used to describe the mechanisms by which living organisms regulate the flow of genetic information. Genes are switched on or off so that cells can conserve resources and respond appropriately to environmental or developmental cues. For example, many bacteria only produce the enzymes that degrade scarce nutrient molecules when the nutrient is actually available. In multicellular organisms, complex programmed patterns of gene expression are responsible for the diverse characteristics of differentiated cells.

KEY CONCEPTS 1.8

The genetic information encoded into the structure of nucleic acids is used to direct the assembly of amino acids in polypeptides. In a process called gene expression, genes are switched on and off so that living cells can conserve resources and respond to environmental or developmental cues.

SUMMARY

1. Biochemistry may be defined as the study of the molecular basis of life. Biochemists have contributed to the following insights into life: (1) life is complex and dynamic, (2) life is organized and self-sustaining, (3) life is cellular, (4) life is information-based, and (5) life adapts and evolves.

2. The molecular evidence concerning the evolutionary relationships of living species is sufficiently compelling that many life scientists now classify all living organisms into three domains: the bacteria, the archaea, and the eukarya.

3. All living things are composed of either prokaryotic cells or eukaryotic cells. Prokaryotes, which include bacteria and the archaea, lack a membrane-bound cellular organelle called a nucleus. The eukaryotes consist of all the remaining species. These cells contain a nucleus and complex structures that are not observed in prokaryotes.

4. Many eukaryotes are multicelluar. Multicellular organisms have several advantages over unicellular ones. These include (1) the provision of a relatively stable environment for most of the organism's cells, (2) the capacity for greater complexity in an organism's form and function, and (3) the ability to exploit environmental resources more effectively than individual single-celled organisms can.

5. Animal and plant cells contain thousands of molecules. Water constitutes 50% to 90% of a cell's content by weight, and ions such as Na^+, K^+, and Ca^{2+} may account for another 1%. Almost all the other kinds of biomolecules are organic. The properties of the element carbon are responsible for the vast variety of organic molecules. The chemical properties of organic molecules are determined by specific arrangements of atoms called functional groups. Different families of organic molecules result when hydrogen atoms on hydrocarbon molecules are replaced by functional groups. Most biomolecules contain more than one functional group.

6. Many of the biomolecules found in cells are relatively small, with molecular weights of less than 1,000 D. Cells contain four families of small molecules: amino acids, sugars, fatty acids, and nucleotides. Members of each group serve several functions: (1) they are used in the synthesis of larger molecules, (2) some small molecules have special biological functions, and (3) many small molecules are components in complex reaction pathways.

7. All life processes consist of chemical reactions catalyzed by enzymes. The reactions of a living cell, which are known collectively as metabolism, result in highly coordinated and purposeful activity. Among the most common reaction types encountered in biochemical processes are (1) nucleophilic substitution, (2) elimination, (3) addition, (4) isomerization, and (5) oxidation-reduction.

8. Living cells are inherently unstable. Only a constant flow of energy prevents them from becoming disorganized. One of the means by which cells obtain energy is oxidation of biomolecules or certain minerals.

9. Metabolism is the sum total of all the enzyme-catalyzed reactions in a living organism. Many of these reactions are organized into pathways. There are two major types of biochemical pathways: anabolic and catabolic.

10. The complex structure of cells requires a high degree of internal order. This is accomplished by four primary means: (1) synthesis of biomolecules, (2) transport of ions and molecules across cell membranes, (3) production of movement, and (4) removal of metabolic waste products and other toxic substances.

11. In genetic information flow the shape and chemical properties of the bases in the nucleotide residues of DNA direct the assembly of polypeptides from amino acids. There are two phases of gene expression: transcription and translation. In transcription, RNA polymerases use the capacity of nucleotide bases to form base pairs to copy the base sequence of genes to synthesize RNA molecules. During translation, ribosomes use the base sequence information in mRNA to construct polypeptides.

SUGGESTED READINGS

Goodsell, D. S., *The Machinery of Life*, Springer-Verlag, New York, 1993.

Lewis, R., *Life*, 3rd ed., WCB/McGraw-Hill, Dubuque, Iowa, 1998.

Mader, S., *Biology*, 6th ed., WCB/McGraw-Hill, Dubuque, Iowa, 1998.

Raven, P. H., and Johnson, G. B., *Biology*, 5th ed., McGraw-Hill, Dubuque, Iowa, 1999.

Tudge, C., *The Variety of Life: A Survey and a Celebration of All the Creatures that Have Ever Lived*, Oxford University Press, New York, 2000.

KEY WORDS

active transport, *22*

addition reaction, *19*

amino acid, *10*

anabolic pathway, *21*

anticodon, *25*

archaea, *5*

autotroph, *20*

bacteria, *5*

biomolecule, *2*

bioremediation, *6*

catabolic pathway, *21*

chemoautotroph, *20*

chemoheterotroph, *20*

codon, *25*

electrophile, *17*

elimination, *18*

energy, *20*

enzyme, *2*

eukarya, *5*

eukaryotic cell, *4*

extremophile, *6*

extremozyme, *6*

fatty acid, *13*

functional group, *9*

gene, *3*

gene expression, *25*

heterotroph, *20*

homeostasis, *2*

hydration, *19*

hydrocarbon, *8*

hydrolysis, *17*

hydrophilic, *10*

hydrophobic, *9*

isomerization, *19*

leaving group, *17*

lipid, *13*

macromolecule, *2*

metabolism, *2*

monosaccharide, *12*

mutation, *4*

neurotransmitter, *10*

nucleic acid, *14*

nucleophile, *17*

nucleophilic substitution, *17*

nucleotide, *14*

organelle *7*

oxidation-reduction (redox), *19*

oxidize, *19*

oxidizing agent, *19*

peptide, *11*

peptide bond, *11*

photoautotroph, *20*

photoheterotroph, *20*

polypeptide, *11*

polysaccharide, *12*

prokaryotic cell, *4*

protein, *11*

purine, *15*

pyrimidine, *15*

reduce, *19*

reducing agent, *19*

reductionism, *25*

saturated, *13*

sugar, *12*

transcription, *23*

transduction, *8*

translation, *24*

translocation, *25*

unsaturated, *13*

REVIEW QUESTIONS

1. Describe the insights that biochemical research has given us about life and living processes.

2. There are three domains of living organisms. What are they? What unique features do the organisms in each domain possess?

3. Describe the major differences between prokaryotic and eukaryotic cells.

4. Identify the functional groups in the following molecules.

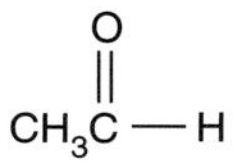

a.

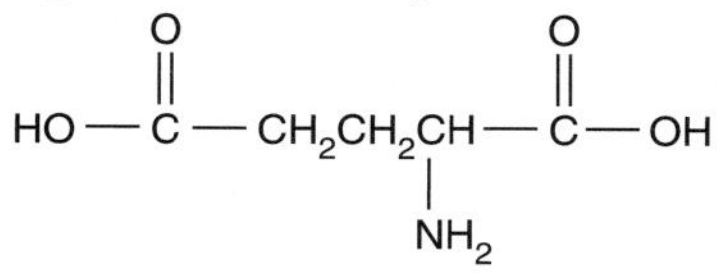

b.

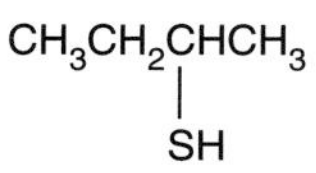

c.

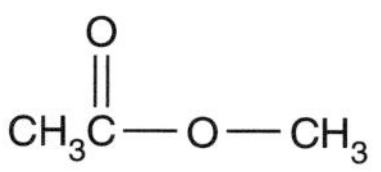

d.

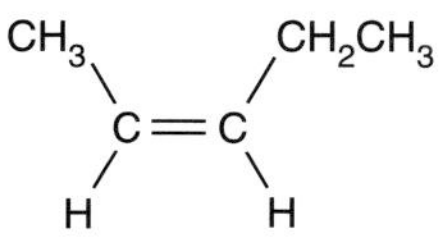

e.

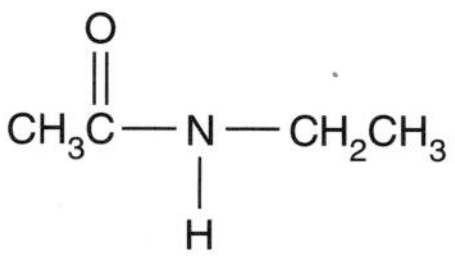

f.

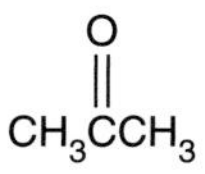

g.

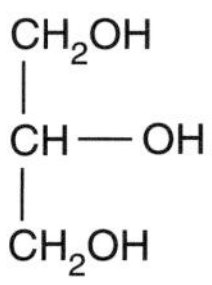

h.

5. Name four classes of small biomolecules. In what larger biomolecules are they found?

6. Define the following terms:
   a. biochemistry
   b. oxidation
   c. reduction
   d. active transport

   e. leaving group
   f. elimination
   g. isomerization
   h. nucleophilic substitution
   i. reducing agent
   j. oxidizing agent

7. List two functions for each of the following biomolecules:
   a. fatty acids
   b. sugars
   c. nucleotides

8. What are the roles of DNA and RNA?

9. How do cells obtain energy from chemical bonds?

10. How do plants dispose of waste products?

11. What is the difference between an unsaturated and a saturated hydrocarbon?

12. What advantages do multicellular organisms have over unicellular organisms?

13. Assign each of the following compounds to one of the major classes of biomolecule:

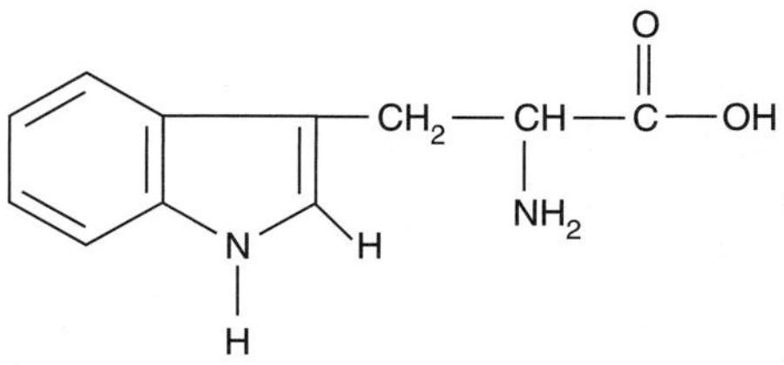

a.

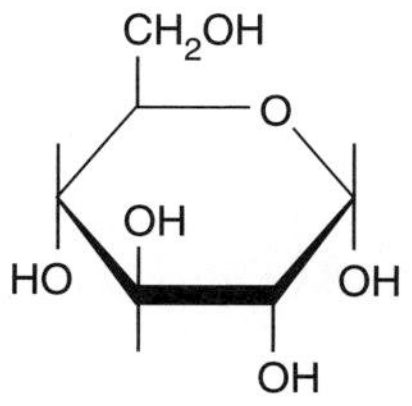

b.

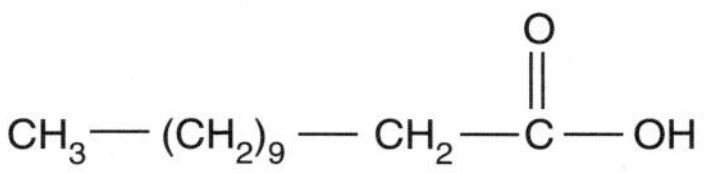

c.

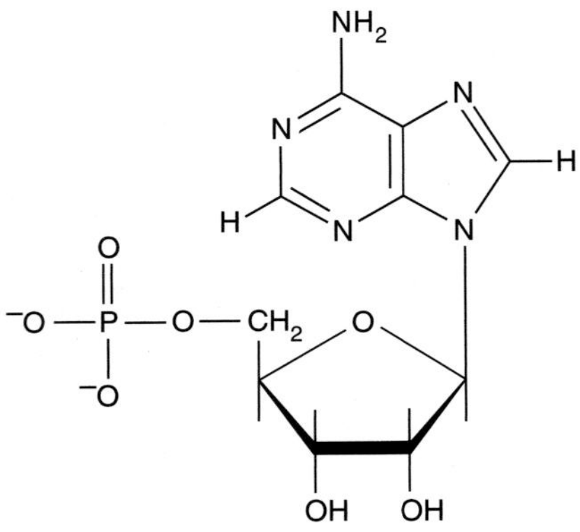

d.

14. Define the following terms:
    a. metabolism
    b. nucleophile
    c. reductionism
    d. electrophile
    e. energy

15. What are organelles? In general, what advantages do they provide to eukaryotes?

16. What are the primary functions of metabolism?

17. Give an example of each of the following reaction processes:
    a. nucleophilic substitution
    b. elimination
    c. oxidation-reduction
    d. addition

18. List several important ions that are found in living organisms.

19. What are the common types of chemical reactions found in living cells?

20. Describe several functions of polypeptides.

21. Carbohydrates are widely recognized as sources of metabolic energy. What are two other critical roles that carbohydrates play in living organisms?

22. What are the largest biomolecules? What functions do they serve in living organisms?

23. Nucleotides have roles in addition to being components of DNA and RNA. Give an example.

24. How is order maintained within living cells?

25. Name several waste products that animal cells produce.

26. What are emergent properties? Provide several examples.

27. Compare the functions of mRNA, rRNA, and tRNA in protein synthesis.

THOUGHT QUESTIONS

1. Biochemical reactions have been viewed as exotic versions of organic reactions. Can you suggest any problems with this assumption?

2. It is often assumed that biochemical processes in prokaryotes and eukaryotes are basically similar. Is this a safe assumption?

3. Much of what is known about biochemical processes is a direct result of research using prokaryotic organisms. Most organisms, however, are eukaryotic. Can you suggest any reasons why so many research efforts have used prokaryotes? Why not use eukaryotes directly?

4. Carbon-nitrogen single bonds are freely rotating; however, amide carbon-nitrogen bonds are much more rigid. Why is this true? What effect does this property of amide bonds have on the shapes that proteins can assume?

5. Why are fatty acids the principal long-term energy reserve of the body?

6. When a substance such as sodium chloride is dissolved in water, the ions that form become completely surrounded by water molecules, which form structures called hydration spheres. When the sodium salt of a fatty acid is mixed with water, the carboxylate group of the molecule becomes hydrated but the hydrophobic hydrocarbon portion of the molecule is poorly hydrated, if at all. The hydrocarbon chains from numerous fatty acids tend to clump together in spherical structures called micelles or, if large numbers are present, into bilayer sheets. Using a circle to represent the carboxylate group and an attached squiggly line to represent the hydrocarbon chain of a fatty acid, draw a picture of a micelle and a bilayer.

7. The bases of two complementary DNA chains pair with each other because of hydrogen bonding; that is,

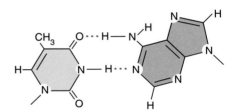

Thymine - Adenine base pair

A new nucleotide has been isolated containing the following purine:

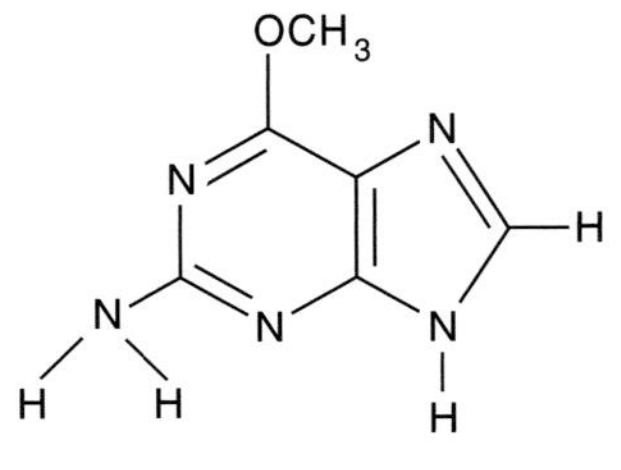

2-Amino-6-methoxypurine

Which of the normal purines and pyrimidines (adenine, guanine, cytosine, or thymine) would you expect to pair with it?

8. In most healthy humans, a diet high in cholesterol results in the inhibition of cholesterol synthesis in the body's cells. What property of living organisms does this phenomenon illustrate?

9. Describe the possible benefits of the production of biomolecules such as insulin by biotechnology.

CHAPTER TWO
Living Cells

Membranes in Living Cells Membranes are an essential feature of living cells. Most biochemical processes occur in or near these dynamic and complex supramolecular structures.

Cells are the structural units of all living organisms. One remarkable feature of cells is their diversity. For example, the human body contains about 200 types of cells. This great variation reflects the variety of functions that cells can perform. However, no matter what their shape, size, or species, cells are also amazingly similar. They are all surrounded by a membrane that separates them from their environment. They are all composed of the same types of molecules.

Cells are the fundamental units of life. They are functional entities, each of which is enclosed in a semipermeable membrane that varies in composition and function both over a single cell surface and between different cell types. There are two basic forms of cell: prokaryotic and eukaryotic. Prokaryotes are most noted for their small sizes and relatively simple structures. Presumably because of these traits, in addition to their remarkably rapid reproduction rates and biochemical diversity, various prokaryotic species occupy virtually every ecological niche in the biosphere. In contrast, the most conspicuous feature of the eukaryotes is their extraordinarily complex internal structure. Because eukaryotes carry out their various metabolic functions in a variety of membrane-bound organelles, they are capable of a more sophisticated intracellular metabolism. The diverse metabolic regulatory mechanisms made possible by this complexity promote two important lifestyle features required by multicellular organisms: cell specialization and intercellular cooperation. Consequently, it is not surprising that the majority of eukaryotes are multicellular organisms composed of numerous types of specialized cells.

Despite their immense diversity of sizes, shapes, and capacities, living cells are also remarkably similar. In fact, all modern cells are believed to have evolved from primordial cells over three billion years ago (see Special Interest Box 2.2). The common features of prokaryotic and eukaryotic cells include their similar chemical composition and the universal use of DNA as genetic material. The objective of this chapter is to provide an overview of cell structure. This review is a valuable exercise because biochemical reactions do not occur in isolation. It is becoming increasingly obvious that our understanding of living processes is incomplete without knowledge of their cellular context. After a brief discussion of some basic themes in cellular structure and function, the essential structural features of prokaryotic and eukaryotic cells will be described in relation to their biochemical roles.

2.1 BASIC THEMES

Within each living cell are hundreds of millions of densely packed biomolecules that perform at a frenetic pace the thousands of tasks that together constitute life. The application of biochemical techniques to investigations of living processes has provided significant insights into the unique chemical and structural properties of biomolecules that make their functional properties possible. A review of the following biochemical themes serves as an introduction to cellular structure and function:

1. Water, the biological solvent
2. Biological membranes
3. Self-assembly
4. Molecular machines

Water

Water dominates living processes. Its chemical and physical properties (described in Chapter 3) that result from its unique polar structure and its high concentration make it an indispensable component of living organisms. Among water's most important properties is its capacity to dissolve a wide range of substances. In fact, the behavior of all other molecules in living organisms is defined by the nature of their interactions with water. **Hydrophilic** molecules, that is, those that possess positive or negative charges or contain relatively large numbers of electronegative oxygen or nitrogen atoms, dissolve easily in water. Examples of simple hydrophilic molecules include salts such as sodium chloride and sugars such as glucose. In contrast, **hydrophobic** molecules, those that possess few electronegative atoms, do not dissolve in water. Instead, water excludes them and they end up confined to nonaqueous regions similar to the droplets of oil that form when oil and water are mixed. The behavior of fatty acids, for example, is dominated by their long hydrocarbon chains. When mixed with water they spontaneously form clusters so contact between the hydrocarbon chains and water molecules is minimized (Figure 2.1). In between these two extremes is an enormous group of both large and small biomolecules, each of which possesses its own unique pattern of hydrophilic and hydrophobic functional groups. Living organisms exploit the distinctive molecular structure of each of these biomolecules. Phospholipids and proteins are excellent examples. In an aqueous environment phospholipid molecules, the principal structural component of membranes, form spontaneously into a membranelike bilayer. Similarly, the proportion and precise placement of the hydrophilic and hydrophobic side chains in each polypeptide largely determine its structural and functional properties. The folding of each newly synthesized polypeptide into its three-dimensional and biologically active form is a process that is driven largely by the action of water to force hydrophobic portions of the polypeptide chain to internal regions of the folded protein.

Biological Membranes

Biological membranes are thin, flexible, and relatively stable sheetlike structures that enclose all living cells and organelles. These membranes can be thought of as noncovalent two-dimensional polymers creating chemically reactive surfaces and possessing unique transport functions between the extracellular and intracellular compartments. They are also versatile and dynamic cellular components that are intricately integrated into all living processes. Among the numerous crucial functions that have been assigned to membranes, the most basic is that of selective physical barriers. Membranes prevent the indiscriminate leakage of molecules and ions out of cells or organelles into their surroundings and allow the timely intake of nutrients and export of waste products. In addition, membranes have significant roles in information processing and energy generation.

Most biological membranes have the same basic structure: a lipid bilayer composed of phospholipids and other lipid molecules into which various proteins are

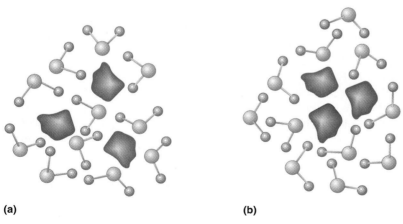

(a)   **(b)**

FIGURE 2.1

Hydrophobic Interactions Between Water and a Nonpolar Substance

As soon as nonpolar substances (e.g., hydrocarbons) are mixed with water (a), they coalesce into droplets (b). Hydrophobic interactions between nonpolar molecules take effect only when the cohesiveness of water and other polar molecules forces nonpolar molecules or regions of molecules close together.

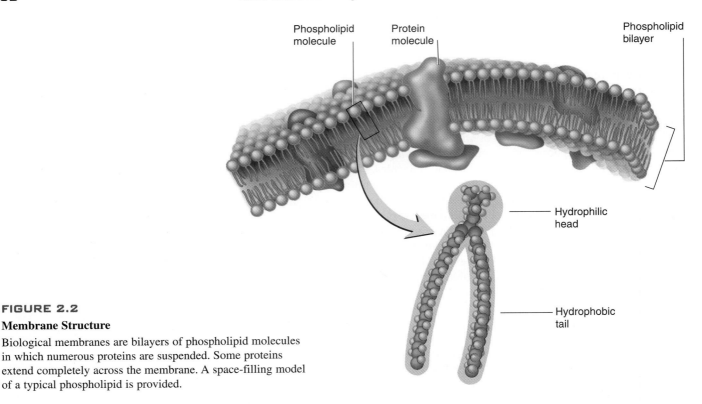

FIGURE 2.2

Membrane Structure

Biological membranes are bilayers of phospholipid molecules in which numerous proteins are suspended. Some proteins extend completely across the membrane. A space-filling model of a typical phospholipid is provided.

KEY CONCEPTS 2.2

Each biological membrane is composed of a lipid bilayer into which proteins are inserted or attached indirectly. Biological membranes are inextricably integrated into all living processes.

KEY CONCEPTS 2.3

In living organisms the molecules in supramolecular structures assemble spontaneously. Biomolecules are able to self-assemble because of the steric information they contain.

embedded or attached indirectly (Figure 2.2). Phospholipids have two features that make them ideally suited to their structural role: a hydrophilic charged group (referred to as a "head group") and a hydrophobic group composed of two fatty acid chains (often called hydrocarbon "tails"). Membrane proteins confer special abilities to membranes such as molecular and ion transport, energy generation, and signal transduction (cellular responsiveness to external stimuli). The amount and types of proteins in the specific cell membrane depend on the environment in which the cell operates.

Self-Assembly

As described in Chapter 1, living organisms are hierarchical systems. Recall that polymeric molecules such as nucleic acids and proteins are constructed from monomers. After their synthesis polymers aggregate (perhaps with smaller molecules) to form specific heterogeneous higher level assemblies, sometimes referred to as *supramolecular structures*. Prominent examples include ribosomes (the protein-synthesizing units that are formed from several different types of protein and RNA), and large protein complexes such as the sarcomeres in muscle cells or proteasomes (a large protein complex that degrades certain proteins).

According to the principle of self-assembly, most molecules that interact to form stable and functional supramolecular complexes are able to do so spontaneously because they inherently possess the steric information required. They have intricately shaped surfaces with complementary structures, charge distributions, and/or hydrophobic regions that allow the formation of numerous relatively weak noncovalent interactions (Figure 2.3). Self-assembly of such molecules is a balance between the tendency of hydrophilic groups to interact with water and for water to exclude hydrophobic groups from the aqueous regions of the cell. In some cases self-assembly processes need assistance. For example, the folding of some proteins requires the aid of molecular chaperones, protein molecules that prevent inappropriate interactions during the folding process. The assembly of certain supramolecular structures (e.g., chromosomes and membranes) requires preexisting information, that is, creation of new structure on a template of an existing structure.

Molecular Machines

In recent years researchers have recognized that many of the multisubunit complexes involved in cellular processes function as molecular machines. Examples include assembly devices such as the ribosomes that rapidly and accurately incorporate amino acids into polypeptides and the sarcomeres, the contractile units of skeletal muscle. Machines can be defined as mechanical devices with moving parts that perform work, the product of force and distance. The optimal functioning of each machine ensures that precisely the correct amount of applied force creates precisely the appropriate amount and direction of movement required for a specific task to be completed successfully. When machines are working properly they permit the accomplishment of tasks that often would be impossible without them. Although biological machines are composed of relatively fragile proteins that cannot withstand the physical conditions associated with human-made machines (e.g., heat and friction), the two do share important features. In addition to being composed of moving parts, both types of devices require energy-transducing mechanisms; that is, they both convert energy into directed motion. Despite the wide diversity of types of work performed by biological machines, they all share one key feature: energy-driven changes in the three-dimensional shapes of proteins. One or more components of biological machines bind nucleotide molecules such as ATP or GTP. The binding of nucleotide molecules to these protein subunits, referred to as **motor proteins**, and the subsequent release of energy that occurs when the nucleotide is hydrolyzed result in a precisely targeted change in the subunit's shape (Figure 2.4). Subsequently, this wave of change is transmitted to nearby subunits in a process that resembles a series of falling dominoes. Biological machines are relatively efficient because the hydrolysis of nucleotides

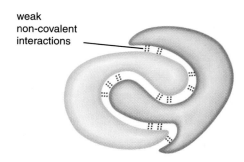

weak
non-covalent
interactions

FIGURE 2.3

Self-Assembly

The information that permits the self-assembly of biomolecules consists of the complementary shapes and distributions of charges and hydrophobic groups in the interacting molecules. Large numbers of weak interactions are required for supramolecular structures to form. In this diagrammatic illustration, several weak noncovalent interactions stabilize the binding of two molecules that possess complementary shapes.

KEY CONCEPTS 2.4

Many molecular complexes in living organisms function as molecular machines; that is, they are mechanical devices with moving parts that perform work.

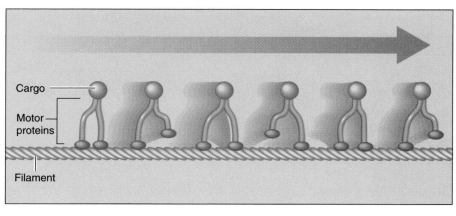

FIGURE 2.4

Biological Machines

Proteins perform work when motor protein subunits bind and hydrolyze nucleotides such as ATP. The energy-induced change in the shape of a motor protein subunit causes an orderly change in the shapes of adjacent subunits. In this diagrammatic illustration, a motor protein complex moves attached cargo (e.g., a vesicle) as it "walks" along a cytoskeletal filament.

is essentially irreversible; therefore, the functional changes that occur in each machine occur in one direction only.

2.2   STRUCTURE OF PROKARYOTIC CELLS

The prokaryotes are an immense and heterogeneous group that includes the bacteria and the archaea. Most prokaryotes are similar in their external appearances. For example, the most commonly observed shapes among prokaryotes are cylindrical or rodlike (bacillus), spheroidal (cocci), and helically coiled (spirilla). Prokaryotes are also characterized by their relatively small size (a typical rod-shaped bacterial cell has a diameter of 1 μm and a length of 2 μm), their capacity to move (i.e., whether they have flagella, whiplike appendages that propel them), and their retention of specific dyes. Because these traits are insufficient to differentiate the thousands of species that are currently known, most are identified on the basis of more subtle characteristics. Among the most useful of these are nutritional requirements, energy sources, chemical composition, and biochemical capacities. Despite their diversity most prokaryotes possess the following common features: cell walls, a plasma membrane, circular DNA molecules, and no internal membrane-enclosed organelles. Because the bacteria are better known, the following discussion will primarily focus on their structural features. The anatomical features of a typical bacterial cell are illustrated in Figures 2.5 and 2.6.

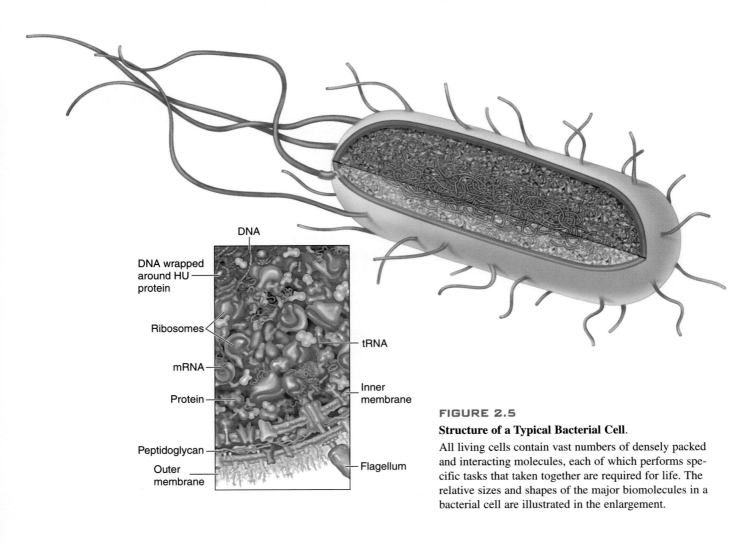

FIGURE 2.5

Structure of a Typical Bacterial Cell.

All living cells contain vast numbers of densely packed and interacting molecules, each of which performs specific tasks that taken together are required for life. The relative sizes and shapes of the major biomolecules in a bacterial cell are illustrated in the enlargement.

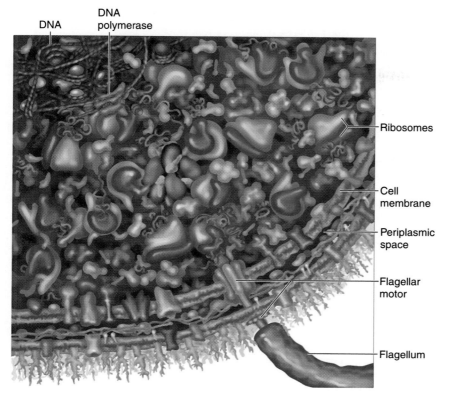

DNA

DNA
polymerase

Ribosomes

Cell
membrane

Periplasmic
space

Flagellar
motor

Flagellum

FIGURE 2.6

Bacterial Cell

Bacterial cells are not the bags of protoplasm they were once envisioned to be. Considering
that they have no membrane compartments, their internal structure is surprisingly well
organized. Note, for example, the spatial separation of the supercoiled DNA molecule
(upper left) from other biomolecules. (Also refer to Figure 2.5.)

Cell Wall

The prokaryotic cell wall (Figure 2.7) is a complex semi-rigid structure that serves
as the primary source of support. It both maintains the organism's shape and protects
it from mechanical injury. The cell wall's strength is largely due to the presence of
complex peptide- and carbohydrate-containing polymers. In the cell walls of many
bacteria the network composed of these polymers is called *peptidoglycan*. Some bacteria secrete substances such as polysaccharides and proteins, collectively known
as the *glycocalyx*, that accumulate on the outside of the cell. Depending on the structure and composition of this material, the glycocalyx may also be referred to as a
slime layer or a *capsule*. Slime layers are disorganized accumulations of gelatinous material that are only loosely attached. In contrast, the material in capsules is
highly organized and firmly attached to the cell wall. Some bacterial species are especially pathogenic (disease-causing) because the capsule allows them to avoid detection or damage by host immune systems, attach to host cells to facilitate colonization,
and slough endotoxins that cause host cell damage.

The thickness and chemical composition of the cell wall and its adjacent
structures determine how avidly a cell wall takes up and/or retains specific dyes.
Most cells can be differentiated on the basis of whether they retain crystal violet stain during the Gram stain procedure. Those that can retain the dye, called
Gram positive, do so because their cell walls consist of a thick peptidoglycan
layer. In contrast, Gram-negative cells possess a thin peptidoglycan layer. Often

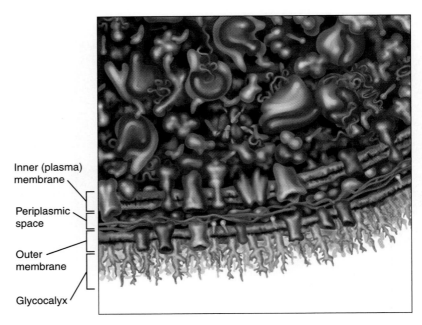

Inner (plasma) membrane

Periplasmic space

Outer membrane

Glycocalyx

FIGURE 2.7

Prokaryotic Cell Wall

The cell wall from the Gram negative organism *Escherichia coli* is complex. As with many Gram negative bacteria, the cell wall of *E. coli* possess a periplasmic space, a gelatinous layer between the inner and outer membrane. It often contains some peptidoglycan molecules.

this thin layer is surrounded by a thick outer lipid bilayer with embedded proteins and attached polysaccharides. The cell walls of the archaea are quite variable in their composition, and some archaea do not contain cell walls. Although archaean cell walls are also different from those of the bacteria, for example, they lack certain sugars and amino acids typically found in bacterial peptidoglycan, they too can stain Gram positive or Gram negative. One consequence of the different cell wall composition of the archaea is that none are susceptible to penicillin, an antibiotic that inhibits the cell wall synthesis of Gram positive bacteria.

Plasma Membrane

Directly inside the cell wall is the **plasma membrane** (Figure 2.8). In addition to acting as a selective permeability barrier, the bacterial plasma membrane possesses receptor proteins that detect nutrients and toxins in their environment. Numerous types of transport proteins involved in nutrient uptake and waste product disposal also occur here. Depending on the species of organism, there may also be proteins involved in energy transduction processes such as **photosynthesis** (the conversion of light energy into chemical energy) and **respiration** (a process whereby fuel molecules are oxidized and their electrons are used to generate ATP).

The composition of membranes in the archaea is remarkably different from both bacteria and the eukaryotes. Instead of the straight-chain fatty acids linked to glycerol through ester bonds that are typically found in the lipids that comprise membranes, the hydrocarbon chains in the membranes of the archaea are linked by ether linkages. In addition, archaean membrane lipids also contain some branched-chain hydrocarbons.

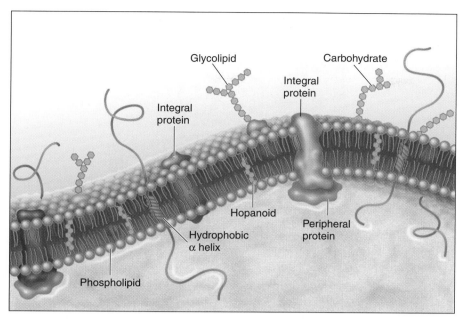

FIGURE 2.8

The Bacterial Plasma Membrane

In this simplified view of the plasma membrane, several classes of protein and lipid are illustrated. Many of these proteins and certain lipids are covalently bound to carbohydrate molecules. (Glycolipids contain carbohydrate groups.) Hopanoids are complex lipid molecules that stabilize bacterial membranes.

Cytoplasm

Despite the absence of internal membranes, prokaryotic cells do appear to have functional compartments (Figure 2.9). The most obvious of these is the **nucleoid**, a spacious, irregularly shaped region that contains a long, circular DNA molecule called a **chromosome**. The bacterial chromosome is attached to the plasma membrane. It typically has numerous regions of highly coiled structure and others that are uncoiled. Protein complexes involved in DNA synthesis and regulation of gene expression are also found within the nucleoid. Many bacteria also contain additional small circular DNA molecules called **plasmids** that exist separately from the cell's chromosome elsewhere in the cytoplasm. Although they are not required for growth or cell division, plasmids usually provide the cell with a biochemical advantage over cells that lack plasmids. For example, DNA segments coding for antibiotic resistance are often found on plasmids. In the presence of the antibiotic, resistant cells synthesize a protein that inactivates the antibiotic before it can damage the cell. Such cells continue to grow and reproduce, whereas susceptible cells die.

Under low magnification the cytoplasm of prokaryotes has a uniform, grainy appearance except for inclusion bodies, large granules that contain organic or inorganic substances. Examples of inclusion bodies are reservoirs of glycogen, or fats (energy storage molecules), or polyphosphate. The remaining portions of the cytoplasm are filled with ribosomes and many types of enzymes and molecular complexes that perform routine tasks such as the synthesis and degradation of biomolecules.

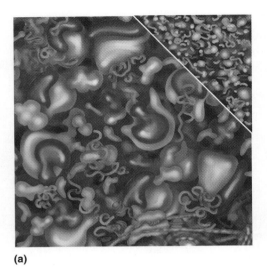

(a)

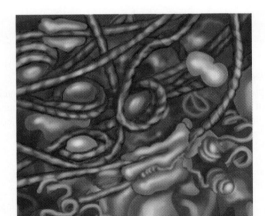

(b)

FIGURE 2.9

Bacterial Cytoplasm

(a) Cytoplasm is a complex mixture of proteins, nucleic acids, and an enormous variety of ions and small molecules. For clarity the small molecules are only drawn in the upper right corner. (b) Close-up view of the nucleoid. Note that DNA is coiled and folded around protein molecules.

Pili and Flagella

Many bacterial cells have external appendages. *Pili* (singular: pilus) are fine, hair-like structures that may allow cells to attach to food sources and host tissues. Sex pili are used by some bacteria to transfer genetic information from donor cells to recipients, a process called *conjugation* (Figure 2.10). In bacteria, the *flagellum* (plural: flagella) is a flexible corkscrew-shaped protein filament that is used for locomotion (Figure 2.11). Cells are pushed forward when flagella rotate in a counterclockwise direction, whereas clockwise rotation results in a stop and tumble motion, allowing the cell to reorient for a subsequent forward run. The filament of the flagellum is anchored into the cell by a protein complex. Among the many components of this complex are motor proteins that convert chemical energy into rotational motion.

FIGURE 2.10

Bacterial Pili

In this electron micrograph, a sex pilus connects two conjugating *E. coli* cells. Note the numerous smaller pili covering the surface of one of the cells.

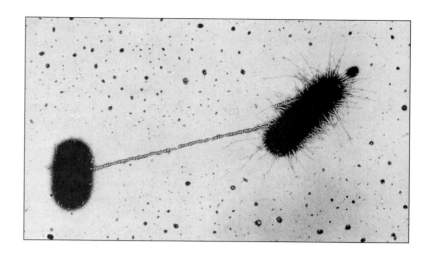

FIGURE 2.11

Structure of a Prokaryotic Flagellum.

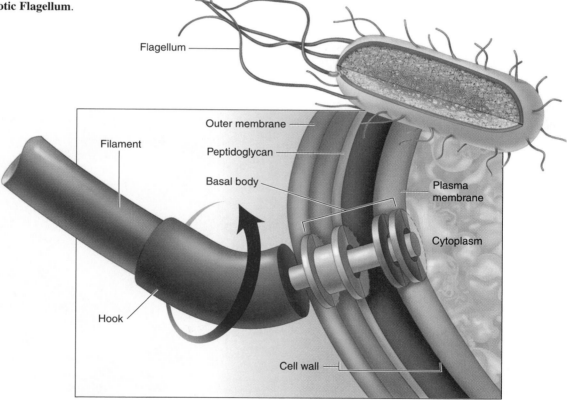

QUESTION 2.1

A typical human hepatocyte (liver cell), a widely studied eukaryotic cell, has a diameter of about 20 *μ*m. Calculate the volume of both a prokaryotic and a eukaryotic cell. To appreciate the magnitude of the size difference between the two cell types, estimate how many bacterial cells would fit inside the liver cell. (*Hint*: Use the expression $V = \pi r^2 h$ for the volume of a cylinder and $V = 4\pi r^3/3$ for the volume of a sphere.)

2.3 STRUCTURE OF EUKARYOTIC CELLS

The structural complexity of eukaryotic cells allows vastly more sophisticated regulation of living processes than is possible in the prokaryotes. The organelles and other structures within these cells are organized into efficient and highly integrated dynamic entities. Although most eukaryotic cells possess similar structural features, there is no "typical" eukaryotic cell. Each cell type has its own characteristic structural and functional properties. They are sufficiently similar, however, that a discussion of the basic components is useful. The generalized structures of cells from animals and plants, the major forms of multicellular eukaryotic organisms, are illustrated in Figures 2.12 and 2.13. The structure and functional properties of each cellular component are briefly described in the sections that follow.

Plasma Membrane

The plasma membrane, like all cell membranes, is composed of a lipid bilayer in which embedded and attached proteins perform many of its functional roles. On the external surface of many eukaryotic cells is a structure called the **glycocalyx** (Figure 2.14), which is composed largely of carbohydrate molecules that are attached to membrane proteins and certain lipid molecules.

FIGURE 2.12
Animal Cell Structure.

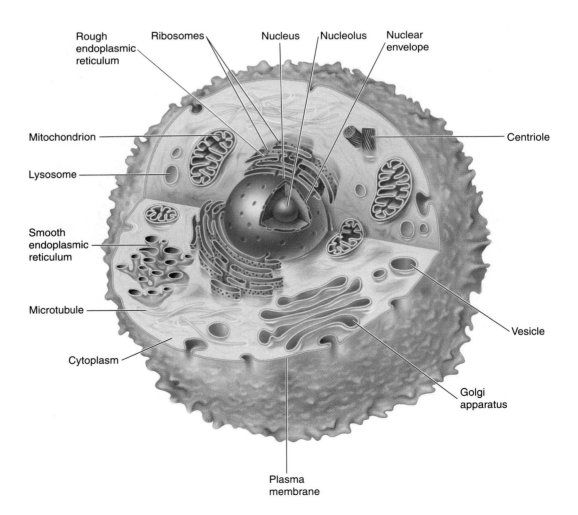

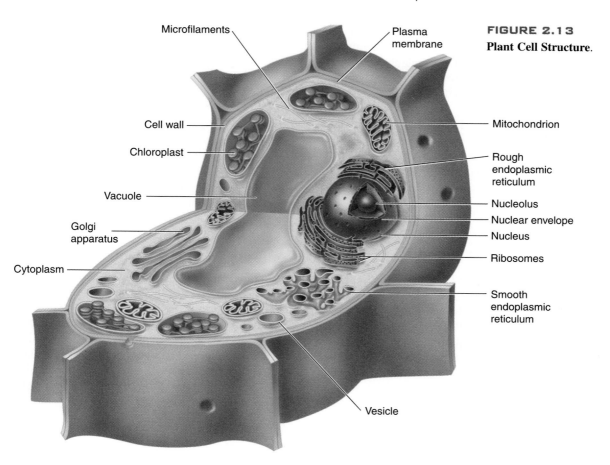

FIGURE 2.13

Plant Cell Structure.

Microfilaments

Plasma membrane

Cell wall

Chloroplast

Vacuole

Golgi apparatus

Cytoplasm

Mitochondrion

Rough endoplasmic reticulum

Nucleolus

Nuclear envelope

Nucleus

Ribosomes

Smooth endoplasmic reticulum

Vesicle

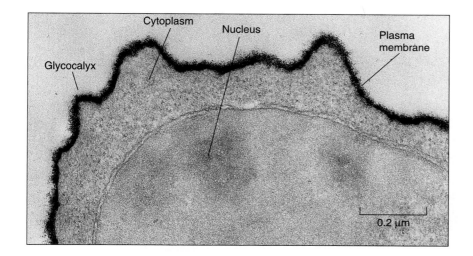

Cytoplasm

Nucleus

Plasma membrane

Glycocalyx

0.2 µm

FIGURE 2.14

The Glycocalyx

Electron micrograph of the surface of a lymphocyte stained to reveal the glycocalyx (cell coat).

The plasma membrane of eukaryotic cells performs several vital functions. As with all plasma membranes, it provides shape and some mechanical strength and protection, as well as a permeability barrier, to the cell. The relatively thick glycocalyx assists in these functions. The plasma membrane, which possesses a diversity of channel complexes that transport ions and molecules and **receptors** that bind signal molecules, is also involved in various types of transport and signaling processes.

The plasma membranes of multicellular eukaryotes have structural properties that allow them to function within groups of cells. Specialized portions of plasma membrane contain molecular complexes that allow the formation of tight contacts between cells to facilitate the transport of metabolites between cells and the integrated functioning of cells within tissues and organs. Within animal tissues, cells typically secrete protein and carbohydrates that form the **extracellular matrix**, a gelatinous material that binds cells and tissues together. In plant cells, protection of the cell is provided mainly by a thick cell wall composed primarily of fibers of the polysaccharide cellulose, which is synthesized by enzyme complexes on the plasma membrane's surface. Newly made cellulose fibers become embedded in a matrix that contains other polysaccharides and some protein. The rigid structure of the cell wall protects it from being damaged by the enormous pressure exerted by water in vascular plants.

Nucleus

The **nucleus** (Figure 2.15) consists of nucleoplasm surrounded by a nuclear envelope. The **nucleoplasm** is a material rich in DNA in which proteins called *lamins* form a fibrous network that provides structural support. A prominent feature of the nucleoplasm is a network of **chromatin fibers** composed of DNA and DNA packaging proteins known as the *histones*. The DNA is believed to be attached to the lamins. The **nuclear envelope** is composed of two membranes that fuse at structures called **nuclear pores**. The outer nuclear membrane is continuous with the rough endoplasmic reticulum. The nuclear pores (Figure 2.16) are relatively

KEY CONCEPTS 2.6

Besides providing mechanical strength and shape to the cell, the plasma membrane is actively involved in selecting which molecules can enter or exit the cell. Receptors on the plasma membrane's surface allow the cell to respond to external stimuli.

FIGURE 2.15

The Eukaryotic Nucleus

The nucleus is an organelle surrounded by a double membrane, the nuclear envelope.

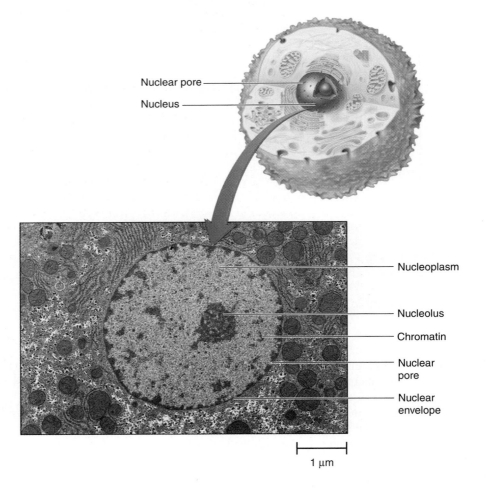

Nuclear pore

Nucleus

Nucleoplasm

Nucleolus

Chromatin

Nuclear pore

Nuclear envelope

1 μm

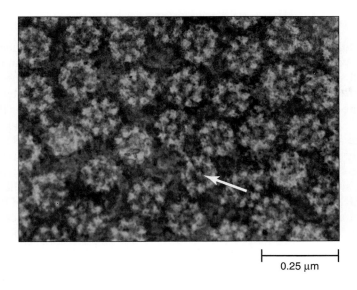

0.25 µm

FIGURE 2.16

The Nuclear Pore Complex

The nuclear envelope is studed with nuclear pore complex structures, one of which is indicated by the arrow in the photo at top.

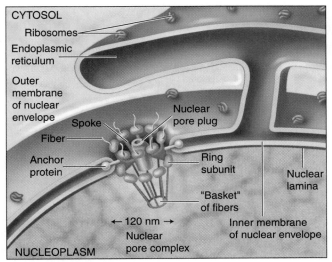

large and complex structures through which pass most of the molecules that enter and leave the nucleus. Many of these substances such as ions, small proteins, and other molecules diffuse through the nuclear pore complex. The movement of larger molecules such as RNA and large proteins into and out of the nucleus is believed to be regulated by protein components of a centrally located transport complex, sometimes referred to as the *nuclear pore plug* (see Figure 2.16).

The importance of the nucleus in regulating cell function has long been appreciated. However, its mechanism of control was not understood until the significance of its major component, DNA, was discovered (see Chapter 16). The nucleus is now known to perform two critical functions. First, it contains the cell's hereditary information. Second, the nucleus exerts a profound influence over all cellular metabolic activities. This influence, which is exerted by directing the synthesis of protein cell components, is in turn affected by the passage of molecules back and forth between the cytoplasm and the nucleus.

When nuclei are stained with certain dyes, one or more spherical structures called nucleoli (singular: nucleolus) become visible. The **nucleolus** is the site of ribosomal RNA synthesis. Its high content of RNA makes it stain differently than the rest of the nucleus.

Endoplasmic Reticulum

The **endoplasmic reticulum (ER)** is a system of interconnected membranous tubules, vesicles, and large flattened sacs. A hint of its importance in cell function is that it often constitutes more than half of a cell's total membrane. The repeatedly folded, continuous sheets of ER membrane enclose an internal space called the ER *lumen*. This compartment, which is often referred to as the *cisternal space*, is entirely separated from the cytoplasm by ER membrane.

There are two forms of ER (Figure 2.17). The **rough ER (RER)**, which is primarily involved in the synthesis of membrane proteins and protein for export from the cell, is so named because of the numerous ribosomes that stud its cytoplasmic surface. The second form lacks attached ribosomes and is called **smooth ER (SER)**. Although the SER membranes are continuous with those of RER, their physical appearances may be significantly different. In hepatocytes (the predominant cell type in liver), for example, SER consists of a tubular network that penetrates large regions of cytoplasm. Functions of SER include lipid synthesis and **biotransformation**, a process in which water-insoluble organic molecules are prepared for excretion.

Ribosomes

The cytoplasmic **ribosomes** of eukaryotes are RNA/protein complexes (20 nm in diameter), whose function is the biosynthesis of proteins. Composed of a variety of proteins and a type of RNA called ribosomal RNA (rRNA), ribosomes

FIGURE 2.17

The Endoplasmic Reticulum

There are two forms of endoplasmic reticulum: RER, the rough endoplasmic reticulum and SER, the smooth endoplasmic reticulum.

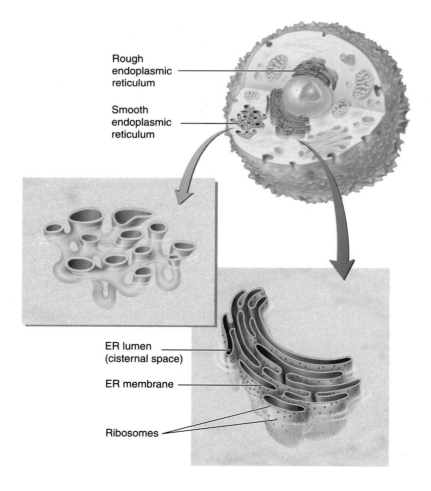

Rough endoplasmic reticulum

Smooth endoplasmic reticulum

ER lumen (cisternal space)

ER membrane

Ribosomes

are complex structures containing two irregularly shaped subunits of unequal size (Figure 2.18). They come together to form whole ribosomes when protein synthesis is initiated; when not in use, the ribosomal subunits separate. The number and distribution of ribosomes in any cell depend on the relative metabolic activity and the proteins being synthesized. Although eukaryotic ribosomes are larger and more complex than those of prokaryotes, they are similar in overall shape and function.

Golgi Apparatus

The **Golgi apparatus** (also known as the **Golgi complex**) is named for the Italian cell biologist Count Camillo Golgi, who first described it in 1898. Formed from relatively large, flattened, saclike membranous vesicles that resemble a stack of plates, the Golgi apparatus (called **dictyosomes** in plants) is involved in the packaging and distribution of cell products to internal and external compartments (Figure 2.19).

The Golgi apparatus has two faces. The plate (or *cisterna*) positioned closest to the ER is on the forming (*cis*) face, whereas the one on the maturing (*trans*) face is typically close to the portion of the cell's plasma membrane that is engaged in secretion. Small membranous vesicles containing newly synthesized protein and lipid bud off from the ER and fuse with the *cis* Golgi membrane. These molecules are transported from one Golgi sac to the next by vesicles, where they are further processed by enzymes. Once the products reach the *trans* face, they are then targeted to other parts of the cell. Secretory products, such as digestive enzymes or hormones, are concentrated within *secretory vesicles* (also known as *secretory granules*) that bud off from the *trans* face. Secretory granules remain

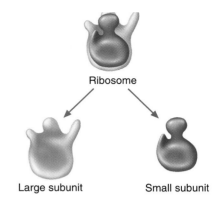

Ribosome

Large subunit   Small subunit

FIGURE 2.18
The Eukaryotic Ribosome.

FIGURE 2.19
The Golgi Apparatus.

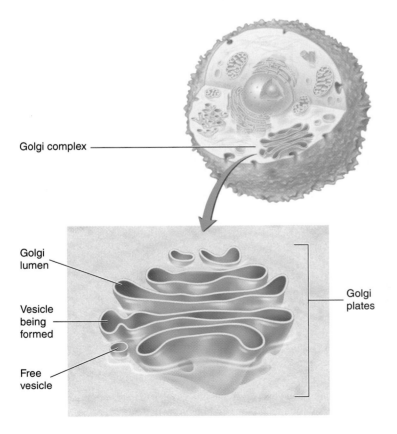

Golgi complex

Golgi lumen

Vesicle being formed

Free vesicle

Golgi plates

KEY CONCEPTS 2.8

Formed from relatively large, flattened, saclike membranous vesicles, the Golgi apparatus is involved in packaging and secretion of cell products.

in storage in the cytoplasm until the cell is stimulated to secrete them. The secretory process, referred to as **exocytosis**, consists of the fusion of the membrane-bound granules with the plasma membrane (Figure 2.20). The contents of the granules are then released into the extracellular space. In plants, the functions of the Golgi apparatus include transport of substances into the cell wall and expansion of the plasma membrane during cell growth.

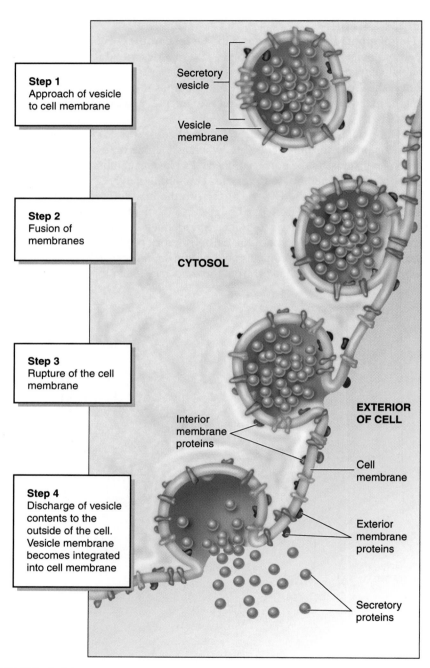

Step 1
Approach of vesicle to cell membrane

Secretory vesicle

Vesicle membrane

Step 2
Fusion of membranes

CYTOSOL

Step 3
Rupture of the cell membrane

EXTERIOR OF CELL

Interior membrane proteins

Cell membrane

Step 4
Discharge of vesicle contents to the outside of the cell. Vesicle membrane becomes integrated into cell membrane

Exterior membrane proteins

Secretory proteins

FIGURE 2.20

Exocytosis

Proteins produced in the ER and processed by the Golgi apparatus are packaged into vesicles that migrate to the plasma membrane and merge with it.

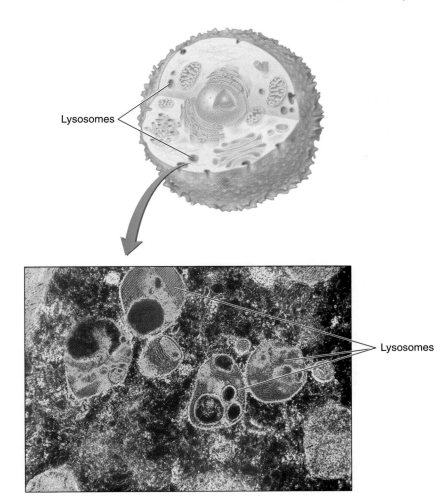

Lysosomes

Lysosomes

FIGURE 2.21

Lysosomes

Lysosomes are membranous sacs that contain hydrolytic enzymes.

Lysosomes

Although the appearance of **lysosomes** differs from one cell type to another, they are typically spherical, saclike organelles with an average diameter of 500 nm (Figure 2.21). Bounded by a single membrane, lysosomes contain granules that are aggregates of digestive enzymes. These proteins are often referred to as *acid hydrolases* because they require an acidic environment to function properly and they use water molecules to split large molecules into fragments. (Because the plant vacuole contains acid hydrolases, it is considered to function to a certain extent like a lysosome. Plant vacuoles are membranous sacs that store a wide variety of substances.)

Lysosomes function in intracellular and extracellular digestion. They are capable of degrading most biomolecules. Lysosomes participate in the life of a cell in three fundamental ways: (1) digestion of food molecules or other substances taken into the cell by **endocytosis** (a process illustrated in Figure 2.22), (2) digestion of worn out or unnecessary cell components, and (3) breakdown of extracellular material.

Two properties of lysosomal membrane are especially interesting. First, certain membrane proteins transport protons across the membrane, thus creating the required acidic environment within the lysosomes. Second, under certain circumstances lysosomal enzymes leak into other parts of the cell. Such an occurrence

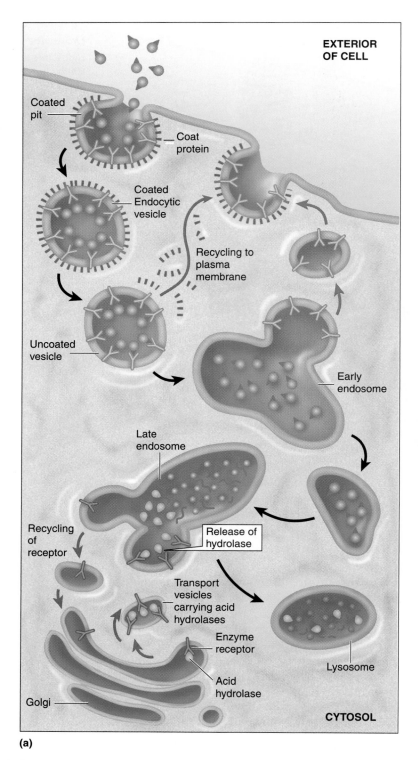

EXTERIOR OF CELL

Coated pit

Coat protein

Coated Endocytic vesicle

Recycling to plasma membrane

Uncoated vesicle

Early endosome

Late endosome

Recycling of receptor

Release of hydrolase

Transport vesicles carrying acid hydrolases

Enzyme receptor

Lysosome

Acid hydrolase

Golgi

CYTOSOL

(a)

FIGURE 2.22

Receptor-Mediated Endocytosis

(a) Extracellular substances may enter the cell during endocytosis, a process in which receptor molecules in the plasma membrane bind to the specific molecules or molecular complexes called ligands. Specialized regions of plasma membrane called coated pits progressively invaginate to form closed vesicles. After the coat proteins are removed, the vesicle fuses with an early endosome, the precursor of lysosomes. The coat proteins are then recycled to the plasma membrane. During endosomal maturation, the proton concentration rises and the ligands are released from their receptors which are subsequently also recycled back to the plasma membrane. As endosomal maturation continues, lysosomal hydrolases are delivered from the Golgi apparatus. Lysosomal formation is complete when all the hydrolases have been transferred to the late endosome and the Golgi membrane has been recycled back to the Golgi apparatus. (b) Electron micrographs illustrating the initial events in endocytosis.

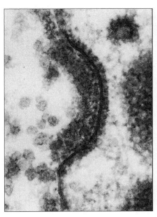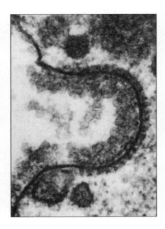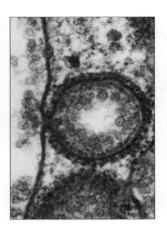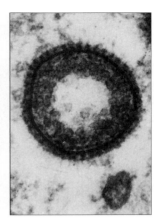

(b)

would ordinarily have devastating consequences, because all of the cell contents would eventually be degraded. In several pathological conditions such as rheumatoid arthritis and gout, there is a release of lysosomal enzymes by macrophages (a white blood cell that plays an important role in inflammatory responses). The release of these enzymes into the affected tissue contributes to further inflammation and tissue destruction.

Although lysosomal function has common features in various tissues, its specific role differs. For example, macrophage lysosomes are prominent components in the normal immunological process by which damaged cells and foreign organisms are degraded. Lysosomal enzymes secreted from osteoclast cells are largely responsible for the resorption phase of bone remodeling.

> **KEY CONCEPTS 2.9**
>
> The function of lysosomes is intracellular and extracellular digestion. These spherical, membranous organelles contain a group of enzymes called acid hydrolases, which degrade most biomolecules.

QUESTION 2.2

In many genetic disorders, a lysosomal enzyme required to degrade a specific molecule is missing or defective. One example of these maladies, often referred to as *lysosomal storage diseases*, is Tay-Sachs disease. Afflicted individuals inherit a defective gene from each parent that codes for an enzyme that degrades a complex lipid molecule. Symptoms include severe mental retardation and death before the age of 5 years. What is the nature of the process that is destroying the patient's cells? (*Hint:* Synthesis of the lipid molecule continues at a normal rate.)

Peroxisomes

Peroxisomes are small spherical membranous organelles that contain oxidative enzymes (proteins that catalyze the transfer of electrons). These organelles, whose enzymatic composition varies among species and cells within individual organisms, are most noted for their involvement in the generation and breakdown of toxic molecules known as *peroxides*. For example, hydrogen peroxide (H_2O_2) is generated when molecular oxygen (O_2) is used to remove hydrogen atoms from specific organic molecules. Once formed, H_2O_2 must be immediately destroyed before it damages the cell. This process is especially important in liver and kidney cells, which have an important detoxifying role in animal bodies. For example, peroxisomes are involved in the oxidation of ingested ethanol.

Two types of peroxisomes have been identified in plants. One, found in leaves, is responsible for an oxygen-consuming process known as *photorespiration* in which carbon dioxide (CO_2) is produced. The other type of peroxisome (often

> **KEY CONCEPTS 2.10**
>
> Peroxisomes contain oxidative enzymes. They are most noted for their involvement in the generation and breakdown of toxic molecules known as peroxides.

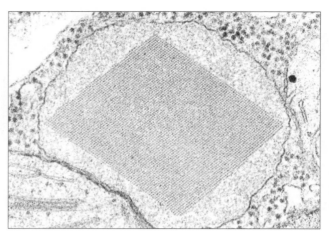

FIGURE 2.23

Peroxisome in a Tobacco Leaf Cell

The granular substance surrounding the crystal-like core is called the matrix.

called **glyoxysomes**) is found in germinating seed. In these structures lipid molecules are converted into carbohydrate, which provides energy for growth and development (Figure 2.23).

Mitochondria

Aerobic metabolism, the mechanism by which the chemical bond energy of food molecules is captured and used to drive the oxygen-dependent synthesis of adenosine triphosphate (ATP), the cell's energy storage molecule, takes place within mitochondria.

Each **mitochondrion** (plural: mitochondria) is bounded by two membranes (Figure 2.24a). The smooth **outer membrane** is relatively porous, because it is permeable to most molecules with masses less than 10,000 D. The **inner membrane,** which is impermeable to ions and a variety of organic molecules, projects inward into folds that are called *cristae* (singular: crista). Embedded in this membrane are structures composed of molecular complexes and called *respiratory assemblies* (described in Chapter 10) that are responsible for the synthesis of ATP. Also present are a series of proteins that are responsible for the transport of specific molecules and ions.

Together, both membranes create two separate compartments: (1) the *intermembrane space* and (2) the *matrix*. The intermembrane space contains several enzymes involved in nucleotide metabolism, whereas the gel-like matrix consists of high concentrations of enzymes and ions and a myriad of small organic molecules. The matrix also contains several circular DNA molecules and all components required for protein synthesis. It should be noted that mitochondria are capable of independent fission and the number of mitochondria per cell will vary with the activity of the cell. (See Special Interest Box 2.1.)

Mitochondria are often depicted as sausage-shaped structures (Figure 2.24b), but their appearance varies considerably among different species and cell types. Their configuration also changes with the physiological status of the cell. For example, the internal appearance of liver mitochondria has been observed to change dramatically during active respiration (Figure 2.25). Additionally, the fragmentation or inordinate swelling of mitochondria is a very sensitive indicator of cell injury.

KEY CONCEPTS 2.11

Aerobic respiration, the process that generates most of the energy required in eukaryotes, takes place in mitochondria.
Embedded in the inner membrane of the mitochondrion are respiratory assemblies, where ATP is synthesized.

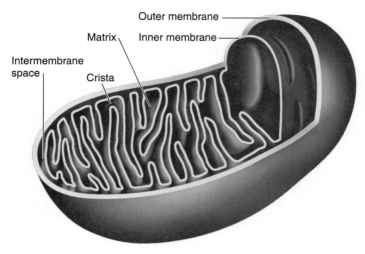

Outer membrane

Inner membrane

Matrix

Intermembrane space

Crista

(a)

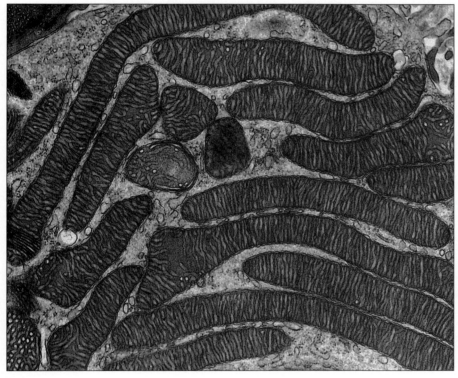

(b)

FIGURE 2.24

The Mitochondrion

(a) Membranes and crista. (b) Mitochondria from adrenal cortex.

Plastids

Plastids, structures that are found only in plants, algae, and some protists, are bounded by a double membrane. Although the inner membrane is not folded as in mitochondria, another separate intricately arranged internal membrane is often

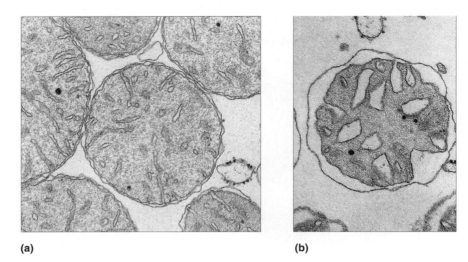

(a)   (b)

FIGURE 2.25

Rat liver Mitochondria in the (a)Low-Energy (Orthodox), and (b) High-Energy (Condensed) Conformations.

present. In plants, all plastids develop from *proplastids*, which are small, nearly colorless structures found in the meristem (a special region in plants made up of undifferentiated cells from which new tissues arise). Proplastids develop according to the requirements of each differentiated cell. Mature plastids are of two types: (1) *leucoplasts,* which store substances such as starch or proteins in storage organs (e.g., roots or tubers), and (2) **chromoplasts**, which accumulate the pigments that are responsible for the colors of leaves, flower petals, and fruits.

Chloroplasts are a type of chromoplast that are specialized for the conversion of light energy into chemical energy. In this process, called **photosynthesis,** which will be described in Chapter 13, light energy is used to drive the synthesis of carbohydrate from CO_2. The structure of chloroplasts (Figure 2.26) is similar in several respects to that of mitochondria. For example, the outer membrane is highly permeable, whereas the relatively impermeable internal membrane contains special carrier proteins that control molecular traffic into and out of the organelle.

An intricately folded internal membrane system, called the **thylakoid membrane**, is responsible for the metabolic function of chloroplasts. For example, chlorophyll molecules, which capture light energy during photosynthesis, are bound to thylakoid membrane proteins. Certain portions of thylakoid membrane form tightly stacked structures called **grana** (singular: granum), whereas the entire membrane encloses a compartment known as the *thylakoid lumen* (or channel). Surrounding the thylakoid membrane is a dense enzyme-filled substance, analogous to the mitochondrial matrix, called the **stroma**. In addition to enzymes, the stroma contains DNA, RNA, and ribosomes. Membrane segments that connect adjacent grana are referred to as *stroma lamellae* (singular: lamella).

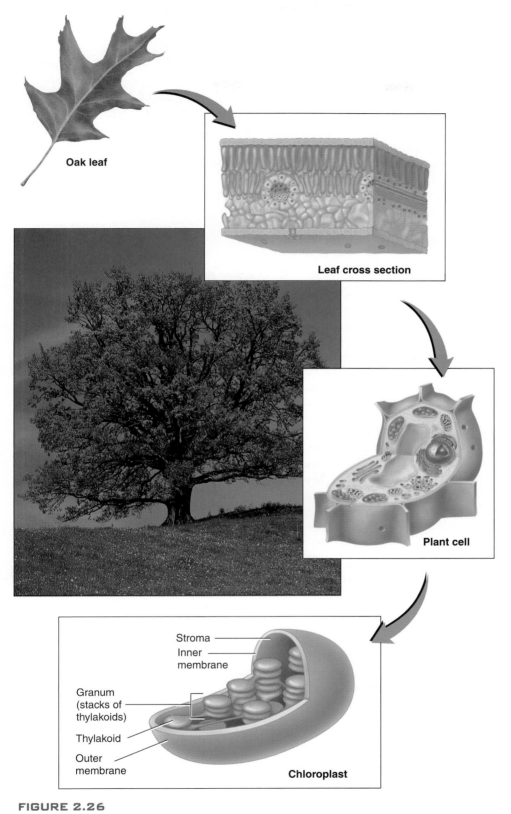

Oak leaf

Leaf cross section

Plant cell

Stroma

Inner membrane

Granum (stacks of thylakoids)

Thylakoid

Outer membrane

Chloroplast

FIGURE 2.26

The Chloroplast

Chloroplasts are one type of organelle found in multicellular plants.

The evolutionary origins of some eukaryotic organelles have long been of interest to life scientists. In her book *The Origin of Eukaryotic Cells*, published in 1970, Lynn Margulis uses the concept of symbiosis to explain this major unresolved biological problem. **Symbiosis**, defined as the living together of two dissimilar organisms in an intimate relationship, is a common biological phenomenon. Such associations vary from the parasitic, in which one organism derives benefit at the other's expense, to the mutualistic, in which both organisms benefit. One example of the former involves humans and trypanosomes, the protozoa that cause African sleeping sickness. The relationship between humans and the intestinal bacterium *Lactobacillus acidophilus* is an example of the latter. A variety of *Lactobacillus* species, obtained from the consumption of fermented milk products, exchange protection, warmth, and nutrition for several beneficial effects to humans. These include protection against pathogenic microorganisms such as *Clostridium difficile*, the lowering of serum cholesterol, and perhaps some anticancer effects.

Margulis proposed that the mitochondria and chloroplasts, as well as cilia and flagella of eukaryotic cells, evolved from prokaryotic cells. According to the *endosymbiotic hypothesis*, eukaryotic cells began as large anaerobic organisms. (The term *anaerobic* indicates that oxygen is not used to generate energy.) Mitochondria arose when small aerobic (oxygen-using) bacteria were ingested by the larger cells. In exchange for benefits such as protection and a constant nutrient supply, the smaller cell provided its host with energy generated by a process known as *aerobic respiration*. As time passed, the bacteria lost their independence because of the transfer of several genes (genetic coding units) to the host cell nucleus. Similarly, chloroplasts are believed to descend from cells that were similar to modern cyanobacteria, whereas cilia and flagella derive from ancient spiral prokaryotes:

The endosymbiotic hypothesis is supported by a considerable amount of indirect evidence.

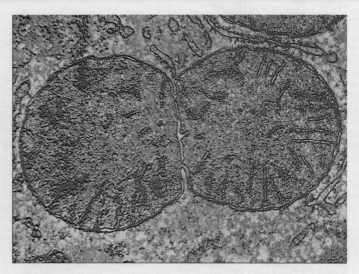

FIGURE 2A

Replication of a Mitochondrion by Binary Fission

1. Mitochondria and chloroplasts are similar in size to many modern prokaryotes.
2. These two organelles reproduce by binary fission, as do bacteria and the archaea (Figure 2A).
3. The genetic information (DNA) and the protein-synthesizing capability of mitochondria and chloroplasts are similar to those of prokaryotes. For example, both mitochondrial and chloroplast DNA are circular and "naked" (i.e., not complexed with histone proteins as nuclear DNA is). (There is insufficient genetic information on these chromosomes to account for all organelle components. However, the nuclear genes that are responsible for synthesis of mitochondrial components resemble prokaryotic genes.)

QUESTION 2.3

Cyanophora paradoxa is a eukaryotic organism that incorporates cyanobacteria, aerobic photosynthesizing prokaryotic organisms, into its cells. Describe the benefits that both species derive from this relationship.

Cytoskeleton

Cytoplasm was once believed to be a structureless solution in which the nucleus was suspended. Experiments have revealed not only the extensive membrane system and the membranous organelles just described, but also an intricate supportive network of proteinaceous fibers and filaments called the **cytoskeleton** (Figure 2.27). Components of the cytoskeleton include **microtubules**, **microfilaments**, and **intermediate fibers**.

Microtubules (diameter = 25 nm), composed of the protein tubulin, are the largest constituent of the cytoskeleton. Although they are found in many cellu-

4. The ribosomes of chloroplasts and mitochondria are similar in size and function to those of prokaryotes. For example, drugs such as the antibiotic chloramphenicol, which kill certain bacteria by inhibiting the protein-synthesizing activities of ribosomes, also inhibit chloroplast and mitochondrial ribosomal function.

5. Traces of the other nucleic acid, RNA, have been found in the basal bodies of cilia and flagella. This evidence is considered by some researchers to support the idea that such eukaryotic structures arose by a symbiotic union.

6. Many modern organisms contain intracellular symbiotic bacteria, cyanobacteria, or algae; that is, such associations are not difficult to establish. For example, a primitive freshwater animal called *Chlorohydra* owes its green color to endosymbiont algae. The hydra's nutrition is supplemented by the photosynthetic activity of the algae.

The endosymbiotic hypothesis is illustrated in Figure 2B.

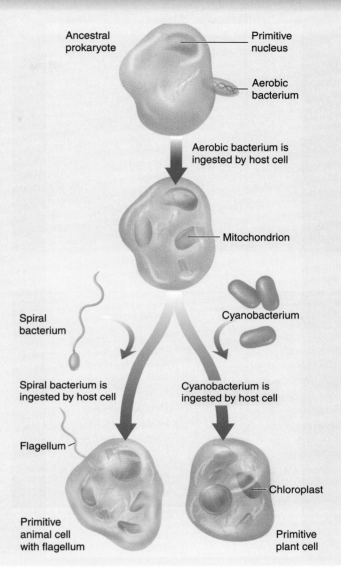

FIGURE 2B

The Endosymbiotic Hypothesis.
The first eukaryotes are believed to have evolved at least 1.5 billion years ago. The transition from ancient prokaryotic to eukaryotic cell structure is arguably the most important one in evolution, except for the origin of life itself. The endosymbiotic hypothesis is an interesting and compelling view of this transition.

lar regions, microtubules are most prominent in long, thin structures that require support (e.g., the extended axons and dendrites of nerve cells). They are also found in the *mitotic spindle* (the structure formed in dividing cells that is responsible for the equal dispersal of chromosomes into daughter cells) and the slender, hair-like organelles of locomotion known as cilia and flagella (Figure 2.28).

Microfilaments, which are small (5–7 nm in diameter) fibers composed of the protein *actin*, perform their functions by interacting with certain cross-linking proteins. Important roles of microfilaments include involvement in cytoplasmic streaming (a process that is most easily observed in plant cells in which cytoplasmic currents rapidly displace organelles such as chloroplasts) and ameboid movement (a type of locomotion created by the formation of temporary cytoplasmic protrusions).

Intermediate fibers (10 nm in diameter) are proteinaceous structures with a heterogeneous composition. Despite this variation, their structural organization is similar in many cell types. Involved primarily in the maintenance of cell shape,

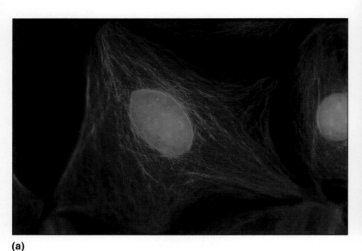

(a)

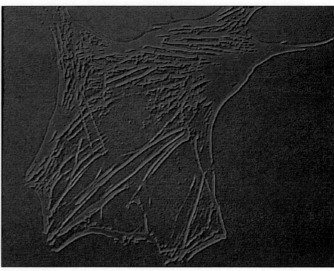

(b)

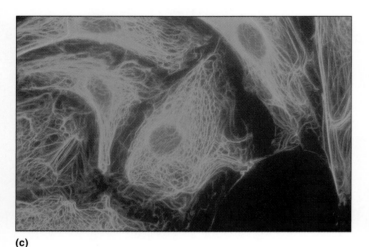

(c)

FIGURE 2.27

The Cytoskeleton.

The major components of the cytoskeleton are microtubles (a) microfilaments (b), and intermediate filaments (c). The intracellular distribution of each type of cytoskeletal component is visualized by staining with fluorescent dyes.

intermediate fibers are especially prominent in cells that are subjected to mechanical stress. For example, one type, known as *keratin filaments*, is found in the outermost cell layers of the skin.

This highly developed framework of the cytoskeleton contributes to living processes in several ways:

1. **Maintenance of overall cell shape**. Eukaryotic cells come in a vast variety of shapes including the bloblike amoeba, columnar epithelial cells, and neurons with complex branching architecture.

2. **Facilitation of coherent cellular movement**. Large-scale cellular movement such as the cytoplasmic streaming that occurs in plant cells and the ameboid movement seen in some animal cells is made possible by a dynamic cytoskeleton that can rapidly assemble and disassemble its structural elements according to the cell's immediate needs.

3. **Provision of a supporting structure that guides the movements of organelles within the cell**. Organelles are moved around within cells by

KEY CONCEPTS 2.12

The cytoskeleton, a highly structured network of proteinaceous filaments, is responsible for maintenance of overall cell shape and facilitation of cell movements. Components of the cytoskeleton include microtubules, microfilaments, and intermediate fibers.

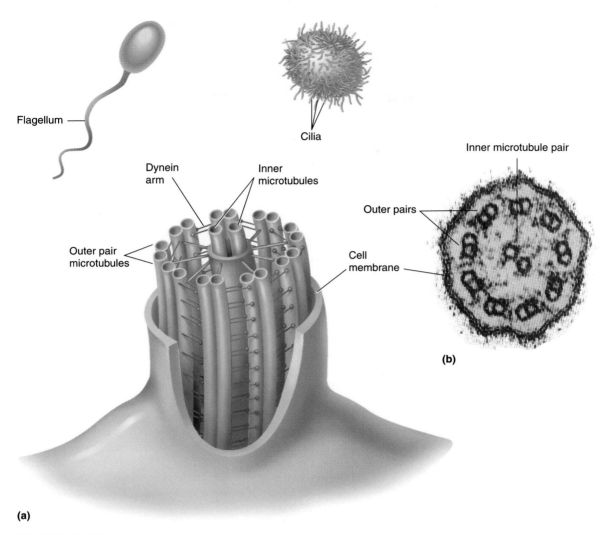

FIGURE 2.28

Cilia and Flagella

(a)The microtubules of eukaryotic cells are arranged in the classic 9+2 pattern. Two central microtubules are surrounded by an outer ring of nine pairs of microtubules. (b) A transmission electron micrograph of a cross section of a flagellum.

being attached to cytoskeletal structures. For example, after cell division, the extension of the endoplasmic reticulum membrane from the newly formed nuclear membrane out to the cell's periphery and the re-formation of the Golgi complex is accomplished by attachment to microtubules. Movement occurs as specific motor proteins linked to microtubules and to the membrane cargo undergo ATP hydrolysis–dependent conformational changes.

Without referring to this chapter, can you draw diagrams of a prokaryotic and a eukaryotic cell and label each structural component? Describe the function of each structure in one sentence.

QUESTION 2.4

The Earth was formed from a cloud of condensing cosmic dust and gas about 4.5 billion years ago. Life arose soon thereafter. Fossil evidence in the form of stromatolites (compressed layers of bacterial remains) indicates that the earliest organisms existed at least 3.6 billion years ago. Although analyses of geological, biological, and chemical evidence have allowed scientists to slowly piece together an overview of the history of life, the precise mechanisms by which life originated are still the subject of considerable speculation. Two basic strategies in origin of life studies have been used: the "top-down" and the "bottom-up" approaches. In the top-down approach, the phylogenetic (evolutionary) history of modern organisms has been traced back through time with phylogenetic analysis, that is, by investigating the similarities and differences among organisms that are clues to their evolutionary past. In the bottom-up approach, researchers have focused on reconstructing the mechanisms by which inanimate matter was transformed on the early Earth into the first primitive living organisms, a process called **abiogenesis**. Investigators of abiogenesis are especially concerned with determining the physical and chemical conditions that facilitated the origin of life. They also analyze biomolecules in modern species that are considered to be vestigial remnants of the prebiotic world. Both strategies have yielded valuable information. Phylogenetic analysis is described in Special Interest Box 17.2. Current views of abiogenesis are outlined below.

Abiogenesis

Currently, one of the most perplexing problems in origin of life studies is to provide a plausible explanation for abiogenesis: How did the first living organism arise from inanimate matter under the physical and chemical conditions that existed during the Earth's early history? In other words, during this primordial period, how did lifeless molecules transform into primitive, yet information-rich, self-organizing life forms? Because the events by which life arose cannot be reproduced in the laboratory, abiogenesis mechanisms are still speculative. Among the most essential issues that must be addressed in abiogenesis investigations are:

1. How were simple organic molecules (e.g., sugars, amino acids, and nucleotides) originally formed?
2. How did these primordial organic molecules link up to form information-rich macromolecules such as proteins and nucleic acids?
3. How did the first cells originate?

There are, as yet, no convincing answers to these questions. As scientists have investigated the enigma of life's origin, several intriguing hypotheses have been developed. The assumptions upon which these explanations of abiogenesis are based include the following:

1. The first life forms were very simple in both structural and functional capabilities when compared with modern organisms.
2. The basic requirement of any life form is the presence of one or more molecules that are able to duplicate themselves using the raw materials available in their environment.

The essential features of most modern views of abiogenesis (Figure 2C) include an early phase in which organic molecules were formed. During a later phase, it is presumed that primitive cell-like structures, called *protocells*, enclosed by lipid precursor molecules possessed a richer diversity of organic molecules. Within the confines of protocells, certain monomer molecules polymerized to form polypeptides and nucleic acids.

Scientific inquiries into the mechanisms of abiogenesis began with Charles Darwin, who suggested that life might have arisen in a "warm little pond," which he speculated contained ammonia, phosphate, and other molecules. Over a long time, in the presence of energy sources such as light and lightning, these molecules would have given rise eventually to the first living organism. In the early twentieth century, J. B. S. Haldane (1892–1964) introduced the term *primordial soup*. He and other scientists speculated that life arose in hot ocean water into which had washed diverse types of inorganic and organic molecules formed in volcanic eruptions and asteroids newly arrived from outer space. In 1924 Alexandr Oparin (1894–1980) proposed that the atmosphere on early Earth was significantly different from today's. In his view the Earth's primitive atmosphere consisted largely of hydrogen, methane, ammonia, and water vapor, but with no oxygen. In other words, this early atmosphere was a reducing atmosphere. In today's oxidizing atmosphere with its relatively high oxygen content, organic molecules do not spontaneously link to form polymers. Instead the opposite happens; that is, such molecules degrade to form inorganic molecules. Oparin also focused his attention on the formation of the first cells. He observed that when hydrophobic molecules and water are mixed, vesicles form that trap other molecules inside. For protocells to have been formed in this manner, the vesicular "membrane" must have been able to prevent vital molecules from escaping, while allowing raw material to enter.

It was not until several decades later that Harold Urey (1893–1981), an American chemist, realized that some of the speculations of Oparin and Haldane could be tested under laboratory conditions. In 1953 Urey and his graduate student Stanley Miller (1930–) subjected a mixture of ammonia, methane, hydrogen gas, and water to conditions that were presumed to simulate those on earth 4 billion years ago. The mixture was placed in a flask in a sealed apparatus and then heated and stirred for several days. Energy was supplied in the form of electric sparks. Analysis of the resulting tarry residue revealed the presence of the amino acids alanine and glycine among lesser amounts of several other organic molecules. Considered a remarkable breakthrough at the time, the Urey-Miller experiment has been criticized recently for several reasons. First, it now appears that the Earth's early atmosphere contained primarily carbon dioxide, nitrogen, and very small amounts of hydrogen, but little methane. Without methane the Urey-Miller experiment produces only glycine. An equally serious problem is the failure of this methodology to produce nucleotides, the building block molecules of the nucleic acids.

In recent years several scientists have provided insights into possible mechanisms concerning the formation of macromolecules. The principal issue in this phase of life's origins concerns the interdependence of DNA and proteins in modern organisms. In other words, what came first, DNA molecules that contain the information that codes for proteins, or proteins, several of which are required to duplicate and transcribe DNA? The most prominent hypothesis that addresses this paradox is the *RNA world concept*. According to this view, neither DNA or proteins were the first information-carrying macromolecules in primitive life. Instead, RNA was the first informational molecule. This concept is based on evidence concerning the functional properties of RNA. RNA not only possesses genetic information; it also can behave as an enzyme. For example, recent experimental and structural evidence have proven conclusively that the formation of peptide bonds during protein synthesis is catalyzed by

an RNA component of ribosomes. In addition, under certain circumstances in living cells, DNA molecules can be synthesized from an RNA molecule in a process that involves an enzyme called *reverse transcriptase*. In a hypothetical scenario, short RNA segments may have originally encoded short peptides. Eventually as protocells became more complex and a more stable form of genetic information provided a selective advantage, a reverse transcriptase started copying RNA sequences into DNA. Eventually this process resulted in the roles of the major informational macromolecules in all modern organisms: DNA, the genetic blueprint; proteins, the devices that perform the tasks of living processes; and RNA, the carrier of information used to manufacture protein.

FIGURE 2C

A Hypothetical Scenario of Abiogenesis.

During the likely stages in the origin of life, energy in the form of light, lightning and heat promoted the formation of organic molecules from inorganic precursors. Later certain molecules polymerized to form polypeptides and the nucleic acids DNA and RNA. Once these macromolecules became enclosed within a membrane-like barrier, their evolution proceded over time.

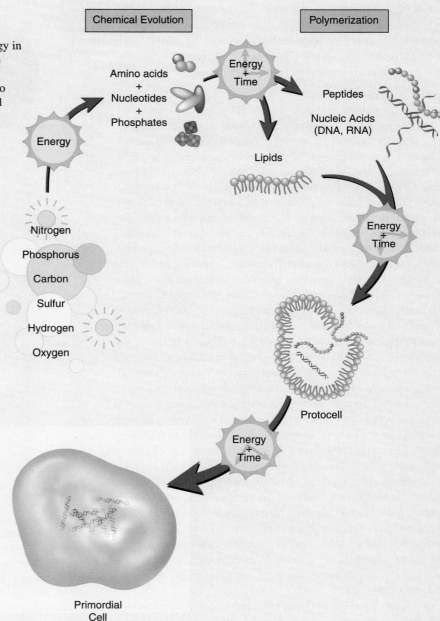

During the past 50 years, our understanding of the functioning of living organisms has undergone a revolution. Much of our current knowledge of biochemical processes is due directly to technological innovations. For example, the development of the electron microscope (EM) as a biological instrument by Keith Porter and his colleagues in the 1940s was responsible for the resolution of the fine structure of organelles. Structures such as the mitochondria and lysosomes were not discovered until this time; under the light microscope they appeared as mere granules. The electron microscope also revealed that the membranes of the Golgi apparatus are often continuous with those of the ER. This discovery is especially significant because of the role both organelles play in protein synthesis. In this box three of the most important cellular techniques used in biochemical research are briefly described: cell fractionation, electron microscopy, and autoradiography.

Cell Fractionation

Cell fractionation techniques (Figure 2D) allow the study of cell organelles in a relatively intact form outside of cells. For example, functioning mitochondria can be used to study cellular energy generation. In these techniques, cells are gently disrupted and separated into several organelle-containing fractions. Cells may be disrupted by several methods, but homogenization is the most commonly used. In this process a cell suspension is placed either in a glass tube fitted with a specially designed glass pestle, or into an electric blender. The resulting homogenate is then separated into several fractions during a procedure called **differential centrifugation**. A refrigerated instrument called the *ultracentrifuge* generates enormous centrifugal forces that separate cell components on the basis of size, surface area, and relative density. (Forces as large as 500,000 times the force of gravity, or 500,000 g, can be generated in unbreakable test tubes placed in the rotor of an ultracentrifuge.) Initially, the homogenate is spun in the ultracentrifuge at low speed (700–1,000 g) for 10 to 20 minutes. The heavier particles, such as the nuclei, form a sediment, or pellet. Lighter particles, such as mitochondria and lysosomes, remain suspended in the *supernatant*, the liquid above the pellet. The supernatant is then transferred to another centrifuge tube and spun at a higher speed (15,000 to 20,000 g) for 10–20 minutes. The resulting pellet contains mitochondria, lysosomes, and peroxisomes. The supernatant, which contains **microsomes** (small closed vesicles formed from ER during homogenization), is transferred to another tube and spun at 100,000 g for 60–120 minutes. Microsomes are deposited in the pellet, and the supernatant contains ribosomes, various cellular membranes, and granules such as glycogen, a carbohydrate polymer. After this latest supernatant is recentrifuged at 200,000 g for 2 to 3 hours, ribosomes and large macromolecules are recovered from the pellet.

Often, the organelle fractions obtained with this technique are not sufficiently pure for research purposes. One method that is often

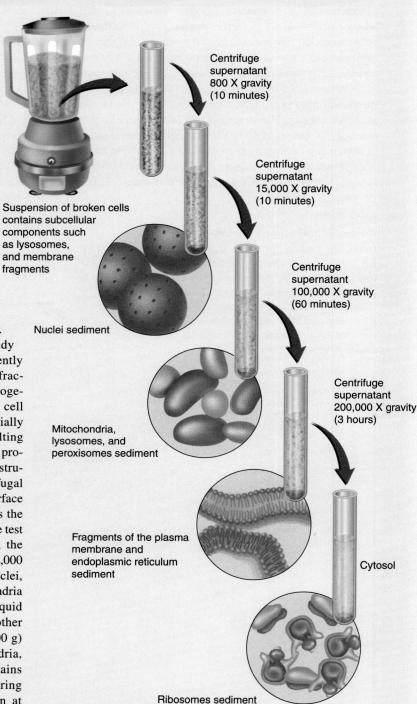

Suspension of broken cells contains subcellular components such as lysosomes, and membrane fragments

Centrifuge supernatant 800 X gravity (10 minutes)

Centrifuge supernatant 15,000 X gravity (10 minutes)

Centrifuge supernatant 100,000 X gravity (60 minutes)

Centrifuge supernatant 200,000 X gravity (3 hours)

Nuclei sediment

Mitochondria, lysosomes, and peroxisomes sediment

Fragments of the plasma membrane and endoplasmic reticulum sediment

Cytosol

Ribosomes sediment

FIGURE 2D

Cell Fractionation.

After the homogenization of cells in a blender, cell components are separated in a series of centrifugations at increasing speeds. As each centrifugation ends, the supernatant is removed, placed into a new centrifuge tube and then subjected to greater centrifugal force. The collected pellet can be resuspended in liquid and then examined by microscopy or biochemical tests.

employed to further purify cell fractions is **density-gradient centrifugation** (Figure 2E). In this procedure the fraction of interest is layered on top of a centrifuge tube containing a solution that consists of a dense substance such as sucrose. (In such a tube the concentration of the sucrose increases from the top to the bottom of the tube.) During centrifugation at high speed for several hours, particles move downward in the gradient until they reach a level that has a density equal to their own. Cell components are then collected by puncturing the plastic centrifuge tube and collecting drops from the bottom. The purity of the individual fractions can be assessed by visual inspection using the electron microscope. However, assays for **marker enzymes** (enzymes that are known to be present in especially high concentration in specific organelles) are more commonly used. For example, glucose-6-phosphatase, the enzyme responsible for converting glucose-6-phosphate to glucose in the liver, is a marker for liver microsomes. Likewise, DNA polymerase, which is involved in DNA synthesis, is a marker for nuclei.

Electron Microscopy

The electron microscope (EM) permits a view of cell ultrastructure not possible with the more common light microscope. Direct magnifications as high as 1,000,000× have been obtained with the EM. Electron micrographs may be enlarged photographically to 10,000,000×. The light microscope, in contrast, magnifies an image to about 1,000×. This difference is due to the greater resolving power of the EM. The **limit of resolution**, defined as the minimum distance between two points that allows for their discrimination as two separate points, is 0.2 μm using the light microscope. The limit of resolution for the EM is approximately 0.5 nm. The lower resolving power of the light microscope is related to the wavelength of visible light. In general, shorter wavelengths allow greater resolution. The EM uses a stream of electrons instead of light to illuminate specimens. Because this electron stream has a much shorter wavelength than visible light, more detailed images can be obtained.

There are two types of EM: the transmission electron microscope (TEM) and the scanning electron microscope (SEM). Like the light microscope, the TEM is used for viewing thin specimens. Because the image in the TEM depends on variations in the absorption of electrons by the specimen (rather than on variations in light absorption), heavy metals such as osmium or uranium are used to increase contrast among cell components. The SEM is used to obtain three-dimensional views of cellular structure. Unlike the TEM, which uses electrons that have passed through a specimen to form an image, the SEM uses electrons that are emitted from the specimen's surface. The specimen is coated with a thin layer of heavy metal and then scanned with a narrow stream of electrons. The electrons emitted from the specimen's surface, sometimes referred to as *secondary electrons*, form an image on a television screen. Although only surface features can be examined with the SEM, this form of microscopy provides very useful information about cell structure and function.

Autoradiography

Autoradiography is used to study the intracellular location and behavior of cellular components. It has been an invaluable tool in biochemistry. For example, it was used to determine the precise sites of DNA, RNA, and protein synthesis within eukaryotic cells. In this procedure, living cells are briefly exposed to radioactively labeled precursor molecules. Tritium (^{3}H) is the most commonly used isotope. The tritiated nucleotide thymidine, for example, is used to study DNA synthesis, because thymidine is incorporated only into DNA molecules. After exposure to the radioactive precursor, the cells are processed for light or electron microscopy. The resulting slides are then dipped in photographic emulsion. After storage in the dark, the emulsion is developed by standard photographic techniques. The location of radioactively labeled molecules is indicated by the developed pattern of silver grains.

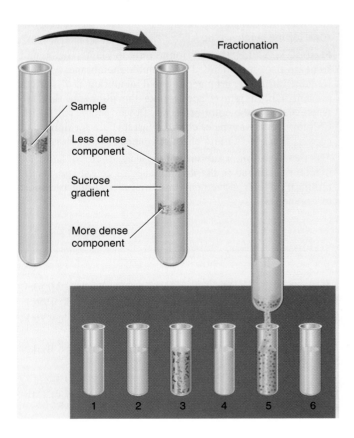

FIGURE 2E

Density Gradient Centrifugation.
The sample is gently layered onto the top of a preformed gradient of an inert substance such as sucrose. As centrifugal force is applied, particles in the sample migrate through the gradient bands according to their densities. After centrifugation, the bottom of the tube is punctured and the individual bands are collected in separate tubes.

SUMMARY

1. Cells are the structural units of all living organisms. Within each living cell are hundreds of millions of densely packed biomolecules. Insights into their structure and functional behavior have been gained through biochemical research. Among the most significant of these are the following: The unique chemical and physical properties of water, the biological solvent, are a crucial determining factor in the behavior of all other biomolecules. Biological membranes are thin, flexible, and relatively stable sheetlike structures that enclose cells and organelles. They are formed from biomolecules such as phospholipids and proteins that together form a selective physical barrier. Membranes also serve as a chemically reactive surface that is integrated into all living processes within the organism. Self-assembly of supramolecular structures occurs within living cells because of the steric information encoded into the intricate shapes of biomolecules that allows numerous weak, noncovalent interactions between complementary surfaces. Many of the multisubunit complexes involved in cellular processes are now known to function as molecular machines; that is, they are mechanical devices composed of moving parts that convert energy into directed motion.

2. There are two types of cells found in all currently existing organisms: prokaryotic and eukaryotic. Prokaryotes are simpler in structure than eukaryotes. They also have a vast biochemical diversity across species lines, because almost any organic molecule can be used as a food source by some species of prokaryote. Unlike the prokaryotes, the eukaryotes carry out their metabolic functions in membrane-bound compartments called organelles.

3. Although the prokaryotic cell lacks a nucleus, a circular DNA molecule called a chromosome is located in an irregularly shaped region called the nucleoid. Many bacteria contain additional small circular DNA molecules called plasmids. Plasmids may carry genes for special function proteins that provide protection, metabolic specialization, or reproductive advantages to the organism. The nucleus of eukaryotes contains DNA, the cell's genetic information. Ribosomal RNA is synthesized in the nucleolus, found within the nucleus.

4. The plasma membrane of both prokaryotes and eukaryotes performs several vital functions. The most important of these is controlled molecular transport, which is facilitated by carrier and channel proteins.

5. The endoplasmic reticulum (ER) is a system of interconnected membranous tubules, vesicles, and large flattened sacs found in eukaryotic cells. There are two forms of ER. The rough ER, which is primarily involved in protein synthesis, is so named because of the numerous ribosomes that stud its cytoplasmic surface. The second form lacks attached ribosomes and is called smooth ER. Functions of the smooth ER include lipid synthesis and biotransformation.

6. The cytoplasmic ribosomes of eukaryotes are relatively small organelles that synthesize proteins. Ribosomes are complex structures composed of a variety of proteins and a type of RNA called ribosomal RNA.

7. Formed from relatively large, flattened, saclike membranous vesicles that resemble a stack of plates, the Golgi apparatus is involved in the packaging and secretion of cell products.

8. Lysosomes are saclike organelles that function in intracellular and extracellular digestion. They contain digestive enzymes that can degrade most biomolecules.

9. Peroxisomes are small spherical membranous organelles that contain a variety of oxidative enzymes. These organelles are most noted for their involvement in the generation and breakdown of peroxides.

10. Aerobic respiration, a process by which cells use O_2 to generate energy, takes place in mitochondria. Each mitochondrion is bounded by two membranes. The smooth outer membrane is permeable to most molecules with masses less than 10,000 D. The inner membrane, which is impermeable to ions and a variety of organic molecules, projects inward into folds that are called cristae. Embedded in this membrane are structures called respiratory assemblies that are responsible for the synthesis of ATP.

11. Plastids, structures that are found only in plants, algae, and some protists, are bounded by a double membrane. Another separate intricately arranged internal membrane is also often present. Chromoplasts accumulate the pigments that are responsible for the color of leaves, flower petals, and fruits. Chloroplasts are a type of chromoplast that are specialized to convert light energy into chemical energy.

12. The cytoskeleton, a supportive network of fibers and filaments, is involved in the maintenance of cell shape, facilitation of cellular movement, and the intracellular transport of organelles.

SUGGESTED READINGS

Alberts, B., Bray, D., Johnson, A., Lewis, J., Raff, M., Roberts, K., and Walter, P., *Essential Cell Biology: An Introduction to the Molecular Biology of the Cell*, Garland, New York, 1998.

Becker, W. M., Kleinsmith, L. J., and Hardin, J., *The World of the Cell*, 4th ed., Benjamin Cummings, New York, 2000.

Davis, P. *The Fifth Miracle: The Search for the Origin and Meaning of Life*, Simon and Schuster, New York, 1999.

deDuve, C., The Birth of Complex Cells, *Scientific American* 274(4):50–57, 1996.

Goodsell, D. S., *Our Molecular Nature: The Body's Motors, Machines and Messages*, Springer-Verlag, New York, 1996.

Goodsell, D. S. *The Machinery of Life*, Springer-Verlag, New York, 1998.

Margulis, L., *What is Life?* University of California Press, Berkeley, 2000.

Margulis, L., and Sagan, D., *Microcosmos: Four Billion Years of Evolution from Our Microbial Ancestors*, University of California Press, Berkeley, 1997.

KEY WORDS

abiogenesis, *58*

aerobic metabolism, *50*

biotransformation, *44*

cell fractionation, *60*

chloroplast, *52*

chromatin fiber, *42*

chromoplast, *52*

chromosome, *37*

cytoskeleton, *54*

density-gradient centrifugation, *60*

dictyosome, *45*

differential centrifugation, *60*

endocytosis, *47*

endoplasmic reticulum (ER), *44*

exocytosis, *46*

extracellular matrix, *42*

glycocalyx, *40*

glyoxysome, *50*

Golgi apparatus (Golgi complex), *45*

granum, *52*

hydrophilic, *30*

hydrophobic, *31*

inner membrane, *50*

intermediate fiber, *54*

limit of resolution, *61*

lysosome, *47*

marker enzyme, *61*

microfilament, *54*

microsome, *60*

microtubule, *54*

mitochondrion, *50*

motor protein, *33*

nuclear envelope, *42*

nuclear pore, *42*

nucleoid, *37*

nucleolus, *43*

nucleoplasm, *42*

nucleus, *42*

outer membrane, *50*

peroxisome, *49*

photosynthesis, *36, 52*

plasma membrane, *36*

plasmid, *37*

plastid, *51*

receptor, *41*

respiration, *36*

ribosome, *44*

rough ER (RER), *44*

smooth ER (SER), *44*

stroma, *52*

symbiosis, *54*

thylakoid membrane, *52*

REVIEW QUESTIONS

1. Define the term *cell*.
2. What evidence is there that all cells have a common ancestor?
3. Draw a diagram of a bacterial cell. Label and explain the function of each of the following components:
   a. nucleoid
   b. plasmid
   c. cell wall
   d. pili
   e. flagella
4. The outer boundary of most eukaryotic cells is a cell membrane, whereas the outer boundary of a prokaryotic cell is a cell wall. How do these structures differ in function?
5. Indicate whether the following structures are present in prokaryotic or eukaryotic cells:
   a. nucleus
   b. plasma membrane
   c. endoplasmic reticulum
   d. mitochondria
   e. nucleolus
6. Briefly define the following terms:
   a. exocytosis
   b. biotransformation
   c. grana

   d. symbiosis
   e. self-assembly
   f. hydrophobic
   g. hydrophilic
   h. motor protein
   i. endosymbiosis
   j. proplastids
   k. thylakoid
7. How do lysosomes participate in the life of a cell?
8. Plastids, structures found only in _____, are of two types. These are _____, which are used to store starch and protein, and _____, which accumulate pigments.
9. List six pieces of evidence that support the endosymbiotic hypothesis.
10. What functions does the cytoskeleton perform in living cells?
11. Many eukaryotic cells lack a cell wall. Suggest several reasons why this is an advantage.
12. What are the two essential functions of the nucleus?
13. What roles do plasma membrane proteins play in cells?
14. Name the two forms of endoplasmic reticulum. What functions do they serve in the cell?
15. Describe the functions of the Golgi apparatus.

THOUGHT QUESTIONS

1. Several pathogenic bacteria (e.g., *Bacillus anthracis*, the cause of anthrax) produce an outermost mucoid layer called a capsule. Capsules may be composed of polysaccharide or protein. What effect do you think this "coat" would have on a bacterium's interactions with an animal's immune system?
2. Eukaryotic cells are more highly specialized than prokaryotic cells. Can you suggest some advantages and disadvantages of specialization?

3. The endosymbiotic hypothesis proposes that mitochondria and chloroplasts are derived from aerobic bacteria. Is there any structural feature of these organelles that precludes their having been developed by eukaryotic cells?
4. In addition to providing support, the cytoskeleton also immobilizes enzymes and organelles in the cytoplasm. What advantage does this immobilization have over allowing the cell contents to freely diffuse in the cytoplasm?

5. A particular organelle found in a eukaryote is thought to have arisen from a free-living organism. The finding of what type of molecule in the organelle would strongly support this hypothesis?

6. Mycoplasmas are unusual bacteria that lack cell walls. With a diameter of 0.3 μm, they are believed to be the smallest known free-living organisms. Some species are pathogenic to humans. For example, *Mycoplasma pneumoniae* causes a very serious form of pneumonia. Assuming that mycoplasmas are spherical, calculate the volume of an individual cell. Compare the volume of a mycoplasma with that of *Escherichia coli*.

7. The dimensions of prokaryotic ribosomes are approximately 14 nm by 20 nm. If ribosomes occupy 20% of the volume of a bacterial cell, calculate how many ribosomes are in a typical cell such as *E. coli*. Assume that the shape of a ribosome is approximately that of a cylinder.

Water: The Medium of Life

The Water Planet. Unique among the planets in the solar system, the earth is an oceanic world. Water's properties make life on earth possible.

The most important factor in the evolution of life on earth is the abundant liquid water that occurs on the planet's surface. The distinctive chemical and physical characteristics of water are so crucial for living systems that life undoubtedly could not have arisen in its absence. Among these characteristics are water's chemical stability and its remarkable solvent properties. Wherever there is water, there are living organisms. Although large bodies of water such as lakes and oceans support diverse and often abundant populations of organisms, certain specially adapted organisms also occur where water is present but in short supply. In deserts, for example, brief and infrequent rainstorms immediately trigger the blooming of certain plants. These plants then race through their entire life cycle before the rainwater has completely evaporated. Even when other conditions are unfavorable, (e.g., extremely hot or cold climates), water makes life possible. Certain prokaryotic species thrive in hot springs;

others flourish in the freezing waters of the Arctic and the Antarctic oceans. Similarly, algae grow on the melting edges of glaciers. Why is water so vital for life? Understanding the critical role of water in living processes requires a review of its molecular structure and the physical and chemical properties that are the consequences of that structure.

Earth is unique among the planets in our solar system primarily because of its vast oceans of water. Formed over billions of years, water was produced during high-temperature interactions between atmospheric hydrocarbons and the silicate and iron oxides in the Earth's mantle. Moisture reached the planet's surface as steam emitted during volcanic eruptions. Oceans formed as the steam condensed and fell back to Earth as rain. This first rain may have lasted more than 60,000 years.

Over millions of years, water has profoundly affected our planet. Whether falling as rain or flowing in rivers, water has eroded the hardest rocks and transformed the mountains and continents.

Life is now believed by many scientists to have arisen in the ancient seas. It is not an accident that life arose in association with water, because this substance has several unusual properties that suit it to be the medium of life. Among these are its thermal properties and unusual solvent characteristics. Water's properties are directly related to its molecular structure.

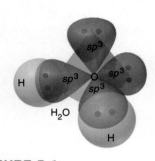

FIGURE 3.1

Tetrahedral Structure of Water.

In water, two of the four sp^3 orbitals of oxygen are occupied by two lone pairs of electrons. Each of the other two half-filled sp^3 orbitals is filled by the addition of an electron from hydrogen.

3.1 MOLECULAR STRUCTURE OF WATER

The water molecule (H_2O) is composed of two atoms of hydrogen and one of oxygen. Water has a tetrahedral geometry because its oxygen atom is sp^3 hybridized. At the center of the tetrahedron is the oxygen atom. Two of the corners are occupied by hydrogen atoms, each of which is linked to the oxygen atom by a single covalent bond (Figure 3.1). The other two corners are occupied by the unshared electron pairs of the oxygen. Oxygen is more electronegative than hydrogen (i.e., oxygen has a greater capacity to attract electrons when bonded to hydrogen). Consequently, the larger oxygen atom bears a partial negative charge (δ^-) and each of the two hydrogen atoms bears a partial positive charge (δ^+) (Figure 3.2). The electron distribution in oxygen-hydrogen bonds is displaced toward the oxygen and, therefore, the bond is **polar**. If water molecules were linear, then the bond polarities would balance each other and water would be nonpolar. However, water molecules are bent; the bond angle is 104.5° (Figure 3.3). In contrast, carbon dioxide (O=C=O), which also possesses polar covalent bonds, is nonpolar because the molecule is linear.

Molecules such as water, in which charge is separated, are called **dipoles**. When molecular dipoles are subjected to an electric field, they orient themselves in the direction opposite to that of the field (Figure 3.4).

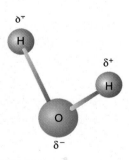

FIGURE 3.2

Charges on a Water Molecule.

The two hydrogen atoms in each molecule carry partial positive charges. The oxygen atom carries a partial negative charge.

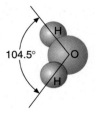

FIGURE 3.3

Space-Filling Model of a Water Molecule.

Because the water molcule has a bent geometry, the distribution of charge within the molecule is asymmetric. Water is therefore polar.

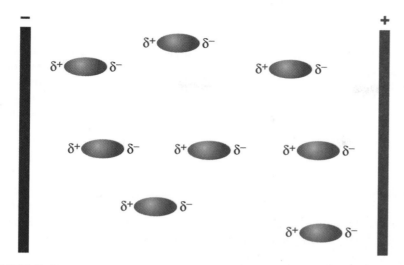

FIGURE 3.4

Molecular Dipoles in an Electric Field.

When polar molecules are placed between charged plates, they line up in opposition to the field.

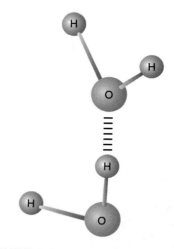

FIGURE 3.5

The Hydrogen Bond.

A hydrogen bond is a weak attraction between an electronegative atom in one molecule and a hydrogen atom in another molecule. The hydrogen bonds between water molecules are represented by short parallel lines.

Because of the large difference in electronegativity of hydrogen and oxygen, the electron-deficient hydrogens of one water molecule are attracted to the unshared pairs of electrons of another water molecule. (Hydrogens attached to nitrogen, sulfur, and fluorine behave the same way.) This interaction is called a **hydrogen bond** (Figure 3.5), which has both electrostatic (ionic) and covalent character. **Electrostatic interactions** between polar molecules play a signifi-cant role in living systems.

3.2 NONCOVALENT BONDING

Noncovalent interactions are usually electrostatic; that is, they occur between the positive nucleus of one atom and the negative electron clouds of another nearby atom. Unlike the stronger covalent bonds, individual noncovalent inter-actions are relatively weak and are therefore easily disrupted (Table 3.1). Nev-ertheless, they play a vital role in determining the physical and chemical properties of water and the structure and function of biomolecules because the cumulative

TABLE 3.1

Bond Strengths of Bonds Typically Found in Living Organisms

Bond Type	Bond Strength	
	kcal/mol	kJ/mol*
Covalent	>50	>210
Noncovalent		
Ionic interactions**	1–20	4–80
Hydrogen bonds	3–7	12–30
van der Waals forces	<1–2.7	0.3–9
Hydrophobic interactions	<1–3	3–12

*1 cal = 4.184 J.

**The actual strength varies considerably with the identity of the interacting species.

effect of many weak interactions can be considerable. Large numbers of noncovalent interactions stabilize macromolecules and supramolecular structures, whereas the capacity of these bonds to be rapidly formed and broken endows biomolecules with the flexibility required for the rapid information flow that occurs in dynamic living processes. In living organisms, the most important noncovalent interactions include ionic interactions, the hydrogen bond, van der Waals forces, and hydrophobic interactions. The first three types are briefly described in this section. Hydrophobic interactions are described in Section 3.4.

Ionic Interactions

Ionic interactions occur between charged atoms or groups. Oppositely charged ions such as sodium (Na^+) and chloride (Cl^-) are attracted to each other. In contrast, ions with similar charges, such as Na^+ and K^+ (potassium), repel each other. In proteins, certain amino acid side chains contain ionizable groups. For example, the side chain of the amino acid glutamic acid ionizes at physiological pH as $-CH_2CH_2COO^-$. The side chain group of the amino acid lysine ($-CH_2CH_2CH_2CH_2-NH_2$) ionizes as $-CH_2CH_2CH_2CH_2NH_3^+$ at physiological pH. The attraction of positively and negatively charged amino acid side chains forms **salt bridges** ($-COO^-{}^+H_3N-$) and the repulsive forces created when similarly charged species come into close proximity are an important feature in many biological processes, such as protein folding, enzyme catalysis, and molecular recognition. It should be noted that stable salt bridges rarely form between biomolecules in the presence of water because the hydration of ions is preferred and the attraction between the biomolecules decreases significantly. Most salt bridges in biomolecules occur in relatively water-free depressions or at biomolecular interfaces where water is excluded.

Hydrogen Bonds

Covalent bonds between hydrogen and oxygen, nitrogen, or sulfur are sufficiently polar that the hydrogen nucleus is weakly attracted to the lone pair of electrons of an oxygen, nitrogen, or sulfur on a neighboring molecule (Figure 3.6). In the water molecule each of oxygen's unshared electron pairs can form a hydrogen bond with nearby water molecules. The resulting intermolecular "bonds" act as a bridge between water molecules. Each hydrogen bond is not especially strong (about 20 kJ/mol) when compared to covalent bonds (e.g., 393 kJ/mol for N—H bonds and 460 kJ/mol for O—H bonds). However, when large numbers of intermolecular hydrogen bonds can be formed (e.g., in the liquid and solid states of water), the molecules involved effectively become large, dynamic, three-dimensional aggregates. In water, the substantial amounts of energy that are required to break up this aggregate explain the high values for its boiling and melting points, heat of vaporization, and heat capacity. Other properties of water, such as surface tension and viscosity, are also due largely to its capacity to form large numbers of hydrogen bonds.

van der Waals Forces

Van der Waals forces are relatively weak, transient electrostatic interactions. They occur between permanent and/or induced dipoles. They may be attractive or repulsive, depending on the distance between the atoms or groups involved. The attraction between molecules is greatest at a distance called the *van der Waals radius*. If molecules approach each other more closely, a repulsive force develops. The magnitude of van der Waals forces depends on how easily an atom is polarized. Electronegative atoms with unshared pairs of electrons are easily polarized.

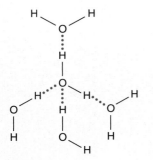

FIGURE 3.6

Hydrogen Bonding Between Water Molecules.

In water, each molecule can form hydrogen bonds with four other water molecules.

(a) Dipole – dipole interactions

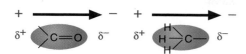

(b) Dipole – induced dipole interactions

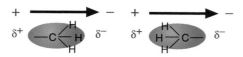

(c) Induced dipole – induced dipole interactions

FIGURE 3.7

Dipolar Interactions.
There are three types of electrostatic interactions involving dipoles: (a) dipole-dipole interactions, (b) dipole–induced dipole interactions, and (c) induced dipole–induced dipole interactions. The relative ease with which electrons respond to an electric field determines the magnitude of van der Waals forces. Dipole-dipole interactions are the strongest; induced dipole–induced dipole interactions are the weakest.

There are three types of van der Waals forces:

1. **Dipole-dipole interactions**. These forces, which occur between molecules containing electronegative atoms, cause molecules to orient themselves so that the positive end of one molecule is directed toward the negative end of another (Figure 3.7a). Hydrogen bonds (described in the previous section) are an especially strong type of dipole-dipole interaction.

2. **Dipole-induced dipole interactions.** A permanent dipole induces a transient dipole in a nearby molecule by distorting its electron distribution (Figure 3.7b). For example, a carbonyl-containing molecule is weakly attracted to a hydrocarbon. Dipole–induced dipole interactions are weaker than dipole-dipole interactions.

3. **Induced dipole-induced dipole interactions**. The motion of electrons in nearby nonpolar molecules results in transient charge imbalance in adjacent molecules (Figure 3.7c). A transient dipole in one molecule polarizes the electrons in a neighboring molecule. This attractive interaction, often called **London dispersion forces**, is extremely weak. The stacking of the base rings in a DNA molecule is a classic example of this type of interaction. Although individually weak, these interactions extending over the length of the DNA molecule provide significant stability.

KEY CONCEPTS 3.1

Noncovalent bonds (i.e., hydrogen bonds, ionic interactions, van der Waals forces, and hydrophobic interactions) play important roles in determining the physical and chemical properties of water. They also have a significant effect on the structure and function of biomolecules.

3.3 THERMAL PROPERTIES OF WATER

Perhaps the oddest property of water is that it is a liquid at room temperature. If water is compared with related molecules of similar molecular weight, it becomes apparent that water's melting and boiling points are exceptionally high (Table 3.2). If water followed the pattern of compounds such as hydrogen sulfide, it would melt at −100°C and boil at −91°C. Under these conditions, most of the Earth's water would be steam, making life unlikely. However, water actually melts at 0°C and boils at +100°C. Consequently, it is a liquid over most of the wide range of temperatures typically found on the Earth's surface. Hydrogen bonding is responsible for this anomalous behavior.

Because of its molecular structure, each water molecule can form hydrogen bonds with four other water molecules. Each of the latter molecules can form hydrogen bonds with other water molecules. The maximum number of hydrogen bonds form when water has frozen into ice (Figure 3.8). Energy is required to break these bonds. When ice is warmed to its melting point, approximately 15%

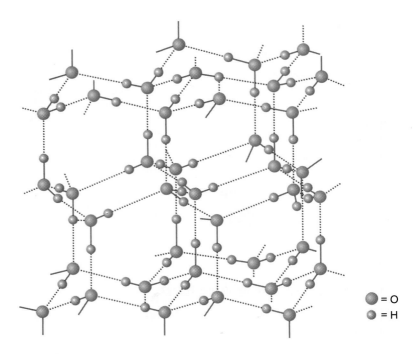

FIGURE 3.8

Hydrogen Bonding Between Water Molecules in Ice.

Hydrogen bonding in ice produces a very open structure. Ice is less dense than water in its liquid state.

of the hydrogen bonds break. Liquid water consists of icelike clusters of molecules whose hydrogen bonds are continuously breaking and forming. As the temperature rises, the movement and vibrations of the water molecules accelerate, and additional hydrogen bonds are broken. When the boiling point is reached, the water molecules break free from one another and vaporize. The energy required to raise water's temperature is substantially higher than expected (Table 3.3). In addition to the energy absorbed in increasing molecular agitation, a significant amount of energy is dissipated by the rapid vibration of shared hydrogens back and forth between oxygen atoms.

One consequence of water's high *heat of vaporization* (the energy required to vaporize one mole of a liquid at a pressure of one atmosphere) and high *heat capacity* (the energy that must be added or removed to change the temperature by one degree Celsius) is that water acts as an effective modulator of climatic temperature. Water can absorb and store solar heat and release it slowly. Consider,

TABLE 3.2

Melting and Boiling Points of Water and Other Group VI Hydrogen-Containing Compounds

Name	Formula	Molecular Weight (daltons*)	Melting Point (°C)	Boiling Point (°C)
Water	H_2O	18	0	100
Hydrogen sulfide	H_2S	34	−85.5	−260.7
Hydrogen selenide	H_2Se	81	−50.4	−241.5
Hydrogen telluride	H_2Te	129.6	−49	−2

*1 dalton = 1 atomic mass unit.

TABLE 3.3

Heat of Fusion of Water and Other Group VI Hydrogen-Containing Compounds

Name	Formula	Molecular Weight (daltons)	Heat of Fusion (cal/g)	Heat of Fusion (J/g)
Water	H_2O	18	80	335
Hydrogen sulfide	H_2S	34	16.7	69.9
Hydrogen selenide	H_2Se	81	7.4	31

The heat of fusion is the amount of heat required to change 1 g of a solid into a liquid at its melting point. 1 cal = 4.184 J.

for example, the relatively moderate temperature transitions when seasons change on land masses near the oceans. In contrast, in the interior of large land masses such as the American Midwest, seasons begin abruptly, and seasonal extremes of temperature may be dramatic. In exceptionally dry areas such as the Sahara desert, daily changes in temperature may equal as much as 100°F. Heat absorbed during the day in a desert is quickly lost by reradiation at night.

Not surprisingly, water plays an important role in the thermal regulation of living organisms. Water's high heat capacity, coupled with the high water content found in most organisms (between 50% and 95%, depending on species), helps maintain an organism's internal temperature. The evaporation of water is used as a cooling mechanism, because it permits large losses of heat. For example, an adult human may eliminate as much as 1200 g of water daily in expired air, sweat, and urine. The associated heat loss may amount to approximately 20% of the total heat generated by metabolic processes.

KEY CONCEPTS 3.2

Hydrogen bonding is responsible for water's unusually high freezing and boiling points. Because water has a high heat capacity, it can absorb and release heat slowly. Water plays an important role in regulating heat in living organisms.

QUESTION 3.1

The ammonia molecule is similar to water. It also has a higher boiling point than would be expected from its molecular weight, a high heat of fusion (the energy required for a substance to change between the solid and liquid states), and a high heat capacity. Draw the structure for solid ammonia. Would you expect this "ice" to be more or less dense than liquid ammonia?

3.4 SOLVENT PROPERTIES OF WATER

Water is the ideal biological solvent. It easily dissolves a wide variety of the constituents of living organisms. Examples include ions (e.g., Na^+, K^+, and Cl^-), sugars, and many of the amino acids. Its inability to dissolve other substances, such as lipids and certain other amino acids, makes supramolecular structures (e.g., membranes) and numerous biochemical processes (e.g., protein folding) possible. In this section the behavior of hydrophilic and hydrophobic substances in water is described. This discussion is followed by a brief review of osmotic pressure, one of the colligative properties of water. Colligative properties are physical properties that are affected not by the specific structure of dissolved solutes, but rather by their numbers.

Hydrophilic Molecules

Water's dipolar structure and its capacity to form hydrogen bonds with electronegative atoms enable water to dissolve both ionic and polar substances. Salts such as sodium chloride (NaCl) are held together by ionic forces. An important aspect of all ionic interactions in aqueous solution is the hydration of ions. Because water molecules are polar, they are attracted to charged ions such as Na^+ and Cl^-. Shells of water molecules, referred to as **solvation spheres**, cluster around both positive and negative ions (Figure 3.9). As ions become hydrated, the attractive force between them is reduced, and the charged species dissolves in the water. The capacity of a solvent to reduce the attractive forces between ions is indicated by its dielectric constant. Water, sometimes referred to as the *universal solvent* because of the large variety of ionic and polar substances it can dissolve, has a very large dielectric constant. Because biomolecules recognize and bind to each other primarily through intermolecular forces such as ionic bonds and other electrostatic interactions, the weakening of these interactions is prevented by excluding water from interacting surfaces. Rapid catalysis of substrate molecules within enzyme active sites, which usually involves interactions between charged species, for example, is made possible by excluding water from the catalytic surface.

Organic molecules with ionizable groups and many neutral organic molecules with polar functional groups also dissolve in water, primarily because of the solvent's hydrogen bonding capacity. Such associations form between water and the carbonyl groups of aldehydes and ketones and the hydroxyl groups of alcohols.

Hydrophobic Molecules

When mixed with water, small amounts of nonpolar substances are excluded from the solvation network of the water; that is, they coalesce into droplets. This process is called the *hydrophobic effect*. Hydrophobic ("water-hating") molecules, such as the hydrocarbons, are virtually insoluble in water. Their association into droplets (or, in larger amounts, into a separate layer) results from the solvent properties of water, not from the relatively weak attraction between the associating nonpolar molecules. When nonpolar molecules enter an aqueous environment, hydrogen-bonded water molecules attempt to form a cagelike structure around

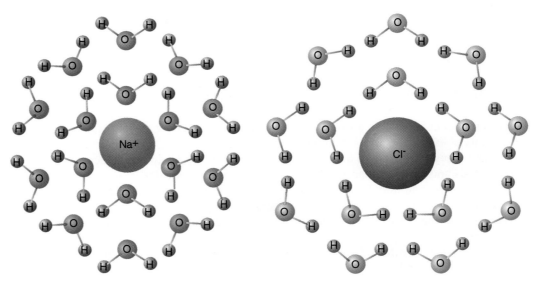

FIGURE 3.9

Solvation Spheres of Water Molecules Around Na^+ and Cl^- Ions.

When an ionic compound such as NaCl is dissolved in water, its ions separate because the polar water molecules attract the ions more than the ions attract each other.

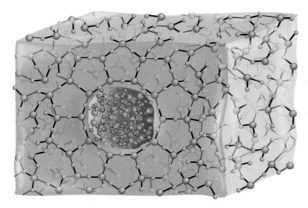

FIGURE 3.10

The Hydrophobic Effect.

When nonpolar molecules and water are mixed, a solvation sphere composed of many layers of highly ordered hydrogen-bonded water molecules forms around the hydrophobic molecules. Although nonpolar molecules, when in close proximity, are attracted to each other
by van der Waals forces, the driving force in the formation of the solvation spheres is the
strong tendency of water molecules to form hydrogen bonds among themselves. Nonpolar
molecules are excluded because they cannot form hydrogen bonds.

them (Figure 3.10). Sufficient energy is not available in the surroundings to form
this cagelike structure, and nonpolar molecules are expelled. The droplets that
form result from the most energetically favorable configuration of the surrounding water molecules. **Hydrophobic interactions** between these excluded nonpolar substances have a profound effect on living cells. For example, they are
primarily responsible for the structure of membranes and the stability of proteins.

Proteins are amino acid polymers. Noncovalent bonding plays an important role
in determining the three-dimensional structures of proteins. Typical examples
of noncovalent interactions (shaded areas) between amino acid side chains are
shown below.

QUESTION 3.2

Which noncovalent bond is primarily responsible for the interactions indicated
in the figure?

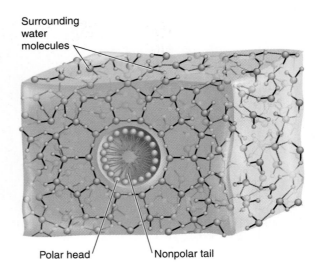

Surrounding
water
molecules

Polar head　　　　　Nonpolar tail

FIGURE 3.11

Formation of Micelles.

The polar heads of amphipathic molecules orient themselves so that they are in contact with water molecules. The nonpolar tails aggregate in the center, away from water.

KEY CONCEPTS 3.3

Water's dipolar structure and its capacity to form hydrogen bonds enable water to dissolve many ionic and polar substances. Because nonpolar molecules cannot form hydrogen bonds, they cannot dissolve in water. Amphipathic molecules, such as fatty acid salts, spontaneously rearrange themselves in water to form micelles.

Amphipathic Molecules

A large number of biomolecules, referred to as **amphipathic**, contain both polar and nonpolar groups. This property significantly affects their behavior in water. For example, ionized fatty acids are amphipathic molecules because they contain hydrophilic carboxylate groups and hydrophobic hydrocarbon groups. When they are mixed with water, amphipathic molecules form structures called *micelles* (Figure 3.11). In **micelles**, the charged species (the carboxylate groups), called *polar heads*, orient themselves so that they are in contact with water. The nonpolar hydrocarbon "tails" become sequestered in the hydrophobic interior. The tendency of amphipathic biomolecules to spontaneously rearrange themselves in water is an important feature of numerous cell components. For example, a group of bilayer-forming phospholipid molecules is the basic structural feature of biological membranes (see Chapter 11).

Osmotic Pressure

Osmosis is the spontaneous process in which solvent molecules pass through a semipermeable membrane from a solution of lower solute concentration to a solution of higher solute concentration. Pores in the membrane are wide enough to allow solvent molecules to pass through in both directions but too narrow for the larger solute molecules or ions to pass. Figure 3.12 illustrates the movement of solvent across a membrane. As the process begins, there are fewer water molecules on the high solute concentration side of the membrane. Over time, more water moves from side A (lower solute concentration) to side B (higher solute concentration). The higher the concentration of water in a solution (i.e., the lower

FIGURE 3.12

Osmotic Pressure.

Over time water diffuses from side A (more dilute) to side B (more concentrated). Equilibrium between the solutions on both sides of a semipermeable membrane is attained when there is no net movement of water molecules from side A to side B. Osmotic pressure stops the net flow of water across the membrane.

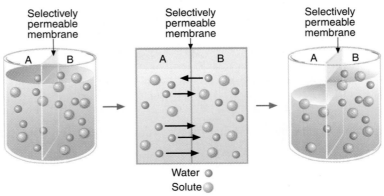

Selectively permeable membrane

Selectively permeable membrane

Selectively permeable membrane

A　B　　A　B　　A　B

Water
Solute

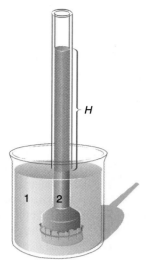

FIGURE 3.13

The Measurement of Osmosis Using an Osmometer.

Volume 1 contains pure water. Volume 2 contains a solution of sucrose. The membrane is permeable to water but not to the sucrose. Therefore there will be a net movement of water into the osmometer. The osmotic pressure is proportional to the height H of the solution in the tube.

the solute concentration), the faster the movement through the membrane. **Osmotic pressure** is the pressure required to stop the net flow of water across the membrane. The force generated by osmosis can be considerable. The principal cause of water flow across cellular membranes, osmotic pressure is a driving force in numerous living processes. For example, osmotic pressure appears to be a significant factor in the formation of sap in trees.

Osmotic pressure depends on solute concentration. A device called an *osmometer* (Figure 3.13) measures osmotic pressure. Osmotic pressure can also be calculated using the following equation, keeping in mind that the final osmotic pressure reflects the contribution of all solutes present.

$$\pi = iMRT$$

where π = osmotic pressure (atm)

   i = van't Hoff factor (reflects the extent of ionization of solutes)

   M = molarity (moles/liter)

   R = gas constant (0.082 L·atm/K·mole)

   T = Kelvin temperature

The concentration of a solution can be expressed in terms of *osmolarity*. The unit of osmolarity is osmol/liter. From the equation above, the osmolarity is equal to iM where i (the van't Hoff factor) represents the degree of ionization of the solute species. The degree of ionization of a 1 M NaCl solution is 90% with 10% of the NaCl existing as ion pairs.

$$i = [Na^+] + [Cl^-] + [NaCl]_{unionized} = 0.9 + 0.9 + 0.1 = 1.9$$

The value of i for this solution is therefore 1.9. The value of i approaches 2 for NaCl solutions as they become increasingly more dilute. The value of i for a 1 M solution of a weak acid that undergoes a 10% ionization is 1.1. The value of i for a nonionizable solute is always 1.0. Several osmotic pressure problems appear on page 76.

Osmotic pressure creates some critical problems for living organisms. Cells typically contain fairly high concentrations of solutes, that is, small organic molecules and ionic salts, as well as lower concentrations of macromolecules. Consequently, cells may gain or lose water because of the concentration of solute in their environment. If cells are placed in an **isotonic solution** (i.e., the concentration of solute and water is the same on both sides of the selectively permeable plasma membrane) there is no net movement of water in either direction across the membrane (Figure 3.14). For example, red blood cells are isotonic to

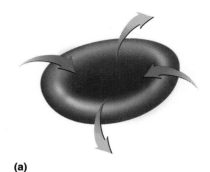

(a)

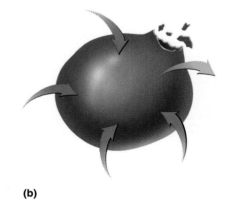

(b)

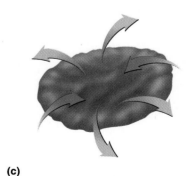

(c)

FIGURE 3.14

The Effect of Hypertonic and Hypotonic Solutions on Animal Cells.

(a) Isotonic solutions do not change cell volume because water is entering and leaving the cell at the same rate; (b) hypotonic solutions cause cell rupture; (c) hypertonic solutions cause cells to shrink (crenation).

a 0.9% NaCl solution. When cells are placed in a solution with a lower solute concentration (i.e., a **hypotonic solution**), water moves into the cells. Red blood cells, for example, swell and rupture in a process called *hemolysis* when they are immersed in pure water. In **hypertonic solutions**, those with higher solute concentrations, cells shrivel because there is a net movement of water out of the cell. The shrinkage of red blood cells in hypertonic solution (e.g., a 3% NaCl solution) is referred to as *crenation*.

Because of their relatively low cellular concentration, macromolecules have little direct effect on cellular osmolarity. However, macromolecules such as the proteins contain a large number of ionizable groups. The large number of ions of opposite charge that are attracted to these groups have a substantial effect on intracellular osmolarity. Unlike most ions, the ionizable groups of proteins cannot penetrate cell membranes. (Cell membranes are not, strictly speaking, osmotic membranes, because they allow various ions, nutrients, and waste products to pass through. The term *dialyzing membrane* gives a more accurate description of their function.) The consequence of this is that, at equilibrium, ion distributions are not equal on the two sides of cell membrane. Instead, the intracellular concentrations of inorganic ions are higher than outside the cell. The existence of this asymmetry on the surfaces of cell membrane results in the establishment of an electrical gradient, called a **membrane potential**, which provides the means for electrical conduction, active transport, and even passive transport. Because of this phenomenon, called the **Donnan effect**, there is a constant tendency toward cellular swelling caused by water entry resulting from osmotic pressure. Cells must, therefore, constantly regulate their osmolarity. Many cells, for example, animal and bacterial cells, "pump out" certain inorganic ions such as Na^+. This process, which requires a substantial proportion of cellular energy, controls cell volume (Special Interest Box 3.1). Several species, such as some protozoa and algae, periodically expel water from special contractile vacuoles. Because plant cells have rigid cell walls, plants use the Donnan effect to create an internal hydrostatic pressure, called *turgor pressure* (Figure 3.15). This process drives cellular growth and expansion and makes many plant structures rigid.

PROBLEM 3.1

When 0.1 g of urea (M.W = 60) is diluted to 100 mL with water, what is the osmotic pressure of the solution? (Assume that the temperature is 25°C.)

Solution
Calculate the osmolarity of the urea solution.

$$\text{Molarity} = \frac{0.1 \text{ g urea} \times 1 \text{ mol}}{60 \text{ g}} \times \frac{1}{0.1 \text{ L}}$$

The number of particles produced per mole of solute is 1. The osmotic pressure at 25°C (298 K) is given by the equation

$$\pi = iMRT$$

Urea is a nonelectrolyte, so $i = 1$

$$\pi = (1)\frac{1.7 \times 10^{-2} \text{ mol}}{L} \frac{0.0821 \text{ L·atm}}{K\cdot mol} (298 \text{ K})$$

$$\pi = 0.42 \text{ atm}$$

PROBLEM 3.2

Estimate the osmotic pressure of a solution of 0.1 M NaCl at 25°C. Assume 100% ionization of solute.

Solution
A solution of 0.1 M NaCl produces 0.2 mol of particles per liter (0.1 mol of Na^+ and 0.1 mol of Cl^-). The osmotic pressure at 25°C (298 K) is

$$\pi = \frac{2 \times 0.1 \text{ mol}}{L} \times \frac{0.0821 \text{ L·atm}}{K\cdot mol} \times 298 \text{ K}$$

$$\pi = 4.9 \text{ atm}$$

QUESTION 3.3

Osmotic pressure may be used to determine the molecular weight of pure substances. It is an especially useful technique because low concentrations of solute produce relatively large osmotic pressures. Although more sophisticated techniques are now available to determine molecular weight, osmotic pressure is still used occasionally because of its simplicity. Determine the molecular weight of myoglobin (an oxygen storage protein that gives muscle its red color). The osmotic pressure of a 1.0-mL water solution containing 1.5×10^{-3} g of the protein was measured at 2.06×10^{-3} atm at 25°C.

3.5 IONIZATION OF WATER

Liquid water molecules have a limited capacity to ionize to form a hydrogen ion (H^+) and a hydroxide ion (OH^-). H^+ does not actually exist in aqueous solution. In water a proton combines with a water molecule to form H_3O^+, commonly referred to as *hydronium ion*. For convenience, H^+ will be used in representing the ionization reactions of water.

The disassociation of water

$$H_2O(l) \rightleftharpoons H^+ + OH^-$$

may be expressed as

$$K_{eq} = \frac{[H^+][OH^-]}{[H_2O]}$$

where K_{eq} is the equilibrium constant for the reaction. Since the molar concentration of pure water (55.5M) is considerably larger than of any solutes, it too is considered a constant. (The concentration of water $[H_2O]$, is obtained by dividing the number of grams in 1 liter of water, 1000g, by the molecular weight of water, 18g/mole.) After this value is substituted into this equation, it may be rewritten as follows:

$$K_{eq} \times 55.5 = [H^+][OH^-]$$

The term $K_{eq} \times 55.5$ is called the ion product of water (K_w). Since the equilibrium constant for the reversible ionization of water is equal to 1.8×10^{-16} M (at 25°C), the above relationship yields

$$K_w = (1.8 \times 10^{-16})(55.5) = [H^+][OH^-]$$
$$= 1.0 \times 10^{-14}$$

This means that the product of $[H^+]$ and $[OH^-]$ in any water solution (at 25°C) is always 1×10^{-14}. Since $[H^+]$ is equal to $[OH^-]$ when pure water dissociates,

$$[H^+] = [OH^-] = 1 \times 10^{-7}\,M$$

Thus the hydrogen ion concentration of pure water is equal to $1 \times 10^{-7}\,M$.

When a solution contains equal amounts of H^+ and OH^-, it is said to be *neutral*. When an ionic or polar substance is dissolved in water, it may change the relative numbers of H^+ and OH^-. Solutions with an excess of H^+ are *acidic*, whereas those with a greater number of OH^- are *basic*. Hydrogen ion concentration varies over a very wide range: commonly between 10^0 and 10^{-14} M, which provides the basis of the pH scale ($pH = -\log [H^+]$).

Acids, Bases, and pH

The hydrogen ion is one of the most important ions in biological systems. The concentration of this ion affects most cellular and organismal processes. For example, the structure and function of proteins and the rates of most biochemical reactions are strongly affected by hydrogen ion concentration. Additionally, hydrogen ions play a major role in processes such as energy generation (see Chapter 10) and endocytosis.

Many biomolecules have acidic or basic properties. There are several possible definitions of these important classes of ionic compounds. For our purposes,

KEY CONCEPTS 3.4

Osmosis is the movement of water across a semipermeable membrane from a dilute solution to a more concentrated solution. Osmotic pressure is the pressure exerted by water on a semipermeable membrane due to a difference in the concentration of solutes on either side of the membrane.

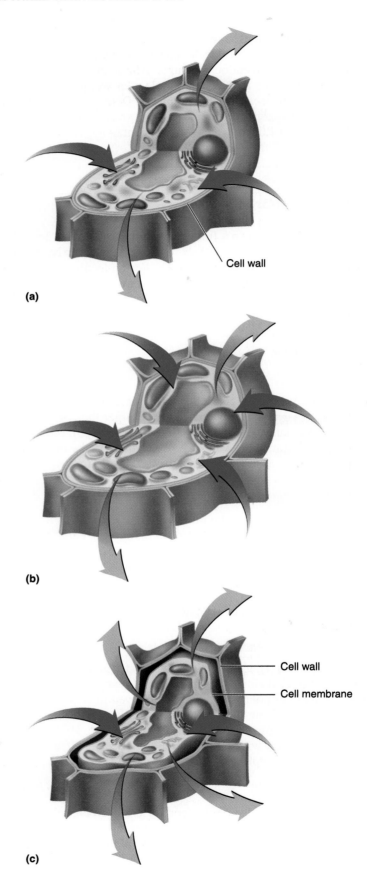

(a)

Cell wall

(b)

Cell wall

Cell membrane

(c)

FIGURE 3.15

Osmotic Pressure and Plant Cells.

(a) Isotonic solutions do not change cell volume. (b) Plant cells typically exist in a hypotonic environment. When water enters these cells they become swollen. Rigid cell walls prevent the cells from bursting. (c) In a hypertonic environment the cell membrane pulls away from the cell wall because of water loss and the plant wilts.

Living cells are in constant danger. Even the smallest changes in the balance of solutes between their interiors and their surroundings make cells vulnerable to potentially damaging changes in osmotic pressure. Any inability to manage osmotic balance can lead to distortions in shape and volume that compromise cell function. In multicellular organisms such as animals, however, individual cells are usually not exposed to significant fluctuations in the osmolarity of their surroundings. Instead, it is now realized, they are continuously challenged by internal variations that are created by normal metabolic processes. Routine tasks such as the uptake of nutrients (e.g., sugars, fatty acids, and amino acids), the excretion of waste products (e.g., H^+ and CO_2), and metabolic processes such as the synthesis and degradation of macromolecules (e.g., proteins and glycogen) cause osmotic imbalances.

Research efforts have revealed that cells possess several sophisticated mechanisms that together rapidly correct even the most minor changes in osmolarity. The best understood of these is the exchange of inorganic ions across membranes (Figure 3A). For example, when a cell is engaged in the synthesis of protein, the resulting reduction in the concentration of amino acids causes water to flow out of the cell. The cell responds by importing K^+, Na^+, Cl^- (in exchange for HCO_3^-) through specialized membrane channel complexes. The osmotic gradient created by this process results in the flow

of water into the cell, thus restoring the cell's normal volume. When protein is degraded, the opposite process occurs. The increased concentration of osmotically active amino acids causes the cell to swell. Ions (e.g., K^+, Cl^-, and HCO_3^-) followed by water then move across the plasma membrane out of the cell, and cell volume is restored.

Cell volume can also be controlled by the synthesis of quantities of osmotically active substances called **osmolytes**. For example, when confronted with osmotic stress some cells produce large amounts of alcohols (e.g., sorbitol, see p. 208), amino acids, or amino acid derivatives such as taurine (see p. 413). Cells have also been observed to restore osmotic balance by synthesizing or degrading macromolecules such as glycogen. The precise means by which cells manage osmotic balance are not yet resolved. It is known that cell volume changes signaled by distortions of the cytoskeleton cause alterations in the expression of genes, some of which code for the synthesis of membrane channel proteins and osmolytes.

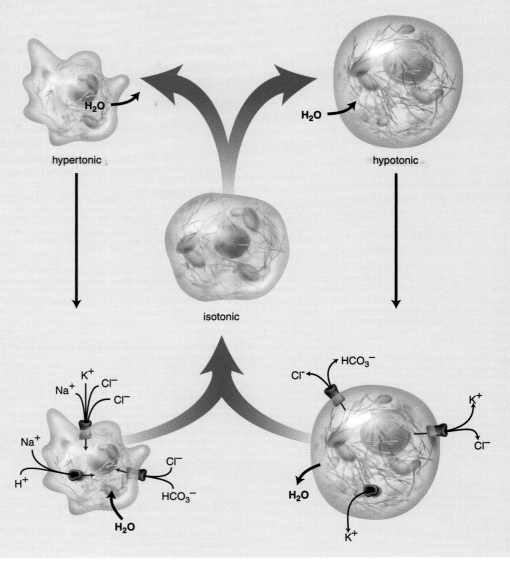

FIGURE 3A

Osmotic Pressure and Cell Volume Changes. Cells shrink when they are exposed to a hypertonic medium or when biochemical processes reduce the number of osmotically active particles. The cell's osmotic balance is restored when inorganic ions such as Na^+, K^+, and Cl^- enter. In certain cells Na^+ and Cl^- may enter in exchange for H^+ and HCO_3^-, respectively. The cell returns to its normal volume as water then flows back into the cell. Cells swell when they are placed in a hypotonic medium or they increase their concentration of osmotically active particles through transport or the degradation of macromolecules. Osmotic balance is restored with the expulsion of inorganic ions, followed by the outflow of water.

however, an **acid** may be defined as a hydrogen ion donor and a **base** as a hydrogen ion acceptor. Strong acids (e.g., HCl) and bases (e.g., NaOH) ionize almost completely in water:

$$HCl \longrightarrow H^+ + Cl^-$$

$$NaOH \longrightarrow Na^+ + OH^-$$

Many acids and bases, however, do not dissociate completely. Organic acids (compounds with carboxyl groups) do not completely dissociate in water. They are referred to as **weak acids**. Organic bases have a small but measurable capacity to combine with hydrogen ions. Many common **weak bases** contain amino groups.

The dissociation of an organic acid is described by the following reaction:

$$\underset{\text{weak acid}}{HA} \;\;\rightleftharpoons\;\; \underset{\text{Conjugate Base of HA}}{H^+ + A^-}$$

Note that the deprotonated product of the dissociation reaction is referred to as a **conjugate base**. For example, acetic acid (CH_3COOH) dissociates to form the conjugate base acetate (CH_3COO^-).

The strength of a weak acid (i.e., its capacity to release hydrogen ions released) may be determined by using the following expression:

$$K_a = \frac{[H^+][A^-]}{[HA]}$$

where K_a is the acid dissociation constant. The larger the value of K_a, the stronger the acid is. Because K_a values vary over a wide range, they are expressed by using a logarithmic scale:

$$pK_a = -\log K_a$$

The lower the pK_a, the stronger the acid. Dissociation constants and pK_a values for several common weak acids are given in Table 3.4.

The **pH scale** (Figure 3.16) conveniently expresses hydrogen ion concentration. pH is defined as the negative logarithm of the concentration of hydrogen ions:

$$pH = -\log[H^+]$$

On the pH scale, neutrality is defined as pH 7 (i.e., $[H^+]$ is equal to 1×10^{-7} M). Solutions with pH values less than 7 (i.e., $[H^+]$ greater than 1×10^{-7} M) are acidic. Those with pH values greater than 7 are basic or alkaline.

It is important to note that although pK_a and pH appear to be similar mathematical expressions, they are in fact different. At constant temperature, the pK_a value of a substance is a constant. In contrast, the pH values of a system may vary.

TABLE 3.4

Dissociation Constants and pKa Values for Common Weak Acids

Acid	HA	A$^-$	K_a	pK_a
Acetic acid	CH_3COOH	CH_3COO^-	1.76×10^{-5}	4.76
Carbonic acid	H_2CO_3	HCO_3^-	$4.30\ 3\ 10^{-7}$	6.37
Bicarbonate	HCO_3^-	CO_3^{2-}	5.61×10^{-11}	10.25
Lactic acid	$CH_3\underset{\mid}{C}HCOOH$ $\;\;OH$	$CH_3\underset{\mid}{C}HCOO^-$ $\;\;OH$	1.38×10^{-4}	3.86
Phosphoric acid	H_3PO_4	$H_2PO_4^-$	7.25×10^{-3}	2.14
Dihydrogen phosphate	$H_2PO_4^-$	HPO_4^{2-}	6.31×10^{-8}	7.20

Equilibrium constants should be expressed in terms of activities rather than concentrations (activity is the effective concentration of a substance in a solution). However, in dilute solutions, concentrations may be substituted for activities with reasonable accuracy.

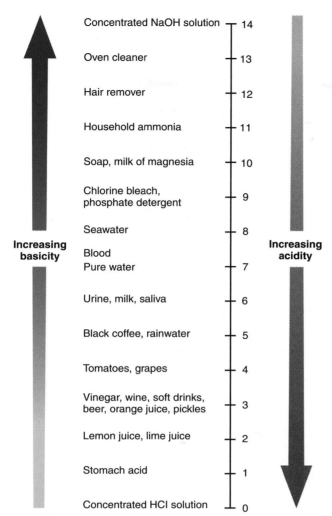

FIGURE 3.16

The pH Scale and the pH of Common Fluids.

Buffers

The regulation of pH is a universal and essential activity of living organisms. Hydrogen ion concentration must typically be kept within very narrow limits. For example, normal human blood has a pH of 7.4. It may vary between 7.35 and 7.45, although blood normally contains many acidic or basic waste products dissolved in it. Certain disease processes cause pH changes that, if not corrected, can be disastrous. **Acidosis**, a condition that occurs when human blood pH falls below 7.35, results from an excessive production of acid in the tissues, loss of base from body fluids, or the failure of the kidneys to excrete acidic metabolites. Acidosis occurs in certain diseases (e.g., diabetes mellitus) and during starvation. If blood pH drops below 7, the central nervous system becomes depressed. This results in coma and eventually death. When blood pH rises above 7.45, **alkalosis** results. This condition, brought on by prolonged vomiting or by ingestion of excessive amounts of alkaline drugs, overexcites the central nervous system. Muscles then go into a state of spasm. If uncorrected, convulsions and respiratory arrest develop.

Buffers help maintain a relatively constant hydrogen ion concentration. The most common buffers consist of weak acids and their conjugate bases. A buffered solution can resist pH changes because an equilibrium between the buffer's components

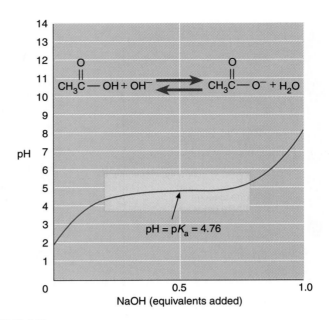

FIGURE 3.17

Titration of Acetic Acid with NaOH.

The shaded band indicates the pH range over which acetate buffer functions effectively. A buffer is most effective at or near its pK_a value.

is established. Therefore buffers obey **Le Chatelier's principle**, which states that if a stress is applied to a reaction at equilibrium, the equilibrium will be displaced in the direction that relieves the stress. Consider a solution containing acetate buffer, which consists of acetic acid and sodium acetate (Figure 3.17). The buffer is created by mixing a solution of sodium acetate with a solution of acetic acid to create an equilibrium mixture of the correct pH and ionic strength.

If hydrogen ions are added, the equilibrium shifts toward the formation of acetic acid with the $[H^+]$ changing little:

$$H^+ + CH_3COO^- \longrightarrow CH_3COOH$$

If hydroxide ions are added, they react with the free hydrogen ions to form water and the equilibrium shifts to the acetate ion and the pH changes little.

BUFFERING CAPACITY The capacity of a buffer to maintain a specific pH depends on two factors: (1) the molar concentration of the acid–conjugate base pair and (2) the ratio of their concentrations. Buffering capacity is directly proportional to the concentration of the buffer components. In other words, the more molecules of buffer present, the more H^+ and OH^- ions can be absorbed without changing the pH. The concentration of the buffer is defined as the sum of the concentration of the weak acid and its conjugate base. For example, a 0.2 M acetate buffer may contain 0.1 mol of acetic acid and 0.1 mol of sodium acetate in 1 L of H_2O. Such a buffer may also consist of 0.05 mol of acetic acid and 0.15 mol of sodium acetate in 1 L of H_2O. The most effective buffers are those that contain equal concentrations of both components. However, there are exceptions. Bicarbonate buffer (p. 86), one of the most important physiological buffers, significantly deviates from such a ratio.

HENDERSON-HASSELBALCH EQUATION In choosing or making a buffer, the concepts of both pH and pK_a are useful. The relationship between these two quantities is expressed in the Henderson-Hasselbalch equation, which is derived from the equilibrium expression below:

$$K_a = \frac{[H^+][A^-]}{[HA]}$$

Solving for $[H^+]$ results in

$$[H^+] = K_a \frac{[HA]}{[A^-]}$$

Taking the negative logarithm of each side, we obtain

$$-\log [H^+] = -\log K_a - \log \frac{[HA]}{[A^-]}$$

Defining $-\log [H^+]$ as pH and $-\log K_a$ as pK_a gives

$$pH = pK_a - \log \frac{[HA]}{[A^-]}$$

If the log term is inverted, thereby changing its sign, the following relationship called the *Henderson-Hasselbalch equation,* is obtained:

$$pH = pK_a + \log \frac{[A^-]}{[HA]}$$

Notice that when $[A^-] = [HA]$, the equation becomes

$$pH = pK_a + \log 1$$
$$= pK_a + 0$$

Under this circumstance, pH is equal to pK_a. The graph in Figure 3.17 illustrates that buffers are most effective when they are composed of equal amounts of weak acid and conjugate base. The most effective buffering occurs in the portion of the titration curve that has a minimum slope, that is, 1 pH unit above and below the value of pK_a.

Typical buffer problems along with their solutions are given next.

KEY CONCEPTS 3.5

Liquid water molecules have a limited capacity to ionize to form H^+ and OH^- ions. The concentration of hydrogen ions is a crucial feature of biological systems primarily because of their effects on biochemical reaction rates and protein structure. Buffers, which consist of weak acids and their conjugate bases, prevent changes in pH (a measure of $[H^+]$).

PROBLEM 3.3

Calculate the pH of a mixture of 0.25 M acetic acid and 0.1 M sodium acetate. The pK_a of acetic acid is 4.76.

Solution

$$pH = pK_a + \log \frac{[\text{acetate}]}{[\text{acetic acid}]}$$

$$pH = 4.76 + \log \frac{0.1}{0.25} = 4.76 - 0.398 = 4.36$$

PROBLEM 3.4

What is the pH in the preceding problem if the mixture consists of 0.1 M acetic acid and 0.25 M sodium acetate?

Solution

$$pH = 4.76 + \log \frac{0.25}{0.1} = 4.76 + 0.398 = 5.16$$

PROBLEM 3.5

Calculate the ratio of lactic acid and lactate required in a buffer system of pH 5.00. The pK_a of lactic acid is 3.86.

Solution

The equation

$$pH = pK_a + \log \frac{[\text{lactate}]}{[\text{lactic acid}]}$$

can be rearranged to

$$\log \frac{[\text{lactate}]}{[\text{lactic acid}]} = pH - pK_a$$
$$= 5.00 - 3.86 = 1.14$$

Therefore the required ratio is

$$\frac{[\text{lactate}]}{[\text{lactic acid}]} = \text{antilog } 1.14$$
$$= 13.8$$

For a lactate buffer to have a pH of 5, the lactate and lactic acid components must be present in a ratio of 13.8 to 1. Because a good buffer contains a mixture of a weak acid and its conjugate base present in near equal concentrations and when the buffered pH is within 1 pH unit of the pK_a, lactate buffer is a poor choice in this situation. A better choice would be the acetate buffer.

PROBLEM 3.6

During the fermentation of wine, a buffer system consisting of tartaric acid and potassium hydrogen tartrate is produced by a biochemical reaction. Assuming that at some time the concentration of potassium hydrogen tartrate is twice that of tartaric acid, calculate the pH of the wine. The pK_a of tartaric acid is 2.96.

Solution

$$pH = pK_a + \log \frac{[\text{hydrogen tartrate}]}{[\text{tartaric acid}]}$$
$$= 2.96 + \log 2$$
$$= 2.96 + 0.30 = 3.26$$

PROBLEM 3.7

What is the pH of a solution prepared by mixing 150 mL of 0.1 M HCl with 300 mL of 0.1 M sodium acetate (NaOAc) and diluting the mixture of 1 L? The pK_a of acetic acid is 4.76.

Solution

The amount of acid present in the solution is given by the equation:

$$mL \times M = mM$$

$$150 \text{ mL} \times 0.1 \text{ M} = 15 \text{ mM of acid}$$

The amount of sodium acetate is found using the same equation:

$$300 \text{ mL} \times 0.1 \text{ M} = 30 \text{ mM}$$

Each mole of HCl will consume 1 mol of sodium acetate and produce 1 mol of acetic acid. This will give 15 mM of acetic acid with 15 mM remaining of sodium acetate (i.e., 30 mM − 15 mM). Substituting these values into the Henderson-Hasselbalch equation gives

$$pH = 4.76 + \log \frac{15}{15}$$
$$= 4.76 + \log 1$$
$$= 4.76$$

Because the log term is a ratio of two concentrations, the volume factor can be eliminated and the molar amounts can be used directly.

What would be the effect of adding an additional 50 mL of 0.1 M HCl to the solution in Problem 3.7 before dilution to 1 L?

Solution

Using the same equation as in Problem 3.7, the amount of HCl would be

$$200 \text{ mL} \times 0.1 \text{ M} = 20 \text{ mM}$$

which is also equal to the concentration of acetic acid.
The amount of sodium acetate would be

$$30 \text{ mM} - 20 \text{ mM} = 10 \text{ mM}$$

Substituting into the Henderson-Hasselbalch equation gives

$$\text{pH} = 4.76 + \log \frac{10}{20}$$
$$= 4.76 + \log 0.5$$
$$= 4.76 - 0.3$$
$$= 4.46$$

WEAK ACIDS WITH MORE THAN ONE IONIZABLE GROUP Some molecules contain more than one ionizable group. Phosphoric acid (H_3PO_4) is a weak polyprotic acid; that is, it can donate more than one hydrogen ion (in this case, three hydrogen ions). During titration with NaOH (Figure 3.18), these ionizations occur in a stepwise fashion with 1 proton being released at a time:

$$H_3PO_4 \underset{pK_1 = 2.1}{\rightleftharpoons} H^+ + H_2PO_4^- \underset{pK_2 = 6.7}{\rightleftharpoons} H^+ + HPO_4^{2-} \underset{pK_3 = 12.3}{\rightleftharpoons} H^+ + PO_4^{3-}$$

(The pK_a for the most acidic group is referred to as pK_1. The pK_a for the next most acidic group is pK_2. The third most acidic pK_a value is pK_3.) At low pH most molecules are fully protonated. As NaOH is added protons are released in the order of decreasing acidity with the least acidic proton (with the largest pK_a value) ionizing last. When the pH is equal to pK_1, equal amounts of H_3PO_4 and $H_2PO_4^-$ exist in the solution.

Amino acids are biomolecules that contain several ionizable groups. Like all amino acids, alanine contains both a carboxyl group and an amino group. At low pH both of these groups are protonated. As the pH rises during a titration with NaOH, the acidic carboxyl group (COOH) loses its proton to form a carboxylate group (COO⁻). The addition of more NaOH eventually causes the ionized amino group to release its proton:

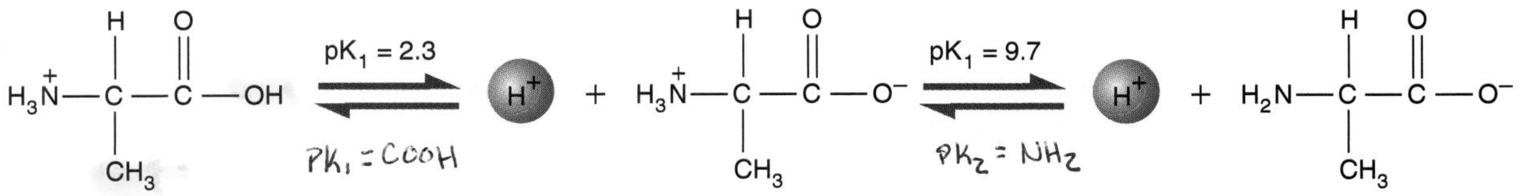

pK₁ = 2.3 pK_1 = CCOH   pK₁ = 9.7 pK_2 = NH₂

Certain amino acids also possess side chains with ionizable groups. For example, the side chain of lysine possesses an ionizable amino group. Because of their structures, alanine, lysine, and the other amino acids can act as effective

$$\frac{pK_1 + pK_2}{2} = PI (6.0)$$

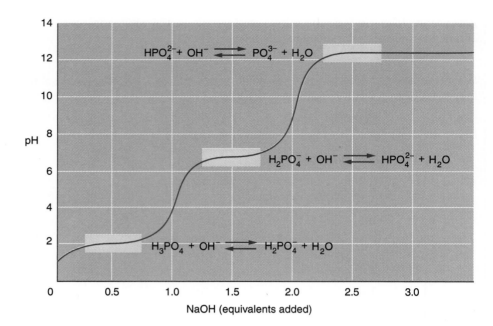

FIGURE 3.18

Titration of Phosphoric Acid with NaOH.

Phosphoric acid (H_3PO_4) is a polyprotic acid that sequentially releases 3 protons when titrated with NaOH.

buffers. The titration and buffering capacity of amino acids are described further in Chapter 5.

Physiological Buffers

The three most important buffers in the body are the bicarbonate buffer, the phosphate buffer, and the protein buffer. Each is adapted to solve specific physiological problems in the body.

BICARBONATE BUFFER Bicarbonate buffer, one of the more important buffers in blood, has three components. The first of these, carbon dioxide, reacts with water to form carbonic acid:

$$CO_2 + H_2O \rightleftharpoons H_2CO_3$$
$$\text{Carbonic acid}$$

Carbonic acid then rapidly dissociates to form H^+ and HCO_3^- ions:

$$H_2CO_3 \rightleftharpoons H^+ + HCO_3^-$$
$$\text{Bicarbonate}$$

Because the concentration of H_2CO_3 is very low in blood, the above equations may be simplified to:

$$CO_2 + H_2O \rightleftharpoons H^+ + HCO_3^-$$

Recall that buffering capacity is greatest at or near the pK_a of the acid–conjugate base pair. Bicarbonate buffer is clearly unusual in that its pK_a is 6.37. It would appear at first glance that bicarbonate buffer is ill-suited to buffer blood. The ratio of HCO_3^- to CO_2 required to maintain a pH of 7.4 is approximately 11 to 1. In other words, the bicarbonate buffer operates in blood near the limit of its buffering power. In addition, the concentrations of CO_2 and HCO_3^- are not exceptionally high. Despite these liabilities, the bicarbonate buffering system is important because both components can be regulated. Carbon dioxide concentration is adjusted by changes in the rate of respiration.

$$CO_2 + H_2O \rightleftharpoons H_2CO_3 \rightleftharpoons HCO_3^- + H^+$$

Bicarbonate concentration is regulated by the kidneys. If bicarbonate concentration ($[HCO_3^-]$) decreases, the kidneys remove H^+ from the blood, shifting the equilibrium to the right and increasing $[HCO_3^-]$. Carbonic acid lost in this process is quickly replenished by hydrating CO_2, a waste product of cellular metabolism. When excess amounts of HCO_3^- are produced, they are excreted by the kidney. As acid is added to the body's bicarbonate system, $[HCO_3^-]$ decreases and CO_2 is formed. Because the excess CO_2 is exhaled, the ratio of HCO_3^- to CO_2 remains essentially unchanged.

PHOSPHATE BUFFER Phosphate buffer consists of the weak acid–conjugate base pair $H_2PO_4^-/HPO_4^{2-}$ (Figure 3.19):

$$H_2PO_4^- \rightleftharpoons H^- + HPO_4^{2-}$$

Dihydrogen   Hydrogen
phosphate   phosphate

With pK_a 7.2, it would appear that phosphate buffer is an excellent choice for buffering the blood. Although the blood pH of 7.4 is well within this buffer system's capability, the concentrations of $H_2PO_4^-$ and HPO_4^{2-} in blood are too low to have a major effect. Instead, the phosphate system is an important buffer in intracellular fluids where its concentration is approximately 75 mEq/L. Phosphate concentration in extracellular fluids such as blood is about 4 mEq/L. (An equivalent is defined as the mass of acid or base that can furnish or accept 1 mol of H^+ ions. The abbreviation mEq indicates a milliequivalent.) Because the normal pH of

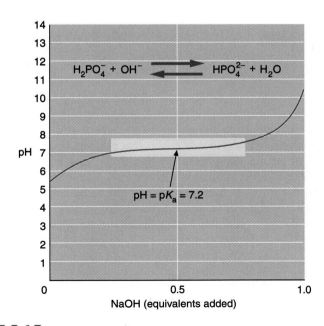

FIGURE 3.19

Titration of $H_2PO_4^-$ by Strong Base.

The shaded band indicates the pH range over which the weak acid–conjugate base pair $H_2PO_4^-/HPO_4^{2-}$ functions effectively as a buffer.

Artificial semipermeable membranes are routinely used in biochemical laboratories to separate small solutes from larger solutes. For example, this technique (referred to as **dialysis**) is used as an important early step in protein purification. An impure protein-containing specimen is placed in a cellophane dialysis bag (Figure 3B), which is then suspended in flowing distilled water or in a buffered solution. After a certain time, all the small solutes leave the bag. The protein solution, which may contain many high-molecular-weight impurities, is then ready for further purification.

Hemodialysis is a clinical application of dialysis that removes toxic waste from the blood of patients suffering from temporary or permanent renal failure. All constituents of blood except blood cells and the plasma proteins move freely between blood and the dialyzing fluid. Because dialysis tubing allows passage of nutrients (glucose, amino acids) and essential electrolyes (Na^+, K^+), these and other vital substances are included in the dialyzing fluid. Their inclusion prevents

a net loss of these materials from the blood. Because no waste products such as urea and uric acid are in the dialyzing fluid, these substances are lost from the blood in large quantities.

Although hemodialysis is very effective in removing toxic waste from the body, it does not solve all the problems brought on by renal failure. For example, until recently, patients suffering from renal failure often became anemic because they lacked a protein hormone called *erythropoietin*, which is normally secreted by the kidney. (Erythropoietin stimulates red blood cell synthesis.) Because of DNA technology (Chapter 18), erythropoietin is now readily administered to dialysis patients.

FIGURE 3B
Dialysis.

Proteins are routinely separated from low-molecular-weight impurities by dialysis. When a dialysis bag containing a cell extract is suspended in water or a buffered solution, small molecules pass out through the membrane's pores. If the solvent outside the bag is continually renewed, all low-molecular-weight impurities are removed from the inside.

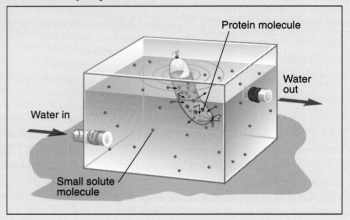

Protein molecule
Water out
Water in
Small solute molecule

KEY CONCEPTS 3.6

The most important buffers in the body are the bicarbonate buffer (blood), the phosphate buffer (intracellular fluids), and the protein buffer.

cell fluids is approximately 7.2 (the range is from 6.9 to 7.4), an equimolar mixture of $H_2PO_4^-$ and HPO_4^{2-} is typically present. Although cells contain other weak acids, these substances are unimportant as buffers. Their concentrations are quite low, and their pK_a values are significantly lower than intracellular pH. For example, lactic acid has a pK_a of 3.86.

PROTEIN BUFFER Proteins are a significant source of buffering capacity. Composed of amino acids linked together by peptide bonds, proteins contain several types of ionizable groups in side chains that can donate or accept protons. Because protein molecules are present in significant concentration in living organisms, they are powerful buffers. For example, the oxygen-carrying protein hemoglobin is the most abundant biomolecule in red blood cells. Because of its structure and high cellular concentration, hemoglobin plays a major role in maintaining blood pH. Also present in high concentrations and buffering the blood are the serum albumins and other proteins.

QUESTION 3.4

Severe diarrhea is one of the most common causes of death in young children. One of the principal effects of diarrhea is the excretion of large quantities of sodium bicarbonate. In which direction does the bicarbonate buffer system shift under this circumstance? What is the resulting condition called?

SUMMARY

1. Water molecules (H_2O) are composed of two atoms of hydrogen and one of oxygen. Each hydrogen atom is linked to the oxygen atom by a single covalent bond. The oxygen-hydrogen bonds are polar and water molecules are dipoles. One consequence of water's polarity is that water molecules are attracted to each other by the electrostatic force between the oxygen of one molecule and the hydrogen of another. This attraction is called a hydrogen bond.

2. Noncovalent bonds are relatively weak and, therefore, easily disrupted. They play a vital role in determining the physical and chemical properties of water and biomolecules. Ionic interactions occur between charged atoms or groups. Although each hydrogen bond is not especially strong when compared to covalent bonds, large numbers of them have a significant effect on the molecules involved. Van der Waals forces occur between permanent and/or induced dipoles. They may be attractive or repulsive.

3. Water has an exceptionally high heat capacity. Its boiling and melting points are significantly higher than those of compounds of comparable structure and molecular weight. Hydrogen bonding is responsible for this anomalous behavior.

4. Water is also a remarkable solvent. Water's dipolar structure and its capacity to form hydrogen bonds enable it to dissolve many ionic and polar substances.

5. Liquid water molecules have a limited capacity to ionize to form a hydrogen ion (H^+) and a hydroxide ion (OH^-). When a solution contains equal amounts of H^+ and OH^- ions, it is said to be neutral. Solutions with an excess of H^+ are acidic, whereas those with a greater number of OH^- are basic. Because organic acids do not completely dissociate in water, they are referred to as weak acids. The acid dissociation constant K_a is a measure of the strength of a weak acid. Because K_a values vary over a wide range, pK_a values ($-\log K_a$) are used instead.

6. The hydrogen ion is one of the most important ions in biological systems. The pH scale conveniently expresses hydrogen ion concentration. pH is defined as the negative logarithm of the hydrogen ion concentration.

7. Because hydrogen ion concentration affects living processes so profoundly, it is not surprising that regulating pH is a universal and essential activity of living organisms. Hydrogen ion concentration is typically kept within narrow limits. Because buffers combine with H^+ ions, they help maintain a relatively constant hydrogen ion concentration. The ability of a solution to resist pH changes is called buffering capacity. Most buffers consist of a weak acid and its conjugate base.

8. Several physical properties of liquid water change when solute molecules are dissolved. The most important of these for living organisms is osmotic pressure, the pressure that prevents the flow of water across cellular membranes.

SUGGESTED READINGS

Lang, F., and Waldegger, S., Regulating Cell Volume, *Am. Sci.* 85:456–463, 1997.

Montgomery, R., and Swenson, C. A., *Quantitative Problems in Biochemical Sciences*, 2nd ed., W. A. Freeman, New York, 1976.

Pennisi, E., Water, Water Everywhere, *Sci. News* 143:121–125, 1993.

Rand, R. P., Raising Water to New Heights, *Science* 256:618, 1992.

Segel, I. H., *Biochemical Calculations*, 2nd ed., John Wiley and Sons, New York, 1976.

Stewart, P. A., *How to Understand Acid-Base: A Quantitative Acid-Base Primer for Biology and Medicine*, Elsevier, New York, 1981.

Stillinger, F. H., Water Revisited, *Science* 209:451–457, 1980.

Wiggins, P. M., Role of Water in Some Biological Processes, *Microbiol. Rev.* 54:432–449, 1990.

KEY WORDS

acid, *80*

acidosis, *81*

alkalosis, *81*

amphipathic molecule, *74*

base, *80*

buffer, *81*

conjugate base, *80*

dialysis, *88*

dipole, *66*

Donnan effect, *76*

electrostatic interaction, *67*

hydrogen bond, *67*

hydrophobic interaction, *73*

hypertonic solution, *76*

hypotonic solution, *76*

isotonic solution, *75*

Le Chatelier's principle, *82*

London dispersion force, *69*

membrane potential, *76*

micelle, *74*

osmolytes, *79*

osmosis, *74*

osmotic pressure, *75*

pH scale, *80*

polar, *66*

salt bridges, *68*

solvation sphere, *72*

van der Waals force, *68*

weak acid, *80*

weak base, *80*

REVIEW QUESTIONS

1. Which of the following are acid–conjugate base pairs?
   a. H_2CO_3, CO_3^{2-}
   b. $H_2PO_4^-$, PO_4^{3-}
   c. HCO_3^-, CO_3^{2-}
   d. H_2O, OH^-

2. What is the hydrogen ion concentration in a solution at pH 8.3?

3. Consider the following titration curve. Estimate the effective buffer range.

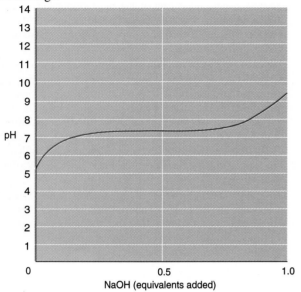

4. Describe how you would prepare a 0.1 M phosphate buffer with a pH of 7.2. What ratio of salt to acid would you use?

5. Which of the following compounds can form hydrogen bonds with like molecules or with water?

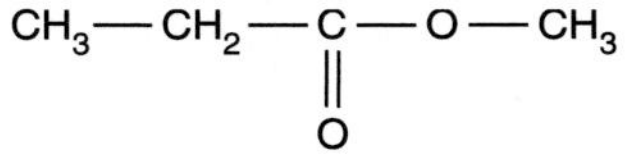

   CH_3—CH_2—$\overset{\underset{\|}{O}}{C}$—$O$—$CH_3$

   (a)

   CH_3—$\overset{\underset{\|}{O}}{C}$—$\overset{\overset{H}{|}}{N}$—$CH_3$

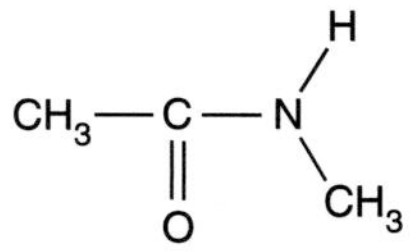

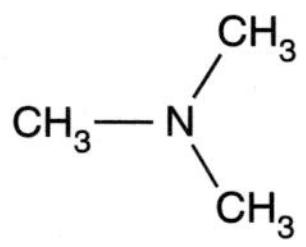

   (b)

   CH_3—$N\overset{CH_3}{\underset{CH_3}{\diagup}}$

   (c)

   CH_3—O—CH_2—O—H

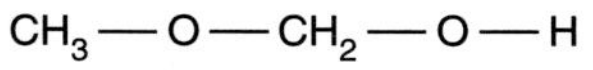

   (d)

6. What is the osmolarity of a 1.3 M solution of sodium phosphate (Na_3PO_4)? Assume 85% ionization for this solution.

7. What direction does water flow when a dialysis bag containing a 3 M solution of the sugar fructose is placed in each of the following solutions?
   a. 1 M sodium lactate
   b. 3 M sodium lactate
   c. 4.5 M sodium lactate

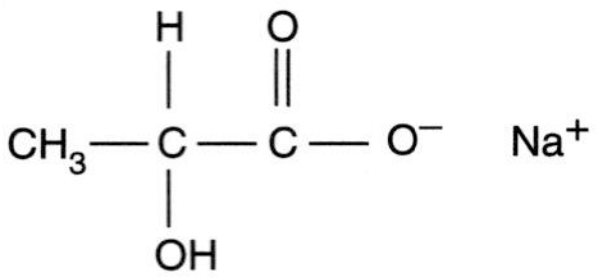

   CH_3—$\overset{\overset{\displaystyle H}{|}}{\underset{\underset{\displaystyle OH}{|}}{C}}$—$\overset{\overset{\displaystyle O}{\|}}{C}$—$O^-$ Na^+

8. What interactions occur between the following molecules and ions?
   a. water and ammonia
   b. lactate and ammonium ion
   c. benzene and octane
   d. carbon tetrachloride and chloroform
   e. chloroform and diethylether

9. A solution containing 56 mg of a protein in 30 mL of distilled water exerts an osmotic pressure of 0.01 atm at $T = 25°C$. Determine the molecular weight of the unknown protein.

10. Tyrosine is an amino acid.

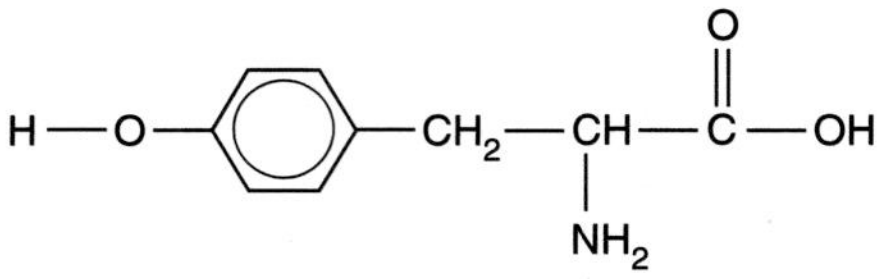

    H—O—⬡—CH_2—$\overset{\underset{\underset{\displaystyle NH_2}{|}}{}}{CH}$—$\overset{\overset{\displaystyle O}{\|}}{C}$—$OH$

    Which atoms in this molecule can form hydrogen bonds?

11. Briefly define the following terms:
    a. hydrogen bond
    b. pH
    c. buffer
    d. osmotic pressure
    e. osmolytes
    f. isotonic
    g. amphipathic
    h. hydrophobic interactions
    i. dipole
    j. induced dipole

12. Which of the following molecules would you expect to have a dipole moment?
    a. CCl_4
    b. $CHCl_3$
    c. H_2O
    d. CH_3OCH_3
    e. CH_3CH_3
    f. H_2

13. Which of the following molecules would you expect to form micelles?
    a. NaCl
    b. CH_3COOH
    c. $CH_3COO^-NH_4^+$
    d. $CH_3(CH_2)_{10}COO^-Na^+$
    e. $CH_3(CH_2)_{10}CH_3$

14. Bicarbonate is one of the main buffers of the blood, and phosphate is the main buffer of the cells. Suggest a reason why this observation is true.

15. Describe how you can increase the buffering capacity of a 0.1 M acetate buffer.

16. Which of the following molecules or ions are weak acids? Explain.
    a. HCl
    b. $H_2PO_4^-$
    c. CH_3COOH
    d. HNO_3
    e. HSO_4^-

17. Which of the following species can form buffer systems?
    a. $NH_4^+ Cl^-$
    b. CH_3COOH, HCl
    c. CH_3COOH, $CH_3COO^-Na^+$
    d. H_3PO_4, PO_4^{3-}

18. What effect does hyperventilation have on blood pH?

19. What is the relationship between osmolarity and molarity?

20. Is it possible to prepare a buffer consisting of only carbonic acid and sodium carbonate?

21. Calculate the ratio of dihydrogen phosphate/hydrogen phosphate in blood at pH 7.4. The K_a is 6.3×10^{-8}.

22. Calculate the pH of a solution prepared by mixing 300 mL of 0.25 M sodium hydrogen ascorbate and 150 mL of 0.2 M HCl. The pK_{a1} of ascorbic acid is 5×10^{-5}.

23. What is the pH of a solution that is 1×10^{-8} in HCl?

THOUGHT QUESTIONS

1. Many fruits can be preserved by candying. The fruit is immersed in a highly concentrated sugar solution, then the sugar is allowed to crystallize. How does the sugar preserve the fruit?

2. Explain why ice is less dense than water. If ice were not less dense than water, how would the oceans be affected? How would the development of life on earth be affected?

3. Why can't seawater be used to water plants?

4. Explain how the acids produced in metabolism are transported to the liver without greatly affecting the pH of the blood.

5. The pH scale is valid only for water. Why is this so?

6. Gelatin is a mixture of protein and water that is mostly water. Explain how the water-protein mixture becomes a solid.

7. Many molecules are polar, yet they do not form significant hydrogen bonds. What is so unusual about water that hydrogen bonding becomes possible?

8. Water has been described as the universal solvent. If this statement were strictly true, could life have arisen in a water medium? Explain.

9. Alcohols (ROH) are structurally similar to water. Why are alcohols not as powerful a solvent as water for ionic compounds? (*Hint*: Methanol is a better solvent for ionic compounds than is propanol.)

10. During stressful situations, some cells in the body convert glycogen to glucose. What effect does this conversion have on cellular osmotic balance? Explain how cells handle this situation.

11. Would you expect a carboxylic acid group within the water-free interior of a protein to have a higher or lower K_a than if it occurs on the protein's surface where it is hydrated?

12. The strength of ionic interactions is weaker in water than in an anhydrous medium. Explain how water weakens these interactions.

Energy

Energy Transformation. The cheetah, the fastest land animal on earth, transforms the chemical bond energy in food to the energy that sustains its living processes.

All living organisms have an unrelenting requirement for energy. The flow of energy that sustains life on earth originates in the sun, where thermonuclear reactions generate radiant energy. A small amount of the solar energy that reaches the earth is captured by organisms such as plants and certain microorganisms. During photosynthesis, organisms called photoautotrophs convert solar energy into the chemical bond energy of sugar molecules. This bond energy is used to produce the vast array of organic molecules found in living organisms and to drive processes such as active transport, cell division, endocytosis, and muscle contraction. Ultimately, chemical bond energy is converted into heat, which is then dissipated into the environment.

Every event in the universe from the collisions among individual atoms in laboratory test tubes to the explosions of stars in deep space involves energy. The energy in the universe comes in many interconvertible forms: gravitational, nuclear, radiant, heat, mechanical, electrical, and chemical. Despite its obvious importance, however, a precise definition of energy remains elusive. According to modern scientific theory, energy is *the* basic constituent of the universe. The relationship between matter and its energy equivalent is defined by Einstein's famous equation $E = mc^2$. The total energy (E) in joules (kg m^2/s^2) in a particle is equal to the mass (m) in kilograms of the particle multiplied by the speed of light ($c = 3.0 \times 10^8$ m/s) squared. Energy is more commonly defined, however, as the capacity to do work. Work is organized molecular motion that involves the movement of an object caused by the use of force. Cellular work is performed by thousands of molecular machines. Each machine repetitively performs a single task powered directly or indirectly by the energy provided by adenosine triphosphate (ATP). Typical examples of work in living cells include the creation of concentration gradients across membranes and the synthesis of biomolecules. The most common types of molecular machines are contractile proteins, transporter complexes, and enzymes.

The investigation of energy transformations that accompany physical and chemical changes in matter is called **thermodynamics**. The principles of thermodynamics are used to evaluate the flow and interchanges of matter and energy. **Bioenergetics**, a branch of thermodynamics, is the study of energy transformations in living organisms. It is especially useful in determining the direction and extent to which specific biochemical reactions occur. These reactions are affected by three factors. Two of these, **enthalpy** (total heat content) and **entropy** (disorder), are related to the first and second laws of thermodynamics, respectively. The third factor, called **free energy** (energy available to do chemical work), is derived from a mathematical relationship between enthalpy and entropy.

The chapter begins with a brief description of thermodynamic concepts and their relationship to biochemical reactions. This is followed by a discussion of free energy, a useful measure of the spontaneity of chemical reactions. The chapter ends with a description of the structure and function of ATP and other high-energy compounds, and a discussion of a remarkable form of autotrophic energy generation that occurs on the ocean floor (Special Interest Box 4.1). The discussion of oxidation-reduction (redox) reactions, the cell's primary mechanism for generating energy, is deferred to Chapter 9 where their roles in metabolism will be described.

4.1 THERMODYNAMICS

The modern concept of energy is an invention of the industrial revolution. Investigations of the relationship between mechanical work and heat by engineers, physicists, mathematicians, physiologists, and physicians in the nineteenth century led eventually to the discovery of a set of rules, called the *laws of thermodynamics*, that describe energy transformations. The three laws of thermodynamics are as follows:

1. **The First Law of Thermodynamics:** The total amount of energy in the universe is constant. Energy can neither be created nor destroyed, but it can be transformed from one form into another.

2. **The Second Law of Thermodynamics:** The disorder of the universe always increases. In other words, all chemical and physical processes occur spontaneously only when disorder is increased.

3. **The Third Law of Thermodynamics:** As the temperature of a perfect crystalline solid approaches absolute zero (0 K), disorder approaches zero.

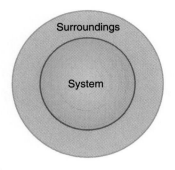

FIGURE 4.1

A Thermodynamic Universe.

A universe consists of a system and its surroundings.

The first two laws are powerful tools that biochemists use to investigate the energy transformations in living systems.

Thermodynamics is concerned with heat and energy transformations. Such transformations are considered to take place in a "universe" composed of a system and its surroundings (Figure 4.1). A system is defined according to the interests of the investigator. For example, a system may be defined as an entire organism or a single cell or a reaction occurring in a flask. In an *open system* matter and energy are exchanged between the system and its surroundings. If only energy can be exchanged with the surroundings, then the system is said to be *closed*. Living organisms, which consume nutrients from their surroundings and release waste products into it, are open systems.

Knowledge of thermodynamic functions such as enthalpy, entropy, and free energy enables biochemists to predict whether a process is spontaneous (thermodynamically favorable). This does not indicate that a reaction will occur (is kinetically favorable), but that it can occur under the right set of conditions. Reactions are kinetically favorable only if there is sufficient energy available to the system undergoing change.

Several thermodynamic properties are state functions. The values of state functions depend only on their initial and final states and are independent of the pathway taken to get from the initial state to the final state. For example, the energy content of a glucose molecule is the same regardless of whether it was synthesized via photosynthesis or the breakdown of lactose (milk sugar). How the energy of a reaction is distributed, however, is not fixed but is governed by the system or pathway undergoing change. For example, living cells use some of the energy in glucose molecules to perform cellular work such as muscle contraction. The remainder is released as disordered heat energy. **Work** (the displacement or movement of an object by a force) and heat are not state functions; that is, their values vary with the pathway. If glucose molecules are instead ignited in a laboratory dish, the overall reaction is the same but all of the chemical bond energy in the glucose is transformed directly into heat and little or no measurable work is performed. The energy content of the glucose molecules is the same in each process. The work accomplished by each process is different.

The exchange of energy between a system and its surroundings can occur in only two ways: Heat (q), random molecular motion, may be transferred to or from the system, or the system may do work (w) on its surroundings or have work done on it by its surroundings. Energy is transferred as heat when the system and its surroundings are at different temperatures. When heat is absorbed by a system, q is a positive number. A negative value for q indicates that heat has been lost from a system. Energy is transferred as work when an object is moved by force. When w is positive, work is done on the system by the surroundings. Work by the system on the surroundings is signified by a negative w value.

The First Law of Thermodynamics

The first law states that energy cannot be created or destroyed. In other words, the total amount of energy in a system and its surroundings, referred to as internal energy (E), must be the same before and after any process occurs. When work is being done and heat is being transferred, the first law can be stated as

$$\Delta E = q + w \qquad (1)$$

where

ΔE = the change in energy of the system
q = the heat absorbed by the system from the surroundings
w = the work done on the system by the surroundings

Chemists have defined the term enthalpy (H), which is related to internal energy by the equation

$$H = E + PV \qquad (2)$$

where PV = pressure-volume work, that is, the work done on or by a system that involves changes in pressure and volume

In biochemical systems in which pressure is constant, enthalpy changes (ΔH) are equal to the heat gained or lost by the system ($\Delta E = q$). When the volume change in such a process is relatively negligible, as occurs in living cells, enthalpy change is essentially equal to the internal energy:

$$\Delta H = \Delta E \qquad (3)$$

If ΔH is negative ($\Delta H < 0$), the reaction or process gives off heat and is referred to as **exothermic**. If ΔH is positive ($\Delta H > 0$), heat is absorbed from the surroundings. Such processes are called **endothermic**. In **isothermic** processes ($\Delta H = 0$), heat is not exchanged with the surroundings.

Equation (3) indicates that the total energy change of a biological system is equivalent to the heat evolved or absorbed by the system. Because the enthalpy of a reactant or product is a state function (independent of pathway), then the enthalpy change for any reaction forming that substance can be used to calculate the ΔH of a reaction involving that substance. If the sum of the ΔH values ($\Sigma\Delta H$) for both the reactants and the products is known, then the enthalpy change for the reaction can be calculated by using the following equation:

$$\Delta H_{reaction} = \Sigma\Delta H_{products} - \Sigma\Delta H_{reactants} \qquad (4)$$

The standard enthalpy of formation per mole, ΔH_f, is commonly used in enthalpy calculations. H_f is the energy evolved or absorbed when 1 mol of a substance is formed from its most stable elements. Note that equation (4) cannot predict the direction of any chemical reaction. It can determine only the heat flow. An enthalpy calculation for a reaction at constant pressure is given below.

KEY CONCEPTS 4.1

At constant pressure, a system's enthalpy change ΔH is equal to the flow of heat energy. If ΔH is negative, the reaction or process is exothermic. If ΔH is positive, the reaction or process is endothermic. In isothermic processes no heat is exchanged with the surroundings.

PROBLEM 4.1

Given the following ΔH values, calculate ΔH for the reaction

$$6\ CO_2 + 6\ H_2O \longrightarrow C_6H_{12}O_6 + 6\ O_2$$

	ΔH_f^*	
	kcal[†]/mol	kJ[‡]/mol
$C_6H_{12}O_6$	− 304.7	− 1274.9
CO_2	− 94.0	− 393.3
H_2O	− 68.4	− 286.2
O_2	0	0

* ΔH_f is the energy evolved to produce a compound from its elements.

† 1 kcal is the energy required to raise the temperature of 1000 g of water 1°C.

‡ The joule (J) is a unit of energy that is gradually replacing the calorie (cal) in scientific usage. One calorie is equal to 4.184 joules.

Solution

The total enthalpy for a reaction is equal to the sum of enthalpy values of the products minus those of the reactants.

$$6\ CO_2\ +\quad 6\ H_2O\quad \longrightarrow\quad C_6H_{12}O_6\ +\ 6\ O_2$$

$$6(-393.3)\ +\ 6(-286.2) \qquad -1274.9\ +\ 6(0)$$

$$-2359.8\ +\ (-1717.2) \qquad -1274.9$$

$$\Delta H = -1274.9 - (-4077.0) = 2802.1\ kJ/mol$$

The positive ΔH indicates that the reaction is endothermic.

Second Law of Thermodynamics

The first law accounts for the energy changes that can occur during a process, but it cannot be used to predict whether and to what extent a specific process will occur. In some circumstances, whether certain processes occur appears to be obvious, for example, the behavior of ice at room temperature or gasoline in an internal combustion engine. Experience tells us that ice melts at temperatures above 0°C and that gasoline molecules can be converted in the presence of oxygen to CO_2 and H_2O. When physical or chemical changes occur with the release of energy, such changes are said to be **spontaneous**. When a constant input of energy is required to support a change, a nonspontaneous process is occurring. Experience convinces us that certain processes will not occur: ice will not form at temperatures above 0°C and gasoline molecules are not formed from an engine's exhaust fumes. In other words, we intuitively understand that there is directionality to these processes and that predictions about their outcome are easily made. When experience cannot be relied on to allow us to make predictions concerning spontaneity and direction, the second law can be used. According to the second law, all spontaneous processes occur in the direction that increases the total disorder of the universe (a system and its surroundings) (Figure 4.2). As a result of spontaneous processes, matter and energy become more disorganized. Gasoline molecules, for example, are hydrocarbons in which carbon atoms are linked in an orderly arrangement. When gasoline burns, the carbon atoms in the gaseous products are randomly dispersed (Figure 4.3). Similarly, the energy that is released as gasoline burns becomes more disordered; it becomes less concentrated and less useful. In a car engine, increased gas pressure in the cylinders drives the pistons and causes the car to move. When the chemical energy in the gasoline molecules and the kinetic energy that moves the car are compared, it becomes apparent that a significant amount of energy does no useful work. It is dissipated, instead, into the surroundings, as evidenced by the hot engine and exhaust fumes.

The degree of disorder of a system is measured by the state function called **entropy** (S). The more disordered a system is, the greater is its entropy value. According to the second law, the entropy change of the universe is positive for every spontaneous process. The increase may take place in any part of the universe (ΔS_{sys} or ΔS_{surr}):

$$\Delta S_{univ} = \Delta S_{sys} + \Delta S_{surr}$$

FIGURE 4.2

A Living Cell as a Thermodynamic System.

(a) The molecules of the cell and its surroundings are in a relatively disordered state. (b) Heat is released from the cell as a consequence of reactions that create order among the molecules inside the cell. This energy increases the random motion, and therefore the disorder, of the molecules outside the cell. This process causes a net positive entropy change. The cell's decrease in entropy is more than offset by an increase in the entropy of the surroundings.

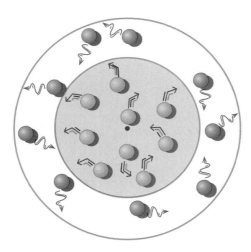
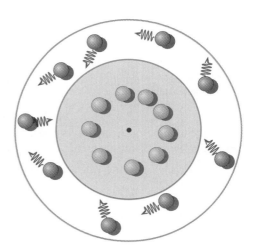

(a)   **(b)**

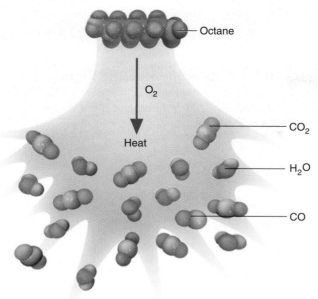

FIGURE 4.3

Gasoline Combustion.

When hydrocarbons such as octane are burned, the release of energy is accompanied by the conversion of highly ordered reactant molecules into relatively disorganized gaseous products such as CO_2 and H_2O. Unfortunately, gasoline combustion is inefficient; that is, other substances such as the poisonous molecule carbon monoxide (CO) that pollute the environment are also released.

Living cells do not increase their internal disorder when they consume and metabolize nutrients. The organism's surroundings increase in entropy instead. For example, the food molecules that humans consume to provide the energy and structural material needed to maintain their complex bodies are converted into vast amounts of disordered waste products (e.g., CO_2, H_2O, and heat) that are discharged into their surroundings.

Although entropy may be considered to be unusable energy, the formation of entropy is not a useless activity. In spontaneous processes, such as the melting of ice and the burning of wood, entropy always increases. Some reactions are said to be entropy driven because the increase in entropy in the system overrides a gain in enthalpy to result in a spontaneous reaction. (By definition, a spontaneous process will occur. The rate at which it occurs, however, may be very rapid or very slow.) In irreversible processes, processes that proceed in only one direction, entropy is now believed to be a driving force. Entropy directs a system toward equilibrium with its surroundings. Once a process reaches equilibrium (i.e., there is no net change in either direction), there is no longer any driving force to propel it. To predict whether a process is spontaneous, the sign of ΔS_{univ} must be known. For example, if the value of ΔS_{univ} for a process is positive (i.e., the entropy of the universe increases), then the process is spontaneous. If ΔS_{univ} is negative, the process does not occur, but the reverse process takes place. The opposite process is spontaneous. If ΔS_{univ} is equal to zero, neither process tends to occur. Organisms that are at equilibrium with their surroundings are dead.

KEY CONCEPTS 4.2

The second law of thermodynamics states that the universe tends to become more disorganized. Entropy increases may take place anywhere in the system's universe. For processes in living organisms, the increase in entropy takes place in their surroundings.

4.2 FREE ENERGY

Although the entropy of the universe always increases in a spontaneous process, measuring it is often impractical because both the ΔS_{sys} and ΔS_{surr} must be known.

A more convenient thermodynamic function for predicting the spontaneity of a process is **free energy**, which can be derived from the expression for ΔS_{univ},

$$\Delta S_{univ} = \Delta S_{surr} + \Delta S_{sys}$$

The ΔS_{surr} is defined as the quantity of heat exchanged per K temperature in the course of a specific chemical or physical change. For an exothermic reaction, heat is released and the value of ΔH is a negative number. Therefore,

$$\Delta S_{surr} = -\Delta H/T$$

By substitution

$$\Delta S_{univ} = -\Delta H/T + \Delta S_{sys}$$

Multiply both sides by $-T$:

$$-T\Delta S_{univ} = \Delta H - T\Delta S_{sys}$$

Josiah Gibbs defined $-T\Delta S_{univ}$ as the state function called the *Gibbs free energy change* or ΔG:

$$\Delta G = \Delta H - T\Delta S_{sys}$$

The change in free energy is negative when ΔS_{univ} is positive, which reflects a spontaneous reaction said to be **exergonic** (Figure 4.4). If the ΔG is positive, the process is said to be **endergonic** (nonspontaneous). When the ΔG is zero, the process is at equilibrium. As with other thermodynamic functions, ΔG provides no information about reaction rates. Reaction rates depend on the precise mechanism by which a process occurs and are dealt with under the study of kinetics (Chapter 6).

Standard Free Energy Changes

A convention known as the *standard state* provides a uniform basis for free energy calculations. The standard free energy, $\Delta G°$, is defined for reactions at 25°C (298 K) and 1.0 atm pressure with all solutes at a concentration of 1.0 M.

The standard free energy change is related to the reaction's equilibrium constant, K_{eq}, the value of the reaction quotient at equilibrium when the forward and reverse reaction rates are equal.

The change in free energy for a reaction

$$aA + bB \rightleftharpoons cC + dD$$

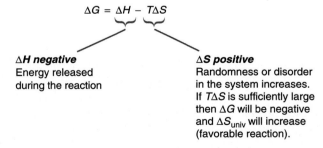

FIGURE 4.4

The Gibbs Free Energy Equation.

At constant pressure, enthalpy (H) is essentially equal to the total energy content of the system. A process is spontaneous if it decreases free energy. Free energy changes ΔG are negative if enthalpy decreases or if the entropy term $T\Delta S$ is sufficiently large.

is related to the reaction's equilibrium constant:

$$K_{eq} = \frac{[C]^c[D]^d}{[A]^a[B]^b}$$

Based on the observation that the free energy of an ideal gas is dependent on its pressure (concentration) and that the state function G can be manipulated in the same way as the state function H, the following equation was derived:

$$\Delta G = \Delta G° + RT \ln \frac{[C]^c[D]^d}{[A]^a[B]^b}$$

If the reaction is allowed to go to equilibrium, the $\Delta G = 0$ and the expression reduces to

$$\Delta G° = -RT \ln K_{eq}$$

This equation allows the calculation of $\Delta G°$ if the K_{eq} is known. Because most biochemical reactions take place at or near pH 7 ($[H^+] = 1.0 \times 10^{-7}$ M), this exception is made in the 1.0 M solute rule in bioenergetics and the free energy change is expressed as $\Delta G°'$. Problem 4.2 is a free energy problem.

PROBLEM 4.2

For the reaction $HC_2H_3O_2 \rightleftharpoons C_2H_3O_2^- + H^+$, calculate $\Delta G°$ and $\Delta G°'$. Assume that $T = 25°C$. The ionization constant for acetic acid is 1.8×10^{-5}. Is this reaction spontaneous? Recall that

$$K_{eq} = \frac{[C_2H_3O_2^-][H^+]}{[HC_2H_3O_2]}$$

Solution

1. Calculate $\Delta G°$.

   $\Delta G° = -RT \ln K_{eq}$

   $= -(8.315 \text{ J/mol·K})(298 \text{ K})\ln(1.8 \times 10^{-5})$

   $= 27{,}071 \text{ J} = 27.1 \text{ kJ/mol}$

   The $\Delta G°$ indicates that under these conditions the reaction is not spontaneous.

2. Calculate $\Delta G°'$. Use the relation between free energy change and standard free energy change.

   $$\Delta G = \Delta G° + RT \ln \frac{[C]^c[D]^d}{[A]^a[B]^b}$$

   For this example the expression becomes

   $$\Delta G°' = \Delta G° + RT \ln \frac{[C_2H_3O_2^-][H^+]}{[HC_2H_3O_2]}$$

   Recall that under standard conditions, [acetate] and [acetic acid] are both 1 M. Substituting values, we have

   $\Delta G°' = 27071 \text{ J/mol} + (8.315 \text{ J/mol·K})(298\text{K})(\ln 10^{-7})$

   $= 27071 - 39{,}939$

   $= -12867.54 = -12.9 \text{ kJ/mol}$

Under the conditions specified for $\Delta G°$ (i.e., 1 M concentrations for all reactants including H^+) the ionization of acetic acid is not spontaneous, as indicated by the positive $\Delta G°$. When the pH value is 7, however, the reaction becomes spontaneous. A low $[H^+]$ makes the ionization of a weak acid such as acetic acid a more likely process, as indicated by the negative $\Delta G°'$.

Coupled Reactions

Many chemical reactions within living organisms have positive $\Delta G^{\circ\prime}$ values. Fortunately, free energy values are additive in any reaction sequence.

$$A + B \rightleftharpoons C + D \qquad\qquad \Delta G^{\circ\prime}_{\text{reaction 1}} \qquad\qquad (1)$$

$$C + E \rightleftharpoons F + G \qquad\qquad \Delta G^{\circ\prime}_{\text{reaction 2}} \qquad\qquad (2)$$

$$A + B + E \rightleftharpoons D + F + G \quad \Delta G^{\circ\prime}_{\text{overall}} = \Delta G^{\circ\prime}_{\text{reaction 1}} + \Delta G^{\circ\prime}_{\text{reaction 2}} \quad (3)$$

Note that reactions (1) and (2) have a common intermediate, C. If the net $\Delta G^{\circ\prime}$ value ($\Delta G^{\circ\prime}_{\text{overall}}$) is sufficiently negative, forming the products F and G is an exergonic process.

The conversion of glucose-6-phosphate to fructose-1,6-bisphosphate illustrates the principle of coupled reactions (Figure 4.5). The common intermediate in this reaction sequence is fructose-6-phosphate. Because the formation of fructose-6-phosphate from glucose-6-phosphate is endergonic ($\Delta G^{\circ\prime}$ is $+1.7$ kJ/mol), the reaction is not expected to proceed as written (at least under standard conditions). The conversion of fructose-6-phosphate to fructose-1,6-bisphosphate is strongly exergonic because it is coupled to the cleavage of the phosphoanhydride bond of ATP. (The cleavage of ATP's phosphoanhydride bond to form ADP yields approximately -30.5 kJ/mol. ATP in living organisms is discussed in Section 4.3). Because $\Delta G^{\circ\prime}_{\text{overall}}$ for the coupled reactions is negative, the reactions do proceed in the direction written at standard conditions.

PROBLEM 4.3

Glycogen is synthesized from glucose-1-phosphate. To be incorporated into glycogen, glucose-1-phosphate is converted to a derivative of the nucleotide uridine disphosphate (UDP). The UDP serves as an excellent leaving group in the condensation reaction to form the glycogen polymer. The reaction is

$$\text{Glucose-1-phosphate} + \text{UTP} \longrightarrow \text{UDP-glucose} + \text{PP}_i$$

If the $\Delta G^{\circ\prime}$ value for this reaction is approximately zero, is this reaction favorable? If PP_i (pyrophosphate) is hydrolyzed,

$$\text{PP}_i + \text{H}_2\text{O} \longrightarrow 2\,\text{P}_i$$

The loss in free energy ($\Delta G^{\circ\prime}$) is -33.5 kJ. How does this second reaction affect the first one? What is the overall reaction? Determine the $\Delta G^{\circ\prime}_{\text{overall}}$ value.

Solution

The overall reaction is

$$\text{Glucose-phosphate} + \text{UTP} \longrightarrow \text{UDP-glucose} + 2\,\text{P}_i$$

$$\begin{aligned}
\Delta G^{\circ\prime}_{\text{overall}} &= \Delta G^{\circ\prime}_{\text{reaction 1}} + \Delta G^{\circ\prime}_{\text{reaction 2}} \\
&= 0 + (-33.5 \text{ kJ}) \\
&= -33.5 \text{ kJ}
\end{aligned}$$

The hydrolysis of PP_i drives the formation of UDP-glucose to the right.

QUESTION 4.1

Within living cells the concentrations of ATP and the products of its hydrolysis (ADP and P_i) are significantly lower than the standard 1 M concentrations. Therefore the actual free energy of hydrolysis of ATP ($\Delta G'$) differs from the standard free energy ($\Delta G^{\circ\prime}$). Unfortunately, it is difficult to obtain an accurate measure of the concentrations of cellular components. For this reason, only estimates

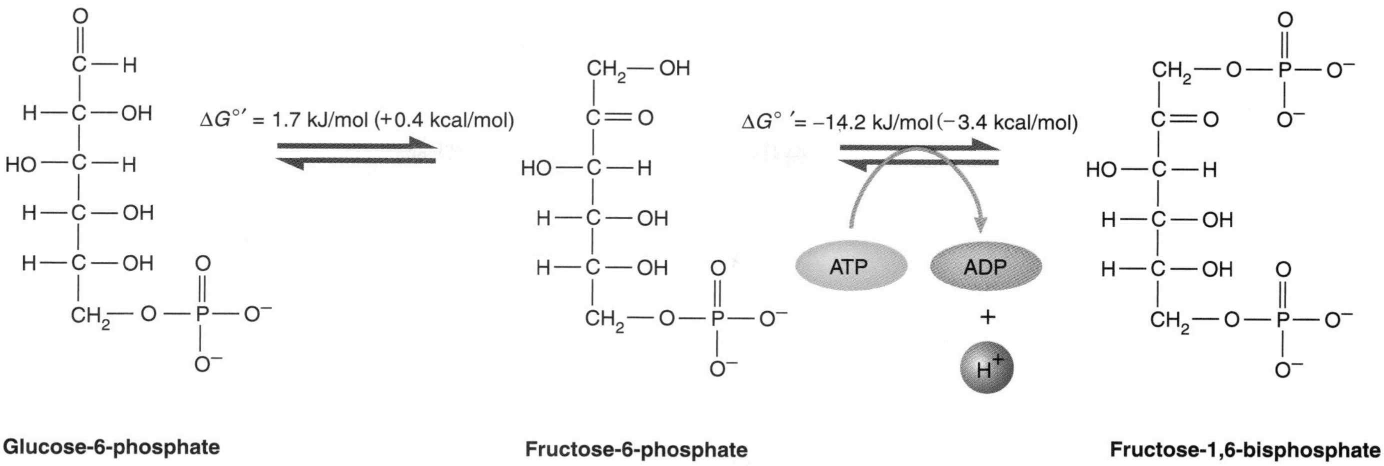

Glucose-6-phosphate   Fructose-6-phosphate   Fructose-1,6-bisphosphate

FIGURE 4.5

A Coupled Reaction.

The net $\Delta G^{\circ\prime}$ value for the two reactions is -12.6 kJ/mol (-3.0 kcal/mol).

can be made. The following equation includes a correction for nonstandard concentrations:

$$\Delta G' = \Delta G^{\circ\prime} + RT \ln \frac{[ADP][P_i]}{[ATP]}$$

The temperature is 37°C. Assume that the pH is 7. Calculate the free energy of hydrolysis of ATP if the concentrations (mM) within a liver cell are as follows:

ATP = 4.0,   ADP = 1.35, P_i = 4.65

$\Delta G^{\circ\prime} = -30.5$ kJ/mol

The Hydrophobic Effect Revisited

Understanding the spontaneous aggregation of nonpolar substances in water is enhanced by consideration of thermodynamic principles. When nonpolar molecules are mixed with water, they disrupt water's energetically favorable hydrogen-bonded interactions. The water molecules then form highly ordered cagelike structures around each nonpolar molecule. The hydrogen bonds that stabilize these structures restrict the motion of the water molecules, thus resulting in a decrease in entropy. Consequently, the free energy of dissolving nonpolar molecules is unfavorable (i.e., ΔG is positive because ΔH is positive and $-T\Delta S$ is strongly positive). The decrease in entropy, however, is proportional to the surface area of contact between nonpolar molecules and water. The aggregation of nonpolar molecules significantly decreases the surface area of their contact with water and thus the water becomes less ordered (i.e., the entropy change, ΔS, is now positive). Because $-T\Delta S$ becomes negative, the free energy of the process is negative, and therefore, it proceeds spontaneously.

4.3 THE ROLE OF ATP

Adenosine triphosphate (ATP) plays an extraordinarily important role within living cells. The hydrolysis of ATP (Figure 4.6) immediately and directly provides the free energy to drive an immense variety of endergonic biochemical

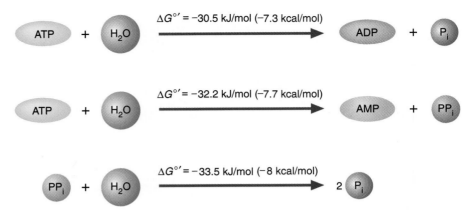

$\Delta G^{\circ\prime} = -30.5$ kJ/mol (-7.3 kcal/mol)

ATP + H_2O $\longrightarrow$ ADP + P_i

$\Delta G^{\circ\prime} = -32.2$ kJ/mol (-7.7 kcal/mol)

ATP + H_2O $\longrightarrow$ AMP + PP_i

$\Delta G^{\circ\prime} = -33.5$ kJ/mol (-8 kcal/mol)

PP_i + H_2O $\longrightarrow$ 2 P_i

FIGURE 4.6

Hydrolysis of ATP.

ATP may be hydrolyzed to form ADP and P_i (orthophosphate) or AMP (adenosine monophosphate) and PP_i (pyrophosphate). Pyrophosphate may be subsequently hydrolyzed to orthophosphate, releasing additional free energy. The hydrolysis of ATP to form AMP and pyrophosphate is often used to drive reactions with high positive $\Delta G^{\circ\prime}$ values or to ensure that a reaction goes to completion.

reactions. Produced from ADP and P_i with energy released by the breakdown of food molecules and the light reactions of photosynthesis, ATP drives several types of processes (Figure 4.7). These include (1) biosynthesis of biomolecules, (2) active transport of substances across cell membranes, and (3) mechanical work such as muscle contraction.

ATP is ideally suited to its role as universal energy currency because of its structure. ATP is a nucleotide composed of adenine, ribose, and a triphosphate unit (Figure 4.8). The two terminal phosphoryl groups ($-PO_3^{2-}$) are linked by phosphoanhydride bonds. Although anhydrides are easily hydrolyzed, the phosphoanhydride bonds of ATP are sufficiently stable under mild intracellular conditions. Specific enzymes facilitate ATP hydrolysis.

The tendency of ATP to undergo hydrolysis, also referred to as its **phosphate group transfer potential**, is not unique. A variety of biomolecules can transfer phosphate groups to other compounds. Table 4.1 lists several important examples.

FIGURE 4.7

The Role of ATP.

ATP is an intermediate in the flow of energy from food molecules to the biosynthetic reactions of metabolism.

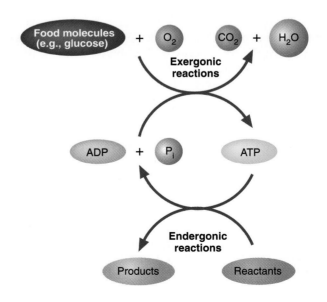

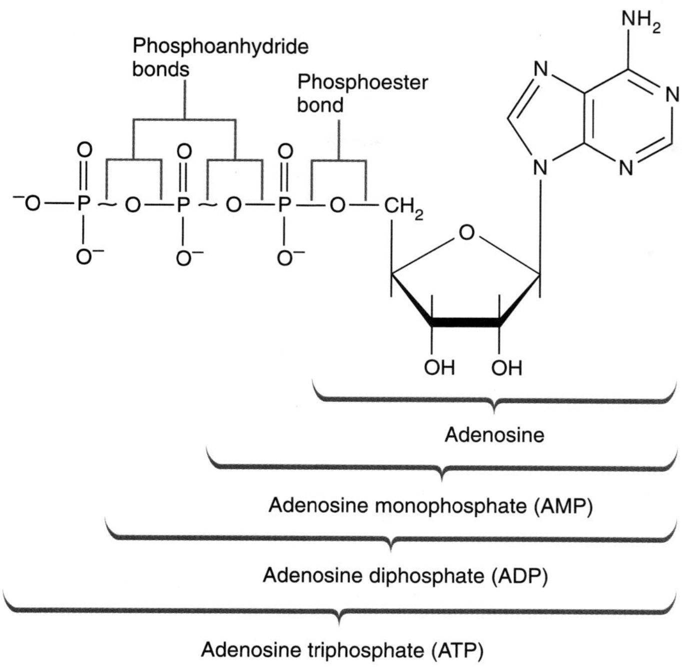

FIGURE 4.8

The Structure of ATP.

The squiggle (~) in ATP indicates that the bonds are easily hydrolyzed.

Phosphorylated compounds with high negative $\Delta G^{\circ\prime}$ values of hydrolysis have larger phosphate group transfer potentials than those compounds with smaller, negative values. Because ATP has an intermediate phosphate group transfer potential, it can be an intermediate carrier of phosphoryl groups from higher energy compounds such as phosphoenolpyruvate to low-energy compounds (Figure 4.9). ATP is therefore the "energy currency" for living systems, because cells usually transfer phosphate by coupling reactions to ATP hydrolysis. The two phosphoanhydride bonds of ATP are often referred to as "high energy." However, the term *high-energy bond* is now considered inappropriate. The term denotes instability of the bond and, therefore, its ability to participate in reactions rather than the quantitative value of the bond energy. To understand why ATP hydrolysis is so exergonic, several factors must be considered.

TABLE 4.1
Standard Free Energy of Hydrolysis of Selected Phosphorylated Biomolecules

Molecule	$\Delta G^{\circ\prime}$	
	kcal/mol	**kJ/mol**
Glucose-6-phosphate	−3.3	−13.8
Fructose-6-phosphate	−3.8	−15.9
Glucose-1-phosphate	−5	−20.9
ATP $\longrightarrow$ ADP + P_i	−7.3	−30.5
ATP $\longrightarrow$ AMP + PP_i	−7.7	−32.2
PP_i	−8.0	−33.5
Phosphocreatine	−10.3	−43.1
Glycerate-1,3-bisphosphate	−11.8	−49.4
Carbamoyl phosphate	−12.3	−51.5
Phosphoenolpyruvate	−14.8	−61.9

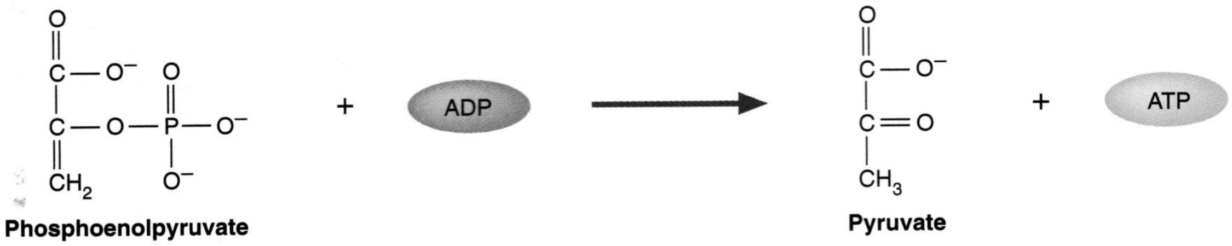

(a)

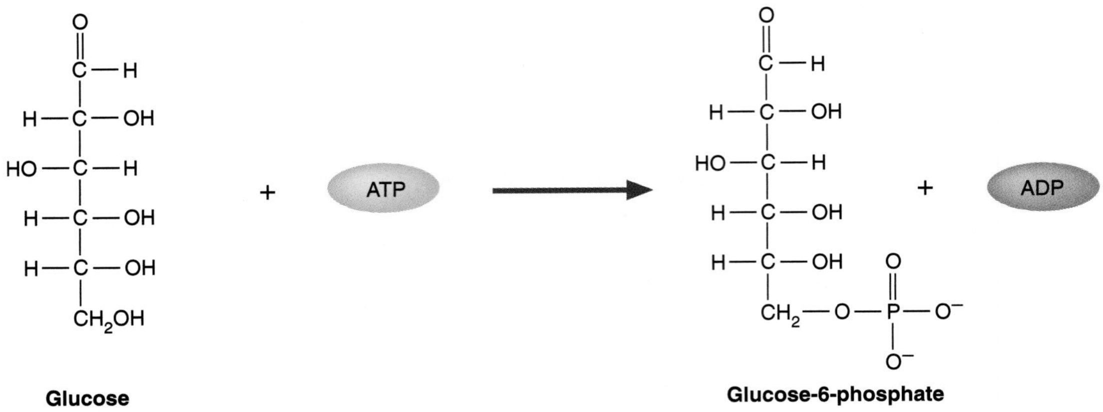

(b)

FIGURE 4.9

Transfer of Phosphoryl Groups.

(a) Transfer of a phosphoryl group from phosphoenolpyruvate to ADP. As discussed in Chapter 8, this reaction is one of two steps that form ATP during glycolysis, a reaction pathway that breaks down glucose. (b) Transfer of a phosphoryl group from ATP to glucose. The product of this reaction, glucose-6-phosphate, is the first intermediate formed during glycolysis.

KEY CONCEPTS 4.4

The hydrolysis of ATP immediately and directly provides the free energy to drive an immense variety of endergonic biochemical reactions. Because ATP has an intermediate phosphate group transfer potential, it can carry phosphoryl groups from higher energy compounds to lower energy compounds. ATP is the energy currency for living systems.

1. At typical intracellular pH values, the triphosphate unit of ATP carries three or four negative charges that repel each other. Hydrolysis of ATP reduces electrostatic repulsion.

2. Because of *resonance hybridization*, the products of ATP hydrolysis are more stable than ATP. When a molecule has two or more alternative structures that differ only in the position of electrons, the result is called a **resonance hybrid**. The electrons in a resonance hybrid with several contributing structures possess much less energy than those with fewer contributing structures. The contributing structures of the phosphate resonance hybrid are illustrated in Figure 4.10.

FIGURE 4.10

Contributing Structures of the Resonance Hybrid of Phosphate.

At physiological pH, orthophosphate is HPO_4^{2-}. In this illustration, H^+ is not assigned permanently to any of the four oxygen atoms.

Living organisms use redox reactions (the addition or removal of electrons from molecules or parts of molecules, see p. 19) as the most important means by which they capture energy from energy sources in their environment and convert it to the chemical bond energy of biomolecules. Organisms are classified according to the energy source they exploit and the redox reactions they use in energy-generating mechanisms.

Most autotrophic ("self-feeding") organisms trap light energy during photosynthesis. (Photosynthesizing organisms, referred to as *photoautotrophs*, include green plants, algae, and cyanobacteria.) A small group of bacterial species use specific inorganic reactions to generate energy to drive their metabolic processes. These organisms are often referred to as **lithotrophs** or **chemolithotrophs** (*lithos* = stone). Found in a diverse set of habitats, including soil and fresh and marine waters, the chemolithotrophs recycle elements in the biosphere. For example, the nitrifying bacteria convert both ammonia and nitrites to nitrates. As an important component of the nitrogen cycle, nitrifying bacteria contribute to soil fertility.

In 1977, scientists discovered a new and unexpected habitat in the Pacific Ocean northeast of the Galápagos Islands. In areas surrounding underwater hot springs, a large number of previously unknown animal species were found living in total darkness. In other marine and terrestrial habitats the photoautotrophs produce the majority of the organic food required to sustain animal life. Yet these hot water springs, 2600 m below the ocean's surface, are teeming with life. Two especially prominent examples are the giant white clam (*Calyptogena magnifica*) and the giant tube worm (*Riftia pachyptila*).

What conditions could account for these oceanic oases? Researchers made a closer inspection of the hot springs, called *hydrothermal vents* or *calderas*, where columns of dark, cloudy, mineral-laden hot water pour through fissures in the sea floor. (Cold water seeps downward through cracks in the earth's crust and is heated by molten lava. Later, as the hydrothermal water is forced upward it mixes with seawater causing mineral precipitates to form that create the chimneylike structures called *black smokers*.) The water in the hot springs is rich in hydrogen sulfide (H_2S). (In the extreme heat and pressure deep within the crust, sulfate is reduced to form the high-energy H—S bond.) The water surrounding the vents contains large numbers of sulfur bacteria, which generate energy by oxidizing H_2S. These organisms are the primary organic food for the hydrothermal vent community. As they oxidize H_2S to form H_2SO_4, the sulfur bacteria use the energy of the earth's interior captured by H_2S formation to convert CO_2 into organic nutrients.

To benefit from this process, several vent animals have established endosymbiotic relationships with the sulfur bacteria. (See Special Interest Box 2.1 for a discussion of endosymbiosis.) One of the best-researched examples is *R. pachyptila*. The giant tube worm, which consists primarily of a long, thin sac attached to a gill plume, has neither a mouth nor a digestive tract. H_2S, O_2, and some CO_2 molecules are absorbed through the gill plume and carried to the tissues through the blood bound to the transport protein hemoglobin. The trophosome, the site of the redox (energy-generating) reactions, is the animal's most prominent organ. It is colonized by a large number of sulfur-oxidizing bacteria. In return for a steady supply of H_2S, O_2, and CO_2 provided by the tube worms' circulatory system, the sulfur bacteria provide the organic nutrients required for the worm's growth and development.

3. The hydrolyzed products of ATP, either ADP and P_i or AMP and PP_i, are more easily solvated than ATP. Recall that the water molecules that form the solvation spheres around ions shield them from one another. The resulting decrease in the repulsive force between phosphate groups drives the hydrolytic reaction.

Walking consumes approximately 100 kcal/mi. For ATP hydrolysis (ATP ⟶ ADP + P_i), the reaction that drives muscle contraction, $\Delta G^{\circ\prime}$ is −7.3 kcal/mol (−30.5 kJ/mol). Calculate how many grams of ATP must be produced to walk a mile. ATP synthesis is coupled to the oxidation of glucose ($\Delta G^{\circ\prime} = -686$ kcal/mol). How many grams of glucose are actually metabolized to produce this amount of ATP? (Assume that only glucose oxidation is used to generate ATP and that 40% of the energy generated from this process is used to phosphorylate ADP. The gram molecular weight of glucose is 180 g, and that of ATP is 507 g.)

QUESTION 4.2

SUMMARY

1. All living organisms unrelentingly require energy. By using bioenergetics, the study of energy transformations, the direction and extent to which biochemical reactions proceed can be determined. Enthalpy (a measure of heat content) and entropy (a measure of disorder) are related to the first and second laws of thermodynamics, respectively. Free energy (the portion of total energy that is available to do work) is related to a mathematical relationship between enthalpy and entropy.

2. Energy and heat transformations take place in a "universe" composed of a system and its surroundings. In an open system, matter and energy are exchanged between the system and its surroundings. If energy but not matter can be exchanged with the surroundings, then the system is said to be closed. Living organisms are open systems.

3. Several thermodynamic quantities are state functions; that is, their value does not depend on the pathway used to make or degrade a specific substance. Examples of state functions are total energy, free energy, enthalpy, and entropy. Some quantities, such as work and heat, are not state functions, because they depend on the pathway.

4. Free energy, a state function that relates the first and second laws of thermodynamics, represents the maximum useful work obtainable from a process. Exergonic processes, that is, processes in which free energy decreases ($\Delta G < 0$), are spontaneous. If the free energy change is positive ($\Delta G > 0$), the process is called endergonic. A system is at equilibrium when the free energy change is zero. The standard free energy ($\Delta G°$) is defined for reactions at 25°C, 1 atm pressure, and 1 M solute concentrations. The standard pH in bioenergetics is 7. The standard free energy change $\Delta G°'$ at pH 7 is used in this textbook.

5. ATP hydrolysis provides most of the free energy required for living processes. ATP is ideally suited to its role as universal energy currency because its phosphoanhydride structure is easily hydrolyzed.

SUGGESTED READINGS

Bergethon, P. R., *The Physical Basis of Biochemistry: The Foundations of Molecular Biophysics,* Springer, New York, 1998.

Hanson, R. W., The Role of ATP in Metabolism, *Biochem. Educ.* 17:86–92, 1989.

Harold, F. M., *The Vital Force: A Study of Bioenergetics*, W. H. Freeman, New York, 1986.

Ho, M. W., *The Rainbow and the Worm: The Physics of Organisms*, World Scientific Publishing, Singapore, 1993.

Schrödinger, E., *What Is Life?*, Cambridge University Press, Cambridge, England, 1944.

KEY WORDS

bioenergetics, *93*

chemolithotrophs, *105*

endergonic process, *98*

endothermic reaction, *95*

enthalpy, *93*

entropy, *93, 96*

exergonic process, *98*

exothermic reaction, *95*

free energy, *93, 98*

isothermic reaction, *95*

lithotrophs, *105*

phosphate group transfer potential, *102*

resonance hybrid, *104*

spontaneous changes, *96*

thermodynamics, *93*

work, *94*

REVIEW QUESTIONS

1. Define each of the following terms:
   a. thermodynamics
   b. endergonic
   c. enthalpy
   d. free energy
   e. high-energy bond
   f. redox reaction
   g. chemolithotroph
   h. phosphate group transfer potential

2. Which of the following thermodynamic quantities are state functions? Explain.
   a. work
   b. entropy
   c. enthalpy
   d. free energy

3. Which of the following reactions could be driven by coupling to the hydrolysis of ATP? (The $\Delta G°'$ value in kJ/mol for each reaction is indicated in parentheses.)

$$\text{ATP} + H_2O \longrightarrow \text{ADP} + P_i \ (-30.5)$$

   a. glucose-1-phosphate $\longrightarrow$ glucose-6-phosphate (-7.1)
   b. glucose + P_i $\longrightarrow$ glucose-6-phosphate ($+13.8$)
   c. acetic acid $\longrightarrow$ acetic anhydride ($+99.6$)
   d. glucose + fructose $\longrightarrow$ sucrose ($+29.3$)
   e. glycylglycine + water $\longrightarrow$ 2 glycine (-9.2)

4. The K_a for the ionization of formic acid is 1.8×10^{-4}. Calculate $\Delta G°$ for this reaction at 25°C.

5. The following reaction is catalyzed by the enzyme glutamine synthetase:

$$\text{ATP} + \text{glutamate} + NH_3 \longrightarrow \text{ADP} + P_i + \text{glutamine}$$

Use the following equations with $\Delta G°'$ values given in kJ/mol to calculate $\Delta G°'$ for the overall reaction.

$$\text{ATP} + H_2O \longrightarrow \text{ADP} + P_i \ (-30.5)$$

$$\text{glutamine} + H_2O \longrightarrow \text{glutamate} + NH_3 \ (-5.0)$$

6. $\Delta G^{\circ\prime}$ values (kJ/mol) for the following reactions are indicated in parentheses.

   Ethyl acetate + water $\longrightarrow$

   $$\text{ethyl alcohol + acetic acid } (-19.7) \quad (i)$$

   Glycose-6-phosphate + water

   $$\text{Glucose} + P_i \ (-13.8) \quad (ii)$$

   Which statements are true?
   a. The rate of reaction (i) is greater than the rate of reaction (ii).
   b. The rate of reaction (ii) is greater than the rate of reaction (i).
   c. Neither reaction is spontaneous.
   d. Reaction rates cannot be determined from energy values.

7. Under standard conditions, which statements are true?
   a. $\Delta G = \Delta G^{\circ}$
   b. $\Delta H = \Delta G$
   c. $\Delta G = \Delta G^{\circ} + RT \ln K_{eq}$
   d. $\Delta G^{\circ} = \Delta H - T\Delta S$
   e. $P = 1$ atm
   f. $T = 273$ K
   g. [reactants] = [products] = 1 molar

8. Which statements concerning free energy change are true?
   a. Free energy change is a measure of the rate of a reaction.
   b. Free energy change is a measure of the maximum amount of work available from a reaction.

   c. Free energy change is a constant for a reaction under any conditions.
   d. Free energy change is related to the equilibrium constant for a specific reaction.
   e. Free energy change is equal to zero at equilibrium.

9. Consider the following reaction:

   Glucose-1-phosphate $\longrightarrow$ glucose-6-phosphate

   $$\Delta G^{\circ} = -7.1 \text{ kJ/mol}$$

   What is the equilibrium constant for this reaction at 25°C?

10. Which of the following compounds would you expect to liberate the least free energy when hydrolyzed? Explain.
    a. ATP
    b. ADP
    c. AMP
    d. phosphoenolpyruvate
    e. phosphocreatine

11. Which statements are true and which are false?
    a. In a closed system, neither energy nor matter is exchanged with the surroundings.
    b. State functions are independent of the pathway.
    c. A process is isothermic if $\Delta H = 0$.
    d. The sign and magnitude of ΔG give important information about the direction and rate of a reaction.
    e. At equilibrium, $\Delta G = \Delta G^{\circ}$.
    f. For two reactions to be coupled, they must have a common intermediate.

THOUGHT QUESTIONS

1. Pyruvate oxidizes to form carbon dioxide and water and liberates 1142.2 kJ/mol. If electron transport also occurs, approximately 12.5 ATP molecules are produced. The free energy of hydrolysis for ATP is -30.5 kJ/mol. What is the apparent efficiency of ATP production?

2. In the reaction

   ATP + glucose $\longrightarrow$ ADP + glucose-6-phosphate

   ΔG° is -16.7 kJ/mol. Assume that the concentration of ATP and ADP are each 1 M and $T = 25°C$. What ratio of glucose-6-phosphate to glucose would allow the reverse reaction to begin?

3. Thermodynamics is based on the behavior of large numbers of molecules. Yet within a cell there may only be a few molecules of a particular type at a time. Do the laws of thermodynamics apply under these circumstances?

4. Frequently, when salts dissolve in water, the solution becomes warm. Such a process is exothermic. When other salts, such as ammonium chloride, dissolve in water, the solution becomes cold, indicating an endothermic process. Because endothermic processes are usually not spontaneous, why does the latter process proceed?

5. Of the three thermodynamic quantities ΔH, ΔG, and ΔS, which provides the most useful criterion of spontaneity in a reaction? Explain.

6. What factors make ATP suitable as an "energy currency" for the cell?

7. To determine the $\Delta G^{\circ\prime}$ of a reaction within a cell, what information would you need?

8. Given the following data, calculate K_{eq} for the denaturation reaction of β-lactoglobin at 25°C:

   $$\Delta H = -88 \text{ kJ/mol}$$

   $$\Delta S = 0.3 \text{ kJ/mol}$$

9. Which of the following compounds would have the higher phosphate group transfer potential? Explain your answer;

10. The free energy of hydrolysis of ATP in systems free of Mg^{2+} is -35.7 kJ/mol. When $[Mg^{2+}]$ is 5 mM, $\Delta G^{\circ}_{observed}$ is approximately -31 kJ/mol at pH 7 and 38°C. Suggest a possible reason for this effect.

Amino Acids, Peptides, and Proteins

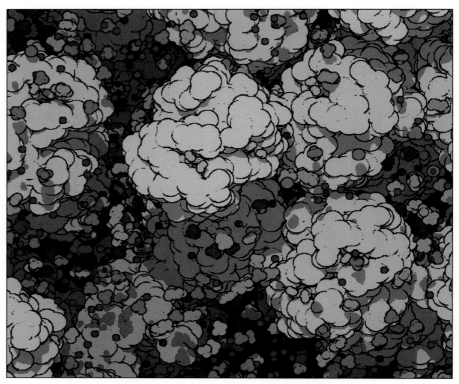

Hemoglobin Within a Red Blood Cell Human red blood cells are filled almost to bursting with the oxygen-carrying protein hemoglobin. The large pink structures are hemoglobin molecules. Sugar and amino acids are shown in green. Positive ions are blue. Negative ions are red. The large blue molecule is an enzyme.

Proteins are essential constituents of all organisms. Most tasks performed by living cells require proteins. The variety of functions that they perform is astonishing. In animals, for example, proteins are the primary structural components of muscle, connective tissue, feathers, nails, and hair. In addition to serving as structural materials in all living organisms, proteins are involved in such diverse functions as metabolic regulation, transport, defense, and catalysis. The functional diversity exhibited by this class of biomolecules is directly related to the combinatorial possibilities of the monomeric units, the 20 amino acids.

As crucial as an uninterrupted flow of energy is to living systems, it is insufficient to maintain the organized complexity of life. Also required is a continuous flow of staggering amounts of timely, precise, and accurate information. Information is a measure of order and is often referred to as *negative entropy*. In general terms, information specifies the instructions required to create a specific organization. In living organisms, information is inherent in the three-dimensional atomic configuration of biomolecules. The information in genes is the instructions for making the proteins and ribonuclear proteins required to sustain life. The proteins and ribonuclear proteins constitute the machinery and structure of the cell. Proteins themselves are informational, each with a unique shape (Figure 5.1), allowing selective interactions with only one or a few other molecules. The enzyme glucokinase will only accept glucose as a substrate, whereas hexokinase will accept glucose, galactose, or mannose despite the fact that the two enzymes catalyze the same reaction in the same way. Glucokinase has high specificity; hexokinase has low specificity. In general, the larger the protein, the greater the potential for multifunctional capacities, so that an enzyme might bind modulating ligands in addition to the substrate.

Proteins may be composed of as many as 20 different amino acids. In each protein the precise types and amounts of each amino acid are covalently linked in the

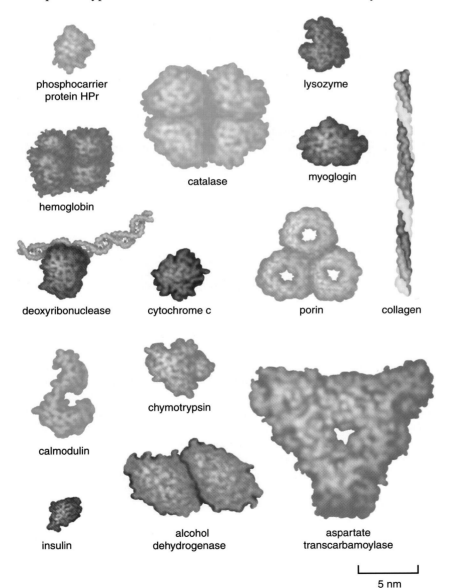

FIGURE 5.1

Protein Diversity.

Proteins occur in an enormous diversity of sizes and shapes.

phosphocarrier protein HPr

lysozyme

catalase

myoglogin

hemoglobin

deoxyribonuclease

cytochrome c

porin

collagen

chymotrypsin

calmodulin

insulin

alcohol dehydrogenase

aspartate transcarbamoylase

5 nm

linear sequence specified by the base sequence of the DNA-generated mRNA for that protein. The ability of each type of the tens of thousands of different proteins to perform its functions is specified by its unique amino acid sequence. During synthesis each polypeptide molecule bends in three-dimensional space as its amino acid components (called **amino acid residues**) interact with each other, largely through noncovalent interactions. The subsequent folding of a protein molecule into its own unique, complex, three-dimensional, and biologically active structure is a process dictated by information inherent in the structures of its amino acids.

Amino acid polymers are often differentiated according to their molecular weights or the number of amino acid residues they contain. Molecules with molecular weights ranging from several thousand to several million daltons (D) are called **polypeptides**. Those with low molecular weights, typically consisting of fewer than 50 amino acids, are called **peptides**. The term **protein** describes molecules with more than 50 amino acids. Each protein consists of one or more polypeptide chains. The distinction between proteins and peptides is often imprecise. For example, some biochemists define oligopeptides as polymers consisting of two to ten amino acids and polypeptides as having more than ten residues. Proteins, in this view, have molecular weights greater than 10,000 D. In addition, the terms *protein* and *polypeptide* are often used interchangeably. Throughout this textbook the terms *peptide* and *protein* will be used as defined above. The term *polypeptide* will be used whenever the topic under discussion applies to both peptides and proteins.

Polypeptides may be broken into their constituent monomer molecules by hydrolysis. The amino acid products of the reaction constitute the *amino acid composition* of the polypeptide.

This chapter begins with a review of the structures and chemical properties of the amino acids. This is followed by descriptions of the structural features of peptides and proteins. The chapter ends with an examination of the structural and functional properties of several well-researched proteins. The emphasis throughout the chapter is on the intimate relationship between the structure and function of polypeptides. In Chapter 6 the functioning of the enzymes, an especially important group of proteins, is discussed. Protein synthesis and the folding process are covered in Chapter 19.

5.1 AMINO ACIDS

The structures of the 20 amino acids that are commonly found in proteins are shown in Figure 5.2. These amino acids are referred to as *standard* amino acids. Common abbreviations for the standard amino acids are listed in Table 5.1. Note that 19 of the standard amino acids have the same general structure (Figure 5.3. These molecules contain a central carbon atom (the α-carbon) to which an amino group, a carboxylate group, a hydrogen atom, and an R (side chain) group are attached.

The exception, proline, differs from the other standard amino acids in that its amino group is secondary, formed by ring closure between the R group and the amino nitrogen. Proline confers rigidity to the peptide chain because rotation about the α-carbon is not possible. This structural feature has significant implications in the structure and, therefore, the function of proteins with a high proline content.

Nonstandard amino acids consist of amino acid residues that have been chemically modified after they have been incorporated into a polypeptide or amino acids that occur in living organisms but are not found in proteins.

At a pH of 7, the carboxyl group of an amino acid is in its conjugate base form ($-COO^-$) and the amino group is in its conjugate acid form ($-NH_3^+$). Thus each amino acid can behave as either an acid or a base. The term **amphoteric** is used to describe this property. Neutral molecules that bear an equal number of positive and negative charges simultaneously are called **zwitterions**. The R group, however, gives each amino acid its unique properties.

KEY CONCEPTS 5.1

Each protein is composed of building blocks called amino acids. Amino acids are amphoteric molecules; that is, they can act as either an acid or a base. In addition to their primary function as components of proteins, amino acids have several important biological roles.

Neutral Nonpolar Amino Acids Hydrophobic

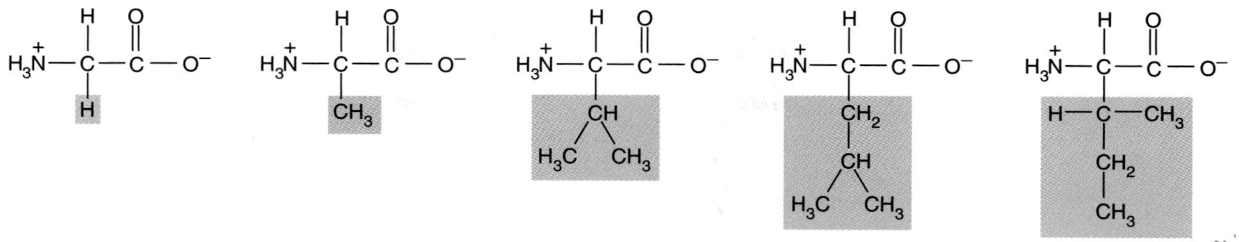

Glycine   Alanine   Valine   Leucine   Isoleucine — Amine

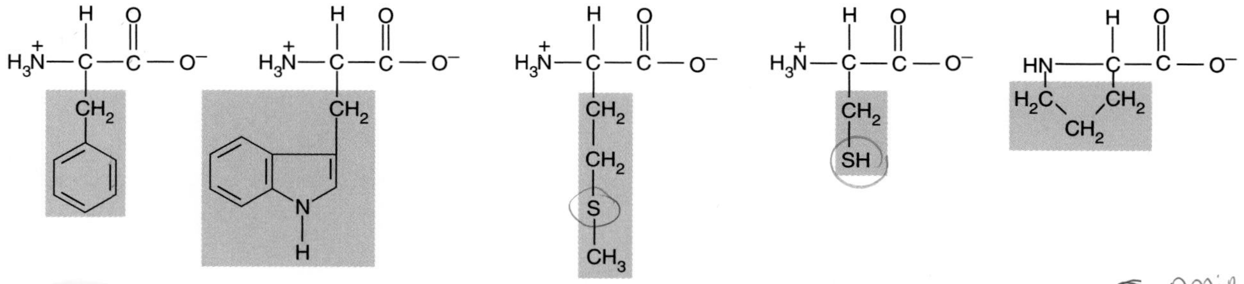

Phenylalanine   Tryptophan   Methionine   Cysteine   Proline — Amine

Neutral Polar Amino Acids Hydrophilic

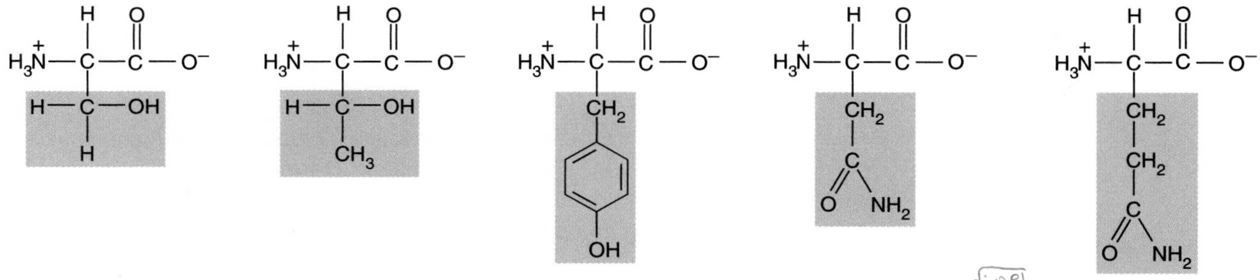

Serine   Threonine   Tyrosine   Asparagine — Amine   Glutamine

Acidic Amino Acids   **Basic Amino Acids**

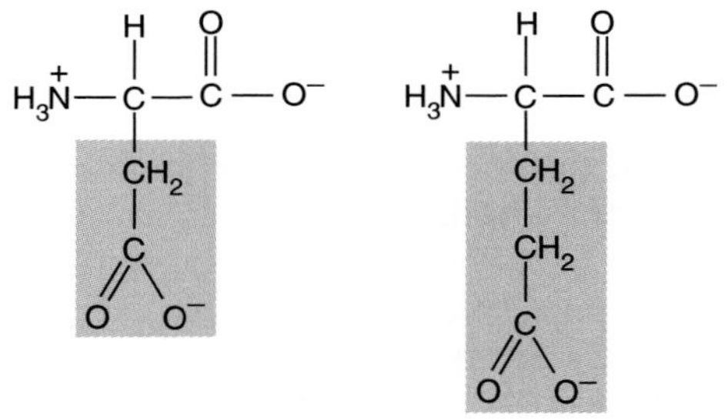

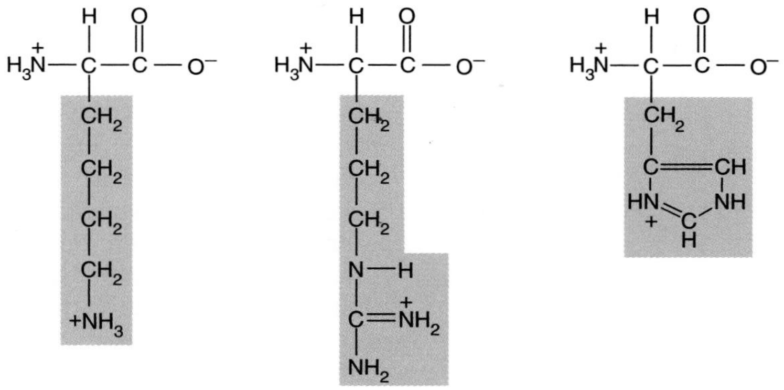

Aspartate   Glutamate   Lysine   Arginine   Histidine

FIGURE 5.2

The Standard Amino Acids.

The side chain is indicated by the shaded box.

TABLE 5.1
Names and Abbreviations of the Standard Amino Acids

Amino Acid	Three-Letter Abbreviation	One-Letter Abbreviation
Alanine	Ala	A
Arginine	Arg	R
Asparagine	Asn	N
Aspartic acid	Asp	D
Cysteine	Cys	C
Glutamic acid	Glu	E
Glutamine	Gln	Q
Gycine	Gly	G
Histidine	His	H
Isoleucine	Ile	I
Leucine	Leu	L
Lysine	Lys	K
Methionine	Met	M
Phenylalanine	Phe	F
Proline	Pro	P
Serine	Ser	S
Threonine	Thr	T
Tryptophan	Trp	W
Tyrosine	Tyr	Y
Valine	Val	V

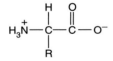

FIGURE 5.3
General Structure of the α-Amino Acids.

FIGURE 5.4
Benzene.

Amino Acid Classes

Because the sequence of amino acids determines the final three-dimensional configuration of each protein, their structures are examined carefully in the next four subsections. Amino acids are classified according to their capacity to interact with water. By using this criterion, four classes may be distinguished: (1) nonpolar and neutral, (2) polar and neutral, (3) acidic, and (4) basic.

NEUTRAL NONPOLAR AMINO ACIDS The neutral nonpolar amino acids contain mostly hydrocarbon R groups. The term *neutral* is used because these R groups do not bear positive or negative charges. Because they interact poorly with water, nonpolar (i.e., hydrophobic) amino acids play an important role maintaining the three-dimensional structure of proteins. Two types of hydrocarbon side chains are found in this group: aromatic and aliphatic. (Recall that **aromatic** hydrocarbons contain cyclic structures that constitute a class of unsaturated hydrocarbons with unique properties. Benzene is one of the simplest aromatic hydrocarbons (Figure 5.4). The term **aliphatic** refers to nonaromatic hydrocarbons such as methane and cyclohexane.) Phenylalanine and tryptophan contain aromatic ring structures. Glycine, alanine, valine, leucine, isoleucine, and proline have aliphatic R groups. A sulfur atom appears in the aliphatic side chains of methionine and cysteine. In methionine the nonbonding electrons of the sulfur atom can form bonds with electrophiles such as metal ions. Although the sulfhydryl (—SH) group of cysteine is nonpolar, it can form weak hydrogen bonds with oxygen and nitrogen. Sulfhydryl groups, which are highly reactive, are important components of many enzymes. Additionally, the sulfhydryl groups of two cysteine molecules may oxidize spontaneously to form a disulfide compound called cystine. (See p. 121 for a discussion of this reaction.)

NEUTRAL POLAR AMINO ACIDS Because polar amino acids have functional groups capable of hydrogen bonding, they easily interact with water. (Polar amino acids are described as "hydrophilic" or "water-loving.") Serine, threonine, tyrosine, asparagine, and glutamine belong to this category. Serine, threonine, and tyrosine contain a polar hydroxyl group, which enables them to participate in hydrogen bonding, an important factor in protein structure. The hydroxyl groups serve other functions in proteins. For example, the formation of the phosphate ester of tyrosine is a common regulatory mechanism. Additionally, the —OH groups of serine and threonine are points for attaching carbohydrates. Asparagine and glutamine are amide derivatives of the acidic amino acids aspartic acid and glutamic acid, respectively. Because the amide functional group is highly polar, the hydrogen-bonding capability of asparagine and glutamine has a significant effect on protein stability.

ACIDIC AMINO ACIDS Two standard amino acids have side chains with carboxylate groups. Because the side chains of aspartic acid and glutamic acid are negatively charged at physiological pH, they are often referred to as aspartate and glutamate.

BASIC AMINO ACIDS Basic amino acids bear a positive charge at physiological pH. They can therefore form ionic bonds with acidic amino acids. Lysine, which has a side chain amino group, accepts a proton from water to form the conjugate acid ($-NH_3^+$). When lysine's side chain in proteins such as collagen is oxidized, strong intramolecular and intermolecular cross-linkages are formed. Because the guanidino group of arginine has a pK_a range of 11.5–12.5 in proteins, it is permanently protonated at physiological pH and, therefore, does not function in acid-base reactions. Histidine, on the other hand, is a weak base, because it is only partially ionized at pH 7. Consequently, histidine residues act as a buffer. They also play an important role in the catalytic activity of numerous enzymes.

KEY CONCEPTS 5.2

Amino acids are classified according to their capacity to interact with water. By using this criterion, four classes may be distinguished: nonpolar, polar, acidic, and basic.

QUESTION 5.1

Shown are the structures of several standard amino acids. Classify them according to whether they are neutral nonpolar, neutral polar, acidic, or basic.

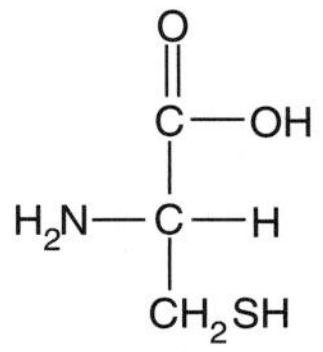

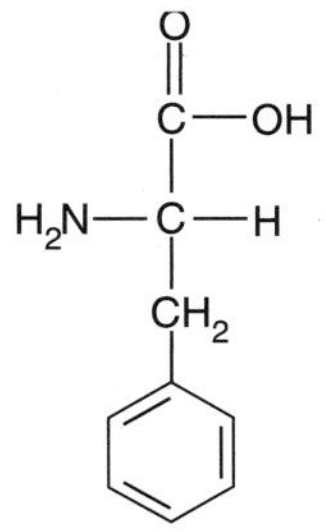

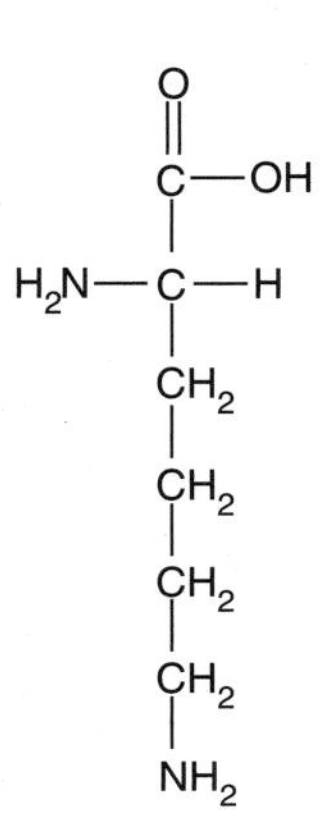

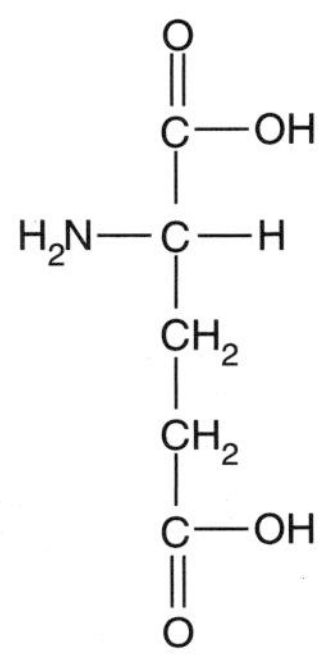

Biologically Active Amino Acids

In addition to their primary function as components of protein, amino acids have several other biological roles.

1. Several α-amino acids or their derivatives act as chemical messengers (Figure 5.5). For example, glycine, γ-amino butyric acid (GABA, a derivative of glutamate), and serotonin and melatonin (derivatives of tryptophan) are **neurotransmitters**, substances released from one nerve cell that influence the function of a second nerve cell or a muscle cell. Thyroxine (a tyrosine derivative produced in the thyroid gland of animals) and indole acetic acid (a tryptophan derivative found in plants) are two examples of hormones. **Hormones** are chemical signal molecules produced in one cell that regulate the function of other cells.

FIGURE 5.5
Some Derivatives of Amino Acids.

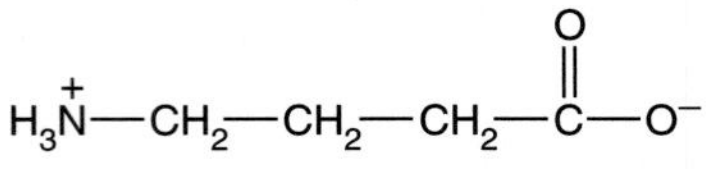

GABA

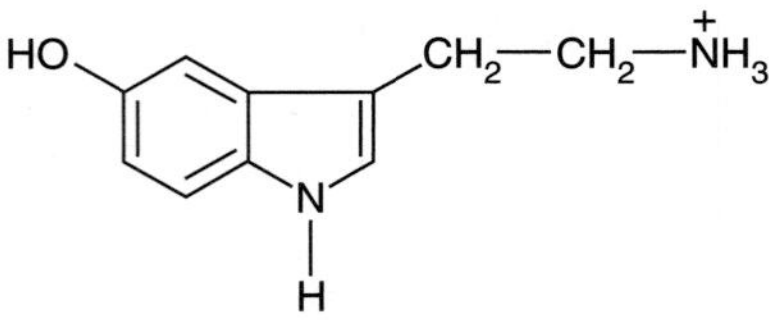

Serotonin

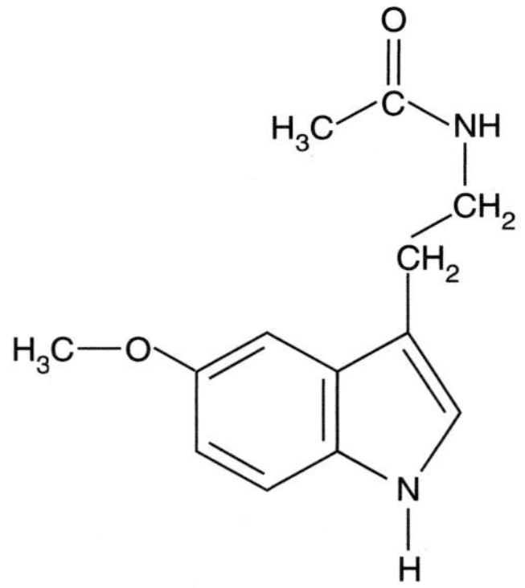

Melatonin

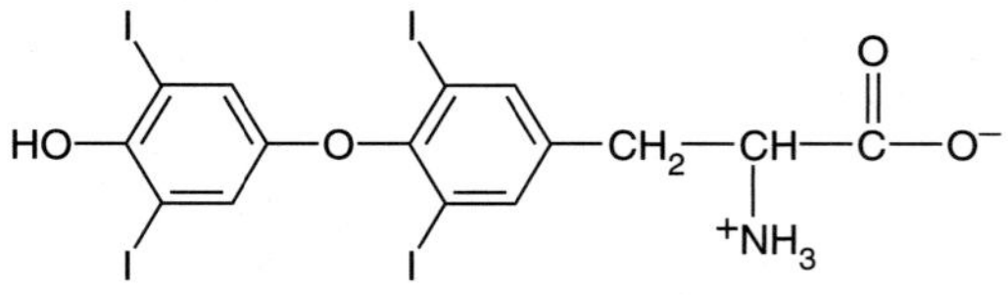

Thyroxine

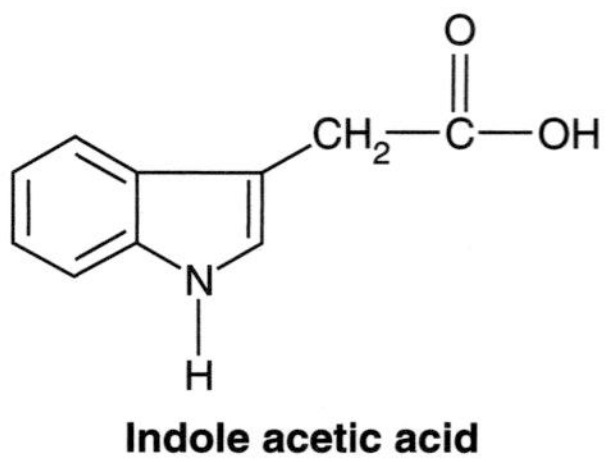

Indole acetic acid

2. Amino acids are precursors of a variety of complex nitrogen-containing molecules. Examples include the nitrogenous base components of nucleotides and the nucleic acids, heme (the iron-containing organic group required for the biological activity of several important proteins), and chlorophyll (a pigment of critical importance in photosynthesis).

3. Several standard and nonstandard amino acids act as metabolic intermediates. For example, arginine, citrulline, and ornithine (Figure 5.6) are components of the urea cycle (Chapter 15). The synthesis of urea, a molecule formed in vertebrate livers, is the principal mechanism for the disposal of nitrogenous waste.

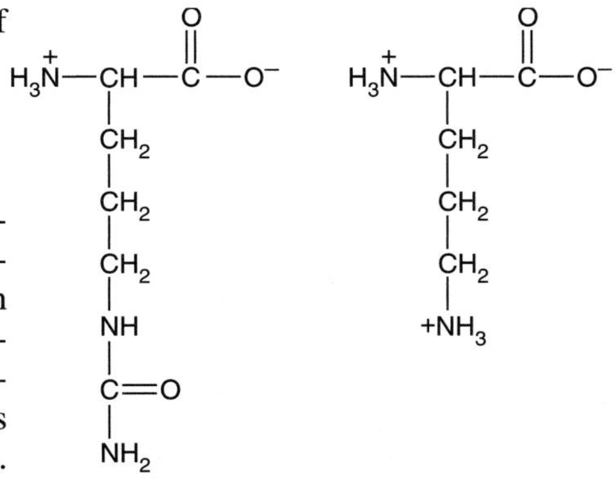

Modified Amino Acids in Proteins

Several proteins contain amino acid derivatives that are formed after a polypeptide chain has been synthesized. Among these modified amino acids is γ-carboxyglutamic acid (Figure 5.7), a calcium-binding amino acid residue found in the blood-clotting protein prothrombin. Both 4-hydroxyproline and 5-hydroxylysine are important structural components of collagen, the most abundant protein in connective tissue. Phosphorylation of the hydroxyl-containing amino acids serine, threonine, and tyrosine is often used to regulate the activity of proteins. For example, the synthesis of glycogen is significantly curtailed when the enzyme glycogen synthase is phosphorylated.

Amino Acid Stereoisomers

Because the α-carbons of 19 of the 20 standard amino acids are attached to four different groups (i.e., a hydrogen, a carboxyl group, an amino group, and an R group), they are referred to as **asymmetric** or **chiral carbons**. Glycine is a symmetrical molecule because its α-carbon is attached to two hydrogens. Molecules with chiral carbons can exist as **stereoisomers**, molecules that differ only in the spatial arrangement of their atoms. Three-dimensional representations of amino acid stereoisomers are illustrated in Figure 5.8. Notice in the figure that the atoms of the two isomers are bonded together in the same pattern except for the position of the ammonium group and the hydrogen atom. These two isomers are mirror images of each other. Such molecules, called **enantiomers**, cannot be superimposed on each other. The physical properties of enantiomers are identical except that they rotate plane-polarized light in opposite directions. (In plane-polarized light, produced by passing unpolarized light through a special filter, the light waves vibrate in only one plane.) Molecules that possess this property are called **optical isomers**.

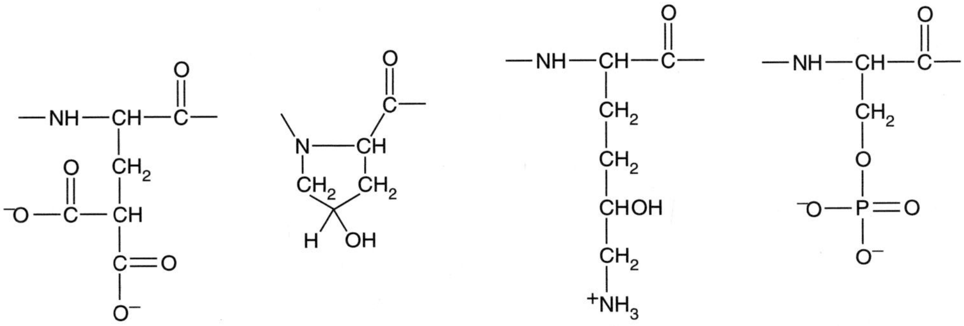

Citrulline **Ornithine**

FIGURE 5.6
Citrulline and Ornithine.

γ-Carboxyglutamate **4-Hydroxyproline** **5-Hydroxylysine** **o-Phosphoserine**

FIGURE 5.7
Some Modified Amino Acid Residues Found in Polypeptides.

FIGURE 5.8

Two Enantiomers.

L-Alanine and D-alanine are mirror images of each other.

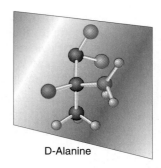

L-Alanine   D-Alanine

D-**Glyceraldehyde**   L-**Glyceraldehyde**

FIGURE 5.9

D- **and** L-**Glyceraldehyde.**

These molecules are mirror images of each other.

KEY CONCEPTS 5.3

Molecules that differ only in the spatial arrangement of some of their atoms are called stereoisomers. Stereoisomers with an asymmetric carbon atom have two non-superimposable mirror-image forms called enantiomers. Most asymmetric molecules in living organisms have only one stereoisomeric form.

Glyceraldehyde is the reference compound for optical isomers (Figure 5.9). One glyceraldehyde isomer rotates the light beam in a clockwise direction and is said to be dextrorotary (designated by +). The other glyceraldehyde isomer, referred to as levorotary (designated by −), rotates the beam in the opposite direction to an equal degree. Optical isomers are often designated as D or L; for example, D-glucose and L-alanine. The D or L indicates the similarity of the arrangement of atoms around a molecule's asymmetric carbon to the asymmetric carbon in either of the glyceraldehyde isomers.

Because many biomolecules have more than one chiral carbon, the letters D and L refer only to a molecule's structural relationship to either of the glyceraldehyde isomers, not to the direction in which it rotates plane-polarized light. Most asymmetric molecules found in living organisms occur in only one stereoisomeric form, either D or L. For example, with few exceptions, only L-amino acids are found in proteins.

Chirality has had a profound effect on the structural and functional properties of biomolecules. For example, the right-handed helices observed in proteins result from the exclusive presence of L-amino acids. Polypeptides synthesized in the laboratory from both D- and L- amino acids do not form helices. In addition, because the enzymes are chiral molecules, they only bind substrate molecules in one enantiomeric form. Proteases, enzymes that degrade proteins by hydrolyzing peptide bonds, cannot degrade artificial polypeptides composed of D-amino acids.

QUESTION 5.2

Certain bacterial species have outer layers composed of polymers made of D-amino acids. Immune system cells, whose task is to attack and destroy foreign cells, cannot destroy these bacteria. Suggest a reason for this phenomenon.

Titration of Amino Acids

Because amino acids contain ionizable groups (Table 5.2), the predominant ionic form of these molecules in solution depends on the pH. Titration of an amino acid illustrates the effect of pH on amino acid structure (Figure 5.10a). Titration is also a useful tool in determining the reactivity of amino acid side chains. Consider alanine, a simple amino acid, which has two titratable groups. During titration with a strong base such as NaOH, alanine loses two protons in a stepwise fashion. In a strongly acidic solution (e.g., at pH 0), alanine is present mainly in the form in which the carboxyl group is uncharged. Under this circumstance the molecule's net charge is +1, because the ammonium group is protonated. Lowering of the H⁺ concentration results in the carboxyl group losing its proton to become a negatively charged carboxylate group. (In a polyprotic acid, the protons are first lost from the group with the lowest pK_a.) At this point, alanine has no net charge and is electrically neutral. The pH at which this occurs is called the **isoelectric point**

TABLE 5.2

pK_a Values for the Ionizing Groups of the Amino Acids

Amino Acid	pK_1 *of COOH*	pK_2 *of NH$_2$*	pK_R
Glycine	2.34	9.6	
Alanine	2.34	9.69	
Valine	2.32	9.62	
Leucine	2.36	9.6	
Isoleucine	2.36	9.6	
Serine	2.21	9.15	
Threonine	2.63	10.43	
Methionine	2.28	9.21	
Phenylalanine	1.83	9.13	
Tryptophan	2.83	9.39	
Asparagine	2.02	8.8	
Glutamine	2.17	9.13	
Proline	1.99	10.6	
Cysteine	1.71	10.78	8.33
Histidine	1.82	9.17	6.0
Aspartic acid	2.09	9.82	3.86
Glutamic acid	2.19	9.67	4.25
Tyrosine	2.2	9.11	10.07
Lysine	2.18	8.95	10.79
Arginine	2.17	9.04	12.48

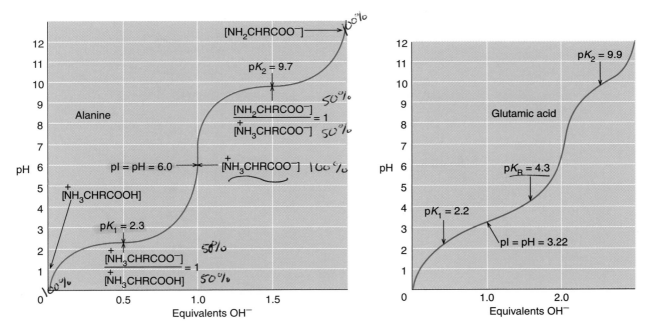

FIGURE 5.10

Titration of (a) Alanine and (b) Glutamic Acid.

(pI). Because there is no net charge at the isoelectric point, amino acids are least soluble at this pH. (Zwitterions crystallize relatively easily.) The isoelectric point for alanine may be calculated as follows:

$$pI = \frac{pK_1 + pK_2}{2}$$

The pK_1 and pK_2 values for alanine are 2.34 and 9.7 respectively (see Table 5.2). The pI value for alanine is therefore

$$pI = \frac{2.34 + 9.7}{2} = 6.0$$

As the titration continues, the ammonium group loses its proton, leaving an uncharged amino group. The molecule then has a net negative charge because of the carboxylate group.

Amino acids with ionizable side chains have more complex titration curves (Figure 5.10b). Glutamic acid, for example, has a carboxyl side chain group. At low pH, glutamic acid has net charge +1. As base is added, the α-carboxyl group loses a proton to become a carboxylate group. Glutamate now has no net charge. As more base is added, the second carboxyl group loses a proton, and the molecule has a −1 charge. Adding additional base results in the ammonium ion losing its proton. At this point, glutamate has a net charge of −2. The pI value for glutamate is the pH halfway between the pK_a values for the two carboxyl groups:

$$pI = \frac{2.19 + 4.25}{2} = 3.22$$

The isoelectric point for histidine is the pH value halfway between the pK values for the two nitrogen-containing groups. The pK_a and pI values of amino acids in peptides and proteins differ somewhat from those of free amino acids, principally because most of the α-amino and α-carboxyl groups are not ionized but are covalently joined in peptide bonds.

Problems 5.1 and 5.2 are sample titration problems, given with their solutions.

PROBLEM 5.1

Consider the following amino acid and its pK_a values:

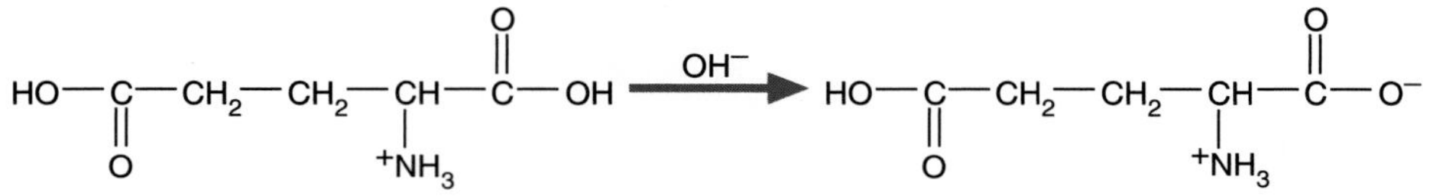

$$pK_{a1} = 2.19 \qquad pK_{a2} = 9.67, \qquad pK_{aR} = 4.25$$

a. Draw the structure of the amino acid as the pH of the solution changes from highly acidic to strongly basic.

Solution

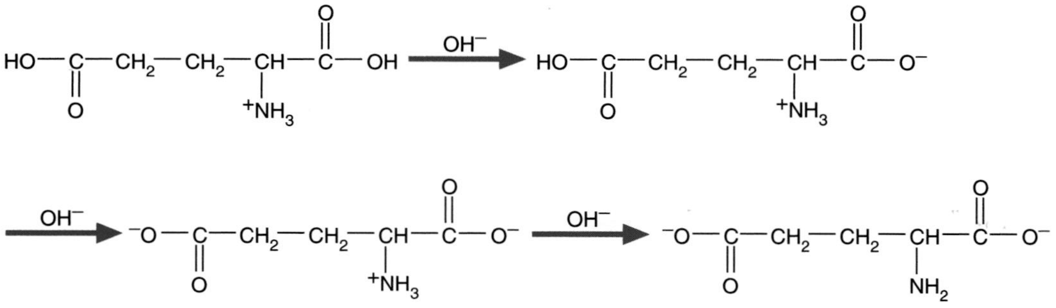

The ionizable hydrogens are lost in order of acidity, the most acidic ionizing first.

b. Which form of the amino acid is present at the isoelectric point?

Solution

The form present at the isoelectric point is electrically neutral:

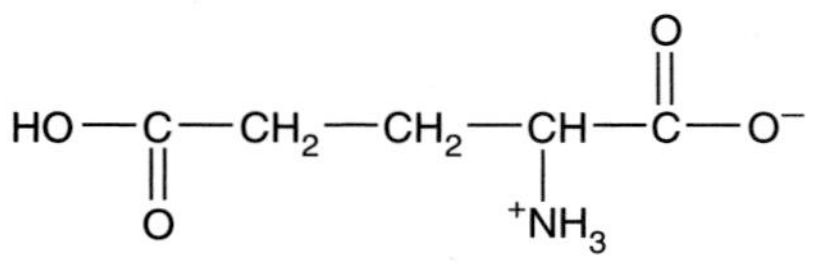

$$HO-\underset{\underset{O}{\|}}{C}-CH_2-CH_2-\underset{\underset{{}^+NH_3}{|}}{CH}-\overset{\overset{O}{\|}}{C}-O^-$$

c. Calculate the isoelectric point.

Solution

The isoelectric point is the average of the two pK_a's bracketing the isoelectric structure:

$$pI = \frac{pK_{a1} + pK_{aR}}{2} = \frac{2.19 + 4.25}{2} = 3.22$$

d. Sketch the titration curve for the amino acid.

Solution

Plateaus appear at the pK_a's and are centered about 0.5 equivalents (Eq), 1.5 Eq, and 2.5 Eq of base. There is a sharp rise at 1 Eq, 2 Eq, and 3 Eq. The isoelectric point is midway on the sharp rise between pK_{a1} and pK_{aR}.

e. In what direction does the amino acid move when placed in an electric field at the following pH values: 1, 3, 5, 7, 9, 12?

Solution

At pH values below the pI, the amino acid is positively charged and moves to the cathode (negative electrode). At pH values above the pI, the amino acid is negatively charged and moves toward the anode (positive electrode). At the isoelectric point, the amino acid has no net charge and therefore does not move in the electric field.

Consider the following tetrapeptide:

Lys—Ser—Asp—Ala

a. Determine the pI for the peptide.

PROBLEM 5.2

Solution

The structure of the tetrapeptide in its most acidic form is shown below.

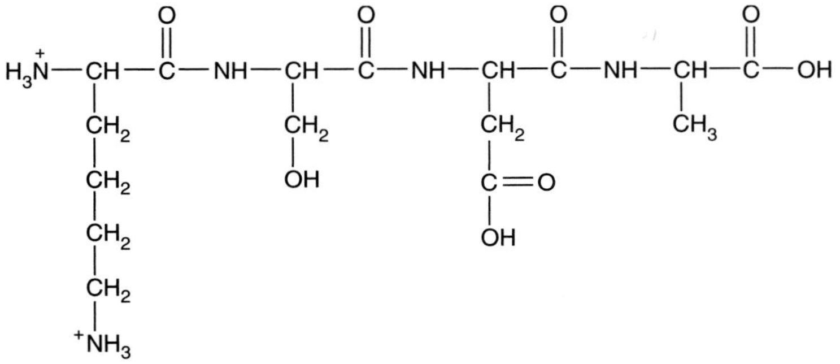

$$H_3\overset{+}{N}-CH-\overset{\overset{O}{\|}}{C}-NH-CH-\overset{\overset{O}{\|}}{C}-NH-CH-\overset{\overset{O}{\|}}{C}-NH-CH-\overset{\overset{O}{\|}}{C}-OH$$

with side chains:

Lys	Ser	Asp	Ala
CH_2	CH_2	CH_2	CH_3
CH_2	OH	$C{=}O$	
CH_2		OH	
CH_2			
${}^+NH_3$			

Refer to Table 5.2 for the pK_a values for lysine and aspartic acid, both of which have ionizable side chains. Lysine also contains a terminal α-amino and alanine a terminal α-carboxyl group. These values are as follows:

Lysine: α-amino = 8.95,   amino side chain = 10.79
Aspartic acid: carboxyl side chain = 3.86
Alanine: α-carboxyl = 2.34

(These values are approximations, because the behavior of amino acids is affected by the presence of other groups.) The electrically neutral peptide is formed after both carboxyl groups have lost their protons but before either ammonium group has lost any protons. The isoelectric point is calculated as follows:

$$pI = \frac{3.86 + 8.95}{2} = \frac{12.81}{2} = 6.4$$

b. In what direction does the peptide move when placed in an electric field at the following pHs: 4 and 9?

Solution

At pH = 4 the peptide is positively charged and moves toward the negative electrode (cathode). At pH = 9 the peptide is negatively charged and will move toward the positive electrode (anode).

Amino Acid Reactions

The functional groups of organic molecules determine which reactions they may undergo. Amino acids with their carboxyl groups, amino groups, and various R groups can undergo numerous chemical reactions. However, two reactions (i.e, peptide bond and disulfide bridge formation) are of special interest because of their effect on protein structure.

PEPTIDE BOND FORMATION Polypeptides are linear polymers composed of amino acids linked together by peptide bonds. **Peptide bonds** (Figure 5.11) are amide linkages formed when the unshared electron pair of the α-amino nitrogen atom of one amino acid attacks the α-carboxyl carbon of another in a nucleophilic acyl substitution reaction. A generalized acyl substitution reaction is shown:

$$
\begin{array}{ccc}
\overset{\displaystyle O}{\underset{\displaystyle \|}{}} & & \overset{\displaystyle O}{\underset{\displaystyle \|}{}} \\
R-C-X + Y^- & \longrightarrow & R-C-Y + X^-
\end{array}
$$

Because this reaction is a dehydration (i.e., a water molecule is removed) the linked amino acids are referred to as *amino acid residues*. When two amino acid molecules are linked, the product is called a dipeptide. For example, glycine and serine can form the dipeptides glycylserine or serylglycine. As amino acids are added and the chain lengthens, the prefix reflects the number of residues. For example, a tripeptide contains three amino acid residues, a tetrapeptide four, and so on. By convention the amino acid residue with the free amino group is called the *N-terminal* residue and is written to the left. The free carboxyl group on the *C-terminal* residue appears on the right. Peptides are named by using their amino acid sequences, beginning from their N-terminal residue. For example,

$$H_2N-Tyr-Ala-Cys-Gly-COOH$$

is a tetrapeptide named tyrosylalanylcysteinylglycine.

QUESTION 5.3

Considering only the 20 standard amino acids, calculate the total number of possible tetrapeptides.

Large polypeptides have well-defined three-dimensional structures. This structure, referred to as the molecule's native conformation, is a direct consequence of its *amino acid sequence* (the order in which the amino acids are linked together). Because all the linkages connecting the amino acid residues consist of single bonds, it might be expected that each polypeptide undergoes constant conformational changes caused by rotation around the single bonds. However, most polypeptides spontaneously fold into a single biologically active form. In the early

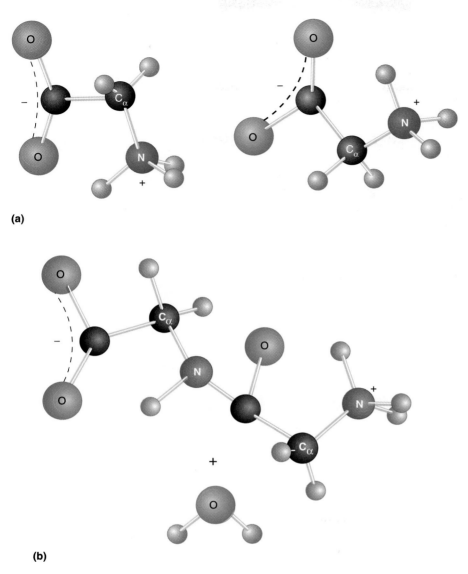

(a)

(b)

FIGURE 5.11

Formation of a Dipeptide.

(a) A peptide bond forms when the α-carboxyl group of one amino acid reacts with the amino group of another. (b) A water molecule is lost in the reaction.

1950s, Linus Pauling and his colleagues proposed an explanation. Using X-ray diffraction studies, they determined that the peptide bond is rigid and planar (flat) (Figure 5.12). Having discovered that the C—N bonds joining each two amino acids are shorter than other types of C—N bonds, Pauling deduced that peptide bonds have a partial double bond character. (This indicates that peptide bonds are resonance hybrids.) The rigidity of the peptide bond has several consequences. Because fully one-third of the bonds in a polypeptide backbone chain cannot rotate freely, there are limits on the number of conformational possibilities. Another consequence is that in extended segments of polypeptide, successive R groups appear on opposite sides (Figure 5.13).

CYSTEINE OXIDATION The sulfhydryl group of cysteine is highly reactive. The most common reaction of this group is a reversible oxidation that forms a disulfide. Oxidation of two molecules of cysteine forms cystine, a molecule that contains a disulfide bond (Figure 5.14). When two cysteine residues form such

FIGURE 5.12

The Peptide Bond.

(a) Resonance forms of the peptide bond. (b) Dimensions of a dipeptide. Because peptide bonds are rigid, the conformational degrees of freedom of a polypeptide chain are limited to rotations around the C_α—C and C_α—N bonds. The corresponding rotations are represented by ψ and ϕ, respectively.

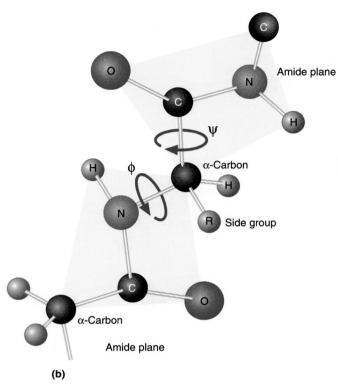

(a)

(b)

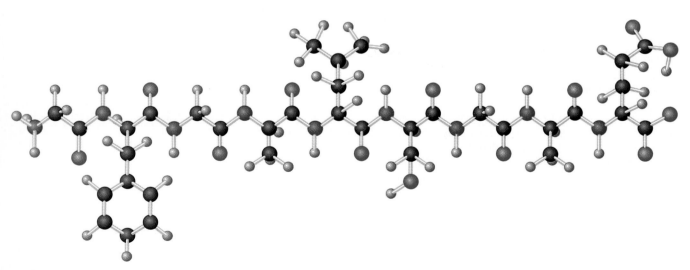

FIGURE 5.13

A Polypeptide Chain.

In polypeptides, successive R groups occur on alternate sides of the peptide bonds. Note that this illustration is a diagrammatic view of an extended polypeptide chain, not a representation of native structure.

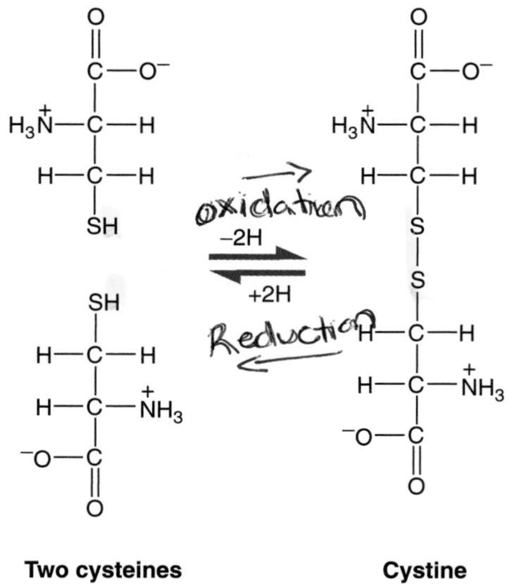

Two cysteines **Cystine**

FIGURE 5.14

Oxidation of Two Cysteine Molecules to form Cystine.

The disulfide bond in a polypeptide is called a disulfide bridge.

KEY CONCEPTS 5.5

Polypeptides are polymers composed of amino acids linked by peptide bonds. The order of the amino acids in a polypeptide is called the amino acid sequence. Disulfide bridges, formed by the oxidation of cysteine residues, are an important structural element in polypeptides and proteins.

a bond, it is referred to as a **disulfide bridge**. This bond can occur in a single chain to form a ring or between two separate chains to form an intermolecular bridge. Disulfide bridges help stabilize many polypeptides and proteins.

QUESTION 5.4

In extracellular fluids such as blood, the sulfhydryl groups of cysteine are quickly oxidized to form cystine. Unfortunately, cystine is the least soluble of the amino acids. In a genetic disorder known as *cystinuria,* defective membrane transport of cystine results in excessive excretion of cystine into the urine. Crystallization of the amino acid results in formation of calculi (stones) in the kidney, ureter, or urinary bladder. The stones may cause pain, infection, and blood in the urine. Cystine concentration in the kidney is reduced by massively increasing fluid intake and administering D-penicillamine. It is believed that penicillamine (Figure 5.15) is effective because penicillamine-cysteine disulfide, which is substantially more soluble than cystine, is formed. What is the structure of the penicillamine-cysteine disulfide?

5.2 PEPTIDES

Although their structures are less complex than the larger protein molecules, peptides have significant biological activities. The structure and function of several interesting examples, presented in Table 5.3, are now discussed.

The tripeptide glutathione (γ-glutamyl-L-cysteinylglycine) contains an unusual γ-amide bond. (Note that the γ-carboxyl group of the glutamic acid residue, not the α-carboxyl group, contributes to the peptide bond.) Found in almost all organisms, glutathione is involved in many important biological processes. Among these are protein and DNA synthesis, drug and environmental toxin metabolism, and amino acid transport. One group of glutathione's functions exploits its effectiveness as a reducing agent. (Because the reducing component of the molecule is the —SH group of the cysteine residue, the abbreviation for glutathione is GSH.) Glutathione protects cells from the destructive effects of oxidation by

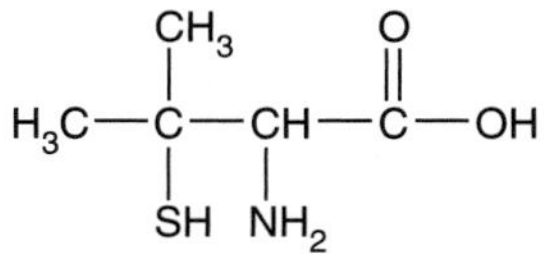

FIGURE 5.15

Structure of Penicillamine.

TABLE 5.3
Selected Biologically Important Peptides

Name	Amino Acid Sequence
Glutathione	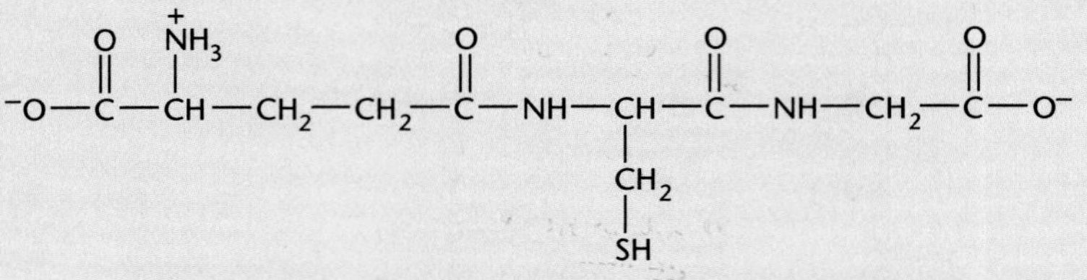
Oxytocin	Cys—Tyr—Ile—Gln—Asn—Cys—Pro—Leu—Gly—NH$_2$ (with S—S disulfide bond between the two Cys)
Vasopressin ↑ BP	Cys—Tyr—Phe—Gln—Asn—Cys—Pro—Arg—Gly—NH$_2$ (with S—S disulfide bond between the two Cys)
Met-enkephalin	Tyr—Gly—Gly—Phe—Met
Leu-enkephalin	Tyr—Gly—Gly—Phe—Leu
Atrial natriuretic factor	Ser[1]—Leu—Arg—Arg—Ser—Ser—Cys—Phe—Gly—Gly[10]—Arg—Met—Asp— Arg—Ile—Gly—Ala—Gln—Ser—Gly—Leu—Gly—Cys—Asn—Ser—Phe—Arg—Tyr[28]
Substance P	Arg—Pro—Lys—Pro—Gln—Phe—Phe—Gly—Leu—Met—NH$_2$
Bradykinin	Arg—Pro—Pro—Gly—Phe—Ser—Pro—Phe—Arg
α-Melanocyte stimulating hormone	Ser—Tyr—Ser—Met—Glu—His—Phe—Arg—Trp—Gly—Lys—Pro—Val
Cholecystokinin	Lys—Ala—Pro—Ser—Gly—Arg—Met—Ser—Ile—Val—Lys—Asn—Leu—Gln— Asn—Lys—Asp—Pro—Ser—His—Arg—Ile—Ser—Asp—Arg—Asp—Tyr—(SO$_3$)— Met—Gly—Trp—Met—Asp—Phe—NH$_2$
Galanin	Gly—Trp—Thr—Leu—Asn—Ser—Ala—Gly—Tyr—Leu—Leu—Gly— Pro—His—Ala—Val—Gly—Asn—His—Arg—Ser—Phe—Ser—Asp—Lys—Asn—G Gly—Leu—Thr—Ser
Neuropeptide Y	Tyr—Pro—Ser—Lys—Pro—Asp—Asn—Pro—Gly—Glu—Asp—Ala—Pro— Ala—Glu—Asp—Met—Ala—Arg—Tyr—Tyr—Ser—Ala—Leu—Arg—His—Tyr— Ile—Asn—Leu—Ile—Thr—Arg—Gln—Arg—Tyr—C—NH$_2$ (C bears a ║O)

reacting with substances such as peroxides, R—O—O—R, byproducts of O$_2$ metabolism. For example, in red blood cells, hydrogen peroxide (H$_2$O$_2$) oxidizes the iron of hemoglobin to its ferric form (Fe^{3+}). Methemoglobin, the product of this reaction, is incapable of binding O$_2$. Glutathione protects against the formation of methemoglobin by reducing H$_2$O$_2$ in a reaction catalyzed by the enzyme glutathione peroxidase. In the oxidized product GSSG, two tripeptides are linked by a disulfide bond:

$$2\ GSH + H_2O_2 \longrightarrow GSSG + 2H_2O$$

Because of the high GSH:GSSG ratio normally present in cells, glutathione is an important intracellular agent.

QUESTION 5.5

Write out the complete structure of oxytocin. What would be the net charge on this molecule at pH 4? At pH 9? Indicate which atoms in oxytocin can potentially form hydrogen bonds with water molecules.

Peptides are one class of signal molecules that multicellular organisms use to regulate their complex activities. Recall that multicellular organisms, consisting of several hundred cell types, must coordinate a huge number of biochemical processes. A stable internal environment is maintained by the dynamic interplay between opposing processes, called *homeostasis*. Peptide molecules with opposing functions are now known to affect the regulation of numerous processes. Examples include feeding behavior, blood pressure and pain perception. The roles of selected peptides in each of these processes are briefly described.

The regulation of food intake and body weight has proven to be considerably more complicated than previously thought. Research into the causes of obesity (excessive body weight), a major health problem in industrialized countries, has revealed that a variety of signal molecules in the brain have an effect on feeding behavior. Among these are appetite-stimulating peptides such as neuropeptide Y (NPY) and galanin, and appetite-inhibiting peptides such as cholecystokinin and α-melanocyte stimulating hormone (α-MSH). Recent evidence suggests that leptin, a protein released primarily by adipocytes (fat cells), reduces food intake by decreasing the expression of genes coding for NPY, galanin, and several other appetite-stimulating signal molecules.

Blood pressure, the force exerted by blood against the walls of blood vessels, is influenced by several factors such as blood volume and viscosity. Two peptides known to affect blood volume are vasopressin and atrial natriuretic factor. Vasopressin, also called antidiuretic hormone (ADH), contains nine amino acid residues. It is synthesized in the hypothalamus, a small structure in the brain that regulates a wide variety of functions including water balance and appetite. The ADH is transported down nerve tracts to the pituitary gland at the base of the brain and released in response to low blood pressure or a high blood Na^+ concentration. ADH stimulates the kidneys to retain water. The structure of ADH is remarkably similar to another peptide produced in the hypothalamus called oxytocin, a signal molecule that stimulates the ejection of milk by mammary glands during lactation and influences sexual, maternal, and social behavior. Oxytocin produced in the uterus stimulates the contraction of uterine muscle during childbirth. Because ADH and oxytocin have similar structures, it is not surprising that the functions of the two molecules overlap. Oxytocin has mild antidiuretic activity and vasopressin has some oxytocin-like activity. Atrial natriuretic factor (ANF), a peptide produced by specialized cells in the heart in response to stretching and in the nervous system, stimulates the production of a dilute urine, an effect opposite to that of vasopressin. ANF exerts its effect, in part, by increasing the excretion of Na^+, a process that causes increased excretion of water, and by inhibiting the secretion of renin by the kidney. (Renin is an enzyme that catalyzes the formation of angiotensin, a hormone that constricts blood vessels.)

Met-enkephalin and leu-enkephalin belong to a group of peptides called the **opioid peptides**, found predominantly in nerve tissue cells. Opioid peptides are molecules that relieve pain (a protective mechanism in animals that warns of tissue damage) and produce pleasant sensations. They were discovered after researchers suspected that the physiological effects of opiate drugs such as morphine resulted from their binding to nerve cell receptors for endogenous molecules. Leu-enkephalin and met-enkephalin are pentapeptides that differ only in their C-terminal amino acid residues. Substance P and bradykinin stimulate the perception of pain, an effect opposed by the opioid peptides.

5.3 PROTEINS

Of all the molecules encountered in living organisms, proteins have the most diverse functions, as the following list suggests:

1. **Catalysis**. *Enzymes* are proteins that direct and accelerate thousands of biochemical reactions in such processes as digestion, energy capture, and

biosynthesis. These molecules have remarkable properties. For example, they can increase reaction rates by factors of between 10^6 and 10^{12}. They can perform this feat under mild conditions of pH and temperature because they can induce or stabilize strained reaction intermediates. Examples include ribulose bisphosphate carboxylase, an important enzyme in photosynthesis, and nitrogenase, a protein complex that is responsible for nitrogen fixation.

2. **Structure**. Some proteins provide protection and support. Structural proteins often have very specialized properties. For example, collagen (the major components of connective tissues) and fibroin (silk protein) have significant mechanical strength. Elastin, the rubberlike protein found in elastic fibers, is found in several tissues in the body (e.g., blood vessels and skin) that must be elastic to function properly.

3. **Movement**. Proteins are involved in all cell movements. For example, actin, tubulin, and other proteins comprise the cytoskeleton. Cytoskeletal proteins are active in cell division, endocytosis, exocytosis, and the ameboid movement of white blood cells.

4. **Defense**. A wide variety of proteins are protective. Examples found in vertebrates include keratin, the protein found in skin cells that aids in protecting the organism against mechanical and chemical injury. The bloodclotting proteins fibrinogen and thrombin prevent blood loss when blood vessels are damaged. The immunoglobulins (or antibodies) are produced by lymphocytes when foreign organisms such as bacteria invade an organism. Binding antibodies to an invading organism is the first step in its destruction.

5. **Regulation**. Binding a hormone molecule or a growth factor to cognate receptors on its target cell changes cellular function. Examples of peptide hormones include insulin and glucagon, both of which regulate blood glucose levels. Growth hormone stimulates cell growth and division. Growth factors are polypeptides that control animal cell division and differentiation. Examples include platelet-derived growth factor (PDGF) and epidermal growth factor (EGF).

6. **Transport**. Many proteins function as carriers of molecules or ions across membranes or between cells. Examples of membrane proteins include the Na^+-K^+ ATPase and the glucose transporter. Other transport proteins include hemoglobin, which carries O_2 to the tissues from the lungs, and the lipoproteins LDL and HDL, which transport lipids from the liver and intestines to other organs. Transferrin and ceruloplasmin are serum proteins that transport iron and copper, respectively.

7. **Storage**. Certain proteins serve as a reservoir of essential nutrients. For example, ovalbumin in bird eggs and casein in mammalian milk are rich sources of organic nitrogen during development. Plant proteins such as zein perform a similar role in germinating seed.

8. **Stress response**. The capacity of living organisms to survive a variety of abiotic stresses is mediated by certain proteins. Examples include cytochrome P_{450}, a diverse group of enzymes found in animals and plants that usually convert a variety of toxic organic contaminants into less toxic derivatives, and metallothionein, a cysteine-rich intracellular protein found in virtually all mammalian cells that binds to and sequesters toxic metals such as cadmium, mercury, and silver. Excessively high temperatures and other stresses result in the synthesis of a class of proteins called the **heat shock proteins** (hsps) that promote the correct refolding of damaged proteins. If such proteins are severely damaged, hsps promote their degradation. (Certain hsps function in the normal process of protein folding.) Cells are protected from radiation by DNA repair enzymes.

Because of their diversity, proteins are often classified in two additional ways: (1) shape and (2) composition. Proteins are classified into two major groups based on their shape. As their name suggests, **fibrous proteins** are long, rod-shaped molecules that are insoluble in water and physically tough. Fibrous proteins, such as the keratins found in skin, hair, and nails, have structural and protective functions. **Globular proteins** are compact spherical molecules that are usually water soluble. Typically, globular proteins have dynamic functions. For example, nearly all enzymes have globular structures. Other examples include the immunoglobulins and the transport proteins hemoglobin and albumin (a carrier of fatty acids in blood).

On the basis of composition, proteins are classified as simple or conjugated. Simple proteins, such as serum albumin and keratin, contain only amino acids. In contrast, each **conjugated protein** consists of a simple protein combined with a nonprotein component. The nonprotein component is called a **prosthetic group**. (A protein without its prosthetic group is called an **apoprotein**. A protein molecule combined with its prosthetic group is referred to as a **holoprotein**.) Prosthetic groups typically play an important, even crucial, role in the function of proteins. Conjugated proteins are classified according to the nature of their prosthetic groups. For example, **glycoproteins** contain a carbohydrate component, **lipoproteins** contain lipid molecules, and **metalloproteins** contain metal ions. Similarly, **phosphoproteins** contain phosphate groups, and **hemoproteins** possess heme groups (p. 144).

Protein Structure

Proteins are extraordinarily complex molecules. Complete models depicting even the smallest of the polypeptide chains are almost impossible to comprehend. Simpler images that highlight specific features of a molecule are useful. Two methods of conveying structural information about proteins are presented in Figure 5.16. Another structural representation, referred to as a ball-and-stick model, can be seen in Figures 5.29 and 5.31 (pp. 145 and 148).

Biochemists have distinguished several levels of the structural organization of proteins. **Primary structure**, the amino acid sequence, is specified by genetic information. As the polypeptide chain folds, it forms certain localized arrangements of adjacent amino acids that constitute **secondary structure**. The overall three-dimensional shape that a polypeptide assumes is called the **tertiary structure**. Proteins that consist of two or more polypeptide chains (or subunits) are said to have a **quaternary structure**.

PRIMARY STRUCTURE Every polypeptide has a specific amino acid sequence. The interactions between amino acid residues determine the protein's three-dimensional structure and its functional role and relationship to other proteins. Polypeptides that have similar amino acid sequences and functions are said to be **homologous**. Sequence comparisons among homologous polypeptides have been used to trace the genetic relationships of different species. For example, the sequence homologies of the mitochondrial redox protein cytochrome c have been used extensively in the study of evolution. Sequence comparisons of cytochrome c among numerous species reveal a significant amount of sequence conservation. The amino acid residues that are identical in all homologues of a protein, referred to as *invariant*, are presumed to be essential for the protein's function. (In cytochrome c the invariant residues interact with heme, a prosthetic group, or certain other proteins involved in energy generation.)

PRIMARY STRUCTURE, EVOLUTION, AND MOLECULAR DISEASES. Because of the essential role of cytochrome c in energy production, individual organisms with amino acid substitutions at invariant positions are not viable. Mutations (alterations in the DNA sequences that code for a protein's amino acid

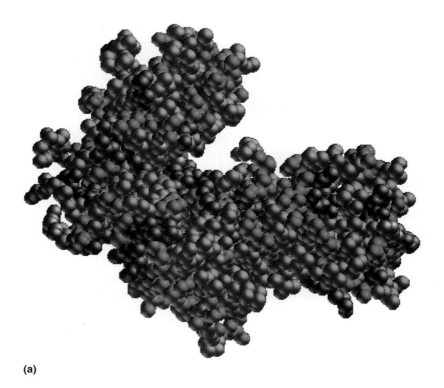

(a)

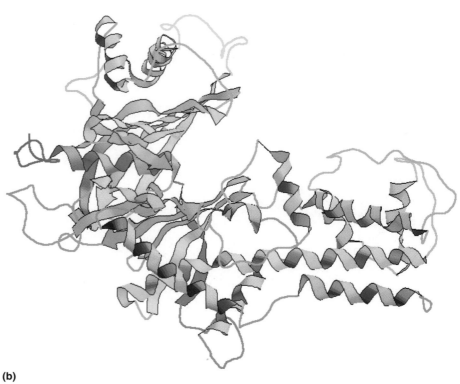

(b)

FIGURE 5.16

The Enzyme Adenylate Kinase.

(a) A space-filling model illustrates the volume occupied by molecular components and overall shape. (b) In a ribbon model β-pleated segments are represented by flat arrows. The α-helices appear as spiral ribbons.

sequence) are random and spontaneous events. Therefore, a significant number of primary sequence changes that do not affect a polypeptide's function occur over time. Some of these substitutions are said to be *conservative* because an amino acid with a chemically similar side chain is substituted. For example, at certain sequence positions leucine and isoleucine, which both contain hydrophobic side chains, may be substituted for each other without affecting function. Some sequence positions are significantly less stringent. These residues, referred to as *variable*, apparently perform nonspecific roles in the polypeptide's function.

Substitutions at conservative and variable sites have been used to trace evolutionary relationships. These studies assume that the longer the time since two species diverged from each other, the larger the number of differences in a certain polypeptide's primary structure. For example, humans and chimpanzees are believed to have diverged relatively recently (perhaps only four million years ago). This presumption, based principally on fossil and anatomical evidence, is supported by cytochrome c primary sequence data, because the protein is identical in both species. Animals such as kangaroos, whales, and sheep, whose cytochrome c molecules each differ by 10 residues from the human protein, are all believed to have evolved from a common ancestor that lived over 50 million years ago.

Some mutations are deleterious without being immediately lethal. Sickle-cell anemia, which is caused by mutant hemoglobin, is a classic example of a group of maladies that Linus Pauling and his colleagues referred to as **molecular diseases**. (Dr. Pauling first demonstrated that sickle-cell patients have a mutant hemoglobin through the use of electrophoresis.) Human adult hemoglobin (HbA) is composed of two identical α-chains and two identical β-chains. Sickle-cell anemia results from a single amino acid substitution in the β-chain of HbA. Analysis of the hemoglobin molecules of sickle-cell patients reveals that the only difference between HbA and sickle-cell hemoglobin (HbS) is at amino acid residue 6 in the β-chain (Figure 5.17). Because of the substitution of a hydrophobic valine for a negatively charged glutamic acid, HbS molecules aggregate to form rigid rodlike structures in the oxygen-free state. The patient's red blood cells become sickle shaped and are susceptible to hemolysis resulting in severe anemia. These red blood cells have a lower than normal oxygen-binding capacity. Intermittent clogging of capillaries by sickled cells also causes tissues to be deprived of oxygen. Sickle-cell anemia is characterized by excruciating pain, eventual organ damage, and earlier death.

Until recently, because of the debilitating nature of sickle-cell disease, affected individuals rarely survived beyond childhood. It might be predicted that the deleterious mutational change that causes this affliction would be rapidly eliminated from human populations. However, the sickle-cell gene is not as rare as would be expected. Sickle-cell disease occurs in individuals who have inherited two copies of the sickle-cell gene. Such individuals, who are said to be *homozygous*, inherit one copy of the defective gene from each parent. Each of the parents, referred to as *heterozygous* because they have one normal HbA gene and one defective HbS gene, is said to have the *sickle-cell trait*. Such people

KEY CONCEPTS 5.6

The primary structure of a polypeptide is its amino acid sequence. The amino acids are connected by peptide bonds. Amino acid residues that are essential for the molecule's function are referred to as invariant. Proteins with similar amino acid sequences and functions are said to be homologous.

Hb A   Val—His—Leu—Thr—Pro—Glu—Glu—Lys—

Hb S   Val—His—Leu—Thr—Pro—Val—Glu—Lys—

        1   2   3   4   5   6   7   8

FIGURE 5.17

Segments of β-Chain in HbA and HbS.

Individuals possessing the gene for sickle-cell hemoglobin produce β-chains with valine instead of glutamic acid at residue 6.

lead normal lives even though about 40% of their hemoglobin is HbS. The incidence of sickle-cell trait is especially high in some regions of Africa. In these regions the disease malaria, caused by the *Anopheles* mosquito–borne parasite *Plasmodium,* is a serious health problem. Individuals with the sickle-cell trait are less vulnerable to malaria because their red blood cells are a less favorable environment for the growth of the parasite than are normal cells. Because sickle-cell trait carriers are more likely to survive malaria than normal individuals, the incidence of the sickle-cell gene has remained high. (In some areas, as many as 40% of the native populations have the sickle-cell trait.)

QUESTION 5.6

A genetic disease called *glucose-6-phosphate dehydrogenase deficiency* is inherited in a manner similar to that of sickle-cell anemia. The defective enzyme cannot keep erythrocytes supplied with sufficient amounts of the antioxidant molecule NADPH (Chapter 8). NADPH protects cell membranes and other cellular structures from oxidation. Describe in general terms the inheritance pattern of this molecular disease. Why do you think that the antimalarial drug primaquine, which stimulates peroxide formation, results in devastating cases of hemolytic anemia in carriers of the defective gene? Does it surprise you that this genetic anomaly is commonly found in African and Mediterranean populations?

SECONDARY STRUCTURE The secondary structure of polypeptides consists of several repeating patterns. The most commonly observed types of secondary structure are the α-helix and the β-pleated sheet. Both α-helix and β-pleated sheet patterns are stabilized by hydrogen bonds between the carbonyl and N—H groups in the polypeptide's backbone. These patterns occur when all the ϕ (phi) angles (rotation angle about C_α—C) in a polypeptide segment are equal and all the ψ (psi) bond angles (rotation angle about C_α—N) are equal (see Figure 5.12b). Because peptide bonds are rigid, the α-carbons are swivel points for the polypeptide chain. Several properties of the R groups (e.g., size and charge, if any) attached to the α-carbon influence the ϕ and ψ angles. Certain amino acids foster or inhibit specific secondary structural patterns. Many fibrous proteins are composed almost entirely of secondary structural patterns.

The *α-helix* is a rigid, rodlike structure that forms when a polypeptide chain twists into a right-handed helical conformation (Figure 5.18). Hydrogen bonds form between the N—H group of each amino acid and the carbonyl group of the amino acid four residues away. There are 3.6 amino acid residues per turn of the helix, and the pitch (the distance between corresponding points per turn) is 54 nm. Amino acid R groups extend outward from the helix. Because of several structural constraints (i.e., the rigidity of peptide bonds and the allowed limits on the values of the ϕ and ψ angles), certain amino acids do not foster α-helical formation. For example, glycine's R group (a hydrogen atom) is so small that the polypeptide chain may be too flexible. Proline, on the other hand, contains a rigid ring that prevents the N—C_α bond from rotating. In addition, proline has no N—H group available to form the intrachain hydrogen bonds that are crucial in α-helix structure. Amino acid sequences with large numbers of charged amino acids (e.g., glutamate and aspartate) and bulky R groups (e.g., tryptophan) are also incompatible with α-helix structures.

β-Pleated sheets form when two or more polypeptide chain segments line up side by side (Figure 5.19). Each individual segment is referred to as a *β-strand.* Rather than being coiled, each β-strand is fully extended. β-Pleated sheets are stabilized by hydrogen bonds that form between the polypeptide backbone N—H and carbonyl groups of adjacent chains. There are two β-pleated sheets: parallel and antiparallel. In *parallel* β-pleated sheet structures, the polypeptide chains

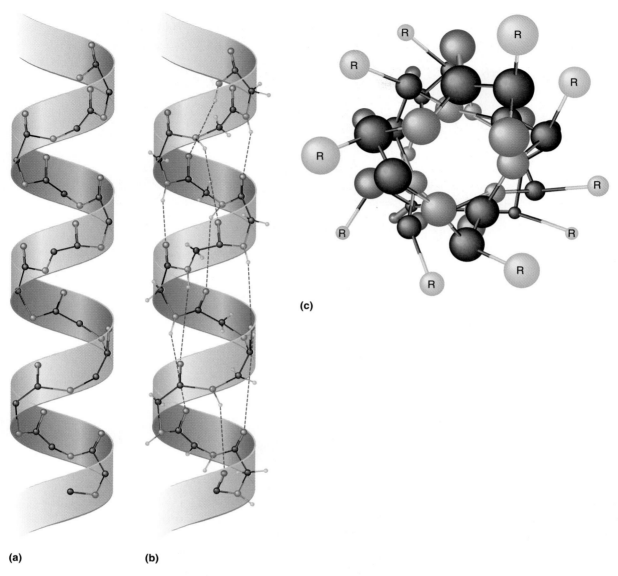

(c)

FIGURE 5.18

The α-Helix.

(a) The helical backbone. (b) A more complete model. Hydrogen bonds form between carbonyl and N—H groups along the long axis of the helix. (c) A top view of the α-helix. The R groups point away from the long axis of the helix.

are arranged in the same direction. *Antiparallel* chains run in opposite directions. Antiparallel β-sheets are more stable than parallel β-sheets because fully colinear hydrogen bonds form. Occasionally, mixed parallel-antiparallel β-sheets are observed.

Many globular proteins contain combinations of α-helix and β-pleated sheet secondary structures (Figure 5.20). These patterns are called **supersecondary structures**. In the *βαβ unit*, two parallel β-pleated sheets are connected by an α-helix segment. In the *β-meander* pattern, two antiparallel β-sheets are connected by polar amino acids and glycines to effect an abrupt change in direction of the polypeptide chain called *reverse* or *β-turns*. In *αα-units*, two successive α-helices separated by a loop or nonhelical segment become enmeshed because of compatible side chains. Several *β-barrel* arrangements are formed when various

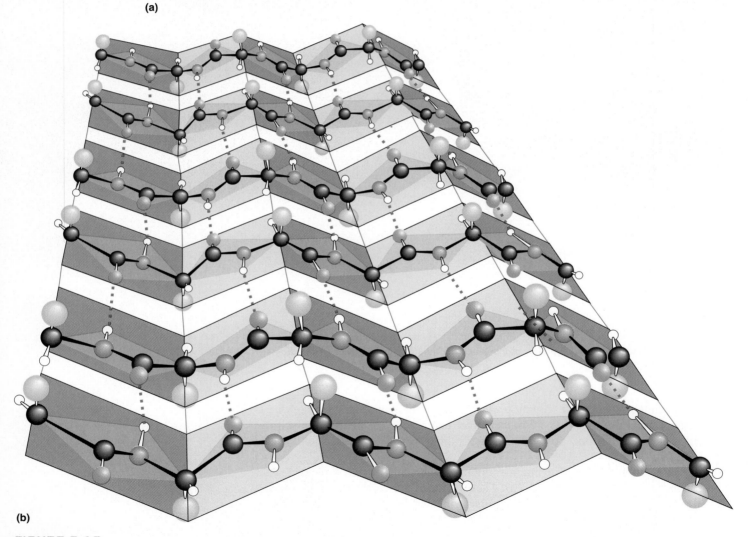

(a)

(b)

FIGURE 5.19

β-Pleated Sheet.

(a) Two forms of β-pleated sheet: Antiparallel and parallel. Hydrogen bonds are represented by dotted lines. (b) A more detailed view of antiparallel β-pleated sheet.

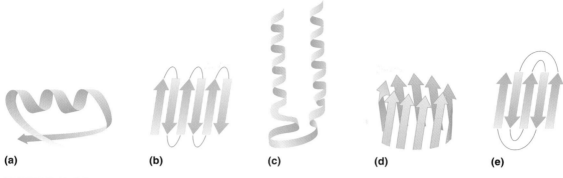

FIGURE 5.20

Selected Supersecondary Structures.
(a) $\beta\alpha\beta$ units, (b) β-meander, (c) $\alpha\alpha$-unit, (d) β-barrel, and (e) Greek key.

β-sheet configurations fold back on themselves. When an antiparallel β-sheet doubles back on itself in a pattern that resembles a common Greek pottery design, the motif is called the *Greek key*.

TERTIARY STRUCTURE Although globular proteins often contain significant numbers of secondary structural elements, several other factors contribute to their structure. The term *tertiary structure* refers to the unique three-dimensional conformations that globular proteins assume as they fold into their native (biologically active) structures. **Protein folding**, a process in which an unorganized, *nascent* (newly synthesized) molecule acquires a highly organized structure, occurs as a consequence of the interactions between the side chains in their primary structure. Tertiary structure has several important features:

1. Many polypeptides fold in such a fashion that amino acid residues that are distant from each other in the primary structure come into close proximity.

2. Because of efficient packing as the polypeptide chain folds, globular proteins are compact. During this process, most water molecules are excluded from the protein's interior making interactions between both polar and nonpolar groups possible.

3. Large globular proteins (i.e., those with more than 200 amino acid residues) often contain several compact units called domains. Domains (Figure 5.21) are typically structurally independent segments that have specific functions (e.g., binding an ion or small molecule).

The following types of interactions stabilize tertiary structure (Figure 5.22):

1. **Hydrophobic interactions**. As a polypeptide folds, hydrophobic R groups are brought into close proximity because they are excluded from water. Then the highly ordered water molecules in solvation shells are released from the interior, increasing the disorder (entropy) of the water molecules. The favorable entropy change is a major driving force in protein folding.

2. **Electrostatic interactions**. The strongest electrostatic interaction in proteins occurs between ionic groups of opposite charge. Referred to as **salt bridges**, these noncovalent bonds are significant only in regions of the protein where water is excluded because of the energy required to remove water molecules from ionic groups near the surface. Salt bridges have been observed to contribute to the interactions between adjacent subunits in complex proteins. The same is true for the weaker electrostatic interactions (ion-dipole, dipole-dipole, van der Waals). They are significant in the interior of the folded protein and between subunits or in protein-ligand interactions. (In proteins that consist of more than one polypeptide

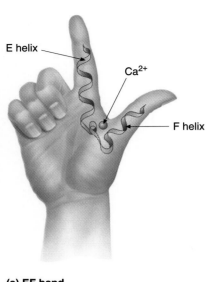

(a) EF hand

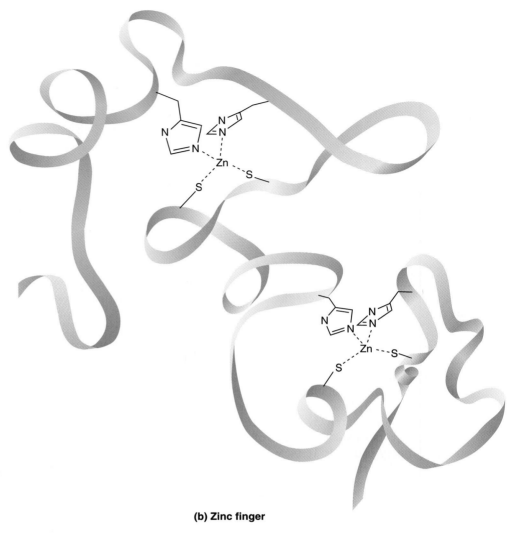

(b) Zinc finger

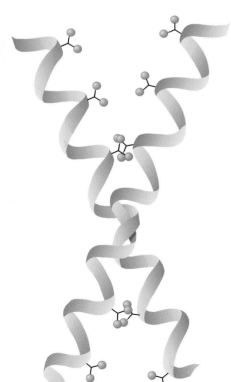

(c) Leucine zipper

FIGURE 5.21

Three Domains Found in Several Proteins.

(a) The EF hand, which consists of a helix-loop-helix configuration, binds specifically to Ca²⁺. (The hand motif helps the viewer to comprehend the three-dimensional structure of the calcium-binding domain.) (b) The zinc finger motif is commonly found in DNA-binding proteins. DNA-binding proteins often possess several zinc fingers, each of which promotes protein-DNA interactions. (See Chapter 18.) Cysteine residues provide the binding sites for the zinc ions. (c) The leucine zipper is another DNA-binding domain. The knobs on the leucine zipper represent leucine side chains.

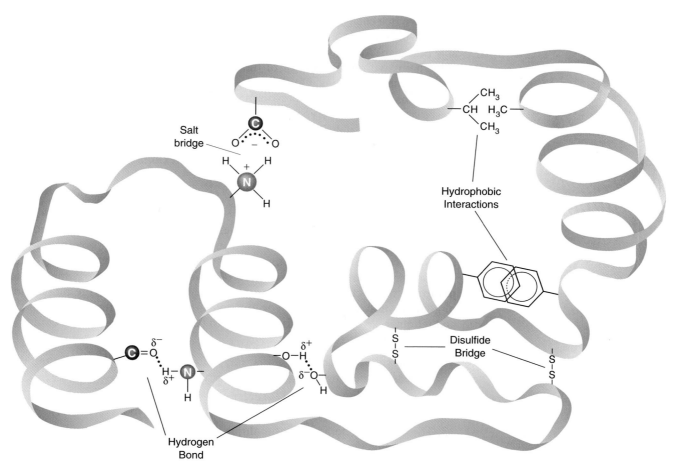

FIGURE 5.22
Interactions That Maintain Tertiary Structure.

chain, each polypeptide is called a **subunit**. **Ligands** are molecules that bind to specific sites on larger molecules such as proteins.) Ligand binding pockets are water-depleted regions of the protein.

3. **Hydrogen bonds**. A significant number of hydrogen bonds form within a protein's interior and on its surface. In addition to forming hydrogen bonds with one another, the polar amino acid side chains may interact with water or with the polypeptide backbone. Again, the presence of water precludes the formation of hydrogen bonds with other species.

4. **Covalent bonds**. Covalent linkages are created by chemical reactions that alter a polypeptide's structure during or after its synthesis. (Examples of these reactions, referred to as posttranslational modifications, are described in Section 19.2.) The most prominent covalent bonds in tertiary structure are the disulfide bridges found in many extracellular proteins. In extracellular environments these strong linkages partly protect protein structure from adverse changes in pH or salt concentrations. Intracellular proteins do not contain disulfide bridges because of high cytoplasmic concentrations of reducing agents.

The precise nature of the forces that promote the folding of proteins (described in detail in Chapter 19) has not yet been completely resolved. It is clear, however, that protein folding is a thermodynamically favorable process with an overall negative free energy change. According to the free energy equation:

$$\Delta G = \Delta H - T\Delta S$$

a negative free energy change in a process is the result of a balance between favorable and unfavorable enthalpy and entropy changes. As a polypeptide folds, favorable (negative) ΔH values are the result in part of the sequestration of hydrophobic side chains within the interior of the molecule and the optimization of other noncovalent interactions. Opposing these factors is the unfavorable decrease in entropy that occurs as the disorganized polypeptide folds into its highly organized native state. For most polypeptide molecules the net free energy change between the folded and unfolded state is relatively modest (the energy equivalent of several hydrogen bonds). The precarious balance between favorable and unfavorable forces, described on p. 138, allows proteins the flexibility they require for biological function.

QUATERNARY STRUCTURE Many proteins, especially those with high molecular weights, are composed of several polypeptide chains. As mentioned, each polypeptide component is called a subunit. Subunits in a protein complex may be identical or quite different. Multisubunit proteins in which some or all subunits are identical are referred to as **oligomers**. Oligomers are composed of **protomers**, which may consist of one or more subunits. A large number of oligomeric proteins contain two or four subunit protomers, referred to as dimers and tetramers, respectively. There appear to be several reasons for the common occurrence of multisubunit proteins:

1. Synthesis of separate subunits may be more efficient than substantially increasing the length of a single polypeptide chain.
2. In supramolecular complexes such as collagen fibers, replacement of smaller worn-out or damaged components can be managed more effectively.
3. The complex interactions of multiple subunits help regulate a protein's biological function.

Polypeptide subunits assemble and are held together by noncovalent interactions such as the hydrophobic effect, electrostatic interactions, and hydrogen bonds, as well as covalent cross-links. As with protein folding, the hydrophobic effect is clearly the most important because the structures of the complementary interfacing surfaces between subunits are similar to those observed in the interior of globular protein domains. Although they are less numerous, covalent cross-links significantly stabilize certain multisubunit proteins. Prominent examples include the disulfide bridges in the immunoglobulins, and the desmosine and lysinonorleucine linkages in certain connective tissue proteins. *Desmosine* (Figure 5.23) cross-links connect four polypeptide chains in the rubberlike connective tissue protein elastin. They are formed as a result of a series of reactions involving the oxidation of lysine side chains. A similar process results in the formation of *lysinonorleucine*, a cross-linking structure that is found in elastin and collagen.

Quite often the interactions between subunits are affected by the binding of ligands. In **allostery**, the control of protein function through ligand binding, binding a ligand to a specific site in a protein triggers a conformational change that alters its affinity for other ligands. Ligand-induced conformational changes in such proteins are called **allosteric transitions** and the ligands that trigger them are called **effectors** or **modulators**. Allosteric effects can be positive or negative, depending on whether effector binding increases or decreases the protein's affinity for other ligands. One of the best understood examples of allosteric effects, the reversible binding of O_2 and other ligands to hemoglobin, is described on pp. 145, 148–151. (Because allosteric enzymes play a key role in the control of metabolic processes, allostery is discussed further in Sections 6.3 and 6.5.)

KEY CONCEPTS 5.7

Biochemists distinguish four levels of the structural organization of proteins. In primary structure, the amino acid residues are connected by peptide bonds. The secondary structure of polypeptides is stabilized by hydrogen bonds. Prominent examples of secondary structure are α-helices and β-pleated sheets. Tertiary structure is the unique three-dimensional conformation that a protein assumes because of the interactions between amino acid side chains. Several types of interactions stabilize tertiary structure: the hydrophobic effect, electrostatic interactions, hydrogen bonds, and certain covalent bonds. Proteins that consists of several separate polypeptide subunits exhibit quaternary structure. Both noncovalent and covalent bonds hold the subunits together.

Desmosine

Lysinonorleucine

FIGURE 5.23

Desmosine and Lysinonorleucine Linkages.

Review the following illustrations of globular proteins. Identify examples of secondary and suprasecondary structure.

QUESTION 5.7

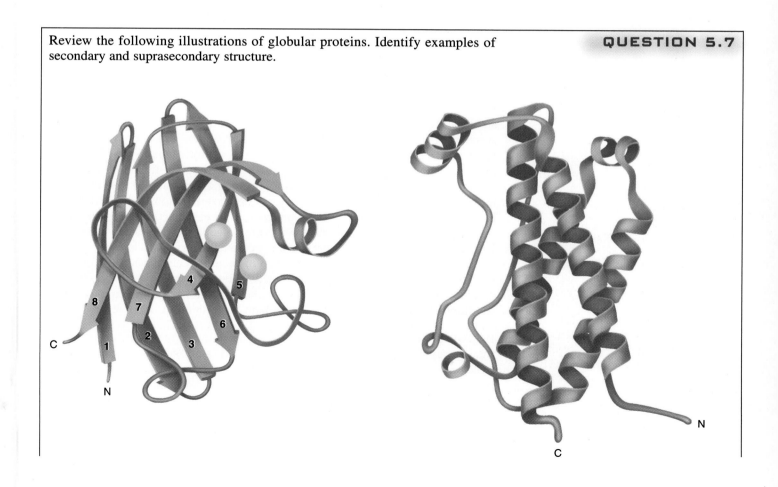

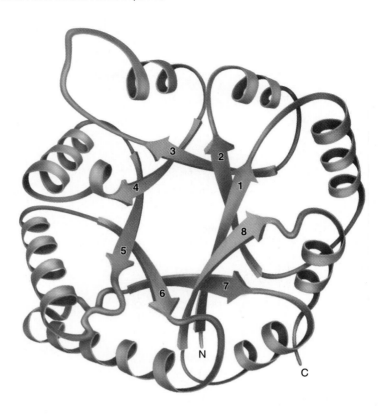

QUESTION 5.8

Illustrate the noncovalent interactions that can occur between the following side chain groups in folded polypeptides: (a) serine and glutamate; (b) arginine and aspartic acid; (c) threonine and serine; (d) glutamine and aspartate; (e) phenylalanine and tryptophan.

PROTEIN DYNAMICS Despite the emphasis so far placed on the forces that stabilize protein structure, it should be recognized that protein function requires some degree of flexibility. The significance of conformational flexibility (continuous, rapid fluctuations in the precise orientation of the atoms in proteins) has been revealed as researchers have investigated protein-ligand interactions. Protein function often involves the rapid opening and closing of cavities in the molecule's surface. The rate that enzymes catalyze reactions is limited in part by how fast product molecules can be released from the active site. Also recall that the transfer of information between biomolecules occurs when molecules with precise complementary surfaces interact in a process involving noncovalent interactions. Information transfer between molecules is always accompanied by modifications in three-dimensional structure. For example, the conformations of the subunits of the O_2-binding hemoglobin molecules undergo specific structural changes as oxygen molecules bind and unbind (p. 149).

LOSS OF PROTEIN STRUCTURE Considering the small differences in the free energy of folded and unfolded proteins, it is not surprising that protein structure is especially sensitive to environmental factors. Many physical and chemical agents can disrupt a protein's native conformation. The process of structure disruption is called **denaturation**. (Denaturation is not usually considered to include the breaking of peptide bonds.) Depending on the degree of denaturation, the molecule may partially or completely lose its biological activity. Denaturation often results in easily observable changes in the physical properties of proteins. For example, soluble and transparent egg albumin (egg white) becomes insoluble

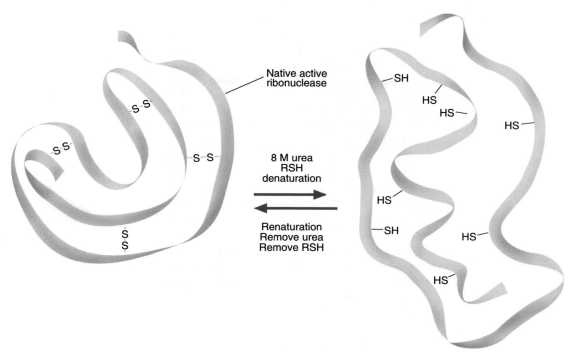

FIGURE 5.24

The Anfinsen Experiment.

Ribonuclease denatured by 8 M urea and a mercaptan (RSH; a reagent that reduces disulfides to sulfhydryl groups) can be renaturated by removing the urea and RSH and air oxidizing the reduced disulfides.

and opaque upon heating. Like many denaturations, cooking eggs is an irreversible process. The following example of a reversible denaturation was demonstrated by Christian Anfinsen in the 1950s. Bovine pancreatic ribonuclease (a digestive enzyme from cattle that degrades RNA) is denatured when treated with β-mercaptoethanol and 8 M urea (Figure 5.24). During this process, ribonuclease, composed of a single polypeptide with four disulfide bridges, completely unfolds and loses all biological activity. Careful removal of the denaturing agents with dialysis results in a spontaneous and correct refolding of the polypeptide and re-formation of the disulfide bonds. The fact that Anfinsen's experiment resulted in a full restoration of the enzyme's catalytic activity served as early evidence that three-dimensional structure is determined by a polypeptide's amino acid sequence. However, most proteins treated similarly do not renature.

Denaturing conditions include the following:

1. **Strong acids or bases**. Changes in pH result in protonation of some protein side groups, which alters hydrogen bonding and salt bridge patterns. As a protein approaches its isoelectric point, it becomes insoluble and precipitates from solution.

2. **Organic solvents**. Water-soluble organic solvents such as ethanol interfere with hydrophobic interactions because they interact with nonpolar R groups and form hydrogen bonds with water and polar protein groups. Nonpolar solvents also disrupt hydrophobic interactions.

3. **Detergents**. These amphipathic molecules disrupt hydrophobic interactions, causing proteins to unfold into extended polypeptide chains. (**Amphipathic molecules** contain both hydrophobic and hydrophilic components.)

4. **Reducing agents**. In the presence of reagents such as urea, reducing agents such as β-mercaptoethanol convert disulfide bridges to sulfhydryl groups. Urea disrupts hydrogen bonds and hydrophobic interactions.

5. Salt concentration. The binding of salt ions to a protein's ionizable groups decreases interaction between oppositely charged groups on the protein molecule. Water molecules then can form solvation spheres around these groups. When large amounts of salt are added to a protein in solution, a precipitate forms. The large number of salt ions can effectively compete with the protein for water molecules, that is, the solvation spheres surrounding the protein's ionized groups are removed. The protein molecules aggregate and then precipitate. This process is referred to as *salting out*. Because salting out is usually reversible, it is often used as an early step in protein purification.

6. Heavy metal ions. Heavy metals such as mercury (Hg^{2+}) and lead (Pb^{2+}) affect protein structure in several ways. They may disrupt salt bridges by forming ionic bonds with negatively charged groups. Heavy metals also bond with sulfhydryl groups, a process that may result in significant changes in protein structure and function. For example, Pb^{2+} binds to sulfhydryl groups in two enzymes in the heme synthetic pathway (Chapter 14). The resultant decrease in hemoglobin synthesis causes severe anemia. (In anemia the number of red blood cells or the hemoglobin concentration is lower than normal.) Anemia is one of the most easily measured symptoms of lead poisoning (Special Interest Box 14.4).

7. Temperature changes. As the temperature increases, the rate of molecular vibration increases. Eventually, weak interactions such as hydrogen bonds are disrupted and the protein unfolds. Some proteins are more resistant to heat denaturation and this fact can be used in purification procedures.

8. Mechanical stress. Stirring and grinding actions disrupt the delicate balance of forces that maintain protein structure. For example, the foam formed when egg white is beaten vigorously contains denatured protein.

Fibrous Proteins

Fibrous proteins typically contain high proportions of regular secondary structures, such as α-helices or β-pleated sheets. As a consequence of their rodlike or sheetlike shapes, many fibrous proteins have structural rather than dynamic roles. Examples of fibrous proteins include α-keratin, collagen, and silk fibroin.

α-**KERATIN** In fibrous proteins, bundles of helical polypeptides are commonly twisted together into larger bundles. The structural unit of the α-keratins, a class of proteins found in hair, wool, skin, horns, and fingernails, is an α-helical polypeptide. Each polypeptide has three domains: an amino terminal "head" domain, a central rodlike α-helical domain, and a carboxyl terminal "tail." Two keratin polypeptides associate to form a coiled coil dimer (Figure 5.25). Two staggered antiparallel rows of these dimers form a left-handed supercoiled structure called a protofilament. Hydrogen bonds and disulfide bridges are the principal interactions between protofilament subunits. Hundreds of filaments, each containing four protofilaments, are packed together to form a macrofibril. Each hair cell, also called a *fiber*, contains several macrofibrils. A strand of hair therefore consists of numerous dead cells packed with keratin molecules.

Many of the physical properties of the α-keratins are reflected in their amino acid compositions. They have a regular α-helical structure because they lack helix-breaking amino acids such as proline and have helix-promoting residues such as alanine and leucine. Because R groups are on the outside of the α-helices, α-keratin's high hydrophobic amino acid content makes this group of proteins very insoluble in water. Its cysteine residues and the formation of interhelix disulfide bridges make α-keratin relatively resistant to stretching. "Hard" keratins, such as those found in horns and nails, have considerably more disulfide bridges than their softer counterparts found in skin. Humans take advantage of the disulfide bridge content of hair during the permanent waving process. After the hair strands are arranged in the desired shape, disulfide bonds are broken with a reducing agent. New disulfide bonds are then formed by an oxidizing agent, thus creating curled hair.

α-Helix

Coiled coil of two α- helices

Protofilament (pair of coiled coils)

Filament (four right-hand twisted protofilaments)

FIGURE 5.25

Keratin.

The α-helical rodlike domains of two keratin polypeptides form a coiled coil. Two staggered antiparallel rows of these dimers form a supercoiled protofilament. Hundreds of filaments, each containing four protofilaments, form a macrofibril.

COLLAGEN Collagen is the most abundant protein in vertebrates. Synthesized by connective tissue cells, collagen molecules are secreted into the extracellular space to become part of the connective tissue matrix. Collagen includes many closely related proteins that have diverse functions. The genetically distinct collagen molecules in skin, bones, tendons, blood vessels, and corneas impart to these structures many of their special properties (e.g., the tensile strength of tendons and the transparency of corneas).

Collagen is composed of three left-handed polypeptide helices that are twisted around each other to form a right-handed triple helix (Figure 5.26). Type I collagen molecules, found in teeth, bone, skin, and tendons, are about 300 nm long and approximately 1.5 nm wide. (There are 20 major families of collagen molecules. Approximately 90% of the collagen found in humans is type I.)

The amino acid composition of collagen is distinctive. Glycine constitutes approximately one-third of the amino acid residues. Proline and 4-hydroxyproline may account for as much as 30% of a collagen molecule's amino acid composition. Small amounts of 3-hydroxyproline and 5-hydroxylysine also occur. (Specific proline and lysine residues in collagen's primary sequence are hydroxylated within the rough ER after the polypeptides have been synthesized. These reactions, which are discussed in Chapter 19, require the antioxidant ascorbic acid.)

Collagen's amino acid sequence primarily consists of large numbers of repeating triplets with the sequence of Gly—X—Y, in which X and Y are often proline and hydroxyproline. Hydroxylysine is also found in the Y position. Simple carbohydrate groups are often attached to the hydroxyl group of hydroxylysine

FIGURE 5.26

Collagen Fibrils.

The bands are formed by staggered collagen molecules. Cross striations are about 680 Å apart. Each collagen molecule is about 3000 Å long.

From Voet and Voet, *Biochemistry*, 2nd edition. Copyright © John Wiley and Sons, Inc., New York, NY.

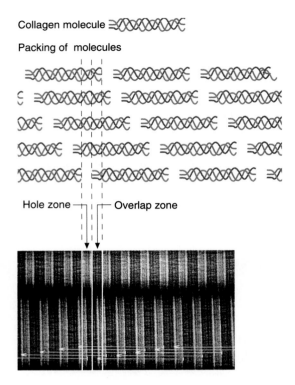

Collagen molecule

Packing of molecules

Hole zone — Overlap zone

residues. It has been suggested that collagen's carbohydrate components are required for *fibrilogenesis*, the assembly of collagen fibers in their extracellular locations, such as tendons and bone. The enzyme lysyl oxidase converts some of the lysine and hydroxylysine side groups to aldehydes through oxidative deamination and this facilitates the spontaneous non-enzymatic formation of strengthening aldimine, and aldol crosslinks. (*Aldimines* are Schiff bases that are formed by the reversible reaction of an amine with an aldehyde functional group.) Cross-linkages between hydroxylysine-linked carbohydrates and the amino group of other lysine and hydroxylysine residues on adjacent molecules also occurs. Increased crosslinking with age leads to the brittleness and breakage of the collagen fibers that occurs with aging.

Glycine is prominent in collagen sequences because the triple helix is formed by interchain hydrogen bonding involving the glycine residues. Therefore every third residue is in close contact with the other two chains. Glycine is the only amino acid with an R group sufficiently small for the space available. Larger R groups would destabilize the superhelix structure. The triple helix is further strengthened by hydrogen bonding between the polypeptides (caused principally by the large number of hydroxyproline residues) and lysinonorleucine linkages that stabilize the orderly arrays of triple helices in the final collagen fibril.

SILK FIBROIN Several insects and spiders produce silk, a substance that consists of the fibrous protein fibroin embedded in an amorphous matrix. In fibroin, which is considered to be a β-keratin, the polypeptide chains are arranged in antiparallel β-pleated sheet conformations (Figure 5.27). β-Pleated sheets form because of fibroin's large content of amino acids with relatively small R groups such as glycine and alanine or serine. (Bulky R groups would distort the almost crystalline regularity of silk protein.) Silk is a strong fabric because the chains are fully extended. Stretching them further requires breaking strong covalent bonds in the polypeptide backbones. Because the pleated sheets are loosely bonded to each other (primarily with weak van der Waals forces), they slide over each other easily. This arrangement gives silk fibers their flexibility.

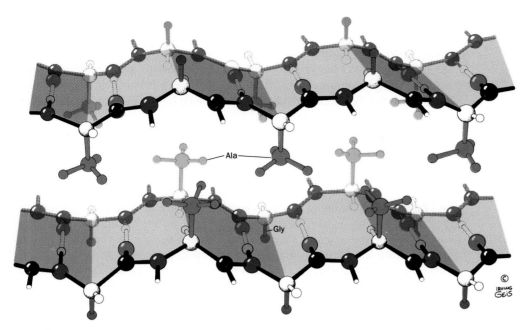

FIGURE 5.27

Molecular Model of Silk Fibroin.

Note that the R groups of alanine on one side of each β-pleated sheet interdigitate with similar residues on the adjacent sheet.

© Irving Geis (illustration) from *Biochemistry*, 2nd ed. by Donald Voet and Judith Voet, published by John Wiley & Sons, Inc.

QUESTION 5.9

Covalent cross-links contribute to the strength of collagen. The first reaction in cross-link formation is catalyzed by the copper-containing enzyme lysyl oxidase, which converts lysine residues to the aldehyde allysine:

Lysine residue $\xrightarrow{\text{Lysyl oxidase}}$ Allysine residue

Allysine then reacts with other side chain aldehyde or amino groups to form cross-linkages. For example, two allysine residues react to form an aldol cross-linked product:

Allysine residue + Allysine residue → Aldol cross-link

(A reaction in which aldehydes form an α, β-unsaturated aldehyde linkage is referred to as an **aldol condensation**. In condensation reactions a small molecule, in this case H_2O, is removed.)

In a disease called *lathyrism*, which occurs in humans and several other animals, a toxin (β-aminopropionitrile) found in sweet peas (*Lathyrus odoratus*) inactivates lysyl oxidase. Consider the abundance of collagen in animal bodies and suggest some likely symptoms of this malady.

Globular Proteins

The biological functions of globular proteins usually involve the precise binding of small ligands or large macromolecules such as nucleic acids or other proteins. Each protein possesses a unique and complex surface that contains cavities or clefts whose structure is complementary to specific ligands. After ligand binding, a conformational change occurs in the protein that is linked to a biochemical event. For example, the binding of ATP to myosin in muscle cells is a critical event in muscle contraction.

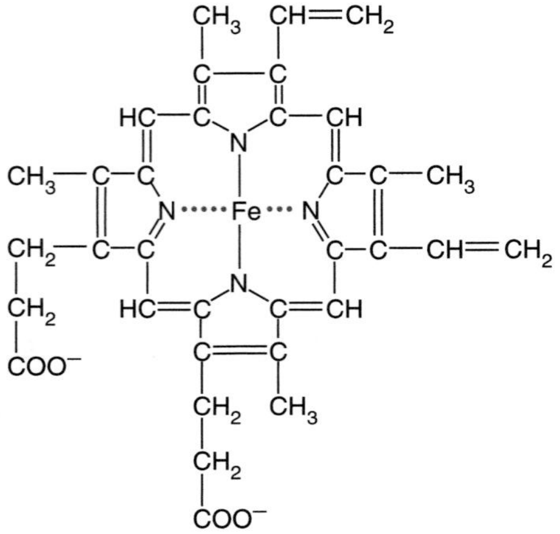

The oxygen-binding proteins myoglobin and hemoglobin are interesting and well-researched examples of globular proteins. They are both members of the hemoproteins, a specialized group of proteins that contain the prosthetic group heme. Although the heme group (Figure 5.28) in both proteins is responsible for the reversible binding of molecular oxygen, the physiological roles of myoglobin and hemoglobin are significantly different. The chemical properties of heme are dependent on the Fe^{2+} ion in the center of the prosthetic group. Fe^{2+} forms six coordinate bonds. The iron atom is bound to the four nitrogens in the center of the protoporphyrin ring. Two other coordinate bonds are available, one on each side of the planar heme structure. In myoglobin and hemoglobin, the fifth coordination bond is to the nitrogen atom in a histidine residue, and the sixth coordination bond is available for binding oxygen. In addition to serving as a reservoir for oxygen within muscle cells, myoglobin facilitates the diffusion of oxygen in metabolically active cells. The role of hemoglobin, the primary protein of red blood cells, is to deliver oxygen to cells throughout the body. A comparison of the structures of these two proteins illustrates several important principles of protein structure, function, and regulation.

FIGURE 5.28

Heme.

Heme consists of a porphyrin ring (composed of four pyrroles) with Fe^{2+} in the center.

MYOGLOBIN Myoglobin, found in high concentration in skeletal and cardiac muscle, gives these tissues their characteristic red color. Diving mammals such as whales, which remain submerged for long periods, possess high myoglobin concentrations in their muscles. Because of the extremely high concentrations of myoglobin, such muscles are typically brown. The protein component of myoglobin, called globin, is a single polypeptide chain that contains eight sections of α-helix (Figure 5.29). The folded globin chain forms a crevice that almost completely encloses a heme group. Free heme [Fe^{2+}] has a high affinity for O_2 and is irreversibly oxidized to form hematin [Fe^{3+}]. Hematin cannot bind O_2. Noncovalent interactions between amino acid side chains and the nonpolar porphyrin ring within the oxygen-binding crevice decrease heme's affinity for O_2. The decreased affinity protects Fe^{2+} from oxidation and allows for the reversible binding of O_2. All of the heme-interacting amino acids are nonpolar with the exception of two histidines, one of which (the proximal histidine) binds directly to the heme iron atom (Figure 5.30). The other (the distal histidine) stabilizes the oxygen-binding site.

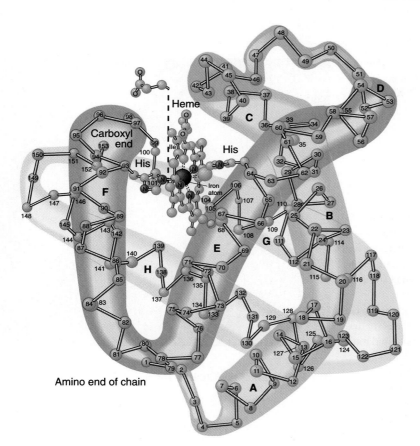

FIGURE 5.29

Myoglobin.

With the exception of the side chain groups of two histidine residues only the α-carbon atoms of the globin polypeptide are shown. Myoglobin's eight helices are designated A through H. The heme group has an iron atom that binds reversibly with oxygen. To improve clarity one of heme's propionic acid side chains has been displaced.
© Irving Geis.

HEMOGLOBIN Hemoglobin is a roughly spherical molecule found in red blood cells, where its primary function is to transport oxygen from the lungs to every tissue in the body. Recall that HbA is composed of two α-chains and two β-chains (Figure 5.31). The HbA molecule is commonly designated $\alpha_2\beta_2$. (There is another type of adult hemoglobin. Approximately 2% of human hemoglobin is HbA$_2$, which contains δ (delta)-chains instead of β-chains.) Before birth, several additional hemoglobin polypeptides are synthesized. The ε (epsilon)-chain, which appears in early embryonic life, and the γ-chain found in the fetus closely resemble the β-chain. Because both $\alpha_2\varepsilon_2$ and $\alpha_2\gamma_2$ hemoglobins have a greater affinity for oxygen than does $\alpha_2\beta_2$ (HBA), the fetus can preferentially absorb oxygen from the maternal bloodstream.

Although the three-dimensional configurations of myoglobin and the α- and β-chains of hemoglobin are very similar, their amino acid sequences have many differences. Comparison of these molecules from dozens of species has revealed nine invariant amino acid residues. Several invariant residues directly affect the oxygen-binding site, whereas others stabilize the α-helical peptide segments. The remaining residues may vary considerably. However, most substitutions are conservative. For example, each polypeptide's interior remains nonpolar.

The four chains of hemoglobin are arranged in two identical dimers, designated as $\alpha_1\beta_1$ and $\alpha_2\beta_2$. Each globin polypeptide has a heme-binding unit similar to that described for myoglobin. Although both myoglobin and hemoglobin bind oxygen reversibly, the latter molecule has a complex structure and more complicated binding properties. The numerous noncovalent interactions (mostly hydrophobic) between the subunits in each $\alpha\beta$-dimer remain largely unchanged when hemoglobin interconverts between its oxygenated and deoxygenated forms (Figure 5.32). In contrast, the relatively small number of interactions between the

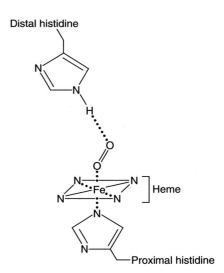

FIGURE 5.30

The Oxygen-Binding Site of Heme Created by a Folded Globin Chain.

From H.R. Horton, et al., *Principles of Biochemistry*,
© 1992, p. 432, fig. 4.27. Reprinted by permission of Prentice-Hall, Inc., Upper Saddle River, NJ.

Some pathogenic organisms damage humans by producing poisonous substances called toxins. *Toxins*, many of which are proteins, exert their effects by several methods:

1. damage to cell membranes,
2. disruption of various intracellular functions, and
3. inhibition of function at nerve cell synapses.

Toxins that act directly on cell membranes, called cytolytic toxins, disturb and ultimately kill the target cells. Produced by many organisms (e.g., bacteria, fungi, plants, fish, and snakes), cytolytic toxins may cause damage in several ways. For example, streptolysin O (67,000 D), produced by the bacterium *Streptococcus pyogenes*, causes pores to form in the target cell membranes. Affected cells are rapidly lysed because the cell membrane is much more permeable to ions such as Na^+. Streptolysin O is believed to cause some of the damage in rheumatic fever.

Cell membrane destruction may also be caused by toxic enzymes. For example, many organisms secrete enzymes, called phospholipases, which cause the hydrolysis of membrane lipid molecules. Phospholipase A2 is found in the venom of several snakes.

Many toxins interfere with intracellular functions. The best-characterized of these are diphtheria toxin and cholera toxin, produced by the bacteria *Corynebacterium diptheriae* and *Vibrio cholerae*, respectively. Both of these toxins contain two subunits, called A and B. The A subunit is responsible for the toxic effect, whereas the B subunit binds to the target cell. Once diphtheria toxin has entered the target cell, the A and B subunits split apart. The A subunit, which is an enzyme, catalyzes a reaction that prevents protein synthesis. The cell dies because it cannot synthesize proteins. The host organisms dies because cardiac, kidney, and nervous tissue are destroyed.

The B subunit of cholera toxin, which is made up of five identical subunits (Figure 5A), attaches to the membranes of intestinal cells. The A subunit is then inserted into these cells, where it activates an enzyme that increases the concentration of a nucleotide called cyclic AMP (cAMP). High sustained concentrations of cAMP, a molecule that opens membrane chloride channels, causes severe diarrhea. (Loss of chloride results in water loss because of osmotic pressure. Several gallons of fluid per day may be lost.) If left untreated, severe dehydration may cause death within 48 hours, by circulatory shock brought on by low blood volume.

Several toxic proteins act as neurotoxins by disrupting the activity of synapses. (A synapse is a junction between two neurons or between a neuron and a muscle cell.) The pain, tremors, and irritability that result from black widow spider bites are caused by α-latrotoxin (125,000 D). This molecule, a single polypeptide, stimulates a massive release of the neurotransmitter acetylcholine (ACh). In contrast, ACh release is inhibited by botulinum toxin, a mixture of several proteins produced by the bacterium *Clostridium botulinum*. Botulism, a malady most commonly caused by eating contaminated canned food, is characterized by vomiting, dizziness, and sometimes paralysis and death. A related species,

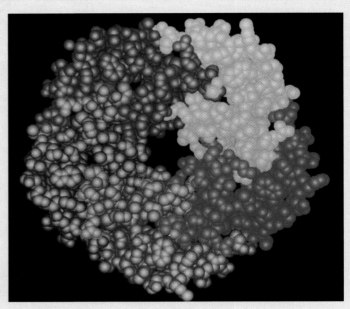

FIGURE 5A

The B-Subunit of the Cholera Toxin.

Clostridium tetani, produces another deadly neurotoxin. Tetanus toxin causes severe paralysis by blocking neurotransmitter release (primarily glycine and γ-aminobutyric acid) in the central nervous system.

Cholera: A Short History

Toxin-producing organisms, such as *V. cholerae*, do not kill only individual humans. Under certain circumstances they can affect an entire civilization. Cholera has had lasting effects on the Western world. Because of improved transportation, the 1817 cholera epidemic in India reached Europe. Traveling at an average speed of five miles a day, the disease left millions dead. Cholera claimed its first British victim in the port city of Sunderland in October 1831. The 22,000 deaths that followed during the next 2 years were due largely to horrendous living conditions during the Industrial Revolution (Figure 5B). Despite intense but misdirected efforts, cholera epidemics often occurred during the decades that followed. How the disease spread was not discovered until the newly emerging science of statistics revealed that poor sanitation and polluted water were responsible. Public pressure, driven by the seemingly endless deaths caused by cholera, eventually forced the British government to assume some responsibility for public health, a relatively modern concept. In 1859 the British Parliament contracted to build an elaborate sewer system in the city of London, at that time the largest municipal project of its kind ever undertaken. Cholera never returned to that city, and the relationship between sanitation and infectious disease was firmly established. In 1883 the German researcher Robert Koch identified the causative agent.

CHOLERA.

THE DUDLEY BOARD OF HEALTH,

HEREBY GIVE NOTICE, THAT IN CONSEQUENCE OF THE

Church-yards at Dudley

Being so full, no one who has died of the CHOLERA will be permitted to be buried after *SUNDAY* next, (To-morrow) in either of the Burial Grounds of St. *Thomas's*, or St. *Edmund's*, in this Town.

All Persons who die from CHOLERA, must for the future be buried in the Church-yard at Netherton.

BOARD of HEALTH, DUDLEY.
September 1st, 1832.

W. MAURICE, PRINTER, HIGH STREET, DUDL.

Royal Society of Medicine

FIGURE 5B

Devastating Effect of Cholera on Britain in the Mid-Nineteenth Century.

(a) A poster attests to the severity of the epidemic. (b) Burning tar to kill infection in the air in Exeter. Other "methods" for preventing the spread of cholera included tobacco fumes, and cleaning with turpentine vinegar.

Bioterrorism and Anthrax

Bioterrorism is the use of bacteria, viruses, or toxins to intimidate or coerce human populations. Bioterrorists are criminals who seek to achieve political goals that are unobtainable by legitimate, nonviolent means. In its latest incarnation, bioterrorism has taken the form of anthrax-contaminated letters sent to prominent individuals in government and the media in the weeks after the World Trade Center disaster. Anthrax is a disease, primarily affecting livestock (sheep, cattle, and horses), that is caused by the Gram-positive bacterium *Bacillus anthracis.* Known since biblical times (it is believed to be one of the ten plagues mentioned in the Book of Exodus, Chapter 9), anthrax has played an important role in the history of medicine and microbiology. In 1876 Robert Koch demonstrated that *B. anthracis* is the causative agent of anthrax. The techniques he developed while investigating anthrax and later tuberculosis led to the development of Koch's postulates, the method still currently used to identify the causes of newly discovered infectious diseases. Louis Pasteur developed the first artificial vaccine against the disease, and in 1881, he conclusively demonstrated to a skeptical scientific community that vaccinated sheep and cattle could withstand the otherwise fatal injection of the live bacterium.

Anthrax is initiated by exposure to heat-resistant endospores of *B. anthracis* (a dormant form of the organism) that can exist in soil or animal products for decades. The spores enter the body through skin abrasions (cutaneous anthrax), the lungs (inhalation anthrax), or the ingestion of contaminated food (gastrointestinal anthrax). After inhalation of the endospores (the most deadly form of the disease) they are absorbed by macrophages, immune system cells that ordinarily ingest and destroy invading bacteria and other foreign material. The macrophages, however, are unable to destroy the endospores because their capsular coating is made of an indigestible polymer composed of D-glutamic acid residues. Instead the endospores germinate into vegetative (actively dividing), disease-causing bacteria that divide until the macrophage bursts. If the exposure to the endospores is sufficient, the rapidly dividing bacteria can overwhelm the immune system and spread throughout the body. Systemic anthrax causes, within days after the first flulike symptoms appear, severe hypotension (low blood pressure), shock, and (in some cases) meningitis. The organism's capacity to inflict such massive damage is made possible by three toxins, which together inactivate critical immune defenses, break into cells, and disrupt normal signaling mechanisms. Once the bacterial cells are released into the blood, they secrete their toxins. Protective antigen (PA), named before its role was discovered, binds to cell-surface receptors. Once on the cell surface, seven PA toxin molecules undergo proteolytic activation and assemble into a doughnut-shaped structure. This complex then binds the toxic enzymes lethal factor (LF) and edema factor (EF) and inserts them into the cell in an endocytosis-like process. LF, the principal cause of death, is a zinc-dependent protease (an enzyme that breaks peptide bonds in proteins) that disrupts the intricate intracellular signaling system. Its most damaging effect is to cause macrophages to release massive amounts of inflammatory molecules that induce shock. EF causes massive swelling (edema) in affected tissues. If the infection is not arrested by antibiotics such as penicillin, the combined effects of these toxins cause death.

FIGURE 5.31

Hemoglobin.

The protein contains four subunits, designated α and β. Each subunit contains a heme group that binds reversibly with oxygen.

© Irving Geiss

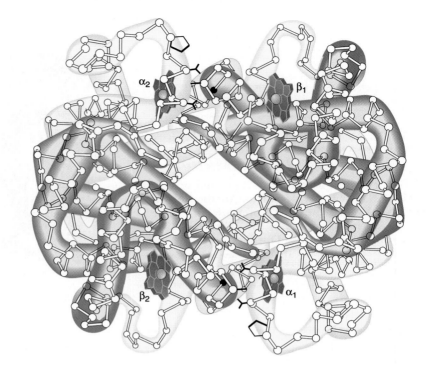

FIGURE 5.32

Three-Dimensional Structure of (a) Oxyhemoglobin and (b) Deoxyhemoglobin.

The β-chains are on top. In the oxy-deoxy transformation, the $\alpha_1\beta_1$ and $\alpha_2\beta_2$ dimers move as units relative to each other. This allows 2,3-bisphosphoglycerate (discussed on p. 151) to bind to the larger central cavity in the deoxy conformation.

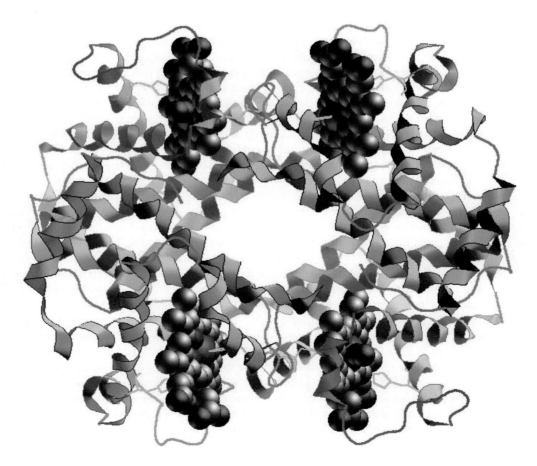

(a)

two dimers change substantially during this transition. When hemoglobin is oxygenated, the salt bridges and certain hydrogen bonds are ruptured as the $\alpha_1\beta_1$ and $\alpha_2\beta_2$ dimers slide by each other and rotate 15° (Figure 5.33). The deoxygenated conformation of hemoglobin (deoxyHb) is often referred to as the T(taut) state and oxygenated hemoglobin (oxyHb) is said to be in the R(relaxed) state. The oxygen-induced readjustments in the interdimer contacts are almost simultaneous. In other words, a conformational change in one subunit is rapidly propagated to the other subunits. Consequently, hemoglobin alternates between two stable conformations, the T and R states.

Because of subunit interactions, the oxygen dissociation curve of hemoglobin has a sigmoidal shape (Figure 5.34). As the first O_2 binds to hemoglobin, the binding of additional O_2 to the same molecule is enhanced. This binding pattern, called **cooperative binding**, results from changes in hemoglobin's three-dimensional structure that are initiated when the first O_2 binds. The binding of the first O_2 facilitates the binding of the remaining three O_2 molecules to the tetrameric hemoglobin molecules. In the lungs, where O_2 tension is high, hemoglobin is quickly saturated (converted to the R state). In tissues depleted of O_2, hemoglobin gives up about half of its oxygen. In contrast to hemoglobin, myoglobin's oxygen dissociation curve has a hyperbolic shape. This simpler binding pattern, a consequence of myoglobin's simpler structure, reflects several aspects of this protein's role in oxygen storage. Because its dissociation curve is well to the left of the hemoglobin curve, myoglobin gives up oxygen only when

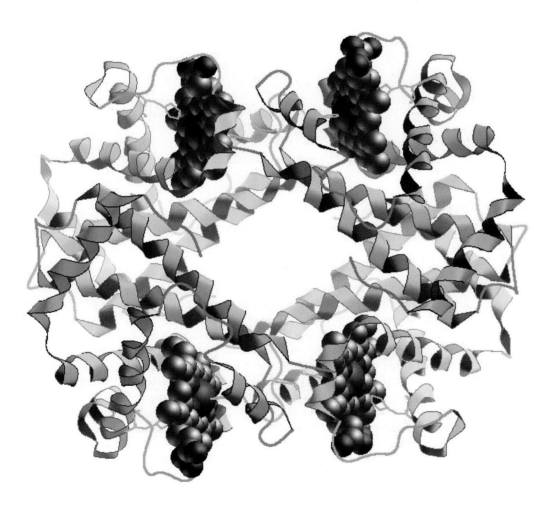

(b)

FIGURE 5.33

The Hemoglobin Allosteric Transition.

When hemoglobin is oxygenated, the $\alpha_1\beta_1$ and $\alpha_2\beta_2$ dimers slide by each other and rotate 15°. (a) Deoxyhemoglobin, (b) oxyhemoglobin.

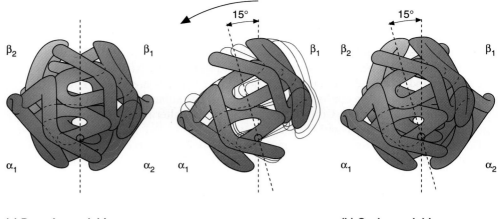

(a) Deoxyhemoglobin   **(b) Oxyhemoglobin**

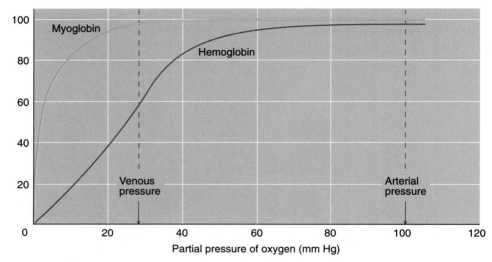

Partial pressure of oxygen (mm Hg)

FIGURE 5.34

Equilibrium Curves Measure the Affinity of Hemoglobin and Myoglobin for Oxygen.

the muscle cell's oxygen concentration is very low (i.e., during strenuous exercise). In addition, because myoglobin has a greater affinity for oxygen than does hemoglobin, oxygen moves from blood to muscle.

The binding of ligands other than oxygen affects hemoglobin's oxygen-binding properties. For example, the dissociation of oxygen from hemoglobin is enhanced if pH decreases. By this mechanism called the *Bohr effect*, oxygen is delivered to cells in proportion to their needs. Metabolically active cells, which require large amounts of oxygen for energy generation, also produce large amounts of the waste product CO_2. As CO_2 diffuses into blood, it reacts with water to form HCO_3^- and H^+. (The bicarbonate buffer was discussed on p. 86.) The subsequent binding of H^+ to several ionizable groups on hemoglobin molecules enhances the dissociation of O_2 by converting hemoglobin to its T state. (Hydrogen ions bind preferentially to deoxyHb. Any increase in H^+ concentration sta-

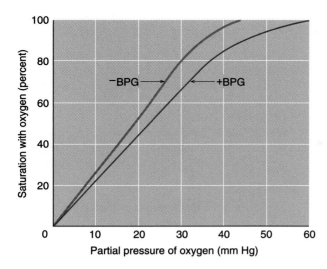

FIGURE 5.35

The Effect of 2,3-Bisphosphoglycerate (BPG) on the Affinity Between Oxygen and Hemoglobin.

In the absence of BPG (–BPG) hemoglobin has a high affinity for O_2; where BPG is present and binds to hemoglobin (+BPG) its affinity for O_2 decreases.

bilizes the deoxy conformation of the protein and therefore speeds its formation.) When a small number of CO_2 molecules bind to terminal amino groups on hemoglobin (forming carbamate or —$NHCOO^-$ groups) the deoxy form (T state) of the protein is additionally stabilized.

2,3-Bisphosphoglycerate (BPG) (also called glycerate-2,3-bisphosphate) is also an important regulator of hemoglobin function. Although most cells contain only trace amounts of BPG, red blood cells contain a considerable amount. BPG is derived from glycerate-1,3-bisphosphate, an intermediate in the breakdown of the energy-rich compound glucose. In the absence of BPG, hemoglobin has a very high affinity for oxygen (Figure 5.35). As with H^+ and CO_2, binding BPG stabilizes deoxyHb. A negatively charged BPG molecule binds in a central cavity within hemoglobin that is lined with positively charged amino acids.

In the lungs the process is reversed. A high oxygen concentration drives the conversion from the deoxyHb configuration to that of oxyHb. The change in the protein's three-dimensional structure initiated by the binding of the first oxygen molecule releases bound CO_2, H^+, and BPG. The H^+ recombine with HCO_3^- to form carbonic acid, which then dissociates to form CO_2 and H_2O. CO_2 subsequently diffuses from the blood into the alveoli.

Fetal hemoglobin (HbF) binds to BPG to a lesser extent than does HbA. Why do you think HbF has a greater affinity for oxygen than does maternal hemoglobin?

QUESTION 5.10

The muscle protein myoglobin and the erythrocyte protein hemoglobin are both oxygen transport proteins. Describe the structural features that allow these molecules to perform their separate functions.

QUESTION 5.11

Because most genetic information is expressed through proteins it is not surprising that considerable time, effort, and funding have been devoted to investigating their properties. Since the amino acid sequence of bovine insulin was determined by Frederick Sanger in 1953, the structures of several thousand proteins have been elucidated. In addition to providing insight into the molecular mechanisms of various proteins, information about protein structure has led to a deeper understanding of the evolutionary relationships between species. Because some inherited diseases are now known to be caused by alterations in the amino acid sequence of specific proteins, new diagnostic tests and clinical therapies have been developed.

The technology used to determine protein structure has changed significantly since the 1950s. For example, the determination of insulin's structure required the efforts of many scientists over 10 years. Currently, a well-funded research team can determine the primary structure of a newly discovered protein in less than a year. Much of this time is devoted to devising efficient methods for the protein's extraction and purification. Because of automated technology, the amino acid sequence determination may require only a few days of work.

A substantial portion of an investigator's time is devoted to the extraction and purification of proteins because of several formidable problems. The most prominent of these are the following:

1. Cells contain thousands of different substances. With rare exceptions (e.g., hemoglobin in red blood cells) the protein of interest exists in extremely small amounts. Separating a specific protein from a cell extract in sufficient quantities for research purposes is often a challenge to the investigator's endurance and ingenuity.

2. Proteins are often unstable and may require special handling. For example, they may be especially sensitive to pH, temperature, or salt concentration, among other factors. Problems in handling a protein may become apparent only after considerable time and effort have been expended. For example, the investigation of nitrogenase, the enzyme that catalyzes the reduction of N_2 to form NH_3, was hindered for years until it was discovered that the enzyme's activity is destroyed by contact with O_2.

The techniques for the isolation, purification, and initial characterization of proteins, which are outlined below, exploit differences of charge, molecular weight, and binding affinities. Many of these techniques apply to the investigation of other biomolecules.

Isolating Techniques

The first step in any project is to develop an assay for the protein of interest. Because the protein is typically extracted from source material that contains hundreds of similar molecules, the assay must be specific. In addition, the assay must be convenient to perform, because it will be used frequently during the investigation. If the protein is an enzyme, the disappearance of the substrate (reactant) or the formation of product can be measured. (This is usually accomplished by using a spectrophotometer, a machine that measures differences in the absorption of a specific wavelength of light.) Nonenzymatic proteins are often detected by employing antibodies.

(Antibodies are proteins produced by an animal's immune system in response to a foreign substance.) Often antibodies, which bind only to specific structures called **antigens**, are linked to radioactive or fluorescent "tags" to enhance their visibility.

Extraction of a protein begins with cell disruption and homogenization (see Biochemical Methods 2.1). This process is often followed by differential centrifugation and, if the protein is a component of an organelle, by density gradient centrifugation.

Purification

After the protein-containing fraction has been obtained, several relatively crude methods may be used to enhance purification. **Salting out** is a technique in which high concentrations of salts such as ammonium sulfate $[(NH_4)_2SO_4]$ are used to precipitate proteins. Because each protein has a characteristic salting-out point, this technique removes many impurities. (Unwanted proteins that remain in solution are then discarded when the liquid is decanted.) When proteins are tightly bound to membrane, organic solvents or detergents often aid in their extraction. Dialysis is routinely used to remove low-molecular-weight impurities such as salts, solvents, and detergents.

As a protein sample becomes progressively more pure, more sophisticated methods are used to achieve further purification. The most commonly used techniques include chromatography and electrophoresis.

Chromatography

Originally devised to separate low-molecular-weight substances such as sugars and amino acids, chromatography has become an invaluable tool in protein purification. There are a wide variety of chromatographic techniques. They can be used to separate protein mixtures on the basis of molecular properties such as size, shape, and weight, or certain binding affinities. Often several techniques must be used sequentially to obtain a demonstrably pure protein.

In all chromatographic methods the protein mixture is dissolved in a liquid known as the **mobile phase**. As the protein molecules pass across the **stationary phase** (a solid matrix), they separate from each other because of their different distributions between the two phases. The relative movement of each molecule results from its capacity to remain associated with the stationary phase while the mobile phase continues to flow.

Three chromatographic methods commonly used in protein purification are gel-filtration chromatography, ion-exchange chromatography and affinity chromatography. In **gel-filtration chromatography** (Figure 5C) a column packed with a gelatinous polymer separates molecules according to their size and shape. Molecules that are larger than the gel pores are excluded and therefore move through the column quickly. Molecules that are smaller than the gel pores diffuse in and out of the pores, so their movement through the column is retarded. The smaller their molecular weight, the slower they move. Differences in these rates separate the protein mixture into bands, which are then collected separately.

Ion-exchange chromatography separates proteins on the basis of their charge. Anion-exchange resins, which consist of positively charged materials, bind reversibly with a protein's negatively charged groups. Similarly, cation-exchange resins bind positively charged groups. After proteins that do not bind to the resin are removed, the protein of interest is recovered by an appropriate change in the solvent pH and/or salt concentration. (A change in pH alters the protein's net charge.)

Affinity chromatography uses the unique biological properties of proteins. That is, it uses a special noncovalent binding affinity between the protein and a special molecule (the ligand). The ligand is covalently bound to an insoluble matrix, which is placed in a column. After nonbinding protein molecules have passed through the column, the protein of interest is removed by altering the conditions that affect binding (i.e., pH or salt concentration).

Electrophoresis

Because proteins are electrically charged, they move in an electric field. In this process, called **electrophoresis**, molecules separate from each other because of differences in their net charge. For example, molecules with a positive net charge migrate toward the negatively charged electrode (cathode). Molecules with a net negative charge will move toward the positively charged electrode (anode). Molecules with no net charge will not move at all.

Electrophoresis, one of the most widely used techniques in biochemistry, is nearly always carried out by using gels such as polyacrylamide or agarose. The gel, functioning much as it does in gel-filtration chromatography, also acts to separate proteins on the basis of their molecular weight and shape. Consequently, gel electrophoresis is highly effective at separating complex mixtures of proteins or other molecules.

Bands resulting from a gel electrophoretic separation may be treated in several ways. During purification, specific bands may be excised from the gel after visualization with ultraviolet light. Each protein-containing slice is then eluted with buffer and prepared for the next step. Because of its high resolving power, gel electrophoresis is often used to assess the purity of protein samples. Staining gels with a dye such as Coomassie brilliant blue is a commonly used method for quickly assessing the success of a purification step.

A variation of gel electrophoresis called SDS gel electrophoresis, used primarily for characterizing proteins, is discussed on p. 156.

Initial Characterization of Proteins

Because protein function is determined by its physical structure, a vast array of analytical techniques have been devised. A few basic biophysical and biochemical techniques are described below.

Amino Acid Composition

The first step in characterizing any protein is the determination of the number of each type of amino acid residue present in the molecule. The process for obtaining this information, referred to as the amino acid composition, begins with the complete hydrolysis of all peptide bonds. Hydrolysis is typically accomplished with 6 N HCl for 10–100 hours. (Long reaction times are required because of difficulties in the hydrolysis of the aliphatic amino acids, Leu, Ile, and Val.) Hydrolysis is

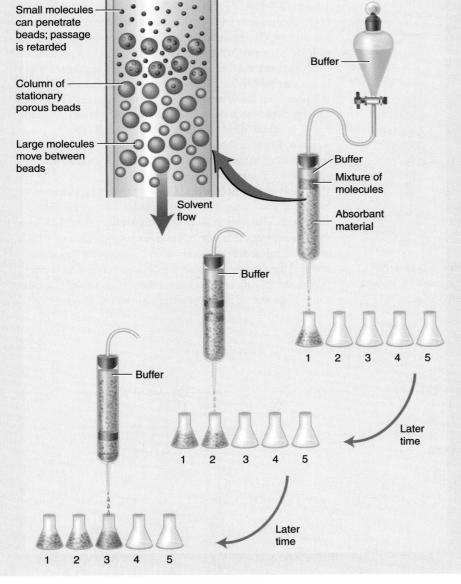

FIGURE 5C

Gel-Filtration Chromatography.

In gel-filtration chromatography the stationary phase is a gelatinous polymer with pore sizes selected by the experimenter to separate molecules according to their sizes. The sample is applied to the top of the column and is eluted with buffer (the mobile phase). As elution proceeds, larger molecules travel faster through the gel than smaller molecules, whose progress is slowed because they can enter the pores. If fractions are collected, the larger molecules appear in the earlier fractions and later fractions contain smaller molecules.

followed by analysis of the resulting amino acid mixture, referred to as the *hydrolysate*. Two automated methods are now commonly used to analyze protein hydrolysates: (1) ion-exchange chromatography and (2) high-pressure liquid chromatography (HPLC).

In the ion-exchange method, the hydrolysate is applied to a column that contains a cation-exchange resin. Initially, an acidic buffer (pH 3) is allowed to flow through the column. Because most amino acids have net positive charges at pH 3, they displace the bound positive ions and bind with the negatively charged resin. Amino acids with net negative charges are the first to be eluted. As the pH and ionic strength of the buffer are increased, the other amino acids are released sequentially from the column. As the amino acids emerge from the column, they are analyzed quantitatively. Heating the amino acids with ninhydrin yields a purple product, known as Ruhemann's purple:

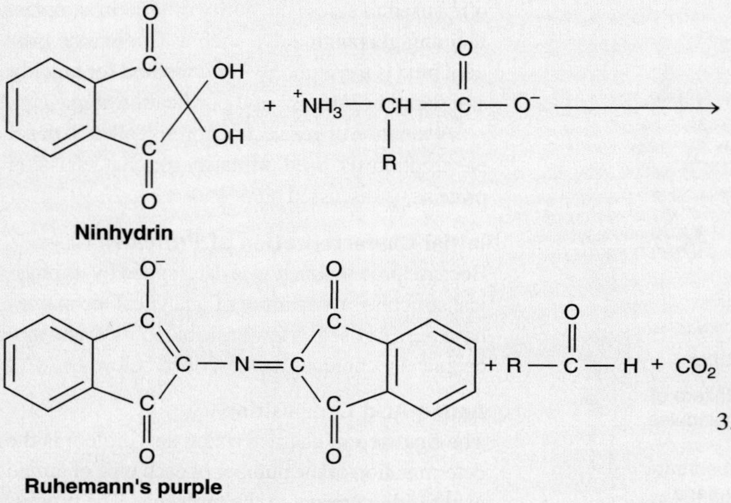

Ninhydrin

Ruhemann's purple

The reaction with the amino acid proline results in a yellow derivative. The amount of each amino acid is then determined by measuring the absorption of light by the ninhydrin derivatives.

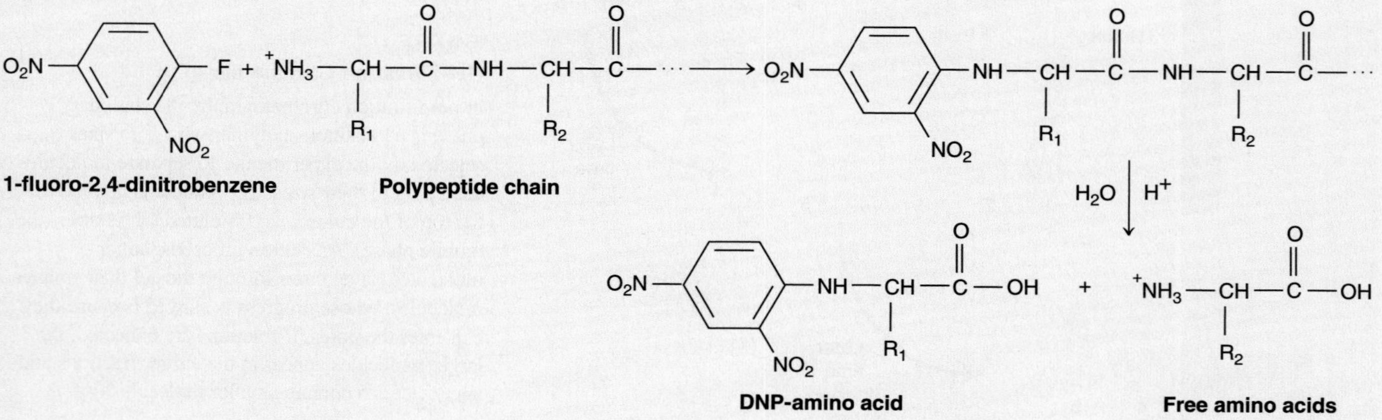

In HPLC, after the hydrolysate is treated with compounds such as Edman's reagent, the products are forced at high pressure through a stainless steel column packed with a stationary phase. Each amino acid derivative is identified according to its retention time on the column. Because the time required for amino acid analysis by HPLC (about 1 hour) is significantly shorter than that of other methods, it is becoming the method of choice.

Amino Acid Sequencing

Determining a protein's primary structure is similar to solving a complex puzzle. Several steps are involved in solving the amino acid sequence of any protein.

1. **Cleavage of all disulfide bonds.** Oxidation with performic acid is commonly used.
2. **Determination of the N-terminal and C-terminal amino acids.** Several methods are available to determine the N-terminal amino acid. In Sanger's method, the polypeptide chain is reacted with 1-fluoro-2,4-dinitrobenzene. The dinitrophenyl (DNP) derivative of the N-terminal amino acid can be isolated and identified by ion-exchange chromatography after the polypeptide is hydrolyzed. A group of enzymes called the carboxypeptidases are used to identify the C-terminal residue. Carboxypeptidases A and B, both secreted by the pancreas, hydrolyze peptides one residue at a time from the C-terminal end. Carboxypeptidase A preferentially cleaves peptide bonds when an aromatic amino acid is the C-terminal residue. Carboxypeptidase B prefers basic residues. Because these enzymes sequentially cleave peptide bonds starting at the C-terminal residue, the first amino acid liberated is the C-terminal residue.
3. **Cleavage of the polypeptide into fragments.** Because of technical problems, long polypeptides cannot be directly sequenced. For this reason the polypeptide is broken into smaller peptides. The use of several reagents, each of which cuts the chain at different sites, creates overlapping sets of fragments. After the amino acid sequence of each fragment is determined, the investigator uses this information to work out the entire sequence of the polypeptide. Of all the enzymes com-

Sanger method

1-fluoro-2,4-dinitrobenzene   **Polypeptide chain**

DNP-amino acid   **Free amino acids**

monly used, the pancreatic enzyme trypsin is the most reliable. It cleaves peptide bonds on the carboxy side of either lysine or arginine residues. The peptide fragments, referred to as *tryptic peptides*, have lysine or arginine carboxyl terminal residues. Chymotrypsin, another pancreatic enzyme, is also often used. It breaks peptide bonds on the carboxyl side of phenylalanine, tyrosine, or tryptophan. Treating the polypeptide with the reagent cyanogen bromide also generates peptide fragments. Cyanogen bromide specifically cleaves peptide bonds on the carboxyl side of methionine residues.

4. **Determination of the sequences of the peptide fragments**. Each fragment is sequenced through repeated cycles of a procedure called the *Edman degradation*. In this method phenylisothiocyanate (PITC), often referred to as Edman's reagent, reacts with the N-terminal residue of each fragment.

    Treatment of the product of this reaction with acid cleaves the N-terminal residue as a phenylthiohydantoin derivative. The derivative is then identified by comparing it with known standards, using electrophoresis or various chromatographic methods. (HPLC is most commonly used.) Because of the large number of steps involved in sequencing peptide fragments, Edman degradation is usually carried out by using a computer-programmed machine called a sequenator.

5. **Ordering the peptide fragments**. The amino acid sequence information derived from two or more sets of polypeptide fragments is next examined for overlapping segments. Such segments make it possible to piece together the overall sequence.

Several typical examples of primary sequence determination problems along with their solutions are given below.

Problem 1
Consider the following peptide:

Gly—Ile—Glu—Trp—Thr—Pro—Tyr—Gln—Phe—Arg—Lys

What amino acids and peptides are produced when the above peptide is treated with each of the following reagents?

a. Carboxypeptidase   b. Chymotrypsin
c. Trypsin   d. DNFB

Solution

a. Because carboxypeptidase cleaves at the carboxyl end of peptides, the products are
   Gly—Ile—Glu—Trp—Thr—Pro—Tyr—Gln—Phe—Arg
   and   Lys

b. Because chymotrypsin cleaves peptide bonds in which aromatic amino acids (i.e., Phe, Tyr, and Trp) contribute a carboxyl group, the products are:
   Gly—Ile—Glu—Trp,   Thr—Pro—Tyr,
   Gln—Phe, and Arg—Lys

c. Trypsin cleaves at the carboxyl end of lysine and arginine. The products are
   Gly—Ile—Glu—Trp—Thr—Pro—Tyr—Gln—Phe—Arg
   and   Lys

d. DNFB tags the amino terminal amino acid. The product is

   DNP—Gly—Ile—Glu—Trp—Thr—Pro—
   Tyr—Gln—Phe—Arg—Lys

    Hydrolysis then cleaves all the peptide bonds. DNP—Gly can then be identified by a chromatographic method.

Problem 2
From the following analytical results, deduce the structure of a peptide isolated from the Alantian orchid that contains 14 amino acids.

Edman degradation

PITC

Dilute H⁺

Phenylthiohydantoin derivative of N-terminal amino acid

Peptide minus N-terminal residue

Complete hydrolysis produces the following amino acids: Gly (3), Leu (3), Glu (2), Pro, Met, Lys (2), Thr, Phe. Treatment with carboxypeptidase releases glycine. Treatment with DNFB releases DNP-glycine. Treatment with a nonspecific proteolytic enzyme produces the following fragments:

Gly—Leu—Glu, Gly—Pro—Met—Lys,
Lys—Glu, Thr—Phe—Leu—Leu—Gly,
Lys—Glu—Thr—Phe—Leu,
Leu—Leu—Gly,
Glu—Thr—Phe, Glu—Gly—Pro,
Pro—Met—Lys—Lys,
and   Gly—Leu

Solution

The amino acid analysis provides information concerning the kind and number of amino acids in the peptide. The carboxypeptidase and DNFB results show that the carboxy and amino terminal amino acids are both glycine. Finally, by overlapping the fragments, the sequence of amino acids can be determined. Remember to start with a fragment that ends with the N-terminal residue, in this case, glycine.

Gly—Leu—Glu, Gly—Pro—Met—Lys, Lys—Glu,
Thr—Phe—Leu—Leu—Gly, Gly—Leu,
Gly—Gly—Pro, Pro—Met—Lys—Lys,
Lys—Glu—Thr—Phe—Leu, Leu—Leu—Gly

The overall structure then becomes

Gly—Leu—Glu—Gly—Pro—Met—Lys—Lys—
Glu—Thr—Phe—Leu—Leu—Gly

Review each piece of analytical data to ensure that it supports the final amino acid sequence.

Molecular Weight Determination

Several methods are available for determining the molecular weight of proteins. Gel-filtration column chromatography, SDS-PAGE, and ultracentrifugation are among the most commonly used.

Because gel acts as a molecular sieve, there is a direct correlation between elution volume (V_e, the volume of solvent required to elute the protein from the column since it first contacted the gel) and molecular weight. When gel-filtration chromatography is used to determine molecular weight, the gel column must be carefully calibrated. This is accomplished by careful measurement of several quantities. The total column volume V_t is equal to the sum of the volume occupied by the gel V_g and the void volume V_o, which is the volume occupied by the solvent molecules:

$$V_t = V_g + V_o$$

The molecular weight of the protein is determined by comparing its relative elution volume ($V_e - V_o/V_g$) to those of several standard molecules (Figure 5D). V_e is the solvent volume required for the elution of a solvent from the column.

A widely used variation of electrophoresis that determines molecular weight employs the powerful negatively charged detergent sodium dodecyl sulfate (SDS) (Figure 5E). In **SDS polyacrylamide gel electrophoresis** (SDS-PAGE) the detergent binds to the hydrophobic regions of protein molecules. As a result of binding SDS

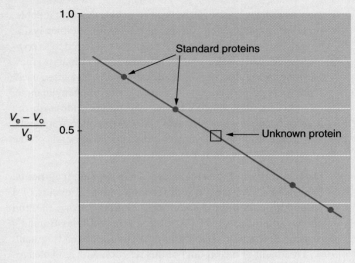

FIGURE 5D

Molecular Weight Determination Using Cell Filtration Chromatography.

The molecular weight of known standards is plotted against their elution values. This standard curve is then used to determine the molecular weight of the unknown protein.

molecules, proteins denature and assume rodlike shapes. (This effect is also achieved by adding mercaptoethanol, which cleaves disulfide bridges.) Because most molecules bind SDS in a ratio roughly proportional to their molecular weights, during electrophoresis SDS-treated proteins migrate toward the anode (+pole) only in relation to their molecular weight because of gel-filtration effects.

Estimates of molecular weight can also be obtained by using ultracentrifugation. Recall that ultracentrifugation separates components on the basis of size, surface area, and relative density. By using high centrifugal forces, the molecular masses of macromolecules such as proteins can be determined. By using an analytical ultracentrifuge, the rate at which such molecules sediment because of the influence of centrifugal force is optically measured.

X-Ray Crystallography

Much of the three-dimensional structural information about proteins was obtained by X-ray crystallography. Because the bond distances in proteins are approximately 15 nm, the electromagnetic radiation used to resolve protein structure must have a short wavelength. Visible light, whose wavelengths λ are 400–700 nm, clearly does not have sufficient resolving power for biomolecules. X-rays, however, have very short wavelengths (0.07–0.25 nm).

In X-ray crystallography, highly ordered crystalline specimens are exposed to an X-ray beam (Figure 5F). As the X-rays hit the crystal, they are scattered by the atoms in the crystal. The diffraction pattern that results is recorded on a photographic plate. The diffraction patterns are used to construct an electron density map. Because there is no objective lens to recombine the scattered X-rays, the three-dimensional image is reconstructed mathematically. Computer programs now perform these extremely complex and laborious computations.

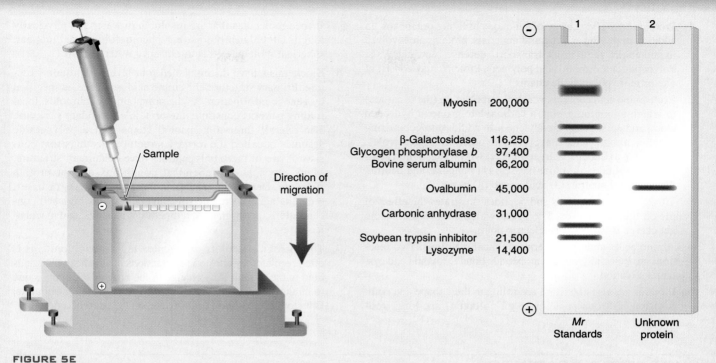

	Mr
Myosin	200,000
β-Galactosidase	116,250
Glycogen phosphorylase *b*	97,400
Bovine serum albumin	66,200
Ovalbumin	45,000
Carbonic anhydrase	31,000
Soybean trypsin inhibitor	21,500
Lysozyme	14,400

FIGURE 5E

Gel Electrophoresis

(a) Gel apparatus. The samples are loaded into wells. After an electric field is applied, the proteins move into the gel. (b) Molecules separate and move in the gel as a function of molecular weight and shape.

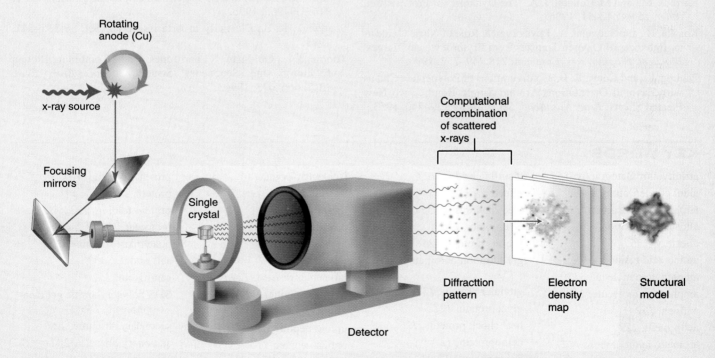

FIGURE 5F

Schematic Diagram of X-Ray Crystallography.

X-rays are useful in the analysis of biomolecules because their wavelength range is sufficiently similar to the magnitude of chemical bonds. Consequently, the resolving power of X-ray crystallography is equivalent to interatomic distances.

SUMMARY

1. Proteins have a vast array of functions in living organisms. In addition to serving as structural materials, proteins are involved in metabolic regulation, transport, defense, and catalysis. Polypeptides are amino acid polymers. Proteins may consist of one or more polypeptide chains.

2. Each amino acid contains a central carbon atom (the α-carbon) to which an amino group, a carboxylate group, a hydrogen atom, and an R group are attached. In addition to comprising protein, amino acids have several other biological roles. According to their capacity to interact with water, amino acids may be separated into four classes: (1) nonpolar and neutral, (2) polar and neutral, (3) acidic, and (4) basic.

3. Titration of amino acids and peptides illustrates the effect of pH on their structures. The pH at which a molecule has no net charge is called its isoelectric point.

4. Amino acids undergo several chemical reactions. Two reactions are especially important: peptide bond formation and cysteine oxidation.

5. Proteins are also classified according to their shape and composition. Fibrous proteins (e.g., collagen) are long, rod-shaped molecules that are insoluble in water and physically tough. Globular proteins (e.g., hemoglobin) are compact, spherical molecules that are usually water soluble.

6. Biochemists have distinguished four levels of protein structure. Primary structure, the amino acid sequence, is specified by genetic information. As the polypeptide chain folds, local folding patterns constitute the protein's secondary structure. The overall three-dimensional shape that a polypeptide assumes is called the tertiary structure. Proteins that consists of two or more polypeptides have quaternary structure. Many physical and chemical conditions disrupt protein structure. Denaturing agents include strong acids or bases, reducing agents, organic solvents, detergents, high salt concentrations, heavy metals, temperature changes, and mechanical stress.

7. The biological activity of complex multisubunit proteins is often regulated by allosteric interactions in which small ligands bind to the protein. Any change in the protein's activity is due to changes in the interactions among the protein's subunits. Effectors can increase or decrease the function of a protein.

SUGGESTED READINGS

Branden, C. and Tooze, J., *Introduction to Protein Structure*, 2nd ed., Garland, New York, 1999.

Doolittle, R. F., Proteins, *Sci. Amer.* 253(10):88–96, 1985.

Karplus, M., and McCannon, J. A., The Dynamics of Proteins, *Sci. Amer.* 254(4):42–51, 1986.

Kosaka, H., and Seiyama, A., Physiological Role of Nitric Oxide as an Enhancer of Oxygen Transfer from Erythrocytes to Tissues, *Biochem. Biophys. Res. Commun.* 218:749–752, 1996.

Pauling, L., and Corey, R. B., Configurations of Polypeptide Chains with Favored Orientations Around Single Bonds: Two New Pleated Sheets, *Proc. Nat. Acad. Sci. USA* 37:729–740, 1953.

Petruzzelli, R., Aureli, G., Lania, A., Galtieri, A., Desideri, A., and Giardina, B., Diving Behavior and Haemoglobin Function: The Primary Structure of the α- and β-Chains of the Sea Turtle (*Caretta caretta*) and Its Functional Implications, *Biochem. J.* 316:959–965, 1996.

Shadwick, R. E., Elasticity in Arteries, *Amer. Sci.* 86:535–541, 1998.

Thorne, J. L., Goldman, N., and Jones, D. T., Combining Protein Evolution and Secondary Structure, *Mol. Biol. Evol.* 13(5):666–673, 1996.

KEY WORDS

affinity chromatography, *153*

aldol condensation, *144*

aliphatic hydrocarbon, *112*

allosteric transition, *136*

allostery, *136*

amino acid residue, *110*

amphipathic molecule, *139*

amphoteric molecule, *110*

antigen, *152*

apoprotein, *127*

aromatic hydrocarbon, *112*

asymmetric carbon, *115*

chiral carbon, *115*

conjugated protein, *127*

cooperative binding, *149*

denaturation, *138*

disulfide bridge, *123*

effector, *136*

electrophoresis, *153*

enantiomer, *115*

fibrous protein, *126*

gel filtration chromatography, *152*

globular protein, *127*

glycoprotein, *127*

heat shock proteins, *126*

hemoprotein, *127*

holoprotein, *127*

homologous polypeptide, *127*

hormones, *114*

ion-exchange chromatography, *152*

isoelectric point, *116*

ligand, *135*

lipoprotein, *127*

metalloprotein, *127*

mobile phase, *152*

modulator, *136*

molecular disease, *129*

neurotransmitter, *114*

oligomer, *136*

opioid peptide, *125*

optical isomer, *115*

peptide, *110*

peptide bond, *120*

phosphoprotein, *127*

polypeptide, *110*

primary structure, *127*

prosthetic group, *127*

protein, *110*

protein folding, *133*

protomer, *136*

quaternary structure, *127*

salt bridges, *133*

salting out, *152*

SDS polyacrylamide gel electrophoresis, *156*

secondary structure, *127*

stationary phase, *152*

stereoisomer, *115*

subunit, *135*

supersecondary structure, *131*

tertiary structure, *127*

zwitterions, *110*

REVIEW QUESTIONS

1. Distinguish between proteins, peptides, and polypeptides.
2. Indicate which of the following amino acids are polar, non-polar, acidic, or basic:
   a. glycine
   b. tyrosine
   c. glutamic acid
   d. histidine
   e. proline
   f. lysine
   g. cysteine
   h. asparagine
   i. valine
   j. leucine
3. Arginine has the following pK_a values:

   pK_1 = 2.17,   pK_2 = 9.04,   pK_R = 12.48

   What is the structure and net charge of arginine at the following pH values? 1, 4, 7, 10, 12
4. Shown is the titration curve for histidine:

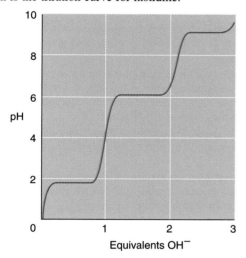

   a. What species are present at each plateau?
   b. Using the titration curve, determine the pK_a of each ionization of histidine.
   c. What is the isoelectric point of histidine?
5. Consider the following molecule:

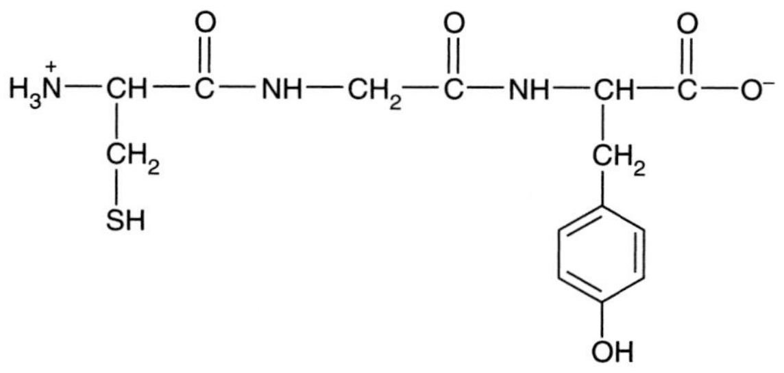

   a. Name it.
   b. Using the three-letter symbols for the amino acids, how would this molecule be represented?
6. Rotation about the peptide bond in glycylglycine is hindered. Draw the resonance forms of the peptide bond and explain why.
7. List six functions of proteins in the body.

8. Differentiate the terms in each pair below:
   a. globular and fibrous proteins
   b. simple and conjugated proteins
   c. apoprotein and holoprotein
9. Define the following terms:
   a. α-carbon
   b. isoelectric point
   c. peptide bond
   d. hydrophobic amino acid
10. Indicate the level(s) of protein structure to which each of the following contributes:
    a. amino acid sequence
    b. β-pleated sheet
    c. hydrogen bond
    d. disulfide bond
11. What type of secondary structure would the following amino acid sequence be *most* likely to have?
    a. polyproline
    b. polyglycine
    c. Ala—Val—Ala—Val—Ala—Val—
    d. Gly—Ser—Gly—Ala—Gly—Ala
12. List three factors that do not foster α-helix formation.
13. Denaturation is the loss of protein function from structural change or chemical reaction. At what level of protein structure or through what chemical reaction does each of the following denaturation agents act?
    a. heat
    b. strong acid
    c. saturated salt solution
    d. organic solvents (e.g., alcohol or chloroform)
14. A polypeptide has a high pI value. Suggest which amino acids might comprise it.
15. Outline the steps to isolate typical protein. What is achieved at each step?
16. Outline the steps to purify a protein. What criteria are used to evaluate purity?
17. List the types of chromatography used to purify proteins. Describe how each separation method works.
18. In sequencing a protein using carboxypeptidase, the protein is first broken down into smaller fragments, which are then separated from one another. Each fragment is then individually sequenced. If this initial fragmentation were not carried out, amino acid residues would build up in the reaction medium. How would these residues inhibit sequencing?
19. In an amino acid analysis, a large protein is broken down into overlapping fragments by using specific enzymes. Why must the sequences be overlapping?
20. Hydrolysis of β-endorphin (a peptide containing 31 amino acid residues) produces the following amino acids:

    Tyr (1), Gly (3), Phe (2), Met, Thr (3), Ser (2), Lys (5), Gln (2), Pro, Leu (2), Val (2), Asn (2), Ala (2), Ile, His, and Glu

    Treatment with carboxypeptidase liberates Gln. Treatment with DNFB liberates DNP-Tyr. Treatment with trypsin produces the following peptides:

    Lys, Gly—Gln, Asn—Ala—Ile—Val—Lys,
    Tyr—Gly—Gly—Phe—Met—Thr—Ser—Glu—Lys,

Asn—Ala—His—Lys, Ser—Gln—Thr—Pro—Leu—
Val—Thr—Leu—Phe—Lys

Treatment with chymotrypsin produces the following peptides:

Lys—Asn—Ala—Ile—Val—Lys—Asn—Ala—
His—Lys—Lys—Gly—Gln
Tyr—Gly—Gly—Phe
Met—Thr—Ser—Glu—Lys—Ser—Gln—Thr—Pro—
Leu—Val—Thr—Leu—Phe

What is the primary sequence of β-endorphin?

21. Consider the following tripeptide:

Gly—Ala—Val

a. What is the approximate isoelectric point?

b. In which direction will the tripeptide move when placed in an electric field at the following pH values? 1, 5, 10, 12

22. The following is the amino acid sequence of bradykinin, a peptide released by certain organisms in response to wasp stings:

Arg—Pro—Pro—Gly—Phe—Ser—Pro—Phe—Arg

What amino acids or peptides are produced when bradykinin is treated with each of the following reagents?
a. carboxypeptidase
b. chymotrypsin
c. trypsin
d. DNFB

THOUGHT QUESTIONS

1. Residues such as valine, leucine, isoleucine, methionine, and phenylalanine are often found in the interior of proteins, whereas arginine, lysine, aspartic acid, and glutamic acid are often found on the surface of proteins. Suggest a reason for this observation. Where would you expect to find glutamine, glycine, and alanine?

2. Proteins that are synthesized by living organisms adopt a biologically active conformation. Yet when such molecules are prepared in the laboratory, they usually fail to spontaneously adopt their active conformations. Can you suggest why?

3. The active site of an enzyme contains sequences that are conserved because they participate in the protein's catalytic activity. The bulk of an enzyme, however, is not part of the active site. Because a substantial amount of energy is required to assemble enzymes, why are they usually so large?

4. Structural protein may incorporate large amounts of immobilized water as part of its structure. Can you suggest how protein molecules "freeze" the water in place and make it part of the protein structure?

5. The peptide bond is a stronger bond than that of esters. What structural feature of the peptide bond gives it additional bond strength?

6. Because of their tendency to avoid water, nonpolar amino acids play an important role in forming and maintaining the three-dimensional structure of proteins. Can you suggest how these molecules accomplish this feat?

7. Chymotrypsin is an enzyme that cleaves other enzymes during sequencing. Why don't chymotrypsin molecules attack each other?

8. Most amino acids appear bluish purple when treated with ninhydrin reagent. Proline and hydroxyproline appear yellow. Suggest a reason for the difference.

Enzymes

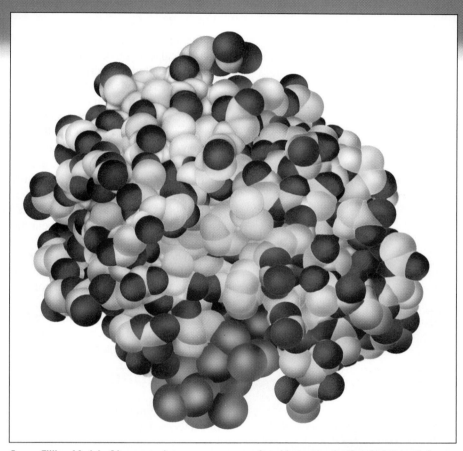

Space–Filling Model of Lysozyme Lysozyme, an enzyme found in tears and saliva, destroys certain bacteria by hydrolyzing cell wall polysaccharides. In this space-filling model the polypeptide is shown with a bound segment of polysaccharide (green).

Life is inconceivable without enzymes. Most of the thousands of biochemical reactions that sustain living processes would occur at imperceptible rates without enzymes. Enzymes are enormously powerful catalysts exhibiting high specificity. Their catalytic activities can be precisely regulated.

One of the most important functions of proteins is their role as catalysts. (Until recently, all enzymes were considered to be proteins. Several examples of catalytic RNA molecules have now been verified. See Chapter 18.) Recall that living processes consist almost entirely of biochemical reactions. Without catalysts these reactions would not occur fast enough to sustain life.

To proceed at a viable rate, most chemical reactions require an initial input of energy. In the laboratory this energy is usually supplied as heat. At temperatures above absolute zero ($-273.1°C$), all molecules possess vibrational energy, which increases as the molecules are heated. Consider the following reaction:

$$A + B \longrightarrow C$$

As the temperature rises, vibrating molecules (A and B) are more likely to collide. A chemical reaction occurs when the colliding molecules possess a minimum amount of energy called the **activation energy** (E_a) or, more commonly in biochemistry, the free energy of activation ($\Delta G^{\ddagger}$). Not all collisions result in chemical reactions because only a fraction of the molecules have sufficient energy or the correct orientation to react (i.e., to break bonds or rearrange atoms into the product molecules). Another way of increasing the likelihood of collisions, thereby increasing the formation of product, is to increase the concentration of the reactants.

In living systems, however, elevated temperatures may harm delicate biological structures and reactant concentrations are usually quite low. Living organisms circumvent these problems by using enzymes.

Enzymes have several remarkable properties. First, the rates of enzymatically catalyzed reactions are often phenomenally high. (Rate increases by factors of 10^6 or greater are common.) Second, in marked contrast to inorganic catalysts, the enzymes are highly specific to the reactions they catalyze. Side products are rarely formed. Finally, because of their complex structures, enzymes can be regulated. This is an especially important consideration in living organisms, which must conserve energy and raw materials.

Because enzymes are involved in so many aspects of living processes, any understanding of biochemistry depends on an appreciation of these remarkable catalysts. This chapter examines their structure and function.

6.1   PROPERTIES OF ENZYMES

How do enzymes work? The answer to this question requires a review of the role of catalysts. By definition a catalyst is a substance that enhances the rate of a chemical reaction but is not permanently altered by the reaction. Catalysts perform this feat because they decrease the activation energy required for a chemical reaction. In other words, catalysts provide an alternative reaction pathway that requires less energy (Figure 6.1). A **transition state** occurs at the apex of both reaction pathways in Figure 6.1. The free energy of activation, $\Delta G^{\ddagger}$, is defined as the amount of energy required to convert 1 mol of **substrate** (reactant) molecules from the ground state (the stable, low-energy form of a molecule) to the transition state. In the reaction in which ethanol is oxidized to form acetaldehyde

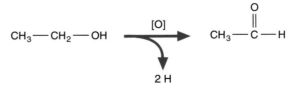

this transition state might look like

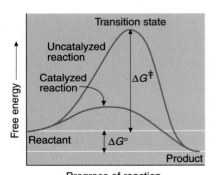

FIGURE 6.1

A Catalyst Reduces the Activation Energy of a Reaction.

A catalyst alters the free energy of activation $\Delta G^{\ddagger}$ and not the standard free energy $\Delta G°$ of the reaction.

Even with an inorganic catalyst, most laboratory reactions require an input of energy, usually in the form of heat. In addition, most of these catalysts are non-specific; that is, they accelerate a wide variety of reactions. Enzymes perform their work at moderate temperatures and are quite specific in the reactions that each one catalyzes. The difference between inorganic catalysts and enzymes is directly related to their structures. In contrast to inorganic catalysts, each type of enzyme molecule contains a unique, intricately shaped binding surface called an **active site**. Substrates bind to the enzyme's active site, which is typically a small cleft or crevice on a large protein molecule. The active site is not just a binding site, however. Several of the amino acid side chains that line the active site actively participate in the catalytic process.

The information within an enzyme's active site (its shape and charge distribution) constrains the motions and allowed conformations of the substrate, making it appear more like the transition state. In other words, the information in the structure of the enzyme is used to optimally orient the substrate. As a result of this information transfer, the energy of the enzyme-substrate complex becomes closer to the $\Delta G^{\ddagger}$, which means that the energy needed for the reaction to proceed to product is reduced. Consequently there is an increase in the rate of the enzyme-catalyzed reaction. Other factors, such as electrostatic effects, general acid-base catalysis, and covalent catalysis (discussed on pp. 177–180), also contribute to the increased rates of enzyme-catalyzed reactions over non–enzyme catalyzed reactions.

Enzymes, like all catalysts, cannot alter the equilibrium of the reaction, but they can increase the rate toward equilibrium. Consider the following reversible reaction:

$$A \rightleftharpoons B$$

Without a catalyst, the reactant A is converted into the product B at a certain rate. Because this is a reversible reaction, B is also converted into A. The rate expression for the forward reaction is $k_F[A]^n$, and that for the reverse reaction $k_R[B]^m$. The superscripts n and m represent the order of a reaction. Reaction order reflects the mechanism by which A is converted to B and vice versa. A reaction order of 2 for the conversion of A to B indicates that it is a bimolecular process and the molecules of A must collide for the reaction to occur (Section 6.3). At equilibrium, the rates for the forward and reverse reactions must be equal:

$$k_F[A]^n = k_R[B]^m \tag{1}$$

which rearranges to

$$\frac{k_F}{k_R} = \frac{[B]^m}{[A]^n} \tag{2}$$

The ratio of the forward and reverse constants is the equilibrium constant:

$$K_{eq} = \frac{[B]^m}{[A]^n} \tag{3}$$

For example, in equation (3), $m = n = 1$ and $k_F = 1 \times 10^{-3}$ s^{-1} and $k_R = 1 \times 10^{-6}$ s^{-1}, then

$$K_{eq} = \frac{10^{-3}}{10^{-6}} = 10^3$$

At equilibrium, therefore, the ratio of products to reactants is 1000 to 1.

In a catalyzed reaction, both the forward rate and the backward rate are increased, but the K_{eq} (in this case 1000) remains unchanged. If the catalyst increases both the forward and reverse rates by a factor of 100, then the forward rate becomes 100,000 and the reverse rate becomes 100. Because of the dramatic increase in the rate of the forward reaction made possible by the catalyst, equilibrium is approached in seconds or minutes instead of hours or days.

The *lock-and-key model* of enzyme action, introduced by Emil Fischer in 1890, partially accounts for enzyme specificity. Each enzyme binds to a single type

FIGURE 6.2

The Induced Fit Model.

Substrate binding causes a conformational change in the enzyme. The conformations of the enzyme hexokinase (a single polypeptide with two domains) (a) before glucose binding and (b) after glucose binding (glucose molecule is not shown). The domains move relative to each other to close around the glucose molecule.

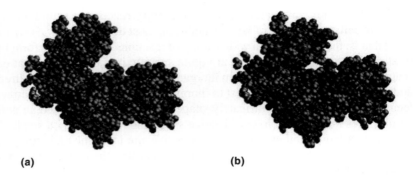

(a)   (b)

of substrate because the active site and the substrate have complementary structures. The substrate's overall shape and charge distribution allow it to enter and interact with the enzyme's active site. In a modern variation by Daniel Koshland of the lock-and-key model, called the *induced-fit model*, the flexible structure of proteins is taken into account (Figure 6.2). In this model, substrate does not fit precisely into a rigid active site. Instead, noncovalent interactions between the enzyme and substrate change the three-dimensional structure of the active site, conforming the shape of the active site to the shape of the substrate in its transition state conformation.

Although the catalytic activity of some enzymes depends only on interactions between active site amino acids and the substrate, other enzymes require nonprotein components for their activities. Enzyme **cofactors** may be ions, such as Mg^{2+} or Zn^{2+}, or complex organic molecules, referred to as **coenzymes**. The protein component of an enzyme that lacks an essential cofactor is called an **apoenzyme**. Intact enzymes with their bound cofactors are referred to as **holoenzymes**.

The activities of some enzymes can be regulated to maintain a stable intracellular environment. For example, adjustments in the rates of enzyme-catalyzed reactions allow cells to respond effectively to changes in the concentrations of nutrients. Organisms may control enzyme activities directly, principally through the binding of activators or inhibitors, the covalent modification of enzyme molecules, or indirectly, by regulating enzyme synthesis. (Control of enzyme synthesis requires gene regulation, a topic covered in Chapters 18 and 19.)

KEY CONCEPTS 6.1

Enzymes are catalysts. Catalysts modify the rate of a reaction because they provide an alternative reaction pathway that requires less energy than the uncatalyzed reaction. Most enzymes are proteins.

QUESTION 6.1

The hexokinases are a class of enzymes that catalyze the ATP-dependent phosphorylation of hexoses (sugars with six carbons). The hexokinases will bind only D-hexose sugars and not their L-counterparts. In general terms, describe the features of enzyme structure that make this discrimination possible.

6.2 CLASSIFICATION OF ENZYMES

In the early days of biochemistry, enzymes were named at the whim of their discoverers. Often enzyme names provided no clue to their function (e.g., trypsin), or several names were used for the same enzyme. Enzymes were often named by adding the suffix *-ase* to the name of the substrate. For example, urease catalyzes the hydrolysis of urea. To eliminate confusion, the International Union of Biochemistry (IUB) instituted a systematic naming scheme for enzymes. Each enzyme is now classified and named according to the type of chemical reaction it catalyzes. In this scheme an enzyme is assigned a four-number classification and a two-part name called a *systematic name*. In addition, a shorter version of the systematic name, called the *recommended name*, is suggested by the IUB for everyday use. For example, alcohol:NAD$^+$ oxidoreductase (E.C. 1.1.1.1) is usually referred to as alcohol dehydrogenase. (The letters E.C. are an abbreviation for enzyme commission.) Because many enzymes were discovered before the institution of the systematic nomenclature, many of the old well-known names have been retained.

The following are the six major enzyme categories:

  1. **Oxidoreductases**. **Oxidoreductases** catalyze oxidation-reduction reactions. Subclasses of this group include the dehydrogenases, oxidases, oxygenases, reductases, peroxidases, and hydroxylases.
  2. **Transferases**. **Transferases** catalyze reactions that involve the transfer of groups from one molecule to another. Examples of such groups include amino, carboxyl, carbonyl, methyl, phosphoryl, and acyl (RC=O). Common trivial names for the transferases often include the prefix *trans*. Examples include the transcarboxylases, transmethylases, and transaminases.
  3. **Hydrolases**. **Hydrolases** catalyze reactions in which the cleavage of bonds is accomplished by adding water. The hydrolases include the esterases, phosphatases, and peptidases.
  4. **Lyases**. **Lyases** catalyze reactions in which groups (e.g., H_2O, CO_2, and NH_3) are removed to form a double bond or are added to a double bond. Decarboxylases, hydratases, dehydratases, deaminases, and synthases are examples of lyases.
  5. **Isomerases**. This is a heterogeneous group of enzymes. **Isomerases** catalyze several types of intramolecular rearrangements. The epimerases catalyze the inversion of asymmetric carbon atoms. Mutases catalyze the intramolecular transfer of functional groups.
  6. **Ligases**. **Ligases** catalyze bond formation between two substrate molecules. The energy for these reactions is always supplied by ATP hydrolysis. The names of many ligases include the term *synthetase*. Several other ligases are called carboxylases.

An example from each enzyme class is illustrated in Table 6.1.

TABLE 6.1
Selected Examples of Enzymes

Enzyme Class	Example	Reaction Catalyzed
Oxidoreductase	Alcohol dehydrogenase	CH_3—CH_2—OH + NAD$^+$ ⟶ CH_3—CH(=O) + NADH + H$^+$
Transferase	Hexokinase	Glucose + ATP ⟶ Glucose-6-phosphate + ADP
Hydrolase	Chymotrypsin	Polypeptide + H_2O ⟶ Peptides
Lyase	Pyruvate decarboxylase	$^-$O—C(=O)—C(=O)—CH_3 (Pyruvate) + H$^+$ ⟶ H—C(=O)—CH_3 + CO_2
Isomerase	Alanine racemase	D–Alanine ⇌ L–Alanine
Ligase	Pyruvate carboxylase	CH_3—C(=O)—C(=O)—O$^-$ (Pyruvate) + HCO$_3^-$ →(ATP → ADP + P$_i$)→ $^-$O—C(=O)—CH_2—C(=O)—C(=O)—O$^-$ (Oxaloacetate)

QUESTION 6.2   Which type of enzyme catalyzes each of the following reactions?

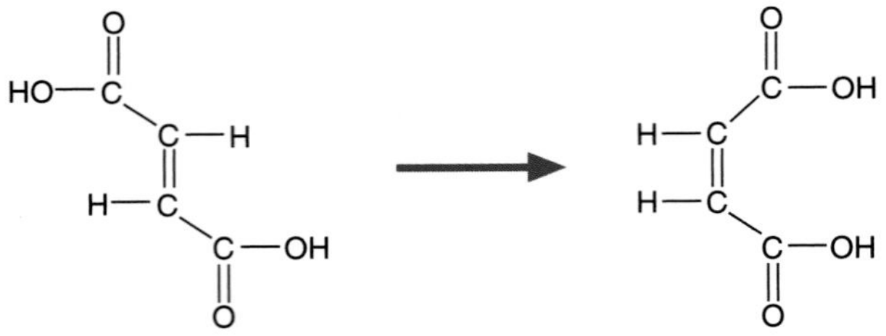

a.

b.

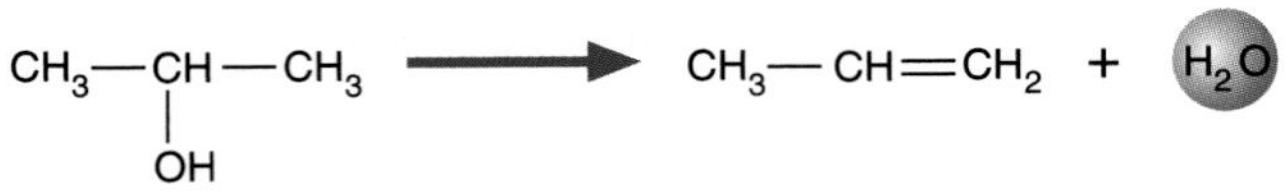

c.

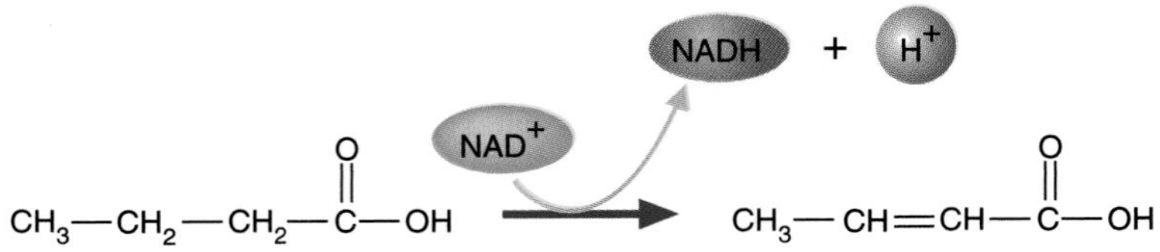

d.

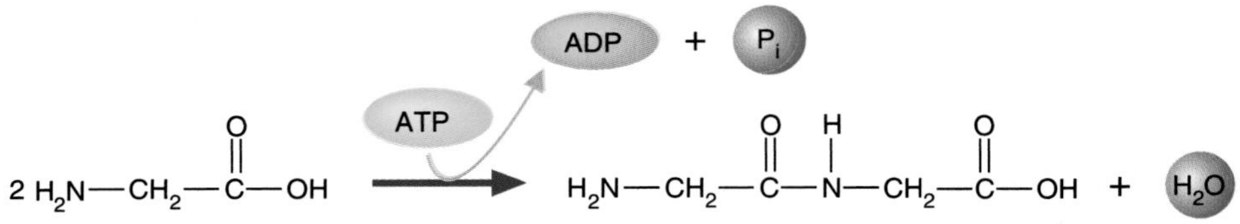

e.

f.

QUESTION 6.3   Aspartame, an artificial sweetener, has the following structure:

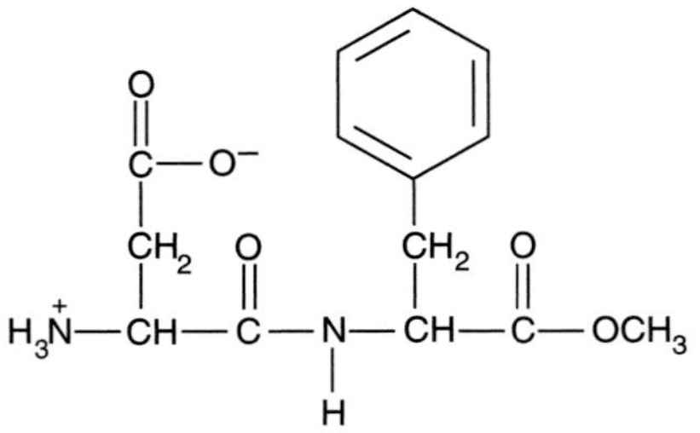

Once it is consumed in food or beverages, aspartame is degraded in the digestive tract to its component molecules. Predict what the products of this process are. What classes of enzymes are involved?

6.3 ENZYME KINETICS

Recall from Chapter 4 that the principles of thermodynamics can predict whether a reaction is spontaneous. However, thermodynamic quantities do not provide any information regarding reaction rates. To be useful to an organism, biochemical reactions must occur at reasonable rates. The rate or **velocity** of a biochemical reaction is defined as the change in the concentration of a reactant or product per unit time. The initial velocity v_0 of the reaction $A \longrightarrow P$ where A = substrate and P = product is

$$v_0 = \frac{-\Delta[A]}{\Delta t} = \frac{\Delta[P]}{\Delta t} \tag{4}$$

where

$[A]$ = concentration of substrate

$[P]$ = concentration of product

t = time

Initial velocity (v_0), the velocity when $[A]$ greatly exceeds the concentration of enzyme $[E]$ and the reaction time is very short, is the rate of the reaction immediately after mixing the enzyme and substrate. It is measured because it can be assumed that the reverse reaction (i.e., conversion of product into substrate), if possible, has not yet occurred to any appreciable extent.

The quantitative study of enzyme catalysis, referred to as **enzyme kinetics**, provides information about reaction rates. Kinetic studies also measure the affinity of enzymes for substrates and inhibitors and provide insight into reaction mechanisms. Enzyme kinetics has several practical applications. These include a greater comprehension of the forces that regulate metabolic pathways and the design of improved therapies.

The rate of the above reaction is proportional to the frequency with which the reacting molecules form product. The reaction rate is

$$v_0 = k[A]^x \tag{5}$$

where

v_0 = rate

k = a rate constant that depends on the reaction conditions (e.g., temperature, pH, and ionic strength)

x = the order of the reaction

Combining equations (4) and (5), we have

$$\frac{\Delta[A]}{\Delta t} = -k[A]^x \tag{6}$$

Another term that is useful in describing a reaction is the reaction's *order*. Order is determined empirically, that is, by experimentation (Figure 6.3). Order is defined as the sum of the exponents on the concentration terms in the rate expression. Determining the order of a reaction allows an experimenter to draw certain conclusions regarding the reaction's mechanism. A reaction is said to follow *first-order kinetics* if the rate depends on the first power of the concentration of a single reactant and suggests that the rate-limiting step is a unimolecular reaction (i.e., no molecular collisions are required). In the reaction $A \longrightarrow P$ the experimental rate equation becomes

$$\text{Rate} = k[A]^1 \tag{7}$$

KEY CONCEPTS 6.2

Enzyme kinetics is the quantitative study of enzyme catalysis. Kinetic studies measure reaction rates and the affinity of enzymes for substrates and inhibitors. Kinetics also provides insight into reaction mechanisms.

FIGURE 6.3

Enzyme Kinetic Studies.

(a) Plot of initial velocity v versus substrate concentration [S]. The rate of the reaction is directly proportional to substrate concentration only when [S] is low. When [S] becomes sufficiently high that the enzyme is saturated, the rate of the reaction is zero-order with respect to substrate. At intermediate substrate concentrations, the reaction has a mixed order (i.e., the effect of substrate on reaction velocity is in transition). (b) Conversion of substrate to product per unit time. The slope of the curve at $t = 0$ equals the initial rate of the reaction.

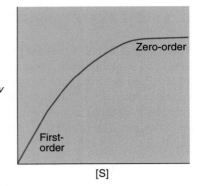

(a)

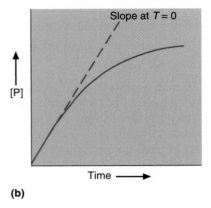

(b)

If [A] is doubled, the rate is observed to double. Reducing [A] by half results in halving the observed reaction rate. In first-order reactions the concentration of the reactant is a function of time, so k is expressed in units of s^{-1}. In any reaction the time required for one-half of the reactant molecules to be consumed is called a *half-life* ($t_{1/2}$).

In the reaction $A + B \longrightarrow P$, if the order of A and B is 1 each, then the reaction is said to be *second-order* and A and B must collide for product to form (a biomolecular reaction):

$$\text{Rate} = k[\text{A}]^1[\text{B}]^1 \tag{8}$$

In this circumstance the reaction rate depends on the concentrations of the two reactants. In other words, both A and B take part in the reaction's rate-determining step. Second-order rate constants are measured in units of $M^{-1}s^{-1}$. Sometimes second-order reactions involve reactants such as water that are present in great excess:

$$A + H_2O \longrightarrow P$$

The second-order rate expression is

$$\text{Rate} = k[\text{A}]^1[\text{H}_2\text{O}]^1$$

Because water is present in excess, however, the reaction appears to be first-order. Such reactions are said to be *pseudo-first-order*. Hydrolysis reactions in biochemical systems are assumed to be pseudo-first-order because of the ready availability of the second reactant, H_2O, in aqueous environments.

Another possibility is that only one of the two reactants is involved in the rate-determining step and it alone appears in the rate expression. For the above example, if rate $= k[\text{A}]^2$, then the rate-limiting step involves collisions between

A molecules. The water is involved in a fast, non–rate limiting step in the reaction mechanism.

When the addition of a reactant does not alter a reaction rate, the reaction is said to be *zero-order* for that reactant. For the reaction A $\longrightarrow$ P the experimentally determined rate expression is

$$\text{Rate} = k[A]^0 = k \tag{10}$$

The rate is constant because the reactant concentration is high enough to saturate all the catalytic sites on the enzyme molecules.

An example of an order determination is given in Problem 6.1.

PROBLEM 6.1

Consider the following reaction:

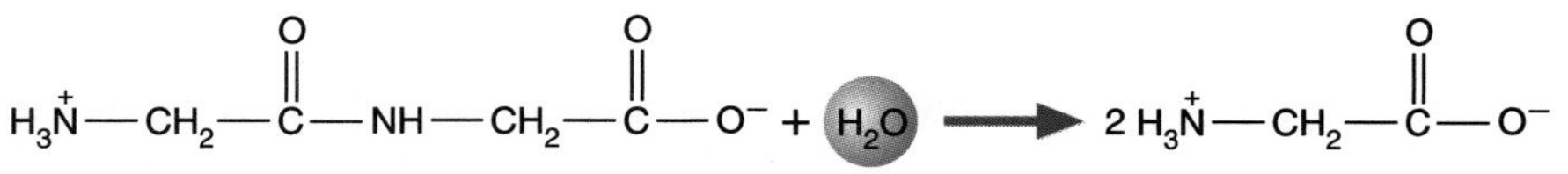

Given the following rate data, determine the order in each reactant and the overall order of the reaction. The concentrations are in moles per liter. Rates are measured in millimoles per second.

[Glycylglycine]	[H$_2$O]	Rate
0.1	0.1	1×10^2
0.2	0.1	2×10^2
0.1	0.2	2×10^2
0.2	0.2	4×10^2

Solution

The overall rate expression is

$$\text{Rate} = k[\text{Glycylglycine}]^x[\text{H}_2\text{O}]^y$$

To evaluate x and y, determine the effect on the rate of the reaction of increasing the concentration of one reactant while keeping the concentration of the other constant.

For this experiment, doubling the concentration of glycylglycine doubles the rate of the reaction; therefore, x is equal to 1. Doubling the concentration of water doubles the rate of the reaction. So y is also equal to 1. The rate expression then is

$$\text{Rate} = k[\text{Glycylglycine}]^1[\text{H}_2\text{O}]^1$$

The reaction is first-order in both reactants and second-order overall.

Michaelis–Menten Kinetics

One of the most useful models in the systematic investigation of enzyme rates was proposed by Leonor Michaelis and Maud Menten in 1913. The concept of the enzyme-substrate complex, first enunciated by Victor Henri in 1903, is central to Michaelis-Menten kinetics. When the substrate S binds in the active site of an enzyme E, an intermediate complex (ES) is formed. During the transition state, the substrate is converted into product. After a brief time, the product dissociates from the enzyme. This process can be summarized as follows:

$$E + S \underset{k_2}{\overset{k_1}{\rightleftharpoons}} ES \overset{k_3}{\longrightarrow} E + P \tag{11}$$

where

k_1 = rate constant for ES formation
k_2 = rate constant for ES dissociation
k_3 = rate constant for product formation and release from the active site

Equation (11) ignores the reversibility of the step in which the ES complex is converted into enzyme and product. This simplifying assumption is allowed if the reaction rate is measured while [P] is still very low. Recall that initial velocities are measured in most kinetic studies.

According to the Michaelis-Menten model, as currently conceived, it is assumed that (1) k_2 is negligible when compared with k_1 and (2) the rate of formation of ES is equal to the rate of its degradation over most of the course of the reaction. (The latter premise is referred to as the *steady-state assumption*.)

$$\text{Rate} = \frac{\Delta P}{\Delta t} = k_3[\text{ES}] \tag{12}$$

To be useful, a reaction rate must be defined in terms of [S] and [E]. The rate of formation of ES is equal to $k_1[\text{E}][\text{S}]$, whereas the rate of ES dissociation is equal to $(k_2 + k_3)[\text{ES}]$. The steady-state assumption equates these two rates:

$$k_1[\text{E}][\text{S}] = (k_2 + k_3)[\text{ES}] \tag{13}$$

$$[\text{ES}] = \frac{[\text{E}][\text{S}]}{(k_2 + k_3)/k_1} \tag{14}$$

Michaelis and Menten introduced a new constant, K_m (now referred to as the *Michaelis constant*):

$$K_m = \frac{k_2 + k_3}{k_1} \tag{15}$$

They also derived the equation

$$v = \frac{V_{max}[\text{S}]}{[\text{S}] + K_m} \tag{16}$$

where V_{max} = maximum velocity that the reaction can attain. This equation, now referred to as the *Michaelis-Menten equation*, has proven to be very useful in defining certain aspects of enzyme behavior. For example, when [S] is equal to K_m, the denominator in equation (16) is equal to 2[S], and v is equal to $V_{max}/2$ (Figure 6.4). The experimentally determined value K_m is considered a constant that is characteristic of the enzyme and the substrate under specified conditions. It may reflect the affinity of the enzyme for its substrate. (If k_3 is much smaller than k_2, that is, $k_3 \ll k_2$, then the K_m value approximates k_2/k_1. In this circumstance, K_m is the dissociation constant for the ES complex.) The lower the value of K_m, the greater the affinity of the enzyme for ES complex formation.

An enzyme's kinetic properties can also be used to determine its catalytic efficiency. The **turnover number** (k_{cat}) of an enzyme is defined as

$$k_{cat} = \frac{V_{max}}{[\text{E}_t]} \tag{17}$$

where

k_{cat} = the number of substrate molecules converted to product per unit time by an enzyme molecule under optimum conditions, that is, saturated with substrate

$[\text{E}_t]$ = total enzyme concentration

Under physiological conditions [S] is usually significantly lower than K_m. A more useful measure of catalytic efficiency is obtained by rearranging equation (17) as

$$V_{max} = k_{cat}[\text{E}_t] \tag{18}$$

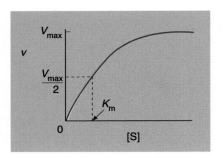

FIGURE 6.4

Reaction Velocity v and Substrate Concentration [S] for a Typical Enzyme-Catalyzed Reaction.

K_m is the substrate concentration at which the enzyme has half-maximal velocity.

and substituting this function in the Michaelis-Menten equation (equation 16):

$$v = \frac{k_{cat}[E_t][S]}{K_m + [S]} \qquad (19)$$

When [S] is very low, [E_t] is approximately equal to [E] and equation (19) reduces to

$$v = (k_{cat}/K_m)[E][S] \qquad (20)$$

In equation (20) the term k_{cat}/K_m is the rate constant for a reaction in which $[S] \ll K_m$. In this reaction the [S] is sufficiently low that the value of k_{cat}/K_m reflects the combined impact of binding and catalysis and is, therefore, a reliable gauge of catalytic efficiency. Examples of k_{cat}, K_m, and k_{cat}/K_m values of selected enzymes are provided in Table 6.2. It should be noted that the upper limit for an enzyme's k_{cat}/K_m value cannot exceed the maximal value of the rate at which the enzyme can bind to substrate molecules (k_1). This limit is imposed by the rate of diffusion of substrate into an enzyme's active site. The *diffusion control limit* on enzymatic reactions is approximately 10^8 to 10^9 M^{-1}s^{-1}. Several enzymes, for example those listed in Table 6.2, have k_{cat}/K_m values that approach the diffusion control limit.

KEY CONCEPTS 6.3

The Michaelis-Menten kinetic model explains several aspects of the behavior of many enzymes. Each enzyme has a K_m value characteristic of that enzyme under specified conditions.

TABLE 6.2
The Values of k_{cat}, K_m, and k_{cat}/K_m for Selected Enzymes*

Enzyme	Reaction Catalyzed	k_{cat} (s^{-1})	K_m (M)	k_{cat}/K_m (M^{-1}s^{-1})
Acetylcholinesterase	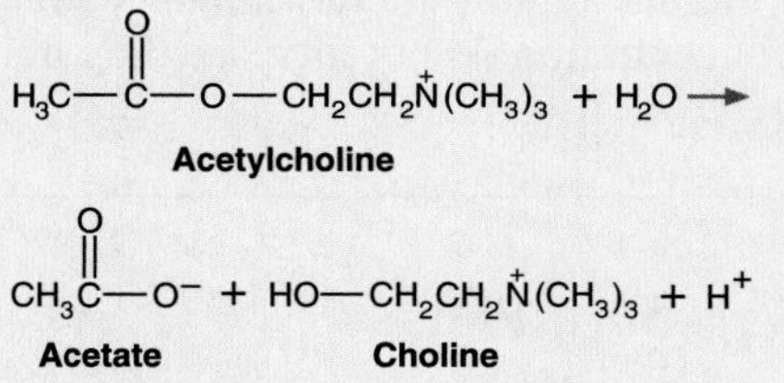 Acetylcholine + H_2O → Acetate + Choline + H^+	1.4×10^4	9×10^{-5}	1.6×10^8
Carbonic anhydrase	$HCO_3^- + H^+ \rightleftharpoons CO_2 + H_2O$	4×10^5	0.026	1.5×10^7
Catalase	$2 H_2O_2 \longrightarrow 2 H_2O + O_2$	4×10^7	1.1	4×10^7
Fumarase	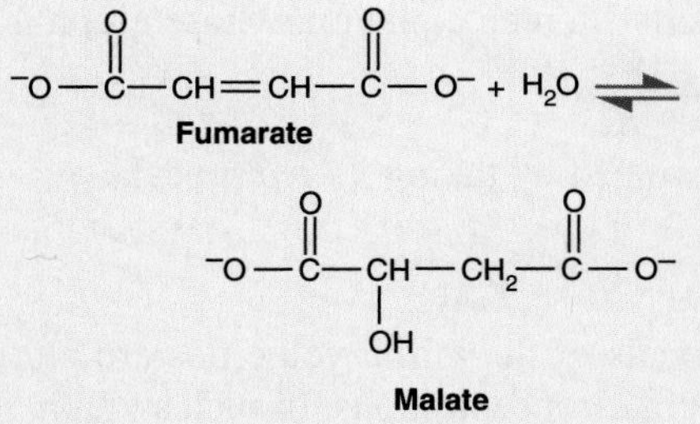 Fumarate + $H_2O \rightleftharpoons$ Malate	8×10^2	5×10^{-6}	1.6×10^8
Triosephosphate isomerase	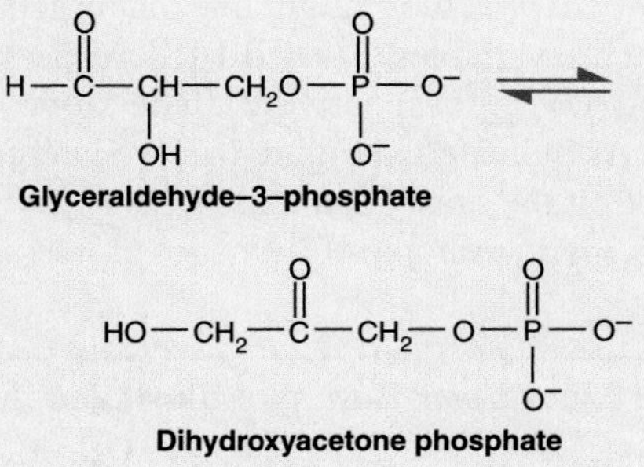 Glyceraldehyde–3–phosphate $\rightleftharpoons$ Dihydroxyacetone phosphate	4.3×10^3	4.7×10^{-4}	2.4×10^8

*Adapted from Fersht, A., *Structure and Mechanism in Protein Science: A Guide to Enzyme Catalysis and Protein Folding*, 2nd ed., W. H. Freeman, New York, 1999.

Because such enzymes convert substrate to product virtually every time the substrate diffuses into the active site, they are said to have achieved *catalytic perfection*. Living organisms overcome the diffusion control limit for the enzymes in biochemical pathways that do not achieve this high degree of catalytic efficiency by organizing them into multienzyme complexes. In these complexes, the active sites of the enzymes are in such close proximity to each other that diffusion is not a factor in the transfer of substrate and product molecules.

Enzyme activity is measured in *international units* (I.U.). One I.U. is defined as the amount of enzyme that produces 1 μmol of product per minute. An enzyme's **specific activity**, a quantity that is used to monitor enzyme purification, is defined as the number of international units per milligram of protein. (A new unit for measuring enzyme activity called the *katal* has recently been introduced. One katal (kat) indicates the amount of enzyme that transforms 1 mole of substrate per second. One katal is equal to 6×10^7 I.U.)

An example of a kinetic determination is given in Problem 6.2.

PROBLEM 6.2

Consider the Michaelis-Menten plot illustrated in Figure 6.5). Identify the following points on the curve:

a. V_{max}

b. K_m

Solution

a. V_{max} is the maximum rate the enzyme can attain. Further increases in substrate concentration do not increase the rate.

b. $K_m = $ [S] at $V_{max}/2$

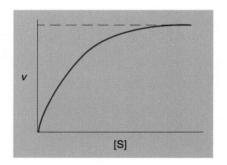

FIGURE 6.5
A Michaelis-Menten Plot.

Lineweaver–Burk Plots

K_m and V_{max} values for an enzyme are determined by measuring initial reaction velocities at various substrate concentrations. Approximate values of K_m and V_{max} can be obtained by constructing a graph, as shown in Figure 6.4. A more accurate determination of these values results from an algebraic transformation of the data. The Michaelis-Menten equation, whose graph is a hyperbola,

$$v = \frac{V_{max}[S]}{[S] + K_m}$$

can be rearranged by taking its reciprocal:

$$\frac{1}{v} = \frac{K_m}{V_{max}} \frac{1}{[S]} + \frac{1}{V_{max}}$$

The reciprocals of the initial velocities are plotted as functions of the reciprocals of substrate concentrations. In such a graph, referred to as a *Lineweaver-Burk double-reciprocal plot*, the straight line that is generated has the form $y = mx + b$, where y and x are variables ($1/v$ and $1/[S]$, respectively) and m and b are constants (K_m/V_{max} and $1/V_{max}$, respectively). The slope of the straight line is K_m/V_{max} (Figure 6.6). As indicated in Figure 6.6, the intercept on the vertical axis is $1/V_{max}$. The intercept on the horizontal axis is $-1/K_m$.

Problem 6.3 is an example of a kinetics problem using the Lineweaver-Burk plot.

PROBLEM 6.3

Consider the Lineweaver-Burk plot illustrated in Figure 6.7. Identify:

a. $-1/K_m$

b. $1/V_{max}$

c. K_m/V_{max}

Solution

a. $A = -1/K_m$

b. $B = 1/V_{max}$

c. $K_m/V_{max} = $ slope

Enzyme Inhibition

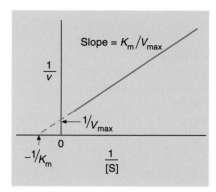

FIGURE 6.6

Lineweaver-Burk Plot.

If an enzyme obeys Michaelis-Menten kinetics, a plot of the reciprocal of the reaction velocity $1/v$ as a function of the reciprocal of the substrate concentration $1/[S]$ will fit a straight line. The slope of the line is K_m/V_{max}. The intercept on the vertical axis is $1/V_{max}$. The intercept on the horizontal axis is $-1/K_m$.

The activity of enzymes can be inhibited. Molecules that reduce an enzyme's activity, called **inhibitors**, include many drugs, antibiotics, food preservatives, and poisons. The investigations of enzyme inhibition and inhibitors carried out by biochemists are important for several reasons. First and foremost, in living systems enzyme inhibition is an important means by which metabolic pathways are regulated. Numerous small biomolecules are used routinely to modulate the rates of specific enzymatic reactions so that the needs of the organism are consistently met. Secondly, numerous clinical therapies are based on enzyme inhibition. For example, many antibiotics and drugs reduce or eliminate the activity of specific enzymes. Currently the most effective AIDS treatment is a multidrug therapy that includes protease inhibitors, molecules that disable a viral enzyme required to make new virus. Finally, investigations of enzyme inhibition have enabled biochemists to develop techniques used to probe the physical and chemical architecture, as well as the functional properties, of enzymes.

Enzyme inhibition can occur when a compound competes with substrate for the active site of the free enzyme, binds to the ES complex at a site removed from the active site, or binds to the free enzyme at a site removed from the active site. Three classes of enzyme inhibitors are described: competitive, uncompetitive, and noncompetitive inhibitors.

COMPETITIVE INHIBITORS Competitive inhibitors bind reversibly to free enzyme, not the ES complex, to form an enzyme-inhibitor (EI) complex.

$$E + S \underset{k_2}{\overset{k_1}{\rightleftharpoons}} ES \overset{k_3}{\longrightarrow} P + E$$
$$+$$
$$I$$
$$k_{I2} \updownarrow k_{I1}$$
$$EI + S \longrightarrow \text{No Reaction}$$

Often the substrate and inhibitor compete for the same site on the enzyme. The concentration of EI complex depends on the concentration of free inhibitor and on the dissociation constant K_I:

$$K_i = \frac{[E][I]}{[EI]} = \frac{k_{I2}}{k_{I1}}$$

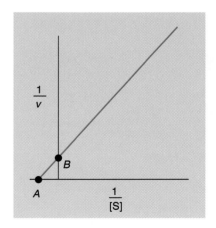

FIGURE 6.7

A Lineweaver-Burk Plot.

Because the EI complex readily dissociates, the enzyme is again available for substrate binding. The enzyme's activity declines (Figure 6.8) because no productive reaction occurs during the limited time that the EI complex exists. The effect of a competitive inhibitor on activity is reversed by increasing the concentration of substrate. At high [S], all the active sites are filled with substrate, and reaction velocity reaches the value observed without an inhibitor.

Substances that behave as competitive inhibitors, that is, they reduce an enzyme's apparent affinity for substrate, are often similar in structure to the substrate. Such molecules include reaction products or unmetabolizable analogs or derivatives of substrate molecules. Succinate dehydrogenase, an enzyme in the Krebs

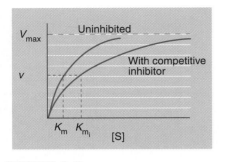

FIGURE 6.8

Michaelis-Menten Plot of Uninhibited Enzyme Activity Versus Competitive Inhibition.

Initial velocity v is plotted against substrate concentration [S]. With competitive inhibition, V_{max} stays constant and K_m increases.

Malonate

FIGURE 6.9

Malonate.

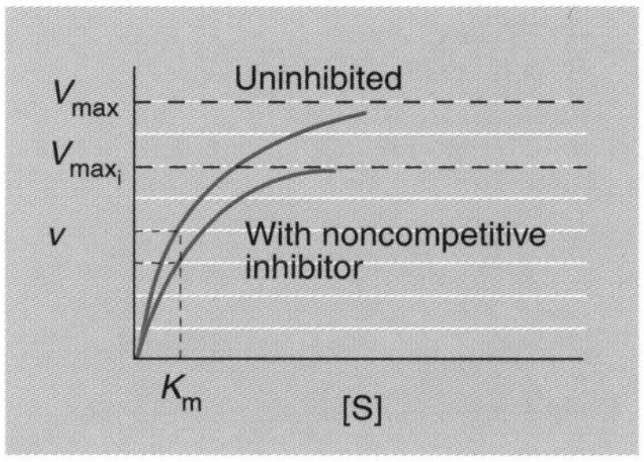

FIGURE 6.10

Michaelis-Menten Plot of Uninhibited Enzyme Activity Versus Noncompetitive Inhibition.

Initial velocity v is plotted against substrate concentration [S]. With noncompetitive inhibition V_{max} decreases and K_m stays constant.

citric acid cycle (Chapter 9), catalyzes a redox reaction that converts succinate to fumarate.

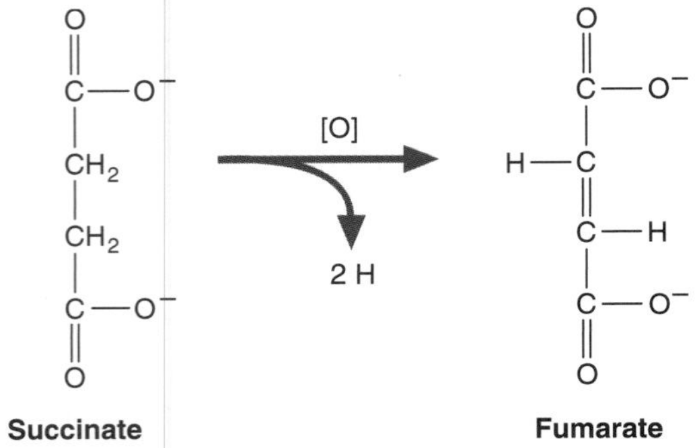

Succinate   **Fumarate**

This reaction is inhibited by malonate (Figure 6.9). Malonate binds to the enzyme's active site but cannot be converted to product.

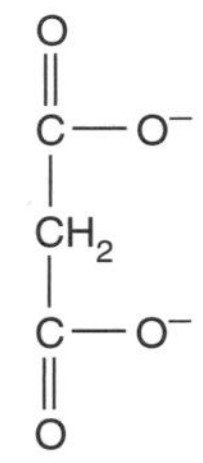

UNCOMPETITIVE INHIBITORS In uncompetitive inhibition, the inhibitor binds only to the enzyme-substrate complex, and not the free enzyme:

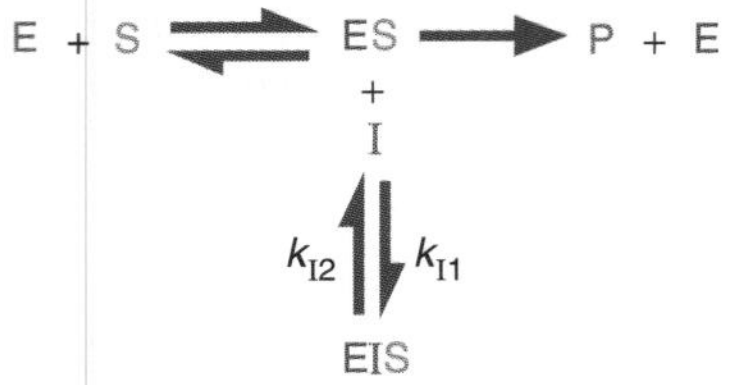

$$E + S \rightleftharpoons ES \longrightarrow P + E$$

The dissociation constant for the binding step of an uncompetitive inhibitor to an enzyme is

$$K_i = \frac{[ES][I]}{[EIS]}$$

Adding more substrate to the reaction results in an increase in reaction velocity, but not to the extent observed in the uninhibited reaction. Uncompetitive inhibition is most commonly observed in reactions in which enzymes bind more than one substrate.

NONCOMPETITIVE INHIBITORS In some enzyme-catalyzed reactions an inhibitor can bind to both the enzyme and the enzyme-substrate complex:

$$E + S \rightleftharpoons ES \longrightarrow P + E$$
$$EI + S \rightleftharpoons EIS \longrightarrow \text{No Reaction}$$

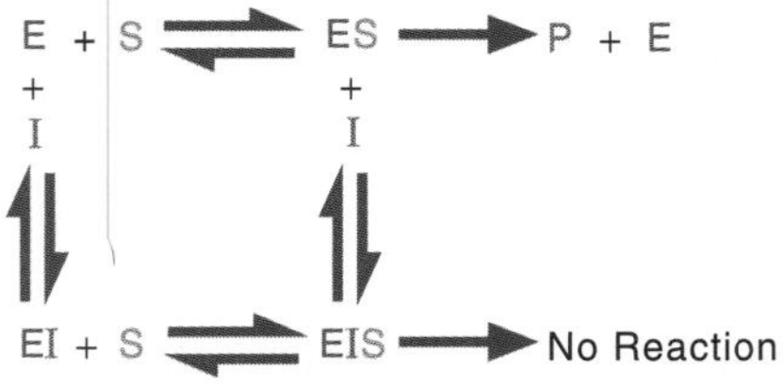

In such circumstances, referred to as **noncompetitive** inhibition, the inhibitor binds to a site other than the active site. Inhibitor binding results in a modification of the enzyme's conformation that prevents product formation (Figure 6.10. Usually, noncompetitive inhibitors do not affect substrate binding and they have little or no structural resemblance to substrate. As with uncompetitive inhibition, noncompetitive inhibition is only partially reversed by increasing the substrate concentration. Analysis of reactions inhibited by noncompetitive inhibitors is often complex for several reasons. As with uncompetitive inhibition, noncompetitive inhibition usually involves two or more substrates. Thus the characteristics of the inhibition that are observed may depend in part on factors such as the order in which the different substrates bind. In addition, there are two determinations of K_I for the binding of noncompetitive inhibitors

$$K_I = \frac{[E][I]}{[EI]}$$

and

$$K_I' = \frac{[E][S][I]}{[EIS]}$$

Depending on the inhibitor being considered, the values of these dissociation constants may or may not be equivalent. There are two forms of noncompetitive inhibition: pure and mixed. In pure noncompetitive inhibition, a rare phenomenon, both K_I values are equivalent. Mixed noncompetitive inhibition is typically more complicated because the K_I values are different.

KINETIC ANALYSIS OF ENZYME INHIBITION Competitive, uncompetitive, and noncompetitive inhibition can be easily distinguished with double-reciprocal plots (Figure 6.11). In two sets of rate determinations, enzyme concentration is held constant. In the first experiment, the velocity of the uninhibited enzyme is established. In the second experiment, a constant amount of inhibitor is included in each enzyme assay. Figure 6.11 illustrates the different effects that inhibitors have on enzyme activity. Competitive inhibition increases the K_m of the enzyme but the V_{max} is unchanged. (This is shown in the double-reciprocal plot as a shift in the horizontal intercept.) In uncompetitive inhibition both K_m and V_{max} are changed, although their ratio remains the same. In noncompetitive inhibition, V_{max} is lowered (i.e., the vertical intercept is shifted), and the K_m is increased (the K_m will be unchanged in the rare case when the k values for binding to E and ES are the same).

IRREVERSIBLE INHIBITION Inhibition may be reversible or irreversible. In reversible inhibition (i.e., competitive, uncompetitive, and noncompetitive inhibition), the inhibitor can dissociate from the enzyme because it binds through noncovalent bonds. Irreversible inhibitors usually bond covalently to the enzyme, often to a side chain group in the active site. For example, enzymes containing free sulfhydryl groups can react with alkylating agents such as iodoacetate:

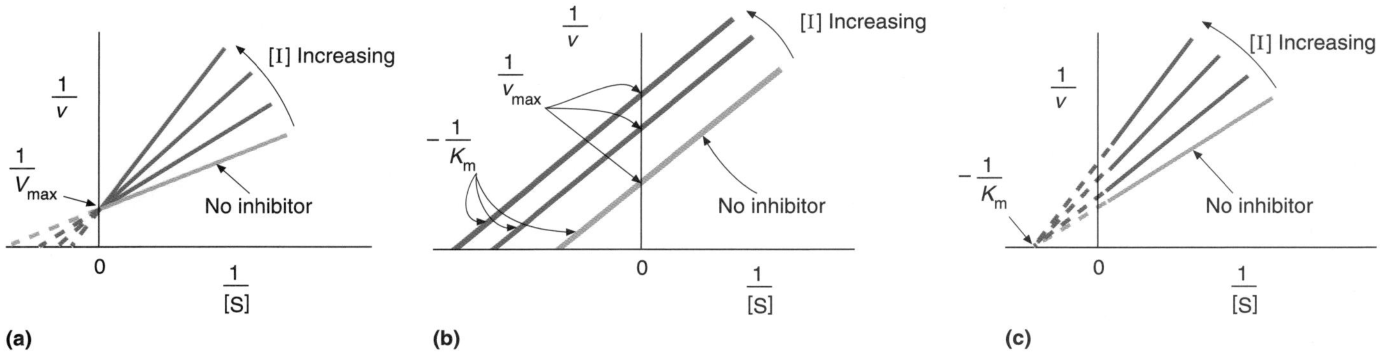

$$\text{Enzyme}-CH_2-SH \;+\; I-CH_2-\overset{\overset{\textstyle O}{\|}}{C}-O^- \;\longrightarrow\; \text{Enzyme}-CH_2-S-CH_2-\overset{\overset{\textstyle O}{\|}}{C}-O^- \;+\; HI$$

KEY CONCEPTS 6.4

Most inhibition of enzymes is competitive, uncompetitive, or noncompetitive. Competitive inhibitors reversibly compete with substrate for the same site on free enzyme. Uncompetitive inhibitors bind only to the enzyme-substrate complex and not the free enzyme. Noncompetitive inhibitors can bind to both the enzyme and the enzyme-substrate complex.

(a)   **(b)**   **(c)**

FIGURE 6.11

Kinetic Analysis of Enzyme Inhibition.

(a) Competitive inhibition. Plots of $1/v$ versus $1/[S]$ in the presence of several concentrations of the inhibitor intersect at the same point on the **vertical** axis, $1/V_{max}$. (b) Uncompetitive inhibition. Plots of $1/v$ versus $1/[S]$ in the presence of several concentrations of the inhibitor do not have a common intersection on either the horizontal or vertical axis. (c) Noncompetitive inhibition. Plots of $1/v$ versus $1/[S]$ in the presence of several concentrations of the inhibitor intersect at the same point on the **horizontal** axis, $-1/K_m$. In noncompetitive inhibition the dissociation constants for ES and EIS are assumed to stay the same.

Glyceraldehyde-3-phosphate dehydrogenase, an enzyme in the glycolytic pathway (Chapter 8), is inactivated by alkylation with iodoacetate. Enzymes that use sulfhydryl groups to form covalent bonds with metal cofactors are often irreversibly inhibited by heavy metals (e.g., mercury and lead). The anemia in lead poisoning is caused in part because of lead binding to a sulfhydryl group of ferrochelatase. Ferrochelatase catalyzes the insertion of Fe^{2+} into heme.

Problem 6.4 is concerned with enzyme inhibition.

PROBLEM 6.4

Consider the Lineweaver-Burk plot illustrated in Figure 6.12.

Line *A* = Normal enzyme-catalyzed reaction
Line *B* = Compound B added
Line *C* = Compound C added
Line *D* = Compound D added

Identify the type of inhibitory action shown by compounds B, C, and D.

Solution
Compound B is a competitive inhibitor because the K_m only has changed. Compound C is a pure noncompetitive inhibitor because the V_{max} only has changed. Compound D is an uncompetitive inhibitor because both K_m and V_{max} have changed.

QUESTION 6.4

Iodoacetamide is an irreversible inhibitor of several enzymes that have a cysteine residue in their active sites. After examining its structure predict the products of the reaction of iodoacetamide with such an enzyme.

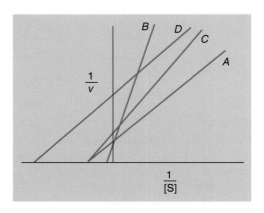

ALLOSTERIC ENZYMES Although the Michaelis-Menten model is an invaluable tool, it does not explain the kinetic properties of many enzymes. For example, plots of reaction velocity versus substrate concentration for many enzymes with multiple subunits are often sigmoidal rather than hyperbolic, as predicted by the Michaelis-Menten model (Figure 6.13). Such effects are seen in an important group of enzymes called the **allosteric enzymes**. The substrate-binding curve in Figure 6.13 resembles the oxygen-binding curve of hemoglobin. There are several other similarities between allosteric enzymes and hemoglobin.

FIGURE 6.12
A Lineweaver-Burk Plot.

FIGURE 6.13

The Kinetic Profile of an Allosteric Enzyme.

The sigmoidal binding curve displayed by many allosteric enzymes resembles the curve for the cooperative binding of O_2 to hemoglobin.

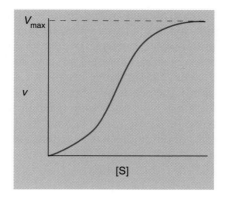

Most allosteric enzymes are multisubunit proteins. The binding of substrate to one protomer in an allosteric enzyme affects the binding properties of adjacent protomers. In addition, the activity of allosteric enzymes is affected by effector molecules that bind to additional sites called *allosteric* or *regulatory sites*. Allosteric enzymes generally catalyze key regulatory steps in biochemical pathways. (Regulation of allosteric enzymes is discussed on p. 190.)

QUESTION 6.5

Drinking methanol can cause blindness in humans, as well as a severe acidosis that may be life-threatening. Methanol is toxic because it is converted in the liver to formaldehyde and formic acid by the enzymes alcohol dehydrogenase and aldehyde dehydrogenase. Methanol poisoning is treated with dialysis and infusions of bicarbonate and ethanol. Explain why each treatment is used.

6.4 CATALYSIS

However valuable kinetic studies are, they reveal little about how enzymes catalyze biochemical reactions. Biochemists use other techniques to investigate the catalytic mechanisms of enzymes. (A *mechanism* is a set of steps in a chemical reaction by which a substrate is changed into a product.) Enzyme mechanism investigations seek to relate enzyme activity to the structure and function of the active site. X-ray crystallography, chemical inactivation of active site side chains, and studies using simple model compounds as substrates and as inhibitors are used.

Catalytic Mechanisms

Despite extensive research, the mechanisms of only a few enzymes are known in significant detail. However, it has become increasingly clear that enzymes use the same catalytic mechanisms as nonenzymatic catalysts. Enzymes achieve significantly higher catalytic rates because their active sites possess structures that are uniquely suited to promote catalysis.

Several factors contribute to enzyme catalysis. The most important of these are (1) proximity and strain effects, (2) electrostatic effects, (3) acid-base catalysis, and (4) covalent catalysis. Each factor will be described briefly.

PROXIMITY AND STRAIN EFFECTS For a biochemical reaction to occur, the substrate must come into close proximity to catalytic functional groups (side chain groups involved in a catalytic mechanism) within the active site. In addition, the substrate must be precisely oriented to the catalytic groups. Once the substrate is correctly positioned, a change in the enzyme's conformation may result in a strained enzyme-substrate complex. This strain helps to bring the enzyme-substrate complex into the transition state. In general, the more tightly the active site can bind the substrate while it is in its transition state, the greater the rate of the reaction. When an enzyme and substrate are in very close proximity, they behave as if they are part of the same molecule. The rates of intramolecular reactions are much higher than intermolecular reactions in which diffusion of the substrate reduces the time spent in the transition state and slows the reaction.

ELECTROSTATIC EFFECTS Recall that the strength of electrostatic interactions is related to the capacity of surrounding solvent molecules to reduce the attractive forces between chemical groups (Chapter 3). Because water is largely excluded from the active site as the substrate binds, the local dielectric constant is often low. The charge distribution in the relatively anhydrous active site may influence the chemical reactivity of the substrate. In addition, weak electrostatic interactions, such as those between permanent and induced dipoles in both the active site and the substrate, are believed to contribute to catalysis. A more efficient binding of substrate lowers the free energy of the transition state, which accelerates the reaction.

ACID-BASE CATALYSIS Chemical groups can often be made more reactive by adding or removing a proton. Enzyme active sites contain side chain groups that act as proton donors or acceptors. These groups are referred to as general acids or general bases. (*General acids* and *general bases* are substances that can release a proton or accept a proton, respectively.) For example, the side chain of histidine (referred to as an imidazole group) often participates in concerted acid-base catalysis because its pK_a range is close to physiological pH. The protonated imidazole ring can serve as a general acid and the deprotonated imidazole ring can serve as a general base:

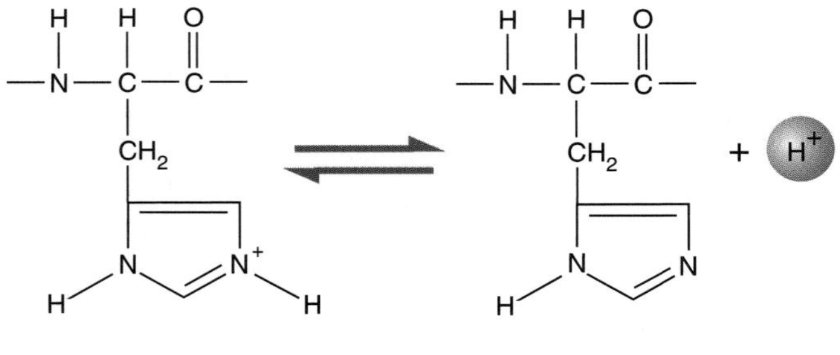

General acid   General base

Histidine

Transfer of protons is a common feature of chemical reactions. For example, consider the hydrolysis of an ester:

Because water is a weak nucleophile, ester hydrolysis is relatively slow in neutral solution. Ester hydrolysis takes place much more rapidly if the pH is raised. As hydroxide ion attacks the polarized carbon atom of the carbonyl group

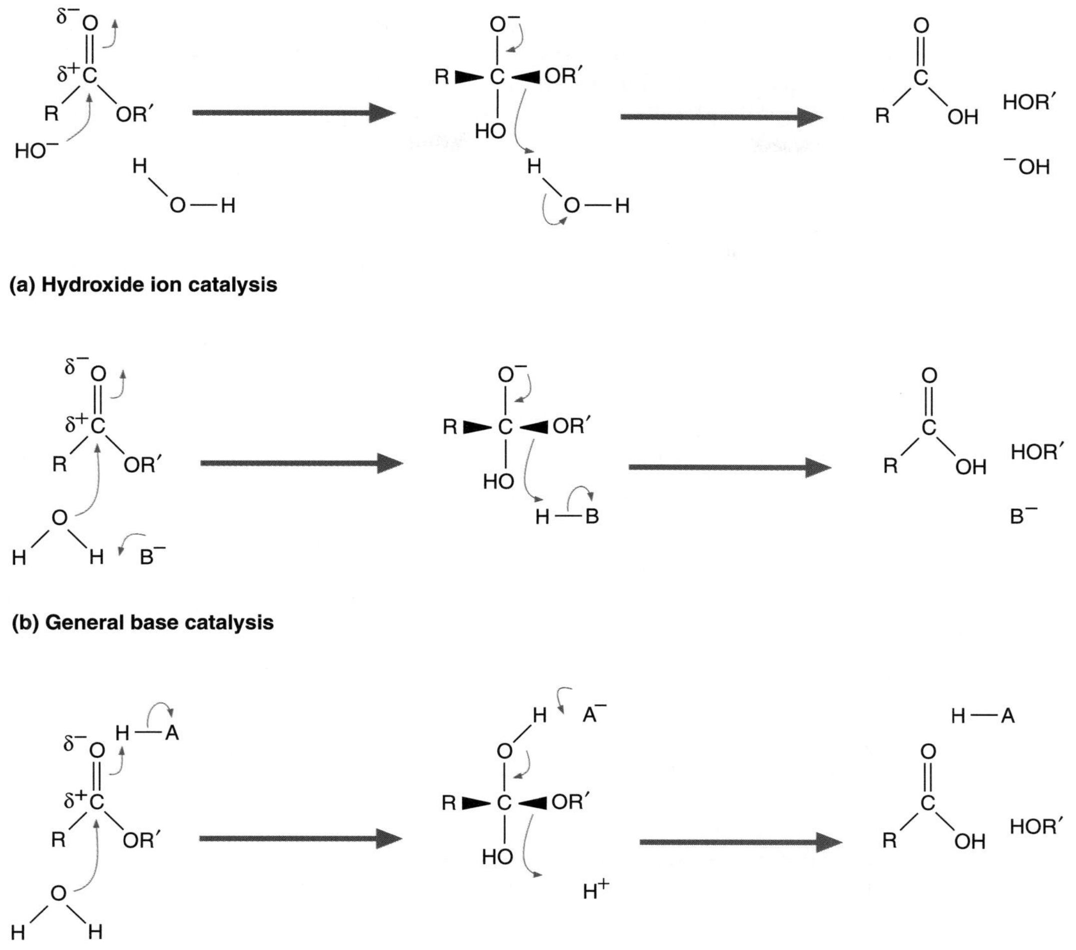

(a) Hydroxide ion catalysis

(b) General base catalysis

(c) General acid catalysis

FIGURE 6.14

Ester Hydrolysis.

Esters can be hydrolyzed in several ways: (a) catalysis by free hydroxide ion, (b) general base catalysis, and (c) general acid catalysis. A colored arrow represents the movement of an electron pair during each mechanism.

(Figure 6.14a), a tetrahedral intermediate is formed. As the intermediate breaks down, a proton is transferred from a nearby water molecule. The reaction is complete when the alcohol is released. However, hydroxide ion catalysis is not practical in living systems. Enzymes use several functional groups that behave as general bases to transfer protons efficiently. Such groups can be precisely positioned in relation to the substrate (Figure 6.14b). Ester hydrolysis can also be catalyzed by a general acid (Figure 6.14c). As the oxygen of the ester's carbonyl group binds to the proton, the carbon atom becomes more positive. The ester then becomes more susceptible to the nucleophilic attack of a water molecule.

COVALENT CATALYSIS In some enzymes a nucleophilic side chain group forms an unstable covalent bond with the substrate. The enzyme-substrate complex then forms product. A class of enzymes called the **serine proteases**

uses the —CH$_2$—OH group of serine as a nucleophile to hydrolyze peptide bonds. (Examples of serine proteases include the digestive enzymes trypsin and chymotrypsin and the blood-clotting enzyme thrombin.) During the first step, the nucleophile attacks the carbonyl group. As the ester bond is formed, the peptide bond is broken. The resulting acyl-enzyme intermediate is hydrolyzed by water in a second reaction:

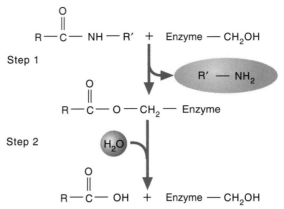

Several other amino acid side chains may act as nucleophiles. The sulfhydryl group of cysteine, the carboxylate groups of aspartate and glutamate, and the imidazole group of histidine can play this role.

Several metal cations and coenzymes also aid catalysis. Their role in facilitating enzyme function is described next.

QUESTION 6.6   Review the structures of the standard amino acids in proteins. Which ones do you think can be involved in acid-base reactions?

The Role of Cofactors in Enzyme Catalysis

Active site amino acid side chains are primarily responsible for catalyzing proton transfers and nucleophilic substitutions. To catalyze other reactions, enzymes require nonprotein cofactors, that is, metal cations and the coenzymes. The structural properties and chemical reactivities of each group are briefly discussed.

METALS The important metals in living organisms fall into two classes: the transition metals (e.g., Fe^{2+} and Cu^{2+}) and the alkali and alkaline earth metals (e.g., Na$^+$, K$^+$, Mg^{2+}, and Ca^{2+}). Because of their electronic structures, the transition metals are most often involved in catalysis. Although they have important functions in living organisms, the alkali and alkaline earth metals are only rarely found in tight complexes with proteins. This discussion is therefore concerned with the properties of the transition metals.

Several properties of transition metals make them useful in catalysis. Metal ions provide a high concentration of positive charge that is especially useful in binding small molecules. Because transition metals act as *Lewis acids* (electron pair acceptors), they are effective electrophiles. (Amino acid side chains are poor electrophiles because they cannot accept unshared pairs of electrons.) Because their directed valences allow them to interact with two or more ligands, metal ions help orient the substrate within the active site. As a consequence, the substrate–metal ion complex polarizes the substrate and promotes catalysis. For example, when the Zn^{2+} cofactor of carbonic anhydrase polarizes a water molecule, a Zn^{2+}-bound OH group forms. The OH group (acting as a nucleophile) attacks CO_2, converting it into HCO_3^-:

Enzyme—Zn^{2+} + H_2O ⇌ Enzyme—Zn^{2+}... ^-OH + H^+

Enzyme—Zn^{2+}... ^-OH + H^+ ⇌ Enzyme—Zn^{2+}... $O—\overset{\overset{\displaystyle O}{\|}}{\underset{\underset{\displaystyle H}{|}}{C}}—OH$

$O\!=\!C\!=\!O$

⇌ Enzyme—Zn^{2+} + H_2CO_3 ⇌ Enzyme—Zn^{2+} + HCO_3^- + H^+

Finally, because transition metals have two or more valence states, they can mediate oxidation-reduction reactions. For example, the reversible oxidation of Fe^{2+} to form Fe^{3+} is important in the function of cytochrome P_{450}. Cytochrome P_{450} is a microsomal enzyme found in animals that processes toxic substances (Chapter 10).

QUESTION 6.7

Copper is a cofactor in several enzymes, including lysyl oxidase and superoxide dismutase. Ceruloplasmin, a deep-blue glycoprotein, is the principal copper-containing protein in blood. It is used to transport Cu^{2+} and maintain appropriate levels of Cu^{2+} in the body's tissues. Ceruloplasmin also catalyzes the oxidation of Fe^{2+} to Fe^{3+}, an important reaction in iron metabolism. Because the metal is widely found in foods, copper deficiency is rare in humans. Deficiency symptoms include anemia, leukopenia (reduction in blood levels of white blood cells), bone defects, and weakened arterial walls. The body is partially protected from exposure to excessive copper (and several other metals) by metallothionein, a small, metal-binding protein that possesses a large proportion of cysteine residues. Certain metals (most notably zinc and cadmium) induce the synthesis of metallothionein in the intestine and liver.

In *Menkes' syndrome* intestinal absorption of copper is defective. How can affected infants be treated to avoid the symptoms of the disorder, which include seizures, retarded growth, and brittle hair?

In another rare inherited disorder, called *Wilson's disease*, excessive amounts of copper accumulate in liver and brain tissue. A prominent symptom of the disease is the deposition of copper in greenish-brown layers surrounding the cornea, called Kayser-Fleischer rings. Wilson's disease is now known to be caused by a defective ATP-dependent protein that transports copper across cell membranes. Apparently, the copper transport protein is required to incorporate copper into ceruloplasmin and to excrete excess copper. In addition to a low copper diet, Wilson's disease is treated with zinc sulfate and the chelating agent penicillamine (p. 123). Describe how these treatments work. (*Hint:* Metallothionein has a greater affinity for copper than for zinc.)

COENZYMES Most coenzymes are derived from vitamins. **Vitamins** (organic nutrients required in small amounts in the human diet) are divided into two classes: water-soluble and lipid-soluble. In addition there are certain vitamin-like nutrients (e.g., lipoic acid, carnitine, and *p*-aminobenzoic acid) that can be

TABLE 6.3
Vitamins and Their Coenzyme Forms

Vitamin	Coenzyme Form	Reaction or Process Promoted
Water-Soluble Vitamins		
Thiamine (B_1)	Thiamine pyrophosphate	Decarboxylation, aldehyde group transfer
Riboflavin (B_2)	FAD and FMN	Redox
Pyridoxine (B_6)	Pyridoxal phosphate	Amino group transfer
Nicotinic acid (niacin)	NAD and NADP	Redox
Pantothenic acid	Coenzyme A	Acyl transfer
Biotin	Biocytin	Carboxylation
Folic acid	Tetrahydrofolic acid	One-carbon group transfer
Vitamin B_{12}	Deoxyadenosylcobalamin, methylcobalamin	Intramolecular rearrangements
Ascorbic acid (vitamin C)	Unknown	Hydroxylation
Lipid-Soluble Vitamins		
Vitamin A	Retinal	Vision, growth, and reproduction
Vitamin D	1,25-Dihydroxycholecalciferol	Calcium and phosphate metabolism
Vitamin E	Unknown	Lipid antioxidant
Vitamin K	Unknown	Blood clotting

synthesized in small amounts and that facilitate enzyme-catalyzed processes. Table 6.3 lists the vitamins, their coenzyme forms, and the reactions they promote. The structure and function of the coenzyme forms of nicotinic acid (niacin) and riboflavin are described in this chapter. The other coenzymes and vitamin-like nutrients are discussed in Chapters 10, 12, 14 and 15.

There are two coenzyme forms of nicotinic acid: nicotinamide adenine dinucleotide (NAD) and nicotinamide adenine dinucleotide phosphate (NADP). These coenzymes occur in oxidized forms (NAD^+ and $NADP^+$) and reduced forms (NADH and NADPH). The structures of NAD^+ and $NADP^+$ both contain adenosine and the N-ribosyl derivative of nicotinamide, which are linked together through a pyrophosphate group (Figure 6.15a). $NADP^+$ has an additional phosphate attached to the $2'-OH$ group of adenosine. (The ring atoms of the sugar in a nucleotide are designated with a prime to distinguish them from atoms in the base.) Both NAD^+ and $NADP^+$ carry electrons for several enzymes in a group known as the dehydrogenases. (Dehydrogenases catalyze hydride ($H:^-$) transfer reactions. Many dehydrogenases that catalyze reactions involved in energy generation use the coenzyme NADH. Those enzymes that require NADPH usually catalyze biosynthetic reactions. A small number of dehydrogenases can use either NADH or NADPH.)

Alcohol dehydrogenase catalyzes the reversible oxidation of ethanol to form acetaldehyde:

$$CH_3-CH_2-OH \ + \ NAD^+ \ \rightleftharpoons \ CH_3-\overset{\overset{\displaystyle O}{\|}}{C}-H \ + \ NADH \ + \ H^+$$

During this reaction NAD^+ accepts a hydride ion (a proton with two electrons) from ethanol, the substrate molecule undergoing oxidation. Note that the equivalent of two hydrogen atoms are removed from the substrate molecules, so an H^+ is produced in addition to the hydride ion. The reversible reduction of NAD^+ is illustrated in Figure 6.15b.

Catalysis

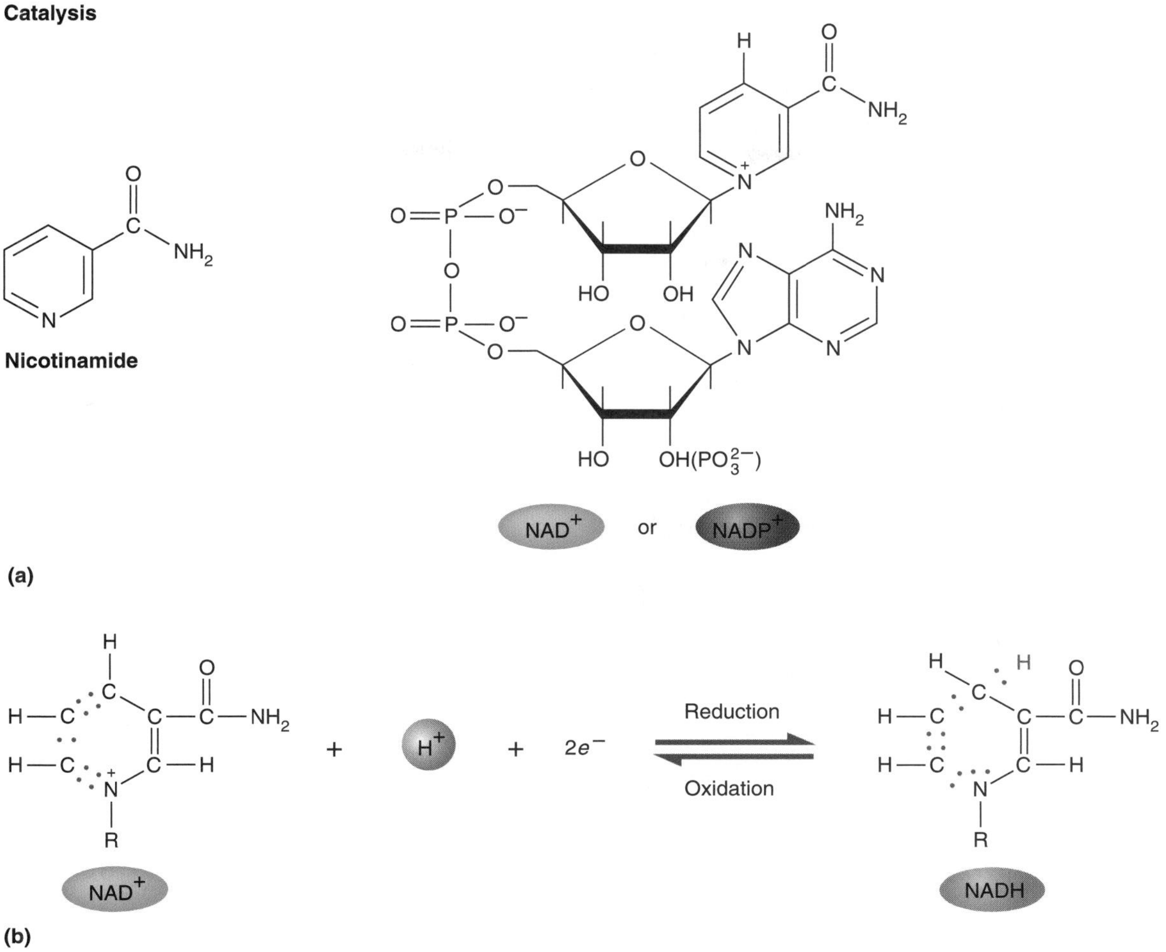

Nicotinamide

(a)

(b)

FIGURE 6.15

Nicotinamide Adenine Dinucleotide (NAD).

(a) Nicotinamide and NAD(P)$^+$. (b) Reversible reduction of NAD$^+$ to NADH$^+$. To simplify the equation, only the nicotinamide ring is shown. The rest of the molecule is designated R.

In most reactions catalyzed by dehydrogenases, the NAD$^+$ (or NADP$^+$) is bound only transiently to the enzyme. After the reduced version of the coenzyme is released from the enzyme, it donates the hydride ion to another molecule, called an *electron acceptor*. The high-energy bond between the hydrogen and the nicotinamide ring provides the energy for the enzyme-mediated transfer of the hydride ion.

Riboflavin (vitamin B$_2$) is a component of two coenzymes: flavin mononucleotide (FMN) and flavin adenine dinucleotide (FAD) (Figure 6.16). FMN and FAD function as tightly bound prosthetic groups in a class of enzymes known as the **flavoproteins**. Flavoproteins are a diverse group of catalysts; they function as dehydrogenases, oxidases, and hydroxylases. These enzymes, which catalyze oxidation-reduction reactions, use the isoalloxazine group of FAD or FMN as a donor or acceptor of two hydrogen atoms. Succinate dehydrogenase is a prominent example of a flavoprotein. It catalyzes the oxidation of succinate to form fumarate, an important reaction in energy production.

KEY CONCEPTS 6.5

The amino acid side chains in the active site of enzymes catalyze proton transfers and nucleophilic substitutions. Other reactions require a group of nonprotein cofactors, that is, metal cations and the coenzymes. Metal ions are effective electrophiles, and they help orient the substrate within the active site. In addition, certain metal cations mediate redox reactions. Coenzymes are organic molecules that have a variety of functions in enzyme catalysis.

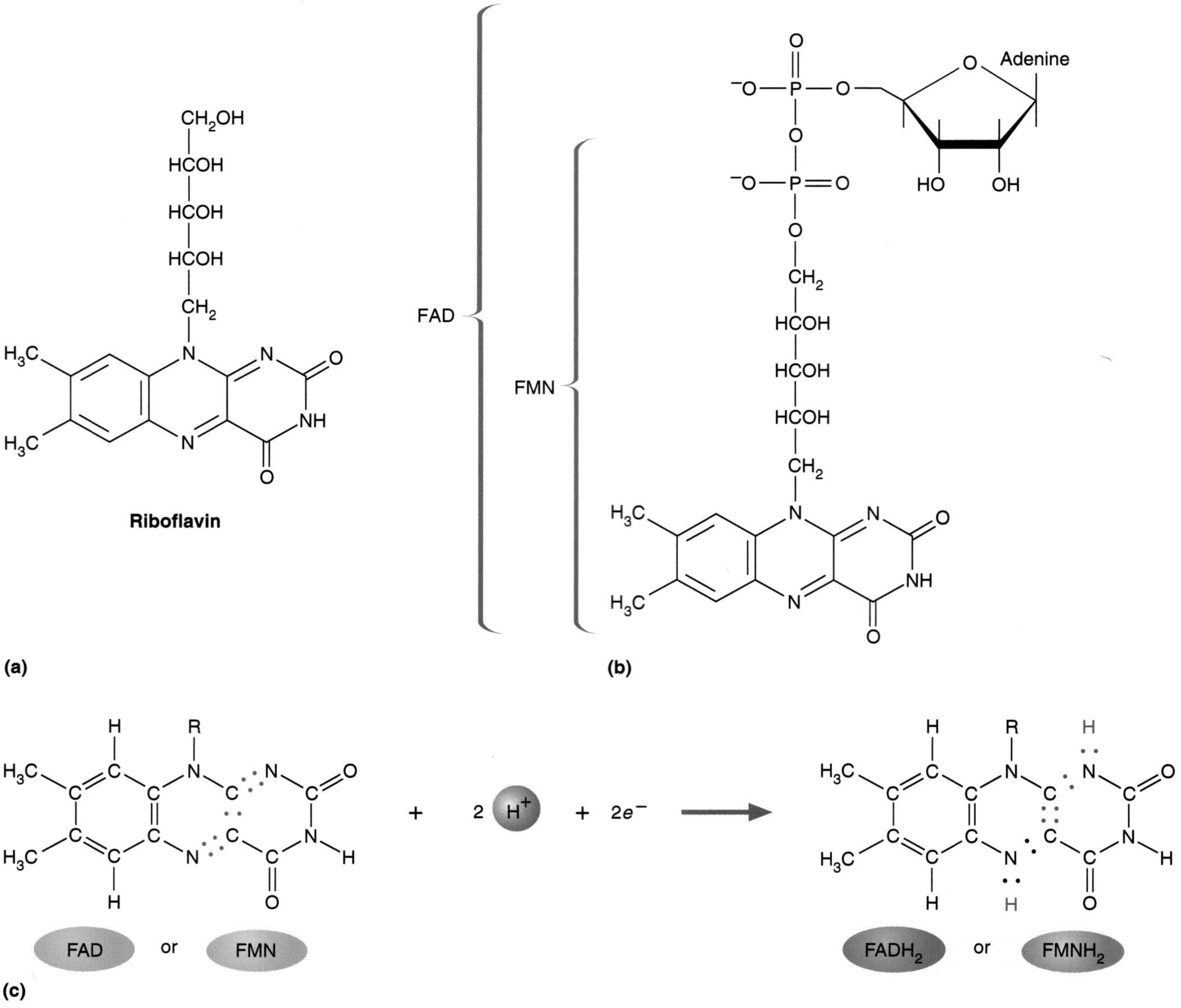

(a)

(b)

(c)

FIGURE 6.16

Flavin Coenzymes.

(a) The vitamin riboflavin. (b) FAD and FMN. (c) Reversible reduction of flavin coenzymes. To simplify the equation, only the isoalloxazine ring system is shown. The rest of the coenzyme is designated R.

QUESTION 6.8   Identify each of the following as a cofactor, coenzyme, apoenzyme, or holoenzyme:

a. Zn^{2+}

b. active alcohol dehydrogenase

c. alcohol dehydrogenase lacking Zn^{2+}

d. FMN

e. NAD^+

Effects of Temperature and pH on Enzyme-Catalyzed Reactions

Any environmental factor that disturbs protein structure may change enzymatic activity. Enzymes are especially sensitive to changes in temperature and pH.

TEMPERATURE All chemical reactions are affected by temperature. The higher the temperature, the higher the reaction rate. The reaction velocity increases because more molecules have sufficient energy to enter into the transition state. The rates of enzyme-catalyzed reactions also increase with increasing temperature. However, enzymes are proteins that become denatured at high temperatures. Each enzyme has an *optimum temperature* at which it operates at maximal efficiency (Figure 6.17). Because enzymes are proteins, optimum temperature values depend on pH and ionic strength. If the temperature is raised beyond the optimum temperature, the activity of many enzymes declines abruptly. An enzyme's optimum temperature is usually close to the normal temperature of the organism it comes from. For example, the temperature optima of most human enzymes are close to 37°C.

pH Hydrogen ion concentration affects enzymes in several ways. First, catalytic activity is related to the ionic state of the active site. Changes in hydrogen ion concentration can affect the ionization of active site groups. For example, the catalytic activity of a certain enzyme requires the protonated form of a side chain amino group. If the pH becomes sufficiently alkaline that the group loses its proton, the enzyme's activity may be depressed. In addition, substrates may also be affected. If a substrate contains an ionizable group, a change in pH may alter its capacity to bind to the active site. Second, changes in ionizable groups may change the tertiary structure of the enzyme. Drastic changes in pH often lead to denaturation.

Although a few enzymes tolerate large changes in pH, most enzymes are active only within a narrow pH range. For this reason, living organisms employ buffers to closely regulate pH. The pH value at which an enzyme's activity is maximal is called the **pH optimum** (Figure 6.18). The pH optima of enzymes vary considerably. For example, the optimum pH of pepsin, a proteolytic enzyme produced in the stomach, is approximately 2. For chymotrypsin, which digests protein in the small intestine, the optimum pH is approximately 8.

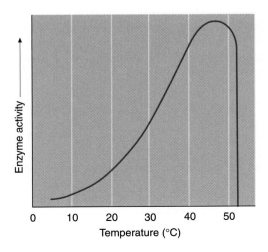

FIGURE 6.17

Effect of Temperature of Enzyme Activity.

Modest increases in temperature increase the rate of enzyme-catalyzed reactions because of an increase in the number of collisions between enzyme and substrate. Eventually, increasing the temperature decreases the reaction velocity. Catalytic activity is lost because heat denatures the enzyme.

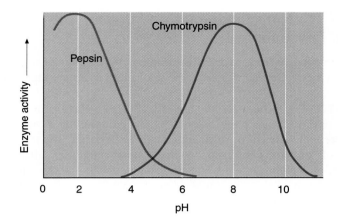

FIGURE 6.18

Effect of pH on Two Enzymes.

Each enzyme has a certain pH at which it is most active. A change in pH can alter the ioniz-able groups within the active site or affect the enzyme's conformation.

Detailed Mechanisms of Enzyme Catalysis

More than 2000 enzymes have been studied, each of which has a unique structure, substrate specificity, and reaction mechanism. Each reaction mechanism is affected by the catalysis-promoting factors of temperature and pH. The mechanisms of a variety of enzymes have been investigated intensively over the past several decades. The catalytic mechanisms of two well-characterized enzymes follow.

CHYMOTRYPSIN Chymotrypsin is a 27,000 D protein that belongs to the serine proteases. The active sites of all serine proteases contain a characteristic set of amino acid residues. In the chymotrypsin numbering system these are His 57, Asp 102, and Ser 195. Studies of crystallized enzyme bound to substrate analogs reveal that these residues are close to each other in the active site. The active site serine residue plays an especially important role in the catalytic mechanisms of this group of enzymes. Serine proteases are irreversibly inhibited by diisopropylfluorophosphate (DFP). In DFP-inhibited enzymes, the inhibitor is covalently bound only to Ser 195 and not to any of the other 29 serines. The special reactivity of Ser 195 is attributed to the proximity of His 57 and Asp 102. The imidazole ring of His 57 lies between the carboxyl group of Asp 102 and the —CH$_2$OH group of Ser 195. The carboxyl group of Asp 102 polarizes His 57, thus allowing it to act as a general base (i.e., the abstraction of a proton by the imidazole group is promoted):

$$-CH_2-\overset{\overset{\displaystyle O}{\|}}{C}-O^- \cdots HN \diagdown N: \quad HO-CH_2-$$

Removing the proton from the serine OH group converts it into a more effective nucleophile.

Chymotrypsin catalyzes the hydrolysis of peptide bonds adjacent to aromatic amino acids. The probable mechanism for this reaction is illustrated in Figure 6.19. Step (a) of the figure shows the initial enzyme-substrate complex. Asp 102, His 57, and Ser 195 are aligned. In addition, the aromatic ring of the substrate's phenylalanine residue is seated in a hydrophobic binding pocket, and

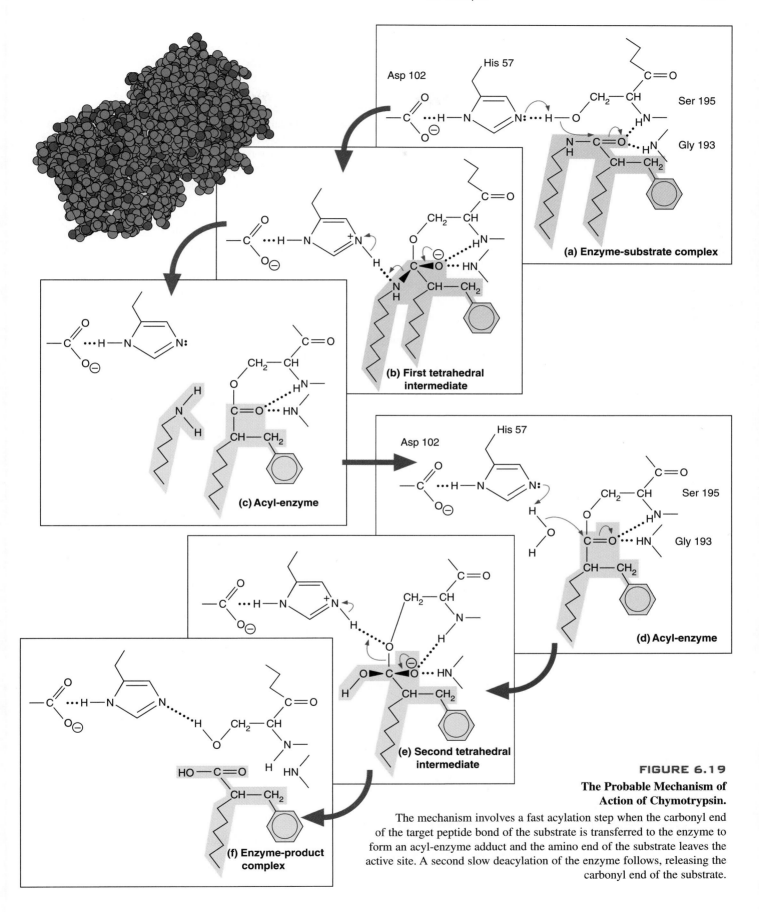

(a) Enzyme-substrate complex

(b) First tetrahedral intermediate

(c) Acyl-enzyme

(d) Acyl-enzyme

(e) Second tetrahedral intermediate

(f) Enzyme-product complex

FIGURE 6.19

The Probable Mechanism of Action of Chymotrypsin.

The mechanism involves a fast acylation step when the carbonyl end of the target peptide bond of the substrate is transferred to the enzyme to form an acyl-enzyme adduct and the amino end of the substrate leaves the active site. A second slow deacylation of the enzyme follows, releasing the carbonyl end of the substrate.

the substrate's peptide bond is hydrogen-bonded to the amide NH groups of Ser 195 and Gly 193. The nucleophilic hydroxyl oxygen of Ser 195 launches a nucleophilic attack on the carbonyl carbon of the substrate. The **oxyanion** that forms as negative charge moves to the carbonyl oxygen is stabilized by hydrogen bonds to the amide NH of Ser 195 and Gly 193. The tetrahedral intermediate, illustrated in Step (b), decomposes to form the covalently bound acyl-enzyme intermediate (Step c). His 57, acting as a general acid, is believed to facilitate this decomposition. In Steps (d) and (e), the two previous steps are reversed. With water acting as an attacking nucleophile, a tetrahedral (oxyanion) intermediate is formed. By Step (f) (the final enzyme-product complex) the bond between the serine oxygen and the carbonyl carbon has been broken. Serine is again hydrogen-bonded to His 57.

ALCOHOL DEHYDROGENASE Recall that alcohol dehydrogenase, an enzyme found in many eukaryotic cells (e.g., animal liver, plant leaves, and yeast), catalyzes the reversible oxidation of an alcohol to form an aldehyde:

$$CH_3-CH_2-OH + NAD^+ \rightleftharpoons CH_3-\overset{O}{\overset{\|}{C}}-H + NADH + H^+$$

In this reaction, which involves the removal of two electrons and two protons, the coenzyme NAD acts as the electron acceptor.

The active site of alcohol dehydrogenase contains two cysteine residues (Cys 48 and Cys 174) and a histidine residue (His 67), all of which are coordinated to a zinc ion (Figure 6.20a). After NAD$^+$ binds to the active site, the substrate ethanol enters and binds to the Zn^{2+} as the alcoholate anion (Figure 6.20b). The electrostatic effect of Zn^{2+} stabilizes the transition state. As the intermediate

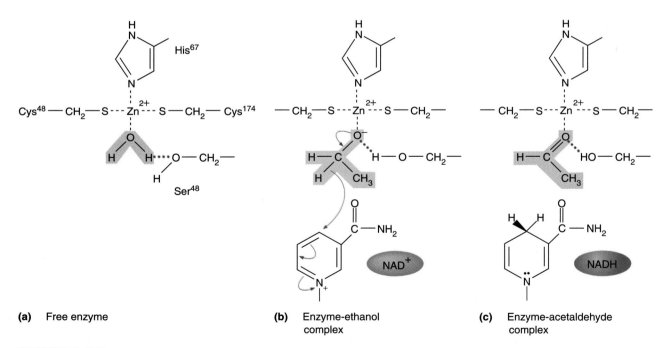

(a) Free enzyme (b) Enzyme-ethanol complex (c) Enzyme-acetaldehyde complex

FIGURE 6.20

Functional Groups of the Active Site of Alcohol Dehydrogenase.

(a) Without a substrate, a molecule of water is one of the ligands of the Zn^{2+} ion. (b) The substrate ethanol probably binds to the Zn^{2+} as the alcoholate anion, displacing the water molecule. (c) NAD$^+$ accepts a hydride ion from the substrate and the aldehyde product is formed.

decomposes, the hydride ion is transferred from the substrate to the nicotinamide ring of NAD^+ (Figure 6.20c). After the aldehyde product is released from the active site, NADH also dissociates.

KEY CONCEPTS 6.6

Each enzyme has a unique structure, substrate specificity, and reaction mechanism. Each mechanism is affected by catalysis-promoting factors that are determined by the structure of the substrate and the enzyme's active site.

6.5 ENZYME REGULATION

The thousands of enzyme-catalyzed chemical reactions in living cells are organized into a series of *biochemical* (or *metabolic*) *pathways*. Each pathway consists of a sequence of catalytic steps. The product of the first reaction becomes the substrate of the next and so on. The number of reactions varies from one pathway to another. For example, animals form glutamine from α-ketoglutarate in a pathway that has two sequential steps, whereas the synthesis of tryptophan by *Escherichia coli* requires 13 steps. Frequently, biochemical pathways have branch points. For example, chorismate, a metabolic intermediate in tryptophan biosynthesis, is also a precursor of phenylalanine and tyrosine.

Living organisms have evolved sophisticated mechanisms for regulating biochemical pathways. Regulation is essential for several reasons:

1. **Maintenance of an ordered state**. Regulation of each pathway results in the production of the substances required to maintain cell structure and function in a timely fashion and without wasting resources.
2. **Conservation of energy**. Cells constantly control energy-generating reactions so that they consume just enough nutrients to meet their energy requirements.
3. **Responsiveness to environmental changes**. Cells can make relatively rapid adjustments to changes in temperature, pH, ionic strength, and nutrient concentrations because they can increase or decrease the rates of specific reactions.

The regulation of biochemical pathways is complex. It is achieved primarily by adjusting the concentrations and activities of certain enzymes. Control is accomplished by (1) genetic control, (2) covalent modification, (3) allosteric regulation, and (4) compartmentation.

Genetic Control

The synthesis of enzymes in response to changing metabolic needs, a process referred to as **enzyme induction**, allows cells to respond efficiently to changes in their environment. For example, *E. coli* cells grown without the sugar lactose initially cannot metabolize this nutrient when it is introduced into the bacterium's growth medium. After its introduction in the absence of glucose, the genes that code for enzymes needed to utilize lactose as an energy source are activated. After all of the lactose is consumed, synthesis of these enzymes is terminated.

The synthesis of certain enzymes may also be specifically inhibited. In a process called *repression*, the end product of a biochemical pathway may inhibit the synthesis of a key enzyme in the pathway. For example, in *E. coli* the products of some amino acid synthetic pathways regulate the synthesis of key enzymes. The mechanism of this form of metabolic control is discussed in Chapter 18.

Covalent Modification

Some enzymes are regulated by the reversible interconversion between their active and inactive forms. Several covalent modifications of enzyme structure cause these changes in function. Many such enzymes have specific residues that may be phosphorylated and dephosphorylated. For example, glycogen phosphorylase

(Chapter 8) catalyzes the first reaction in the degradation of glycogen, a carbohydrate energy storage molecule. In a process controlled by hormones, the inactive form of the enzyme (glycogen phosphorylase b) is converted to the active form (glycogen phosphorylase a) by adding a phosphate group to a specific serine residue. Other types of reversible covalent modification include methylation, acetylation, and nucleotidylation (the covalent addition of a nucleotide).

Several enzymes are produced and stored as inactive precursors called **proenzymes** or **zymogens**. Zymogens are converted into active enzymes by the irreversible cleavage of one or more peptide bonds. For example, chymotrypsinogen is produced in the pancreas. After chymotrypsinogen is secreted into the small intestine, it is converted to its active form in several steps (Figure 6.21). Initially, trypsin (another proteolytic enzyme) cleaves the peptide bond between Arg 15 and Ile 16. Later, chymotrypsin cleaves other peptide bonds, creating the catalytically active enzyme that assists in the digestion of dietary protein. Other enzymes activated by partial proteolysis include pepsin, trypsin, elastase, collagenase, and the blood-clotting enzyme thrombin.

Allosteric Regulation

In each biochemical pathway there are one or more enzymes whose catalytic activity can be modulated in response to cellular needs. Such **regulatory enzymes** usually catalyze the first unique (or "committed") step in a pathway. Another typical control point is the first step of a branch in a pathway that leads to an alternate product. There are two major strategies for controlling regulatory enzymes: covalent modification and allosteric regulation. It should be noted that the control of metabolic processes is complex. There are no simple rules that explain all aspects of the regulation of metabolic pathways. Attempts to increase the pace of pathways such as glycolysis in bioengineered bacteria by increasing the concentrations of specific regulatory enzymes have usually failed. In some circumstances increased synthesis of all the enzymes in a pathway has been observed to be required for a substantial increase in the output of product.

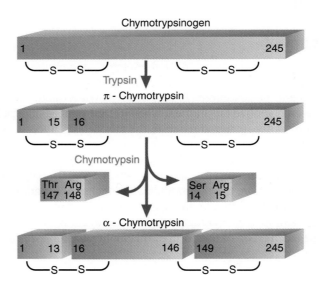

FIGURE 6.21

The Activation of Chymotrypsinogen.

The inactive zymogen chymotrypsinogen is activated in several steps. After its secretion into the small intestine, chymotrypsinogen is converted into π-chymotrypsin when trypsin, another proteolytic enzyme, cleaves the peptide bond between Arg 15 and Ile 16. Later chymotrypsin cleaves several other peptide bonds and conformational changes cause the formation of α-chymotrypsin.

Cells use allosteric regulation to respond effectively to certain changes in intracellular conditions. Recall that allosteric enzymes are usually composed of several protomers whose properties are affected by effector molecules. The binding of an effector to an allosteric enzyme can increase or decrease the binding of substrate to that enzyme. If the ligands are identical (e.g., the binding of a substrate influences the binding of additional substrate), then the allosteric effects are referred to as *homotropic*. Homotropic allostery is also referred to as *cooperativity*. *Heterotropic effects* are those that involve modulating ligands, which are different from the substrate.

Allosteric effects may be positive or negative. Recall that the binding curves for allosteric enzymes are sigmoidal. The binding of an effector shifts the curve to a higher or lower activity, depending on whether it is an activator or inhibitor (Figure 6.22). For example, aspartate transcarbamoylase (ATCase) in *E. coli* is an allosteric enzyme that catalyzes the first step in a reaction pathway that leads to the synthesis of the pyrimidine nucleotide cytidine triphosphate (CTP). CTP acts as an inhibitor of ATCase activity. This is an example of **negative feedback** inhibition, a process in which the product of a pathway inhibits the activity of the pacemaker enzyme (Figure 6.23). The purine nucleotide ATP acts as an activator. ATP activation of ATCase makes sense because nucleic acid biosynthesis requires relatively equal amounts of purine and pyrimidine nucleotides. When ATP concentration is higher than that of CTP, ATCase is activated. When

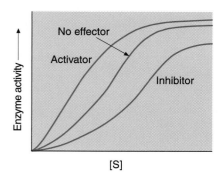

FIGURE 6.22

The Rate of an Enzyme-Catalyzed Reaction as a Function of Substrate Concentration.

The activity of allosteric enzymes is affected by positive effectors (activators) and negative inhibitors.

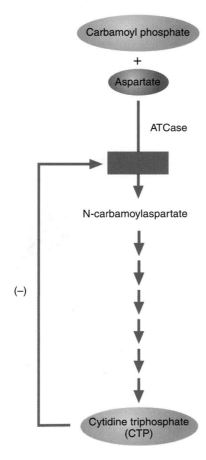

FIGURE 6.23

Feedback Inhibition.

ATCase (aspartate transcarbamoylase) catalyzes the committed step in the synthesis of CTP. The binding of CTP, the product of the pathway, to ATCase, inhibits the enzyme.

ATP concentration is lower than that of CTP, the net effect on ATCase is inhibitory. The inhibitor CTP shifts the curve to the right, indicating an increase in the apparent K_m of ATCase. The activator ATP shifts the curve to the left, indicating a lower K_m.

Two theoretical models that attempt to explain the behavior of allosteric enzymes are the concerted model and the sequential model. In the *concerted* (or *symmetry*) model, it is assumed that the enzyme exists in only two states: T(aut) and R(elaxed). Substrates and activators bind more easily to the R conformation, whereas inhibitors favor the T conformation. The term *concerted* is applied to this model because the conformations of all the protein's protomers are believed to change simultaneously when the first effector binds. (This rapid concerted change in conformation maintains the protein's overall symmmetry.) The binding of an activator shifts the equilibrium in favor of the R form. An inhibitor shifts the equilibrium toward the T conformation.

The behavior of the enzyme phosphofructokinase appears to be consistent with the concerted model. Phosphofructokinase catalyzes the transfer of a phosphate group from ATP to the OH group on C-1 of fructose-6-phosphate.

$$\text{Fructose-6-phosphate + ATP} \longrightarrow \text{fructose-1,6-bisphosphate + ADP}$$

This reaction is the most important regulatory control point in glycolysis, an important energy-generating biochemical pathway. Phosphofructokinase contains four identical subunits, each of which has an active site and an allosteric site. The enzyme is stimulated by ADP, AMP, and other metabolites. It is inhibited by phosphoenolpyruvate (an intermediate in glycolysis), citrate (an intermediate in the citric acid cycle, a related biochemical pathway), and ATP. (ATP is both a substrate and an inhibitor; that is, it can bind to both the active site and the allosteric site.) The binding of the allosteric effectors alters the rate of glycolysis in response to changes in the cell's energy needs and the availability of other fuels. Kinetic data suggest that phosphofructokinase has two conformations: T and R. When fructose-6-phosphate binds, all four subunits convert from the T conformation to the R conformation.

Although the concerted model explains several aspects of allosteric enzymes, it has certain limitations. First of all, the concerted model is too simple to account for the complex behavior of many enzymes. For example, it cannot account for **negative cooperativity**, a phenomenon observed in a few enzymes in which the binding of the first ligand reduces the affinity of the enzyme for similar ligands. The concerted model accounts only for **positive cooperativity**, in which the first ligand increases subsequent ligand binding. In addition, the concerted model makes no allowances for hybrid conformations.

In the sequential model, it is assumed that proteins are flexible. The binding of a ligand to one protomer in a multisubunit protein prompts a conformational change that is sequentially transmitted to adjacent protomers. The more sophisticated sequential model (Figure 6.24) allows for the intermediate conformations that are believed to be a closer approximation than the conformations in the simpler concerted model. Negative cooperativity has also been observed. A ligand binding to one protomer can induce conformation changes in adjacent protomers that might make ligand binding less likely.

It should be emphasized that the concerted and sequential models are theoretical models, that is, the behavior of many allosteric proteins appears to be more complex than can be accounted for by either model. For example, the cooperative binding of O_2 by hemoglobin (the most thoroughly researched allosteric protein) appears to exhibit features of both models. The binding of the first O_2 initiates a concerted T $\longrightarrow$ R transition that involves small changes in the conformation of each subunit (a feature of the sequential model). In addition, hemoglobin species with only one or two bound O_2 have been observed.

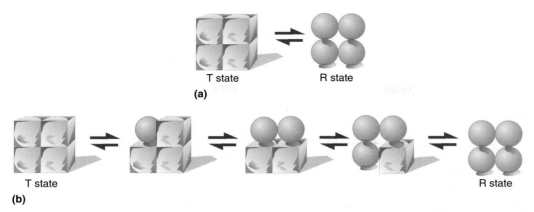

(a)

(b)

FIGURE 6.24

Allosteric Interaction Models.

(a) In the concerted model, the enzyme exists in only two conformations. Substrates and activators have a greater affinity for the R state. Inhibitors favor the T state. (b) In the sequential model, one protomer assumes an R conformation as it binds to substrate. As the first protomer changes its conformation, the affinity of nearby protomers for ligand is affected.

Compartmentation

In recent years the long-held assumption that cells are essentially enzyme-filled, membrane-bound bags has increasingly become implausible. Based on early research efforts that elucidated the major biochemical pathways, this premise resulted in part from the relative ease with which certain enzymes can be isolated from extracts of homogenized cells. Although some scientists continue to view much of metabolism as involving the actions of independent enzymes dispersed in aqueous or lipid phases of cells, it is at least generally recognized that reactions that can be demonstrated *in vitro* (in the laboratory) have little in common with their *in vivo* ("in life") counterparts. For example, substrate concentrations and rates of reaction differ widely in the two environments. In fact, significant evidence indicates that in living organisms, biochemical reactions only take place in a supramolecular context; that is, the precise spatial relationship of enzymes in a pathway to each other and to substrates, energy sources, and regulatory elements is a critical functional feature. Compartmentation, a mechanism that living cells use to control biochemical reactions, plays a key role in metabolic processes. At any given time, hundreds, if not thousands, of different chemical reactions are occurring simultaneously. Many of these reactions occur in incompatible processes; for example, the synthesis and degradation of macromolecules such as proteins and nucleic acids. The spatial separation of enzymes, substrates, and regulatory molecules into different regions or compartments allows cells to efficiently utilize relatively scarce resources. The confinement of certain biomolecules within a compartment is crucial for the coupling of unfavorable reactions with energetically favorable reactions. Cells use several strategies to compartmentalize biochemical processes. These range from the association of enzyme molecules into multiprotein complexes to the use of easily identified membrane-bound compartments.

Multiprotein complexes are molecular machines. Used by both prokaryotes and eukaryotes, each type of complex efficiently performs a specific biological task. Prominent examples of processes that require the coordinated functioning of large numbers of associated proteins include DNA synthesis, transcription, fatty acid synthesis, and protein degradation. Each element in a multiprotein complex is spatially and temporally positioned to perform a specific operation so that a biological task can be accomplished. In addition to the catalysis of certain biochemical reactions, examples of such operations include the generation of force by motor proteins and the integration of the complex into its appropriate position in the cell's organizational framework. Other common constituents of multiprotein complexes are subunits that facilitate the binding of regulatory molecules or cellular structural elements (e.g., specific membrane or cytoskeletal components). In other words, catalytically competent enzymes are insufficient for proper function. Each enzymatic activity must be oriented so that it can perform its task in an appropriate and timely manner.

Within eukaryotic cells, compartmentation is also facilitated by segregation of biochemical pathways into membrane-bound organelles. In addition to physically separating opposing pathways, the use of a complex system of organelles allows the efficiency of cellular processes to be maximized. Recall that all cellular membranes are semipermeable. Each type possesses receptor and transport molecules that control the passage of substrates, products, and regulatory molecules from one compartment to another. The flux of each biochemical pathway can be precisely regulated by altering the rates at which such molecules enter and leave. For example, fatty acid biosynthesis occurs in the cytoplasm, whereas the energy-generating reactions of fatty acid oxidation occur within the mitochondria. By using hormones and other mechanisms, metabolic control over fatty acid metabolism is exerted by regulating the transport of specific molecules (e.g., fatty acids or their degradation products) across the mitochondrial membrane. Close coordination between the two processes prevents excess waste of energy caused by significant overlap of synthesis and oxidation of fatty acids. Net synthesis occurs when the cell's energy is high and net oxidation occurs when the cell's energy is low.

Another factor related to cellular compartmentation is that special microenvironments are often created within organelles. For example, lysosomes contain hydrolytic enzymes that require a relatively high concentration of hydrogen ions for optimum activity. (Optimal lysosomal enzyme activity occurs at pH 5. The cell's cytoplasmic pH is approximately 7.2.) Lysosomes can concentrate H^+ because the lysosomal membrane, which is itself impervious to H^+, contains an energy-driven H^+ pump.

KEY CONCEPTS 6.7

All biochemical pathways are regulated to maintain the ordered state of living cells. Regulation is accomplished by genetic control, covalent modification of enzymes, allosteric regulation, and cell compartmentation.

QUESTION 6.9

Drugs are chemicals that alter or enhance physiological processes. For example, aspirin suppresses pain, and antibiotics kill infectious organisms. Once a drug is consumed, it is absorbed and distributed to the tissues, where it performs its function. Eventually, drug molecules are processed (primarily in the liver) and excreted. The dosage of each drug that physicians prescribe is based on the amount required to achieve a therapeutic effect and the drug's average rate of excretion from the body. Various reactions prepare drug molecules for excretion. Examples include oxidation, reduction, and conjugation reactions. (In conjugation reactions, small polar or ionizable groups are attached to a drug molecule to improve its solubility.) Not surprisingly, enzymes play an important role in drug metabolism. The amount of certain enzymes directly affects a patient's ability to metabolize a specific drug. For example, isoniazid is an antituberculosis agent. (Tuberculosis is a highly infectious, chronic, debilitating disease caused by *Mycobacterium tuberculosis*.) It is metabolized by N-acetylation. (In N-acetylation an amide bond forms between the substrate and an acetyl group.) The rate at which isoniazid is acetylated determines its clinical effectiveness.

Two tuberculosis patients with similar body weights and symptoms are given the same dose of isoniazid. Although both take the drug as prescribed, one patient fails to show a significant clinical improvement. The other patient is cured. Because genetic factors appear to be responsible for the differences in drug metabolism, can you suggest a reason why these patients reacted so differently to isoniazid? How can physicians improve the percentage of patients who are cured?

As knowledge concerning enzymes has grown, they have increasingly been used to solve problems. The earliest uses of enzymes were in food processing. For example, renin, a proteolytic enzyme obtained from calf stomachs, has been used for thousands of years to produce cheese. More recently, certain enzymes have become invaluable tools in medicine. A new era in medicine began in 1954 when researchers discovered that aspartate aminotransferase (ASAT, also known as serum glutamate-oxaloacetate transaminase, or SGOT) is elevated in the blood serum of patients with myocardial infarction (heart attack). Shortly thereafter, it was discovered that the serum levels of both ASAT and alanine aminotransferase (ALAT, also known as serum glutamate-pyruvate transaminase, or SGPT) become elevated after the liver is damaged. In the ensuing years, medical scientists have investigated dozens of enzymes. Enzyme technology currently plays a role in two aspects of medical practice: diagnosis and therapy. Examples of each are briefly discussed.

Diagnostic Uses of Enzymes

Enzymes are useful in modern medical practice for several reasons. Enzyme assays provide important information concerning the presence and severity of disease. In addition, enzymes often provide a means of monitoring a patient's response to therapy. Genetic predispositions to certain diseases may also be determined by measuring specific enzyme activities.

In the clinical laboratory, enzymes are used in two ways. First the activity of certain enzymes may be measured directly. For example, the measurement of blood levels of acid phosphatase activity is used to diagnose prostatic carcinoma (a urinary tract tumor that occurs in males). Second, several enzymes are used as reagents. Because purified enzymes are available, detecting certain metabolites is more accurate and cost-effective. For example, the enzyme urate oxidase is used to measure blood levels of uric acid, a metabolite whose concentration is usually high in patients suffering from gout.

For an enzyme to be used in diagnosis, several conditions must be met.

1. **Ease of measurement**. An enzyme assay should be both accurate and convenient. The development of such an assay involves determining the saturating levels of substrate and cofactors, as well as good temperature and pH control. (Recall that under zero-order conditions, velocity is proportional to the enzyme's concentration.)
2. **Convenience of method for obtaining clinically useful specimens**. Specimens of blood, urine, and (to a lesser extent) cerebrospinal fluid are readily available. However, blood, which contains a variety of enzymes and metabolites, is most often used in clinical diagnoses. Enzymes are measured by using blood *plasma* (the liquid remaining after the blood cells have been removed) or blood *serum* (the straw-colored liquid that results when blood has been allowed to clot).

Blood plasma contains two types of enzymes. The enzymes that are in the highest concentrations are specific to plasma. They include the enzymes involved in the blood-clotting process (e.g., thrombin and plasmin) and lipoprotein metabolism (Chapter 12). The nonspecific plasma enzymes have no physiological role in plasma and are normally present in low concentrations. In the normal turnover of cells, intracellular enzymes are released. An organ damaged by disease or trauma may elevate nonspecific plasma enzymes. For useful assays of these enzymes, measured enzyme activity should be commensurate with the damage. Because few enzymes are specific to one organ, the activities of several enzymes must often be measured. The procedure to confirm a diagnosis of myocardial infarction illustrates the use of enzymes in diagnosis.

Myocardial Infarction

Interruption of the heart's blood supply leads to the death of cardiac muscle cells. The symptoms of myocardial infarction include pain in the left side of the chest that may radiate to the neck, left shoulder, and arm, and irregular breathing. The initial diagnosis is based on these and other symptoms. Therapy is instituted immediately. Physicians then use several enzyme assays to confirm the diagnosis and to monitor the course of treatment. The enzymes most commonly assayed are creatine kinase (CK) and lactate dehydrogenase (LDH). Each enzyme's activity shows a characteristic time profile in terms of its release from damaged cardiac muscle cells and rate of clearance from blood (Figure 6A).

The blood levels of both enzymes must be monitored. Neither enzyme concentration gives sufficient information for the duration of treatment. For example, CK is the first enzyme detected during a myocardial infarction. CK's serum concentration rises and falls so rapidly that it is of little clinical use after several days. LDH, whose serum concentration rises later, is used to monitor the later stages of heart damage. Monitoring the activity of several enzymes also prevents misdiagnosis. Recall that few enzymes are specific

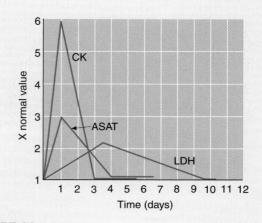

FIGURE 6A

Characteristic Pattern of Serum Cardiac Enzyme Concentrations Following a Myocardial Infarction.

to a particular organ. Until recently, serum levels of ASAT and ALAT were used to distinguish between heart and liver damage. (The ratios of the activities of the two enzymes are different in the two organs.) However, a technique employing variants of CK and LDH is now considered to be more reliable.

Both CK and LDH occur in multiple forms called isozymes. **Isozymes** are active forms of an enzyme with slightly different amino acid sequences. Isozymes can be distinguished from each other because they migrate differently during electrophoresis. CK, which occurs as a dimer, has two types of protomer: muscle type (M) and brain type (B). Heart muscle contains CK_2 (MB) and CK_3 (MM). Only the MB isozyme is found exclusively in heart muscle. Its concentration in blood reaches a maximum within a day following the infarction. CK_3 (MM), which is also found in other tissues, peaks a day after CK_2 (Figure 6B). LDH is a tetramer composed of two protomers: heart type (H) and muscle type (M). There are five different LDH isozymes. LDH 1 (H_4) and LDH 2 (H_3M) are found only in heart muscle and red blood cells. LDH 5 (M_4) occurs in both liver and skeletal muscle. LDH_3 and LDH_4 are found in other organs. The differences in the migration patterns of LDH isozymes of a normal individual and a patient suffering from myocardial infarction are illustrated in Figure 6C. Information generated from measuring the blood levels and migration patterns of both CK and LDH virtually guarantees that a correct diagnosis will be made.

Therapeutic Uses of Enzymes

The use of enzymes in medical therapy has been limited. When administered to patients, enzymes are often rapidly inactivated or degraded. The large amounts of enzyme that are often required to sustain a therapy may provoke allergic reactions. There are, however, several examples of successful enzyme therapies.

Streptokinase is a proteolytic enzyme produced by *Streptococcus pyogenes* (a bacterium that causes throat and skin infec-

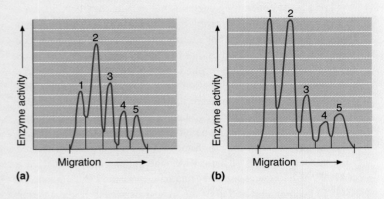

(a)　　　　　　　　　　　　(b)

FIGURE 6C

Electrophoresis Pattern of Lactate Dehydrogenase Isozymes.
(a) LDH isozymes from a normal individual. (b) LDH isozymes from a myocardial infarction patient.

tions). Streptokinase promotes the growth of the bacterium in tissue because it digests blood clots. It is currently used with significant success in the treatment of myocardial infarction, which results from the occlusion of the coronary arteries. If administered soon after the beginning of a heart attack, streptokinase can often prevent or significantly reduce further damage to the heart. Streptokinase catalyzes the conversion of plasminogen to plasmin, the trypsinlike enzyme that digests fibrin (the primary component of blood clots). Human tissue plasminogen activator (tPA), a product of recombinant DNA technology (Biochemical Methods 18.1) that acts in a similar fashion, is also used to treat myocardial infarction.

The enzyme asparaginase is used to treat several types of cancer. It catalyzes the following reaction:

$$\text{L-asparagine} + H_2O \longrightarrow \text{L-aspartate} + NH_3$$

Asparaginase occurs in plants, vertebrates, and bacteria but not in human blood. Unlike most normal cells, the cells in certain kinds of tumor, such as in several forms of adult leukemia, cannot synthesize asparagine. Infusing asparaginase reduces the blood's concentration of asparagine and often causes tumor regression. (The regression of these tumors is not completely understood. Presumably, lack of asparagine inhibits protein synthesis, causing cell death.) Asparaginase therapy has several serious side effects, such as allergic reactions and liver damage. Unfortunately, after several asparaginase treatments some patients may develop resistance. (*Resistance* is a condition in which tumor cells grow despite the presence of a toxic substance that previously caused tumor regression.) Apparently, some tumor cells can induce the synthesis of asparagine synthetase, an enzyme that converts aspartate to asparagine.

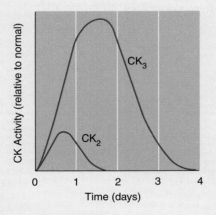

FIGURE 6B

Characteristic Pattern of Serum Creatine Kinase Concentrations Following a Myocardial Infarction.

SUMMARY

1. Enzymes are biological catalysts. They enhance reaction rates because they provide an alternative reaction pathway that requires less energy than an uncatalyzed reaction. In contrast to some inorganic catalysts, most enzymes catalyze reactions at mild temperatures. In addition, enzymes are specific to the types of reactions they catalyze. Each type of enzyme has a unique, intricately shaped binding surface called an active site. Substrate binds to the enzyme's active site, which is a small cleft or crevice in a large protein molecule. In the lock-and-key model of enzyme action, the structures of the enzyme's active site and the substrate transition state are complementary. In the induced-fit model, the protein molecule is assumed to be flexible.

2. Each enzyme is currently classified and named according to the type of reaction it catalyzes. There are six major enzyme categories: oxidoreductases, transferases, hydrolases, lyases, isomerases, and ligases.

3. Enzyme kinetics is the quantitative study of enzyme catalysis. According to the Michaelis-Menten model, when the substrate S binds in the active site of an enzyme E, an ES transition state complex is formed. During the transition state, the substrate is converted into product. After a time the product dissociates from the enzyme. In the Michaelis-Menten equation,

$$v = \frac{V_{max}[S]}{[S] + K_m}$$

V_{max} is the maximal velocity for the reaction, and K_m is a rate constant. Experimental determinations of K_m and V_{max} are made with Lineweaver-Burk double-reciprocal plots.

4. The turnover number (k_{cat}) is a measure of the number of substrate molecules converted to product per unit time by an enzyme when it is saturated with substrate. Because [S] is rel-atively low under physiological conditions ([S] $\ll K_m$), the term k_{cat}/K_m is a more reliable gauge of the catalytic efficiency of enzymes.

5. Enzyme inhibition may be reversible or irreversible. Irreversible inhibitors usually bind covalently to enzymes. In reversible inhibition, the inhibitor can dissociate from the enzyme. The most common types of reversible inhibition are competitive, uncompetitive, and noncompetitive.

6. The kinetic properties of allosteric enzymes are not explained by the Michaelis-Menten model. Most allosteric enzymes are multisubunit proteins. The binding of substrate or effector to one subunit affects the binding properties of other protomers.

7. Enzymes use the same catalytic mechanisms as nonenzymatic catalysts. Several factors contribute to enzyme catalysis: proximity and strain effects, electrostatic effects, acid-base catalysis, and covalent catalysis. Combinations of these factors affect enzyme mechanisms.

8. Active site amino acid side chains are primarily responsible for catalyzing proton transfers and nucleophilic substitutions. Nonprotein cofactors (metals and coenzymes) are used by enzymes to catalyze other types of reactions.

9. Enzymes are sensitive to environmental factors such as temperature and pH. Each enzyme has an optimum temperature and an optimum pH.

10. The chemical reactions in living cells are organized into a series of biochemical pathways. The pathways are controlled primarily by adjusting the concentrations and activities of enzymes through genetic control, covalent modification, allosteric regulation, and compartmentation.

SUGGESTED READINGS

Copeland, R. A., *Enzymes: A Practical Introduction to Structure, Mechanism, and Data Analysis*, Wiley-VCH, New York, 1996.

Cornish-Bowden, A., *Fundamentals of Enzyme Kinetics*, Portland Press, London, 1995.

Dische, Z., The Discovery of Feedback Inhibition, *Trends Biochem. Sci.*, 1:269–270, 1976.

Fersht, A., *Structure and Mechanism in Protein Science: A Guide to Enzyme Catalysis and Protein Folding*, W. H. Freeman, New York, 1999.

Gutfreund, H., *Kinetics for the Life Sciences: Receptors, Transmitters and Catalysts*, Cambridge University Press, Cambridge, 1995.

Kraut, J., How Do Enzymes Work? *Science*, 242:533–540, 1998.

Miller, J. A., Women in Chemistry, in Kass-Simon, G., and Fannes, P. (Eds.), *Women in Science: Righting the Record*, pp. 300–334, Indiana University Press, Bloomington, 1990.

Monod, J., Changeux, J. P., and Jacob, F., Allosteric Proteins and Cellular Control Systems, *J. Mol. Biol.*, 6:306–329, 1963.

Perutz, M. F., Mechanisms of Cooperativity and Allosteric Regulation in Proteins, *Quart. Rev. Biophys.*, 22:139–151, 1989.

Schultz, P. G., The Interplay Between Chemistry and Biology in the Design of Enzymatic Catalysts, *Science*, 240:426–433, 1988.

Segel, I. R., *Biochemical Calculations: How to Solve Mathematical Problems in General Biochemistry*, 2nd ed., John Wiley & Sons, New York, 1976.

KEY WORDS

activation energy, *162*	cofactor, *164*	hydrolase, *165*	lyase, *165*
active site, *163*	enzyme induction, *189*	inhibitor, *173*	negative cooperativity, *192*
allosteric enzyme, *176*	enzyme kinetics, *167*	isomerase, *165*	negative feedback, *191*
apoenzyme, *164*	flavoprotein, *183*	isozyme, *196*	noncompetitive, *174*
coenzyme, *164*	holoenzyme, *164*	ligase, *165*	oxidoreductase, *165*

REVIEW QUESTIONS

1. Clearly define the following terms:
   a. activation energy
   b. catalyst
   c. active site
   d. coenzyme
   e. velocity of a chemical reaction
   f. half-life
   g. turnover number
   h. katal
   i. noncompetitive inhibitor
   j. repression

2. What are four important properties of enzymes?

3. Living things must regulate the rate of catalytic processes. Explain how the cell regulates enzymatic reactions.

4. Determine the class of enzyme that is most likely to catalyze each of the following reactions, see bottom Figure 6D:

5. Several factors contribute to enzyme catalysis. What are they? Briefly explain the effect of each.

6. List three reasons why the regulation of biochemical processes is important.

7. Describe negative feedback inhibition.

8. Describe the two models that explain the binding of allosteric enzymes. Use either model to explain the binding of oxygen to hemoglobin.

9. What are the major coenzymes? Briefly describe the function of each.

10. What properties of transition metals make them useful as enzyme cofactors?

11. The ΔH for the following reaction is -28.2 kJ/mol:

$$C_6H_{12}O_6 + 6\ O_2 \longrightarrow 6\ CO_2 + 6\ H_2O$$
glucose

   Explain why glucose is stable in an oxygen atmosphere for appreciable periods of time.

12. Enzymes act by reducing the activation energy of a reaction. Describe several ways in which this is accomplished.

13. In enzyme kinetics, why are measurements made at the start of a reaction?

14. Histidine is frequently used as a general acid or general base in enzyme catalysis. Consider the pK_a's of the side groups of the amino acids listed in Table 5.2 to suggest a reason why this is so.

15. Enzymes are stereochemically specific; that is, they often convert only one stereoisomeric form of substrate into product. Why is such specificity inherent in their structure?

FIGURE 6D

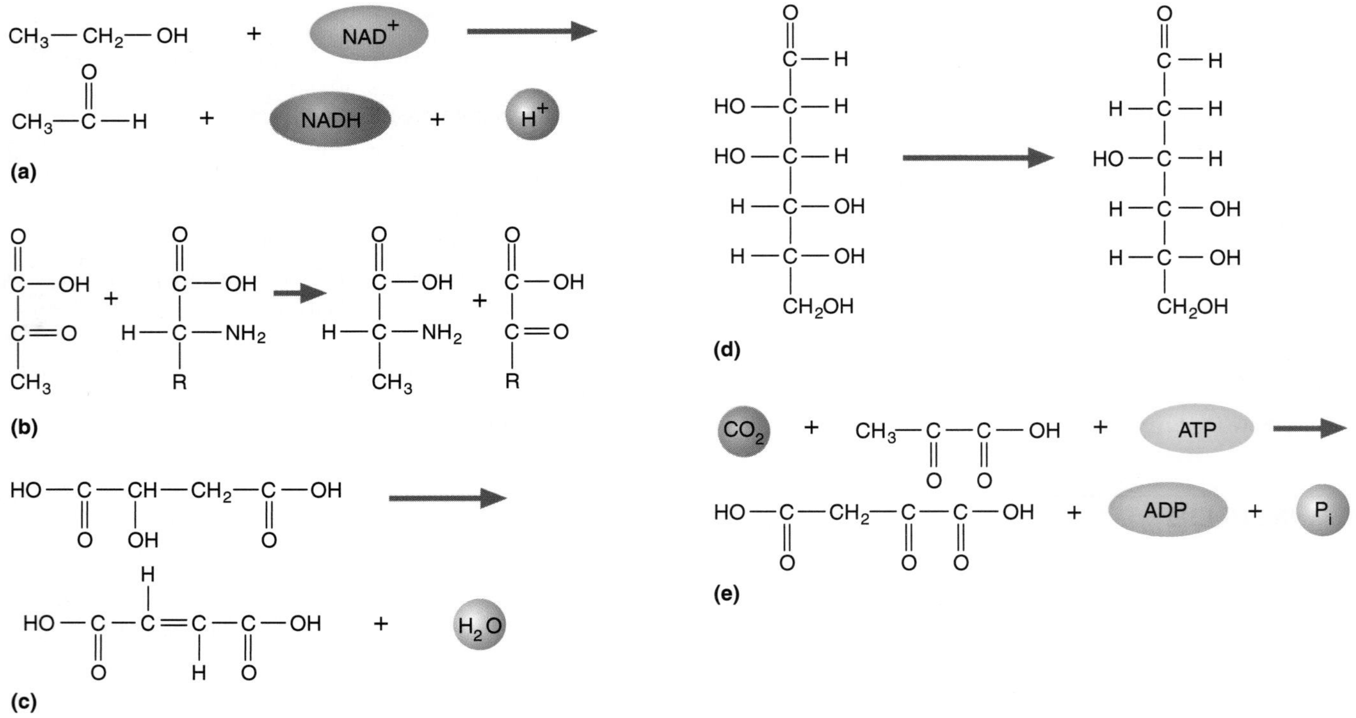

(a)

(b)

(c)

(d)

(e)

THOUGHT QUESTIONS

1. Consider the following reaction:

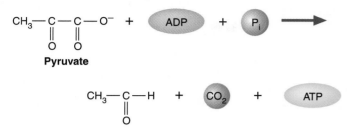

Pyruvate

Using the following data, determine the order of the reaction for each substrate and the overall order of the reaction.

Experiment	[Pyruvate]*	[ADP]	[P$_1$]	[Rate]
1	0.1	0.1	0.1	8×10^{-4}
2	0.2	0.1	0.1	1.6×10^{-3}
3	0.2	0.2	0.1	3.2×10^{-3}
4	0.1	0.1	0.2	3.2×10^{-3}

*Concentrations are in moles per liter. Enzyme rates are measured in moles per liter per second.

2. Consider the following data for an enzyme-catalyzed hydrolysis reaction by the inhibitor I:

[Substrate] (M)	v (μmol/min)	v_1 (μmol/min)
6×10^{-6}	20.8	4.2
1×10^{-5}	29	5.8
2×10^{-5}	45	9
6×10^{-5}	67.6	13.6
1.8×10^{-4}	87	16.2

Using a Michaelis-Menten plot, determine K_m for the uninhibited reaction and the inhibited reaction.

3. Using the data in Question 2:
   a. Generate a Lineweaver-Burk plot for the data.
   b. Explain the significance of the (i) horizontal intercept, (ii) vertical intercept, (iii) slope.
   c. What type of inhibition is being measured?

4. What are the two types of enzyme inhibitors? Give an example of each.

5. Two experiments were performed with the enzyme ribonuclease. In experiment 1 the effect of increasing substrate concentration on reaction velocity was measured. In experiment 2 the reaction mixtures were identical to those in experiment 1 except for the addition of 0.1 mg of an unknown compound to each tube. Plot the data according to the Lineweaver-Burk method. Determine the effect of the unknown compound on the enzyme's activity. (Substrate concentration is measured in millimoles per liter. Velocity is measured in the change in optical density per hour.)

Experiment 1		Experiment 2	
[S]	v	[S]	v
0.5	0.81	0.5	0.42
0.67	0.95	0.67	0.53
1	1.25	1	0.71
2	1.61	2	1.08

6. Suggest a reason why enzymes can be partially protected from thermal denaturation by high concentrations of substrate.

7. Describe the mechanism for chymotrypsin. Show how the amino acid residues of the active site participate in the reaction.

8. Alcohol dehydrogenase is inhibited by numerous alcohols. Using the data given in the following table, calculate the k_{cat}/K_m values for each of the alcohols. Which of the listed alcohols is most easily metabolized by alcohol dehydrogenase?

Kinetic parameters for hamster testes ADH

Substrate	K_m (μM)	k_{cat} (min^{-1})
Ethanol	960	480
1-butanol	440	450
1-hexanol	69	182
12-hydroxy-dodecanoate	50	146
all-trans-retinol	20	78
benzyl alcohol	410	82
2-butanol	250,000	285
cyclohexanol	31,000	122

9. 4-Methyl pyrazole has been developed as a long-acting and less toxic alternative to ethanol in the treatment of ethylene glycol poisoning. Shown is a Lineweaver-Burk plot of the inhibition of alcohol dehydrogenase by various concentrations of 4-methyl pyrazole. What type of inhibition appears to be exhibited by this molecule?

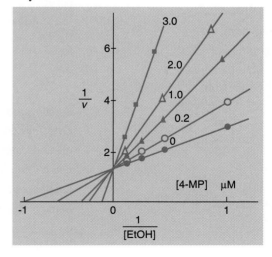

Carbohydrates

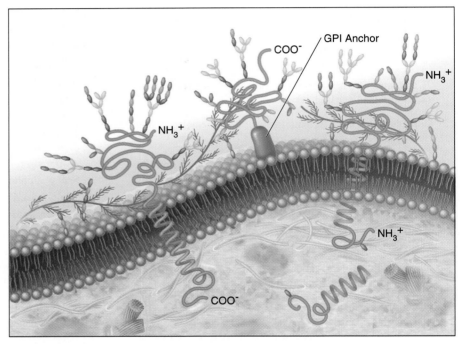

The Cell Surface Significant amounts of carbohydrate are attached to membrane protein and lipids on the external surface of cells. (Colored balls = sugar residues; GPI anchor = a complex lipid molecule that connects many cell surface proteins to the phospholipid bilayer of the plasma membrane)

Carbohydrates are not just an important source of rapid energy production for living cells. They are also structural building blocks of cells and components of numerous metabolic pathways. A broad range of cellular phenomena, such as cell recognition and binding (e.g., by other cells, hormones, and viruses), depend on carbohydrates. Chapter 7 describes the structures and chemistry of typical carbohydrate molecules found in living organisms.

Carbohydrates, the most abundant biomolecules in nature, are a direct link between solar energy and the chemical bond energy of living organisms. (More than half of all '"organic" carbon is found in carbohydrates.) They are formed during *photosynthesis* (Chapter 13), a biochemical process in which light energy is captured and used to drive the biosynthesis of energy-rich organic molecules from the energy-poor inorganic molecules CO_2 and H_2O. Most carbohydrates contain carbon, hydrogen, and oxygen in the ratio $(CH_2O)_n$, hence the name "hydrate of carbon." They have been adapted for a wide variety of biological functions, which include energy sources (e.g., glucose), structural elements (e.g., cellulose and chitin in plants and insects, respectively), and precursors in the production of other biomolecules (e.g., amino acids, lipids, purines, and pyrimidines). Carbohydrates are classified as monosaccharides, disaccharides, oligosaccharides, and polysaccharides according to the number of simple sugar units they contain. Carbohydrate moieties also occur as components of other biomolecules. A vast array of *glycoconjugates* (protein and lipid molecules with covalently linked carbohydrate groups) are distributed among all living species, most notably among the eukaryotes. Certain carbohydrate molecules (the sugars ribose and deoxyribose) are structural elements of nucleotides and nucleic acids.

In recent years it has become increasingly apparent that carbohydrate provides living organisms with enormous informational capacities. Investigations of biological processes such as signal transduction, cell-cell interactions, and endocytosis have revealed that they typically involve the binding of glycoconjugates such as glycoproteins and glycolipids or free carbohydrate molecules with complementary receptors. Chapter 7 provides a foundation for understanding the complex processes in living organisms by reviewing the structure and function of the most common carbohydrates and glycoconjugates. The chapter ends with a discussion of the *sugar code*, the mechanism by which carbohydrate structure is used to encode biological information (Special Interest Box 7.3).

7.1 MONOSACCHARIDES

Monosaccharides or simple sugars are polyhydroxy aldehydes or ketones. Recall from Chapter 1 that monosaccharides with an aldehyde functional group are called **aldoses**, whereas those with a ketone group are called *ketoses* (Figure 7.1). The simplest aldose and ketose are glyceraldehyde and dihydroxyacetone, respectively (Figure 7.2). Sugars are also classified according to the number of carbon atoms they contain. For example, the smallest sugars, called *trioses*, contain three carbon atoms. Four-, five-, and six-carbon sugars are called *tetroses*, *pentoses*, and *hexoses*, respectively. The most abundant monosaccharides found in living cells are the pentoses and hexoses. Often, class names such as aldohexoses and ketopentoses, which combine information about carbon number and functional groups, describe monosaccharides. For example, glucose, a six-carbon aldehyde-containing sugar, is referred to as an aldohexose.

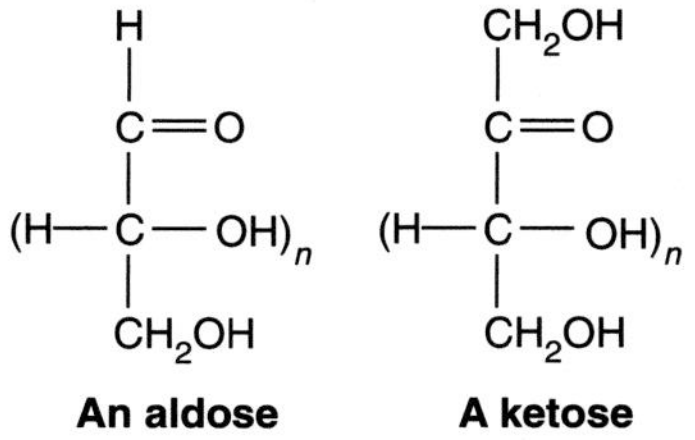

The sugar structures shown in Figures 7.1 and 7.2 are known as Fischer projections (in honor of the great Nobel prize–winning German chemist Emil Fischer).

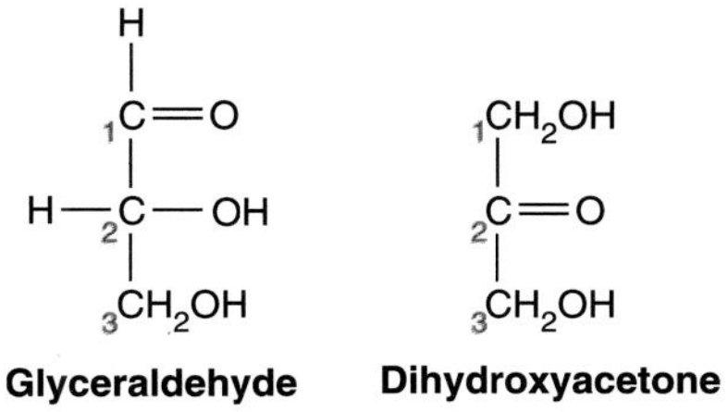

FIGURE 7.1

General Formula for Aldose and Ketose Forms of Monosaccharides.

FIGURE 7.2

Glyceraldehyde (an aldotriose) and Dihydroxyacetone (a ketotriose).

In these structures the carbohydrate backbone is drawn vertically with the most highly oxidized carbon usually shown at the top. The horizontal lines are understood to project toward the viewer, and the vertical lines recede from the viewer.

QUESTION 7.1

Identify the class of each of the following sugars. For example, glucose is an aldohexose.

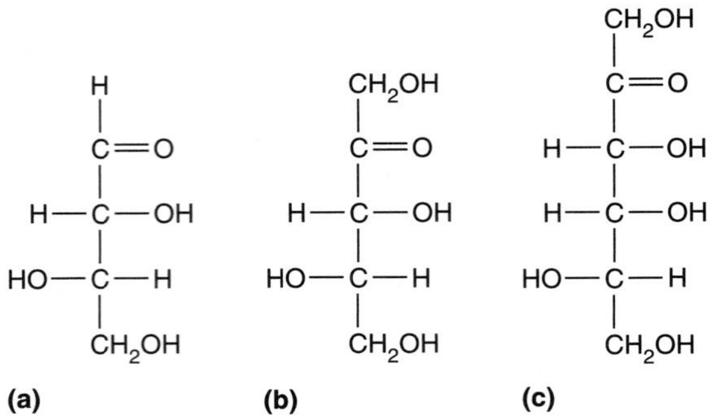

(a)   (b)   (c)

Monosaccharide Stereoisomers

When the number of chiral carbon atoms increases in optically active compounds, the number of possible optical isomers also increases. The total number of possible isomers can be determined by using van't Hoff's rule: A compound with n chiral carbon atoms has a maximum of 2^n possible stereoisomers. For example, when n is equal to 4, there are 2^4 or 16 stereoisomers (8 D-stereoisomers and 8 L-stereoisomers).

In optical isomers the reference carbon is the asymmetric carbon that is most remote from the carbonyl carbon. Its configuration is similar to that of the asymmetric carbon in either D- or L-glyceraldehyde. Almost all naturally occurring sugars have the D- configuration. They can be considered to be derived from either the triose D-glyceraldehyde (the aldoses) or the nonchiral triose dihydroxyacetone (the ketoses). (Note that although dihydroxyacetone does not have an asymmetric carbon, it clearly is the parent compound for the ketoses.) In the D-aldose family of sugars (Figure 7.3), which contains most biologically important monosaccharides, the hydroxyl group is to the right on the chiral carbon atom farthest from the most oxidized carbon (in this case the aldehyde group) in the molecule (e.g., carbon 5 in a six-carbon sugar).

Stereoisomers that are not enantiomers (mirror-image isomers) are called **diastereomers**. For example, the aldopentoses D-ribose and L-ribose are enantiomers, as are D-arabinose and L-arabinose (Figure 7.4. The sugars D-ribose and D-arabinose are diastereomers because they are isomers but not mirror images.

Diastereomers that differ in the configuration at a single asymmetric carbon atom are called **epimers**. For example, D-glucose and D-galactose are epimers because their structures differ only in the configuration of the OH group at carbon 4 (Figure 7.3). D-Mannose and D-galactose are not epimers because their configurations differ at more than one carbon.

QUESTION 7.2

When viewed in two dimensions (as on the printed page), the structural differences between optical isomers may appear trivial. However, many biomolecules are optically active, and the capacity of enzymes to distinguish between D- and L-substrate molecules is an important feature of the chemistry of living cells. For example, most of the enzymes that break down and use dietary carbohydrates can bind to D-sugars but not to their L-isomers. Can you convince yourself that

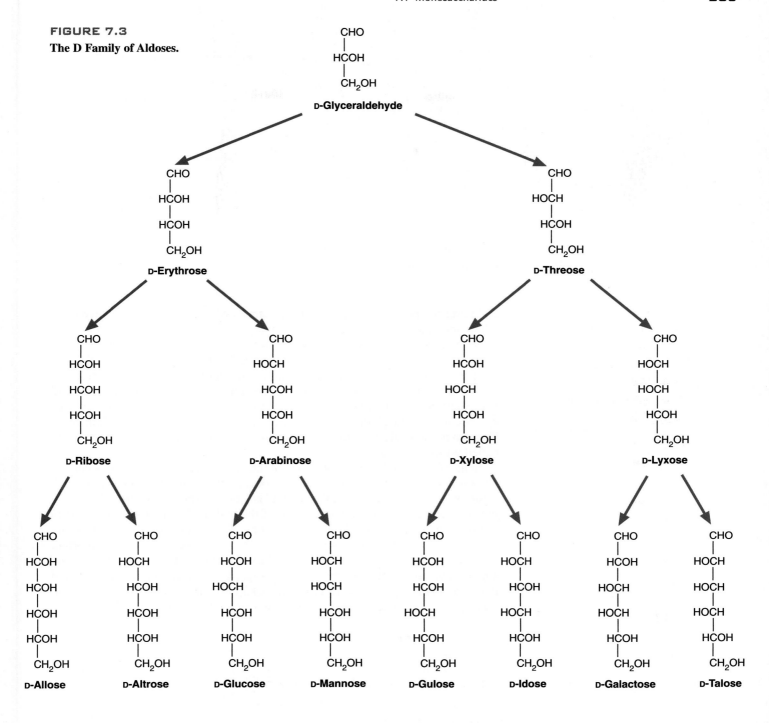

FIGURE 7.3

The D Family of Aldoses.

the D- and L-isomers of an optically active molecule are indeed different in three-dimensional space? Make models of D- and L-glyceraldehyde with an organic chemistry model kit or with colored styrofoam balls and toothpicks.

Cyclic Structure of Monosaccharides

Sugars that contain four or more carbons exist primarily in cyclic forms. Ring formation occurs in aqueous solution because aldehyde and ketone groups react reversibly with hydroxyl groups present in the sugar to form cyclic **hemiacetals**

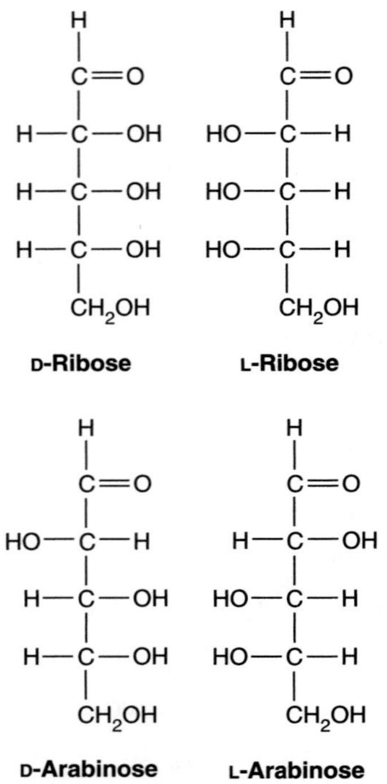

FIGURE 7.4

The Optical Isomers D- and L-Ribose and D- and L-Arabinose.

D-Ribose and D-arabinose are diastereomers; that is, they are not mirror images.

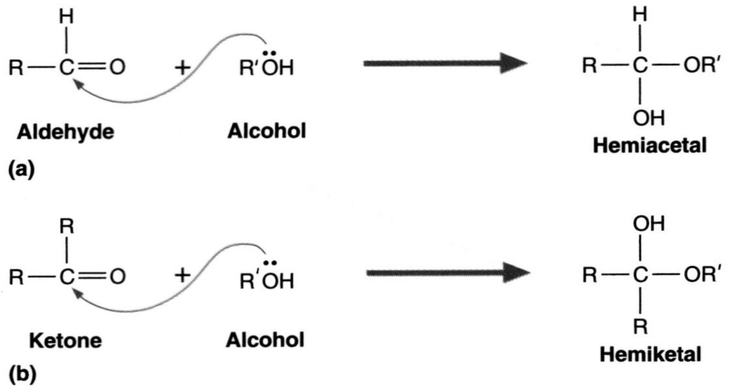

FIGURE 7.5

Formation of Hemiacetals and Hemiketals.

(a) From an aldehyde. (b) From a ketone.

and **hemiketals**, respectively. Ordinary hemiacetals and hemiketals, which form when molecules containing an aldehyde or ketone functional group react with an alcohol, are unstable and easily revert to the aldehyde or ketone forms (Figure 7.5). When the aldehyde or ketone group and the alcohol functional group are part of the same molecule, however, an intramolecular cyclization reaction occurs that can form stable products. The most stable cyclic hemiacetal and hemiketal rings contain five or six atoms. As cyclization occurs, the carbonyl carbon becomes a new chiral center. This carbon is called the *anomeric carbon atom*. The two possible diastereomers that may form during the cyclization reaction are called **anomers**.

In aldose sugars the hydroxyl group of the newly formed hemiacetal occurs on carbon 1 (the anomeric carbon) and may occur either above the ring (in the "up" position) or below the ring (in the "down" position). When the hydroxyl is down, the structure is in the α-anomeric form. If the hydroxyl is up, the structure is in the β-anomeric form. In Fischer projections, the α-anomeric hydroxyl occurs on the right and the β-hydroxyl occurs on the left (Figure 7.6). It is important to note that the anomers are defined relative to the D- and L-classification of sugars. The above rules apply only to D-sugars, the most common ones found in nature. In the L-sugars the α-anomeric OH group is above the ring. The cyclization of sugars is more easily visualized by using Haworth structures.

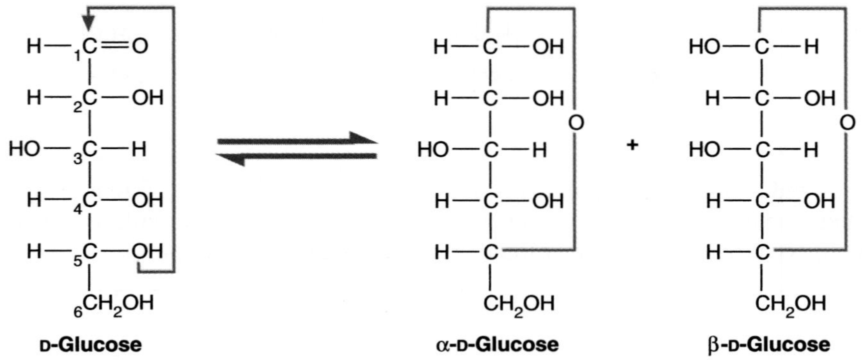

FIGURE 7.6

Monosaccharide Structure.

Formation of the hemiacetal structure of glucose. Both the α- and β-anomers of glucose form.

HAWORTH STRUCTURES Fischer representations of cyclic sugar molecules use a long bond to indicate ring structure. A more accurate picture of carbohydrate structure was developed by the English chemist W. N. Haworth (Figure 7.7).

Haworth structures more closely depict proper bond angles and lengths than do Fischer representations. To convert from the traditional Fischer formula of a D-pentose or D-hexose to a Haworth formula, the following steps should be followed:

1. Draw a five- or six-membered ring with the oxygen placed as shown below:

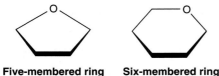

Five-membered ring   Six-membered ring

2. Starting with the anomeric carbon to the right of the ring oxygen, place hydroxyl groups either above or below the plane of the ring. Groups that are pointing to the left in the Fischer projection formula should go above the plane of the ring, and those that are pointing to the right in the Fischer projection formula should go below the ring.

3. In D-sugars, the last carbon position (e.g., C-6 glucose) is always up.

Five-membered hemiacetal rings are called *furanoses* because of their structural similarity to furan (Figure 7.8). For example, the cyclic form of fructose depicted in Figure 7.9 is called fructofuranose. Six-membered rings are called *pyranoses* because of their similarity to pyran. Glucose, in the pyranose form, is called glucopyranose.

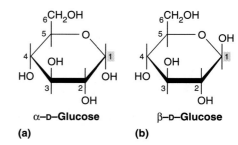

FIGURE 7.7

Haworth Structures of the Anomers of Glucose.

(a) α-D-Glucose. (b) β-D-Glucose.

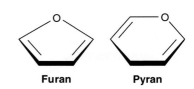

Furan   Pyran

FIGURE 7.8

Furan and Pyran.

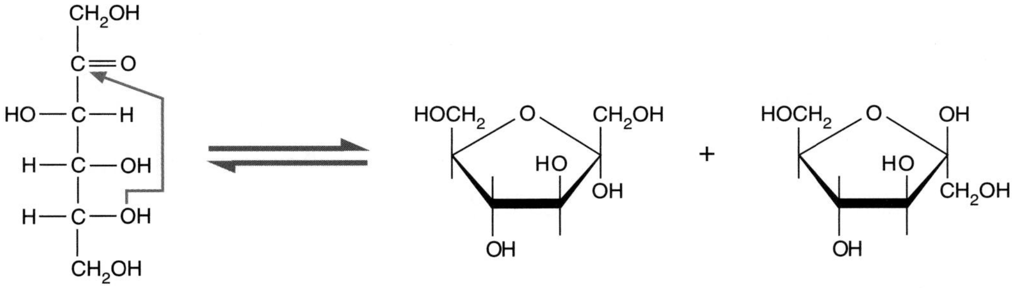

D-Fructose   α-D-Fructofuranose   β-D-Fructofuranose

FIGURE 7.9

Fischer and Haworth Forms of D-Fructose.

Convert the following Fischer structures into cyclic Haworth structures: **QUESTION 7.3**

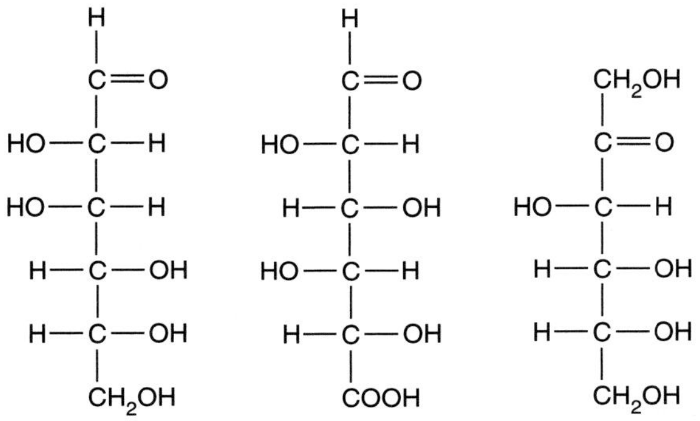

Name the anomers that result from the cyclization of these molecules.

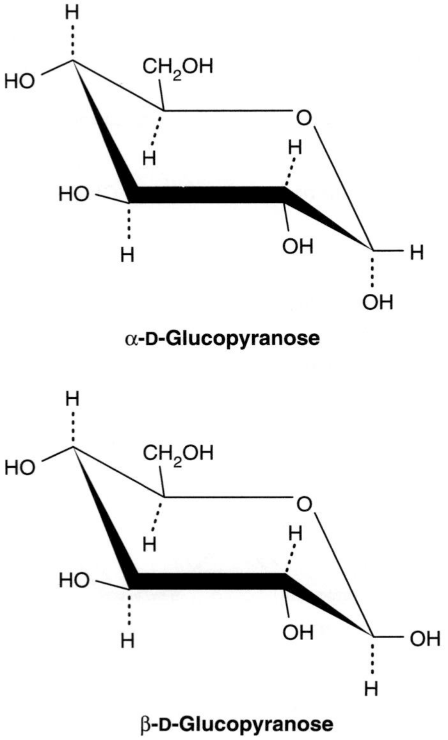

α-D-Glucopyranose

β-D-Glucopyranose

FIGURE 7.10

α- and β-Glucose.

CONFORMATIONAL STRUCTURES Although Haworth projection formulas are often used to represent carbohydrate structure, they are oversimplifications. Bond angle analysis and X-ray analysis demonstrate that *conformational formulas* are more accurate representations of monosaccharide structure (Figure 7.10). Conformational structures are more accurate because they illustrate the puckered nature of sugar rings.

Space-filling models, whose dimensions are proportional to the radius of the atoms, also give useful structural information. (See Figures 7.19, 7.20, and 7.21.) Monosaccharides undergo most of the reactions that are typical of aldehydes, ketones, and alcohols. The most important of these reactions in living organisms are described.

MUTAROTATION The α- and β-forms of monosaccharides are readily interconverted when dissolved in water. This spontaneous process, called **mutarotation**, produces an equilibrium mixture of α- and β-forms in both furanose and pyranose ring structures. The proportion of each form differs with each sugar type. Glucose, for example, exists primarily as a mixture of α- (38%) and β- (62%) pyranose forms (Figure 7.11). Fructose is predominantly found in the α- and β-furanose forms. The open chain formed during mutarotation can participate in oxidation-reduction reactions.

OXIDATION-REDUCTION REACTIONS In the presence of oxidizing agents, metal ions such as Cu^{2+}, and certain enzymes, monosaccharides readily undergo

KEY CONCEPTS 7.1

Monosaccharides, polyhydroxy aldehydes or ketones, are either aldoses or ketoses. Sugars that contain four or more carbons primarily have cyclic forms. Cyclic aldoses or ketoses are hemiacetals and hemiketals, respectively.

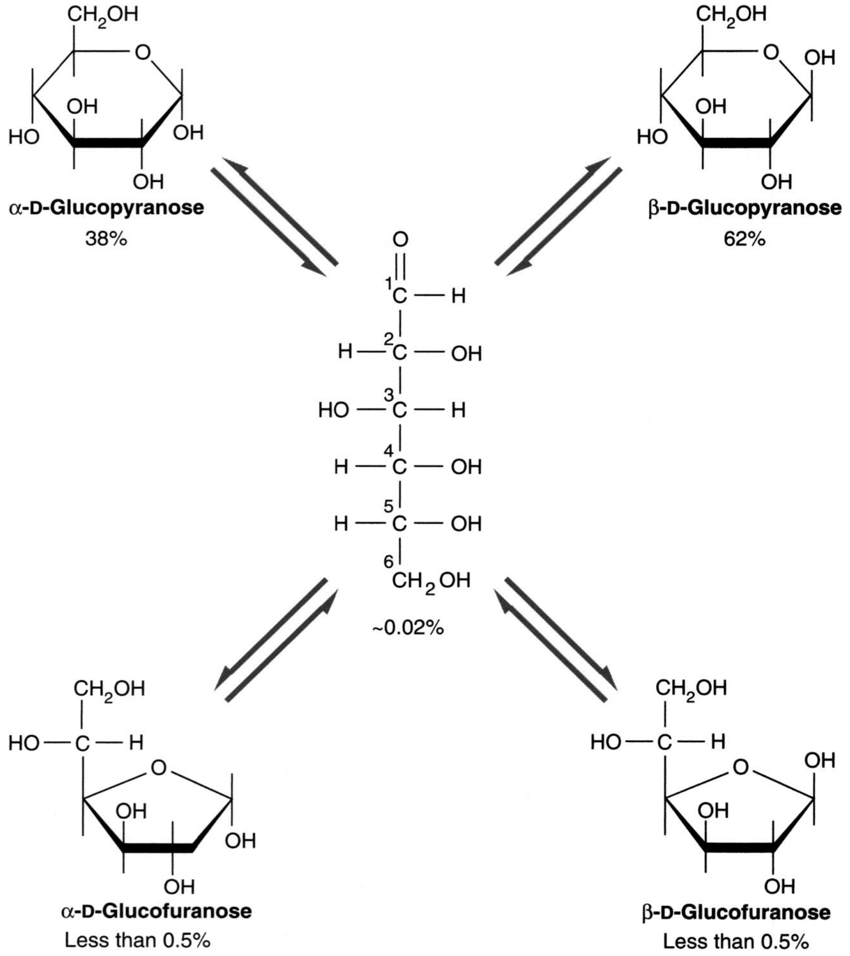

FIGURE 7.11

Equilibrium Mixture of D-Glucose.

When glucose is dissolved in water at 25°C, the anomeric forms of the sugar undergo very rapid interconversions. When equilibrium is reached (i.e., there is no net change in the occurrence of each form), the glucose solution contains the percentages shown.

α-D-Glucopyranose
38%

β-D-Glucopyranose
62%

~0.02%

α-D-Glucofuranose
Less than 0.5%

β-D-Glucofuranose
Less than 0.5%

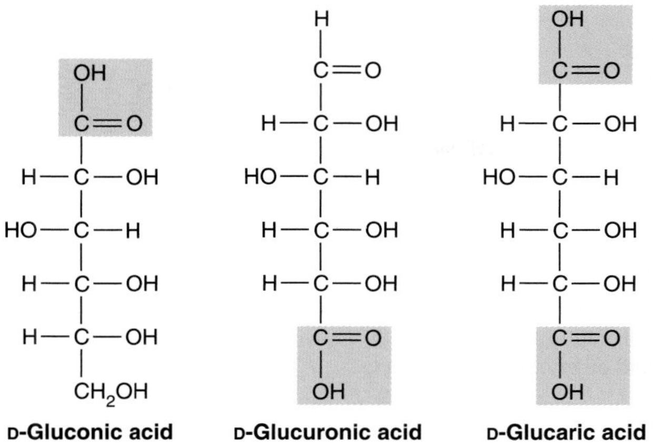

FIGURE 7.12
Oxidation Products of Glucose.
The newly oxidized groups are highlighted.

several oxidation reactions. Oxidation of an aldehyde group yields an **aldonic acid**, whereas oxidation of a terminal CH_2OH group (but not the aldehyde group) gives a **uronic acid**. Oxidation of both the aldehyde and CH_2OH gives an **aldaric acid** (Figure 7.12).

The carbonyl groups in both aldonic and uronic acids can react with an OH group in the same molecule to form a cyclic ester known as a **lactone**:

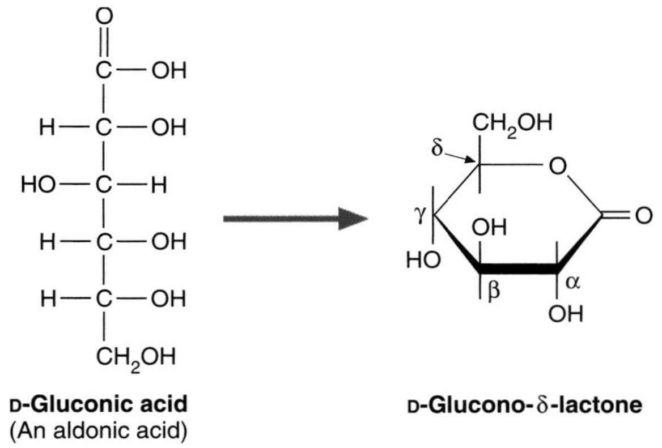

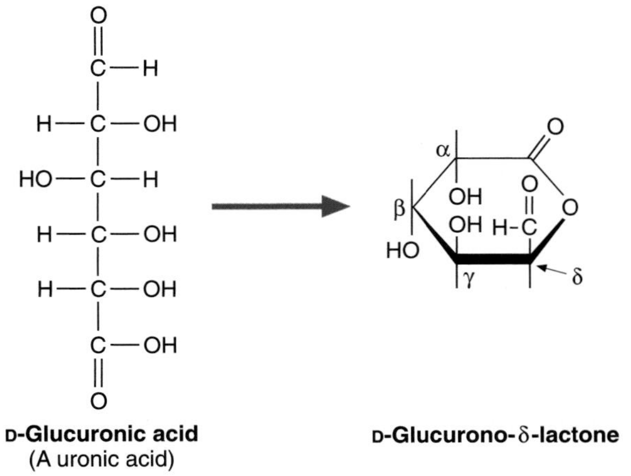

Lactones are commonly found in nature. For example, L-ascorbic acid (vitamin C) is a lactone derivative of D-glucuronic acid (Special Interest Box 7.1).

Sugars that can be oxidized by weak oxidizing agents such as Benedict's reagent are called **reducing sugars** (Figure 7.13). Because the reaction occurs

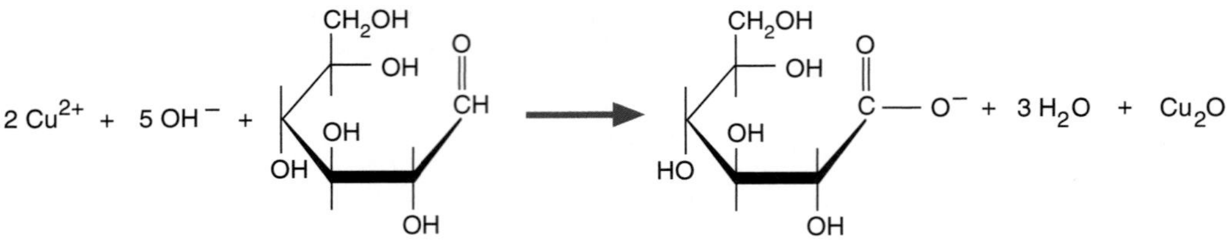

$$2\,Cu^{2+} + 5\,OH^- + \text{[glucose ring structure]} \longrightarrow \text{[gluconate ring structure]} + 3\,H_2O + Cu_2O$$

FIGURE 7.13

Reaction of Glucose with Benedict's Reagent.

Benedict's reagent, copper(II) sulfate in a solution of sodium carbonate and sodium citrate, is reduced by the monosaccharide glucose. Glucose is oxidized to form the salt of gluconic acid. The reaction also forms the reddish-brown precipitate Cu_2O and other oxidation products.

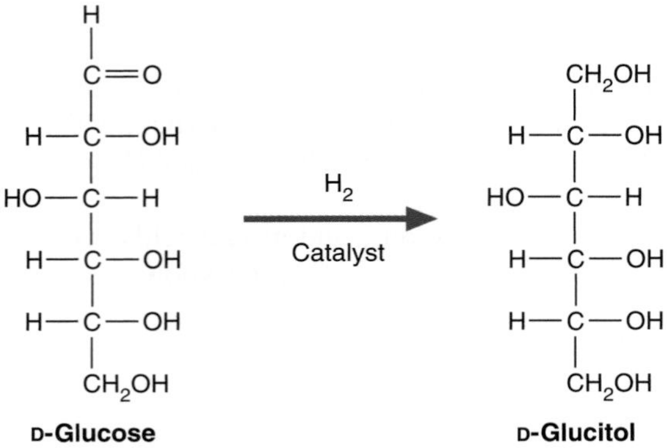

D-Glucose   D-Glucitol

FIGURE 7.14

Laboratory Reduction of Glucose to Form D-Glucitol (Sorbitol).

only with sugars that can revert to the open chain form, all monosaccharides are reducing sugars.

REDUCTION Reduction of the aldehyde and ketone groups of monosaccharides yields the sugar alcohols (**alditols**). Reduction of D-glucose, for example, yields D-glucitol, also known as D-sorbitol (Figure 7.14). Sugar alcohols are used commercially in processing foods and pharmaceuticals. Sorbitol, for example, improves the shelf life of candy because it helps prevent moisture loss. Adding sorbitol syrup to artificially sweetened canned fruit reduces the unpleasant aftertaste of the artificial sweetener saccharin. Once consumed, sorbitol is converted into fructose in the liver.

ISOMERIZATION Monosaccharides undergo several types of isomerization. For example, after several hours an alkaline solution of D-glucose also contains D-mannose and D-fructose. Both isomerizations involve an intramolecular shift of a hydrogen atom and a relocation of a double bond (Figure 7.15). The intermediate formed is called an **enediol**. The reversible transformation of glucose to fructose is an example of an aldose-ketose interconversion. Because the configuration at a single asymmetric carbon changes, the conversion of glucose to mannose is referred to as an **epimerization**. Several enzyme-catalyzed reactions involving enediols occur in carbohydrate metabolism (Chapter 8).

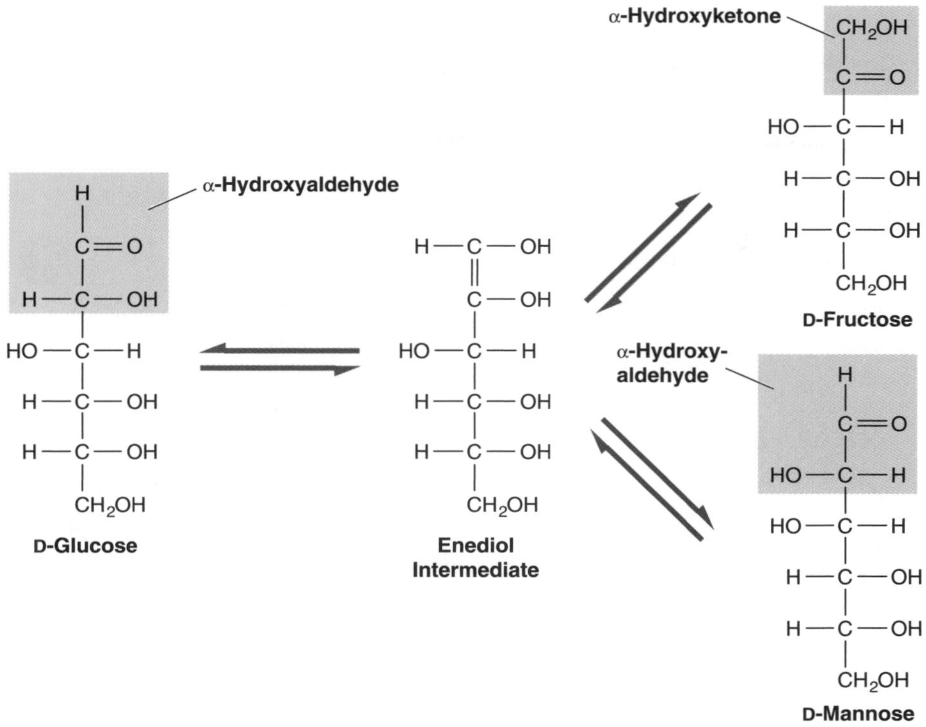

FIGURE 7.15

Isomerization of D-Glucose to Form D-Mannose and D-Fructose.

An enediol intermediate is formed in this process.

ESTERIFICATION Like all free OH groups, those of carbohydrates can be converted to esters by reactions with acids. Esterification often dramatically changes a sugar's chemical and physical properties. Phosphate and sulfate esters of carbohydrate molecules are among the most common ones found in nature.

Phosphorylated derivatives of certain monosaccharides are important metabolic components of living cells. They are frequently formed during reactions with ATP. They are important because many biochemical transformations use nucleophilic substitution reactions. Such reactions require a leaving group. In a carbohydrate molecule this group is most likely to be an OH group. However, because OH groups are poor leaving groups, any substitution reaction is unlikely. The problem is solved by converting an appropriate OH group to a phosphate ester, which can then be displaced by an incoming nucleophile. As a consequence, a slow reaction now occurs much more rapidly.

Sulfate esters of carbohydrate molecules are found predominantly in the proteoglycan components of connective tissue. Because sulfate esters are charged, they bind large amounts of water and small ions. They also participate in forming salt bridges between carbohydrate chains.

Draw the following compounds:

(a) α- and β-anomers of D-galactose

(b) aldonic acid, uronic acid, and aldaric acid derivatives of galactose

(c) galactitol

(d) δ-lactone of galactonic acid

QUESTION 7.4

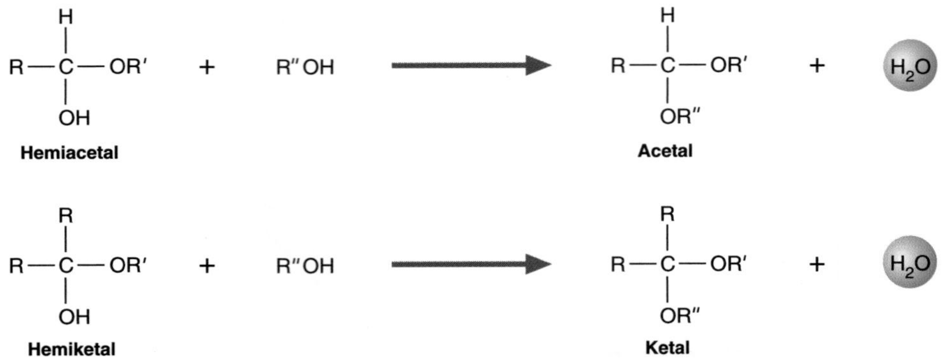

FIGURE 7.16

Formation of Acetals and Ketals.

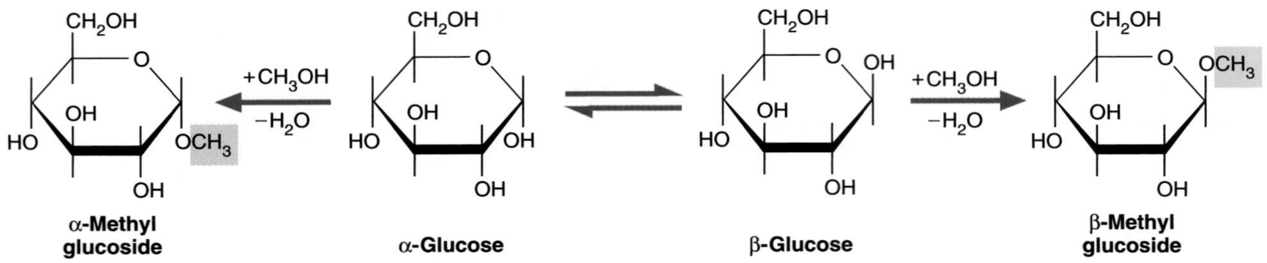

FIGURE 7.17

Methyl Glucoside Formation.

Noncarbohydrate components of glycosides are called aglycones. The highlighted methyl groups are aglycones.

FIGURE 7.18

Salicin.

GLYCOSIDE FORMATION Hemiacetals and hemiketals react with alcohols to form the corresponding **acetal** or **ketal** (Figure 7.16). When the cyclic hemiacetal or hemiketal form of the monosaccharide reacts with an alcohol, the new linkage is called a **glycosidic linkage**, and the compound is called a **glycoside**. The name of the glycoside specifies the sugar component. For example, the acetals of glucose and the ketals of fructose are called *glucoside* and *fructoside*, respectively. Additionally, glycosides derived from sugars with five-membered rings are called *furanosides*; those from six-membered rings are called *pyranosides*. A relatively simple example shown in Figure 7.17 illustrates the reaction of glucose with methanol to form two anomeric types of methyl glucosides. Because glycosides are acetals, they are stable in basic solutions. Carbohydrate molecules that contain only acetal groups do not test positive with Benedict's reagent. (Acetal formation "locks" a ring so it cannot undergo oxidation or mutarotation.) Only hemiacetals act as reducing agents.

If an acetal linkage is formed between the hemiacetal hydroxyl group of one monosaccharide and a hydroxyl group of another monosaccharide, the resulting glycoside is called a **disaccharide**. A molecule containing a large number of monosaccharides linked by glycosidic linkages is called a **polysaccharide**.

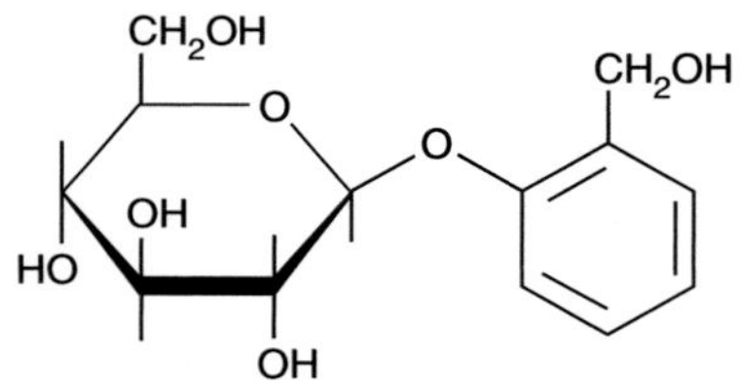

QUESTION 7.5

Draw the structure of a D-glucosamine molecule linked to threonine via a β-glycosidic linkage.

QUESTION 7.6

Glycosides are commonly found in nature. One example is salicin (Figure 7.18), a compound found in willow tree bark that has antipyretic (fever-reducing) and analgesic properties. Can you identify the carbohydrate and aglycone (noncarbohydrate) components of salicin?

(a)　　　　　　　　　　　　　　**(b)**

FIGURE 7.19

α-D-**Glucopyranose.**

Compare the information provided by (a) the space-filling model and (b) the Haworth structure. Carbon atoms are green, oxygen atoms are red, and hydrogen atoms are white.

Important Monosaccharides

Among the most important monosaccharides that occur in living organisms are glucose, fructose, and galactose. The principal functional roles of these molecules are briefly described.

GLUCOSE D-Glucose, originally called dextrose, is found in large quantities throughout the living world (Figure 7.19). It is the primary fuel for living cells. In animals, glucose is the preferred energy source of brain cells and cells that have few or no mitochondria, such as erythrocytes. Cells that have a limited oxygen supply, such as those in the eyeball, also use large amounts of glucose to generate energy. Dietary sources include plant starch and the disaccharides lactose, maltose, and sucrose.

FRUCTOSE D-Fructose, originally called levulose, is often referred to as fruit sugar because of its high content in fruit. It is also found in some vegetables and in honey (Figure 7.20). This molecule is an important member of the ketose family of sugars. On a per gram basis, fructose is twice as sweet as sucrose. It can therefore be used in smaller amounts. For this reason, fructose is often used as a sweetening agent in processed food products. Large amounts of fructose are used in the male reproductive tract. It is synthesized in the seminal vesicles and then incorporated into semen. Sperm use the sugar as an energy source.

GALACTOSE Galactose is necessary to synthesize a variety of biomolecules (Figure 7.21). These include lactose (in lactating mammary glands), glycolipids,

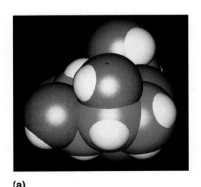

(a)　　　　　　　　**(b)**

FIGURE 7.20

β-D-**Fructofuranose.**

(a) Space-filling model and (b) Haworth structure.

Ascorbic acid, a lactone with a molecular weight of 176.1 D, is a powerful reducing agent:

OH
|
HO—CH₂—CH
Ascorbic acid

It is synthesized by all mammals except guinea pigs, apes, fruit-eating bats, and, of course, humans. These species must obtain ascorbic acid (also called *vitamin C*) in their diet. The three enzymes necessary to synthesize ascorbic acid have been isolated from microsomal fractions of liver tissue from species that can synthesize the molecule. In humans and guinea pigs, however, one of the enzymes (gulonolactone oxidase) has not been detected. (Antibodies that bind specifically to gulonolactone oxidase do not bind to human or guinea pig hepatic microsomes.) Presumably, lack of this enzyme prevents these species from synthesizing ascorbic acid.

Scurvy (Figure 7A), the disease caused by a significant depletion of the body's stores of ascorbic acid, was common in Europe before the introduction of root crops at the end of the Middle Ages. However, as Europeans took increasingly longer voyages to search for a trade route to the Far East, outbreaks of the malady were dramatic (and often documented). For example, a ship's log written during Vasco da Gama's expedition (1498) reveals an outbreak of scurvy on the return journey across the Arabian Sea. The sailors "again suffered from their gums, their legs also swelled and other parts of the body and these swellings spread until the sufferer died, without exhibiting symptoms of any other disease." During the expedition the sailors became convinced that oranges were a curative for scurvy, whose symptoms did not appear until 10 weeks after setting sail.

When James Lind, a surgeon in the British Royal Navy, performed his famous experiment (perhaps the first controlled clinical trial) in 1746, he compared the curative effects of several treatments on 12 sailors who had scurvy. Two men were assigned to each of six daily treatments for two weeks. Briefly, the treatments

FIGURE 7A

Scurvy Grass Promoted as a Cure for the Disease
The title page of the English version of Moellenbrok's seventeenth-century book and his illustration.

were (1) 1 quart of hard cider, (2) 25 mL. of elixir of vitriol three times a day, (3) 18 mL of vinegar three times a day, (4) 0.3 L of seawater, (5) two oranges and one lemon, and (6) 4 mL of a medicinal paste containing garlic and mustard seed, among other herbs. Contrary to popular opinion, Lind's work did not prove the efficacy of citrus fruits in the treatment of scurvy; these had already been an accepted treatment for several hundred years. Actually, citrus fruits were Lind's positive controls. He was apparently interested in testing the effectiveness of the official treatment of the Royal Navy: vinegar (treatment 3). The results of the trial: Those who received the oranges and lemons were fit for duty after 6 days. The other treatments varied widely in outcome but were all significantly less effective than citrus fruits. Lind subsequently reported his results to the British Admiralty, and in 1754 his work *A Treatise on Scurvey* was published. Eventually (1795), the Royal Navy approved the daily ration of 1-1/2 ounces of lime juice on all ships, giving rise to the nickname of "limeys" for British sailors.

KEY CONCEPTS 7.2

Glucose, fructose, and galactose are among the most important monosaccharides in living organisms.

certain phospholipids, proteoglycans, and glycoproteins. Synthesis of these substances is not diminished by diets that lack galactose or the disaccharide lactose (the principal dietary source of galactose), because the sugar is readily synthesized from glucose-1-phosphate. As was mentioned previously, galactose and glucose are epimers at carbon 4. The interconversion of galactose and glucose is catalyzed by an enzyme called an epimerase.

 In *galactosemia*, a genetic disorder, an enzyme required to metabolize galactose is missing. Galactose, galactose-1-phosphate, and galactitol (a sugar alcohol derivative) accumulate and cause liver damage, cataracts, and severe

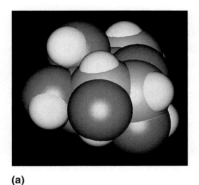

(a)   **(b)**

mental retardation. The only effective treatment is early diagnosis and a diet free of galactose.

Monosaccharide Derivatives

Simple sugars may be converted to closely related chemical compounds. Several of these are important metabolic and structural components of living organisms.

URONIC ACIDS Recall that uronic acids are formed when the terminal CH_2OH group of a monosaccharide is oxidized. Two uronic acids are important in animals: D-glucuronic acid and its epimer, L-iduronic acid (Figure 7.22). In liver cells, glucuronic acid is combined with molecules such as steroids, certain drugs, and bilirubin (a degradation product of the oxygen-carrying protein hemoglobin) to improve water solubility. This process helps remove waste products from the body. Both D-glucuronic acid and L-iduronic acid are abundant in connective tissue carbohydrate components.

AMINO SUGARS In amino sugars a hydroxyl group (most commonly on carbon 2) is replaced by an amino group (Figure 7.23). These compounds are common

(a)

(b)

FIGURE 7.22
α-D-**Glucuronate (a) and**
β-L-**Iduronate (b).**

FIGURE 7.23
Amino Sugars.
(a) *α*-D-Glucosamine, (b) *α*-D-galactosamine, (c) N-acetyl-*α*-D-glucosamine, and (d) N-acetylneuraminic acid (sialic acid).

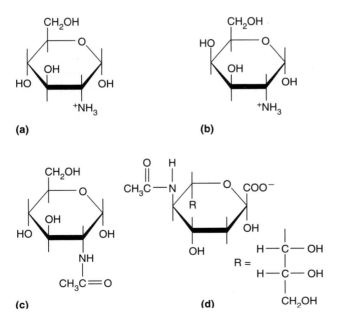

(a)   **(b)**

(c)   **(d)**

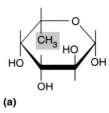

(a)

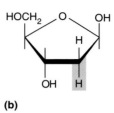

(b)

FIGURE 7.24

Deoxy Sugars.

(a) β-L-Fucose (6-deoxygalactose) and
(b) 2-deoxy-β-D-ribose. The carbon atoms
that have —OH groups replaced by —H
are highlighted.

constituents of the complex carbohydrate molecule found attached to cellular proteins and lipids. The most common amino sugars of animal cells are D-glucosamine and D-galactosamine. Amino sugars are often acetylated. One such molecule is N-acetyl-glucosamine. N-acetyl-neuraminic acid (the most common form of sialic acid) is a condensation product of D-mannosamine and pyruvic acid, a 2-ketocarboxylic acid. Sialic acids are ketoses containing nine carbon atoms that may be amidated with acetic or glycolic acid (hydroxyacetic acid). They are common components of glycoproteins and glycolipids.

DEOXYSUGARS Monosaccharides in which an —H has replaced an —OH group are known as *deoxysugars*. Two important deoxysugars found in cells are L-fucose (formed from D-mannose by reduction reactions) and 2-deoxy-D-ribose (Figure 7.24). Fucose is often found among the carbohydrate components of glycoproteins, such as those of the ABO blood group determinates on the surface of red blood cells. As was mentioned in Section 1.2, 2-deoxyribose is the pentose sugar component of DNA.

7.2 DISACCHARIDES AND OLIGOSACCHARIDES

When they are linked together by glycosidic bonds, monosaccharides form a variety of molecules that perform diverse biological functions. Disaccharides are glycosides composed of two monosaccharides. The term **oligosaccharide** is often used for relatively small sugar polymers that consist of two to ten or more monosaccharide units.

If one monosaccharide molecule is linked through its anomeric carbon atom to the hydroxyl group on carbon 4 of another monosaccharide, the glycosidic linkage is designated as 1,4. Because the anomeric hydroxyl group may potentially be in either the α- or β-configuration, two possible disaccharides may form when two sugar molecules are linked: $\alpha(1,4)$ or $\beta(1,4)$. Other varieties of glycosidic linkages (i.e., α or β(1,1), (1,2), (1,3), and (1,6) linkages) also occur (Figure 7.25).

Digestion of disaccharides and other carbohydrates is mediated by enzymes synthesized by cells lining the small intestine. Deficiency of any of these enzymes causes unpleasant symptoms when the undigestible disaccharide sugar is ingested. Because carbohydrates are absorbed principally as monosaccharides, any undigested dissacharide molecules pass into the large intestine, where osmotic pressure draws water from the surrounding tissues (diarrhea). Bacteria in the colon digest the disaccharides (fermentation), thus producing gas (bloating and cramps). The most commonly known deficiency is *lactose intolerance*, which may occur in most human adults except those with ancestors from northern Europe and/or certain African groups. Caused by the greatly reduced synthesis of the enzyme lactase following childhood, lactose intolerance is treated by eliminating the sugar from the diet or (in some cases) by treating food with the enzyme lactase.

FIGURE 7.25

Glycosidic Bonds.

Several types of glycosidic bonds can form between monosaccharides. The sugar α-D-glucopyranose (shown on the left) can theoretically form glycosidic linkages with any of the alcoholic functional groups of another monosaccharide, in this case another molecule of α-D-glucopyranose.

Lactose (milk sugar) is a disaccharide found in milk. It is composed of one molecule of galactose linked through the hydroxyl group on carbon 1 in a β-glycosidic linkage to the hydroxyl group of carbon 4 of a molecule of glucose (Figure 7.26). Because the anomeric carbon of galactose is in the β-configuration, the linkage between the two monosaccharides is designated as β(1,4). The inability to hydrolyze the β(1,4) linkage (caused by lactase deficiency) is common among animals, so carbohydrates with such linkages (such as cellulose) cannot be digested. Because the glucose component contains a hemiacetal group, lactose is a reducing sugar.

Maltose, also known as malt sugar, is an intermediate product of starch hydrolysis and does not appear to exist freely in nature. Maltose is a disaccharide with an α(1,4) glycosidic linkage between two D-glucose molecules. In solution the free anomeric carbon undergoes mutarotation, which results in an equilibrium mixture of α- and β-maltoses (Figure 7.27).

Cellobiose, a degradation product of cellulose, contains two molecules of glucose linked by a β(1,4) glycosidic bond (Figure 7.28). Like maltose, whose structure is identical except for the direction of the glycosidic bond, cellobiose does not freely exist in nature.

Sucrose (common table sugar: cane sugar or beet sugar) is produced in the leaves and stems of plants. It is a transportable energy source throughout the entire plant. Containing both α-glucose and β-fructose residues, sucrose differs from the previously described disaccharides in that the monosaccharides are linked through a glycosidic bond between both anomeric carbons (Figure 7.29). Because neither monosaccharide ring can revert to the open chain form, sucrose is a nonreducing sugar.

Oligosaccharides are small polymers most often found attached to polypeptides in glycoproteins (pp. 225–228) and some glycolipids (Chapter 11). Among the best-characterized oligosaccharide groups are those attached to membrane and secretory proteins found in the endoplasmic reticulum and Golgi complex of various cells. There are two broad classes of oligosaccharides: N-linked and O-linked. N-linked oligosaccharides are attached to polypeptides by an N-glycosidic bond with the side chain amide group of the amino acid asparagine. There are three major types of asparagine-linked oligosaccharides: high-mannose, hybrid, and complex (Figure 7.30). O-linked oligosaccharides are attached to

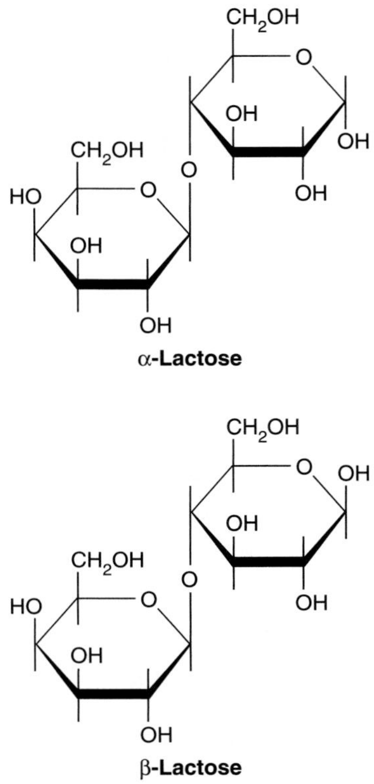

FIGURE 7.26

α- and β-Lactose.

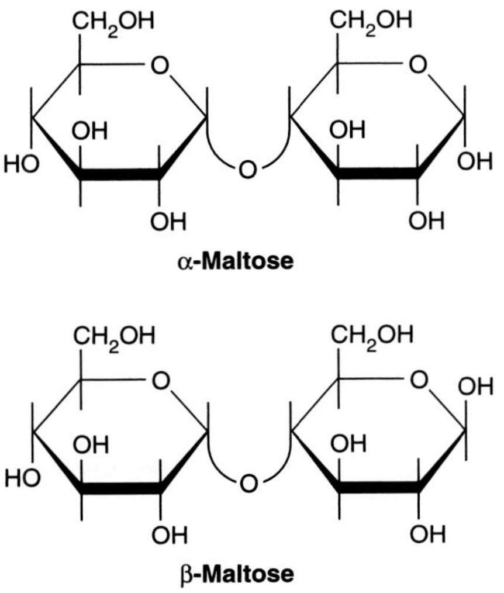

FIGURE 7.27

α- and β-Maltose.

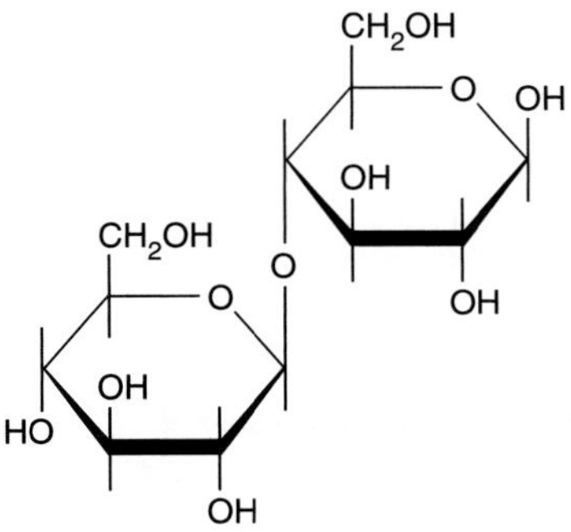

FIGURE 7.28

β-Cellobiose.

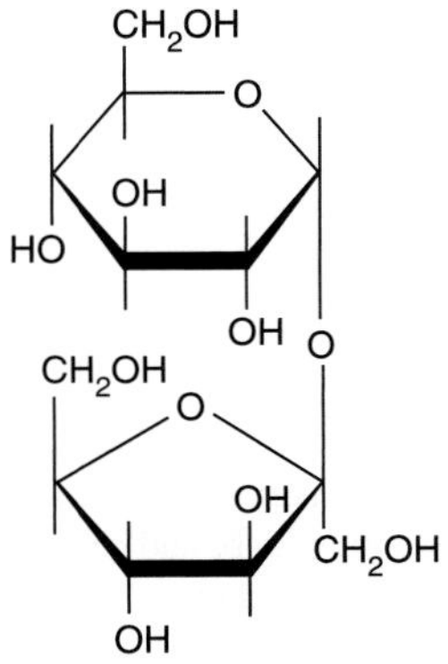

FIGURE 7.29

Sucrose.

The glucose and fructose residues are linked by an α, β(1,2) glycosidic bond.

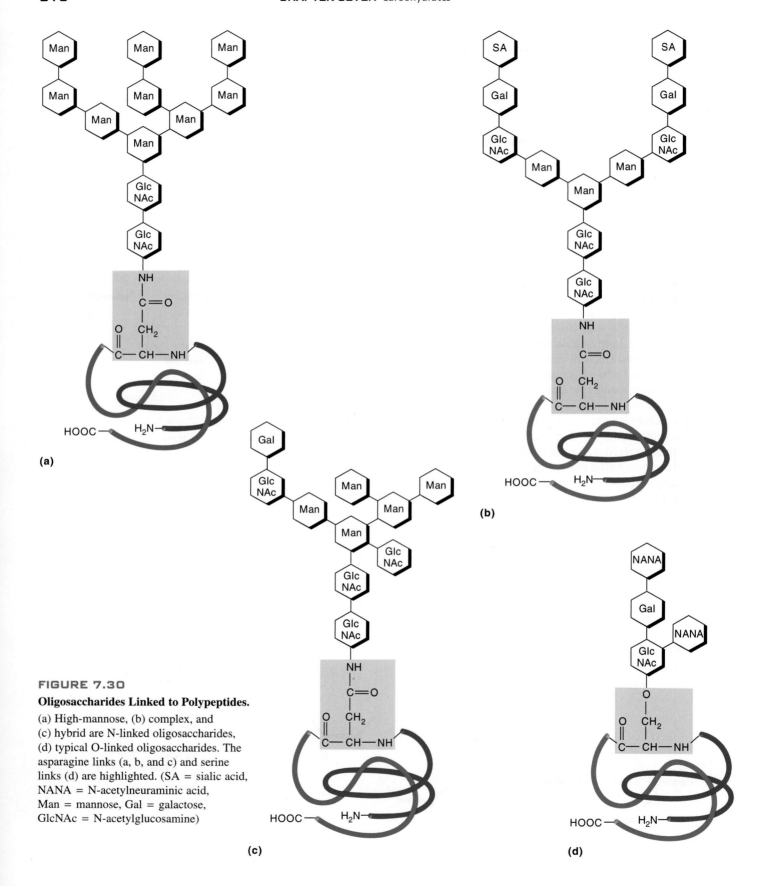

FIGURE 7.30

Oligosaccharides Linked to Polypeptides.

(a) High-mannose, (b) complex, and
(c) hybrid are N-linked oligosaccharides,
(d) typical O-linked oligosaccharides. The
asparagine links (a, b, and c) and serine
links (d) are highlighted. (SA = sialic acid,
NANA = N-acetylneuraminic acid,
Man = mannose, Gal = galactose,
GlcNAc = N-acetylglucosamine)

polypeptides by the side chain hydroxyl group of the amino acids serine or threonine in polypeptide chains or the hydroxyl group of membrane lipids.

QUESTION 7.7

Which of the following sugars or sugar derivatives are reducing sugars?

(a) glucose
(b) fructose
(c) α-methyl-D-glucoside
(d) sucrose

Which of the above compounds are capable of mutarotation?

7.3 POLYSACCHARIDES

KEY CONCEPTS 7.3

Disaccharides are glycosides composed to two monosaccharide units. Examples of disaccharides are maltose, lactose, cellobiose, and sucrose. Oligosaccharides contain two to ten monosaccharide units.

Polysaccharide molecules are used as storage forms of energy or as structural materials. They are composed of large numbers of monosaccharide units connected by glycosidic linkages. Most common polysaccharides are large molecules containing from hundreds to thousands of sugar units. These molecules may have a linear structure like that of cellulose or amylose, or they may have branched shapes like those found in glycogen and amylopectin. Unlike nucleic acids and proteins, which have specific molecular weights, the molecular weights of many polysaccharides have no fixed values. The size of such molecules is a reflection of the metabolic state of the cell producing them. For example, when blood sugar levels are high (e.g., after a meal), the liver synthesizes glycogen. Glycogen molecules in a well-fed animal may have molecular weights as high as 2×10^7 D. When blood sugar levels fall, the liver enzymes begin breaking down the glycogen molecules, releasing glucose into the blood stream. If the animal continues to fast, the process continues until glycogen reserves are almost used up.

Polysaccharides may be divided into two classes: *homopolysaccharides*, which are composed of one type of monosaccharide, and *heteropolysaccharides*, which contain two or more types of monosaccharides.

Homopolysaccharides

The homopolysaccharides found in abundance in nature are starch, glycogen, cellulose, and chitin. Starch, glycogen, and cellulose all yield D-glucose when they are hydrolzyed. Starch and glycogen are glucose storage molecules in plants and animals, respectively. Cellulose is the primary structural component of plant cells. **Chitin**, the principal structural component of the exoskeletons of arthropods such as insects and crustaceans and the cell walls of many fungi, yields the glucose derivative N-acetyl glucosamine when it is hydrolyzed.

STARCH Starch, the energy reservoir of plant cells, is a significant source of carbohydrate in the human diet. Much of the nutritional value of the world's major foodstuffs (e.g., potatoes, rice, corn, and wheat) comes from starch. Two polysaccharides occur together in starch: amylose and amylopectin.

Amylose is composed of long, unbranched chains of D-glucose residues that are linked with α(1,4) glycosidic bonds (Figure 7.31). A number of polysaccharides, including both types of starch, have one *reducing end* in which the ring can open to form a free aldehyde group with reducing properties. The internal anomeric carbons in these molecules are involved in acetal linkages and are not free to act as reducing agents.

Amylose molecules, which typically contain several thousand glucose residues, vary in molecular weight from 150,000 to 600,000 D. Because the linear amylose

FIGURE 7.31

Amylose.

(a) The D-glucose residues of amylose are linked through $\alpha(1,4)$ glycosidic bonds. (b) The amylose polymer forms a left-handed helix.

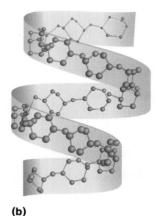

(a)   **(b)**

molecule forms long, tight helices, its compact shape is ideal for its storage function. The common iodine test for starch works because molecular iodine inserts itself into these helices. (The intense blue color of a positive test comes from electronic interactions between iodine molecules and the helically arranged glucose residues of the amylose.)

The other form of starch, **amylopectin**, is a branched polymer containing both $\alpha(1,4)$ and $\alpha(1,6)$ glycosidic linkages. The $\alpha(1,6)$ branch points may occur every 20–25 glucose residues and prevent helix formation (Figure 7.32a). The number of glucose units in amylopectin may vary from a few thousand to a million.

Starch digestion begins in the mouth, where the salivary enzyme α-amylase initiates hydrolysis of the glycosidic linkages. Digestion continues in the small intestine, where pancreatic α-amylase randomly hydrolyses all the $\alpha(1,4)$ glycosidic bonds except those next to the branch points. The products of α-amylase are maltose, the trisaccharide maltotriose, and the α-limit dextrins (oligosaccharides that typically contain eight glucose units with one or more $\alpha(1,6)$ branch points). Several enzymes secreted by cells that line the small intestine convert these intermediate products into glucose. Glucose molecules are then absorbed into intestinal cells. After passage into the bloodstream, they are transported to the liver and then to the rest of the body.

GLYCOGEN Glycogen is the carbohydrate storage molecule in vertebrates. It is found in greatest abundance in liver and muscle cells. (Glycogen may make up as much as 8–10% of the wet weight of liver cells and 2–3% of that of muscle cells.) Glycogen (Figure 7.32b) is similar in structure to amylopectin except that it has more branch points, possibly at every fourth glucose residue in the core of the molecule. In the outer regions of glycogen molecules, branch points are not so close together (approximately every 8–12 residues). Because the molecule is more compact than other polysaccharides, it takes up little space, an important consideration in mobile animal bodies. The many nonreducing ends in glycogen molecules allow for stored glucose to be rapidly mobilized when the animal demands much energy.

QUESTION 7.8

It has been estimated that two high-energy phosphate bonds must be expended to incorporate one glucose molecule into glycogen. Why is glucose stored in muscle and liver in the form of glycogen, not as individual glucose molecules? In other words, why is it advantageous for a cell to expend metabolic energy to polymerize glucose molecules? (*Hint*: Besides the reasons given above, refer to Chapter 3 for another problem that glucose polymerization solves.)

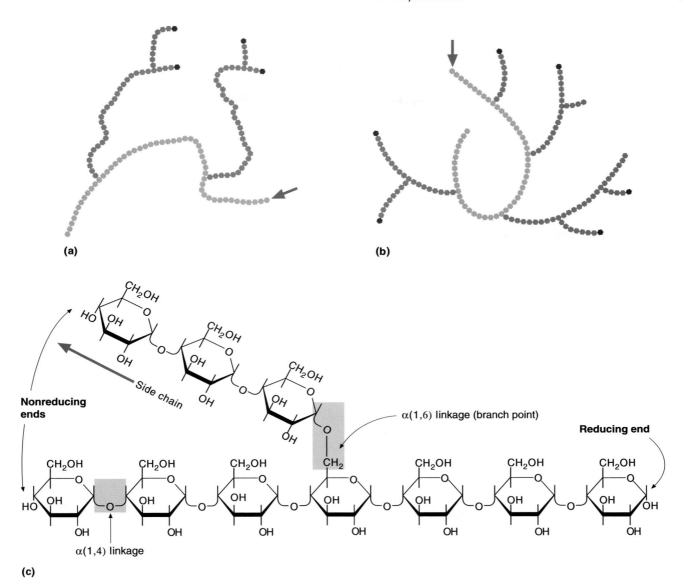

FIGURE 7.32

Amylopectin (a) and Glycogen (b).

Each hexagon represents a glucose molecule. Notice that each molecule has only one reducing end (arrow) and numerous nonreducing ends. (c) Detail from (a) or (b).

CELLULOSE **Cellulose** is a polymer composed of D-glucopyranose residues linked by $\beta(1,4)$ glycosidic bonds (Figure 7.33). It is the most important structural polysaccharide of plants. Because cellulose comprises about one-third of plant biomass, it is the most abundant organic substance on earth. Approximately 100 trillion kg of cellulose are produced each year.

Pairs of unbranched cellulose molecules, which may contain as many as 12,000 glucose units each, are held together by hydrogen bonding to form sheetlike strips called *microfibrils* (Figure 7.34). Each bundle of microfibrils contains approximately 40 of these pairs. These structures are found in both plant primary and secondary cell walls, where they provide a structural framework that both protects and supports cells.

The ability to digest cellulose is found only in microorganisms that possess the enzyme cellulase. Certain animal species (e.g., termites and cows) use such organisms in their digestive tracts to digest cellulose. The breakdown of the cellulose

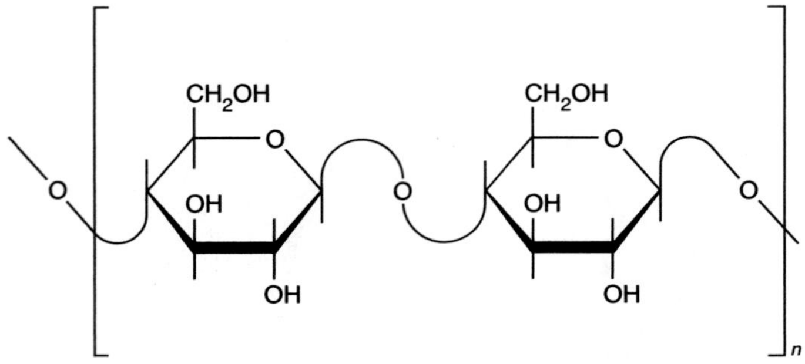

FIGURE 7.33
Cellulose.

FIGURE 7.34

Cellulose Microfibrils.

Intermolecular hydrogen bonds between adjacent cellulose molecules are largely responsible for the great strength of cellulose.

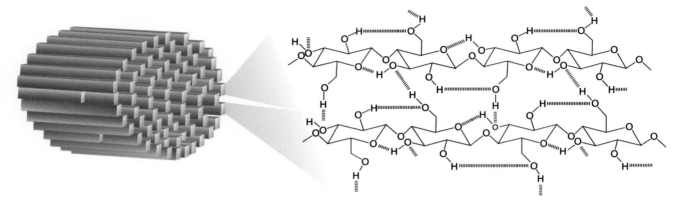

makes glucose available to both the micoorganisms and their host. Although many animals cannot digest cellulose-containing plant materials, these substances play a vital role in nutrition. Cellulose is one of several plant products that make up the dietary fiber that is now believed to be important for good health.

Because of its structural properties, cellulose has enormous economic importance. Products such as wood, paper, and textiles (e.g., cotton, linen, and ramie) owe many of their unique characteristics to their cellulose content.

CHITIN The structure (Figure 7.35) and function of chitin are similar to those of cellulose. As with cellulose, the monomeric units (in this case N-acetyl-glucosamine) are linked in unbranched chains by $\beta(1,4)$ glycosidic bonds. Microfibrils are formed from adjacent chitin molecules that are strongly hydrogen-bonded together. Unlike cellulose, in which the chains are arranged

FIGURE 7.35

Chitin.

Chitin is composed of N-acetylglu-cosamine residues.

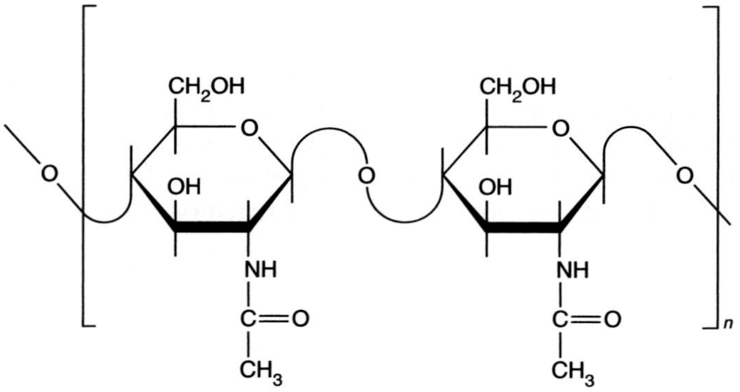

Weaving plant fibers is a relatively recent development in human history, apparently for technological reasons. Before our ancestors learned to spin or weave, they discovered which plants contained useful fibers and how these fibers could be extracted. According to archaeological evidence, one of the earliest plants used for fiber was flax (*Linum usitatissimum*). It was woven into linen at least 8000 years ago. The cultivation and uses of flax were beautifully illustrated on the walls of Egyptian tombs (Figure 7B).

The chemical properties of cellulose contribute to the qualities that make linen such an attractive fabric (i.e., its smoothness, strength, and water absorbency). Bast fibers, found in the phloem (a component of the plant's vascular system) are used to make linen. They contain thicker cell walls than most of the other plant tissues. In a chemical process called *retting*, bast fibers are harvested after the rest of the plant is decomposed by bacteria. Because of the large amount of cellulose within their cell walls, bast fibers can withstand the numerous corrosive chemical reactions of decomposition.

FIGURE 7B

Harvesting Flax.

From a wall painting in the tomb of Sennedjem, Deir-el-Medina, twentieth dynasty, about 1150 B.C.

in parallel bundles, chitin occurs in three types of hydrogen-bonded microfibrils: α-chitin, β-chitin, and γ-chitin. In the most abundant and stable form, α-chitin, the chains are arranged in antiparallel arrays. In the more flexible β- and γ-forms, the chains are parallel or a mixture of parallel and antiparallel, respectively.

Heteropolysaccharides

Heteropolysaccharides are high-molecular-weight carbohydrate polymers that contain more than one kind of monosaccharide. Important examples include the **glycosaminoglycans** (GAGs), the principal components of proteoglycans (Section 7.4), and murein, a major component of bacterial cell walls.

GLYCOSAMINOGLYCANS GAGs are linear polymers with disaccharide repeating units (Table 7.1). Many of the sugar residues are amino derivatives.

The repeating units contain a hexuronic acid (a uronic acid containing six carbon atoms), except for keratan sulfate, which contains galactose. Usually an N-acetylhexosamine sulfate is also present, except in hyaluronic acid, which contains N-acetylglucosamine. Many disaccharide units contain both carboxyl and sulfate functions groups. GAGs are classified according to their sugar residues, the linkages between these residues, and the presence and location of sulfate groups. Five classes are distinguished: *hyaluronic acid, chondroitin sulfate, dermatan sulfate, heparin* and *heparan sulfate*, and *keratan sulfate*.

TABLE 7.1

Disaccharide Repeating Units in Selected Glycosaminoglycans

Name	Repeating Unit	Molecular Weight (D)	Comments
Chondroitin sulfate	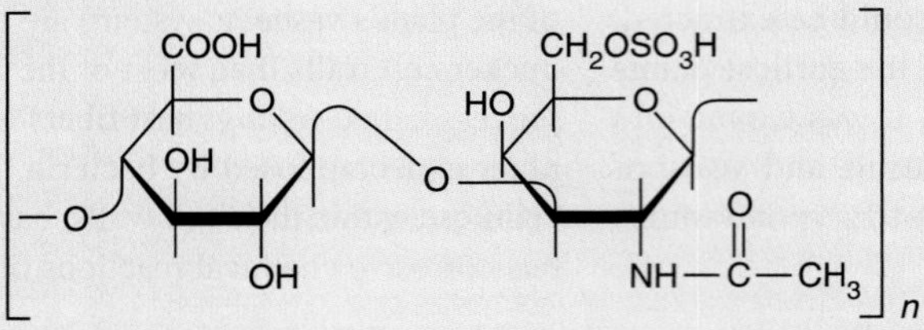 (1,4)-O- β-D-Glucopyranosyluronic acid-(1,3)-2-acetamido-2-deoxy-6-O-sulfo-β-D-galactopyranose	5000–50,000	May also have sulfate on carbon 6. Important component of cartilage.
Dermatan sulfate	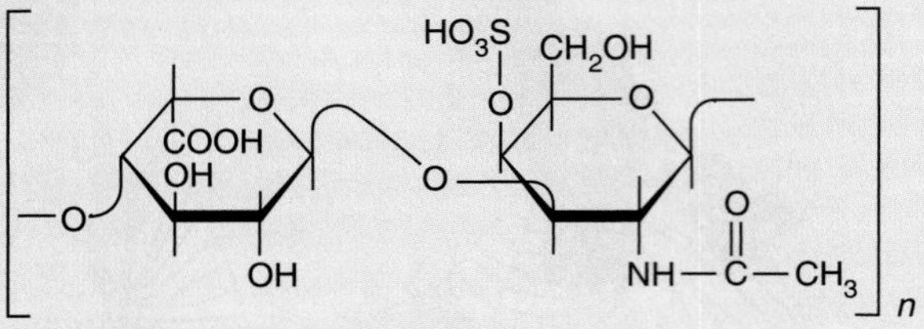 (1,4)-O-α-L-Idopyranosyluronic acid-(1,3)-2-acetamido-2-deoxy-4-O-sulfo-β-D-galactopyranose	15,000–40,000	Varying amounts of D-glucuronic acid may be present. Concentration increases during aging process.
Heparin	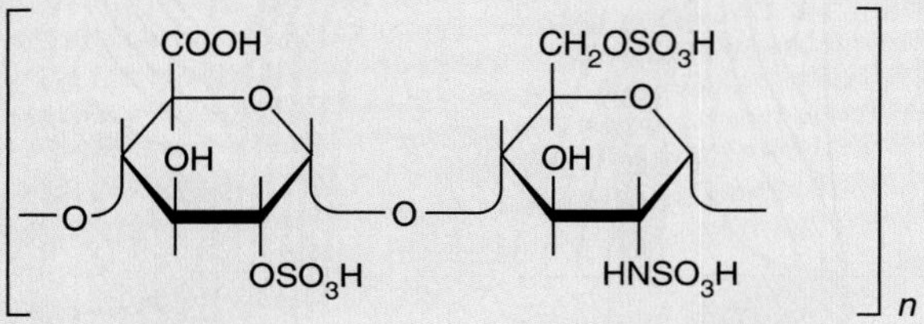 (1,4)-O-α-D-Glucopyranosyluronic acid-2-sulfo-(1,4)-2-sulfamido-2-deoxy-6-O-sulfo-α-D-glucopyranose	6000–25,000	Anticoagulant activity. Found in mast cells. Also contains D-glucuronic acid.
Keratan sulfate	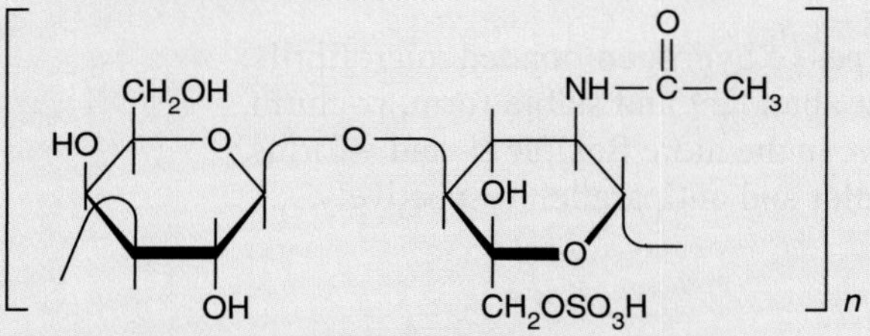 (1,3)-O-β-D-Galactopyranose-(1,4)-2-acetamido-2-deoxy-6-O-sulfo-β-D-glucopyranose	4000–19,000	Minor constituent of proteoglycans. Found in cornea, cartilage, and intervertebral disks.
Hyaluronic acid	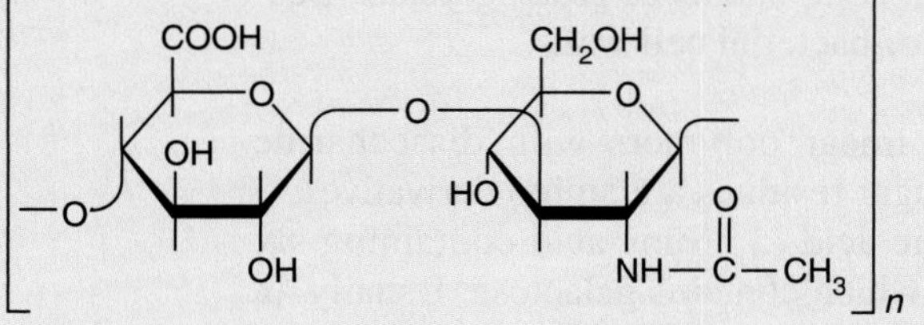 (1,4)-O-β-D-Glucopyranosyluronic acid-(1,3)-2-acetamido-2-deoxy-β-D-glucopyranose	4000	Most abundant GAG in the vitreous humor of the eye and the synovial fluid of joints.

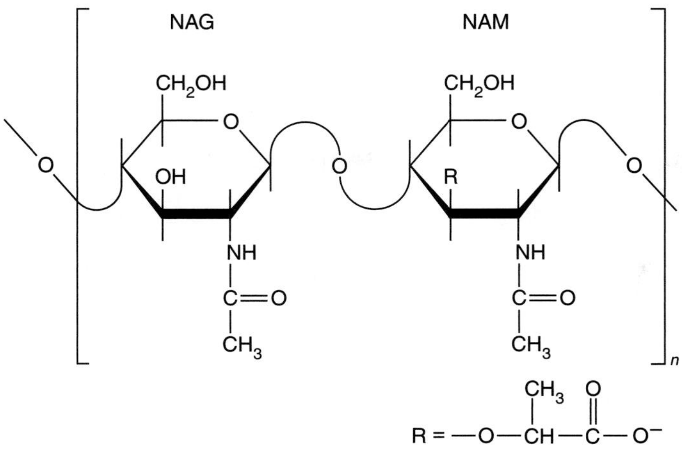

FIGURE 7.36
Disaccharide Repeating Unit of Murein (Peptidoglycan).
(NAG = N-acetylglucosamine, NAM = N-acetyl-muramic acid)

GAGs have many negative charges at physiological pH. The charge repulsion keeps GAGs separated from each other. Additionally, the relatively inflexible polysaccharide chains are strongly hydrophilic. GAGs occupy a huge volume relative to their mass because they attract large volumes of water. For example, hydrated hyaluronic acid may occupy a volume 1000 times greater than its dry state.

MUREIN **Murein** (Figure 7.36), also referred to as peptidoglycan, is a complex polymer that is the major structural feature of the cell walls of all bacteria. It contains two sugar derivatives: N-acetyl-glucosamine and N-acetyl-muramic acid [N-acetyl-glucosamine bonded to lactic acid ($CH_3CH(OH)COOH$) by an ether linkage], and several different amino acids (some of which are D-isomers). Murein consists of three basic components: a backbone composed of disaccharide repeating units linked by $\beta(1,4)$ glycosidic bonds, parallel tetrapeptide chains, each of which is attached to N-acetyl-muramic acid, and a series of peptide cross-bridges that form between the tetrapeptide chains of parallel polysaccharide backbones (Figure 7.37). Murein is largely responsible for the shape and the rigidity of bacterial cell walls.

KEY CONCEPTS 7.4

Polysaccharide molecules, composed of large numbers of monosaccharide units, are used in energy storage and as structural materials.

7.4 GLYCOCONJUGATES

The compounds that result from the covalent linkages of carbohydrate molecules to both proteins and lipids are collectively known as the **glycoconjugates**. These substances have profound effects on the function of individual cells, as well as the cell-cell interactions of multicellular organisms. There are two classes of carbohydrate-protein conjugate: proteoglycans and glycoproteins. Although both molecular types contain carbohydrate and protein, their structures and functions appear, in general, to be substantially different. The *glycolipids*, which are oligosaccharide-containing lipid molecules, are found predominantly on the outer surface of plasma membranes. A discussion of their structure is deferred until Chapter 11.

Proteoglycans

Proteoglycans are distinguished from the more common glycoproteins by their extremely high carbohydrate content, which may constitute as much as 95% of the dry weight of such molecules. These molecules are found predominantly in the extracellular matrix (intercellular material) of tissues. All proteoglycans contain GAG chains. The GAG chains are linked to protein molecules (known as *core proteins*) by N- and O-glycosidic linkages. The diversity of proteoglycans results

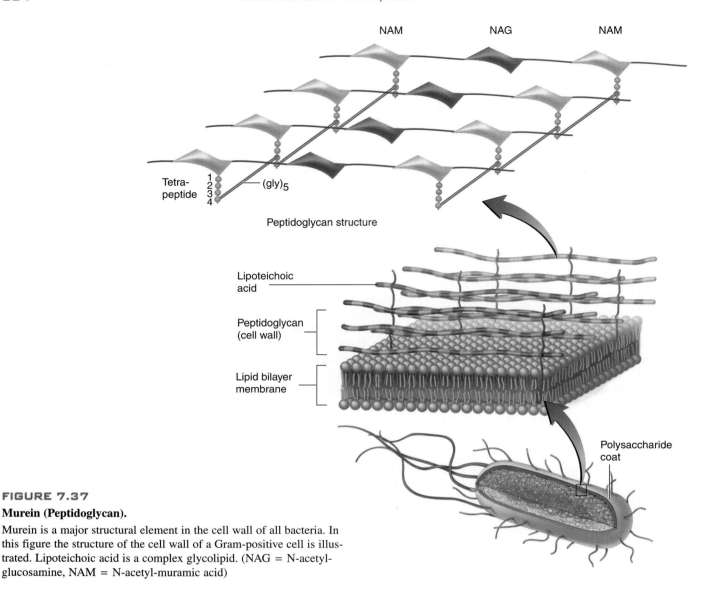

NAM   NAG   NAM

Tetra-peptide
1
2
3
4
——(gly)₅

Peptidoglycan structure

Lipoteichoic acid

Peptidoglycan (cell wall)

Lipid bilayer membrane

Polysaccharide coat

FIGURE 7.37

Murein (Peptidoglycan).

Murein is a major structural element in the cell wall of all bacteria. In this figure the structure of the cell wall of a Gram-positive cell is illustrated. Lipoteichoic acid is a complex glycolipid. (NAG = N-acetyl-glucosamine, NAM = N-acetyl-muramic acid)

from both the number of different core proteins and the large variety of classes and lengths of the carbohydrate chains (Figure 7.38).

Because proteoglycans contain large numbers of GAGs, which are polyanions, large volumes of water and cations are trapped within their structure. Proteoglycan molecules occupy thousands of times as much space as a densely packed molecule of the same mass. Proteoglycans contribute support and elasticity to tissues. Consider, for example, the strength, flexibility, and resilience of cartilage. The structural diversity of proteoglycans allows them to play a variety of structural and functional roles in living organisms. Together with matrix proteins such as collagen, fibronectin, and laminin, they form an organized meshwork that provides strength and support to multicellular tissues. Proteoglycans are also present at the surface of cells, where they are directly bound to the plasma membrane. Although the function of these latter molecules is not yet clear, they may play an important role in membrane structure and cell-cell interactions.

A number of genetic diseases associated with proteoglycan metabolism, known as *mucopolysaccharidoses*, have been identified. Because proteoglycans are con-

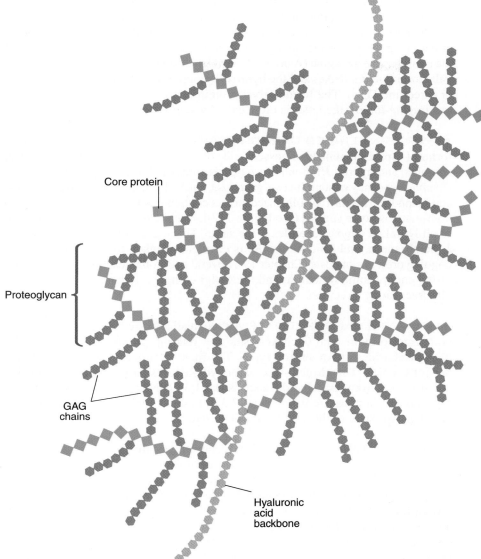

Core protein

Proteoglycan

GAG
chains

Hyaluronic
acid
backbone

FIGURE 7.38

Proteoglycan.

Proteoglycan aggregates are typically found in the extracellular matrix of connective tissue. The noncovalent attachment of each proteoglycan to hyaluronic acid via the core protein is mediated by two linker proteins (not shown). Proteoglycans interact with numerous fibrous proteins in the extracellular matrix such as collagen, elastin, and fibronectin (a glycoprotein involved in cell adhesion).

stantly being synthesized and degraded, their excessive accumulation (caused by missing or defective lysosomal enzymes) has very serious consequences. For example, in *Hurler's syndrome*, an autosomal recessive disorder (a disease in which one copy of the defective gene is inherited from each parent), deficiency of a specific enzyme causes dermatan sulfate to accumulate. Symptoms include mental retardation, skeletal deformity, and death in early childhood.

Glycoproteins

Glycoproteins are commonly defined as proteins that are covalently linked to carbohydrate through N- or O-linkages. The carbohydrate composition of glycoprotein varies from 1% to more than 85% of total weight. The carbohydrates found include monosaccharides and disaccharides such as those attached to the structural protein collagen and the branched oligosaccharides on plasma glycoproteins. Although the glycoproteins are sometimes considered to include the proteoglycans, structural reasons seem to be sufficient to examine them separately, such as the relative absence in glycoproteins of uronic acids, sulfate groups, and the disaccharide repeating units that are typical of proteoglycans.

The carbohydrate groups of glycoproteins are linked to the polypeptide by either (1) an N-glycosidic linkage between N-acetylglucosamine (GlcNAc) and the amino acid asparagine (Asn) or (2) an O-glycosidic linkage between N-acetyl-galactosamine (GalNAc) and the hydroxyl group of the amino acids serine (Ser) or threonine (Thr). The former glycoprotein class is sometimes referred to as *asparagine-linked*; the latter is often called *mucin-type*.

ASPARAGINE-LINKED CARBOHYDRATE As mentioned, three structural forms of asparagine-linked oligosaccharide occur in glycoproteins: high-mannose, complex, and hybrid. *High-mannose type* is composed of GlcNAc and mannose. *Complex type* may contain fucose, galactose, and sialic acid in addition to GlcNAc and mannose. *Hybrid-type* oligosaccharides contain features of both complex and high-mannose type species (see Figure 7.30). Despite these differences, the core of all N-linked oligosaccharides is the same. This core, which is constructed on a membrane-bound lipid molecule, is covalently linked to asparagine during ongoing protein synthesis (Chapter 19). Several additional reactions, in the lumen of the endoplasmic reticulum and the Golgi complex, form the final N-linked oligosaccharide structures.

MUCIN-TYPE CARBOHYDRATE Although all N-linked oligosaccharides are bound to protein via GlcNAc-Asn, the linking groups of O-glycosidic oligosaccharides are of several types. The most common of these is GalNAc-Ser (or GalNAc-Thr). Mucin-type carbohydrate units vary considerably in size and structure, from disaccharides such as Gal-1,3-GalNAc, found in the antifreeze glycoprotein of antarctic fish (Figure 7.39), to the complex oligosaccharides of blood groups such as those of the ABO system.

GLYCOPROTEIN FUNCTIONS Glycoproteins are a diverse group of molecules that are ubiquitous constituents of most living organisms (Table 7.2). They occur in cells, in both soluble and membrane-bound form, and in extracellular fluids. Vertebrate animals are particularly rich in glycoproteins. Examples of such substances include the metal-transport proteins transferrin and ceruloplasmin, the blood-clotting factors, and many of the components of complement (proteins involved in cell destruction during immune reactions). A number of hormones (chemicals produced by certain cells that are transported by blood to other cells, where they exert regulatory effects) are glycoproteins.

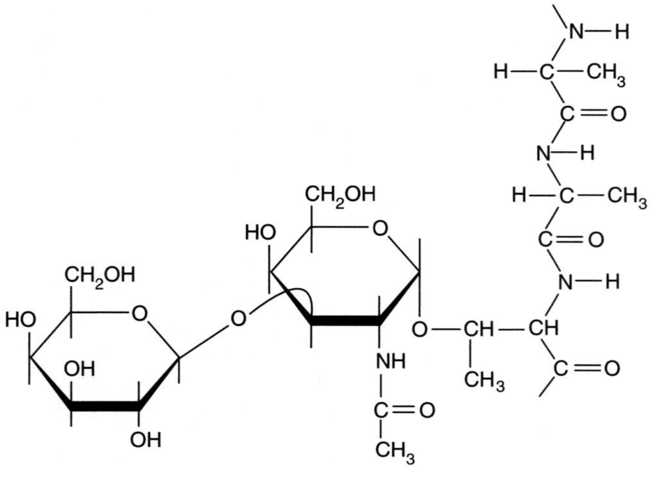

FIGURE 7.39

Antifreeze Glycoprotein.

This segment is a recurring glyco-tripeptide unit. Each disaccharide unit, composed of β-1,3 linked residues of galactose and N-acetylgalactosamine, is attached to the polypeptide chain by a glycosidic linkage to a threonine residue.

TABLE 7.2
Glycoproteins

Type	Example	Source	Molecular Weight (D)	Percent Carbohydrate
Enzyme	Ribonuclease B	Bovine	14,700	8
Immunoglobulin	IgA	Human	160,000	7
	IgM	Human	950,000	10
Hormone	Chorionic gonadotropin	Human placenta	38,000	31
	FSH	Human	34,000	20
Membrane protein	Glycophorin	Human RBC	31,000	60
Lectin (carbohydrate-binding proteins)	Potato lectin	Potato	50,000	50

Consider, for example, follicle-stimulating hormone (FSH), produced by the anterior pituitary gland. FSH stimulates the development of both eggs and sperm. Additionally, many enzymes are glycoproteins. Ribonuclease (RNase), the enzyme that degrades ribonucleic acid, is a well-researched example. Other glycoproteins are integral membrane proteins (Chapter 11). Of these, Na+-K+-ATPase (an ion pump found in the plasma membrane of animal cells) and the major histocompatibility antigens (cell surface markers used to cross-match organ donors and recipients) are especially interesting examples.

Although many glycoproteins have been studied, the role of carbohydrate is still not clearly understood. Despite challenging technical problems, some progress has been made in discerning how the carbohydrate component contributes to biological activity. Recent research has focused on how carbohydrate stabilizes protein molecules and functions in recognition processes in multicellular organisms.

The presence of carbohydrate on protein molecules protects them from denaturation. For example, bovine RNase A is more susceptible to heat denaturation than its glycosylated counterpart RNase B. Several other studies have shown that sugar-rich glycoproteins are relatively resistant to proteolysis (splitting of polypeptides by enzyme-catalyzed hydrolytic reactions). Because the carbohydrate is on the molecule's surface, it may shield the polypeptide chain from proteolytic enzymes.

The carbohydrates in glycoproteins seem to affect biological function. In some glycoproteins this contribution is more easily discerned than in others. For example, a large content of sialic acid residues is responsible for the high viscosity of salivary mucins (the lubricating glycoproteins of saliva). Another interesting example is the antifreeze glycoproteins of antarctic fish. Apparently, their disaccharide residues form hydrogen bonds with water molecules. This process retards the growth of ice crystals.

Glycoproteins are now known to be important in complex recognition phenomena such as cell-molecule, cell-virus, and cell-cell interactions. Prime examples of glycoprotein involvement in cell-molecule interactions include the insulin receptor, whose binding to insulin facilitates the transport of glucose into numerous cell types. It does so, in part, by recruiting glucose transporters to the plasma membrane. In addition, the glucose transporter that is directly responsible for transporting the sugar into cells is also a glycoprotein. The interaction between gp120, the target cell binding glycoprotein of HIV (the AIDS virus), and host cells is a fascinating example of cell-virus interaction. The attachment of gp120 to the CD4 receptor found on the surface of several human cell types is now considered to be an early step in the infective process. Removal of carbohydrate from purified gp120 significantly reduces the binding of the viral protein to the CD4 receptor. Paradoxically, the oligosaccharides attached to the glycoprotein on the surface of vesicular stomatitis virus are not required for viral infectivity.

Cell structure glycoproteins, components of the *glycocalyx* (also known as the *cell coat*), are now recognized as players in cellular adhesion. This process is a critical event in the cell-cell interactions of growth and differentiation (Figure 7.40). The best-characterized of these substances are called cell adhesion molecules (CAMs). CAMs are now believed to be involved in the embryonic development of the mouse nervous system. Sialic acid residues in the N-linked oligosaccharides of several CAMs have been shown to be important in this phenomenon.

Recent improvements in technology have led to an increased appreciation of the importance of carbohydrate in glycoproteins. Consequently, there is currently a heightened interest in studying cellular glycosylation patterns. Carbohydrate structure is now used to investigate normal processes such as nerve development and certain disease processes. For example, changes in the galactose content of the antibody class IgG have recently been shown to be directly related to the severity (i.e., the degree of inflammation) of juvenile arthritis. Additionally, recent evidence indicates that changes in glycosylation patterns accompany changes in the behavior of cancer cells. This knowledge is currently making tumor detection and the metastatic process (the spread of cancerous cells from a tumor to other body parts) more accessible to investigation. The role of glycoconjugates in living processes is explored further in Special Interest Box 7.3.

KEY CONCEPTS 7.5

Glycoconjugates are biomolecules in which carbohydrate is covalently linked to either proteins or lipids. Proteoglycans are composed of relatively large amounts of carbohydrate (GAG units) covalently linked to small polypeptide components. Glycoproteins are proteins covalently linked to carbohydrate through N- or O-linkages.

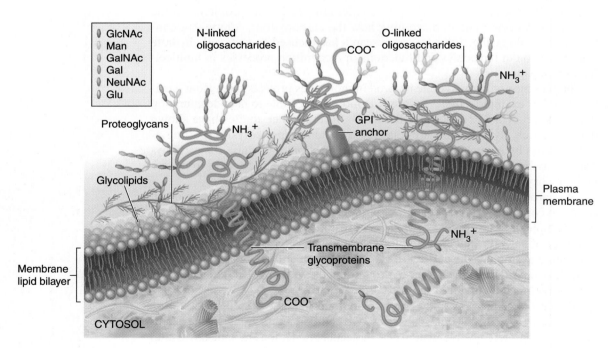

FIGURE 7.40

The Glycocalyx.

A GPI (glycosylphosphatidylinositol) anchor is a specialized structure that attaches several diverse types of oligosaccharide to the plasma membrane of some eukaryotic cells.

Biological information transfer occurs as one molecule, typically a protein, binds to another molecule with a complementary shape, thereby changing both of their three-dimensional conformations. The complex signaling pathways involved in biological processes are made possible by serial information transfer; that is, information is transferred from molecule to molecule to achieve a specific result. For example, in a process called *chemotaxis*, a motile cell such as the bacterium *E. coli* senses a certain nutrient molecule when the latter binds to a plasma membrane receptor. Once the receptor is activated by the ligand, the signal is transferred across the cell membrane by conformational changes. A rapid cascade of intracellular reactions is then initiated that alters the rotation direction of the cell's flagellar motors so that the bacterium can efficiently move toward higher concentrations of the nutrient. Living organisms require extraordinarily large coding capacities because each information transfer event, whether it is the conversion of a substrate to product within an enzyme's active site, the transduction of a hormonal signal, or the engulfment of a bacterial cell by a macrophage, is initiated by the specific binding of one unique molecule by another that has been selected from millions of other nearby molecules. In other words, the functioning of systems as profoundly complicated as living organisms requires a correspondingly large repertoire of molecular codes. To succeed as a coding mechanism, a class of molecules must provide a large capacity for variations in shape because of the tremendous number of different messages that must be quickly and unambiguously deciphered. Recall, for example, that the enormous diversity of proteins observed in living organisms is composed of just 20 amino acids. The total number of hexapeptides that can be synthesized from these amino acids is an impressive 6.4×10^7. Despite their longstanding reputation as information-poor and repetitive structures such as glycogen and cellulose, carbohydrates, as a class of biomolecules, have structural properties (e.g., glycosidic linkage variations, branching, and anomeric isomers) that provide them with significant coding capacity. In contrast to the peptide linkages that form exclusively between the amino and carboxyl groups in amino acid residues to create a linear peptide molecule, the glycosidic linkages between monosaccharides can be considerably more variable. Consequently, the potential number of permutations in oligosaccharides is substantially higher than that predicted for peptides. For example, the total number of possible linear and branched hexasaccharides that can form from 20 simple or modified monosaccharides is 1.44×10^{15}.

Once information has been encoded it must be translated. The translation of the sugar code is accomplished by lectins. **Lectins** are carbohydrate binding proteins that are not antibodies and have no enzymatic activity. Originally discovered in plants, they are now known to exist in all organisms. Lectins, which usually consist of two or four subunits, possess recognition domains that bind to specific carbohydrate groups via hydrogen bonds, van der Waals forces, and hydrophobic interactions. Biological processes that involve lectin binding include a vast array of cell-cell interactions (Figure 7C). Prominent examples include infections by microorganisms, the mechanisms of many toxins, and physiological processes such as leukocyte rolling. The essential features of each are briefly described.

Infection by many bacteria is initiated when they become firmly attached to host cells. Often, attachment is mediated by the binding of bacterial lectins to oligosaccharides on the cell's surface. *Helicobacter pylori*, the causative agent of gastritis and stomach ulcers, possesses several lectins that allow it to establish a chronic infection in the mucous lining of the stomach. One of these lectins binds with high affinity to a portion of the type O blood group determinant, an oligosaccharide, a circumstance that explains the observation that humans with type O blood are at considerably greater risk of developing ulcers than those with other blood types. Those with types A or B blood, however, are not immune to infection because the bacterium can also use other lectins to achieve adhesion.

The damaging effects of many bacterial toxins occur only after endocytosis into the host cell, a process that is initiated by lectin-ligand binding. The binding of the B subunit of cholera toxin (Special Interest Box 5.1) to a glycolipid on the surface of intestinal cells results in the uptake of the toxic A subunit. Once the A subunit is internalized it proceeds to disrupt the mechanism that regulates chloride transport, a process that results in a life-threatening diarrhea.

Leukocyte rolling is a well-known example of cell-cell interaction mediated by lectin binding. When a tissue becomes damaged in an animal either by infection with a pathogenic organism or by physical trauma, it emits signal molecules that create an inflammation. In response to certain of these molecules, the endothelial cells that line nearby blood vessels produce and insert a protein called selectin into their plasma membranes. The selectins are a family of lectins that act as cell adhesion molecules. Once selectin is displayed on the surface of the endothelial cells, it binds transiently to the *selectin* ligand (an oligosaccharide) on white blood cells such as neutrophils. This relatively weak binding serves to slow the rapid motion of the neutrophils as they flow in blood so that they appear to roll along the lumenal surface of the blood vessel. Once rolling has been initiated, and white

continued on page 230

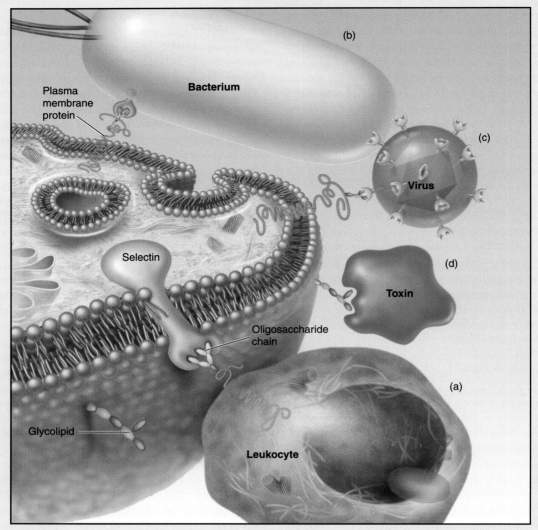

FIGURE 7C

Role of Oligosaccharides in Biological Recognition.

The specific binding of lectins (carbohydrate binding proteins) to the oligosaccharide groups (colored ovals) of glycoconjugate molecules is an essential feature of many biological phenomena. (a) Cell-cell interactions (e.g., leukocyte rolling), (b and c) cell-pathogen infections, and (d) the binding of toxins (e.g., cholera toxin) to cells.

blood cells approach the inflammation site, they encounter other signal molecules that cause them to express another lectin called *integrin* on their surfaces. The binding of integrin with its oligosaccharide ligand on the endothelial surface of the blood vessel causes the neutrophils to stop rolling. Subsequently, the neutrophils undergo changes that allow them to squeeze between the cells of the endothelium and migrate to the infected site where they proceed to consume and degrade bacteria and cellular debris.

SUMMARY

1. Carbohydrates, the most abundant organic molecules in nature, are classified as monosaccharides, disaccharides, oligosaccharides, and polysaccharides according to the number of simple sugar units they contain. Carbohydrate moieties also occur as components of other biomolecules. Glycoconjugates are protein and lipid molecules with covalently linked carbohydrate groups. They include proteoglycans, glycoproteins, and glycolipids.

2. Monosaccharides with an aldehyde functional group are called aldoses; those with a ketone group are known as ketoses. Simple sugars belong to either the D or L family, according to whether the configuration of the asymmetric carbon farthest from the aldehyde or ketone group resembles the D- or L-isomer of glyceraldehyde. The D family contains most biologically important sugars.

3. Sugars containing five or six carbons exist in cyclic forms that result from reactions between hydroxyl groups and either aldehyde (hemiacetal product) or ketone groups (hemiketal product). In both five-membered rings (furanoses) and six-membered rings (pyranoses), the hydroxyl group attached to the anomeric carbon lies either below (α) or above (β) the plane of the ring. The spontaneous interconversion between α- and β-forms is called mutarotation.

4. Simple sugars undergo a variety of chemical reactions. Derivatives of these molecules, such as uronic acids, amino sugars, deoxy sugars, and phosphorylated sugars, have important roles in cellular metabolism.

5. Hemiacetals and hemiketals react with alcohols to form acetals and ketals, respectively. When the cyclic hemiacetal or hemiketal form of a monosaccharide reacts with an alcohol, the new linkage is called a glycosidic linkage, and the compound is called a glycoside.

6. Glycosidic bonds form between the anomeric carbon of one monosaccharide and one of the free hydroxyl groups of another monosaccharide. Disaccharides are carbohydrates composed of two monosaccharides. Oligosaccharides, carbohydrates that typically contain as many as 10 monosaccharide units, are often attached to proteins and lipids. Polysaccharide molecules, which are composed of large numbers of monosaccharide units, may have a linear structure like cellulose and amylose or a branched structure like glycogen and amylopectin. Polysaccharides may consist of only one sugar type (homopolysaccharides) or multiple types (heteropolysaccharides).

7. The three most common homopolysaccharides found in nature (starch, glycogen, and cellulose) all yield D-glucose when hydrolyzed. Cellulose is a plant structural material; starch and glycogen are storage forms of glucose in plant and animal cells, respectively. Chitin, the principal structural material in insect exoskeletons, is composed of N-acetyl-glucosamine residues linked in unbranched chains. Glycosaminoglycans, the principal components of proteoglycan, and murein, a major component of bacterial cell walls, are examples of the heteropolysaccharides, carbohydrate polymers that contain more than one kind of monosaccharide.

8. The enormous heterogeneity of proteoglycans, which are found predominantly in the extracellular matrix of tissues, allows them to play diverse, but as yet poorly understood, roles in living organisms. Glycoproteins occur in cells, in both soluble and membrane-bound forms, and in extracellular fluids. Because of their diverse structures the glycoconjugates, which include proteoglycans, glycoproteins, and glycolipids, play important roles in information transfer in living organisms.

SUGGESTED READINGS

Dwek, R. A., Glycobiology: More Functions for Oligosaccharides, *Science*, 269:1234–1235, 1995.

Gabius, H.-J., Biological Information Transfer Beyond the Genetic Code: The Sugar Code, *Natur Wissenschaften*, 87(3):108–121, 2000.

Lehmann, J., *Carbohydrates: Structure and Biology*, Thieme, New York, 1998.

Robyt, J. F., *Essentials of Carbohydrate Chemistry*, Springer, New York, 1998.

Ruoslahti, E., Structure and Biology of Proteoglycans, *Ann. Rev. Cell Biol.*, 4:229–255, 1988.

Sharon, N., Carbohydrates, *Sci. Amer.*, 243(5):90–116, 1980.

Sharon, N., and Halina, L., Carbohydrates in Cell Recognition, *Sci. Amer.*, 268(1):82–89, 1993.

Yeh, O., and Feeney, R. E., Antifreeze Proteins: Structures and Mechanisms of Function, *Chem. Rev.*, 96:601–617, 1996.

KEY WORDS

acetal, *210*

aldaric acid, *207*

alditol, *208*

aldonic acid, *207*

aldose, *201*

amylopectin, *218*

amylose, *217*

anomer, *204*

cellobiose, *215*

cellulose, *219*

chitin, *217*

diastereomers, *202*

disaccharide, *210*

enediol, *208*

epimer, *202*

epimerization, *208*

glycoconjugate, *201*

glycogen, *218*

glycosaminoglycan, *221*

glycoside, *210*

glycosidic linkage, *210*

hemiacetal, *203*

hemiketal, *204*

ketal, *210*

lactone, *207*

lactose, *215*

lectin, *229*

maltose, *215*

monosaccharide, *201*

murein, *223*

mutarotation, *206*

oligosaccharide, *214*

polysaccharide, *210*

proteoglyan, *223*

reducing sugars, *207*

sucrose, *215*

uronic acid, *207*

REVIEW QUESTIONS

1. Write Haworth structures for the following compounds:
   a. α-D-glucopyranose and β-D-glucofuranose
   b. sucrose
   c. D-lactose
   d. sialic acid
   e. pyranose form of D-mannose
   f. chondroitin sulfate, repeating unit

2. Give an example of each of the following:
   a. epimer
   b. glycosidic linkage
   c. reducing sugar
   d. monosaccharide
   e. anomer
   f. diastereomer

3. What structural relationship is indicated by the term D-sugar? Why are (+) glucose and (−) fructose both classified as D-sugars?

4. Name an example of each of the following classes of compounds:
   a. glycoprotein
   b. proteoglycan
   c. disaccharide
   d. glycosaminoglycan

5. What is the difference between a heteropolysaccharide and a homopolysaccharide? Give examples.

6. Convert each of the following Fischer representations to a Haworth formula:

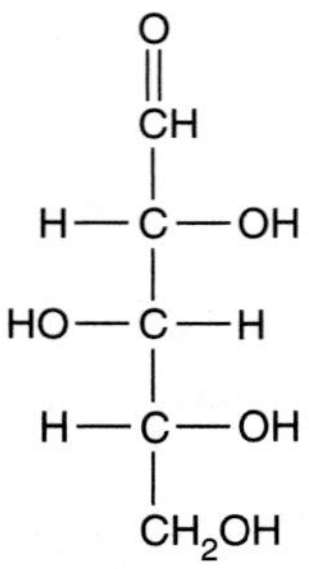

a.

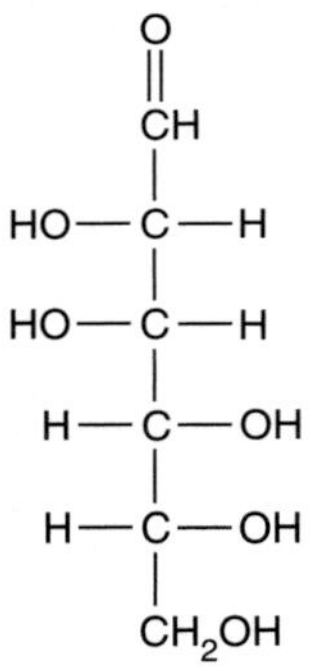

$$
\begin{array}{c}
O \\
\parallel \\
CH \\
| \\
H - C - OH \\
| \\
HO - C - H \\
| \\
H - C - OH \\
| \\
CH_2OH
\end{array}
$$

b.

$$
\begin{array}{c}
O \\
\parallel \\
CH \\
| \\
HO - C - H \\
| \\
HO - C - H \\
| \\
H - C - OH \\
| \\
H - C - OH \\
| \\
CH_2OH
\end{array}
$$

c.

$$
\begin{array}{c}
O \\
\parallel \\
CH \\
| \\
H - C - OH \\
| \\
HO - C - H \\
| \\
HO - C - H \\
| \\
H - C - OH \\
| \\
CH_2OH
\end{array}
$$

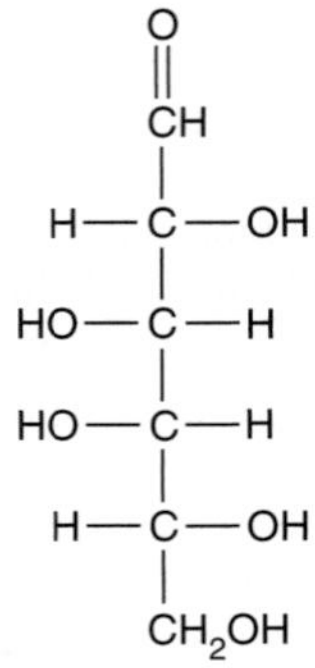

7. Which of the following carbohydrates are reducing and which nonreducing?
   a. starch
   b. cellulose
   c. fructose
   d. sucrose
   e. ribose

8. What structural differences characterize starch, cellulose, and glycogen?

9. What shape do carbohydrate chains linked with $\alpha(1,4)$ glycosidic bonds generally have?

10. Determine the number of possible stereoisomers for the following compounds:

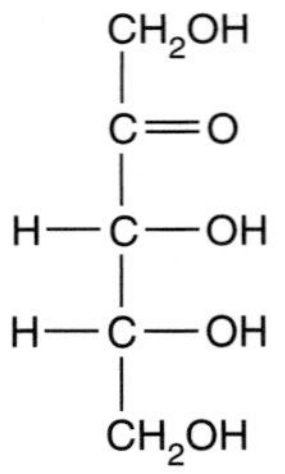

Ribulose

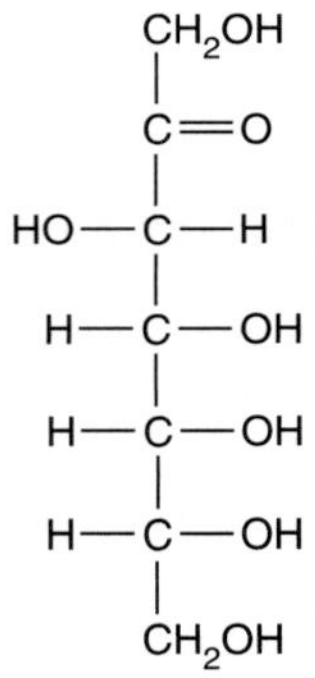

$$
\begin{array}{c}
CH_2OH \\
| \\
C = O \\
| \\
H - C - OH \\
| \\
H - C - OH \\
| \\
CH_2OH
\end{array}
$$

Sedoheptulose

$$
\begin{array}{c}
CH_2OH \\
| \\
C = O \\
| \\
HO - C - H \\
| \\
H - C - OH \\
| \\
H - C - OH \\
| \\
H - C - OH \\
| \\
CH_2OH
\end{array}
$$

11. Raffinose is the most abundant trisaccharide found in nature.

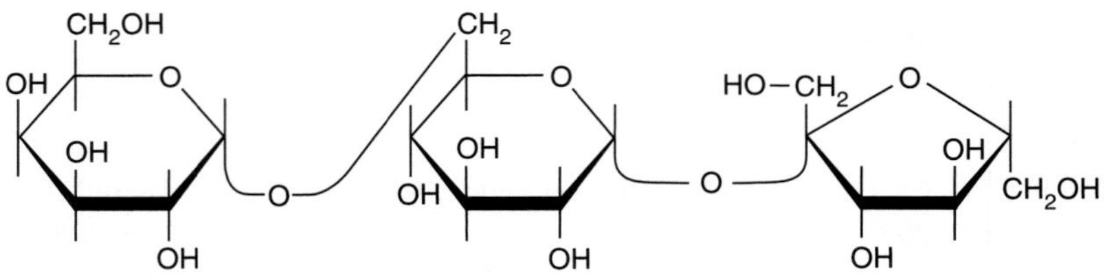

a. Name the three monosaccharide units of raffinose.

b. There are two glycosidic linkages. Are they α or β?

c. Is raffinose a reducing or nonreducing sugar?

d. Is raffinose capable of mutarotation?

12. Give at least one function of each of the following:
    a. glycogen
    b. glycosaminoglycans
    c. glycoconjugates
    d. proteoglycans
    e. glycoproteins
    f. polysaccharides

13. The polymer chains of glycosaminoglycans are widely spread apart and bind large amounts of water.
    a. What two functional groups of the polymer make this binding of water possible?
    b. What type of bonding is involved?

14. In glycoproteins, what are the three amino acids to which the carbohydrate groups are most frequently linked?

15. Chondroitin sulfate chains have been likened to a large fishnet, passing small molecules through their matrix but excluding large ones. Use the structure of chondroitin sulfate and proteoglycans to explain this analogy.

16. Define the term *reducing sugar*. What structural feature does a reducing sugar have?

17. Compare the structures of proteoglycans and glycoproteins. How are structural differences related to their functions?

18. What role is carbohydrate thought to play in maintaining glycoprotein stability?

19. How does the structure of cellulose differ from starch and glycogen?

20. Determine which of the following sugar pairs are enantiomers, diastereomers, epimers, or an aldose-ketose pair.
    a. D-erythrose and D-threose
    b. D-glucose and D-mannose
    c. D-ribose and L-ribose
    d. D-allose and D-galactose
    e. D-glyceraldehyde and dihydroxyacetone

THOUGHT QUESTIONS

1. β-Galactosidase is an enzyme that hydrolyses only $\beta(1,4)$ linkages of lactose. An unknown trisaccharide is converted by β-galactosidase into maltose and galactose. Draw the structure of the trisaccharide.

2. Steroids are large, polycyclic complex, lipid-soluble molecules that are very insoluble in water. Reaction with glucuronic acid makes a steroid much more water-soluble and enables transport through the blood. What structural feature of the glucuronic acid increases the solubility?

3. Many bacteria are surrounded by a proteoglycan coat. Use your knowledge of the properties of this substance to suggest a function for such a coat.

4. It has long been recognized that breast milk protects infants from infectious diseases, especially those that affect the digestive tract. The main reason for this protection appears to be a large group of oligosaccharides that are components of human milk. Suggest a rationale for the protective effect of these oligosaccharides. (*Hint*: Recall that the damage inflicted by many pathogenic organisms and toxins is initiated when they adhere to target cells via the binding of glycoconjugates to lectins.)

5. Alginic acid, isolated from seaweed and used as a thickening agent for ice cream and other foods, is a polymer of D-mannuronic acid with $\beta(1,4)$ glycosidic linkages.
   a. Draw the structure of alginic acid.
   b. Why does this substance act as a thickening agent?

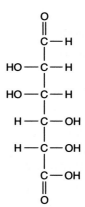

D-Mannuronic acid

6. What is the maximum number of stereoisomers for mannuronic acid?

7. A polysaccharide is found in the shells of arthropods (e.g., lobsters and grasshoppers) and of mollusks (e.g., oysters and snails). It can be obtained from these sources by soaking the shells in cold dilute hydrochloric acid to dissolve the calcium carbonate. The threadlike substance formed is composed of linear long-chain molecules. Hydrolysis with boiling acid gives D-glucosamine and acetic acid in equimolar amounts. Milder enzymatic hydrolysis gives N-acetyl-D-glucosamine as the sole product. The polysaccharide's linkages are identical to those of cellulose. What is the structure of this polymer?

8. Proteoglycan aggregates in tissues form hydrated, viscous gels. Can you think of any obvious mechanical reason why their capacity to form gels is important to cell function? (*Hint*: Liquid water is virtually incompressible.)

Carbohydrate Metabolism

Products of Fermentation Humans use certain microorganisms to metabolize sugar in the absence of oxygen to produce cheese, wine, and bread.

Carbohydrates play several crucial roles in the metabolic processes of living organisms. They serve as energy sources and as structural elements in living cells. Chapter 8 focuses on the role of carbohydrates in energy production. Because the monosaccharide glucose is a prominent energy source in almost all living cells, major emphasis is placed on its synthesis, degradation, and storage. The use of other sugars is also discussed.

Living cells are in a state of ceaseless activity. To maintain its "life" each cell depends on highly coordinated and complex biochemical reactions. In Chapter 8, several major reaction pathways of the carbohydrate metabolism of animals are discussed. During **glycolysis**, an ancient pathway found in almost all organisms, a small amount of energy is captured as a glucose molecule is converted to two molecules of pyruvate. Glycogen, a storage form of glucose in vertebrates, is synthesized by **glycogenesis** when glucose levels are high and degraded by **glycogenolysis** when glucose is in short supply. Glucose can also be synthesized from noncarbohydrate precursors by reactions referred to as **gluconeogenesis**. The **pentose phosphate pathway** enables cells to convert glucose-6-phosphate, a derivative of glucose, to ribose-5-phosphate (the sugar used to synthesize nucleotides and nucleic acids) and other types of monosaccharides. NADPH, an important cellular reducing agent, is also produced by this pathway. In Chapters 9 and 13, other related pathways are discussed. *Photosynthesis*, a process in which light energy is captured to drive carbohydrate synthesis, is described in Chapter 13. In Chapter 9 the glyoxylate cycle is considered. In the *glyoxylate cycle* some organisms (primarily plants) manufacture carbohydrate from fatty acids.

The synthesis and usage of glucose, the major fuel of most organisms, are the focus of any discussion of carbohydrate metabolism. In vertebrates, glucose is transported throughout the body in the blood. If cellular energy reserves are low, glucose is degraded by the glycolytic pathway. Glucose molecules not required for immediate energy production are stored as glycogen in liver and muscle. Depending on a cell's metabolic requirements, glucose can also be used to synthesize, for example, other monosaccharides, fatty acids, and certain amino acids. For this reason, glycolysis is an example of an **amphibolic pathway**. (Amphibolic pathways function in both anabolic and catabolic processes.) Figure 8.1 summarizes the major pathways of carbohydrate metabolism in animals.

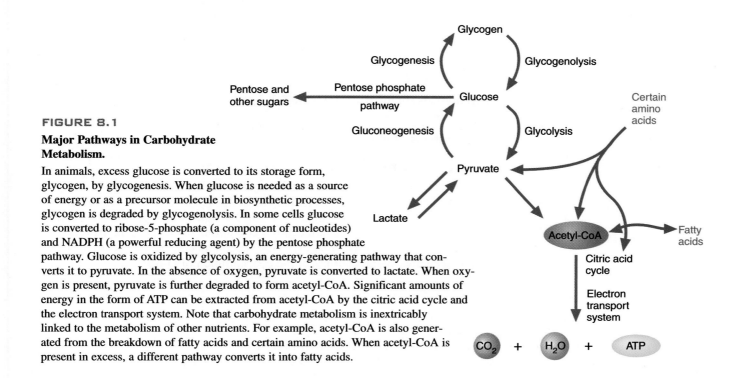

FIGURE 8.1

Major Pathways in Carbohydrate Metabolism.

In animals, excess glucose is converted to its storage form, glycogen, by glycogenesis. When glucose is needed as a source of energy or as a precursor molecule in biosynthetic processes, glycogen is degraded by glycogenolysis. In some cells glucose is converted to ribose-5-phosphate (a component of nucleotides) and NADPH (a powerful reducing agent) by the pentose phosphate pathway. Glucose is oxidized by glycolysis, an energy-generating pathway that converts it to pyruvate. In the absence of oxygen, pyruvate is converted to lactate. When oxygen is present, pyruvate is further degraded to form acetyl-CoA. Significant amounts of energy in the form of ATP can be extracted from acetyl-CoA by the citric acid cycle and the electron transport system. Note that carbohydrate metabolism is inextricably linked to the metabolism of other nutrients. For example, acetyl-CoA is also generated from the breakdown of fatty acids and certain amino acids. When acetyl-CoA is present in excess, a different pathway converts it into fatty acids.

8.1   GLYCOLYSIS

Glycolysis, a series of reactions that occurs in almost every living cell, is believed to be among the oldest of all the biochemical pathways. Both the enzymes and the number and mechanisms of the steps in the pathway are highly homologous in prokaryotes and eukaryotes. Also, glycolysis is an anaerobic process, which it would have had to be in the oxygen-poor atmosphere of pre-eukaryotic Earth.

In glycolysis, also referred to as the *Embden-Meyerhof-Parnas pathway*, each glucose molecule is split and converted to two three-carbon units (pyruvate). During this process several carbon atoms are oxidized. The small amount of energy captured during glycolytic reactions (about 5% of the total available) is stored temporarily in two molecules each of ATP and NADH. The subsequent metabolic fate of pyruvate depends on the organism being considered and its metabolic circumstances. In **anaerobic organisms** (those that do not use oxygen to generate energy), pyruvate may be converted to waste products. Examples include ethanol, lactic acid, acetic acid, and similar molecules. Using oxygen as a terminal electron acceptor, aerobic organisms such as animals and plants completely oxidize pyruvate to form CO_2 and H_2O in an elaborate stepwise mechanism known as **aerobic respiration**.

Glycolysis, which consists of 10 reactions, occurs in two stages:

1. Glucose is phosphorylated twice and cleaved to form two molecules of glyceraldehyde-3-phosphate (G-3-P). The two ATP molecules consumed during this stage are like an investment, because this stage creates the actual substrates for oxidation in a form that is trapped inside the cell.

2. Glyceraldehyde-3-phosphate is converted to pyruvate. Four ATP molecules and two NADH are produced. Because two ATP were consumed in stage 1, the net production of ATP per glucose molecule is 2.

The glycolytic pathway can be summed up in the following equation:

$$\text{D-Glucose} + 2\text{ ADP} + 2\text{ P}_i + 2\text{ NAD}^+ \longrightarrow$$
$$2\text{ pyruvate} + 2\text{ ATP} + 2\text{ NADH} + 2\text{ H}^+ + 2\text{ H}_2\text{O}$$

The Reactions of the Glycolytic Pathway

Glycolysis is summarized in Figure 8.2. The 10 reactions of the glycolytic pathway are as follows:

1. Synthesis of glucose-6-phosphate. Immediately after entering a cell, glucose and other sugar molecules are phosphorylated. Phosphorylation prevents transport of glucose out of the cell and increases the reactivity of the oxygen in the resulting phosphate ester. Several enzymes, called the hexokinases, catalyze the phosphorylation of hexoses in all cells in the body. ATP, a cosubstrate in the reaction, is complexed with Mg^{2+}. (ATP-Mg^{2+} complexes are common in kinase-catalyzed reactions.) Under intracellular conditions the reaction is irreversible; that is, the enzyme has no ability to retain or accommodate the product of the reaction in its active site, regardless of the concentration of G-6-P.

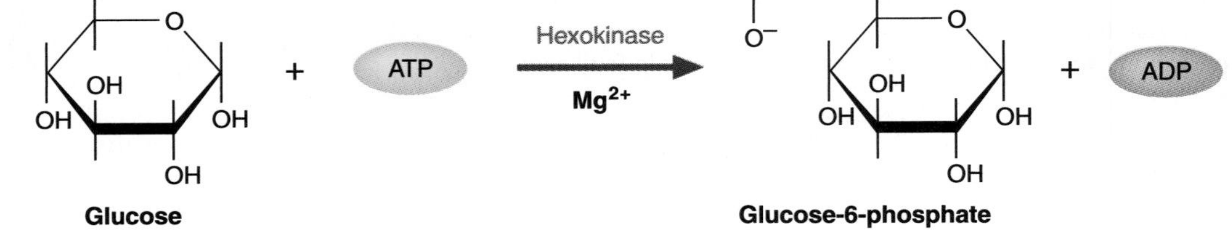

Glucose   **Glucose-6-phosphate**

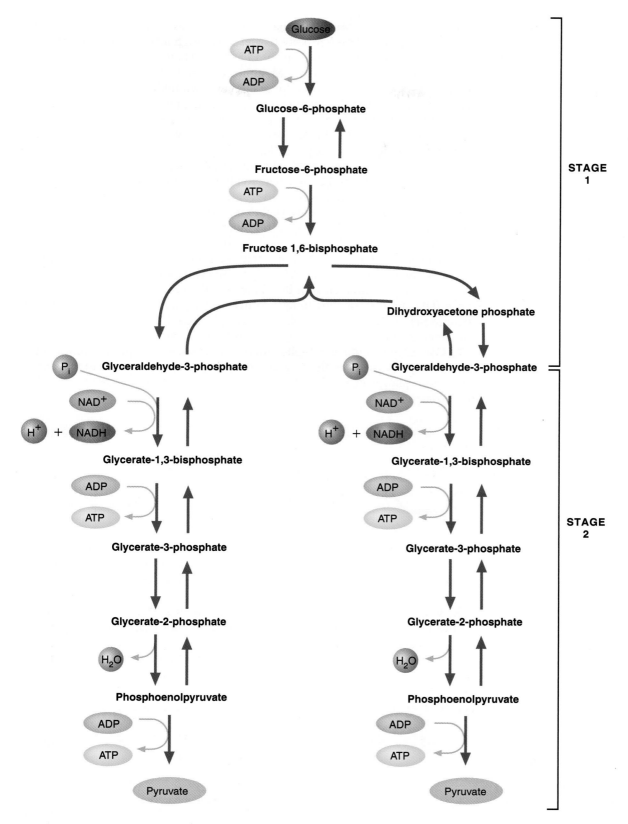

FIGURE 8.2

The Glycolyic Pathway.

In glycolysis each glucose molecule is converted into two pyruvate molecules. In addition, two molecules each of ATP and NADH are produced. Reactions with double arrows are reversible reactions and those with single arrows are irreversible reactions that serve as control points in the pathway.

The animal liver has four hexokinases. Three of these enzymes, found in varying concentrations in other body tissues, have high affinities for glucose relative to its concentration in blood (i.e., they are half-saturated at concentrations of less than 0.1 mM, although blood glucose levels are approximately 4–5 mM). In addition, these enzymes are inhibited from phosphorylating glucose molecules by glucose-6-phosphate, the product of the reaction. When blood glucose levels are low, these properties allow cells such as those in brain and muscle to obtain sufficient glucose. When blood glucose levels are high, cells do not phosphorylate more glucose molecules than required to meet their immediate needs. The fourth enzyme, called hexokinase D (or glucokinase), catalyzes the same reaction but has significantly different kinetic properties that permit the liver to divert glucose into storage as glycogen. This capacity provides the resources used to maintain blood glucose levels, a major role of the liver. Glucokinase requires much higher glucose concentrations for optimal activity (about 10 mM), and it is not inhibited by glucose-6-phosphate. Consequently, after a carbohydrate meal the liver does not remove large quantities of glucose from the blood for glycogen synthesis until other tissues have satisfied their requirements for this molecule. Between meals, when blood glucose falls, another enzyme unique to liver cells (and kidney under starvation conditions) called glucose-6-phosphatase (Section 8.2) facilitates the release of sugar mobilized from glycogen stores into the blood.

    2. Conversion of glucose-6-phosphate to fructose-6-phosphate. During reaction 2 of glycolysis, the aldose glucose-6-phosphate is converted to the ketose fructose-6-phosphate by phosphoglucose isomerase (PGI) in a readily reversible reaction:

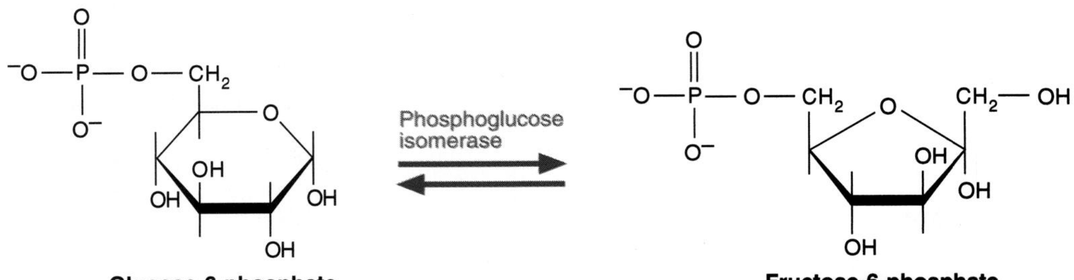

Glucose-6-phosphate   **Fructose-6-phosphate**

Recall that the isomerization reaction of glucose and fructose involves an enediol intermediate (Figure 7.15). This transformation makes C-1 of the fructose product available for phosphorylation.

    3. The phosphorylation of fructose-6-phosphate. Phosphofructokinase-1 (PFK-1) irreversibly catalyzes the phosphorylation of fructose-6-phosphate to form fructose-1,6-bisphosphate:

Fructose-6-phosphate   **Fructose-1,6-bisphosphate**

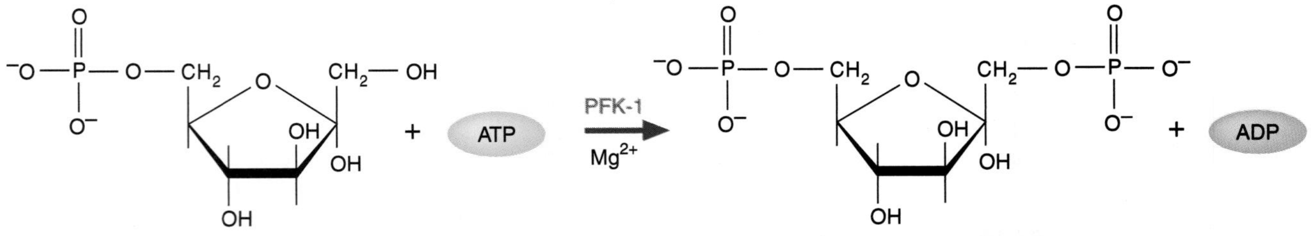

Investing a second molecule of ATP serves several purposes. First of all, because ATP is used as the phosphorylating agent, the reaction proceeds with a large decrease in free energy. After fructose-1,6-bisphosphate has been synthesized, the cell is committed to glycolysis. Because fructose-1,6-bisphosphate eventually splits into two trioses, another purpose for phosphorylation is to prevent any later product from diffusing out of the cell.

PFK-1 is a major regulatory enzyme in glycolysis. Its activity is allosterically inhibited by high levels of ATP and citrate, which are indicators that the cell's energy charge is high and that the citric acid cycle, a major component of the cell's energy-generating capacity, has slowed down. AMP levels increase when the energy charge of the cell is low and are a better predictor of energy deficit than ADP levels. AMP is an allosteric activation of PFK-1. Fructose-2,6-bisphosphate is an allosteric activator of PFK-1 activity in the liver and is synthesized by phosphofructokinase-2 (PFK-2) in response to hormonal signals correlated to blood glucose levels. When serum glucose levels are high, hormone-stimulated increase in fructose-2,6-bisphosphate coordinately increases the activity of PFK-1 (activates glycolysis) and decreases the activity of the enzyme that catalyzes the reverse reaction, fructose-1,6-bisphosphatase (inhibits gluconeogenesis, Section 8.2). AMP is an allosteric inhibiton of fructose-1,6-bisphosphatase. PFK-2 is a bifunctional enzyme that behaves as a phosphatase when phosphorylated in response to the hormone glucagon (low blood sugar). It functions as a kinase when dephosphorylated in response to the hormone insulin (high blood sugar).

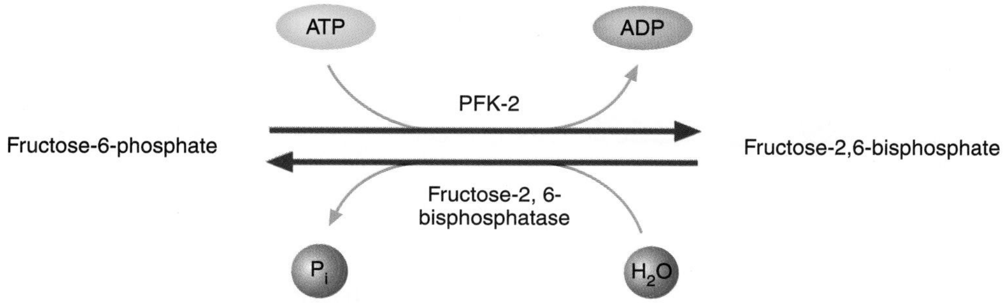

4. Cleavage of fructose-1,6-bisphosphate. Stage 1 of glycolysis ends by cleaving fructose-1,6-bisphosphate into two three-carbon molecules: glyceraldehyde-3-phosphate (G-3-P) and dihydroxyacetone phosphate (DHAP). This reaction is an **aldol cleavage**, hence the name of the enzyme: aldolase. Aldol cleavages are the reverse of aldol condensations, described on p. 144. In aldol cleavages an aldehyde and a ketone are products.

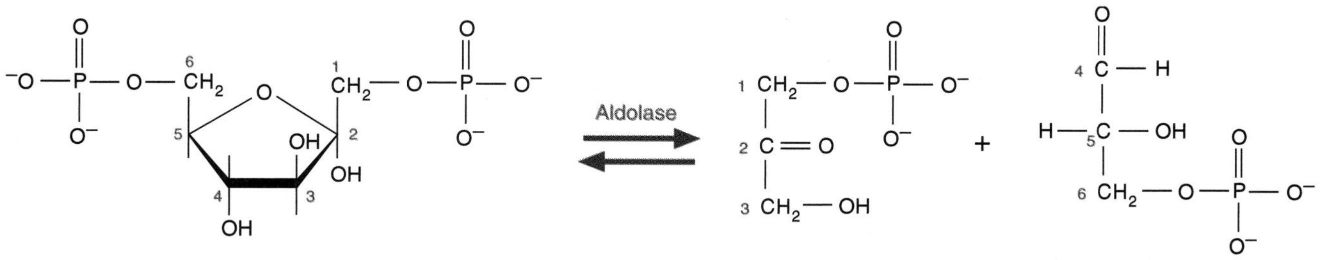

Fructose-1,6-bisphosphate → (Aldolase) → **Dihydroxyacetone phosphate** + **Glyceraldehyde-3-phosphate**

Although the cleavage of fructose-1,6-bisphosphate is frequently unfavorable ($\Delta G^{\circ\prime} = +23.8$ kJ/mol), the reaction proceeds because the products are rapidly removed.

5. The interconversion of glyceraldehyde-3-phosphate and dihydroxyacetone phosphate. Of the two products of the aldolase reaction, only G-3-P serves as a substrate for the next reaction in glycolysis. To prevent the loss of the other three-carbon unit from the glycolytic pathway, triose phosphate isomerase catalyzes the interconversion of DHAP and G-3-P:

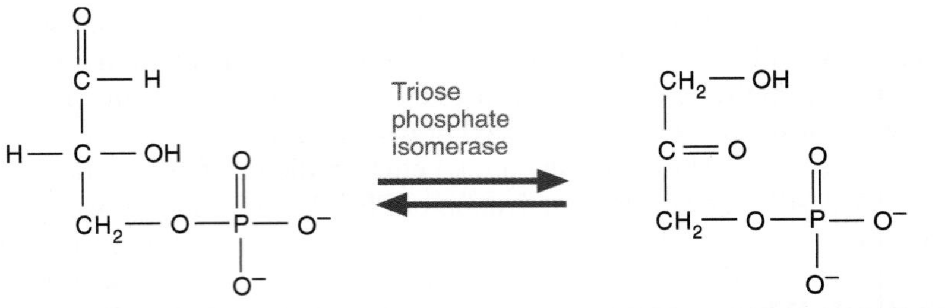

Glyceraldehyde-3-phosphate   **Dihydroxyacetone phosphate**

After this reaction, the original molecule of glucose has been converted to two molecules of G-3-P.

6. Oxidation of glyceraldehyde-3-phosphate. During reaction 6 of glycolysis, G-3-P undergoes oxidation and phosphorylation. The product, glycerate-1,3-bisphosphate, contains a high-energy bond, which may be used in the next reaction to generate ATP:

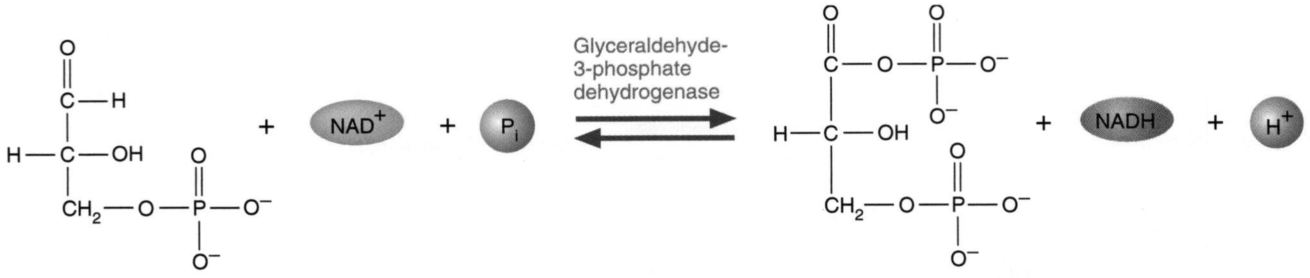

Glyceraldehyde-3-phosphate   **Glycerate-1,3-bisphosphate**

This complex process is catalyzed by glyceraldehyde-3-phosphate dehydrogenase, a tetramer composed of four identical subunits. Each subunit contains one binding site for G-3-P and another for NAD$^+$.

As the enzyme forms a covalent thioester bond with the substrate (Figure 8.3), a hydride ion (H:$^-$) is transferred to NAD$^+$ in the active site. The NADH then leaves the active site and is replaced by NAD$^+$. The acyl enzyme adduct is attacked by inorganic phosphate and the product leaves the active site.

7. Phosphoryl group transfer. In this reaction ATP is synthesized as phosphoglycerate kinase catalyzes the transfer of the high-energy phosphoryl group of glycerate-1,3-bisphosphate to ADP:

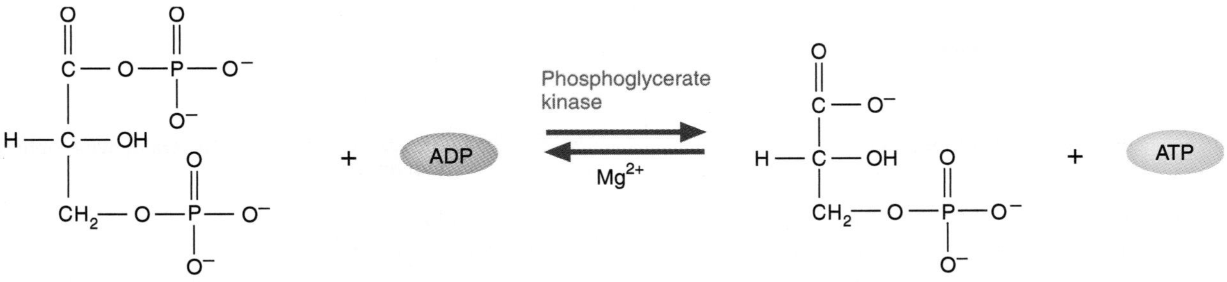

Glycerate-1,3-bisphosphate   **Glycerate-3-phosphate**

Reaction 7 is an example of a substrate-level phosphorylation. Because the synthesis of ATP is endergonic, it requires an energy source. In **substrate-level phosphorylations**, ATP is produced because of the transfer of a phosphoryl group from a substrate with a high phosphoryl group transfer potential. Because two molecules of glycerate-1,3-bisphosphate are formed for every glucose molecule, this reaction

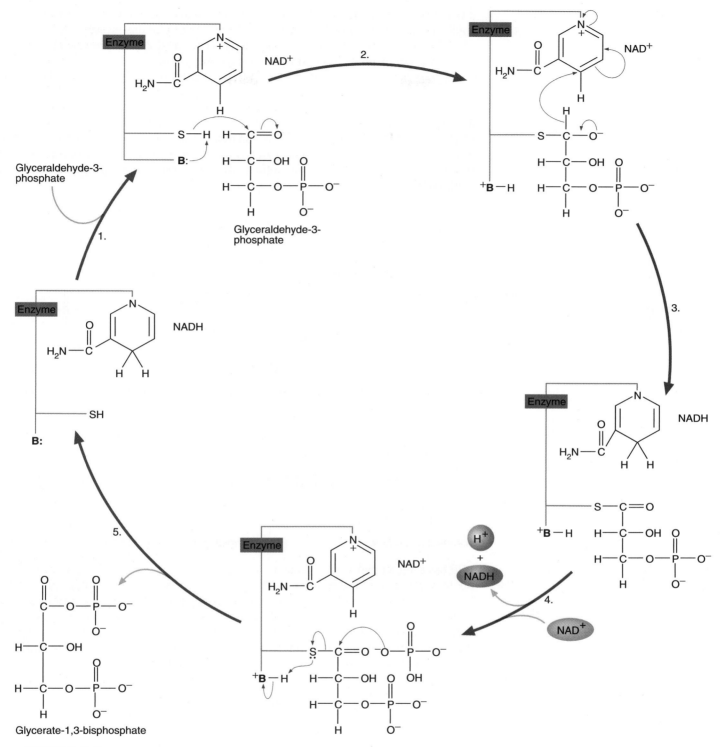

FIGURE 8.3

Glyceraldehyde-3-Phosphate Dehydrogenase Reactions.

In the first step the substrate, glyceraldehyde-3-phosphate, enters the active site. As the enzyme catalyzes the reaction of the substrate with a sulfhydryl group within the active site (Step 2), the substrate is oxidized (Step 3). The bound NADH is then reoxidized by the transfer of a hydride ion to a cytoplasmic NAD⁺ (Step 4). Displacement of the enzyme by inorganic phosphate (Step 5) liberates the product, glycerate-1,3-bisphosphate, thus returning the enzyme to its original form.

produces two ATP molecules and the investment of phosphate bond energy is recovered. Any later ATP synthesis may be considered interest on this investment.

8. The interconversion of 3-phosphoglycerate and 2-phosphoglycerate. Glycerate-3-phosphate has a low phosphoryl group transfer potential. As such, it is a poor candidate for further ATP synthesis. Cells convert glycerate-3-phosphate with its energy-poor phosphate ester to phosphoenolpyruvate (PEP), which has an exceptionally high phosphoryl group transfer potential. (The standard free energies of hydrolysis of glycerate-3-phosphate and PEP are −12.6 kJ/mol and −58.6 kJ/mol, respectively.) In the first step in this conversion (reaction 8), phosphoglycerate mutase catalyzes the conversion of a C-3 phosphorylated compound to a C-2 phosphorylated compound through a two-step addition/elimination cycle.

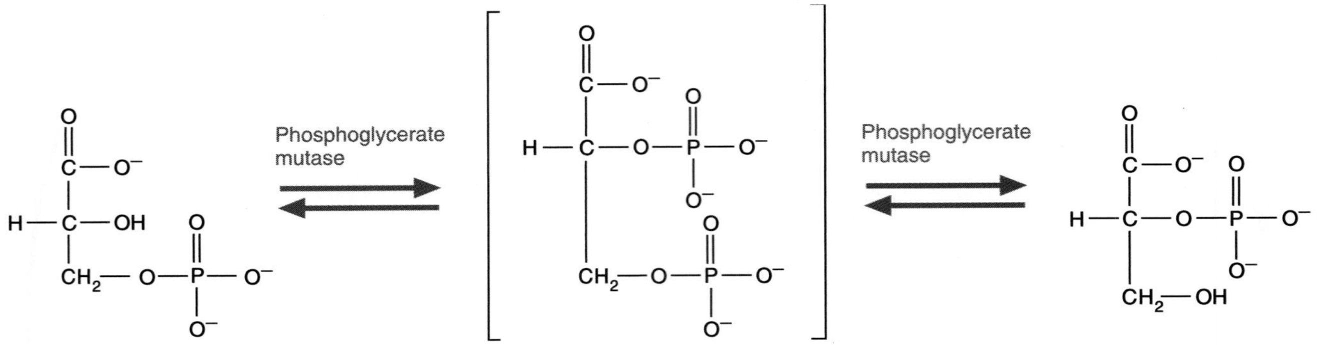

9. Dehydration of 2-phosphoglycerate. Enolase catalyzes the dehydration of glycerate-2-phosphate to form PEP:

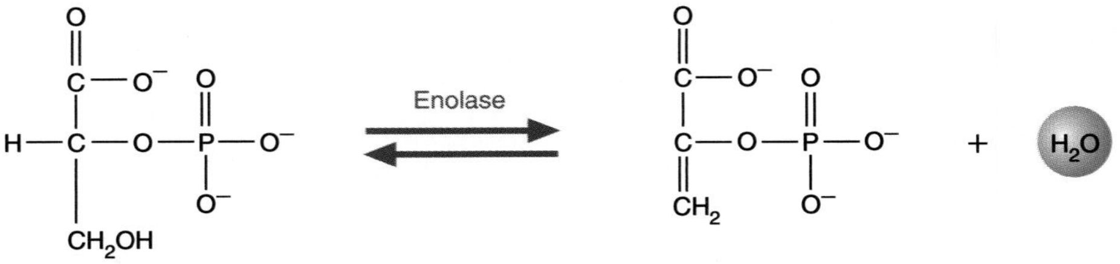

Glycerate-2-phosphate   Phosphoenolpyruvate (PEP)

PEP has a higher phosphoryl group transfer potential than does glycerate-2-phosphate because it contains an enol-phosphate group instead of a simple phosphate ester. The reason for this difference is made apparent in the next reaction. Aldehydes and ketones have two isomeric forms. The *enol* form contains a carbon-carbon double bond and a hydroxyl group. Enols exist in equilibrium with the more stable carbonyl-containing *keto* form. The interconversion of keto and enol forms, also called **tautomers**, is referred to as **tautomerization**:

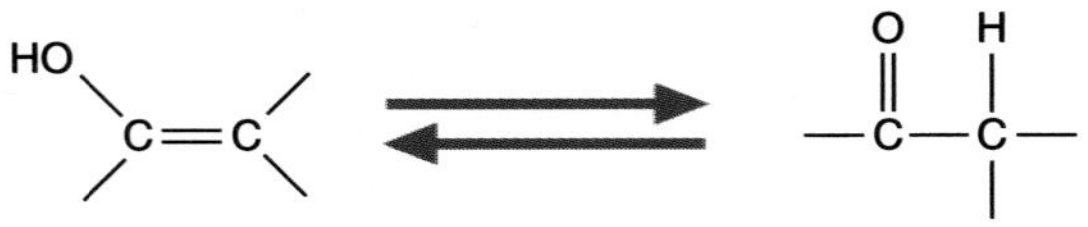

Enol form   Keto form

This tautomerization is restricted by the presence of the phosphate group, as is the resonance stabilization of the free phosphate ion. As a result, phosphoryl transfer to ADP in reaction 10 is highly favored.

10. Synthesis of pyruvate. In the final reaction of glycolysis, pyruvate kinase catalyzes the transfer of a phosphoryl group from PEP to ADP. Two molecules of ATP are formed for each molecule of glucose.

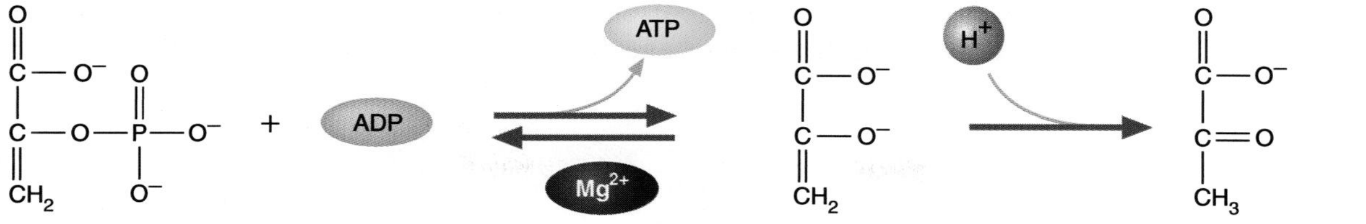

PEP   **Pyruvate (enol form)**   **Pyruvate (keto form)**

Because the free energy of hydrolysis is exceptionally large, PEP is irreversibly converted to pyruvate. The free energy loss, which makes the reaction irreversible, is associated with the spontaneous conversion (tautomerization) of the enol form of pyruvate to the more stable keto form.

The Fates of Pyruvate

In terms of energy, the result of glycolysis is the production of two ATPs and two NADHs per molecule of glucose. Pyruvate, the other product of glycolysis, is still an energy-rich molecule, which can produce a substantial amount of ATP. Before this can happen, however, an intermediate transitional molecule is formed through decarboxylation. This molecule is acetyl-CoA, which is the entry-level substrate for the **citric acid cycle**, an amphibolic pathway that completely oxidizes two carbons to CO_2 and NADH. In the presence of oxygen, this cycle is kept going as the electrons of NADH (and $FADH_2$, another electron carrier) produced in the citric acid cycle are delivered to oxygen via the **electron transport system** to produce water. The electron transport system is a linked series of oxidation-reduction reactions that transfer electrons from donors such as NADH to acceptors such as O_2. Coupled to this process is the generation of a proton gradient that drives the synthesis of ATP. Under anaerobic conditions, further oxidation of pyruvate is impeded. A number of cells and organisms compensate by converting this molecule to a more reduced organic compound and regenerating the NAD^+ required for glycolysis to continue (Figure 8.4). This process of NAD^+ regeneration is referred to as **fermentation**. Muscle cells and certain bacterial species (e.g., *Lactobacillus*) produce NAD^+ by transforming pyruvate into lactate:

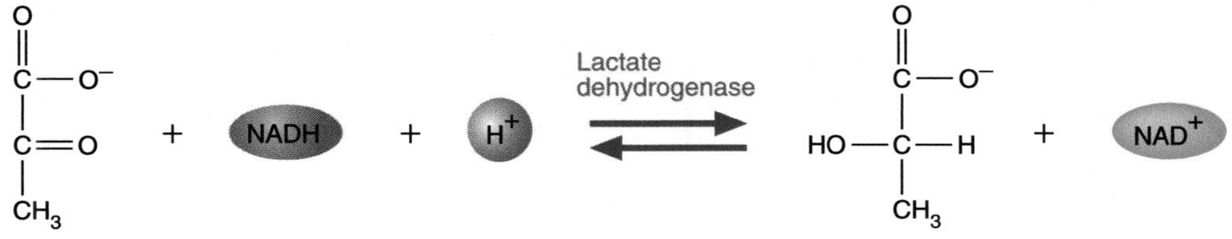

Pyruvate   **Lactate**

In rapidly contracting muscle cells the demand for energy is high. After the O_2 supply is depleted, *lactic acid fermentation* provides sufficient NAD^+ to allow glycolysis (with its low level of ATP production) to continue for a short time (Figure 8.5).

Most molecules of ethanol are detoxified in the liver by two reactions. In the first, ethanol is oxidized to form acetaldehyde. This reaction, catalyzed by alcohol dehydrogenase, produces large amounts of NADH:

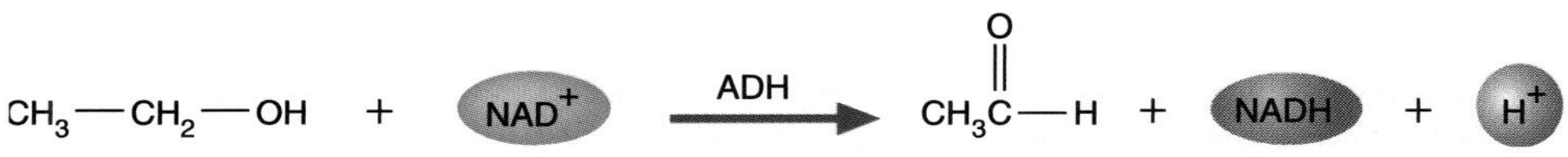

QUESTION 8.1

Soon after its production, acetaldehyde is converted to acetate by aldehyde dehydrogenase, which catalyzes a reaction that also produces NADH:

$$CH_3-\underset{\underset{O}{\|}}{C}-H \;+\; NAD^+ \;+\; H_2O \xrightarrow{\text{Aldehyde dehydrogenase}}$$

$$CH_3-\underset{\underset{O}{\|}}{C}-O^- \;+\; NADH \;+\; 2\,H^+$$

One common effect of alcohol intoxication is the accumulation of lactate in the blood. Can you explain why this effect occurs?

Microorganisms that use lactic acid fermentation to generate energy can be separated into two groups. *Homolactic fermenters* produce only lactate. For example, several species of lactic acid bacteria sour milk. *Heterolactic* or *mixed acid fermenters* produce several organic acids. Mixed acid fermentation, for example, occurs in the rumen of cattle. Symbiotic organisms, some of which digest cellulose, synthesize organic acids (e.g., lactic, acetic, propionic, and butyric acids). The organic acids are absorbed from the rumen

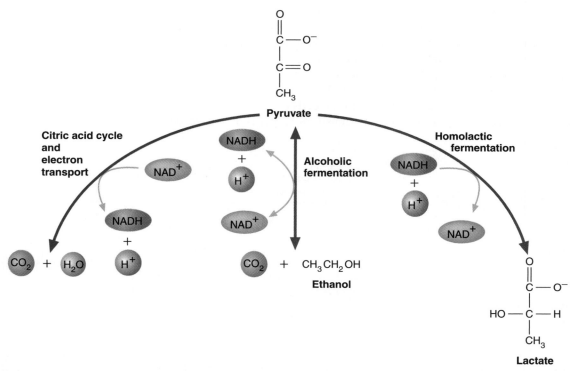

FIGURE 8.4

The Fates of Pyruvate.

When oxygen is available (*left*), aerobic organisms completely oxidize pyruvate to CO_2 and H_2O. In the absence of oxygen, pyruvate can be converted to several types of reduced molecules. In some cells (e.g., yeast), ethanol and CO_2 are produced (*middle*). In others (e.g., muscle cells), homolactic fermentation occurs in which lactate is the only organic product (*right*). Some microorganisms use heterolactic fermentation reactions (not shown) that produce other acids or alcohols in addition to lactate. In all fermentation processes the principal purpose is to regenerate NAD^+ so that glycolysis can continue.

and used as nutrients. Gases such as methane and carbon dioxide are also produced.

Alcoholic fermentation occurs in yeast and several bacterial species. In yeast, pyruvate is decarboxylated to form acetaldehyde, which is then reduced by NADH to form ethanol. (In a **decarboxylation** reaction, an organic acid loses a carboxyl group as CO_2.)

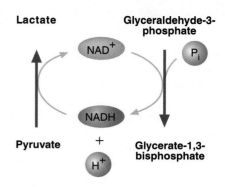

FIGURE 8.5

Recycling of NADH During Anaerobic Glycolysis.

The NADH produced during the conversion of glyceraldehyde-3-phosphate to glycerate-1,3-bisphosphate is oxidized when pyruvate is converted to lactate. This process allows the cell to continue producing ATP for a short time until O_2 is again available.

Alcoholic fermentation by certain yeasts is used commercially to produce wine, beer, and bread (Special Interest Box 8.1). Certain bacterial species produce alcohols other than ethanol. For example, *Clostridium acetobutylicum*, an organism related to the causative agents of botulism and tetanus, produces butanol. Until recently, this organism was used commercially to synthesize butanol, an alcohol used to produce detergents and synthetic fibers. A petroleum-based synthetic process has now replaced microbial fermentation.

The Energetics of Glycolysis

During glycolysis, the decrease in the free energy of glucose is coupled to the synthesis of a net yield of two ATP. However, comparison of the standard free energy changes of the individual reactions (Figure 8.6) reveals no discernible pattern that explains the efficiency of this pathway. A more useful method for evaluating free energy changes takes into account the conditions (e.g., pH and metabolite concentrations) under which cells actually operate. As illustrated in Figure 8.6, free energy changes measured in red blood cells indicate that only three reactions (1, 3, and 10, see pp. 236–243) have significantly negative ΔG values. These reactions, catalyzed by hexokinase, PFK-1, and pyruvate kinase, respectively, are for all practical purposes irreversible; that is, each goes to completion as written. The values for the remaining reactions (2, 4–9) are so close to zero that they operate near equilibrium. Consequently, these latter reactions are easily reversible; small changes in substrate or product concentrations can alter the direction of each reaction. Not surprisingly, in gluconeogenesis (Section 8.2), the pathway by which glucose can be generated from pyruvate and certain other substrates, all of the glycolytic enzymes are involved except for those that catalyze reactions 1, 3, and 10. Gluconeogenesis uses different enzymes to bypass the irreversible steps of glycolysis.

Regulation of Glycolysis

The regulation of glycolysis is complex because of the crucial role of glucose in energy generation and in the synthesis of numerous metabolites. The rate at which the glycolytic pathway operates is controlled primarily by allosteric

Alcoholic beverage production has a long and colorful history. Humans probably began making fermented beverages at least 10,000 years ago. However, the archaeological evidence is about 5500 years old. Ancient wine-stained pottery demonstrates that wine making was a flourishing trade in Sumer (now western Iran) by 3500 B.C. By that time, cultivation of the grape vine (*Vitis vinifera*), which originated in central Asia, had spread throughout the Middle East, especially in Mesopotamia (modern Iraq) and Egypt (Figure 8A). Wines were also made from sweet dates and the sap of palm trees.

These ancient peoples also knew how to produce beer by fermenting barley, a starchy grain. (A Sumerian tablet dated approximately 1750 B.C., which contains directions for brewing beer, is probably one of the oldest known recipes.) Beer making was probably a profitable occupation because Sumerian soldiers received a portion of their pay as beer. Beer was also popular in ancient Egypt. Numerous references to beer have been found on the walls of ancient tombs. Beer produced in ancient China, Japan, and central Africa was made from millet.

In addition to their intoxicating properties, both wine and beer were valued in the ancient world because of their medicinal properties. Wine was especially esteemed by ancient physicians. For example, Hippocrates (460–370 B.C.), the Greek physician who gave the medical profession its ethical ideals, prescribed wine as a diuretic, as a wound dressing, and (in moderate amounts) as a nourishing beverage.

Although humans have been making alcoholic beverages for thousands of years, fermentation was understood only relatively recently. As their businesses became more competitive in the nineteenth century, commercial producers of wine and beer in Europe provided substantial financial support for scientific investigations of fermentation. For example, Louis Pasteur was working for the French wine industry when he discovered that wine fermentation is caused by yeast and that wine spoilage (i.e., vinegar formation) is caused by microbial contamination. Pasteur was credited with saving the French wine industry after he discovered that briefly heating wine to 55°C kills unwanted organisms without affecting the wine's taste. This process is now called pasteurization.

Wine Making

Grapes are well suited to the fermentative process because they contain enough sugar to reach a fairly high alcohol content (about 10%). In addition, because the pH of wine is about 3, it is acidic enough to suppress the growth of most other microorganisms.

FIGURE 8A
Egyptian Wall Painting Illustrating Wine Production.

regulation of three enzymes: hexokinase, PFK-1, and pyruvate kinase. The reactions catalyzed by these enzymes are irreversible and can be switched on and off by allosteric effectors. In general, allosteric effectors are molecules whose cellular concentrations are sensitive indicators of a cell's metabolic state. Some allosteric effectors are product molecules. For example, hexokinase is inhibited by excess glucose-6-phosphate. Several energy-related molecules also act as allosteric effectors. For example, a high AMP concentration (an indicator of low energy production) activates PFK-1 and pyruvate kinase. In contrast, a high ATP

The distinctive flavor and bouquet (aroma) of each wine are determined by many factors. Prominent among these are the strain of grape vine used and its growing conditions (e.g., the mineral content and drainage of soil and the amount and intensity of sunlight).

Wine making begins when ripe grapes are crushed into juice. The crush contains grape skins, seeds, and a liquid referred to as *must*. The must contains sugars (primarily glucose and fructose) in variable amounts (from 12% to 27%), and small amounts of several organic acids (e.g., tartaric, malic, and citric acids). White wines are made by using grapes with unpigmented skins or musts from which pigmented grape skins have been removed before fermentation. Red wines result when pigmented grape skins remain in the must throughout fermentation. During fermentation, yeasts not only convert sugar to alcohol but also produce volatile and aromatic molecules that are not present in the original must. Among these are as many as 10,000 different types of molecules such as complex esters, long-chain alcohols, various acids, glycerol, and other substances that contribute to the wine's unique character. Some of these molecules, referred to as *congeners*, may contribute to hangovers. Examples include ethyl acetate and amyl alcohol. Tyramine, derived from the amino acid tyrosine and found in red wine, is especially well known for this effect.

$$CH_3C \overset{\overset{\textstyle O}{\|}}{} - O - CH_2CH_3 \qquad CH_3CH_2CH_2CH_2CH_2 - OH$$

Ethyl Acetate **Amyl Alcohol**

$$HO - \langle \bigcirc \rangle - CH_2CH_2NH_2$$

Tyramine

In commercial wine production, both temperature and oxygen concentration are carefully controlled. At lower temperatures, yeasts produce more of the molecules that enhance flavor and aroma. In addition, other organisms are less likely to flourish during a cool fermentation. A high oxygen concentration at the beginning of a fermentation causes rapid cell division, so more yeast cells are available to ferment sugar. Later, as the oxygen concentration is reduced, the yeast cells excrete larger and larger amounts of alcohol. After fermentation, yeast cells and other particulates are allowed to settle before the wine is carefully decanted. The new wine is then placed in wooden barrels, where it slowly ages. Controlled oxidation results in the complex flavors and aromas typical of fine wines.

Beer Brewing

Beers are made from starchy grains. Although wheat and oats and other grains have been used in beer production, barley is the preferred grain. In addition to their high starch content and large amounts of relevant enzymes, barley seeds (called *corns*) have several structural layers that protect them during storage and the early stages of brewing.

The first step in brewing is a process referred to as *malting*, in which starch is broken down into glucose and maltose. During malting, the grain, steeped in water, is allowed to begin germination. As germination proceeds gibberellin, a plant hormone, stimulates enzyme production. Large quantities of enzymes such as amylase and other enzymes (e.g., protease, ribonuclease, and phosphatase) make the wort (the malt extract that will eventually become beer) a suitable food for yeast. After germination is terminated by drying out the grain, the resulting *malt* is cured at 100°C. (The color and flavor of beer develop, to a significant extent, during curing.) Curing reduces the malt's moisture content to about 2–5% and arrests enzymatic activity. (Amylase is resistant to high temperatures; its optimum temperature is 70°C.)

Brewing proceeds with *mashing*, in which finely crushed malt is mixed with water and enough enzymes to further degrade any remaining starch and protein. After mashing, the dissolved product (now called the *wort*) is separated from an insoluble residue (referred to as the spent grain) by filtration. (Spent grain is usually sold as cattle fodder.) Afterward, the wort is boiled with hops, the dried cones of the vine *Humulus lupulus*, which give beer its bitter taste. After cooling and removal of the hops, fermentation begins as the wort is inoculated with a pure strain of yeast. (A strain of *Saccharomyces cerevisiae*, sometimes referred to as *brewer's yeast*, is often used.) Fermentation is carefully controlled by varying the temperature and other parameters. Fermentation continues until the desired level of alcohol is reached. (In the United States, the amount of alcohol in beer varies between 3.6% and 4.9% by weight.) After the new beer is filtered to remove yeast, it is stored (or *lagered*) for several months to permit sedimentation. Beer production ends with filtration and pasteurization.

concentration (an indicator that the cell's energy requirements are being met) inhibits both enzymes. Citrate and acetyl-CoA, which accumulate when ATP is in rich supply, inhibit PFK-1 and pyruvate kinase, respectively. Fructose-2,6-bisphosphate, produced via hormone-induced covalent modification of PFK-2, is an indicator of high levels of available glucose and allosterically activates PFK-1. Accumulated fructose-1,6-bisphosphate activates pyruvate kinase, providing a feedforward mechanism of control (i.e., fructose-1,6-bisphosphate is an allosteric activator). The regulation of glycolysis is summarized in Table 8.1.

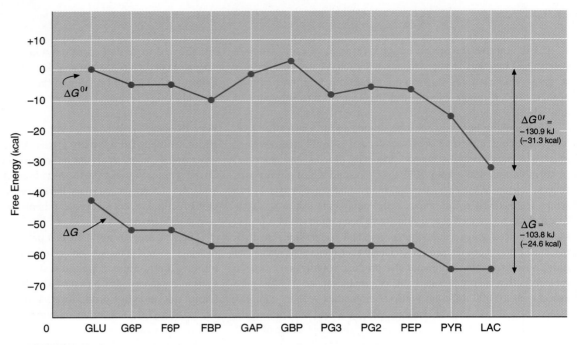

FIGURE 8.6

Free Energy Changes During Glycolysis in Red Blood Cells.

Note that the standard free energy changes ($\Delta G^{\circ\prime}$) for the reactions in glycolysis show no consistent pattern. In contrast, actual free energy values (ΔG) based on metabolite concentrations measured in red blood cells clearly illustrate why reactions 1, 3, and 10 (the conversions of glucose to glucose-6-phosphate, fructose-6-phosphate to fructose-1,6-bisphosphate, and phosphoenolpyruvate to pyruvate, respectively) are irreversible. The ready reversibility of the remaining reactions is indicated by their near zero ΔG values. (GLU = glucose, G6P = glucose-6-phosphate, F6P = fructose-6-phosphate, FBP = fructose-1,6-bisphosphate, GAP = glyceraldehyde phosphate, PG3 = glycerate-3-phosphate, PG2 = glycerate-2-phosphate, PEP = phosphoenolpyruvate, PYR = pyruvate, LAC = lactate)

TABLE 8.1

Allosteric Regulation of Glycolysis

Enzyme	Activator	Inhibitor
Hexokinase		Glucose-6-phosphate, ATP
PFK-1	Fructose-2,6-bisphosphate, AMP	Citrate, ATP
Pyruvate kinase	Fructose-1,6-bisphosphate, AMP	Acetyl-CoA, ATP

Glucagon, present when serum glucose is low, activates the phosphatase function of PFK-2, reducing the level of fructose-2,6-bisphosphate in the cell. Insulin, present when serum glucose is high, activates the kinase function of PFK-2, increasing the level of fructose-2,6-bisphosphate in the cell.

QUESTION 8.2

Insulin is a hormone secreted by the pancreas when blood sugar increases. Its most easily observable function is to reduce the blood sugar level to normal. The binding of insulin to most body cells promotes the transport of glucose across the plasma membrane. The capacity of an individual to respond to a carbohydrate meal by reducing blood glucose concentration quickly is referred to as *glucose tolerance*. Chromium-deficient animals show a decreased glucose tolerance; that is, they cannot remove glucose from blood quickly enough. The metal is believed to facilitate the binding of insulin to cells. Do you think the chromium is acting as an allosteric activator or cofactor?

Louis Pasteur, the great nineteenth century French chemist and microbiologist, was the first scientist to make the following observation. Cells that can oxidize glucose completely to CO_2 and H_2O use glucose more rapidly in the absence of O_2 than in its presence. O_2 seems to inhibit glucose consumption. Explain in general terms the significance of this finding, now referred to as the **Pasteur effect**.

QUESTION 8.3

8.2 GLUCONEOGENESIS

Gluconeogenesis, the formation of new glucose molecules from noncarbohydrate precursors, occurs primarily in the liver. These precursors include lactate, pyruvate, glycerol, and certain α-keto acids (molecules derived from amino acids). In certain situations (i.e., metabolic acidosis or starvation) the kidney can make new glucose. Between meals adequate blood glucose levels are maintained by the hydrolysis of liver glycogen. When liver glycogen is depleted (e.g., prolonged fasting or vigorous exercise), the gluconeogenesis pathway provides the body with adequate glucose. Brain and red blood cells rely exclusively on glucose as their energy source. Under exceptional circumstances, brain cells can also use certain fatty acid derivatives to generate energy. Exercising skeletal muscle uses glucose stored as glycogen in the muscle cell in combination with fatty acids stored in micellar form in the muscle cell.

Gluconeogenesis Reactions

The reaction sequence in gluconeogenesis is largely the reverse of glycolysis. Recall, however, that three glycolytic reactions (the reactions catalyzed by hexokinase, PFK-1, and pyruvate kinase) are irreversible. In gluconeogenesis, alternate reactions catalyzed by different enzymes are used to bypass these obstacles. The reactions unique to gluconeogenesis are summarized below. The entire gluconeogenic pathway and its relationship to glycolysis are illustrated in Figure 8.7. The bypass reactions of gluconeogenesis are as follows:

1. Synthesis of PEP. PEP synthesis from pyruvate requires two enzymes: pyruvate carboxylase and PEP carboxykinase. Pyruvate carboxylase, found within mitochondria, converts pyruvate to oxaloacetate (OAA):

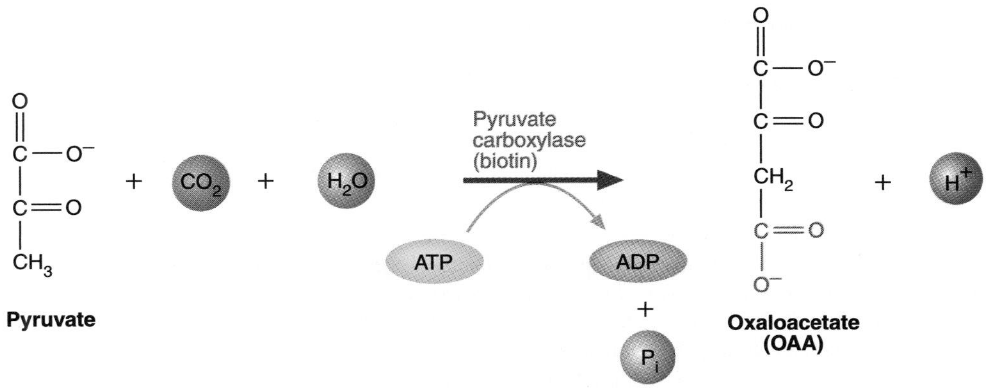

The coenzyme *biotin*, which functions as a CO_2 carrier, is covalently bound to the enzyme through the side chain amino group of a lysine residue. OAA

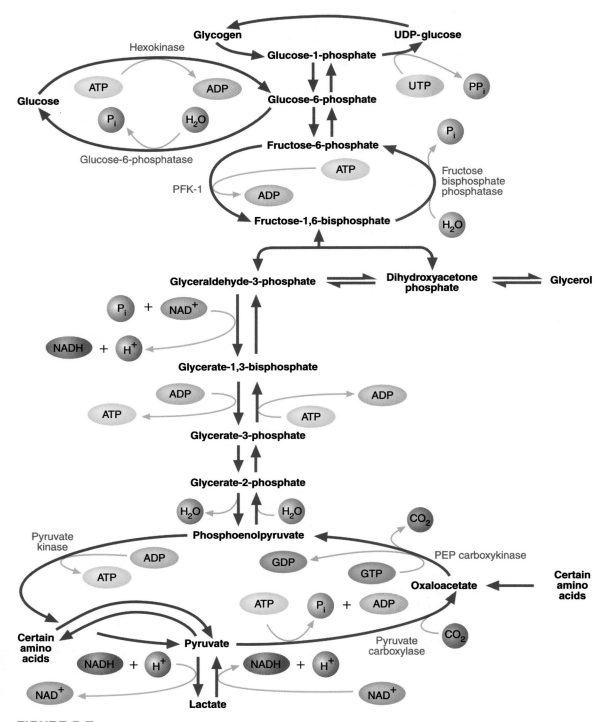

FIGURE 8.7

Carbohydrate Metabolism: Gluconeogenesis and Glycolysis.

In gluconeogenesis, which occurs when blood sugar levels are low and liver glycogen is depleted, 7 of the 10 reactions of glycolysis are reversed. Three irreversible glycolytic reactions are bypassed by alternative reactions. The major substrates for gluconeogenesis are certain amino acids (derived from muscle), lactate (formed in muscle and red blood cells), and glycerol (produced from the degradation of triacylglycerols). In contrast to the reactions of glycolysis, which occur only in cytoplasm, several reactions of gluconeogenesis occur within the mitochondria (the reactions catalyzed by pyruvate carboxylase and, in some species, PEP carboxykinase) and the endoplasmic reticulum (the reaction catalyzed by glucose-6-phosphatase).

is decarboxylated and phosphorylated by PEP carboxykinase in a reaction driven by the hydrolysis of guanosine triphosphate (GTP):

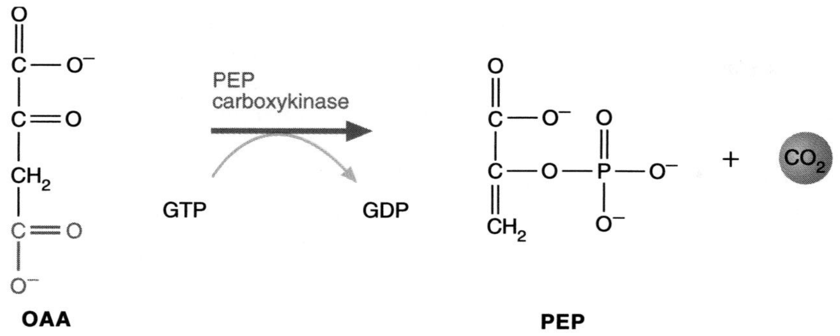

OAA   **PEP**

PEP carboxykinase is found within the mitochondria of some species and in the cytoplasm of others. In humans this enzymatic activity is found in both compartments. Because the inner mitochondrial membrane is impermeable to OAA, cells that lack mitochondrial PEP carboxykinase transfer OAA into the cytoplasm by using, for example, the **malate shuttle**. In this process, OAA is converted into malate by mitochondrial malate dehydrogenase. After the transport of malate across mitochondrial membrane, the reverse reaction is catalyzed by cytoplasmic malate dehydrogenase.

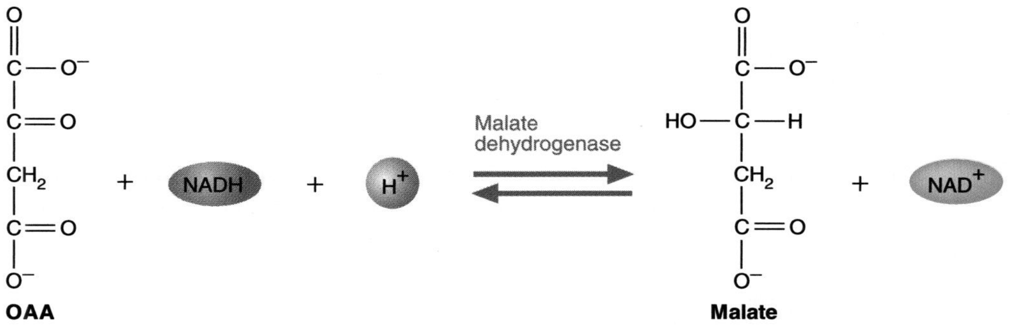

OAA   **Malate**

2. Conversion of fructose-1,6-bisphosphate to fructose-6-phosphate. The irreversible PFK-1–catalyzed reaction in glycolysis is bypassed by fructose-1,6-bisphosphatase:

Fructose-1,6-bisphosphate   **Fructose-6-phosphate**

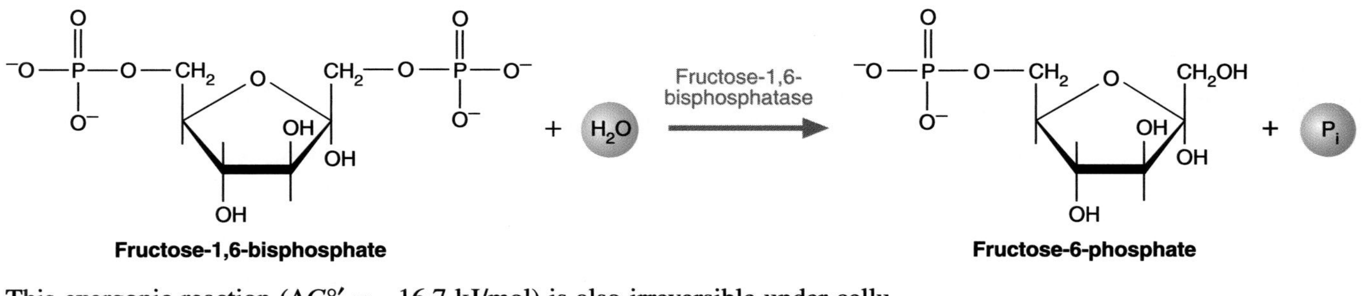

This exergonic reaction ($\Delta G^{\circ\prime} = -16.7$ kJ/mol) is also irreversible under cellular conditions. ATP is not regenerated. Fructose-1,6-bisphosphatase is an allosteric enzyme. Its activity is stimulated by citrate and inhibited by AMP and fructose-2,6-bisphosphate:

3. Formation of glucose from glucose-6-phosphate. Glucose-6-phosphatase, found only in liver and kidney, catalyzes the irreversible hydrolysis of glucose-6-phosphate to form glucose and P_i. Glucose is subsequently released into the blood.

As stated, each of the above reactions is matched by an opposing irreversible reaction in glycolysis. Each set of such paired reactions is referred to as a *substrate cycle*. Because they are coordinately regulated (an activator of the enzyme catalyzing the forward reaction serves as an inhibitor of the enzyme catalyzing the reverse reaction), very little energy is wasted despite the fact that both enzymes may be operating at some level at the same time. *Flux control* (regulation of the flow of substrate and removal of product) is more effective if transient accumulation of product is funneled back through the cycle. The catalytic velocity of the forward enzyme will remain high if the concentration of the substrate is maximized. The gain in catalytic efficiency more than makes up for the small energy loss in recycling the product.

Gluconeogenesis is an energy-consuming process. Instead of generating ATP (as in glycolysis), gluconeogenesis requires the hydrolysis of six high-energy phosphate bonds.

QUESTION 8.4

Malignant hyperthermia is a rare, inherited disorder triggered during surgery by certain anesthetics. A dramatic (and dangerous) rise in body temperature (as high as 112°F) is accompanied by muscle rigidity and acidosis. The excessive muscle contraction is initiated by a large release of calcium from the sarcoplasmic reticulum. (The sarcoplasmic reticulum is a calcium-storing organelle in muscle cells.) Acidosis results from excessive lactic acid production. Prompt treatment to reduce body temperature and to counteract the acidosis is essential to save the patient's life. A probable contributing factor to this disorder is wasteful cycling between glycolysis and gluconeogenesis. Explain why this is a reasonable explanation.

QUESTION 8.5

The reaction summary for gluconeogenesis is shown below. After examining the gluconeogenic pathway, account for each component in the equation. (*Hint*: The hydrolysis of each nucleotide releases a proton.)

$$2\ C_3H_4O_3 + 4\ \text{ATP} + 2\ \text{GTP} + 2\ \text{NADH} + 2\ H^+ + 6\ H_2O \longrightarrow$$

Pyruvic acid

$$1\ C_6H_{12}O_6 + 4\ \text{ADP} + 2\ \text{GDP} + 2\ \text{NAD}^+ + 6\ HPO_4^{2-} + 6\ H^+$$

Glucose

QUESTION 8.6

Patients with *Von Gierke's disease* (a glycogen storage disease) lack glucose-6-phosphatase activity. Two prominent symptoms of this disorder are fasting hypoglycemia and lactic acidosis. Can you explain why these symptoms occur?

Gluconeogenesis Substrates

As was previously mentioned, several metabolites are gluconeogenic precursors. Three of the most important substrates are described briefly.

Lactate is released by red blood cells and other cells that lack mitochondria or have low oxygen concentrations. In the **Cori cycle**, lactate is released by skeletal muscle during exercise (Figure 8.8). After lactate is transferred to the liver, it is reconverted to pyruvate by lactate dehydrogenase and then to glucose by gluconeogenesis.

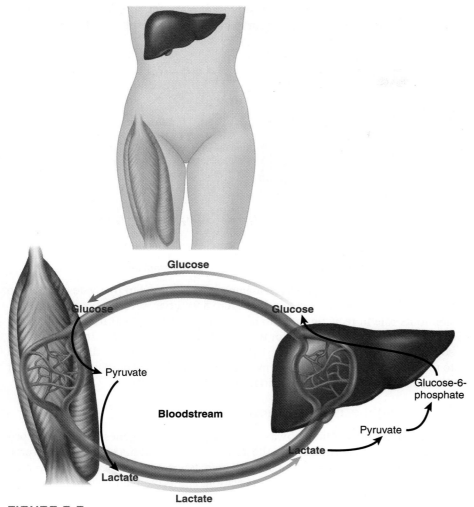

FIGURE 8.8

The Cori Cycle.

During strenuous exercise, lactate is produced in muscle cells under anaerobic conditions. After passing through blood to the liver, lactate is converted to glucose by gluconeogenesis.

Glycerol, a product of fat metabolism in adipose tissue, is transported to the liver in the blood, and then converted to glycerol-3-phosphate by glycerol kinase. (Glycerol kinase is found only in liver.) Oxidation of glycerol-3-phosphate to form DHAP occurs when cytoplasm NAD^+ concentration is relatively high.

Of all the amino acids that can be converted to glycolytic intermediates (molecules referred to as *glucogenic*), alanine is perhaps the most important. (The metabolism of the glucogenic amino acids is described in Chapter 15.) When exercising muscle produces large quantities of pyruvate, some of these molecules are converted to alanine by a transamination reaction involving glutamate:

$$CH_3-\overset{\overset{O}{\|}}{C}-\overset{\overset{O}{\|}}{C}-O^- \quad + \quad {}^-O-\overset{\overset{O}{\|}}{C}-CH_2-CH_2-\overset{\overset{H}{|}}{\underset{\overset{|}{{}^+NH_3}}{C}}-\overset{\overset{O}{\|}}{C}-O^-$$

Pyruvate   L-Glutamate

⇅ Alanine transaminase

$$CH_3-\overset{\overset{H}{|}}{\underset{\overset{|}{{}^+NH_3}}{C}}-\overset{\overset{O}{\|}}{C}-O^- \quad + \quad {}^-O-\overset{\overset{O}{\|}}{C}-CH_2-CH_2-\overset{\overset{O}{\|}}{C}-\overset{\overset{O}{\|}}{C}-O^-$$

L-Alanine   α-Ketoglutarate

After it is transported to the liver, alanine is reconverted to pyruvate and then to glucose. The **glucose-alanine cycle** (Figure 8.9) serves several purposes. In addition to its role in recycling α-keto acids between muscle and liver, the glucose-alanine cycle is a mechanism for transporting NH_4^+ to the liver. In α-keto acids, sometimes referred to as carbon skeletons, a carbonyl group is directly attached to the carboxyl group. Examples include pyruvate and α-ketoglutarate. The liver then converts NH_4^+, a very toxic ion, to urea (Chapter 15).

Gluconeogenesis Regulation

As with other metabolic pathways, the rate of gluconeogenesis is affected primarily by substrate availability, allosteric effectors, and hormones. Not surprisingly, gluconeogenesis is stimulated by high concentrations of lactate, glycerol, and amino acids. A high-fat diet, starvation, and prolonged fasting make large quantities of these molecules available.

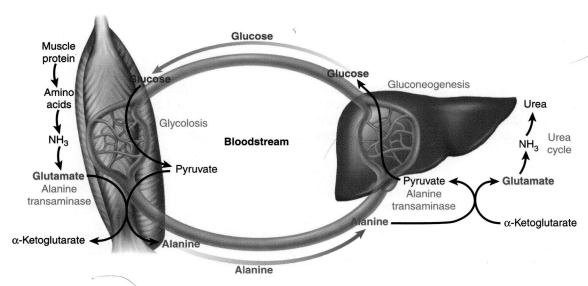

FIGURE 8.9

The Glucose-Alanine Cycle.

Alanine is formed from pyruvate in muscle. After it is transported to the liver, alanine is reconverted to pyruvate by alanine transaminase. Eventually pyruvate is used in the synthesis of new glucose. Because muscle cannot synthesize urea from amino nitrogen, the glucose-alanine cycle is used to transfer amino nitrogen to the liver.

The human brain is composed of an estimated 100 billion neurons that together integrate all of the functions of the body. Despite its relatively small size (about 1.5 kg or 2% of the body weight of an average adult), the human brain uses, under resting conditions, between 15% and 20% of the body's cardiac output. This profoundly complex organ requires such a rich blood supply because of its high metabolic rate. Even a brief interruption in the continuous flow of oxygen, nutrients, and energy, in the form of glucose, can cause unconsciousness. The metabolic processes in the brain are so complicated that investigations of brain function have been very limited until the relatively recent development of computerized radiographic imaging technologies such as PET (positron emission tomography). PET studies allow noninvasive investigations of brain functioning. PET scans detect radioactivity emitted by compounds within the brain. The most frequently used radiotracer molecule for PET scan studies of brain is 2-deoxy-2-[^{18}F]fluoro-β-D-glucose, referred to as ^{18}F-deoxyglucose. Because ^{18}F-deoxyglucose is short lived, the individual undergoing the procedure is exposed to very small amounts of radiation.

2-Deoxy-2-[^{18} F] fluoro-β-D-Glucose

After a person is injected with a small amount of ^{18}F-deoxyglucose, PET scans of the brain are obtained by photographing the γ-rays that are emitted after the radiolabeled glucose molecules are transported into cells. Once in a cell, both glucose and its analogue ^{18}F-deoxyglucose are converted to phosphate esters by hexokinase. Unlike glucose, however, the phosphorylated product of the radiotracer, ^{18}F-deoxyglucose-6-phosphate, is neither an inhibitor of hexokinase nor a substrate for phosphoglucose isomerase. Consequently,

the "trapped" phosphorylated radiotracer molecule accumulates within the cell. As the radioactive isotope decays by emitting positrons, these particles encounter nearby electrons and γ-rays are formed. The PET scanner converts the emitted rays into color-coded images that reveal the intensity of the metabolic activity of the structures being examined. PET scans show changes in metabolic activity because as a brain region becomes more active, it requires more nutrients and energy, hence its uptake of glucose increases and the scan "lights up." PET scans (Figure 8B) have been used to study the brains of healthy volunteers to identify brain areas involved in tasks such as reading, memory retrieval, and problem solving. They have also been used to detect, diagnose, and investigate several types of brain tumor, degenerative diseases such as Alzheimer's and Parkinson's, and psychiatric illnesses such as schizophrenia.

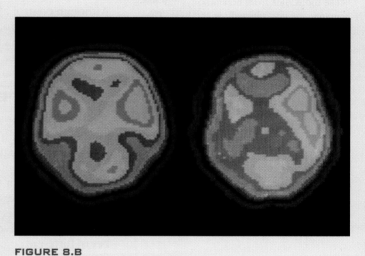

FIGURE 8.B

PET Images of Uptake of ^{18}F-Deoxyglucose in the Brains of a Normal Adult (left) and a Depressed Adult (right).

Regions of lower glucose metabolism are depicted in blue and green, and those of high glucose uptake in red and yellow.

The four key enzymes in gluconeogenesis (pyruvate carboxylase, PEP carboxykinase, fructose-1,6-bisphosphatase, and glucose-6-phosphatase) are affected to varying degrees by allosteric modulators. For example, fructose-1,6-bisphosphatase is activated by ATP and inhibited by AMP and fructose-2,6-bisphosphate. Acetyl-CoA activates pyruvate carboxylase. (The concentration of acetyl-CoA, a product of fatty acid degradation, is especially high during starvation.)

As with other biochemical pathways, hormones affect gluconeogenesis by altering the concentrations of allosteric effectors and the rate key enzymes are synthesized. As mentioned previously, glucagon depresses the synthesis of fructose-2,6-bisphosphate, activating the phosphatase function of PFK-2. The lowered concentration of fructose-2,6-bisphosphate reduces activation of PFK-1 and releases the inhibition of fructose-1,6-bisphosphatase.

Another effect of glucagon binding to liver cells is the inactivation of the glycolytic enzyme pyruvate kinase. (Protein kinase C, an enzyme activated by cAMP, converts pyruvate kinase to its inactive phosphorylated conformation.) Hormones also influence gluconeogenesis by altering enzyme synthesis. For example, the synthesis of gluconeogenic enzymes is stimulated by cortisol (a steroid hormone produced in the cortex of the adrenal gland). (*Cortisol* facilitates the body's adaptation to stressful situations. Its actions affect carbohydrate, protein, and lipid metabolism.) Finally, insulin action leads to the synthesis of new molecules of glucokinase, PFK-1, and PFK-2. Glucagon action leads to the synthesis of new molecules of PEP carboxykinase, fructose-1,6-bisphosphatase, and glucose-6-phosphatase.

These hormones accomplish this feat by altering the phosphorylation state of certain target proteins in the liver cell, which in turn modifies gene expression. The key point to remember is that it is the insulin/glucagon ratio that exerts the major regulatory effects on carbohydrate metabolism. After a carbohydrate meal, the insulin/glucagon ratio is high and glycolysis in the liver predominates over gluconeogenesis. After a period of fasting or following a high-fat, low-carbohydrate meal, the insulin/glucagon ratio is low and gluconeogenesis in the liver predominates over glycolysis. The availability of ATP is the second important regulator in the reciprocal control of glycolysis and gluconeogenesis in that high levels of AMP, the low-energy hydrolysis product of ATP, increase the flux through glycolysis at the expense of gluconeogenesis and low levels of AMP increase the flux through gluconeogenesis at the expense of glycolysis. Although control at the PFK-1/fructose-1,6-bisphosphatase cycle would appear to be sufficient for this pathway, control at the pyruvate kinase step is key because it permits the maximal retention of PEP, a molecule with a very high phosphate transfer potential.

KEY CONCEPTS 8.2

Gluconeogenesis, the synthesis of new glucose molecules from noncarbohydrate precursors, occurs primarily in the liver. The reaction sequence is the reverse of glycolysis except for three reactions that bypass irreversible steps in glycolysis.

8.3   THE PENTOSE PHOSPHATE PATHWAY

The pentose phosphate pathway is an alternative metabolic pathway for glucose oxidation in which no ATP is generated. Its principal products are NADPH, a reducing agent required in several anabolic processes, and ribose-5-phosphate, a structural component of nucleotides and nucleic acids. The pentose phosphate pathway occurs in the cytoplasm in two phases: oxidative and nonoxidative. In the oxidative phase of the pathway, the conversion of glucose-6-phosphate to ribulose-5-phosphate is accompanied by the production of two molecules of NADPH.

The nonoxidative phase involves the isomerization and condensation of a number of different sugar molecules. Three intermediates in this process that are useful in other pathways are ribose-5-phosphate, fructose-6-phosphate, and glyceraldehyde-3-phosphate.

The oxidative phase of the pentose phosphate pathway consists of three reactions (Figure 8.10a). In the first reaction, glucose-6-phosphate dehydrogenase (G-6-PD) catalyzes the oxidation of glucose-6-phosphate. 6-Phosphogluconolactone and NADPH are products in this reaction. 6-Phosphogluconolactone is then hydrolyzed to produce 6-phosphogluconate. A second molecule of NADPH is produced during the oxidative decarboxylation of 6-phosphogluconate, a reaction that yields ribulose-5-phosphate.

A substantial amount of the NADPH required for reductive processes (i.e., lipid biosynthesis) and antioxidant mechanisms is supplied by these reactions. For this reason this pathway is most active in cells in which relatively large amounts of lipids are synthesized, for example, adipose tissue, adrenal cortex, mammary glands, and the liver.

NADPH is also a powerful antioxidant. (**Antioxidants** are substances that prevent the oxidation of other molecules. Their roles in living processes are described in Chapter 10.) Consequently, the oxidative phase of the pentose phosphate pathway is also quite active in cells that are at high risk for oxidative damage, such as red blood cells. The nonoxidative phase of the pathway begins with the conversion of ribulose-5-phosphate to ribose-5-phosphate by ribulose-5-phosphate isomerase or to xylulose-5-phosphate by ribulose-5-phosphate epimerase. During the remaining reactions of the pathway (Figure 8.10b), transketolase and transaldolase catalyze the interconversions of trioses, pentoses, and hexoses. Transketolase is a TPP-requiring enzyme that transfers two-carbon units from

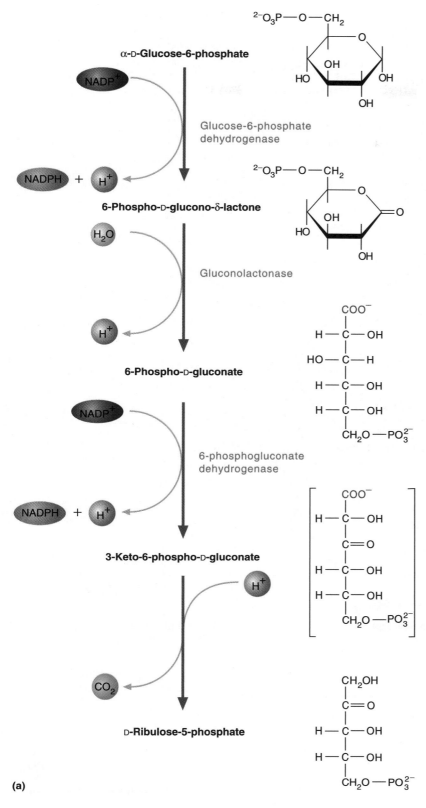

(a)

FIGURE 8.10

The Pentose Phosphate Pathway.

(a) The oxidative phase. NADPH is an important product of these reactions. (b) The nonoxidative phase. When cells require more NADPH than pentose phosphates, the enzymes in the nonoxidative phase convert ribose-5-phosphate into the glycolytic intermediates fructose-6-phosphate and glyceraldehyde-3-phosphate.

a ketose to an aldose. (TPP, thiamine pyrophosphate, is the coenzyme form of thiamine, also known as vitamin B_1.) Two reactions are catalyzed by transketolase. In the first reaction, the enzyme transfers a two-carbon unit from xylulose-5-phosphate to ribose-5-phosphate, yielding glyceraldehyde-3-phosphate

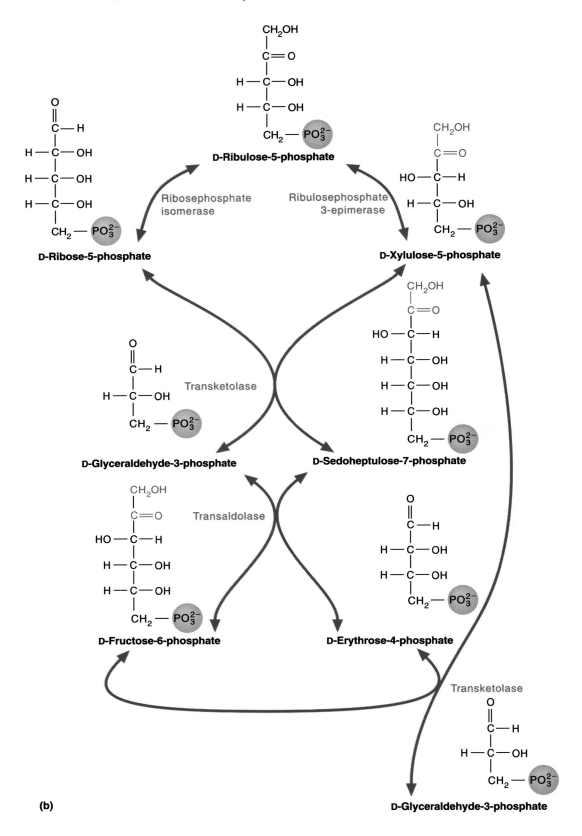

(b)

and sedoheptulose-7-phosphate. In the second transketolase-catalyzed reaction, a two-carbon unit from another xylulose-5-phosphate molecule is transferred to erythrose-4-phosphate to form a second molecule of glyceraldehyde-3-phosphate and fructose-6-phosphate. (Erythrose-4-phosphate is used by some organisms to synthesize aromatic amino acids.) Transaldolase transfers three-carbon units from a ketose to an aldose. In the reaction catalyzed by transaldolase, a three-carbon unit is transferred from sedoheptulose-7-phosphate to glyceraldehyde-3-phosphate. The products formed are fructose-6-phosphate and erythrose-4-phosphate. The result of the nonoxidative phase of the pathway is the synthesis of ribose-5-phosphate and the glycolytic intermediates glyceraldehyde-3-phosphate and fructose-6-phosphate.

When pentose sugars are not required for biosynthetic reactions, the metabolites in the nonoxidative portion of the pathway are converted into glycolytic intermediates that can then be further degraded to generate energy or converted into precursor molecules for biosynthetic processes (Figure 8.11). For this reason the pentose phosphate pathway is also referred to as the *hexose monophosphate shunt*. In plants, the pentose phosphate pathway is involved in the synthesis of glucose during the dark reactions of photosynthesis (Chapter 13).

The pentose phosphate pathway is regulated to meet the cell's moment-by-moment requirements for NADPH and ribose-5-phosphate. The oxidative phase is very active in cells such as red blood cells or hepatocytes in which demand for NADPH is high. In contrast, the oxidative phase is virtually absent in cells (e.g., muscle cells) that synthesize little or no lipid. G-6-PD catalyzes a key regulatory step in the pentose phosphate pathway. Its activity is inhibited by NADPH and stimulated by GSSG (GSSG is the oxidized form of glutathione, an important cellular antioxidant that is discussed in Chapter 10) and glucose-6-phosphate. In addition, diets high in carbohydrate increase the synthesis of both G-6-PD and phosphogluconate dehydrogenase.

KEY CONCEPTS 8.3

The pentose phosphate pathway produces NADPH, ribose-5-phosphate, and several glycolytic intermediates.

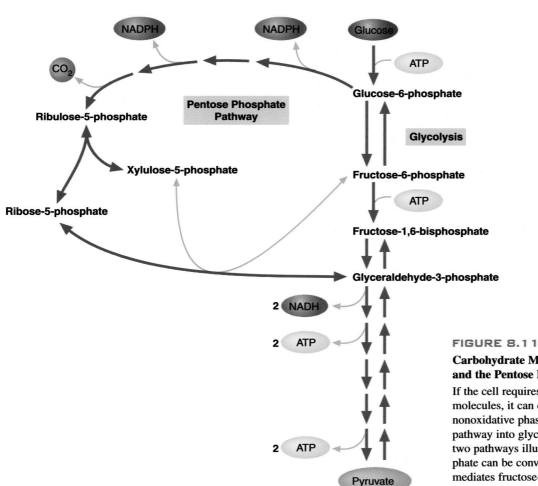

FIGURE 8.11

Carbohydrate Metabolism: Glycolysis and the Pentose Phosphate Pathway.

If the cell requires more NADPH than ribose molecules, it can channel the products of the nonoxidative phase of the pentose phosphate pathway into glycolysis. As this overview of the two pathways illustrates, excess ribose-5-phosphate can be converted into the glycolytic intermediates fructose-6-phosphate and glyceraldehyde-3-phosphate.

8.4 METABOLISM OF OTHER IMPORTANT SUGARS

Several sugars other than glucose are important in vertebrates. The most notable of these are fructose, galactose, and mannose. Besides glucose, these molecules are the most common sugars found in oligosaccharides and polysaccharides. They are also energy sources. The reactions by which these sugars are converted into glycolytic intermediates are illustrated in Figure 8.12.

Fructose Metabolism

Dietary sources of fructose include fruit, honey, and the disaccharide sucrose. Fructose, a significant source of carbohydrate in the human diet (second only

FIGURE 8.12
Carbohydrate Metabolism: Other Important Sugars.
Galactose, mannose, and fructose can be converted to glycolytic intermediates.

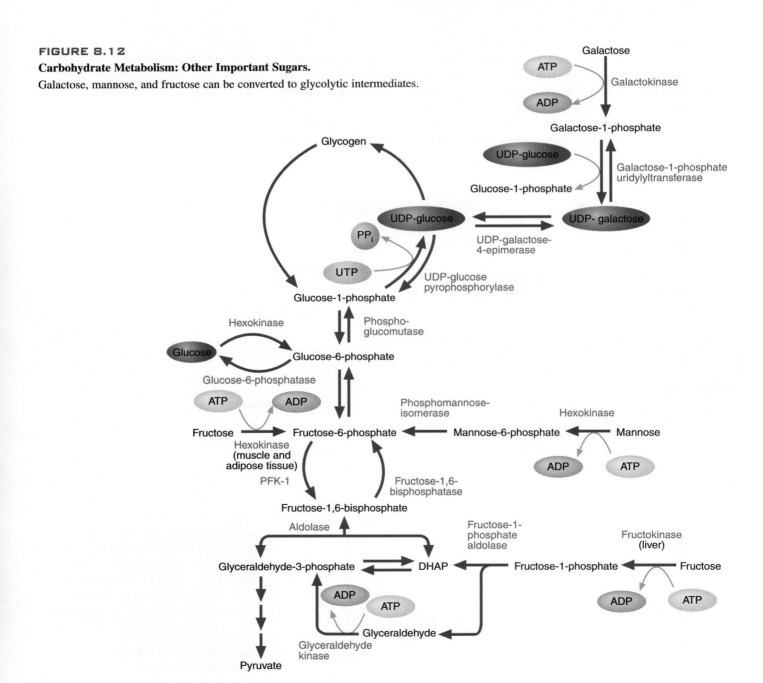

to glucose), can enter the glycolytic pathway by two routes. In the liver, fructose is converted to fructose-1-phosphate by fructokinase:

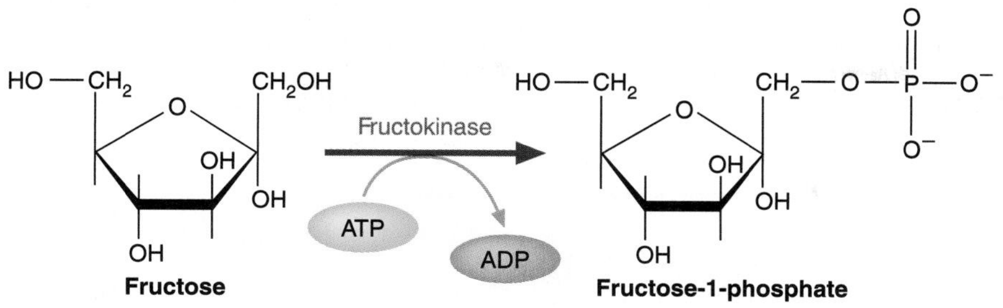

Fructose → **Fructose-1-phosphate** (Fructokinase, ATP → ADP)

When fructose-1-phosphate enters the glycolytic pathway, it is first split into dihydroxyacetone phosphate (DHAP) and glyceraldehyde by fructose-1-phosphate aldolase. DHAP is then converted to glyceraldehyde-3-phosphate by triose phosphate isomerase. Glyceraldehyde-3-phosphate is generated from glyceraldehyde and ATP by glyceraldehyde kinase.

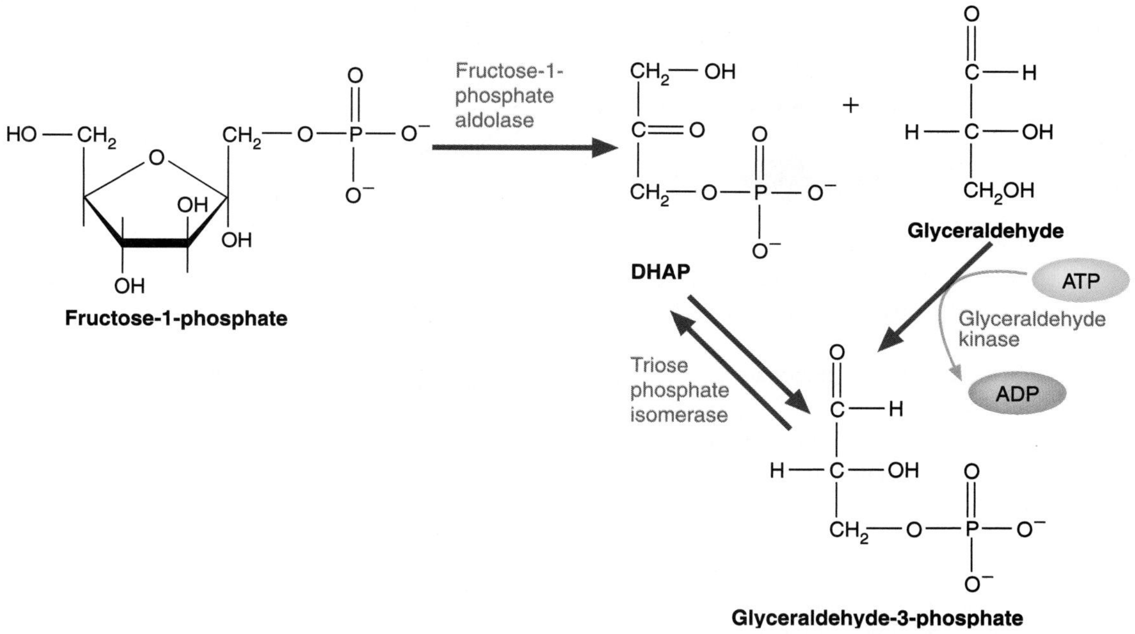

The conversion of fructose-1-phosphate into glycolytic intermediates bypasses two regulatory steps (the reactions catalyzed by hexokinase and PFK-1); thus fructose is metabolized more quickly than glucose.

In muscle and adipose tissue, fructose is converted to the glycolytic intermediate fructose-6-phosphate by hexokinase. Because the hexokinases have a low affinity for fructose, this reaction is of minor importance unless fructose consumption is exceptionally high.

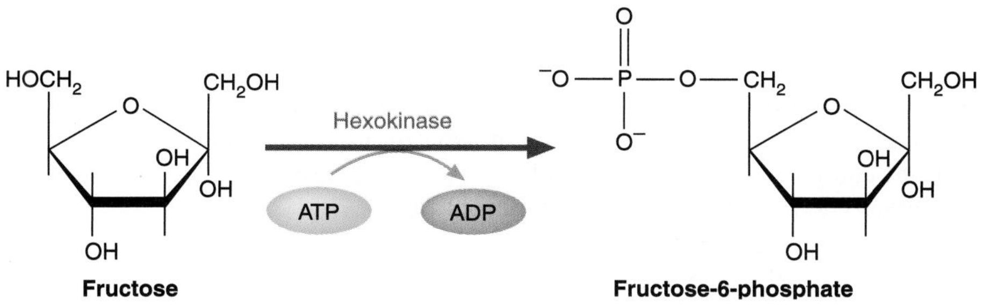

Fructose → **Fructose-6-phosphate** (Hexokinase, ATP → ADP)

Galactose Metabolism

Although galactose and glucose have similar structures (i.e., they are epimers), several reactions are required for this sugar to enter the glycolytic pathway. Galactose is initially converted to galactose-1-phosphate by galactokinase:

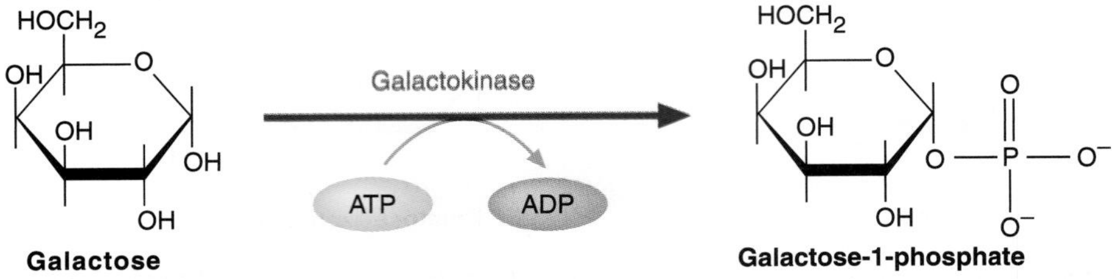

Galactokinase

ATP → ADP

Galactose → **Galactose-1-phosphate**

Then galactose-1-phosphate is transformed into the nucleotide derivative UDP-galactose. During fetal development and childhood the first step in this conversion is catalyzed by galactose-1-phosphate uridyltransferase. (The hereditary disorder galactosemia, described on p. 212, is caused by the absence of this enzyme.)

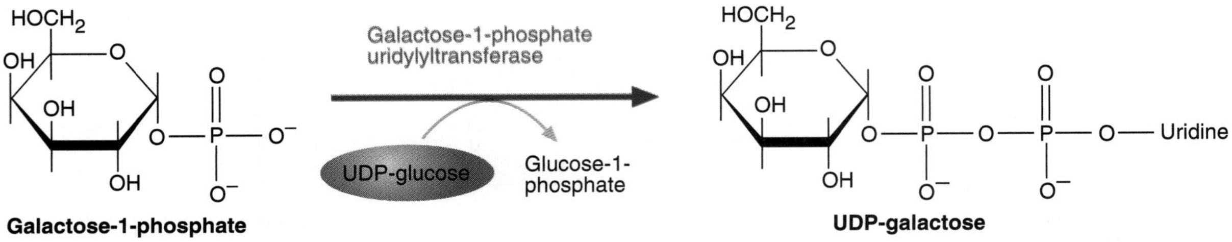

Galactose-1-phosphate uridylyltransferase

UDP-glucose → Glucose-1-phosphate

Galactose-1-phosphate → **UDP-galactose**

Beginning in adolescence, UDP-galactose is produced in a reaction catalyzed by UDP-galactose pyrophosphorylase:

Galactose-1-phosphate + UTP ⇌ UDP-galactose + PP$_i$

Then UDP-glucose is formed by the isomerization of galactose catalyzed by UDP-glucose-4-epimerase:

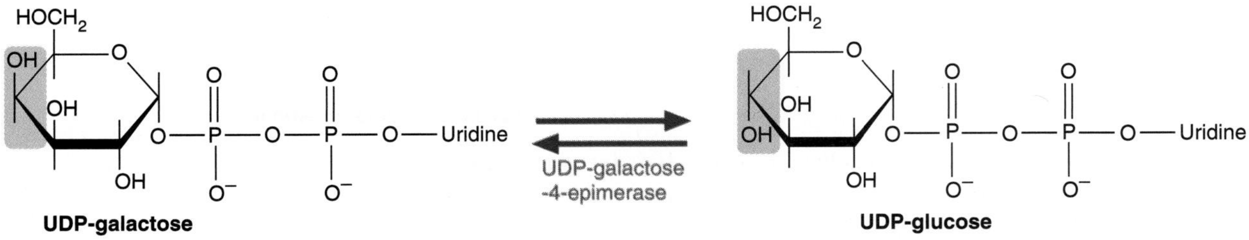

UDP-galactose ⇌ UDP-glucose

UDP-galactose-4-epimerase

UDP-galactose → **UDP-glucose**

Depending on the cell's metabolic needs, UDP-glucose is used directly in glycogen synthesis or is converted to glucose-1-phosphate by UDP-glucose pyrophosphorylase. Glucose-1-phosphate enters the glycolytic pathway after its conversion to glucose-6-phosphate by phosphoglucomutase.

Mannose Metabolism

Mannose is an important component of the oligosaccharides that are found in glycoproteins. Because it is a minor component in the diet, mannose is an unimportant energy source. After phosphorylation by hexokinase, mannose enters the glycolytic pathway as fructose-6-phosphate.

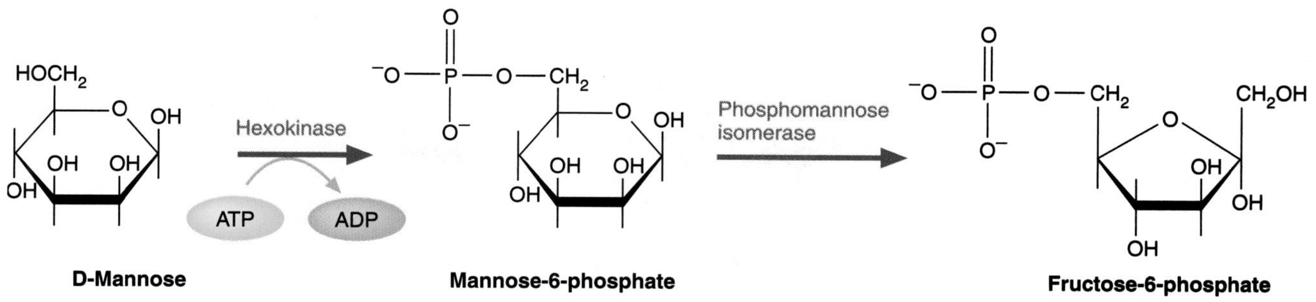

D-Mannose → Mannose-6-phosphate (Hexokinase, ATP → ADP) → Fructose-6-phosphate (Phosphomannose isomerase)

8.5 GLYCOGEN METABOLISM

The synthesis and degradation of glycogen are carefully regulated so that sufficient glucose is available for the body's energy needs. Both glycogenesis and glycogenolysis are controlled primarily by three hormones: insulin, glucagon, and epinephrine.

Glycogenesis

Glycogen synthesis occurs after a meal, when blood glucose levels are high. It has long been recognized that the consumption of a carbohydrate meal is followed promptly by liver glycogenesis. Until recently, it was presumed that blood glucose is the sole direct precursor in this process. However, it now appears that under physiological conditions, a portion of the new glycogen is formed by a mechanism involving the following sequence: dietary glucose $\longrightarrow$ C_3-molecule $\longrightarrow$ liver glycogen. Lactate and alanine are believed to be the most likely C_3-molecules in this process. As indicated in Figure 8.7, both of these molecules are easily converted to glucose in the liver. The following discussion traces the synthesis of glycogen from glucose-6-phosphate.

Glycogenesis involves the following set of reactions:

1. Synthesis of glucose-1-phosphate. Glucose-6-phosphate is reversibly converted to glucose-1-phosphate by phosphoglucomutase, an enzyme that contains a phosphoryl group attached to a reactive serine residue:

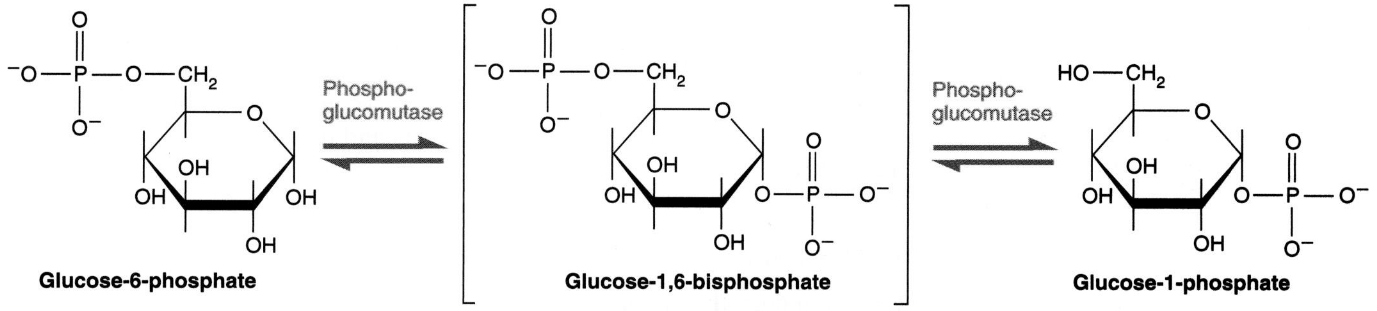

Glucose-6-phosphate ⇌ (Phosphoglucomutase) Glucose-1,6-bisphosphate ⇌ (Phosphoglucomutase) Glucose-1-phosphate

The enzyme's phosphoryl group is transferred to glucose-6-phosphate, forming glucose-1,6-bisphosphate. As glucose-1-phosphate forms, the phosphoryl group attached to C-6 is transferred to the enzyme's serine residue.

2. Synthesis of UDP-glucose. Glycosidic bond formation is an endergonic process. Derivatizing the sugar with a good leaving group provides the driving force for most sugar transfer reactions. For this reason, sugar-nucleotide synthesis is a common reaction preceding sugar transfer and polymerization processes. Uridine diphosphate glucose (UDP-glucose) is more reactive than glucose and is held more securely in the active site of the enzymes catalyzing transfer reactions (referred to as a group as glycosyl transferases). Because UDP-glucose contains two phosphoryl bonds, it is a highly energized molecule. Formation of UDP-glucose, whose $\Delta G^{\circ\prime}$ value is near zero, is a reversible reaction catalyzed by UDP-glucose pyrophosphorylase:

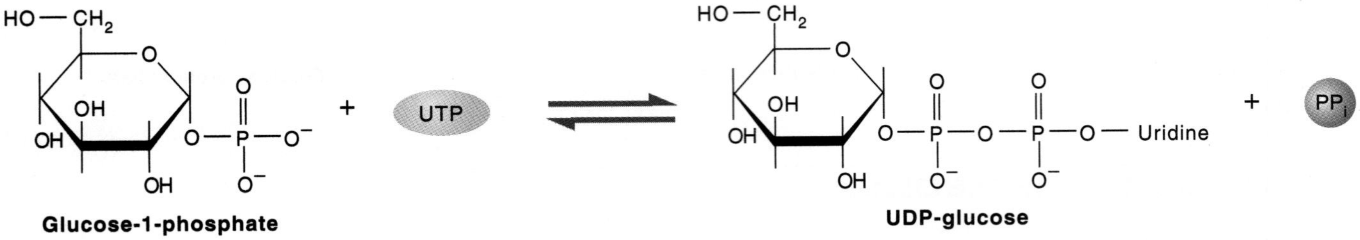

Glucose-1-phosphate   **UDP-glucose**

However, the reaction is driven to completion because pyrophosphate (PP_i) is immediately and irreversibly hydrolyzed by pyrophosphatase with a large loss of free energy ($\Delta G^{\circ\prime} = -33.5$ kJ/mol):

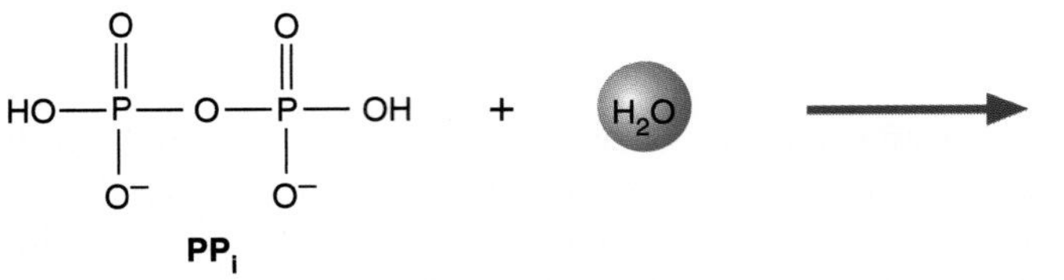

(Recall that removing product shifts the reaction equilibrium to the right. This cellular strategy is common.)

3.` Synthesis of glycogen from UDP-glucose. The formation of glycogen from UDP-glucose requires two enzymes:

a. Glycogen synthase, which catalyzes the transfer of the glucosyl group of UDP-glucose to the nonreducing ends of glycogen (Figure 8.13a), and

b. Amylo-$\alpha(1,4 \rightarrow 1,6)$-glucosyl transferase (branching enzyme), which creates the $\alpha(1,6)$ linkages for branches in the molecule (Figure 8.13b).

Glycogen synthesis requires a glycogen chain. Glycogen synthesis is now believed to be initiated by the transfer of glucose from UDP-glucose to a specific tyrosine residue in a "primer" protein called *glycogenin*. Large glycogen granules, each consisting of a single highly branched glycogen molecule, can be observed in the cytoplasm of liver and muscle cells of well-fed animals. The enzymes responsible for glycogen synthesis and degradation coat each granule's surface.

Glycogenolysis

Glycogen degradation requires the following two reactions:

1. Removal of glucose from the nonreducing ends of glycogen. Using inorganic phosphate (P_i), glycogen phosphorylase cleaves the $\alpha(1,4)$ linkages on the outer branches of glycogen to yield glucose-1-phosphate. Glycogen phosphorylase stops when it comes within four glucose residues of a branch point (Figure 8.14). (A glycogen molecule that has been degraded to its branch points is called a *limit dextrin*.)

2. Hydrolysis of the $\alpha(1,6)$ glycosidic bonds at branch points of glycogen. Amylo-$\alpha(1,6)$-glucosidase, also called debranching enzyme, begins the removal of $\alpha(1,6)$ branch points by transferring the outer three of the four glucose residues attached to the branch point to a nearby nonreducing end. It then

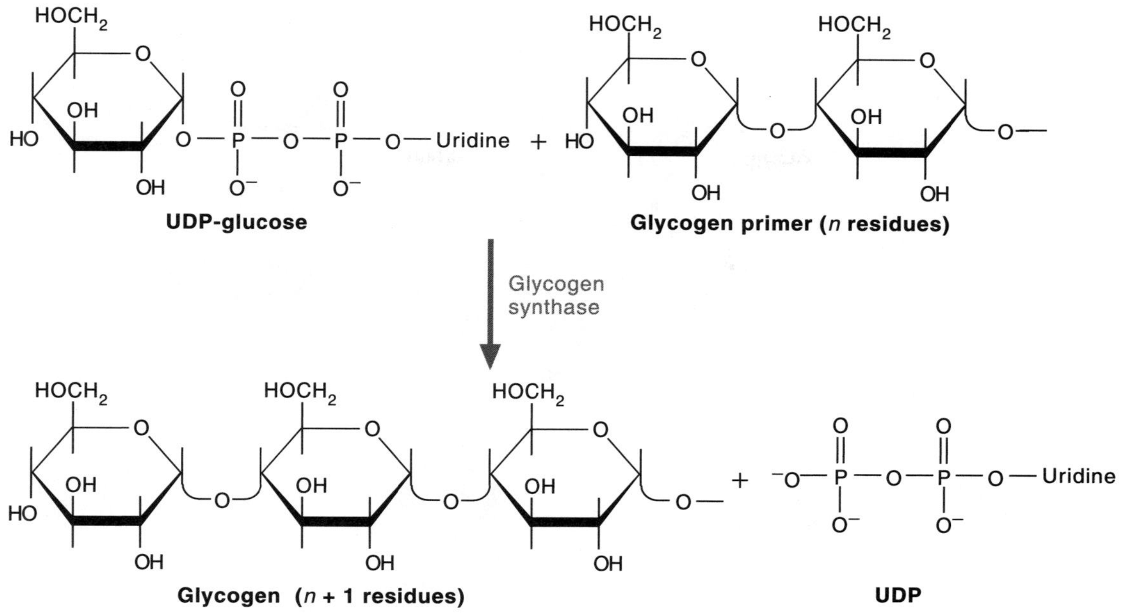

FIGURE 8.13

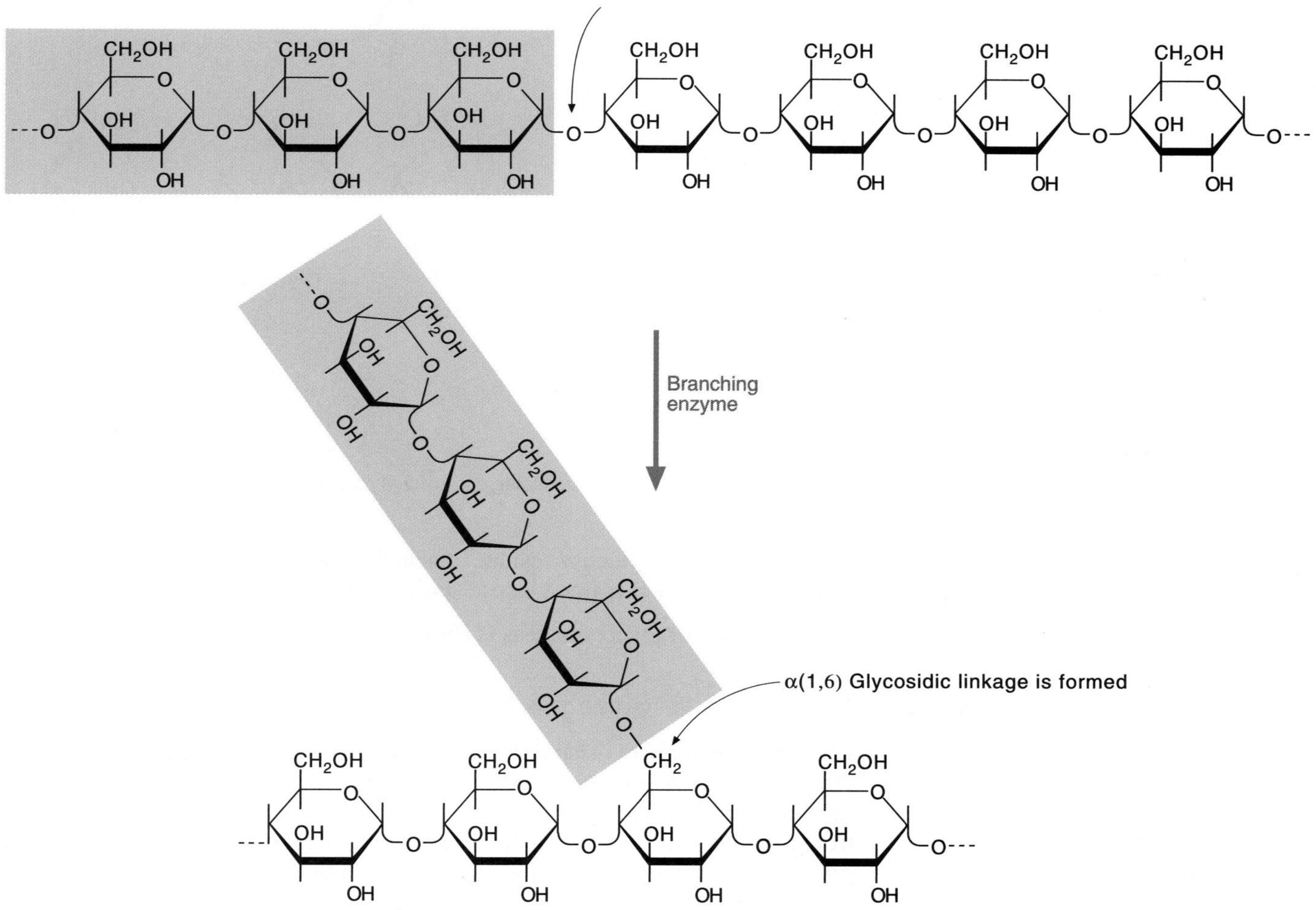

Glycogen Synthesis.

(a) The enzyme glycogen synthase breaks the ester linkage of UDP-glucose and forms an $\alpha(1,4)$ glycosidic bond between glucose and the growing glycogen chain. (b) Branching enzyme is responsible for the synthesis of $\alpha(1,6)$ linkages in glycogen.

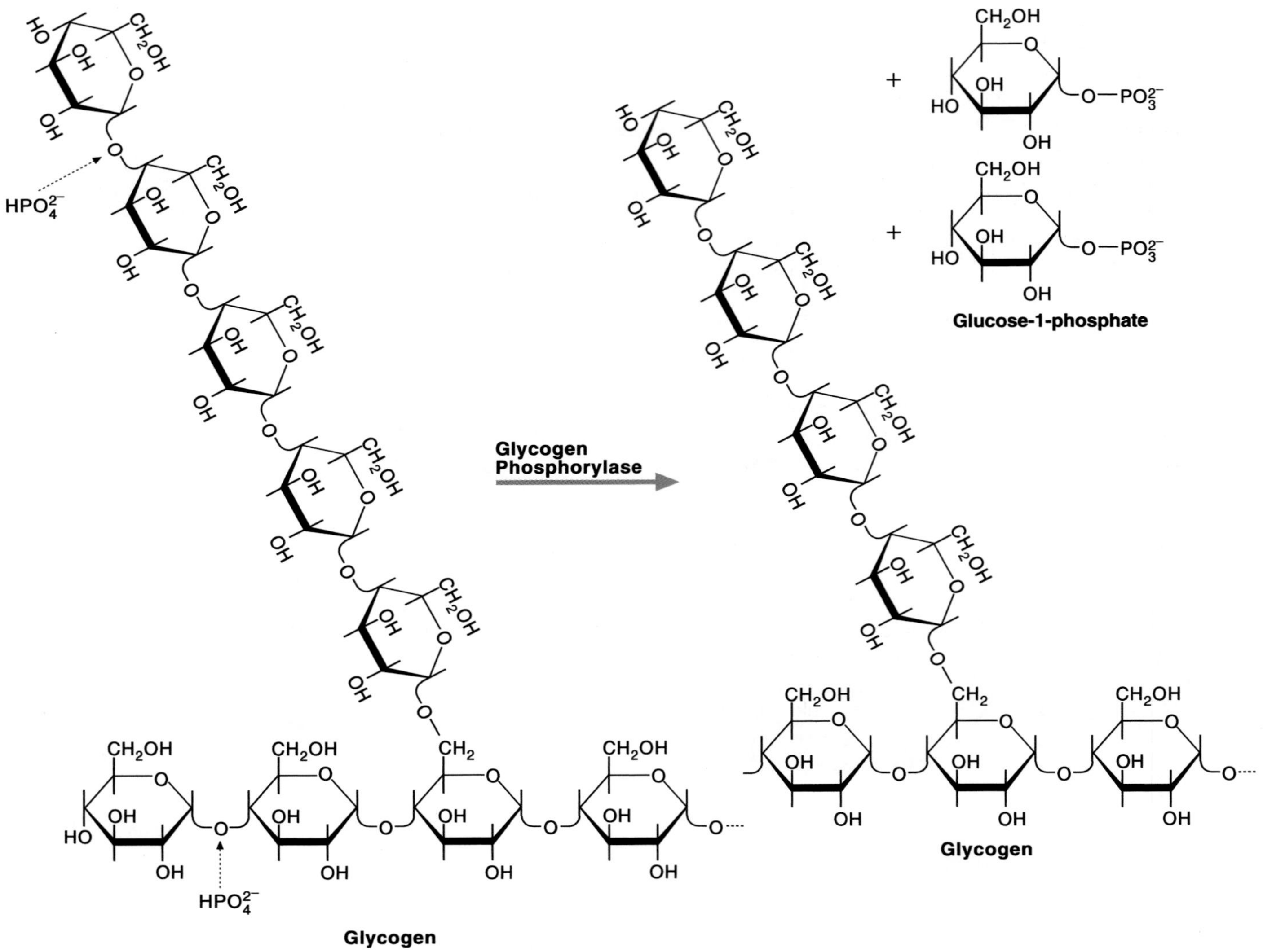

Glucose-1-phosphate

Glycogen Phosphorylase →

Glycogen

Glycogen

FIGURE 8.14

Glycogen Degradation.

Glycogen phosphorylase catalyzes the removal of glucose residues from the nonreducing ends of a glycogen chain.

removes the single glucose residue attached at each branch point. The product of this latter reaction is free glucose (Figure 8.15).

A summary of glycogenolysis is shown in Figure 8.16.

Regulation of Glycogen Metabolism

Glycogen metabolism is carefully regulated to avoid wasting energy. Both synthesis and degradation are controlled through a complex mechanism involving insulin, glucagon, and epinephrine. These hormones initiate processes that control several sets of enzymes. The binding of glucagon to liver cells stimulates glycogenolysis and inhibits glycogenesis. As blood glucose levels drop in the hours after a meal, glucagon ensures that glucose will be released into the bloodstream. After glucagon binds to its receptor, adenylate cyclase (a cell membrane enzyme) is stimulated to convert ATP to the intracellular signal molecule 3'-5' cyclic AMP, usually referred to as cAMP. Then cAMP initiates a reaction cascade (described in Chapter 16) that amplifies the original signal. Within seconds a few glucagon molecules have initiated release of thousands of glucose molecules.

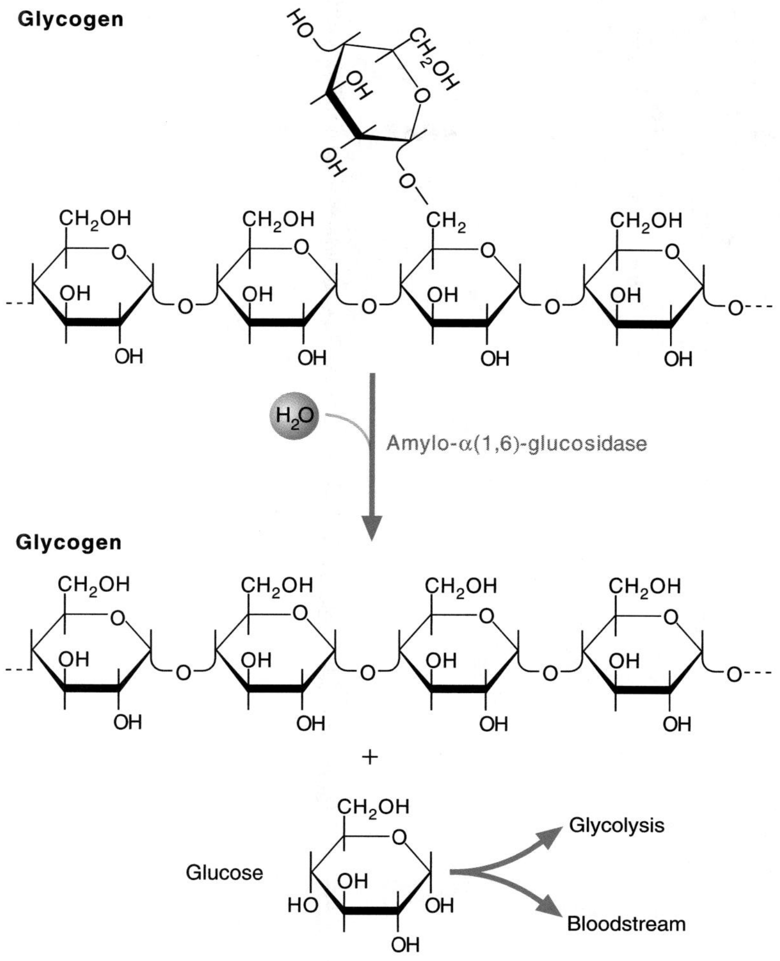

Glycogen

Amylo-α(1,6)-glucosidase

Glycogen

Glucose → Glycolysis

Glucose → Bloodstream

FIGURE 8.15

Glycogen Degradation.

Branch points in glycogen are removed by debranching enzyme (amylo-α(1,6)-glucosidase).

When occupied, the insulin receptor becomes an active tyrosine kinase enzyme that causes a phosphorylation cascade that ultimately has the opposite effect of the glucagon/cAMP system: the enzymes of glycogenolysis are inhibited and the enzymes of glycogenesis are activated. Insulin also increases the rate of glucose uptake into several types of target cells, but not liver or brain cells.

Emotional or physical stress releases epinephrine from the adrenal medulla. Epinephrine promotes glycogenolysis and inhibits glycogenesis. In emergency situations, when epinephrine is released in relatively large quantities, massive production of glucose provides the energy required to manage the situation. This effect is referred to as the flight-or fight response. Epinephrine initiates the process by activating adenylate cyclase in liver and muscle cells. Two other second messengers, calcium ions and inositol trisphosphate (Chapter 16), are also believed to be involved in epinephrine's action.

Glycogen synthase and glycogen phosphorylase have both active and inactive conformations that are interconverted by covalent modification. The active form of glycogen synthase, known as the I (independent) form, is converted to the inactive or D (dependent) form by phosphorylation. In contrast, the inactive form of glycogen phosphorylase (phosphorylase b) is converted to the active form (phosphorylase a) by the phosphorylation of a specific serine residue. The phosphorylating enzyme

KEY CONCEPTS 8.4

During glycogenesis, glycogen synthase catalyzes the transfer of the glucosyl group of UDP-glucose to the nonreducing ends of glycogen, and glycogen branching enzyme catalyzes the formation of branch points. Glycogenolysis requires glycogen phosphorylase and debranching enzyme. Glycogen metabolism is regulated by the actions of three hormones: glucagon, insulin, and epinephrine.

FIGURE 8.16

Glycogen Degradation.

Glycogen phosphorylase cleaves the $\alpha(1,4)$ linkages of glycogen to yield glucose-1-phosphate until it comes within four glucose residues of a branch point. Debranching enzyme transfers three of these residues to a nearby nonreducing end and releases the fourth residue as free glucose. The repeated actions of both enzymes can lead to the complete degradation of glycogen.

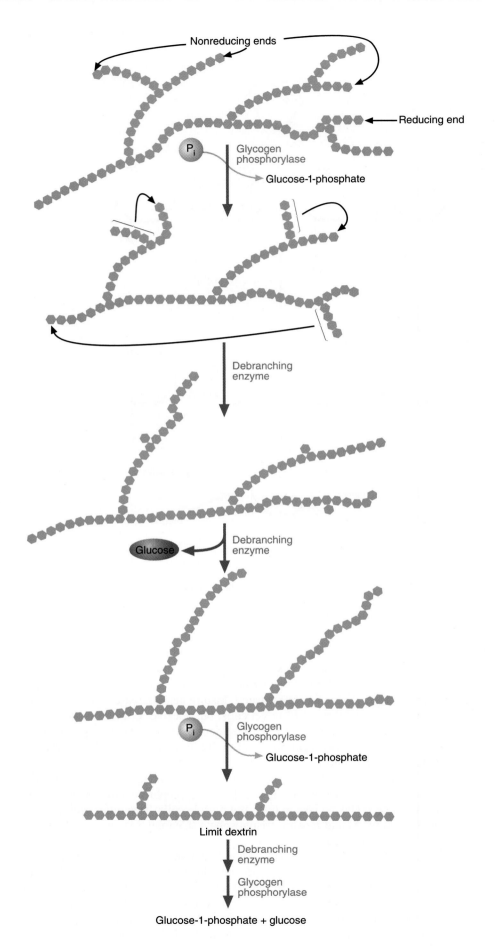

is called phosphorylase kinase. Phosphorylation of both glycogen synthase and phosphorylase kinase is catalyzed by a protein kinase, which is activated by cAMP. Glycogen synthesis occurs when glycogen synthase and glycogen phosphorylase have been dephosphorylated. This conversion is catalyzed by phosphoprotein phosphatase. Phosphoprotein phosphatase also inactivates phosphorylase kinase. The major factors in this complex process are summarized in Figure 8.17.

Glycogen storage diseases are caused by inherited defects of one or more enzymes involved in glycogen synthesis or degradation. Patients with *Cori's disease*, caused by a deficiency of debranching enzyme, have enlarged livers (*hepatomegaly*) and low blood sugar concentrations (**hypoglycemia**). Can you suggest what causes these symptoms?

QUESTION 8.7

FIGURE 8.17

Major Factors Affecting Glycogen Metabolism.

The binding of glucagon (released from the pancreas in response to low blood sugar) and/or epinephrine (released from the adrenal glands in response to stress) to their cognate receptors on the surface of target cells initiates a reaction cascade that converts glycogen to glucose-1-phosphate and inhibits glycogenesis. Insulin inhibits glycogenolysis and stimulates glycogenesis in part by decreasing the synthesis of cAMP and activating phosphoprotein phosphatase.

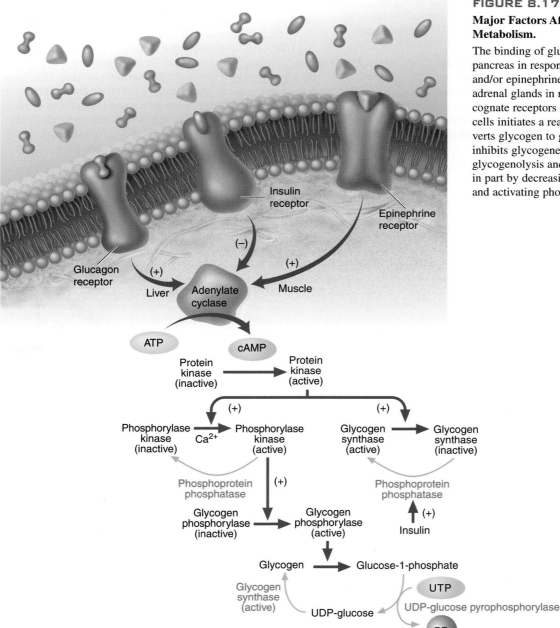

SUMMARY

1. The metabolism of carbohydrates is dominated by glucose because this sugar is an important fuel molecule in most organisms. If cellular energy reserves are low, glucose is degraded by the glycolytic pathway. Glucose molecules that are not required for immediate energy production are stored as either glycogen (in animals) or starch (in plants).

2. During glycolysis, glucose is phosphorylated and cleaved to form two molecules of glyceraldehyde-3-phosphate. Each glyceraldehyde-3-phosphate is then converted to a molecule of pyruvate. A small amount of energy is captured in two molecules each of ATP and NADH. In anaerobic organisms, pyruvate is converted to waste products. During this process, NAD^+ is regenerated so that glycolysis can continue. In the presence of O_2, aerobic organisms convert pyruvate to acetyl-CoA and then to CO_2 and H_2O. Glycolysis is controlled primarily by allosteric regulation of three enzymes—hexokinase, PFK-1, and pyruvate kinase—and by the hormones glucagon and insulin.

3. During gluconeogenesis, molecules of glucose are synthesized from noncarbohydrate precursors (lactate, pyruvate, glycerol, and certain amino acids). The reaction sequence in gluconeogenesis is largely the reverse of glycolysis. The three irreversible glycolytic reactions (the synthesis of pyruvate, the conversion of fructose-1,6-bisphosphate to fructose-6-phosphate, and the formation of glucose from glucose-6-phosphate) are bypassed by alternate energetically favorable reactions.

4. The pentose phosphate pathway, in which glucose-6-phosphate is oxidized, occurs in two phases. In the oxidative phase, two molecules of NADPH are produced as glucose-6-phosphate is converted to ribulose-5-phosphate. In the nonoxidative phase, ribose-5-phosphate and other sugars are synthesized. If cells need more NADPH than ribose-5-phosphate, a component of nucleotides and the nucleic acids, then metabolites of the nonoxidative phase are converted into glycolytic intermediates.

5. Several sugars other than glucose are important in vertebrate carbohydrate metabolism. These include fructose, galactose, and mannose.

6. The substrate for glycogen synthesis is UDP-glucose, an activated form of the sugar. UDP-glucose pyrophosphorylase catalyzes the formation of UDP-glucose from glucose-1-phosphate and UTP. Glucose-6-phosphate is converted to glucose-1-phosphate by phosphoglucomutase. To form glycogen requires two enzymes: glycogen synthase and branching enzyme. Glycogen degradation requires glycogen phosphorylase and debranching enzyme. The balance between glycogenesis (glycogen synthesis) and glycogenolysis (glycogen breakdown) is carefully regulated by several hormones (insulin, glucagon, and epinephrine).

SUGGESTED READINGS

Fothergill-Gilmore, L.A., and Michels, P.A., Evolution of Glycolysis, *Prog. Biophys. Mol. Biol.*, 59:105–135, 1993.

Hallfrisch, J., Metabolic Effects of Dietary Fructose, *FASEB J.*, 4:2652–2660, 1990.

Lehmann, J., *Carbohydrates: Structure and Biology*, Thieme, New York, 1998.

Pilkus, S. J., Mahgrabi, M. R., and Claus, T. A., Hormonal Regulation of Hepatic Gluconeogenesis and Glycolysis, *Ann. Rev. Biochem.*, 57:755–783, 1988.

Shulman, G. I., and Landau, B. R., Pathways of Glycogen Repletion, *Physiol. Rev.*, 72(4):1019–1035, 1992.

VanSchaftingen, E., Fructose-2,6-Bisphosphate, *Adv. Enzymol.*, 59:315–395, 1987.

KEY WORDS

aerobic respiration, *236*
aldol cleavage, *239*
amphibolic pathway, *235*
anaerobic organisms, *236*
antioxidants, *256*
citric acid cycle, *243*
Cori cycle, *252*
decarboxylation, *245*
electron transport system, *243*
fermentation, *243*
gluconeogenesis, *235*
glucose-alanine cycle, *254*
glycogenesis, *235*
glycogenolysis, *235*
glycolysis, *235*
hypoglycemia, *269*
malate shuttle, *251*
Pasteur effect, *249*
pentose phosphate pathway, *235*
substrate-level phosphorylation, *240*
tautomer, *242*
tautomerization, *242*

REVIEW QUESTIONS

1. Upon entering a cell, glucose is phosphorylated. Give two reasons why this reaction is required.
2. Describe the functions of the following molecules:
   a. insulin
   b. glucagon
   c. fructose-2, 6-bisphosphate
   d. congeners
   e. glutathione
   f. GSSG
   g. NADPH
3. Describe the structural differences between ribose-5-phosphate and ribulose-5-phosphate.

4. In which locations in the eukaryotic cell do the following processes occur?
   a. gluconeogenesis
   b. glycolysis
   c. pentose phosphate pathway
5. Compare the entry-level substrates, products, and metabolic purposes of glycolysis and gluconeogenesis.
6. Define substrate-level phosphorylation. Which two reactions in glycolysis are in this category?
7. What is the principal reason that organisms such as yeast produce alcohol?
8. Why is pyruvate not oxidized to CO_2 and H_2O under anaerobic conditions?
9. Describe how epinephrine promotes the conversion of glycogen to glucose.
10. Glycolysis occurs in two stages. Describe what is accomplished in each stage.

11. What effects do the following molecules have on gluconeogenesis?
    a. lactate
    b. ATP
    c. pyruvate
    d. glycerol
    e. AMP
    f. acetyl-CoA
12. Describe the physiological conditions that activate gluconeogenesis.
13. The following two reactions constitute a wasteful cycle:

$$\text{Glucose} + \text{ATP} \longrightarrow \text{glucose-6-phosphate}$$

$$\text{Glucose-6-phosphate} + H_2O \longrightarrow \text{glucose} + P_i$$

Suggest how such wasteful cycles are prevented or controlled.

THOUGHT QUESTIONS

1. An individual has a genetic deficiency that prevents the production of glucokinase. Following a carbohydrate meal, do you expect blood glucose levels to be high, low, or about normal? What organ accumulates glycogen under these circumstances?
2. Glycogen synthesis requires a short primer chain. Explain how new glycogen molecules are synthesized given this limitation.
3. Why is fructose metabolized more rapidly than glucose?
4. What is the difference between an enol-phosphate ester and a normal phosphate ester that gives PEP such a high phosphate group transfer potential?

5. In aerobic oxidation, oxygen is the ultimate oxidizing agent (electron acceptor). Name two common oxidizing agents in anaerobic fermentation.
6. Why is it important that gluconeogenesis is not the exact reverse of glycolysis?
7. Compare the structural formula of ethanol, acetate, and acetaldehyde. Which molecule is the most oxidized? Which is the most reduced? Explain your answers.

CHAPTER NINE

Aerobic Metabolism I: The Citric Acid Cycle

In aerobic cells most energy is generated within the mitochondrion. Dioxygen (O_2) is the final electron acceptor in the oxidation of nutrient molecules.

Over two billion years ago prokaryotes such as the cyanobacteria began creating an oxygenated atmosphere. The oxygen they produced as a waste product of photosynthesis triggered a revolution in the living world. Many organisms were metabolically incapable of dealing with the growing amounts of this highly reactive molecule. Although many species either became extinct or were forced into isolated oxygen-free habitats, others evolved molecular mechanisms that allowed them to exploit dioxygen (O_2, often referred to as oxygen) as a means of capturing energy. Modern aerobic organisms transduce the chemical bond energy of food molecules into the bond energy of ATP by using oxygen as the terminal acceptor of the electrons extracted from food molecules. The capacity to use oxygen to oxidize nutrients such as glucose and fatty acids yields a substantially greater amount of energy than does fermentation.

As the first primordial life forms emerged on the Earth, they were sustained by preformed, simple organic molecules such as carboxylic and amino acids. The source of these substances, which were used as building block and fuel molecules by primitive living systems, is believed to have been chemical reactions driven by electrical discharges, solar radiation, and thermal forces deep within the planet. In addition, untold tons of chemical matter arrived from outer space as the Earth was bombarded by meteors and other cosmic debris. Early organisms (primordial prokaryotic cells) eventually became so abundant that they consumed organic molecules faster than those molecules were formed by natural forces. As supplies of preformed molecules dwindled, some organisms evolved new mechanisms for obtaining food. Some organisms evolved the capacity to synthesize photo-sensitive pigments that captured light energy and converted it to chemical bond energy. This mechanism, photosynthesis, had a dramatic and far-reaching effect on the global environment. Over three billion years ago, photosynthetic cells began to produce their own food by using light energy to transform CO_2 and H_2O into organic molecules. Dioxygen (O_2) is a by-product of this process. As photosynthesis occurred on an ever increasing scale, the oxygen content of the atmosphere increased. Because O_2 combines readily with other molecules (e.g., $4 NH_3 + 3 O_2 \longrightarrow 2 N_2 + 6 H_2O$), the earth's atmosphere was gradually converted (over a billion-year time span) to one consisting principally of dinitrogen, water vapor, carbon dioxide, and oxygen. Most living organisms that arose under the reducing conditions of the primitive Earth were unprepared for living in an oxi-dizing atmosphere. The species that survived the transition did so because they developed methods to protect themselves from oxygen's toxic effects. Their descendants, today's organisms, use one of the following strategies. **Obligate anaerobes**, organisms that grow only in the absence of oxygen, avoid the gas by living in highly reduced environments such as soil. They use fermentative processes to satisfy their energy requirements. **Aerotolerant anaerobes**, which also depend on fermentation for their energy needs, possess detoxifying enzymes and antioxidant molecules that protect against oxygen's toxic products. **Facul-tative anaerobes** not only possess the mechanisms needed for detoxifying oxy-gen metabolites, they can also generate energy by using oxygen as an electron acceptor when the gas is present. Finally, **obligate aerobes** are highly depen-dent on oxygen for energy production. They protect themselves from the poten-tially dangerous consequences of exposure to oxygen with elaborate mechanisms composed of enzymes and antioxidant molecules.

Facultative anaerobes and obligate aerobes that use oxygen to generate energy employ the following biochemical processes: the citric acid cycle, the electron transport pathway, and oxidative phosphorylation. In eukaryotes these processes occur within the mitochondrion (Figure 9.1). The citric acid cycle is a meta-bolic pathway in which two carbon fragments derived from organic fuel mole-cules are oxidized to form CO_2 and the coenzymes NAD^+ and FAD are reduced to form NADH and $FADH_2$, which act as electron carriers. The electron transport pathway, also referred to as the electron transport chain (ETC), is a mechanism by which electrons are transferred from reduced coenzymes to an acceptor (usu-ally O_2). In oxidative phosphorylation, the energy released by electron transport is captured in the form of a proton gradient that drives the synthesis of ATP, the energy currency of living organisms. Chapter 9 begins with a review of oxidation-reduction reactions and the relationship between electron flow and energy trans-duction. This is followed by a detailed discussion of the citric acid cycle, the central pathway in aerobic metabolism, and its roles in energy generation and biosynthesis. In Chapter 10, the discussion of aerobic metabolism continues with an examination of electron transport and oxidative phosphorylation, the means by which aerobic organisms use oxygen to generate significant amounts of ATP. It ends with a review of *oxidative stress*, a series of reactions in which toxic oxygen

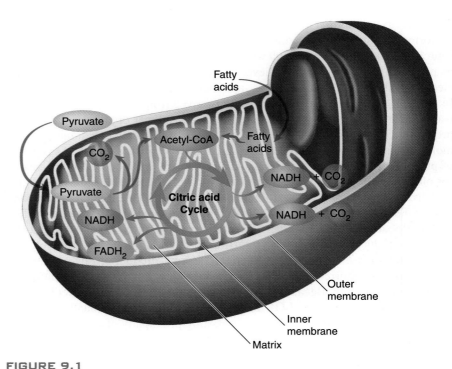

FIGURE 9.1

Aerobic Metabolism in the Mitochondrion.

In eukaryotic cells aerobic metabolism occurs within the mitochondrion. Acetyl-CoA, the oxidation product of pyruvate, fatty acids, and certain amino acids (not shown), is oxidized by the reactions of the citric acid cycle within the mitochondrial matrix. The principal products of the cycle are the reduced coenzymes NADH and $FADH_2$ and CO_2. The high-energy electrons of NADH and $FADH_2$ are subsequently donated to the electron transport chain (ETC), a series of electron carriers in the inner membrane. The terminal electron acceptor for the ETC is O_2. The energy derived from the electron transport mechanism drives ATP synthesis by creating a proton gradient across the inner membrane. The large folded surface of the inner membrane is studded with ETC complexes, numerous types of transport proteins, and ATP synthase, the enzyme complex responsible for ATP synthesis.

metabolites damage cell structure and function, and the methods that living organisms use to protect themselves.

9.1 OXIDATION-REDUCTION REACTIONS

In living organisms, both energy-capturing and energy-releasing processes consist largely of redox reactions. Recall that redox reactions occur when electrons are transferred between an electron donor (reducing agent) and an electron acceptor (oxidizing agent). In some redox reactions, only electrons are transferred. For example, in the reaction

$$Cu^+ + Fe^{3+} \rightleftharpoons Cu^{2+} + Fe^{2+}$$

an electron is transferred from Cu^+ to Fe^{3+}. Cu^+, the reducing agent, is oxidized to form Cu^{2+}. Meanwhile, Fe^{3+} is reduced to Fe^{2+}. In many reactions, however, both electrons and protons are transferred. For example, in the reaction catalyzed by lactate dehydrogenase, 2 protons (H^+) and 2 electrons are transferred as pyruvate is reduced to form lactate and NAD^+ (Figure 9.2).

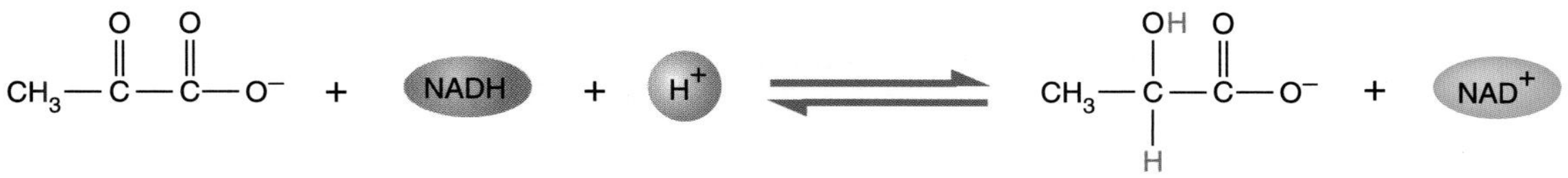

$$CH_3-\overset{\overset{O}{\|}}{C}-\overset{\overset{O}{\|}}{C}-O^- \ + \ NADH \ + \ H^+ \ \rightleftharpoons \ CH_3-\overset{\overset{OH}{|}}{\underset{H}{C}}-\overset{\overset{O}{\|}}{C}-O^- \ + \ NAD^+$$

FIGURE 9.2

Reduction of Pyruvate by NADH.

In this redox reaction, both protons and electrons are transferred.

Redox reactions are more easily understood if they are separated into half-reactions. For example, in the reaction between copper and iron the Cu^+ ion lost an electron to become Cu^{2+}:

$$Cu^+ \rightleftharpoons Cu^{2+} + e^-$$

This equation indicates that Cu^+ is the electron donor. (Together Cu^+ and Cu^{2+} constitute a **conjugate redox pair**.) As Cu^+ loses an electron, Fe^{3+} gains an electron to form Fe^{2+}:

$$Fe^{3+} + e^- \rightleftharpoons Fe^{2+}$$

In this half-reaction, Fe^{3+} is an electron acceptor. The separation of redox reactions emphasizes that electrons are always the common intermediates between half-reactions.

The constituents of half-reactions may be observed in an electrochemical cell (Figure 9.3). Each half-reaction takes place in a separate container or *half-cell*. The movement of electrons generated in the half-cell undergoing oxidation (e.g., $Cu^+ \longrightarrow Cu^{2+} + e^-$) generates a voltage (or potential difference) between the two half-cells. The sign of the voltage (measured by a voltmeter) is positive or negative according to the direction of the electron flow. The magnitude of the potential difference is a measure of the energy that drives the reaction.

The tendency for a specific substance to lose or gain electrons is called its **redox** or **reduction potential**. The redox potential of a conjugate redox pair is measured in an electrochemical cell against a reference standard, usually a standard hydrogen electrode. The redox potential of the standard hydrogen electrode is 0.0 V at 1 atm, by definition. Substances with a more negative

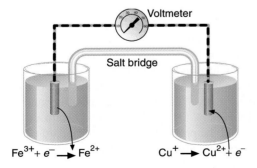

FIGURE 9.3

An Electrochemical Cell.

Electrons flow from the copper electrode through the voltmeter to the iron electrode. The salt bridge containing KCl completes the electrical circuit. The voltmeter measures the electrical potential as electrons flow from one half-cell to the other.

TABLE 9.1
Standard Reduction Potentials*

Redox Half-Reaction	Standard Reduction Potentials ($E^{0'}$) (V)
$2\ H^+ + 2\ e^- \rightarrow H_2$	−0.42
α-Ketoglutarate + CO_2 + 2 H^+ + 2 e^- → isocitrate	−0.38
$NAD^+ + H^+ + 2\ e^- \rightarrow NADH$	−0.32
$S + 2\ H^+ + 2\ e^- \rightarrow H_2S$	−0.23
$FAD + 2\ H^+ + 2\ e^- \rightarrow FADH_2$	−0.22
Acetaldehyde + 2 H^+ + 2 e^- → ethanol	−0.20
Pyruvate + 2 H^+ + 2 e^- → lactate	−0.19
Oxaloacetate + 2 H^+ + 2 e^- → malate	−0.166
$Cu^+ \rightarrow Cu^{2+} + e^-$	−0.16
Fumarate + 2 H^+ + 2 e^- → succinate	−0.031
Cytochrome b (Fe^{3+}) + e^- → cytochrome b (Fe^{2+})	+0.075
Cytochrome c_1 (Fe^{3+}) + e^- → cytochrome c_1 (Fe^{2+})	+0.22
Cytochrome c (Fe^{3+}) + e^- → cytochrome c (Fe^{2+})	+0.235
Cytochrome a (Fe^{3+}) + e^- → cytochrome a (Fe^{2+})	+0.29
$NO_3^- + 2\ H^+ + 2\ e^- \rightarrow NO_2^- + H_2O$	+0.42
$NO_2^- + 8\ H^+ + 6\ e^- \rightarrow NH_4^+ + 2\ H_2O$	+0.44
$Fe^{3+} + e^- \rightarrow Fe^{2+}$	+0.77
$\frac{1}{2}O_2 + 2\ H^+ + 2\ e^- \rightarrow H_2O$	+0.82

*By convention, redox reactions are written with the reducing agent to the right of the oxidizing agent and the number of electrons transferred. In this table the redox pairs are listed in order of increasing E_0' values. The more negative the E_0' value is for a redox pair, the lower its affinity for electrons. The more positive the E_0' value is, the greater the affinity of the redox pair for electrons. Under appropriate conditions, a redox half-reaction reduces any of the half-reactions below it in the table.

reduction potential will transfer electrons to a substance with a more positive reduction potential and the $\Delta E^{0'}$ will be positive. In biochemistry the reference half-reaction is

$$2\ H^+ + 2\ e^- \rightleftharpoons H_2$$

when

$$pH = 7$$
$$Temperature = 25°C$$
$$Pressure = 1\ atm$$

Under these conditions the reduction potential of the hydrogen electrode is −0.42 V when measured against the standard hydrogen electrode in which the [H^+] is 1 M. Substances with reduction potentials lower than −0.42 V (i.e., those with more negative values) have a lower affinity for electrons than does H^+. Substances with higher reduction potentials (i.e., those with more positive values) have a greater affinity for electrons (Table 9.1). (The pH in the test electrode is 7.0 for each of the redox half-reactions and the pH of the reference standard electrode is 0.)

Electrons flow spontaneously from a species with a more negative $E^{0'}$ value to a species with a more positive $E^{0'}$, so that the $\Delta E^{0'}$ is positive. The relationship between $\Delta E^{0'}$ and $\Delta G^{\circ'}$ is

$$\Delta G^{\circ'} = -nF\ \Delta E^{0'}$$

where

$\Delta G^{\circ'}$ = the standard free energy

n = the number of electrons transferred

F = the Faraday constant (96,485 J/V·mol)

$\Delta E^{0\prime}$ = the difference in reduction potential between the electron donor and the electron acceptor under standard conditions

Most of the aerobic cell's free energy is captured by the mitochondrial electron transport system (Chapter 10). During this process, electrons are transferred from a redox pair with a more negative reduction potential ($NADH/NAD^+$) to those with more positive reduction potentials. The last component in the system is the $H_2O/\frac{1}{2} O_2$ pair:

$$\frac{1}{2} O_2 + NADH + H^+ \longrightarrow H_2O + NAD^+$$

QUESTION 9.1

Using Table 9.1 determine which of the following reactions will proceed as written:

$$CH_3CH_2OH + 2 \text{ cyt b } (Fe^{3+}) \longrightarrow CH_3CHO + 2 \text{ cyt b } (Fe^{2+}) + 2 H^+$$

$$NO_2^- + H_2O + 2 \text{ cyt b } (Fe^{3+}) \longrightarrow 2 \text{ cyt b } (Fe^{2+}) + NO_3^- + 2 H^+$$

QUESTION 9.2

Which of the following reactions are redox reactions? For each redox reaction identify the oxidizing and reducing agents.

1. Glucose + ATP $\longrightarrow$ glucose-1-phosphate + ADP

2.

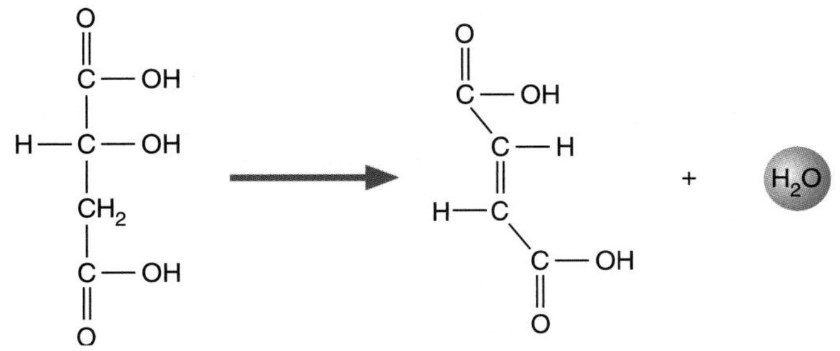

3. Lactate + NAD^+ $\longrightarrow$ pyruvate + $NADH$ + H^+

4. $NO_2^- + 8 H^+ + 6 \text{ cyt b } (Fe^{2+}) \longrightarrow NH_4^+ + 2 H_2O + 6 \text{ cyt b } (Fe^{3+})$

5. $CH_3CHO + NADH + H^+ \longrightarrow CH_3CH_2OH + NAD^+$

The free energy released as a pair of electrons passes from NADH to O_2 under standard conditions is calculated as follows:

$$\begin{aligned}
\Delta G^{0\prime} &= -mF\Delta E^{0\prime} \\
&= -2(96.5 \text{ kJ/V·mol})[0.815 - (-0.32)] \\
&= -220 \text{ kJ/mol}
\end{aligned}$$

A significant portion of the free energy generated as electrons move from NADH to O_2 in the electron transport system is used to synthesize ATP.

In several metabolic processes, electrons move from redox pairs with more positive reduction potentials to those with more negative reduction potentials. Of course, energy is required. The most prominent example of this phenomenon is photosynthesis (Chapter 13). Photosynthetic organisms use captured light energy to drive electrons from electron donors, such as water, to electron acceptors with more negative reduction potentials (Figure 9.4). The energized electrons eventually flow back to acceptors with more positive reduction potentials, thereby providing energy for ATP synthesis and CO_2 reduction to form carbohydrate.

KEY CONCEPTS 9.1

In living organisms, both energy-capturing and energy-releasing processes consist primarily of redox reactions. In redox reactions electrons move between an electron donor and an electron acceptor. In many reactions, both electrons and protons are transferred.

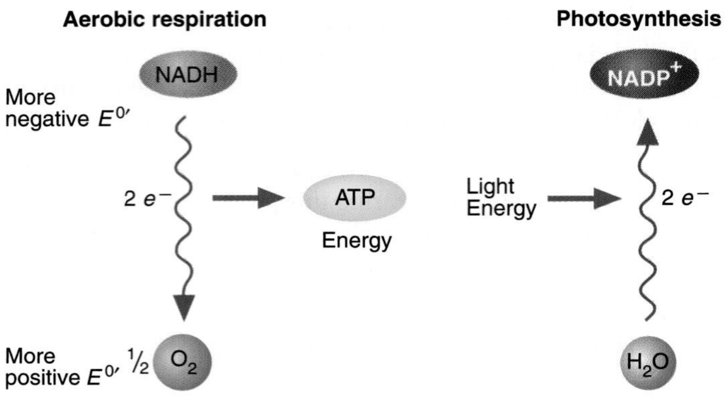

FIGURE 9.4

Electron Flow and Energy.

Electron flow may be used to generate and capture energy in aerobic respiration. Energy may also be used to drive electron flow in photosynthesis. ($NADP^+$ is a more phosphorylated version of NAD^+.)

Aerobic respiration   Photosynthesis

PROBLEM 9.1

Use the following half-cell potentials to calculate (a) the overall cell potential and (b) $\Delta G^{\circ\prime}$.

$$\text{Succinate} + \tfrac{1}{2} O_2 \longrightarrow \text{fumarate} + H_2O$$

The half-reactions are

$$\text{Fumarate} + 2\,H^+ + 2\,e^- \longrightarrow \text{succinate} \qquad (E'_0 = -0.031 \text{ V})$$

$$\tfrac{1}{2} O_2 + 2\,H^+ + 2\,e^- \longrightarrow H_2O \qquad (E_0 = +0.82 \text{ V})$$

Solution

Write one of the half-reactions as an oxidation (i.e., reverse the equation) and add the two half-reactions:

$$\text{Succinate} \longrightarrow \text{fumarate} + 2\,H^+ + 2\,e^- \qquad (E'_0 = +0.031 \text{ V}) \text{ (oxidation)}$$

$$\tfrac{1}{2} O_2 + 2\,H^+ + 2\,e^- \longrightarrow H_2O \; (E'_0 = +0.82\text{V}) \text{ (reduction)}$$

The net reaction is therefore

$$\text{Succinate} + \tfrac{1}{2} O_2 \longrightarrow \text{fumarate} + H_2O$$

a. The overall potential is the sum of the potentials for each half-cell:

$$\Delta E'_0 = E'_0 \text{ (electron acceptor)} - E'_0 \text{ (electron donor)}$$

$$\Delta E'_0 = (+0.82 \text{ V}) - (-0.031 \text{ V})$$

$$= +0.82 + 0.031 = +0.85 \text{ V}$$

b. Use the formula to find $\Delta G^{\circ\prime}$.

$$\Delta G^{\circ\prime} = nF\Delta E^{0\prime}$$

$$= -(2)(96.5 \text{ kJ/V·mol})(0.85 \text{ V})$$

$$= -164.05 \text{ kJ/mol}$$

$$= -164 \text{ kJ/mol}$$

QUESTION 9.3

Because redox reactions play an important role in living processes, biochemists need to determine the oxidation state of the atoms in a molecule. In one method, the oxidation state of an atom is determined by assigning numbers to carbon atoms based on the type of groups attached to them. For example, a bond to a hydrogen is assigned the value -1. A bond to another carbon atom is valued at 0, and a bond to an electronegative atom such as oxygen or nitrogen is valued at $+1$. The values of a single carbon atom in a molecule may range from -4 (e.g., CH_4) to $+4$ (CO_2). Note that methane is a high-energy molecule and carbon dioxide is a low-energy molecule. As carbon changes its oxidation state from -4 (methane) to $+4$ (carbon dioxide), a large amount of energy is released. This process is therefore highly exothermic.

Ethanol is degraded in the liver by a series of redox reactions. Identify the oxidation state of the indicated carbon atom in each molecule in the following reaction sequence:

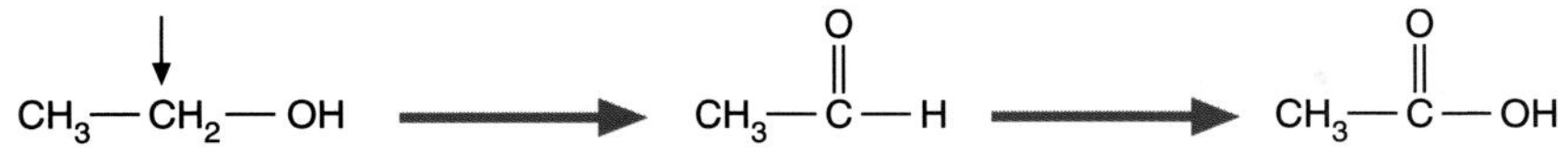

As CO_2 is incorporated into organic molecules during photosynthesis, is it being oxidized or reduced?

QUESTION 9.4

In Section 9.2 the citric acid cycle is examined. In this pathway, which is the first phase of aerobic metabolism, the energy released by the oxidation of two-carbon fragments derived from glucose, fatty acids, and some amino acids is conserved in the reduced coenzymes NADH and $FADH_2$.

9.2 CITRIC ACID CYCLE

The citric acid cycle (Figure 9.5) is a series of biochemical reactions aerobic organisms use to release chemical energy stored in the two-carbon acetyl group in acetyl-CoA. Acetyl-CoA is composed of an acetyl group derived from the breakdown of carbohydrates, lipids, and some amino acid that is linked to the acyl carrier molecule **coenzyme A** (Figure 9.6). Acetyl-CoA is synthesized from pyruvate (a partially oxidized product of the degradation of sugars and certain amino acids) in a series of reactions. Acetyl-CoA is also the product of fatty acid catabolism (described in Chapter 11) and certain reactions in amino acid metabolism (Chapter 15). In the citric acid cycle, the carbon atoms are oxidized to CO_2 and the high-energy electrons are transferred to NAD^+ and FAD to form the reduced coenzymes NADH and $FADH_2$, respectively.

In the first reaction of the citric acid cycle, a two-carbon acetyl group condenses with a four-carbon molecule (oxaloacetate) to form a six-carbon molecule (citrate) (Figure 9.7). During the subsequent seven reactions, in which two CO_2 molecules are produced and four pairs of electrons are removed from carbon compounds, citrate is reconverted to oxaloacetate. During one step in the cycle, the high-energy molecule guanosine triphosphate (GTP) is produced during a substrate-level phosphorylation. The net reaction for the citric acid cycle is as follows:

$$\text{Acetyl-CoA} + 3 \text{ NAD}^+ + \text{FAD} + \text{GDP} + P_i + 2 \text{ H}_2\text{O} \longrightarrow$$
$$2 \text{ CO}_2 + 3 \text{ NADH} + \text{FADH}_2 + \text{CoASH} + \text{GTP} + 3 \text{ H}^+$$

In addition to its role in energy production, the citric acid cycle plays another important role in metabolism. Cycle intermediates are substrates in a variety of biosynthetic reactions. Table 9.2 provides a summary of the roles of coenzymes in the citric acid cycle.

Conversion of Pyruvate to Acetyl–CoA

After its transport into the mitochondrial matrix, pyruvate is converted to acetyl-CoA in a series of reactions catalyzed by the enzymes in the pyruvate dehydrogenase complex. The net reaction, an oxidative decarboxylation, is as follows:

$$\text{Pyruvate} + \text{NAD}^+ + \text{CoASH} \longrightarrow \text{Acetyl-CoA} + \text{NADH} + \text{CO}_2 + \text{H}_2\text{O} + \text{H}^+$$

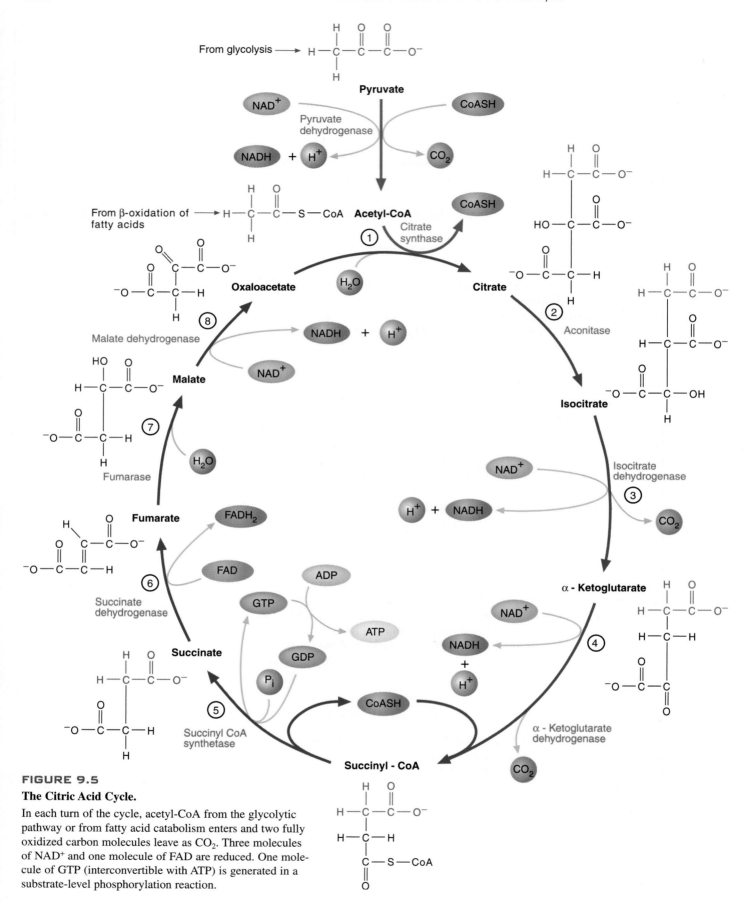

FIGURE 9.5

The Citric Acid Cycle.

In each turn of the cycle, acetyl-CoA from the glycolytic pathway or from fatty acid catabolism enters and two fully oxidized carbon molecules leave as CO_2. Three molecules of NAD^+ and one molecule of FAD are reduced. One molecule of GTP (interconvertible with ATP) is generated in a substrate-level phosphorylation reaction.

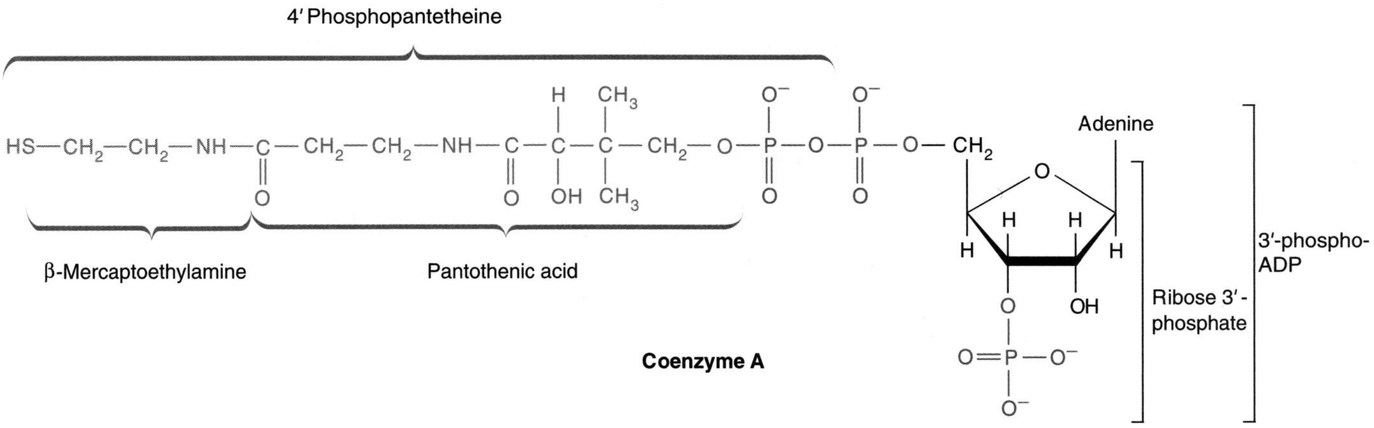

4′ Phosphopantetheine

β-Mercaptoethylamine   Pantothenic acid

Adenine

3′-phospho-ADP

Ribose 3′-phosphate

Coenzyme A

FIGURE 9.6

Coenzyme A.

In coenzyme A a 3′ phosphate derivative of ADP is linked to pantothenic acid via a phosphate ester bond. The β-mercaptoethylamine group of coenzyme A is attached to pantothenic acid by an amide bond. Coenzyme A is a carrier of acyl groups that range in size from the acetyl group to long chain fatty acids. Because the reactive SH group forms a thioester bond with acyl groups, coenzyme A is often abbreviated as CoASH. The carbon-sulfur bond of a thioester is more easily cleaved than the carbon-oxygen bond of an ester. Because the thioester linkage is more easily hydrolyzed than a simple ester linkage, the transfer of the acyl group is highly favored.

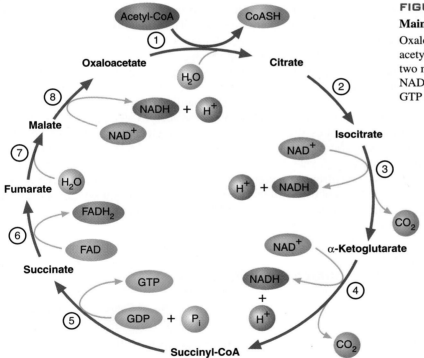

FIGURE 9.7

Main Reactions of the Citric Acid Cycle.

Oxaloacetate, a four-carbon molecule, condenses with acetyl-CoA to form citrate, a six-carbon molecule. Then two molecules of CO_2 are formed. Three molecules of NADH, one molecule of $FADH_2$, and one molecule of GTP are also formed.

Despite the apparent simplicity of this highly exergonic reaction ($\Delta G^{\circ\prime} = -33.5$ kJ/mol), its mechanism is one of the most complex known. The pyruvate dehydrogenase complex is a large multienzyme structure that contains three enzyme activities: pyruvate dehydrogenase (E_1), also known as pyruvate decarboxylase, dihydrolipoyl transacetylase (E_2), and dihydrolipoyl dehydrogenase (E_3). Each enzyme activity is present in multiple copies. Table 9.3 summarizes the number

TABLE 9.2
Summary of the Coenzymes in the Citric Acid Cycle

Coenzyme	Functions
Thiamine pyrophosphate (TPP)	Decarboxylation and aldehyde group transfer
Lipoic acid	Carrier of hydrogens or acetyl groups
NADH	Electron carrier
$FADH_2$	Electron carrier
Coenzyme A (CoASH)	Acetyl group carrier

of copies of each enzyme and the required coenzymes of *E. coli* pyruvate dehydrogenase.

In the first step, pyruvate dehydrogenase catalyzes the decarboxylation of pyruvate. A nucleophile is formed when a basic residue of the enzyme extracts a proton from the thiazole ring of thiamine pyrophosphate (TPP). The intermediate, hydroxyethyl-TPP (HETPP), forms after the nucleophilic thiazole ring attacks the carbonyl group of pyruvate with the resulting loss of CO_2 (Figure 9.8).

In the next several steps, the hydroxyethyl group of HETPP is converted to acetyl-CoA by dihydrolipoyl transacetylase. Lipoic acid (Figure 9.9) plays a crucial role in this transformation. Lipoic acid is bound to the enzyme through an amide linkage with the ε-amino group of a lysine residue. It reacts with HETPP to form an acetylated lipoic acid and free TPP. The acetyl group is then transferred to the sulfhydryl group of coenzyme A. Subsequently, the reduced lipoic acid is reoxidized by dihydrolipoyl dehydrogenase. The $FADH_2$ is reoxidized by NAD^+ (with its more negative reduction potential) to form the FAD required for the oxidation of the next reduced lipoic acid residue.

The activity of pyruvate dehydrogenase is regulated by two mechanisms: product inhibition and covalent modification (Section 6.5). The enzyme complex is allosterically activated by NAD^+, CoASH, and AMP. It is inhibited by high concentrations of ATP and the reaction products acetyl-CoA and NADH. In vertebrates these molecules also activate a kinase, which converts the active pyruvate dehydrogenase complex to an inactive phosphorylated form. High concentrations of the substrates pyruvate, CoASH, and NAD^+ inhibit the activity of the kinase. The pyruvate dehydrogenase complex is reactivated by a dephosphorylation reaction catalyzed by a phosphoprotein phosphatase. The phosphoprotein phosphatase is activated when the mitochondrial ATP concentration is low.

KEY CONCEPTS 9.2

Pyruvate is converted to acetyl-CoA by the enzymes in the pyruvate dehydrogenase complex. TPP, FAD, NAD^+, and lipoic acid are required coenzymes.

TABLE 9.3
E. Coli Pyruvate Dehydrogenase Complex

Enzyme Activity	Function	No. of Copies per Complex*	Coenzymes
Pyruvate dehydrogenase (E_1)	Decarboxylates pyruvate	24 (20–30)	TPP
Dihydrolipoyl transacetylase (E_2)	Catalyzes transfer of acetyl group to CoASH	24 (60)	Lipoic acid, CoASH
Dihydrolipoyl dehydrogenase (E_3)	Reoxidizes dihydrolipoamide	12 (20–30)	NAD^+, FAD

*The number of molecules of each enzyme activity found in mammalian pyruvate dehydrogenase is shown in parentheses.

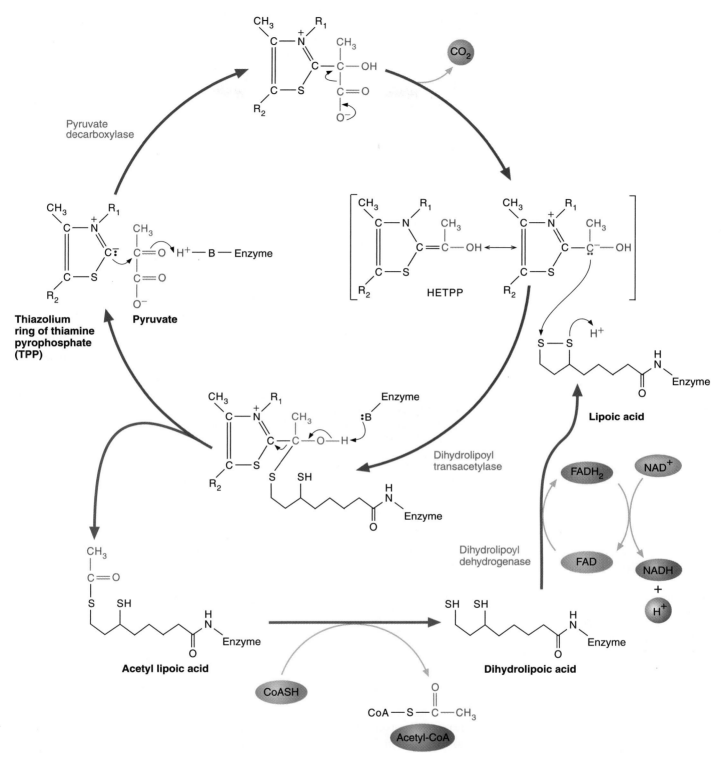

FIGURE 9.8

The Reactions Catalyzed by the Pyruvate Dehydrogenase Complex.

Pyruvate decarboxylase, which contains TPP, catalyzes the formation of the HETPP. Using lipoic acid as a cofactor, dihydrolipoyl transacetylase converts the hydroxyethyl group of HETPP to acetyl-CoA. Dihydrolipoyl dehydrogenase reoxidizes the reduced lipoic acid. (Refer to Figure 9.9 for the structure of lipoic acid.)

FIGURE 9.9

Lipoamide.

Lipoic acid is covalently bonded to the enzyme through an amide linkage with the ε-amino group of a lysine residue.

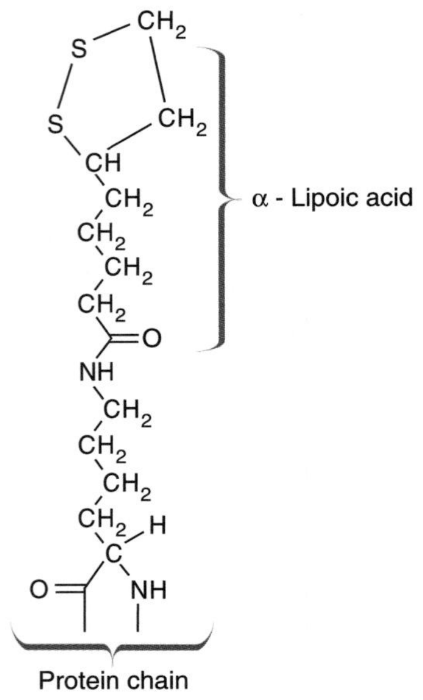

Reactions of the Citric Acid Cycle

The citric acid cycle is composed of eight reactions that occur in two stages:

1. The two-carbon acetyl group of acetyl-CoA enters the cycle by reacting with the four-carbon compound oxaloacetate (reactions 1–4). Two molecules of CO_2 are subsequently liberated.
2. Oxaloacetate is regenerated so it can react with another acetyl-CoA (reactions 5–8).

The reactions of the citric acid cycle are as follows:

1. Introduction of two carbons as acetyl-CoA. The citric acid cycle begins with the condensation of acetyl-CoA with oxaloacetate to form citrate:

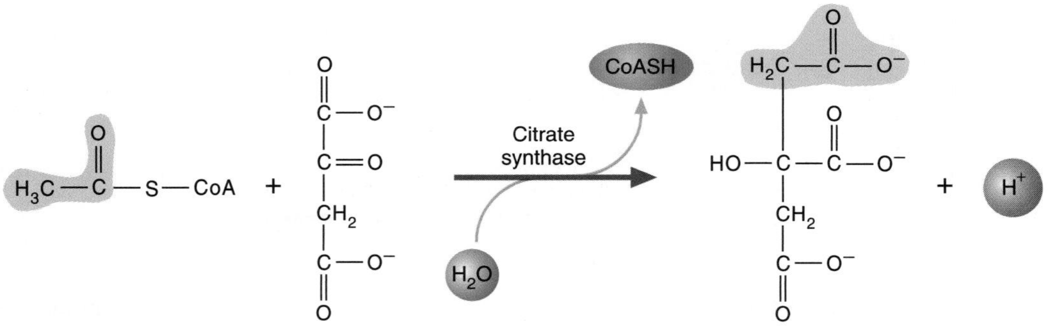

Note that this reaction is an aldol condensation. In this reaction the enzyme removes a proton from the methyl group of acetyl-CoA, thereby converting it to a carbanion. The nucleophilic methyl carbanion subsequently attacks the carbonyl carbon of oxaloacetate. The product, citroyl-CoA, rapidly hydrolyzes to form citrate and CoASH. Because of the hydrolysis of the high-energy thioester bond, the overall standard free energy change is equal to −33.5 kJ/mol, and citrate formation is highly exergonic.

2. Citrate is isomerized to form a secondary alcohol that can be easily oxidized. In the next reaction of the cycle, citrate, which contains a tertiary alcohol, is reversibly converted to isocitrate by aconitase. During this isomerization reaction, an intermediate called *cis*-aconitate is formed by dehydration. The carbon-carbon double bond of *cis*-aconitate is then rehydrated to form the more reactive secondary alcohol, isocitrate.

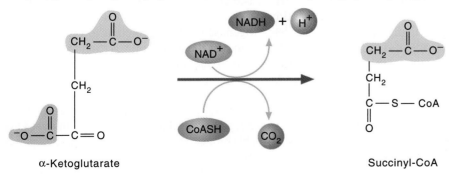

Citrate   *cis*-Aconitate   Isocitrate

3. Isocitrate is oxidized to form NADH and CO$_2$. The oxidative decarboxylation of isocitrate, catalyzed by isocitrate dehydrogenase, occurs in two steps. First, isocitrate is oxidized to form oxalosuccinate, a transient intermediate:

Isocitrate   Oxalosuccinate   α-Ketoglutarate

Immediate decarboxylation of oxalosuccinate results in the formation of α-keto-glutarate, an α-keto acid. There are two forms of isocitrate dehydrogenase in mammals. The NAD$^+$-requiring isozyme is found only within mitochondria. The other isozyme, which requires NADP$^+$, is found in both the mitochondrial matrix and the cytoplasm. In some circumstances the latter enzyme is used within both compartments to generate NADPH, which is required in biosynthetic processes. Note that the NADH produced in the conversion of isocitrate to α-ketoglutarate is the first link between the citric acid cycle and the ETC and oxidative phosphorylation.

4. α-Ketoglutarate is oxidized to form a second molecule each of NADH and CO$_2$. The conversion of α-ketoglutarate to succinyl-CoA is catalyzed by the enzyme activities in the α-ketoglutarate dehydrogenase complex: α-ketoglutarate dehydrogenase, dihydrolipoyl transsuccinylase, and dihydrolipoyl dehydrogenase.

α-Ketoglutarate   Succinyl-CoA

This highly exergonic reaction ($\Delta G^{\circ\prime} = -33.5$ kJ/mol), an oxidative decarboxylation, is analogous to the conversion of pyruvate to acetyl-CoA catalyzed by pyruvate dehydrogenase. In both reactions, energy-rich thioester molecules are products, that is, acetyl-CoA and succinyl-CoA. Other similarities between the two multienzyme complexes are that the same cofactors (TPP, CoASH, lipoic acid, NAD^+, and FAD) are required and the same or similar allosteric effectors are inhibitors. α-Ketoglutarate dehydrogenase is inhibited by succinyl-CoA, NADH, ATP, and GTP. An important difference between the two complexes is that the control mechanism of the α-ketoglutarate dehydrogenase complex does not involve covalent modification.

5. The cleavage of succinyl-CoA is coupled to a substrate-level phosphorylation. The cleavage of the high-energy thioester bond of succinyl-CoA to form succinate, catalyzed by succinate thiokinase, is coupled in mammals to the substrate-level phosphorylation of GDP. In many other organisms, ADP is phosphorylated instead.

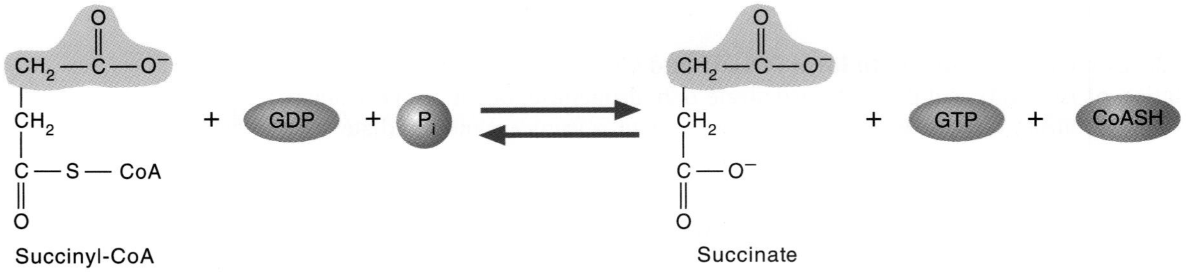

Succinyl-CoA + GDP + P_i ⇌ Succinate + GTP + CoASH

ATP is synthesized in the following reaction catalyzed by nucleoside diphosphate kinase:

GTP + ADP ⇌ GDP + ATP

6. The four-carbon molecule succinate is oxidized to form fumarate and FADH$_2$. Succinate dehydrogenase catalyzes the oxidation of succinate to form fumarate:

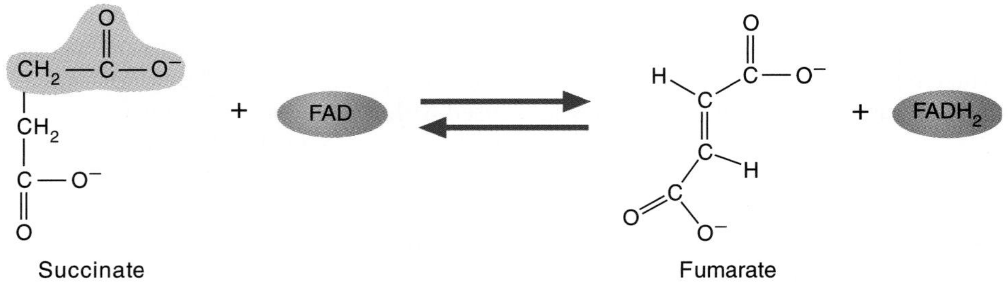

Succinate + FAD ⇌ Fumarate + FADH$_2$

Unlike the other citric acid cycle enzymes, succinate dehydrogenase is not found within the mitochondrial matrix. Instead, it is tightly bound to the inner mitochondrial membrane. The oxidation of an alkane requires a stronger oxidizing agent than NAD^+. Succinate dehydrogenase is a flavoprotein using FAD to drive the oxidation of succinate to fumarate. (In the complete reaction the electrons donated to the FAD covalently bound to succinate dehydrogenase are subsequently passed to coenzyme Q, a component of the ETC. The $\Delta G^{\circ\prime}$ for this process is -5.6 kJ/mol.) Succinate dehydrogenase is activated by high concentrations of succinate, ATP, and P_i and inhibited by oxaloacetate. Recall that the enzyme is also inhibited by malonate, a structural analogue of succinate. (This inhibitor was used by Hans Krebs in his pioneering work on the citric acid cycle.)

7. Fumarate is hydrated. Fumarate is converted to L-malate in a reversible stereospecific hydration reaction catalyzed by fumarase (also referred to as fumarate hydratase):

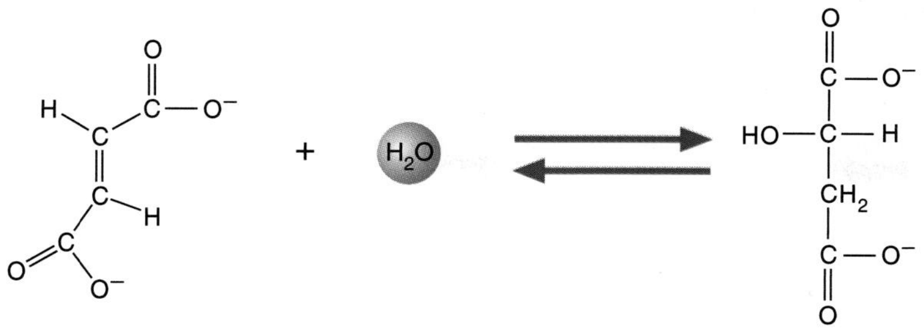

Fumarate L-Malate

8. Malate is oxidized to form oxaloacetate and a third NADH. Finally, oxaloacetate is regenerated with the oxidation of L-malate:

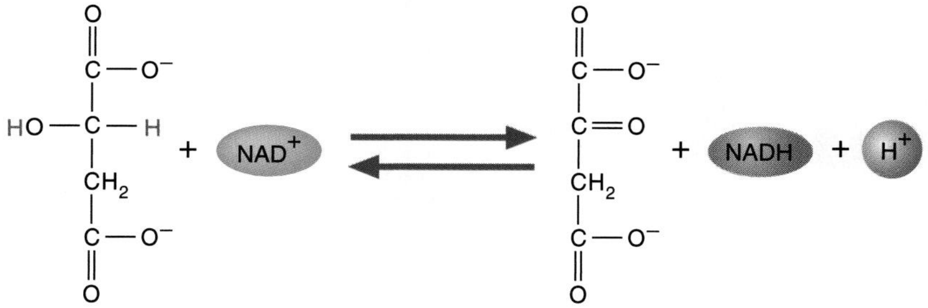

Malate dehydrogenase uses NAD^+ as the oxidizing agent in a highly endergonic reaction ($\Delta G^{\circ\prime} = +29$ kJ/mol). The reaction is pulled to completion because of the removal of oxaloacetate in the next round of the cycle.

Fate of Carbon Atoms in the Citric Acid Cycle

In each turn of the citric acid cycle, two carbon atoms enter as the acetyl group of acetyl-CoA and two molecules of CO_2 are released. A careful review of Figure 9.5 reveals that the two carbon atoms released as CO_2 molecules are not the same two carbons that just entered the cycle. Instead, the released carbon atoms are derived from oxaloacetate that reacted with the incoming acetyl-CoA. The incoming carbon atoms subsequently form one-half of succinate. Because of the symmetric structure of succinate, the carbon atoms derived from the incoming acetyl group become evenly distributed in all of the molecules derived from succinate. Consequently, incoming carbon atoms are released as CO_2 only after two or more turns of the cycle.

Trace the labeled carbon in $CH_3{}^{14}C$-SCoA through one round of the citric acid cycle. After examining Figure 9.5, suggest why more than two turns of the cycle are required before all the labeled carbon atoms are released as ${}^{14}CO_2$.

QUESTION 9.5

The Amphibolic Citric Acid Cycle

Amphibolic pathways can function in both anabolic and catabolic processes. The citric acid cycle is obviously catabolic, because acetyl groups are oxidized to form CO_2 and energy is conserved in reduced coenzyme molecules. The citric acid cycle is also anabolic, because several citric acid cycle intermediates are precursors in biosynthetic pathways (Figure 9.10). For example,

KEY CONCEPTS 9.3

The citric acid cycle begins with the condensation of a molecule of acetyl-CoA with oxaloacetate to form citrate, which is eventually reconverted to oxaloacetate. During this process, two molecules of CO_2, three molecules of NADH, one molecule of $FADH_2$, and one molecule of GTP are produced.

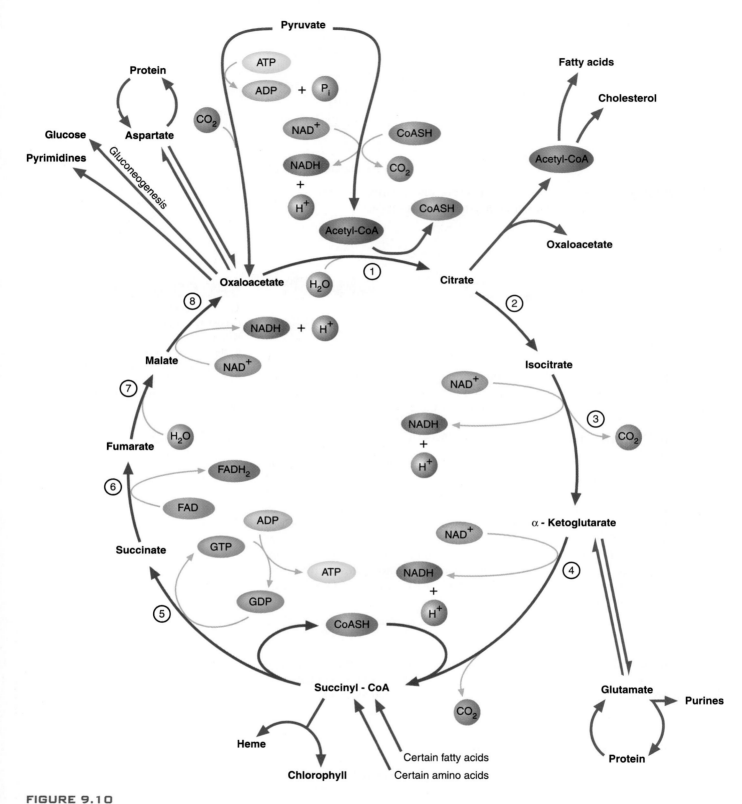

FIGURE 9.10

The Amphibolic Citric Acid Cycle.

The citric acid cycle operates in both anabolic processes (e.g., the synthesis of fatty acids, cholesterol, heme, and glucose) and catabolic processes (e.g., amino acid degradation and energy production).

oxaloacetate is used in both gluconeogenesis (Chapter 8) and amino acid synthesis (Chapter 14). α-Ketoglutarate also plays an important role in amino acid synthesis. The synthesis of porphyrins such as heme uses succinyl-CoA (Chapter 14). Finally, the synthesis of fatty acids and cholesterol in the cytoplasm requires acetyl-CoA (Chapter 12).

Anabolic processes drain the citric acid cycle of the molecules required to sustain its role in energy generation. Several reactions, referred to as **anaplerotic** reactions, replenish them. One of the most important anaplerotic reactions is catalyzed by pyruvate carboxylase. A high concentration of acetyl-CoA, an indicator of an insufficient oxaloacetate concentration, activates pyruvate carboxylase. As a result, oxaloacetate concentration increases. Any excess oxaloacetate that is not used within the citric acid cycle is used in gluconeogenesis (Chapter 8). Other anaplerotic reactions include the synthesis of succinyl-CoA from certain fatty acids (Chapter 12) and the α-keto acids α-ketoglutarate and oxaloacetate from the amino acids glutamate and aspartate, respectively, via transamination reactions (Chapter 14).

KEY CONCEPTS 9.4

The citric acid cycle is an amphibolic pathway; that is, it plays a role in both anabolism and catabolism. The citric acid cycle intermediates used in anabolic processes are replenished by several anaplerotic reactions.

QUESTION 9.6

Pyruvate carboxylase deficiency is a usually fatal disease caused by a missing or defective enzyme that converts pyruvate to oxaloacetate. It is characterized by varying degrees of mental retardation and disturbances in several metabolic pathways, especially those involving amino acids and their degradation products. A prominent symptom of this malady is *lactic aciduria* (lactic acid in the urine). After reviewing the function of pyruvate carboxylase, explain why this symptom occurs.

Citric Acid Cycle Regulation

The citric acid cycle is precisely regulated so that the cell's energy and biosynthetic requirements are constantly met. Regulation is achieved primarily by the modulation of key enzymes and the availability of certain substrates. Because of its prominent role in energy production, the cycle also depends on a continuous supply of NAD$^+$, FAD, and ADP.

The citric acid cycle enzymes citrate synthase, isocitrate dehydrogenase, and α-ketoglutarate dehydrogenase are closely regulated because they catalyze reactions that represent important metabolic branch points (Figure 9.11).

Citrate synthase, the first enzyme in the cycle, catalyzes the formation of citrate from acetyl-CoA and oxaloacetate. Because the concentrations of acetyl-CoA and oxaloacetate are low in mitochondria in relation to the amount of the enzyme, any increase in substrate availability stimulates citrate synthesis. (Under these conditions the reaction is first order with respect to substrate. Therefore the rate of citrate synthesis is influenced by changes in concentrations of acetyl-CoA and oxaloacetate.) High concentrations of succinyl-CoA (a "downstream" intermediate product of the cycle) and citrate inhibit citrate synthase by acting as allosteric inhibitors. Other allosteric regulators of this reaction are NADH and ATP, whose concentrations reflect the cell's current energy status. A resting cell has high NADH/NAD$^+$ and ATP/ADP ratios. As a cell becomes metabolically active, NADH and ATP concentrations decrease. Consequently, key enzymes such as citrate synthase become more active.

Isocitrate dehydrogenase catalyzes the second closely regulated reaction in the cycle. Its activity is stimulated by relatively high concentrations of ADP and NAD$^+$ and inhibited by ATP and NADH. Isocitrate dehydrogenase is closely regulated because of its important role in citrate metabolism (Figure 9.12). As previously described, the conversion of citrate to isocitrate is reversible. An

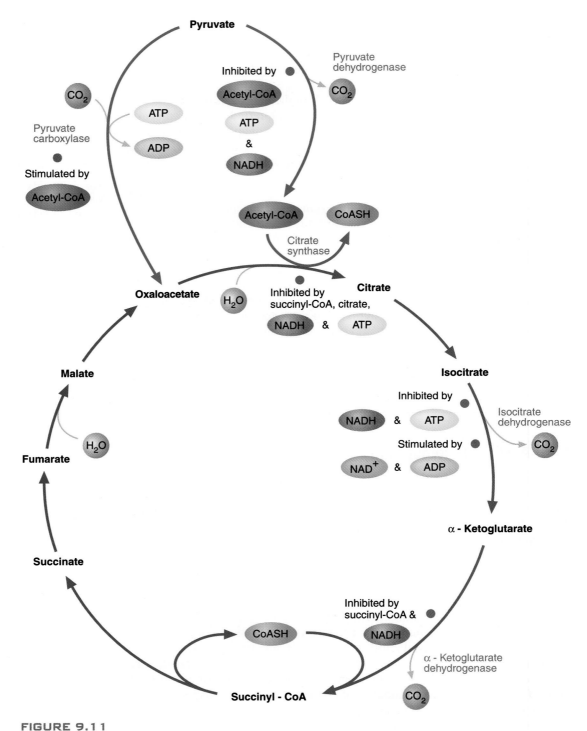

FIGURE 9.11

Control of the Citric Acid Cycle.

The major regulatory sites of the cycle are indicated. Activators and inhibitors of regulated enzymes are shown in color.

equilibrium mixture of the two molecules consists largely of citrate. (The reaction is driven forward because isocitrate is rapidly transformed to α-ketoglutarate.) Of the two molecules, only citrate can penetrate the mitochondrial inner membrane. (When a substantial number of citrate molecules move into cyto-

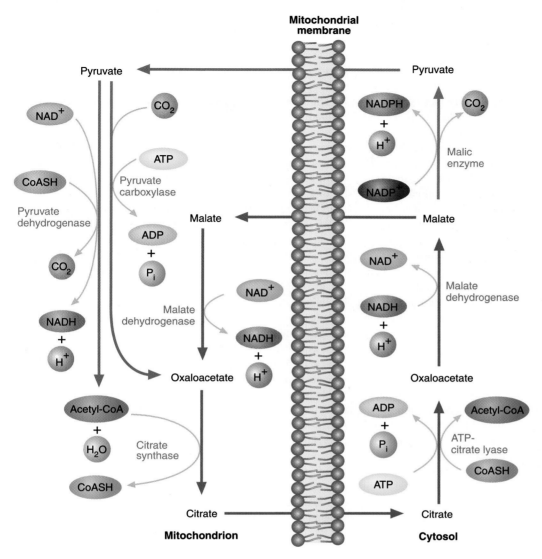

FIGURE 9.12

Citrate Metabolism.

When citrate, a citric acid cycle intermediate, moves from the mitochondrial matrix into the cytoplasm, it is cleaved to form acetyl-CoA and oxaloacetate by citrate lyase. The citrate lyase reaction is driven by ATP hydrolysis. Most of the oxaloacetate is reduced to malate by malate dehydrogenase. Malate may then be oxidized to pyruvate and CO_2 by malic enzyme. The NADPH produced in this reaction is used in cytoplasmic biosynthetic processes, such as fatty acid synthesis. Pyruvate enters the mitochondria, where it may be converted to oxaloacetate or acetyl-CoA. Malate may also reenter the mitochondria, where it is reoxidized to form oxaloacetate.

plasm, the cell's current requirement for energy is low.) Citrate transport is used to transfer acetyl-CoA out of mitochondria because acetyl-CoA cannot penetrate the inner mitochondrial membrane. Once in the cytoplasm, citrate is cleaved by citrate lyase. The acetyl-CoA formed is used in several biosynthetic processes, such as fatty acid synthesis. Oxaloacetate is used in biosynthetic reactions, or it can be converted to malate. Malate either reenters the mitochondrion, where it is reconverted to oxaloacetate, or is converted in the cytoplasm to pyruvate by malic enzyme. Pyruvate then reenters the mitochondrion. In addition to being a precursor of acetyl-CoA and oxaloacetate in the cytoplasm, citrate also acts directly to regulate several cytoplasmic processes. Citrate is an allosteric activator

Cancer is a group of diseases in which genetically damaged cells proliferate autonomously. Such cells cannot respond to normal regulatory mechanisms that ensure the intercellular cooperation required in multicellular organisms. Consequently, they continue to proliferate, thereby robbing nearby normal cells of nutrients and eventually crowding surrounding healthy tissue.

It has long been recognized that cancerous tumors have abnormal energy metabolism. For example, in the 1930s, Otto Warburg (1883–1970), the German biochemist who developed the first reliable methods for studying energy metabolism in animal tissues, observed that tumor cells convert glucose to lactic acid at an abnormally high rate regardless of their supply of oxygen. As has been described previously, in the presence of oxygen the flux of glucose through glycolysis is dramatically lower in normal aerobic cells than when oxygen is absent. Until the 1980s it was assumed that both the intracellular and extracellular pH values of tumor cells are acidic. At that time technological advances that allowed the noninvasive measurement of the pH values of tumors revealed that malignant cells regulate their internal pH so that it is either neutral or slightly alkaline. It is now recognized that both circumstances (the low external pH and the relatively high internal pH values) benefit tumor cells. A low external pH promotes *metastasis* (the migration of rapidly dividing malignant cells to other body tissues); a high internal pH promotes the processes involved in cell growth and division.

For many years molecular oncologists (biochemists and molecular biologists who investigate the molecular basis of cancer) have attempted to discover how malignant cells override the regulation of energy metabolism. Although the precise mechanisms remain uncertain, it is now believed that changes in energy metabolism are caused by alterations in gene expression (Chapter 18), many of which are promoted by chronic *hypoxia* (inadequate oxygen concentrations). Recent research has revealed that in normal tissue, hypoxia triggers gene expression changes that produce a cascade of physiological responses that allow cells to survive during temporarily adverse conditions. Many of these changes in gene expression result from the activation of the transcription factor hypoxia inducible factor-1 (HIF-1). (**Transcription factors** are proteins that regulate or initiate RNA synthesis by binding to specific DNA sequences called response elements.) Prominent among the genes whose expression is upregulated by HIF-1 are those that code for glucose transporter proteins, most glycolytic enzymes, and LDH. In the early stages of **carcinogenesis** (the process whereby cells become genetically unstable and eventually cancerous) the cells deep within the tumor are deprived of both oxygen and nutrients. Those cells that can adapt to these hostile conditions have a survival advantage over those that cannot. Consequently, increased glycolytic flux and the resulting overproduction of lactate are common features of tumors. As carcinogenesis proceeds and the tumor develops its own blood supply (a process referred to as *angiogenesis*) by inducing the synthesis of proteins such as VEGF (vascular endothelial growth factor), its cells eventually become supplied with adequate amounts of oxygen and nutrients. The glycolytic rate of these cells, however, remains high. In this phenomenon, referred to as **aerobic glycolysis**, tumor cells derive the energy required to support their rapid cell divisions from a mixed metabolism that involves a high rate of glycolysis and some level of oxidative phosphorylation. In many tumor cells the production of lactate is so high that little acetyl-CoA is produced. In this circumstance, the amino acid glutamine is often the major oxidizable substrate for an abnormal and incomplete citric acid cycle. (Glutamine is converted to glutamate, the precursor of α-ketoglutarate. The subsequent oxidation of α-ketoglutarate produces only two molecules of NADH instead of the normal three molecules.) In other words, the normally transient upregulation of hypoxia inducible genes and the high levels of glucose consumption that accompany it are now permanent features of the growing tumor.

of the first reaction of fatty acid synthesis. In addition, citrate metabolism provides some of the NADPH used in fatty acid synthesis. Finally, because citrate is an inhibitor of PFK-1, it inhibits glycolysis.

The activity of α-ketoglutarate dehydrogenase is strictly regulated because of the important role of α-ketoglutarate in several metabolic processes (e.g., amino acid metabolism). When a cell's energy stores are low, α-ketoglutarate dehydrogenase is activated and α-ketoglutarate is retained within the cycle at the expense of biosynthetic processes. As the cell's supply of NADH rises, the enzyme is inhibited, and α-ketoglutarate molecules become available for biosynthetic reactions.

Two enzymes outside the citric acid cycle profoundly affect its regulation. The relative activities of pyruvate dehydrogenase and pyruvate carboxylase determine the degree to which pyruvate is used to generate energy and biosynthetic precursors. For example, if a cell is using a cycle intermediate such as α-ketoglutarate in biosynthesis, the concentration of oxaloacetate falls and acetyl-CoA accumulates. Because acetyl-CoA is an activator of pyruvate carboxylase (and an inhibitor of pyruvate dehydrogenase), more oxaloacetate is produced from pyruvate, thus replenishing the cycle.

KEY CONCEPTS 9.5

The citric acid cycle is closely regulated, thus ensuring that the cell's energy and biosynthetic needs are met. Allosteric effectors and substrate availability primarily regulate the enzymes citrate synthase, isocitrate dehydrogenase, α-ketoglutarate dehydrogenase, pyruvate dehydrogenase, and pyruvate carboxylase.

Fluoroacetate, F—CH$_2$—COO$^-$, is a toxic substance found in certain plants that grow in Australia and South Africa. Animals that ingest these plants die. Research indicates, however, that fluoroacetate is not poisonous by itself. Once it is consumed, fluoroacetate is converted into a toxic metabolite, fluorocitrate. In affected cells, citrate accumulates. Can you suggest how fluoroacetate is converted to fluorocitrate? Why are animals killed whereas the plants are unaffected by fluoroacetate?

QUESTION 9.7

The Glyoxylate Cycle

Plants and some fungi, algae, protozoans, and bacteria can grow by using two-carbon compounds. (Molecules such as ethanol, acetate, and acetyl-CoA, derived from fatty acids, are the most common substrates.) The series of reactions responsible for this capability, referred to as the **glyoxylate cycle**, is a modified version of the citric acid cycle. In plants the glyoxylate cycle occurs in organelles called glyoxysomes (p. 50). In the absence of photosynthesis, for example, growth in germinating seed is supported by the conversion of oil reserves (triacylglycerol) to carbohydrate). In other eukaryotic organisms and in bacteria, glyoxylate enzymes occur in cytoplasm.

The glyoxylate cycle (Figure 9.13) consists of five reactions. The first two reactions (the synthesis of citrate and isocitrate) are familiar ones, because they also occur in the citric acid cycle. However, the formation of citrate from

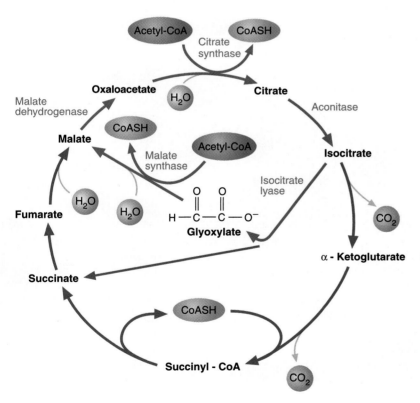

FIGURE 9.13

The Glyoxylate Cycle.

Using some of the enzymes of the citric acid cycle, the glyoxylate cycle converts two molecules of acetyl-CoA to one molecule of oxaloacetate. Both decarboxylation reactions of the citric acid cycle are bypassed.

oxaloacetate and acetyl-CoA and the isomerization of citrate to form isocitrate are catalyzed by glyoxysome-specific isozymes. The next two reactions are unique to the glyoxylate cycle. Isocitrate is split into two molecules (succinate and glyoxylate) by isocitrate lyase. (This reaction is an aldol cleavage.) Succinate, a four-carbon molecule, is eventually converted to malate by mitochondrial enzymes (Figure 9.14). The two-carbon molecule glyoxylate reacts with a second molecule of acetyl-CoA to form malate in a reaction catalyzed by malate synthase. The cycle is completed as malate is converted to oxaloacetate by malate dehydrogenase.

The glyoxylate cycle allows for the net synthesis of larger molecules from two-carbon molecules for the following reason. The decarboxylation reactions of the citric acid cycle, in which two molecules of CO_2 are lost, are bypassed. By using two molecules of acetyl-CoA, the glyoxylate cycle produces one molecule each of succinate and oxaloacetate. The succinate product is used in the synthesis of metabolically important molecules, the most notable of which is glucose. (In organisms, such as animals, that do not possess isocitrate lyase and malate synthase, gluconeogenesis always involves molecules with at least three carbon atoms. In these organisms there is no net synthesis of glucose from fatty acids.) The oxaloacetate product is used to sustain the glyoxylate cycle.

KEY CONCEPTS 9.6

Organisms that possess the glyoxylate cycle can use two-carbon molecules to sustain growth. In plants the glyoxylate cycle occurs in organelles called glyoxysomes.

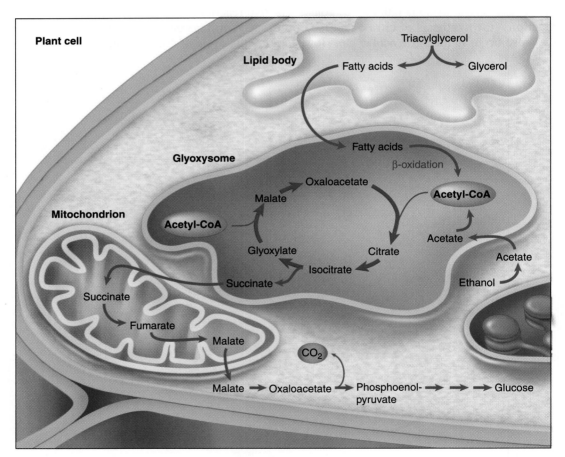

FIGURE 9.14

Role of the Glyoxylate Cycle in Gluconeogenesis.

The acetyl-CoA used in the glyoxylate cycle is derived from the breakdown of fatty acids (β-oxidation, see Chapter 12). In organisms with the appropriate enzymes, glucose can be produced from two-carbon compounds such as ethanol and acetate. In plants the reactions are localized within lipid bodies, glyoxysomes, mitochondria, and the cytoplasm.

Hans Krebs (1900–1981), a German-born British biochemist, is best known for his discovery of the citric acid cycle, arguably one of the most important contributions to biochemistry in the twentieth century. Krebs's efforts to elucidate the details of oxidative metabolism (the means by which nutrients are converted to energy) were made possible by many factors. As in most scientific research, Krebs was the beneficiary of discoveries by many other scientists that provided information about substrates (e.g., succinate, fumarate, and malate) and reactions (e.g., dehydrogenations, hydrations, and dehydrations) that were known to be involved in cellular respiration. The most prominent of these other scientists were Otto Warburg (1883–1970) and Albert Szent-Györgyi (1893–1986). Krebs was also fortunate to have spent the first few years of his career working in Otto Warburg's laboratory in the Kaiser Wilhelm Institut in Berlin. There Krebs learned techniques, developed by Warburg, that later proved to be crucial in his own research projects. Among these were *manometry* (a method for determining the concentration of a specific substance by using an instrument that measures the uptake of O_2 or the release of CO_2), *spectrophotometry* (a technique that measures the concentration of a substance by determining: the proportion of incident light that is absorbed by the substance at a specific wavelength, see p. 424 for a discussion of light), and the preparation of experimental tissue slices. Over several years, beginning in 1933, Krebs, aided by the prodigious efforts of his research student William A. Johnson, slowly pieced together the elements of the oxidative pathway that he eventually realized was a cycle. It was only some years after Krebs and Johnson reported their work in 1937 that the citric acid cycle was recognized as the principal means by which carbohydrates are oxidized in living cells. In addition to the controversial nature of their work (e.g., Szent-Györgyi believed that molecules such as succinate and fumarate were acting as shuttles for hydrogen atoms to O_2), one of the principal problems that delayed recognition was the identity of "active acetate," the intermediate in the conversion of pyruvate to citrate. Krebs observed that when pyruvate was added to tissue slices, large amounts of citrate were produced. However, the addition of acetate, the expected product of the reaction, had no effect. It was not until much later that Fritz Lipmann and Nathan Kaplan discovered acetyl-CoA (1945) and Severo Ochoa and Feodor Lynen established that acetyl-CoA reacts with oxaloacetate to form citrate (1951). For his efforts Krebs received a knighthood and the Nobel prize in physiology and medicine in 1953 (shared with Fritz Lipmann). Other important contributions of Hans Krebs include the discovery of the urea cycle and the glyoxylate cycle. The *urea cycle* (Section 15.1), the mechanism by which some animals convert toxic waste nitrogen to urea (a water-soluble product that can then be excreted), was discovered in 1932 by Krebs and Kurt Heinsleit, a medical student. In 1957 in a joint publication Hans Kornberg and Hans Krebs reported the discovery of the *glyoxylate cycle* (see p. 293).

Many years after his discovery of the citric acid cycle, when asked to reflect upon why he had succeeded when so many other brilliant scientists had failed to elucidate this mechanism, Krebs replied in part,

"My outlook was that of a biologist trying to elucidate chemical events in living cells. I was thus accustomed to correlating chemical reactions in living matter with the activities of the cell as a whole. By putting together pieces of information in jigsaw-puzzle manner, and by attempting to find missing links, I tried to arrive at a coherent picture of metabolic processes. So my mind was prepared to make use of any piece of information which might have a bearing on the intermediary stages of the combustion of foodstuffs. This difference in outlook was, I believe, an important factor in determining who first stumbled on the concept of the tricarboxylic cycle."*

*Krebs, H. A., The History of the Tricarboxylic Acid Cycle, *Perspect. Biol. Med.,* 14:166–167, 1970.

SUMMARY

1. Aerobic organisms have an enormous advantage over anaerobic organisms, that is, a greater capacity to obtain energy from organic food molecules. To use oxygen to generate energy requires the following biochemical pathways: the citric acid cycle, the electron transport pathway, and oxidative phosphorylation.

2. Most reactions that capture or release energy are redox reactions. In these reactions, electrons are transferred between an electron donor (reducing agent) and an electron acceptor (oxidizing agent). In some reactions, only electrons are transferred. In other reactions, both electrons and protons are transferred. The tendency for a specific conjugate redox pair to lose an electron is called its redox potential. Electrons flow spontaneously from electronegative redox pairs to those that are more positive. In favorable redox reactions $\Delta E^{\circ\prime}$ is positive and $\Delta G^{\circ\prime}$ is negative.

3. The citric acid cycle is a series of biochemical reactions that eventually completely oxidize organic substrates, such as glucose and fatty acids, to form CO_2, H_2O, and the reduced coenzymes NADH and $FADH_2$. Pyruvate, the product of the glycolytic pathway, is converted to acetyl-CoA, the citric acid cycle substrate.

4. In addition to its role in energy generation, the citric acid cycle also plays an important role in several biosynthetic processes, such as gluconeogenesis, amino acid synthesis, and porphyrin synthesis.

5. The glyoxylate cycle, found in plants and some fungi, algae, protozoans, and bacteria, is a modified version of the citric acid cycle in which two-carbon molecules, such as acetate, are converted to precursors of glucose.

SUGGESTED READINGS

Gibble, G. W., Fluoroacetate Toxicity, *J. Chem. Ed.*, 50:460–462, 1973.

Graham, T .E, and Gibala, M. J., Anaplerosis of the Tricarboxylic Acid Cycle in Human Skeletal Muscle during Exercise: Magnitude, Sources, and Potential Physiological Significance, *Adv. Exp. Med. Biol.*, 441:271–286, 1998.

Huynen, M. A., Dandekar, T., and Bork, P., Variation and Evolution of the Citric Acid Cycle: A Genomic Perspective, *Trends Microbiol.*, 7:281–291, 1999.

Krebs, H. A., The History of the Tricarboxylic Cycle, *Perspect. Biol. Med.*, 14:154–170, 1970.

Kornberg, H. L., Tricarboxylic Acid Cycles, *Bioessays*, 7:236–238, 1987.

Kornberg, H., Krebs and His Trinity of Cycles, *Nat. Rev. Mol. Cell Biol.*, 1(3):225–227, 2000.

Masters, C., and Crane, D., *The Peroxisome: A Vital Organelle*, Cambridge University Press, Cambridge, 1995.

Scheffler, I. E., *Mitochondria*, Wiley-Liss, New York, 1999.

KEY WORDS

aerobic glycolysis, *292*	coenzyme A, *279*	obligate aerobe, *273*	transcription factor, *292*
aerotolerant anaerobe, *273*	conjugate redox pair, *275*	obligate anaerobe, *273*	
anaplerotic, *289*	facultative anaerobe, *273*	redox potential, *275*	
carcinogenesis, *292*	glyoxylate cycle, *293*	reduction potential, *275*	

REVIEW QUESTIONS

1. Define the following terms:
   a. anaerobes
   b. anaplerotic reactions
   c. glyoxysomes
   d. reduction potential
   e. conjugate redox pair
   f. coenzyme A
   g. amphibolic
   h. lactic aciduria
   i. carcinogenesis
   j. aerobic glycolysis

2. Describe in general terms how the appearance of molecular oxygen in the Earth's atmosphere about three billion years ago affected the history of living organisms.

3. Describe the lifestyles of the following organisms:
   a. obligate anaerobes
   b. aerotolerant anaerobes
   c. facultative anaerobes
   d. obligate aerobes

4. A runner needs a tremendous amount of energy during a race. Explain how the use of ATP by contracting muscle affects the citric acid cycle.

5. Describe two important roles of the citric acid cycle.

6. Acetyl-CoA is manufactured in the mitochondria and used in the cytoplasm to synthesize fatty acids. However, acetyl-CoA cannot penetrate the mitochondrial membrane. How is this problem solved?

7. Describe the glyoxylate cycle. How is it used to synthesize complex molecules from two-carbon molecules?

8. Which of the following conditions indicates a low cell energy status? What biochemical reaction(s) does each condition affect?
   a. high NADH/NAD$^+$ ratio
   b. high ATP/ADP ratio
   c. high acetyl-CoA concentration
   d. low citrate concentration
   e. high succinyl-CoA concentration

9. Shown are the structures of flavin mononucleotide (FMN) and its reduced form FMNH$_2$. Identify each molecule.

THOUGHT QUESTIONS

1. What is the significance of substrate-level phosphorylation reactions? Which of the reactions in the citric acid cycle involve a substrate-level phosphorylation? Name another example from a biochemical pathway with which you are familiar.

2. You have just consumed a piece of fruit. Trace the carbon atoms in the fructose in the fruit through the biochemical pathways between their uptake into tissue cells and their conversion to CO_2.

3. Outline the biosynthesis of the amino acid glutamate from pyruvate.

4. Outline the degradation of glutamate to form CO_2.

5. Determine the standard free energy ($\Delta G°'$)for the following reactions:

   a. $NADH + H^+ + \frac{1}{2} O_2 \longrightarrow NAD^+ + H_2O$

   b. Cytochrome c (Fe^{2+}) $+ \frac{1}{2} O_2 \longrightarrow$ cytochrome c (Fe^{3+}) $+ H_2O$

CHAPTER TEN

Aerobic Metabolism II: Electron Transport and Oxidative Phosphorylation

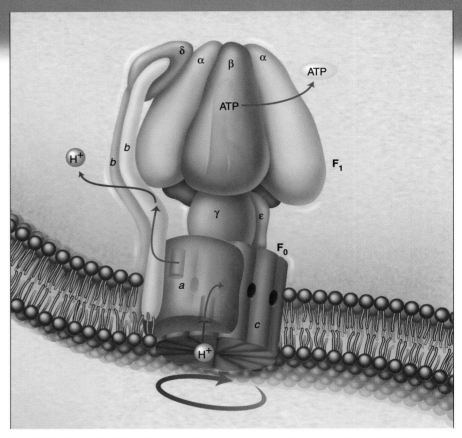

The ATP Synthase The ATP synthase is a rotating molecular machine that synthesizes ATP. This multi-protein complex is composed of two principal domains: the membrane-spanning F_0 component and the ATP-synthesizing F_1 component. The flow of protons through F_0, made possible by a gradient created by electron transport, generates a torque that forces the shaft (the γ subunit) to rotate. Rotational force within F_1 then triggers conformational changes that result in ATP synthesis.

The use of oxygen by aerobic organisms provides enormous advantages in comparison to an anaerobic lifestyle because the aerobic oxidation of nutrients such as glucose and fatty acids yields a substantially greater amount of energy than does fermentation. Oxygen also facilitates reactions such as hydroxylations. Because oxygen is a highly reactive molecule, however, an inescapable price is paid for its use: toxic metabolites are formed. Research efforts reveal that aerobic cells have an array of mechanisms that protect against oxygen's deleterious effects. Numerous enzymes and antioxidant molecules usually prevent oxidative cell damage with extraordinary precision. Despite this high level of protection, however, oxygen metabolites sometimes cause serious damage to living cells. Abnormally stressful conditions can overwhelm an organism's antioxidant mechanisms. Disorders such as cancer, heart disease, and the nerve damage following spinal cord injuries are now known to be caused in part by oxygen metabolites.

The complexity of modern aerobic cells, especially those in multicellular organisms, is made possible by thousands of small and intricate molecular machines. As described previously (see p. 33), molecular machines are proteins that use the chemical bond energy of nucleotides to drive the coordinated operation of their component parts. The structure of each machine, whether it is a pump or an enzyme or a signal transducer, is designed so that it can bind ATP (in some cases GTP) in a specific orientation. When the terminal phosphoanhydride bond of the nucleotide hydrolyses, the resulting repulsion that occurs between the adjacent phosphate groups is used to accomplish work. Typically, the energy released by nucleotide hydrolysis initiates a series of ordered shape changes in the protein machine that culminate in the accomplishment of its biological task. In many instances, shape change is caused when nucleotide hydrolysis results in the covalent attachment of phosphate to a specific amino acid residue in the protein machine. In fact, protein function is often controlled (i.e., turned on and off) by reversible phosphorylation of specific amino acid residues. Although all living cells possess molecular machines, aerobic eukaryotic cells possess by far the largest variety. The complex structure and functioning of these cells is sustained by the extraordinarily large amounts of ATP they can generate. This capacity is made possible by their ability to use O_2 as the terminal acceptor of the electrons extracted from fuel molecules. Other organisms can generate energy with anaerobic electron transport pathways (e.g., the anaerobic chemoorganotrophs synthesize ATP by using terminal electron acceptors such as sulfur, sulfate, nitrate, ferric iron, or CO_2), but they are relatively simple in structure and occupy only isolated ecological niches. In contrast, aerobic organisms are found in great abundance throughout the biosphere.

The vast difference in energy-generating capacity between aerobic and anaerobic organisms is directly related to the chemical and physical properties of oxygen. Oxygen has several properties that in combination have made possible a highly favorable mechanism for extracting energy from organic molecules. First of all, oxygen is found almost everywhere on the Earth's surface. Recall that obligate anaerobes, those organisms poisoned by oxygen, are found in very isolated habitats (see p. 273). In contrast, most other electron acceptors are relatively rare. Second, oxygen diffuses easily across cell membranes. This is not true of several other electron acceptors. For example, charged species such as sulfate and nitrate do not readily diffuse across cell membranes. Finally, oxygen is highly reactive so it readily accepts electrons. This capacity, however, is responsible for another property of oxygen, its tendency to form highly destructive metabolites.

In Chapter 10 the basic principles of oxidative phosphorylation, the complex mechanism by which modern aerobic cells manufacture ATP, are described. The discussion begins with a review of the electron transport system in which electrons are donated by reduced coenzymes to the electron transport chain (ETC). The ETC is a series of electron carriers in the inner membrane of the mitochondria of eukaryotes and the plasma membrane of aerobic prokaryotes. This is followed by a description of chemiosmosis, the means by which the energy extracted from electron flow is captured and used to synthesize ATP. Chapter 10 ends with a discussion of the formation of toxic oxygen products and the strategies that cells use to protect themselves.

10.1 ELECTRON TRANSPORT

The mitochondrial electron transport chain (ETC), also referred to as the electron transport system, is a series of electron carriers arranged in the inner membrane in order of increasing electron affinity that transfer the electrons derived from reduced coenzymes to oxygen. (There are other electron transport systems within

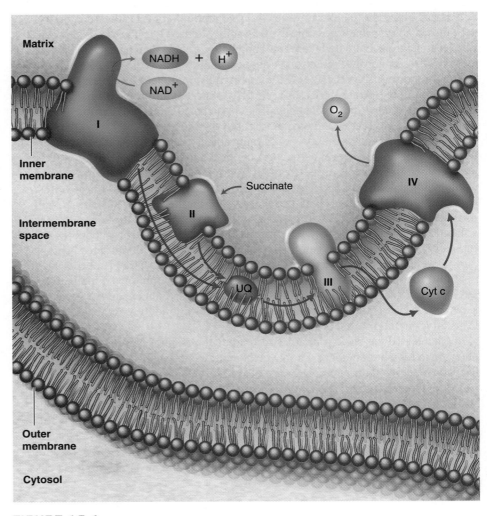

cells. Several examples are discussed in Special Interest Box 10.1 and Chapter 13.) During this transfer, a decrease in oxidation-reduction potential ($E^{\circ\prime}$) occurs. When NADH is the electron donor and oxygen is the electron acceptor, the change in potential is +1.14 V (i.e., +0.82V−(−0.32V), see Table 9.1). This process, in which oxygen is used to generate energy from food molecules, is sometimes referred to as **aerobic respiration**. The energy released during electron transfer is coupled to several endergonic processes, the most prominent of which is ATP synthesis. Other processes driven by electron transport pump Ca^{2+} into the mitochondrial matrix and generate heat in brown adipose tissue (described on p. 319). Reduced coenzymes, derived from glycolysis, the citric acid cycle, and fatty acid oxidation, are the principal sources of electrons.

Electron Transport and Its Components

The components of the ETC in eukaryotes are located in the inner mitochondrial membrane (Figure 10.1). Most ETC components are organized into four

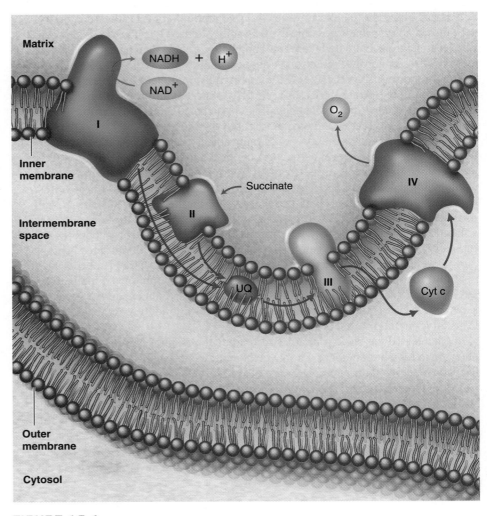

FIGURE 10.1

The Electron Transport Chain.

Complexes I and II transfer electrons from NADH and succinate, respectively, to UQ. Complex III transfers electrons from UQH_2 to cytochrome c. Complex IV transfers electrons from cytochrome c to O_2. The arrows represent the flow of electrons.

complexes, each of which consists of several proteins and prosthetic groups. Each complex is briefly described. The roles of two other molecules, coenzyme Q (ubiquinone, UQ) and cytochrome c (cyt c) are also described.

Complex I, also referred to as the NADH dehydrogenase complex, catalyzes the transfer of electrons from NADH to UQ. The major sources of NADH include several reactions of the citric acid cycle (see pp. 284–287), and fatty acid oxidation (Chapter 12). Composed of at least 25 different polypeptides, complex I is the largest protein component in the inner membrane. In addition to one molecule of FMN, the complex contains seven iron-sulfur centers (Figure 10.2). Iron-sulfur centers, which may consist of two or four iron atoms complexed with an equal number of sulfide ions, mediate 1-electron transfer reactions. Proteins that contain iron-sulfur centers are often referred to as *nonheme iron proteins.* Although the structure and function of complex I are still poorly understood, it is believed that NADH reduces FMN to $FMNH_2$. Electrons are then transferred from $FMNH_2$ to an iron-sulfur center, 1 electron at a time. After transfer from one iron-sulfur center to another, the electrons are eventually donated to UQ (Figure 10.3).

Figure 10.4 illustrates the transfer of electrons through complex I. Electron transport is accompanied by the movement of protons from the matrix across the inner membrane and into the intermembrane space. The significance of this phenomenon for ATP synthesis will be discussed.

The succinate dehydrogenase complex (complex II) consists primarily of the citric acid cycle enzyme succinate dehydrogenase and two iron-sulfur proteins. Complex II mediates the transfer of electrons from succinate to UQ. The

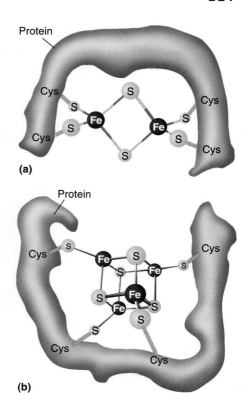

(a)

(b)

FIGURE 10.2

(a) 2 Fe, 2 S and (b) 4 Fe, 4 S Iron-Sulfur Centers.

The cysteine residues are part of a polypeptide.

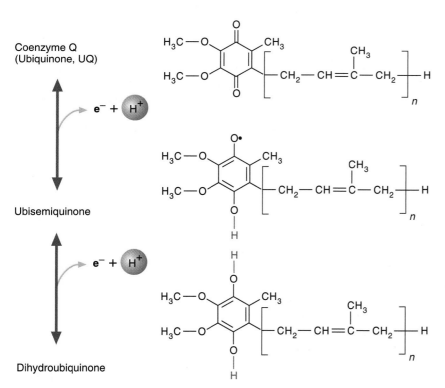

FIGURE 10.3

Structure and Oxidation States of Coenzyme Q.

The length of the side chain varies among species. For example, some bacteria have six isoprene units, although mammals have ten.

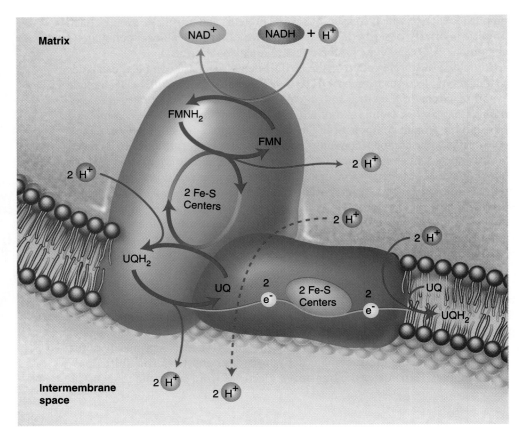

FIGURE 10.4

Transfer of Electrons Through Complex I of the Mitochondrial Electron Transport Chain.
Electron transfer begins with the reduction of FMN by NADH, a process that requires 1 proton from the matrix. $FMNH_2$ subsequently transfers a pair of electrons to six to eight Fe-S centers. (Because the path of the electrons is not known, only four Fe-S centers are shown.) The sequential transfer of the 2 electrons to the first Fe-S center eventually releases 4 protons into the intermembrane space. The mechanism by which these protons are transferred across the membrane is still unclear. However, an internal UQ is believed to be involved in the transfer of two of the protons. The second pair of protons is transferred as the 2 electrons are sequentially transferred from the internal UQ through a series of Fe-S centers to an external UQ.

oxidation site for succinate is located on the larger of the iron-sulfur proteins. This molecule also contains a covalently bound FAD. Other flavoproteins also donate electrons to UQ (Figure 10.5). In some cell types glycerol-3-phosphate dehydrogenase, an enzyme located on the outer face of the inner mitochondrial membrane, transfers electrons from cytoplasmic NADH to the ETC (see p. 316). Acyl-CoA dehydrogenase, the first enzyme in fatty acid oxidation (Chapter 12), transfers electrons to UQ from the matrix side of the inner membrane.

Complex III transfers electrons from reduced coenzyme Q (UQH_2) to cytochrome c. Because it contains two b-type cytochromes, one cytochrome c_1 (cyt c_1), and one iron-sulfur center, complex III is sometimes referred to as the cytochrome bc_1 complex. The cytochromes (Figure 10.6) are a series of electron transport proteins that contain a heme prosthetic group similar to those found in hemoglobin and myoglobin. Electrons are transferred one at a time as each oxidized iron atom (Fe^{3+}) is reversibly reduced to Fe^{2+}. The movement of electrons from UQH_2 to cytochrome c is a complex, multistep process. Because UQ is lipid-soluble, it diffuses within the inner membrane between the electron donors in complex I or II and the electron acceptor in complex

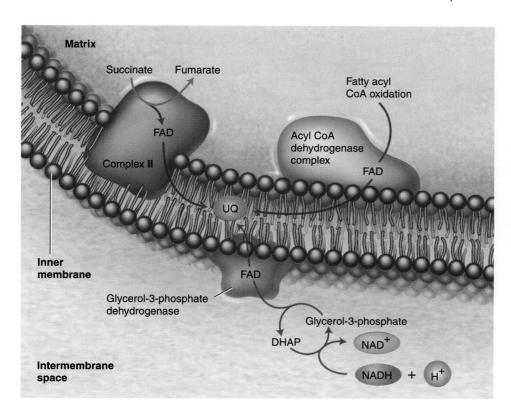

FIGURE 10.5

Path of Electrons From Succinate, Glycerol-3-Phosphate, and Fatty Acids to UQ.

Electrons from succinate are transferred to FAD in complex II and several Fe-S centers and then to UQ. Electrons from cytoplasmic NADH are transferred to UQ via a pathway involving glycerol-3-phosphate and the flavoprotein glycerol-3-phosphate dehydrogenase (see p. 316). Fatty acids are oxidized as coenzyme A derivatives. Acyl-CoA dehydrogenase, one of several enzymes in fatty acid oxidation, transfers 2 electrons to FAD. They are then donated to UQ.

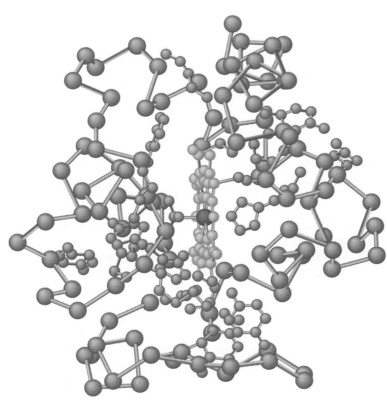

FIGURE 10.6

Structure of Cytochrome c.

Cytochrome c is an example of the cytochromes, a class of small proteins, each of which possesses a heme prosthetic group. During the electron transport process, the iron in the heme is alternately oxidized and reduced.

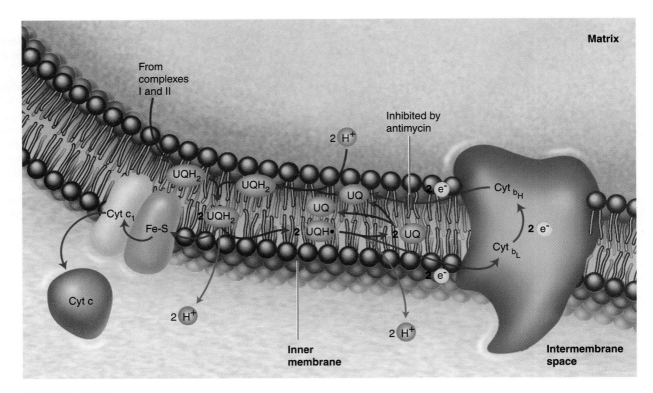

FIGURE 10.7

Electron Transport Through Complex III.

The green and blue arrows represent the flow of electrons. The blue arrows represent the path of UQ in its various oxidation states (the Q cycle) and of protons. UQH_2 is oxidized to UQ in two steps at an enzyme site adjacent to the intermembrane space. The first electron is transferred to the Fe-S protein. The second electron is transferred to cyt b. (Two molecules of UQH_2 undergo those reactions.) One of the two molecules of UQ produced diffuses to the site on the matrix side where it is reduced to form UQH_2. (The transfer of electrons from the two b cytochromes is inhibited by antimycin.) Once formed, the UQH_2 diffuses back to the oxidation site, where it joins the pool of UQH_2 coming from complexes I and II. The electrons transferred from UQH_2 to the Fe-S center then reduce cyt c. Four protons are released on the cytoplasmic side of the inner membrane.

III. Electron transfer begins with the oxidation of UQH_2 by the iron-sulfur protein in complex III, which generates ubisemiquinone ($UQH\cdot$). Then the reduced iron-sulfur protein transfers an electron to cyt c_1, which transfers it to cyt c. See Figure 10.7 for additional details.

Cytochrome oxidase (complex IV) is a protein complex that catalyzes the 4-electron reduction of O_2 to form H_2O. The membrane-spanning complex (Figure 10.8) in mammals may contain between six and thirteen subunits, depending on species. It also contains two atoms of copper in addition to the heme iron atoms of cytochromes a and a_3. (The copper atoms alternate between the +1 and +2 oxidation states, Cu^{1+} and Cu^{2+}.) The iron atom of cyt a_3 is closely associated with a copper atom referred to as Cu_B. The other copper atom (Cu_A) is a short distance from the heme of cyt a. Cytochrome c, a protein that is loosely attached to the inner membrane on its outer surface, transfers electrons one at a time to cyt a and Cu_A. The electrons are then donated to cyt a_3 and Cu_B, which occur on the matrix (inner) side of the membrane. This electron shuttle allows 4 electrons and 4 protons to be delivered to the dioxygen molecule bound to cyt a_3-Fe^{2+}. Two water molecules are formed and leave the site.

$$O_2 + 4\,H^+ + 4\,e^- \longrightarrow 2\,H_2O$$

During each sequential redox reaction in the ETC, an electron loses energy. During the oxidation of NADH there are three steps in which the change in reduc-

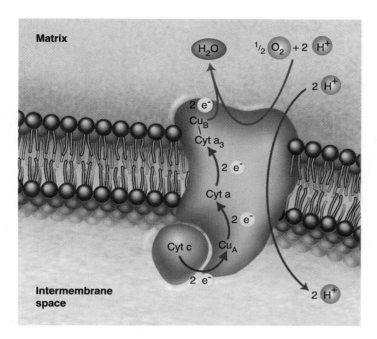

FIGURE 10.8

Electron Transport Through Complex IV.

Each of two reduced cyt c molecules donates 2 electrons one at a time to Cu_A. The electrons are subsequently transferred to cyt a, cyt a_3, and Cu_B. The transfer of a total of 4 electrons from cyt c converts O_2 and 4 protons to two molecules of water. (The half reaction is illustrated in the figure.) A total of 10 protons are transferred across the membrane per electron pair donated by NADH to each oxygen atom.

tion potential ($\Delta E^{\circ\prime}$) is sufficient for ATP synthesis. These steps, which occur within complexes I, III, and IV, are referred to as sites I, II, and III, respectively (Figure 10.9). Recent experimental evidence indicates that approximately 2.5 molecules of ATP are synthesized for each pair of electrons transferred between NADH and O_2 in the ETC. Approximately 1.5 molecules of ATP result from the transfer of each pair donated by the $FADH_2$ produced by succinate oxidation. The mechanism by which ATP synthesis is believed to be coupled to electron transport is described on page 307.

The mechanism by which ATP synthesis is believed to be coupled to electron transport is described on page 307.

Electron Transport Inhibitors

Several molecules specifically inhibit the electron transport process (Figure 10.10). Used in conjunction with reduction potential measurements, inhibitors have been invaluable in determining the correct order of ETC components. In such experiments, electron transport is measured with an oxygen electrode. (Oxygen consumption is a sensitive measure of electron transport.) When electron transport is inhibited, oxygen consumption is reduced or eliminated. Oxidized ETC components accumulate on the oxygen side of the site of inhibition. Reduced ETC components accumulate on the nonoxygen side of the site of inhibition. For example, antimycin A inhibits cyt b. If this inhibitor is added to a suspension of mitochondria, NAD^+, the flavins, and cyt b molecules become more reduced. The cytochromes c_1, c, and a become more oxidized. Other prominent examples of ETC inhibitors include rotenone and amytal, which inhibit NADH dehydrogenase (complex I). Carbon monoxide (CO), azide (N_3^-), and cyanide (CN^-) inhibit cytochrome oxidase.

KEY CONCEPTS 10.1

The electron transport chain is a series of complexes consisting of electron carriers located in the inner mitochondrial membrane in eukaryotic cells.

FIGURE 10.9

The Energy Relationships in the Mito-chondrial Electron Transport Chain.

Relatively large decreases in free energy occur in three steps. During each of these steps (i.e., at sites I, II, and III), sufficient energy is released to account for the synthesis of ATP.

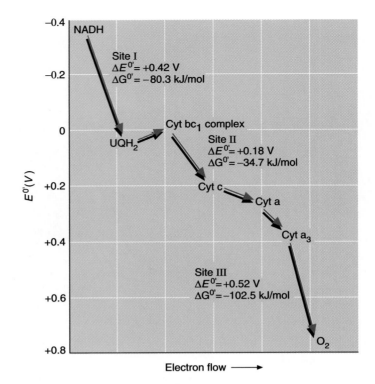

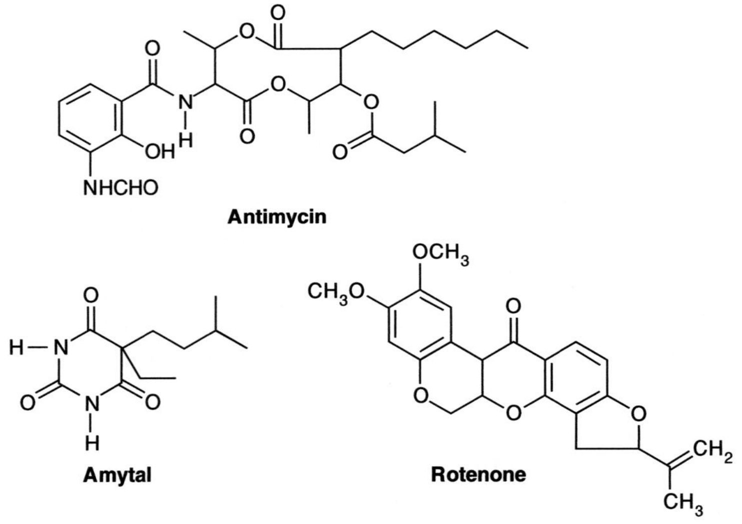

FIGURE 10.10

Several Inhibitors of the Mitochondrial Electron Transport Chain.

Antimycin blocks the transfer of electrons from the b cytochromes. Amytal and rotenone block NADH dehydrogenase.

Antimycin

Amytal

Rotenone

QUESTION 10.1

Which compound in each of the following pairs is the better reducing agent?

a. NADH/H_2O

b. UQ/$FADH_2$

c. Cyt c (oxidized)/cyt c (reduced)

d. $FADH_2$/NADH

e. NADH/$FMNH_2$

TABLE 10.1

Supramolecular Components of the Electron Transport Chain

Enzyme Complex	Prosthetic Groups
Complex I (NADH dehydrogenase)	FMN, FeS
Complex II (Succinate dehydrogenase)	FAD, FeS
Complex III (Cytochrome bc_1 complex)	Hemes, FeS
Cytochrome c	Heme
Complex IV (Cytochrome oxidase)	Hemes, Cu

10.2 OXIDATIVE PHOSPHORYLATION

Oxidative phosphorylation, the process whereby the energy generated by the ETC is conserved by the phosphorylation of ADP to yield ATP, has been studied since the 1940s. The only type of phosphorylation reaction with which biochemists were then familiar was substrate-level phosphorylation (e.g., the oxidation and subsequent phosphorylation of glyceraldehyde-3-phosphate to form bisphosphoglycerate, p. 240). It is not surprising, therefore, that the first mechanism proposed to explain the coupling of electron transport and ATP synthesis involved a high-energy intermediate. According to the *chemical coupling hypothesis*, a high-energy intermediate generated by the electron transport process is used in a second reaction to drive the formation of ATP from ADP and P_i. Despite research efforts that spanned several decades, the proposed intermediate has never been identified. In addition, the hypothesis could not account for several experimental findings. For example, it failed to explain how certain molecules, called uncouplers, prevent ATP synthesis during electron transport. More important, however, the chemical coupling hypothesis did not explain why the entire inner mitochondrial membrane must be intact during ATP synthesis.

The Chemiosmotic Theory

In 1961, Peter Mitchell, a British biochemist, proposed a mechanism by which the free energy generated during electron transport drives ATP synthesis. Now widely accepted, Mitchell's model, referred to as the **chemiosmotic coupling theory** (Figure 10.11), has the following principal features:

1. As electrons pass through the ETC, protons are transported from the matrix and released into the intermembrane space. As a result, an electrical potential Ψ and a proton gradient ΔpH arise across the inner membrane. The electrochemical proton gradient is sometimes referred to as the **protonmotive force** Δp.

2. Protons, which are present in the intermembrane space in great excess, can pass through the inner membrane and back into the matrix down their concentration gradient only through special channels. (The inner membrane itself is impermeable to ions such as protons.) As the thermodynamically favorable flow of protons occurs through a channel, each of which contains an ATP synthase activity, ATP synthesis occurs.

Mitchell suggested that the free energy release associated with electron transport and ATP synthesis is coupled by the protonmotive force created by the ETC. (The term *chemiosmotic* emphasizes that chemical reactions can be coupled to osmotic gradients.) An overview of the chemiosmotic model as it operates in the mitochrondion is illustrated in Figure 10.12.

Protein-mediated electron transfer is a device used in a diverse array of biological transformations. Well-known electron transfer processes include the mitochondrial electron transport system, photosynthesis (Chapter 13), and nitrogen fixation (Chapter 15). Less well known biochemical reactions in which electron transfer plays a crucial role include nitric oxide synthesis and the cytochrome P_{450} electron transport systems. Each of these mechanisms is briefly outlined.

Nitric Oxide Synthesis

Nitric oxide (NO) is a highly reactive gas. Because of its free radical structure NO has been regarded, until recently, primarily as a contributing factor in the destruction of the ozone layer in the Earth's atmosphere and as a precursor of acid rain. (A **radical** is an atom or molecule with an unpaired electron.) Recent research has revealed, however, that NO is an important signal molecule that is produced throughout the mammalian body. Especially high concentrations are found in the central nervous system. Physiological functions in which NO is now believed to play a role include the regulation of blood pressure, the inhibition of blood clotting, and the destruction of foreign, damaged, or cancerous cells by macrophages. The disruption of the normally precise regulation of NO synthesis has been linked to numerous pathological conditions that include stroke, migraine headache, male impotence, septic shock, and several neurodegenerative diseases such as Parkinson's disease.

NO is synthesized by NO synthase (NOS), a heme-containing metalloenzyme that catalyzes a two-step oxidation of L-arginine to L-citrulline (Figure 10A). In this complex reaction electrons are transferred from NADPH to O_2 by an electron transport chain that involves several redox components. The functional enzyme is a homodimer (Figure 10B). Each monomer has two major domains. The reductase domain possesses binding sites for NADPH, FAD, and FMN. The other domain, which has oxygenase activity, binds tetrahydrobiopterin (BH_4) and the substrates arginine and O_2. (BH_4 is a redox cofactor originally identified as an essential component in the hydroxylation of aromatic amino acids. Its structure and properties are described in Chapter 14.) Between the two major domains is the binding site for calmodulin (CAM), a small calcium-binding protein that regulates a variety of enzymes. The formation of the dimer requires the binding of an Fe^{3+} heme group to the oxygenase domain of each monomer. During NO syn-

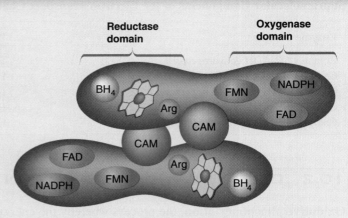

FIGURE 10B Diagrammatic Structure of NOS.
The catalytically active NOS is a homodimer. Each monomer binds NADPH, FAD, FMN, BH_4, and CAM in addition to the substrates arginine and O_2.

thesis CAM accelerates the rate of electron transfer from the reductase domain to the heme group.

The biosynthesis of NO begins with the hydroxylation of L-arginine. The electron source for this reaction is NADPH, which donates 2 electrons to FAD, which in turn reduces FMN. The role of BH_4 is still unclear although it is known to be essential for the activation of O_2 by the electrons donated by NADPH. The prod-

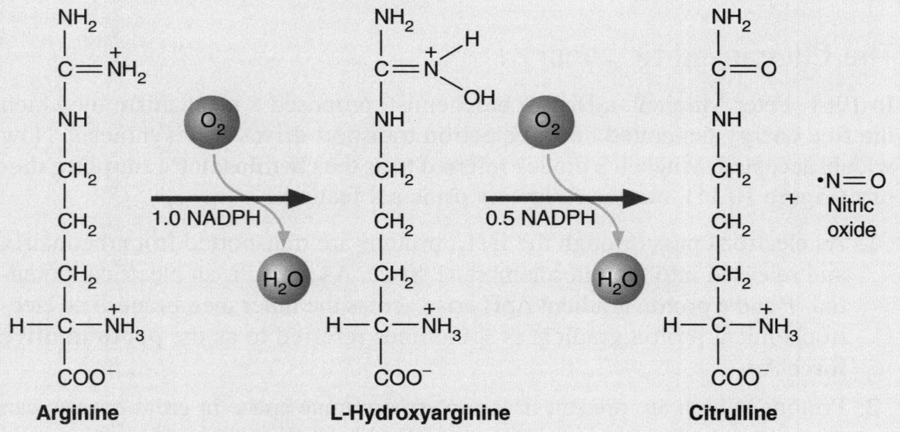

FIGURE 10A The NOS-Catalyzed Reaction.
The biosynthesis of NO is a two-step oxidation of arginine to form citrulline. During the reaction 2 mol of O_2 and 1.5 mol of NADPH are consumed per mol of citrulline formed.

Examples of the evidence that supports the chemiosmotic theory include the following:

1. Actively respiring mitochondria expel protons. The pH of a weakly buffered suspension of mitochondria measured by an electrode drops when O_2 is added.

uct of this reaction, L-hydroxyarginine, remains bound to NOS. The steps in the subsequent reaction have not yet been resolved. It is believed that L-hydroxyarginine reacts with a heme-peroxy complex (R—O—OH) to give citrulline and NO.

Cytochrome P_{450} Electron Transport Systems

Cytochrome P_{450} electron transport systems are an important feature of biotransformation in animal bodies. **Biotransformation** is a series of enzyme-catalyzed processes in which potentially toxic and usually hydrophobic substances are converted into less toxic water-soluble derivatives that can then be more easily excreted. Substrates for biotransformation include endogenous substances, such as cholesterol, and foreign molecules, called xenobiotics, such as drugs and nonnutritive components of food (e.g., glycosides and numerous fatty acid and amino acid derivatives).

The cytochrome P_{450} system, found in microsomal and inner mitochondrial membranes (Figure 10C), consists of two enzymes: NADPH-cytochrome P_{450} reductase and cytochrome P_{450}. In each reaction two electrons are transferred one at a time from NADPH to a cytochrome P_{450} protein by NADPH cytochrome P_{450} reductase. The latter enzyme is a flavoprotein that contains both FAD and FMN in a ratio of 1:1 per mole of enzyme.

The hemoproteins referred to as cytochrome P_{450} are so named because of the complexes they form with carbon monoxide. In the presence of the gas, light is strongly absorbed at a wavelength of 450 nm. Well over 100 cytochrome P_{450} genes have been identified so far. Each gene codes for an enzyme with a unique speci-

ficity range. Cytochrome P_{450} proteins found in the liver have broad and overlapping specificities. For example, molecules as diverse as alkanes, aromatics, ethers, and sulfides are routinely oxidized. In contrast, cytochrome P_{450} proteins in the adrenal glands, ovaries, and testes that add hydroxyl groups to steroid molecules have narrow specificities. Despite this diversity, all cytochrome P_{450} isozymes contain 1 mol of heme. In addition, the cytochrome P_{450} proteins are similar in their physical properties and catalytic mechanisms.

Despite an enormous variety of substrates, all of the oxidative reactions catalyzed by cytochrome P_{450} may be viewed as hydroxylation reactions (i.e., an OH group appears in each reaction) (Figure 10D). The general reaction is as follows:

$$R-H + O_2 + NADPH + H^+ \longrightarrow ROH + H_2O + NADP^+$$

Where R—H is the substrate

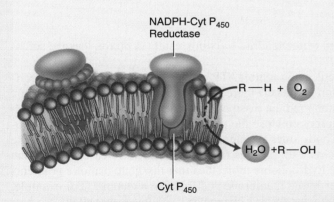

(a)

(b)

(c)

(d)

(e)

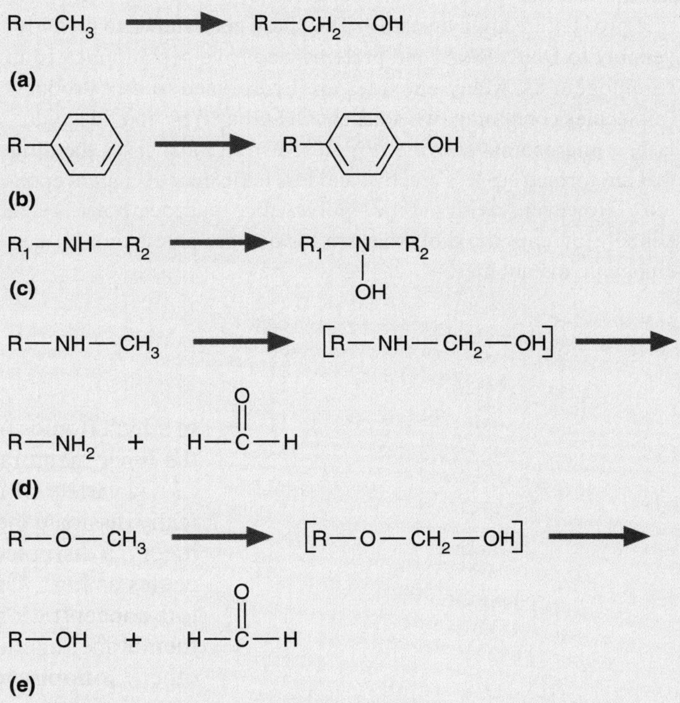

FIGURE 10D Diverse Substrates Oxidized by Cytochrome P_{450} Isozymes.
Among the reactions catalyzed by cytochrome P_{450} are (a) aliphatic oxidation, (b) aromatic hydroxylation, (c) N-hydroxylation, (d) N-dealkylation, and (e) O-dealkylation.

FIGURE 10C The Cytochrome P_{450} Electron Transport System.
Cytochrome P_{450} and cytochrome P_{450} reductase are components of an electron transport system used to oxidize both endogenous and exogenous molecules.

The typical pH gradient across the inner membrane is approximately 0.05 pH unit.

2. ATP synthesis stops when the inner membrane is disrupted. For example, although electron transport continues, ATP synthesis stops in mitochondria placed

The oxygenation reaction is initiated when the substrate binds to oxidized cytochrome P_{450} (Fe^{3+}). This binding promotes a reduction of the enzyme-substrate complex by an electron transferred from NADPH via cytochrome P_{450} reductase (Fe^{2+}—substrate). After reduction, cytochrome P_{450} can bind O_2. Then the electron from heme iron is transferred to the bound O_2, thus forming a transient $[Fe^{3+}(O_2^-)]$—substrate species. (If the bound substrate is easily oxidized, it can be converted into a peroxy radical.) A second electron transferred from the flavoprotein results in the generation of a $[Fe^{3+}(O_2^{2-})]$—substrate complex. This brief association ends when the oxygen-oxygen bond is broken. One oxygen atom is released in a water molecule, while the other remains bound to heme. After a hydrogen atom or electron is abstracted from the substrate, the oxygen species (now a powerful oxidant) is transferred to the substrate. The cycle ends by releasing the product from the active site. Depending on the nature of the substrate, the product is either an epoxide (three-membered ring containing an ether group) or an alcohol.

Epoxides are highly reactive. They have been shown to bind irreversibly to DNA, RNA, and proteins, and have been implicated in carcinogenesis. Many epoxides are hydrolyzed to diol products (molecules containing two adjacent OH groups) by epoxide hydrolase, a microsomal enzyme (Figure 10E). In most cases the diols that are formed are less reactive and less toxic than the parent epoxide. However, with some polycyclic hydrocarbons (e.g., benzo[a]pyrene) the diols that are formed are precursors for carcinogenic metabolites.

Examples of endogenous biotransformation reactions catalyzed by cytochrome P_{450} electron transport systems include the synthesis of steroid hormones from cholesterol (Chapter 12) and the conversion of vitamin D_3 to its biologically active form 1,25-dihydroxyvitamin D_3.

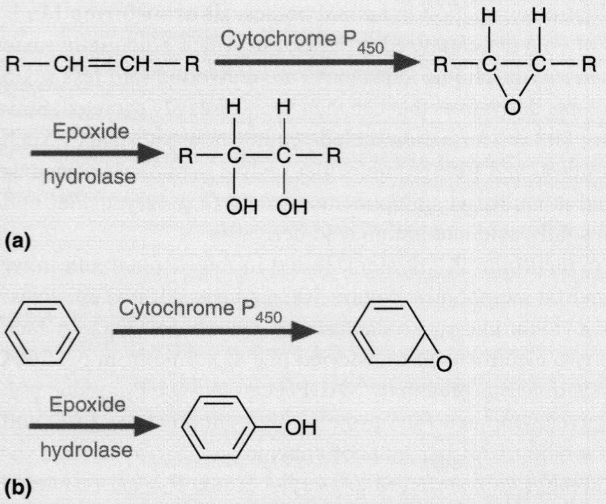

(a)

(b)

FIGURE 10E Conversion of Substrates to Alcohols Via an Epoxide Intermediate.
(a) Aliphatic substrates, such as hydrocarbons. (b) Aromatic substrates, such as benzene.

in a hypertonic solution. Mitochondrial swelling results in proton leakage across the inner membrane.

3. A variety of molecules that inhibit ATP synthesis are now known to specifically dissipate the proton gradient (Figure 10.13). According to the chemiosmotic theory, a disrupted proton gradient dissipates the energy derived from food molecules as heat. **Uncouplers** collapse the proton gradient by equalizing the proton concentration on both sides of membranes. (As they diffuse across the membrane, uncouplers pick up protons from one side and release them on the other.) **Ionophores** are hydrophobic molecules that dissipate osmotic gradients by inserting themselves into a membrane and forming a channel. For example, gramicidin is an antibiotic that forms a channel in membranes that allows the passage of H^+, K^+, and Na^+.

Proton gradients that are generated by electron transport systems are dissipated for two general purposes: ATP is synthesized as protons flow through the ATP synthase, and regulated proton leakage is used to drive several types of biological work. ATP synthesis and its regulation within mitochondria are described, followed by a brief overview of energy generation from glucose. Section 10.2 ends with a discussion of nonshivering thermogenesis, a mechanism in which the mitochondrial proton gradient in certain animal cells is used to regulate body temperature.

KEY CONCEPTS 10.2

In aerobic organisms the energy used to drive the synthesis of most ATP molecules is the protonmotive force. The protonmotive force is generated as free energy is released when electrons flow through the electron transport chain.

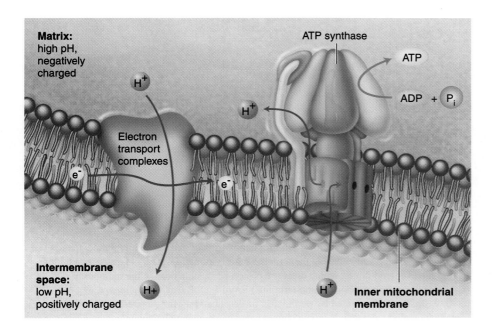

FIGURE 10.11

The Chemiosmotic Theory.

The flow of electrons through the electron transport complexes is coupled to the flow of protons across the inner membrane from the matrix to the intermembrane space. This process raises the matrix pH. In addition, the matrix becomes negatively charged with respect to the intermembrane space. Protons flow passively into the matrix through a channel in the ATP synthase. This flow is coupled to ATP synthesis.

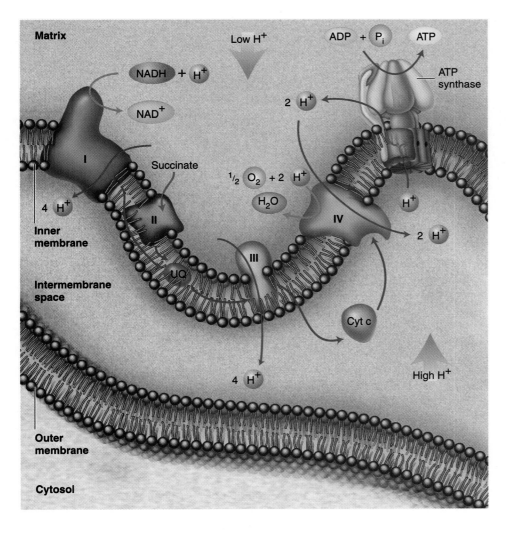

FIGURE 10.12

Overview of the Chemiosmotic Model.

In Mitchell's model protons are driven from the mitochondrial matrix across the inner membrane and into the intermembrane space by the electron transport mechanism. The energy captured from electron transport is used to create an electrical potential and a proton gradient. Because the inner membrane is impermeable to protons, they can only traverse the membrane by flowing through specific proton channels. The flow of protons through the ATP synthase drives the synthesis of ATP. (See Figure 10.1 for brief descriptions of the roles of complexes I, II, III, and IV in electron transport.)

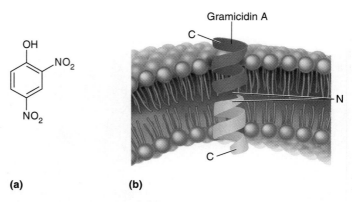

(a) (b)

FIGURE 10.13

Uncouplers.

(a) Dinitrophenol, (b) gramicidin A. Dinitrophenol diffuses across the membrane, picking up protons on one side and releasing them on the other. Gramicidin A, an 11-residue peptide, forms an end-to-end dimer, which creates a proton permeable pore in the membrane. (C = carboxy terminus; N = amino terminus) From J. D. Rawn, *Biochemistry*, © 1989, p. 1039, fig. 31.18. Reprinted by permission of Prentice-Hall, Inc., Upper Saddle River, NJ.

QUESTION 10.2   Dinitrophenol (DNP) is an uncoupler used as a diet aid in the 1920s, until several deaths occurred. Suggest why DNP consumption results in weight loss. The deaths caused by DNP were a result of liver failure. Explain. (*Hint*: Liver cells contain an extraordinarily large number of mitochondria.)

ATP Synthesis

Early electron microscopic studies of mitochondria revealed the presence of numerous lollipop-shaped structures studding the inner membrane on its inner surface (Figure 10.14). Experiments begun in the early 1960s, using submitochondrial particles, revealed that each lollipop is a proton translocating ATP synthase. (*Submitochondrial particles*, or SMP, are small membranous vesicles formed when mitochondria are subjected to sonication. Figure 10.14a illustrates that SMP are inside out, that is, the lollipops project to the outside.) Further work showed that the ATP synthase (Figure 10.15) consists of two major components. The F_1 unit, the active ATPase, possesses five different subunits present in the ratio α_3, β_3, γ, δ and ε. There are three nucleotide-binding catalytic sites on F_1. The F_0 unit, a transmembrane channel for protons, has three subunits present in the ratio a, b_2, and c_{12}.

It is currently believed that the translocation of 3 protons through the ATP synthase is required for each ATP molecule synthesized. (The transfer of an additional proton is required for the transport of ATP and OH^- out of the matrix in exchange for ADP and P_i.) It now appears that the effect of protonmotive force is to induce a three-step rotation of 120° each of the F_0 channel unit. The rotor (or revolving) component of this molecular machine consists of subunits ε, γ, and c_{12}, whereas subunits a, b_2, δ, α_3, and β_3 comprise the stator (or stationary) component.

As protons flow through F_0, the rotation of c_{12} (the proton channel) is transferred to the γ subunit that projects into the core of the F_1 component. The rotation of the γ subunit puts it in three possible positions relative to each $\alpha\beta$-dimer. The binding affinity of the catalytic β subunit changes with the alternating position of the γ subunit. Conformation A binds ADP and P_i weakly, conformation

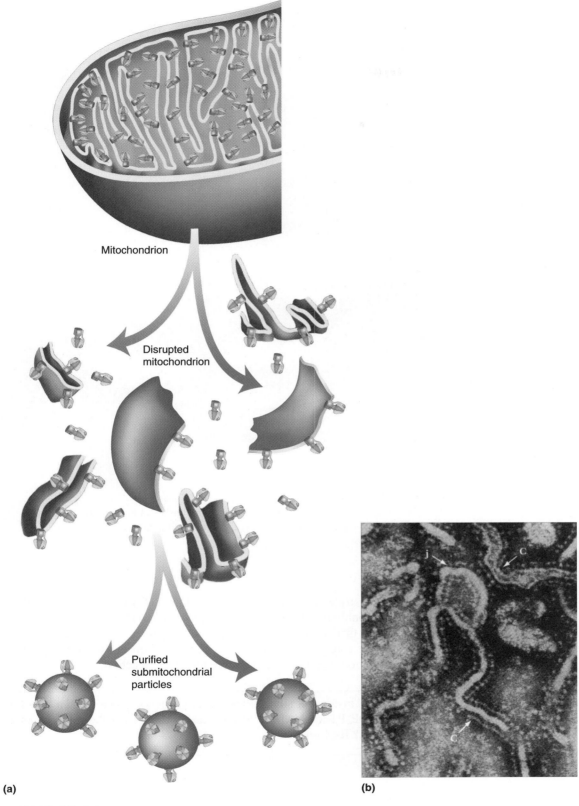

(a)

(b)

FIGURE 10.14

The ATP Synthase.

(a) After disruption of mitochondria, the inner membrane fragments reseal to form inverted submitochondrial particles.
(b) An early electron micrograph of submitochondrial particles revealing the "lollipop"-like structures that would
eventually be identified as ATP synthase.

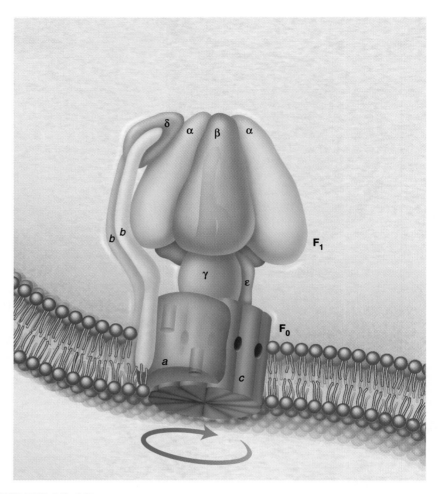

FIGURE 10.15

ATP Synthase from *Escherichia coli.*

The rotor consists of ε, γ, and c_{12} subunits. The stator consists of a, b_2, δ, α_3, and β_3 subunits. The molecular components of ATP synthase are well conserved among bacteria, plants, and animals.

B drives the substrates close together to facilitate ATP formation, and conformation C is a nonbinding form that effectively expels ATP from the active site. It is the last step that serves as the driving force for the reaction. Without a proton gradient, the rotor does not function. The direction of proton flow determines rotor rotation direction and the direction of the reaction.

QUESTION 10.3

A suspension of inside-out submitochondrial particles is placed in a solution that contains ADP, P_i, and NADH. Will increasing the [H^+] of the solution result in ATP synthesis? Explain.

Control of Oxidative Phosphorylation

Control of oxidative phosphorylation allows a cell to produce only the amount of ATP immediately required to sustain its current activities. Recall that under normal circumstances, electron transport and ATP synthesis are tightly coupled.

The value of the *P/O ratio* (the number of moles of P_i consumed for each oxygen atom reduced to H_2O) reflects the degree of coupling observed between electron transport and ATP synthesis. The measured maximum ratio for the oxidation of NADH is 2.5. The maximum P/O ratio for $FADH_2$ is 1.5.

The control of oxidative phosphorylation by ADP concentration is illustrated by the fact that mitochondria can oxidize NADH and $FADH_2$ only when ADP and P_i are present in sufficient concentration. If isolated mitochondria are provided with an oxidizable substrate (e.g., succinate), all of the ADP is eventually converted to ATP. At this point, oxygen consumption becomes greatly depressed. Oxygen consumption increases dramatically when ADP is supplied. The control of aerobic respiration by ADP is referred to as **respiratory control**. The formation of ATP appears to be strongly related to the ATP mass action ratio ($[ATP]/[ADP][P_i]$). In other words, ATP synthase is inhibited by a high concentration of its product (ATP) and activated when ADP and P_i concentrations are high. The relative amounts of ATP and ADP within mitochondria are controlled largely by the two transport proteins in the inner membrane: the ADP-ATP translocator and the phosphate carrier.

The *ADP-ATP translocator* (Figure 10.16) is a dimeric protein responsible for the 1:1 exchange of intramitochondrial ATP for the ADP produced in the

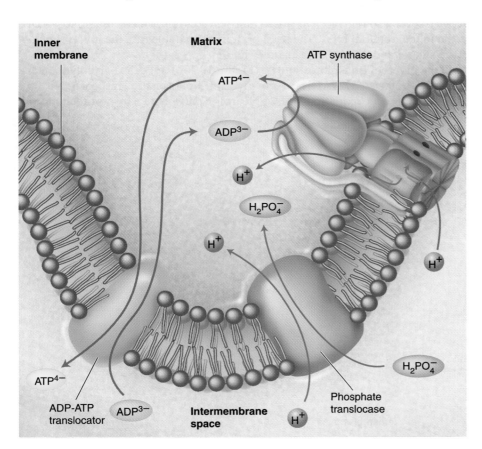

FIGURE 10.16

The ADP-ATP Translocator and the Phosphate Translocase.

The transport of $H_2PO_4^-$ across the inner mitochondrial membrane by the phosphate translocase is driven by the proton gradient. For every 4 protons that are transported out of the matrix, 3 drive the ATP synthase rotor and 1 drives the inward transport of phosphate. The simultaneous exchange of ADP^{3-} and ATP^{4-}, required for continuing ATP synthesis and mediated by the ADP-ATP translocator, is driven by the potential difference across the inner membrane.

cytoplasm. As previously described, there is a potential difference across the inner mitochondrial membrane (negative inside). Because ATP molecules have one more negative charge than ADP molecules, the outward transport of ATP and inward transport of ADP are favored. The transport of H$_2$PO$_4^-$ along with a proton is mediated by the phosphate translocase, also referred to as the H$_2$PO$_4^-$/H$^+$ symporter. (*Symporters* are transmembrane transport proteins that move solutes across a membrane in the same direction. See Section 11.2 for a discussion of membrane transport mechanisms.) The inward transport of 4 protons is required for the synthesis of each ATP molecule; 3 to drive the ATP synthase rotor and 1 to drive the inward transport of phosphate.

The Complete Oxidation of Glucose

A summary of the sources of ATP produced from one molecule of glucose is provided in Table 10.2. ATP production from fatty acids, the other important energy source, is discussed in Chapter 12. Several aspects of this summary require further discussion. Recall that two molecules of NADH are produced during glycolysis. When oxygen is available, the oxidation of this NADH by the ETC is preferable (in terms of energy production) to lactate formation. The inner mitochondrial membrane, however, is impermeable to NADH. Animal cells have evolved several shuttle mechanisms to transfer electrons from cytoplasmic NADH to the mitochrondrial ETC. The most prominent examples are the glycerol phosphate shuttle and the malate-aspartate shuttle.

In the **glycerol phosphate shuttle** (Figure 10.17a), DHAP, a glycolytic intermediate, is reduced by NADH to form glycerol-3-phosphate. This reaction is followed by the oxidation of glycerol-3-phosphate by mitochondrial

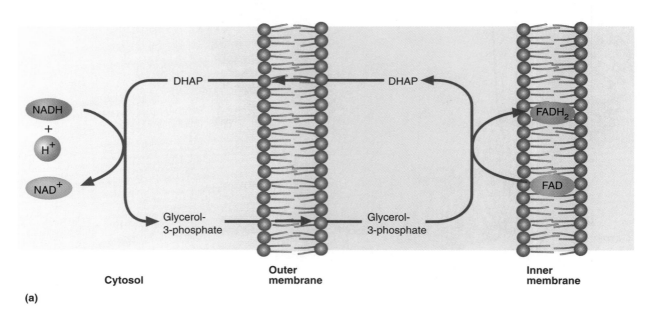

(a)

FIGURE 10.17

Shuttle Mechanisms that Transfer Electrons from Cytoplasmic NADH to the Respiratory Chain.

(a) The glycerol-3-phosphate shuttle dihydroxyacetone phosphate (DHAP) is reduced to form glycerol-3-phosphate. Glycerol-3-phosphate (G-3-P) is reoxidized by mitochondrial glycerol-3-phosphate dehydrogenase. (b) The aspartate-malate shuttle. Oxaloacetate is reduced by NADH to form malate. Malate is transported into the mitochondrial matrix, where it is reoxidized to form oxaloacetate and NADH. Because oxaloacetate cannot penetrate the inner membrane, it is converted to aspartate in a transamination involving glutamate. Two inner membrane carriers are required for this shuttle mechanism: the glutamate-aspartate transport protein and the malate-α-ketoglutarate transport protein.

glycerol-3-phosphate dehydrogenase. (The mitochondrial enzyme uses FAD as an electron acceptor.) Because glycerol-3-phosphate interacts with the mitochondrial enzyme on the outer face of the inner membrane, the substrate does not actually enter the matrix. The $FADH_2$ produced in this reaction is then oxidized by the ETC. FAD as an electron acceptor produces only 1.5 ATP per molecule of cytoplasmic NADH.

Although the **malate-aspartate shuttle** (Figure 10.17b) is a more complicated mechanism than the glycerol phosphate shuttle, it is more energy efficient. The shuttle begins with the reduction of cytoplasmic oxaloacetate to malate by NADH. After its transport into the mitochondrial matrix, malate is reoxidized. The NADH produced is then oxidized by the ETC. For the shuttle to continue, oxaloacetate

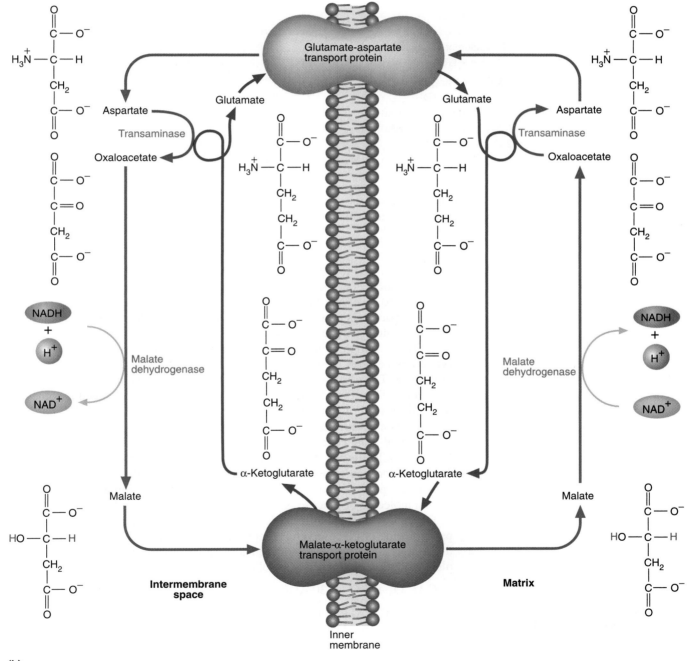

(b)

TABLE 10.2

Summary of ATP Synthesis from the Oxidation of One Molecule of Glucose

	NADH	FADH$_2$	ATP
Glycolysis (cytoplasm)			
Glucose $\longrightarrow$ glucose-6-phosphate			−1
Fructose-6-phosphate $\longrightarrow$ fructose-1,6-bisphosphate			−1
Glyceraldehyde-3-phosphate $\longrightarrow$ glycerate-1,3-bisphosphate	+2		
Glycerate-1,3-bisphosphate $\longrightarrow$ glycerate-3-phosphate			+2
Phosphoenolpyruvate $\longrightarrow$ pyruvate			+2
Mitochondrial Reactions			
Pyruvate $\longrightarrow$ acetyl-CoA	+2		
Citric acid cycle			
Oxidation of isocitrate, α-ketoglutarate, and malate	+6		
Oxidation of succinate		+2	
GDP $\longrightarrow$ GTP			+1.5*
Oxidative Phosphorylation			
2 Glycolytic NADH			+4.5† (3) ‡
2 NADH (pyruvate to acetyl-CoA)			+5
6 NADH (citric acid cycle)			+15
2 FADH$_2$ (citric acid cycle)			+3
			31 (29.5)

*This number reflects the price of transport into the cytoplasm.

† Assumes the malate-aspartate shuttle.

‡ Assumes the glycerol phosphate shuttle.

must be returned to the cytoplasm. Because the inner membrane is impermeable to oxaloacetate, it is converted to aspartate in a transamination reaction (Chapter 15) involving glutamate.

The aspartate is transported to the cytoplasm in exchange for glutamate (via the glutamate-aspartate transport protein) where it can be converted to oxaloacetate. The α-ketoglutarate is transported to the cytoplasm in exchange for malate (via the malate-α-ketoglutarate transport protein) where it can be converted to glutamate. The glutamate-aspartate transporter requires moving a proton into the matrix. Therefore, the net ATP synthesis using this mechanism is somewhat reduced. Instead of generating 2.5 molecules of ATP for each NADH molecule, the yield is approximately 2.25 molecules of ATP.

One final issue concerned with ATP synthesis from glucose remains. Recall that two molecules of ATP are produced in the citric acid cycle (from GTP). The price for their transport into the cytoplasm, where they will be used, is the uptake of 2 protons into the matrix. Therefore the total amount of ATP produced from a molecule of glucose is reduced by about 0.5 molecule of ATP.

Depending on the shuttle used, the total number of molecules of ATP produced per molecule of glucose varies (approximately) from 29.5 to 31. Assuming that the average amount of ATP produced is 30 molecules, the net reaction for the complete oxidation of glucose is as follows:

$$C_6H_{12}O_6 + 6\ O_2 + 30\ ADP + 30\ P_i \longrightarrow 6\ CO_2 + 6\ H_2O + 30\ ATP$$

The number of ATP molecules generated during the complete oxidation of glucose is in sharp contrast to the two molecules of ATP formed by glycolysis. Quite obviously, organisms that use oxygen to oxidize glucose have a substantial advantage.

KEY CONCEPTS 10.4

The aerobic oxidation of glucose yields between 29.5 and 31 ATP molecules.

Traditionally, the oxidation of each NADH and $FADH_2$ by the ETC was believed to result in the synthesis of three molecules of ATP and two molecules of ATP, respectively. As noted, recent measurements, which have considered such factors as proton leakage across the inner membrane, have reduced these values somewhat. Using the earlier values, calculate the number of ATP molecules generated by the aerobic oxidation of a glucose molecule. First assume that the glycerol phosphate shuttle is operating, then assume that the malate-aspartate shuttle is transferring reducing equivalents into the mitochondrion.

Calculate the maximum number of ATP that can be generated from a mole of sucrose.

QUESTION 10.4

QUESTION 10.5

Uncoupled Electron Transport and Heat Generation

Newborn babies, hibernating animals, and cold-adapted animals all require more heat production than is normally generated by metabolism. Recall from Chapter 4 that heat is a consequence of cellular reactions that create an ordered state. (The loss of heat, the most disorganized form of energy, increases the entropy of the surroundings.) Warm-blooded animals use this heat to maintain their body temperature. Under normal conditions, electron transport and ATP synthesis are tightly coupled, so heat production is kept to a minimum. In a specialized form of adipose tissue called *brown fat*, most of the energy produced by the mitochondrial ETC is not used to produce ATP. Instead, it is dissipated as heat. (This tissue has a brown appearance because of the large number of mitochondria it contains.) About 10% of the protein in the mitochondrial inner membrane of brown fat cells is a unique 33-kD protein called **uncoupling protein (UCP)** or *thermogenin*. When the uncoupling protein is active, it dissipates the proton gradient by translocating protons. Uncoupling protein is activated when it binds to fatty acids. As the proton gradient decreases, large amounts of energy are dissipated as heat.

The entire process of heat generation from brown fat, called *nonshivering thermogenesis*, is regulated by norepinephrine. (In shivering thermogenesis, heat is produced by nonvoluntary muscle contraction.) Norepinephrine, a neurotransmitter released from specialized neurons that terminate in brown adipose tissue, initiates a cascade mechanism that ultimately hydrolyzes fat molecules. The fatty acid products of fat hydrolysis activate the uncoupler protein. Fatty acid oxidation continues until the norepinephrine signal is terminated or the cell's fat reserves are depleted.

10.3 OXIDATIVE STRESS

Oxygen is not essential to generate energy; many living organisms (all of which are anaerobic prokaryotes) use glycolysis to provide all their energy needs. Why then is oxygen used by the vast majority of living organisms to extract energy from organic molecules? In addition to the large amounts of energy generated by using this gaseous substance, it is readily available and easily distributed within organisms. (Oxygen diffuses rapidly into and out of cells because it is soluble in the nonpolar lipid core of membranes.) As previously mentioned (see p. 299), however, the advantages of using oxygen are linked to a dangerous property it possesses. Oxygen can accept single electrons to form unstable derivatives, referred to as **reactive oxygen species (ROS)**. Examples of ROS include the superoxide radical, hydrogen peroxide, the hydroxyl radical, and singlet oxygen. Because ROS are so reactive, they can seriously damage living cells if formed in significant amounts. In living organisms, ROS formation is usually kept to a minimum by antioxidant defense mechanisms.

(**Antioxidants** are substances that inhibit the reaction of molecules with oxygen radicals. Often, antioxidants are effective because they are more easily oxidized than the atom or molecule being protected.)

Under certain conditions, referred to collectively as **oxidative stress**, antioxidant mechanisms are overwhelmed and some damage may occur. Damage results primarily from enzyme inactivation, polysaccharide depolymerization, DNA breakage, and membrane destruction. Examples of circumstances that may cause serious oxidative damage include certain metabolic abnormalities, the overconsumption of certain drugs or exposure to intense radiation, or repeated contact with certain environmental contaminants (e.g., tobacco smoke).

In addition to contributing to the aging process, oxidative damage has been linked to at least 100 human diseases. Examples include cancer, cardiovascular disorders such as atherosclerosis, myocardial infarction, and hypertension, and neurological disorders such as amyotrophic lateral sclerosis (ALS or Lou Gehrig's disease), Parkinson's disease, and Alzheimer's disease. Several types of cells are now known to deliberately produce large quantities of ROS. For example, in animal bodies scavenger cells such as macrophages and neutrophils continuously search for microorganisms and damaged cells. In an oxygen-consuming process referred to as the **respiratory burst**, ROS are generated and used to kill and dismantle these cells.

Reactive Oxygen Species

The properties of oxygen are of course directly related to its molecular structure. The diatomic oxygen molecule is a diradical. A **radical** is an atom or group of atoms that contains one or more unpaired electrons. Dioxygen is a diradical because it possesses two unpaired electrons. For this and other reasons, when it reacts, dioxygen can accept only 1 electron at a time.

Recall that during mitochondrial electron transport, H_2O is formed as a consequence of the sequential transfer of 4 electrons to O_2. During this process, several ROS are formed. Cytochrome oxidase (and other oxygen-activating proteins) traps these reactive intermediates within its active site until all 4 electrons have been transferred to oxygen. However, electrons may leak out of the electron transport pathway and react with O_2 to form ROS (Figure 10.18).

Under normal circumstances, cellular antioxidant defense mechanisms minimize any damage. ROS are also formed during nonenzymatic processes. For example, exposure to UV light and ionizing radiation cause ROS formation.

The first ROS formed during the reduction of oxygen is the superoxide radical O_2^-. Most superoxide radicals are produced by electrons derived from the Q cycle in complex III and by the flavoprotein NADH dehydrogenase (complex I). O_2^- acts as a nucleophile and (under specific circumstances) as either an oxidizing agent or a reducing agent. Because of its solubility properties, O_2^- causes considerable damage to the phospholipid components of membranes. When it is generated in an aqueous environment, O_2^- reacts with itself to produce O_2 and hydrogen peroxide (H_2O_2):

$$2\ H^+ + 2\ O_2^- \longrightarrow O_2 + H_2O_2$$

H_2O_2 is not a radical because it does not have any unpaired electrons. The limited reactivity of H_2O_2 allows it to cross membranes and become widely dispersed. The subsequent reaction of H_2O_2 with Fe^{2+} (or other transition metals) results in the production of the hydroxyl radical ($\cdot OH$), a highly reactive species.

$$Fe^{2+} + H_2O_2 \longrightarrow Fe^{3+} + \ \cdot OH + OH^-$$

The hydroxyl radical, which is highly reactive, diffuses only a short distance before it reacts with whatever biomolecule it collides with. Radicals such as the hydroxyl radical are especially dangerous because they can initiate an autocatalytic radical chain reaction (Figure 10.19). Singlet oxygen ($^1O_2$), a highly excited

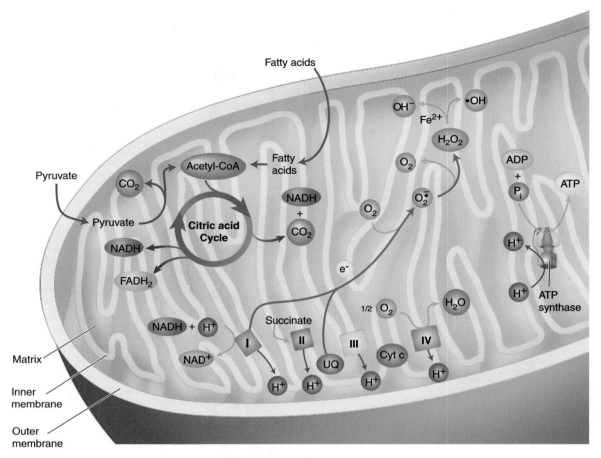

FIGURE 10.18

Overview of Oxidative Phosphorylation and ROS Formation in the Mitochondrion.

Oxidative phosphorylation involves five multiprotein complexes: complexes I, II, III, and IV (the principal components of the ETC) and the ATP synthase. Pyruvate and fatty acids, the major fuel molecules, are transported into the mitochondrion where they are oxidized by the citric acid cycle. The hydrogen atoms liberated during this process are carried by NADH and $FADH_2$ to the ETC. The energy that is released by the electron transport system is used to pump protons from the matrix into the intermembrane space. The electrochemical gradient created by proton pumping is used to synthesize ATP as protons flow through the ATP synthase. However, no system is perfect. Electrons inadvertently leak from the ETC and react with O_2 to form superoxide (O_2^-). In the presence of Fe^{2+}, superoxide is converted into the hydroxyl radical (•OH). Superoxide is also converted to hydrogen peroxide.

state created when dioxygen absorbs sufficient energy to shift an unpaired electron to a higher orbital, can be formed from superoxide:

$$2\ O_2^- + 2\ H^+ \longrightarrow H_2O_2 + {}^1O_2$$

or from peroxides:

$$2\ ROOH \longrightarrow 2\ ROH + {}^1O_2$$

Because it is a powerful oxidant, singlet oxygen is even more reactive than the hydroxyl radical, although it is not a radical.

As mentioned (see p. 320), ROS are generated during several other cellular activities besides the reduction of O_2 to form H_2O. These include the biotransformation of xenobiotics and the respiratory burst (Figure 10.20) in white blood cells. In addition, electrons often leak from the electron transport pathways in the endoplasmic reticulum (e.g., the cytochrome P_{450} electron transport system) to form superoxide by combining with O_2.

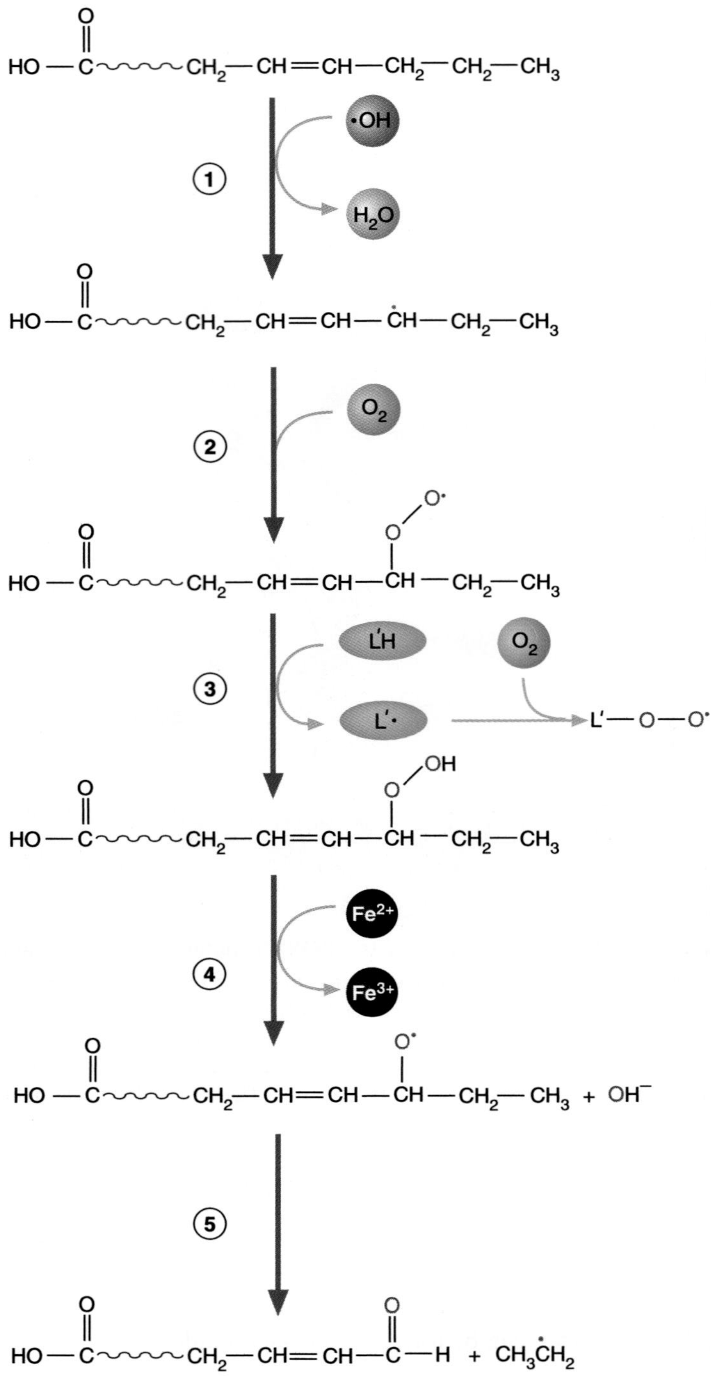

FIGURE 10.19

Radical Chain Reaction.

Step 1: Lipid peroxidation reactions begin after a hydrogen atom is extracted from an unsaturated fatty acid (LH ⟶ L•). Step 2: The lipid radical (L•) then reacts with O_2 to form a peroxyl radical (L• + O_2 ⟶ L—O—O•). Step 3: The radical chain reaction begins when the peroxyl radical extracts a hydrogen atom from another fatty acid molecule (L—O—O• + L′H ⟶ L—O—OH + L′•). Step 4: The presence of a transition metal such as Fe^{2+} initiates further radical formation (L—O—O—H + Fe^{2+} ⟶ LO• + HO⁻ + Fe^{3+}). Step 5: One of the most serious consequences of lipid peroxidation is the formation of aldehydes, which involves a radical cleavage reaction. The chain reaction continues as the free radical product then reacts with a nearby molecule.

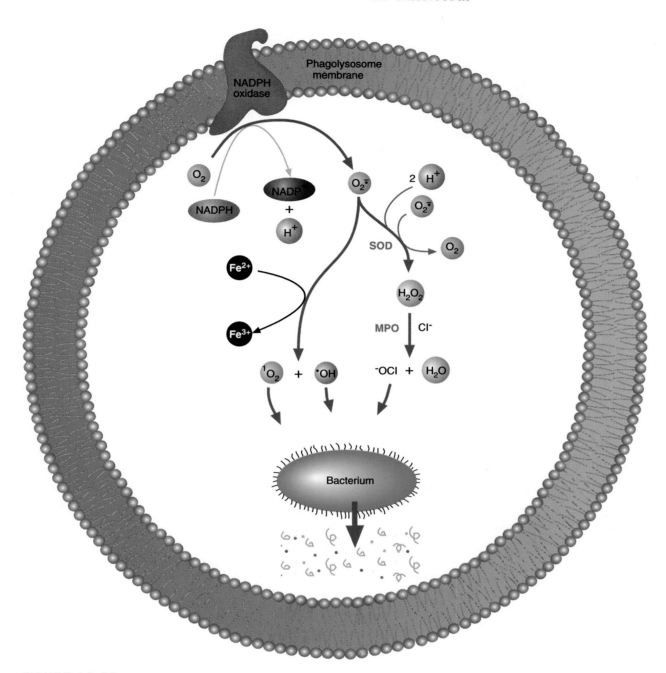

FIGURE 10.20

The Respiratory Burst.

The respiratory burst provides a dramatic example of the destructiveness of ROS. Within seconds after a phagocytic cell binds to a bacterium (or other foreign structure), its oxygen consumption increases nearly 100-fold. During endocytosis the bacterium is incorporated into a large vesicle called a phagosome. Phagosomes then fuse with lysosomes to form phagolysosomes. Two destructive processes then ensue: the respiratory burst and digestion by lysosomal enzymes. The respiratory burst is initiated when NADPH oxidase converts O_2 to O_2^-. Two molecules of O_2^- combine in a reaction catalyzed by SOD (superoxide dismutase) to form H_2O_2. H_2O_2 is next converted to several types of bacteriocidal (bacteria-killing) molecules by myeloperoxidase (MPO), an enzyme found in abundance in phagocytes. For example, MPO catalyzes the oxygenation of halide ions (e.g., Cl^-) to form hypohalides. Hypochlorite (the active ingredient in household bleach) is extremely bacteriocidal. In the presence of Fe^{2+}, O_2^- and H_2O_2 react to form $\cdot OH$ and $^1O_2$ (singlet oxygen), both of which are extremely reactive. After the bacterial cell disintegrates, lysosomal enzymes digest the fragments that remain.

To protect themselves from oxidative stress, living organisms have developed several antioxidant defense mechanisms. These mechanisms employ several metalloenzymes and antioxidant molecules.

Antioxidant Enzyme Systems

The major enzymatic defenses against oxidative stress are provided by superoxide dismutase, glutathione peroxidase, and catalase. The wide distribution of these enzymatic activities underscores the ever present problem of oxidative damage.

The superoxide dismutases (SOD) are a class of enzymes that catalyze the formation of H_2O_2 and O_2 from the superoxide radical:

$$2\ O_2^{-} + 2\ H^{+} \longrightarrow H_2O_2 + O_2$$

There are two major forms of SOD. In humans the Cu-Zn isoenzyme occurs in cytoplasm. A Mn-containing isozyme is found in the mitochondrial matrix. Lou Gehrig's disease, a fatal degenerative condition in which motor neurons are destroyed, is now known to be caused by a mutation in the gene that codes for the cytosolic Cu-Zn isozyme of SOD.

Glutathione peroxidase, a selenium-containing enzyme, is a key component in an enzymatic system most responsible for controlling cellular peroxide levels. Recall that this enzyme catalyzes the reduction of a variety of substances by the reducing agent GSH (Section 5.2). In addition to reducing H_2O_2 to form water, glutathione peroxidase transforms organic peroxides into alcohols:

$$2\ GSH + R\text{—}O\text{—}O\text{—}H \longrightarrow G\text{—}S\text{—}S\text{—}G + R\text{—}OH + H_2O$$

Several ancillary enzymes support glutathione peroxidase function (Figure 10.21). GSH is regenerated from GSSG by glutathione reductase:

$$G\text{—}S\text{—}S\text{—}G + NADPH + H^{+} \longrightarrow 2\ GSH + NADP^{+}$$

The NADPH required in the reaction is provided primarily by several reactions of the pentose phosphate pathway (Chapter 8). Recall that NADPH is also produced by the reactions catalyzed by isocitrate dehydrogenase and malic enzyme.

Catalase is a heme-containing enzyme that uses H_2O_2 to oxidize other substrates:

$$RH_2 + H_2O_2 \longrightarrow R + 2\ H_2O$$

Copious amounts of catalase are found in peroxisomes, where relatively large amounts of H_2O_2 are generated as a by-product in various oxidative reactions:

$$RH_2 + O_2 \longrightarrow R\bullet + H_2O_2$$

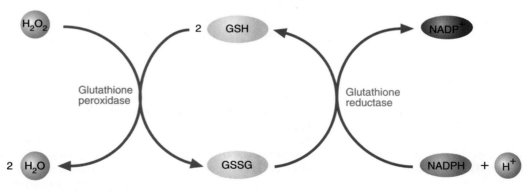

FIGURE 10.21

The Glutathione Redox Cycle.

Glutathione peroxidase utilizes GSH to reduce the peroxides generated by cellular aerobic metabolism. GSH is regenerated from its oxidized form, GSSG, by glutathione reductase. NADPH, the reducing agent in this reaction, is supplied by the pentose phosphate pathway and several other reactions.

Because of their role in oxygen transport, red blood cells are especially prone to oxidative stress. The millions of hemoglobin molecules within each cell are potential prooxidants, that is, the heme group promotes the production of ROS. Recall that oxygen attaches to the sixth coordination bond of the heme group (see p. 145):

$$[Fe^{2+}(heme)(His)(O_2)]$$

When it does so, an intermediate structure forms in which an electron is delocalized between the iron atom and the oxygen:

$$[Fe^{2+}(O_2)] \rightleftharpoons [Fe^3(O_2^-)]$$

Occasionally, oxyhemoglobin decomposes and releases O_2^-. Under normal conditions, a few percent of hemoglobin molecules become oxidized at any one time. Consequently, red blood cells are constantly exposed to O_2^- and the oxidized product of hemoglobin, called methemoglobin, with its heme- Fe^{3+} group is no longer able to bind oxygen. Red blood cells become fragile because the lipid peroxidation caused by H_2O_2 damages the cell's plasma membrane. When such cells pass through narrow blood vessels, they may rupture. If the oxidative stress is severe, hemolytic anemia results. Fortunately, red blood cells are usually well protected. They possess high concentrations of Cu-Zn SOD, catalase, and glutathione peroxidase, and a very active pentose phosphate pathway. The NADPH produced by the oxidative phase of the pentose phosphate pathway is used to reduce GSSG to GSH (Figure 10.21). However, red blood cells have a specific vulnerability to oxidative stress because they derive NADPH only from the pentose phosphate pathway.

In *glucose-6-phosphate dehydrogenase deficiency*, the red blood cell's capacity to protect itself from oxidative stress is reduced. Affected individuals produce low amounts of NADPH because they possess a defective enzyme. (There are over 100 known variants of the G-6-PD gene. The capacity to produce NADPH therefore varies widely among G-6-PD–deficient individuals.) A lower than normal NADPH concentration impairs the individual's capacity to generate GSH.

Under normal conditions, many carriers of the mutant gene are asymptomatic. However, any additional oxidative stress can have serious consequences. For example, administration of the antimalarial drug primaquine to G-6-PD–deficient individuals results in hemolytic anemia. The drug kills the malarial parasite *Plasmodium* because it stimulates the production of hydrogen peroxide. The resultant lowering of NADPH and GSH levels in red blood cells (that already have lower than normal amounts) causes the lysis of the red cell membrane. G-6-PD–deficient individuals are resistant to malaria. (*Plasmodium* is especially sensitive to oxidizing conditions, so any circumstance that lowers cellular antioxidant capacity inhibits the infection.) It is not surprising, therefore, that G-6-PD deficiency is one of the most common human genetic anomalies. In geographic areas in which malaria is endemic (e.g., the Mediterranean and Middle East regions), individuals who possess the defective enzyme are less likely to die of the disease than those who do not. (Recall that sickle-cell trait also confers resistance to malaria.)

When H_2O_2 is present in excessive amounts, catalase converts it to water:

$$2\ H_2O_2 \longrightarrow 2\ H_2O + O_2$$

Selenium is generally considered a toxic element. (It is the active component of loco weed.) However, there is growing evidence that selenium is also an essential trace element. Because glutathione peroxidase activity is essential to protect red blood cells against oxidative stress, a selenium deficiency can damage red blood cells. Although sulfur is in the same chemical family as selenium, it cannot be substituted. Can you explain why? (*Hint*: Selenium is more easily oxidized than sulfur.) Is sulfur or selenium a better scavenger for oxygen when this gas is present in trace amounts?	**QUESTION 10.6**
Ionizing radiation is believed to damage living tissue by producing hydroxyl radicals. Drugs that protect organisms from radiation damage usually have —SH groups. Unfortunately, they must be taken *before* radiation exposure. How do such drugs protect against radiation damage? Can you suggest any type of non-sulfhydryl group–containing molecule that would protect against hydroxyl radical–induced damage?	**QUESTION 10.7**

QUESTION 10.8

In some regions where malaria is endemic (e.g., the Middle East), fava beans are a staple food. Fava beans are now known to contain two β-glycosides called vicine and convicine:

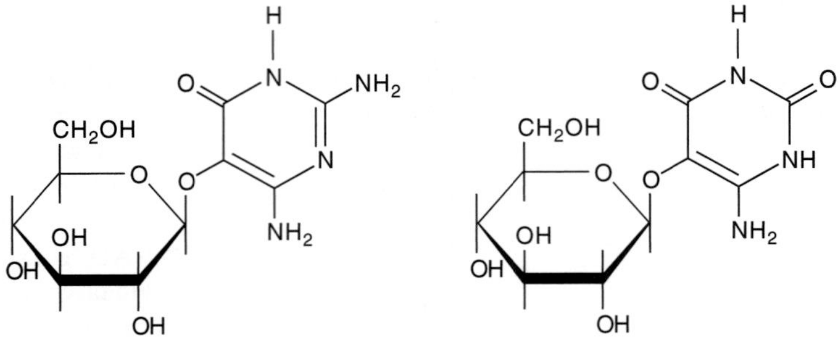

Vicine Convicine

It is believed that the aglycone components of these substances, called divicine and isouramil, respectively, can oxidize GSH. Individuals who eat fresh fava beans are protected to a certain extent from malaria. A condition known as *favism* results when some glucose-6-phosphate dehydrogenase–deficient individuals develop a severe hemolytic anemia after eating the beans. Explain why.

Antioxidant Molecules

Living organisms use antioxidant molecules to protect themselves from radicals. Some prominent antioxidants include GSH, α-tocopherol (vitamin E), ascorbic acid (vitamin C), and β-carotene (Figure 10.22).

α-Tocopherol, a potent radical scavenger, belongs to a class of compounds referred to as *phenolic antioxidants.* Phenols are effective antioxidants because the radical products of these molecules are resonance stabilized and thus relatively stable:

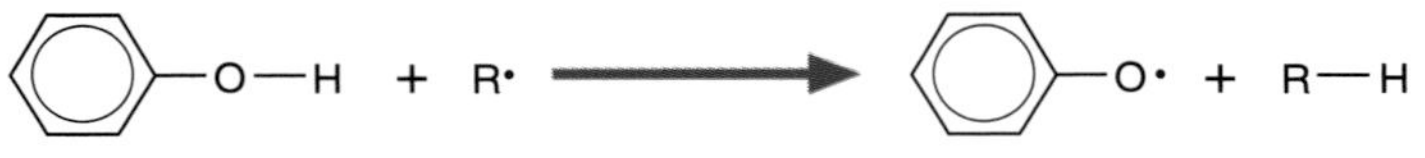

Phenol

Resonance - stabilized phenolic radical

Because vitamin E (found in vegetable and seed oils, whole grains, and green, leafy vegetables) is lipid-soluble, it plays an important role in protecting membranes from lipid peroxyl radicals.

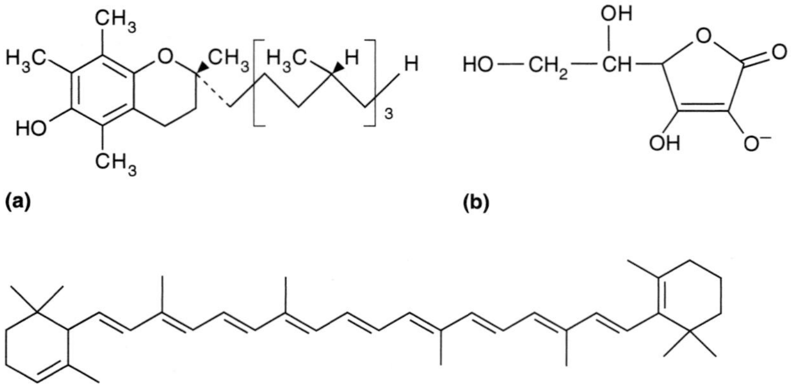

FIGURE 10.22

Selected Antioxidant Molecules.
(a) α-Tocopherol (vitamin E), (b) ascorbate (vitamin C), and (c) β-carotene.

β-Carotene, found in yellow-orange and dark green fruits and vegetables such as carrots, sweet potatoes, broccoli, and apricots, is a member of a class of plant pigment molecules referred to as the *carotenoids*. In plant tissue the carotenoids absorb some of the light energy used to drive photosynthesis and protect against the ROS that form at high light intensities. In animals, β-carotene is a precursor of retinol (vitamin A) and an important antioxidant in membranes. (Retinol is a precursor of retinal, the light-absorbing pigment found in the rod cells of the retina.)

KEY CONCEPTS 10.7

Antioxidant molecules protect cell components from oxidative damage. Prominent antioxidants include GSH and the dietary components α-tocopherol, β-carotene, and ascorbic acid.

Ascorbic acid has been shown to be an efficient antioxidant. Present largely as ascorbate, this water-soluble molecule scavenges a variety of ROS within the aqueous compartments of cells and in extracellular fluids. Ascorbate is reversibly oxidized as shown:

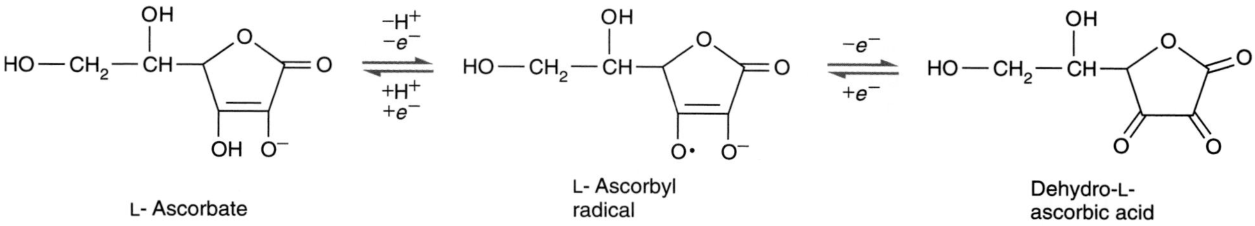

Ascorbate protects membranes through two mechanisms. First, ascorbate reacts with peroxyl radicals formed in the cytoplasm before they can reach the membrane, thereby preventing lipid peroxidation. Second, ascorbate enhances the antioxidant activity of vitamin E by regenerating reduced α-tocopherol from the α-tocopheroxyl radical (Figure 10.23). Ascorbate is then regenerated by reacting with GSH.

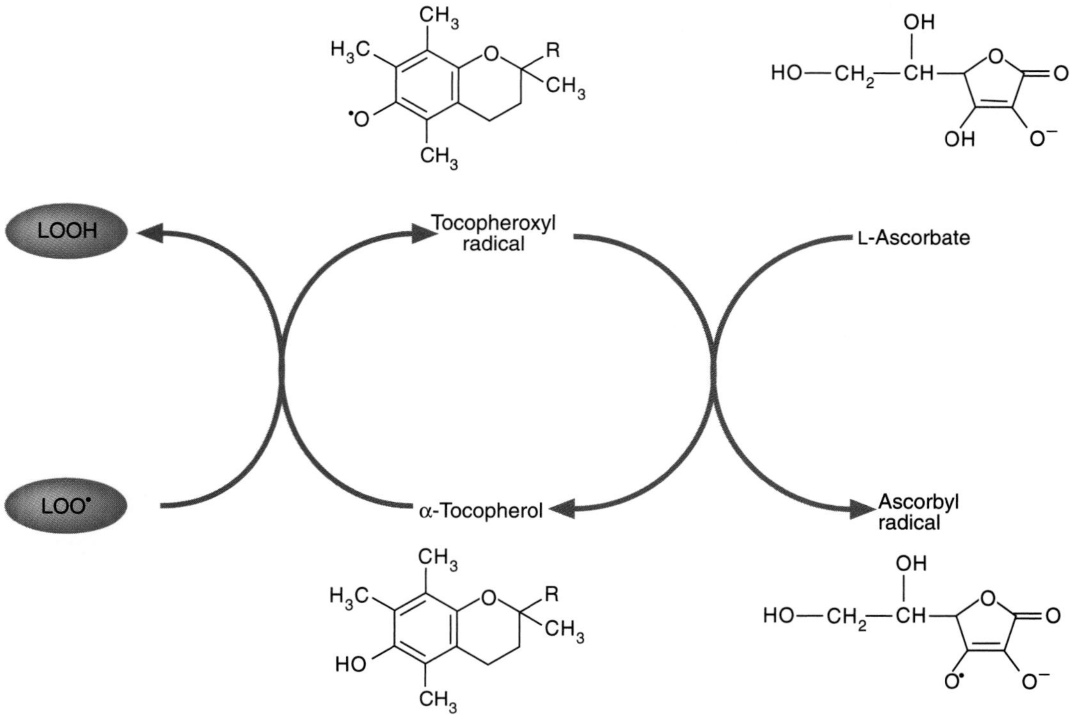

FIGURE 10.23

Regeneration of α-Tocopherol by L-Ascorbate.

L-Ascorbate, a water-soluble molecule, protects membranes from oxidative damage by regenerating α-tocopherol from α-tocopheroxyl radical. The ascorbyl radical formed in this process is reconverted to L-ascorbate during a reaction with GSH.

The tissue damage that occurs during a myocardial infarct or a cerebrovascular accident (stroke) is caused by *ischemia*, a condition in which there is inadequate blood flow. Heart attacks and strokes are usually caused by atherosclerosis accompanied by blood clot formation in an essential artery. (In atherosclerosis, soft masses of fatty material called *plaques* are formed in the linings of blood vessels.) Unlike skeletal muscle, which is fairly resistant to ischemic injury, heart and brain are extremely sensitive to hypoxic (low-oxygen) conditions. For example, significant brain damage occurs if the brain is deprived of oxygen for more than a few minutes. The stimulation of anaerobic glycolysis, which leads to lactate production and acidosis, is an early response of cells to ischemia. Because energy production by glycolysis is inefficient, ATP levels begin to fall. As they do so, adenine nucleotides are degraded to form hypoxanthine (Chapter 15). Without sufficient ATP, cells cannot maintain appropriate intracellular ion concentrations. For example, cytoplasmic calcium levels rise. One of the consequences of this circumstance is the activation of calcium-dependent enzymes such as proteases and phospholipases (enzymes that degrade the phospholipids in membranes). As osmotic pressure increases, affected cells swell and leak their contents. (Recall that the leakage of specific enzymes into blood is used to diagnose heart and liver damage (Special Interest Box 6.1).) The blood supply is further compromised as neutrophils, attracted to the damaged site via chemotaxis, clog the blood vessels. Eventually, lysosomal enzymes begin to leak from lysosomes. Because lysosomal enzymes are active only at low pH values, their presence in an increasingly acidic cytoplasm eventually results in the hydrolysis of cell components. If an oxygen supply is not soon reestablished, affected cells are irreversibly damaged.

The reoxygenation of an ischemic tissue, a process referred to as *reperfusion*, can be a life-saving therapy. For example, using streptokinase to remove artery-occluding clots in heart attack patients, accompanied by administration of oxygen, has been a remarkably successful life-saving strategy. However, depending on the duration of the hypoxic episode, the reintroduction of oxygen to ischemic tissue may also result in further damage. Recent research with antioxidants reveals that ROS are largely responsible for reperfusion-initiated cell damage. The exact mechanism by which reperfusion causes ROS production is still unclear. However, there are several likely possibilities. For example, the leakage of electrons from swollen mitochondria may result in ROS formation. In addition, the release of iron from cell components such as myoglobin, which can result from ROS-inflicted damage, can cause additional production of $\cdot OH$. Other reperfusion-caused damage is the result of the conversion of hypoxanthine to uric acid, which involves the formation of O_2^- and $\cdot OH$, and ROS synthesis in neutrophils. Finally, the acidosis caused by lactate accumulation in compromised heart muscle cells unloads abnormally high amounts of oxygen from hemoglobin. This latter condition greatly facilitates further ROS synthesis. Currently, ROS-quenching antioxidant molecules are being investigated for clinical use. Examples include α-tocopherol and mannitol (the sugar alcohol derived from mannose).

QUESTION 10.9

BHT (butylated hydroxytoluene) is an antioxidant that is widely used as a food preservative. Quercitin is an example of a large group of potent antioxidants found in fruits and vegetables called the flavonoids.

BHT

Quercitin

What structural characteristic of these molecules is responsible for their antioxidant properties?

SUMMARY

1. Dioxygen (O_2), sometimes referred to as oxygen, is used by aerobic organisms as a terminal electron acceptor in energy generation. Several physical and chemical properties of oxygen make it suitable for this role. In addition to its ready availability (it occurs almost everywhere on the Earth's surface), oxygen diffuses easily across cell membranes and it is highly reactive so that it readily accepts electrons.

2. The NADH and $FADH_2$ molecules produced in glycolysis, the β-oxidation pathway, and the citric acid cycle generate usable energy in the electron transport pathway. The pathway

consists of a series of redox carriers that receive electrons from NADH and FADH$_2$. At the end of the pathway the electrons, along with protons, are donated to oxygen to form H$_2$O.

3. During the oxidation of NADH, there are three steps in which the energy loss is sufficient to account for ATP synthesis. These steps, which occur within complexes I, III, and IV, are referred to as sites I, II, and III, respectively.

4. Oxidative phosphorylation is the mechanism by which electron transport is coupled to the synthesis of ATP. According to the chemiosmotic theory, the creation of a proton gradient that accompanies electron transport is coupled to ATP synthesis.

5. The complete oxidation of a molecule of glucose results in the synthesis of 29.5 to 31 molecules of ATP, depending on whether the glycerol phosphate shuttle or the malate-aspartate shuttle transfers electrons from cytoplasmic NADH to the mitochondrial ETC.

6. The use of oxygen by aerobic organisms is linked to the production of ROS. ROS form because the diradical oxygen molecule accepts electrons one at a time. Examples of ROS include the superoxide radical, hydrogen peroxide, the hydroxyl radical, and singlet oxygen. The danger from the highly reactive ROS is usually kept to a minimum by cellular antioxidant defense mechanisms.

SUGGESTED READINGS

Beal, M. F., Oxidative Metabolism, *Ann. N.Y. Acad. Sci.*, 924:164–169, 2000.

Chans, S. I., and Li, P. M., Cytochrome Oxidase: Understanding Nature's Design of a Proton Pump, *Biochem.*, 29(1):1–12, 1990.

Hinkle, P. C., Kumar, A., Resetar, A., and Harris, D. L., Mechanistic Stoichiometry of Mitochondrial Oxidative Phosphorylation, *Biochem.*, 30:3576–3582, 1991.

Hinkle, P. C., and McCarty, R. E., How Cells Make ATP, *Sci. Amer.*, 238(3):104–123, 1978.

Junge, W., Lill, H., and Engelbrecht, S., ATP Synthase: An Electrochemical Transducer with Rotatory Mechanics, *Trends Biochem. Sci.*, 22(11):420–423, 1997.

Mitchell, P., Keilin's Respiratory Chain Concept and Its Chemiosmotic Consequences, *Science*, 206:1148–1159, 1979.

Nicholls, D. G., and Ferguson, S. J., *Bioenergetics 2*, Academic Press, London, 1992.

Nicholls, D. G., and Rial, E., Brown Fat Mitochondria, *Trends Biochem. Sci.*, 9:489–491, 1984.

Nicholls, D. G., and Rial, E., A History of the First Uncoupling Protein, UCP1, *J. Bioenerg. Biomembr.*, 31(5):399–406, 1999.

Rice-Evans, C. A., Miller, N. J., and Paganga, G., Antioxidant Properties of Phenolic Compounds, *Trends Plant Sci.*, 2(4):152–159, 1997.

Sies, H., (Ed.), *Oxidative Stress*, Academic Press, London, 1985.

Stock, D., Gibbons, C., Arechaga, I., Leslie, A. G., and Walker, J. E., The Rotary Mechanism of ATP Synthase, *Curr. Opin. Struc. Biol.*, 10(6):672–679, 2000.

KEY WORDS

aerobic respiration, *300*	glycerol phosphate shuttle, *316*	protonmotive force, *307*	α-tocopherol, *326*
antioxidant, *320*	ionophore, *310*	radical, *320*	uncoupler, *310*
biotransformation, *309*	malate-aspartate shuttle, *317*	reactive oxygen species (ROS), *319*	uncoupling protein, *319*
β-carotene, *327*	oxidative phosphorylation, *307*	respiratory burst, *320*	
chemiosmotic coupling theory, *307*	oxidative stress, *320*	respiratory control, *315*	

REVIEW QUESTIONS

1. Define the following terms:
   a. chemical coupling hypothesis
   b. chemiosmotic coupling theory
   c. ionophore
   d. respiratory control
   e. ischemia
   f. aerobic respiration
   g. radical
   h. biotransformation
   i. epoxide
   j. protonmotive force
   k. proton gradient
   l. uncoupler
   m. ROS

2. What are the principal sources of electrons for the electron transport pathway?

3. Describe the processes that are believed to be driven by mitochondrial electron transport.

4. Describe the principal features of the chemiosmotic theory.

5. The chemical coupling hypothesis failed to explain why mitochondrial membrane must be intact during ATP synthesis. How does the chemiosmotic theory account for this phenomenon?

6. How does dinitrophenol inhibit ATP synthesis?

7. Four protons are required to drive the phosphorylation of ADP. Account for the function of each proton in this process.

8. List several reasons why oxygen is widely used in energy production.

9. Which of the following are reactive oxygen species? Why is each ROS dangerous?
   a. O$_2$
   b. OH$^-$

    c. $RO\cdot$
    d. O_2^-
    e. CH_3OH
    f. $^1O_2$

10. Describe the types of cellular damage produced by ROS.

11. Describe the enzymatic activities used by cells to protect themselves from oxidative damage.

12. Explain how a defect in the gene for glucose-6-phosphate dehydrogenase can provide a survival advantage.

THOUGHT QUESTIONS

1. $^{14}CH_3$—COOH is fed to microorganisms during an experiment. Trace the ^{14}C label through the citric acid cycle. How many ATP molecules can be generated from 1 mol of this substance? (The conversion of acetate to acetyl CoA requires the consumption of 2ATP.)

2. Ethanol is oxidized in the liver to form acetate, which is converted to acetyl-CoA. Determine how many molecules of ATP are produced from 1 mol of ethanol. Note that 2 mol of NADH are produced when ethanol is oxidized to form acetate.

3. Glutamine is degraded to form NH_4^+, CO_2, and H_2O. How many ATP can be generated from 1 mol of this amino acid?

4. Consumption of dinitrophenol by animals results in an immediate increase in body temperature. Explain this phenomenon. Why is this practice a very bad idea?

CHAPTER ELEVEN
Lipids and Membranes

OUTLINE

Biological Membrane A biological membrane is a dynamic compartmental barrier composed of a lipid bilayer noncovalently complexed with proteins, glycoproteins, glycolipids, and cholesterol.

Lipids are naturally occurring substances that dissolve in hydrocarbons but not in water. They perform a stunning array of functions in living organisms. Some lipids are vital energy reserves. Others are the primary structural components of biological membranes. Still other lipid molecules act as hormones, antioxidants, pigments, or vital growth factors and vitamins. This chapter describes the structures and properties of the major lipid classes found in living organisms.

Lipids are a diverse group of biomolecules. Molecules such as fats and oils, phospholipids, steroids, and the carotenoids, which differ widely in both structure and function, are all considered lipids. Because of this diversity, the term **lipid** has an operational rather than a structural definition. Lipids are defined as those substances from living organisms that dissolve in nonpolar solvents such as ether, chloroform, and acetone but not appreciably in water. The functions of lipids are also diverse. Several types of lipid molecules (e.g., phospholipids and sphingolipids) are important structural components in cell membranes. Another type, the fats and oils (both of which are triacylglycerols), store energy efficiently. Other types of lipid molecules are chemical signals, vitamins, or pigments. Finally, some lipid molecules that occur in the outer coatings of various organisms have protective or waterproofing functions.

In Chapter 11 the structure and function of each major type of lipid is described. The lipoproteins, complexes of protein and lipid that transport lipids in animals, are discussed. Chapter 11 ends with an overview of membrane structure and function. In Chapter 12 the metabolism of several major lipids is described.

11.1 LIPID CLASSES

Lipids may be classified in many different ways. For this discussion, lipids can be subdivided into the following classes:

1. Fatty acids and their derivatives
2. Triacylglycerols
3. Wax esters
4. Phospholipids (phosphoglycerides and sphingomyelin)
5. Sphingolipids (molecules other than sphingomyelin that contain the amino alcohol sphingosine)
6. Isoprenoids (molecules made up of repeating isoprene units, a branched five-carbon hydrocarbon)

Each class is discussed.

Fatty Acids and Their Derivatives

As previously described (see p. 13), fatty acids are monocarboxylic acids that typically contain hydrocarbon chains of variable lengths (between 12 and 20 carbons). The structures and names of several common fatty acids are illustrated in Table 11.1. Fatty acids are important components of several types of lipid molecules. They occur primarily in triacylglycerols and several types of membrane-bound lipid molecules.

Most naturally occurring fatty acids have an even number of carbon atoms that form an unbranched chain. (Unusual fatty acids with branched or ring-containing chains are found in some species.) Fatty acid chains that contain only carbon-carbon single bonds are referred to as *saturated*. Those molecules that contain one or more double bonds are said to be *unsaturated*. Because double bonds are rigid structures, molecules that contain them can occur in two isomeric forms: *cis* and *trans*. In **cis-isomers**, for example, similar or identical groups are on the same side of a double bond (Figure 11.1a). When such groups are on opposite sides of a double bond, the molecule is said to be a ***trans*-isomer** (Figure 11.1b).

The double bonds in most naturally occurring fatty acids are in a *cis* configuration. The presence of a *cis* double bond causes an inflexible "kink" in a fatty acid chain (Figure 11.2). Because of this structural feature, unsaturated fatty acids

(a)

(b)

FIGURE 11.1

Isomeric Forms of Unsaturated Molecules.

In *cis*-isomers (a) both R groups are on the same side of the carbon-carbon double bond. *Trans*-isomers (b) have R groups on different sides.

TABLE 11.1
Examples of Fatty Acids

Common Name	Structure	Abbreviation
Saturated Fatty Acids		
Myristic acid	$CH_3(CH_2)_{12}COOH$	14:0
Palmitic acid	$CH_3(CH_2)_{12}CH_2CH_2COOH$	16:0
Stearic acid	$CH_3(CH_2)_{12}CH_2CH_2CH_2CH_2COOH$	18:0
Arachidic acid	$CH_3(CH_2)_{12}CH_2CH_2CH_2CH_2CH_2CH_2COOH$	20:0
Lignoceric acid	$CH_3(CH_2)_{12}CH_2CH_2CH_2CH_2CH_2CH_2CH_2CH_2CH_2CH_2COOH$	24:0
Cerotic acid	$CH_3(CH_2)_{12}CH_2CH_2CH_2CH_2CH_2CH_2CH_2CH_2CH_2CH_2CH_2CH_2COOH$	26:0
Unsaturated Fatty Acids		
Palmitoleic acid	$CH_3(CH_2)_5C{=}C(CH_2)_7COOH$	$16:1^{\Delta 9}$
Oleic acid	$CH_3(CH_2)_7C{=}C(CH_2)_7COOH$	$18:1^{\Delta 9}$
Linoleic acid	$CH_3(CH_2)_4C{=}C{-}CH_2{-}C{=}C(CH_2)_7COOH$	$18:2^{\Delta 9,12}$
α-Linolenic acid	$CH_3CH_2C{=}C{-}CH_2{-}C{=}C{-}CH_2{-}C{=}C(CH_2)_7COOH$	$18:3^{\Delta 9,12,15}$
Arachidonic acid	$CH_3(CH_2)_3{-}(CH_2{-}C{=}C)_4{-}(CH_2)_3COOH$	$20:4^{\Delta 5,8,11,14}$

do not pack as closely together as saturated fatty acids. Less energy is required to disrupt the intermolecular forces between unsaturated fatty acids. Therefore, they have lower melting points and are liquids at room temperature. For example, a sample of palmitic acid (16:0), a saturated fatty acid, melts at 63°C, whereas palmitoleic acid ($16:1^{\Delta 9}$) melts at 0°C. (In the abbreviations for specific fatty acids, the number to the left of the colon is the total number of carbon atoms, and the number to the right is the number of double bonds. A superscript denotes the placement of a double bond. For example, $\Delta 9$ signifies that there are eight carbons between the carboxyl group and the double bond, i.e., the double bond occurs between carbons 9 and 10.) Interestingly, fatty acids with *trans* double bonds have three-dimensional structures similar to those of unsaturated fatty acids. It should also be noted that the presence of one or more double bonds in a fatty acid makes it susceptible to oxidative attack (Figure 10.19). Consequences of this phenomenon include the effects of oxidative stress on cell membranes (Special Interest Box 10.2) and the tendency of oils (see p. 322) to become rancid.

Fatty acids with one double bond are referred to as **monounsaturated** molecules. When two or more double bonds occur in fatty acids, usually separated by methylene groups ($-CH_2-$), they are referred to as **polyunsaturated**. The monounsaturated fatty acid oleic acid ($18:1^{\Delta 9}$) and the polyunsaturated linoleic acid ($18:2^{\Delta 9,12}$) are among the most abundant fatty acids in living organisms.

Organisms such as plants and bacteria can synthesize all the fatty acids they require from acetyl-CoA (Chapter 10). Mammals obtain most of their fatty acids from dietary sources. However, these organisms can synthesize saturated fatty acids and some monounsaturated fatty acids. They can also modify some dietary

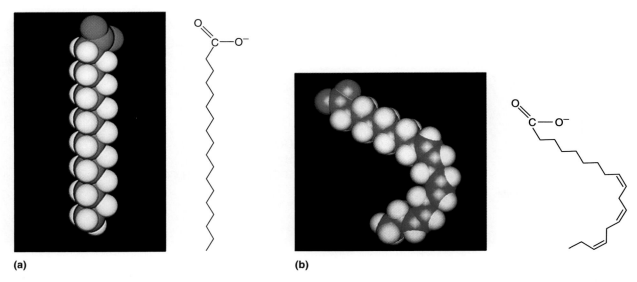

FIGURE 11.2

Space-Filling and Conformational Models.

(a) A saturated fatty acid (stearic acid), and (b) an unsaturated fatty acid (α-linolenic acid). (Green spheres = carbon atoms; white spheres = hydrogen atoms; red spheres = oxygen atoms)

fatty acids by adding two-carbon units and introducing some double bonds. Fatty acids that can be synthesized are called **nonessential fatty acids**. Because mammals do not possess the enzymes required to synthesize linoleic ($18:2^{\Delta 9,12}$) and linolenic ($18:2^{\Delta 9,12,15}$) acids, these **essential fatty acids** must be obtained from the diet. Rich sources of essential fatty acids, which have several critical physiological functions, include some vegetable oils, nuts, and seeds. In addition to contributing to proper membrane structure, linoleic and linolenic acids are precursors of several important metabolites. The most-researched examples of fatty acid derivatives are the eicosanoids (Special Interest Box 11.1). Dermatitis (scaly skin) is an early symptom in individuals on low-fat diets that are deficient in essential fatty acids. Other signs of this deficiency include poor wound healing, reduced resistance to infection, alopecia (hair loss), and thrombocytopenia (reduction in the number of platelets, the blood component involved in the clotting process).

Fatty acids have several important chemical properties. The reactions that they undergo are typical of short-chain carboxylic acids. For example, fatty acids react with alcohols to form esters:

$$R-\overset{\overset{\textstyle O}{\|}}{C}-OH \ + \ R'-OH \ \rightleftharpoons \ R-\overset{\overset{\textstyle O}{\|}}{C}-O-R' \ + \ H_2O$$

This reaction is reversible; that is, under appropriate conditions a fatty acid ester can react with water to produce a fatty acid and an alcohol. Unsaturated fatty acids with double bonds can undergo hydrogenation reactions to form saturated fatty acids. Finally, unsaturated fatty acids are susceptible to oxidative attack. (This feature of fatty acid chemistry is described in Chapter 12.)

Certain fatty acids (primarily myristic and palmitic acids) are covalently attached to a wide variety of eukaryotic proteins. Such proteins are referred to as *acylated* proteins. Fatty acid groups (called **acyl groups**) clearly facilitate the interactions between membrane proteins and their hydrophobic environment. Fatty acids are transported from fat cells to body cells esterified to serum proteins and enter cells via acyl transfer reactions. Some of the acylated proteins in cells

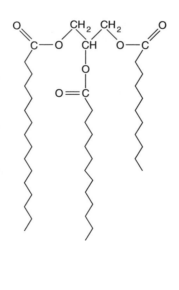

FIGURE 11.3

Triacylglycerol.

Each triacylglycerol molecule is composed of glycerol esterified to three (usually different) fatty acids.

are representative of this transfer process. However, many additional aspects of the role of protein acylation are still not understood.

Triacylglycerols

Triacylglycerols are esters of glycerol with three fatty acid molecules (Figure 11.3). (Glycerides with one or two fatty acid groups, called monoacylglycerols and diacylglycerols, respectively, are metabolic intermediates. They are normally present in small amounts.) Because triacylglycerols have no charge (i.e., the carboxyl group of each fatty acid is joined to glycerol through a covalent bond), they are sometimes referred to as **neutral fats**. Most triacylglycerol molecules contain fatty acids of varying lengths, which may be unsaturated, saturated, or a combination (Figure 11.4). Depending on their fatty acid compositions, triacylglycerol mixtures are referred to as fats or oils. *Fats*, which are solid at room

KEY CONCEPTS 11.1

Fatty acids are monocarboxylic acids, most of which are found in triacylglycerol molecules, several types of membrane-bound lipid molecules, or acylated membrane proteins.

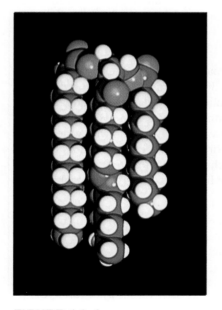

FIGURE 11.4

Space-Filling and Conformational Models of a Triacylglycerol.

Triacylglycerols are highly reduced molecules that serve as a rich source of chemical bond energy.

temperature, contain a large proportion of saturated fatty acids. *Oils* are liquid at room temperature because of their relatively high unsaturated fatty acid content. (Recall (see p. 333) that unsaturated fatty acids do not pack together as closely as do saturated fatty acids.)

In animals, triacylglycerols (usually referred to as *fat*) have several roles. First, they are the major storage and transport form of fatty acids. Triacylglycerol molecules store energy more efficiently than glycogen for several reasons:

1. Because triacylglycerols are hydrophobic, they coalesce into compact, anhydrous droplets within cells. A specialized type of cell called the *adipocyte*, found in adipose tissue, stores triacylglycerols. Because glycogen (the other major energy storage molecule) binds a substantial amount of water, the anhydrous triacylglycerols store an equivalent amount of energy in about one-eighth of glycogen's volume.

2. Triacylglycerol molecules are less oxidized than carbohydrate molecules. Therefore, triacylglycerols release more energy (38.9 kJ/g of fat compared with 17.2 kJ/g of carbohydrate) when they are degraded.

A second important function of fat is to provide insulation in low temperatures. Fat is a poor conductor of heat. Because adipose tissue, with its high triacylglycerol content, is found throughout the body (especially underneath the skin), it prevents heat loss. Finally, in some animals fat molecules secreted by specialized glands make fur or feathers water-repellent.

In plants, triacylglycerols constitute an important energy reserve in fruits and seeds. Because these molecules contain relatively large amounts of unsaturated fatty acids (e.g., oleic and linoleic), they are referred to as plant oils. Seeds rich in oil include peanut, corn, palm, safflower, and soybean. Avocados and olives are fruits with a high oil content.

KEY CONCEPTS 11.2

Triacylglycerols are molecules consisting of glycerol esterified to three fatty acids. In both animals and plants they are a rich energy source.

QUESTION 11.1

Oils can be converted to fats in a commercial nickel-catalyzed process referred to as *partial hydrogenation*. Under relatively mild conditions (180°C and pressures of about 1013 torr or 1.33 atm) enough double bonds are hydrogenated for liquid oils to solidify. This solid material, oleomargarine, has a consistency like butter. Propose a practical reason why oils are not completely hydrogenated during commercial hydrogenation processes.

QUESTION 11.2

Soapmaking is an ancient process. The Phoenicians, a seafaring people who dominated trade in the Mediterranean Sea about 3000 years ago, are believed to have been the first to manufacture soap. Traditionally, soap has been made by heating animal fat with potash. (Potash is a mixture of potassium hydroxide (KOH) and potassium carbonate (K_2CO_3) obtained by mixing wood ash with water.) Currently, soap is made by heating beef tallow or coconut oil with sodium or potassium hydroxide. During this reaction, which is a *saponification* (the reverse of esterification), triacylglycerol molecules are hydrolyzed to give glycerol and the sodium or potassium salts of fatty acids:

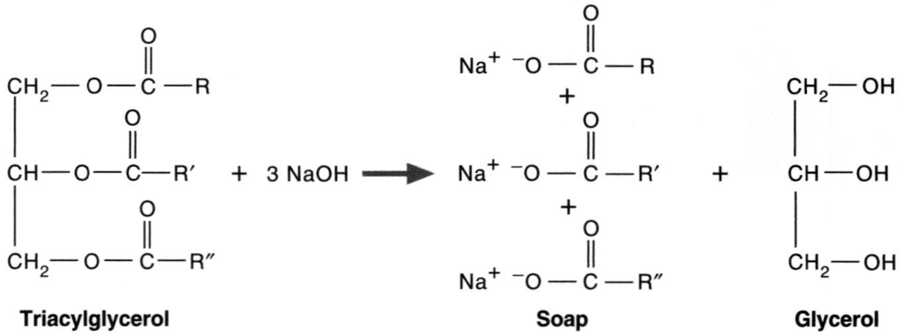

Triacylglycerol   Soap   Glycerol

Fatty acid salts (soaps) are amphipathic molecules, that is, they possess polar and nonpolar domains, that spontaneously form into micelles (Figure 3.11). Soap micelles have negatively charged surfaces that repel each other. Because soap can act as an emulsifying agent it is a cleansing agent. (*Emulsifying agents* promote the formation of an emulsion, that is, the dispersal of one substance in another.) When soap and grease are mixed together, an emulsion forms. Complete the following diagram, and explain how this process occurs. (*Hint*: Recall that "like dissolves like.")

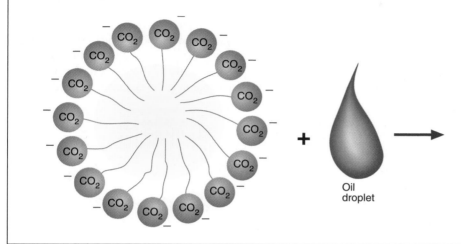

Wax Esters

Waxes are complex mixtures of nonpolar lipids. They are protective coatings on leaves, stems, and fruits of plants and the skin and fur of animals. Esters composed of long-chain fatty acids and long-chain alcohols are prominent constituents of most waxes. Well-known examples of waxes include carnauba wax, produced by the leaves of the Brazilian wax palm, and beeswax. The predominant constituent of carnauba wax is the wax ester melissyl cerotate (Figure 11.5). Triacontyl hexadecanoate is one of several important wax esters in beeswax. Waxes also contain hydrocarbons, alcohols, fatty acids, aldehydes, and sterols (steroid alcohols).

Phospholipids

Phospholipids have several roles in living organisms. They are first and foremost structural components of membranes. In addition, several phospholipids are emulsifying agents and surface active agents. (A *surface active agent* is a substance that lowers the surface tension of a liquid, usually water, so that it spreads out over a surface.) Phospholipids are suited to these roles because they are amphipathic molecules. Despite their structural differences, all phospholipids have hydrophobic and hydrophilic domains. The hydrophobic domain is composed largely of the hydrocarbon chains of fatty acids; the hydrophilic domain, called a **polar head group**, contains phosphate and other charged or polar groups.

When phospholipids are suspended in water, they spontaneously rearrange into ordered structures (Figure 11.6). As these structures form, phospholipid hydrophobic groups are buried in the interior to exclude water. Simultaneously, hydrophilic polar head groups are oriented so that they are exposed to water. When phospholipid molecules are present in sufficient concentration, they form bimolecular layers. This property of phospholipids (and other amphipathic lipid molecules) is the basis of membrane structure (see pp. 353–360).

There are two types of phospholipids: phosphoglycerides and sphingomyelins. **Phosphoglycerides** are molecules that contain glycerol, fatty acids, phosphate,

$$CH_3-(CH_2)_{24}-\overset{\displaystyle O}{\overset{\displaystyle \|}{C}}-O-(CH_2)_{29}-CH_3$$

FIGURE 11.5

The Wax Ester Melissyl Cerotate.

Found in carnauba wax, melissyl cerotate is an ester formed from melissyl alcohol and cerotic acid.

The **eicosanoids** are a diverse group of extremely powerful hormonelike molecules produced in most mammalian tissues. They mediate a wide variety of physiological processes. Examples include smooth muscle contraction, inflammation, pain perception, and blood flow regulation. Eicosanoids are also implicated in several diseases such as myocardial infarct and rheumatoid arthritis. Because they are generally active within the cell in which they are produced, the eicosanoids are called **autocrine** regulators instead of hormones. Most eicosanoids are derived from arachidonic acid ($20:4^{\Delta 5,8,11,14}$), which is also called 5,8,11,14-eicosatetraenoic acid. (Arachidonic acid is synthesized from linoleic acid by adding a three-carbon unit followed by decarboxylation and desaturation.)

Production of eicosanoids begins after arachidonic acid is released from membrane phospholipid molecules by the enzyme phospholipase A_2. The eicosanoids, which include the prostaglandins, thromboxanes, and leukotrienes (Figure 11A), are extremely difficult to study because they are active for short periods (often measured in seconds or minutes). In addition, they are produced only in small amounts.

Prostaglandins are arachidonic acid derivatives that contain a cyclopentane ring with hydroxy groups at C-11 and C-15. Molecules belonging to the E series of prostaglandins have a carbonyl group at C-9, whereas the F series molecules have an OH group at the same position. The subscript number in a prostaglandin name indicates

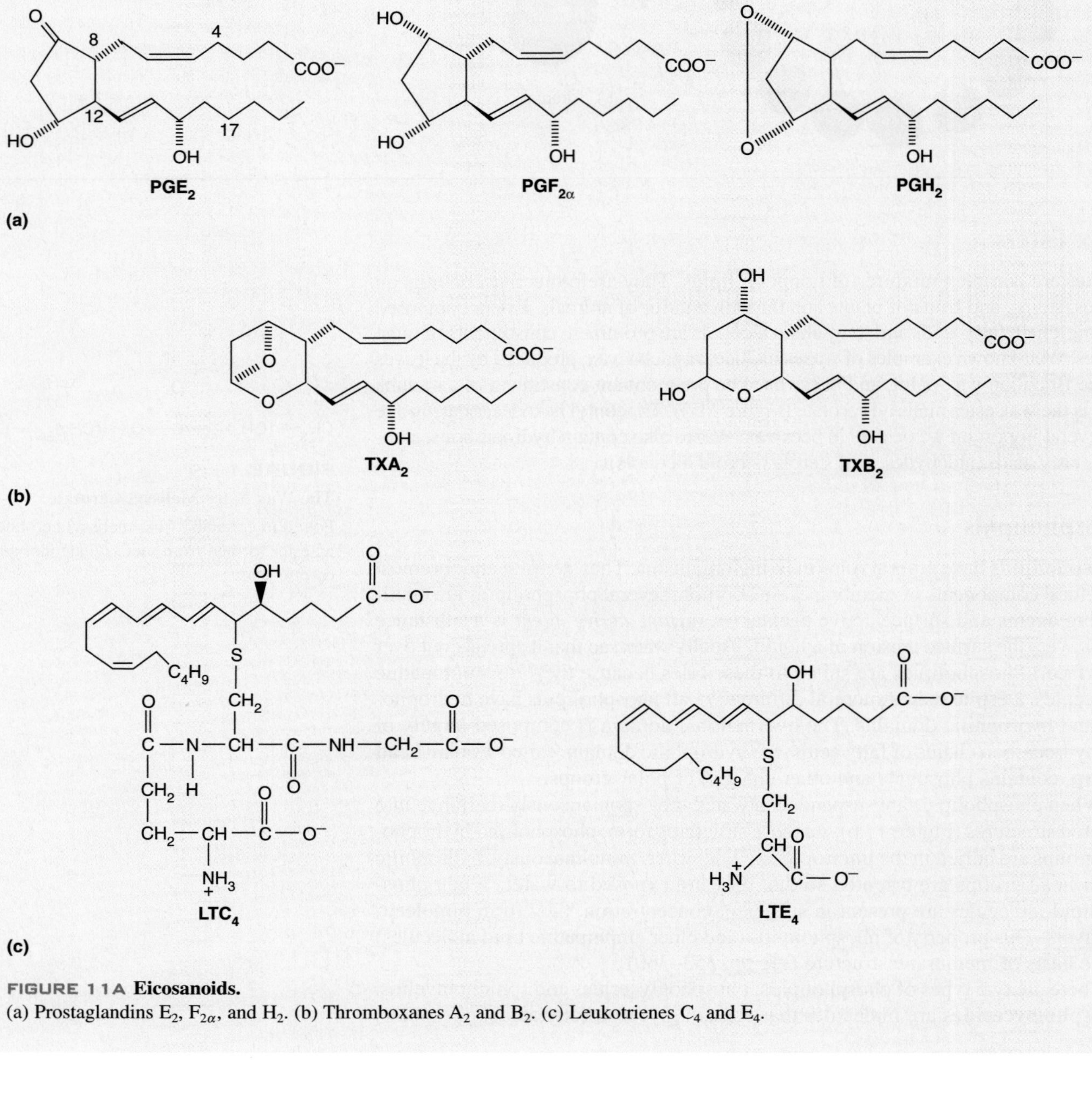

FIGURE 11A Eicosanoids.
(a) Prostaglandins E_2, $F_{2\alpha}$, and H_2. (b) Thromboxanes A_2 and B_2. (c) Leukotrienes C_4 and E_4.

the number of double bonds in the molecule. The 2-series, derived from arachidonic acid, appears to be the most important group of prostaglandins in humans. Prostaglandins are involved in a wide range of regulatory functions. For example, prostaglandins promote inflammation, an infection-fighting process that produces pain and fever. They are also involved in reproductive processes (e.g., ovulation and uterine contractions during conception and labor) and digestion (e.g., inhibition of gastric secretion). Additional biological actions of selected prostaglandins are shown in Figure 11B. Prostaglandin metabolism is complex for the following reasons:

1. There are many types of prostaglandins.
2. The types and amounts of prostaglandins are different in each tissue or organ.
3. Certain prostaglandins have opposite effects in different organs, that is, their receptors are tissue-specific. (For example, several E-series prostaglandins cause smooth muscle relaxation in organs such as the intestine and uterus. The same molecules promote contraction of the smooth muscle in the cardiovascular system.)

The **thromboxanes** are also derivatives of arachidonic acid. They differ from other eicosanoids in that their structures have a cyclic ether. Thromboxane A_2 (TXA_2), the most prominent member of this group of eicosanoids, is primarily produced by platelets (cell fragments in the blood that initiate blood clot formation).

Once it is released, TXA_2 promotes platelet aggregation and vasoconstriction.

The **leukotrienes** are linear (noncyclic) derivatives of arachidonic acid whose synthesis is initiated by a peroxidation reaction. The leukotrienes differ in the position of this peroxidation step and the nature of the thioether group attached near the site of peroxidation. The name *leukotrienes* stems from their early discovery in white blood cells (leukocytes) and the presence of a *triene* (three conjugated double bonds) in their structures. (The term *conjugated* indicates that carbon-carbon double bonds are separated by one carbon-carbon single bond.) Leukotrienes LTC_4, LTD_4, and LTE_4 have been identified as components of slow-reacting substance of anaphylaxis (SRS-A). (The subscript in a leukotriene name indicates the total number of double bonds in the molecule.) *Anaphylaxis* is an unusually severe allergic reaction that results in respiratory distress, low blood pressure, and shock. During inflammation (a normal response to tissue damage) these molecules increase fluid leakage from blood vessels into affected areas. LTB_4, a potent chemotactic agent, attracts infection-fighting white blood cells to damaged tissue. (Chemotactic agents are also referred to as chemoattractants.) Other effects of leukotrienes include vasoconstriction and bronchoconstriction (both caused by the contraction of smooth muscle in blood vessels and the air passages in the lungs) and edema (increased capillary permeability that causes fluid to leak out of blood vessels).

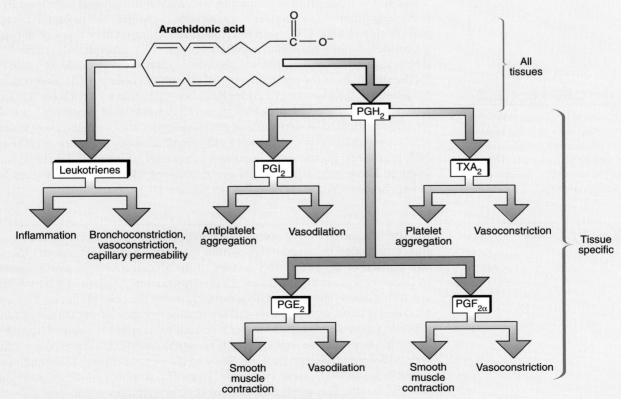

FIGURE 11B **Biological Actions of Selected Eicosanoid Molecules.**
The synthesis of these molecules is discussed in Chapter 12.

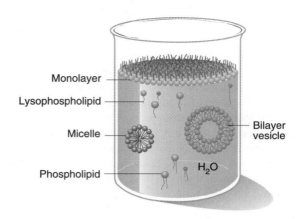

FIGURE 11.6

Phospholipid Molecules in Aqueous Solution.

Each molecule is represented as a polar head group attached to one or two fatty acyl chains. (Lysophospholipid molecules possess only one fatty acyl chain.) The monolayer on the surface of the water forms first. As the phospholipid concentration increases, bilayer vesicles begin to form. Because of their smaller size, lysophospholipid molecules form micelles.

and an alcohol (e.g., choline). **Sphingomyelins** differ from phosphoglycerides in that they contain sphingosine instead of glycerol. Because sphingomyelins are also classified as sphingolipids, their structures and properties are discussed in that section.

Phosphoglycerides are the most numerous phospholipid molecules found in cell membranes. The simplest phosphoglyceride, phosphatidic acid, is the precursor for all other phosphoglyceride molecules. Phosphatidic acid is composed of glycerol-3-phosphate that is esterified with two fatty acids. Phosphoglyceride molecules are classified according to which alcohol becomes esterified to the phosphate group. For example, if the alcohol is choline, the molecule is called phosphatidylcholine (PC) (also referred to as lecithin). Other types of phosphoglycerides include phosphatidylethanolamine (PE), phosphatidylserine (PS), diphosphatidylglycerol (dPG), and phosphatidylinositol (PI). (Refer to Table 11.2 for the structures of the common types of phosphoglycerides.) The most common fatty acids in the phosphoglycerides have between 16 and 20 carbons. Saturated fatty acids usually occur at C-1 of glycerol. The fatty acid substituent at C-2 is usually unsaturated. A derivative of phosphatidylinositol, namely, phosphatidyl-4,5-bisphosphate (PIP_2), is found in only small amounts in plasma membranes. PIP_2 is now recognized as an important component of intracellular signal transduction. The *phosphatidylinositol cycle*, initiated when certain hormones bind to membrane receptors, is described in Section 16.4.

KEY CONCEPTS 11.3

Phospholipids are amphipathic molecules that play important roles in living organisms as membrane components, emulsifying agents, and surface active agents. There are two types of phospholipids: phosphoglycerides and sphingomyelins.

QUESTION 11.3

Dipalmitoylphosphatidylcholine is the major component of *surfactant*, a substance produced by certain cells within the lungs. Surfactant reduces the surface tension of the moist inner surface of the alveoli. (Alveoli, also referred to as alveolar sacs, are the functional units of respiration. Oxygen and carbon dioxide diffuse across the walls of alveolar sacs, which are one cell thick.) The water on alveolar surfaces has a high surface tension because of the attractive forces between the molecules. If the water's surface tension is not reduced, the alveolar sac tends to collapse, making breathing extremely difficult. If premature infants lack sufficient surfactant, they are likely to die of suffocation. This condition is called *respiratory distress syndrome*. Draw the structure of dipalmitoylphosphatidylcholine. Considering the general structural features of phospholipids, propose a reason why surfactant is effective in reducing surface tension.

TABLE 11.2

Major Classes of Phosphoglycerides

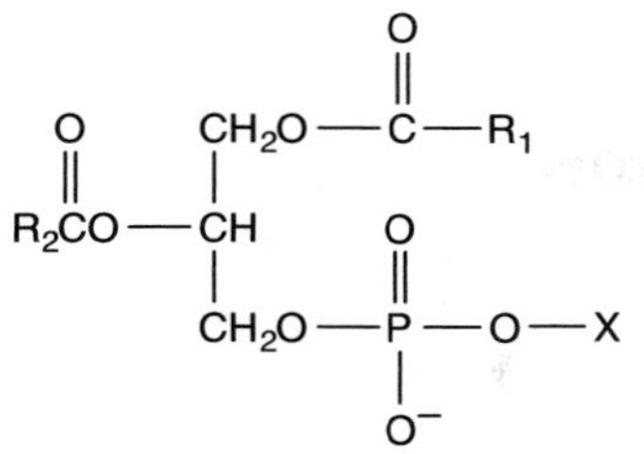

	X Substituent	
Name of X-OH	**Formula of X**	**Name of Phospholipid**
Water	—H	Phosphatidic acid
Choline	—$CH_2CH_2\overset{+}{N}(CH_3)_3$	Phosphatidylcholine (lecithin)
Ethanolamine	—$CH_2CH_2\overset{+}{N}H_3$	Phosphatidylethanolamine (cephalin)
Serine	—CH_2—CH with $\overset{+}{N}H_3$ and COO^-	Phosphatidylserine
Glycerol	—CH_2CHCH_2OH with OH	Phosphatidylglycerol
Phosphatidylglycerol	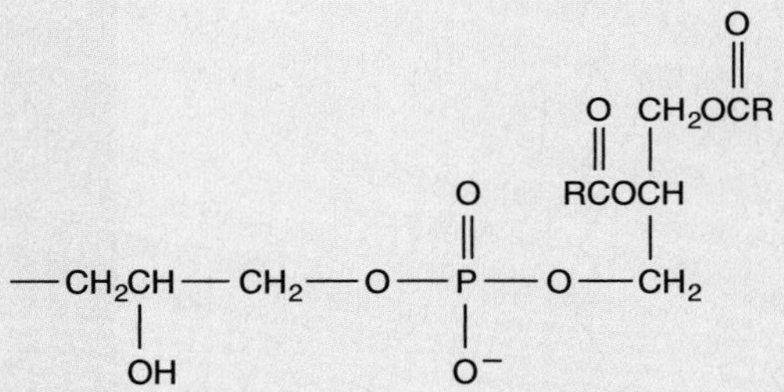 —CH_2CH—CH_2—O—P—O—CH_2 with OH, O^-, $RCOCH$, CH_2OCR	Diphosphatidylglycerol (cardiolipin)
Inositol	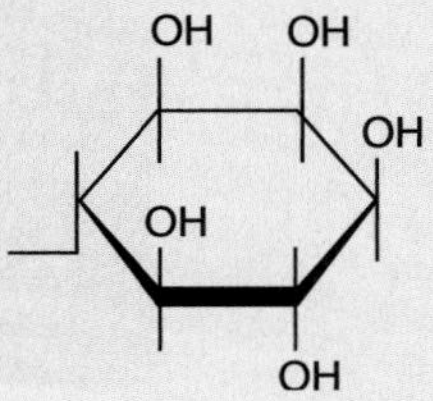 (inositol ring with OH groups)	Phosphatidylinositol

Sphingolipids

Sphingolipids are important components of animal and plant membranes. All sphingolipid molecules contain a long-chain amino alcohol. In animals this alcohol is primarily sphingosine (Figure 11.7). Phytosphingosine is found in plant sphingolipids. The core of each type of sphingolipid is *ceramide*, a fatty acid amide derivative of sphingosine. In *sphingomyelin*, the 1-hydroxyl group of ceramide is esterified to the phosphate group of phosphorylcholine or phosphorylethanolamine (Figure 11.8). Sphingomyelin is found in most animal cell membranes. However, as its name suggests, sphingomyelin is found in greatest abundance in the myelin sheath of nerve cells. (The myelin sheath is formed by successive wrappings of the cell

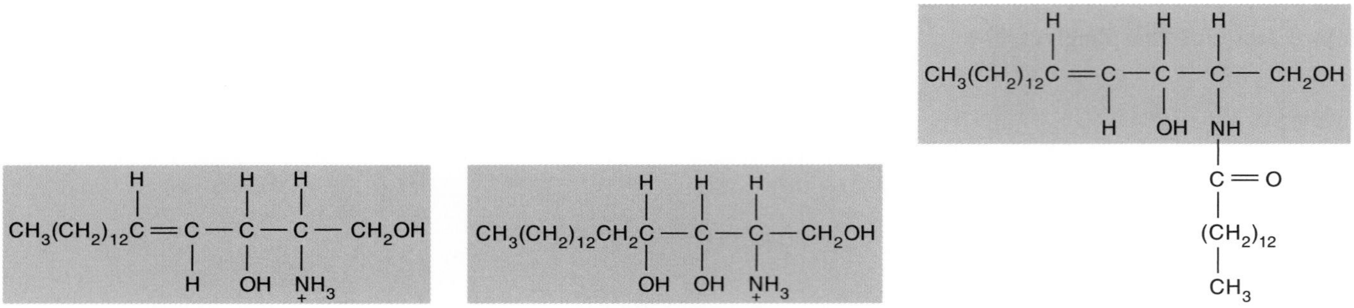

Sphingosine Phytosphingosine A Ceramide

FIGURE 11.7

Sphingolipid Components.

Note that the *trans*-isomer of sphingosine occurs in sphingolipids.

FIGURE 11.8

Conformational and Space-Filling Models of Sphingomyelin.

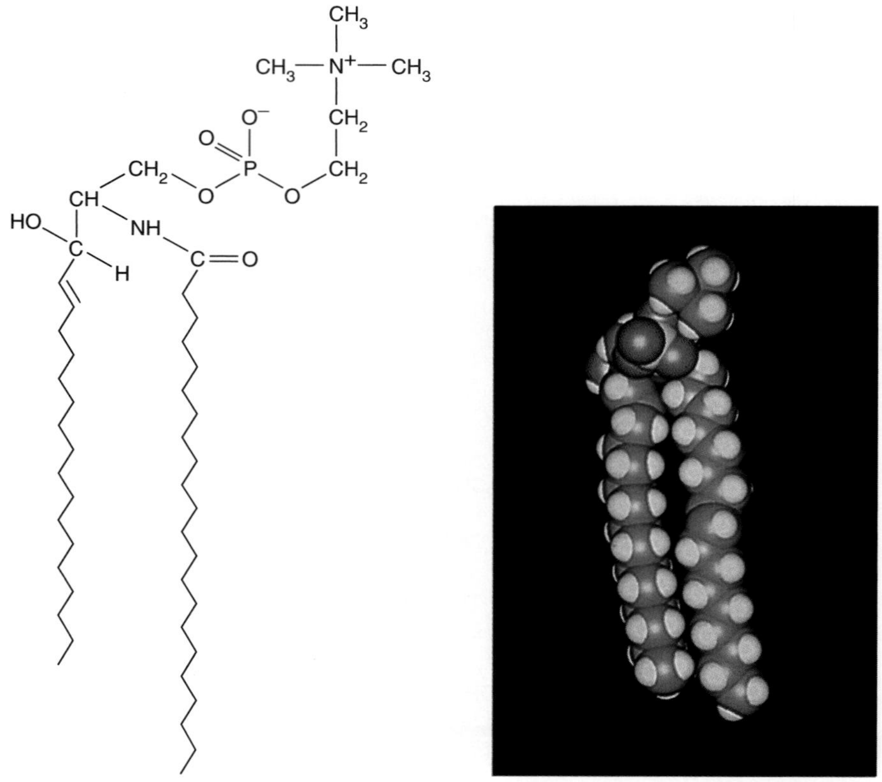

membrane of a specialized myelinating cell around a nerve cell axon. Its insulating properties facilitate the rapid transmission of nerve impulses.

The ceramides are also precursors for the **glycolipids**, sometimes referred to as the *glycosphingolipids* (Figure 11.9). In glycolipids a monosaccharide, disaccharide, or oligosaccharide is attached to a ceramide through an O-glycosidic linkage. Glycolipids also differ from sphingomyelin in that they contain no phosphate. The most important glycolipid classes are the cerebrosides, the sulfatides, and the gangliosides. *Cerebrosides* are sphingolipids in which the head group is a monosaccharide. (These molecules, unlike phospholipids, are nonionic.) Galactocerebrosides, the most common example of this class, are almost entirely found in the cell membranes of the brain. If a cerebroside is sulfated, it is referred to as a *sulfatide*. Sulfatides are negatively charged at physiological pH. Sphingolipids

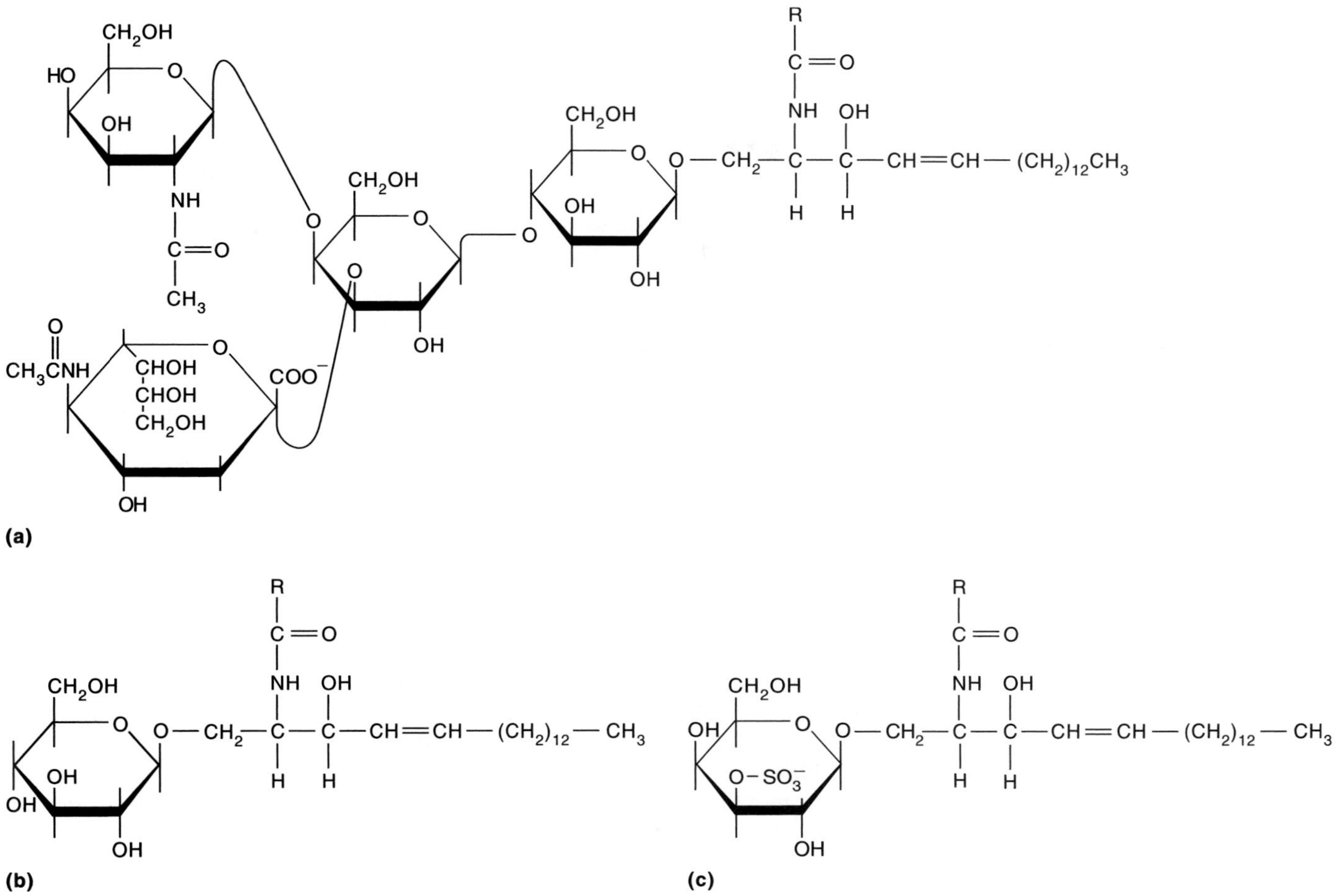

FIGURE 11.9
Selected Glycolipids.
(a) Tay-Sachs ganglioside (GM$_2$), (b) glucocerebroside, and (c) galactocerebroside sulfate (a sulfatide).

that possess oligosaccharide groups with one or more sialic acid residues are called *gangliosides*. Although gangliosides were first isolated from nerve tissue, they also occur in most other animal tissues. The names of gangliosides include subscript letter and numbers. The letters M, D, and T indicate whether the molecule contains one, two, or three sialic acid residues (see Fig. 7.23d), respectively. The numbers designate the sequence of sugars that are attached to ceramide. The Tay-Sachs ganglioside G$_{M2}$ is illustrated in Figure 11.9.

The role of glycolipids is still unclear. Certain glycolipid molecules may bind bacterial toxins, as well as bacterial cells, to animal cell membranes. For example, the toxins that cause cholera, tetanus, and botulism bind to glycolipid cell membrane receptors. Bacteria that have been shown to bind to glycolipid receptors include *E. coli, Streptococcus pneumoniae,* and *Neisseria gonorrhoeae,* the causative agents of urinary tract infections, pneumonia, and gonorrhea, respectively.

KEY CONCEPTS 11.4

Sphingolipids, important membrane components of animals and plants, contain a complex long-chain amino alcohol (either sphingosine or phytosphingosine). The core of each sphingolipid is ceramide, a fatty acid amide derivative of the alcohol molecule. Glycolipids are derivatives of ceramide that possess a carbohydrate component.

Sphingolipid Storage Diseases

Each lysosomal storage disease (see p. 49) is caused by a hereditary deficiency of an enzyme required for the degradation of a specific metabolite. Several lysosomal storage diseases are associated with sphingolipid metabolism. Most of

TABLE 11.3

Selected Sphingolipid Storage Diseases*

Disease	Symptom	Accumulating Sphingolipid	Enzyme Deficiency
Tay-Sachs disease	Blindness, muscle weakness, seizures, mental retardation	Ganglioside G_{M2}	β-Hexosaminidase A
Gaucher's disease	Mental retardation, liver and spleen enlargement, erosion of long bones	Glucocerebroside	β-Glucosidase
Krabbe's disease	Demyelination, mental retardation	Galactocerebroside	β-Galactosidase
Niemann-Pick disease	Mental retardation	Sphingomyelin	Sphingomyelinase

*Many diseases are named for the physicians who first described them. Tay-Sachs disease was reported by Warren Tay (1843–1927), a British ophthalmologist, and Bernard Sachs (1858–1944), a New York neurologist. Phillipe Gaucher (1854–1918), a French physician, and Knud Krabbe (1885–1961), a Danish neurologist, first described Gaucher's disease and Krabbe's disease, respectively. Niemann-Pick disease was first characterized by the German physicians Albert Niemann (1880–1921) and Ludwig Pick (1868–1944).

 these diseases, also referred to as the *sphingolipidoses*, are fatal. The most common sphingolipid storage disease, Tay-Sachs disease, is caused by a deficiency of β-hexosaminidase A, the enzyme that degrades the ganglioside G_{M2}. As cells accumulate this molecule, they swell and eventually die. Tay-Sachs symptoms (i.e., blindness, muscle weakness, seizures, and mental retardation) usually appear several months after birth. Because there is no therapy for Tay-Sachs disease or for any other of the sphingolipidoses, the condition is always fatal (usually by age 3). Examples of the sphingolipidoses are summarized in Table 11.3.

Isoprenoids

The **isoprenoids** are a vast array of biomolecules that contain repeating five-carbon structural units known as *isoprene units* (Figure 11.10). Isoprenoids are not synthesized from isoprene (methylbutadiene). Instead, their biosynthetic pathways all begin with the formation of isopentenyl pyrophosphate from acetyl-CoA (Chapter 12).

The isoprenoids consist of terpenes and steroids. **Terpenes** are an enormous group of molecules that are found largely in the essential oils of plants. (Essential oils are plant extracts used for thousands of years in perfumes and medicines.) Steroids are derivatives of the hydrocarbon ring system of cholesterol.

TERPENES The terpenes are classified according to the number of isoprene residues they contain (Table 11.4). *Monoterpenes* are composed of two isoprene

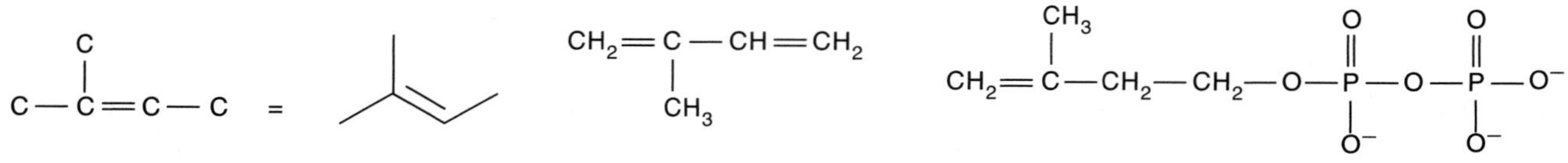

(a)

Isoprene unit

(b)

Isoprene

(c)

Isopentenyl pyrophosphate

FIGURE 11.10

Isoprene.

(a) Basic isoprene structure, (b) the organic molecule isoprene, (c) isopentenyl pyrophosphate.

TABLE 11.4
Examples of Terpenes

Type	Number of Isoprene Units	Example	
		Name	Structure
Monoterpene	2	Geraniol	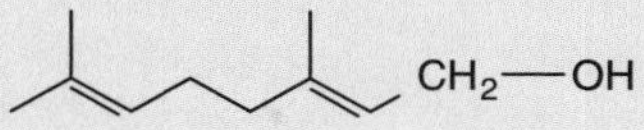
Sesquiterpene	3	Farnesene	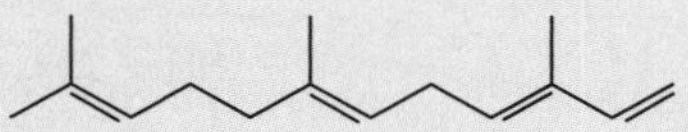
Diterpene	4	Phytol	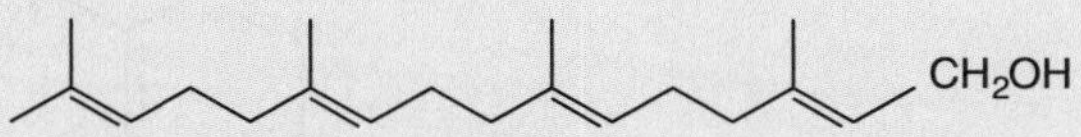
Triterpene	6	Squalene	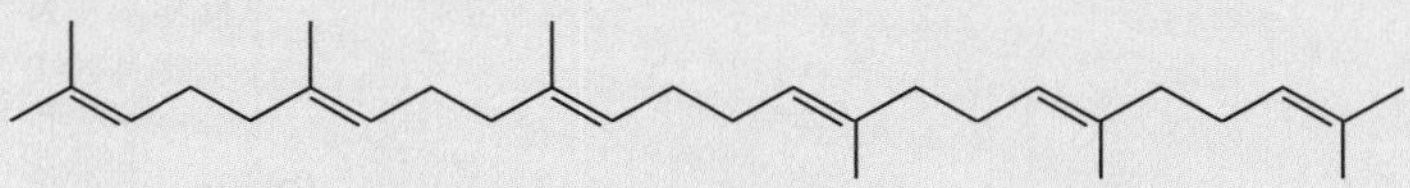
Tetraterpene	8	β-Carotene	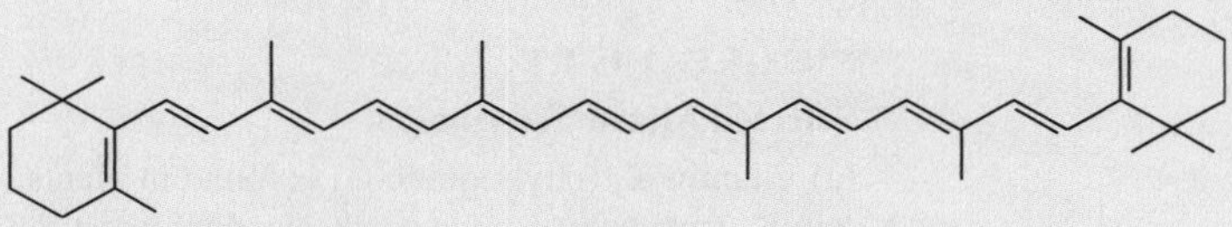
Polyterpene	Thousands	Rubber	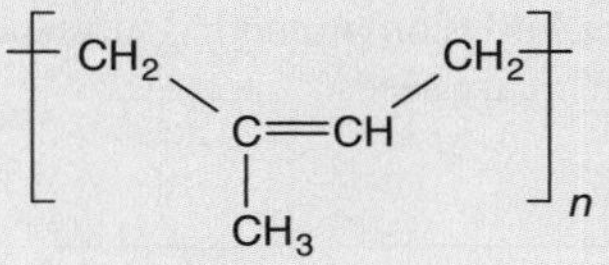

units (10 carbon atoms). Geraniol is a monoterpene found in oil of geranium. Terpenes that contain three isoprenes (15 carbons) are referred to as *sesquiterpenes*. Farnesene, an important constituent of oil of citronella (a substance used in soap and perfumes), is a sesquiterpene. Phytol, a plant alcohol, is an example of the *diterpenes*, molecules composed of four isoprene units. Squalene, which is found in large quantities in shark liver oil, olive oil, and yeast, is a prominent example of the *triterpenes*. (Squalene is an intermediate in the synthesis of the steroids.) **Carotenoids**, the orange pigments found in most plants, are the only *tetraterpenes* (molecules composed of eight isoprene units). The *carotenes* are hydrocarbon members of this group. The *xanthophylls* are oxygenated derivatives of the carotenes. *Polyterpenes* are high-molecular-weight molecules composed of hundreds or thousands of isoprene units. Natural rubber is a polyterpene composed of between 3000 and 6000 isoprene units.

Several important biomolecules are composed of nonterpene components attached to isoprenoid groups (often referred to as *prenyl* or *isoprenyl* groups). Examples of these biomolecules, referred to as **mixed terpenoids**, include vitamin E (α-tocopherol) (Figure 10.22a), ubiquinone (Figure 10.3), vitamin K, and some cytokinins (plant hormones) (Figure 11.11).

A variety of proteins in eukaryotic cells are now known to be covalently attached to prenyl groups after their biosynthesis on ribosomes. The prenyl groups most often involved in this process, referred to as **prenylation**, are farnesyl and geranylgeranyl groups (Figure 11.12). The function of protein prenylation is not clear. There is some evidence that it plays a role in the control of cell growth. For example, *Ras proteins*, a group of cell growth regulators, are activated by prenylation reactions.

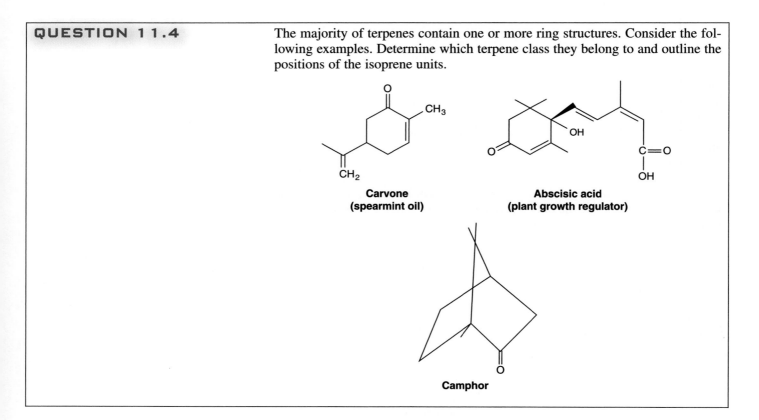

FIGURE 11.11

Selected Mixed Terpenoids.

(a) Vitamin K_1 (phylloquinone) is found in plants, where it acts as an electron carrier in photosynthesis. Vitamin K_2 (menaquinone) is synthesized by intestinal bacteria and plays an important role in blood coagulation. (b) Cytokinins are cell-division-promoting substances in plants. Some cytokinins are mixed terpenoids. Zeatin is found in immature maize seeds.

QUESTION 11.4

The majority of terpenes contain one or more ring structures. Consider the following examples. Determine which terpene class they belong to and outline the positions of the isoprene units.

Carvone
(spearmint oil)

Abscisic acid
(plant growth regulator)

Camphor

STEROIDS **Steroids** are complex derivatives of triterpenes. They are found in all eukaryotes and a small number of bacteria. Each type of steroid is composed of four fused rings. Steroids are distinguished from each other by the placement

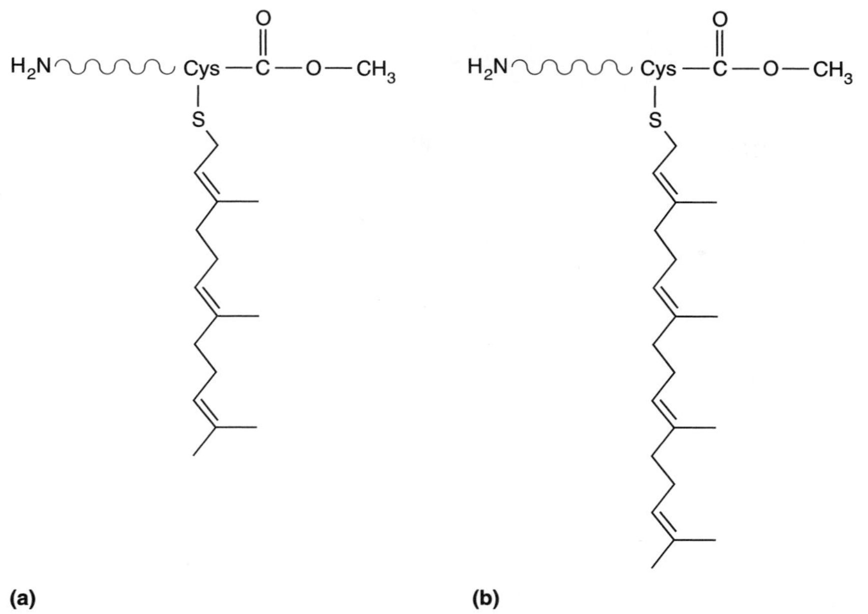

$$H_2N\text{\raisebox{0pt}{$\sim\!\!\sim\!\!\sim$}} Cys-\overset{\displaystyle O}{\overset{\|}{C}}-O-CH_3$$

(a)   (b)

FIGURE 11.12

Prenylated Proteins.

Prenyl groups are covalently attached at the SH group of C-terminal cysteine residues. Many prenylated proteins are also methylated at this residue. (a) Farnesylated protein, (b) geranylgeranylated protein.

of carbon-carbon double bonds and various substituents (e.g., hydroxyl, carbonyl, and alkyl groups).

Cholesterol, an important molecule in animals, is an example of a steroid (Figure 11.13). In addition to being an essential component in animal cell membranes, cholesterol is a precursor in the biosynthesis of all steroid hormones, vitamin D, and bile salts (Figure 11.14). Cholesterol possesses two methyl substituents (C-18 and C-19), which are attached to C-13 and C-10, respectively, and a Δ5 double bond. A branched hydrocarbon side chain is attached to C-17. Because this molecule has a hydroxyl group (attached to C-3), it is classified as a *sterol*. (Although the term *steroid* is most properly used to designate molecules that contain one or more carbonyl or carboxyl groups, it is often used to describe all derivatives of the steroid ring structure.) Cholesterol is usually stored within cells as a fatty acid ester. The esterification reaction is catalyzed by the enzyme *acyl-CoA:cholesterol acyltransferase* (ACAT), located on the cytoplasmic face of the endoplasmic reticulum.

QUESTION 11.5

Bile salts are emulsifying agents; that is, they promote the formation of mixtures of hydrophobic substances and water. Produced in the liver, bile salts assist in the digestion of fats in the small intestine. They are formed by linking bile acids to hydrophilic substances such as the amino acid glycine. After reviewing the structure of cholic acid in Figure 11.14, suggest how the structural features of bile salts contribute to their function.

Practically all plant steroid molecules are sterols. The function of plant sterols is still relatively unclear. They undoubtedly play an important role in membrane structure and function. Certain sterol derivatives, such as the cardiac glycosides, are known to protect plants that produce them from predators. Most plant sterols

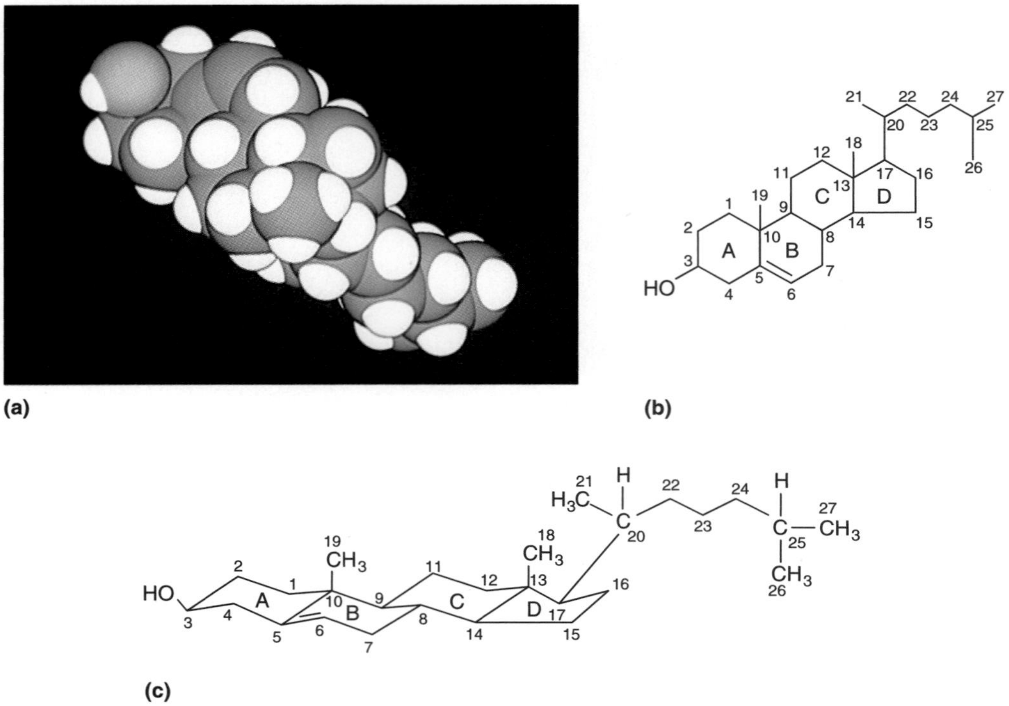

(a)   **(b)**

(c)

FIGURE 11.13

Structure of Cholesterol.

(a) Space-filling model, (b) conventional view, and (c) conformational model. Space-filling models and conformational models (see p. 206) are more accurate representations of molecular structure than is the conventional view.

KEY CONCEPTS 11.5

Isoprenoids are a large group of biomolecules with repeating units derived from isopentenyl pyrophosphate. There are two types of isoprenoids: terpenes and steroids.

possess a one- or two-carbon substituent attached to C-24. The most abundant sterols in green algae and higher plants are β-sitosterol and stigmasterol (Figure 11.15).

Cardiac glycosides, molecules that increase the force of cardiac muscle contraction, are among the most interesting steroid derivatives. Glycosides are carbohydrate-containing acetals (see p. 210). Although several cardiac glycosides are extremely toxic (e.g., *ouabain*, obtained from the seeds of the plant *Strophanthus gratus*), others have valuable medicinal properties (Figure 11.16). For example, *digitalis*, an extract of the dried leaves of *Digitalis purpurea* (the foxglove plant), is a time-honored stimulator of cardiac muscle contraction. *Digitoxin*, the major "cardiotonic" glycoside in digitalis, is used to treat congestive heart failure, an illness in which the heart is so damaged by disease processes (e.g., myocardial infarct) that pumping is impaired. In higher than therapeutic doses, digitoxin is extremely toxic. Both ouabain and digitoxin inhibit Na^+–K^+ ATPase (see p. 363).

Lipoproteins

Although the term *lipoprotein* can describe any protein that is covalently linked to lipid groups (e.g., fatty acids or prenyl groups), it is most often used for a group of molecular complexes found in the blood plasma of mammals (especially humans). Plasma lipoproteins transport lipid molecules (triacylglycerols, phospholipids, and cholesterol) through the bloodstream from one organ to another. Lipoproteins also contain several types of lipid-soluble antioxidant molecules (e.g., α-tocopherol and several carotenoids). (The function of *antioxidants*, substances that protect biomolecules from free radicals, is described in Chapter 10.) The protein components of lipoproteins are called *apolipoproteins* or *apoproteins*.

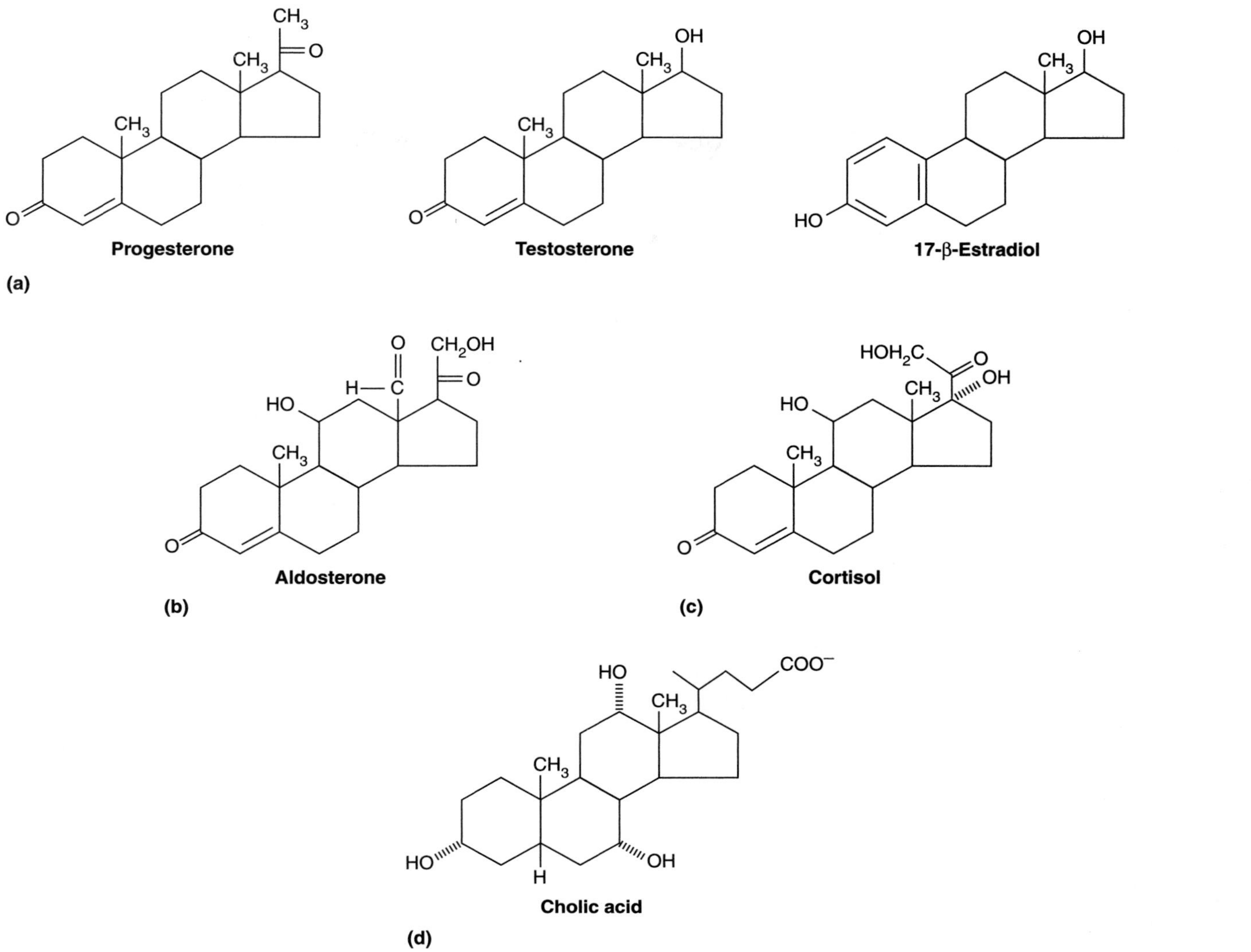

FIGURE 11.14

Animal Steroids.

(a) Sex hormones (molecules that regulate the development of sexual structures and various reproductive behaviors). (b) A mineralocorticoid (a molecule produced in the adrenal cortex that regulates plasma concentrations of several ions, especially sodium). (c) A glucocorticoid (a molecule that regulates the metabolism of carbohydrates, fats, and proteins). (d) A bile acid (a molecule produced in the liver that aids the absorption of dietary fats and fat-soluble vitamins in the intestine).

A generalized lipoprotein is shown in Figure 11.17. The relative amounts of lipid and protein components of the major types of lipoprotein are summarized in Figure 11.18.

Lipoproteins are classified according to their density. **Chylomicrons**, which are large lipoproteins of extremely low density, transport dietary triacylglycerols and cholesteryl esters from the intestine to muscle and adipose tissues. **Very low density lipoproteins** (VLDL) ($0.95-1.006$ g/cm^3), synthesized in the liver, transport lipids to tissues. As VLDL are transported through the body, they become depleted of triacylglycerols and some apoproteins and phospholipids. Eventually, the triacylglycerol-depleted VLDL remnants are either picked up by the liver or converted to **low-density lipoproteins** (LDL) ($1.006-1.063$ g/cm^3). LDL carry cholesterol to tissues. In a complex process (Section 11.2) elucidated by Michael Brown and Joseph Goldstein (recipients of the 1985 Nobel prize for medicine or physiology), LDL are engulfed by cells after binding to LDL

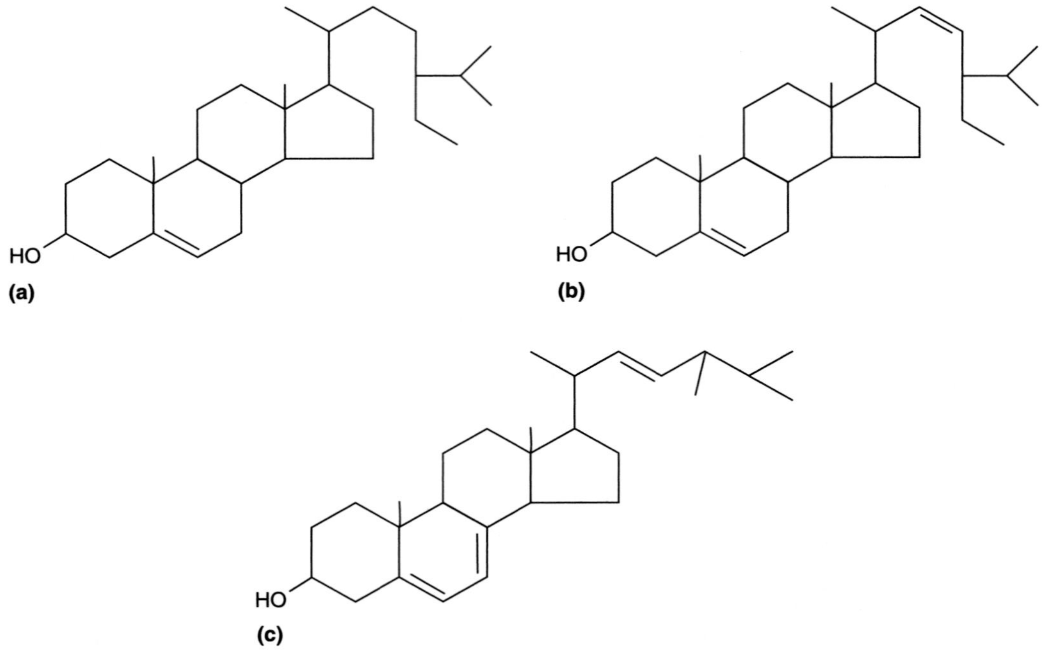

FIGURE 11.15

Plant Steroids.

(a) β-Sitosterol, (b) stigmasterol, and (c) ergosterol (found in fungi). One of the most important roles of plant sterols is the stabilization of cell membranes.

FIGURE 11.16

Cardiac Glycosides.

Each cardiac glycoside possesses a glycone (carbohydrate) and an aglycone component. (a) In ouabain the glycone is one rhamnose residue. The steroid aglycone of ouabain is called ouabagenin. (b) The glycone of digitoxin is composed of three digitoxose residues. The aglycone of digitoxin is called digitoxigenin.

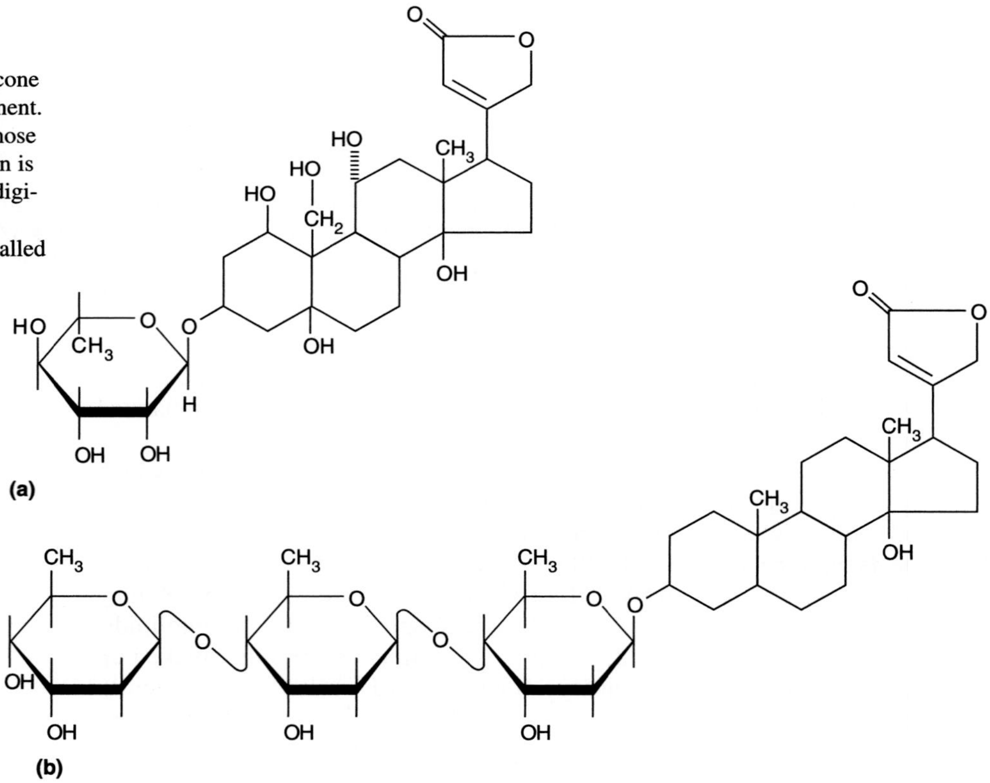

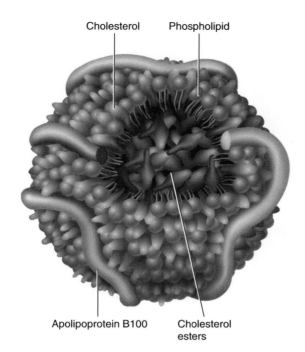

Cholesterol   Phospholipid

Apolipoprotein B100   Cholesterol esters

FIGURE 11.17

Plasma Lipoproteins.

Lipoproteins vary in diameter from 5 to 1000 nm. Each type of lipoprotein contains a neutral lipid core composed of cholesteryl esters and/or triacylglycerols. This core is surrounded by a layer of phospholipid, cholesterol, and protein. Charged and polar residues on the surface of a lipoprotein enable it to dissolve in blood. In LDL (low-density lipoprotein), the example illustrated in this figure, each particle is composed of a core of cholesteryl esters surrounded by a monolayer that consists of hundreds of cholesterol and phospholipid molecules and one molecule of apolipoprotein B-100.

receptors. The role of **high-density lipoprotein** (HDL) (1.063–1.210 g/cm^3), also produced in the liver, appears to be the scavenging of excessive cholesterol from cell membranes. Cholesteryl esters are formed when the plasma enzyme lecithin:cholesterol acyltransferase (LCAT) transfers a fatty acid residue from lecithin to cholesterol (Figure 11.19). It is now believed that HDL transport these cholesteryl esters to the liver. The liver, the only organ that can dispose of excess cholesterol, converts most of it to bile acids (Chapter 12).

Lipoproteins and Atherosclerosis

Atherosclerosis is a chronic disease in which soft masses, called *atheromas*, accumulate on the inside of arteries. These deposits are also referred to as *plaque*. During plaque formation, which is a progressive process, smooth muscle cells, macrophages, and various cell debris build up. As macrophages fill with lipid (predominantly cholesterol and cholesteryl esters derived from LDL deposits in a mechanically damaged artery wall), they take on a foam-like appearance, hence the name *foam cells*. Eventually, atherosclerotic plaque may calcify and protrude sufficiently into arterial lumens (the interior cavity) that blood flow is impeded. Disruption of vital organ functions, especially those of the brain, heart, and lungs caused by oxygen and nutrient deprivation, usually ensues. In *coronary artery disease*, one of the most common consequences of atherosclerosis, this deprivation damages heart muscle (Special Interest Box 10.3).

KEY CONCEPTS 11.6

Plasma lipoproteins transport lipids through the bloodstream. On the basis of density, lipoproteins are classified into four major classes: chylomicrons, VLDL, LDL, and HDL.

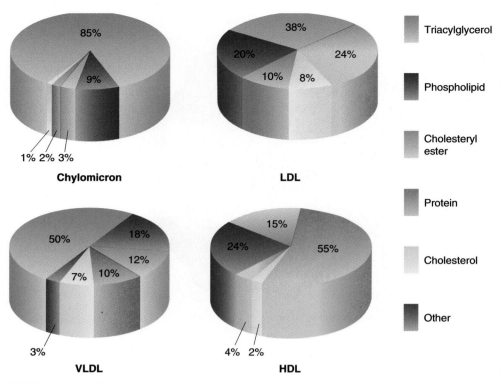

FIGURE 11.18

Proportional (Relative) Number of Cholesterol, Cholesteryl Ester, Phospholipid, and Protein Molecules in Four Major Classes of Plasma Lipoproteins.

Chylomicrons are the largest plasma lipoproteins because they contain the largest percentage of lipid molecules. In contrast, HDL have a relatively small diameter because their high protein content makes them dense.

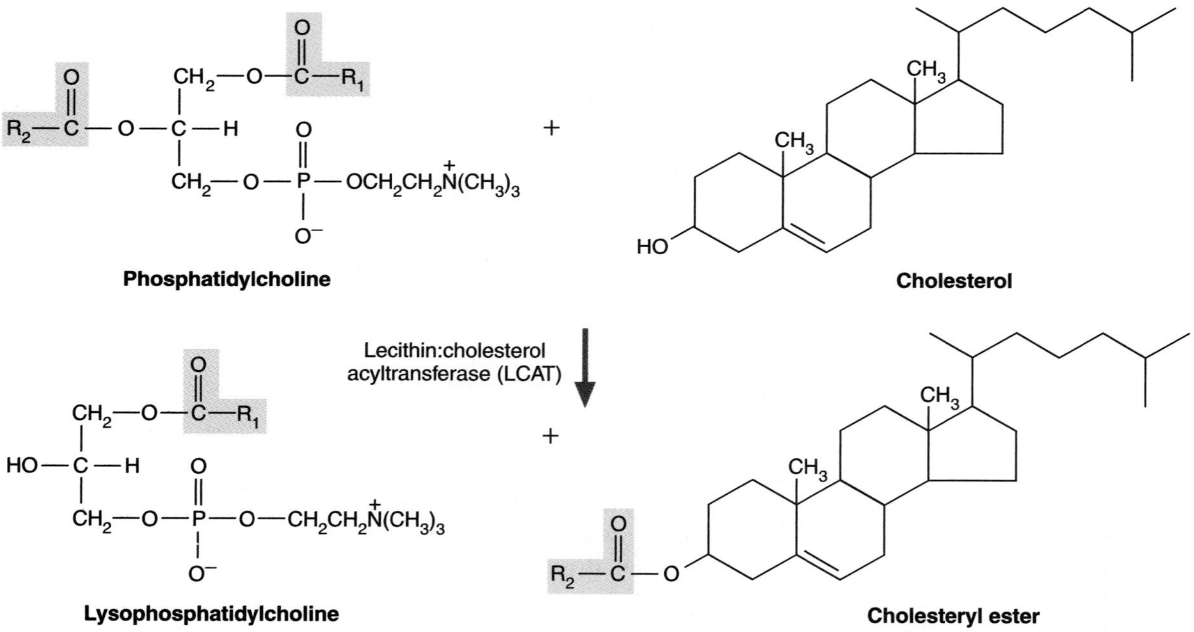

Phosphatidylcholine

Cholesterol

Lecithin:cholesterol acyltransferase (LCAT)

Lysophosphatidylcholine

Cholesteryl ester

FIGURE 11.19

Reaction Catalyzed by Lecithin:Cholesterol Acyltransferase (LCAT).

Cholesteryl ester transfer protein, a protein associated with the LCAT-HDL complex, transfers cholesteryl esters from HDL to VLDL and LDL. The acyl groups are highlighted in color.

Because most of the cholesterol found in plaque is obtained by the ingestion of LDL by foam cells, it is not surprising that high plasma LDL levels are directly correlated with high risk for coronary artery disease. (Recall that LDL have a high cholesterol and cholesteryl ester content.) In contrast, a high plasma HDL level is considered to be associated with a low risk for coronary artery disease. Liver cells (hepatocytes) are the only cells that possess HDL receptors. (Other high risk factors include a high-fat diet, smoking, stress, and a sedentary lifestyle.)

LDL play a significant role in atherosclerosis because cells that become foam cells possess LDL receptors. The binding of LDL to LDL receptors initiates endocytosis (Figure 2.22). Under normal circumstances the cholesterol and other lipids released after LDL enter a cell are used to meet the cell's structural and metabolic needs. Because LDL receptor function is usually highly regulated, the intake of a relatively large number of LDL particles is followed by decreased synthesis of LDL receptors. Macrophages, unlike other cell types, do not exhibit this decrease in LDL receptor synthesis. Although the cause of atherosclerosis is still not understood, recent research has illuminated several aspects. For example, the macrophages found within atherosclerotic plaque possess high levels of LDL receptors with an affinity for oxidized (i.e., damaged) LDL. Clinical trials and animal studies reveal that diets supplemented with ascorbic acid and α-tocopherol, two powerful antioxidants, can retard or arrest plaque formation.

11.2 MEMBRANES

Most of the properties attributed to living organisms (e.g., movement, growth, reproduction, and metabolism) depend, either directly or indirectly, on membranes. All biological membranes have the same general structure. As previously mentioned (Chapter 2), membranes contain lipid and protein molecules. In the currently accepted concept of membranes, referred to as the **fluid mosaic model**, membrane is a bimolecular lipid layer (**lipid bilayer**). The proteins, most of which float within the lipid bilayer, largely determine a membrane's biological functions. Because of the importance of membranes in biochemical processes, the remainder of Chapter 11 is devoted to a discussion of their structure and functions.

Membrane Structure

Because each type of living cell has its own functions, it follows that the structure of its membranes is also unique. Not surprisingly, the proportion of lipid and protein varies considerably among cell types and among organelles within each cell (Table 11.5). The types of lipid and protein found in each membrane also vary.

TABLE 11.5
Chemical Composition of Some Cell Membranes

Membrane	Protein %	Lipid %	Carbohydrate %
Human erythrocyte plasma membrane	49	43	8
Mouse liver cell plasma membrane	46	54	2–4
Amoeba plasma membrane	54	42	4
Mitochondrial inner membrane	76	24	1–2
Spinach chloroplast lamellar membrane	70	30	6
Halobacterium purple membrane	75	25	0

Source: G. Guidotti, Membrane Proteins, *Ann. Rev. Biochem.*, 41:731, 1972.

Most risk factors for atherosclerosis can be reduced by changes in behavior. Exercise, a low-fat diet, and quitting tobacco use usually decrease the likelihood of coronary artery disease. However, an individual's genetic inheritance can sometimes play a decisive role. For example, some people with poor health habits (e.g., smoking and high-fat diets) do not have high plasma LDL and cholesterol values. Similarly, some individuals with high blood cholesterol never have a heart attack. In contrast, other people, apparently at low risk because of healthy lifestyles, do have heart attacks. Recently, researchers have proposed that a variant of LDL, called lipoprotein (a) (Lp(a)), may be at least partially responsible for these anomalous cases. Like LDL, Lp(a) consists of phospholipids, cholesterol, and the apoprotein B-100. An additional apoprotein, a glycoprotein called apoprotein (a) (apo(a)), is attached to Lp(a) by a disulfide linkage to apoprotein B-100. The amount of Lp(a) in an individual's blood plasma is genetically determined and does not vary markedly throughout life. Lp(a) levels, which range from 10 mg/dL to 100 mg/dL, are also not influenced by exercise, diet, or lipid-lowering medications.

Both Lp(a) and apo(a) are polymorphic, that is, their physical properties differ among individuals. The size and density of Lp(a) differ depending on lipid composition and the variant of apo(a) they possess. For unknown reasons the molecular weight of apo(a) is inversely related to Lp(a) plasma concentration. A high blood plasma Lp(a) concentration is correlated with a high risk of coronary artery disease.

Although Lp(a) presumably functions, like other LDL, as a means of delivery of lipids from the liver to other organs, its precise function and its role in cardiovascular disease remain unresolved. Several mechanisms have been proposed. For example, it has been observed that atherosclerosis is promoted when macrophages that consume Lp(a) migrate into the lining of blood vessels where they become transformed into foam cells. A more interesting mechanism links atherosclerosis to a structural property of apo(a).

The structure of apo(a) closely resembles that of plasminogen. (*Plasminogen* is a blood plasma zymogen (see p. 190). When it is activated by a protease, most notably *plasminogen activator*, the fibrin-dissolving protease *plasmin* is produced.) Because the formation of a blood clot often triggers heart attacks, plasminogen is important in dissolving clots. Some researchers have proposed that under certain conditions, apo (a) may promote coronary artery disease by binding to either fibrin or plasminogen activator. Because apo(a) does not digest fibrin, blood clot dissolution is undermined. High plasma levels of Lp(a) have also been shown to cause the proliferation of smooth muscle cells within arterial walls. (Recall that the initial formation of plaque is associated with an increase in the number of such cells in the artery wall.) Lp(a) may interfere with the activity of a growth-retarding substance that normally suppresses cell division in arterial walls.

MEMBRANE LIPIDS When amphipathic molecules are suspended in water, they spontaneously rearrange into ordered structures (Figure 11.6). As these structures form, hydrophobic groups become buried in the water-depleted interior. Simultaneously, hydrophilic groups become oriented so that they are exposed to water. Phospholipids form into bimolecular layers when sufficiently concentrated. This property of phospholipids (and other amphipathic lipid molecules) is the basis of membrane structure.

Membrane lipids are largely responsible for several other important features of biological membranes:

1. Membrane fluidity. The term *fluidity* describes the resistance of membrane components to movement. Rapid lateral movement (Figure 11.20) of lipid molecules is apparently responsible for the proper functioning of many membrane proteins. (The movement of lipid molecules from one side of a lipid bilayer to the

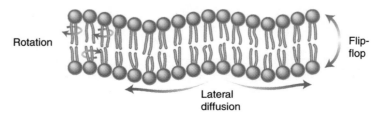

FIGURE 11.20

Lateral Diffusion in Biological Membranes.

Lateral movement of phospholipid molecules is usually relatively rapid. "Flip-flop," the transfer of a lipid molecule from one side of a lipid bilayer to the other, is rare.

other is relatively rare.) A membrane's fluidity is largely determined by the percentage of unsaturated fatty acids in its phospholipid molecules. (Recall that unsaturated hydrocarbon chains pack less densely than saturated ones; see p. 332.) A high concentration of unsaturated chains results in a more fluid membrane. Cholesterol moderates membrane stability without greatly compromising fluidity because it contains both rigid (ring system) and flexible (hydrocarbon tail) structural elements (Figure 11.21).

2. Selective permeability. Because of their hydrophobic nature, the hydrocarbon chains in lipid bilayers provide a virtually impenetrable barrier to the transport of ionic and polar substances. Specific membrane proteins regulate the movement of such substances into and out of cells. To cross a lipid bilayer, a polar substance must shed some or all of its hydration sphere and bind to a carrier protein for membrane translocation or pass through an aqueous protein channel. Both methods shield the hydrophilic molecule from the hydrophobic core

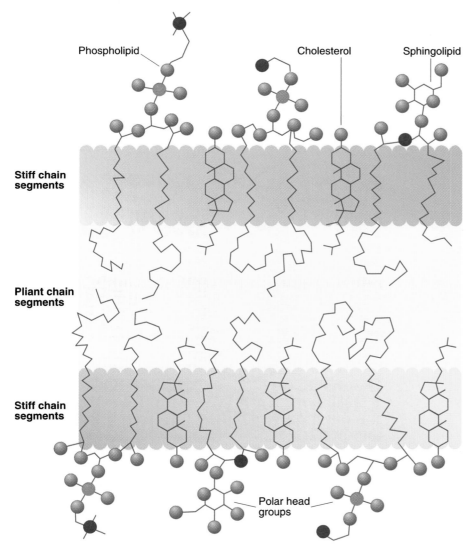

FIGURE 11.21

Lipid Bilayer.

The flexible hydrocarbon chains in the hydrophobic core (lightly shaded area in the middle) make the membrane fluid. The steroid ring system of cholesterol molecules, positioned in the outer surfaces, stiffens the membrane. (Nitrogen atoms are red, oxygen atoms are blue, and phosphorus atoms are orange.) Cell membranes are about 7 to 9 nm thick.

of the membrane. Most transmembrane water movement accompanies ion transport. Nonpolar substances simply diffuse through the lipid bilayer down their concentration gradients. Each membrane exhibits its own transport capability or selectivity based on its protein component.

3. Self-sealing capability. When lipid bilayers are disrupted, they immediately and spontaneously reseal (Figure 11.22) because a break in a lipid bilayer exposes the hydrophobic hydrocarbon chains to water. Because breaches in cell membranes can be lethal, this resealing property is critical. (In living cells, certain pro-

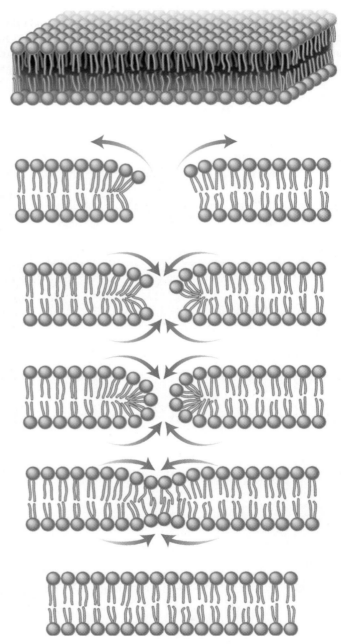

FIGURE 11.22

Membrane Self-Sealing.

Disruptions in a lipid bilayer are rapidly resealed. When the hydrophobic tails of lipid molecules are suddenly exposed to polar water molecules, lipids respond by forming hydrophilic edges consisting of polar head groups. As membrane edges draw closer to each other, they fuse and re-form the bilayer.

tein components of membrane and the cytoskeleton, as well as calcium ions, also assist in membrane resealing.)

4. Asymmetry. Biological membranes are asymmetric; that is, the lipid composition of each half of a bilayer is different. For example, the human red blood cell membrane possesses substantially more phosphatidylcholine and sphingomyelin on its outside surface. Most of the membrane's phosphatidylserine and phosphatidylethanolamine are on the inner side. Membrane asymmetry is not unexpected, because each side of a membrane is exposed to a different environment. Asymmetry originates during membrane synthesis, because phospholipid biosynthesis occurs on only one side of a membrane (Special Interest Box 12.3). The protein components of membranes (discussed below) also exhibit considerable asymmetry with distinctly different functional domains within membrane and on the cytoplasmic and extracellular faces of membrane.

MEMBRANE PROTEINS Most of the functions associated with biological membranes require protein molecules. Membrane proteins are often classified by the function they perform. Most of these molecules are structural components, enzymes, hormone receptors, or transport proteins.

Membrane proteins are also classified according to their structural relationship to membrane. Proteins that are embedded in and/or extend through a membrane are referred to as *integral proteins* (Figure 11.23). Such molecules can be extracted only by disrupting the membrane with organic solvents or detergents (Figure 11.24). *Peripheral proteins* are bound to membrane primarily through interactions with integral membrane proteins. Some peripheral proteins interact directly with the lipid bilayer. Typically, peripheral proteins can be released from membrane by relatively gentle methods (e.g., concentrated salt solutions or pH changes alter noncovalent interactions between amino acid side chains). For several technical reasons (Biochemical Methods 11.1), red blood cell plasma membrane proteins have been widely studied.

The two major integral proteins in red blood cell membrane are glycophorin and the anion channel protein (Figure 11.25). *Glycophorin* is a 31-kD glycoprotein with 131 amino acid residues. Approximately 60% of its weight is carbohydrate. Certain oligosaccharide groups on glycophorin constitute the ABO and MN blood group antigens. (These and other blood group antigens are markers used to classify blood for transfusions and organ transplantation.) However, despite intense research efforts, the function of glycophorin is still unknown. *Anion channel protein* (also referred to as band 3) is composed of two identical subunits, each consisting of 929 amino acids. This protein channel plays an important role in CO_2 transport in

(continued on p. 360)

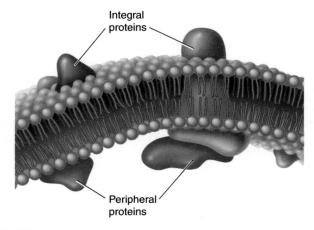

Integral
proteins

Peripheral
proteins

FIGURE 11.23

Integral and Peripheral Membrane Proteins.

Integral membrane proteins are released only if the membrane is disrupted. Many peripheral proteins can be removed with mild reagents.

Recognition of the vast array of functions performed by biological membranes has led to the development of numerous techniques for their investigation. Research accomplished over the past few decades reveals substantial information about membrane structure and function. One of the most important principles revealed by these research efforts is that the membranes of most living organisms have many structural similarities. These common features allow biochemists to apply (with caution) information gained from one membrane system to solving structural problems of other membranes. For example, the structural features of the red blood cell (rbc) membrane have proved to be valuable in studies of other membranes.

Membrane research requires reasonably pure specimens. Most membranes are obtained by a stepwise process that begins with cell homogenization, followed by several centrifugation steps. (Refer to p. 60 for details.) There are several reasons why rbc membrane is a popular choice in membrane research. First, because the rbc plasma membrane is the only membrane in the cell, it cannot be contaminated with intracellular membranes (a common problem in membrane research). Second, rbc membrane preparation is relatively simple. Rbc membrane is prepared by exposing red blood cells to a hypotonic solution. After washing with an isotonic solution the rbc membrane fragments can be resealed to form "ghosts." Ghosts are, in effect, empty red blood cells. Rbc membrane can also be converted to numerous small vesicles by disruption and resealing. (Both inside-out and right-side-out vesicles can be produced.) Finally, rbc are easily obtained in large quantities from blood banks.

Membrane Composition

Once they are isolated, membranes are analyzed biochemically for lipid and protein composition. After they are extracted from membranes with organic solvents (e.g., chloroform), lipids are separated into classes by column chromatography. Phospholipids, the major lipid membrane component, are often resolved and identified by thin-layer chromatography. Extracting and purifying membrane proteins requires detergents. Commonly used detergents are triton X-100 and SDS (sodium dodecyl sulfate). Because membrane proteins require a hydrophobic environment to maintain their structure and biological activity, they are also investigated in the presence of detergent. (In the absence of detergent, integral membrane proteins often aggregate and precipitate.) For example, SDS-PAGE (see p. 156) is often used to resolve membrane proteins.

Membrane Morphology

The arrangement of proteins within biological membranes can be directly observed by using freeze-fracture electron microscopy. In this technique, a rapidly frozen membrane is struck with a microtome knife. (A microtome is an instrument used for cutting thin sections of biological specimens for microscopic study.) The membrane often splits along the inner surfaces of the two lipid layers. In preparation for viewing in the electron microscope, the delicate membrane's inner surfaces are shadowed with a thin layer of heavy metal (usually platinum). Numerous intramembranous particles are commonly observed in freeze-fractured membranes.

Although X-ray crystallography (see p. 156) has had limited use in determining membrane morphology, it usually provides low-resolution structural detail. High-resolution information requires highly ordered crystalline samples. Most membranes, however, contain an assortment of different proteins. In a few instances (i.e., biological membranes that possess only one type of protein), X-ray crystallography has provided high-resolution structural information. The "purple membrane" of the bacterium *Halobacterium halobium* is an excellent example. In the presence of O_2 this halophilic (salt-loving) organism depends on aerobic metabolism for energy production. If O_2 concentration is low and light intensity is high, the organism produces crystalline membrane patches called purple membrane. The pigment bacteriorhodopsin, the sole protein component of purple membrane, acts as a light-driven proton pump. (The pH gradient that results from the pumping activity of bacteriorhodopsin is used to synthesize ATP.) Relatively pure specimens of purple membrane are easily obtained by decreasing the salt concentration in the medium. The organism's ordinary membrane disintegrates, leaving the purple membrane intact.

Initial protein structure studies indicated that bacteriorhodopsin is a 248-residue polypeptide. Its amino-terminal residue is on the membrane's outside surface, and its carboxyl residue projects into the cytoplasm. Careful analysis of bacteriorhodopsin's primary sequence revealed seven peptide segments with amino acid sequences typical of α-helices. Using electron microscopy and X-ray crystallography, researchers determined that bacteriorhodopsin possesses seven α-helices, which are roughly perpendicular to the membrane (Figure 11C).

Membrane Fluidity

Membrane fluidity is one of the most important assumptions of the fluid mosaic model of membrane structure. One measure of membrane fluidity, the ability of membrane components to diffuse laterally, can be demonstrated when cells from two different species are fused to form a **heterokaryon** (Figure 11D). (Certain viruses or chemicals are used to promote cell-cell fusion.) The plasma membrane proteins of each cell type can be tracked because they are labeled with different fluorescent markers. Initially, the proteins are confined to their own side of the heterokaryon membrane. As time passes, the two fluorescent markers intermix, indicating that proteins move freely in the lipid bilayer.

Another technique, referred to as **fluorescence recovery after photobleaching** (FRAP), is also used to observe lateral diffusion. Cell plasma membranes are uniformly labeled with a fluorescent marker. Using a laser beam, the fluorescence in a small area is destroyed (or "bleached"). Using video equipment, the lateral movement of membrane components into and out of the bleached area can be tracked as a function of time.

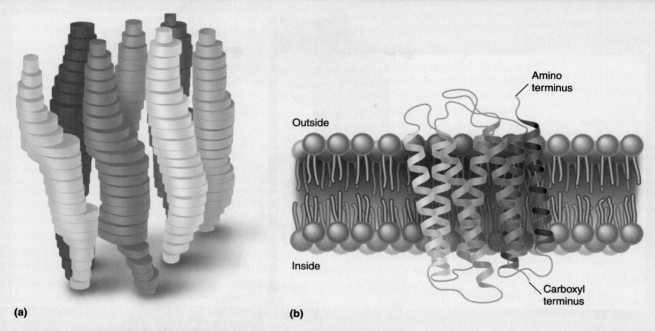

(a)

(b)

FIGURE 11C Bacteriorhodopsin Structure.
An early electron density map of bacteriorhodopsin (a) created by using electron micrographs of the crystallized protein contributed eventually to more detailed models (b) made possible by more sophisticated methods.

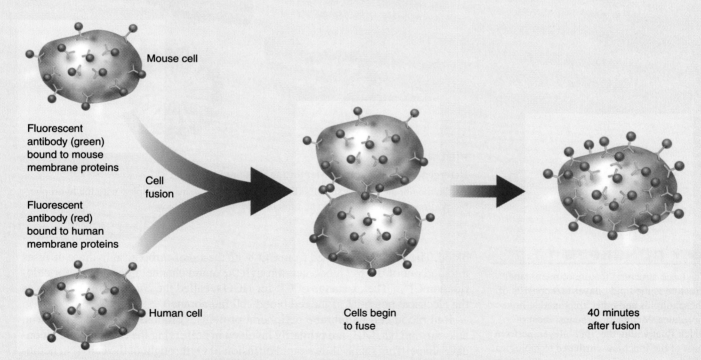

FIGURE 11D Fusion of Fluorescence-Labeled Mouse and Human Cells to Form a Heterokaryon.
This experiment demonstrates that membranes are fluid and that proteins can move freely within the lipid bilayer. Metabolic inhibitors do not slow down protein movement, but lowering the temperature below 15°C does.

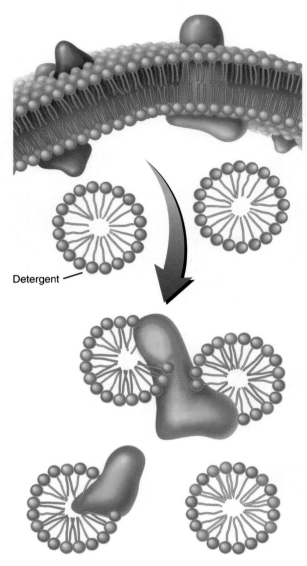

FIGURE 11.24

Detergent Solubilization of Membrane Proteins.

As the membrane is mechanically disrupted, detergent forms a complex with the hydrophobic portions of integral membrane proteins and membrane lipids.

blood. The HCO_3^- ion formed from CO_2 with the aid of carbonic anhydrase diffuses into and out of the red blood cell through the anion channel in exchange for chloride ion Cl^-). (The exchange of Cl^- for HCO_3^-, called the *chloride shift*, preserves the electrical potential of the red blood cell membrane.)

Red blood cell membrane peripheral proteins, composed largely of spectrin, ankyrin, and band 4.1, are primarily involved in preserving the cell's unique biconcave shape. This shape allows rapid diffusion of O_2 throughout the cell. (No hemoglobin molecule is greater than 1 μm away from the cell's surface.) *Spectrin* is a tetramer, composed of two $\alpha\beta$-dimers, that binds to ankyrin and band 4.1. *Ankyrin* is a large globular polypeptide (215 kD) that links spectrin to the anion channel protein. (This is a connecting link between the red blood cell's cytoskeleton and its plasma membrane.) *Band 4.1* binds to both spectrin and *actin filaments* (a cytoskeletal component found in many cell types). Because band 4.1 also binds to glycophorin, it too links the cytoskeleton and the membrane.

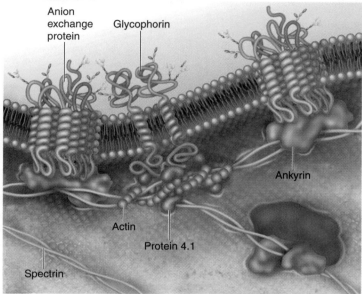

FIGURE 11.25

Red Blood Cell Integral Membrane Proteins.

The integral membrane proteins glycophorin and anion exchange protein are components in a network of linkages that connect the plasma membrane to structural elements of the cytoskeleton (e.g., actin, spectrin, protein 4.1, and ankyrin).

Membrane Function

Membranes are involved in a bewildering array of functions in living organisms. Among the most important of these are the transport of molecules and ions into and out of cells and organelles and the binding of hormones and other biomolecules. Each of these topics is discussed briefly. A description of receptor-mediated endocytosis follows.

MEMBRANE TRANSPORT Membrane transport mechanisms are vital to living organisms. Ions and molecules constantly move across cell plasma membranes and across the membranes of organelles. This flux must be carefully regulated to meet each cell's metabolic needs. For example, a cell's plasma membrane regulates the entrance of nutrient molecules and the exit of waste products. Additionally, it regulates intracellular ion concentrations. Because lipid bilayers are generally impenetrable to ions and polar substances, specific transport components must be inserted into cellular membranes. Several examples of these structures, referred to as transport proteins or permeases, are discussed.

Biological transport mechanisms are classified according to whether they require energy. Major types of biological transport are illustrated in Figure 11.26. In **passive transport**, there is no direct input of energy. In contrast, **active transport** requires energy to transport molecules against a concentration gradient.

In **simple diffusion**, each solute, propelled by random molecular motion, moves down its concentration gradient (i.e., from an area of high concentration

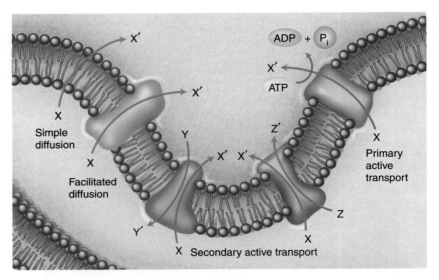

FIGURE 11.26

Transport Across Membranes.

The major transport processes are simple and facilitated diffusion and primary and secondary active transport. In simple diffusion the spontaneous transport of a specific solute is driven by its concentration gradient. Facilitated diffusion is the movement of a solute down its concentration gradient across a membrane that occurs through protein channels or carriers. In both primary and secondary active transport energy is required to move solutes across a membrane against their concentration gradients. In primary active transport, this energy is usually provided directly by ATP hydrolysis. In secondary active transport, solutes are moved across a membrane by energy stored in a concentration gradient of a second substance that has been created by ATP hydrolysis or other energy-generating mechanisms.

to an area of low concentration). In this spontaneous process, there is a net movement of solute until an equilibrium is reached. As Figure 11.27 illustrates, a system reaching equilibrium becomes more disordered, that is, entropy increases. Because there is no input of energy, transport occurs with a negative change in free energy. In general, the higher the concentration gradient, the faster the rate of solute diffusion. The diffusion of gases such as O_2 and CO_2 across membranes is proportional to their concentration gradients. The diffusion of organic molecules also depends on molecular weight and lipid solubility.

In **facilitated diffusion**, the second type of passive transport, the transport of certain large or charged molecules occurs through special channels or carriers. *Channels* are tunnel-like transmembrane proteins. Each type is designed for

FIGURE 11.27

Simple Diffusion.

By time t, one-fourth of the molecules present on one side of a semipermeable membrane have diffused by random thermal motion to the other side. Eventually, equilibrium is approached.

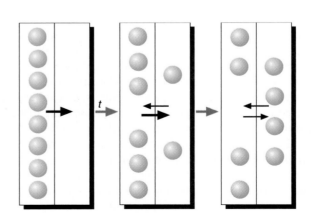

the transport of a specific solute. Many channels are chemically or voltage-regulated. Chemically regulated channels open or close in response to a specific chemical signal. For example, a *chemically gated Na+ channel* in the nicotinic acetylcholine receptor complex (found in muscle cell plasma membranes) opens when acetylcholine binds. Na+ rushes into the cell and the membrane potential falls. Because membrane potential is an electrical gradient across the membrane (see p. 76), a decrease in membrane potential is membrane *depolarization*. Local depolarization caused by acetylcholine leads to the opening of nearby Na+ channels (these are referred to as *voltage-gated Na+ channels*). *Repolarization*, the reestablishment of the membrane potential, begins with the diffusion of K+ ions out of the cell through *voltage-gated K+ channels*. (The diffusion of *K+* ions out of the cell makes the inside less positive, that is, more negative.

Another form of facilitated diffusion involves membrane proteins called *carriers* (sometimes referred to as *passive transporters*). In carrier-mediated transport, a specific solute binds to the carrier on one side of a membrane and causes a conformational change in the carrier. The solute is then translocated across the membrane and released. The red blood cell *glucose transporter* is the best-characterized example of passive transporters. It allows D-glucose to diffuse across the red blood cell membrane for use in glycolysis and the pentose phosphate pathway. Facilitated diffusion increases the rate at which certain solutes move down their concentration gradients. This process cannot cause a net increase in solute concentration on one side of the membrane.

The two forms of active transport are primary and secondary. In *primary active transport*, energy is provided by ATP. Transmembrane ATP-hydrolyzing enzymes use the energy derived from ATP to drive the transport of ions or molecules. The Na+-K+ pump (also referred to as the Na+-K+ ATPase) is a prominent example of a primary transporter. (The Na+ and K+ gradients are required for maintaining normal cell volume and membrane potential. Refer to Special Interest Box 3.1.) In *secondary active transport*, concentration gradients generated by primary active transport are harnessed to move substances across membranes. For example, the Na+ gradient created by the Na+-K+ ATPase pump is used in kidney tubule cells and intestinal cells to transport D-glucose (Figure 11.28).

Impaired membrane transport mechanisms can have very serious consequences. One of the best understood examples of dysfunctional transport occurs in cystic fibrosis. **Cystic fibrosis** (CF), a fatal autosomal recessive disease, is

KEY CONCEPTS 11.8

Membrane transport mechanisms are classified as passive or active according to whether they require energy. In passive transport, solutes moving across membranes move down their concentration gradient. In active transport, energy derived directly or indirectly from ATP hydrolysis or other energy sources is required to move an ion or molecule against its concentration gradient.

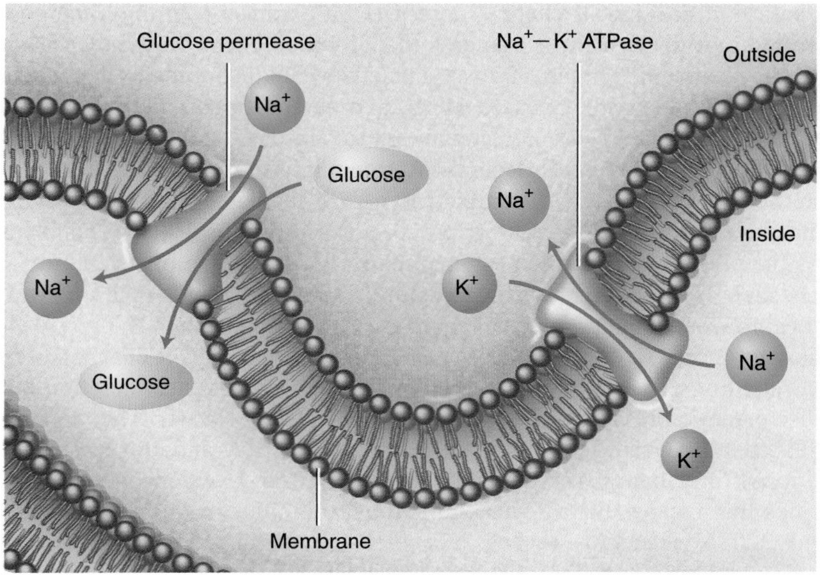

FIGURE 11.28

The Na+-K+ ATPase and Glucose Transport.

The Na+-K+ ATPase maintains the Na+ gradient essential to maintain membrane potential. In certain cells, glucose transport depends on the Na+ gradient. Glucose permease transports both Na+ and glucose. Only when both substrates are bound does the protein change its conformation, thus initiating transport.

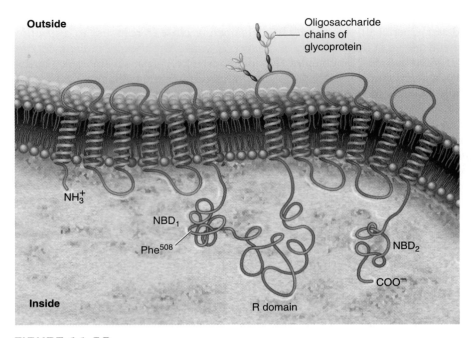

FIGURE 11.29

The Cystic Fibrosis Transmembrane Regulator (CFTR).

CFTR is a chloride channel composed of two domains (each with six membrane-spanning helices) that constitute the Cl^- pore, two nucleotide-binding domains (NBD), and a regulatory (R) domain. Transport of Cl^- through the pore, driven by ATP hydrolysis, occurs when specific amino acid residues on the R domains are phosphorylated. The most commonly observed CF-causing mutation is a deletion of Phe^{508} in NBD_1. The precise structural relationships among the pore-forming helices remain unclear.

caused by a missing or defective plasma membrane glycoprotein called **cystic fibrosis transmembrane conductance regulator (CFTR)**. CFTR (Figure 11.29, which functions as a chloride channel in epithelial cells, is a member of a family of proteins referred to as ABC transporters. (ABC transporters are so named because each contains a polypeptide segment called an *ATP binding cassette*.) CFTR contains five domains. Two domains, each containing six membrane-spanning helices, form the Cl^- channel pore. Chloride transport through the pore is controlled by the other three domains (all of which occur on the cytoplasmic side of the plasma membrane). Two are nucleotide-binding domains that bind and hydrolyze ATP and use the released energy to drive conformational changes in the pore. The regulatory (R) domain contains several amino acid residues that must be phosphorylated by cAMP-dependent protein kinase for chloride transport to occur. (See Section 16.4 for a discussion of cAMP-mediated signal transduction.)

The chloride channel is vital for proper absorption of salt (NaCl) and water across the plasma membranes of the epithelial cells that line ducts and tubes in tissues such as lungs, liver, small intestine, and sweat glands. Chloride transport occurs when signal molecules open CFTR Cl^- channels in the apical (top) membrane surface of epithelial cells. In CF the failure of CFTR channels results in the retention of Cl^- within the cells. A thick mucus or other secretion forms because osmotic pressure causes the excessive uptake of water. The most obvious features of CF are lung disease (obstructed air flow and chronic bacterial infections), and pancreatic insufficiency (impaired production of digestive enzymes that can result in severe nutritional deficits). In the majority of CF patients, CFTR is defective because of a deletion mutation at Phe^{508}, which causes protein misfolding that prevents the processing and insertion of the mutant protein into the plasma membrane. Less common causes of CF (over 100 have been described) include defective formation of CFTR mRNA molecules, mutations in the nucleotide-binding domains that result in ineffective binding or hydrolysis of ATP, and mutations in the pore-forming domains that cause reduced chloride transport.

CF patients rarely survived childhood before the development of modern therapies. It is only because of antibiotics (used principally in the treatment of lung infections) and commercially available digestive enzymes (replacing the enzymes normally produced in the pancreas) that many CF patients can now expect to live into their 30s. Yet as with the sickle-cell gene (see p. 129), defective CF genes are not rare. With an approximate incidence of 1 in every 2500 Caucasians, CF is the most common fatal genetic disorder in this population. Recent experiments with "knock-out" mice indicate that carriers of the mutant gene are protected from diseases that kill because of diarrhea. (Knock-out animals are bred so that they contain one copy of a defective gene in all of their cells.) Such animals lose significantly less body fluid because they have a reduced number of functional chloride channels. It is suspected that CF carriers (individuals that have only one copy of a defective CF gene) are also less susceptible to fatal diarrhea (e.g., cholera) for the same reason. The CF gene did not spread beyond Western Europe (e.g., the incidence among East Asians is approximately 1 in 100,000) because CF carriers secrete slightly more salt in their sweat than noncarriers. (The epithelial cells that line sweat gland ducts cannot reabsorb chloride efficiently.) In warmer climates where sweating is a common feature of daily life, chronic excessive salt loss is far more dangerous than intermittent exposure to diarrhea-causing microorganisms.

QUESTION 11.6

Suggest the mechanism(s) by which each of the following substances is (are) transported across cell membranes:

a. CO_2

b. Glucose

c. Cl^-

d. K^+

e. Fat molecules

f. α-Tocopherol

QUESTION 11.7

Describe the types of noncovalent interactions that promote the stability and the functional properties of biological membranes.

QUESTION 11.8

Transport mechanisms are often categorized according to the number of transported solutes and the direction of solute transport.

1. Uniporters transport one solute.

2. Symporters transport two different solutes simultaneously in the same direction.

3. Antiporters transport two different solutes simultaneously in opposite directions.

After examining the examples of transport discussed in this chapter, determine which of the above categories each one belongs to.

A basic characteristic of living cells is the capacity to rapidly move water across cell membranes in response to changes in osmotic pressure. It was long assumed by many researchers that simple diffusion was responsible for most water flow. In a wide variety of cell types such as red blood cells and certain kidney cells, it became apparent that water flow is extraordinarily rapid. In the early 1990s the first of a series of water channel proteins, now called the **aquaporins**, was characterized. Found initially in red blood cell membrane and then in kidney tubule cells, aquaporin-1 (AQP-1) is an intrinsic membrane protein complex that facilitates water flow, which has been measured at more than 10^9/molecules/s/channel. Aquaporins have been found in almost all living organisms, with at least 10 different forms in mammals. Recent experimental evidence suggests that water flow through aquaporin channels is regulated. For example, three mammalian aquaporins appear to be regulated by pH. Others are regulated by phosphorylation reactions or by the binding of specific signal molecules. In 1993, the cause of a rare inherited form of **nephrogenic diabetes insipidis** (an autosomal recessive disease in which the kidneys of affected individuals cannot produce concentrated urine), was discovered to be a mutation in the gene for AQP-2. The mutated AQP-2 does not respond to the antidiuretic hormone vasopressin (Table 16.1, p. 544).

Of all the aquaporins, AQP-1, a homotetramer, is the best characterized. Each subunit is a polypeptide containing 269 amino acid residues that form a water-transporting pore with six α-helical membrane-spanning domains connected by five loops (Figure 11E). In the functional monomer, the two loops that both possess an Asn-Pro-Ala (NPA) sequence meet in the middle to form the water-binding site. The pore, which has been measured at 3 Å, is only slightly larger than a water molecule (2.8 Å). As illustrated in Figure 11F, the NPA sequences are juxtaposed in the channel. The movement of only water molecules, and not smaller species such as H⁺, through the channel is believed to be made possible by the formation of hydrogen bonds between water molecules and the Asn residues of the two NPA sequences. The hydrophobic environment created by the amino acid residues on the other helices that comprise the pore causes the breakage of the hydrogen bonds between water molecules as they move in single file toward the narrowest portion of the pore. It also forces the oxygen atom of each water molecule to orient itself toward the Asn residues. When the water molecule approaches the 3 Å constriction in the pore, its oxygen atom sequentially forms and breaks hydrogen bonds with the side chains of the two Asn residues. The absence of other hydrogen bonding partners prevents the transport of protons.

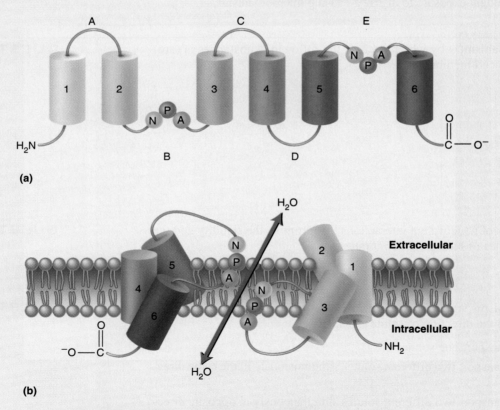

FIGURE 11E Diagrammatic View of the Aquaporin Monomer.
Each AQP-1 monomer possesses six membrane-spanning α-helices that are connected by five loops (a). In the functional monomer, the two loops containing the NPA sequences form the water-selective site (b).

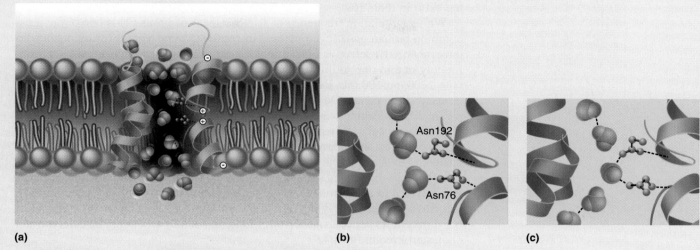

(a)   **(b)**   **(c)**

FIGURE 11F Water Transport Through the AQP-1 Monomer.
Water molecules move through the pore in single file. As each molecule approaches the constriction in the pore it is forced to orient its oxygen atom so it can then form and break hydrogen bonds with the side chains of the two Asn residues. (a) Within the pore of the aquaporin monomer the B and E loops (see Figure 11D) create a positive electrostatic environment in which the oxygen atom of each water molecule is oriented toward the two Asn residues. (b) and (c) The sequential formation and breakage of hydrogen bonds between the oxygen of water molecules and the two Asn residue side chains mediate the movement of water through the pore.

MEMBRANE RECEPTORS Membrane receptors play a vital role in the metabolism of all living organisms. They provide mechanisms by which cells monitor and respond to changes in their environment. In multicellular organisms the binding of chemical signals, such as the hormones and neurotransmitters of animals, to membrane receptors is a vital link in intracellular communication. Other receptors are engaged in cell-cell recognition or adhesion. For example, lymphocytes perform a critical role in the immune system function of identifying and then destroying foreign or virus-infected cells when they transiently bind to cell surfaces throughout the body. Similarly, the capacity of cells to recognize and adhere to other appropriate cells in a tissue is of crucial importance in many organismal processes, such as embryonic and fetal development.

The binding of a ligand to a membrane receptor results in a conformational change, which then causes a specific programmed response. Sometimes, receptor responses appear to be relatively straightforward. For example, the binding of acetylcholine to an acetylcholine receptor opens a cation channel. However, most responses are complex. Currently, the most intensively researched example of membrane receptor function is LDL receptor–mediated endocytosis, which is discussed next.

The low-density lipoprotein receptor is responsible for the uptake of cholesterol-containing lipoproteins into cells. As often happens, this receptor was discovered during an investigation of an inherited disease, in this case, *familial hypercholesterolemia* (FH). The LDL receptor was discovered as Brown and Goldstein were investigating the uptake of LDL into the fibroblasts of FH patients. The biochemical defect that causes FH was then identified as mutations in the LDL receptor gene.

Patients with FH have elevated levels of plasma cholesterol because they have missing or defective LDL receptors. (Recall that LDL transport cholesterol to tissues.) Heterozygous individuals (also referred to as *heterozygotes*) inherit one defective LDL receptor gene. Consequently, they possess half the number of functional LDL receptors. With blood cholesterol values of 300–600 mg/100 mL it is not surprising that heterozygotes have heart attacks as early as the age of 40. They also develop disfiguring *xanthomas* (cholesterol deposits in the skin) in their 30s.

With a population frequency of 1 in 500, heterozygous FH is one of the most common human genetic anomalies. In contrast, *homozygotes* (individuals who have inherited a defective LDL receptor gene from both parents) are rare (approximately one in one million). These patients have plasma cholesterol values of 650–1200 mg/100 mL. Both xanthomas and heart attacks occur during childhood or early adolescence. Death usually occurs before the age of 20.

The LDL receptor is a complex glycoprotein (Figure 11.30) found on the surface of many cells. When cells need cholesterol for the synthesis of membrane or steroid hormones, they produce LDL receptors and insert them into discrete coated regions of plasma membrane. (Coated membrane regions usually constitute about 2% of a cell's surface. The protein *clathrin*, which has a unique structure referred to as *triskelion*, is the major protein component of coated regions. It forms a latticelike polymer during the initial stages of endocytosis.) The number of receptors per cell varies from 15,000 to 70,000, depending on cell type and cholesterol requirements.

The process of LDL receptor–mediated endocytosis, as observed in fibroblasts, occurs in several steps (Figure 2.22a). It begins within several minutes after

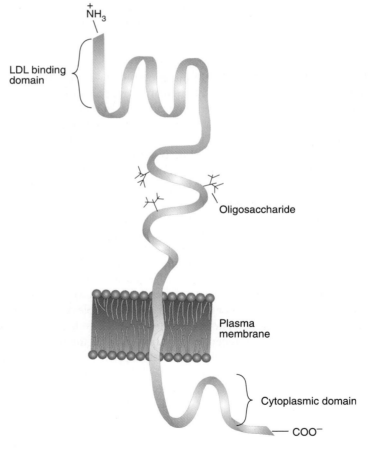

FIGURE 11.30

The LDL Receptor.

The cytoplasmic domain of the LDL receptor plays a critical role in forming of coated pits, an important feature of receptor-mediated endocytosis. Once LDL binds to a receptor, both are rapidly internalized.

LDL have bound to LDL receptors. The coated region surrounding the bound receptor, referred to as a *coated pit*, pinches off and becomes a *coated vesicle*. Subsequently, *uncoated vesicles* are formed as clathrin depolymerizes. Before uncoated vesicles fuse with lysosomes, LDL are uncoupled from LDL receptors as the pH changes from 7 to 5. (This change is created by ATP-driven proton pumps in the vesicle membrane.) LDL receptors are recycled to the plasma membrane, and LDL-containing vesicles fuse with lysosomes. Subsequently, LDL proteins are degraded to amino acids, and cholesteryl esters are hydrolyzed to cholesterol and fatty acids. Under normal circumstances, LDL receptor–mediated endocytosis is a highly regulated process. For example, cholesterol (or a derivative) suppresses the activity of HMG CoA reductase, the enzyme (Chapter 12) that catalyzes the rate-controlling step in cholesterol synthesis. Additionally, cholesterol stimulates ACAT activity and depresses the synthesis of LDL receptors. The genetic defects that cause FH prevent affected cells from obtaining sufficient cholesterol from LDL. The most common defect is failure to synthesize the receptor. Other defects include ineffective intracellular processing of newly synthesized receptor, defects in the receptor's binding of LDL, and the inability of receptors to cluster in coated pits.

SUMMARY

1. Lipids are a diverse group of biomolecules that dissolve in non-polar solvents. They can be separated into the following classes: fatty acids and their derivatives, triacylglycerols, wax esters, phospholipids, lipoproteins, sphingolipids, and the isoprenoids.

2. Fatty acids are monocarboxylic acids that occur primarily in triacylglycerols, phospholipids, and sphingolipids. The eicosanoids are a group of powerful hormonelike molecules derived from long chain fatty acids. The eicosanoids include the prostaglandins, thromboxanes, and the leukotrienes.

3. Triacylglycerols are esters of glycerol with three fatty acid molecules. Triacylglycerols that are solid at room temperature (i.e., they possess mostly saturated fatty acids) are called fats. Those that are liquid at room temperature (i.e., they possess a high unsaturated fatty acid content) are referred to as oils. Triacylglycerols, the major storage and transport form of fatty acids, are an important energy storage form in animals. In plants they store energy in fruits and seeds.

4. Phospholipids are structural components of membranes. There are two types of phospholipids: phosphoglycerides and sphingomyelins.

5. Sphingolipids are also important components of animal and plant membranes. They contain a long-chain amino alcohol. In animals this alcohol is sphingosine. Phytosphingosine is found in plant sphingolipids. Glycolipids are sphingolipids that possess carbohydrate groups and no phosphate.

6. Isoprenoids are molecules that contain repeating five-carbon isoprene units. The isoprenoids consist of the terpenes and the steroids.

7. Plasma lipoproteins transport lipid molecules through the bloodstream from one organ to another. They are classified according to their density. Chylomicrons are large lipoproteins of extremely low density that transport dietary triacylglycerols and cholesteryl esters from the intestine to adipose tissue and skeletal muscle. VLDL, which are synthesized in the liver, transport lipids to tissues. As VLDL travel through the bloodstream, they are converted to LDL. LDL are engulfed by cells after binding to LDL receptors on the plasma membrane. HDL, also produced in the liver, scavenge cholesterol from cell membranes and other lipoprotein particles. LDL play an important role in the development of atherosclerosis.

8. According to the fluid mosaic model, the basic structure of membranes is a lipid bilayer in which proteins float. Membrane lipids (the majority of which are phospholipids) are primarily responsible for the fluidity, selective permeability, and self-sealing properties of membranes. Membrane proteins usually define the biological functions of specific membranes. Depending on their location, membrane proteins can be classified as integral or peripheral. Examples of functions in which membranes are involved include transport and the binding of hormones and other extracellular metabolic signals.

SUGGESTED READINGS

Brown, M. S., and Goldstein, J. L., How LDL Receptors Influence Cholesterol and Atherosclerosis, *Sci. Amer.*, 251(5):58–66, 1984.

Gennis, R. B., *Biomembranes: Molecular Structure and Function,* Springer-Verlag, New York, 1989.

Glomset. J. A., Gelb, M. H., and Farnworth, C. C., Prenyl Proteins in Eukaryotic Cells: A New Type of Membrane Anchor, *Trends Biochem. Sci.*, 15:139–142, 1990.

Gounaris, K., and Barber, J., Monogalactosyldiacylglycerol: The Most Abundant Polar Lipid in Nature, *Trends Biochem. Sci.*, 8:378–381, 1983.

King, L. S., Yasui, M. and Agre, P., Aquaporins in Health and Disease, *Mol. Med. Today*, 6:60–65, 2000.

Lawn, R. M., Lipoprotein (a) in Heart Disease, *Sci. Amer.*, 266(6):54–60, 1992.

McNeil, P. L., Cell Wounding and Healing, *Amer. Sci.*, 79:222–235, 1991.

Nicholls, D. G., and Ferguson, S. J., *Bioenergetics 2*, Academic Press, New York, 1992.

Scanu, A. M., Lawn, R. M., and Berg, K., Lipoprotein (a) and Atherosclerosis, *Ann. Intern. Med.*, 115(3):209–218, 1991.

Souter, A. K., Familial Hypercholesterolaemia: Mutations in the Gene for the Low Density Lipoprotein Receptor, *Mol. Med. Today*, 1(2):90–97, 1995.

Superko, H. R., The Atherogenic Lipoprotein Profile. *Sci. Med.*, 4(5):36–45, 1997.

Welsh, M. J, and Smith, A. E., Cystic Fibrosis, *Sci. Amer.*, 273(6):52–59, 1995.

Wine, J. T., Cystic Fibrosis Lung Disease, *Sci. Med.*, 6(3):34–43, 1999.

KEY WORDS

active transport, *361*

acyl group, *334*

aquaporin, *366*

atherosclerosis, *351*

autocrine, *338*

carotenoid, *345*

chylomicron, *349*

cis-isomer, *332*

cystic fibrosis, *363*

cystic fibrosis transmembrane conductance regulator, *364*

eicosanoid, *338*

essential fatty acid, *334*

facilitated diffusion, *362*

fluid mosaic model, *353*

fluorescence recovery after photobleaching, *358*

glycolipid, *342*

heterokaryon, *358*

high-density lipoprotein, *351*

isoprenoid, *344*

leukotriene, *339*

lipid, *332*

lipid bilayer, *353*

low-density lipoprotein, *349*

mixed terpenoid, *345*

monounsaturated, *333*

nephrogenic diabetes insipidis, *366*

neutral fat, *335*

nonessential fatty acid, *334*

passive transport, *361*

phosphoglyceride, *337*

phospholipid, *337*

polar head group, *337*

polyunsaturated, *333*

prenylation, *345*

prostaglandin, *338*

simple diffusion, *361*

sphingolipid, *341*

sphingomyelin, *340*

steroid, *346*

terpene, *344*

thromboxane, *339*

trans-isomer, *332*

very low density lipoprotein, *349*

wax, *337*

REVIEW QUESTIONS

1. Clearly define the following terms:
   a. lipid
   b. autocrine regulator
   c. amphipathic
   d. sesquiterpene
   e. lipid bilayer
   f. prenylation
   g. fluidity
   h. chylomicron
   i. voltage-gated channel
   j. terpene
   k. CFTR
   l. aquaporin

2. List a major function of each of the following classes of lipid:
   a. phospholipids
   b. sphingolipids
   c. oils
   d. waxes
   e. steroids
   f. carotenoids

3. To which lipid class does each of the following molecules belong?

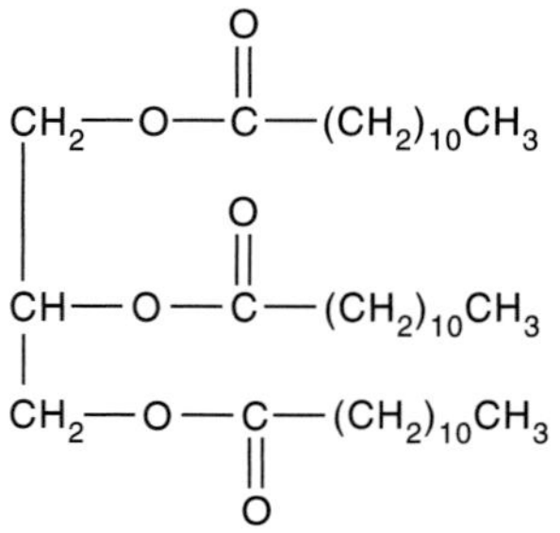

a.

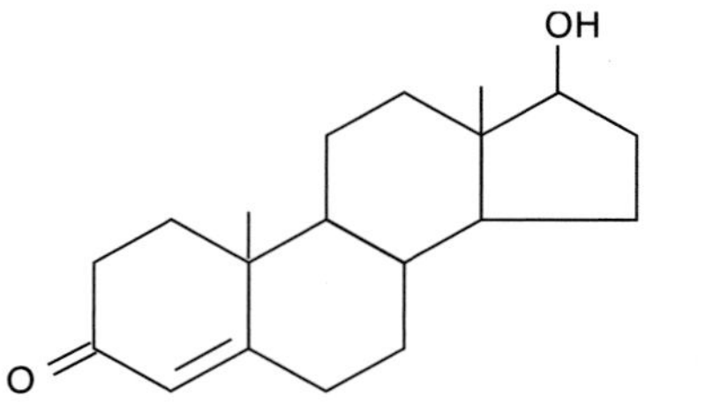

b.

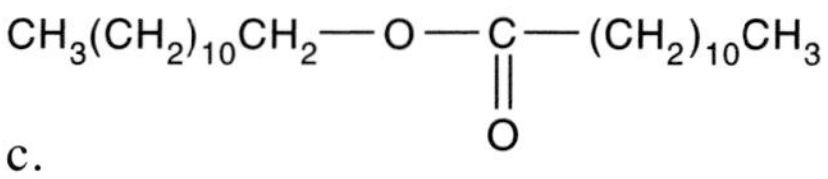

c. $CH_3(CH_2)_{10}CH_2-O-\overset{\overset{\displaystyle O}{\|}}{C}-(CH_2)_{10}CH_3$

d. $CH_3(CH_2)_7CH=CH(CH_2)_7COOH$

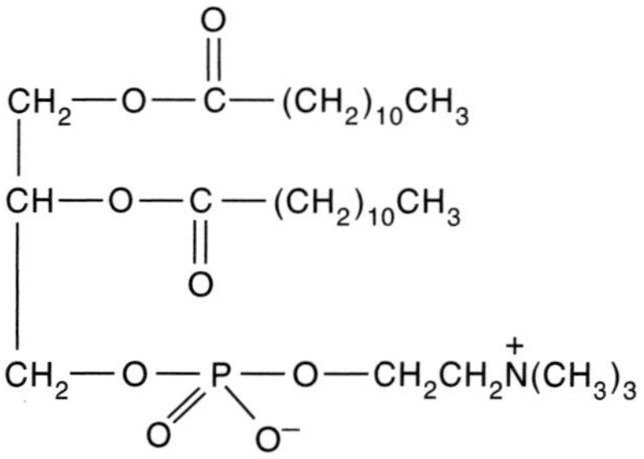

e.

f.

4. Sphingomyelins are amphipathic molecules. Draw the structure of a typical sphingomyelin. Identify which regions are hydrophilic and which are hydrophobic.

5. What role do plasma lipoproteins play in the human body? Why do plasma lipoproteins require a protein component to accomplish their role?

6. Describe several factors that influence membrane fluidity.

7. The fluid mosaic model of membrane structure has been very useful in explaining membrane behavior. However, the description of membrane as proteins floating in a phospholipid sea is oversimplified. Describe some components of membrane that are restricted in their lateral motion.

8. Which of the following statements or phrases concerning ionophores are true?
   a. form channels through which ions flow
   b. require energy
   c. ions may diffuse in either direction
   d. may cause voltage gates
   e. transport all ions with equal ease

9. Explain the differences in the ease of lateral movement and bilayer translocation movement of phospholipids.

10. Explain how potassium moves across a membrane. How are the channels opened? What other ions flow during this process?

11. From what fatty acid are most of the eicosanoids derived? List several medical conditions in which it may appear advantageous to suppress their synthesis.

12. In which of the following processes do the prostaglandins not have a major recognized role?
   a. reproduction
   b. digestion
   c. respiration
   d. inflammation
   e. smooth muscle contraction and relaxation

13. Classify each of the following as a monoterpene, diterpene, triterpene, tetraterpene, sesquiterpene, or polyterpene:

a.

b.

c.

d.

e.

f.

14. Which of the following is not a function of triacylglycerols?
   a. energy storage
   b. insulation
   c. shock absorption
   d. membrane structure

15. What molecules perform the roles (listed in question 14) that are not attributed to triacylglycerols?

16. How does the function of HDL promote the reduction of coronary artery disease risk?

17. For which of the following properties of membranes are lipids not directly responsible? In what features are lipids directly involved?
   a. selective permeability
   b. self-sealing capability
   c. fluidity
   d. asymmetry
   e. active transport of ions

18. Describe how glucose is transported across membranes in the kidney. What type of transport is involved?

19. Describe how carrier-mediated transport operates. Give an example.

20. Compare and contrast the following processes: active transport, passive transport, diffusion, and facilitated diffusion.

21. How do the following substances move across the plasma membranes of animal cells?
   a. CO_2
   b. H_2O
   c. glucose
   d. Cl^-
   e. Na^+

THOUGHT QUESTIONS

1. Animal cells are enclosed in a cell membrane. According to the fluid mosaic model, this membrane is held together by hydrophobic interactions. Consider the shear forces involved. Why does this membrane not break every time an animal moves?

2. Species-specific antigens are located on the surfaces of human and canine cells. If a heterokaryon is formed from red blood cell membrane from both species, what will happen to each set of antigens? What does this suggest about the nature of membranes?

3. Suggest a reason why elevated LDL levels are a risk factor for coronary artery disease.

4. Glycolipids are nonionic lipids that can orient themselves into bilayers as phospholipids do. Suggest a reason why they can accomplish this feat although they lack an ionic group like that of the phospholipids.

5. Explain why spontaneous phospholipid translocation (the movement of a molecule from one side of a bilayer to the other) is so slow.

6. As bacteria are exposed to higher temperatures, their membranes become increasingly fluid. What molecule could a researcher use to restore the organism's original fluidity characteristics? Consider the anatomy of bacteria. Why would this procedure not be feasible?

7. Mammals in the Arctic (e.g., reindeer) have higher levels of unsaturated fatty acids in their legs and hooves than in the rest of their bodies. Suggest a reason for this phenomenon. Does it have a survival advantage?

8. Explain why entropy increases when a lipid bilayer forms from phospholipid molecules.

CHAPTER TWELVE
Lipid Metabolism

OUTLINE

FATTY ACIDS AND TRIACYLGLYC-EROLS

Fatty Acid Degradation

The Complete Oxidation of a Fatty Acid

SPECIAL INTEREST BOX 12.1
FATTY ACID OXIDATION: DOUBLE BONDS AND ODD CHAINS

Fatty Acid Biosynthesis

SPECIAL INTEREST BOX 12.2
EICOSANOID METABOLISM

Regulation of Fatty Acid Metabolism in Mammals

MEMBRANE LIPID METABOLISM

Phospholipid Metabolism

SPECIAL INTEREST BOX 12.3
MEMBRANE BIOGENESIS

Sphingolipid Metabolism

ISOPRENOID METABOLISM

Cholesterol Metabolism

Sterol Metabolism in Plants

The roles lipids play in living organisms are largely due to their hydrophobic structures. As prominent components of cell membranes, lipids are primarily responsible for the integrity of individual cells and the intracellular compartments that are the hallmark of eukaryotic organisms. The hydrophobic and highly reduced structure of triacylglycerols makes them a compact and rich source of energy for cellular processes. Chapter 12 focuses on the metabolism of the major classes of lipids, that is, how they are synthesized and degraded and how these processes are regulated. A major emphasis is placed on the central metabolite in lipid metabolism: acetyl-coenzyme A. Because of its prominent role in several human diseases, the metabolism of cholesterol is also discussed.

Acetyl-CoA, the energy-rich molecule composed of coenzyme A and an acetyl group, plays a preeminent role in the metabolism of lipids. In most lipid-related metabolic processes, acetyl-CoA is either a substrate or a product. For example, acetyl-CoA that is not immediately required by a cell in energy production is used in fatty acid synthesis. When fatty acids are degraded to generate energy, acetyl-CoA is produced. Similarly, three acetyl-CoA molecules are combined to form isopentenyl pyrophosphate, the building block molecule in isoprenoid synthetic reactions. Therefore molecules as diverse as the terpenes and steroids found in animals and plants are all synthesized from acetyl-CoA. In Chapter 12, the metabolism of the major classes of lipid is discussed: fatty acids, triacylglycerols, phospholipids, sphingolipids, and isoprenoids. In addition, the metabolism of several important fatty acid metabolites, that is, the eicosanoids and the ketone bodies, is reviewed. Because of the major role that lipids play in providing energy and structural materials for living cells, several metabolic control mechanisms are discussed throughout the chapter.

12.1 FATTY ACIDS AND TRIACYLGLYCEROLS

Fatty acids are an important and efficient energy source for many living cells. In animals, most fatty acids are obtained in the diet. For example, in the average U.S. diet, between 30% and 40% of calories ingested are provided by fat. Triacylglycerol molecules are digested within the lumen of the small intestine (Figure 12.1). After being mixed with **bile salts**, amphipathic molecules with detergent properties that are produced in the liver and temporarily stored in the gallbladder (see p. 412), triacylglycerol molecules are digested by pancreatic lipase to form fatty acids and monoacylglycerol. These latter molecules are then transported across the plasma membrane of intestinal wall cells (enter-ocytes), where they are reconverted to triacylglycerols. Enterocytes combine triacylglycerols with dietary cholesterol and newly synthesized phospholipids and protein to form chylomicrons (large, low-density lipoproteins). After their secretion into the lymph (tissue fluid derived from blood), chylomicrons pass from the lymph into the blood. Most chylomicrons are removed from blood by adipose tissue cells (adipocytes), the body's primary lipid storage depot. Lipoprotein lipase, synthesized by cardiac and skeletal muscle, lactating mammary gland, and adipose tissue, is transferred to the endothelial surface of the capillaries, where it converts the triacylglycerol in chylomicrons into fatty acids and glycerol. (Lipoprotein lipase is activated when it binds to one of the apoprotein component of chylomicrons. The triacylglycerols in VLDL are also degraded by lipoprotein lipase.) Because adipose tissue cannot use glycerol, this molecule is carried in the blood to the liver, where the enzyme glycerol kinase converts it to glycerol-3-phosphate. (Adipocytes lack glycerol kinase. They derive glycerol-3-phosphate from dihydroxyacetone phosphate, a glycolytic intermediate.) In liver cells, glycerol-3-phosphate can then be used in the synthesis of triacylglycerols, phospholipids, or glucose.

Depending on an animal's current metabolic needs, fatty acids may be converted to triacylglycerols, degraded to generate energy, or used in membrane synthesis. For example, serum glucose levels are high immediately after a meal. The hormone insulin promotes triacylglycerol storage by inactivating triacylglycerol lipase (an enzyme that hydrolyses the ester bonds of fat molecules) in the adipocyte, increasing triacylglycerol synthesis and transport via VLDL from the liver, and promoting lipoprotein lipase activity and uptake of fatty acids by the adipocytes. Because glycolysis in the adipocytes provides DHAP and, therefore, glycerol-3-phosphate, new triacyglycerols are formed in the adipocyte. In contrast, when blood glucose is low (insulin levels decrease and glucagon levels increase) triacylglycerol lipase inhibition is released and fat will be mobilized from the adipocyte forming glycerol and fatty acids. As discussed, glycerol is a substrate for gluconeogenesis. Fatty acids are degraded by the body's cells to generate energy.

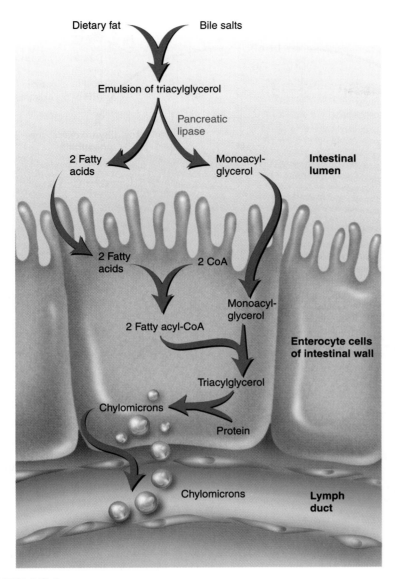

FIGURE 12.1

Digestion and Absorption of Triacylglycerols in the Small Intestine.

After triacylglycerols are emulsified (solubilized) by mixing with bile salts, they are digested by intestinal lipases, the most important of which is pancreatic lipase. The products, fatty acids and monoacylglycerol, are transported into enterocytes and resynthesized to form triacylglycerol. The triacylglycerol molecules, along with newly synthesized phospholipid and protein, are then incorporated into chylomicrons. After the chylomicrons are transported into lymph, via exocytosis, and then blood, they are taken up by peripheral tissues.

Triacylglycerol synthesis (referred to as **lipogenesis**) is illustrated in Figure 12.2. Glycerol-3-phosphate or dihydroxyacetone phosphate reacts sequentially with three molecules of acyl-CoA (fatty acid esters of CoASH). Acyl-CoA molecules are produced in the following reaction:

$$R-\overset{\overset{\textstyle O}{\|}}{C}-O^- + \text{ATP} + \text{CoASH} \longrightarrow R-\overset{\overset{\textstyle O}{\|}}{C}-S-\text{CoA} + \text{PP}_i + \text{AMP}$$

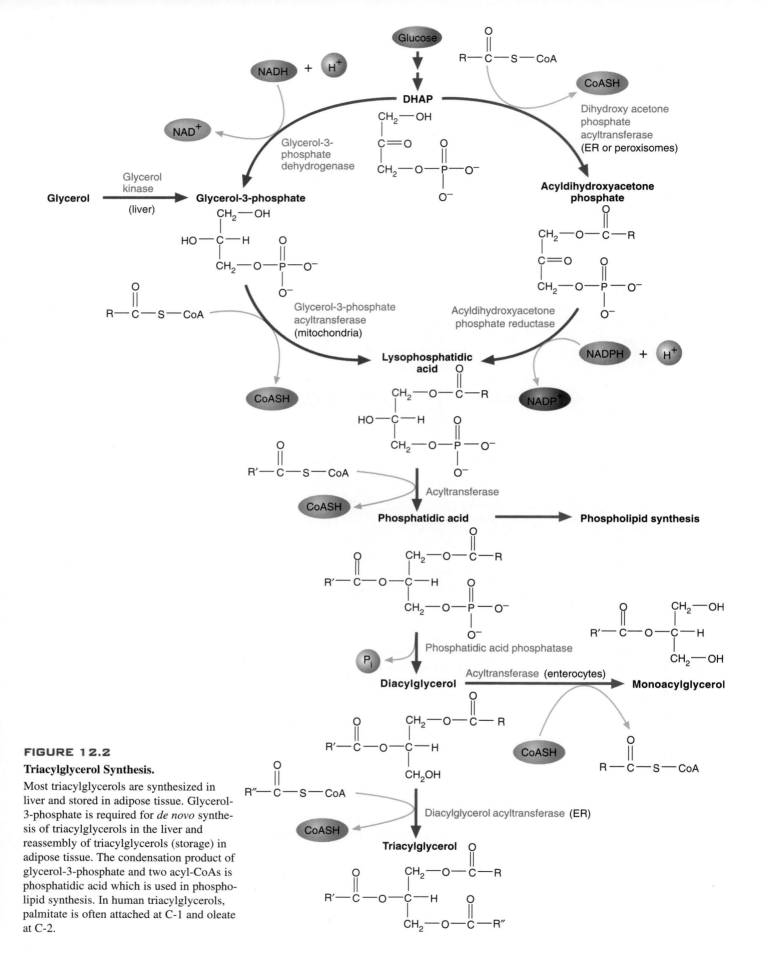

FIGURE 12.2

Triacylglycerol Synthesis.

Most triacylglycerols are synthesized in liver and stored in adipose tissue. Glycerol-3-phosphate is required for *de novo* synthesis of triacylglycerols in the liver and reassembly of triacylglycerols (storage) in adipose tissue. The condensation product of glycerol-3-phosphate and two acyl-CoAs is phosphatidic acid which is used in phospholipid synthesis. In human triacylglycerols, palmitate is often attached at C-1 and oleate at C-2.

(Note that the reaction is driven to completion by the hydrolysis of pyrophosphate by pyrophosphatase.) In the synthesis of triacylglycerols, phosphatidic acid is formed by two sequential acylations of glycerol-3-phosphate or by a pathway involving the direct acylation of dihydroxyacetone phosphate. In the latter pathway, acyldihydroxyacetone phosphate is later reduced to form lysophosphatidic acid. Depending on the pathway used, lysophosphatidic acid synthesis utilizes either an NADH or an NADPH. Phosphatidic acid is produced when lysophosphatidic acid reacts with a second acyl-CoA. Once formed, phosphatidic acid is converted to diacylglycerol by phosphatidic acid phosphatase. A third acylation reaction forms triacylglycerol. Fatty acids derived from both the diet and *de novo* synthesis are incorporated into triacylglycerols. (The term *de novo* is used by biochemists to indicate new synthesis.) *De novo* synthesis of fatty acids is discussed.

When energy reserves are low, the body's fat stores are mobilized in a process referred to as **lipolysis** (Figure 12.3). Lipolysis occurs during fasting, during vigorous exercise, and in response to stress. Several hormones (e.g., glucagon and epinephrine) bind to specific adipocyte plasma membrane receptors and begin a reaction sequence similar to the activation of glycogen phosphorylase. Hormone binding to the receptor elevates cytosolic cAMP levels, which, in turn, activates hormone-sensitive triacylglycerol lipase. (Triacylglycerols are constantly being synthesized and mobilized in adipose tissue. The activation of hormone-sensitive lipase vastly increases the rate of hydrolyzing triacylglycerols.)

KEY CONCEPTS 12.1

When energy reserves are high, triacylglycerols are stored in a process called lipogenesis. When energy reserves are low, triacylglycerols are degraded in a process called lipolysis to form fatty acids and glycerol.

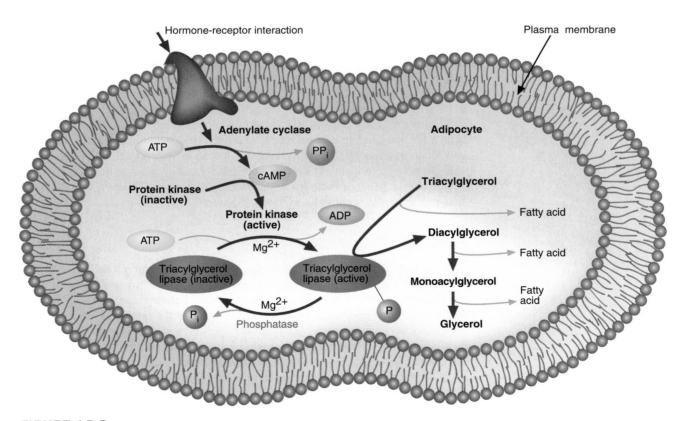

FIGURE 12.3

Diagrammatic View of Lipolysis.

When certain hormones bind to their receptors in adipose tissue, a cascade mechanism releases fatty acids and glycerol from triacylglycerol molecules. Triacylglycerol lipase (sometimes referred to as hormone-sensitive lipase) is activated when it is phosphorylated by protein kinase. Protein kinase is activated by cAMP. After their transport across the plasma membrane, fatty acids are transported in blood to other organs bound to serum albumin.

Both products of lipolysis (i.e., fatty acids and glycerol) are released into the blood. As stated previously (see p. 253), glycerol is transported to the liver where it can be used in lipid or glucose synthesis. After their transport across the adipocyte plasma membrane, fatty acids become bound to serum albumin. (Serum albumin is an abundant protein that transports numerous substances in blood. Other examples of hydrophobic albumin ligands include steroids and eicosanoids.) The albumin-bound fatty acids are carried to tissues throughout the body, where they are oxidized to generate energy. Fatty acids are transported into cells by a protein in the plasma membrane. This process is linked to the active transport of sodium. The amount of fatty acid that is transported depends on its concentration in blood and the relative activity of the fatty acid transport mechanism. Cells vary widely in their capacity to transport and use fatty acids. For example, some cells (e.g., brain and red blood cells) cannot use fatty acids as fuel, although others (e.g., cardiac muscle) rely on them for a significant portion of their energy requirements. Once they enter a cell, fatty acids must be transported to their destinations (i.e., mitochondria, endoplasmic reticulum, and other organelles). Several **fatty acid–binding proteins** (water-soluble proteins whose sole function is to bind and transport hydrophobic fatty acids) may be responsible for this transport.

Most fatty acids are degraded to form acetyl-CoA within mitochondria in a process referred to as β-oxidation. β-Oxidation is also known to occur in peroxisomes. Other oxidative mechanisms are also available to degrade certain nonstandard fatty acids (Special Interest Box 12.1).

Fatty acids are synthesized when an organism has met its energy needs and nutrient levels are high. (Glucose and several amino acids are substrates for fatty acid synthesis.) Fatty acids are synthesized from acetyl-CoA in a process that is similar to the reverse of β-oxidation. Although most fatty acids are supplied in the diet, most animal tissues can synthesize some saturated and unsaturated fatty acids. In addition, animals can elongate and desaturate dietary fatty acids. For example, arachidonic acid is produced by adding a two-carbon unit and introducing two double bonds to linoleic acid.

In plants the fatty acids of triacylglycerols are used predominantly as an energy source for germinating seeds. Once synthesized (in a reaction sequence similar to that found in animals), triacylglycerols are stored in vesicles called *oil bodies*. As seeds begin to germinate, the synthesis of lipases causes a massive breakdown of triacylglycerols. Most of the fatty acids released in this process are used to synthesize carbohydrates within glyoxysomes (Section 9.2).

Fatty Acid Degradation

Most fatty acids are degraded by the sequential removal of two-carbon fragments from the carboxyl end of fatty acids. During this process, referred to as **β-oxidation**, acetyl-CoA is formed as the bond between the α- and β-carbon atoms is broken. (β-Oxidation is so named because the β-carbon of fatty acids, which is two carbons removed from the carboxyl group, is oxidized.) Other mechanisms for degrading fatty acids are known. Most of these degrade unusual fatty acids; for example, odd-chain or branched chain molecules usually require an α-oxidation step in which the fatty acid chain is shortened by one carbon by a stepwise oxidative decarboxylation. In some organisms, the carbon farthest from the carboxyl group may be oxidized by a process called ω-oxidation. The subsequent β-oxidation generates a short chain dicarboxylic acid. Very little is known about the structure, mechanism, or relevance of ω-oxidating enzymes. Once formed, acetyl-CoA and other short chain products can be used to generate energy or provide metabolic intermediates. β-Oxidation is discussed next. The degradation of odd-chain, branched chain, and unsaturated fatty acids is discussed in Special Interest Box 12.1.

You have just consumed a cheeseburger. Trace the fat molecules (triacylglycerol) from the cheeseburger to your adipocytes (fat cells).

VLDL secretion by liver cells depends directly on the intracellular concentration of fatty acids. A cytoplasmic fatty acid–binding protein (FABP) may be responsible for the transport of fatty acids to the SER, the site of triacylglycerol synthesis. Because hepatic VLDL secretion has been found to be greater in female rats than in male rats, there may be a connection between an animal's sex hormone status, FABP concentration, and VLDL secretion rate. If this is so, what do you expect to happen if a male rat is injected with estrogen? In general, what mechanisms are involved? (*Hint*: Steroid hormones exert their metabolic effects by causing changes in gene expression.)

β-Oxidation occurs primarily within mitochondria. (β-Oxidation within peroxisomes is also discussed.) Before β-oxidation begins, each fatty acid is activated in a reaction with ATP and CoASH (see p. 375). The enzyme that catalyzes this reaction, acyl-CoA synthetase, is found in the outer mitochondrial membrane. Because the mitochondrial inner membrane is impermeable to most acyl-CoA molecules, a special carrier called *carnitine* is used to transport acyl groups into the mitochondrion (Figure 12.4). Carnitine-mediated transfer of acyl groups into the mitochondrial matrix is accomplished through the following mechanism (Figure 12.5):

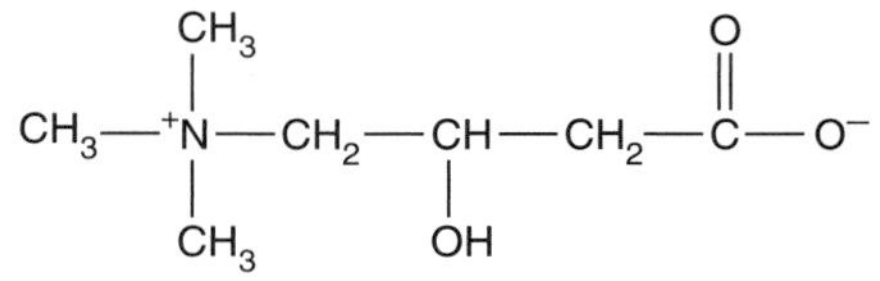

FIGURE 12.4

Structure of Carnitine.

1. Each acyl-CoA molecule is converted to an acylcarnitine derivative:

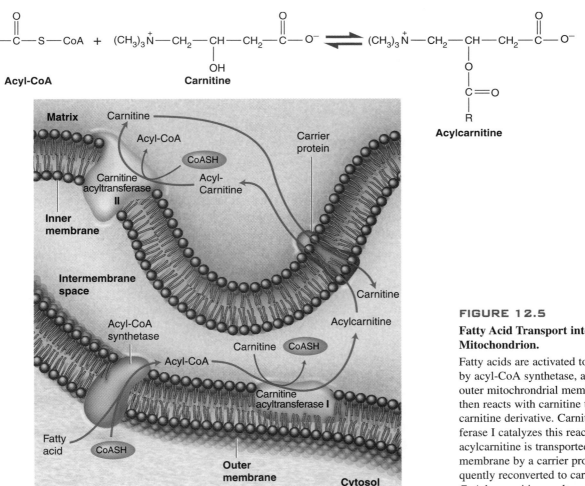

FIGURE 12.5

Fatty Acid Transport into the Mitochondrion.

Fatty acids are activated to form acyl-CoA by acyl-CoA synthetase, an enzyme in the outer mitochrondrial membrane. Acyl-CoA then reacts with carnitine to form an acylcarnitine derivative. Carnitine acyltransferase I catalyzes this reaction. After acylcarnitine is transported across the inner membrane by a carrier protein, it is subsequently reconverted to carnitine and acyl-CoA by carnitine acyltransferase II.

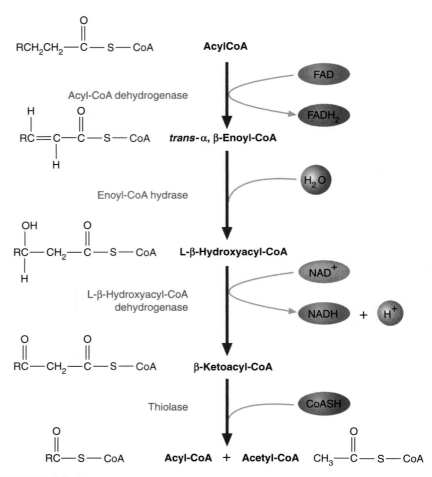

FIGURE 12.6

β-Oxidation of Acyl-CoA.

The β-oxidation of acyl-CoA molecules includes four reactions that occur in the mitochondrial matrix. Each cycle of reactions forms acetyl-CoA and an acyl-CoA that is shorter by two carbons.

This reaction is catalyzed by carnitine acyltransferase I.

2. A carrier protein within the mitochondrial inner membrane transfers acylcarnitine into the mitochondrial matrix.

3. Acyl-CoA is regenerated by carnitine acyltransferase II.

4. Carnitine is transported back into the intermembrane space by the carrier protein. It then reacts with another acyl-CoA.

A summary of the reactions of the β-oxidation of saturated fatty acids is shown in Figure 12.6. The pathway begins with an oxidation-reduction reaction, catalyzed by acyl-CoA dehydrogenase (an inner mitochondrial membrane flavoprotein), in which one hydrogen atom each is removed from the α- and β-carbons and transferred to the enzyme-bound FAD:

The $FADH_2$ produced in this reaction then donates 2 electrons to the mitochondrial electron transport chain (ETC). There are several isozymes of acyl-CoA dehydrogenase, each specific to a different fatty acid chain length. The product of this reaction is *trans*-α,β-enoyl-CoA.

The second reaction, catalyzed by enoyl-CoA hydrase, involves a hydration of the double bond between the α- and β-carbons:

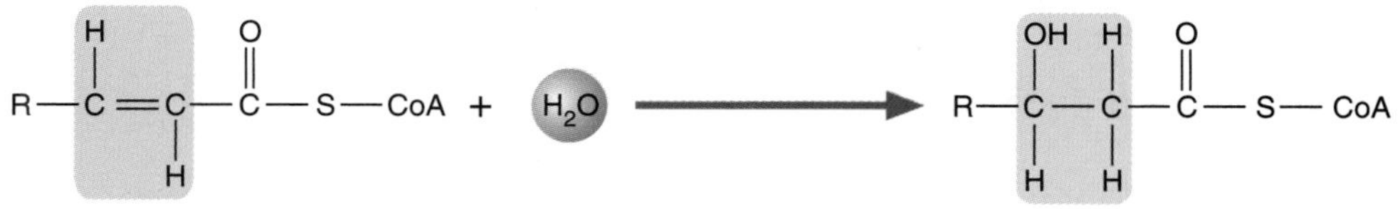

***trans*-α, β-Enoyl-CoA** **β-Hydroxylacyl-CoA**

The β-carbon is now hydroxylated. In the next reaction this hydroxyl group is oxidized. The production of a β-ketoacyl-CoA is catalyzed by β-hydroxyacyl-CoA dehydrogenase:

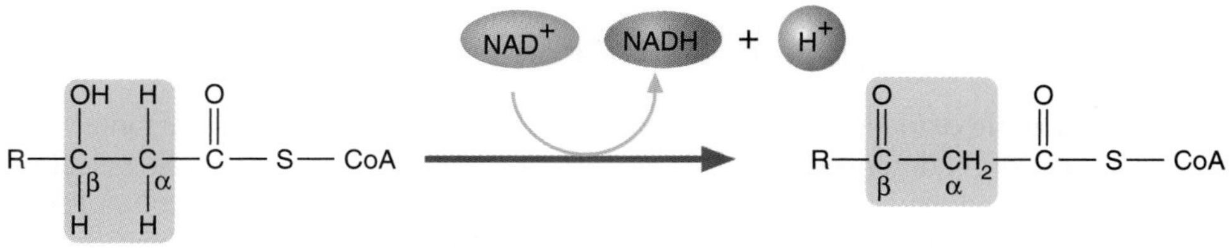

3-Hydroxylacyl-CoA **β-Ketoacyl-CoA**

The electrons transferred to NAD^+ are later donated to Complex I of the ETC. Finally, thiolase (sometimes referred to as β-ketoacyl-CoA thiolase) catalyzes a C_α-C_β cleavage:

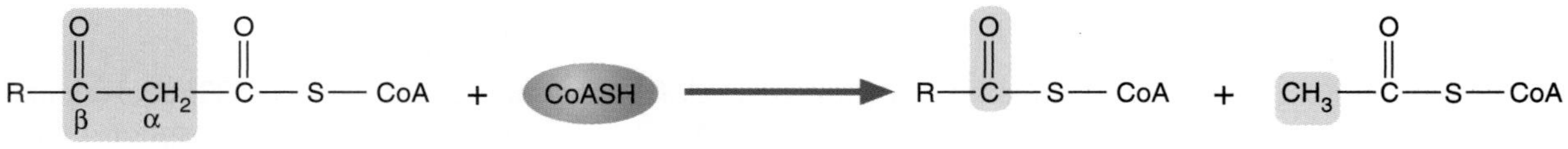

β-Ketoacyl-CoA **Acyl-CoA** **Acetyl-CoA**

In this reaction, sometimes called a **thiolytic cleavage**, an acetyl-CoA molecule is released. The other product, an acyl-CoA, now contains two fewer C atoms.

The four steps just outlined constitute one cycle of β-oxidation. During each later cycle, a two-carbon fragment is removed. This process, sometimes called the *β-oxidation spiral*, continues until, in the last cycle, a four-carbon acyl-CoA is cleaved to form two molecules of acetyl-CoA.

The following equation summarizes the oxidation of palmitoyl-CoA:

$$CH_3(CH_2)_{14}\overset{O}{\overset{\|}{C}}—S—CoA + 7\,FAD + 7\,NAD^+ + 7\,CoASH + 7\,H_2O \longrightarrow$$

$$8\,CH_3\overset{O}{\overset{\|}{C}}—S—CoA + 7\,FADH_2 + 7\,NADH + 7\,H^+$$

The acetyl-CoA molecules produced by fatty acid oxidation are converted via the *citric acid cycle* to CO_2 and H_2O as additional NADH and $FADH_2$ are formed. A portion of the energy released as NADH and $FADH_2$ are oxidized by the ETC is later captured in ATP synthesis via *oxidative phosphorylation*. The complete oxidation of acetyl-CoA is discussed in Chapter 10. The calculation of total number of ATP that can be generated from palmitoyl is reviewed next.

QUESTION 12.3

Identify each of the following biomolecules:

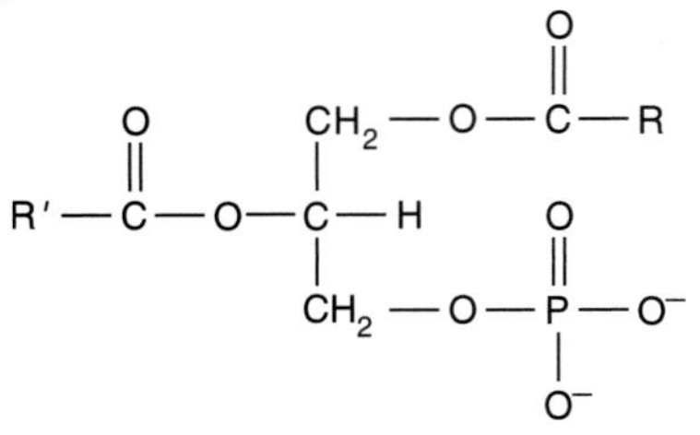

(a) 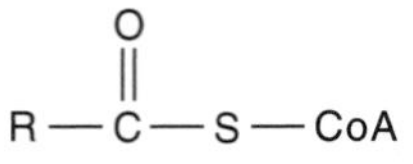   (b) 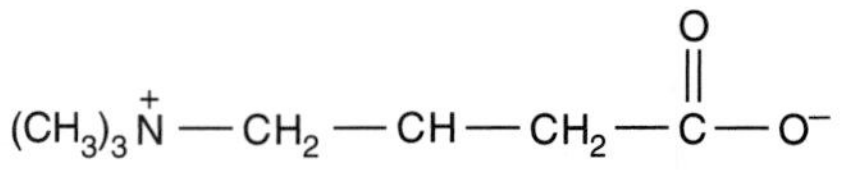   (c)

QUESTION 12.4

In the absence of oxygen, cells can produce small amounts of ATP from the anaerobic oxidation of glucose. This is not true for fatty acid oxidation. Explain.

The Complete Oxidation of a Fatty Acid

The aerobic oxidation of a fatty acid generates a large number of ATP molecules. As previously described (see p. 315), the oxidation of each $FADH_2$ during electron transport and oxidative phosphorylation yields approximately 1.5 molecules of ATP. Similarly, the oxidation of each NADH yields approximately 2.5 molecules of ATP. The yield of ATP from the oxidation of palmitoyl-CoA, which generates 7 $FADH_2$ 7 NADH, and 8 acetyl-CoA molecules to form CO_2 and H_2O is calculated as follows:

$$7\ FADH_2 \times 1.5\ ATP/FADH_2 \quad = 10.5\ ATP$$

$$7\ NADH \times 2.5\ ATP/NADH \quad = 17.5\ ATP$$

$$8\ Acetyl\text{-}CoA \times 10\ ATP/acetyl\text{-}CoA = \frac{80\ ATP}{108\ ATP}$$

The formation of palmitoyl-CoA from palmitic acid uses two ATP equivalents. (The synthesis of ATP from AMP involves two sequential phosphorylation reactions.) The net synthesis of ATP per molecule of palmitoyl-CoA is therefore 106 molecules of ATP.

The yield of ATP from the oxidations of palmitic acid and glucose can be compared. Recall that the total number of ATP molecules produced per glucose molecule is approximately 31. If glucose and palmitic acid molecules are compared in terms of the number of ATP molecules produced per carbon atom, palmitic acid is a superior energy source. The ratio for glucose is 31/6 or 5.2 ATP molecules per carbon atom. Palmitic acid yields 106/16 or 6.6 ATP molecules per carbon atom. The oxidation of palmitic acid generates more energy than that of glucose because palmitic acid is a more reduced molecule. (Glucose with its six oxygen atoms is a partially oxidized molecule.)

QUESTION 12.5

Determine the number of moles of NADH, $FADH_2$, and ATP molecules that can be synthesized from 1 mol of stearic acid.

β-OXIDATION IN PEROXISOMES β-Oxidation of fatty acids also occurs within peroxisomes. In animals peroxisomal β-oxidation appears to shorten very long-chain fatty acids. The resulting medium-chain fatty acids are further degraded within mitochondria. In many plant cells, β-oxidation occurs predominantly in peroxisomes. (Fatty acids are not an important source of energy in most plant tissues. Although some plant mitochondria contain β-oxidation enzymes, this pathway is not considered to contribute to energy generation to any substantial degree.) In some germinating seeds, β-oxidation occurs in glyoxysomes. (Glyoxysomes are specialized peroxisomes that possess the glyoxylate cycle enzymes. See p. 293.) The acetyl-CoA derived from glyoxysomal β-oxidation is converted to carbohydrate by the glyoxylate cycle and gluconeogenesis.

Peroxisomal membrane possesses an acyl-CoA ligase activity that is specific for very long-chain fatty acids. Mitochondria apparently cannot activate long-chain fatty acids such as tetracosanoic (24:0) and hexacosanoic (26:0). Peroxisomal carnitine acyltransferases catalyze the transfer of these molecules into peroxisomes, where they are oxidized to form acetyl-CoA and medium-chain acyl-CoA molecules (i.e., those possessing between 6 and 12 carbons). Medium-chain acyl-CoAs are further degraded via β-oxidation within mitochondria.

Although the reactions of peroxisomal β-oxidation are similar to those in mitochondria, there are some notable differences. First, the initial reaction in the peroxisomal pathway is catalyzed by a different enzyme. This reaction is a dehydration catalyzed by an acyl-CoA oxidase. The reduced coenzyme $FADH_2$ then donates its electrons directly to O_2 instead of UQ (coenzyme Q). This feature of peroxisomal β-oxidation is in sharp contrast to the mitochondrial pathway, which synthesizes ATP. The H_2O_2 produced when $FADH_2$ is oxidized is converted to H_2O by catalase. Second, the next two reactions in peroxisomal β-oxidation are catalyzed by two enzyme activities (enoyl-CoA hydrase and 3-hydroxyacyl CoA dehydrogenase) found on the same protein molecule. Finally, the last enzyme in the pathway (β-ketoacyl-CoA thiolase) has a different substrate specificity than its mitochondrial version. It does not efficiently bind medium-chain acyl-CoAs.

QUESTION 12.6

In addition to β-oxidation, peroxisomes have other vital roles in lipid metabolism. For example, the synthesis of various ether-type lipids occurs within peroxisomes. In a rare autosomal recessive disease called *Zellweger syndrome*, affected individuals lack peroxisomes. Abnormalities in several organs (especially in brain, liver, and kidney) result in death in the first year of life. Because the absence of an organelle cannot be confirmed by microscopic methods, biochemical techniques must be used to diagnose Zellweger syndrome. (The organelle could be so altered by the genetic defect as to be undetectable.) Suggest in general terms several biochemical methods of diagnosing this malady.

THE KETONE BODIES Most of the acetyl-CoA produced during fatty acid oxidation is used by the citric acid cycle or in isoprenoid synthesis (Section 12.3). Under normal conditions, fatty acid metabolism is so carefully regulated that only small amounts of excess acetyl-CoA are produced. In a process called **ketogenesis**, acetyl-CoA molecules are converted to acetoacetate, β-hydroxybutyrate, and acetone, a group of molecules called the **ketone bodies** (Figure 12.7).

Ketone body formation, which occurs within the matrix of liver mitochondria, begins with the condensation of two acetyl-CoAs to form acetoacetyl-CoA. Then acetoacetyl-CoA condenses with another acetyl-CoA to form β-hydroxy-β-methylglutaryl-CoA (HMG-CoA). In the next reaction, HMG-CoA is cleaved to form acetoacetate and acetyl-CoA. Acetoacetate is then reduced to form β-hydroxybutyrate. Acetone is formed by the spontaneous decarboxylation of acetoacetate when the latter molecule's concentration is high. (This condition, referred to as **ketosis**, occurs in uncontrolled diabetes, a metabolic disease discussed in Special Interest Box 16.3, and during starvation. In both of these conditions there is a heavy reliance on fat stores and β-oxidation of fatty acids to supply energy.)

KEY CONCEPTS 12.2

In β-oxidation, fatty acids are degraded by breaking the bond between the α- and β-carbon atoms. The ketone bodies are produced from excess molecules of acetyl-CoA.

The β-oxidation pathway degrades saturated fatty acids with an even number of carbon atoms. Certain additional reactions are required to degrade unsaturated, odd-chain, and branched-chain fatty acids.

Unsaturated Fatty Acid Oxidation

The oxidation of unsaturated fatty acids such as oleic acid requires additional enzymes. They are needed because, unlike the *trans* double bonds introduced during β-oxidation, the double bonds of most naturally occurring unsaturated fatty acids have a *cis* configuration. The enzyme enoyl-CoA isomerase converts the *cis*-β,γ-double bond to a *trans*-α,β double bond. The β-oxidation of oleic acid is illustrated in Figure 12A.

Odd–Chain Fatty Acid Oxidation

Although most fatty acids contain an even number of carbon atoms, some organisms (e.g., some plants and microorganisms) produce odd-chain fatty acid molecules. β-Oxidation of such fatty acids proceeds normally until the last β-oxidation cycle, which yields one acetyl-CoA molecule and one propionyl-CoA molecule. Propionyl-CoA is then converted to succinyl-CoA, a citric acid cycle intermediate (Figure 12B). Ruminant animals such as cattle and sheep

derive a substantial amount of energy from the oxidation of odd-chain fatty acids. These molecules are produced by microorganisms in the rumen (stomach).

α-Oxidation

α-Oxidation is a mechanism for getting around the presence of a branch in a fatty acid molecule such as phytanic acid, a branched 20-carbon fatty acid. (Phytanic acid is an oxidation product of phytol, a diterpene alcohol esterified to chlorophyll, the photosynthetic pigment.) Phytol, found in green vegetables, is converted to phytanic acid after it is ingested. Phytanic acid is a component of foods derived from herbivorous (plant-eating) animals (e.g., dairy products). In some plant tissues (e.g., leaves and seeds), α-oxidation is a major mechanism in the degradation of long-chain fatty acids.

β-Oxidation of phytanic acid is blocked by the methyl group substituent on C-3 (the β-position). Consequently, the first step in phytanic acid catabolism is an α-oxidation in which the molecule is converted to a α-hydroxy fatty acid. (α-Hydroxylating activity has been detected in the ER and in mitochondria.) This reaction is followed by the removal of the carboxyl group (Figure 12C). After activation to a CoA derivative, the product, pristanic acid, can be

FIGURE 12A *β*-**Oxidation of Oleoyl-CoA.**
β-Oxidation of the CoA derivative of oleic acid progresses until Δ³-*cis*-dodecenoyl-CoA is produced. This molecule is not a suitable substrate for β-oxidation, because it contains a *cis* double bond. After conversion of the β,γ *cis* double bond to an α,β *trans* double bond, β-oxidation recommences.

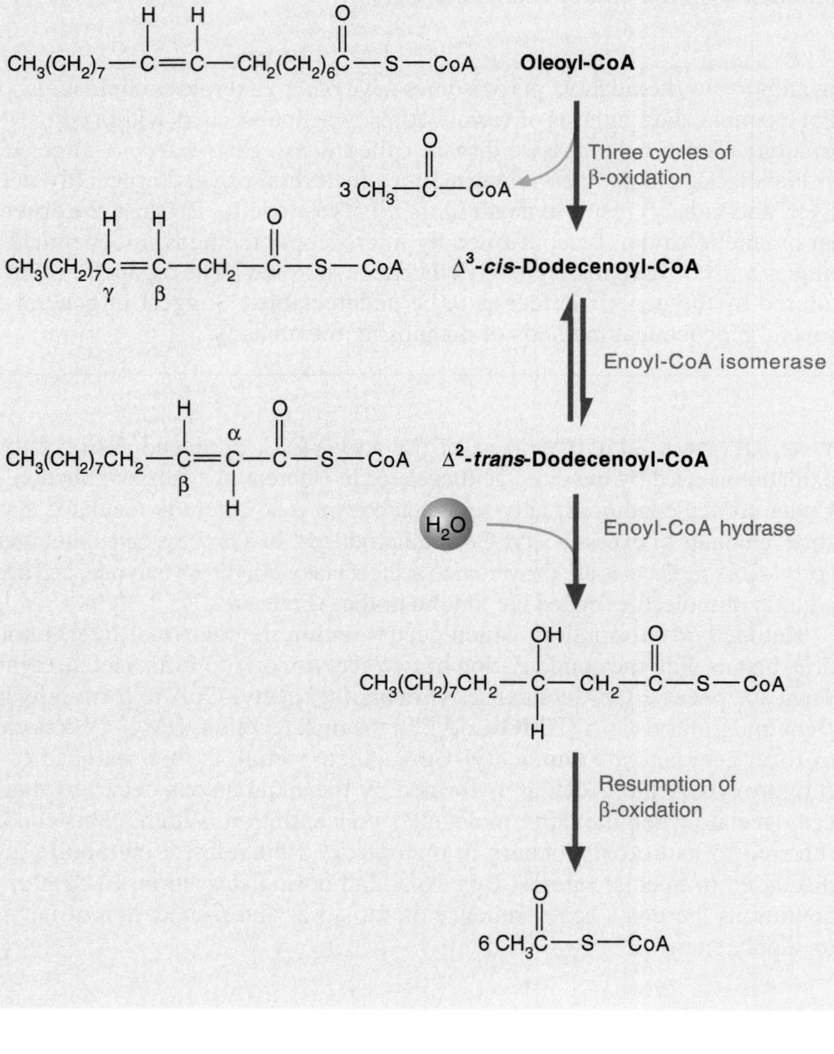

further degraded by β-oxidation. All subsequent side chain methyl groups will now be in the α-position which is not a problem for β-oxidation enzymes

The ability to oxidize phytanic acid is critical because large quantities of this molecule are found in the diet. In *Refsum's disease* (also referred to as *phytanic acid storage syndrome*) a build-up of phytanic acid causes very serious neurological problems. In this rare autosomal recessive condition, nerve damage is caused by a lack of α-hydroxylating activity. The mechanism by which phytanic acid accumulation causes nerve damage is unknown. Eating less phytanic acid–containing foods (i.e., dairy foods) can significantly reduce nerve damage.

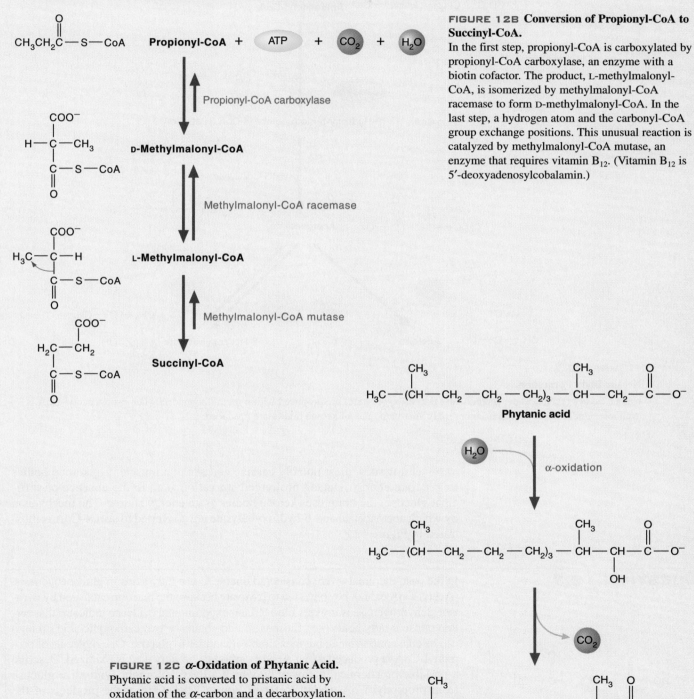

FIGURE 12B Conversion of Propionyl-CoA to Succinyl-CoA.
In the first step, propionyl-CoA is carboxylated by propionyl-CoA carboxylase, an enzyme with a biotin cofactor. The product, L-methylmalonyl-CoA, is isomerized by methylmalonyl-CoA racemase to form D-methylmalonyl-CoA. In the last step, a hydrogen atom and the carbonyl-CoA group exchange positions. This unusual reaction is catalyzed by methylmalonyl-CoA mutase, an enzyme that requires vitamin B_{12}. (Vitamin B_{12} is 5′-deoxyadenosylcobalamin.)

FIGURE 12C α-Oxidation of Phytanic Acid.
Phytanic acid is converted to pristanic acid by oxidation of the α-carbon and a decarboxylation. Pristanic acid is further degraded to generate a mixture of acetyl-CoA, propionyl-CoA, and isobutyl-CoA by the β-oxidation pathway.

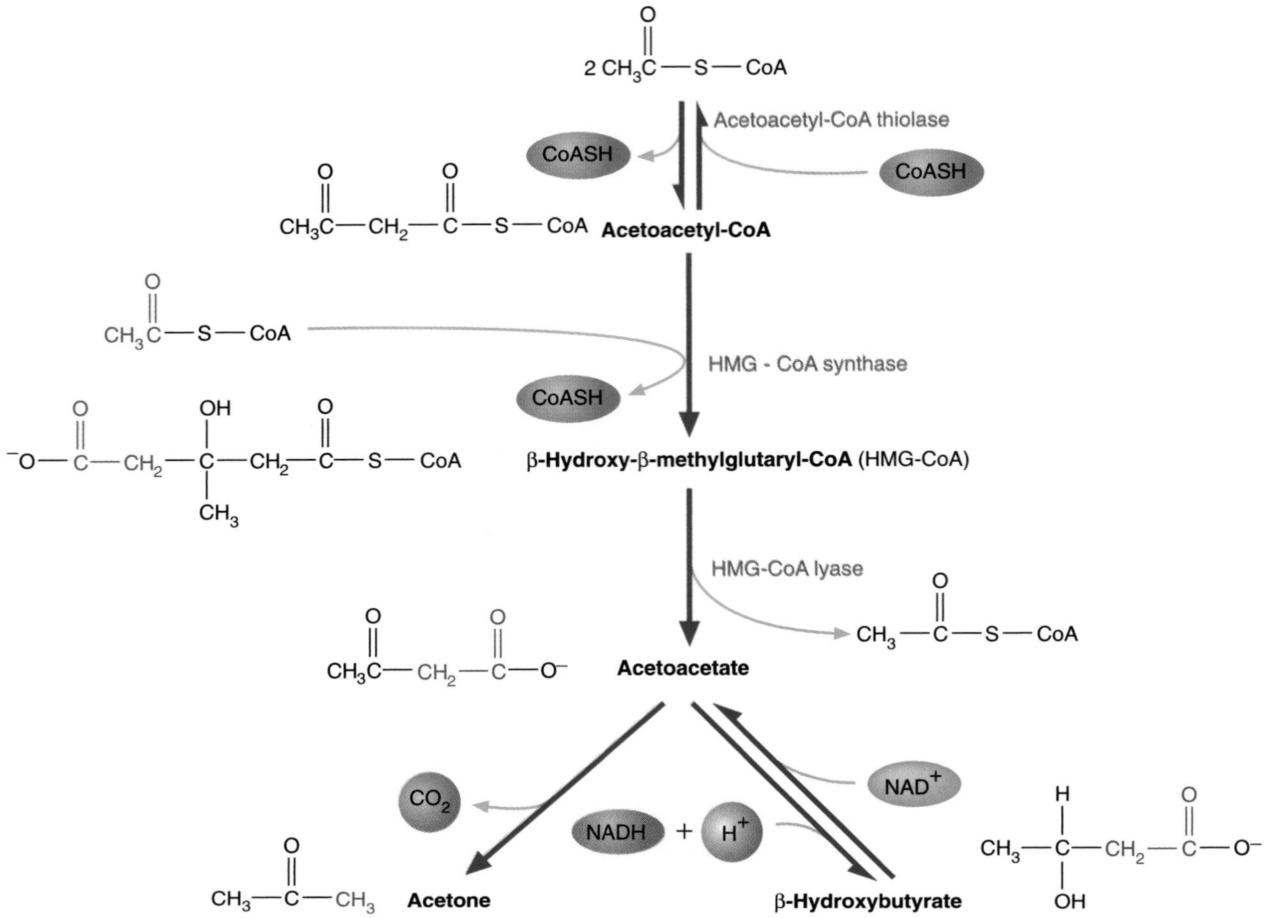

FIGURE 12.7

Ketone Body Formation.

Ketone bodies (acetoacetate, acetone, and β-hydroxybutyrate) are produced within the mitochondria when excess acetyl-CoA is available. Under normal circumstances, only small amounts of ketone bodies are produced.

Several tissues, most notably cardiac and skeletal muscle, use ketone bodies to generate energy. During prolonged starvation (i.e., in the absence of sufficient glucose) the brain uses ketone bodies as an energy source. The mechanism by which acetoacetate and β-hydroxybutyrate are converted to acetyl-CoA is illustrated in Figure 12.8.

QUESTION 12.7

FIGURE 12.9

Adipic Acid.

In the past, mammals were considered unable to use fatty acids in gluconeogenesis. (Acetyl-CoA cannot be converted to pyruvate because the reaction catalyzed by pyruvate dehydrogenase is irreversible.) Recent experimental evidence indicates that certain unusual fatty acids (i.e., those with odd chains or two carboxylic acid groups) can be converted in small but measurable quantities to glucose. One molecule of propionyl-CoA is produced when an odd-carbon chain fatty acid is oxidized. Describe a possible biochemical pathway by which a liver cell might synthesize glucose from propionyl-CoA. (*Hint*: Refer to Figure 12B). One of the products of the β-oxidation of dicarboxylic acids is succinyl-CoA. Propose a biochemical pathway for the conversion of the molecule illustrated in Figure 12.9 to glucose.

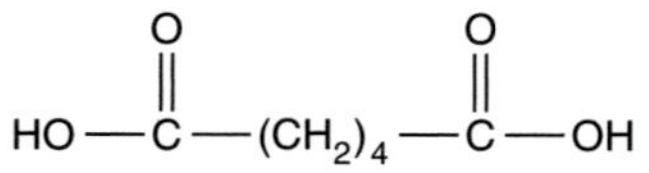

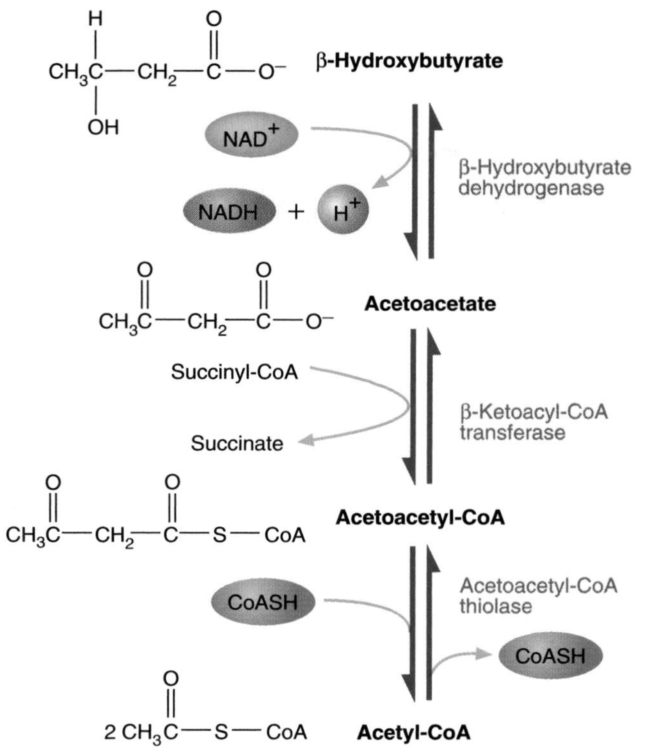

FIGURE 12.8

Conversion of Ketone Bodies to Acetyl-CoA.
Some organs (e.g., heart and skeletal muscle) can use ketone bodies (β-hydroxybutyrate and acetoacetate) as an energy source under normal conditions. During starvation the brain uses them as an important fuel source. Because liver does not have β-oxoacid-CoA transferase, it cannot use ketone bodies as an energy source. These reactions are reversible.

Fatty Acid Biosynthesis

Although fatty acid synthesis occurs within the cytoplasm of most animal cells, liver is the major site for this process. (Recall, for example, that liver produces VLDL. See p. 349.) Fatty acids are synthesized when the diet is low in fat and/or high in carbohydrate or protein. Most fatty acids are synthesized from dietary glucose. As discussed, glucose is converted to pyruvate in the cytoplasm. After entering the mitochondrion, pyruvate is converted to acetyl-CoA, which condenses with oxaloacetate, a citric acid cycle intermediate, to form citrate. When mitochondrial citrate levels are sufficiently high (i.e., cellular energy requirements are low), citrate enters the cytoplasm, where it is cleaved to form acetyl-CoA and oxaloacetate. The net reaction for the synthesis of palmitic acid from acetyl-CoA is as follows:

8 Acetyl-CoA + 14 NADPH + 14 H$^+$ + 7 ATP $\longrightarrow$

Palmitate + 14 NADP$^+$ + 7 ADP + 7 P$_i$ + 8 CoASH + 6 H$_2$O

A relatively large quantity of NADPH is required in fatty acid synthesis. A substantial amount of NADPH is provided by the pentose phosphate pathway (see p. 256). Reactions catalyzed by isocitrate dehydrogenase (see p. 285) and malic enzyme (see p. 291) provide smaller amounts.

The biosynthesis of fatty acids is outlined in Figure 12.10. At first glance, fatty acid synthesis appears to be the reverse of the β-oxidation pathway. For example, fatty acids are constructed by the sequential addition of two-carbon groups supplied by acetyl-CoA. Additionally, the same intermediates are found in both pathways (i.e., β-ketoacyl-, β-hydroxyacyl-, and α,β-unsaturated acyl groups).

As previously discussed (see p. 338), many important eicosanoids are derived from arachidonic acid. Almost all cellular arachidonic acid is stored in cell membranes as esters at C-2 of glycerol in phosphoglycerides. Release of arachidonic acid from membrane, considered to be the rate-limiting step in eicosanoid synthesis (Figure 12D), results from binding an appropriate chemical signal to its receptor on a target cell plasma membrane. For example, the release of arachidonic acid in platelets is caused by binding thrombin, an enzyme that plays an important role in blood clotting. (Thrombin is a proteolytic enzyme that converts the soluble plasma protein fibrinogen into fibrin, which then forms an insoluble meshwork.) Platelet aggregation, triggered by

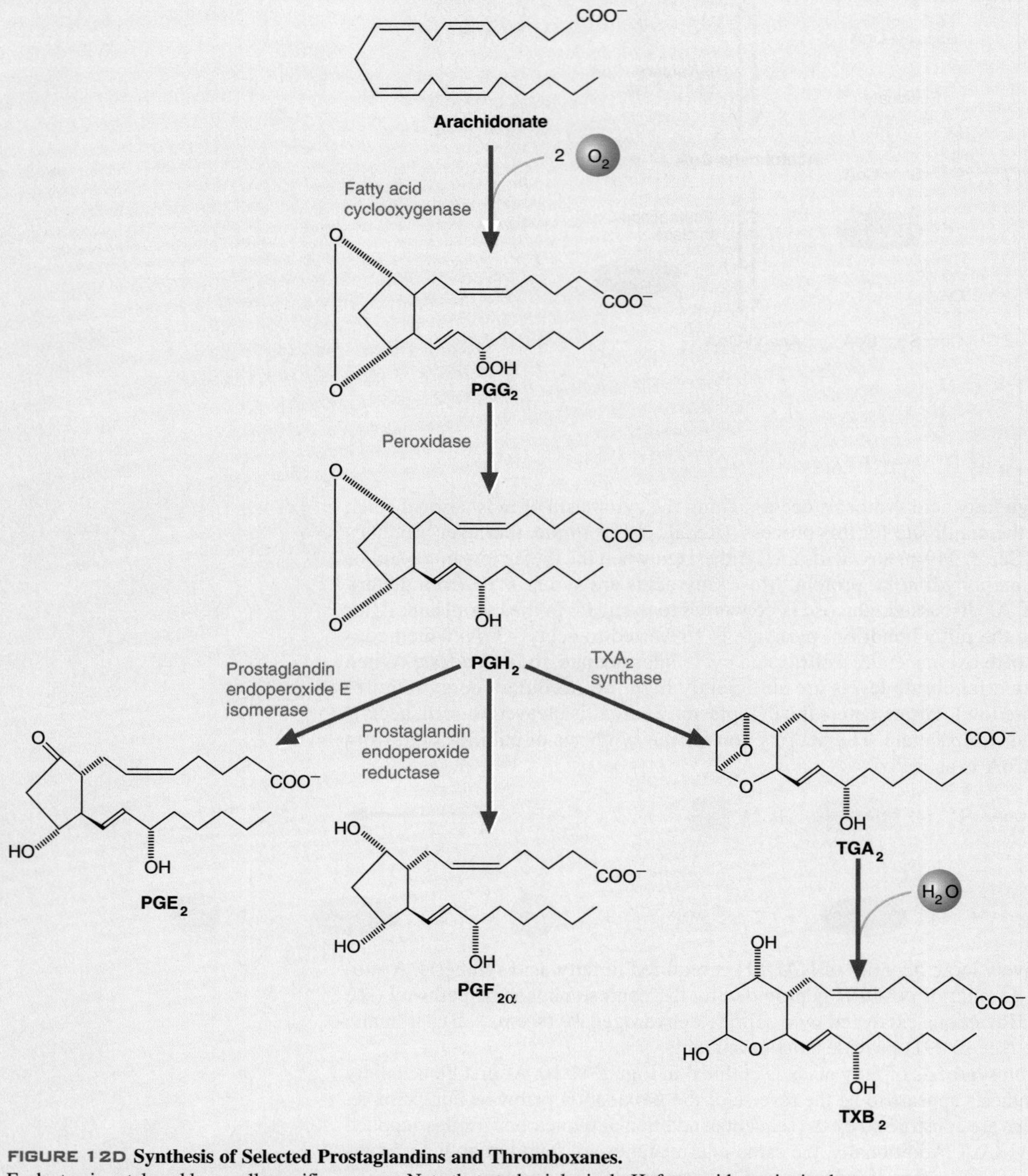

FIGURE 12D Synthesis of Selected Prostaglandins and Thromboxanes.
Each step is catalyzed by a cell-specific enzyme. Note that at physiological pH, fatty acids are ionized.

the eicosanoid TXA, is an early critical step in the blood-clotting process.

Most often, the release of arachidonic acid is catalyzed by phospholipase A_2. Certain steroids that suppress inflammation inhibit phospholipase A_2. Phospholipase A_2 cleaves acyl groups from C-2 of a phosphoglyceride, thus forming a fatty acid and lysophosphoglyceride. Once they are released, arachidonic acid molecules may be converted (depending on both cell type and intracellular conditions) into a variety of eicosanoid molecules. Prostaglandin synthesis begins when cyclooxygenase converts arachidonic acid into PGG_2. (Aspirin inactivates cyclooxygenase by acetylating a critical serine residue in the enzyme.) Then the formation of PGH_2 from PGG_2 is catalyzed by peroxidase. (Prostaglandin endoperoxidase synthase, an ER enzyme, possesses both the cyclooxygenase and the peroxidase activities.) PGH_2 is a precursor of several eicosanoids. For example, PGE_2 and PGF_2 are formed from PGH_2 by the actions of prostaglandin endoperoxide E isomerase and prostaglandin endoperoxide reductase, respectively. In platelets and lung cells, TXA_2 synthase catalyzes the conversion of PGH_2 to TXA_2. Within seconds, TXA_2 is spontaneously hydrolyzed to the inactive molecule TXB_2.

In a separate pathway, arachidonic acid is converted into the leukotrienes. Several enzymes, referred to as lipoxygenases, catalyze the addition of hydroperoxy groups to arachidonic acid. (Lipoxygenases are found in many mammalian tissues and in some plants.) The products of these reactions, called the monohydroperoxyeicosatetraenoic acids (HPETEs), are the direct precursors of the leukotrienes. For example, 5-lipooxygenase catalyzes the synthesis of 5-HPETE (Figure 12E). 5-HPETE is then converted to LTA_4. (Note that LTA_4 possesses an epoxide. **Epoxides** are three-membered rings containing an ether functional group.) LTC_4 is formed by the addition of GSH. LTC_4 is converted to LTD_4 by the removal of glutamic acid. Finally, LTE_4 is formed when glycine is removed from LTD_4. The role of leukotrienes is unclear, although several are believed to act as chemoattractants or as intracellular signals.

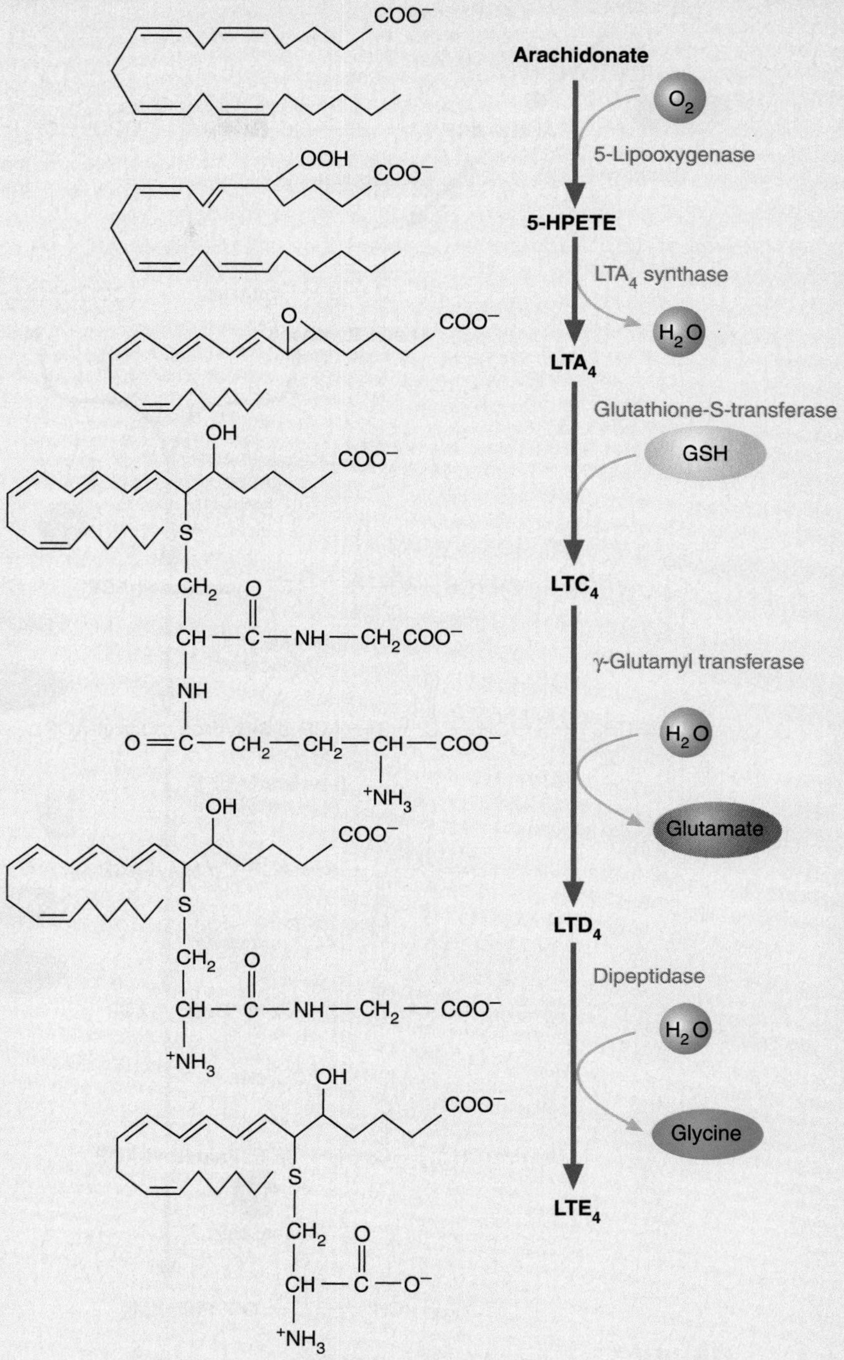

FIGURE 12E Synthesis of Selected Leukotrienes.
LTC_4, LTD_4, and LTE_4 are components of slow-reacting substance of anaphylaxis.

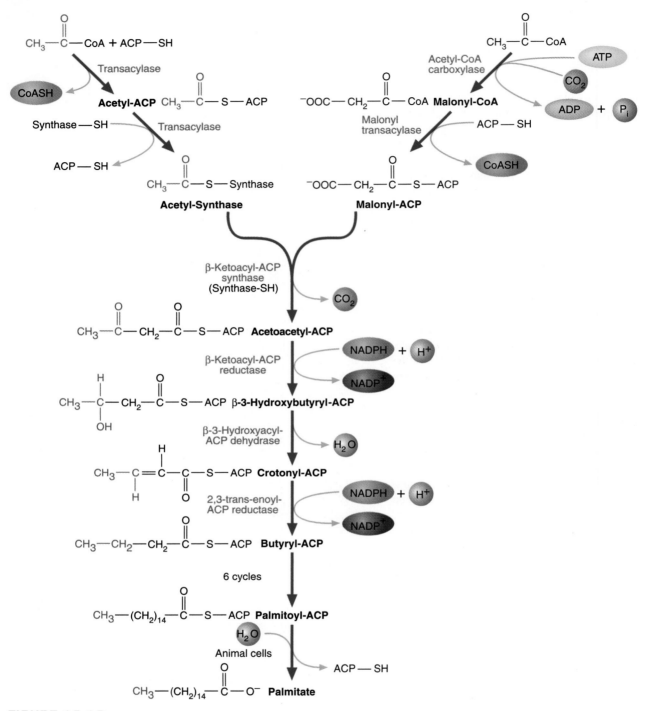

FIGURE 12.10

Fatty Acid Biosynthesis.

During each cycle of fatty acid biosynthesis, the molecule grows by two carbons. Most of the reactions in this pathway occur on a multienzyme complex.

Saturated fatty acids containing up to 16 carbon atoms (palmitate) are assembled in cytoplasm from acetyl-CoA. Depending on cellular conditions, the product of this process (palmitoyl-CoA) can be used directly in the synthesis of several types of lipid (e.g., triacylglycerol or phospholipids), or it can enter the mitochondrion. Several mitochondrial enzymes catalyze elongating and desaturating reactions. Endoplasmic reticulum (ER) possesses similar enzymes. A closer examination, however, reveals several notable differences between fatty acid synthesis and β-oxidation.

1. Location. Fatty acid synthesis occurs predominantly in the cytoplasm. (Recall that β-oxidation occurs within mitochondria and peroxisomes.)

2. Enzymes. The enzymes that catalyze fatty acid synthesis are significantly different in structure than those in β-oxidation. In eukaryotes, most of these enzymes are components of a multienzyme complex referred to as fatty acid synthase.

3. Thioester linkage. The intermediates in fatty acid synthesis are linked through a thioester linkage to **acyl carrier protein** (ACP), a component of fatty acid synthase. (Recall that acyl groups are attached to CoASH through a thioester linkage during β-oxidation.) Acyl groups are attached to both ACP and CoASH via a phosphopantetheine prosthetic group (Figure 12.11).

4. Electron carriers. In contrast to β-oxidation, which produces NADH and FADH$_2$, fatty acid synthesis consumes NADPH.

Fatty acid synthesis begins with the carboxylation of acetyl-CoA to form malonyl-CoA. (Acetyl-CoA carboxylation is considered an activating reaction. Activation is necessary in fatty acid synthesis because the condensation of acetyl groups is an endergonic reaction. As malonyl-CoA is decarboxylated during the condensation reaction, sufficient energy is generated to drive the process.) The carboxylation of acetyl-CoA to form malonyl-CoA (Figure 12.12), catalyzed by acetyl-CoA carboxylase, is the rate-limiting step in fatty acid synthesis. Mammalian acetyl-CoA carboxylase contains two subunits, each with a bound biotin cofactor. (Recall that biotin acts as a CO_2 carrier. See p. 249.) Fatty acid synthesis begins when acetyl-CoA carboxylase dimers aggregate to form high molecular weight filamentous polymers (four million to eight million D). Polymerization begins when cytoplasmic citrate levels rise. Depolymerization occurs when malonyl-CoA or palmitoyl-CoA levels are high. Phosphorylation of acetyl-CoA carboxylase in response to the binding of glucagon or epinephrine also causes depolymerization. In contrast, insulin facilitates dimer aggregation via phosphorylation and activation of an associated

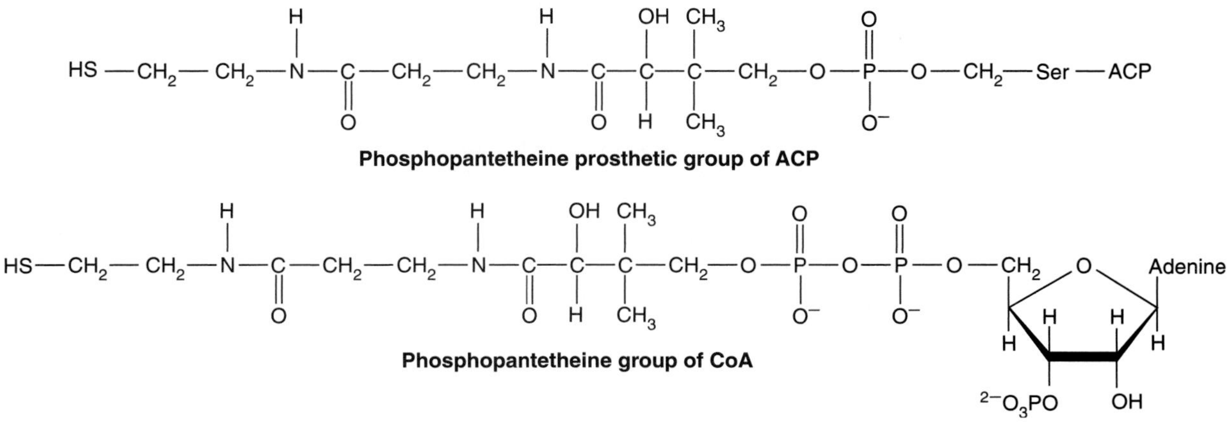

Phosphopantetheine prosthetic group of ACP

Phosphopantetheine group of CoA

FIGURE 12.11

Comparison of the Phosphopantetheine Group in Acyl Carrier Protein (ACP) and in Coenzyme A (CoASH).

Fatty acids are attached to this prosthetic group on ACP during fatty acid biosynthesis and on CoASH during β-oxidation.

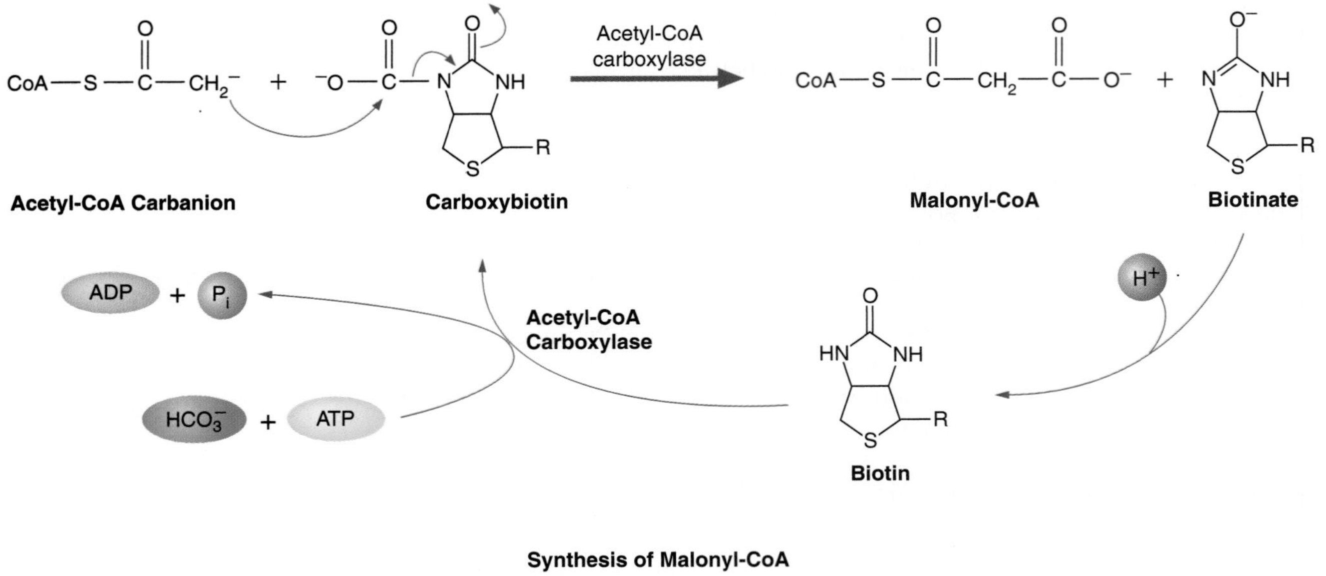

Synthesis of Malonyl-CoA

FIGURE 12.12

Synthesis of Malonyl-CoA.

The reaction begins with the ATP-dependent carboxylation of the biotin cofactor of the enzyme. The carboxylase abstracts a proton from the α-carbon of the acetyl-CoA to generate a reactive carbanion. The carbanion attacks the carbon of carboxybiotin to yield malonyl-CoA and biotinate. The biotinate is protonated by the enzyme to regenerate its biotin form.

phosphatase. The carboxylase dimer aggregates and becomes active only in its dephosphorylated form.

The remaining reactions in fatty acid synthesis take place on the fatty acid synthase multienzyme complex. This complex, the site of seven enzyme activities and ACP, is a 500-kD dimer. Because the enormous polypeptides in the dimer are arranged in a head-to-tail configuration, two fatty acids can be constructed simultaneously. A proposed mechanism for palmitate synthesis is shown in Figure 12.13.

During the first reaction on fatty acid synthase, acetyl transacylase catalyzes the transfer of the acetyl group from an acetyl-CoA molecule to the SH group of a cysteinyl residue of β-ketoacyl-ACP synthase. Malonyl-ACP is formed when malonyl transacylase transfers a malonyl group from malonyl-CoA to the SH group of the pantetheine prosthetic group of ACP (reaction 2). Then β-keto-acyl-ACP synthase catalyzes a condensation reaction (reaction 3) in which acetoacetyl-ACP is formed (Figure 12.14).

During the next three steps, consisting of two reductions and a dehydration, the acetoacetyl group is converted to a butyryl group. (The flexible phosphopantetheine arm acts as a tether so that the substrate does not diffuse away between steps in the cycle. This greatly increases the speed and efficiency of the process.) β-Ketoacyl-ACP reductase catalyzes the reduction of acetoacetyl-ACP to form β-hydroxybutyryl-ACP. β-Hydroxyacyl-ACP dehydrase later catalyzes a dehydration, thus forming crotonyl-ACP. Butyryl-ACP is produced when 2,3-*trans*-enoyl-ACP reductase reduces the double bond in crotonyl-ACP. In the last step of the first cycle of fatty acid synthesis, the butyryl group is transferred from the pantetheine group to the cysteine residue of β-ketoacyl-ACP synthase. The newly freed ACP-SH group now binds to another malonyl group. The process is then repeated. Eventually, palmitoyl-ACP is synthesized. Now the palmitoyl group is released from fatty acid synthase when thioesterase converts it to palmitate.

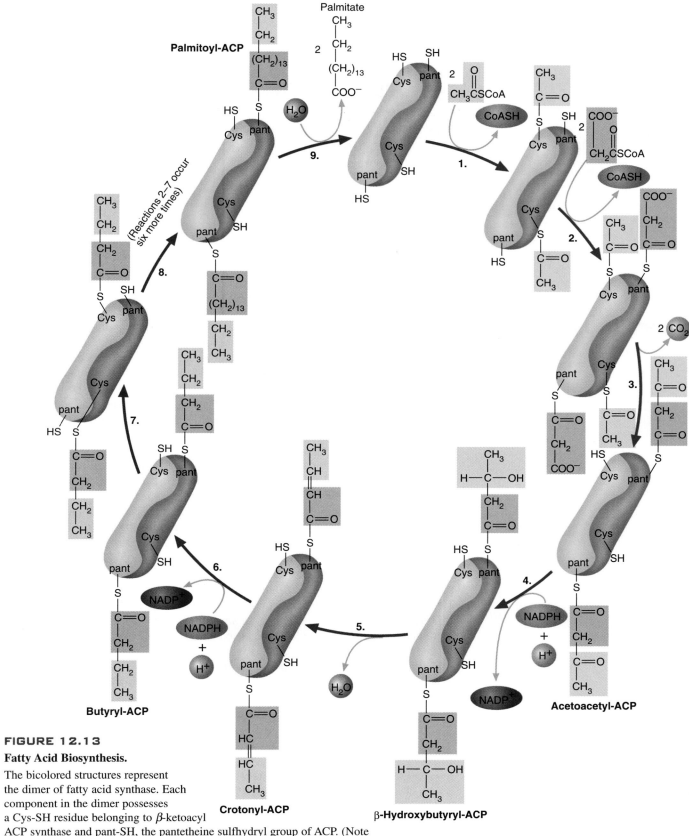

FIGURE 12.13

Fatty Acid Biosynthesis.

The bicolored structures represent
the dimer of fatty acid synthase. Each
component in the dimer possesses
a Cys-SH residue belonging to β-ketoacyl
ACP synthase and pant-SH, the pantetheine sulfhydryl group of ACP. (Note
that fatty acids are attached via a thioester linkage to the thiol terminal on ACP during fatty acid biosynthesis and on CoA during
β-oxidation.) The enzymes that catalyze the reactions in steps 1 through 6 are: (1) acetyl-CoA-ACP transacylase, (2) malonyl-CoA-ACP
transacylase, (3) β-ketoacyl-ACP synthase, (4) β-ketoacyl-ACP reductase, (5) β-hydroxyacyl-ACP dehydrase, and (6) 2,3-*trans*-enoyl-ACP
reductase. In step 7 the butyryl group is transferred from ACP to the cysteine-SH group of β-ketoacyl-ACP synthase by acetyl-CoA-ACP
transacylase. Step 8 in the diagram represents the repeated cycles of condensation and reduction needed to produce palmitoyl-ACP. Step 9,
the release of the product of the process, palmitic acid, from the enzyme complex, is catalyzed by a thioesterase.

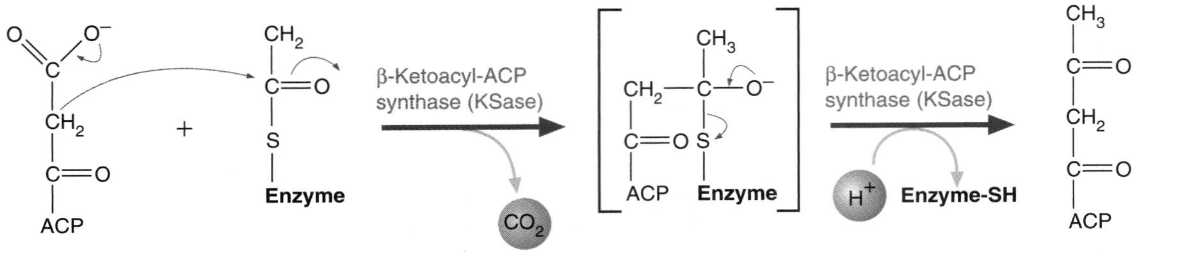

FIGURE 12.14

Formation of Acetoacetyl-ACP.

The acetyl group is shown bound to the enzyme β-ketoacyl-ACP synthase through a cysteine residue. The carbonyl group of the acetyl group is attacked by the central carbon on the malonyl group attached to ACP. Acetoacetyl-ACP is generated as the C—S bond is broken.

QUESTION 12.8

Rheumatoid arthritis is an autoimmune disease in which the joints are chronically inflamed. (In **autoimmune diseases** the immune system fails to distinguish between self and nonself. For reasons that are not understood, specific lymphocytes are stimulated to produce antibodies, referred to as *autoantibodies*. These molecules bind to surface antigens on the patient's own cells as if they were foreign. Then the immune system attacks affected cells.) In rheumatoid arthritis, several types of white blood cells infiltrate joint tissue as part of the inflammatory process. The leakage of lysosomal enzymes from actively phagocytosing cells (neutrophils and macrophages) leads to further tissue damage. The inflammatory response is perpetuated by the release of eicosanoids by white blood cells. A variety of eicosanoids have been implicated. For example, macrophages are known to produce PGE_4, TXA_2, and several leukotrienes.

Currently, the treatment of rheumatoid arthritis consists of suppressing pain and inflammation. (The disease continues to progress despite treatment.) Because of its low cost and relative safety, aspirin plays an important role in the treatment of rheumatoid arthritis and other types of inflammation. Certain steroids are more potent than aspirin in reducing inflammation, that is, they immediately and dramatically reduce painful symptoms. However, steroids have serious side effects. For example, prednisone may depress the immune system, cause fat redistribution to the neck ("buffalo hump"), and cause serious behavioral changes. For these and other reasons, prednisone is used to treat rheumatoid arthritis only when a patient does not respond to aspirin or similar drugs.

Review the effects of aspirin and steroids on eicosanoid metabolism described in Special Interest Box 12.2. Suggest a reason why this information is relevant to the treatment of rheumatoid arthritis. Does, it explain the difference between the effectiveness of aspirin and steroids in treating inflammation?

QUESTION 12.9

Excessive consumption of fructose has been linked to a condition referred to as *hypertriglyceridemia* (high blood levels of triacylglycerols). The most common source of fructose for most Americans is sucrose. (The fructose content of fresh fruits and vegetables is so low in comparison to many processed foods that it would be difficult to consume sufficient quantities to induce hypertriglyceridemia.) Sucrose is digested in the small intestine by the enzyme sucrase to yield one molecule each of fructose and glucose. Digestion is so rapid that the blood concentrations of these sugars become quiet high. Recall that once it reaches the liver, fructose is converted to fructose-1-phosphate (see p. 261). It is now believed that fructose-1-phosphate promotes hexokinase D activity. (Apparently, fructose-1-phosphate binds to and inactivates a protein that depresses hexokinase D activity.) After reviewing fructose metabolism and fatty acid and triacylglycerol synthesis, suggest how hypertriglyceridemia might result from a diet that is rich in sucrose.

FATTY ACID ELONGATION AND DESATURATION Elongation and desaturation of fatty acids synthesized in cytoplasm or obtained from the diet are accomplished primarily by ER enzymes. (These processes occur only when the diet provides an inadequate supply of appropriate fatty acids.) Fatty acid elongation and desaturation (the formation of double bonds) are especially important in the regulation of membrane fluidity and the synthesis of the precursors for a variety of fatty acid derivatives, such as the eicosanoids. For example, myelination (a process in which a myelin sheath is formed around certain nerve cells) depends especially on the ER fatty acid synthetic reactions. Very long chain saturated and monounsaturated fatty acids are important constituents of the cerebrosides and sulfatides found in myelin. Cells apparently regulate membrane fluidity by adjusting the types of fatty acids that are incorporated into membrane lipids. For example, in cold weather, more unsaturated fatty acids are incorporated. (Recall that unsaturated fatty acids have a lower freezing point than do saturated fatty acids. See p. 333.) If the diet does not provide a sufficient number of these molecules, fatty acid synthetic pathways are activated. Although elongation and desaturation are closely integrated processes, for the sake of clarity they are discussed separately.

ER fatty acid chain elongation, which uses two-carbon units provided by malonyl-CoA, is a cycle of condensation, reduction, dehydration, and reduction reactions similar to those observed in cytoplasmic fatty acid synthesis. In contrast to the cytoplasmic process, the intermediates in the ER elongation process are CoA esters. These reactions can lengthen both saturated and unsaturated fatty acids. Reducing equivalents are provided by NADPH.

$$R-\overset{\overset{\displaystyle O}{\|}}{C}-S-CoA \;+\; {}^-O-\overset{\overset{\displaystyle O}{\|}}{C}-CH_2-\overset{\overset{\displaystyle O}{\|}}{C}-S-CoA \longrightarrow$$

$$R-CH_2-CH_2-\overset{\overset{\displaystyle O}{\|}}{C}-S-CoA \;+\; CO_2 \;+\; CoASH$$

Acyl-CoA molecules are desaturated in ER membrane in the presence of NADH and O_2. All components of the desaturase system are integral membrane proteins that are apparently randomly distributed on the cytoplasmic surface of the ER. The association of cytochrome b_5 reductase (a flavoprotein), cytochrome b_5, and oxygen-dependent desaturases constitutes an electron transport system. This system efficiently introduces double bonds into long-chain fatty acids (Figure 12.15). Both the flavoprotein and cytochrome b_5 (found in a ratio of approximately 1:30) have hydrophobic peptides that anchor the proteins into the microsomal membrane. Animals typically have Δ^9, Δ^6, and Δ^5 desaturases that use electrons supplied by NADH via the electron transport system to activate the oxygen needed to create the double bond. Plants contain additional desaturases for the Δ^{12} and Δ^{15} positions.

Because elongation and desaturation systems are in close proximity to each other in microsomal membrane, a variety of long-chain polyunsaturated acids are typically produced. A prominent example of this interaction is the synthesis of arachidonic acid ($20{:}4^{\Delta5,8,11,14}$) from linoleic acid ($18{:}2^{\Delta9,12}$).

FATTY ACID SYNTHESIS IN PLANTS Because of less intensive research efforts and several technical problems, plant fatty acid synthesis is less well

KEY CONCEPTS 12.3

In animals, fatty acids are synthesized in the cytoplasm from acetyl-CoA and malonyl-CoA. Microsomal enzymes elongate and desaturate newly synthesized fatty acids as well as those obtained in the diet.

FIGURE 12.15

Desaturation of Stearoyl-CoA.

The desaturase uses electrons provided by an electron transport system composed of cytochrome b_5 reductase and cytochrome b_5 to activate the oxygen (not shown) needed to create the double bond. NADH is the electron donor.

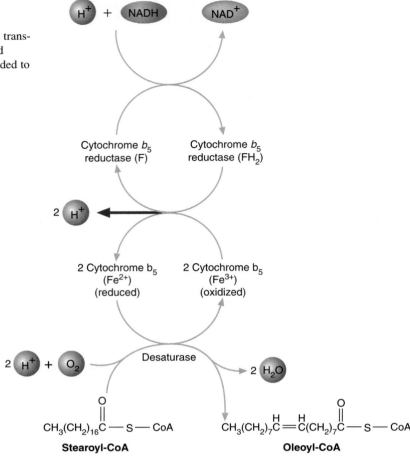

understood than the animal process. It is known, however, that plant fatty acid synthesis has several notable features:

1. Location. Plant fatty acid synthesis appears to be limited to chloroplasts. A chloroplast isozyme of pyruvate dehydrogenase catalyzes the conversion of pyruvate to acetyl-CoA. Pyruvate is also derived from glycerate-3-phosphate, an intermediate in the **Calvin cycle**, a biosynthetic pathway in which plants incorporate CO_2 into sugar molecules. (The Calvin cycle is discussed in Chapter 13.)

2. Metabolic control. The regulation of fatty acid synthesis in plants is poorly understood. It remains unclear whether the reaction catalyzed by acetyl-CoA carboxylase is a rate-limiting step in plants, because malonyl-CoA is used in several other biosynthetic pathways (e.g., bioflavonoid synthesis). (Recall that this reaction is rate-limiting in animal cells. See p. 390.)

3. Enzymes. The structures of both acetyl-CoA carboxylase and fatty acid synthase in plants more closely resemble their counterparts in *Escherichia coli* than those in animal cells. For example, in *E. coli* and plants, each of the enzyme activities of fatty acid synthase is found on a separate protein.

Regulation of Fatty Acid Metabolism in Mammals

Because animals have such varying requirements for energy, the metabolism of fatty acids (*the* major energy source in animals) is carefully regulated (Figure 12.16). Both short- and long-term regulatory mechanisms are used. In most short-term regulation (measured in minutes) the activities of already existing molecules of key regulatory enzymes are modified by hormones. For example, glucagon or epineph-

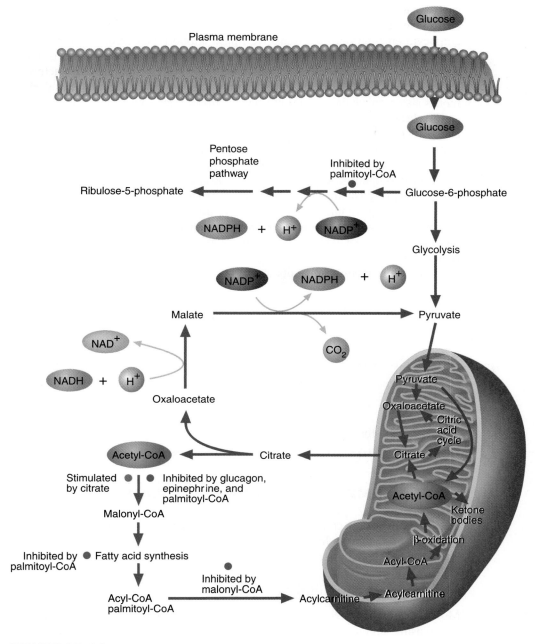

FIGURE 12.16

Intracellular Fatty Acid Metabolism.

Fatty acids are synthesized in cytoplasm from acetyl-CoA, which is formed within the mitochondrion. Because the inner membrane is impermeable to acetyl-CoA, it is transferred out as citrate. Citrate is produced from acetyl-CoA and oxaloacetate in the citric acid cycle, a reaction pathway in the mitochondrial matrix. Citrate is transferred to cytoplasm when β-oxidation is suppressed, that is, when the cell needs little energy. Later it is cleaved to form oxaloacetate and acetyl-CoA. When the cell needs more energy, fatty acids are transported into the mitochondrion as acylcarnitine derivatives. Then acyl-CoA is degraded to acetyl-CoA via β-oxidation. (The further oxidation of acetyl-CoA to generate ATP is described in Chapter 9.) Note that the hormones glucagon and epinephrine and the substrates citrate, malonyl-CoA, and palmitoyl-CoA are important regulators of fatty acid metabolism. Fatty acid metabolism and carbohydrate metabolism are interrelated. Pyruvate, the precursor of acetyl-CoA, is a product of glycolysis. A portion of the NADPH, the reducing agent required in fatty acid synthesis, is generated by several reactions in the pentose phosphate pathway. NADPH is also produced by converting malate, formed by the reduction of oxaloacetate, to pyruvate.

rine (released when the body's energy reserves are low or when there is an increased energy requirement) stimulates the phosphorylation of several enzymes. When the adipocyte enzyme hormone-sensitive lipase is phosphorylated, it catalyzes the hydrolysis of triacylglycerol. (The release of norepinephrine from neurons in the sympathetic nervous system and growth hormone from the pituitary also activates hormone-sensitive lipase.) Subsequently, fatty acids are released into the blood. Hormones also regulate the use of fatty acids within tissues. For example, acetyl-CoA carboxylase is inhibited by glucagon. As cellular malonyl-CoA concentration falls, and fatty acid synthesis decreases. Because malonyl-CoA inhibits carnitine acyltransferase I activity, fatty acid groups may now be transported into the mitochondria, where they are degraded to generate energy. The effect of insulin on fatty acid metabolism is opposite to that of glucagon and epinephrine. The secretion of insulin in response to high blood glucose levels promotes lipogenesis. Insulin promotes fatty acid synthesis by stimulating the phosphorylation of acetyl-CoA carboxylase (by a process that is independent of the cAMP–protein kinase mechanism). Simultaneous lipolysis is prevented by insulin's inhibition of the cAMP-mediated activation of protein kinase. This latter process leads to dephosphorylation (and therefore inactivation) of hormone-sensitive lipase.

QUESTION 12.10   Identify each of the following biomolecules:

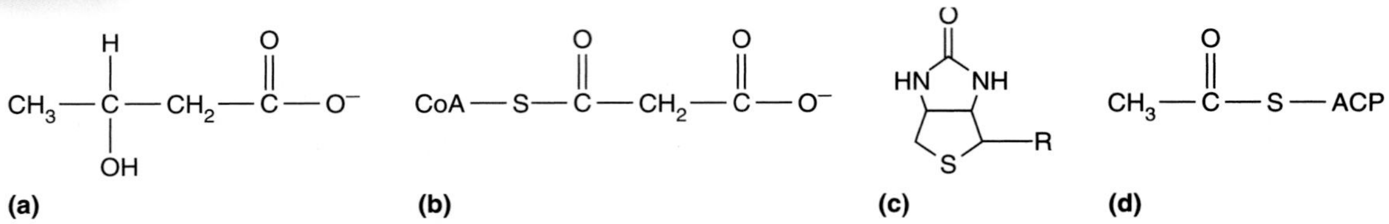

(a)   (b)   (c)   (d)

What is the function of each?

12.2 MEMBRANE LIPID METABOLISM

The lipid bilayer of cell membranes is composed primarily of phospholipids and sphingolipids. After a discussion of the metabolism of these lipid classes, several aspects of membrane biogenesis are briefly described.

Phospholipid Metabolism

KEY CONCEPTS 12.4

Phospholipid synthesis occurs in the membrane of the SER. After phospholipids are synthesized, they are remodeled by altering their fatty acid composition. Phospholipid degradation is catalyzed by several phospholipases.

Most of the reactions involved in lipid biosynthesis appear to be located in the smooth endoplasmic reticulum (SER), although several enzyme activities have also been detected in the Golgi complex. Because each enzyme is a membrane-associated protein with its active site facing the cytoplasm, the biosynthesis of phospholipid occurs at the interface of ER membrane and cytoplasm. The fatty acid composition of phospholipids changes somewhat after their synthesis. (Typically, unsaturated fatty acids replace the original saturated fatty acids incorporated during synthesis.) Most of this remodeling is accomplished by certain phospholipases and acyl transferases. Presumably, this process allows a cell to adjust the fluidity of its membranes.

The syntheses of phosphatidylethanolamine and phosphatidylcholine are similar (Figure 12.17). Phosphatidylethanolamine synthesis begins in the cytoplasm

FIGURE 12.17

Phospholipid Synthesis.

After ethanolamine or choline has entered a cell, it is phosphorylated and converted to a CDP derivative. Then phosphatidylethanolamine or phosphatidylcholine is formed when diacylglycerol reacts with the CDP derivative. Triacylglycerol is produced if diacylglycerol reacts with acyl-CoA. CDP-Diacylglycerol, formed from phosphatidic acid and CTP, is a precursor of several phospholipids, for example, phosphatidylglycerol and phosphatidylinositol.

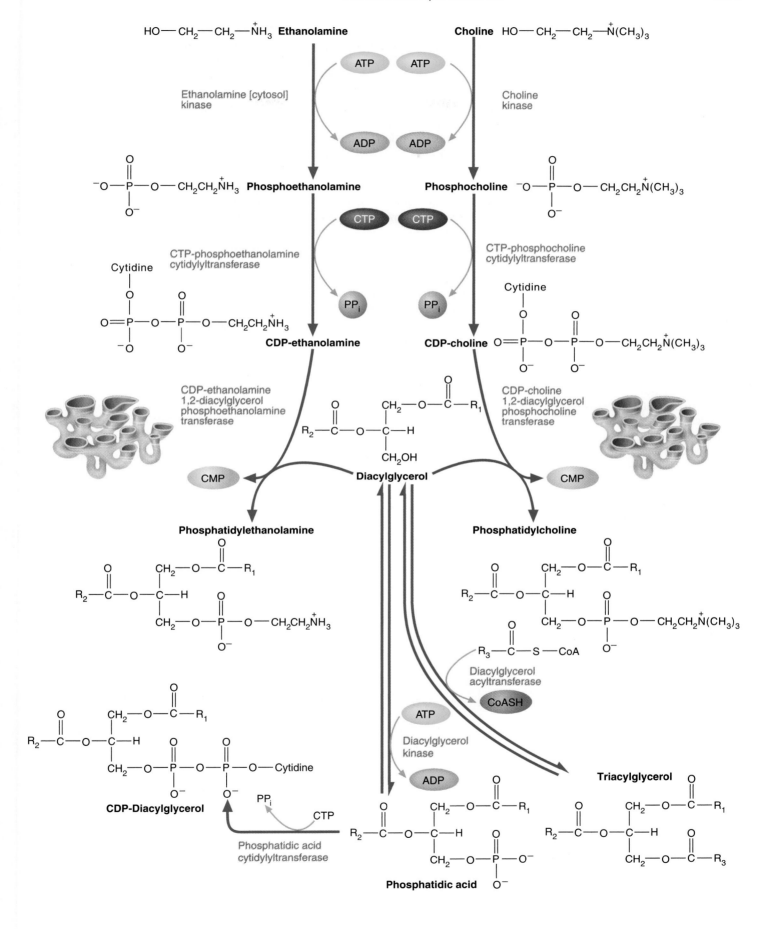

when ethanolamine enters the cell and is then immediately phosphorylated. Subsequently, phosphoethanolamine reacts with CTP (cytidine triphosphate) to form the activated intermediate CDP-ethanolamine. Several nucleotides serve as high-energy carriers of specific molecules. CDP derivatives have an important role in the transfer of polar head groups in phosphoglyceride synthesis. (Recall that UDP plays a similar role in glycogen synthesis. See p. 264.) CDP-ethanolamine is converted to phosphatidylethanolamine when it reacts with diacylglycerol (DAG). This reaction is catalyzed by an enzyme on the endoplasmic reticulum. As noted, the biosynthesis of phosphatidylcholine is similar to that of phosphatidylethanolamine. The choline required in this pathway is obtained in the diet. However, phosphatidylcholine is also synthesized in the liver from phosphatidylethanolamine (Figure 12.18). Phosphatidylethanolamine is methylated in three steps by the enzyme phosphatidylethanolamine-N-methyltransferase to form the trimethylated product phosphatidylcholine. *S-Adenosylmethionine* (SAM) is the methyl donor in this set of reactions. (The role of SAM in cellular methylation processes is discussed in Chapter 14.)

Phosphatidylserine is generated in a reaction in which the ethanolamine residue of phosphatidylethanolamine is exchanged for serine (Figure 12.19). This reaction, which is catalyzed by an ER enzyme, is reversible. In mitochondria, phosphatidylserine is converted to phosphatidylethanolamine in a decarboxylation reaction.

FIGURE 12.18

Conversion of Phosphatidylethanolamine to Phosphatidylcholine.

Methylation reactions that use 5-adenosylmethionine (SAM) are discussed in Chapter 14. (SAH is an abbreviation of S-adenosylhomocysteine.)

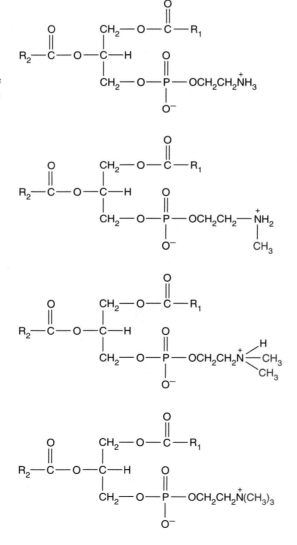

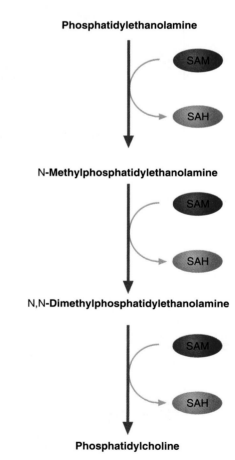

Phosphatidylethanolamine

N-Methylphosphatidylethanolamine

N,N-Dimethylphosphatidylethanolamine

Phosphatidylcholine

Because membranes are dynamic structures, the mechanism by which they are synthesized is complex. Currently, little is known about the synthesis of the membrane bilayer except for the following features: phospholipid translocation across membranes and the intracellular transfer of phospholipids between membranes.

If recently synthesized phospholipid molecules remained only on the cytoplasmic surface of the ER, a monolayer would form. Unassisted bilayer transfer of phospholipid, however, is extremely slow. (For example, half-lives of 8 days have been measured across artificial membrane.) A process known as *phospholipid translocation* is now believed to be responsible for maintaining the bilayer in membranes (Figure 12F). Transmembrane movement of phospholipid molecules (or flip-flop), which may occur in as little as 15 seconds, appears to be mediated by phospholipid translocator proteins. One protein (sometimes referred to as *flippase*) that transfers choline-containing phospholipids across the ER membrane has been identified. Because the hydrophilic polar head group of a phospholipid molecule is probably responsible for the low rate of spontaneous translocation, an interaction between flippase and polar head groups is believed to be involved in phosphatidylcholine transfer. Translocation results in a higher concentration of phosphatidylcholine on the lumenal side of the ER membrane than that

of other phospholipids. This process is therefore partially responsible for the membrane asymmetry discussed in Chapter 11.

Two mechanisms have been proposed to explain the transport of phospholipids from the ER to other cellular membranes: protein-mediated transfer and a vesicular process. Several experiments have demonstrated that water-soluble proteins, known as *phospholipid exchange proteins*, can bind to specific phospholipid molecules and transfer them to another bilayer. Vesicular transport of phospholipids and membrane proteins in structures known as *transition vesicles* from the ER to the Golgi complex is not clearly understood. However, evidence of transfer of luminal material from the ER to the Golgi cisternae clearly supports vesicular transport.

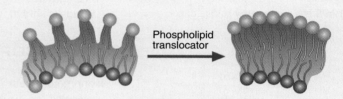

FIGURE 12F **Translocation of Newly Synthesized Phospholipid.** Transfer of selected phospholipid molecules allows balanced growth of the bilayer.

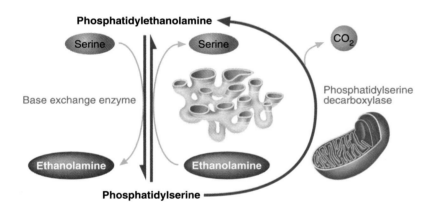

FIGURE 12.19

Phosphatidylserine Synthesis.

Phosphatidylserine can be synthesized from phosphatidylethanolamine in a reaction in which the polar head groups are exchanged. Phosphatidylethanolamine can also be synthesized from phosphatidylserine in a decarboxylation reaction. This reaction is an important source of ethanolamine in many eukaryotes.

Phospholipid turnover is rapid. (**Turnover** is the rate at which all molecules in a structure are degraded and replaced with newly synthesized molecules.) For example, in animal cells, approximately two cell divisions are required for the replacement of one-half of the total number of phospholipid molecules. Phosphoglycerides are degraded by the phospholipases. Each phospholipase, which catalyzes the cleavage of a specific bond in phosphoglyceride molecules, is named according to the bond cleaved. Phospholipases A_1 and A_2, which hydrolyze the ester bonds of phosphoglycerides at C-1 and C-2, respectively, contribute to the phospholipid remodeling described.

Sphingolipid Metabolism

Recall that sphingolipids in animals possess ceramide, a derivative of the amino alcohol sphingosine. The synthesis of ceramide begins with the condensation of palmitoyl-CoA with serine to form 3-ketosphinganine. This reaction is catalyzed by 3-ketosphinganine synthase, a pyridoxal-5′- phosphate–dependent enzyme. (Because

KEY CONCEPTS 12.5

The synthesis of all sphingolipids begins with the production of ceramide. Sphingolipids are degraded within lysosomes by specific hydrolytic enzymes.

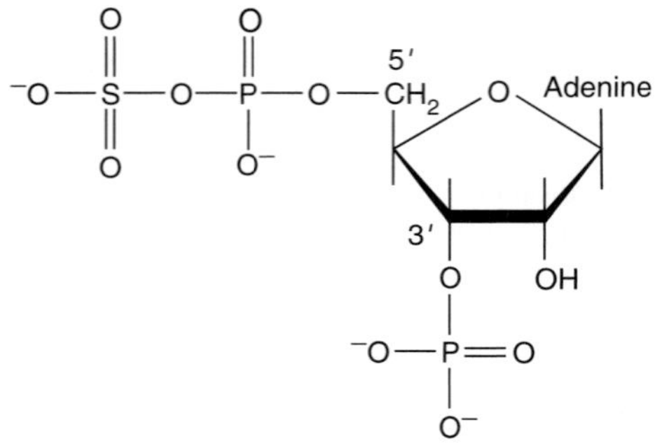

FIGURE 12.20

3′-Phosphoadenosine-5′-Phosphosulfate (PAPS).

PAPS is a high-energy sulfate donor.

pyridoxal-5′-phosphate plays an important role in amino acid metabolism, the biochemical function of this coenzyme is discussed in Chapter 14.) 3-Ketosphinganine is subsequently reduced by NADPH to form sphinganine. In a two-step process involving acyl-CoA and $FADH_2$, sphinganine is converted to ceramide. Sphingomyelin is formed when ceramide reacts with phosphatidylcholine. (In an alternative reaction, CDP-choline is used in place of phosphatidylcholine.) When ceramide reacts with UDP-glucose, glucosylceramide (a common cerebroside, sometimes referred to as glucosylcerebroside) is produced. Galactocerebroside, a precursor of other glycolipids, is synthesized when ceramide reacts with UDP-galactose. The sulfatides are synthesized when the galactocerebrosides react with the sulfate donor molecule **3′-phosphoadenosine-5′-phosphosulfate** (PAPS) (Figure 12.20). The transfer of sulfate groups is catalyzed by the microsomal enzyme sulfotransferase. Sphingolipids are degraded within lysosomes. Recall that specific diseases, called the sphingolipidoses (p. 344), result when enzymes required for degrading these molecules are missing or defective. The synthesis of sphingomyelin and glycosphingolipids is shown in Figure 12.21).

12.3 ISOPRENOID METABOLISM

Isoprenoids occur in all eukaryotes. Despite the astonishing diversity of isoprenoid molecules, the mechanisms by which different species synthesize them are similar. In fact, the initial phase of isoprenoid synthesis (the synthesis of isopentenyl pyrophosphate) appears to be identical in all of the species in which this process has been investigated. Figure 12.22 illustrates the relationships among the isoprenoid classes.

Because of its importance in human biology, cholesterol has received enormous attention from researchers. For this reason the metabolism of cholesterol is better understood than that of any other isoprenoid molecule.

Cholesterol Metabolism

The cholesterol that is used throughout the body is derived from two sources: diet and *de novo* synthesis. When the diet provides sufficient cholesterol, the synthesis of this molecule is depressed. In normal individuals cholesterol delivered by LDL suppresses cholesterol synthesis. Cholesterol biosynthesis is stimulated when the diet is low in cholesterol. As described previously, cholesterol is used as a cell membrane component and in the synthesis of important metabolites. An important mechanism for disposing of cholesterol is conversion to bile acids.

CHOLESTEROL SYNTHESIS Although all tissues can make cholesterol (e.g., adrenal glands, ovaries, testes, skin, and intestine), most cholesterol molecules are synthesized in the liver. Cholesterol synthesis can be divided into three phases:

1. formation of HMG-CoA (β-hydroxy-β-methylglutaryl-CoA) from acetyl-CoA,
2. conversion of HMG-CoA to squalene, and
3. conversion of squalene to cholesterol.

The first phase of cholesterol synthesis is a cytoplasmic process (Figure 12.23). (Recall that the initial substrate, acetyl-CoA, is produced in mitochondria from fatty acids or pyruvate. Also observe the similarity of the first phase of cholesterol synthesis to ketone body synthesis. Refer to Figure 12.8.) The condensation of two acetyl-CoA molecules to form β-ketobutyryl-CoA (also referred to as acetoacetyl-CoA) is catalyzed by thiolase.

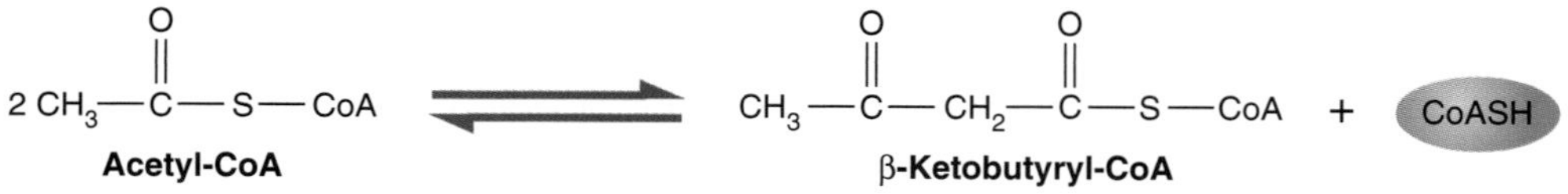

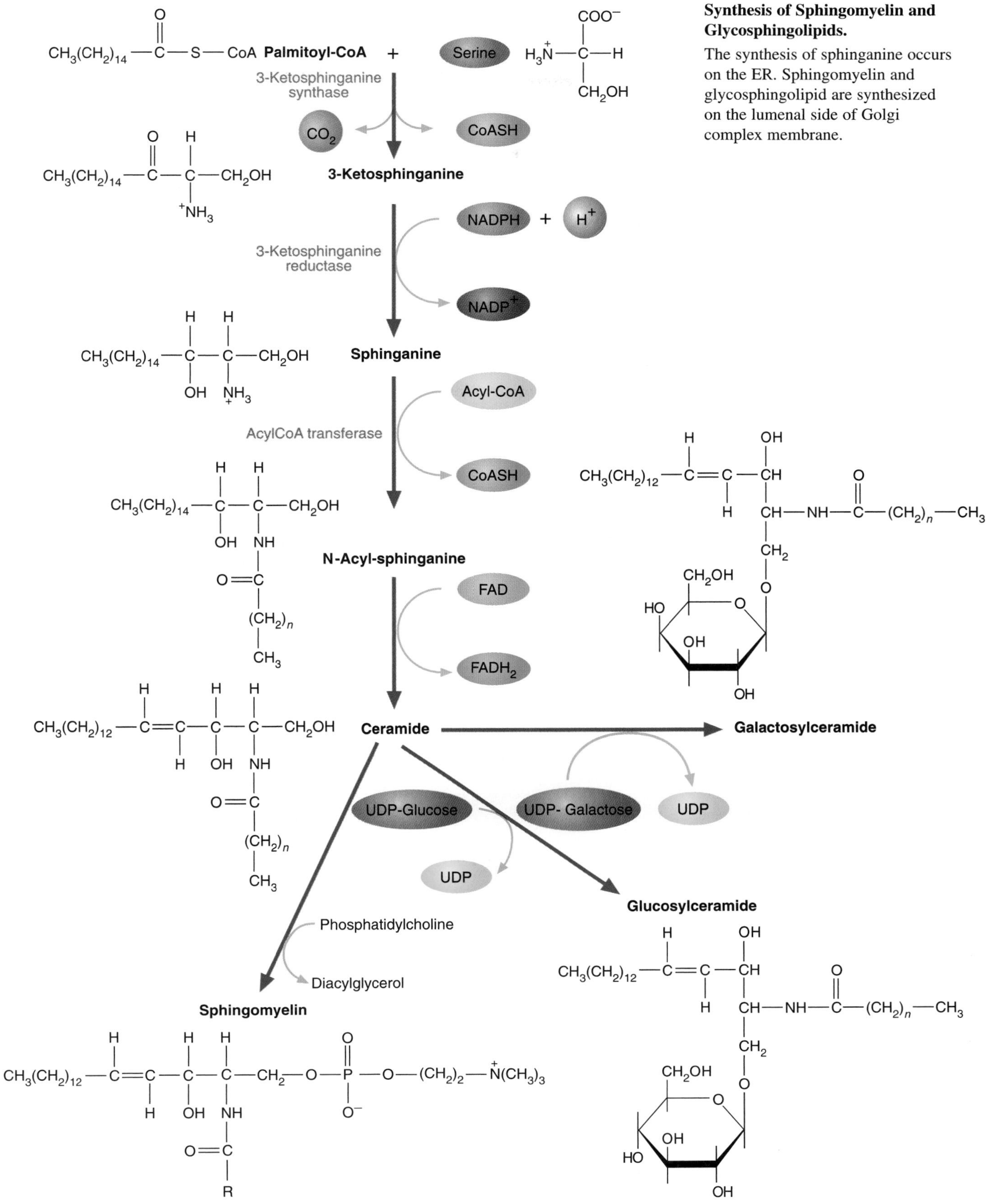

FIGURE 12.21

Synthesis of Sphingomyelin and Glycosphingolipids.

The synthesis of sphinganine occurs on the ER. Sphingomyelin and glycosphingolipid are synthesized on the lumenal side of Golgi complex membrane.

FIGURE 12.22
Isoprenoid Biosynthesis.

Isoprenoid biosynthetic pathways produce an astonishing variety of products in different cell types and different species. Despite this diversity, the beginning of isoprenoid biosynthesis appears to be identical in most of the species investigated (e.g., yeast, mammals, and plants). (HMG-CoA = β-hydroxy-β-methylglutaryl-CoA)

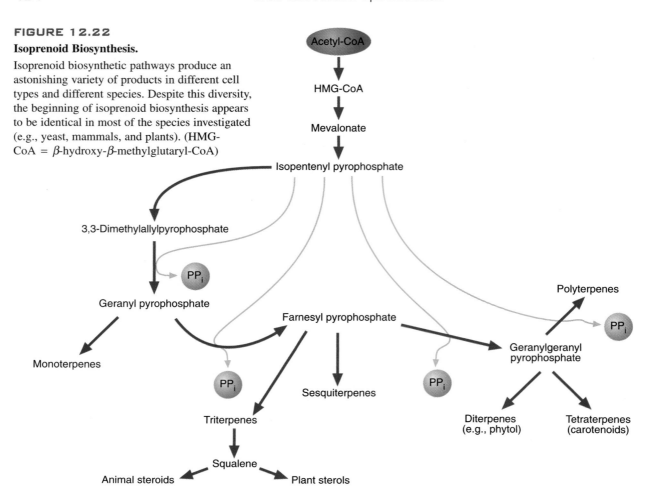

FIGURE 12.23
Cholesterol Synthesis.

Several reactions occur in the cytoplasm, but most enzymes involved in cholesterol biosynthesis occur within ER membrane. The enzymes are indicated by the following numbers: 1 = HMG-CoA reductase, 2 = Squalene synthase, 3 = Squalene monooxygenase, 4 = 2,3-Oxidosqualene lanosterol cyclase, 5 = Enzymes catalyzing 20 separate reactions. Note that squalene and lanosterol are acted upon by ER membrane enzymes while they are bound to carrier proteins in the cytoplasm.

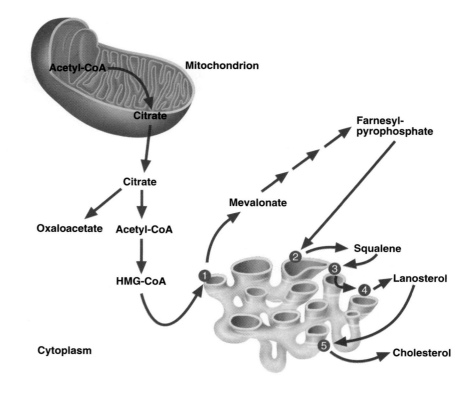

In the next reaction, β-ketobutyryl-CoA condenses with another acetyl-CoA to form β-hydroxy-β-methylglutaryl-CoA (HMG-CoA). This reaction is catalyzed by β-hydroxy-β-methylglutaryl-CoA synthase (HMG-CoA synthase).

β-Ketobutyryl-CoA   **Acetyl-CoA**   **HMG-CoA**

The second phase of cholesterol synthesis begins with the reduction of HMG-CoA to form mevalonate. NADPH is the reducing agent.

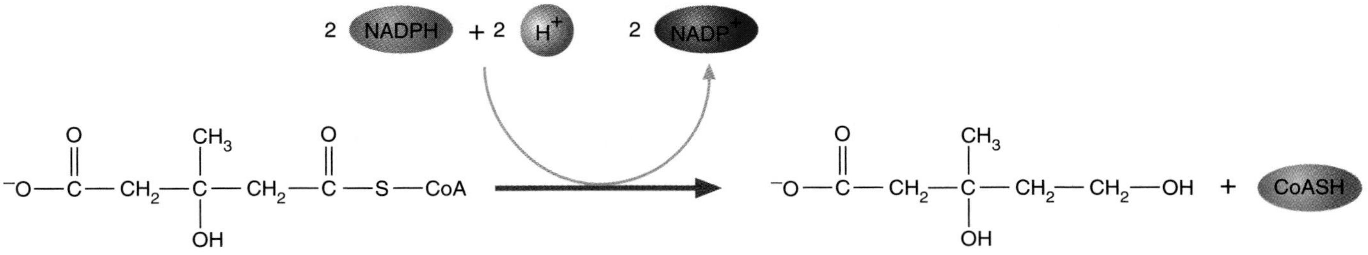

HMG-CoA   **Mevalonate**

HMG-CoA reductase, which catalyzes the latter reaction, is the rate-limiting enzyme in cholesterol synthesis. An accumulation of cholesterol in the cell either from endogenous synthesis or exogenously from LDL intake and degradation reduces the activity of HMG-CoA reductase in two ways: it inhibits HMG-CoA reductase synthesis and enhances degradation of existing enzyme. The activity and cellular concentration of HMG-CoA reductase, located on the cytoplasmic surface of the ER, is affected to varying degrees by the concentrations of intermediate products of the pathway (e.g., mevalonate, farnesol, squalene, and 7-dehydrocholesterol). The precise mechanism by which this strategically important enzyme is regulated, however, remains unresolved.

In a series of cytoplasmic reactions, mevalonate is subsequently converted to farnesylpyrophosphate. Mevalonate kinase catalyzes the synthesis of phosphomevalonate. A second phosphorylation reaction catalyzed by phosphomevalonate kinase produces 5-pyrophosphomevalonate.

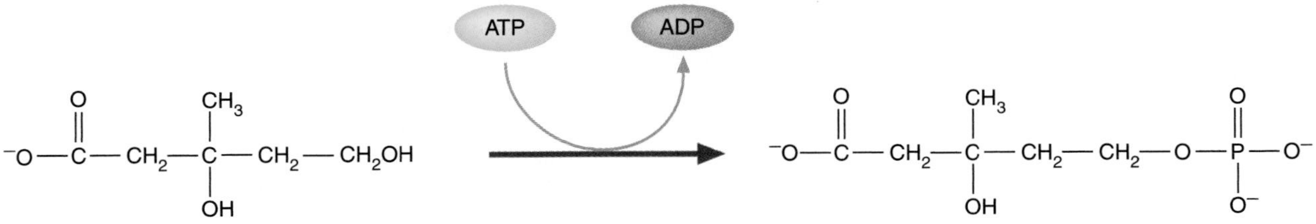

Mevalonate   **Phosphomevalonate**

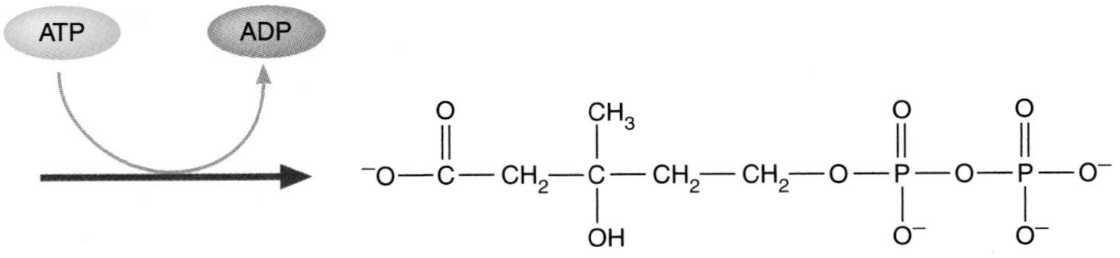

5-Pyrophosphomevalonate

(Phosphorylation reactions significantly increase the solubility of these hydrocarbon molecules in the cytoplasm.) 5-Pyrophosphomevalonate is converted to isopentenyl pyrophosphate in a process involving a decarboxylation and a dehydration:

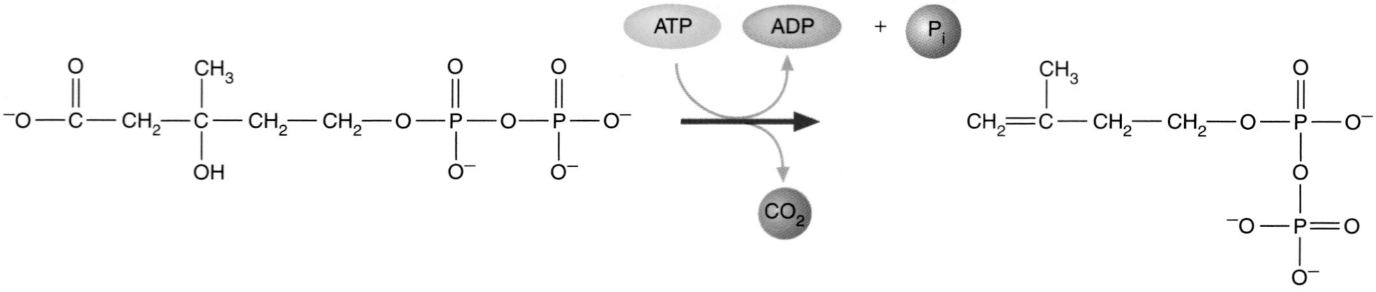

5-Pyrophosphomevalonate   **Isopentenyl pyrophosphate**

Isopentenyl pyrophosphate is next converted to its isomer dimethylallylpyrophosphate by isopentenyl pyrophosphate isomerase. (A $CH_2=CH-CH_2-$ group on an organic molecule is sometimes referred to as an *allyl group*.)

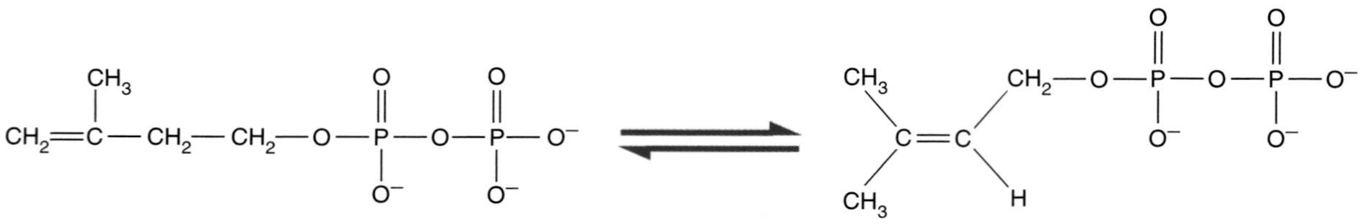

Isopentenylpyrophosphate   **Dimethylallylpyrophosphate**

Geranylpyrophosphate is generated during a condensation reaction between isopentenylpyrophosphate and dimethylallylpyrophosphate (Figure 12.24). Pyrophosphate is also a product of this reaction and two subsequent reactions. (Recall that reactions in which pyrophosphate is released are irreversible because of subsequent pyrophosphate hydrolysis.) Geranyl transferase catalyzes the condensation reaction between geranylpyrophosphate and isopentenylpyrophosphate that forms farnesylpyrophosphate. Squalene is synthesized when farnesyl transferase (a microsomal enzyme) catalyzes the condensation of two farnesylpyrophosphate molecules. (Farnesyl transferase is sometimes referred to as squalene synthase.) This reaction requires NADPH as an electron donor.

The last phase of the cholesterol biosynthetic pathway (Figure 12.25) begins by binding squalene to a specific cytoplasmic protein carrier called **sterol carrier protein**. The conversion of squalene to lanosterol occurs while the intermediates are bound to this protein. The enzyme activities required for the oxygen-dependent epoxide formation (squalene monooxygenase) and subsequent cyclization (2,3-oxidosqualene lanosterol cyclase) that result in lanosterol synthesis have been localized in microsomes. Squalene monooxygenase requires NADPH and FAD for activity. After its synthesis, lanosterol binds to a second carrier protein, to which it remains attached during the remaining reactions. All of the enzyme activities that catalyze the remaining 20 reactions needed to convert lanosterol to cholesterol are embedded in microsomal membranes. In a series of transformations involving NADPH and some oxygen, lanosterol is converted to 7-dehydrocholesterol. This product is then reduced by NADPH to form cholesterol.

Recall that cholesterol is a precursor for all steroid hormones and the bile salts. These syntheses are briefly outlined now. The metabolism of steroids is extremely complex and as yet incompletely understood. A variety of cells and

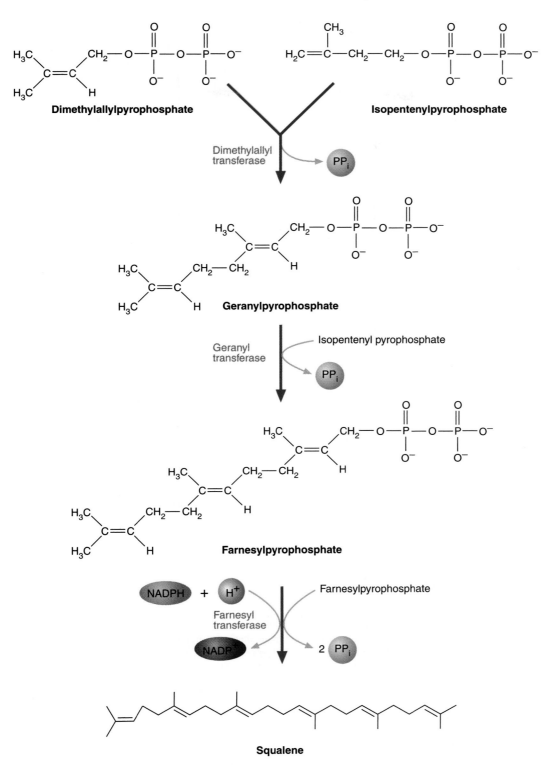

FIGURE 12.24

Synthesis of Squalene from Dimethylallylpyrophosphate and Isopentenylpyrophosphate.

The immediate precursor of squalene is farnesylpyrophosphate, which contains three C_5 isoprenoid groups.

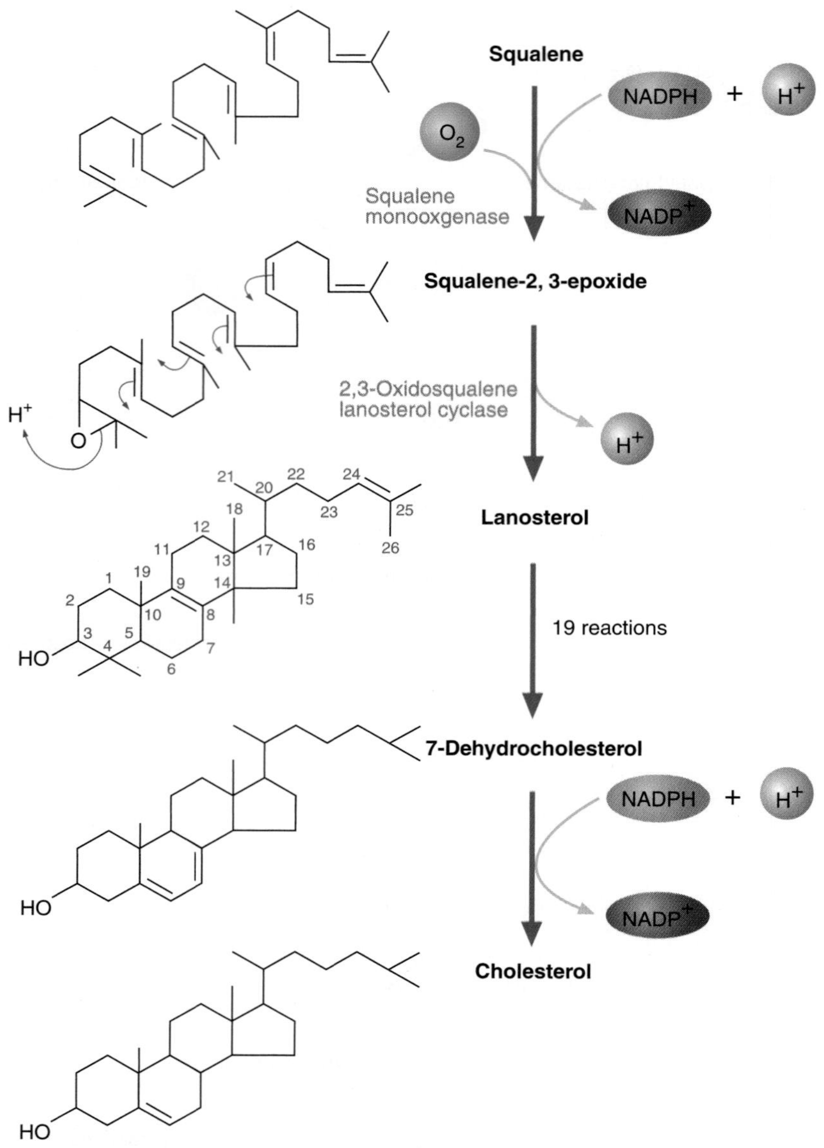

FIGURE 12.25

Synthesis of Cholesterol from Squalene.

This is the major route in mammals. In an alternative minor route, squalene is converted to desmosterol, which is then reduced to form cholesterol. The details of these and many reactions in the major route are poorly understood. (Desmosterol differs from cholesterol because of a C=C double bond between C-24 and C-25.)

organelles figure prominently in the production and processing of these powerful substances. The initial reaction in steroid hormone synthesis, the conversion of cholesterol to pregnenolone (Figure 12.26), is catalyzed by desmolase, a mitochondrial enzyme. Desmolase is an enzyme complex composed of two hydroxylases, one of which is a cytochrome P_{450} enzyme. Cytochrome P_{450} is involved in reactions involving steroid and xenobiotic metabolism (see Special Interest Box 10.1). After its synthesis, pregnenolone is transported to the ER, where it is converted to progesterone. Pregnenolone and progesterone are precursors for all other steroid hormones (Figure 12.27). In addition to its precursor role, progesterone also acts as a hormone. Its primary hormonal role is the

FIGURE 12.26

Progesterone Synthesis.

Pregnenolone is synthesized in the mitochondria. It is transported to the ER, where it is converted to progesterone. This latter process oxidizes the hydroxyl group and isomerizes a C=C double bond.

regulation of several physiological changes in the uterus. During the menstrual cycle, progesterone is produced by specialized cells within the ovary. During pregnancy the progesterone that is produced in large quantities by the placenta prevents uterine smooth muscle contractions.

The amounts and types of steroids synthesized in a specific tissue are carefully regulated. Cells in each tissue are programmed during embryonic and fetal development to respond to chemical signals by inducing the synthesis of a unique set of specific enzymes. The most important chemical signals that are now believed to influence steroid metabolism are peptide hormones secreted from the pituitary (a hormone-producing structure found in the brain) and several prostaglandins. (See Chapter 16 for a discussion of hormones and hormone action.) For example, *adrenocorticotropic hormone* (ACTH) is a peptide hormone secreted by the pituitary gland that stimulates adrenal steroid synthesis. One of the consequences of ACTH binding to adrenal cell receptors is an increased synthesis of 17-α-hydroxylase and 11-β-hydroxylase. In contrast, prostaglandin $F_{2\alpha}$ has been observed to inhibit the induction of progesterone synthesis in the ovaries. This latter process is stimulated by luteinizing hormone (LH), a protein produced by the pituitary. (The functional effect of prostaglandin $F_{2\alpha}$ on progesterone synthesis is still unclear.)

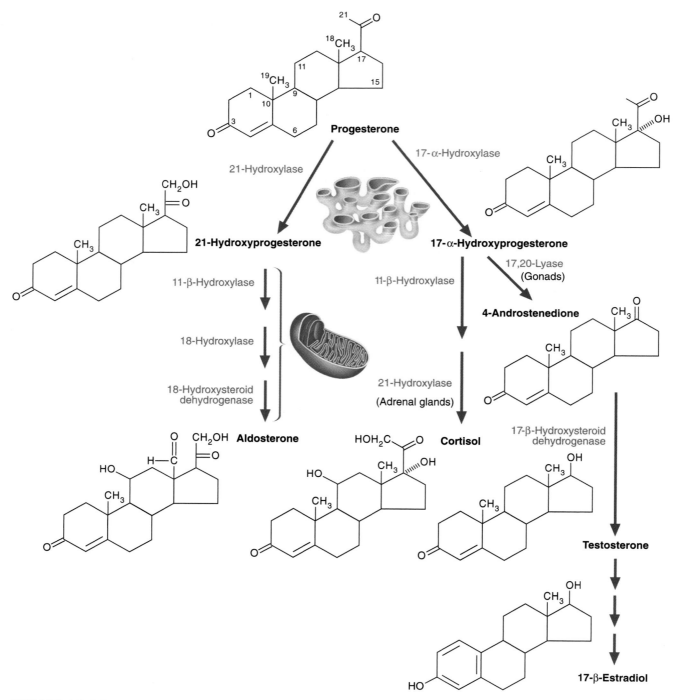

Synthesis of Selected Steroids.

The enzyme 17-α-hydroxylase is found in all steroid-producing cells. Most other enzymes are tissue-specific.

The enzymatic processes by which cholesterol is converted to biologically active steroids, as well as the means by which steroids are inactivated and prepared for excretion, comprise an elaborate mechanism referred to as biotransformation (Special Interest Box 10.1). During biotransformation the same (or in some cases similar) enzymes are also used to solubilize hydrophobic xenobiotics so that they can be more easily excreted. The biotransformation of cholesterol into bile acids is outlined next.

QUESTION 12.11

During an experiment cholesterol is synthesized using

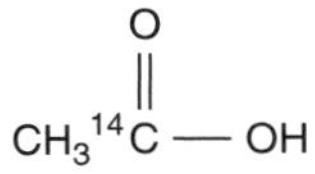

$$CH_3\,^{14}\!C \overset{\displaystyle O}{\overset{\displaystyle \|}{\underset{}{}}} \!\!-\!OH$$

as the substrate. Which atoms in the cholesterol that is recovered will be labeled?

QUESTION 12.12

Cortisol (also referred to as *hydrocortisone*) is a potent glucocorticoid. (**Glucocorticoids** are hormones that promote carbohydrate, protein, and fat metabolism. For example, glucocorticoids stimulate gluconeogenesis, lipolysis, and an increased uptake of amino acids by the liver.) Cortisol also possesses a small amount of mineralocorticoid activity. (**Mineralocorticoids** regulate Na^+ and K^+ metabolism. For example, aldosterone, the most important mineralocorticoid in humans, induces the reabsorption of Na^+ from urine. It also promotes the secretion of K^+ and H^+ into urine.) For steroids to have either glucocorticoid or mineralocorticoid activity, they must possess a hydroxy group at C-11.

In *Addison's disease* an inadequate secretion of glucocorticoids and mineralocorticoids results in hypoglycemia, an imbalance in the body's Na^+ and K^+ concentrations, dehydration, and low blood pressure. In the past, individuals who had undiagnosed Addison's disease found that consumption of large amounts of licorice provided some relief from their symptoms. (Licorice, an extract of the plant *Glycyrrhiza glabra*, is used as a flavoring agent in candy and some medicines.) In these patients, licorice consumption resulted in sodium retention, hypokalemia (low blood K^+ concentration), and a rise in blood pressure. This effect was more pronounced when cortisol was administered.

Recently, it has been discovered that the active ingredient in licorice, called glycyrrhizic acid, inhibits 11-β-hydroxysteroid dehydrogenase, the enzyme that reversibly converts cortisol to cortisone, its inactive metabolite. Refer to the structure of cortisol (Figure 12.27) and deduce the structure of cortisone. Because cortisol is regarded as a glucocorticoid, why does glycyrrhizic acid consumption appear to affect mineral metabolism? Cortisone is often used to treat Addison's disease although it is physiologically inactive. Suggest a reason why its use is justified.

CHOLESTEROL DEGRADATION Unlike many other types of biomolecules, cholesterol and other steroids cannot be degraded to smaller molecules. Instead, they are converted to derivatives whose improved solubility properties allow their excretion. The most important mechanism for degrading and eliminating cholesterol is the synthesis of the bile acids. Bile acid synthesis, which occurs in the liver, is outlined in Figure 12.28. The conversion of cholesterol to 7-α-hydrocholesterol, catalyzed by cholesterol-7-hydroxylase (a microsomal enzyme), is the rate-limiting reaction in bile acid synthesis. In later reactions, the double bond at C-5 is rearranged and reduced, and an additional hydroxyl group is introduced. The products of this process, cholic acid and deoxycholic acid, are converted to bile salts by microsomal enzymes that catalyze conjugation reactions. (In **conjugation reactions** a molecule's solubility is increased by converting it into a derivative that contains a water-soluble group. Amides and esters are

KEY CONCEPTS 12.6

Cholesterol is synthesized from acetyl-CoA in a multistep pathway that occurs primarily in the liver. Small amounts of cholesterol are used to synthesize biologically powerful steroid hormones. Cholesterol is degraded primarily by conversion to the bile salts, which facilitate the emulsification and absorption of dietary fat.

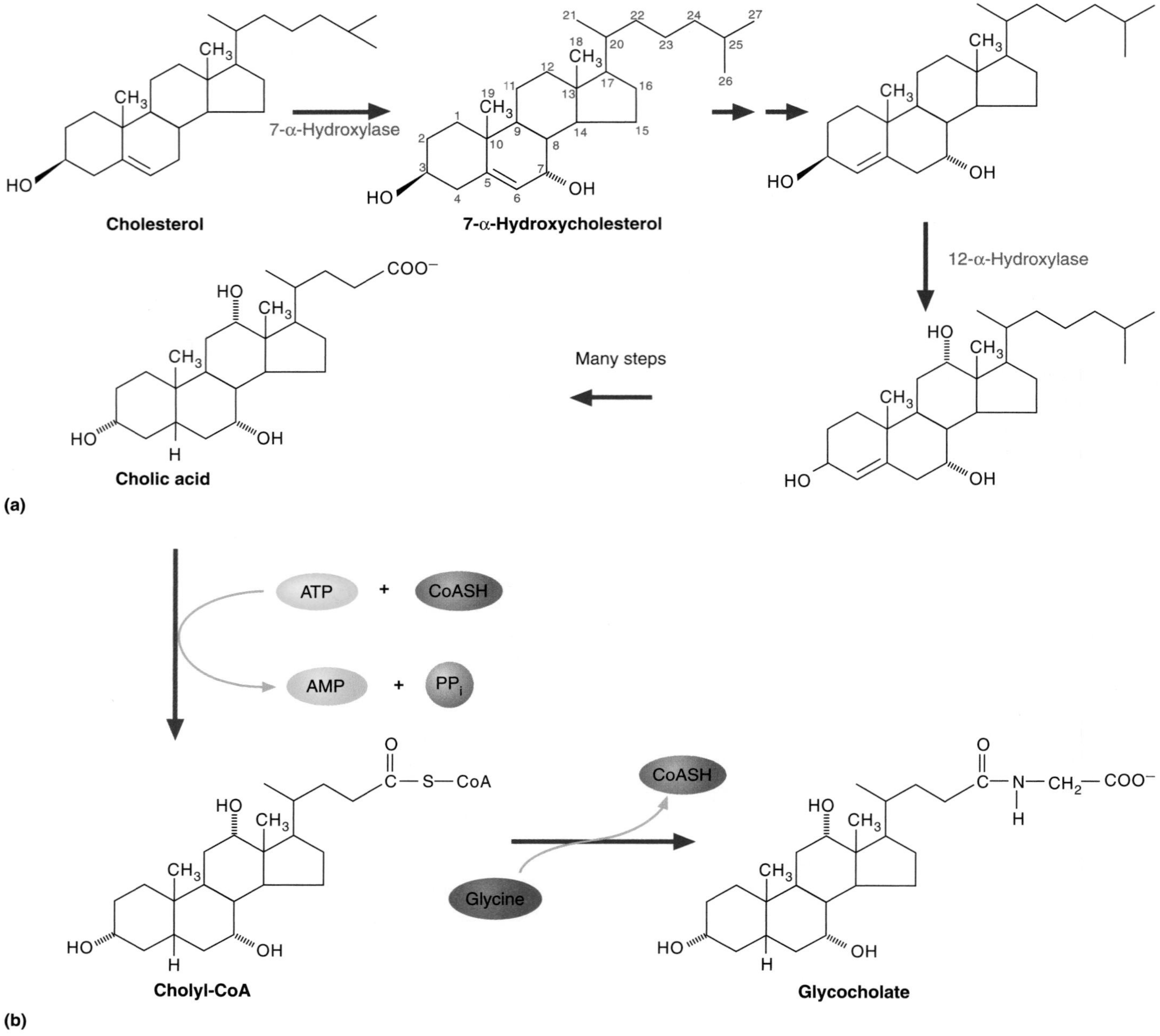

FIGURE 12.28

Synthesis of the Bile Acid Cholic Acid (a) and the Bile Salt Glycocholate (b).

Because of their higher solubility, bile acids primarily act as detergents during dietary fat digestion.

common examples of these conjugated derivatives.) Most bile acids are conjugated with glycine or taurine (Figure 12.29).

The bile salts are important components of *bile*, a yellowish green liquid produced by hepatocytes that aids in the digestion of lipids. In addition to bile salts, bile contains cholesterol, phospholipids, and bile pigments (bilirubin and biliverdin). The bile pigments are degradation products of heme. After it is secreted into the bile ducts and stored in the gallbladder, bile is used in the small intestine to enhance the absorption of dietary fat. Bile acts as an emulsifying agent, that is, it promotes the breakup of large fat droplets into smaller ones. Bile salts are also involved in the formation of so-called biliary micelles, which

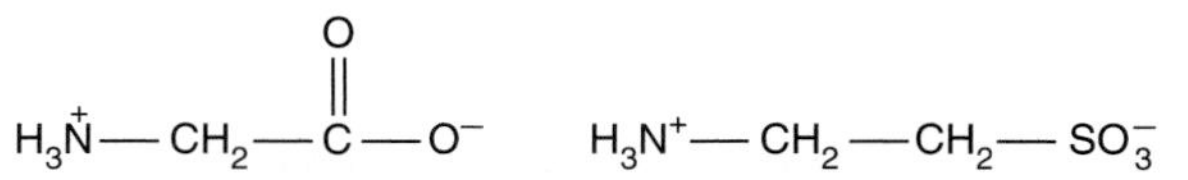

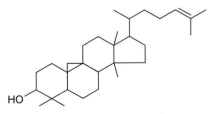

FIGURE 12.29

Structure of Glycine and Taurine.

Most bile acids are conjugated with glycine or taurine in the liver.

aid in absorbing fat and the fat-soluble vitamins (A, D, E, and K). Most bile salts are reabsorbed in the distal ilium (near the end of the small intestine). They enter the blood and are transported back to the liver, where they are resecreted into the bile ducts with other bile components. The biological significance of bile acid conjugation reactions appears to be that the conjugation process prevents premature absorption of bile acids in the biliary tract (the duct system and gallbladder) and small intestine. The reabsorption of bile salts in the distal ilium of the small intestine (necessary for effective recycling) is apparently triggered by the glycine or taurine signal. (It has been estimated that bile salt molecules are recycled 18 times before they are finally eliminated.)

QUESTION 12.13

The formation of gallstones (crystals that are usually composed of cholesterol and inorganic salts) within the gallbladder or bile ducts afflicts millions of people. Predisposing factors for this excruciatingly painful malady include obesity and infection of the gallbladder (*cholecystitis*). Because cholesterol is virtually insoluble in water, it is solubilized in bile by its incorporation into micelles composed of bile salts and phospholipids. Gallstones tend to form when cholesterol is secreted into bile in excessive quantities. Suggest a reason why an obese person is prone to gallstone formation. (*Hint*: HMG-CoA reductase activity is higher in obese individuals.)

Sterol Metabolism in Plants

Relatively little is known about plant sterols. (Most of the research effort in steroid metabolism has been expended in the investigation of steroid-related human diseases.) It appears, however, that the initial phase of plant sterol synthesis is very similar to that of cholesterol synthesis with the following exception. In plants and algae the cyclization of squalene-2,3-epoxide leads to the synthesis of *cycloartenol* (Figure 12.30) instead of lanosterol. Many subsequent reactions in plant sterol pathways involve SAM-mediated methylation reactions. There appear to be two separate isoprenoid biosynthetic pathways in plant cells: the ER/cytoplasm pathway and a separate chloroplast pathway. The roles of these pathways in plant isoprenoid metabolism are still unclear.

Cycloartenol

FIGURE 12.30

Structure of Cycloartenol.

Cycloartenol is an intermediate formed during the synthesis of plant sterols.

SUMMARY

1. Acetyl-CoA plays a central role in most lipid-related metabolic processes. For example, acetyl-CoA is used in the synthesis of fatty acids. When fatty acids are degraded to generate energy, acetyl-CoA is the product.

2. Depending on the body's current energy requirements, newly digested fat molecules are used to generate energy or are stored within adipocytes. When the body's energy reserves are low, fat stores are mobilized in a process referred to as lipolysis. In lipolysis, triacylglycerols are hydrolyzed to fatty acids and glycerol. Glycerol is transported to the liver, where it can be used in lipid or glucose synthesis. Most fatty acids are degraded to form acetyl-CoA within mitochondria in a process referred to as β-oxidation. Peroxisomal β-oxidation appears to shorten very long fatty acids. Other reactions degrade odd-chain fatty acids and unsaturated fatty acids. When the product of fatty acid degradation (acetyl-CoA) is present in excess, ketone bodies are produced.

3. The first step in eicosanoid synthesis is the release of arachidonic acid from C-2 of glycerol in membrane phosphoglyceride molecules. Cyclooxygenase converts arachidonic acid into PGG_2, which is a precursor of the prostaglandins and the thromboxanes. The lipoxygenases convert arachidonic acid to the precursors of the leukotrienes.

4. Fatty acid synthesis begins with the carboxylation of acetyl-CoA to form malonyl-CoA. The remaining reactions of fatty acid synthesis take place on the fatty acid synthase multienzyme complex. Several enzymes are available to elongate and desaturate dietary and newly synthesized fatty acids.

5. After phospholipids are synthesized at the interface of the SER and the cytoplasm, they are often "remodeled," that is, their fatty acid composition is adjusted. The turnover (i.e., the degradation and replacement) of phospholipids, mediated by the phospholipases, is rapid.

6. Synthesis of the ceramide component of sphingolipids begins with the condensation of palmitoyl-CoA with serine to form 3-ketosphinganine. In a two-step process involving acyl-CoA and $FADH_2$, sphinganine (formed when 3-ketosphinganine is reduced by NADPH) is converted to ceramide. Sphingolipids are degraded within lysosomes.

7. Phospholipid translocator proteins, phospholipid exchange proteins, and transition vesicles are involved in the complicated process of membrane synthesis and delivery of membrane components to their cellular destinations.

8. Cholesterol synthesis can be divided into three phases: formation of HMG-CoA from acetyl-CoA, conversion of HMG-CoA to squalene, and conversion of squalene to cholesterol. Cholesterol is the precursor for all steroid hormones and the bile salts. Bile salts are used to emulsify dietary fat. They are the primary means by which the body can rid itself of cholesterol.

SUGGESTED READINGS

Drayr, J.-P., and Vamecq, J., The Gluconeogenicity of Fatty Acids in Mammals, *Trends Biochem. Sci.*, 14:478–479, 1989.

Goldstein, J. L., and Brown, M. S., Regulation of the Mevalonate Pathway, *Nature*, 343:425–430, 1990.

Gurr, M. I., Frayn, K.N., and Harwood, J. L., *Lipid Biochemistry: An Introduction*, 5th ed., Blackwell Scientific Inc., Oxford, 2001.

Hampton, R., Dimster-Denk, D., and Rine, J., The Biology of HMG-CoA Reductase: The Pros of Contra-regulation, *Trends. Biochem. Sci.*, 21:140–145, 1996.

Hashimoto, T., Peroxisomal beta-Oxidation Enzymes, *Cell Biochem. Biophys.*, 32:63–72, 2000.

Johnson, M., Carey, F., and McMillan, R. M., Alternative Pathways of Arachidonate Metabolism: Prostaglandins, Thromboxanes and Leukotrienes, *Essays Biochem.*, 19:40–141, 1983.

Vance, J. E., Eukaryotic Lipid-Biosynthetic Enzymes: The Same but Not the Same, *Trends Biochem. Sci.*, 23(11):423–428, 1998.

Weissman, G., Aspirin, *Sci. Amer.*, 264:84–90, 1991.

KEY WORDS

acyl carrier protein, *387*	epoxides, *389*	ketosis, *383*	3'-phosphoadenosine-5'-phosphosulfate, *401*
autoimmune disease, *392*	fatty acid–binding protein, *378*	lipogenesis, *375*	
bile salts, *374*	glucocorticoid, *411*	lipolysis, *377*	sterol carrier protein, *406*
Calvin cycle, *396*	ketogenesis, *383*	mineralocorticoid, *411*	thiolytic cleavage, *381*
conjugation reaction, *411*	ketone body, *383*	β-oxidation, *378*	turnover, *400*

REVIEW QUESTIONS

1. Define the following terms:
   a. *de novo*
   b. oil bodies
   c. β-oxidation
   d. turnover
   e. thiolytic cleavage
   f. autoantibodies
   g. ketone bodies
   h. biotransformation
   i. conjugation reaction

2. What is the function of each of the following substances?
   a. carnitine

b. flippase
c. thrombin
d. thiolase
e. desmolase
f. phospholipid exchange protein
g. sterol carrier protein
h. ACTH
i. glucocorticoid

3. What are the differences between β-oxidation in mitochondria and in peroxisomes? What similarities are there between these processes?

4. List three differences between fatty acid synthesis and β-oxidation.

5. How does fatty acid synthesis in plants differ from fatty acid synthesis in animals?

6. Explain how hormones act to modify the metabolism of fatty acids in both the short and long term. Give examples.

7. What is the difference between a steroid and a sterol?

8. Show how the following fatty acid is oxidized:

$$CH_3CH_2CH_2\overset{\underset{\displaystyle |}{CH_3}}{CH}-CH_2-\overset{\underset{\displaystyle}{\overset{\displaystyle O}{\|}}}{C}-OH$$

Indicate at which points α-and β-oxidation are carried out.

9. Insulin is released after carbohydrate intake. Describe two ways insulin acts to influence fatty acid metabolism.

10. β-Oxidation of naturally occurring monounsaturated fatty acids requires an additional enzyme. What is this enzyme and how does it accomplish its task?

11. Identify the hydrophobic and hydrophilic regions in the following molecule. How do you think it orients itself in a membrane?

$$CH_3-(CH_2)_{15}CH_2-\overset{\overset{\displaystyle O}{\|}}{C}-O-\overset{\underset{\displaystyle |}{\displaystyle}}{\underset{\displaystyle CH_2}{\overset{\displaystyle CH_2-O-\overset{\overset{\displaystyle O}{\|}}{C}-CH_2(CH_2)_{15}-CH_3}{|}}}C-H$$

$$CH_2-O-\overset{\overset{\displaystyle O}{\|}}{\underset{\underset{\displaystyle O^-}{|}}{P}}-O-CH_2CH_2\overset{+}{N}(CH_3)_3$$

12. Gaucher's disease is an inherited deficiency of β-glucocerebrosidase. Glucocerebroside is deposited in macrophages that die, releasing their contents into the tissues. Some affected individuals may have neurologic disorders while quite young; others may not show ill effects until much later in life. The disease may be detected by assaying white blood cells for the ability to hydrolyze the β-glycosidic bond of artificial substrates. Examine the following glucocerebroside and determine which bond is cleaved by glucocerebrosidase.

$$CH_3-(CH_2)_{12}-CH=CH-\overset{\underset{\displaystyle |}{OH}}{CH}-\overset{\underset{\displaystyle |}{NH}}{CH}-CH_2-O$$

13. Determine the number of moles of ATP that can be generated from the fatty acids in 1 mol of tristearin. (Tristearin is a triacylglycerol composed of glycerol esterified to three stearic acid molecules.) What is the fate of the glycerol?

14. Outline the biosynthesis of bile salts. What are the functions of these substances?

15. How are lipid molecules such as animal steroid molecules like estrogen and β-carotene related to each other? What biosynthetic reactions do these specific molecules have in common?

THOUGHT QUESTIONS

1. A class of pharmaceuticals called the statins inhibit the enzyme HMG-CoA reductase. What is the primary effect of this drug on patients?

2. What are the potential consequences of ineffective regulation of the opposing processes of β-oxidation and fatty acid synthesis?

3. Describe the possible effects of low levels of carnitine on the metabolism of an affected individual.

4. Phospholipases show an enhanced activity for a substrate above the critical micelle concentration. (The critical micelle concentration, or cmc, is that concentration of a lipid above which micelles begin to form.)
   a. What type of noncovalent interactions are possible between the lipid and the enzyme at this stage?
   b. What do these interactions suggest about the structure of phospholipases?

5. When the production of acetyl-CoA exceeds the body's capacity to oxidize it, acetoacetic acid, β-hydroxybutyrate, and acetone accumulate. When generated in large amounts, these substances can exceed the blood's buffering capacity. As the blood pH falls, the ability of red blood cells to carry oxygen is affected. Subsequently, the brain can be starved for oxygen, and a fatal coma can result. Explain how severe dieting can produce this condition.

6. The acyl-CoA dehydrogenase deficiency diseases are a group of inherited defects that impair the β-oxidation of fatty acids. Symptoms of the disease range from nausea and vomiting to frequent comas. Symptoms may be alleviated by eating

regularly and avoiding periods of starvation (12 hours or more). Why does this simple procedure alleviate the symptoms?

7. There is an unusually high concentration of phosphatidylcholine on the lumenal side of the ER. What structural feature of phosphatidylcholine is responsible for this? Explain how this structural feature produces this effect.

8. During periods of stress or fasting, blood glucose levels fall. In response, fatty acids are released by adipocytes. Explain how the drop in blood glucose triggers fatty acid release.

9. Butyric acid, a simple four-carbon fatty acid, is oxidized by β-oxidation. Calculate the number of $FADH_2$ and NADH molecules produced in this oxidation. How many acetyl-CoA molecules are also produced?

Photosynthesis

Evolution of Oxygen by an Aquatic Plant Light is the principal regulator of photosynthesis. Light affects the activities of regulatory enzymes in photosynthetic processes by indirect mechanisms, which include changes in pH, Mg^{2+} concentration, the ferredoxin-thioredoxin system, and phytochrome.

Without question, photosynthesis is the most important biochemical process on the Earth. With a few minor exceptions, photosynthesis is the only mechanism by which an external source of energy is harnessed by the living world. As with other energy-yielding processes, photosynthesis involves oxidation-reduction reactions. Water is the source of electrons and protons that reduce CO_2 to form organic compounds. Chapter 13 is devoted to a discussion of the principles of photosynthetic processes. The relationship between photosynthetic reactions and the structure of chloroplasts and the relevant properties of light are emphasized.

As living organisms became abundant on the primitive Earth, their consumption of the organic nutrients produced by geochemical processes outpaced production. The abundant CO_2 in the Earth's early atmosphere was a natural carbon source for organic synthesis. (Much of this CO_2 was of volcanic origin or was generated during the anaerobic degradation of organic nutrients by living organisms.) However, CO_2 is an oxidized, low-energy molecule. For this reason the processes by which CO_2 is incorporated into organic molecules require energy and reducing power. (The formation of carbon-carbon bonds requires free energy now provided by ATP hydrolysis. Reducing power is required because a strong electron donor must provide the high-energy electrons needed to convert CO_2 to a CH_2O unit once the former has been incorporated into an organic molecule.)

The evolution of photosynthetic mechanisms (referred to as **photosystems**) provided both the energy and the reducing power for organic synthesis. The organisms that possessed them had a definite survival advantage, because they no longer depended on an uncertain supply of preformed organic nutrients. These primitive organisms are believed to have been similar to modern green sulfur bacteria. Green sulfur bacteria possess a photosystem that uses light energy to drive a relatively simple electron transport process. As a pigment molecule in the photosystem absorbs light energy, an electron is energized and then donated to the first of several electron acceptors. Eventually, two light-excited electrons are donated to NAD^+, thus forming the reducing agent NADH. (Modern green sulfur bacteria may also use NADPH as a reducing agent in photosynthesis. In more advanced species, NADPH is used exclusively as the reducing agent.) The membrane component that mediates the conversion of light energy into chemical energy is a pigment-protein complex referred to as a **reaction center**. The electrons removed from the reaction center are replaced when oxidized components of the reaction center strip electrons from H_2S, thus generating S. As electrons flow through the photosystem, protons are pumped across the membrane, thus creating an electrochemical gradient. ATP is synthesized as protons move back into the cell through an ATP synthase. This photosystem therefore provides the bacterial cell with the NADH and ATP that incorporate CO_2 into organic molecules. However, H_2S is not normally produced in large quantities and most H_2S molecules are produced in relatively isolated areas. (Recall, for example, the hydrothermal vents described in Special Interest Box 4.1.)

The next critical step in the evolution of life was a photosystem that removed and used electrons from H_2O. Because water is abundant, photosynthetic organisms now penetrated and occupied vast new areas on the planet. Water-based photosynthesis also had a profound impact on other organisms. As photosynthetic organisms proliferated, they provided a new and richer supply of organic molecules for other life forms. Their most significant contribution, however, was the accumulation of gaseous oxygen in the atmosphere. As previously described, some organisms (i.e., those that survived this period) adapted to these changing conditions by evolving mechanisms that protected them against oxygen's toxic effects. Eventually, organisms began to use oxygen to generate energy.

The electrons removed from H_2O have a more positive redox potential than those in H_2S. A larger input of energy is therefore required to drive the transfer of electrons from H_2O to electron acceptors with more negative reduction potentials (see p. 277). Consequently, more elaborate and sophisticated mechanisms are required to convert water's low-energy electrons into the high-energy electrons needed in ATP and NADPH synthesis. In Chapter 13 the principles of this process, that is, photosynthesis in plants and algae, are described. The discussion begins with a detailed view of chloroplast structure. After a brief review of the relevant properties of light, the reactions that constitute modern photosynthesis will be described. These include the light reactions and the light-independent reactions. During the light reactions, electrons are energized and eventually used in the synthesis of both ATP and NADPH. These molecules are then used in the light-independent reactions (often referred to as the dark reactions) to drive the synthesis of carbohydrate. Several photosynthetic variations, referred to as C4 metabolism and crassulacean acid metabolism, are also discussed. Chapter 13 ends with a discussion of several mechanisms that control photosynthesis in plants.

13.1 CHLOROPHYLL AND CHLOROPLASTS

The essential feature of photosynthesis is the absorption of light energy by specialized pigment molecules (Figure 13.1). The **chlorophylls** are green pigment molecules that resemble heme. *Chlorophyll a* plays the principal role in eukaryotic photosynthesis, because its absorption of light energy directly drives photochemical events. *Chlorophyll b* acts as a light-harvesting pigment by absorbing light energy and passing it on to chlorophyll a. The orange-colored **carotenoids** are isoprenoid molecules that either function as light-harvesting pigments (e.g., lutein, a xanthophyll, see p. 345) or protect against reactive oxygen species (ROS) (e.g., β-carotene).

In plants and algae, photosynthesis takes place within specialized organelles called chloroplasts (Chapter 2, p. 52). Chloroplasts resemble mitochondria in several respects. First, both organelles have an outer and an inner membrane with different permeability characteristics (Figure 13.2). The outer membrane of each organelle is highly permeable, whereas the inner membrane possesses specialized carrier molecules that regulate molecular traffic. Second, the chloroplast inner membrane encloses an inner space, referred to as the **stroma**, that resembles the mitochondrial matrix. The stroma possesses a variety of enzymes (e.g., those that catalyze the light-independent reactions and starch synthesis), DNA, and ribosomes. There are also notable differences between the organelles. For example, chloroplasts are substantially larger than mitochondria. Although their shapes and sizes vary, many plant mitochondria are rod-shaped structures approximately 1500 nm long and 500 nm wide. Many chloroplasts are spheroid-shaped with lengths from 4000 to 6000 nm and widths of approximately 2000 nm. The reasons for this range of sizes are unknown. In addition, chloroplasts possess a distinct third membrane, referred to as the **thylakoid membrane**, that forms an intricate set of flattened vesicles. As described previously, thylakoid membrane is folded into a series of the disklike vesicular structures called **grana**. Each *granum* is a stack of several flattened vesicles. The internal compartment created by the formation of grana is referred to as the **thylakoid lumen** (or *space*). The thylakoid membrane that interconnects the grana is called the **stromal lamellae**. Adjacent layers of membrane that fit closely together within each granum are said to be appressed. The stromal lamellae are nonappressed.

The pigments and proteins responsible for the light-dependent reactions of photosynthesis are found within thylakoid membrane (Figure 13.3). Most of these molecules are organized into the working units of photosynthesis.

1. Photosystem I. Photosystem I (PSI), which energizes and transfers the electrons that eventually are donated to $NADP^+$, is a large membrane-spanning protein-pigment complex composed of several polypeptides. The largest of these are two nearly identical 83 kD subunits designated A and B. Although it possesses over 200 molecules of chlorophyll a, the essential role of photosystem I (the donation of energized electrons to a series of electron carriers within thylakoid membrane) is performed by two special chlorophyll a molecules that reside within the reaction center. These molecules, referred to as a special pair, are located in the core complex of PSI, the AB dimer. Because they absorb light at 700 nm, the special pair within photosystem I is sometimes referred to as P700. In addition to the special pair, the AB dimer contains a series of single electron carriers: A_0, A_1, and

Incorporating CO_2 into organic molecules requires energy and reducing power. In photosynthesis, both of these requirements are provided by a complex process driven by light energy.

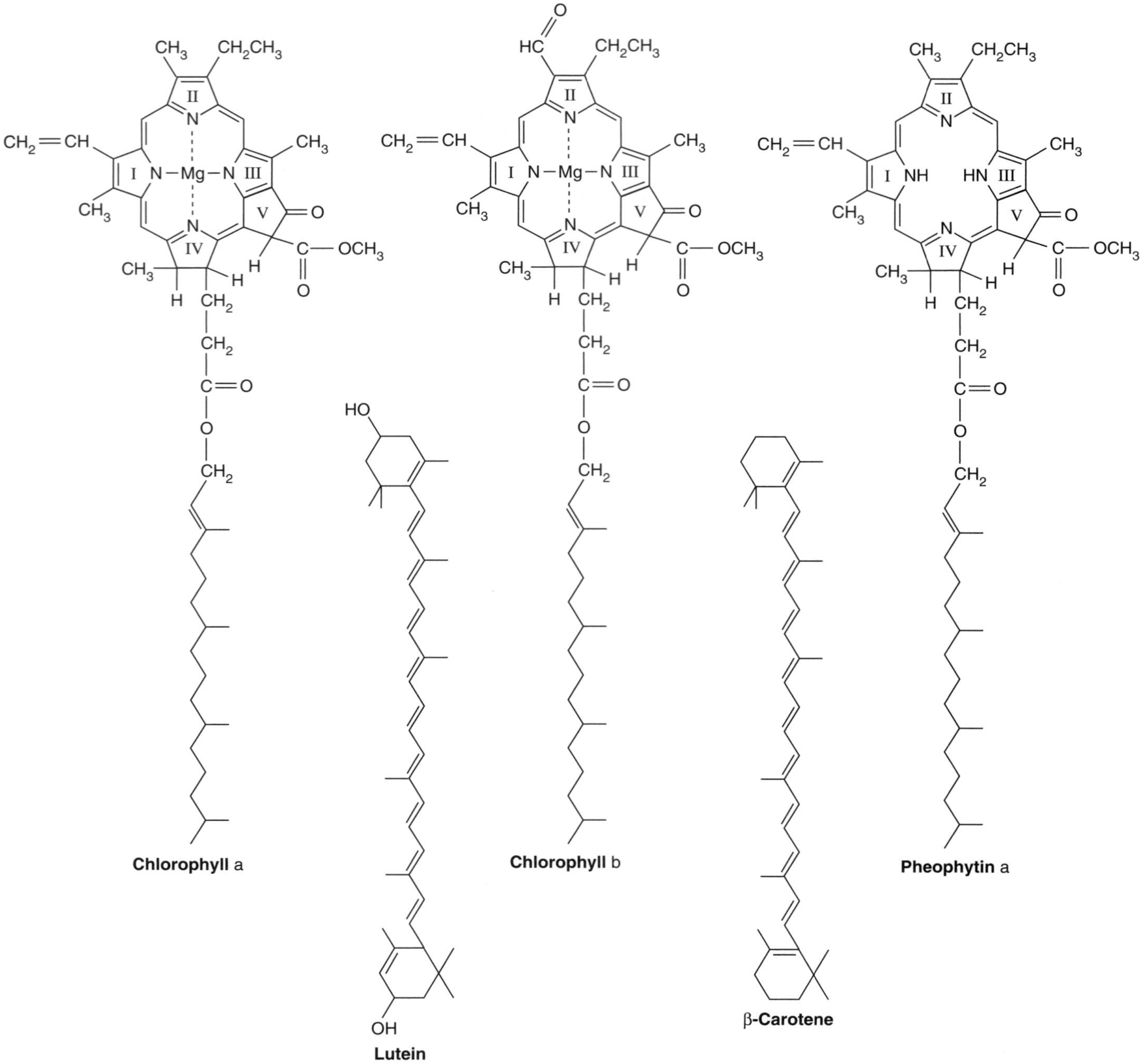

FIGURE 13.1

Pigment Molecules Used in Photosynthesis.

Chlorophylls a and b are found in almost all photosynthesizing organisms. They possess a complex cyclic structure (called a porphyrin) with a magnesium atom at its center. Chlorophyll a possesses a methyl group attached to ring II of the porphyrin, whereas chlorophyll b has an aldehyde group attached to the same site. Pheophytin a is similar in its structure to chlorophyll a. The magnesium atom is replaced by 2 protons. Chlorophylls a and b and pheophytin a all possess a phytol chain esterified to the porphyrin. The phytol chain extends into and anchors the molecule to the membrane. Lutein and β-carotene are the most abundant carotenoids in thylakoid membranes.

F_x. A_0 is a specific chlorophyll a molecule that accepts an energized electron from P700 and transfers it to A_1. A_1, which has been identified as phylloquinone (vitamin K_1), is a molecule similar in structure to ubiquinone (Figure 10.3) that is sometimes abbreviated as Q. The electron is then transferred from A_1 to F_x, a 4Fe-4S center. Subsequently, the electron is donated to F_A and F_B, two 4Fe-4S centers

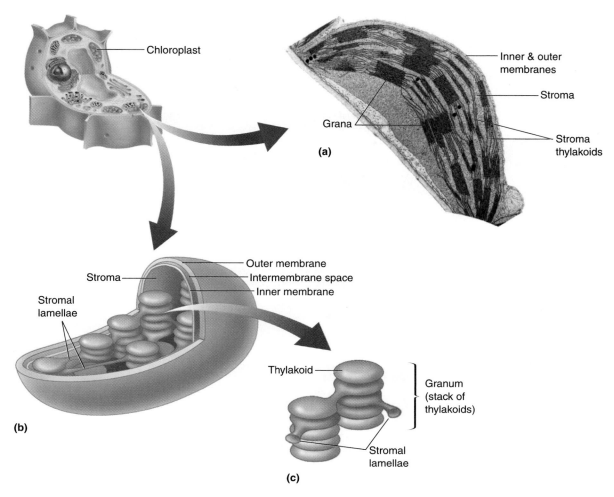

FIGURE 13.2

Chloroplast Structure.

Chloroplasts have inner and outer membranes. A third membrane forms within the aqueous, enzyme-rich stroma into flattened sacs called thylakoids. A stack of thylakoids is called a granum. Unstacked, connecting thylakoid membrane is referred to as stroma lamellae. (a) An electron micrograph of a chloroplast, (b) a diagrammatic view of a chloroplast, (c) a cutaway view of a granum.

in an adjacent 9 kD protein. Chlorophyll a molecules other than the special pair, as well as small amounts of chlorophyll b and carotenoids that occur in photosystem I, act as antenna pigments. **Antenna pigments** absorb light energy and transfer it to the reaction center. This phenomenon is described more fully in Section 13.2. Most PSI complexes are located in nonappressed thylakoid membrane, that is, membrane that is directly exposed to the stroma.

   2. Photosystem II. The function of photosystem II is to oxidize water molecules and donate energized electrons to electron carriers that eventually reduce photosystem I. Photosystem II is a large membrane-spanning protein-pigment complex now believed to possess at least 23 components. The most prominent of these is the reaction center, a protein-pigment complex composed of two polypeptide subunits known as D_1 (33 kD) and D_2 (31 kD) (the D_1/D_2 dimer), cytochrome b_{559}, and a special pair of chlorophyll a molecules (referred to as P680) that absorb light at 680 nm. The oxygen-evolving component of PSII consists of manganese stabilizing protein (MSP), a membrane-associated complex that contains a manganese cluster, and a tyrosine residue (tyr[161]), often referred to as Y_Z, located on D_1. Also associated with the D_1/D_2 dimer are several electron acceptors. Pheophytin a

Cyt b$_6$f   ATP Synthase   LHCII   PSII Complex

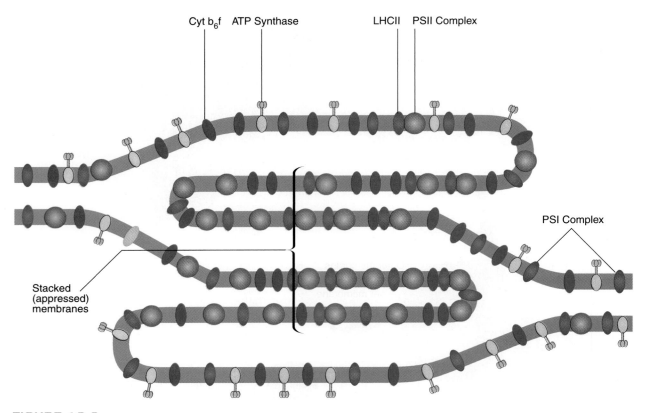

PSI Complex

Stacked
(appressed)
membranes

FIGURE 13.3

The Working Units of Photosynthesis.

PSI are most abundant in the unstacked stromal lamellae. In contrast, PSII are located primarily in the stacked regions of thylakoid membrane. Cytochrome b$_6$f is found in both areas of thylakoid membrane. The ATP synthase is found only in thylakoid membrane that is directly in contact with the stroma.

is a chlorophyll-like pigment that accepts an electron from P680. This electron is donated in turn to two forms of plastoquinone (PQ), a molecule similar to ubiquinone. Q_A is a plastoquinone that is permanently bound to D_2, whereas Q_B is reversibly bound to D_1. Several hundred antenna pigment molecules are also associated with the reaction center. A fraction of these light-harvesting molecules are associated with 43 kD and 47 kD proteins that are components of the core structure of PSII. The preponderance of accessory pigment molecules and several proteins, however, belong to a detachable unit referred to as *light harvesting complex II (LHCII).* LHCII consists of a transmembrane protein that binds numerous chlorophyll a and chlorophyll b molecules and carotenoids. LHCII units are a major component of thylakoid membrane. They possess approximately half of the chloroplast's chlorophyll content and a substantial percentage of photosynthesis-related protein. PSI (P700) and PSII (P680) have absorption properties that overlap and, if they were physically close to each other, most of the light energy would be preferentially transferred to PSI, which has a lower energy absorption requirement (longer wavelength). Balanced absorption is maintained through a reduced plastoquinone (PQH_2)-dependent phosphorylation of LHCII. When PSII absorption exceeds that of PSI, PQH_2 accumulates and LHCII becomes phosphorylated. A conformational change in LHCII releases it from its association with PSII and confinement to the appressed lamellae. Movement of LHCII to regions of the membrane containing PSI leads to an increased oxidation of PQH_2 or its broken association with PSII reduces the production of PQH_2. Regardless of which event predominates, the end result is that balance is restored because when PQH_2 levels drop the LHCII is dephosphorylated and locks its association with PSII and the appressed lamellae. Physical separation of PSI and PSII is required for this excitation balancing function to work efficiently and is absolutely necessary because both the intensity and energy composition of light vary significantly throughout the day and from sunny to shady conditions. The transfer of electrons from PSII to PSI is not compromised by this physical separation because the transfer is mediated through mobile electron carriers. Most PSII units are found in thylakoid membrane within grana, that is, in appressed membrane, not exposed to stroma.

3. Cytochrome b_6f complex. Cytochrome b_6f complex, found throughout the thylakoid membrane, is similar in structure and function to the cytochrome bc_1 complex in mitochondrial inner membrane. (Recall that cytochrome bc_1 complex is involved in transferring electrons from UQ to cytochrome c in the mitochondrial ETC and pumping protons across the inner membrane.) The cytochrome b_6f complex plays a critical role in the transfer of electrons from PSII to PSI. An iron-sulfur site on the complex accepts electrons from the membrane-soluble electron carrier plastoquinone and donates them to a small water-soluble copper-containing protein called plastocyanin. The mechanism that transports electrons from PQH_2 through the cytochrome b_6f complex appears to be similar to the Q cycle in mitochondria (Figure 10.7).

4. ATP synthase. The chloroplast ATP synthase (Figure 13.4), also referred to as CF_0CF_1ATP synthase, is structurally similar to the mitochondrial ATP synthase. The CF_0 component is a membrane-spanning protein complex that contains a proton-conducting channel. The CF_1 head piece, which projects into the stroma, possesses an ATP-synthesizing activity. Although the actual mechanism of chloroplast ATP synthesis is not known, it is clear that a transmembrane proton gradient produced during light-driven electron transport drives ADP phosphorylation.

KEY CONCEPTS 13.2

In chloroplasts a double membrane encloses an inner space called the stroma. The stroma contains the enzymes that catalyze the light-independent reactions of photosynthesis. The third membrane forms into flattened sacs called thylakoids. Thylakoid membrane contains the pigments and proteins of the light-dependent reactions.

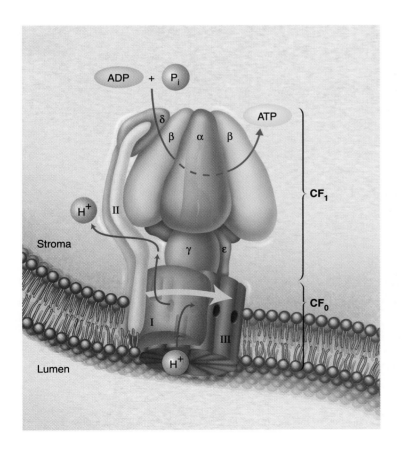

FIGURE 13.4

Diagrammatic View of the Chloroplast ATP Synthase.

The ATP synthase is composed of two components: an integral membrane protein complex (CF_0) that contains a proton pore and an extrinsic protein complex (CF_1) that synthesizes ATP. CF_0 contains four different types of subunits: I, II, III, and IV. The proton pore is composed of multiple copies of subunit II. Subunit IV (not shown) binds CF_0 to CF_1. CF_1 consists of five different subunits: α, β, γ, δ, and ε.

The synthesis of each ATP molecule is believed to require pumping approximately 3 protons across the membrane into the thylakoid space. The ATP synthase is found in thylakoid membrane that is directly in contact with the stroma.

QUESTION 13.1

Describe where each of the following molecules, molecular complexes, or processes is localized in the chloroplast. Explain their functions.

    a. LHCII

    b. lutein

    c. PSII

    d. MSP

    e. CF_0CF_1

    f. P700

    g. P680

    h. O_2 generation

13.2 LIGHT

The sun emits energy in the form of electromagnetic radiation, which propagates through space as waves, some of which impinge on the Earth. Visible light, the energy source that drives photosynthesis, occupies a small part of the electromagnetic radiation spectrum (Figure 13.5). Many of the properties of light

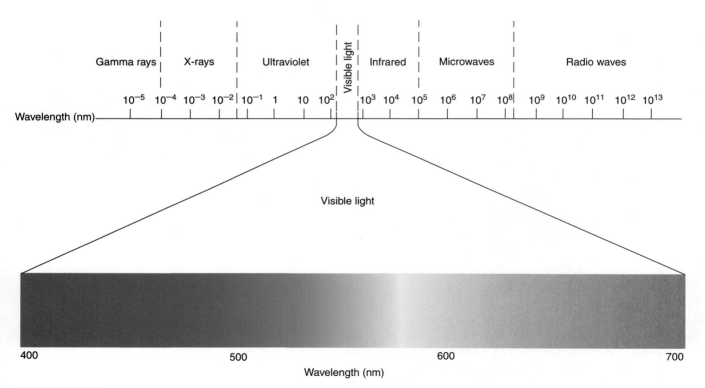

FIGURE 13.5

The Electromagnetic Spectrum.

Gamma rays, which have short wavelengths, have high energy. At the other end of the spectrum, the radio waves (long wavelengths) have low energy. Visible light is the portion of the spectrum to which the visual pigments in the retina of eyes are sensitive. Pigment molecules in chloroplasts are also sensitive to portions of the visible spectrum.

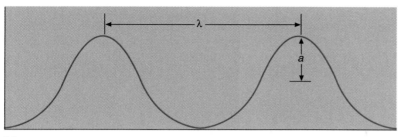

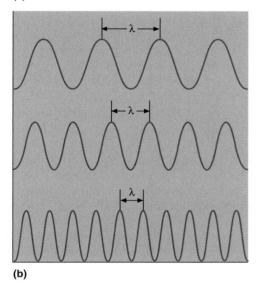

FIGURE 13.6

Properties of Waves.

(a) A wavelength λ is the distance between two consecutive peaks in a wave. The amplitude a or height of a wave is related to the intensity of electromagnetic radiation. (b) Frequency is the number of waves that pass a point in space per second. Radiation with the shortest wavelength has the highest frequency.

are explained by its wave behavior (Figure 13.6). Energy waves are described by the following terms:

1. Wavelength. Wavelength λ is the distance from the crest of one wave to the crest of the next wave.

2. Amplitude. Amplitude a is the height of a wave. The intensity of electromagnetic radiation (e.g., the brightness of light) is proportional to a^2.

3. Frequency. Frequency v is the number of waves that pass a point in space per second.

For each type of radiation the wavelength multiplied by the frequency equals the velocity c of the radiation.

$$\lambda v = c$$

This equation rearranges to

$$\lambda = c/v$$

The wavelength therefore depends on both the frequency and the velocity of the wave.

The wavelengths of visible light range from 400 nm (violet light) to 700 nm (red light). In comparison, highly energetic X-rays and γ-rays have wavelengths that are 10 thousand to 10 million times shorter. On the other end of the spectrum are low-energy radio waves; these have wavelengths on the order of meters to kilometers.

Why do green light waves have less energy than blue light waves?   **QUESTION 13.2**

In addition to behaving as a wave, visible light (and other types of electromagnetic radiation) exhibits the properties of particles such as mass and acceleration (Einstein's observation that energy has mass, or $E = mc^2$, applies to the photon.) When light interacts with matter, it does so in discrete packets of energy called photons. The energy ε of a photon is proportional to the frequency of the radiation.

$$\varepsilon = h\nu$$

where h is Planck's constant (6.63×10^{-34} J•s).

According to the quantum theory, radiant energy can be absorbed or emitted only in specific quantities called quanta. When a molecule absorbs a quantum of energy, an electron is promoted from its ground state orbital (the lowest energy level) to a higher orbital (Figure 13.7). For absorption to occur, the energy difference between the two orbitals must exactly equal the energy of the absorbed photon. Complex molecules often absorb at several wavelengths. For example, chlorophyll produces an absorption spectrum with broad and multiple peaks (blue-violet region and red region). Both of these facts suggest that chlorophyll absorbs photons of many different energies with varying probabilities. Those wavelengths that are not absorbed we see with our eyes and so a chlorophyll solution (or a leaf) appears green. The absorption spectrum of chlorophyll and several other pigment molecules is discussed in Biochemical Methods 13.1. Molecules that absorb electromagnetic energy possess structures called chromophores. In **chromophores** electrons move easily to higher energy levels when energy is absorbed. Visible chromophores typically possess extended chains of conjugated double bonds and aromatic rings. Molecules with a small number of conjugated double bonds or isolated double bonds absorb energy in the ultraviolet portion of the electromagnetic spectrum. In other words, without the resonance stabilization provided by a sufficient number of double bonds, the absorption of considerable energy is required for an electron to reach a higher orbital.

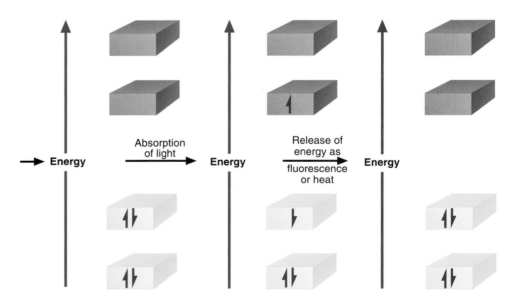

FIGURE 13.7

An Electron Absorbing Light.

If a molecule absorbs a photon, an electron becomes excited and moves to a higher orbital. There is usually no change in the spin of the excited electron. As long as the spins of the two unpaired electrons remain antiparallel, the molecule is said to be in an excited singlet state. An excited molecule can return to its ground state by releasing energy as fluorescence or heat. In addition, the energy may be transferred to another molecule, or the energized electron itself may be donated to another molecule.

Once an electron is excited, it can return to its ground state in several ways:

1. Fluorescence. In **fluorescence** a molecule's excited state decays as it emits a photon. Because the excited electron loses some energy initially by relaxing to a lower vibrational (energy) state, a transition resulting in the emission of a photon has a lower energy (longer wavelength) than that of the photon originally absorbed. Fluorescent decay can occur as quickly as 10^{-15} s. (Although various chlorophylls absorb light energy throughout the visible spectrum, they emit only photons with low energy at or beyond the red end of the visible spectrum.)

2. Resonance energy transfer. In **resonance energy transfer**, the excitation energy is transferred to a neighboring molecule through interaction between adjacent molecular orbitals. A neighboring molecule whose absorption spectrum overlaps the emission spectrum of the target chromophore can absorb photons released when that chromophore returns to its ground state.

3. Oxidation-reduction. An excited electron is transferred to a neighboring molecule. An excited electron occupies a normally unoccupied orbital and is bound less tightly than when it occupies a normally filled orbital. A molecule with an excited electron is a strong reducing agent. It returns to its ground state by reducing another molecule.

4. Radiationless decay. The excited molecule decays to its ground state by converting the excitation energy into heat.

Of all these responses to energy absorption, the most important in photosynthesis are resonance energy transfer and oxidation-reduction. Resonance energy transfer plays a critical role in harvesting light energy by accessory pigment molecules (Figure 13.8). Eventually, the energy absorbed and transmitted by light-harvesting complexes reaches the reaction center chlorophyll molecules. When these molecules become excited, they can lose an electron to a specific

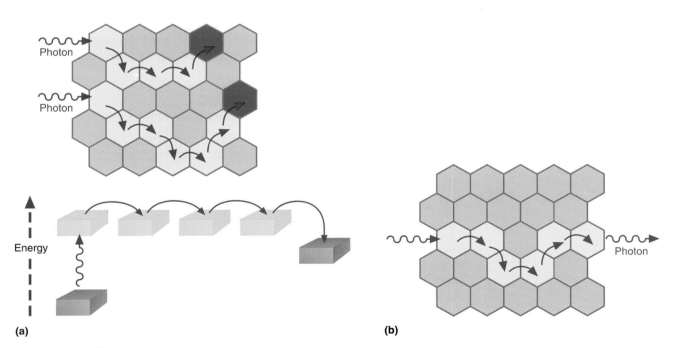

FIGURE 13.8

Resonance Energy Transfer.

Energy flows through a light-harvesting complex. (a) Once a photon is absorbed by a molecule in a light-harvesting complex, it migrates randomly through the complex by resonance energy transfer. The energy is donated from one antenna molecule to another (yellow hexagons) until it is trapped by a reaction center (dark green hexagons) or it is reemitted (b). The reaction center traps the excitation energy because its lowest excited state has a lower energy than the antenna molecules have.

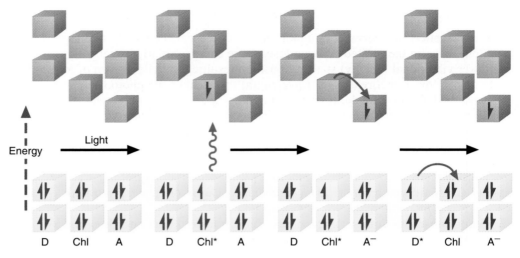

FIGURE 13.9

Electron Transfer.

Electron transfer initiates photosynthesis. When light energy is absorbed by a chlorophyll complex (Chl), an electron is transferred to an acceptor (A). The oxidized chlorophyll complex (Chl*) extracts an electron from a donor (D).

KEY CONCEPTS 13.3

The light energy absorbed by chromophores causes electrons to move to higher energy levels. In photosynthesis energy absorption is used to drive electron flow.

acceptor molecule (Figure 13.9). P700 passes electrons to ferredoxin (an iron-sulfur protein), and P680 passes them to pheophytin a. (Once P700 and P680 have been oxidized, they are referred to as P700* and P680*.) The electron hole left in the reaction center chlorophyll molecules is filled by an electron from a donor molecule. Plastocyanin and water play this role in PSI and PSII, respectively. Fluorescence also plays a role in photosynthesis when light absorption exceeds the capacity of the photosystems to transfer energy. Then photons are reemitted by a protective mechanism.

QUESTION 13.3

Explain the observation that antenna pigment molecules have different absorption spectra than do P680 and P700.

13.3 LIGHT REACTIONS

As described previously, the **light reactions** are a mechanism by which electrons are energized and subsequently used in ATP and NADPH synthesis. In O_2-evolving species both photosystems I and II are required. (In other species only photosystem I is used.) During photosynthesis, the two photosystems couple the light-driven oxidation of water molecules to the reduction of $NADP^+$. The overall reaction is

$$2\ NADP^+ + 2\ H_2O \rightleftharpoons 2\ NADPH + O_2 + 2\ H^+$$

Because the standard reduction potentials for the half-reactions are

$$O_2 + 4\ e^- + 4\ H^+ \rightleftharpoons 2\ H_2O \qquad E^{0'} = +0.816\ V$$

and

$$NADP^+ +\ H^+ + 2\ e^- \rightleftharpoons NADPH \quad E^{0'} = -0.320\ V$$

the coupled process has a standard redox potential of -1.136 V. The minimum free energy change for this process (calculated using the equation $\Delta G'_0 = -nF \Delta E^{0'}$; Section 9.1) is approximately 438 kJ (104.7 kcal) per mole of O_2 generated. In comparison, a mole of photons of 700 nm light provides approximately 170 kJ (40.6 kcal). Experimental observations have revealed that the absorption of 8 or more photons (i.e., 2 photons per electron) are required for each O_2 generated. Consequently, a total of 1360 kJ (325 kcal) (i.e., 8 times 170 kJ) are absorbed for each mole of O_2 produced. This energy is more than sufficient to account for reducing $NADP^+$ and establishing the proton gradient for ATP synthesis.

The process of light-driven photosynthesis begins with the excitation of PSII by light energy. One electron at a time is transferred to a chain of electron carriers that connects the two photosystems. As electrons are transferred from PSII to PSI, protons are pumped across the thylakoid membrane from the stroma into the thylakoid space. ATP is synthesized as protons flow back into the stroma through the ATP synthase. When P700 absorbs an additional photon it releases an energized electron. (This electron is immediately replaced by an electron provided by PSII.) The newly energized electron is passed through a series of iron-sulfur proteins and a flavoprotein to $NADP^+$, the final electron acceptor. This sequence, referred to as the **Z scheme**, is outlined in Figure 13.10. A more detailed view of the Z scheme is shown in Figure 13.11.

Photosystem II and Oxygen Generation

When LHCII absorbs a photon, its energy is transferred as previously described to P680 PSII. Then an energized electron is donated to *pheophytin a* (see Figure 13.1), a molecule similar to chlorophyll in its structure. Reduced pheophytin a passes this electron to Q_A (plastoquinone). When a second electron is transferred from P680 and 2 protons are transferred from the stroma, reduced plastoquinone (Q_AH_2), also referred to as plastoquinol, is formed:

Plastoquinone → (2 H^+, 2 e^-) → **Plastoquinol**

One at a time, Q_AH_2 donates its 2 electrons to Q_B. Two additional stromal protons are used in the reduction of Q_B. Reduced Q_B (Q_BH_2) donates its electrons to the cytochrome b_6f complex. Finally, the cytochrome b_6f complex donates its electrons to plastocyanin (PC), a mobile peripheral membrane protein. PC, a single-electron carrier, then transfers these electrons to P700 in PSI.

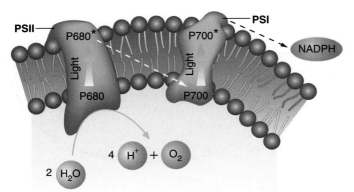

FIGURE 13.10

The Z Scheme.

Two photons must be absorbed to drive an electron from H_2O to $NADP^+$. The arrows represent the flow of electrons. The vertical arrangement of the components is according to their $E^{0'}$ values. The most negative values are at the top of the figure. As electrons move from one carrier to the next, they lose energy, that is, their $E^{0'}$ values become less negative.

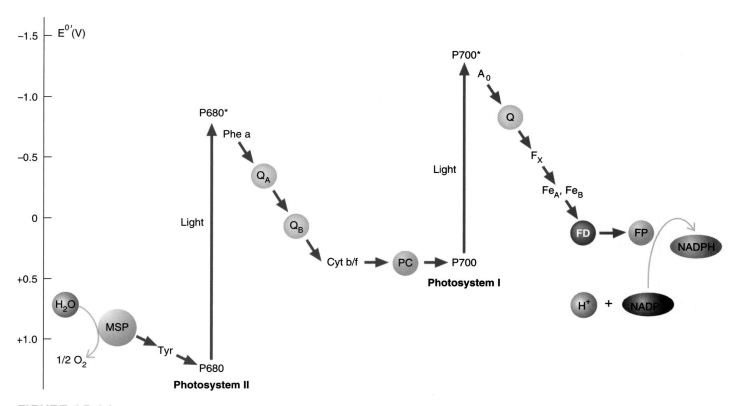

FIGURE 13.11

More Details of the Z Scheme.

The flow of electrons from photosystem II to photosystem I drives the transport of protons into the thylakoid lumen. Electron transfer through the iron-sulfur proteins Fe_A and Fe_B is not understood. $E^{0'}$ values are approximate. MSP contains a manganese cluster. (MSP = manganese stabilizing protein)

Recall that electrons transferred from P680 are replaced when H_2O is oxidized. An *oxygen-evolving complex*, composed in part of the manganese cluster in MSP and the tyrosine residue located on D_1, is responsible for the transfer of electrons from H_2O to P680*. (Tyrosine is effective in electron transfer because the tyrosyl radical formed is resonance stabilized.) The evolution of one O_2 requires splitting two H_2O, which releases 4 protons and 4 electrons. Experimental evidence indicates that H_2O is converted to O_2 by a mechanism referred to as the *water-oxidizing clock* (Figure 13.12). The O_2-evolving complex has five oxidation states: S_0, S_1,

FIGURE 13.12

The Water-Oxidizing Clock.

The O_2-evolving apparatus has five oxidation states. Four electrons are removed for every O_2 that is evolved. Each electron is sequentially transferred to a tyrosine residue (Y_Z) and then to P680. Protons are also released during the cycle.

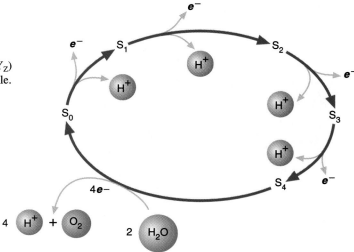

S_2, S_3, and S_4. S_0 is the most reduced state and S_4 is the most oxidized state of the complex. It is now believed that the Mn cluster, a group of four manganese atoms bound to MSP near the PSII reaction center, is mostly responsible for these transitions. As the oxygen-evolving complex abstracts electrons and protons from H_2O, it cycles through the five oxidation states. The electrons are transferred one at a time to a tyrosine residue on the D_1 polypeptide and then to the P680 reaction center. The protons released in the process remain in the thylakoid lumen, where they contribute to the pH gradient that drives ATP synthesis.

QUESTION 13.4

Excessive amounts of light can depress photosynthesis. Recent research indicates that PSII is extremely vulnerable to light damage. Plants often survive this damage because they possess efficient repair systems. It now appears that cells delete and resynthesize damaged components and recycle undamaged ones. For example, the D_1 polypeptide, apparently the most vulnerable component of PSII, is rapidly replaced after it is damaged. After reviewing the role of PSII, suggest the proximate cause of light-induced damage of D_1. (*Hint*: The D_1/D_2 dimer binds two molecules of β-carotene.)

Photosystem I and NADPH Synthesis

As described, the absorption of a photon by P700 leads to the release of an energized electron. This electron is then passed through a series of electron carriers, the first of which is a chlorophyll a molecule (A_0). As the electron is donated sequentially to phylloquinone (Q) and to several iron-sulfur proteins (the last of which is ferredoxin), it is moved from the lumenal surface of the thylakoid membrane to its stromal surface. Ferredoxin, a mobile, water-soluble protein, then donates each electron to a flavoprotein called ferredoxin-NADP oxidoreductase (FNR). The flavoprotein uses a total of 2 electrons and a stromal proton to reduce $NADP^+$ to NADPH. The transfer of electrons from ferredoxin to $NADP^+$ is referred to as the *noncyclic electron transport pathway*. In some species (e.g., algae), electrons can return to PSI by way of a *cyclic electron transport pathway* (Figure 13.13). In this process, which typically occurs when a chloroplast has a high $NADPH/NADP^+$ ratio, no NADPH is produced. Instead, electrons are used to pump additional protons across the thylakoid membrane. Consequently, additional molecules of ATP are synthesized.

QUESTION 13.5

Describe the role of each of the following molecules in photosynthesis:

  a. plastocyanin

  b. β-carotene

  c. ferredoxin

  d. plastoquinone

  e. pheophytin a

  f. lutein

QUESTION 13.6

Because PSII and PSI operate in series, photosynthesis is efficient when they receive light at equal rates. Recent research indicates that phosphorylation helps balance the activities of the two photosystems. After reviewing the structural and functional properties of PSI, PSII, and LHCII, describe the regulatory roles of plastoquinone and the kinase and phosphatase enzymatic activities that make the balanced regulation of PSI and PSII possible. Review the effects of covalent modification of proteins (Chapter 5) and suggest why phosphorylation has this effect on photosynthesis.

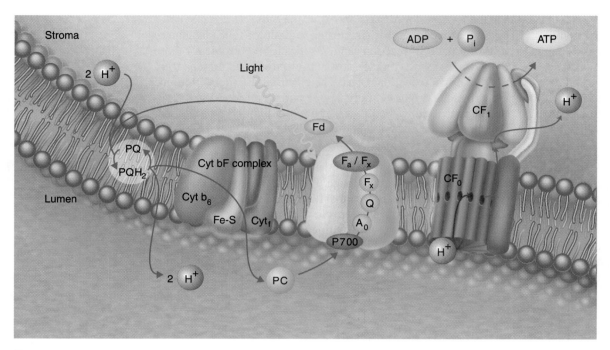

FIGURE 13.13

The Cyclic Electron Transport Pathway.

A Q cycle similar to that observed in the electron transport pathway that links PSI and PSII is believed to be responsible for pumping 2 protons across the thylakoid membrane for each electron transported. The proton flow drives ATP synthesis. No NADPH is produced.

Photophosphorylation

During photosynthesis light energy captured by an organism's photosystems is transduced (i.e., converted from one form to another) into ATP phosphate bond energy. This conversion is referred to as **photophosphorylation**. It is apparent from the preceding discussions that there are many similarities between mitochondrial and chloroplast ATP synthesis. For example, many of the same molecules and terms that are encountered in aerobic respiration (Chapter 10) are also relevant to discussions of photosynthesis. Additionally, in both organelles, electron transport is used to induce a proton gradient, which in turn drives ATP synthesis. Although there are a variety of differences between aerobic respiration and photosynthesis, the essential difference between the two processes is the conversion of light energy into redox energy by chloroplasts. (Recall that mitochondria produce redox energy by extracting high-energy electrons from food molecules.) Another critical difference involves the permeability characteristics of mitochondrial inner membrane and thylakoid membrane. In contrast to the inner membrane, the thylakoid membrane is permeable to Mg^{2+} and Cl^-. Therefore, Mg^{2+} and Cl^- move across the thylakoid membrane as electrons and protons are transported during the light reaction. The electrochemical gradient across the thylakoid membrane that drives ATP synthesis therefore consists mainly of a proton gradient that may be as great as 3.5 pH units.

Although experimental observations of chloroplasts played a critical role in the development of the chemiosmotic theory, several issues related to chloroplast ATP synthesis remain unclear. The most prominent of these is the $H^+/2\ e^-$ ratio. Note that in the light reactions outlined in Figure 13.14, a total of six H^+ ions are transported for each pair of electrons. Because the movement of three H^+ ions through the ATP synthase is required for the synthesis of one ATP molecule, two ATP molecules are produced for every NADPH molecule that is synthesized.

KEY CONCEPTS 13.4

Eukaryotic photosynthesizing cells possess two photosystems, PSI and PSII, which are connected in series in a mechanism referred to as the Z scheme. The water-oxidizing clock component of PSII generates O_2. The protons are used in the synthesis of ATP in a chemiosmotic mechanism. PSI is responsible for the synthesis of NADPH.

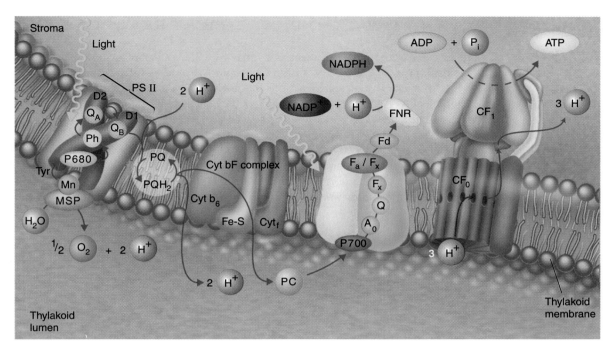

FIGURE 13.14

Membrane Organization of the Light Reactions in Chloroplasts: The Electron Transport Chain and the ATP Synthase Complex.

As 2 electrons move from each water molecule to $NADP^+$ (blue arrows), about two H^+ are pumped from the stroma into the thylakoid lumen. Two additional H^+ are generated within the lumen by the oxygen-evolving complex. The flow of protons through the proton pore in CF_0 drives the synthesis of ATP in CF_1. (MSP = manganese stabilizing protein; ph = pheophytin; Fd = ferredoxin; FNR = ferredoxin-NADP oxidoreductase)

However, in some circumstances (e.g., high light intensity) a ratio closer to 4 H^+/2 e^- has been observed. With this ratio, approximately 1.3 ATP molecules are synthesized for each NADPH molecule. The reason for the ratio reduction is unknown. Some recent experimental evidence indicates that under high light intensity the cytochrome b_6f complex fails to pump the additional 2 protons.

Describe the effect of uncoupler molecules on ATP synthesis in chloroplasts. Compare it to that observed in mitochondria.

QUESTION 13.7

When chloroplasts are first soaked in an acidic solution (pH = 4) and then transferred to a basic solution (pH = 8), there is a quick and brief burst of ATP synthesis. Explain.

QUESTION 13.8

A variety of herbicides kill plants by inhibiting photosynthetic electron transport. Atrazine, a triazine herbicide, blocks electron transport between Q_A and Q_B in PSII. DCMU (3-(3,4-dichlorophenyl)-1,1-dimethylurea) also blocks electron flow between the two molecules of plastoquinone. Paraquat is a member of a family of compounds called bipyridylium herbicides. Paraquat is reduced by PSI but is easily reoxidized by O_2 in a process that produces superoxide and hydroxyl radicals. Plants die because their cell membranes are destroyed by radicals. Of the herbicides just discussed, determine which, if any, are most likely to be toxic to humans and other animals. What specific damage may occur?

QUESTION 13.9

13.4   THE LIGHT-INDEPENDENT REACTIONS

The incorporation of CO_2 into carbohydrate by eukaryotic photosynthesizing organisms, a process that occurs within chloroplast stroma, is often referred to as the **Calvin cycle.** Because the reactions of the Calvin cycle can occur without light if sufficient ATP and NADPH are supplied, they have often been called the dark reactions. The name *dark reactions* is somewhat misleading, however. The Calvin cycle reactions typically occur only when the plant is illuminated, because ATP and NADPH are produced by the light reactions. Therefore **light-independent reactions** is a more appropriate term. Because of the types of reactions that occur in the Calvin cycle, it is also referred to as the *reductive pentose phosphate cycle* (RPP cycle) and the *photosynthetic carbon reduction cycle* (PCR cycle).

The Calvin Cycle

The net equation for the Calvin cycle (Figure 13.15) is

$$3\ CO_2 + 6\ NADPH + 9\ ATP \longrightarrow$$
$$\text{Glyceraldehyde-3-Phosphate} + 6\ NADP^+ + 9\ ADP + 8\ P_i$$

For every three molecules of CO_2 that are incorporated into carbohydrate molecules, there is a net gain of one molecule of glyceraldehyde-3-phosphate. The fixation of six CO_2 into one hexose molecule occurs at the expense of 12 NADPH and 18 ATP. The reactions of the cycle can be divided into three phases:

1. Carbon fixation. Carbon fixation, the mechanism by which inorganic CO_2 is incorporated into organic molecules, consists of a single reaction. Ribulose-1,5-bisphosphate carboxylase catalyzes the carboxylation of ribulose-1,5-bisphosphate to form two molecules of glycerate-3-phosphate. Plants that produce glycerate-3-phosphate as the first stable product of photosynthesis are referred to as **C3 plants.** (Notable exceptions are described in Special Interest Box 13.2.) Ribulose-1,5-bisphosphate carboxylase, a complex molecule composed of eight large subunits (L) (56 kD) and eight small subunits (S) (14 kD), is the pacemaker enzyme of the Calvin cycle. Its activity is regulated by CO_2, O_2, Mg^{2+}, and pH, as well as other metabolites. Each L subunit contains an active site that binds substrate. The catalytic activity of the L subunits is enhanced by the S subunits. Because the CO_2 fixation reaction is extremely slow, plants compensate by producing a large number of copies of the enzyme, which often constitutes approximately half of a leaf's soluble protein. For this reason, ribulose-1,5-bisphosphate carboxylase is often described as the world's most abundant enzyme.

2. Reduction. The next phase of the cycle consists of two reactions. Six molecules of glycerate-3-phosphate are phosphorylated at the expense of six ATP molecules to form glycerate-1,3-bisphosphate. The latter molecules are then reduced by $NADP^+$-glyceraldehyde-3-phosphate dehydrogenase to form six molecules of glyceraldehyde-3-phosphate. These reactions are similar to reactions encountered in gluconeogenesis. Unlike the dehydrogenase in gluconeogenesis, the Calvin cycle enzyme uses NADPH as a reducing agent.

3. Regeneration. As noted previously, the net production of fixed carbon in the Calvin cycle is one molecule of glyceraldehyde-3-phosphate. The other five glyceraldehyde-3-phosphate molecules are processed in the remainder of the Calvin cycle reactions to regenerate three molecules of ribulose-1,5-bisphosphate. Two molecules of glyceraldehyde-3-phosphate are isomerized to form dihydroxyacetone phosphate. One dihydroxyacetone molecule combines with a third glyceraldehyde-3-phosphate molecule to form fructose-1,6-bisphosphate. The latter molecule is then hydrolyzed to fructose-6-phosphate. Fructose-6-phosphate subsequently combines with a fourth molecule of glyceraldehyde-3-phosphate to form xylulose-5-phosphate and erythrose-4-phosphate. Erythrose-4-phosphate

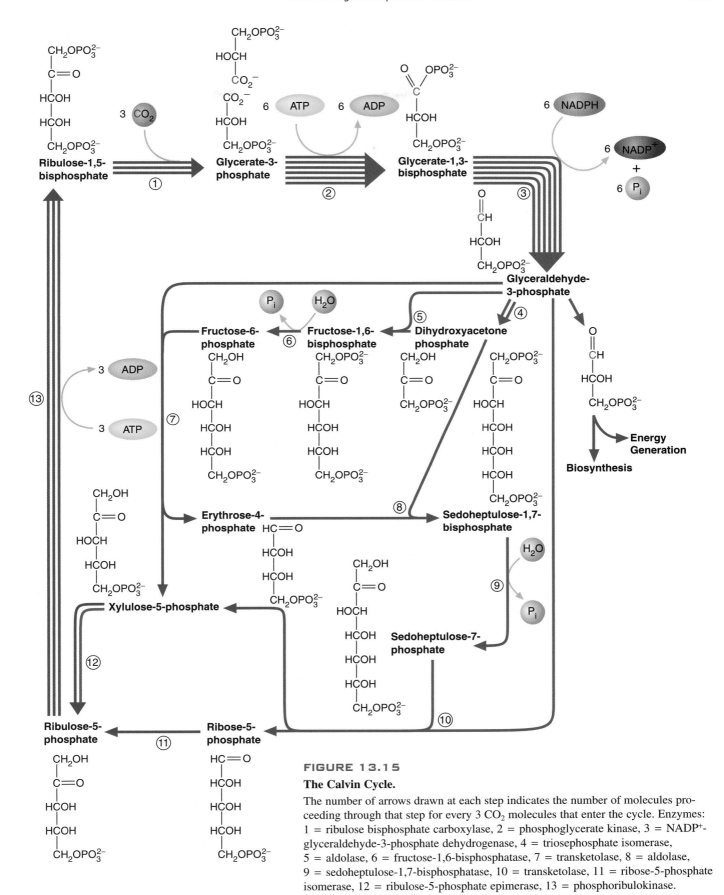

FIGURE 13.15

The Calvin Cycle.

The number of arrows drawn at each step indicates the number of molecules proceeding through that step for every 3 CO_2 molecules that enter the cycle. Enzymes: 1 = ribulose bisphosphate carboxylase, 2 = phosphoglycerate kinase, 3 = NADP$^+$-glyceraldehyde-3-phosphate dehydrogenase, 4 = triosephosphate isomerase, 5 = aldolase, 6 = fructose-1,6-bisphosphatase, 7 = transketolase, 8 = aldolase, 9 = sedoheptulose-1,7-bisphosphatase, 10 = transketolase, 11 = ribose-5-phosphate isomerase, 12 = ribulose-5-phosphate epimerase, 13 = phosphoribulokinase.

combines with dihydroxyacetone phosphate to form sedoheptulose-1,7-bisphosphate, which is then hydrolyzed to form sedoheptulose-7-phosphate. The fifth molecule of glyceraldehyde-3-phosphate combines with sedoheptulose-7-phosphate to form ribose-5-phosphate and a second molecule of xylulose-5-phosphate. Ribose-5-phosphate and both molecules of xylulose-5-phosphate are separately isomerized to ribulose-5-phosphate. In the last step, three molecules of ribulose-5-phosphate are phosphorylated at the expense of three ATP molecules to form three molecules of ribulose-1,5-bisphosphate. The remaining molecule of glyceraldehyde-3-phosphate is either used within the chloroplast in starch synthesis or exported to the cytoplasm, where it may be used in the synthesis of sucrose or other metabolites.

QUESTION 13.10

The overall equation for photosynthesis is

$$6\,CO_2 + 6\,H_2O \xrightarrow{\text{light}} C_6H_{12}O_6 + 6\,O_2$$

From which of the two substrates in this equation are the atoms of molecular oxygen derived?

QUESTION 13.11

Many of the reactions in the regeneration phase of the Calvin cycle are similar to reactions encountered in previous chapters in this textbook. Review the reactions in this phase of the Calvin cycle and determine which reactions they resemble. (*Hint*: Review Chapter 8.)

QUESTION 13.12

When plant cells are illuminated, their cytoplasmic ATP/ADP and NADH/NAD ratios rise significantly. The following shuttle mechanism is believed to contribute to the transfer of ATP and reducing equivalents from the chloroplast into the cytoplasm. Once dihydroxyacetone phosphate is transported from the stroma into the cytoplasm, it is converted to glyceraldehyde-3-phosphate and then to glycerate-1,3-bisphosphate. (This reaction is the reverse of the reaction in which glyceraldehyde-3-phosphate is formed during carbon fixation.) In the cytoplasmic reaction, the reducing equivalents are donated to NAD^+ to form NADH. In a later reaction, glycerate-1,3-bisphosphate is converted to glycerate-3-phosphate with the concomitant production of one molecule of ATP. Glycerate-3-phosphate is then transported back into the chloroplast, where it is reconverted to glyceraldehyde-3-phosphate.

This shuttle somewhat depresses mitochondrial respiration processes. Review the regulation of aerobic respiration (Chapter 9) and suggest how photosynthesis suppresses this aspect of mitochondrial function.

Photorespiration

Photorespiration is perhaps the most curious feature of photosynthesis. In this light-dependent process, oxygen is consumed, and CO_2 is liberated by plant cells that are actively engaged in photosynthesis. Photorespiration is a multistep mechanism initiated by ribulose bisphosphate carboxylase. In addition to its carboxylation function, this enzyme also possesses an oxygenase activity. (For this reason the name *ribulose-1,5-bisphosphate carboxylase-oxygenase*, or *rubisco*, is sometimes used.) Because the enzyme's active site binds to both CO_2 and O_2, these substrates compete.

In the oxygenation reaction, ribulose-1,5-bisphosphate is converted to glycolate-2-phosphate and glycerate-3-phosphate (Figure 13.16). In a complex series

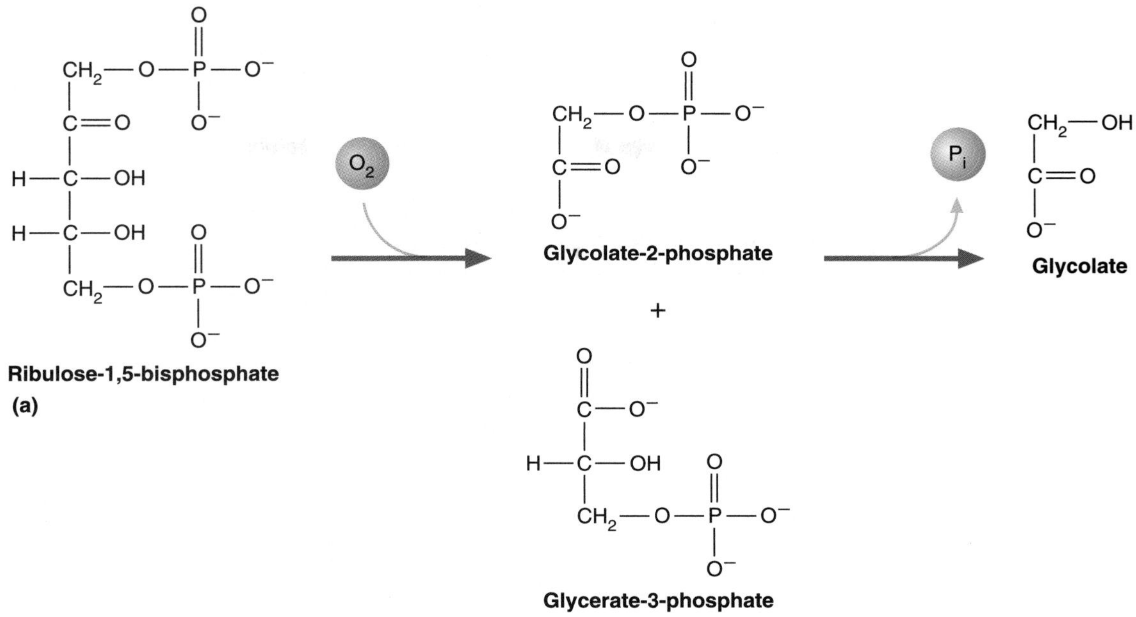

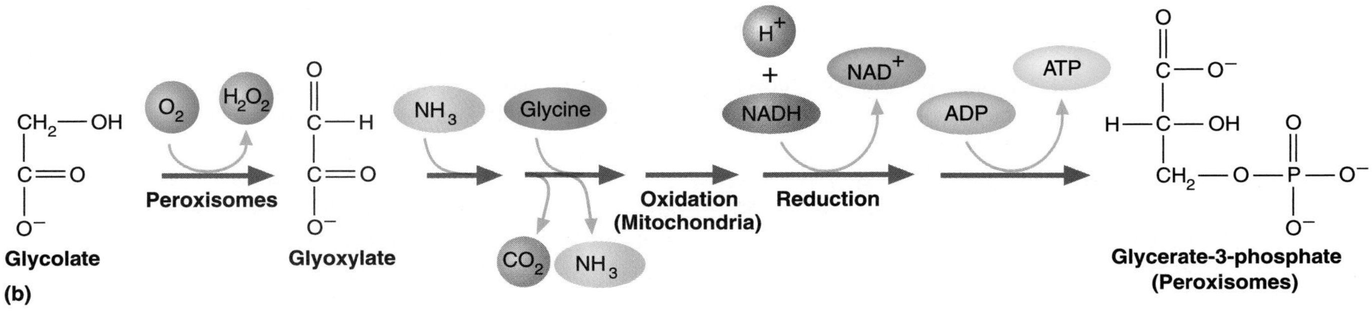

FIGURE 13.16

Photorespiration.

Photorespiration is a complex multistep process catalyzed by enzymes in several cellular compartments. (a) The synthesis of glycolate occurs within the stroma. (b) Glycolate is converted to glyoxylate within peroxisomes. In a complex series of reactions that occur in peroxisomes, mitochondria, and cytoplasm, glyoxylate is converted to glycerate-3-phosphate and CO_2.

of reactions, glycolate-2-phosphate is oxidized by O_2. Ultimately, the glycerate-3-phosphate produced by this pathway is used to produce (via the Calvin cycle) ribulose-1,5-bisphosphate. Photorespiration is a wasteful process. It loses fixed carbon (as CO_2), and consumes both ATP and NADH.

The rate of photorespiration depends on several parameters. These include the concentrations of CO_2 and O_2 to which photosynthesizing cells are exposed. Photorespiration is depressed by CO_2 concentrations above 0.2%. (Because photorespiration and photosynthesis occur concurrently, CO_2 is released during CO_2 fixation. When the rates of CO_2 release and fixation are equal, the *CO_2 compensation point* has been reached. The lower the CO_2 compensation point, the less photorespiration takes place. Many C3 plants have CO_2 compensation points between 0.02% and 0.03% of CO_2 in the air near photosynthesizing cells.) In contrast, high O_2 concentrations and high temperatures promote photorespiration. Consequently, this process is favored when plants are exposed to high temperatures and any condition that causes low CO_2 and/or high O_2 concentrations. For example, photorespiration is a serious problem for C3 plants in hot, dry environments. To conserve water, these plants close their stomata, thus reducing the

KEY CONCEPTS 13.5

The Calvin cycle is a series of light-independent reactions in which CO_2 is incorporated into organic molecules. The Calvin cycle reactions occur in three phases: carbon fixation, reduction, and regeneration. Photorespiration is a wasteful process in which photosynthesizing cells evolve CO_2.

Because glyceraldehyde-3-phosphate and dihydroxyacetone phosphate are readily interconverted, these two molecules (referred to the **triose phosphates**) are both considered to be Calvin cycle products. The synthesis of triose phosphate is sometimes referred to as the *C3 pathway*. Plants that produce triose phosphates during photosynthesis are called *C3 plants*. Triose phosphate molecules are used by plant cells in such biosynthetic processes as the formation of polysaccharides, fatty acids, and amino acids. Initially, most triose phosphate is used in the synthesis of starch and sucrose (Figure 13A). The metabolism of each of these molecules is briefly discussed below.

Starch Metabolism

During very active periods of photosynthesis, triose phosphates are converted to starch. Under normal conditions, approximately 30% of the CO_2 fixed by leaves is incorporated into starch, which is stored as water-insoluble granules. During a subsequent dark period, most chloroplast starch is degraded and converted to sucrose. Sucrose is then exported to storage organs and rapidly growing tissues. In these tissues (e.g, tubers and seeds), most sucrose molecules are used to synthesize starch, which is stored primarily within a specialized plastid called an *amyloplast*.

Triose phosphates retained within the chloroplast are converted to fructose-6-phosphate by aldolase and fructose-1,6-bisphosphatase. Glucose-1-phosphate, the starting material for starch synthesis, is produced from fructose-6-phosphate by phosphoglucoisomerase and phosphoglucomutase. The conversion of glucose-1-phosphate to ADP-glucose by ADP-glucose pyrophosphorylase is the rate-limiting step in starch synthesis. ADP-glucose is incorporated into starch by starch synthase. Like glycogen synthase (p. 264), starch synthase adds each monosaccharide unit to a preexisting polysaccharide chain. The $\alpha(1,6)$ branch points of amylopectin are introduced by branching enzyme.

Several enzymes contribute to starch breakdown. Both α- and β-amylases cleave $\alpha(1,4)$ glycosidic bonds. β-Amylase catalyzes the successive removal of maltose units from the nonreducing ends of starch chains. Maltose is degraded to form glucose by α-glucosidase.

Glucose-1-phosphate is the product when $\alpha(1,4)$ glycosidic bonds at nonreducing ends are broken by starch phosphorylase. Branch points in starch are removed by debranching enzyme. The products of starch digestion, glucose and glucose-1-phosphate, are then converted to triose phosphate and exported to the cytoplasm. In photosynthesizing cells, most triose phosphate is converted to sucrose.

Sucrose imported from leaves is the substrate for most of the starch synthesis in nonphotosynthesizing cells. Sucrose is converted into fructose and UDP-glucose in a reversible reaction catalyzed by sucrose synthase. Fructose is then converted to glucose-1-phosphate by hexokinase and phosphoglucomutase. UDP-glucose is converted to glucose-1-phosphate by UDP-glucose pyrophosphorylase. The conversion of sucrose to two molecules of glucose-1-phosphate is a cytoplasmic process. After its transport into an amyloplast, glucose-1-phosphate is used in starch synthesis. (Smaller amounts of the glycolytic intermediates glyceraldehyde-3-phosphate and dihydroxyacetone phosphate are also transported into amyloplasts and used in starch synthesis.)

Sucrose Metabolism

Sucrose has several important roles in plants. First, sucrose accounts for a large portion of the CO_2 absorbed during photosynthesis. Second, most of the carbon translocated throughout plants is in the form of sucrose. Finally, sucrose is an important energy storage form in many plants.

Sucrose is synthesized in the cytoplasm. After their transport from chloroplasts, triose phosphates are converted to fructose-1,6-bisphosphate and subsequently to glucose-6-phosphate. The latter molecule is converted to glucose-1-phosphate by phosphoglucomutase. UDP-glucose (formed by glucose-1-phosphate uridyltransferase from glucose-1-phosphate) and fructose-6-phosphate combine to form sucrose-6-phosphate. Sucrose-6-phosphate synthesis is catalyzed by sucrose phosphate synthase. Sucrose phosphatase catalyzes the hydrolysis of sucrose-6-phosphate to form sucrose and P_i. The free energy change of the latter reaction ($\Delta G^{\circ'} = -18.4$ kJ/mol) ensures that sucrose production continues in sucrose-storing tissues.

CO_2 concentration within leaf tissue. (*Stomata* are pores on the surface of leaves. When they are open, CO_2, O_2, and H_2O vapor can readily diffuse down the concentration gradients between the leaf's interior and the external environment.) In addition, as photosynthesis continues, O_2 levels increase. Depending on the severity of the circumstances, from 30% to 50% of a plant's yield of fixed carbon may be lost. This effect can be serious because several C3 plants (e.g., soybeans and oats) are major food crops.

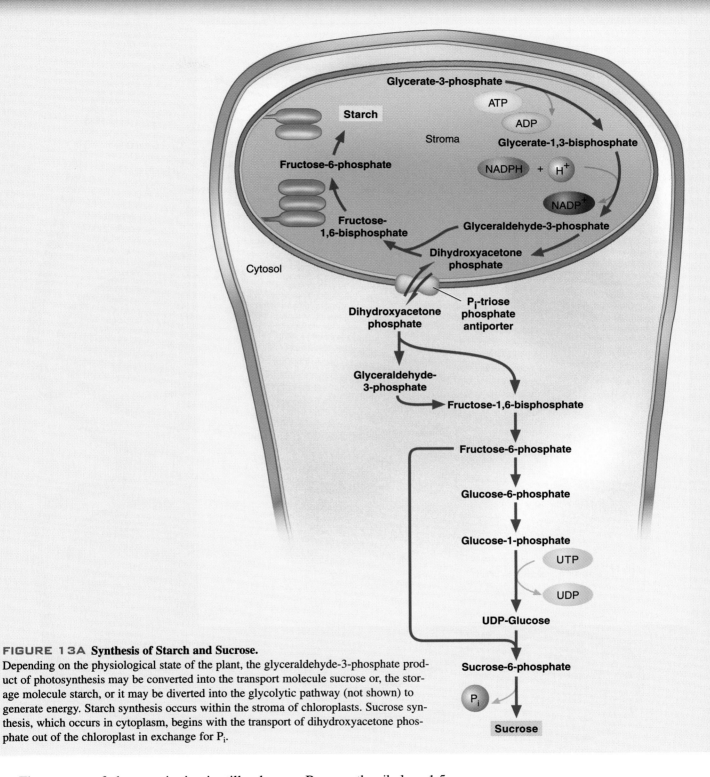

FIGURE 13A Synthesis of Starch and Sucrose.
Depending on the physiological state of the plant, the glyceraldehyde-3-phosphate product of photosynthesis may be converted into the transport molecule sucrose or, the storage molecule starch, or it may be diverted into the glycolytic pathway (not shown) to generate energy. Starch synthesis occurs within the stroma of chloroplasts. Sucrose synthesis, which occurs in cytoplasm, begins with the transport of dihydroxyacetone phosphate out of the chloroplast in exchange for P_i.

The purpose of photorespiration is still unknown. Because the ribulose-1,5-bisphosphate carboxylases of all photosynthetic organisms so far investigated possess oxygenase activity, it is currently believed that the enzyme's structure may make photorespiration necessary. Nevertheless, two types of photosynthesizing plants, collectively called **C4 plants**, have developed elaborate mechanisms to suppress photorespiration. These mechanisms, referred to as C4 metabolism and crassulacean acid metabolism, are described in Special Interest Box 13.2.

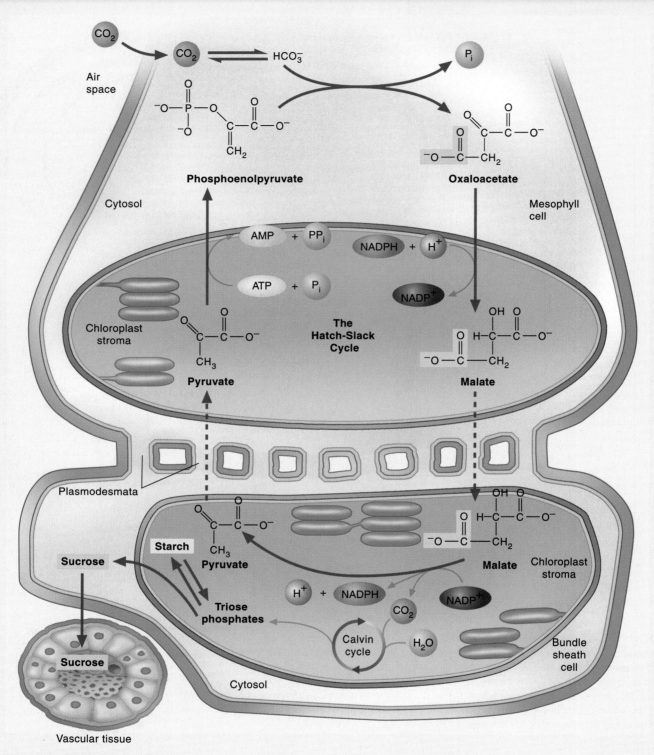

FIGURE 13B C4 Metabolism.
In the C4 pathway mesophyll cells, which are in direct contact with the air space in the leaf, take up CO_2 and use it to synthesize oxaloacetate, which is then reduced to malate. (Some C4 plants synthesize aspartate instead of malate.) Malate then diffuses to bundle sheath cells, where it is reconverted to pyruvate. The CO_2 released in this reaction is used in the Calvin cycle, eventually yielding triose phosphate molecules. Triose phosphate is subsequently converted to starch or sucrose. Pyruvate returns to the mesophyll.

C4 Metabolism

C4 plants are found primarily in the tropics. Such plants include sugar cane and maize (corn). They have been assigned the name *C4 plants* because a four-carbon molecule (oxaloacetate) plays a prominent role in a biochemical pathway that avoids photorespiration. This pathway is called the *C4 pathway* or the *Hatch-Slack pathway* (after its discoverers).

C4 plants possess two types of photosynthesizing cells in their leaves: mesophyll cells and bundle sheath cells. (In C3 plants, photosynthesis occurs in mesophyll cells.) Most mesophyll cells in both plant types are positioned so that they are in direct contact with air when the leaf's stomata are open. In C4 plants, CO_2 is captured in specialized mesophyll cells that incorporate it into oxaloacetate (Figure 13B). Phosphoenolpyruvate carboxylase (PEP carboxylase) catalyzes this reaction. Oxaloacetate is then reduced to malate. Once formed, malate diffuses into bundle sheath cells. (As their name implies, bundle sheath cells form a layer around vascular bundles, which contain phloem and xylem vessels.) Within bundle sheath cells, malate is decarboxylated to pyruvate in a reaction that reduces $NADP^+$ to NADPH. The pyruvate product of this latter reaction diffuses back to a mesophyll cell, where it can be reconverted to PEP. Although this reaction is driven by the hydrolysis of one molecule of ATP, there is a net cost of two ATP molecules. An additional ATP molecule is required to convert the AMP product to ADP so that it can be rephosphorylated during photosynthesis. This circuitous process delivers CO_2 and NADPH to the chloroplasts of bundle sheath cells, where ribulose-1,5-bisphosphate carboxylase and the other enzymes of the Calvin cycle use them to synthesize triose phosphates.

C4 metabolism is important because it assimilates CO_2 when it is advantageous for the plant. In hot environments, C4 plants open their stomata only at night after the air temperature decreases and the risk of water loss is low. The CO_2 enters through the stomata and is immediately incorporated into oxaloacetate. The next morning, when light becomes available to drive photosynthesis, the CO_2 released within bundle sheath cells is fixed into sugar molecules as photosynthesis provides adequate amounts of ATP and NADPH to drive this process. Because the concentration of CO_2 within bundle sheath cells is significantly higher than that of O_2, photorespiration is virtually eliminated. Consequently, the net rate of photosynthesis in C4 plants can be at least one-third higher than that of C3 plants. Agricultural scientists are investigating the use of genetic engineering to introduce the C4 pathway into C3 plants.

Crassulacean Acid Metabolism

Crassulacean acid metabolism (CAM) is another mechanism by which certain plants avoid photorespiration (Figure 13C). CAM plants, most of which are succulents (e.g., cacti), typically grow in regions of high light intensity and very limited water supply. (CAM is named for the Crassulaceae, a group of plants in which this pathway was first investigated.) They employ a strategy similar to that used by C4 plants. At night, when their stomata are open, CO_2 is incorporated into oxaloacetate by PEP carboxylase. After malate is formed, it is stored within the vacuole until photosynthesis begins the next morning, when CO_2 is regenerated.

The most significant difference between C4 metabolism and CAM is the way in which PEP carboxylation is separated from the Calvin cycle. Recall that in C4 metabolism the two processes are spatially separated (i.e., two cell types are used). In CAM the processes are temporally separated within mesophyll cells. In other words, during daylight hours, CO_2 is regenerated from malate that was synthesized during the night.

FIGURE 13C Crassulacean Acid Metabolism.
At night the stomata of CAM plants open to allow CO_2 to enter. Within mesophyll cells PEP carboxylase (1) incorporates CO_2 (as HCO_3^-) into oxaloacetate. Afterward, oxaloacetate is reduced by malate dehydrogenase (2) to form malate. Malate is stored in the cell's vacuole until daylight. Light stimulates the decarboxylation of malate by malic enzyme (3) to form pyruvate and CO_2. As a result of this temporal separation of reactions, CO_2 can be incorporated into sugar molecules via the Calvin cycle during the day when the plant's stomata are closed to avoid water loss.

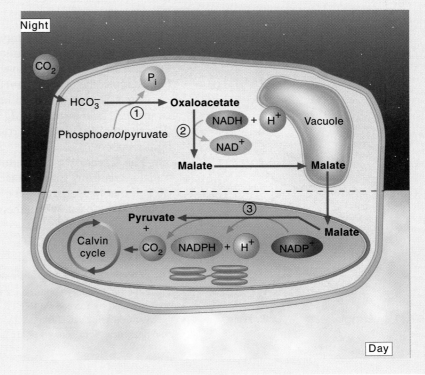

QUESTION 13.13   Briefly explain the significance of photorespiration and the strategies that C4 and CAM plants use to avoid it.

13.5   REGULATION OF PHOTOSYNTHESIS

Because plants must adapt to a wide variety of environmental conditions, the regulation of photosynthesis is complex. Although the control of most photosynthetic processes is far from being completely understood, several control features are well established. Most of these processes are directly or indirectly controlled by light. After a brief description of general light-related effects, the control of the activity of ribulose-1,5-bisphosphate carboxylase, the key regulatory enzyme in photosynthesis, is discussed.

Light Control of Photosynthesis

Investigations of photosynthesis are complicated by several factors. The most prominent of these is that the photosynthetic rate depends on temperature and cellular CO_2 concentration, as well as on light. Nevertheless, numerous investigations have firmly established light as an important regulator of most aspects of photosynthesis. This is not surprising, considering light's role in driving photosynthesis.

Many of the effects of light on plants are mediated by changes in the activities of key enzymes. Because plant cells possess enzymes that operate in several competing pathways (i.e., glycolysis, pentose phosphate pathway, and the Calvin cycle), careful metabolic regulation is critical. Light assists in this regulation by activating certain photosynthetic enzymes and deactivating several enzymes in degradative pathways. Among the light-activated enzymes are ribulose-1,5-bisphosphate carboxylase, $NADP^+$-glyceraldehyde-3-phosphate dehydrogenase, fructose-1,6-bisphosphatase, sedoheptulose-1,7-bisphosphatase, and phosphoribulokinase. Examples of light-inactivated enzymes include phosphofructokinase and glucose-6-phosphate dehydrogenase.

Light affects enzymes by indirect mechanisms. Among the best-researched are the following:

1. pH. Recall that during the light reactions, protons are pumped across the thylakoid membrane from the stroma into the thylakoid lumen. As the pH of the stroma increases from 7 to approximately 8, the activities of several enzymes are affected. For example, the pH optimum of ribulose-1,5-bisphosphate carboxylase is 8.

2. Mg^{2+}. Several photosynthetic enzymes are activated by Mg^{2+}. Light induces an increase in the stromal Mg^{2+} concentration from 1–3 mM to about 3–6 mM. (Recall that Mg^{2+} moves across thylakoid membrane into the stroma during the light reactions.)

3. The ferredoxin-thioredoxin system. Thioredoxins are small proteins that transfer electrons from reduced ferredoxin to certain enzymes (Figure 13.17). (Recall that ferredoxin is an electron donor in PSI.) When exposed to light, PSI reduces ferredoxin, which then reduces ferredoxin-thioredoxin reductase (FTR),

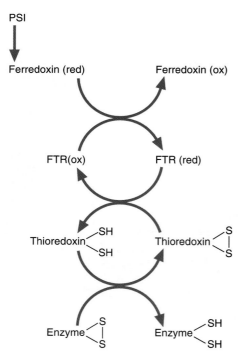

FIGURE 13.17

The Ferredoxin-Thioredoxin System.

Using the light energy captured by PSI, energized electrons are donated to ferredoxin. Electrons donated by ferredoxin to FTR (ferredoxin-thioredoxin reductase) are used to reduce the disulfide bridge of thioredoxin. Thioredoxin then reduces the disulfide bridges of susceptible enzymes. Some enzymes are activated by this process, whereas others are inactivated.

an iron-sulfur protein that mediates the transfer of electrons between ferredoxin and thioredoxin. Reduced thioredoxins activate several enzymes (e.g., fructose-1,6-bisphosphatase, sedoheptulose-1,7-bisphosphatase, and phosphoribulokinase) and inactivate others (e.g., NADP$^+$-glyceraldehyde-3-phosphate dehydrogenase).

4. Phytochrome. Phytochrome is a 120-kD protein that possesses a chromophore (Figure 13.18). Phytochrome exists in two forms: P_r and P_{fr}. P_r, the inactive blue form, absorbs red light (670 nm). The absorption of longer wavelengths (720 nm) (i.e., far red light) converts P_r to P_{fr}, the active green form. (In the dark, P_{fr} decays back to P_r.) Phytochrome apparently mediates hundreds of plant responses to light, many of which are initiated by changes in intracellular Ca^{2+} levels. Examples of phytochrome-mediated processes include seed germination, stem elongation, flower production, and the differentiation of chloroplasts from proplastids. (Several of these processes are also known to be promoted by specific plant hormones.) Phytochrome has specific effects on photosynthetic processes. These include controlling the rate of synthesis of the small subunit of ribulose-1,5-bisphosphate carboxylase and positioning chloroplasts within photosynthesizing cells.

Control of Ribulose-1,5-Bisphosphate Carboxylase

The genes that code for ribulose-1,5-bisphosphate carboxylase (rubisco) are found within the chloroplast (the L subunit) and the nucleus (the S subunit). The activation of these genes is mediated by an increase in light intensity (illumination). Phytochrome also appears to play a role in this activating process. Once the S subunit is transported from the cytoplasm into the chloroplast, both subunits assemble to form the L_8S_8 holoenzyme. A protein called the *large subunit-binding protein* appears to assist in the assembly of the holoenzyme. When illumination is low, the synthesis of both subunits is rapidly depressed.

The activity of rubisco is modified by a number of metabolic signals. When photosynthesis is active the pH in the stroma increases (H$^+$ is being pumped out of the stroma into the thylakoid lumen) and the Mg^{2+} concentration increases (Mg^{2+} moves into the stroma as H$^+$ moves out). Both of these changes increase the activity of rubisco. In other words, when the photosynthetic rate is high, the rate of CO_2 fixation is high. An important consideration in this process is whether the stomata are open or closed (see the discussion of photorespiration on p. 436). Although CO_2 is the preferred substrate for rubisco, under physiological conditions both the carboxylase activity and the oxidase activity compete significantly with each other. If the stomata are closed, as they would be on a hot, dry day, O_2 accumulation in the leaf tissue greatly compromises the proportional participation of the carboxylase activity of rubisco. Some plants use the C4 cycle to diminish this competition by trapping the CO_2 in a four-carbon intermediate and

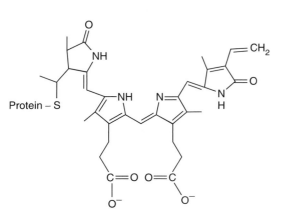

Phytochrome

FIGURE 13.18

Phytochrome.

The absorption of light changes the arrangement of conjugated double bonds in the molecule.

delivering the CO_2 via decarboxylation directly to a rubisco molecule that is protected from exposure to O_2.

Rubisco is subject to covalent modification as a means of regulation. The active site of the L subunit must be carbamoylated at a specific lysine residue to be active. Carbamoylation is the nonenzymatic carboxylation of a free primary amino group, in this case the ε-amino group of a certain lysine in the active site of rubisco. The rate of carbamoylation is dependent on the CO_2 concentration and an alkaline pH. The ribulose-1,5-bisphosphate can and does bind to the active site in both its modified and unmodified forms, but catalysis can only occur when rubisco is carbamoylated. The level of activation is cooperative and increases as more of the eight subunits are modified. A specific protein called rubisco activase mediates an ATP-dependent removal of ribulose-1,5-bisphosphate from the active site so that carbamoylation can occur followed by enzyme activation. When in the dark, photosynthesis is depressed and the ATP required for this activation process is greatly reduced, as is the NADPH required for the Calvin cycle. In the dark, some plants manufacture the substrate analogue carboxyarabinitol-1-phosphate (CA1P), which serves as an effective competitive inhibitor of the activation of rubisco. (A substrate analogue is a molecule that has a structure similar to the transition state intermediate of the enzyme, but which the enzyme cannot act on catalytically.) When the substrate analogue is bound to the rubisco active site, it cannot be dislodged by rubisco activase.

D-Carboxyarabinitol-1-phosphate   **Rubisco Transition State**

This regulatory mechanism ensures that CO_2 fixation only occurs at an appreciable rate when the concentration of CO_2 and available energy are high.

The enzyme fructose-1,6-bisphosphate phosphatase is more active during the day in plants because of the light-dependent increase in pH and Mg^{2+}. The enzyme is also acted upon by the ferredoxin-thioredoxin system (active during photosynthesis) to produce the active free thiol (reduced) form of the enzyme. Other enzymes involved in the Calvin cycle and carbohydrate metabolism are also regulated by light-dependent mechanisms (e.g., phosphoribulokinase, sedoheptulose-1,7-bisphosphatase, and glyceraldehyde-3-phosphate dehydrogenase). In this way, the interdependent activities of photosynthesis, CO_2 fixation, and sucrose synthesis are coordinately regulated.

Considering the discussions related to photosynthesis control, it becomes obvious that the metabolic control systems of plants are extraordinarily sophisticated. Despite considerable research efforts, many aspects of controlling function in plants remain to be elucidated. This endeavor is clearly important, because greater insight into photosynthetic production is critical for improving crop yields and developing plant resources in the future.

Despite the complexity of photosynthesis, many aspects of this life-sustaining process are now understood. Technologies whose use contributed (and continue to contribute) to this research effort include spectroscopy, photochemistry, X-ray crystallography, and radioactive tracers. The principles of these technologies and examples of their use are briefly discussed in Biochemical Methods 13.1.

KEY CONCEPTS 13.6

Light is the principal regulator of photosynthesis. Light affects the activities of regulatory enzymes in photosynthetic processes by indirect mechanisms, which include changes in pH, Mg^{2+} concentration, the ferredoxin-thioredoxin system, and phytochrome.

Most of the technologies used in biochemical research have a variety of applications. This is certainly true of the following techniques used in photosynthesis research.

Spectroscopy

Spectroscopy measures the absorption of electromagnetic radiation by molecules. Instruments that measure this absorption, called spectrophotometers, can scan a wide range of frequencies. A graph of a sample's absorption of electromagnetic radiation is called an **absorption spectrum**.

In photosynthesis research, the relative absorbance of radiation by various plant components has been measured to determine their contribution to light harvesting. This work revealed that most light absorbance is accomplished by the chlorophylls and the carotenoids. The absorption spectra of several plant pigments are shown in Figure 13D. As expected, the chlorophylls absorb little light between 500 and 699 nm (green and yellow-green light). They do absorb strongly between 400 and 500 nm (violet and blue light) and between 600 and 700 nm (orange and red light).

If the effect of wavelength on the rate of photosynthesis is measured, an **action spectrum** is generated. Note in Figure 13D that the action spectrum of a typical leaf suggests that photosynthesis at specific wavelengths (e.g., 650 nm and 680 nm) uses light absorbed by chlorophylls a and b, respectively. Intact leaves absorb light more efficiently than pure pigments because in intact leaves nonabsorbed wavelengths are reflected from chloroplast to chloroplast. Every time an internal reflection occurs, a small percentage of the reflected wavelength is absorbed. Eventually, a significant percentage of the wavelengths that strike a leaf are absorbed.

In the 1950s, Robert Emerson used a more precise version of the action spectrum to investigate photosynthesis. When he measured the number of oxygen molecules produced per quantum of light absorbed over the visible spectrum, he observed that light with wavelengths longer than 690 nm are ineffective in promoting photosynthesis. However, if blue wavelengths are used in addition to the red ones, the photosynthetic rate (i.e., the rate of oxygen evolution) is significantly enhanced. This phenomenon, referred to as the **Emerson enhancement effect**, was later used to support the theory of two separate photosystems (PSI and PSII).

Another type of spectroscopy is known as **electron spin resonance spectroscopy (ESR)**. In molecules that possess unpaired electrons, the energy of such electrons can be measured in a rapidly changing magnetic field. Because each electron generates its own magnetic field, it orients itself with or against an external field. (Electrons are always affected by their molecular environments.) The ESR spectrum is a measure of the difference between these two energy levels. Although ESR is a valuable technique in many areas of biochemistry, it has been especially useful in photosynthesis research. For example, ESR played an important role in determining that the photon-absorbing component of photosynthetic reaction centers is a pair of chlorophyll molecules.

Photochemistry

Photochemistry is the study of chemical reactions that are initiated by light absorption. During photochemical reactions, chemical bonds may be cleaved when ions or radicals are formed. Excited molecules may also be isomerized or converted to oxidizing agents. Several techniques monitor photochemical events. These measure product formation or fluorescence or phosphorescence emission.

One of the more notable uses of photochemistry in photosynthesis research was a study that resulted in the discovery of the water-

(continued on page 446)

FIGURE 13D Light Absorbance Measurements in Photosynthesis Research.
(a) Absorbance spectrum of visible light by photopigments. (b) Action spectrum.

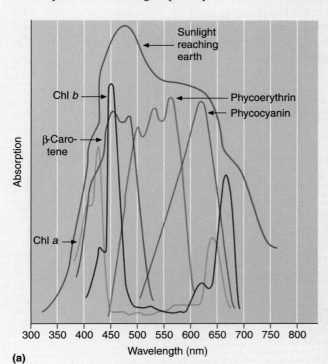

(a)

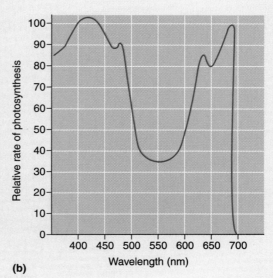

(b)

oxidizing clock. (Refer to Figure 13.12, p. 430.) Pierre Joliot and Bessel Kok studied PSII by measuring the evolution of O_2 when algae or chloroplasts were exposed to brief flashes of light after a period of darkness. (More recently, these experiments have been repeated using membranous vesicles into which PSII reaction centers were inserted.) In 1969, Joliot found that no O_2 is released on the first and second flashes. There is then a burst of O_2 evolution on the third flash. Subsequent O_2 evolution follows an oscillating pattern, with a maximal amount being produced every fourth flash. In 1970, Kok suggested that the oxygen-evolving complex of PSII exists in five transient oxidation states, S_0 through S_4. After the first flash, P680 is converted to P680*. The clock, which provides the electrons that reconvert P680* to P680, releases O_2 when the S_4 state has been reached. It is now believed that the first burst of O_2 comes in the third flash because during a period of darkness the reaction center relaxes into S_1 rather than into S_0. Subsequently, of course, O_2 evolution peaks at every fourth flash. The dampening of the oscillations results from random inefficiencies in the absorption of light by the large number of photosystem complexes being measured.

X–Ray Crystallography

Although X-ray crystallography has been a valuable tool in determining molecular structure (Biochemical Methods 5.1), because investigators cannot crystallize hydrophobic biomolecules it has had limited use. Because many important photosynthetic components are found within membrane, this technique has not been useful in photosynthesis research. However, recently small amphipathic organic molecules have been used during the extraction and purification of membrane proteins, a process referred to as cocrystallization. Using this technique the structures of the reaction center in the rhodopseudomonads (a group of purple nonsulfur bacteria) have been determined. The structural information from X-ray crystallography combined with the knowledge gained from spectroscopy has provided a coherent view of photosynthetic electron transport.

Radioactive Tracers

Because numerous reaction pathways occur simultaneously within living organisms, tracing specific biochemical pathways can be frustrating. However, if biomolecules can be "tagged" (labeled) with a tracer (a substance whose presence can be monitored), reaction pathways become easier to investigate. Radioactive isotopes have been very valuable in tracing the metabolic fate of labeled molecules.

The nucleus of a radioisotope is unstable, i. e., it decays to form a more stable nucleus. This process can be monitored by instruments that measure radiation emissions, such as Geiger counters and scintillation counters, or by autoradiography (Biochemical Methods 2.1).

One of the earliest radioactive tracers was ^{14}C, used by Melvin Calvin and his associates in the 1950s as they investigated carbon fixation in algae. To determine the pathway by which CO_2 is incorporated into carbohydrate, the Calvin team devised an ingenious apparatus (Figure 13E). The labeling of reaction intermediates is limited to the first few stages of the carbon fixation pathway. Unlabeled CO_2 is bubbled into a transparent reservoir that contains a suspension of the algae *Chlorella*. After the reservoir is illuminated and photosynthesis is well underway, a stopcock is opened, and the algae are allowed to flow through a narrow glass tube into a beaker of boiling methanol. (Once algae enter boiling methanol, they are killed, and their metabolism is arrested.) $^{14}CO_2$ can be introduced at specific points along the tube. The exposure of the algae to ^{14}C can then be precisely timed. Photosynthesis continues as the algae flow in the tube, and the organism's processing of the labeled carbon continues until the cells are killed in the methanol. The Calvin team analyzed the alcohol extract with paper chromatography and autoradiography. They determined the pathway by which carbon is assimilated in the algae by varying the exposure time. For example, the team found that, after a 5-second exposure to $^{14}CO_2$, most ^{14}C appears in glycerate-3-phosphate. After a 30-second exposure, most ^{14}C is found in hexose-phosphate.

FIGURE 13E

Calvin Apparatus for Investigations of CO_2 Fixation.

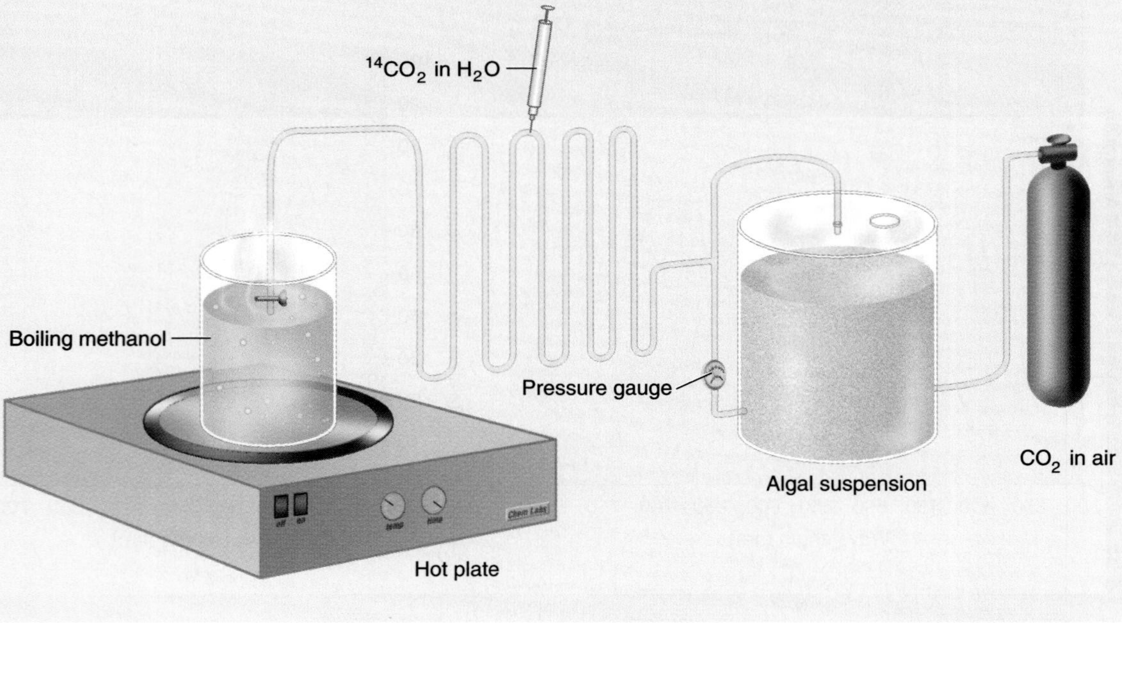

$^{14}CO_2$ in H_2O

Boiling methanol

Pressure gauge

Algal suspension

CO_2 in air

Hot plate

SUMMARY

1. In plants, photosynthesis takes place in chloroplasts. Chloroplasts possess three membranes. The outer membrane is highly permeable, whereas the inner membrane possesses a variety of carrier molecules that regulate molecular traffic into and out of the chloroplast. A third membrane, called the thylakoid membrane, forms an intricate series of flattened vesicles called grana.

2. Photosynthesis consists of two major phases: the light reactions and the light-independent reactions. During the light reactions, water is oxidized, O_2 is evolved, and the ATP and NADPH required to drive carbon fixation are produced. The major working units of the light reactions are photosystems I and II, cytochrome b_6f complex, and the ATP synthase. During the light-independent reactions, CO_2 is incorporated into organic molecules. The first stable product of carbon fixation is glycerate-3-phosphate. The Calvin cycle is composed of three phases: carbon fixation, reduction, and regeneration.

3. Most of the carbon incorporated during the Calvin cycle is used initially to synthesize starch and sucrose, both of which are important energy sources. Sucrose is also important because it is used to translocate fixed carbon throughout the plant.

4. Photorespiration is a process whereby O_2 is consumed and CO_2 is released from plants. Its role in plant metabolism is not understood, because it is apparently a wasteful process. C4 plants and CAM plants, which grow in relatively stringent environments, have developed biochemical and anatomical mechanisms for suppressing photorespiration.

5. Because plants must adapt to ever-changing environmental conditions, the regulation of photosynthesis is complex. Several features of this control are now well established. Light is an important regulator of most aspects of photosynthesis. Many of the effects of light are mediated by changes in the activities of key enzymes. The mechanisms by which light effects these changes include changes in pH, Mg^{2+} concentration, the ferredoxin-thioredoxin system, and phytochrome. The most important enzyme in photosynthesis is ribulose-1,5-bisphosphate carboxylase. Its activity is highly regulated. Light activates the synthesis of both types of the enzyme's subunits. In addition, its activity is affected by allosteric effectors.

SUGGESTED READINGS

Allen, J. F., and Forsberg, J., Molecular Recognition in Thylakoid Structure and Function, *Trends Plant Sci.*, 6(7):317–326, 2001.

Barber, J., and Andersson, B., Too Much of a Good Thing: Light Can Be Bad for Photosynthesis, *Trends Biochem. Sci.*, 17:61–66, 1992.

Blankenship, R. E., and Hartman, H., The Origin and Evolution of Oxygenic Photosynthesis, *Trends Biochem. Sci.*, 23:94–97, 1998.

Furuya, M., and Schafer, E., Photoperception and Signalling of Induction Reactions by Different Phytochromes, *Trends Plant Sci.*, 1(9):301–307, 1996.

Govindjee, and Coleman, W. J., How Plants Make Oxygen, *Sci. Amer.*, 262:50–58, 1990.

Hoganson, C. W., and Babcock, G. T., A Metalloradical Mechanism for the Generation of Oxygen from Water in Photosynthesis, *Science*, 277:1953–1956, 1997.

Huber, R., A Structural Basis of Light Energy and Electron Transfer in Biology, *Eur. J. Biochem.*, 187:283–305, 1990.

Prince, R. C., Photosynthesis: The Z-Scheme Revised, *Trends Biochem. Sci.*, 21(4):121–122, 1996.

Rogner, M., Boekema, E. J., and Barber, J. How Does Photosystem 2 Split Water? The Structural Basis of Efficient Energy Conversion, *Trends Biochem. Sci.*, 21(2):44–49, 1996.

Szalai, V. A., and Brudvig, G. W., How Plants Produce Dioxygen, *Amer. Sci.*, 86:342–351, 1998.

Youvan, D. C., and Marrs, B. L., Molecular Mechanism of Photosynthesis, *Sci. Amer.*, 256(6):42–48, 1987.

KEY WORDS

REVIEW QUESTIONS

1. Define the following terms:
   a. photosystem
   b. reaction center
   c. light reaction
   d. dark reaction
   e. chloroplast
   f. photorespiration
2. What was the most significant contribution of early photosynthetic organisms to the Earth's environment?
3. List the three primary photosynthetic pigments and describe the role each plays in photosynthesis.
4. List five ways in which chloroplasts resemble mitochondria.
5. Excited molecules can return to the ground state by several means. Describe each briefly. Which of these processes are important in photosynthesis? Describe how they function in a living organism.
6. What is the final electron acceptor in photosynthesis?
7. What reactions occur during the light reactions of photosynthesis?
8. What reactions occur during the light-independent reactions of photosynthesis?
9. Why is the oxygen-evolving system referred to as a clock?
10. If the rate of photosynthesis versus the incident wavelength of light is plotted, an action spectrum is obtained. How can the action spectrum provide information about the nature of the light-absorbing pigments involved in photosynthesis?
11. Using the action spectrum for photosynthesis on p.445, determine what wavelengths of light appear to be optimal for photosynthesis.
12. What is the Emerson enhancement effect? How was it used to demonstrate the existence of two different photosystems? (*Hint*: Refer to Biochemical Methods 13.1.)
13. List the types of metals that are components of the photosynthesis mechanism. What functions do they serve?
14. Explain the following observation. When a photosynthetic system is exposed to a brief flash of light, no oxygen is evolved. Only after several bursts of light is oxygen evolved.
15. What is the Z scheme of photosynthesis? How are the products of this reaction used to fix carbon dioxide?
16. Where does carbon dioxide fixation take place in the cell?
17. The chloroplast has a highly organized structure. How does this structure help make photosynthesis possible?

THOUGHT QUESTIONS

1. Without carbon dioxide, chlorophyll fluoresces. How does carbon dioxide prevent this fluorescence?
2. The statement has been made that the more extensively conjugated a chromophore is, the less energy a photon needs to excite it. What is conjugation and how does it contribute to this phenomenon?
3. Increasing the intensity of the incident light but not its energy increases the rate of photosynthesis. Why is this so?
4. Both oxidative phosphorylation and photophosphorylation trap energy in high-energy bonds. How are these processes different? How are they the same?
5: In C3 plants, high concentrations of oxygen inhibit photosynthesis. Why is this so?
6. Generally, increasing the concentration of carbon dioxide increases the rate of photosynthesis. What conditions could prevent this effect?
7. It has been suggested that chloroplasts, like mitochondria, evolved from living organisms. What features of the chloroplast suggest that this is true?
8. Explain why photorespiration is repressed by high concentrations of carbon dioxide.
9. Why does exposing C3 plants to high temperatures raise the carbon dioxide compensation point?
10. Certain herbicides act by promoting photorespiration. These herbicides are lethal to C3 plants but do not affect C4 plants. Why is this so?
11. Corn, a grain of major economic importance, is a C4 plant, and many weeds in temperate climates are C3 plants. Therefore the herbicides described in Question 10 are widely used. What effect is likely if these materials are not degraded before they wash into the ocean?

CHAPTER FOURTEEN

Nitrogen Metabolism I: Synthesis

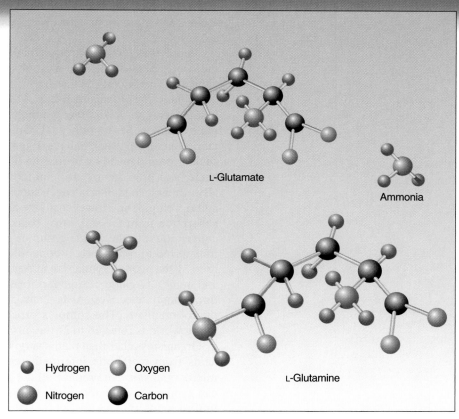

Nitrogen Metabolites. Glutamate, glutamine, and ammonia are among the most important molecules in nitrogen metabolism.

Nitrogen is an essential element found in proteins, nucleic acids, and myriad other biomolecules. Despite the important role that it plays in living organisms, biologically useful nitrogen is scarce. Although nitrogen gas (N_2) is plentiful in the atmosphere, it is almost chemically inert. Therefore, converting N_2 to a useful form requires a major expenditure of energy. Only a few organisms can perform this function, referred to as nitrogen fixation. Certain microorganisms (all of which are bacteria or cyanobacteria) can reduce N_2 to form NH_3 (ammonia). Plants and microorganisms can absorb NH_3 and NO_3^- (nitrate), the oxidation product of ammonia. Both molecules are then used to synthesize nitrogen-containing biomolecules. (The conversion of NH_3 to NO_3^- in a process referred to as nitrification is also performed by certain microorganisms.) Animals cannot synthesize the nitrogen-containing molecules they require from NH_3 and NO_3^-. Instead, *they must acquire "organic nitrogen," primarily amino acids, from dietary sources (i.e., plants and plant-eating animals). In a complex series of reaction pathways, animals then use amino acid nitrogen to synthesize important metabolites. Chapter 14 describes the synthesis of the major nitrogen-containing molecules (e.g., amino acids and nucleotides) and a small selection of molecules that represent the rich diversity of metabolites that contain this critically important element.*

Nitrogen is found in an astonishingly vast array of biomolecules. Examples of major nitrogen-containing metabolites include amino acids, nitrogenous bases, porphyrins, and several lipids. In addition, miscellaneous nitrogen-containing metabolites that are required in smaller amounts (e.g., the biogenic amines and glutathione) are also critically important in the metabolism of many eukaryotes.

As previously mentioned (see p. 6), the incorporation of nitrogen into organic molecules begins with the fixation (reduction) of N_2 by prokaryotic microorganisms. The organisms that fix nitrogen are found in many environments. For example, some organisms live in symbiotic relationships within the roots of certain plants (e.g., *Rhizobium* is found within the root nodule cells of leguminous plants such as peas and beans). Other nitrogen-fixing organisms are found in marine and fresh water, in hot springs, or within the guts of certain animals. Plants such as corn depend on absorbing NH_3 and NO_3^- which are synthesized by soil bacteria or provided by artificial fertilizers. Because little fixed nitrogen is usually available to plants, nitrogen supply is often the limiting factor in plant growth and development. However plants acquire NH_3, whether by nitrogen fixation, absorption from the soil, or by reduction of absorbed NO_3^-, it is assimilated by conversion into the amide group of glutamine. Then this "organic nitrogen" is transferred to other carbon-containing compounds to produce the amino acids used by the plant to synthesize nitrogenous molecules (e.g., proteins, nucleotides, and heme). Organic nitrogen, primarily in the form of amino acids, then flows throughout the ecosystem as plants are consumed by animals and decomposing microorganisms. The complex process that transfers nitrogen throughout the living world is referred to as the *nitrogen cycle*.

Organisms other than plants vary widely in their capacity to synthesize amino acids from metabolic intermediates and fixed nitrogen. For example, many microorganisms can produce all the amino acids they need. In contrast, animals can synthesize only about half the amino acids they require. The **nonessential amino acids (NAA)** are synthesized from readily available metabolites. The amino acids that must be provided in the diet to ensure proper nitrogen balance and adequate growth are referred to as **essential amino acids (EAA)**.

Following the digestion of dietary protein in the body's digestive tract, free amino acids are transported across intestinal mucosal cells and into the blood. Because most diets do not provide amino acids in the proportions that the body requires, their concentrations must be adjusted by metabolic mechanisms. The amino acids released to the blood in the intestine already show some changes in their relative concentrations. For example, alanine levels are higher and glutamate and glutamine levels are lower than those found in the predigested protein. Further changes occur when the blood reaches the liver, where the fate of each amino acid is determined. Excessive amounts of all NAA and most EAA are degraded. The concentrations of certain EAA, referred to as the **branched chain amino acids (BCAA)**, remain unchanged. (The BCAA are leucine, isoleucine, and valine.) Therefore blood leaving the liver after a protein-rich meal is enriched in BCAA because of selective degradation of excessive amounts of other amino acids. Apparently, BCAA represent a major transport form of amino nitrogen from the liver to other tissues, where they are used in the synthesis of the NAA required for protein synthesis, as well as various amino acid derivatives.

Transamination reactions dominate amino acid metabolism. In these reactions, catalyzed by a group of enzymes referred to as the *aminotransferases* or *transaminases*, α-amino groups are transferred from an α-amino acid to an α-keto acid:

Acceptor keto acid	Donor amino acid		New keto acid	New amino acid

(Recall that in *α-keto acids* such as α-ketoglutarate and pyruvate, a carbonyl group is directly adjacent to the carboxyl group. Because transamination reactions are readily reversible, they play important roles in both the synthesis and degradation of the amino acids.

After a discussion of nitrogen fixation, the essential features of amino acid biosynthesis are described. This is followed by descriptions of the biosynthesis of selected nitrogen-containing molecules. A special emphasis is placed on the anabolic pathways of the nucleotides. In the following chapter (Chapter 15) the flow of nitrogen atoms is traced through several catabolic pathways to the nitrogenous waste products excreted by animals.

14.1   NITROGEN FIXATION

Several circumstances limit the amount of usable nitrogen available in the biosphere. Because of the chemical stability of the atmospheric gas dinitrogen (N_2), its reduction to form NH_3 (referred to as **nitrogen fixation**) requires a large energy input. For example, at least 16 ATP are required to reduce one N_2 to two NH_3. In addition, only a few prokaryotic species can "fix" nitrogen. The most prominent of these are several species of free-living bacteria (e.g., *Azotobacter vinelandii* and *Clostridium pasteurianum*), the cyanobacteria (e.g., *Nostoc muscorum* and *Anabaena azollae*), and symbiotic bacteria (e.g., several species of *Rhizobium*). Symbiotic organisms form mutualistic, that is, mutually beneficial, relationships with host plants or animals. *Rhizobium* species, for example, infect the roots of leguminous plants such as soybeans and alfalfa (Special Interest Box 14.1).

All species that can fix nitrogen possess the *nitrogenase complex*. Its structure, similar in all species so far investigated, consists of two proteins called dinitrogenase and dinitrogenase reductase. Dinitrogenase (240 kD), also referred to as *Fe-Mo protein*, is an $\alpha_2\beta_2$-heterotetramer that contains two molybdenum (Mo) atoms, and 30 iron atoms. It catalyzes the reaction $N_2 + 8\ H^+ + 8\ e^- \longrightarrow 2\ NH_3 + H_2$. Dinitrogenase reductase (60 kD) (also referred to as *Fe protein*) is a dimer containing identical subunits.

Despite considerable research, nitrogen fixation is not yet completely understood. However, several aspects of nitrogen fixation have been elucidated (Figure 14.1). NADH (or NADPH) is the ultimate source of the electrons required in the reduction of dinitrogen. The reduced coenzyme molecules donate electrons to the iron-sulfur protein ferredoxin, which then transfers them to dinitrogenase reductase. The hydrolysis of 16 molecules of ATP is required to transfer 8 electrons from dinitrogenase reductase to dinitrogenase to facilitate the reduction of one N_2 to two NH_3 molecules and two hydrogen ions to one H_2 molecule. Once it is synthesized, ammonia is translocated out of the bacterial cells into symbiotic host cells, where it is used in glutamine synthesis (Section 14.2). (In the presence of ATP and low amounts of N_2, significant amounts of H_2 are evolved. It has been

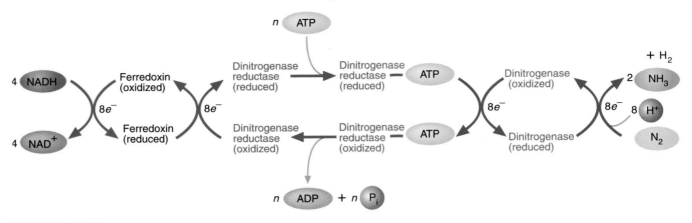

FIGURE 14.1

Schematic Diagram of the Nitrogenase Complex Illustrating the Flow of Electrons and Energy in Enzymatic Nitrogen Fixation.
The high energy of activation of nitrogen fixation is overcome by a large number of ATP molecules (about 16 ATP per N_2 molecule). Both the binding of ATP to dinitrogenase reductase and its subsequent hydrolysis cause conformational changes in the protein that facilitate the transfer of electrons to dinitrogenase.

estimated, for example, that U.S. soybean crops infected with *Rhizobium japonicum* produce billions of cubic feet of H_2 annually.)

Both components of the nitrogenase complex are irreversibly inactivated by O_2. Nitrogen-fixing organisms solve this problem in several ways. Anaerobic organisms such as *Clostridium* grow only in anaerobic soil, whereas many of the cyanobacteria produce specialized nitrogenase-containing cells called heterocysts. The thick cell walls of heterocysts isolate the enzyme from atmospheric oxygen. Legumes produce an oxygen-binding protein called *leghemoglobin*, which traps oxygen before it can interact with the nitrogenase complex.

QUESTION 14.1

Provide the structures for the products of the nitrogenase complex for each of the following substrates (real and hypothetical) that contain triple bonds:

a. hydrogen cyanide
b. dinitrogen
c. acetylene

QUESTION 14.2

What are the substrates and enzyme components in nitrogen fixation? Arrange each in the correct order in which electrons and protons are transferred to dinitrogen.

14.2 AMINO ACID BIOSYNTHESIS

Living organisms differ in their capacity to synthesize the amino acids required for protein synthesis. Although plants and many microorganisms can produce all their amino acids from readily available precursors, other organisms must obtain some preformed amino acids from their environment. For example, mammalian tissues can synthesize NAA (Table 14.1) by relatively simple reaction pathways. In contrast, EAA must be obtained from the diet because mammals lack the long and complex reaction pathways required for their synthesis.

Farmers have used crop rotation utilizing legumes for centuries to preserve soil fertility. In crop rotation, fields are alternately used to grow nitrogen-demanding crops such as corn and the nitrogen-producing legumes. Despite its importance in food production, the origin of this practice remains obscure. It was not until 1888 that bacteria (later called *Rhizobium*) were discovered within the root nodules of leguminous plants. Subsequent patent applications for a legume innoculant, filed by both British and American companies, were an early milestone in commercial attempts to improve agriculture.

As nitrogen's role in soil fertility became known, commercially produced fertilizers came into use. Guano was one of the most popular commercial fertilizers until the early twentieth century, (Guano is a rich source of nitrate. Found in large deposits on islands off the coast of Chile and Peru, guano is essentially the manure of sea birds.) Commercial production of artificial fertilizers became possible in the early twentieth century. Between 1907 and 1909 the German chemist Fritz Haber developed a method of producing ammonia from N_2 and H_2 at high temperature and pressure, referred to as the Haber process. The original Haber process used H_2 obtained from the production of coke (coal heated so intensely that most gases have been removed) and N_2 derived from the fractional distillation of liquid air. The hydrogen and nitrogen gases were heated at 550°C at a pressure of 2 atm in the presence of an iron catalyst. The ammonia produced was oxidized to form nitrate, which was used as a fertilizer. The higher temperatures and pressures (i.e., 700°C and 1000 atm) used in modern industrial nitrogen fixation have significantly improved the efficiency of the Haber process.

Modern improvements in the efficiency of the Haber process have led to the widespread use of commercially produced fertilizer. This factor plus an ever increasing world population (estimated to be 6.9 billion in 2010) and dwindling and inevitably more expensive fossil fuel supplies have resulted in substantially increased agricultural costs. In response to the growing pressure to improve the cost-effectiveness of agricultural crop production, life scientists have explored alternative technologies. One possible answer to future fixed nitrogen requirements is the use of DNA technology to produce plant species with their own nitrogen-fixing capability. Although scientists have been working for almost two decades on this project, success has not yet been achieved. Biological nitrogen fixation is unexpectedly complex. For example, nitrogen fixation by the bacterium *Klebsiella* requires the functions of at least 18 genes (referred to as nif genes). In addition to coding for the protein components of nitrogenase, these genes also code for molecules used in electron transport, metal processing, and coordination of the components into a functioning nitrogen-fixing system. Despite these difficulties, any success in this endeavor (e.g., producing plants that can fix their own nitrogen or develop symbiotic relationships with nitrogen-fixing bacteria) will reduce dependence on expensive artificial fertilizers. The consequences in terms of food production and food prices will be revolutionary.

Amino Acid Metabolism Overview

Amino acids serve a number of functions. Although the most important role of amino acids is the synthesis of proteins, they are also the principal source of the nitrogen atoms required in various synthetic reaction pathways. In addition, the nonnitrogen parts of amino acids (referred to as carbon skeletons) are a source of energy, as well as precursors, in several reaction pathways. Therefore an adequate intake of amino acids, in the form of dietary protein, is essential for an animal's proper growth and development.

Dietary protein sources differ widely in their proportions of the EAA. In general, complete proteins (those containing sufficient quantities of EAA) are of animal origin (e.g., meat, milk, and eggs). Plant proteins often lack one or more EAA. For example, gliadin (wheat protein) has insufficient amounts of lysine, and zein (corn protein) is low in both lysine and tryptophan. Because plant proteins differ in their amino acid compositions, plant foods can provide a high-quality source of essential amino acids only if they are eaten in appropriate combinations. One such combination includes beans (low in methionine) and cereal grains (low in lysine).

The amino acid molecules that are immediately available for use in metabolic processes are referred to as the **amino acid pool**. In animals, amino acids in the pool are derived from the breakdown of both dietary and tissue proteins. Excreted nitrogenous products such as urea and uric acid are output from the pool. Amino acid metabolism is a complex series of reactions in which the amino acid molecules required for the syntheses of proteins and metabolites are continuously being synthesized and degraded. Depending on current metabolic

TABLE 14.1

The Essential and Nonessential Amino Acids in Humans

Essential	Nonessential
Isoleucine	Alanine
Leucine	Arginine*
Lysine	Asparagine
Methionine	Aspartate
Phenylalanine	Cysteine
Threonine	Glutamate
Tryptophan	Glutamine
Valine	Glycine
	Histidine*
	Proline
	Serine
	Tyrosine

*Amino acids that are essential for infants.

requirements, certain amino acids are synthesized or interconverted and then transported to tissue, where they are used. When nitrogen intake (primarily amino acids) equals nitrogen loss, the body is said, to be in *nitrogen balance*. This is the condition of healthy adults. In *positive nitrogen balance*, a condition that is characteristic of growing children, pregnant women, and recuperating patients, nitrogen intake exceeds nitrogen loss. The excess of nitrogen is retained because the amount of tissue proteins being synthesized exceeds the amount being degraded. *Negative nitrogen balance* exists when an individual cannot replace nitrogen losses with dietary sources. *Kwashiorkor* ("the disease the first child gets when the second is on the way") is a form of malnutrition caused by a prolonged insufficient intake of protein. Its symptoms include growth failure, apathy, ulcers, liver enlargement, and diarrhea, as well as decreased mass and function of the heart and kidneys. Prevalent in Africa, Asia, and Central and South America, kwashiorkor can be treated by feeding high-quality protein-rich foods such as milk, eggs, and meat.

Transport of amino acids into cells is mediated by specific membrane-bound transport proteins, several of which have been identified in mammalian cells. They differ in their specificity for the types of amino acids transported and in whether the transport process is linked to the movement of Na^+ across the plasma membrane. (Recall that the gradient created by the active transport of Na^+ can move molecules across membrane. Na^+-dependent amino acid transport is similar to that observed in the glucose transport process illustrated in Figure 11.28.) For example, several Na^+-dependent transport systems have been identified within the lumenal plasma membrane of enterocytes. Na^+-independent transport systems are responsible for transporting amino acids across the portion of enterocyte plasma membrane in contact with blood vessels. The γ-glutamyl cycle (Section 14.3) is believed to assist in transporting some amino acids into specific tissues (i.e., brain, intestine, and kidney).

QUESTION 14.3

If a single essential amino acid is missing from a cell, protein synthesis is blocked. Explain.

Reactions of Amino Groups

Once amino acid molecules enter cells, the amino groups are available for synthetic reactions. This metabolic flexibility is effected primarily by transamination reactions in which amino groups are transferred from an α-amino acid to an α-keto acid. However, another class of reactions, in which NH_4^+ or the amide nitrogen of glutamine is used to supply the amino group or the amide nitrogen of certain amino acids, also occurs. These reaction types are discussed next.

TRANSAMINATION Eukaryotic cells possess a large variety of aminotransferases. Found within both the cytoplasm and mitochondria, these enzymes possess two types of specificity: (1) the type of α-ammo acid that donates the α-amino group and (2) the α-keto acid that accepts the α-amino group. Although the aminotransferases vary widely in the type of amino acids they bind, most of them use glutamate as the amino group donor:

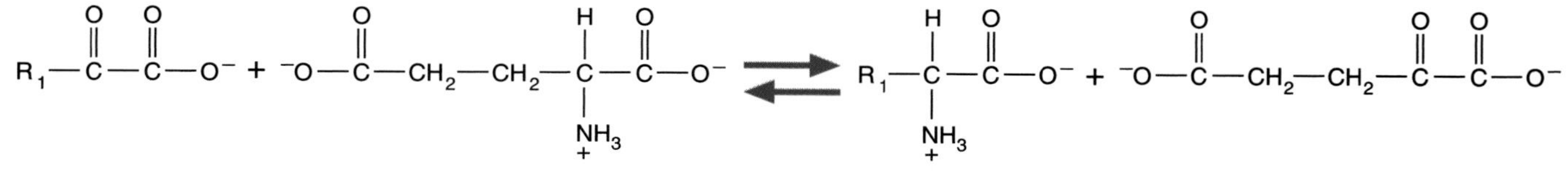

Acceptor α-keto acid   **Glutamate**   **New amino acid**   **α-Ketoglutarate**

Because glutamate is produced when α-ketoglutarate (a citric acid cycle intermediate) accepts an amino group, these two molecules (referred to as the *α-ketoglutarate/glutamate pair*) have a strategically important role in both amino acid metabolism and metabolism in general. Two other such pairs have important functions in metabolism. In addition to its role in transamination reactions, the *oxaloacetate/aspartate pair* is involved in the disposal of nitrogen in the urea cycle (Chapter 15). One of the most important functions of the *pyruvate/alanine pair* is in the alanine cycle (Figure 8.9). Because α-ketoglutarate and oxaloacetate are citric acid cycle intermediates, transamination reactions often represent an important mechanism for meeting the energy requirements of cells.

Transamination reactions require the coenzyme pyridoxal-5'-phosphate (PLP), which is derived from pyridoxine (vitamin B_6). PLP is also required in numerous other reactions of amino acids. Examples include racemizations, decarboxylations, and several side chain modifications. (**Racemizations** are reactions in which mixtures of L- and D-amino acids are formed.) The structures of the vitamin and its coenzyme form are illustrated in Figure 14.2.

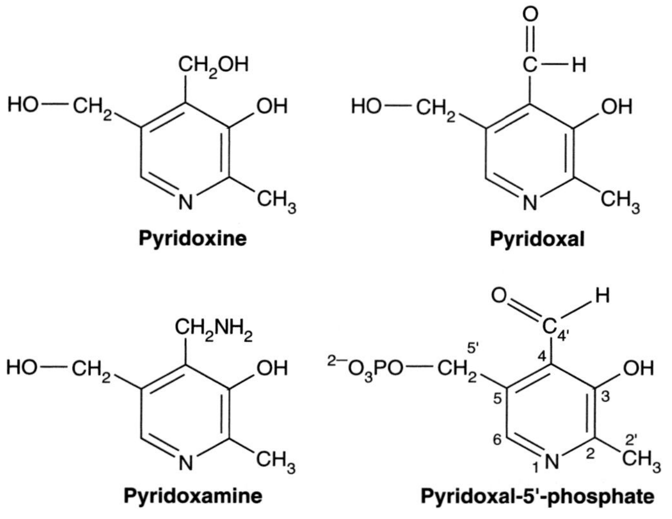

FIGURE 14.2

Vitamin B_6.

Vitamin B_6 includes (a) pyridoxine, (b) pyridoxal, and (c) pyridoxamine. (Pyridoxine is found in leafy green vegetables. Pyridoxal and pyridoxamine are found in animal foods such as fish, poultry, and red meat.) The biologically active form of vitamin B_6 is (d) pyridoxal-5'-phosphate.

PLP is bound in the enzyme active site by noncovalent interactions and a Schiff base (R′—CH=N—R, an aldimine) formed by the condensation of the aldehyde group of PLP and the ε-amino group of a lysine residue.

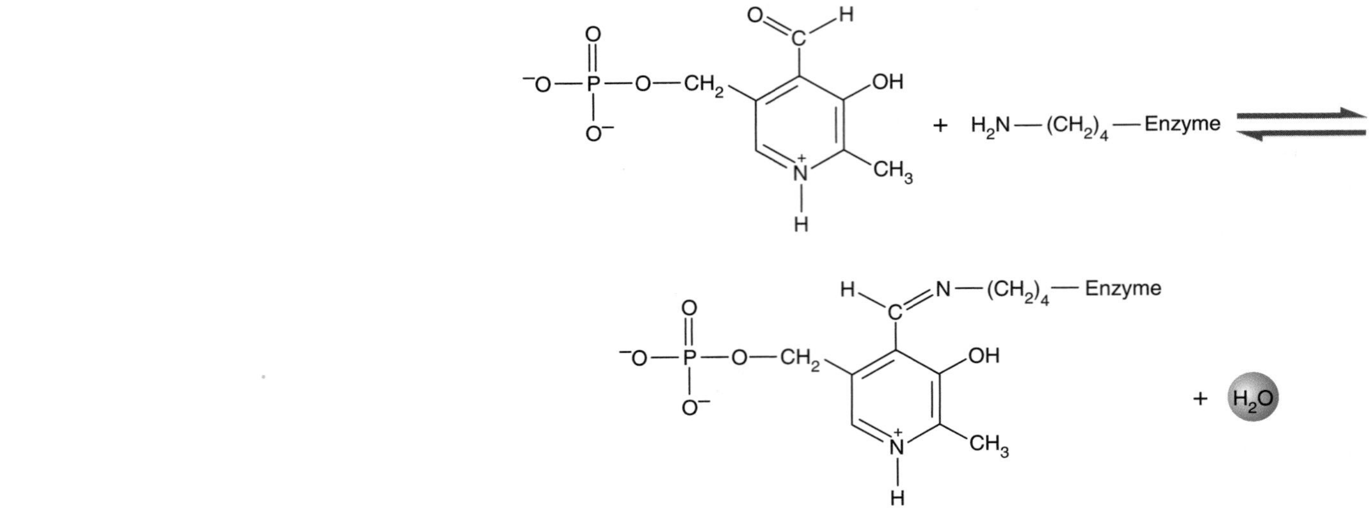

Additional stabilizing forces include ionic interactions between amino acid side chains and PLP's pyridinium ring and phosphate group. The positively charged pyridinium ring also functions as an electron sink, stabilizing negatively charged reaction intermediates.

Amino acid substrates become bound to PLP via the α-amino group in an imine exchange reaction. Then one of three bonds of the α-carbon atom is selectively broken in the active sites in each type of PLP-dependent enzyme.

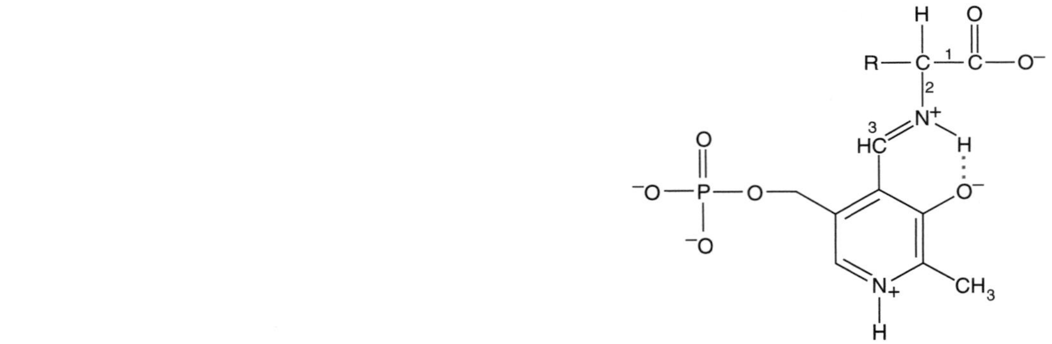

This selectivity is dependent upon the presence or absence of a nearby base catalyst and the orientation of the amino acid in the active site. If an initial deprotonation of the α-carbon of the amino group donor occurs, then transamination (bond 2 broken), or racemization or elimination (bond 3 broken) may occur. If the initial deprotonation does not occur, then decarboxylation results (bond 1 broken).

Despite the apparent simplicity of the transamination reaction, the mechanism is quite complex. The reaction begins with the formation of a Schiff base between PLP and the α-amino group of an α-amino acid (Figure 14.3). When the α-hydrogen atom is removed by a general base in the enzyme active site, a resonance-stabilized intermediate forms. With the donation of a proton from a general acid and a subsequent hydrolysis, the newly formed α-keto acid is released from the enzyme. A second α-keto acid then enters the active site and is converted into an α-amino acid in a reversal of the reaction process that has just been described. Transamination reactions are examples of a reaction mechanism referred to as a

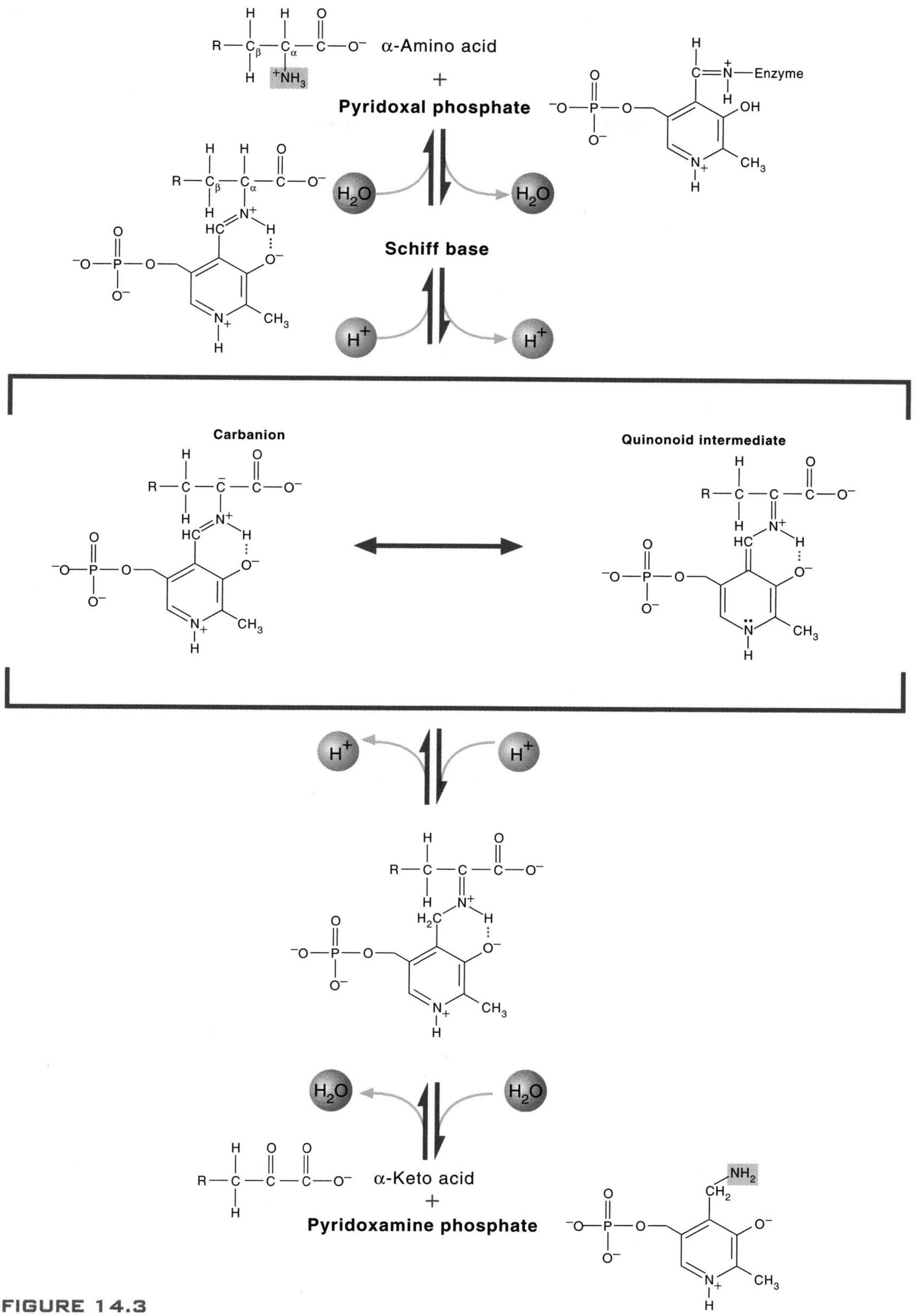

FIGURE 14.3

The Transamination Mechanism.

The donor amino acid forms a Schiff base with pyridoxal phosphate within the enzyme's active site. After a proton is lost, a carbanion forms and is resonance-stabilized by interconversion to a quinonoid intermediate. After an enzyme-catalyzed proton transfer and a hydrolysis, the α-keto product is released. A second α-keto acid then enters the active site. This acceptor α-keto acid is converted to an α-amino acid product as the mechanism just described is reversed.

bimolecular ping-pong reaction. The mechanism is so named because the first substrate must leave the active site before the second one can enter.

QUESTION 14.4

Provide the structure of the α-keto acid product of the transamination of each of the following molecules:

a. glutamine

b. isoleucine

c. phenylalanine

d. aspartate

e. cysteine

Because transamination reactions are reversible, it is theoretically possible for all amino acids to be synthesized by transamination. However, experimental evidence indicates that there is no net synthesis of an amino acid if its α-keto acid precursor is not independently synthesized by the organism. For example, alanine, aspartate, and glutamate are nonessential for animals because their α-keto acid precursors (i.e., pyruvate, oxaloacetate, and α-ketoglutarate) are readily available metabolic intermediates. Because the reaction pathways for synthesizing molecules such as phenylpyruvate, α-keto-β-hydroxybutyrate, and imidazolepyruvate do not occur in animal cells, phenylalanine, threonine, and histidine must be provided in the diet. (Reaction pathways that synthesize amino acids from metabolic intermediates, not only by transamination, are referred to as *de novo* pathways.)

DIRECT INCORPORATION OF AMMONIUM IONS INTO ORGANIC MOLECULES There are two principal means by which ammonium ions are incorporated into amino acids and eventually other metabolites: (1) reductive amination of α-keto acids and (2) formation of the amides of aspartic and glutamic acid with subsequent transfer of the amide nitrogen to form other amino acids.

Glutamate dehydrogenase, an enzyme found in both the mitochondria and cytoplasm of eukaryotic cells and in some bacterial cells, catalyzes the direct amination of α-ketoglutarate:

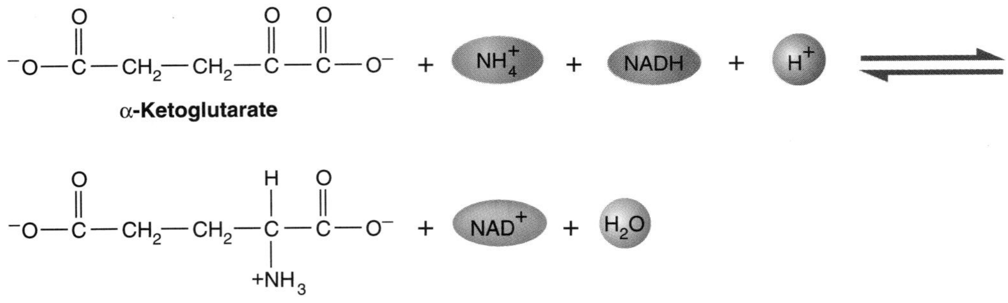

α-**Ketoglutarate**

Glutamate

The primary function of this enzyme in eukaryotes appears to be catabolic (i.e., a means of producing NH_4^+ in preparation for nitrogen excretion). However, the reaction is reversible. When excess ammonia is present, the reaction is driven toward glutamate synthesis.

Ammonium ions are also incorporated into cell metabolites by the formation of glutamine, the amide of glutamate:

$$^-O-\overset{\overset{\displaystyle O}{\|}}{C}-CH_2-CH_2-\overset{\overset{\displaystyle H}{|}}{\underset{\underset{\displaystyle ^+NH_3}{|}}{C}}-\overset{\overset{\displaystyle O}{\|}}{C}-O^- \;+\; \boxed{ATP} \;+\; \boxed{NH_4^+} \longrightarrow$$

Glutamate

$$H_2N-\overset{\overset{\displaystyle O}{\|}}{C}-CH_2-CH_2-\overset{\overset{\displaystyle H}{|}}{\underset{\underset{\displaystyle ^+NH_3}{|}}{C}}-\overset{\overset{\displaystyle O}{\|}}{C}-O^- \;+\; \boxed{ADP} \;+\; \boxed{P_i}$$

Glutamine

The brain, a rich source of the enzyme glutamine synthase, is especially sensitive to the toxic effects of NH_4^+. Brain cells convert NH_4^+ to glutamine, a neutral, nontoxic molecule. Glutamine is then transported to the liver, where the production of nitrogenous waste occurs.

In plants, the pathway by which most NH_4^+ is incorporated into organic molecules requires two enzymes: glutamine synthase and glutamate synthase. After NH_4^+ is incorporated into glutamine by glutamine synthase, the amide nitrogen is transferred to the 2-keto group of α-ketoglutarate by glutamate synthase. The 2 electrons required in this reaction are provided by reduced ferredoxin in some plant tissues (e.g., leaves) and NADPH in other tissues (e.g., roots and germinating seeds).

$$H_2N-\overset{\overset{\displaystyle O}{\|}}{C}-CH_2-CH_2-\overset{\overset{\displaystyle H}{|}}{\underset{\underset{\displaystyle ^+NH_3}{|}}{C}}-\overset{\overset{\displaystyle O}{\|}}{C}-O^- \;+\; ^-O-\overset{\overset{\displaystyle O}{\|}}{C}-CH_2-CH_2-\overset{\overset{\displaystyle O}{\|}}{C}-\overset{\overset{\displaystyle O}{\|}}{C}-O^- \;+\; 2\,e^-$$

Glutamine   **α-Ketoglutarate**

$$\longrightarrow \quad 2\; ^-O-\overset{\overset{\displaystyle O}{\|}}{C}-CH_2-CH_2-\overset{\overset{\displaystyle H}{|}}{\underset{\underset{\displaystyle ^+NH_3}{|}}{C}}-\overset{\overset{\displaystyle O}{\|}}{C}-O^-$$

Glutamate

One of the two glutamate products of this reaction is then used as a substrate in a glutamine synthase–catalyzed reaction. Consequently, in plants, there is a net production of one molecule of glutamate for every NH_4^+ that enters the process:

$$\boxed{NH_4^+} \;+\; ^-O-\overset{\overset{\displaystyle O}{\|}}{C}-CH_2-CH_2-\overset{\overset{\displaystyle O}{\|}}{C}-\overset{\overset{\displaystyle O}{\|}}{C}-O^- \;+\; \boxed{ATP} \;+\; 2\,e^-$$

α-Ketoglutarate

$$\longrightarrow \quad ^-O-\overset{\overset{\displaystyle O}{\|}}{C}-CH_2-CH_2-\overset{\overset{\displaystyle H}{|}}{\underset{\underset{\displaystyle ^+NH_3}{|}}{C}}-\overset{\overset{\displaystyle O}{\|}}{C}-O^- \;+\; \boxed{ADP} \;+\; \boxed{P_i}$$

Glutamate

KEY CONCEPTS 14.1

In transamination reactions, amino groups are transferred from one carbon skeleton to another. In reductive animation, amino acids are synthesized by the incorporating of free NH_4^+ or the amide nitrogen of glutamine or asparagine into α-keto acids. Ammonium ions are also incorporated into cellular metabolites by the animation of glutamate to form glutamine.

Synthesis of the Amino Acids

The amino acids differ from other classes of biomolecules in that each member of this class is synthesized by a unique pathway. Despite the tremendous diversity of amino acid synthetic pathways, they have one common feature. The carbon skeleton of each amino acid is derived from commonly available metabolic intermediates. Thus in animals, all NAA molecules are derivatives of either glycerate-3-phosphate, pyruvate, α-ketoglutarate, or oxaloacetate. Tyrosine, synthesized from the essential amino acid phenylalanine, is an exception to this rule.

On the basis of the similarities in their synthetic pathways, the amino acids can be grouped into six families: glutamate, serine, aspartate, pyruvate, the aromatics, and histidine. The amino acids in each family are ultimately derived from one precursor molecule. In the discussions of amino acid synthesis that follow, the intimate relationship between amino acid metabolism and several other metabolic pathways is apparent. Amino acid biosynthesis is outlined in Figure 14.4.

THE GLUTAMATE FAMILY The glutamate family includes—in addition to glutamate—glutamine, proline, and arginine. As described, α-ketoglutarate may be converted to glutamate by reductive amination and by transamination reactions involving a number of amino acids. Although the relative contribution of these reactions to glutamate synthesis varies with cell type and metabolic circumstances, transamination appears to play a major role in the synthesis of most glutamate molecules in eukaryotic cells. In addition to serving as a component of proteins and as a precursor for other amino acids, glutamate is also used in the central nervous system as an excitatory neurotransmitter. (The binding of *excitatory neurotransmitters* to certain receptors on nerve cell membrane promotes membrane depolarization.)

The conversion of glutamate to glutamine, catalyzed by glutamine synthase, takes place in a number of mammalian tissues (liver, brain, kidney, muscle, and intestine). BCAA (branched chain amino acids) are an important source of amino groups in glutamine synthesis. As mentioned, blood that leaves the liver is selectively enriched in BCAA. Many more BCAA are taken up by peripheral tissues than are needed for protein synthesis. The amino groups of the BCAA may be used primarily for the synthesis of nonessential amino acids. In addition to its role in protein synthesis, glutamine is the amino group donor in numerous biosynthetic reactions (e.g., purine, pyrimidine, and amino sugar syntheses) and, as was previously mentioned, as a safe storage and transport form of NH_4^+. Glutamine is therefore a major metabolite in living organisms. Other functions of glutamine vary, depending on the cell type being considered. For example, in the kidney and small intestine, glutamine is a major source of energy. In the small intestine, approximately 55% of glutamine carbon is oxidized to CO_2.

Proline is a cyclized derivative of glutamate. As shown in Figure 14.5, a γ-glutamyl phosphate intermediate is reduced to glutamate-γ-semialdehyde. The enzyme catalyzing the phosphorylation of glutamate (γ-glutamyl kinase) is regulated by negative feedback inhibition by proline. Glutamate-γ-semialdehyde cyclizes spontaneously to form Δ^1-pyrroline-5-carboxylate. Δ^1-Pyrroline-5-carboxylate reductase catalyzes the reduction of Δ^1-pyrroline-5-carboxylate to form proline. The interconversion of Δ^1-pyrroline-5-carboxylate and proline may act as a shuttle mechanism to transfer reducing equivalents derived from the pentose phosphate pathway into mitochondria. This process may partially explain the high turnover of proline in many cell types. Proline can also be synthesized from ornithine, a urea cycle intermediate. (The urea cycle is a pathway in which urea, the principal mammalian nitrogenous waste product, is produced. Urea synthesis is discussed in Chapter 15.) The enzyme catalyzing ornithine's conversion to glutamate-γ-semialdehyde, ornithine aminotransferase, is found in relatively high concentration in cells (e.g., fibroblasts) where the demand for proline incorporation into collagen is high.

Glutamate is also a precursor of arginine. Arginine synthesis begins with the acetylation of the α-amino group of glutamate. N-acetylglutamate is then

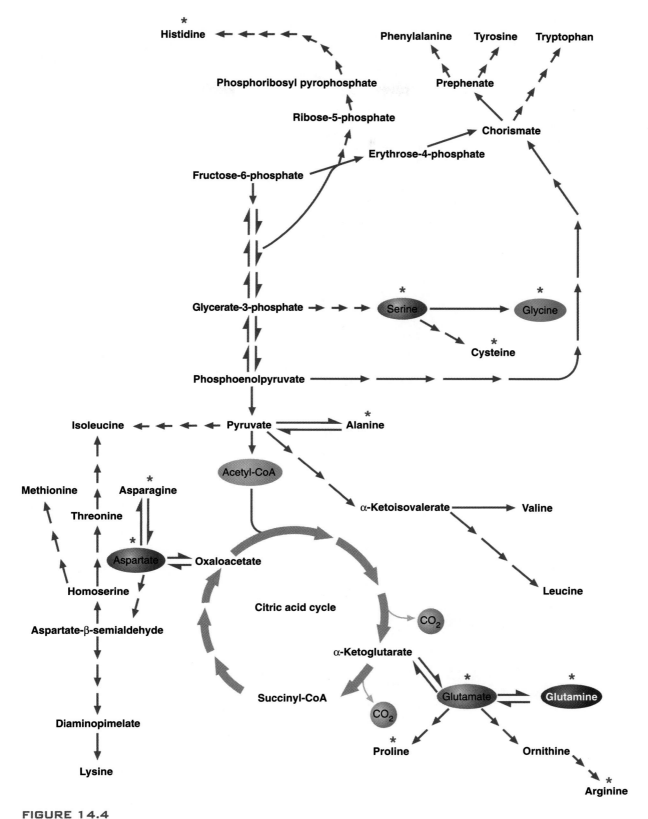

FIGURE 14.4

Biosynthesis of the Amino Acids.

Intermediates in the central metabolic pathways provide the carbon skeleton precursor molecules required for the synthesis of each amino acid. The number of reactions in each pathway is indicated. The nonessential amino acids for mammals are indicated by asterisks. (In mammals, tyrosine can be synthesized from phenylalanine.)

FIGURE 14.5

Biosynthesis of Proline and Arginine from Glutamate.

Proline is synthesized from glutamate in three steps. The second step is a spontaneous cyclization reaction. In arginine synthesis the acetylation of glutamate prevents the cyclization reaction. In mammals the reactions that convert ornithine to arginine are part of the urea cycle.

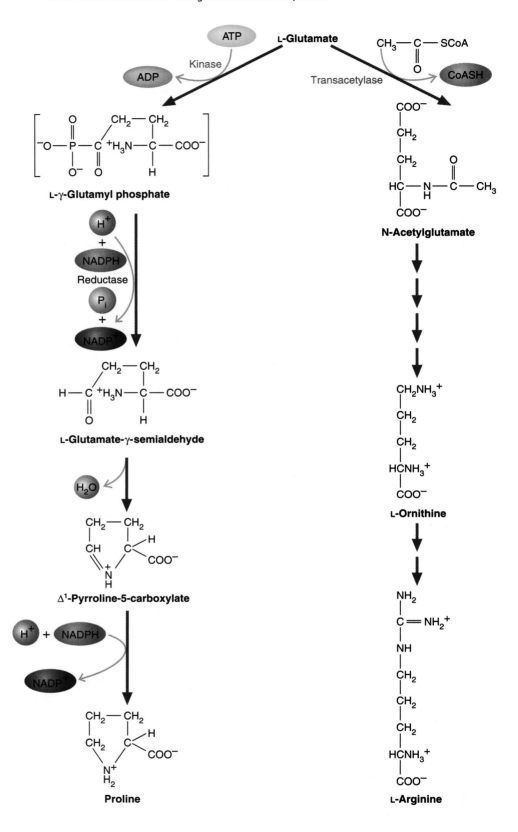

converted to ornithine in a series of reactions that include a phosphorylation, a reduction, a transamination, and a deacetylation (removal of an acetyl group). The subsequent reactions in which ornithine is converted to arginine are part of the urea cycle. In infants, in whom the urea cycle is insufficiently functional, arginine is an essential amino acid.

THE SERINE FAMILY The members of the serine family—serine, glycine, and cysteine—derive their carbon skeletons from the glycolytic intermediate glycerate-3-phosphate. The members of this group play important roles in numerous anabolic pathways. Serine is a precursor of ethanolamine and sphingosine. Glycine is used in the purine, porphyrin, and glutathione synthetic pathways. Together, serine and glycine contribute to a series of biosynthetic pathways that are referred to collectively as one-carbon metabolism (discussed in Section 14.3). Cysteine plays a significant role in sulfur metabolism (Chapter 15).

Serine is synthesized in a direct pathway from glycerate-3-phosphate that involves dehydrogenation, transamination, and hydrolysis by a phosphatase (Figure 14.6). Cellular serine concentration controls the pathway through feedback inhibition of phosphoglycerate dehydrogenase and phosphoserine phosphatase. The latter enzyme catalyzes the only irreversible step in the pathway.

The conversion of serine to glycine consists of a single complex reaction catalyzed by serine hydroxymethyltransferase, a pyridoxal phosphate–requiring enzyme. During the reaction, which is an aldol cleavage, serine binds to pyridoxal phosphate. The reaction yields glycine and a chemically reactive formaldehyde group that is transferred to tetrahydrofolate (THF) to form N^5,N^{10}-methylene tetrahydrofolate. (The coenzyme tetrahydrofolate is discussed in Section 14.3.) Serine is the major source of glycine. Smaller amounts of glycine can be derived from choline, when the latter molecule is present in excess. The synthesis of glycine from choline consists of two dehydrogenations and a series of demethylations. Glycine acts as an inhibitory neurotransmitter within the central nervous system. (When *inhibitory neurotransmitters* bind to nerve cell receptors, which are usually linked to chloride channels, the membrane becomes hyperpolarized. Because the inside of the membrane is more negative in hyperpolarized neurons than it is in resting neurons, action potentials are unlikely.)

Cysteine synthesis is a primary component of sulfur metabolism. The carbon skeleton of cysteine is derived from serine (Figure 14.7). In animals the sulfhydryl group is transferred from methionine by way of the intermediate molecule homocysteine. (Plants and some bacteria obtain the sulfhydryl group by reduction of SO_4^{2-} to S^{2-} as H_2S. A few organisms use H_2S directly from the environment.) Both enzymes involved in the conversion of serine to cysteine (cystathionine synthase and γ-cystathionase) require pyridoxal phosphate.

THE ASPARTATE FAMILY Aspartate, the first member of the aspartate family of amino acids, is derived from oxaloacetate in a transamination reaction:

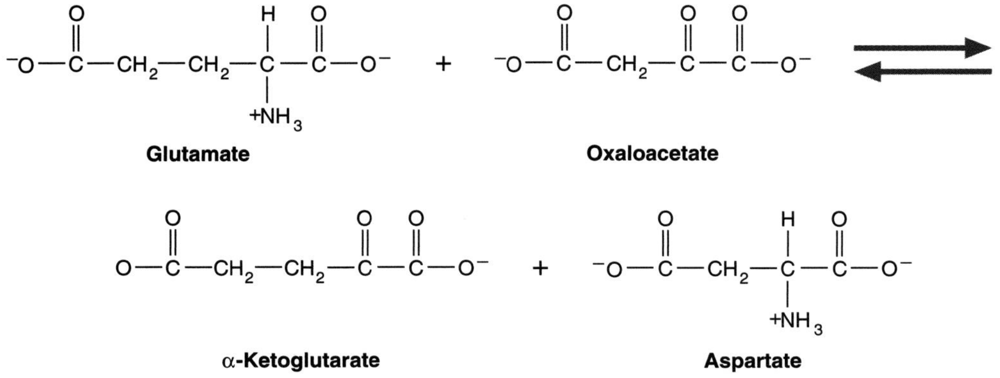

Glutamate   **Oxaloacetate**

α-Ketoglutarate   **Aspartate**

Aspartate transaminase (AST) (also known as glutamic oxaloacetic transaminase, or GOT), the most active of the aminotransferases, is found in most cells. Because AST isozymes occur in both mitochondria and the cytoplasm and the reaction that it catalyzes is reversible, this enzymatic activity significantly influences the flow of carbon and nitrogen within the cell. For example, excess glutamate is converted via AST to aspartate. Aspartate is then used as a source of both nitrogen (for

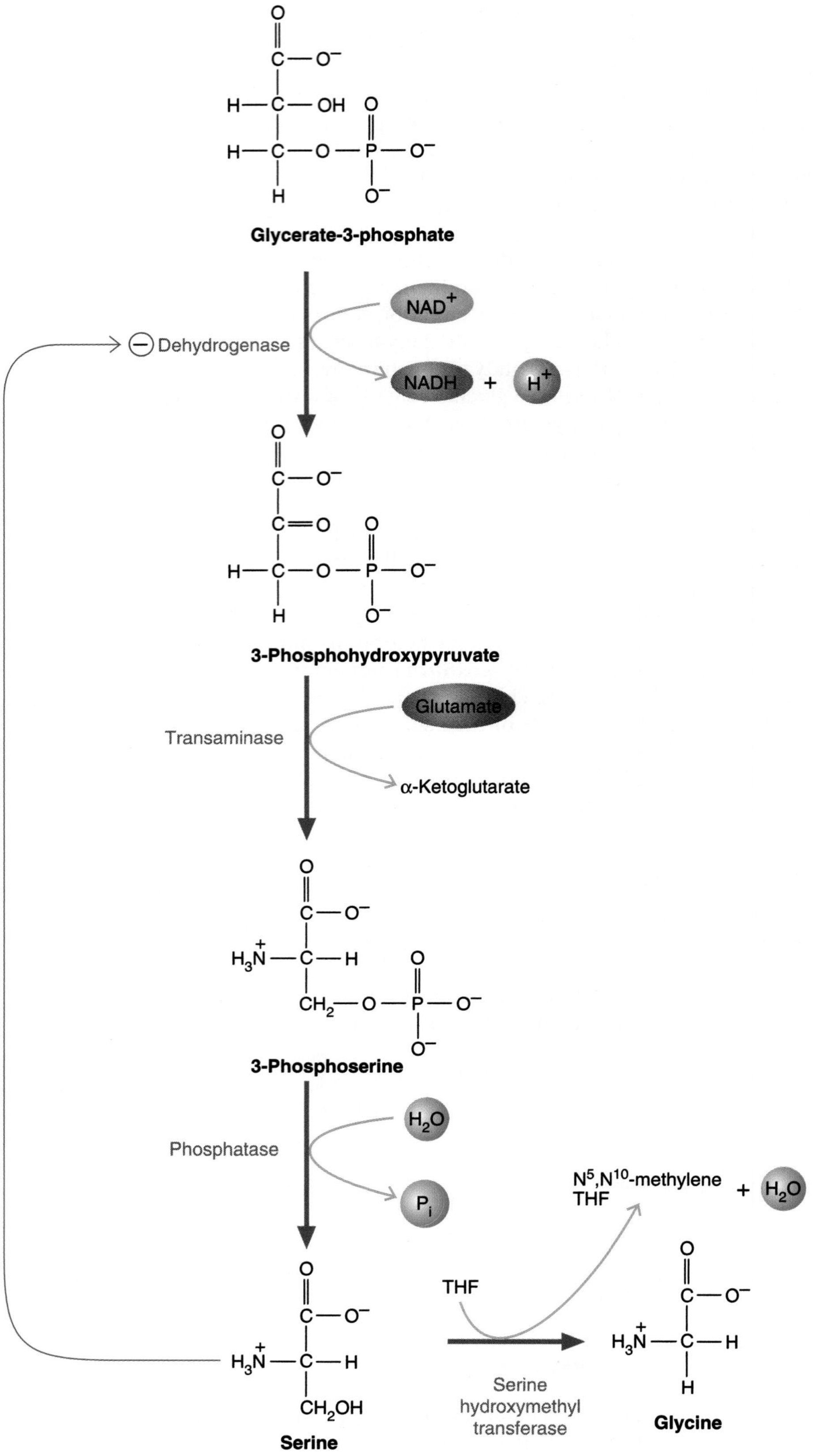

FIGURE 14.6

Biosynthesis of Serine and Glycine.
Serine inhibits glycerate-3-phosphate-dehy-
drogenase, the first reaction in the pathway.

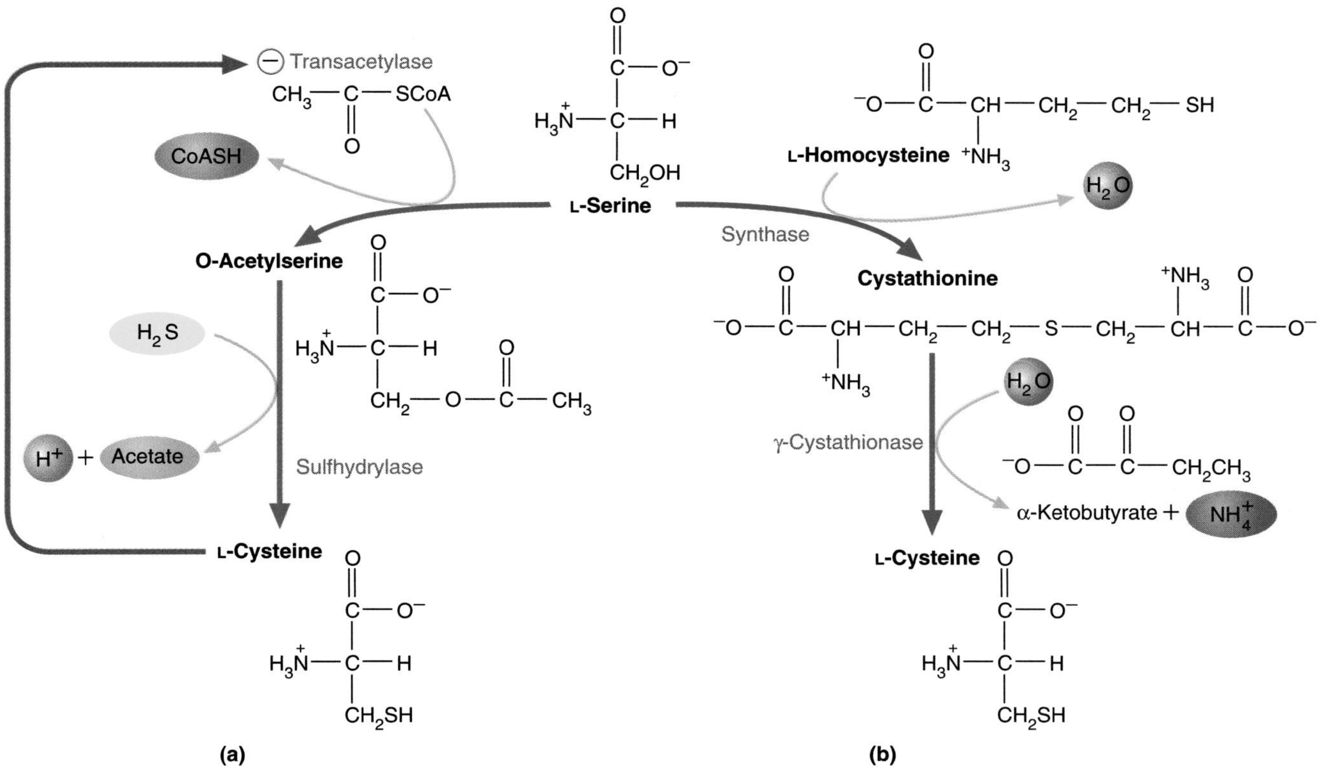

FIGURE 14.7

Biosynthesis of Cysteine.

(a) In plants and some bacteria, cysteine is synthesized in a two-step pathway. Serine is acetylated by serine acetyltransferase. The acetyl group is then displaced in a reaction with H_2S. (b) In animals serine condenses with homocysteine (derived from methionine) to form cystathionine. γ-Cystathionase catalyzes the cleavage of cystathionine to yield cysteine, α-ketobutyrate, and NH_4^+.

urea formation) and the citric acid cycle intermediate fumarate. Aspartate is also an important precursor in nucleotide synthesis.

The aspartate family also contains asparagine, lysine, methionine, and threonine. Threonine contributes to the reaction pathway in which isoleucine is synthesized. The synthesis of isoleucine, often considered to be a member of the pyruvate family, is discussed on p. 467.

Asparagine, the amide of aspartate, is not formed directly from aspartate and NH_4^+. Instead, the amido group of glutamine is transferred by amido group transfer during an ATP-requiring reaction catalyzed by asparagine synthase:

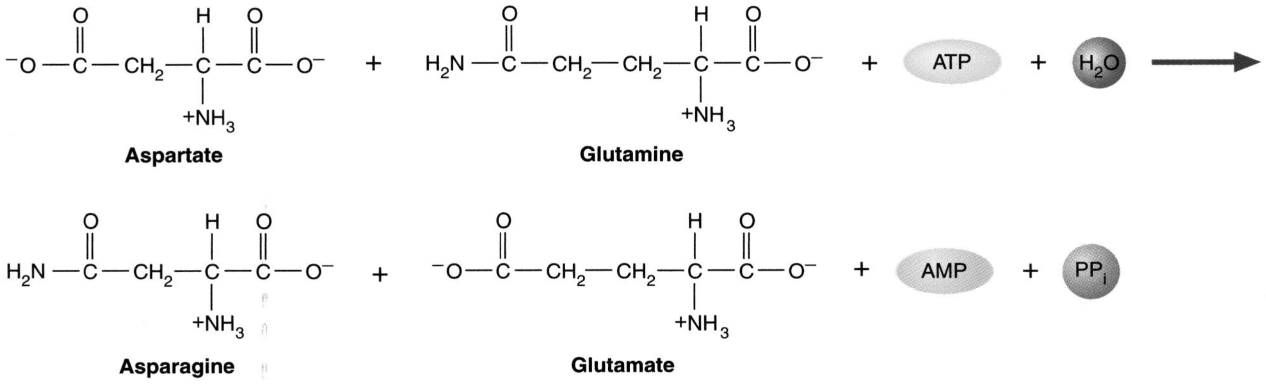

The synthesis of the other members of the aspartate family (Figure 14.8) is initiated by aspartate kinase (often referred to as aspartokinase) in an ATP-requiring reaction in which the side chain carboxyl group is phosphorylated. Aspartate

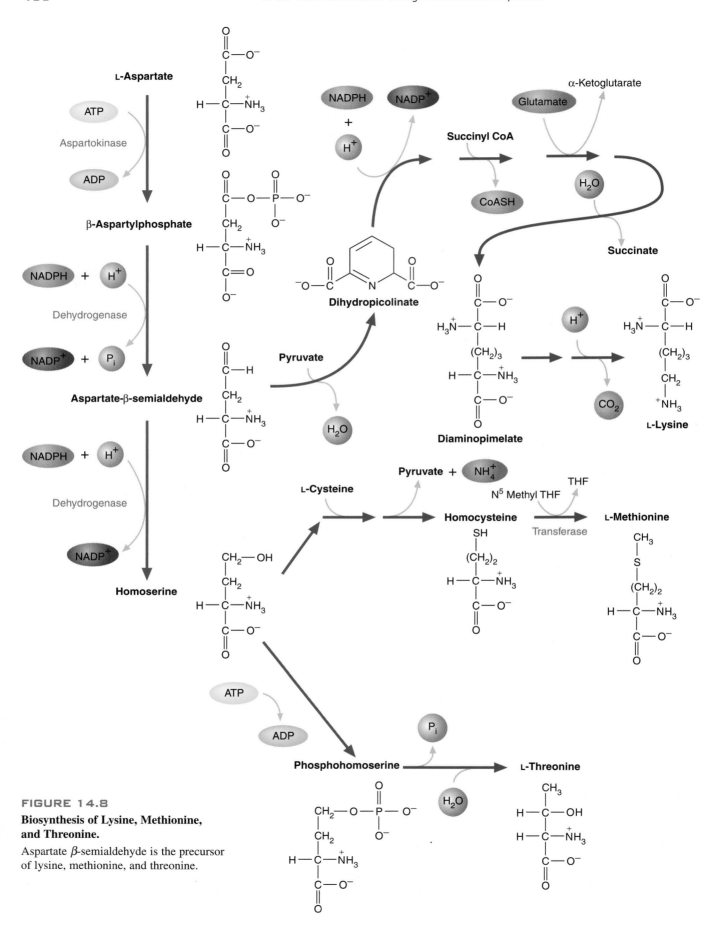

FIGURE 14.8

Biosynthesis of Lysine, Methionine, and Threonine.

Aspartate β-semialdehyde is the precursor of lysine, methionine, and threonine.

β-semialdehyde, produced by the NADPH-dependent reduction of β-aspartylphosphate, represents an important branch point in plant and bacterial amino acid synthesis. The semialdehyde can either react with pyruvate to form dihydropicolinic acid (a precursor of lysine) or be reduced to homoserine. Lysine is synthesized from dihydropicolinic acid in a series of reactions that is still poorly characterized. Homoserine also occurs at a branch point. It is the precursor in the synthesis of both methionine and threonine.

THE PYRUVATE FAMILY The pyruvate family consists of alanine, valine, leucine, and isoleucine. Alanine is synthesized from pyruvate in a single step:

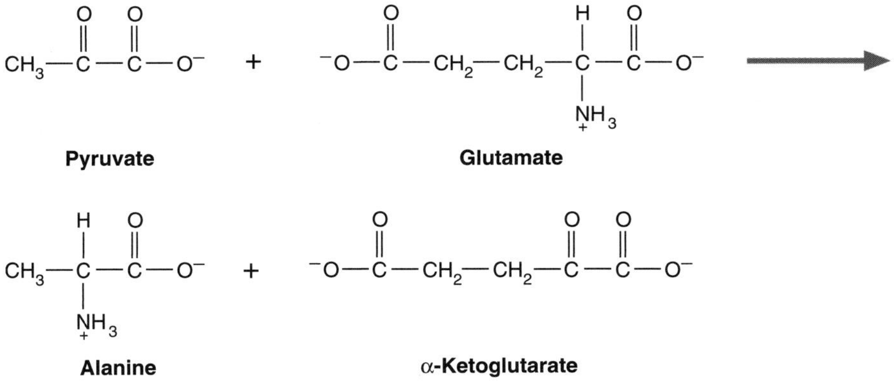

Pyruvate   **Glutamate**

Alanine α-**Ketoglutarate**

Although the enzyme that catalyzes this reaction, alanine aminotransferase, has cytoplasmic and mitochondrial forms, the majority of its activity has been found in the cytoplasm. Recall that the alanine cycle (Chapter 8) contributes to the maintenance of blood glucose. BCAA are the ultimate source of many of the amino groups transferred from glutamate in the alanine cycle (Figure 8.9).

The syntheses of valine, leucine, and isoleucine from pyruvate are illustrated in Figure 14.9. Valine and isoleucine are synthesized in parallel pathways with the same four enzymes. Valine synthesis begins with the condensation of pyruvate with hydroxyethyl-TPP (a decarboxylation product of a pyruvate-thiamine pyrophosphate intermediate) catalyzed by acetohydroxy acid synthase. The α-acetolactate product is then reduced to form α,β-dihydroxyisovalerate followed by a dehydration to α-ketoisovalerate. Valine is produced in a subsequent transamination reaction. (α-Ketoisovalerate is also a precursor of leucine.) Isoleucine synthesis also involves hydroxyethyl-TPP, which condenses with α-ketobutyrate to form α-aceto-α-hydroxybutyrate. (α-Ketobutyrate is derived from L-threonine in a deamination reaction catalyzed by threonine deaminase.) α,β-Dihydroxy-β-methylvalerate, the reduced product of α-aceto-α-hydroxybutyrate, subsequently loses an H_2O molecule, thus forming α-keto-β-methylvalerate. Isoleucine is then produced during a transamination reaction. In the first step of leucine biosynthesis from α-ketoisovalerate, acetyl-CoA donates a two-carbon unit. Leucine is formed after isomerization, reduction, and transamination.

THE AROMATIC FAMILY The aromatic family of amino acids includes phenylalanine, tyrosine, and tryptophan. Of these, only tyrosine is considered to be nonessential in mammals. Either phenylalanine or tyrosine is required for the synthesis of dopamine, epinephrine, and norepinephrine, an important class of biologically potent molecules referred to as the **catecholamines** (Special Interest Box 14.2). Tryptophan is a precursor in the synthesis of NAD+, NADP+, and the neurotransmitter serotonin.

The benzene ring of the aromatic amino acids is formed by the *shikimate pathway*. The carbons in the benzene ring are derived from erythrose-4-phosphate and phosphoenolpyruvate. These two molecules condense to form 2-keto-3-deoxy-arabinoheptulosonate-7-phosphate, a molecule that is subsequently converted to chorismate in a series of reactions that are outlined in Figure 14.10. Chorismate is the branch point in the syntheses of various aromatic compounds.

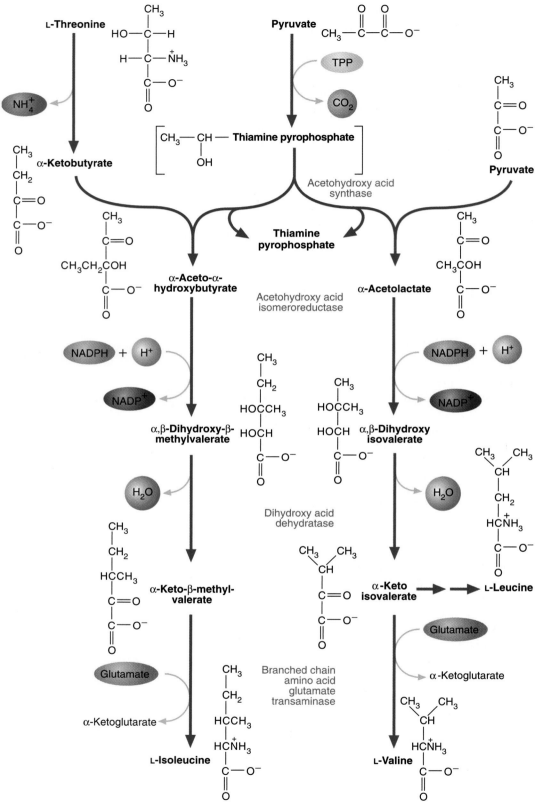

FIGURE 14.9

Biosynthesis of Valine, Leucine, and Isoleucine.

The valine and isoleucine biosynthetic pathways share four enzymes. Isoleucine synthesis begins with the reaction of α-ketobutyrate (a derivative of threonine) with pyruvate. In valine synthesis the condensation of two pyruvate molecules is the first step. Leucine is produced by a series of reactions that begin with α-ketoisovalerate, an intermediate in valine synthesis.

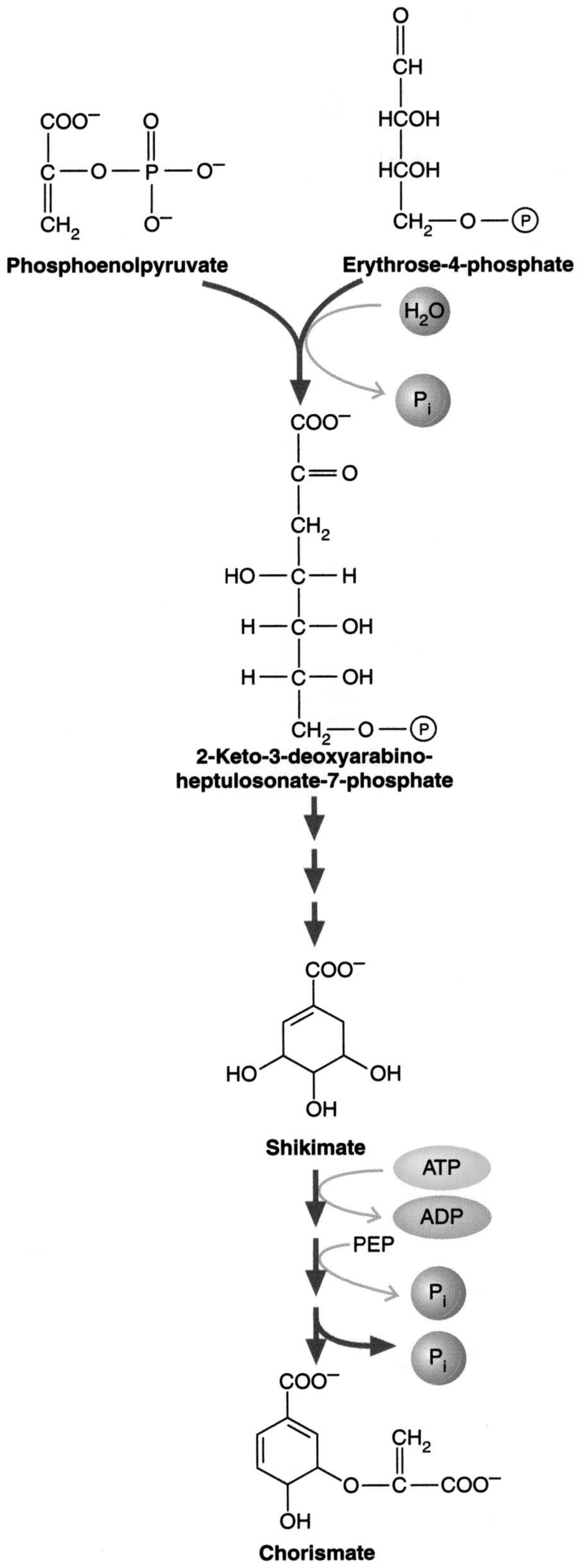

FIGURE 14.10

Chorismate Biosynthesis.

Chorismate is an intermediate in the shikimate pathway. The formation of chorismate involves the ring closure of an intermediate (2-keto-3-deoxyarabino-heptulosonate-7-phosphate) and the subsequent creation of two double bonds. The side chain of chorismate is derived from phosphoenolpyruvate (PEP).

Figure 14.11 illustrates the syntheses of phenylalanine, tyrosine, and tryptophan from chorismate. (Chorismate is also a precursor in the synthesis of the aromatic rings in the mixed terpenoids, e.g., the tocopherols, the ubiquinones, and plastoquinone.)

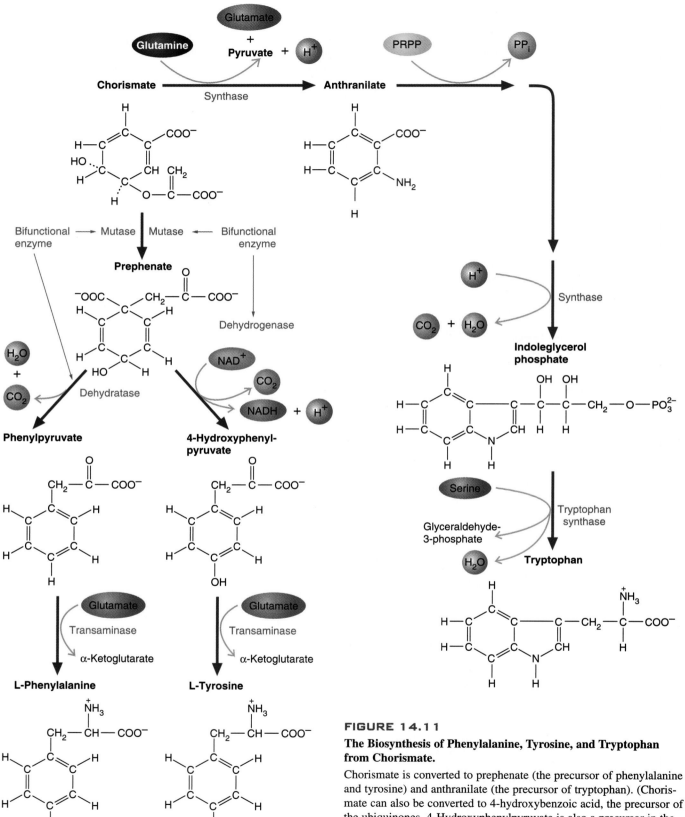

FIGURE 14.11

The Biosynthesis of Phenylalanine, Tyrosine, and Tryptophan from Chorismate.

Chorismate is converted to prephenate (the precursor of phenylalanine and tyrosine) and anthranilate (the precursor of tryptophan). (Chorismate can also be converted to 4-hydroxybenzoic acid, the precursor of the ubiquinones. 4-Hydroxyphenylpyruvate is also a precursor in the synthesis of plastoquinone and various tocopherols.) PRPP is an abbreviation for phosphoribosylpyrophosphate.

Tyrosine is not an essential amino acid in animals because it is synthesized from phenylalanine in a hydroxylation reaction. The enzyme involved, phenylalanine-4-monoxygenase, requires the coenzyme tetrahydrobiopterin (Section 14.3), a folic acid–like molecule derived from GTP. Because this reaction also is a first step in phenylalanine catabolism, it is discussed further in Chapter 15.

HISTIDINE Histidine is considered to be nonessential in healthy human adults. In human infants and many animals, histidine must be provided by the diet. Because of its unique chemical properties, histidine contributes substantially to protein structure and function. Recall, for example, that histidine residues bind heme prosthetic groups in hemoglobin. In addition, histidine often acts as a general acid during enzyme-catalyzed reactions. Of all the amino acids, histidine's biosynthesis is the most unusual. Histidine is synthesized from phosphoribosylpyrophosphate (PRPP), ATP, and glutamine (Figure 14.12). Synthesis begins with the condensation of PRPP with ATP to form phosphoribosyl-ATP. Phosphoribosyl-ATP is then hydrolyzed by phosphoribosyl-ATP pyrophosphorylase to phosphoribosyl-AMP. In the next step, a hydrolytic reaction opens the adenine ring. After an isomerization and the transfer of an amino group from glutamine, imidazole glycerol phosphate is synthesized. (The other product of the latter reaction, 5′-phosphoribosyl-4-carboxamide-5-aminoimidazole, is used in the synthesis of purine nucleotides. See Section 14.3.) Histidine is produced from imidazole glycerol phosphate in a series of reactions that include a dehydration, a transamination, a phosphorolysis, and an oxidation.

KEY CONCEPTS 14.2

There are six families of amino acids: glutamate, serine, aspartate, pyruvate, the aromatics, and histidine. The nonessential amino acids are derived from precursor molecules available in many organisms. The essential amino acids are synthesized from metabolites produced only in plants and some microorganisms.

14.3 BIOSYNTHETIC REACTIONS OF AMINO ACIDS

As described, amino acids are precursors of many physiologically important nitrogen-containing molecules, in addition to serving as building blocks for polypeptides. In the following discussion the syntheses of several examples of these molecules (e.g., neurotransmitters, glutathione, alkaloids, nucleotides, and heme) are described. Because many of these processes involve the transfer of carbon groups, this section begins with a brief description of one-carbon metabolism.

One-Carbon Metabolism

Carbon atoms have several oxidation states. Those of biological interest are found in methanol, formaldehyde, and formate. Table 14.2 lists the equivalent one-carbon groups that are actually involved in synthetic reactions.

The most important carriers of one-carbon groups in biosynthetic pathways are folic acid and S-adenosylmethionine. The metabolism of each is described briefly. (The function of biotin, a carrier of CO_2 groups, is discussed in Section 8.2.)

FOLIC ACID Folic acid, also known as folate or folacin, is a B vitamin. Its structure consists of a pteridine nucleus and *para*-aminobenzoic acid, linked to one or more glutamic acid residues (Figure 14.13). Once it is absorbed by the body, folic acid is converted by dihydrofolate reductase to the biologically active form, tetrahydrofolic acid (THF). The carbon units carried by THF (i.e., methyl, methylene, methenyl, and formyl groups) are bound to N^5 and/or N^{10} of the pteridine ring. Figure 14.14 illustrates the interconversions of the one-carbon units carried by THF, as well as their origin and metabolic fate. A substantial number of one-carbon units enter the THF pool as N^5,N^{10}-methylene THF, produced during the conversion of serine to glycine and the cleavage of glycine (catalyzed by glycine synthase).

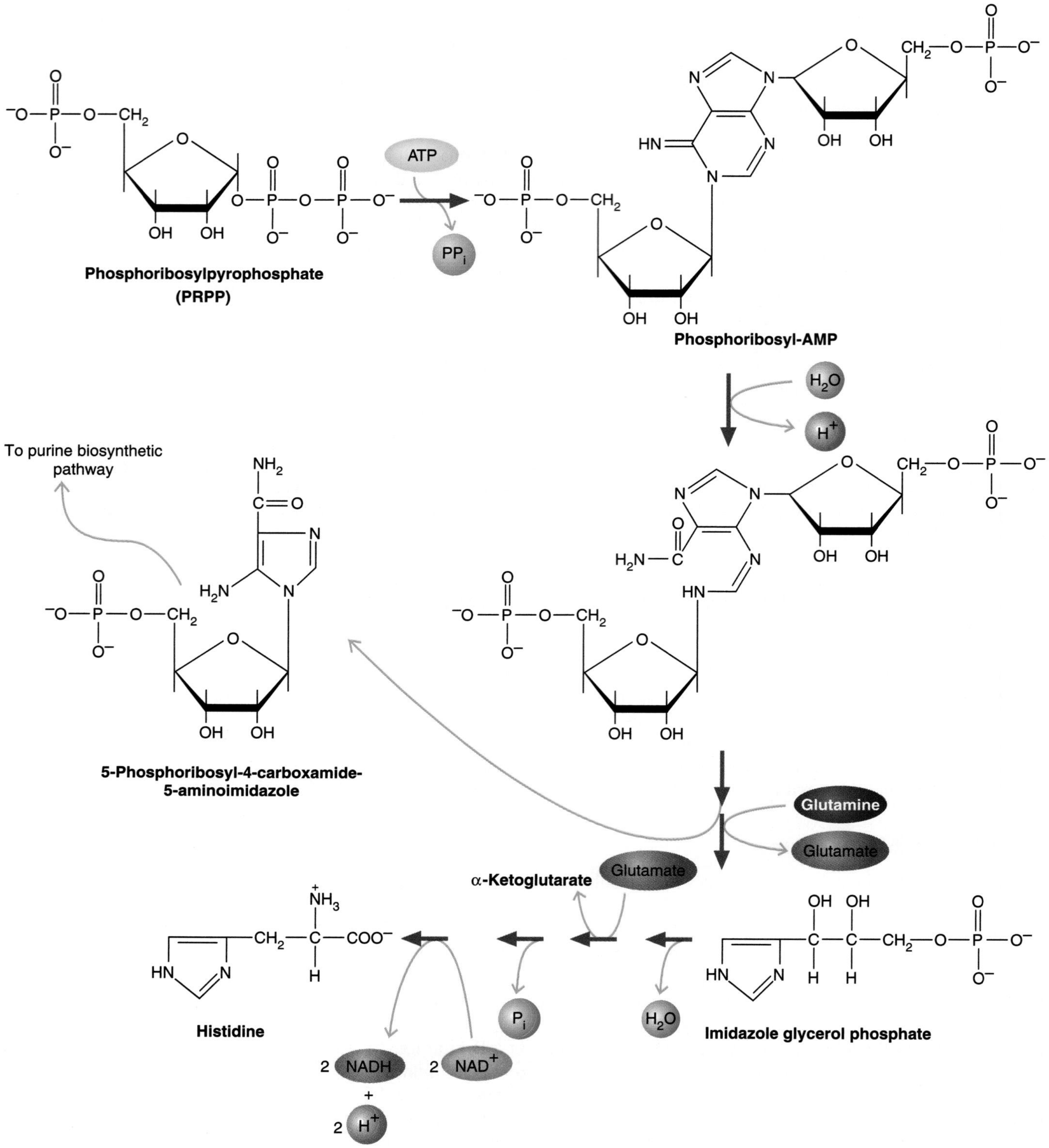

FIGURE 14.12

Histidine Biosynthesis.

Histidine is derived from three biomolecules: PRPP (five carbons), the adenine ring from ATP (one nitrogen and one carbon), and glutamine (one nitrogen). The ATP used in the first reaction in the pathway is regenerated when 5-phosphoribosyl-4-carboxamide-5-aminoimidazole (released in a subsequent reaction) is diverted into the purine nucleotide biosynthetic pathway.

TABLE 14.2
One-Carbon Groups

Oxidation Level	Methanol (most reduced)	Formaldehyde	Formate (most oxidized)
One-carbon group	Methyl ($-CH_3$)	Methylene ($-CH_2-$)	Formyl ($-CHO$) Methenyl ($-CH=$

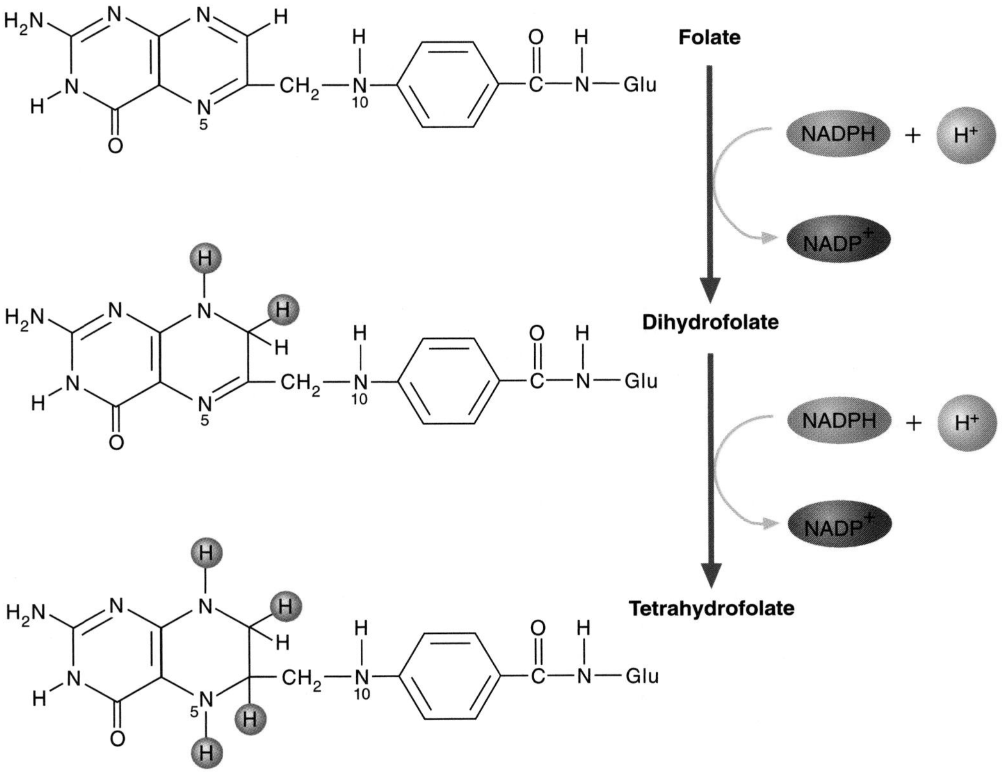

FIGURE 14.13

Biosynthesis of Tetrahydrofolate (THF).

The vitamin folic acid (folate) is converted to its biologically active form by two successive reductions of the pteridine ring. Both reactions are catalyzed by dihydrofolate reductase.

In Figure 14.14, vitamin B_{12} is required for the N^5-methyl THF–dependent conversion of homocysteine to methionine. **Vitamin B_{12}** (cobalamin) is a complex, cobalt-containing molecule synthesized only by microorganisms (Figure 14.15). (During the purification of cobalamin, a cyanide group attaches to cobalt.) Animals obtain cobalamin from intestinal flora and by consuming foods derived from other animals (e.g., liver, eggs, shrimp, chicken, and pork). A deficiency of vitamin B_{12} results in **pernicious anemia**. In addition to low red blood cell counts, the symptoms of this malady include weakness and various neurological disturbances. Pernicious anemia is most often caused by decreased secretion of intrinsic factor, a glycoprotein secreted by stomach cells, which is required for the absorption of the vitamin in the intestine. Vitamin B_{12} absorption can also be inhibited by several gastrointestinal disorders, such as celiac disease or tropical sprue, both of which damage the lining of the intestine.

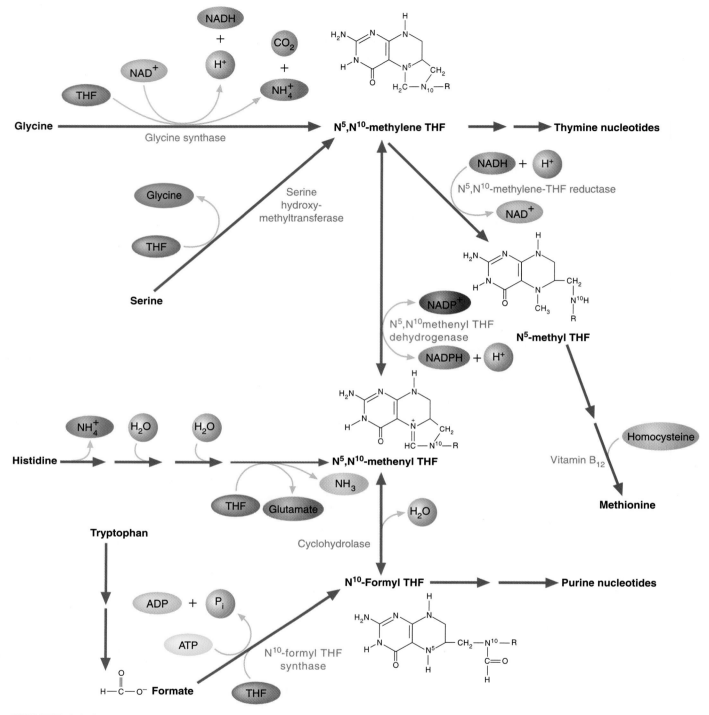

FIGURE 14.14

Structures and Enzymatic Interconversions of THF Coenzymes.

The THF coenzymes play a critical role in one-carbon metabolism. The interconversions of the coenzymes are reversible except for the conversion of N^5,N^{10}-methylene THF to N^5-methyl THF.

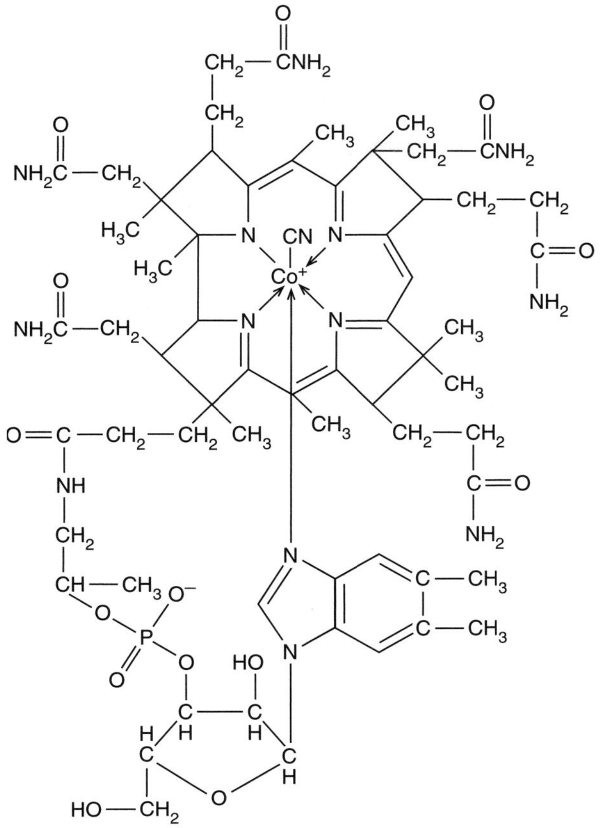

FIGURE 14.15

Structure of Cyanocobalamin, a Derivative of Vitamin B$_{12}$.

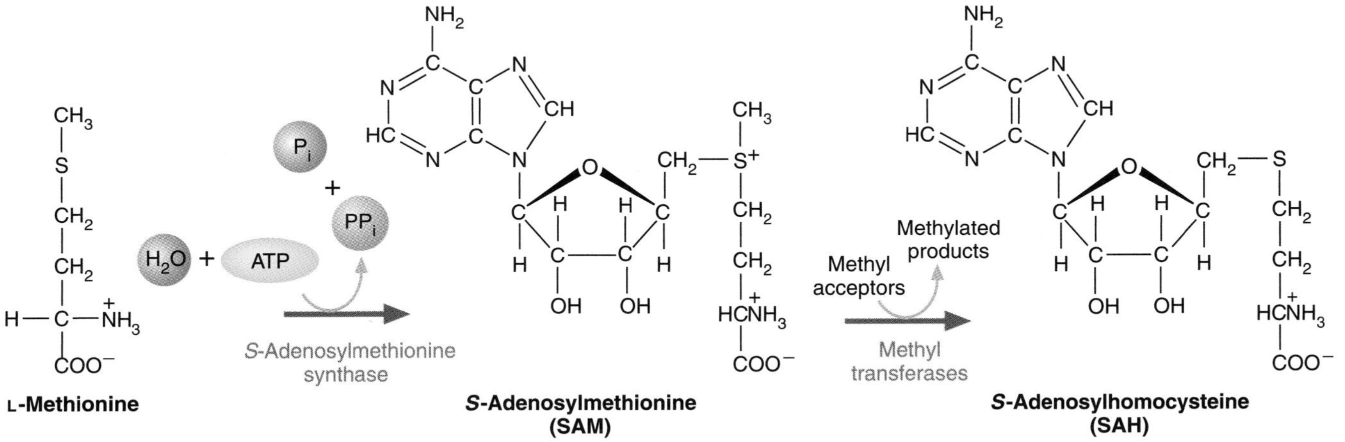

FIGURE 14.16

The Formation of S-Adenosylmethionine.

One of the principal functions of SAM is to serve as a methylating agent.

A reduction in vitamin B$_{12}$ absorption has also been observed in the presence of intestinal overgrowths of microorganisms induced by antibiotic treatments.

S-ADENOSYLMETHIONINE S-Adenosylmethionine (SAM) is the major methyl group donor in one-carbon metabolism. Formed from methionine and ATP (Figure 14.16), SAM contains an "activated" methyl thioether group, which can be transferred to a variety of acceptor molecules (Table 14.3). S-Adenosylhomocysteine

KEY CONCEPTS 14.3

Tetrahydrofolate, the biologically active form of folic acid, and S-adenosylmethionine are important carriers of single carbon atoms in a variety of synthetic reactions.

TABLE 14.3

Examples of Transmethylation Acceptors and Products

Methyl Acceptors	Methylated Product
Phosphatidylethanolamine	Phosphatidylcholine (p. 341)
Norepinephrine	Epinephrine (p. 480)
Guanidinoacetate	Creatine
γ-Aminobutyric acid	Carnitine (p. 379)

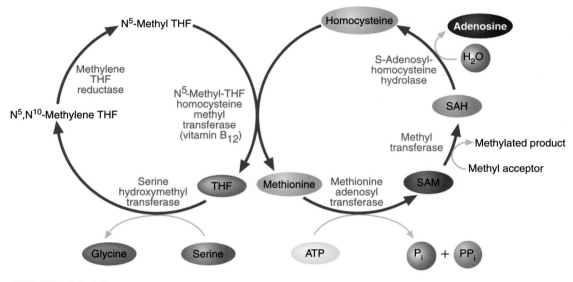

FIGURE 14.17

The Tetrahydrofolate and S-Adenosylmethionine Pathways.

The THF and SAM pathways intersect at the reaction, catalyzed by N^5-methyl THF homocysteine methyltransferase, in which homocysteine is converted to methionine.

(SAH) is a product in these reactions. The loss of free energy that accompanies S-adenosylhomocysteine formation makes the methyl transfer irreversible. SAM acts as a methyl donor in many transmethylation reactions, some of which occur in the synthesis of phospholipids, several neurotransmitters, and glutathione.

The importance of SAM in metabolism is reflected in the several mechanisms that provide for the synthesis of sufficient amounts of its precursor, methionine, when the latter molecule is temporarily absent from the diet. For example, choline is used as source of methyl groups to convert homocysteine into methionine. Homocysteine can also be methylated in a reaction utilizing N^5-methyl THF. This latter reaction is a bridge between the THF and SAM pathways (Figure 14.17).

QUESTION 14.5

The following substances are derived from amino acids. For each, provide the name of the amino acid precursor.

a.

b.

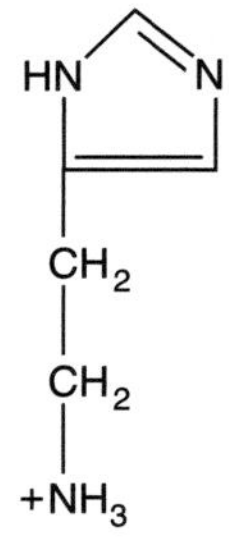

c.

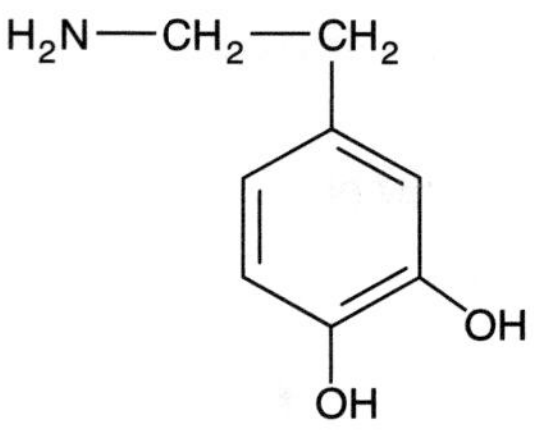

d.

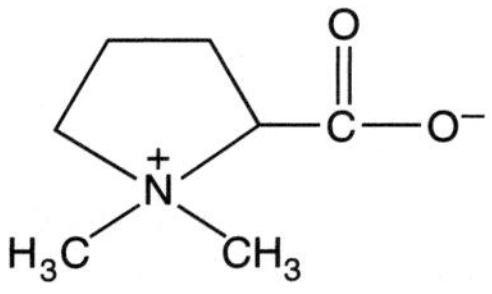

e.

Amethopterin, also referred to as **methotrexate**, is a structural analogue of folate. (**Analogues** are compounds that closely resemble other molecules.) Methotrexate has been used to treat several types of cancer. It has been especially successful in childhood leukemia.

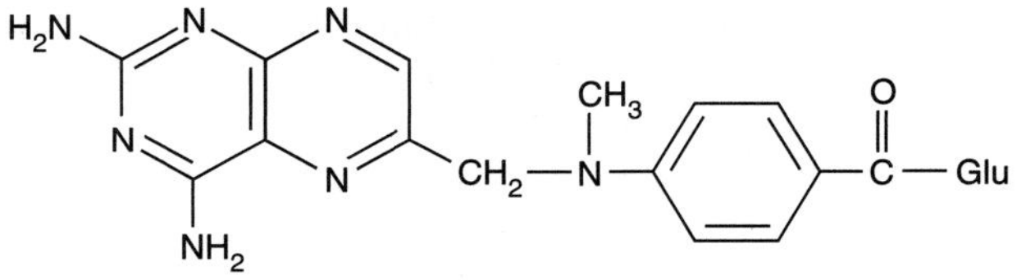

Amethopterin (methotrexate)

Using your knowledge of cell biology and biochemistry, suggest a biochemical mechanism that explains why amethopterin is effective against cancer. (*Hints*: Compare the structures of folate and methotrexate. Review Figure 14.13.) Explain why methotrexate treatment for cancer causes temporary baldness.

QUESTION 14.6

Melatonin is a hormone derived from serotonin. It is produced in the brain's light-sensitive pineal gland. The pineal's secretion of melatonin is depressed by nerve impulses that originate in the retina of the eye and other light-sensitive tissue in the body in response to light. Pineal function is involved in *circadian rhythms*, patterns of activity associated with light and dark, such as sleep/wake cycles. In many mammals the functioning of the pineal gland also regulates seasonal cycles of fertility and infertility. (Melatonin inhibits the secretion of certain hormones from the hypothalamus and pituitary that stimulate the functioning of the ovaries and testes.) For example, in some species (e.g., deer) the males are fertile only in early spring, ensuring that newborn animals will be mature enough to survive the next winter.

After serotonin is produced within the pineal gland, it is converted to 5-hydroxy-N-acetyltryptamine by N-acetyl transferase. 5-Hydroxy-N-acetyltryptamine is then methylated by O-methyl transferase. SAM is the methylating agent. With this information, draw the synthetic pathway of melatonin.

QUESTION 14.7

Auxins are a class of plant growth regulators. Synthesized in meristematic (actively growing) tissue in response to light, auxin molecules diffuse to nearby cells. Depending on several parameters (e.g., concentration and location), auxins may

QUESTION 14.8

either stimulate or inhibit cell growth. For example, auxins stimulate growth in the main shoot but inhibit growth in lateral shoots. Indole-3-acetic acid (IAA), the best-characterized auxin, is derived from tryptophan.

Auxin
(indole acetic acid)

Compare the structure of IAA with that of tryptophan. Suggest two reaction types involved in IAA synthesis.

QUESTION 14.9

It now appears that oxidative stress is a causative factor in the brain damage associated with stroke, head trauma, and several age-related neurological disorders (e.g., Parkinson's disease and Huntington's disease). There is also compelling evidence that excitatory neurotransmitters such as glutamate contribute to the brain damage. In a destructive mechanism that is not yet completely understood, glutamate can act as an excitotoxin; that is, excessive release of glutamate excites nearby neurons to death. Under normal circumstances, neurons are saved from such damage by glutamate transporters, which remove glutamate from the extracellular space. The excitotoxic effect of glutamate apparently results from the disabling of glutamate transporters. NO or its oxidized derivative the peroxynitrite anion ($ONOO^-$) is believed to actually cause the damage. (The peroxynitrite anion is formed when NO reacts with O_2^-. Once it forms, the peroxynitrite anion quickly decomposes to $\cdot OH$ and NO_2^-) Use your knowledge of oxidative stress to suggest how O_2^- might be produced within nerve cells. What defense mechanisms are probably available to protect the brain from oxidative damage? Why might they provide inadequate protection after a stroke or head trauma? Considering NO's role in normal glutamate function in the brain, why is this molecule such a lethal factor in excessive oxidative stress?

Glutathione

The nitrogen-containing molecule glutathione (γ-glutamylcysteinylglycine) is the most common intracellular thiol. (Its concentration in mammalian cells varies from 0.5 to 10 mM.) The functions of glutathione (GSH) include involvement in DNA and RNA synthesis and the synthesis of certain eicosanoids and other biomolecules. (In many of these processes, GSH acts as a reducing agent. As such, it maintains the sulfhydryl groups of enzymes and other molecules in a reduced state.) In addition to protecting cells from radiation, oxygen toxicity, and environmental toxins, GSH also promotes amino acid transport. After a brief discussion of its synthesis, the transport role of GSH is described. This is followed by a discussion of an interesting class of enzymes called the glutathione-S-transferases.

GSH is synthesized in a pathway composed of two reactions. In the first reaction, γ-glutamylcysteine synthase catalyzes the condensation of glutamate with cysteine (Figure 14.18). γ-Glutamylcysteine, the product of this reaction, then combines with glycine to form GSH in a reaction catalyzed by glutathione synthase.

TRANSPORT Transport of GSH out of cells appears to serve several functions. Among these are (1) transfer of the sulfur atoms of cysteine between

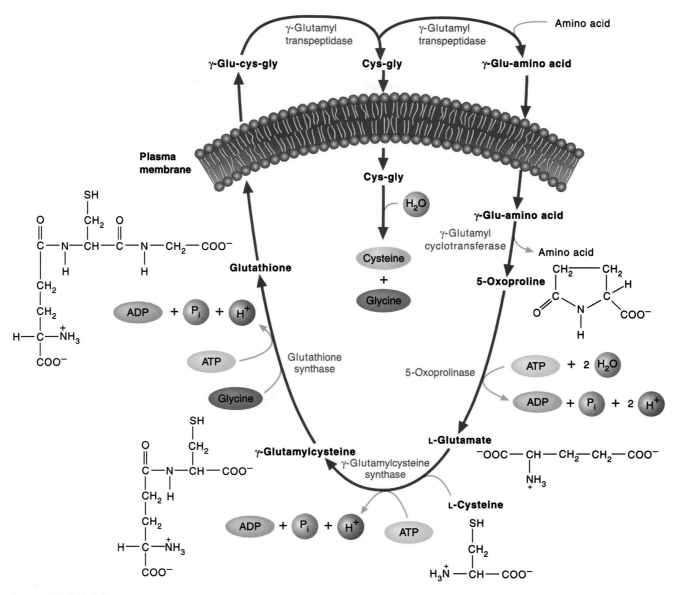

FIGURE 14.18

The γ-Glutamyl Cycle.

The functions of the γ-glutamyl cycle are described in the text. (1) Glutathione is excreted from the cell. γ-Glutamyltranspeptidase converts GSH to γ-Glu-amino acid derivative and Cys-Gly. (2) the γ-glu-amino acid is transported into the cell, where it is converted to 5-oxoproline and the free amino acid. 5-Oxoproline is eventually reconverted to GSH. (3) Cys-Gly is transported into the cell, where (4) it is hydrolyzed to cysteine and glycine.

cells, (2) protection of the plasma membrane from oxidative damage, and (3) transfer to membrane-bound γ-glutamyl transpeptidase leading to the formation of γ-glutamyl amino acid derivatives. The latter process, which initiates the γ-glutamyl cycle, occurs in brain, intestine, pancreas, liver, and kidney cells.

The γ-glutamyl cycle provides for the active transport of several amino acids, especially cysteine and methionine, as well as GSH itself. Some researchers consider the amino acid transport function of GSH controversial. An alternative

(continued on p. 482)

More than 30 different substances have been proven or proposed to act as neurotransmitters. **Neurotransmitters** are either excitatory or inhibitory. As noted, excitatory neurotransmitters (e.g., glutamate and acetylcholine) open sodium channels and promote the depolarization of the membrane in another cell (either another neuron or an effector cell, such as a muscle cell). If the second (postsynaptic) cell is a neuron, the wave of depolarization (referred to as an action potential) triggers the release of neurotransmitter molecules as it reaches the end of the axon. (Most neurotransmitter molecules are stored in numerous membrane-enclosed *synaptic vesicles*.) When the action potential reaches the nerve ending, the neurotransmitter molecules are released by exocytosis into the synapse. If the postsynaptic cell is a muscle cell, sufficient release of excitatory neurotransmitter molecules results in muscle contraction. Inhibitory neurotransmitters (e.g., glycine) open chloride channels and make the membrane potential in the postsynaptic cell even more negative, that is, they inhibit the formation of an action potential.

A significant percentage of neurotransmitter molecules are either amino acids or amino acid derivatives (Table 14.4). The latter class is often referred to as the **biogenic amines**. After a brief discus-

TABLE 14.4
Amino Acid and Amine Neurotransmitters

Amino Acids	Amines
Glycine	Norepinephrine*
Glutamate	Epinephrine*
γ-Aminobutyric acid (GABA)	Dopamine*
	Serotonin
	Histamine

*These molecules are referred to as the catecholamines.

sion of several biogenic amines, the properties of nitric oxide, a newly recognized neurotransmitter, are described.

γ-Aminobutyric Acid
γ-Aminobutyric acid (GABA) acts as an inhibitory neurotransmitter in the central nervous system. The binding of GABA to its receptor increases the nerve cell membrane's permeability to chloride ions. (The benzodiazepines, a class of tranquilizers that alleviate

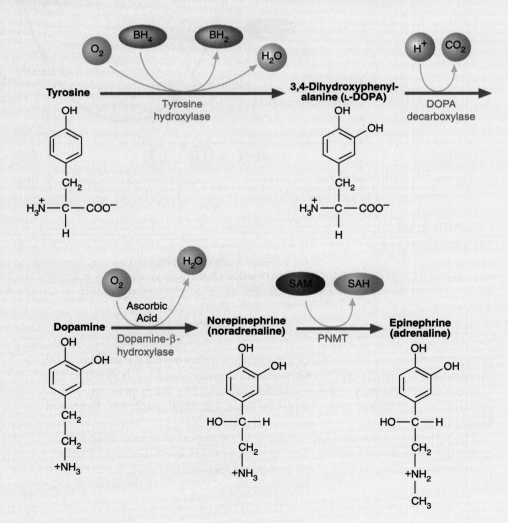

FIGURE 14A Biosynthesis of the Catecholamines.
Dopamine, norepinephrine, and epinephrine act as neurotransmitters and/or hormones. (PNMT is an abbreviation for phenylethanolamine-N-methyltransferase.)

anxiety and aggressive behavior, have been shown to enhance GABA's ability to increase membrane conductance of chloride.)

GABA is produced by the decarboxylation of glutamate. The reaction is catalyzed by glutamate decarboxylase, which is a pyridoxal phosphate–requiring enzyme:

Glutamate   **γ-Aminobutyric acid (GABA)**

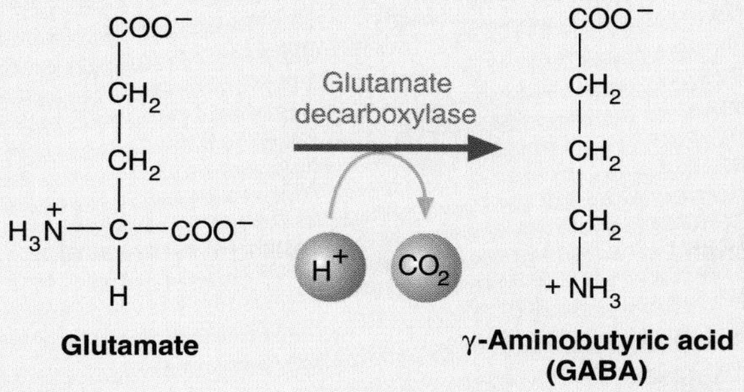

The Catecholamines

The *catecholamines* (dopamine, norepinephrine, and epinephrine) are derivatives of tyrosine. Dopamine (D) and norepinephrine (NE) are used in the brain as excitatory neurotransmitters. Outside the central nervous system, NE and epinephrine (E) are released primarily from the adrenal medulla, as well as the peripheral nervous system. Because both NE and E regulate aspects of metabolism, they are often considered hormones.

The first, and rate-limiting, step in catecholamine synthesis is the hydroxylation of tyrosine to form 3,4-dihydroxyphenyl alanine (DOPA)

(Figure 14A). Tyrosine hydroxylase, the mitochondrial enzyme that catalyzes the reaction, requires a cofactor known as *tetrahydrobiopterin* (BH_4). (BH_4, a folic acid–like molecule, is an essential cofactor in the hydroxylation of aromatic amino acids. BH_4 is regenerated from its oxidized metabolite, BH_2, by reduction with NADPH.)

Tyrosine hydroxylase uses BH_4 to activate O_2. One oxygen atom is attached to tyrosine's aromatic ring, while the other atom oxidizes the coenzyme. DOPA, the product of the reaction, is used in the synthesis of the other catecholamines.

DOPA decarboxylase, a pyridoxal phosphate–requiring enzyme, catalyzes the synthesis of dopamine from DOPA. Dopamine is produced in neurons found in certain structures in the brain. It is believed to exert an inhibitory action within the central nervous system. Deficiency in dopamine production has been found to be associated with Parkinson's disease, a serious degenerative neurological disorder (Special Interest Box 14.3). The precursor L-DOPA is used to alleviate the symptoms of Parkinson's disease because dopamine cannot penetrate the blood-brain barrier. (The blood-brain barrier protects the brain from toxic substances. Many polar molecules and ions cannot move from blood capillaries, although most lipid-soluble substances readily pass across. The blood-brain barrier consists of connective tissue and specialized cells called astrocytes that envelop the capillaries.) Once L-DOPA is transported into appropriate nerve cells, it is converted to dopamine.

Norepinephrine is synthesized from tyrosine in the chromaffin cells of the adrenal medulla in response to fright, cold, and exercise

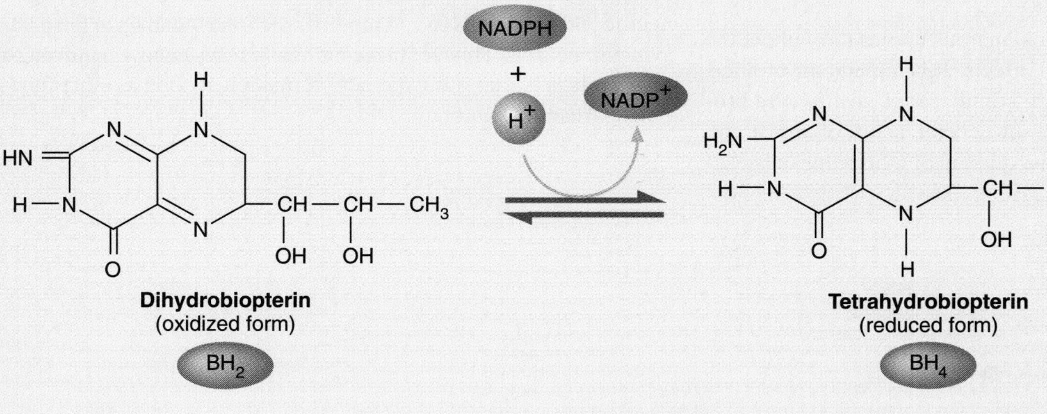

Dihydrobiopterin
(oxidized form)
BH_2

Tetrahydrobiopterin
(reduced form)
BH_4

as well as low levels of blood glucose. NE acts to stimulate the degradation of triacylglycerol and glycogen. It also increases cardiac output and blood pressure. The hydroxylation of dopamine to produce NE is catalyzed by the copper-containing enzyme dopamine-β-hydroxylase, an oxidase that requires the antioxidant ascorbic acid for full activity.

As described, the secretion of epinephrine in response to stress, trauma, extreme exercise, or hypoglycemia causes a rapid mobilization of energy stores, that is, glucose from the liver and fatty acids from adipose tissue. The reaction in which NE is methylated to form E is mediated by the enzyme phenylethanolamine-N-methyltransferase (PNMT). Although the enzyme occurs predominantly in the chromaffin cells of the adrenal medulla, it is also

found in certain portions of the brain where E functions as a neurotransmitter. Recent evidence indicates that both E and NE are present in several other organs (e.g., liver, heart, and lung). Bovine PNMT is a monomeric protein (30 kD) that uses SAM as a source of methyl groups.

Serotonin

Serotonin is found in various cells within the central nervous system, where it inhibits feeding. Serotonin has been implicated in human eating disorders such as anorexia nervosa, bulimia, and the carbohydrate craving associated with seasonal affective disorder (SAD). SAD is a clinical depression triggered

(continued on page 482)

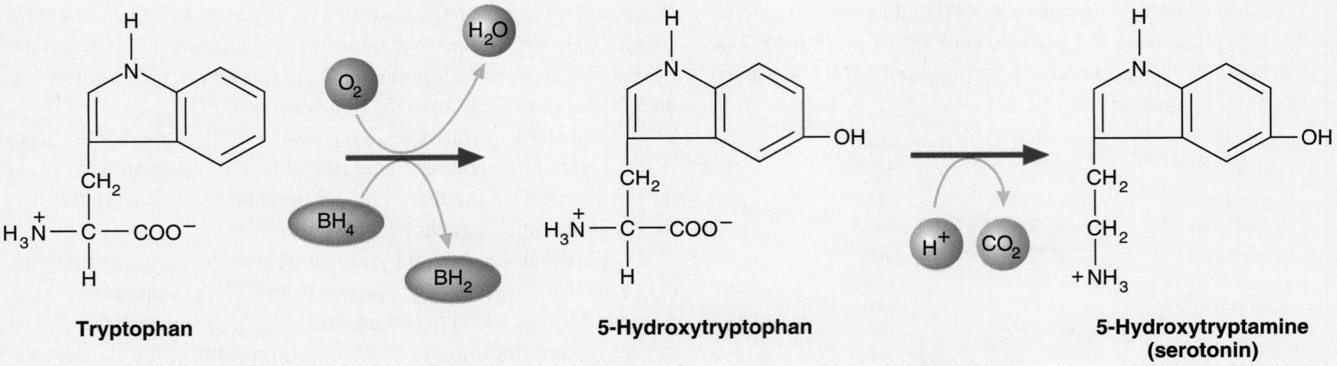

Tryptophan

5-Hydroxytryptophan

**5-Hydroxytryptamine
(serotonin)**

by the decreased daylight in autumn and winter. Additionally, serotonin appears to affect mood, temperature regulation, pain perception, and sleep. The hallucinogenic drug LSD (lysergic acid diethylamide) apparently competes with serotonin for specific brain cell receptors. Serotonin is also found in the gastrointestinal tract, blood platelets, and mast cells.

Tryptophan hydroxylase uses O_2 and the electron donor BH_4 to hydroxylate C-5 of tryptophan. The product, called 5-hyroxytryptophan, then undergoes a decarboxylation catalyzed by 5-hydroxytryptophan decarboxylase, a pyridoxal phosphate–requiring enzyme. Serotonin, often referred to as 5-hydroxytryptamine, is the product of this reaction.

Histamine

Histamine, an amine produced in numerous tissues throughout the body, has complex physiological effects. It is a mediator of allergic and inflammatory reactions, a stimulator of gastric acid production, and a neurotransmitter in several areas of the brain. Histamine is formed by the decarboxylation of L-histidine in a reaction catalyzed by histidine decarboxylase, a pyridoxal phosphate–requiring enzyme.

Nitric Oxide

In addition to the many other functions of nitric oxide (Special Interest Box 10.1), it is also a neurotransmitter. Nitric oxide (NO), which is synthesized from the amino acid arginine by NO synthase (Figure 10A), is produced in many areas of the brain, where its formation has been linked to the neurotransmitter function of glutamate. When glutamate is released from a neuron and binds to a certain class of glutamate receptor, a transient flow of Ca^{2+} through the postsynaptic membrane is triggered that stimulates NO synthesis. Once it is synthesized, NO diffuses back to the presynaptic cell, where it signals further release of glutamate. In other words, NO acts as a so-called retrograde neurotransmitter; that is, it promotes a cycle in which glutamate is released from the presynaptic neuron and then binds to and promotes action potentials in the postsynaptic neuron. This potentiating mechanism is now believed to play a role in learning and memory formation, as well as other functions in mammalian brain.

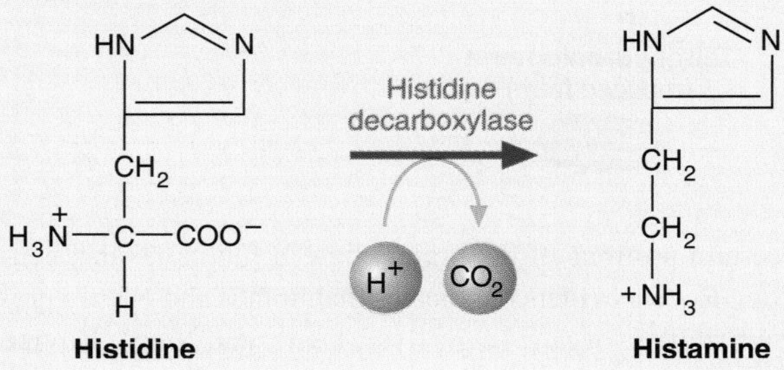

Histidine

**Histidine
decarboxylase**

Histamine

KEY CONCEPTS 14.4

Glutathione (GSH), the most common intracellular thiol, is involved in many cellular activities. In addition to reducing sulfhydryl groups, GSH protects cells against toxins and promotes the transport of some amino acids.

view is that the γ-glutamyl cycle generates a signal that activates amino acid uptake. 5-Oxoproline has been proposed as such a signal.

Several human diseases are associated with deficiencies in GSH metabolism. One of the most notable is *GSH synthase deficiency* (also known as *5-oxoprolinuria*). GSH synthase deficiency is characterized by a severe acidosis, hemolysis (red blood cell destruction), and central nervous system damage. Because of the enzyme deficiency, the concentration of γ-glutamylcysteine increases. This molecule is then converted to 5-oxoproline and cysteine by γ-glutamyl cyclotransferase. The production of 5-oxoproline soon exceeds the capacity of oxoprolinase to convert it to glutamate. Consequently, the concentration of 5-oxoproline in blood and urine begins to rise.

GLUTATHIONE-S-TRANSFERASES As mentioned, GSH contributes to the protection of cells from environmental toxins. GSH does so by reacting with a large variety of foreign molecules to form GSH conjugates (Figure 14.19). The bonding of these substances with GSH, which prepares them for excretion, may be spontaneous, or it may be catalyzed by the GSH-S-transferases (also known as the *ligandins*). Before their excretion in urine, GSH conjugates are usually converted to mercapturic acids by a series of reactions initiated by γ-glutamyltranspeptidase.

Alkaloids

The **alkaloids** are a large, heterogeneous group of basic nitrogen-containing molecules produced in the leaves, seeds, or bark of some plants. They are derived from α-amino acids (or closely related molecules) in complex and poorly understood pathways. Although many alkaloids have profound physiological properties when consumed by animals, their roles in plants are relatively obscure. Because they often have bitter tastes or are poisonous, alkaloids may protect plants against herbivores, insects, and microbes. Several examples of this group of natural products are illustrated in Figure 14.20.

Alkaloids are classified according to their heterocyclic rings. For example, cocaine, a central nervous system stimulant, and atropine, a muscle relaxant, are examples of the *tropane alkaloids* in which a nitrogen appears in a bridge of a seven-membered ring structure. Nicotine, the addictive and toxic component of tobacco, is an example of the *pyridine alkaloids* in which a nitrogen appears as a member of a six atom aromatic ring. (Nicotine is an effective insecticide.) The addictive

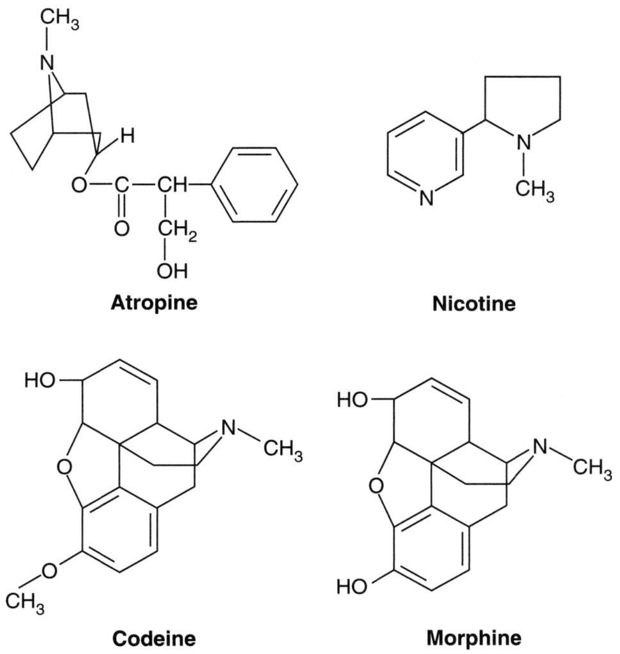

FIGURE 14.20

Structures of Several Alkaloids.

More than 5000 alkaloids have been isolated from plants. Their roles in plants are usually unknown. Alkaloids are often physiologically potent molecules in animals.

FIGURE 14.19

Formation of a Mercapturic Acid Derivative of a Typical Organic Contaminant.

GSH-S-transferase catalyzes the synthesis of a GSH derivative of dichlorobenzene.

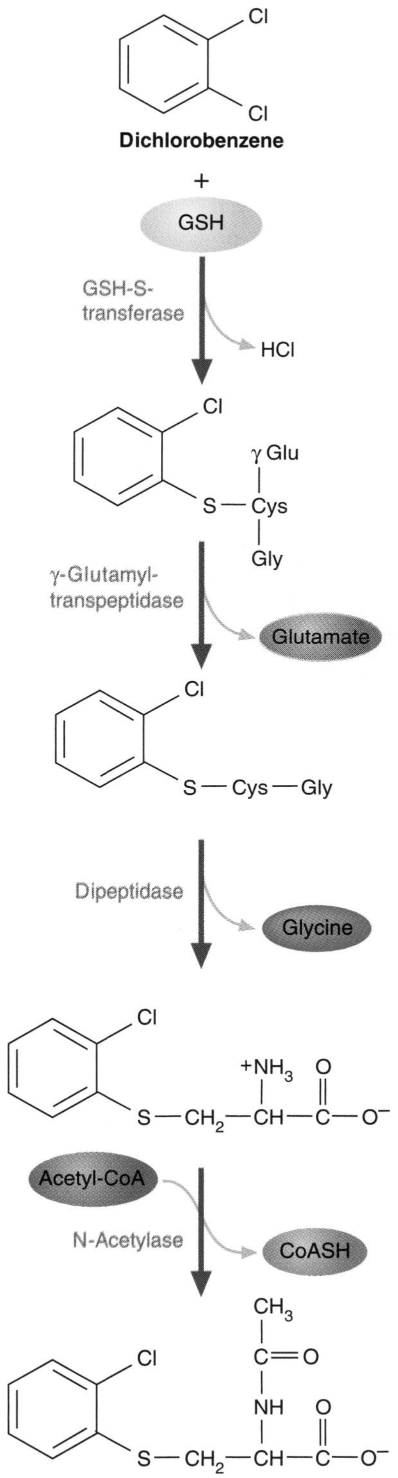

Parkinson's disease, formerly known as *paralysis agitans*, is a movement disorder caused by damage to brain structures called the basal ganglia and substantia nigra. Symptoms of Parkinson's disease, most commonly observed in adults past 40 years of age, include tremor, skeletal muscle rigidity, and difficulty in initiating movement. The inability of certain neurons within the substantia nigra to produce and release dopamine is believed to be the primary cause of Parkinson's disease. (Dopamine produced within the substantia nigra normally acts to inhibit neural activities within the basal ganglia.) As stated, because dopamine does not cross the protective blood-brain barrier, the precursor molecule L-DOPA (also known as levodopa) is used to treat Parkinson's patients.

In the late 1970s a substantial clue to the cause of the nerve cell destruction in Parkinson's disease was provided by young drug addicts using the synthetic heroin substitute MPPP (1-methyl-4-phenyl-4-proprionoxypiperidine) (Figure 14B). Several unfortunate individuals, later found to have consumed MPPP, were diagnosed with Parkinson's disease despite their youth and lack of a family history of the disease. Considerable research revealed that under certain reaction conditions the synthesis of MPPP produces a toxic by-product called MPTP (1-methyl-4-phenyl-1,2,3,6-tetrahydropyridine). Once it has been consumed, MPTP is converted to MPP$^+$ (1-methyl-4-phenylpyridinium) in the brain by the enzyme monoamine oxidase. After its synthesis, MPP$^+$ is transported by a dopamine-specific transport mechanism into certain neurons. Although the mechanism by which nerve cells are destroyed by MPP$^+$ is not completely understood, it appears that one of its effects is to inhibit NADH dehydrogenase, a component of the mitochondrial electron transport complex.

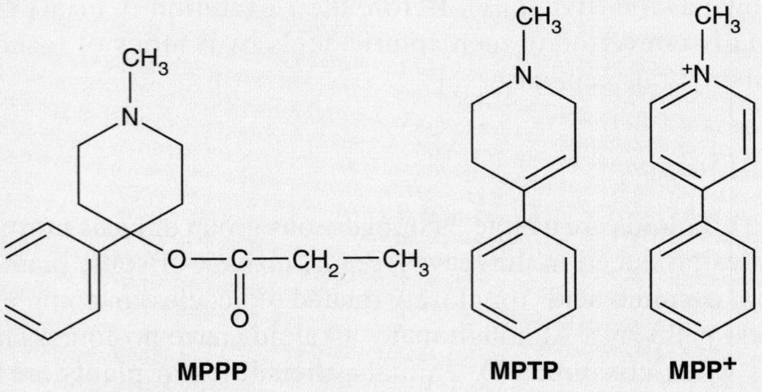

FIGURE 14B **Formation of MPP$^+$, a Neurotoxin.**
MPPP, also referred to as meperidine, is a synthetic analgesic with morphinelike properties. If the chemical reaction used to synthesize MPPP is not carefully regulated, a toxic by-product is also produced. When this latter molecule, MPTP, is inadvertently consumed, it is converted in the brain to MPP$^+$, a neurotoxic agent, in an oxidation reaction catalyzed by monoamine oxidase (see p. 517).

---

components of opium (codeine and morphine) are examples of the *isoquinoline alkaloids* in which a nitrogen appears in one ring of a multiring system.

Nucleotides

Nucleotides are complex nitrogen-containing molecules required for cell growth and development. Not only are nucleotides the building blocks of the nucleic acids, they also play several essential roles in energy transformation and regulate many metabolic pathways. As described, each nucleotide is composed of three parts: a nitrogenous base, a pentose sugar, and one or more phosphate groups. The nitrogenous bases are derivatives of either purine or pyrimidine, which are planar heterocyclic aromatic compounds.

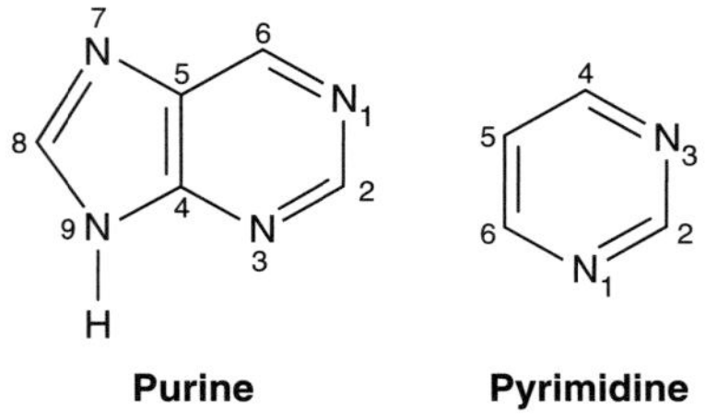

Purine   **Pyrimidine**

Common naturally occurring **purines** include adenine, guanine, xanthine, and hypoxanthine; thymine, cytosine, and uracil are common **pyrimidines** (Figure 14.21). Because of their aromatic structures, the purines and pyrimidines absorb UV light. At pH 7, this absorption is especially strong at 260 nm. Purine and

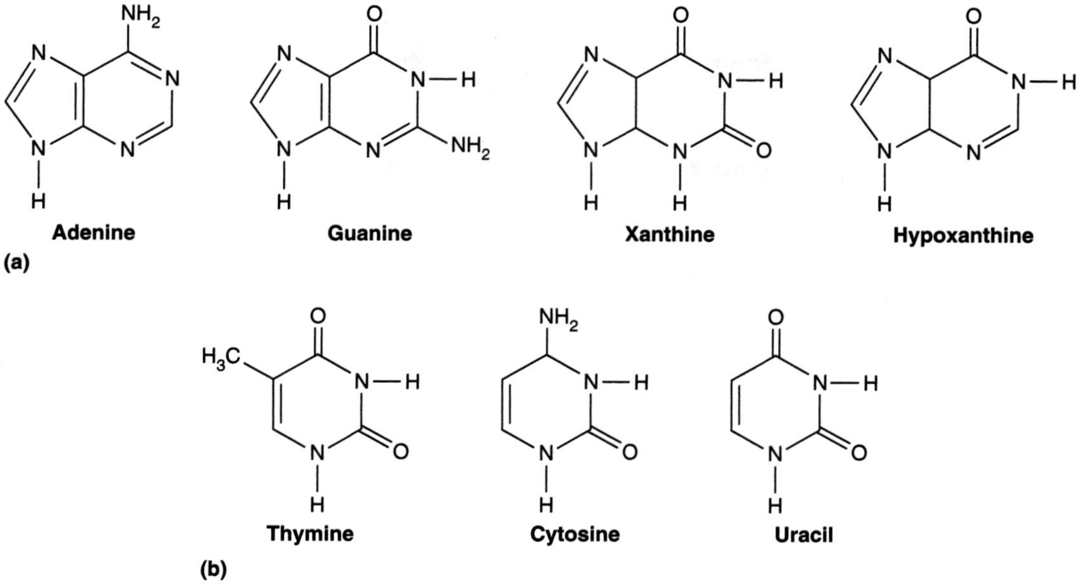

(a)

Adenine   Guanine   Xanthine   Hypoxanthine

(b)

Thymine   Cytosine   Uracil

FIGURE 14.21

The Most Common Naturally Occurring Purines (a) and Pyrimidines (b).

pyrimidine bases have tautomeric forms; that is, they undergo spontaneous shifts in the relative position of a hydrogen atom and a double bond in a three atom sequence involving heteroatoms. This property is especially important because the precise location of hydrogen atoms on the oxygen and nitrogen atoms affects the interaction of bases in nucleic acid molecules. Adenine and cytosine have both amino and imino forms; guanine, thymine, and uracil have both keto (lactam) and enol (lactim) forms (Figure 14.22). At physiological pH the amino and keto forms are the most stable.

When a purine or pyrimidine base is linked through a β-N-glycosidic linkage to C-1 of a pentose sugar, the molecule is called a **nucleoside**. Two types of sugar are found in nucleosides: ribose or deoxyribose. Ribose-containing nucleosides with adenine, guanine, cytosine, and uracil are referred to as adenosine, guanosine, cytidine, and uridine, respectively. When the sugar component is deoxyribose, the prefix *deoxy* is used. For example, the deoxy nucleoside with adenine is called deoxyadenosine. Because the base thymine usually occurs only in deoxyribonucleosides, deoxythymidine is called thymidine. Possible confusion in the identification of atoms in the base and sugar components of nucleosides is avoided by using a superscript prime to denote the atoms in the sugar.

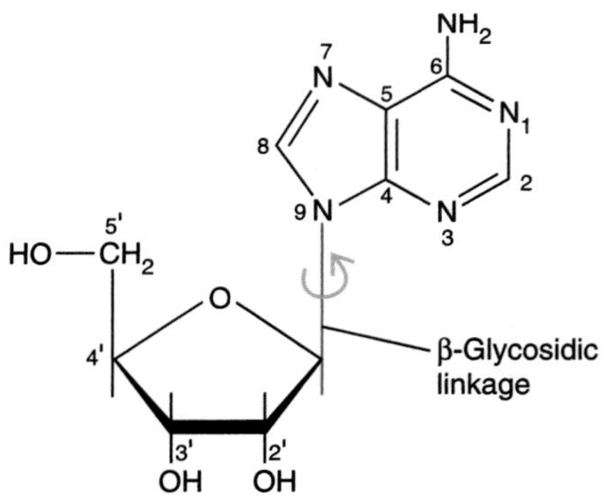

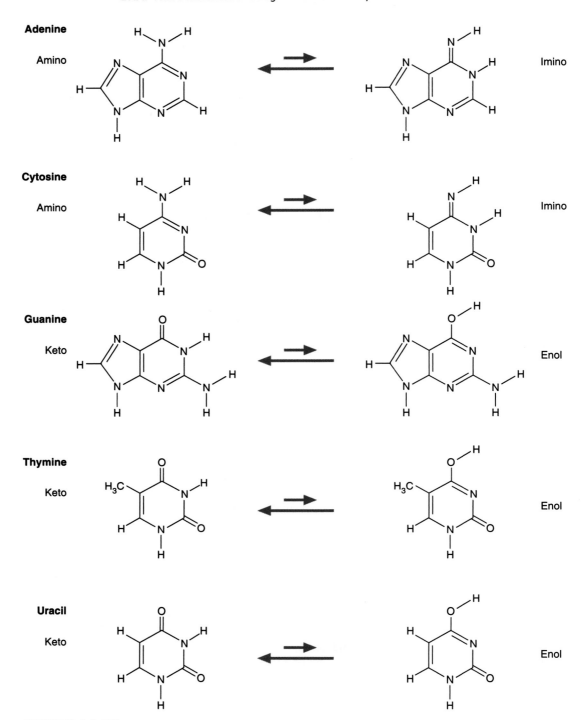

FIGURE 14.22

Tautomers of Adenine, Cytosine, Guanine, Thymine, and Uracil.

At physiological pH the amino and keto tautomers of nitrogenous bases are the predominant forms.

Rotation around the N-glycosidic bond of nucleosides creates two conformations: *syn* and *anti*. Purine nucleosides occur as either *syn* or *anti* forms. In
pyrimidine nucleosides the *anti* conformation predominates because of steric hin-
derance between the pentose sugar and the carbonyl oxygen at C-2.

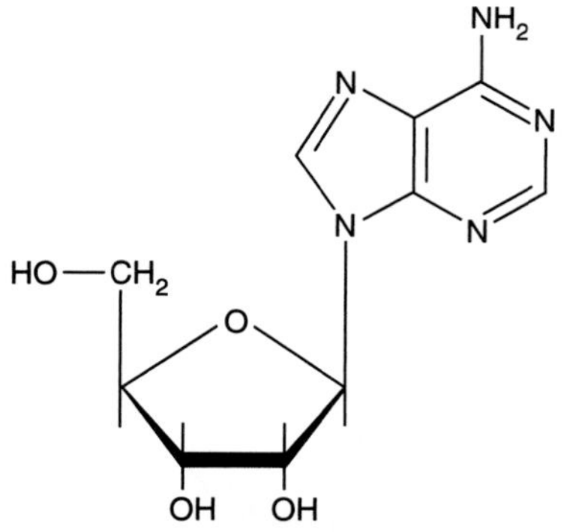

Anti-Adenosine 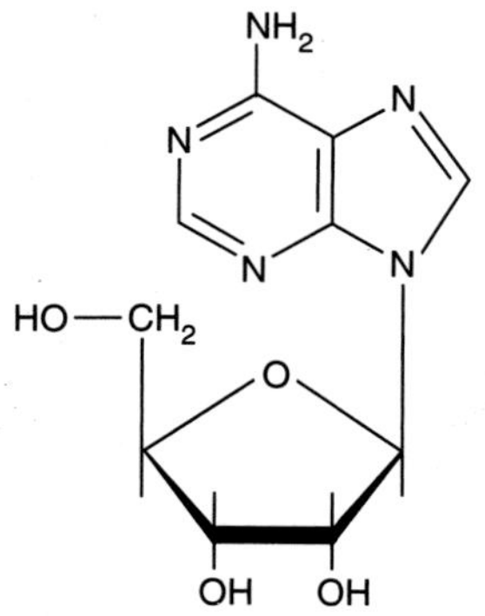   **Syn-Adenosine** 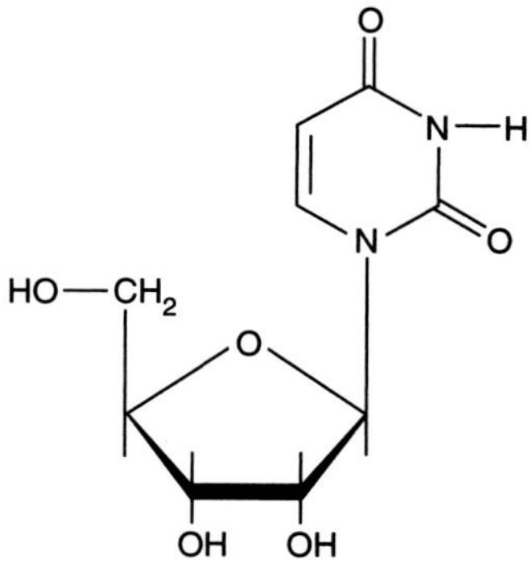   **Anti-Uridine**

Nucleotides are nucleosides in which one or more phosphate groups are bound to the sugar (Figure 14.23). Most naturally occurring nucleotides are 5′-phosphate esters. If one phosphate group is attached at the 5′-carbon of the sugar, the

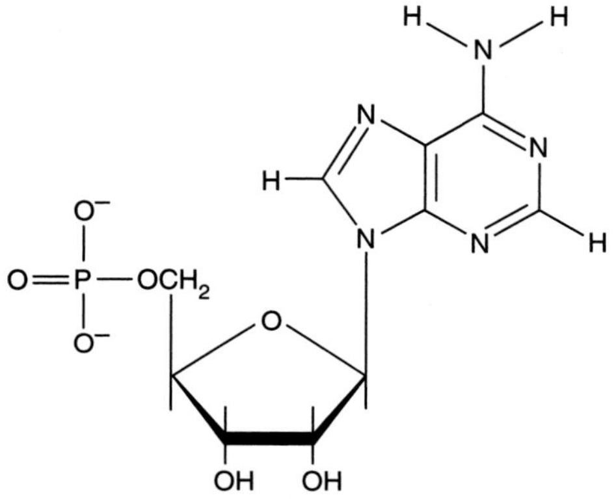

Adenosine-5'-monophosphate
(AMP)

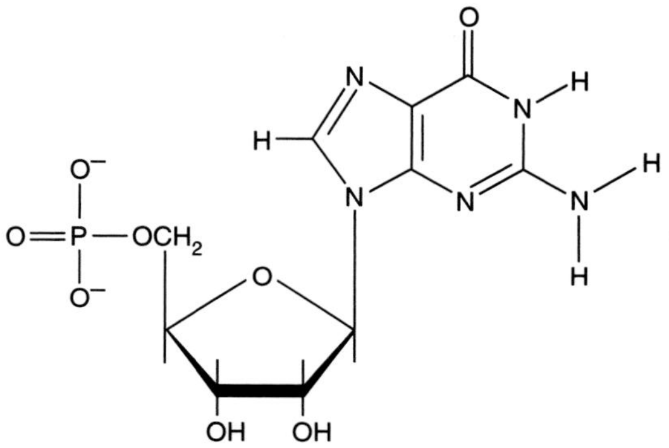

Guanosine-5'-monophosphate
(GMP)

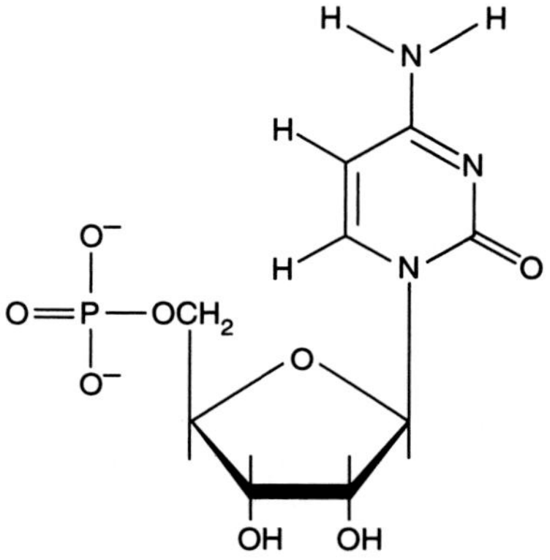

Cytidine-5'-monophosphate
(CMP)

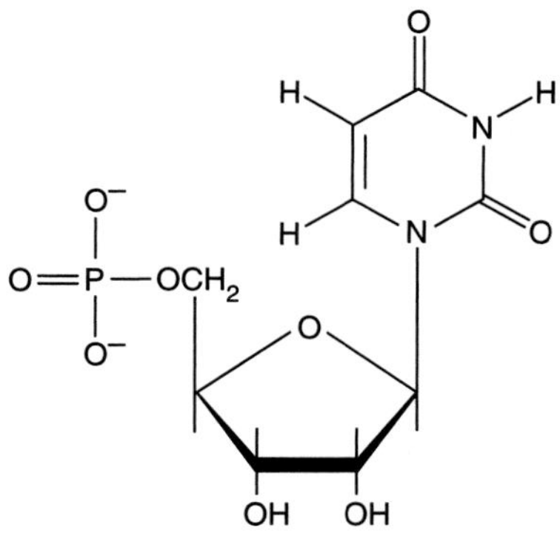

Uridine-5'-monophosphate
(UMP)

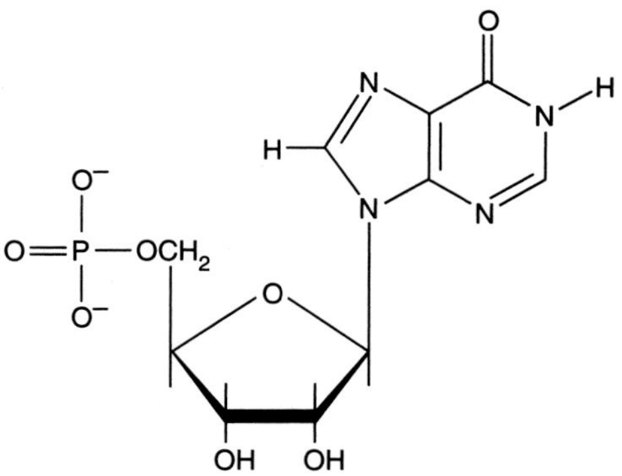

Inosine-5'-monophosphate
(IMP)

(a)

FIGURE 14.23

Common Ribonucleotides (a) and Deoxyribonucleotides (b)

The names of nucleotides containing both deoxyribose and thymine do not have the prefix *deoxy*. Inosine-5′-monophosphate (IMP) is an intermediate in purine nucleotide synthesis. The base component of IMP is hypoxanthine.

FIGURE 14.23
Continued.

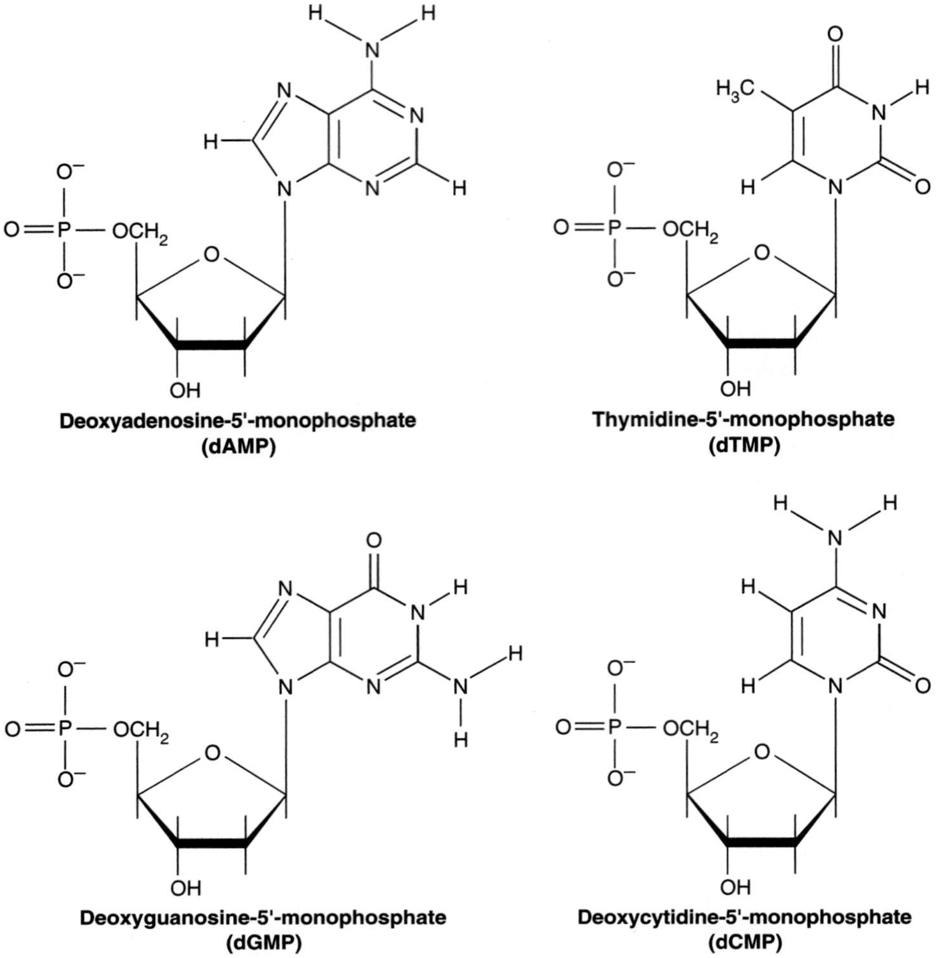

Deoxyadenosine-5'-monophosphate
(dAMP)

Thymidine-5'-monophosphate
(dTMP)

Deoxyguanosine-5'-monophosphate
(dGMP)

Deoxycytidine-5'-monophosphate
(dCMP)

(b)

molecule is named as a nucleoside monophosphate, e.g., adenosine-5′-monophosphate (AMP). Nucleoside di- and triphosphates contain two and three phosphate groups, respectively. Phosphate groups make nucleotides strongly acidic. (Protons dissociate from the phosphate groups at physiological pH.) Because of their acidic nature, nucleotides may also be named as acids. For example, AMP is often referred to as adenylic acid or adenylate. Nucleoside di- and triphosphates form complexes with Mg^{2+}. In nucleoside triphosphates such as ATP, Mg^{2+} can form α,β and β,γ complexes:

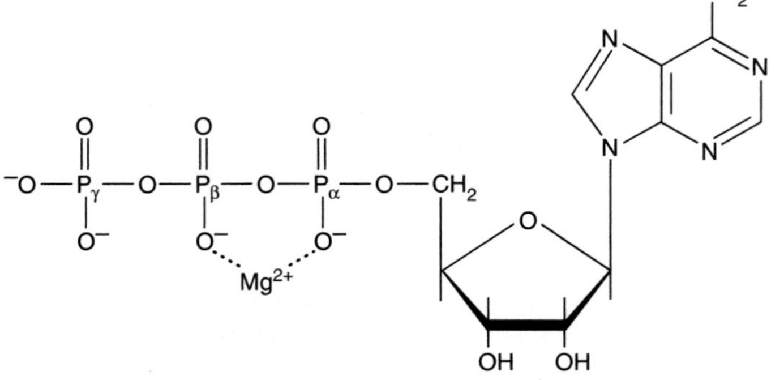

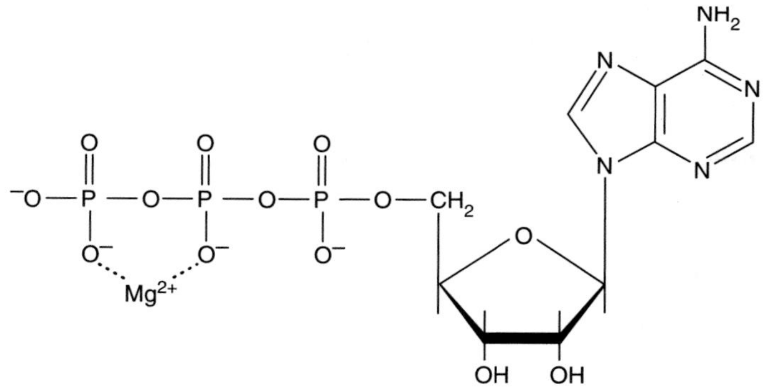

Purine and pyrimidine nucleotides can be synthesized in *de novo* and salvage pathways. These pathways are described.

PURINE NUCLEOTIDES The *de novo* synthesis of purine nucleotides begins with the formation of 5-phospho-α-D-ribosyl-1-pyrophosphate (PRPP) catalyzed by ribose-5-phosphate pyrophosphokinase (PRPP synthetase).

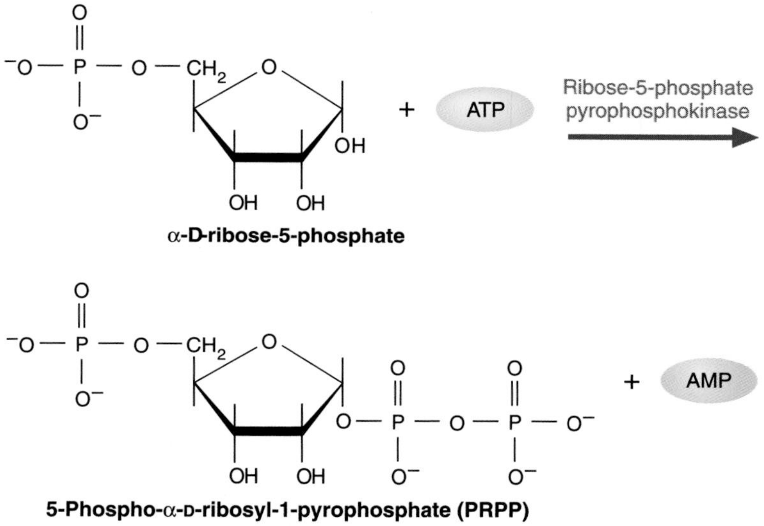

α-**D-ribose-5-phosphate**

5-Phospho-α-D-ribosyl-1-pyrophosphate (PRPP)

(The substrate for this reaction, α-D-ribose-5-phosphate, is a product of the pentose phosphate pathway.) Figure 14.24 illustrates the initial phase in the pathway by which PRPP is converted to inosine monophosphate (inosinate), the first purine nucleotide. The process begins with the displacement of the pyrophosphate group of PRPP by the amide nitrogen of glutamine in a reaction catalyzed by glutamine PRPP amidotransferase. This reaction is the committed step in purine synthesis. The product formed is 5-phospho-β-D-ribosylamine.

Once 5-phospho-β-D-ribosylamine is formed, the building of the purine ring structure begins. Phosphoribosylglycinamide synthase catalyzes the formation of an amide bond between the carboxyl group of glycine and the amino group of 5-phospho-β-D-ribosylamine. In eight subsequent reactions the first purine nucleotide IMP is formed. Other precursors of the base component of IMP (hypoxanthine) include CO_2, aspartate, and N^{10}-formyl THF. This pathway requires the hydrolysis of four ATP molecules.

The conversion of IMP to either adenosine monophosphate (AMP or adenylate) or guanosine monophosphate (GMP or guanylate) requires two reactions (Figure 14.25).

AMP differs from IMP in only one respect: An amino group replaces a keto oxygen. The amino nitrogen provided by aspartate becomes linked to IMP in a

(continued on p. 492)

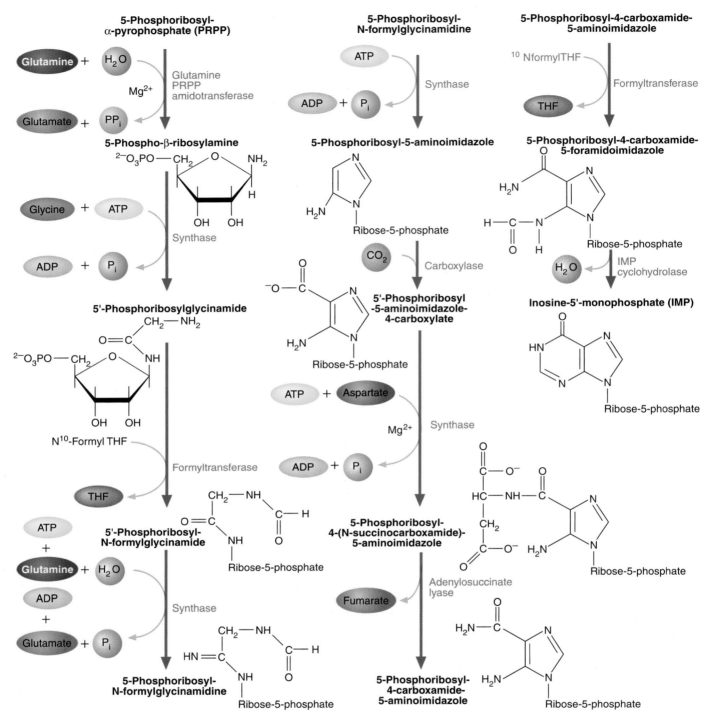

FIGURE 14.24

Synthesis of Inosine-5′-Monophosphate.

The biosynthesis of IMP begins with the reaction between an amino group of glutamine with C-1 of PRPP. The product, 5-phospho-β-ribosyl-amine, subsequently undergoes a series of reactions in which the purine ring is constructed using carbon atoms from formate (via N^{10}-formyl THF) and CO_2, and nitrogen atoms from glycine, glutamine, and aspartate.

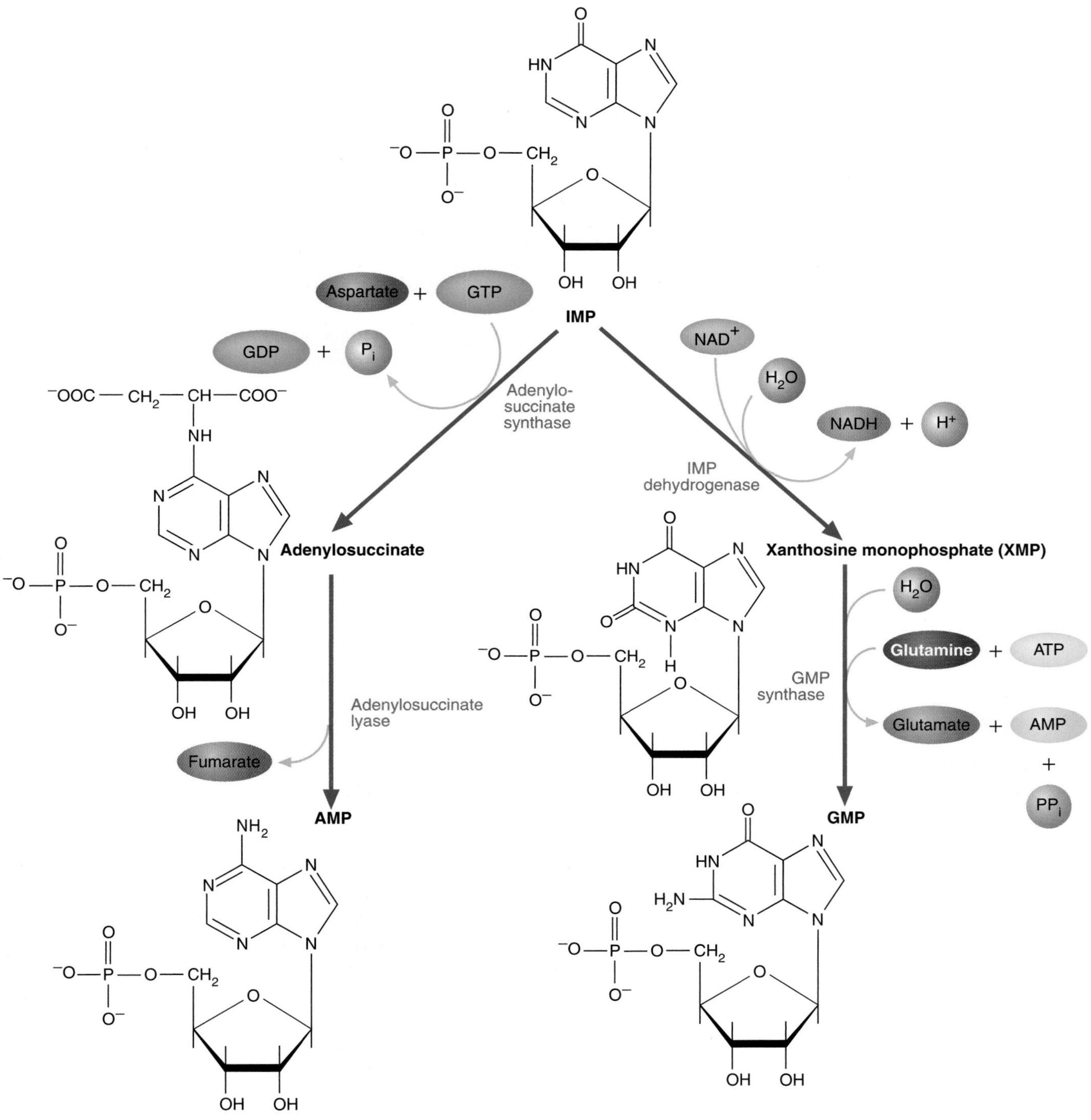

FIGURE 14.25

Biosynthesis of AMP and GMP from IMP.

In the first step of AMP synthesis, the C-6 keto oxygen of the hypoxanthine base moiety of IMP is replaced by the amino group of aspartate. In the second step the product of the first reaction, adenylosuccinate, is hydrolyzed to form AMP and fumarate. GMP synthesis begins with the oxidation of IMP to form XMP. GMP is produced as the amide nitrogen of glutamine replaces the C-2 keto oxygen of XMP. Note that AMP formation requires GTP and GMP formation requires ATP.

GTP-requiring reaction catalyzed by adenylosuccinate synthase. In the step the product adenylosuccinate eliminates fumarate to form AMP. (The enzyme that catalyzes this reaction also catalyzes a similar step in IMP synthesis.) The conversion of IMP to GMP begins with a dehydrogenation utilizing NAD^+ catalyzed by IMP dehydrogenase. The product is referred to as xanthosine monophosphate (XMP). XMP is then converted to GMP by the donation of an amino nitrogen by glutamine in an ATP-requiring reaction catalyzed by GMP synthase.

Nucleoside triphosphates are the most common nucleotide used in metabolism. They are formed in the following manner. Recall that ATP is synthesized from ADP and P_i during certain reactions in glycolysis and aerobic metabolism. ADP is synthesized from AMP in a reaction catalyzed by adenylate kinase:

$$AMP + ATP \longrightarrow 2\,ADP$$

Other nucleoside triphosphate are synthesized in ATP-requiring reactions catalyzed by a series of nucleoside monophosphate kinases:

$$NMP + ATP \rightleftharpoons NDP + ADP$$

Nucleoside diphosphate kinase catalyzes the formation of nucleoside triphosphates:

$$N_1DP + N_2TP \rightleftharpoons N_1TP + N_2DP$$

where N_1 and N_2 are purine or pyrimidine bases.

In the purine salvage pathway, purine bases obtained from the normal turnover of cellular nucleic acids or (to a lesser extent) from the diet are reconverted into nucleotides. Because the *de novo* synthesis of nucleotides is metabolically expensive (i.e., relatively large amounts of phosphoryl bond energy are used), many cells have mechanisms to retrieve purine bases. Hypoxanthine-guaninephosphoribosyltransferase (HGPRT) catalyzes nucleotide synthesis using PRPP and either hypoxanthine or guanine. The hydrolysis of pyrophosphate makes these reactions irreversible.

Hypoxanthine   +   PRPP ⟶ IMP   + PP_i

Guanine   +   PRPP ⟶ GMP   + PP_i

Deficiency of HGPRT causes *Lesch-Nyhan syndrome*, a devastating disease characterized by excessive production of uric acid (Section 15.3) and certain neurological symptoms (self-mutilation, involuntary movements, and mental retardation). Affected children appear normal at birth but begin to deteriorate at about 3 to 4 months of age.

Adenine phosphoribosyltransferase (ARPT) catalyzes the transfer of adenine to PRPP, thus forming AMP:

Adenine   +   PRPP ⟶ AMP   + PP_i

The relative importance of the *de novo* and salvage pathways is unclear. However, the severe symptoms of hereditary HGPRT deficiency indicate that the purine salvage pathway is vitally important. In addition, investigations of purine nucleotide synthesis inhibitors for treating cancer indicate that both pathways must be inhibited for significant tumor growth suppression.

The regulation of purine nucleotide biosynthesis is summarized in Figure 14.26. The pathway is controlled to a considerable degree by PRPP availability. Several products of the pathway inhibit both ribose-5-phosphate pyrophospho-

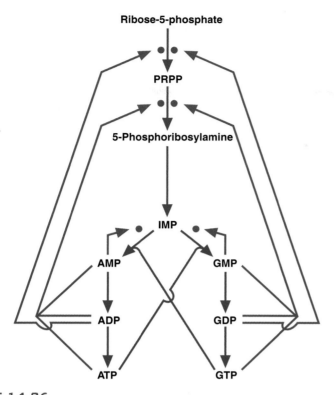

FIGURE 14.26

Purine Nucleotide Biosynthesis Regulation.

Feedback inhibition is indicated by red arrows. The stimulation of AMP synthesis by GTP and GMP synthesis by ATP ensures a balanced synthesis of both families of purine nucleotides.

kinase and glutamine-PRPP amidotransferase. The combined inhibitory effect of the end products is synergistic (i.e., the net inhibition is greater than the inhibition of each nucleotide acting alone). At the IMP branch point, both AMP and GMP regulate their own syntheses by feedback inhibition of adenylosuccinate synthase and IMP dehydrogenase. The hydrolysis of GTP drives the synthesis of adenylosuccinate, whereas ATP drives XMP synthesis. This reciprocal arrangement is believed to facilitate the maintenance of appropriate cellular concentrations of adenine and guanine nucleotides.

PYRIMIDINE NUCLEOTIDES In pyrimidine nucleotide synthesis the pyrimidine ring is assembled first and then linked to ribose phosphate. The carbon and nitrogen atoms in the pyrimidine ring are derived from bicarbonate, aspartate, and glutamine. Synthesis begins with the formation of carbamoyl phosphate in an ATP-requiring reaction catalyzed by the cytoplasmic enzyme carbamoyl phosphate synthetase II (Figure 14.27). (Carbamoyl phosphate synthetase I is a mitochondrial enzyme involved in the urea cycle, described in Chapter 15.) One molecule of ATP provides a phosphate group, while the hydrolysis of another ATP drives the reaction. Carbamoyl phosphate next reacts with aspartate to form carbamoyl aspartate. The closure of the pyrimidine ring is then catalyzed by dihydroorotase. The product, dihydroorotate, is then oxidized to form orotate. Dihydroorotate dehydrogenase, the enzyme that catalyzes this reaction, is a flavoprotein associated with the inner mitochondrial membrane. (The NADH produced in this reaction donates its electrons to the electron transport complex.) Once synthesized on the cytoplasmic face of the inner mitochondrial membrane, orotate is converted by orotate

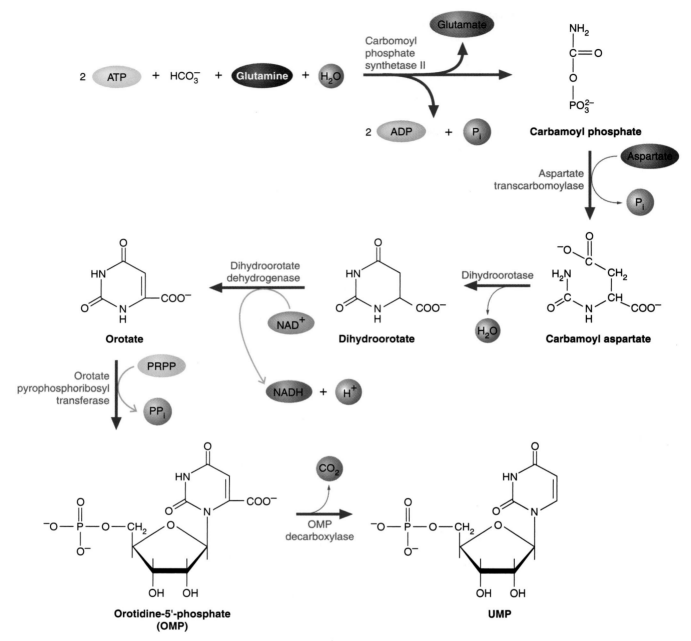

FIGURE 14.27

Pyrimidine Nucleotide Synthesis.

The metabolic pathway in which UMP is synthesized is composed of six enzyme-catalyzed reactions.

pyrophosphoribosyl transferase to orotidine-5′-phosphate (OMP), the first nucleotide in the pathway, by reacting with PRPP. Uridine-5′-phosphate (UMP) is produced when OMP is decarboxylated in a reaction catalyzed by OMP decarboxylase. Both orotate pyrophosphoribosyl transferase and OMP decarboxylase activities occur on a protein referred to as UMP synthase. (In a rare genetic disease called *orotic aciduria*, there is excessive urinary excretion of orotic acid because UMP synthase is defective. Symptoms include retarded growth and anemia. Treatment with a combination of pyrimidine nucleotides, which inhibit the production of orotate and provide the building blocks for nucleic acid synthesis, reverses the disease process.) UMP is a precursor for the other pyrimidine nucleotides. Two sequential phosphorylation reactions form UTP, which then accepts an amide nitrogen from glutamine to form CTP.

DEOXYRIBONUCLEOTIDES All of the nucleotides discussed so far are ribonucleotides, molecules that are principally used as the building blocks of RNA, as nucleotide derivatives of molecules such as sugars, or as energy sources. The nucleotides required for DNA synthesis, the 2′-deoxyribonucleotides, are produced by reducing ribonucleoside diphosphates (Figure 14.28). The electrons used in the synthesis of 2′-deoxyribonucleotides are ultimately donated by NADPH. *Thioredoxin*, a low-molecular-weight protein (13 kD) with two sulfhydryl groups, mediates the transfer of hydrogen atoms from NADPH to ribonucleotide reductase, the enzyme that catalyzes the reduction of ribonucleotides to form deoxyribonucleotides. The regeneration of reduced thioredoxin is catalyzed by thioredoxin reductase.

Regulation of ribonucleotide reductase is complex. The binding of dATP (deoxyadenosine triphosphate) to a regulatory site on the enzyme decreases catalytic activity. The binding of deoxyribonucleoside triphosphates to several other enzyme sites alters substrate specificity so that there are differential increases in the concentrations of each of the deoxyribonucleotides. This latter process balances the production of the 2′-deoxyribonucleotides required for cellular processes, especially that of DNA synthesis.

The deoxyuridylate (dUMP) produced by dephosphorylation of the dUDP product of ribonucleotide reductase is not a component of DNA, but its methylated derivative deoxythymidylate (dTMP) is. The methylation of dUMP is catalyzed by thymidylate synthase, which utilizes N^5,N^{10}-methylene THF. As the methylene group is transferred, it is reduced to a methyl group, while the folate coenzyme is oxidized to form dihydrofolate. THF is regenerated from dihydrofolate by dihydrofolate reductase and NADPH. (This reaction is the site of action of some anticancer drugs, such as methotrexate.) Deoxyuridylate can also be synthesized from dCMP by deoxycytidylate deaminase.

The pyrimidine salvage pathway, which uses preformed pyrimidine bases from dietary sources or from nucleotide turnover, is of minor importance in

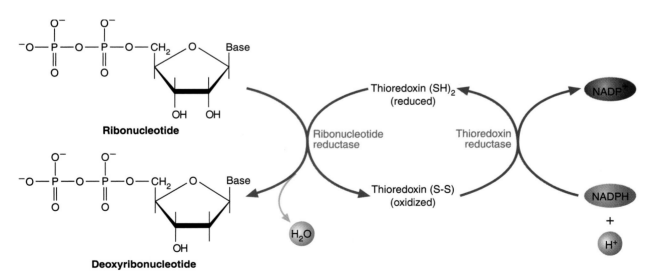

FIGURE 14.28

Deoxyribonucleotide Biosynthesis.

Electrons for the reduction of ribonucleotides ultimately come from NADPH. Thioredoxin, a small protein with two thiol groups, mediates the transfer of electrons from NADPH to ribonucleotide reductase.

mammals. In bacteria, phosphoriboyltransferase catalyzes the synthesis of nucleotides from PRPP and either uracil or thymine. In an alternative pathway found in both bacteria and some higher animals, uridine phosphorylase catalyzes the synthesis of uridine from uracil and ribose-1-phosphate. UMP is produced in a later reaction between uridine and ATP, catalyzed by uridine kinase. Similar enzymes that catalyze salvage reactions for other pyrimidine nucleotides have also been identified.

In mammals, carbamoyl phosphate synthetase II is the key regulatory enzyme in the biosynthesis of pyrimidine nucleotides. The enzyme is inhibited by UTP, the product of the pathway, and stimulated by purine nucleotides. In many bacteria, aspartate carbamoyl transferase is the key regulatory enzyme. It is inhibited by CTP and stimulated by ATP.

QUESTION 14.10

What is the source of the phosphoribosyl group of nucleotides? What pathway is the ultimate source of the ribose in this molecule?

Heme

Heme, one of the most complex molecules synthesized by mammalian cells, has an iron-containing porphyrin ring. As described previously, heme is an essential structural component of hemoglobin, myoglobin, and the cytochromes. Almost all aerobic cells synthesize heme because it is required for the cytochromes of the mitochondrial ETC. The heme biosynthetic pathway is especially prominent in liver, bone marrow, and intestine cells and in reticulocytes (the nucleus-containing precursor cells of red blood cells). Heme is synthesized from the relatively simple components glycine and succinyl-CoA.

In the first step of heme synthesis, glycine and succinyl-CoA condense to form δ-aminolevulinate (ALA) (Figure 14.29). This reaction, which requires pyridoxal phosphate, is the rate-controlling step in porphyrin synthesis. ALA synthase, a mitochondrial enzyme, is allosterically inhibited by *hemin*, an Fe^{3+}-containing derivative of heme. (When a red blood cell's porphyrin concentration exceeds that of globin, heme accumulates and is subsequently oxidized to hemin.) Hemin production also decreases synthesis of ALA synthase. In the next step of porphyrin synthesis, two molecules of ALA condense to form porphobilinogen. Porphobilinogen synthase, which catalyzes this reaction, is a zinc-containing enzyme that is extremely sensitive to heavy-metal poisoning (Special Interest Box 14.4). Uroporphyrinogen I synthase catalyzes the symmetric condensation of four porphobilinogen molecules. An additional protein is also required in this reaction. Uroporphyrinogen III cosynthase alters the specificity of uroporphyrinogen synthase so that the asymmetrical molecule uroporphyrinogen III is produced. When four CO_2 molecules are removed, catalyzed by uroporphyrinogen decarboxylase, coproporphyrinogen is synthesized. This reaction is followed by the removal of two additional CO_2 molecules, thus forming protoporphyrinogen IX. Oxidation of the porphyrin ring's methylene groups forms protophorphyrin IX, the direct precursor of heme. The final step in the synthesis of heme (also called *protoheme IX*) is the insertion of Fe^{2+}, a reaction that occurs spontaneously but is accelerated by ferrochelatase.

Protoporphyrin IX is also a precursor of the chlorophylls. After magnesium (Mg^{2+}) is incorporated, the enzyme Mg-protoporphyrin methylesterase catalyzes the addition of a methyl group to form Mg-protophorphyrin IX monomethylester. This molecule is then converted to chlorophyll in several light-induced reactions.

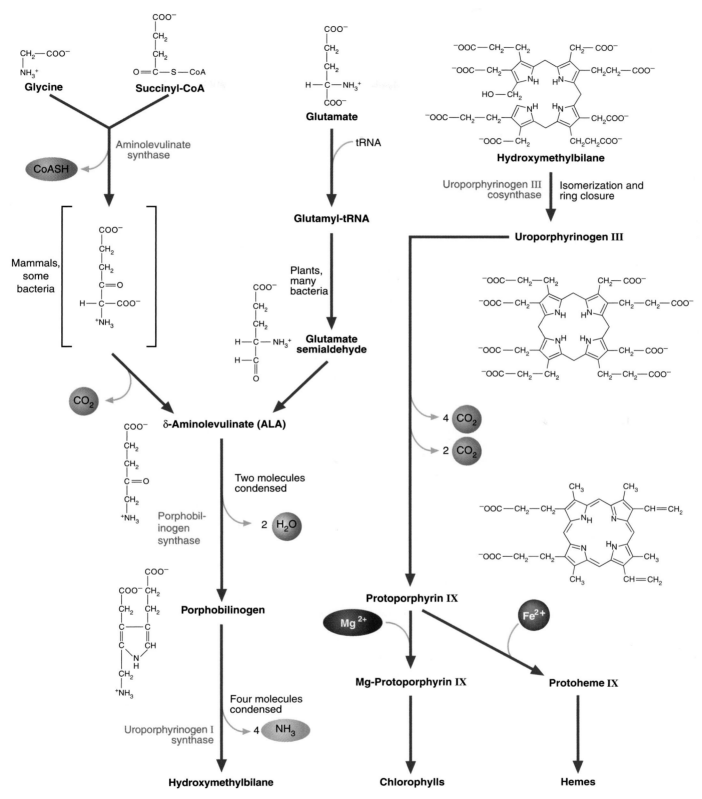

FIGURE 14.29

Heme and Chlorophyll Biosynthesis.

In plants and some bacterial species, ALA is synthesized from glutamate in a process involving glutamyl tRNA. Refer to Chapter 17 for a discussion of tRNA (transfer RNA).

The heavy metal lead is absorbed primarily in the gastrointestinal and respiratory tracts and deposited in soft tissues (e.g., kidney, liver, and central nervous system) and in bone. (In affected individuals, approximately 95% of lead is sequestered in bone.) Symptoms of lead intoxication, which vary according to the degree of exposure, include anorexia, muscle pain and weakness, abdominal pain, infertility, stillbirth, and encephalopathy. (*Lead encephalopathy* is a disorder of the brain's cerebral cortex. It is characterized by clumsiness, headache, irritability, insomnia, mental retardation, and, in extreme cases, hallucinations and paralysis.) Possible renal effects of lead exposure include chronic nephritis (an inflammatory condition), and disturbances in the kidney's capacity to reabsorb nutrients such as amino acids, glucose, and phosphate. Despite intensive research efforts, federal regulations, and increasing public awareness, lead poisoning (also referred to as *plumbism*) remains a serious public health problem. For example, children in impoverished neighborhoods in large cities are still at high risk for high blood levels of lead. Lead-based paint is consumed by children because of its sweet taste. In addition, the soil in these areas often has lead levels substantially above acceptable standards. (These high levels are no doubt due in part to lead compounds formerly used in gasoline.)

Lead and Heme Synthesis

Lead is toxic largely because it forms bonds with the sulfhydryl groups of proteins. Any protein with free sulfhydryl groups is therefore vulnerable. Among the best-researched examples of lead-sensitive biomolecules are several enzymes that catalyze reactions in heme biosynthesis. Inhibition of porphobilinogen synthase by lead occurs with relatively low lead levels. Therefore, detecting its substrate (ALA) in the urine serves as an early warning of lead intoxication. The inhibition of ferrochelatase is a more reliable indicator of a serious lead exposure. In acute lead poisoning (caused by accidental ingestion of relatively large amounts of lead compounds), its substrate (protoporphyrin IX) accumulates in tissues. In chronic lead poisoning (a slow, progressive process), protoporphyrin IX complexed with zinc appears in blood. (Because of its high affinity for zinc, protoporphyrin IX forms complexes with this metal when ferrochelatase is inhibited.) Because zinc protoporphyrin in blood is easily measured, its detection is a valuable diagnostic tool.

Lead Poisoning: An Ancient Heritage

Since ancient times the soft grayish-blue metal called lead has been extremely useful. Because it resists corrosion and can be easily shaped, lead has many commercial and industrial applications. For example, lead alloys have long been used in plumbing and shipbuilding. Additionally, several lead compounds have vibrant colors and have been valued as components of paint and cosmetics. However, lead is highly toxic. First used at least 8000 years ago (probably in areas near the Aegean Sea), lead soon became a source of economic strength in the ancient world. For this reason, lead poisoning may have been one of the earliest occupational diseases. However, plumbism was not limited to artisans and metalworkers. Because lead containers stored and preserved wine and foods, and lead pipes transported water, the wealthy were also at high risk. The decline of the Roman Empire has been blamed in part on the effects of lead-contaminated wine and food (e.g., insanity and infertility) on the Roman aristocracy.

Although several ancient physicians were aware that lead was harmful, it was not until the Industrial Revolution in Europe and America that any sustained attention was paid to lead poisoning. Numerous observations of sterility, miscarriages, stillbirths, and premature delivery in both female leadworkers and the wives of male leadworkers resulted, by the end of the nineteenth century, in the removal of female workers from the industry. In the twentieth century, improvements in testing techniques and an awakening social conscience significantly reduced lead exposure. The most serious (and obvious) effects of lead toxicity are now rarely observed. However, lead is believed to be responsible for more subtle injuries. For example, in one controversial hypothesis, some cases of renal disease and hypertension are linked to mild lead exposure. In addition, several researchers have associated intellectual dullness and lowered IQ scores to relatively low levels of lead exposure.

QUESTION 14.11

Identify each of the following biomolecules. Explain the function of each in biochemical processes.

(a)   (b)

(c) structure: NH_2—C=O—O—PO_3^{2-}

(d) structure: ^-O—$P(=O)(O^-)$—O—CH_2 ... ribose ring with OH OH, O—$P(=O)(O^-)$—O—$P(=O)(O^-)$—O^-

(c)   **(d)**

SUMMARY

1. Nitrogen is an essential element in living systems, because it is found in proteins, nucleic acids, and myriad other biomolecules. Biologically useful nitrogen, a scarce resource, is produced in a process referred to as nitrogen fixation, Only a few organisms are capable of fixing nitrogen to form ammonia and nitrate, the oxidation product of ammonia.

2. Organisms vary widely in their ability to synthesize amino acids. Some organisms (e.g., plants and some microorganisms) can produce all required amino acid molecules from fixed nitrogen. Animals can produce only some amino acids. Nonessential amino acids are produced from readily available precursor molecules, whereas essential amino acids must be acquired in the diet.

3. Two types of reactions play prominent roles in amino acid metabolism. In transamination reactions, new amino acids are produced when α-amino groups are transferred from donor α-amino acids to acceptor α-keto acids. Because transamination reactions are reversible, they play an important role in both amino acid synthesis and degradation. Ammonium ions or the amide nitrogen of glutamine can also be directly incorporated into amino acids and eventually other metabolites.

4. On the basis of the biochemical pathways in which they are synthesized, the amino acids can be divided into six families: glutamate, serine, aspartate, pyruvate, aromatics, and histidine.

5. Amino acids are precursors of many physiologically important biomolecules. Many of the processes that synthesize these molecules involve the transfer of carbon groups. Because many of these transfers involve one-carbon groups (e.g., methyl, methylene, methenyl, and formyl), the overall process is referred to as one-carbon metabolism. S-Adenosylmethionine (SAM) and tetrahydrofolate (THF) are the most important carriers of one-carbon groups.

6. Molecules derived from amino acids include several neurotransmitters (e.g., GABA, the catecholamines, serotonin, histamine, and nitric oxide) and hormones (e.g., indole acetic acid). Glutathione is an example of an amino acid derivative that plays an essential role in cells. Alkaloids are a diverse group of basic nitrogen-containing molecules. The role of alkaloids in the plants that produce them is poorly understood. Several alkaloid molecules have profound physiological effects on animals. The nucleotides, molecules that serve as the building blocks of the nucleic acids (as well as energy sources and metabolic regulators), possess heterocyclic nitrogenous bases as part of their structures. These bases, called the purines and the pyrimidines, are derived from various amino acid molecules. Heme is an example of a complex heterocyclic ring system that is derived from glycine and succinyl-CoA. The biosynthetic pathway that produces heme is similar to the one that produces chlorophyll in plants.

SUGGESTED READINGS

Coyle, J. T., and Puttfarcken, P., Oxidative Stress, Glutamate and Neurodegenerative Disorders, *Science*, 262:689–695, 1993.

Lancaster, J. R., Nitric Oxide in Cells, *Amer. Sci.*, 80(3):248–259, 1992.

Leigh, J. A., Nitrogen Fixation in Methanogens: The Archaeal Perspective, *Curr. Issues Mol. Biol.*, 2(4):125–131, 2000.

Meister, A., and Anderson, M. E., Glutathione, *Ann. Rev. Biochem.*, 52:711–760, 1983.

Orme-Johnson, W. H., Molecular Basis of Biological Nitrogen Fixation, *Ann. Rev. Biophys. Biophys. Chem.*, 14:419–459, 1985.

Reichard, P., and Ehrenberg, A., Ribonucleotide Reductase—A Radical Enzyme, *Science*, 221:514–519, 1983.

Wedein, R. P., *Poison in the Pot: The Legacy of Lead*, Southern Illinois University Press, Carbondale and Edwardville, 1984.

Wendehenne, D., Pugin, A., Klessig, D. F., and Durner, J., Nitric Oxide: Comparative Synthesis and Signaling in Animal and Plant Cells, *Trends Plant Sci.*, 6:177–183, 2001.

KEY WORDS

alkaloid, *483*

amethopterin, *477*

amino acid pool, *453*

analogue, *477*

biogenic amine, *480*

branched chain amino acid, *450*

catecholamine, *467*

essential amino acid, *450*

methotrexate, *477*

neurotransmitter, *480*

nitrogen fixation, *451*

nonessential amino acid, *450*

nucleoside, *485*

pernicious anemia, *473*

purine, *484*

pyrimidine, *484*

racemization, *455*

transamination, *451*

vitamin B$_{12}$, *473*

REVIEW QUESTIONS

1. Define the following terms:
   a. essential amino acid
   b. nitrogen balance
   c. *de novo* pathway
   d. biogenic amine
   e. excitotoxin
   f. retrograde neurotransmitter
   g. analogue
   h. auxin
   i. pernicious anemia

2. Why are transamination reactions important in both the synthesis and degradation of amino acids?

3. Give two reasons why nitrogen compounds are limited in the biosphere.

4. Nitrogenase complexes are irreversibly inactivated by oxygen. Explain how nitrogen-fixing bacteria solve this problem.

5. Using reaction equations, illustrate how α-ketoglutarate is converted to glutamate. Name the enzymes and cofactors required.

6. In PLP-catalyzed reactions, the pyridinium ring acts as an electron sink. Describe this process.

7. The concentration in the blood of which of the following amino acids is not affected by passage through the liver?
   a. alanine
   b. isoleucine
   c. phenylalanine
   d. valine
   e. glycine
   f. proline

8. What are the two major classes of neurotransmitters? How do their modes of action differ? Give an example of each type of neurotransmitter.

9. Illustrate the pathways to synthesize the following amino acids:
   a. glutamine
   b. methionine
   c. threonine
   d. glycine
   e. cysteine

10. Determine the synthetic family to which each of the following amino acids belongs:
    a. alanine
    b. phenylalanine
    c. methionine
    d. tryptophan
    e. histidine
    f. serine

11. What are the two most important carriers in one-carbon metabolism? Give examples of their roles in metabolism.

12. Glutathione is an important intracellular thiol. List five functions of glutathione in the body.

13. List ten essential amino acids in humans. Why are they essential?

14. Shown below are six compounds. Indicate which are nucleosides, nucleotides, or purine or pyrimidine bases.

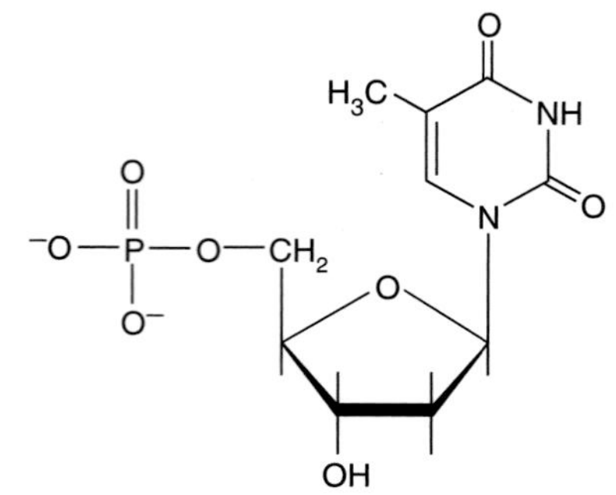

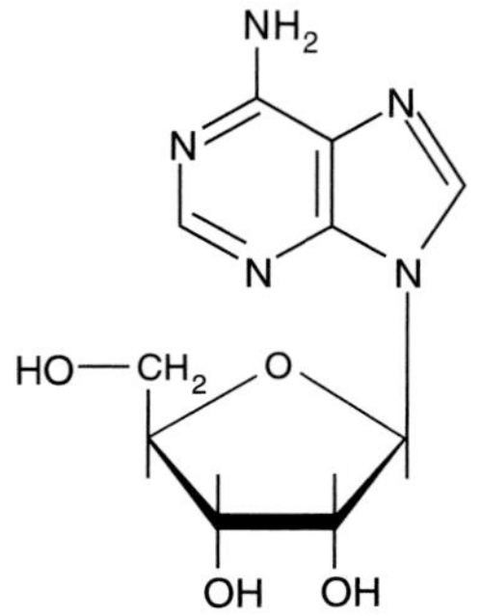

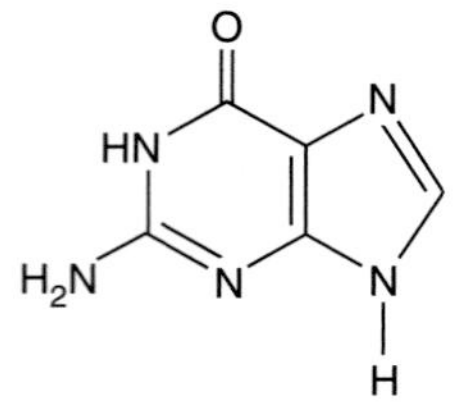

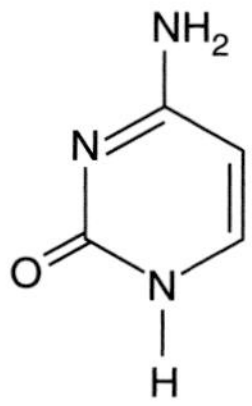

15. In pyrimidine nucleosides the *anti* conformation predominates because of steric interactions with pentose. Do the purine nucleosides have similar interactions?

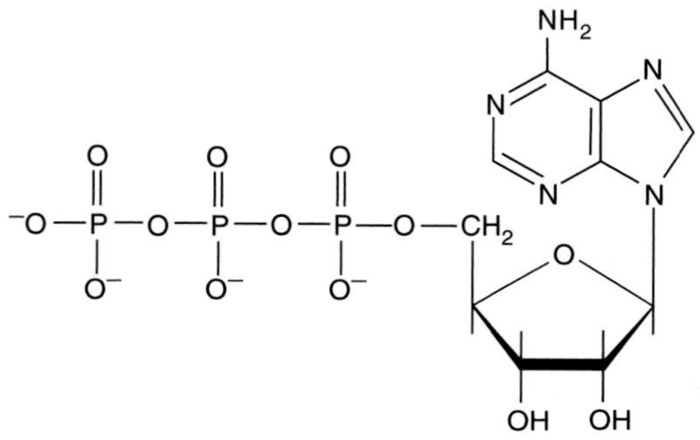

16. Describe the steric interactions that determine the conformations that pyrimidine nucleosides assume.

17. Outline the reactions involved during the assembly of the purine ring. Include structures in your answer.

18. Referring to Question 17, calculate the number of ATP molecules that are required to synthesize a purine. Referring to the purine salvage pathway, calculate the number of ATP molecules that are required to prepare the same molecule. How many ATP molecules are saved by the purine salvage pathway?

19: Explain how the γ-glutamyl cycle acts to transport amino acids across a membrane. How does the location of the γ-glutamyl transpeptidase help drive this process?

20. The amino acids glutamine and glutamate are central to amino acid metabolism. Explain.

21. A mutant bacterium is unable to synthesize glycine. What intermediate in purine biosynthesis will accumulate?

22. Transamination reactions have been described as ping-pong reactions. Using the reaction of alanine with α-ketoglutarate, indicate how this ping-pong reaction works.

23. Pyridoxal phosphate acts as an intermediate carrier of amino groups during transamination reactions. Write a series of reactions to show the role of pyridoxal phosphate in the reaction of alanine and α-ketoglutarate.

24. What is the biologically active form of folic acid? How is it formed?

25. Identify the following biomolecules. Describe the metabolic role of each.

(a)   (b)   (c)

(d)   (e)

THOUGHT QUESTIONS

1. Trace the following radiolabeled molecule through the biosynthetic route to AMP synthesis.

   $H_3N^+—{}^{14}CH_2—COO^-$

2. Although they are consumed by animals in the diet, the purine and pyrimidine bases (unlike fatty acids and sugars) are not used to generate energy. Explain.

3. Why do marathon runners prefer beverages with sugar instead of amino acids during a long run?

4. When susceptible people consume monosodium glutamate, they experience several extremely unpleasant symptoms, such as increased blood pressure and body temperature. Using your knowledge of glutamate activity, explain these symptoms.

5. Radiation exerts part of its damaging effect by causing the formation of hydroxyl radicals. Write a reaction equation to explain how glutathione acts to protect against this form of radiation damage.

6. In pyrimidine nucleotide synthesis, the carbon and nitrogen atoms are derived from bicarbonate, aspartate, and glutamine. Devise a simple experiment to prove the source of the nitrogen atoms. Do not forget to take into account the nitrogen exchange in amino acids.

7. Most amino acids can be readily interconverted with the corresponding α-keto acid. This is not true of lysine. In the multistep process to form the corresponding α-keto acid of lysine, an early step leads to the formation of a Schiff base. Explain how this product might be formed.

8. By definition, essential amino acids are not synthesized by an organism. Arginine is classified as an essential amino acid in children even though it is part of the urea cycle. Explain.

9. If an organism is incubated with ${}^{14}CH_2(OH)CH(NH_2)COOH$, what positions in the purine ring will be labeled?

10. Indicate the source of each carbon and nitrogen in the pyrimidine ring.

11. In PLP-catalyzed reactions, the bond broken in the substrate molecule must be perpendicular to the plane of the pyridinium ring. Considering the bonds present in this ring, describe why this arrangement stabilizes the carbanion.

Nitrogen Metabolism II: Degradation

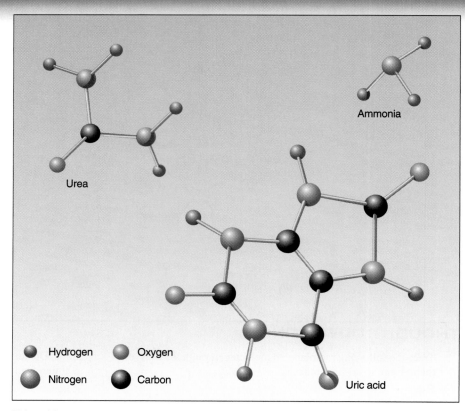

Uric acid, urea, and ammonia are among the most common nitrogenous waste molecules.

The metabolism of nitrogen-containing molecules such as proteins and nucleic acids differs significantly from that of carbohydrates and lipids. Whereas the latter molecules can be stored and mobilized as needed for biosynthetic reactions or for energy generation, there is no nitrogen-storing molecule. (One exception to this rule is storage protein in seeds.) Organisms must constantly replenish their supply of usable nitrogen to replace organic nitrogen that is lost in catabolism. For example, animals must have a steady supply of amino acids in their diets to replace the nitrogen excreted as urea, uric acid, and other nitrogenous waste products.

Despite their apparent stability, most living cells are constantly undergoing renovation. To maintain, repair, and/or reproduce themselves, cells must acquire nutrients from their environment. As described, nutrients are used both as building block molecules and as the energy sources required to drive cellular processes. One of the most obvious aspects of cellular renovation is the turnover of protein and nucleic acids, the macromolecules that are most responsible for the complexity of modern living processes. Additionally, the capacity to both synthesize and degrade these molecules in a timely manner allows organisms to respond to physiological and environmental cues.

Because of the continual synthesis and degradation of proteins and nucleic acids, as well as that of other nitrogen-containing molecules, nitrogen atoms flow through living organisms. Because of the difficulty and expense encountered in fixing nitrogen (and its subsequent scarcity), it is not surprising that living organisms recycle organic nitrogen into a variety of metabolites before it is eventually reconverted to its inorganic form. In some organisms (e.g., plants and microorganisms) this process does not begin until death. Then *decomposers*, microorganisms that inhabit soil and water, convert the organic nitrogen of all dead organisms to ammonia. Ammonia, or its oxidized products nitrate and nitrite, may subsequently be absorbed and used by nearby organisms. Alternatively, nitrate may be converted to atmospheric nitrogen, a process referred to as *denitrification*.

Animals, which typically have a more energetic and aggressive lifestyle, appear to be more wasteful of organic nitrogen. To maintain metabolic flexibility, animals have had to develop mechanisms for disposing of excess and toxic nitrogen-containing molecules (i.e., amino acids and nucleotides) that are not immediately required in cellular processes. Such molecules are converted into nitrogenous waste. Although many variations are observed among species, the following generalizations can be made. The nitrogen in amino acids is removed by deamination reactions and converted to ammonia. The toxic nature of this molecule requires that, it must be detoxified and/or excreted as fast as it is generated. Because of its solubility, many aquatic animals can excrete ammonia itself, which dissolves in the surrounding water and is quickly diluted. (These organisms are referred to as *ammonotelic*.) Terrestrial animals, which must conserve body water, convert ammonia to molecules that can be excreted without a large loss of water. Mammals, for example, convert ammonia to urea. (Urea-producing organisms are referred to as *ureotelic*.) Other animals, such as birds, certain reptiles, and insects, which have even more stringent water conservation problems, are called *uricotelic* because they convert ammonia to uric acid. In many animals (e.g., humans and birds), uric acid is also the nitrogenous waste product of purine nucleotide catabolism.

Because nitrogen catabolic pathways are similar in many organisms and most research efforts in nitrogen catabolism have concentrated on mammals, the mammalian pathways are the focus of this chapter. Chapter 15 begins with a discussion of the pathways that degrade amino acids to form ammonia and the carbon skeletons used in anabolic and catabolic processes. This is followed by a discussion of urea synthesis. Chapter 15 ends with descriptions of the degradation of several amine neurotransmitters, the nucleotides, and the porphyrin heme.

15.1 AMINO ACID CATABOLISM

Despite the complexity of amino acid degradative pathways, the following general statements can be made. The catabolism of the amino acids usually begins by removing the amino group. Amino groups can then be disposed of in urea

The cellular concentration of each type of protein is a consequence of a balance between its synthesis and its degradation. Although it appears to be wasteful, the continuous degradation and resynthesis of proteins, a process referred to as **protein turnover**, serves several purposes. First of all, metabolic flexibility is afforded by relatively quick changes in the concentrations of key regulatory enzymes, peptide hormones, and receptor molecules. Protein turnover also protects cells from the accumulation of abnormal proteins. Finally, numerous physiological processes are just as dependent on timely degradative reactions as they are on synthetic ones. Prominent examples include eukaryotic cell cycle control and antigen presentation. The progression of eukaryotic cells through the phases of the cell cycle (Section 18.1) is regulated by the precisely timed synthesis and degradation of a class of proteins called the *cyclins*. In *antigen presentation* certain immune system cells (e.g., macrophages) engulf foreign or abnormal substances. Most of the molecules that can elicit an immune response, referred to as *antigens*, are polypeptides or proteins. Partially degraded antigen is transferred to the macrophage's plasma membrane where it is used to activate certain T lymphocytes (T cells) via cell-cell interactions. T cells are the principal regulators of the body's immune response.

Proteins differ significantly in their turnover rates, which are measured in half-lives. (A *half-life* is the time required for 50% of a specified amount of a protein to be degraded.) Proteins that play structural roles typically have long half-lives. For example, some connective tissue proteins (e.g., the collagens) often have half-lives that are measured in years. In contrast, the half-lives of regulatory enzymes are typically measured in minutes. Several selected examples are listed in Table 15.1.

Despite considerable research, the mechanisms of protein turnover are still unclear. However, several aspects of this process are now known. Proteins are degraded by proteolytic enzymes found throughout the cell. These include the cytoplasmic Ca^{2+}-activated calpains and the lysosomal cathepsins. In addition, ubiquination is now believed to have a major role in protein turnover. In **ubiquination**, illustrated in Figure 15A, several molecules of a small 76-residue eukaryotic protein called **ubiquitin** are covalently attached to some proteins destined for degradation. Once a protein is ubiquinated, it is degraded by a multisubunit proteolytic complex called a **proteosome**. Because ubiquitin molecules are not degraded in this process, they then become available for new rounds of protein degradation.

Ubiquitin, found in several cellular compartments (e.g., cytoplasm and the nucleus), belongs to a class of proteins referred to as *stress proteins*. Stress proteins, also called **heat shock proteins** (hsp), are so named because their syntheses are accelerated (and in some cases initiated) when cells encounter stress. (The name *heat shock protein* is misleading, because a variety of stressful conditions besides elevated temperature induce their synthesis.) Other stress proteins act as **molecular chaperones**, that is, they promote protein folding (p. 692). Heat shock proteins and molecular chaperones also play significant roles in protein transport and intermolecular interactions.

The mechanisms that target protein for destruction by ubiquination or by other degradative processes are not fully understood. However, the following features of proteins appear to mark them for destruction:

TABLE 15.1
Human Protein Half-Lives

Protein	Approximate Value of Half Life (h)
Ornithine decarboxylase	0.5
Tyrosine aminotransferase	2
Tryptophan oxygenase	2
PEP carboxykinase	5
Arginase	96
Aldolase	118
Glyceraldehyde-3-phosphate dehydrogenase	130
Cytochrome c	150
Hemoglobin	2,880

synthesis. The carbon skeletons produced from the standard amino acids are then degraded to form seven metabolic products: acetyl-CoA, acetoacetyl-CoA, pyruvate, α-ketoglutarate, succinyl-CoA, fumarate, and oxaloacetate. Depending on the animal's current metabolic requirements, these molecules are used to synthesize fatty acids or glucose or to generate energy. Amino acids degraded to form acetyl-CoA or acetoacetyl-CoA are referred to as **ketogenic** because they can be converted to either fatty acids or ketone bodies. The carbon skeletons of the **glucogenic** amino acids, which are degraded to pyruvate or a citric acid cycle intermediate, can then be used in gluconeogenesis. After discussions of deamination pathways and urea synthesis, the pathways that degrade carbon skeletons are described.

1. **N-terminal residues**. The N-terminal residue of a protein is partially responsible for its susceptibility to degradation. For example, proteins with methionine or alanine N-terminal residues have substantially longer half-lives than do those with leucine or lysine.

2. **Peptide motifs**. Proteins with certain homologous sequences are rapidly degraded. For example, proteins that have extended sequences containing proline, glutamate, serine, and threonine have half-lives of less than 2 hours. (PEST sequences are named for the one-letter abbreviations for these amino acids. See Table 5.1.) The *cyclin destruction box* is a set of homologous nine-residue sequences near the N-terminus of cyclins that ensures rapid ubiquination.

3. **Oxidized residues**. Oxidized amino acid residues (i.e., residues that are altered by oxidases or by attack by ROS) promote protein degradation.

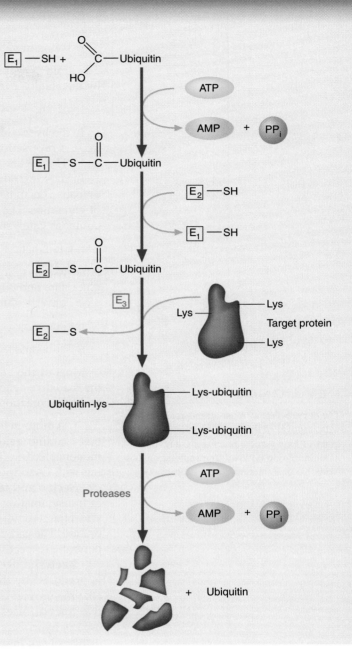

FIGURE 15A Ubiquination of Protein.
Three enzymes are involved in preparing ubiquitin for its role in protein degradation. In the first step, an activating enzyme E_1 forms a thiol ester with ubiquitin. (This reaction is driven by the hydrolysis of ATP to AMP.) Ubiquitin is then transferred from E_1 to E_2. Ubiquitin may be transferred from E_2 (ubiquitin-conjugating enzyme) directly to a target protein. Often E_2 is a substrate for a ubiquitin targeting protein called E_3, which identifies specific proteins to be degraded. In this latter process ubiquitin is transferred from E_2 to E_3 and then to the target protein. Ubiquitin is attached to target proteins via a covalent bond between the C-terminal glycine of ubiquitin and the ε side chain amino group of the lysine residues of the targeted protein. Most cells possess a single type of E_1 and numerous families of E_2 and E_3.

Deamination

The removal of the α-amino group from amino acids involves two types of biochemical reactions: transamination and oxidative deamination. Both reactions have been described (Section 14.2). (Recall that transamination reactions occupy important positions in nonessential amino acid synthesis.) Because these reactions are reversible, amino groups are easily shifted from abundant amino acids and used to synthesize those that are scarce. Amino groups become available for urea synthesis when amino acids are in excess. Urea is synthesized in especially large amounts when the diet is high in protein or when there is massive breakdown of protein, for example, during starvation.

In muscle, excess amino groups are transferred to α-ketoglutarate to form glutamate:

$$\alpha\text{-Ketoglutarate} + \text{L-Amino acid} \rightleftharpoons \text{L-Glutamate} + \alpha\text{-Keto acid}$$

The amino groups of glutamate molecules are transported in blood to the liver by the alanine cycle (Figure 8.9).

$$\text{Pyruvate} + \text{L-Glutamate} \rightleftharpoons \text{L-Alanine} + \alpha\text{-Ketoglutarate}$$

In the liver, glutamate is formed as the reaction catalyzed by alanine transaminase is reversed. The oxidative deamination of glutamate yields α-ketoglutarate and NH_4^+.

In most extrahepatic tissues, the amino group of glutamate is released via oxidative deamination as NH_4^+. Ammonia is carried to the liver as the amide group of glutamine. The ATP-requiring reaction in which glutamate is converted to glutamine is catalyzed by glutamine synthetase:

$$\text{L-Glutamate} + NH_4^+ + \text{ATP} \longrightarrow \text{L-Glutamine}$$

After its transport to the liver, glutamine is hydrolyzed by glutaminase to form glutamate and NH_4^+. An additional NH_4^+ is generated as glutamate dehydrogenase converts glutamate to α-ketoglutarate:

$$\text{L-Glutamine} + H_2O \longrightarrow \text{L-Glutamate} + NH_4^+$$

$$\text{L-Glutamate} + H_2O + NAD^+ \longrightarrow \alpha\text{-Ketoglutarate} + \text{NADH} + H^+ + NH_4^+$$

Most of the ammonia generated in amino acid degradation is produced by the oxidative deamination of glutamate. Additional ammonia is produced in several other reactions catalyzed by the following enzymes:

1. L-Amino acid oxidases. Small amounts of ammonia are generated by various L-amino acid oxidases, found in liver and kidney, that require a flavin mononucleotide (FMN) coenzyme. FMN is regenerated from $FMNH_2$ by reacting with O_2 to form H_2O_2.

2. Serine and threonine dehydrases. Serine and threonine are not substrates in transamination reactions. Their amino groups are removed by the pyridoxal phosphate–requiring hepatic enzymes serine dehydratase and threonine dehydratase. The carbon skeleton products of these reactions are pyruvate and α-ketobutyrate, respectively.

3. Bacterial urease. A major source of ammonia in liver (approximately 25%) is produced by the action of certain bacteria in the intestine that possess the enzyme urease. Urea present in the blood circulating through the lower digestive tract diffuses across cell membranes and into the intestinal lumen. Once urea is hydrolyzed by bacterial urease to form ammonia, the latter substance diffuses back into the blood, which transports it to the liver.

Urea Synthesis

In ureotelic organisms the urea cycle disposes of approximately 90% of surplus nitrogen. As shown in Figure 15.1, urea is formed from ammonia, CO_2, and aspartate in a cyclic pathway referred to as the **urea cycle**. Because the urea cycle was discovered by Hans Krebs and Kurt Henseleit, it is often referred to as the **Krebs urea cycle** or the *Krebs-Henseleit cycle*.

Urea synthesis, which occurs in hepatocytes, begins with the formation of carbamoyl phosphate in the matrix of mitochondria. The substrates for this reaction, catalyzed by carbamoyl phosphate synthetase I, are NH_4^+ and HCO_3^-. (The nitrogen source for carbamoyl phosphate synthetase II, the enzyme involved in pyrimidine synthesis, is glutamine.)

KEY CONCEPTS 15.1

The degradation of most amino acids begins with removal of the α-amino group. Two types of biochemical reactions are involved in amino group removal: transamination and oxidative deamination.

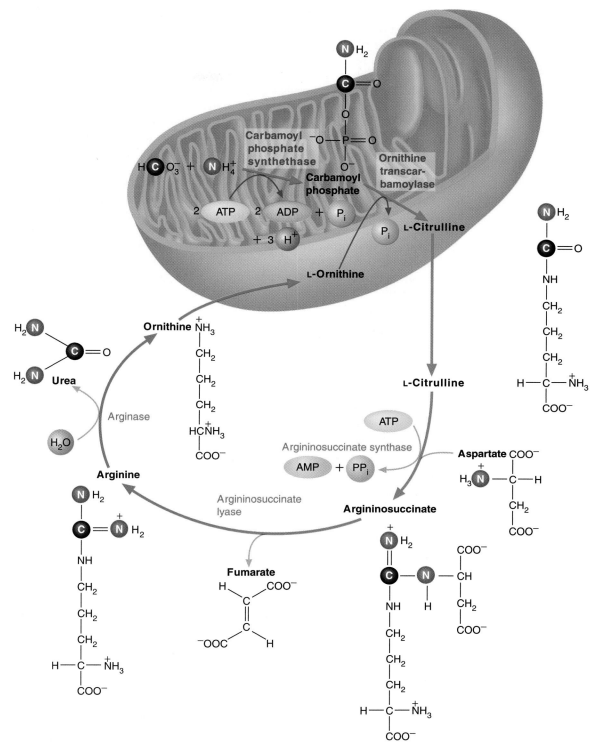

FIGURE 15.1

The Urea Cycle.

The urea cycle converts NH_4^+ to urea, a less toxic molecule. The sources of the atoms in urea are shown in color. Citrulline is transported across the inner membrane by a carrier for neutral amino acids. Ornithine is transported in exchange for H^+ or citrulline. Fumarate is transported back into the mitochondrial matrix (for reconversion to malate) by carriers for α-ketoglutarate or tricarboxylic acids.

Because two molecules of ATP are required in carbamoyl phosphate synthesis, this reaction is essentially irreversible. (One is used to activate HCO_3^-. The second molecule is used to phosphorylate carbamate.) Carbamoyl phosphate subsequently reacts with ornithine to form citrulline. This reaction, catalyzed by ornithine transcarbamoylase, is driven to completion because phosphate is released from carbamoyl phosphate. (Recall from Table 4.2 that carbamoyl phosphate has a high phosphate group transfer potential.) Once it is formed, citrulline is transported to the cytoplasm, where it reacts with aspartate to form argininosuccinate. (The α-amino group of aspartate, formed from oxaloacetate by transamination reactions in the liver, provides the second nitrogen that is ultimately incorporated into urea.) This reaction, which is catalyzed by argininosuccinate synthase, is reversible. It is pulled forward by the cleavage of pyrophosphate by pyrophosphatase. Argininosuccinate lyase subsequently cleaves argininosuccinate to form arginine (the immediate precursor of urea) and fumarate. In the final reaction of the urea cycle, arginase catalyzes the hydrolysis of arginine to form ornithine and urea. Once it forms, urea diffuses out of the hepatocytes and into the bloodstream. It is ultimately eliminated in the urine by the kidney. Ornithine (a basic amino acid) returns to the mitochondria for condensation with carbamoyl phosphate to begin the cycle again. Because arginase is found in significant amounts only in the livers of ureotelic animals, urea is produced only in this organ.

After its transport back into the mitochondrial matrix, fumarate is hydrated to form malate, a component of the citric acid cycle. The oxaloacetate product of the citric acid cycle can be used in energy generation, or it can be converted to glucose or aspartate. The relationship between the urea cycle and the citric acid cycle, often referred to as the **Krebs bicycle**, is outlined in Figure 15.2:

$$CO_2 + NH_4^+ + aspartate + 3\ ATP + 2\ H_2O \longrightarrow$$
$$urea + fumarate + 2\ ADP + 2\ P_i + AMP + PP_i + 5\ H^+$$

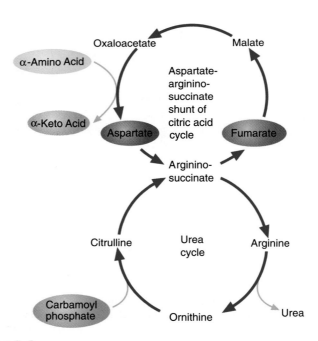

FIGURE 15.2

The Krebs Bicycle.

The aspartate used in urea synthesis is generated from oxaloacetate, a citric acid cycle intermediate.

Hyperammonemia is a condition in which the concentration of NH_4^+ is excessive (i.e., greater than 60 μM) in blood. Elevated concentrations of ammonia are serious; the consequences of **ammonia intoxication** include lethargy, tremors, slurred speech, blurred vision, protein-induced vomiting (vomiting caused by the consumption of dietary protein), coma, and death.

Hyperammonemia may be caused by genetic defects or cirrhosis of the liver. In congenital (inherited) hyperammonemia, a relatively rare condition, one or more of the urea cycle enzymes is missing or defective. Complete absence of a urea cycle enzyme is fatal soon after birth. Brain damage can be minimized in infants who have partial deficiencies in urea synthesis if aggressive therapy is initiated immediately after birth. (Therapy consists of diets with severe restrictions on protein intake.) In cirrhosis, loss of liver function is devastating because of wide spread inflammation and necrosis (cell death). It is most commonly caused by prolonged, excessive consumption of ethanol. Less common causes of cirrhosis include prolonged exposure to toxic chemicals such as carbon tetrachloride, hepatitis (an inflammation of the liver that is often caused by viral infections), and amebiasis (an infection with parasitic amebas).

Because most of the symptoms of ammonia intoxication are manifested in brain tissue, ammonia is considered a neurotoxic agent. Although significant research effort has been devoted to elucidating the effects of ammonia on brain cells, the mechanism of the damage is still unclear. Concentrations of ammonium ions as low as 1–2 μM have been observed to disrupt both inhibitory and excitatory nerve transmission. Inhibitory neurotransmitters such as glycine become ineffective because NH_4^+ inactivates Cl^- channels. NH_4^+ prevents the binding of glutamate, an excitatory neurotransmitter, to its postsynaptic receptors. Glutamate metabolism can also be compromised by its reaction with NH_4^+ catalyzed by glutamine synthase (see p. 459), which may cause nerve tissue to become depleted of glutamate. Depletion of α-ketoglutarate, a citric acid cycle intermediate, has also been implicated. Other toxic effects of ammonia on the brain may include inhibition of amino acid transport and the Na^+-K^+ ATPase.

Four high energy phosphates are consumed in the synthesis of one molecule of urea. Two molecules each of ATP are required to regenerate ATP from AMP and two molecules of ATP from two molecules of ADP.

Control of the Urea Cycle

Because ammonia is so toxic (Special Interest Box 15.2), it is not surprising that the urea cycle is subject to stringent regulation. There are long- and short-term regulatory mechanisms. The levels of all five urea cycle enzymes are altered by variations in dietary protein consumption. Within several days after a significant dietary change, there are twofold to threefold changes in enzyme levels. Several hormones (e.g., glucagon and the glucocorticoids) are believed to be involved in the altered rates of enzyme synthesis.

The urea cycle enzymes are controlled in the short term by the concentrations of their substrates. Carbamoyl phosphate synthetase I is also allosterically activated by *N-acetylglutamate*. This latter molecule is a sensitive indicator of the cell's glutamate concentration. (Recall that a significant amount of NH_4^+ is derived from glutamate.) N-acetylglutamate is produced from glutamate and acetyl-CoA in a reaction catalyzed by N-acetylglutamate synthase.

KEY CONCEPTS 15.2

Urea is synthesized from ammonia, CO_2, and aspartate. The urea cycle is carefully regulated to prevent hyperammonemia.

QUESTION 15.1

Although arginine is an intermediate in the urea cycle it is an essential amino acid in young animals. Suggest a reason for this phenomenon.

QUESTION 15.2

In some clinical circumstances, patients with hyperammonemia are treated with antibiotics. Suggest a rational basis for this therapy.

Catabolism of Amino Acid Carbon Skeletons

The α-amino acids can be grouped into classes according to their end products. As previously mentioned, these are acetyl-CoA, acetoacetyl-CoA, pyruvate, and several citric acid cycle intermediates. Each group is briefly discussed. The

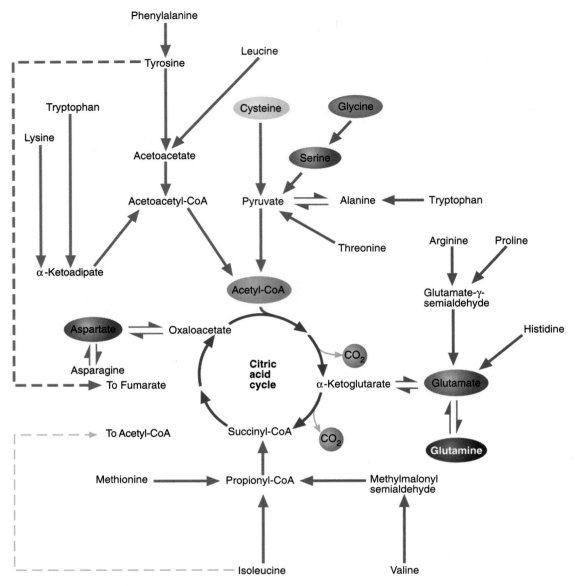

FIGURE 15.3

Degradation of the 20 α-Amino Acids Found in Proteins.

The α-amino groups are removed early in the catabolic pathways. Carbon skeletons are converted to common metabolic intermediates.

degradation pathways for the 20 α-amino acids found in proteins are outlined in Figure 15.3.

AMINO ACIDS FORMING ACETYL-CoA In all, 10 α-amino acids yield acetyl-CoA. This group is further divided according to whether pyruvate is an intermediate in acetyl-CoA formation. (Recall that pyruvate is converted to acetyl-CoA by the pyruvate dehydrogenase complex.) The amino acids whose degradation involves pyruvate are alanine, serine, glycine, cysteine, and threonine. The other five amino acids converted to acetyl-CoA by pathways not involving pyruvate are lysine, tryptophan, tyrosine, phenylalanine, and leucine. The two reaction sequences are outlined in Figures 15.4 and 15.5.

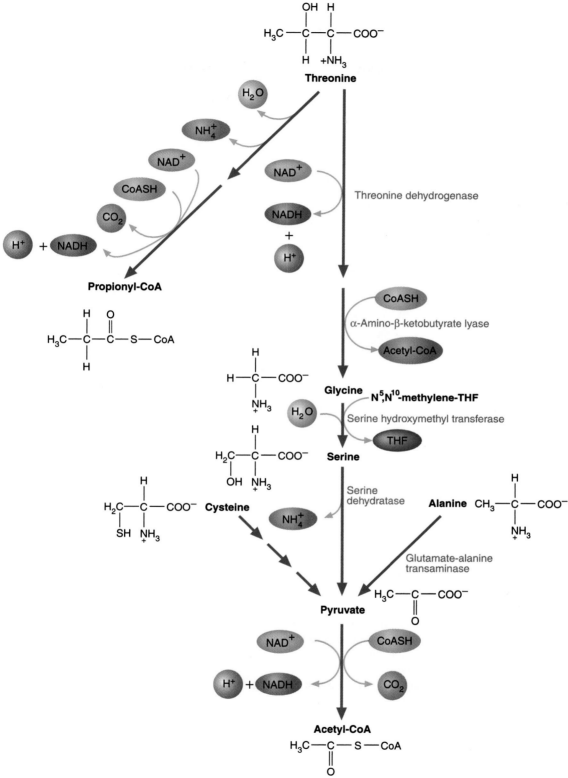

FIGURE 15.4

The Catabolic Pathways of Threonine, Glycine, Serine, Cysteine, and Alanine.

Pyruvate is an intermediate in the conversion of these amino acids to acetyl-CoA. Note that glycine is also degraded by glycine synthase to form CO_2, NH_4^+, and N^5,N^{10}-methylene-THF in an NAD^+-requiring reaction. In primates most threonine molecules are degraded to propionyl-CoA.

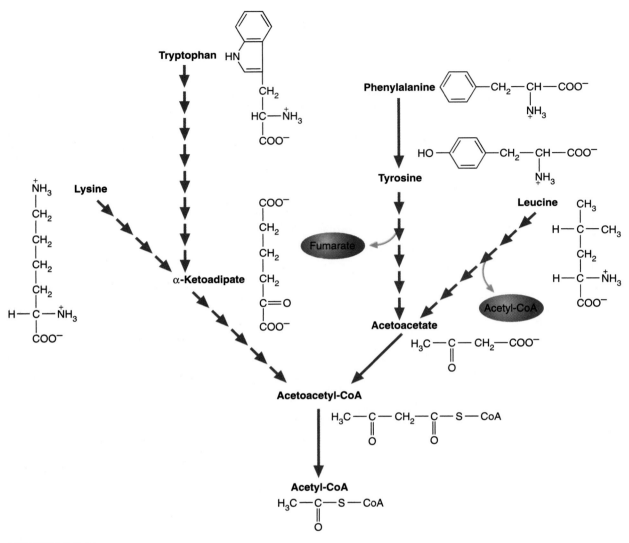

FIGURE 15.5

The Catabolic Pathways of Lysine, Tryptophan, Phenylalanine, Tyrosine, and Leucine.

These pathways are long and complex. The number of reactions in each segment is indicated by the number of arrows.

1. Alanine. Recall that the reversible transamination reaction involving alanine and pyruvate is an important component of the alanine cycle discussed previously (Section 8.2).

2. Serine. As described, serine is converted to pyruvate by serine dehydratase.

3. Glycine. Glycine can be converted to serine by serine hydroxymethyltransferase. (The hydroxymethyl group is donated by N^5,N^{10}-methylene THF as described in Section 14.3.) Then serine is converted to pyruvate, as previously described. Most glycine molecules, however, are degraded to CO_2, NH_4^+, and a methylene group removed by THF. The enzyme involved is glycine synthase (also referred to as glycine cleavage enzyme), which requires NAD^+.

4. Cysteine. In animals, cysteine is converted to pyruvate by several pathways. In the principal pathway, the conversion occurs in three steps. Initially, cysteine is oxidized to cysteine sulfate. Pyruvate is produced after a transamination and a desulfuration reaction.

5. Threonine. In the major degradative pathway, threonine is oxidized by threonine dehydrogenase to form α-amino-β-ketobutyrate. The latter molecule is metabolized further to form lactate via pyruvate, or it can be cleaved by α-amino-β-ketobutyrate lyase to form acetyl-CoA and glycine. As previously discussed, glycine is converted to acetyl-CoA via pyruvate. Alternatively, threonine can be degraded to α-ketobutyrate by threonine dehydratase and subsequently to propionyl-CoA. Propionyl-CoA is then converted to succinyl-CoA (see p. 385).

6. Lysine. Lysine is converted to α-ketoadipate in a series of reactions that include two oxidations, removal of the side chain amino group, and a transamination. Acetoacetyl-CoA is produced in a further series of reactions that involve several oxidations, a decarboxylation, and a hydration. Acetoacetyl-CoA can be converted to acetyl-CoA in a reaction that is the reverse of a step in ketone body formation.

7. Tryptophan. Tryptophan is converted to α-ketoadipate in a long, complex series of eight reactions, which also yield formate and alanine. Acetyl-CoA is synthesized from α-ketoadipate as described for lysine. The alanine produced in this pathway is converted to acetyl-CoA via pyruvate.

8. Tyrosine. Tyrosine catabolism begins with a transamination and a dehydroxylation. Homogentisate is synthesized in the latter reaction, catalyzed by the ascorbate-requiring enzyme parahydroxyphenylpyruvate dioxygenase. Homogentisate is converted to maleylacetoacetate by homogentisate oxidase. Acetoacetate and fumarate are then generated in isomerization and hydration reactions.

9. Phenylalanine. Phenylalanine is converted to tyrosine by phenylalanine-4-monooxygenase in a reaction illustrated in Figure 15.6. Tyrosine is degraded to form acetoacetate and fumarate.

10. Leucine. Leucine, one of the branched chain amino acids, is converted to HMG-CoA in a series of reactions that include a transamination, two oxidations, a carboxylation, and a hydration. HMG-CoA is then converted to acetyl-CoA and acetoacetate by HMG-CoA lyase.

AMINO ACIDS FORMING α-KETOGLUTARATE Five amino acids (arginine, proline, histidine, glutamate, and glutamine) are degraded to α-ketoglutarate. An outline of their catabolism is illustrated in Figure 15.7. Each pathway is briefly described.

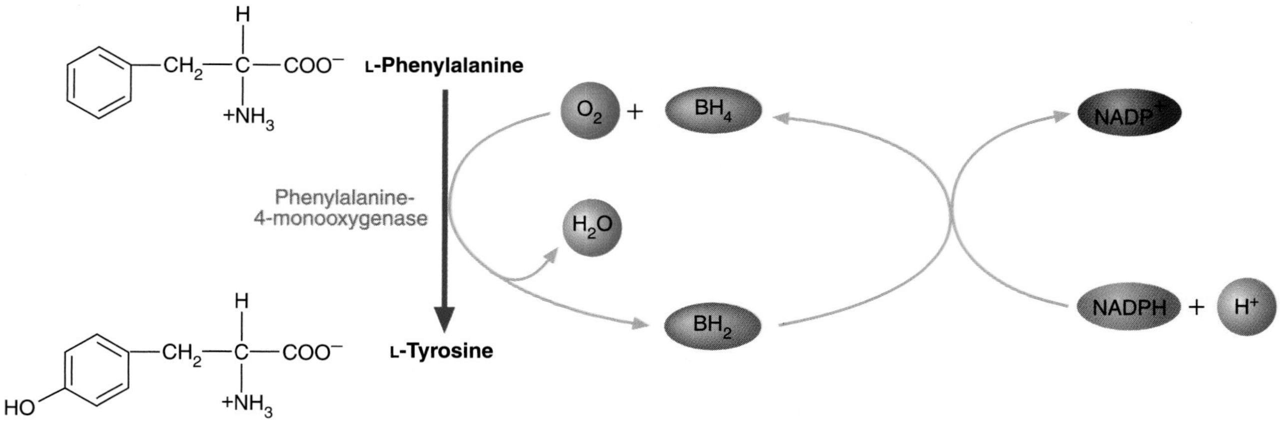

FIGURE 15.6

The Conversion of Phenylalanine to Tyrosine.

The reaction catalyzed by phenylalanine-4-monooxygenase is irreversible. The electrons required for the hydroxylation of phenylalanine are carried to O_2 from NADPH by tetrahydrobiopterin.

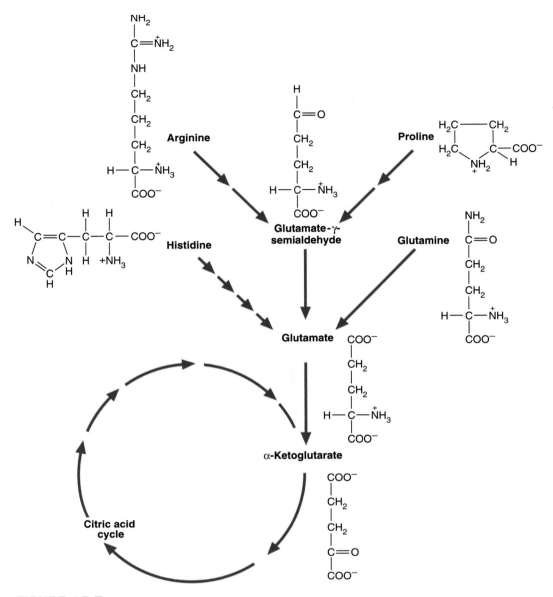

FIGURE 15.7

The Catabolic Pathways of Arginine, Proline, Histidine, Glutamine, and Glutamate.

All these amino acids are eventually converted to α-ketoglutarate.

    1. Glutamate and glutamine. Glutamine is converted to glutamate by gluta-minase. As described previously, glutamate is converted to α-ketoglutarate by glu-tamate dehydrogenase or by transamination.

    2. Arginine. Recall that arginine is cleaved by arginase to form ornithine and urea. In a subsequent transamination reaction, ornithine is converted to glutamate-γ-semialdehyde. Glutamate is then produced as glutamate-γ-semialdehyde is hydrated and oxidized. α-Ketoglutarate is produced by a transamination reac-tion or by oxidative deamination.

    3. Proline. Proline catabolism begins with an oxidation reaction that produces Δ^1-pyrroline. The latter molecule is converted to glutamate-γ-semialdehyde by a hydration reaction. Glutamate is then formed by another oxidation reaction.

    4. Histidine. Histidine is converted to glutamate in four reactions: a nonox-idative deamination, two hydrations, and the removal of a formamino group ($NH{=}CH{-}$) by THF.

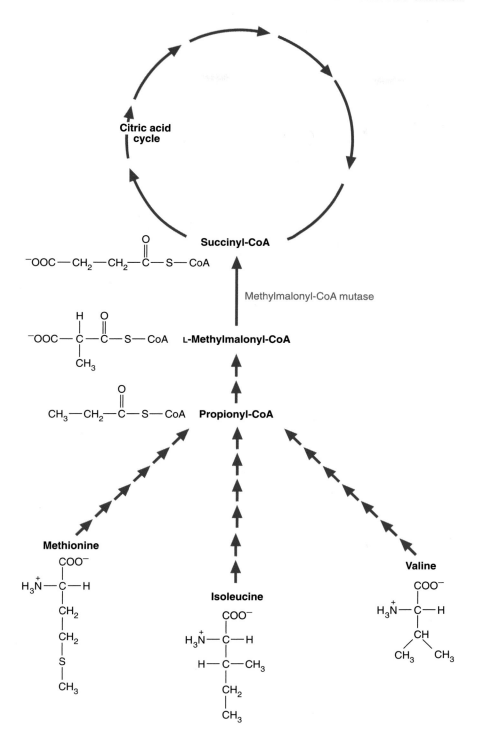

FIGURE 15.8

The Catabolic Pathways of Methionine, Isoleucine, and Valine.

Propionyl-CoA and L-methylmalonyl-CoA are intermediates in the conversion of these amino acids to succinyl-CoA. Methyl-malonyl-CoA mutase is a vitamin B_{12}–requiring enzyme. Note that threonine is also degraded via the propionyl-CoA/suc-cinyl-CoA pathway (see Figure 15.4).

AMINO ACIDS FORMING SUCCINYL-CoA Succinyl-CoA is formed from the carbon skeletons of methionine, isoleucine, valine, and threonine (as already discussed). An outline of the reactions that degrade the first three of these amino acids is illustrated in Figure 15.8.

1. Methionine. Methionine degradation begins with the formation of S-adeno-sylmethionine, which is followed by a demethylation reaction, as described (Figure 14.16). S-Adenosylhomocysteine, the product of the latter reaction, is hydrolyzed to adenosine and homocysteine. Homocysteine then combines with serine to yield cystathionine. Cysteine, α-ketobutyrate, and NH_4^+ result from

FIGURE 15.9

The Transulfuration Pathway.

The sulfur atom of methionine becomes the sulfur atom of cysteine. The sulfate generated in cysteine catabolism is excreted or used in several biosynthetic or catabolic pathways. The transulfuration and methylation pathways are intimately related.

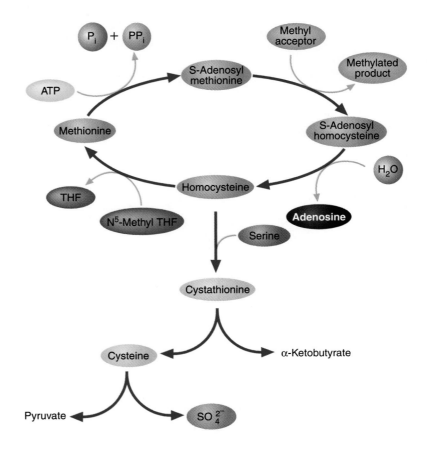

the cleavage of cystathionine. α-Ketobutyrate is then converted to propionyl-CoA by α-ketoacid dehydrogenase. Propionyl-CoA is converted to succinyl-CoA in three steps. The enzyme that catalyzes the last of these, methylmalonyl-CoA mutase, requires methylcobalamin. The conversion of methionine to cysteine is sometimes referred to as the **transulfuration pathway** (Figure 15.9). A substantial amount of the sulfate produced from cysteine degradation is excreted in urine. Sulfate is also used in the synthesis of sulfatides and proteoglycans. Additionally, molecules such as steroids and certain drugs are excreted as sulfate esters.

2. Isoleucine and valine. The first four reactions in the degradation of isoleucine and valine are identical. Initially, both amino acids undergo transamination reactions to form α-keto-β-methylvalerate and α-ketoisovalerate, respectively. This is followed by the formation of CoA derivatives, and oxidative decarboxylation, oxidation, and dehydration reactions. The product of the isoleucine pathway is then hydrated, dehydrogenated, and cleaved to form acetyl-CoA and propionyl-CoA. In the valine degradative pathway the α-keto acid intermediate is converted into propionyl-CoA after a double bond is hydrated and CoA is removed by hydrolysis. After the formation of an aldehyde by the oxidation of the hydroxyl group, propionyl-CoA is produced as a new thioester is formed during an oxidative decarboxylation.

AMINO ACIDS FORMING OXALOACETATE Both aspartate and asparagine are degraded to form oxaloacetate. Aspartate is converted to oxaloacetate with a single transamination reaction. Asparagine is initially hydrolyzed to yield aspartate and NH_4^+ by asparaginase.

KEY CONCEPTS 15.3

Amino acid carbon skeletons can be degraded into one or more of several metabolites. These include acetyl-CoA, acetoacetyl-CoA, α-ketoglutarate, succinyl-CoA, and oxaloacetate.

Taurine is a sulfur-containing amine synthesized from cysteine. With the exception of its incorporation in bile salts, taurine's physiological role is still poorly understood. However, several pieces of information suggest that taurine is an important metabolite. For example, taurine is found in brain tissue in large amounts. In addition, domestic cats have recently been observed to develop congestive heart failure if fed a taurine-free diet. (Cats cannot synthesize taurine. For this reason they must consume meat in their diet. Cats that are fed vegetarian diets soon become listless and eventually die prematurely.) In most animals, taurine is synthesized from cysteine sulfinate (the oxidation product of cysteine) in two reactions: a decarboxylation followed by an oxidation of the sulfinate group ($-SO_2^-$) to form sulfonate ($-SO_3^-$). With this information, determine the biosynthetic pathway for taurine. (*Hint*: The structure of taurine is illustrated in Chapter 12 on p. 413. Also refer to Figure 15.9.)

Some amino acids are classified as both ketogenic and glucogenic. Review the amino acid catabolic pathways and determine which amino acids belong to both categories.

15.2 DEGRADATION OF SELECTED NEUROTRANSMITTERS

The previous discussion of amino acid catabolic disorders indicates that catabolic processes are just as important for the proper functioning of cells and organisms as are anabolic processes. This is no less true for molecules that act as neurotransmitters. To maintain precise information transfer, neurotransmitters are usually quickly degraded or removed from the synaptic cleft. An extreme example of enzyme inhibition illustrates the importance of neurotransmitter degradation. Recall that acetylcholine is the neurotransmitter that initiates muscle contraction. Shortly afterwards, the action of acetylcholine is terminated by the enzyme acetylcholinesterase. (Acetylcholine must be destroyed rapidly so that muscle can relax before the next contraction.) Acetylcholinesterase is a serine esterase that hydrolyzes acetylcholine to acetate and choline. Serine esterases have catalytic mechanisms similar to those of the serine proteases (Section 6.4). Both types of enzymes are irreversibly inhibited by DFP (diisopropylfluorophosphate). Exposure to DFP causes muscle paralysis because acetylcholinesterase is irreversibly inhibited. With each nerve impulse, more acetylcholine molecules enter the neuromuscular synaptic cleft. The accumulating acetylcholine molecules repetitively bind to acetylcholine receptors. The overstimulated muscle cells soon become paralyzed (nonfunctional). Affected individuals suffocate because of paralyzed respiratory muscles.

The catecholamines epinephrine, norepinephrine, and dopamine are inactivated by oxidation reactions catalyzed by monoamine oxidase (MAO) (Figure 15.10). Because MAO is found within nerve endings, catecholamines must be transported out of the synaptic cleft before inactivation. (The process by which neurotransmitters are transported back into nerve cells so that they can be reused or degraded is referred to as *reuptake*.) Epinephrine, released as a hormone from the adrenal gland, is carried in the blood and is catabolized in nonneural tissue (perhaps the kidney). Catecholamines are also inactivated in methylation reactions catalyzed by catechol-O-methyltransferase (COMT). These two enzymes (MAO and COMT) work together to produce a large variety of oxidized and methylated metabolites of the catecholamines.

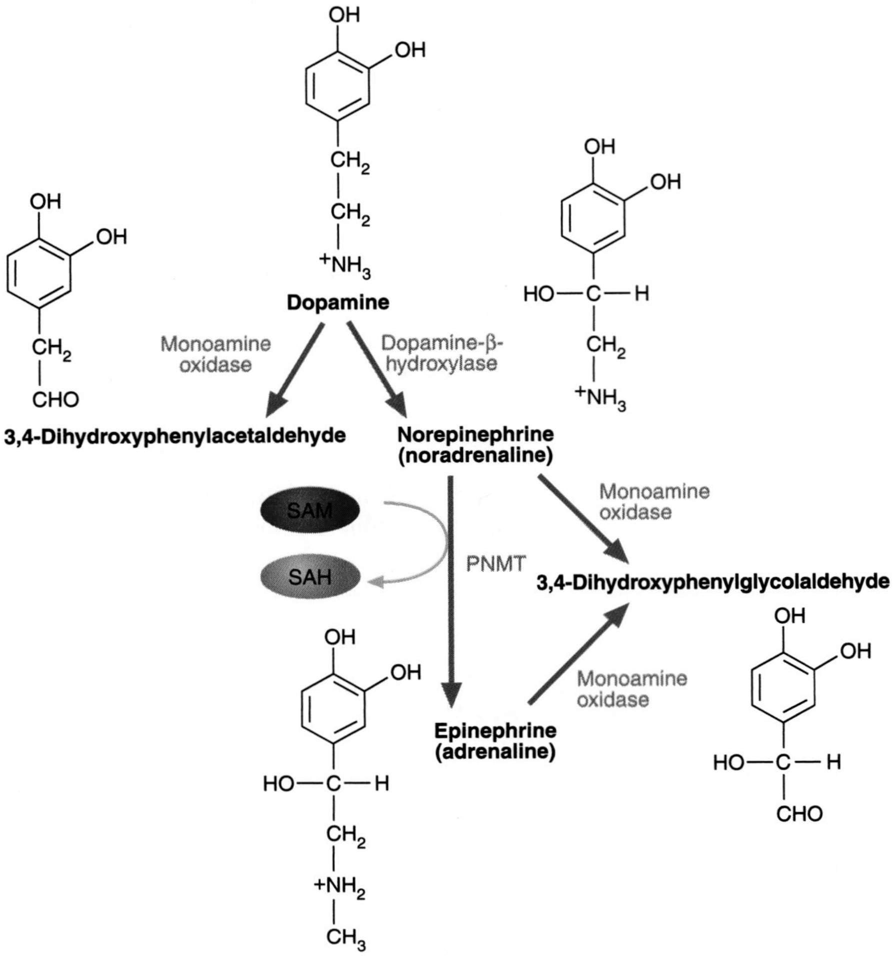

FIGURE 15.10

Inactivation of the Catecholamines.

Monoamine oxidase is a flavoprotein that catalyzes the oxidative deamination of amines to form the corresponding aldehydes. O_2 is the electron acceptor, and NH_3 and H_2O_2 are the other products. (PNMT = phenylethanolamine-N-methyltransferase.)

KEY CONCEPTS 15.4

Information transfer in animals requires that after their release neurotransmitters must be quickly degraded or removed from the synaptic cleft.

After its reuptake into nerve cells, serotonin is degraded in a two-step pathway (Figure 15.11). In the first reaction, serotonin is oxidized by MAO. The product, 5-hydroxyindole-3-acetaldehyde, is then further oxidized by aldehyde dehydrogenase to form 5-hydroxyindole-3-acetate.

QUESTION 15.5

Identify each of the following neurotransmitters. Explain how each is inactivated.

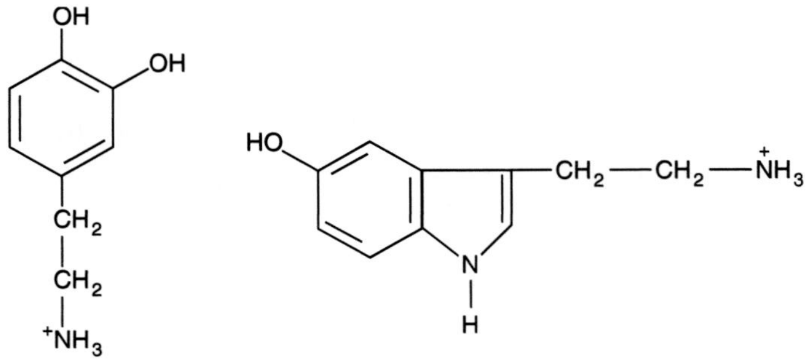

(a)   (b)

Defects in amino acid catabolism were among the first genetic diseases to be recognized and investigated by medical scientists. These "inborn errors of metabolism" result from **mutations** (permanent changes in genetic information, i.e., DNA structure). Most commonly, in the genetic diseases related to amino acid metabolism, the defective gene codes for an enzyme. The metabolic blockage that results from such a deficit disrupts what are ordinarily highly coordinated cellular and organismal processes, producing abnormal amounts and/or types of metabolites. Because these metabolites (or their heightened concentrations) are often toxic, permanent damage or death ensues. Several of the most commonly observed inborn errors of amino acid metabolism are discussed below.

Alkaptonuria, caused by a deficiency of homogentisate oxidase, was the first disease to be linked to genetic inheritance involving a single enzyme. In 1902, Archibald Garrod proposed that a single inheritable unit (later called a gene) was responsible for the urine in alkaptonuric patients turning black. Large quantities of homogentisate, the substrate for the defective enzyme, are excreted in urine. Homogentisate turns black when it is oxidized as the urine is exposed to air. Although black urine appears to be an essentially benign (if somewhat disconcerting) condition, alkaptonuria is not innocuous, because alkaptonuric patients develop arthritis in later life. In addition, pigment accumulates gradually and unevenly darkens the skin.

Albinism is an example of a genetic defect with serious consequences. The enzyme tyrosinase is deficient. Consequently, *melanin*, a black pigment found in skin, hair, and eyes, is not produced. It is formed from tyrosine in several cell types, for example, the melanocytes in skin. In such cells, tyrosinase converts tyrosine to DOPA and DOPA to dopaquinone. A large number of molecules of the latter product, which is highly reactive, condense to form melanin. Because of the lack of pigment, affected individuals (called albinos) are extremely sensitive to sunlight. In addition to their susceptibility to skin cancer and sunburn, they often have poor eyesight.

Phenylketonuria, caused by a deficiency of phenylalanine hydroxylase, is one of the most common genetic diseases associated with amino acid metabolism. If this condition is not identified and treated immediately after birth, mental retardation and other forms of irreversible brain damage occur. This damage results mostly from the accumulation of phenylalanine. (The actual mechanism of the damage is not understood.) When it is present in excess, phenylalanine undergoes transamination to form phenylpyruvate, which is also converted to phenyllactate and phenylacetate. Large amounts of these molecules are excreted in the urine. Phenylacetate gives the urine its characteristic musty odor. Phenylketonuria is treated with a low-phenylalanine diet.

In *maple syrup urine disease*, also called *branched chain ketoaciduria*, the α-keto acids derived from leucine, isoleucine, and valine accumulate in large quantities in blood. Their presence in urine imparts a characteristic odor that gives the malady its name. All three α-keto acids accumulate because of a deficient branched chain α-keto acid dehydrogenase complex. (This enzymatic activity is responsible for the conversion of the α-keto acids to their acyl-CoA derivatives.) If left untreated, affected individuals experience vomiting, convulsions, severe brain damage, and mental retardation. They often die before 1 year of age. As with phenylketonuria, treatment consists of rigid dietary control.

Deficiency of methylmalonyl-CoA mutase results in *methylmalonic acidemia*, a condition in which methylmalonate accumulates in blood. The symptoms are similar to those of maple syrup urine disease. Methylmalonate may also accumulate because of a deficiency of adenosylcobalamin or weak binding of this coenzyme by a defective enzyme. Some affected individuals respond to injections of large daily doses of vitamin B_{12}.

(c)   (d)

A variety of medical conditions are currently being treated with medications that block the biological activity or the metabolism of neurotransmitters. The term *antagonist* is used to describe molecules that block the biological actions of normal neurotransmitters. For example, certain drug molecules used to treat

QUESTION 15.6

hypertension (high blood pressure) antagonize the action of the catecholamines. (The binding of catecholamines to specific receptor molecules in the cardiovascular system constricts blood vessels.) Another interesting example is certain medications that treat *obsessive compulsive disorder* (OCD). OCD is a condition characterized by the persistent intrusion of unwanted and disturbing thoughts and/or the compulsive performance of certain acts such as handwashing. For unknown reasons, serotonin reuptake inhibitors have been remarkably effective in improving patient symptoms. Neurotransmitter reuptake inhibitors have effects similar to **agonists**, substances that boost or amplify the physiological effects of a neurotransmitter.

Myasthenia gravis is treated with drugs that inhibit acetylcholinesterase, the enzyme that degrades acetylcholine. *Myasthenia gravis* is an autoimmune disease in which autoantibodies bind to and initiate the destruction of the acetylcholine receptor in skeletal muscle cell membranes. Gradually, the number of functional acetylcholine receptors is reduced. This condition is characterized by muscle weakness and fatigability. Eventually, patients develop difficulty in speaking and swallowing. However, a short time after consuming reversible cholinesterase inhibitors (e.g., neostigmine or physostigmine), patients experience significant improvement in their symptoms. Based on your knowledge of the action of acetylcholine, can you suggest how anticholinesterase drugs achieve this short-term clinical improvement? (*Hint*: For a muscle cell to contract, a threshold number of acetylcholine receptors must bind acetylcholine. In normal individuals this number of receptors is significantly lower than the number of receptors in muscle cell membrane. Also note that the productive binding and unbinding of a neurotransmitter to its receptor are often rapid.)

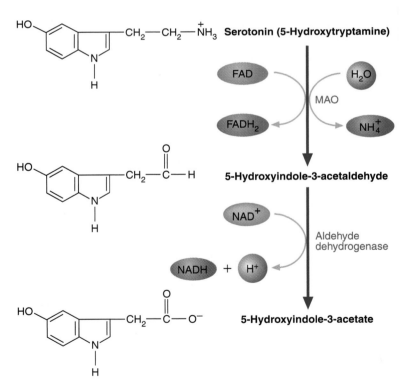

FIGURE 15.11

Degradation of Serotonin.

In the major catabolic pathway, serotonin is deaminated and oxidized to form 5-hydroxyindole-3-acetaldehyde. The latter molecule is then further oxidized to form 5-hydroxyindole-3-acetate.

15.3 NUCLEOTIDE DEGRADATION

In most living organisms, purine and pyrimidine nucleotides are constantly degraded. In animals, degradation occurs because of the normal turnover of nucleic acids and nucleotides and the digestion of dietary nucleic acids. During digestion, nucleic acids are hydrolyzed to oligonucleotides by enzymes called **nucleases**. (**Oligonucleotides** are defined as short nucleic acid segments containing fewer than 50 nucleotides.) Enzymes that are specific for breaking internucleotide bonds in DNA are called *deoxyribonucleases* (DNases); those that degrade RNA are called *ribonucleases* (RNases). Once formed, oligonucleotides are further hydrolyzed by various *phosphodiesterases*, a process that produces a mixture of mononucleotides. *Nucleotidases* remove phosphate groups from nucleotides, yielding nucleosides. These latter molecules are hydrolyzed by *nucleosidases* to free bases and ribose or deoxyribose, which are then absorbed. Alternatively, nucleosides may be absorbed by intestinal enterocytes.

Generally speaking, dietary purine and pyrimidine bases are not used in significant amounts to synthesize cellular nucleic acids. Instead, they are degraded within enterocytes. Purines are degraded to uric acid in humans and birds. Pyrimidines are degraded to β-alanine or β-aminoisobutyric acid, as well as NH_3 and CO_2. In contrast to the catabolic processes for other major classes of biomolecules (e.g., sugars, fatty acids, and amino acids), purine and pyrimidine catabolism does not result in ATP synthesis. The major pathways for the degradation of purine and pyrimidine bases are described next.

Purine Catabolism

Purine nucleotide catabolism is outlined in Figure 15.12. There is some variation in the specific pathways used by different organisms or tissues to degrade AMP. In muscle, for example, AMP is initially converted to IMP by AMP deaminase (also referred to as adenylate aminohydrolase). IMP is subsequently hydrolyzed to inosine by 5′-nucleotidase. In most tissues, however, AMP is hydrolyzed by 5′-nucleotidase to form adenosine. Adenosine is then deaminated by adenosine deaminase (also called adenosine aminohydrolase) to form inosine.

Purine nucleoside phosphorylase converts inosine, guanosine, and xanthosine to hypoxanthine, guanine, and xanthine, respectively. (The ribose-1-phosphate formed during these reactions is reconverted to PRPP by phosphoribomutase.) Hypoxanthine is oxidized to xanthine by xanthine oxidase, an enzyme that contains molybdenum, FAD, and two different Fe-S centers. (Xanthine oxidase–catalyzed reactions produce O_2^- in addition to forming H_2O_2. Guanine is deaminated to xanthine by guanine deaminase (also called guanine aminohydrolase). Xanthine molecules are further oxidized to uric acid by xanthine oxidase.

Several diseases result from defects in purine catabolic pathways. *Gout*, which is often characterized by high blood levels of uric acid and recurrent attacks of arthritis, is caused by several metabolic abnormalities (Special Interest Box 15.4). Two different immunodeficiency diseases are now known to result from defects in purine catabolic reactions. In *adenosine deaminase deficiency*, large concentrations of dATP inhibit ribonucleotide reductase. Consequently, DNA synthesis is depressed. For reasons that are not yet clear, this metabolic distortion is observed primarily in the T and B lymphocytes. (*T lymphocytes*, or **T cells**, bear antibody-like molecules on their surfaces. They bind to and destroy foreign cells in a process referred to as **cellular immunity**. *B lymphocytes*, or **B cells**, produce antibodies that bind to foreign substances, thereby initiating their destruction by other immune system cells. The production of antibodies by B cells is referred to as the **humoral immune response**.) Children with adenosine deaminase deficiency usually die before the age of two because of massive infections. In *purine nucleoside phosphorylase deficiency,* levels of purine nucleotides are high and synthesis of uric acid decreases. High levels of dGTP are apparently responsible for the impairment of T cells that is characteristic of this malady.

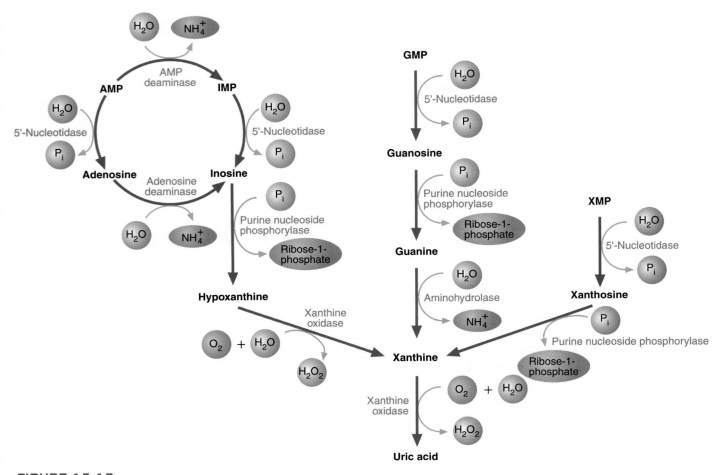

FIGURE 15.12

Purine Nucleotide Catabolism.

Ribose-1-phosphate is released in AMP, GMP, and XMP catabolism. Xanthine oxidase–catalyzed reactions generate O_2^-.

Many animals degrade uric acid further (Figure 15.13). Urate oxidase converts uric acid to allantoin, an excretory product in many mammals. Allantoinase catalyzes the hydration of allantoin to form allantoate, which is excreted by bony fish. Other fish, as well as amphibians, produce allantoicase, which splits allantoic acid into glyoxylate and urea. Finally, marine invertebrates degrade urea to NH_4^+ and CO_2 in a reaction catalyzed by urease.

QUESTION 15.7 Many animals, other than primates and birds, possess the enzyme urate oxidase. Suggest a reason why these organisms do not suffer from gout.

QUESTION 15.8 One of the more fascinating aspects of biochemistry is that living organisms use the same molecule for different purposes. Two interesting examples are allantoin and allantoate. As previously mentioned, these molecules serve as nitrogenous waste in several animal groups. Some plant species (i.e., certain legumes such as soybeans and snapbeans) begin to synthesize allantoin and allantoate once they become infected by nitrogen-fixing bacteria. Both molecules, referred to as the *ureides*, are nitrogen transport compounds. (Other legumes, such as peas and alfalfa, use asparagine for nitrogen transport whether they are infected or not.) Once they are synthesized, the ureides are transported through the xylem to the leaves. In leaves, the nitrogen is released and used primarily in amino acid synthesis. Allantoate is

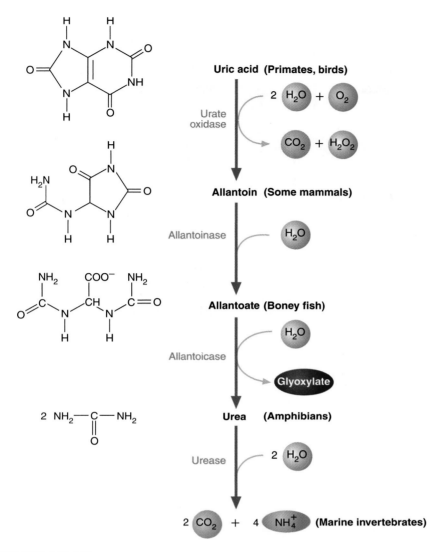

FIGURE 15.13

Uric Acid Catabolism.

Many animals possess enzymes that allow them to convert uric acid to other excretory products. The final excretory products of specific animal groups are indicated.

degraded to glyoxylate, four molecules of NH_4^+, and two molecules of CO_2 by three reactions that are not yet completely characterized. Based on the information provided in Chapter 14 and this chapter, trace the transport of NH_4^+ in root nodules to its incorporation into amino acids in leaves. Assume that the same or similar enzymes are used to synthesize allantoin and allantoate as observed in animals.

Pyrimidine Catabolism

In humans the purine ring cannot be degraded. This is not true for the pyrimidine ring. An outline of the pathway for pyrimidine nucleotide catabolism is illustrated in Figure 15.14.

Before they can be degraded, cytidine and deoxycytidine are converted to uridine and deoxyuridine, respectively, by deamination reactions catalyzed by cytidine deaminase. Similarly, deoxycytidylate (dCMP) is deaminated to form deoxyuridylate (dUMP). The latter molecule is then converted to deoxyuridine by 5′-nucleotidase. Uridine and deoxyuridine are then further degraded by nucleoside

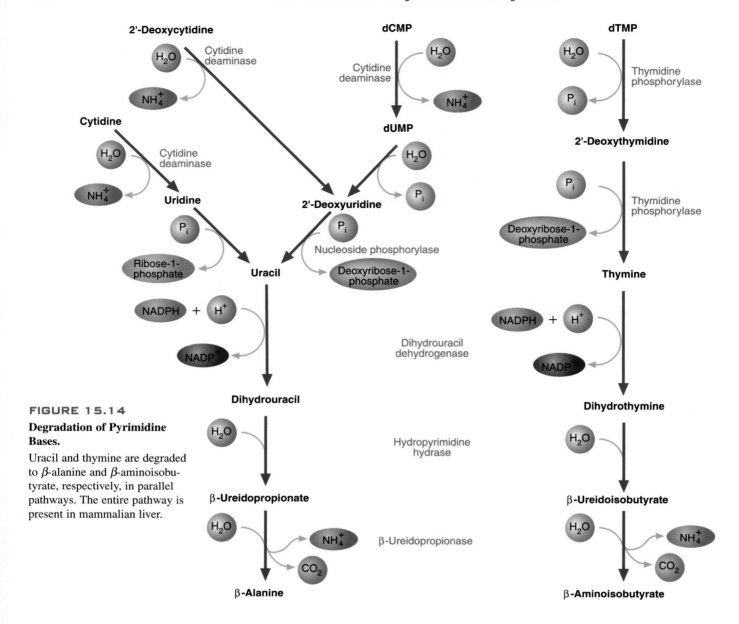

FIGURE 15.14

Degradation of Pyrimidine Bases.

Uracil and thymine are degraded to β-alanine and β-aminoisobutyrate, respectively, in parallel pathways. The entire pathway is present in mammalian liver.

phosphorylase to form uracil. Thymine is formed from thymidylate (dTMP) by the sequential actions of 5′-nucleotidase and nucleoside phosphorylase.

Uracil and thymine are converted to their end products, β-alanine and β-aminoisobutyrate, respectively, in parallel pathways. In the first step, uracil and thymine are reduced by dihydrouracil dehydrogenase to their corresponding dihydro derivatives. As these latter molecules are hydrolyzed, the rings open, yielding β-ureidopropionate and β-ureidoisobutyrate, respectively. Finally, β-alanine and β-aminoisobutyrate are produced in deamination reactions catalyzed by β-ureidopropionase.

In several conditions, β-aminoisobutyrate is produced in such large quantities that it appears in urine. Among these are a genetic predisposition for slow β-aminoisobutyrate conversion to succinyl-CoA and diseases that cause massive cell destruction, such as leukemia. Because it is soluble, excess β-aminoisobutyrate does not cause problems comparable to those observed in gout.

QUESTION 15.9   Identify each of the following biomolecules. Explain how they are produced.

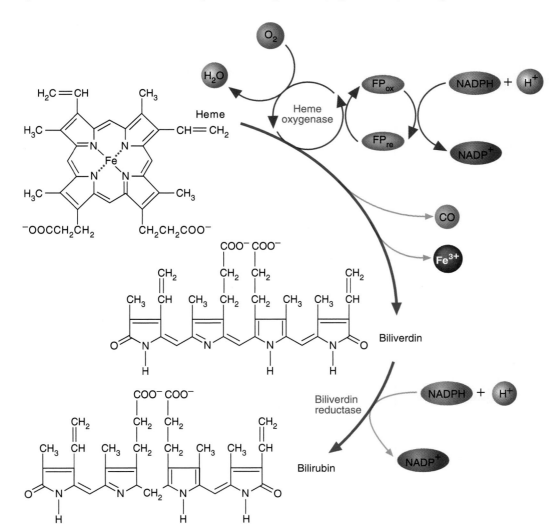

(a) **(b)** **(c)**

The products of pyrimidine base catabolism, β-alanine and β-aminoisobutyrate, can be further degraded to acetyl-CoA and succinyl-CoA, respectively. Can you suggest the type of reactions required to accomplish these transformations?

15.4 HEME BIOTRANSFORMATION

Bilirubin, an orange pigment, is the product of a series of reactions that degrade the heme groups of various hemoproteins. Approximately 80% of the 250–400 mg/dL of bilirubin formed daily is derived from hemoglobin in aging red blood cells. This conversion, which occurs predominantly in the reticuloendothelial cells of liver, spleen, and bone marrow, takes place in two phases. (Figure 15.15) During the

FIGURE 15.15

Bilirubin Synthesis.

Heme oxygenase, which catalyzes the conversion of free heme groups to biliverdin and CO, functions as part of a microsomal electron transport system similar to that of cytochrome P_{450}. (FP = NADPH-cytochrome P_{450} reductase.) Heme oxygenase requires 3 O_2 and 5 NADPH. Biliverdin reductase can use NADPH or NADH as a reductant.

Gout is a disorder in which sodium urate crystals are deposited in and around joints. (The name *gout* is derived from "gutta," the Latin word for "drop." According to ancient belief, a poisonous substance falls drop by drop into joints.) This deposition, which occurs because of **hyperuricemia** (high blood levels of uric acid, greater than 7 mg/dL in men and 6 mg/dL in women), causes a form of arthritis (joint inflammation). The initial attacks of gouty arthritis are usually acute (sudden) and most frequently affect the big toe, although other joints in the foot or leg may also be involved. The inflammation caused by urate crystal deposition attracts white blood cells, which engulf the crystals. Further tissue destruction is caused when urate crystals disrupt the lysosomal membranes in the white blood cells, resulting in the leakage of lysosomal enzymes into the tissues. In addition, visible structures called tophi (urate crystal "stones") may form near joints and cause grotesque deformities. The deposition of urate crystals within the kidney causes impaired renal function. Although hyperuricemia is a necessary predisposing factor to gout, for unknown reasons only a small percentage of individuals with high blood urate levels ever display classic gout symptoms. Circumstances that may provoke gouty arthritis include excessive food and/or alcohol consumption or starvation.

There are two forms of gout: primary and secondary. *Primary gout* is most often caused by genetic defects in purine metabolism. For example, several variants of ribose-5-phosphate pyrophosphokinase are not effectively regulated by allosteric inhibitors (e.g., P_i, GDP, or ADP). Consequently, PRPP concentrations rise, causing the increased synthesis of purine nucleotides. (Recall that PRPP concentration is an important regulator of purine nucleotide synthesis.) The overproduction of purine nucleotides then leads to increased uric acid synthesis. HGPRT deficiency also causes hyperuricemia because of decreased salvage of purine bases. Hyperuricemia can also be caused by genetic defects in other pathways. For example, in glucose-6-phosphatase deficiency, hypoglycemia develops in affected individuals because they cannot produce blood glucose from glucose-6-phosphate. Consequently, high liver concentrations of glucose-6-phosphate stimulate the synthesis of ribose-5-phosphate and PRPP.

Secondary (or acquired) *gout* is caused by seemingly unrelated disorders. These conditions may cause hyperuricemia by either overproduction of uric acid or its undersecretion by the kidneys. For example, leukemia patients overproduce uric acid either because of massive cell destruction or the chemotherapy treatment required to destroy the cancerous cells. Hyperuricemia also results when certain drugs interfere with the renal secretion of uric acid into the urine. Patients with lead poisoning are also likely to develop gout because of renal damage.

Gout is treated with diet and with several drugs. Dietary control (i.e., reduced consumption of food that is rich in nucleic acids such as liver and sardines) depresses uric acid synthesis in some individuals who are susceptible to primary gout. Allopurinol and colchicine are often used in gout therapy. Because allopurinol inhibits xanthine oxidase, it depresses uric acid synthesis. (Allopurinol is converted to alloxanthine by xanthine oxidase. Alloxanthine acts as a competitive inhibitor of the enzyme.) Hypoxanthine and xanthine, whose levels increase with allopurinol treatment, are easily excreted because of their solubility properties. In addition, the conversion of allopurinol to allopurinol ribonucleotide by HGPRT reduces PRPP levels. This circumstance depresses purine nucleotide synthesis. Colchicine, an alkaloid that is known to disrupt microtubules, reduces joint inflammation. It is currently believed that colchicine acts against inflammation by disrupting white blood cell activity.

Saturnine Gout

In years past, gout was associated with rich diets and especially with excessive consumption of alcoholic beverages. In recent years this association has been discounted because so many individuals lead overindulgent lives without developing gout. However, recent clinical research and some historical detective work indicate that the old connection between gout and alcoholic beverages may have been accurate.

Until well into the nineteenth century many bottles of wine and other alcoholic beverages were likely to be contaminated with lead. For example, the large-scale consumption of port wine by the English gentry during the eighteenth century is now believed to have been largely responsible for a gout epidemic that occurred among this population. (Port wines were imported from Portugal. To maximize their profits, Portuguese exporters added lead salts, which are very effective preservatives. In recent years, port wine bottles from this era were tested and found to contain large amounts of lead.) Similarly, in the past, rum was often stored in containers lined with lead-containing glazes.

The term *saturnine gout* reflects the connection made between gout and lead exposure by several nineteenth-century physicians. (The medieval alchemists believed that the planet Saturn had lead-like properties.) Proving the connection has been more difficult. Because bone is the major reservoir for lead (both calcium and lead are divalent), chronic lead exposure may often not be easily diagnosed. Lead can be transferred in small amounts from bone to tissues such as the kidney over long periods. Consequently, tissue damage may continue for years after the original lead exposure. Long before tissue damage becomes obvious, blood lead levels have returned to near normal values. Saturnine gout is now believed to be caused by hyperuricemia from kidney damage. Although the kidney damage is irreversible, further damage can be avoided by removing lead from the body with chelation therapy. A chelating agent such as ethylenediaminetetraacetic acid (EDTA) binds to lead with a higher affinity than it does to calcium. (Chelating agents are molecules with carboxylate groups that bind to metal cations. EDTA binds to metals with two or more positive charges.) Because lead-EDTA chelate is soluble, it is excreted in the urine.

first phase, heme is oxidized by heme oxygenase, an ER enzyme that is a component of an electron transport system similar to that of cytochrome P_{450} (Special Interest Box 10.1). The products of this reaction are the dark green pigment biliverdin and carbon monoxide (CO). During the second phase, biliverdin is converted into bilirubin in a reaction catalyzed by the cytoplasmic enzyme biliverdin reductase. Meanwhile, the CO diffuses out of the cell, is then transported in the blood to the lungs where it leaves the body on exhalation.

The product bilirubin is a very toxic compound. For example, it is known to inhibit RNA and protein synthesis and carbohydrate metabolism in the brain. Mitochondria appear to be especially sensitive to its effects. Bilirubin is also a metabolically expensive molecule to produce. For example, bilirubin is virtually insoluble in water, because of intramolecular hydrogen bonding. Therefore, sophisticated transport mechanisms and conjugation reactions in the liver (Figure 15.16) are required for excretion as components of bile in the gastrointestinal tract. Because bilirubin creates so many problems, considerable effort has been devoted to elucidating its purpose. (Many species, such as amphibians, reptiles, and birds, excrete the water-soluble precursor biliverdin.) Because it reacts with peroxyradicals, bilirubin may act as an antioxidant. During bilirubin transport in blood, the radical scavenging pigment is distributed throughout the circulatory system. (The association of bilirubin with the plasma protein albumin protects cells from the molecule's toxic effects.)

KEY CONCEPTS 15.6

The heme group of hemoproteins is first converted to biliverdin and then to bilirubin. After undergoing conjugation reactions in the liver, bilirubin is excreted in bile.

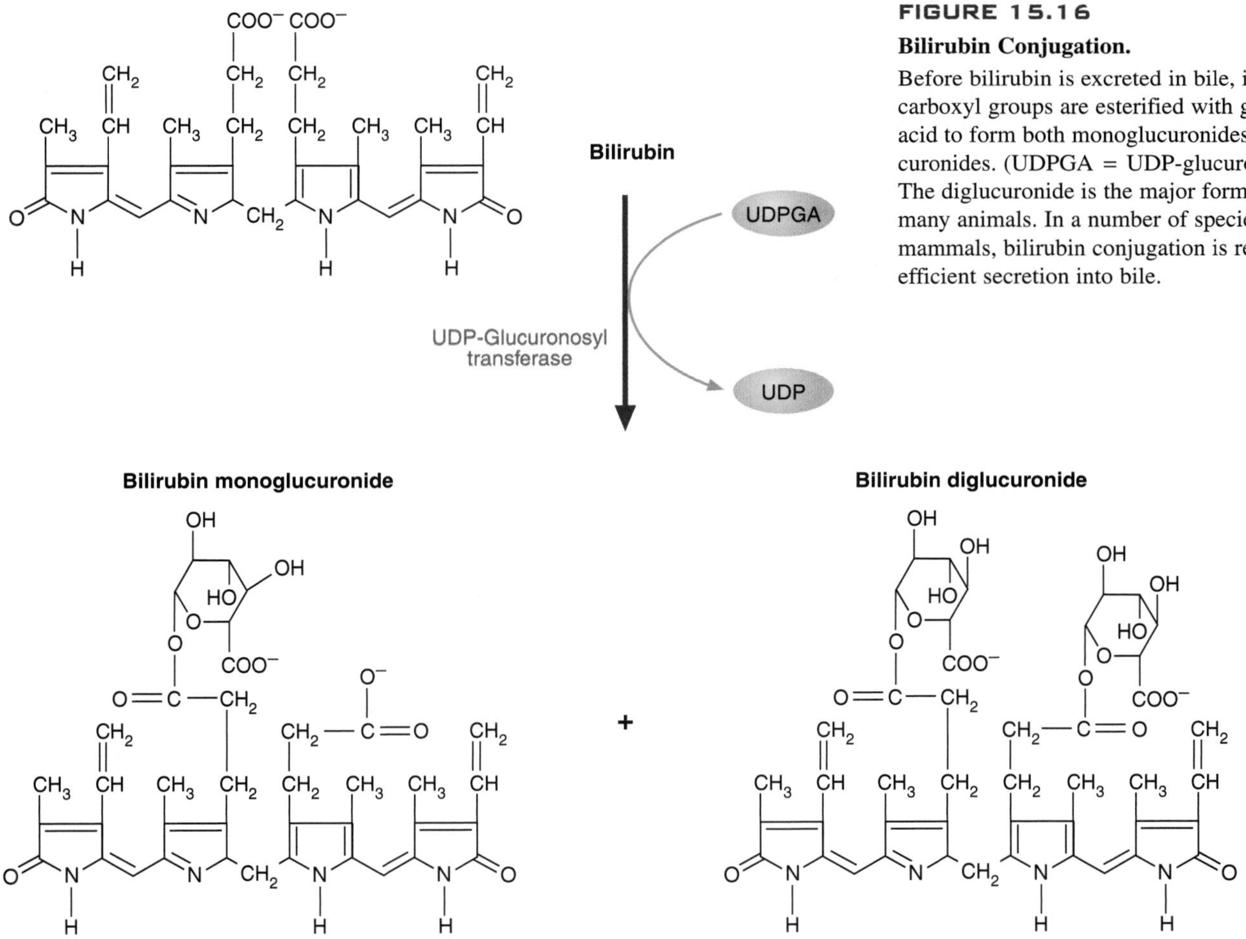

FIGURE 15.16

Bilirubin Conjugation.

Before bilirubin is excreted in bile, its propionyl carboxyl groups are esterified with glucuronic acid to form both monoglucuronides and diglucuronides. (UDPGA = UDP-glucuronic acid.) The diglucuronide is the major form produced in many animals. In a number of species, especially mammals, bilirubin conjugation is required for efficient secretion into bile.

SUMMARY

1. Animals are constantly synthesizing and degrading nitrogen-containing molecules such as proteins and nucleic acids. Protein turnover is believed to provide cells with metabolic flexibility, protection from accumulations of abnormal proteins, and the timely destruction of proteins during developmental processes. Ubiquitin is one stress protein that plays an important role in targeting proteins for destruction.

2. In general, amino acid degradation begins with deamination. Most deamination is accomplished by transamination reactions, which are followed by oxidative deaminations that produce ammonia. Although most deaminations are catalyzed by glutamate dehydrogenase, other enzymes also contribute to ammonia formation. Ammonia is prepared for excretion by the enzymes of the urea cycle. Aspartate and CO_2 also contribute atoms to urea.

3. Amino acids are classified as ketogenic or glucogenic on the basis of whether their carbon skeletons are converted to fatty acids or to glucose. Several amino acids can be classified as both ketogenic and glucogenic because their carbon skeletons are precursors for both fat and carbohydrates.

4. The degradation of neurotransmitters is critical to the proper functioning of information transfer in animals. The amine neurotransmitters such as acetylcholine, the catecholamines, and serotonin are among the best-researched examples.

5. The turnover of nucleic acids is accomplished by several types of enzymes. The nucleases degrade the nucleic acids to oligonucleotides. (The deoxyribonucleases degrade DNA; the ribonucleases degrade RNA.) The phosphodiesterases convert the oligonucleotides to mononucleotides. By removing phosphate groups, the nucleotidases convert nucleotides to nucleosides. The nucleosidases hydrolyze nucleosides to form free bases and ribose or deoxyribose. The nucleoside phosphorylases convert ribonucleosides to free bases and ribose-1-phosphate. Dietary nucleic acids are generally degraded in the intestine and are not used in salvage pathways. Cellular purines are converted to uric acid. Many animals degrade uric acid further because they produce enzymes that are not present in primates. Pyrimidine bases are degraded to either β-alanine (UMP, CMP, dCMP) or β-aminoisobutyrate (dTMP).

6. The porphyrin heme is degraded to form the excretory product bilirubin in a biotransformation process that involves the enzymes heme oxygenase and biliverdin reductase and UDP-glucuronosyltransferase. After undergoing a conjugation reaction, bilirubin is excreted as a component of bile.

SUGGESTED READINGS

Adams, J. D. Jr., Chang, M. L., and Klaidman, L., Parkinson's Disease—Redox Mechanisms, *Curr. Med. Chem.,* 8(7):809–814, 2001.

Bachmair, A., Finley, D., and Varshavsky, A., *In vivo* Half-Life of a Protein Is a Function of Its Amino-Terminal Residue, *Science,* 234:179–186, 1986.

Chain, D. G., Schwartz, J. H., and Hegde, A. N., Ubiquitin-Mediated Proteolysis in Learning and Memory, *Mol. Neurobiol.,* 20(2–3):125–142, 1999.

Hershko, A., The Ubiquitin Pathway for Protein Degradation, *Trends Biochem. Sci.,* 16(7):265–268, 1991.

Hilt, W., and Wolf, D. H., Proteasomes: Destruction as a Programme, *Trends Biochem. Sci.,* 21(3):96–102, 1996.

Holmes, F. L., Hans Krebs and the Discovery of the Ornithine Cycle, *Fed. Proc.,* 39:216–225, 1980.

Rechsteiner, M., and Rogers, S. W., PEST Sequences and Regulation by Proteolysis, *Trends Biochem. Sci.* 21(7):267–271, 1996.

Weissman, A. M., Themes and Variations on Ubiquitylation, *Nat. Rev. Mol. Cell Biol.,* 2(3):169–178, 2001.

Wellner, D., and Meister, A., A Survey of Inborn Errors of Amino Acid Metabolism and Transport in Man, *Ann. Rev. Biochem.,* 50:911–968, 1981.

Winkler, R. G., Blevins, D. G., Polacco, J. C., and Randall, D. D., Ureide Catabolism in Nitrogen-Fixing Legumes., *Trends Biochem. Sci.,* 13:97–100, 1988.

KEY WORDS

agonist, *520*	humoral immune response, *521*	Krebs urea cycle, *506*	proteosome, *504*
ammonia intoxication, *509*	hyperammonemia, *509*	molecular chaperones, *504*	T cell, *521*
B cell, *521*	hyperuricemia, *526*	mutation, *519*	transulfuration pathway, *516*
cellular immunity, *521*	ketogenic, *504*	nuclease, *521*	ubiquination, *504*
glucogenic, *504*	Krebs bicycle, *508*	oligonucleotide, *521*	ubiquitin, *504*
heat shock protein, *504*		protein turnover, *504*	urea cycle, *506*

REVIEW QUESTIONS

1. Define the following terms:
   a. denitrification
   b. ammonotelic
   c. protein turnover
   d. ubiquination
   e. ammonia intoxication
   f. humoral immune response
   g. hyperuricemia

2. What are the major molecules used to excrete nitrogen?

3. What are three purposes served by protein turnover?

4. What are the structural features of proteins that mark them for destruction?

5. What are the seven metabolic products produced by the degradation of amino acids?

6. Indicate which of the following amino acids are ketogenic and which are glucogenic:
   a. tyrosine
   b. lysine
   c. glycine
   d. alanine
   e. valine
   f. threonine

7. Describe how each of the following amino acids is degraded:
   a. lysine
   b. glutamate
   c. glycine
   d. aspartate
   e. tyrosine
   f. alanine

8. In humans the purine ring cannot be degraded. How is it excreted? What reactions are involved?

9. The urea cycle occurs partially in the cytosol and partially in the mitochondria. Discuss the urea cycle reactions with reference to their cellular locations.

10. Describe how the glucose-alanine cycle acts to transport ammonia to the liver.

11. In individuals with PKU, is tyrosine an essential amino acid?

12. Urea formation is energetically expensive, requiring the expenditure of 4 mol of ATP per mole of urea formed. However, NADH is produced when fumarate is reconverted to aspartate. How many ATP molecules are produced by the mitochondrial oxidation of the NADH? What is the net ATP requirement for urea synthesis?

13. Describe the Krebs bicycle. What compound links the citric acid and urea cycles?

14. Most amino acids are degraded in the liver. This is not true of the branched chain amino acids. Where are they primarily degraded?

15. Describe how a protein is targeted for degradation.

16. Provide the names of the organisms that use the following substances as nitrogenous waste molecules:
    a. uric acid
    b. urea
    c. allantoate
    d. NH_4^+
    e. allantoin

17. Which of the following molecules yields uric acid when degraded?
    a. DNA
    b. FAD
    c. CTP
    d. PRPP
    e. β-alanine
    f. urea
    g. NAD^+

THOUGHT QUESTIONS

1. Mammals excrete most nitrogen atoms as urea. The process requires the expenditure of considerable amounts of ATP energy. Why is it not practical to excrete nitrogen as ammonia as some aquatic species do? What toxic effects would this have on a mammal?

2. Describe how increasing concentrations of ammonia stimulate the formation of N-acetylglutamate and turn on the urea cycle.

3. Explain how a defective enzyme in the urea cycle can produce high levels of ammonia.

4. PKU can be caused by deficiencies in phenylalanine hydroxylase and by enzymes catalyzing the formation and regeneration of 5,6,7,8-tetrahydrobiopterin. How can this second defect cause the symptoms of PKU?

5. Individuals who cannot produce 5,6,7,8-tetrahydrobiopterin must be supplied with L-dopa and 5-hydroxytryptophan, metabolic precursors to norepinephrine and serotonin. Why does supplying 5,6,7,8-tetrahydrobiopterin have no effect?

6. In their in vitro studies using liver slices, Krebs and Henseleit observed that urea formation was stimulated by the addition of ornithine, citrulline, and arginine. Other amino acids had no effect. Explain these observations.

7. Specify which type of carbon unit is transferred by each of the following compounds:
   a. N^5,N^{10}-methylene THF
   b. serine
   c. choline
   d. S-adenosylmethionine

8. Caffeine, a methylated xanthine found in chocolate, coffee, and tea, is excreted as uric acid. Using your knowledge of the metabolism of other purine compounds, suggest how caffeine is metabolized.

9. Some animals living in a fluid medium excrete nitrogen as ammonia. Land animals, which conserve water, excrete urea and uric acid. Why does the excretion of these molecules aid in water conservation?

CHAPTER SIXTEEN

Integration of Metabolism

The food that animals consume supplies their bodies with the nourishment required to sustain living processes. Complex regulatory mechanisms ensure that the demands of each cell for energy and metabolites are consistently met.

Previous chapters deal with several important topics, for example, the metabolism of carbohydrates, lipids, and other molecules. However, the whole is not just the sum of its parts. Multicellular organisms are extraordinarily complex, more so than their components would suggest. Chapter 16 takes a wider view of functioning of the mammalian body. Initially, the division of labor that allows the sophisticated functioning of the multicellular body is considered. This is followed by a discussion of the feeding-fasting cycle, a complex multiorgan process. Hormones and growth factors, the major tools of intercellular communication, and their mechanisms of action are then described. Chapter 16 also includes a discussion of diabetes mellitus, a disease that has widespread metabolic effects.

It should now be evident that the maintenance of living processes in multicellular organisms is a complicated business. Recall that, despite changes in their internal and external environments, these organisms must constantly sustain adequate (if not optimal) operating conditions as they simultaneously engage in growth and repair activities. To accomplish these functions, the anabolic and catabolic reaction pathways that use carbohydrates, lipids, and proteins as energy sources and biosynthetic precursors must be precisely regulated. As described, multicellular organisms can efficiently exploit their environment because of the division of labor among their constituent cells, tissues, and organs. Mammals, the most carefully investigated group of multicellular organisms, have a sophisticated and mutually beneficial division of labor. Each organ performs specific functions to serve both the short- and long-term interests of the body.

The operation of such a complex system as the body is maintained by a continuous flow of information among its parts. A simple system for information transfer is composed of a stimulus sent by a sender, a message carrier (or messenger), and a receiver. In such a system, only one response to the signal is possible. Physiological systems, however, are extraordinarily complex and require finely modulated responses to complex stimuli. In addition, for coordinated functioning each body part must also receive information about events in other parts. Because multicellular organisms are hierarchical organizations of cells, tissues, organs, and organ systems, it is not surprising that a large number of signals, message carriers, and receivers are required. In the mammalian body, much information transfer is accomplished by hormones. These messenger molecules are arranged in complex hierarchies that allow for a high degree of sophisticated regulation.

In Chapter 16 the focus of the discussion is the integration of the major metabolic processes in mammals. The chapter begins with an overview of metabolic processes and a description of the metabolic contributions of several major organs. This is followed by a discussion of the feeding-fasting cycle, which illustrates several important control mechanisms. Chapter 16 ends with a brief review of the major mammalian hormones and their mechanisms of action.

16.1 OVERVIEW OF METABOLISM

The central metabolic pathways are common to most organisms. Throughout the life of an organism, a precise balance is struck between anabolic (synthetic) and catabolic (degradative) processes. An overview of the principal anabolic and catabolic pathways in heterotrophs such as animals is illustrated in Figure 16.1. As a young animal grows and matures, the rate of anabolic processes is greater than that of catabolic processes. As healthy adulthood is reached, anabolic processes slow, and growth essentially stops. Throughout the remainder of its life (except during illness or pregnancy) the animal's tissues exist in a metabolic steady state. In a **steady state** the rate of anabolic processes is approximately equal to that of catabolic processes. Consequently, the appearance and functioning of the animal change little from one day to the next. Only over long periods do the inevitable signs of aging appear.

How are animals (or other multicellular organisms) able to maintain a balance between anabolic and catabolic processes as they respond and adapt to changes in their environment? The answer to this question is not fully understood. However, various forms of intercellular communication are believed to play an important role. Most intercellular communication occurs by means of chemical signals. Once released into the extracellular environment, each chemical signal is

FIGURE 16.1

Overview of Metabolism.

In this simplified overview of metabolism, the anabolic and catabolic pathways of the major food molecules in heterotrophs (i.e., those biochemical pathways that synthesize, degrade, or interconvert important biomolecules and generate energy) are illustrated.

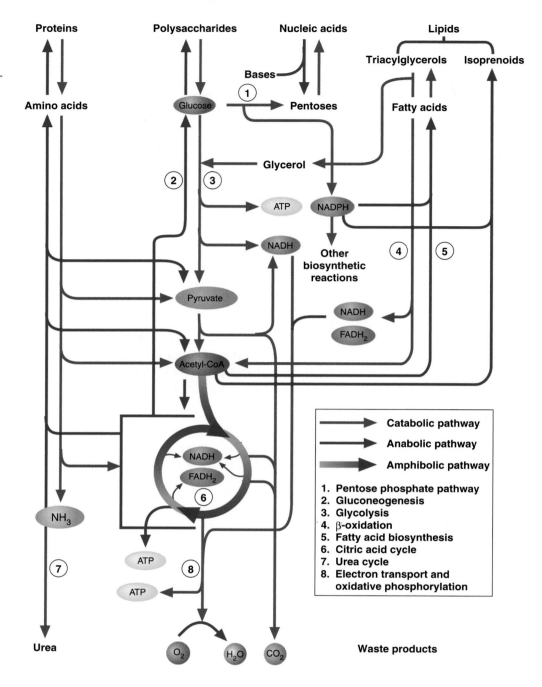

KEY CONCEPTS 16.1

Throughout life, organisms strike a balance between anabolic and catabolic processes so that they can meet their metabolic needs as they respond to environmental changes. The information transfers that sustain living processes are managed largely by hormones.

recognized by specific cells (called **target cells**), which then respond in a specific manner. Most chemical signals are modified amino acids, fatty acids derivatives, peptides, proteins, or steroids.

In animals the nervous and endocrine systems are primarily responsible for coordinating metabolism. The nervous system provides a rapid and efficient mechanism for acquiring and processing environmental information. Nerve cells, called neurons, release neurotransmitters (Section 14.3) at the end of long cell extensions called axons into tiny intercellular spaces called synapses. The neurotransmitter molecules bind to nearby cells, evoking specific responses from those cells.

Metabolic regulation by the endocrine system is achieved by secretion of chemical signals called hormones directly into the blood. The endocrine system is composed of specialized cells, many of which are found in the glands. After hormone molecules are secreted, they travel through the blood until they reach a target cell. Most hormone-induced changes in cell function result from alterations in the activity or concentration of enzymes. Hormones interact with cells by binding to specific receptor molecules. The receptors for most water-soluble hormones (e.g., polypeptides and epinephrine) are located on the surface of target cells. Binding these hormones to membrane-bound receptors triggers an intracellular response. The intracellular actions of many hormones are mediated by a group of molecules referred to as **second messengers**. (The hormone molecule is the first messenger.) Several second messengers have been identified. These include the nucleotides cyclic AMP (cAMP) and cyclic GMP (cGMP), calcium ions, and the inositol phospholipid system. Most second messengers act to modulate enzymes, often by a powerful amplification device called an enzyme cascade. In an *enzyme cascade* (Figure 16.2) enzymes undergo conformational transitions that switch the enzymes from their inactive forms to their active forms, or vice versa, in a sequentially expanding array leading to a substantial amplification of the original signal. This process is often initiated when a second messenger binds to a specific enzyme. For example, binding cAMP to inactive protein kinase A converts it to active protein kinase A which, in turn, modifies the activity of many target enzymes through phosphorylation. The original signal generates an amplified and diversified response, via a second messenger (at the signal level) in some cases and an enzyme cascade (at the catalytic level) in most cases. A cAMP system accomplishes amplification at both levels.

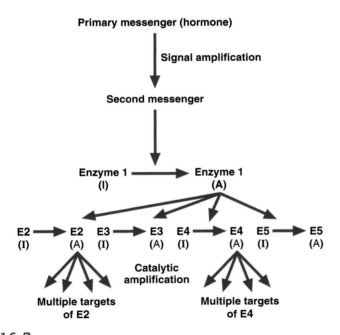

FIGURE 16.2

An Amplified Enzyme Cascade.

And enzyme cascade is a powerful mechanism in which a series of enzymes are sequentially activated. Activation is often initiated by a second messenger molecule (signal amplification). The enzyme activated by the second messenger modifies multiple copies of a number of different target enzymes. Those target enzymes that are activated in the process of modification may also modify multiple copies of a second set of target proteins. These expanded enzymatic responses are referred to as catalytic amplification (I = inactive, A = active.)

Steroid hormones are lipid-soluble molecules that act by a different mechanism. Once a steroid hormone has diffused into a cell, it binds to a specific receptor protein in the cytoplasm. The hormone-receptor complex moves into the nucleus, where it binds to specific sites on DNA. Steroid-receptor complexes alter a cell's pattern and rate of gene transcription and, ultimately, protein synthesis. (This topic is discussed in Chapter 18.) The thyroid hormones act similarly.

Research increasingly shows that the distinction between the nervous and endocrine systems is not as clear as was once thought. For example, certain nerve cells, referred to as neurosecretory cells, synthesize and release hormones into the blood. Oxytocin and vasopressin (see p. 125) are two prominent examples. In addition, several neurotransmitters act through second messengers. Epinephrine, which can function as a neurotransmitter and a hormone, induces Tissue-specific effects dependent upon the nature of the receptor to which it binds.

QUESTION 16.1

Review the epinephrine-stimulated activation of glycogen breakdown (Figure 8.17). Identify the following signal transduction components in this biochemical process: signal, messenger, and receiver.

16.2 THE DIVISION OF LABOR

Each organ in the mammalian body has several roles that contribute to the individual's function. For example, some organs are consumers of energy so that they may perform certain energy-driven tasks (e.g., muscle contraction). Other organs, such as those in the digestive tract, are responsible for efficiently supplying energy-rich nutrient molecules for use elsewhere. The roles of several organs in relation to their metabolic contributions are discussed below.

Small Intestine

The most obvious role of the small intestine is the digestion of nutrients such as carbohydrates, lipids, and proteins into molecules that are small enough to be absorbed (sugars, fatty acids, glycerol, and amino acids). Nutrient absorption by the enterocytes of the small intestine is an extremely vital and complicated process that involves numerous enzymes and transport mechanisms. As described, (p. 374) enterocytes then transport these molecules (and water, minerals, vitamins, and other substances) into the blood and lymph, which carry them throughout the body.

The enterocytes require enormous amounts of energy to support active transport and lipoprotein synthesis. Although some glucose is used, most of the energy is supplied by glutamine. During the digestive process, enterocytes obtain glutamine from degraded dietary protein. Under fasting conditions, glutamine is acquired from arterial blood. Enterocytes also use some glutamine to form Δ^1-pyrroline-5-carboxylate, which is ultimately converted to proline. Other products of glutamine metabolism include lactate, citrate, ornithine, and citrulline. The liver receives blood containing dietary nutrients plus these products of glutamine metabolism. It uses lactate and alanine to synthesize glucose for export and glycogen for storage. The glucose in blood is preferentially delivered to glucose-dependent tissues (e.g., brain, red blood cells, and adrenal medulla).

Liver

The liver performs a stunning variety of metabolic activities. In addition to its key roles in carbohydrate, lipid, and amino acid metabolism, the liver monitors and regulates the chemical composition of blood and synthesizes several plasma proteins. The liver distributes several types of nutrients to other parts of the body. Because of its metabolic flexibility, the liver reduces the fluctuations in nutrient availability caused by drastic dietary changes and intermittent feeding and fasting. For example, a sudden shift from a high-carbohydrate diet to one that is rich in proteins increases (within hours) the synthesis of the enzymes required for amino acid metabolism. Finally, the liver plays a critically important protective role in processing foreign molecules.

Muscle

Skeletal muscle is specialized to perform intermittent mechanical work. As described previously, the energy sources that provide ATP for muscle contraction depend on the degree of muscular activity and the physical status of the individual. During fasting and prolonged starvation, some skeletal muscle protein is degraded to provide amino acids (e.g., alanine) to the liver for gluconeogenesis.

In contrast to skeletal muscle, cardiac muscle must continuously contract to sustain blood flow throughout the body. To maintain its continuous operation, cardiac muscle relies on glucose in the fed state and fatty acids in the fasting state. It is not surprising, therefore, that cardiac muscle is densely packed with mitochondria. It can also use other energy sources, such as glucose, ketone bodies, pyruvate, and lactate. Lactate is produced only in small quantities in cardiac muscle because the isozyme of lactate dehydrogenase found in this tissue is inhibited by large concentrations of its substrate, pyruvate. The limited production of lactate means that glycolysis alone cannot be sustained in cardiac muscle.

Adipose Tissue

The role of adipose tissue is primarily the storage of energy in the form of triacylglycerols (p. 336). Depending on current physiological conditions, adipocytes store fat derived from the diet and liver metabolism or degrades stored fat to supply fatty acids and glycerol to the circulation. Recall that these metabolic activities are regulated by several hormones (i.e., insulin, glucagon, and epinephrine).

Brain

The brain ultimately directs most metabolic processes in the body. Sensory information from numerous sources is integrated in several areas in the brain. These areas then direct the activities of the motor neurons that innervate muscles and glands. Much of the body's hormonal activity is controlled either directly or indirectly by the hypothalamus and the pituitary gland (Section 16.4).

Like the heart, the brain does not provide energy to other organs or tissues. Under normal conditions, the brain uses glucose as its sole fuel. Because it stores little glycogen, the brain is highly dependent on a continuous supply of glucose in the blood. During prolonged starvation, the brain can adapt to using ketone bodies as an energy source.

Kidney

The kidney has several important functions that contribute significantly to maintaining a stable internal environment. These include

1. filtration of blood plasma, which results in the excretion of water-soluble waste products (e.g., urea and certain foreign compounds),

2. reabsorption of electrolytes, sugars, and amino acids from the filtrate,

3. regulation of blood pH, and

4. regulation of the body's water content.

Considering the functions of the kidney, it is not surprising that much of the energy generated in this organ is consumed by transport processes. Energy is provided largely by fatty acids and glucose. Under normal conditions, the small amounts of glucose formed by gluconeogenesis are used only within certain kidney cells. The rate of gluconeogenesis increases during starvation and acidosis. The kidney uses glutamine and glutamate (via glutaminase and glutamate dehydrogenase, respectively) to generate ammonia, which is used in pH regulation. (Recall that NH_3 reversibly combines with H^+ to form NH_4^+.) The carbon skeleton of glutamine and glutamate can then be used by the kidney as a source of energy.

KEY CONCEPTS 16.2

Each organ in the mammalian system contributes to the body's overall function.

QUESTION 16.2

Describe two functions concerned with nutrient metabolism for each of the following organs:

   a. intestine
   b. liver
   c. muscle
   d. adipose tissue
   e. kidney
   f. brain

16.3 THE FEEDING-FASTING CYCLE

Despite their consistent requirements for energy and biosynthetic precursor molecules, mammals consume food only intermittently. This is possible because of elaborate mechanisms for storing and mobilizing energy-rich molecules derived from food (Figure 16.3). The changes in the status of various biochemical pathways during transitions between feeding and fasting illustrate metabolic integration and the profound regulatory influence of hormones. Substrate concentrations are also an important factor in metabolism. In discussions of the feeding-fasting cycle, the terms *postprandial* and *postabsorption* are often used. In the **postprandial** state, which occurs directly after a meal has been digested or absorbed, blood nutrient levels are elevated above those in the fasting phase. During the **postabsorptive** state, for example, after an overnight fast, nutrient levels in blood are low.

The Feeding Phase

As the feeding phase begins, food is propelled along the gastrointestinal tract by muscle contractions. As it moves through the organs, food is broken into smaller particles and exposed to enzymes. Ultimately, the products of digestion (consisting largely of sugars, fatty acids, glycerol, and amino acids) are absorbed by the small intestine and transported into the blood and lymph. This phase is regulated by interactions between enzyme-producing cells of the digestive organs, the nervous system, and several hormones. The nervous system is responsible for the waves of smooth muscle contraction that propel food along the tract, as well as regulating the secretions of several digestive structures (e.g., from salivary and gastric glands). Hormones such as gastrin, secretin, and cholecystokinin also contribute to the digestive process. (See Table 16.1 in Section 16.4.) They do so by stimulating the secretion of enzymes or digestive aids such as bicarbonate and bile.

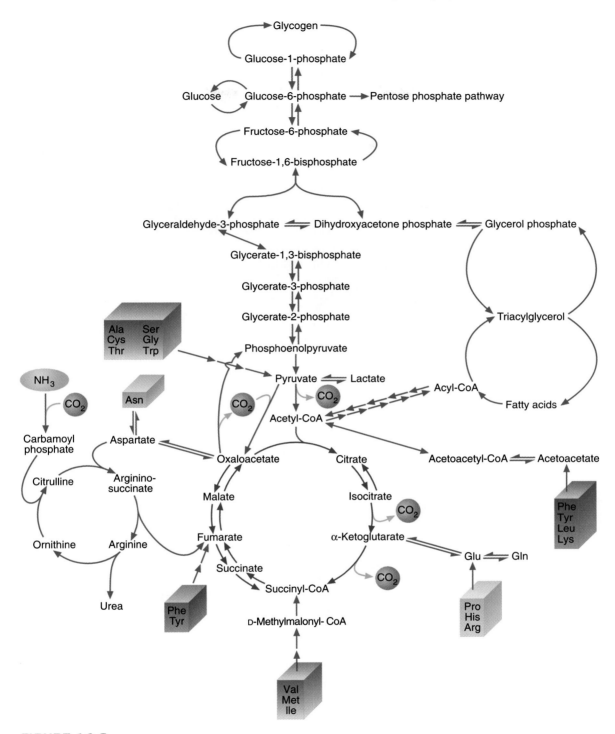

FIGURE 16.3

Nutrient Metabolism in Mammals.

Despite the variability of the mammalian diet, these organisms usually provide their cells with adequate nutrients. Control mechanisms that regulate biochemical pathways are responsible for this phenomenon.

The early postprandial state is illustrated in Figure 16.4. As described, sugars and amino acids are absorbed and transported by the portal blood to the liver. The portal blood also contains a high level of lactate that is a product of enterocyte metabolism. Most lipid molecules are transported from the small intestine in lymph as

FIGURE 16.4

The Early Postprandial State.
The primary substrates for glycogen synthesis in liver are amino acids and lactate (not shown) derived from portal blood. Note that the primary use for glucose in fat cells is as the precursor or glycerol. Fat cells do not carry out significant *de novo* fatty acid synthesis.

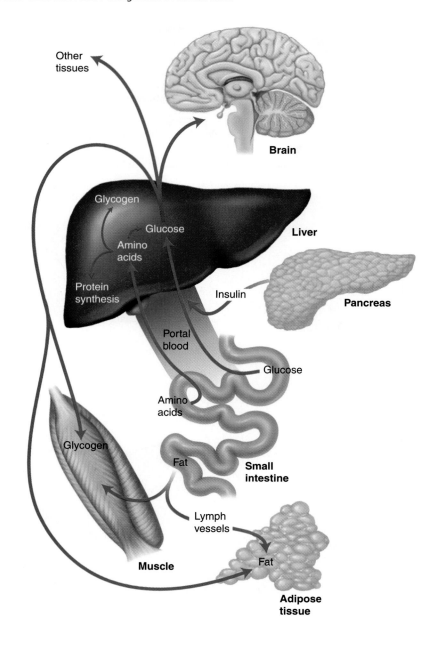

chylomicrons. Chylomicrons pass into the bloodstream, which carries them to tissues such as muscle and adipose tissue. After most triacylglycerol molecules have been removed from chylomicrons, these structures, now referred to as *chylomicron remnants*, are then taken up by the liver. The phospholipid, protein, cholesterol, and few remaining triacylglycerol molecules are then degraded or reused. For example, cholesterol is used to synthesize bile acids, and fatty acids are used in new phospholipid synthesis. Phospholipids, as well as other newly synthesized lipid and protein molecules, are then incorporated into lipoproteins for export to other tissues.

As glucose moves through the blood from the small intestine to the liver, the β-cells within the pancreas are stimulated to release insulin. (High blood glucose and insulin levels depress glucagon secretion by the pancreatic α-cells. The opposing effects of insulin and glucagon on glucose and fat metabolism are illustrated in Figure 16.5.) Insulin release triggers several processes that ensure the storage of nutrients. These include glucose uptake by muscle and adipose tissue, glycogenesis in liver and muscle, fat synthesis in liver, fat storage in

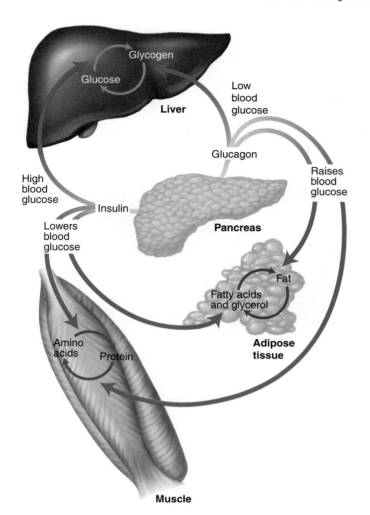

FIGURE 16.5

The Opposing Effects of Insulin and Glucagon on Blood Glucose Levels.

In general, insulin promotes anabolic processes (e.g., fat synthesis, glycogenesis, and protein synthesis). Glucagon raises blood glucose levels by promoting glycogenolysis in liver and protein degradation in muscle. It also promotes lipolysis.

adipocytes, and gluconeogenesis (using excess amino acids and lactate). Recall that in the liver, most glycogen and fatty acids are synthesized from three-carbon molecules such as lactate, not directly from blood glucose. In addition, insulin also influences amino acid metabolism. For example, insulin promotes the transport of amino acids into the cells (especially liver and muscle cells). In general, insulin stimulates protein synthesis in most tissues.

Although the effects of insulin on postprandial metabolism are profound, other factors (e.g., substrate supply and allosteric effectors) also affect the rate and degree to which these processes occur. For example, elevated levels of fatty acids in blood promote lipogenesis in adipose tissue. Regulation by several allosteric effectors further ensures that competing pathways do not occur simultaneously; for example, in many cell types fatty acid synthesis is promoted by citrate (an activator of acetyl-CoA carboxylase), whereas fatty acid oxidation is depressed by malonyl-CoA (an inhibitor of carnitine acyltransferase I activity). The control of fatty acid metabolism is described in Section 12.1.

The Fasting Phase

The early postabsorptive state (Figure 16.6) of the feeding-fasting cycle begins as the nutrient flow from the intestine diminishes. As blood glucose and insulin levels fall back to normal, glucagon is released. Glucagon acts to prevent hypoglycemia by promoting glycogenolysis and gluconeogenesis in liver. Decreased insulin reduces

FIGURE 16.6

The Early Postabsorptive State.

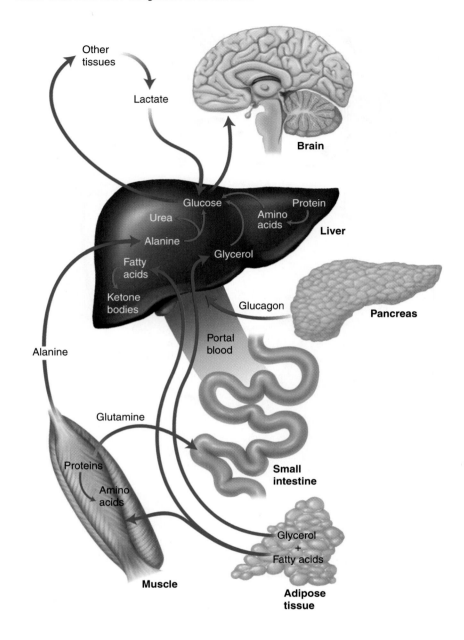

energy storage in several tissues and leads to increased lipolysis and the release of amino acids such as alanine and glutamine from muscle. (Recall that several tissues use fatty acids in preference to glucose. Glycerol and alanine are substrates for gluconeogenesis, and glutamine is an energy source for enterocytes.)

If a fast becomes prolonged (e.g., overnight), several metabolic strategies maintain blood glucose levels. Increased mobilization of fatty acids from adipose tissue during the postabsorptive state is stimulated by norepinephrine. These fatty acids provide an alternative to glucose for muscle. (Reduced skeletal muscle consumption of glucose spares its use for brain. Recall that glucose is normally the only fuel source in brain.) In addition, the action of glucagon increases gluconeogenesis, using amino acids derived from muscle. (During fasting, insulin levels decline significantly.)

Under conditions of extraordinarily prolonged fasting (starvation), the body makes metabolic changes to ensure that adequate amounts of blood glucose are available to sustain energy production in the brain and other glucose-requiring cells. Additionally, fatty acids from adipose tissue and ketone bodies from liver are mobilized to sustain the other tissues. Because glycogen is depleted after sev-

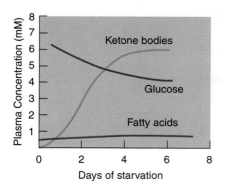

FIGURE 16.7

The Plasma Levels of Fatty Acids, Glucose, and Ketone Bodies During the Early Days of Starvation.

The concentrations of fatty acids and ketones increase. In contrast, glucose levels decrease.

eral hours of fasting, gluconeogenesis plays a critical role in providing sufficient glucose. During early starvation, large amounts of amino acids from muscle are used for this purpose. However, after several weeks, the breakdown of muscle protein declines significantly because the brain is using ketone bodies as a fuel source. Figure 16.7 illustrates the plasma levels of glucose, fatty acids, and ketone bodies as starvation proceeds for several days.

Explain the metabolic changes that occur during starvation. What appears to be the principal purpose for the preferential degradation of muscle tissue during starvation?

QUESTION 16.3

Explain the changes in liver metabolism that occur when blood glucose levels drop after a meal has been digested.

QUESTION 16.4

16.4 INTERCELLULAR COMMUNICATION

Hormones are synthesized and secreted by specialized cells and exert biochemical effects on target cells. When these target cells are distant from the hormone-secreting cells, the hormones are referred to as **endocrine** hormones. (Paracrine hormones exert their effects on nearby cells.) Some hormones exert very specific effects on one type of target cell; other hormones act on a variety of target cells. For example, thyroid-stimulating hormone (TSH) stimulates follicular cells in the thyroid gland to release T_3 (triiodothyronine) and T_4 (thyroxine) (Figure 16.8). In contrast, T_3 (the active form of the hormone) and T_4

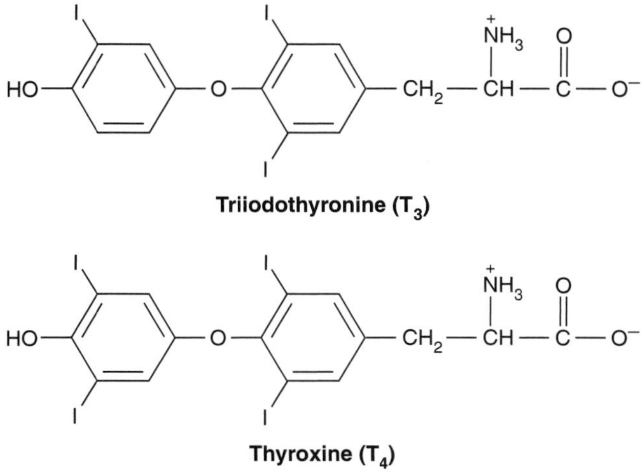

Triiodothyronine (T_3)

Thyroxine (T_4)

FIGURE 16.8

Structure of the Thyroid Hormones, T_3 and T_4.

Regular physical exercise has a profound effect on health. Benefits include a wide range of physiological adaptations that increase cardiovascular fitness, decrease the incidence and/or severity of chronic disease, improve mood, and effectively slow the process of aging. Endurance training, an intense form of exercise in which large skeletal muscles are rapidly contracted, causes the body over time to function more efficiently. Many of the details of the mechanisms by which endurance training effects these changes are now understood. The following description of the molecular basis of the training effect begins with an overview of nutrient metabolism in exercising muscle. Following a description of the practical features of endurance training, the metabolic changes induced by endurance training are discussed.

Energy Sources for Exercising Muscle

There are two metabolic strategies for supplying the energy requirement of contracting muscles: anaerobic metabolism and aerobic metabolism. The term *anaerobic exercise* describes short intense bursts of physical activity. Glycolysis generates the ATP required to drive muscle contraction. At such high levels of metabolism, muscle uses oxygen faster than it can be supplied by the cardiovascular system. As soon as the oxygen supply is depleted, lactate levels begin to increase. Muscle contraction can continue only until muscle lactic acid levels rise to a level that causes muscle fatigue. (When lactic acid levels reach a certain threshold value, muscle cells become unresponsive to neural stimulation.) When muscle contraction occurs at a pace for which adequate amounts of oxygen can be supplied, the exercise is said to be *aerobic*. When exercise is aerobic, the activity can be sustained long enough to induce substantial changes in metabolism. Among these are an increased **basal metabolic rate** (BMR) and resistance to fatigue. The BMR is a measure of the energy required to support essential life-sustaining metabolic processes. BMR is determined when a person is at rest after an all-night fast.

When adequate oxygen is available, muscle uses two primary fuels: glucose and fatty acids. During rest or low-intensity exercise, energy is provided by small amounts of both glucose and fatty acids. A small amount of glucose is required in part because it is converted to oxaloacetate, needed in the citric acid cycle (Figure 16.3). As physical activity intensifies, the release of fatty acids from adipose tissue fat stores increases. (This process is mediated by the sympathetic nervous system, which stimulates the adrenal gland to secrete epinephrine and norepinephrine. Recall that these hormones activate hormone-sensitive lipase.) Eventually, muscle glycogen is depleted, and glucose derived from hepatic glycogen is used. If intense physical activity lasts long enough, the body's glycogen reserves become almost totally depleted. Then muscle cells must depend on fatty acid oxidation to supply the energy for muscle contraction. However, because glucose is primarily responsible for maintaining a large supply of citric acid cycle intermediates, the capacity of the muscle to generate energy then drops to 60% of its previous level. In endurance training, exercise is performed at a submaximal level, that is, when glucose is available to supplement fatty acid oxidation. Following a description of the practical features of endurance training, the metabolic changes induced by endurance training are discussed.

Endurance Training

Endurance training incorporates the following features:

1. The exercise is performed regularly, for example every day or every other day.
2. A large mass of muscle must be used to perform the exercise. Exercises such as jogging, walking, bicycling, and cross-country skiing are excellent choices because a large amount of muscle is used. The more muscle is trained, the larger the amount of fat that can be oxidized.
3. The exercise must be nonstop. To induce a "training effect" (i.e., a change in muscle metabolism), the heart rate must stimulate a variety of cellular reactions in numerous cell types (e.g., it stimulates glycogenolysis in liver cells and glucose absorption in the small intestine).

Control of physiological responses often involves several hormones. In some systems, two or more hormones act in opposition to each other (e.g., insulin and glucagon in the regulation of blood glucose). In other control systems, several hormones act in information hierarchies. Section 16.4 begins with a description of the best-researched example of such a hierarchy, referred to as a hormone cascade mechanism. This is followed by a discussion of growth factors, specialized proteins that stimulate cell division in susceptible cells.

Sensitive techniques are now available to detect and measure hormones. The most common of these, radioimmunoassays and enzyme-linked immunosorbent assays (ELISA), are described in Biochemical Methods 16.1.

The Hormone Cascade System

The list of molecules now recognized as hormones has grown astonishingly large. Table 16.1 contains a small selection of the better-known mammalian hormones.

be elevated above a specific threshold for a minimum time. This threshold value is determined largely by a person's age. For example, a 20-year-old's minimum training heart rate is approximately 130 beats per minute, whereas 104 is sufficient for an individual over 60. These values are 65% of the age-adjusted maximum heart rate. (Maximum heart rate is calculated by subtracting a person's age from 220.)

4. The exercise must be performed for a minimum time. The time required depends on the type of exercise chosen. For example, between 30 and 45 minutes (depending on speed) are required for a walking program, whereas 20 minutes of jogging are sufficient to induce a training effect.

5. The exercise must be aerobic. In other words, it must not be so intense that breathing is difficult. An aerobic workout can also be achieved by monitoring breathing or heart rate. Breathing should not be labored. The exercise should not cause the heart rate to exceed 80% of the age-adjusted maximum heart rate. For a 20-year-old the maximum heart rate during exercise should not exceed 160 (i.e., 80% of 200, the maximum heart rate).

The Training Effect

The endurance training of skeletal muscle is now known to induce a number of metabolic adaptations that enhance the body's capacity to burn fat. (The precise mechanisms by which these changes are effected are not understood.) These adaptations also serve to conserve muscle glycogen, an important consideration if both fatty acids and glucose are to be degraded simultaneously. The most obvious adaptations occur in skeletal muscle.

1. The number of mitochondria per muscle fiber increases. This increase is partially responsible for making muscle cells more responsive to metabolic regulators. For example, muscle glycogen is conserved in part because of the inhibition of PFK-1 by various metabolic regulators produced during fatty acid oxidation, such as citrate. The larger number of mitochondria increase the efficiency with which these regulators can inhibit glycolytic enzymes.

2. Fatty acids are more efficiently degraded because of increased synthesis of molecules that facilitate fatty acid transport and oxidation. The concentration of citric acid cycle and β-oxidation enzymes increases, as well as the components of the ETC. In addition, the capacity of the muscle cell to remove fatty acids from blood and to transport them into mitochondria increases. For example, increases in the synthesis of fatty acid transporter proteins and fatty acid–binding proteins, as well as carnitine and carnitine acyltransferase, have been observed.

3. The vascularization of muscle tissue increases. As the number of capillaries increases, the transit time for blood through muscle increases (i.e., there is increased resistance to flow because of a greater surface area for the exchange of nutrients). The exchange of nutrients and waste products between the blood and muscle fibers is more efficient.

Endurance training also has a noteworthy effect on fat metabolism. In sedentary individuals, especially those who have undergone calorie-restricted diets, adipocytes are resistant to the stimulation by the sympathetic nervous system that accompanies physical exercise. For reasons that are still poorly understood, endurance training results in an increased sensitivity of adipocytes to these hormones. This is an important point, because the amount of fatty acids transported into muscle cells and used to drive muscle contraction is directly related to their concentration in blood. Trained muscle can generate more energy by oxidizing ketone bodies.

In mammals, most metabolic activities are controlled to one degree or another by hormones. The synthesis and secretion of many hormones are regulated by a complex cascade mechanism and ultimately controlled by the central nervous system. In this system, outlined in Figure 16.9, sensory signals are received by the hypothalamus, an area in the brain that links the nervous and endocrine systems. (In addition to regulating hormone production, nerve cells in the hypothalamus monitor and/or regulate vital body functions, such as body temperature, blood pressure, water balance, and body weight. The hypothalamus is also associated with certain behaviors, for example, anger, sexual arousal, and feelings of pain and pleasure.) Once it is appropriately stimulated, the hypothalamus induces the secretion of several hormones produced by the anterior lobe of the pituitary gland. (The pituitary gland, which is attached to the hypothalamus by the pituitary stalk, consists of two distinct parts: the anterior lobe, or adenohypophysis, and the posterior lobe, or neurohypophysis.) The hypothalamus does so by synthesizing a series of specific peptide-releasing hormones. Releasing hormones then pass into a specialized capillary bed referred to as the hypothalamohypophyseal portal system, which transports them directly to the adenohypophysis. Each releasing hormone stimulates specific cells to synthesize and

TABLE 16.1
Selected Mammalian Hormones

Source	Hormone	Function
Hypothalamus	Gonadotropin-releasing hormone* (GnRH)	Stimulates LH and FSH secretion
	Corticotropin-releasing hormone* (CRH)	Stimulates ACTH secretion
	Growth hormone–releasing hormone* (GHRH)	Stimulates GH secretion
	Somatostatin*	Inhibits GH and TSH secretion
	Thyrotropin-releasing hormone* (TRH)	Stimulates TSH and prolactin secretion
Pituitary	Luteinizing hormone* (LH)	Stimulates cell development and synthesis of sex hormones in ovaries and testes
	Follicle-stimulating hormone* (FSH)	Promotes ovulation and estrogen synthesis in ovaries and sperm development in testes
	Corticotropin* (ACTH) (adrenocorticotropic hormone)	Stimulates steroid synthesis in adrenal cortex
	Growth hormone* (GH)	General anabolic effects in many tissues
	Thyrotropin* (TSH) (thyroid-stimulating hormone)	Stimulates thyroid hormone synthesis
	Prolactin*	Stimulates milk production in mammary glands and assists in the regulation of the male reproductive system
	Oxytocin*	Uterine contraction and milk ejection
	Vasopressin*	Blood pressure and water balance
Gonads	Estrogens† (estradiol)	Maturation and function of reproductive system in females
	Progestins† (progesterone)	Implantation of fertilized eggs and maintenance of pregnancy
	Androgens† (testosterone)	Maturation and function of reproductive system in males
Adrenal cortex	Glucocorticoids† (cortisol, corticosterone)	Diverse metabolic effects as well as inhibiting the inflammatory response
	Mineralocorticoids† (aldosterone)	Mineral metabolism
Thyroid	Triiodothyronine‡ (T_3)	General stimulation of many cellular reactions
	Thyroxine‡ (T_4) (after conversion to T_3)	
Gastrointestinal tract	Gastrin*	Stimulates secretion of stomach acid and pancreatic enzymes
	Secretin*	Regulates pancreatic exocrine secretions
	Cholecystokinin*	Stimulates secretion of digestive enzymes and bile
	Somatostatin*	Inhibits secretion of gastrin and glucagon
Pancreas	Insulin*	General anabolic effects including glucose uptake and lipogenesis
	Glucagon*	Glycogenolysis and lipolysis
	Somatostatin*	Inhibits the secretion of glucagon

*Peptide or polypeptide.

†Steroid.

‡Amino acid derivative.

secrete one or more types of hormone. For example, the tripeptide thyrotropin-releasing hormone (TRH) stimulates the secretion of TSH and prolactin. (Prolactin promotes milk production in new mothers. Ordinarily, prolactin release is prevented by other hormonal and neural factors.) The hormones of the anterior pituitary are sometimes referred to as tropic ("to turn" or "to change"), because they stimulate the synthesis and release of hormones from other endocrine glands. For example TSH stimulates the thyroid gland to release the thyroid hormones T_3 and T_4.

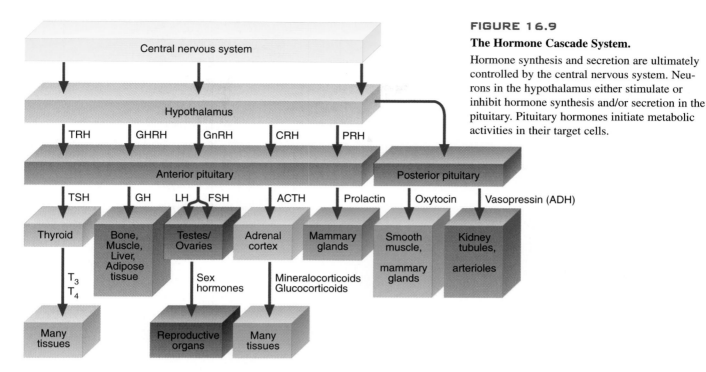

FIGURE 16.9

The Hormone Cascade System.

Hormone synthesis and secretion are ultimately controlled by the central nervous system. Neurons in the hypothalamus either stimulate or inhibit hormone synthesis and/or secretion in the pituitary. Pituitary hormones initiate metabolic activities in their target cells.

The anatomy and function of the posterior pituitary differ from those of the anterior lobe. The hormones secreted by the neurohypophysis (i.e., the peptides vasopressin and oxytocin; see Section 5.2) are actually synthesized in separate types of neurons that originate in the hypothalamus. After their synthesis, both hormones are packaged with associated proteins called neurophysins into secretory granules. The granules then migrate down the axons into the posterior lobe. They are secreted into the bloodstream when an action potential reaches the nerve endings and initiates exocytosis. Oxytocin is secreted as a response to nerve impulses initiated by the stretching of the uterus late in pregnancy and suckling during breast feeding. Vasopressin (antidiuretic hormone) is secreted in response to neural signals from specialized cells called osmoreceptors, which are sensitive to changes in blood osmolality.

Animals employ several mechanisms to prevent excessive hormone synthesis and release. The most prominent of these is feedback inhibition. The hypothalamus and anterior pituitary are controlled by the target cells they regulate. For example, TSH release by the anterior pituitary is inhibited when blood levels of T_3 and T_4 rise (Figure 16.10). The thyroid hormones inhibit the responsiveness of TSH-synthesizing cells to TRH. In addition, several tropic hormones inhibit the synthesis of their releasing factors.

Target cells also possess mechanisms that protect against overstimulation by hormones. In a process referred to as **desensitization**, target cells adjust to changes in stimulation levels by decreasing the number of cell surface receptors or by inactivating those receptors. The reduction in cell surface receptors in response to stimulation by specific hormone molecules is called **down-regulation**. In down-regulation, receptors are internalized by endocytosis. Depending on cell type and several metabolic factors, the receptors may eventually be recycled to the cell surface or be degraded. If degraded, proteins must be synthesized to replace receptors. Some disease states are caused by or associated with target cell insensitivity to specific hormones. For example, some cases of diabetes are associated with **insulin resistance**, caused by a decrease in functional insulin receptors. (Diabetes is discussed in Special Interest Box 16.3.)

KEY CONCEPTS 16.4

Many of the hormones in the mammalian body are controlled by a complex cascade mechanism and ultimately regulated by the central nervous system.

FIGURE 16.10

Feedback Inhibition.

The secretion of thyroxine from the thyroid is stimulated by TSH. The secretion of TSH is in turn stimulated by TRH. As blood levels of thyroxine rise, TRH secretion is inhibited.

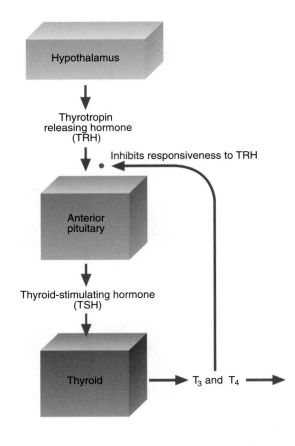

| QUESTION 16.5 | Cortisone is one of several synthetic adrenal glucocorticosteroids that are prescribed for the treatment of inflammatory and allergic disorders. It can be administered orally. In contrast, the hormone insulin, used in the treatment of diabetes mellitus, can only be injected. Explain. |

| QUESTION 16.6 | When a hormone molecule binds reversibly to its receptor in or on a target cell, it does so through noncovalent interactions. Describe the types of interactions that are most likely involved in this binding process. |

| QUESTION 16.7 | The thymus gland is a bilobed organ found just above the heart. It promotes the differentiation of certain lymphocytes to form T cells. (Recall that T cells confer cellular immunity.) In addition, the thymus gland secretes several thymic hormones that stimulate T cell function after these cells have left the thymus gland.

Physical and emotional stress are now known to depress cellular immunity. For example, the risk of developing cancer and severe infections increases with the duration of chronic stress. Although the mechanism by which stress induces this change in function is not clearly understood, elevated blood levels of glucocorticoids (e.g., cortisol) are believed to play an important role. Thymic hormone secretion is depressed when glucocorticoid levels are high. In addition, circadian rhythms alter thymic hormone and glucocorticoid levels. (Thymic hormone levels are higher and glucocorticoid levels are lower at night than during the day.) Assuming that glucocorticoids depress thymic hormone secretion, outline (in general terms) how thymic hormone is affected by stressful situations and by light/dark cycles. What other hormones are involved in both of these processes? (*Hint*: Review melatonin synthesis in Chapter 14.) |

Growth Factors

The survival of multicellular organisms requires that cell growth and cell division (mitosis) be rigorously controlled. The conditions that regulate these processes have not yet been completely resolved. However, a variety of hormonelike polypeptides and proteins, called **growth factors** (or **cytokines**), are now believed to regulate the growth, differentiation, and proliferation of various cells. Often, the actions of several growth factors are required to promote cellular responses. Growth factors differ from hormones in that they are synthesized by a variety of cell types rather than by specialized glandular cells. Examples of mammalian growth factors include epidermal growth factor (EGF), platelet-derived growth factor (PDGF), and the somatomedins. Similar molecules such as the interleukins, which promote cell proliferation and differentiation within the immune system, are also considered cytokines. Several growth-suppressing molecules have also been characterized. The mechanisms by which the cytokines exert their effects are not clearly understood. The few aspects of cytokine action that have been identified resemble some of those observed with hormones (Section 16.5). Although some growth factors are found in circulating blood, most have paracrine and/or autocrine activity.

Epidermal growth factor (EGF) (6.4 kD), one of the first cellular growth factors identified, is a **mitogen** (a stimulator of cell division) for a large number of epithelial cells, such as epidermal and gastrointestinal lining cells. EGF triggers cell division when it binds to plasma membrane EGF receptors, which are transmembrane tyrosine kinases structurally similar to insulin receptors (described on p. 558).

Platelet-derived growth factor (PDGF) (31 kD) is secreted by blood platelets during the clotting reaction. Acting with EGF, PDGF stimulates mitosis in fibroblasts and other nearby cells during wound healing. PDGF also promotes collagen synthesis in fibroblasts.

The **somatomedins** are a group of polypeptides that mediate the growth-promoting actions of GH. Produced in the liver and in a variety of other tissue cells (e.g., muscle, fibroblasts, bone, and kidney) when GH binds to its cell surface receptor, the somatomedins are the major stimulators of growth in animals. The somatomedins are secreted by the liver into the bloodstream, in contrast to the paracrine action exhibited by other somatomedin-producing cells. In addition to stimulating cell division, the somatomedins promote (but to a lesser degree) the same metabolic processes as does the hormone insulin (e.g., glucose transport and fat synthesis). For this reason the two somatomedins found in humans have been renamed **insulin-like growth factors I and II** (IGF-I and IGF-II). Like other polypeptide growth factors, the somatomedins trigger intracellular processes by binding to cell surface receptors. Not surprisingly, somatomedin receptors are also tyrosine kinases.

Interleukin-2 (IL-2) (13 kD) is a member of a group of cytokines that regulate the immune system in addition to promoting cell growth and differentiation. IL-2 is secreted by T cells after they have been activated by binding to a specific antigen-presenting cell. These cells are also stimulated to produce IL-2 receptors. The binding of IL-2 to these receptors stimulates cell division so that numerous identical T cells are produced. This process, as well as other aspects of the immune response, continues until the antigen is eliminated from the body.

Several cytokines are growth inhibitors. The **interferons** are a group of polypeptides produced by a variety of cells in response to several stimuli, such as antigens, mitogens, viral infections, and certain tumors. The type I interferons protect cells from viral infection by stimulating the phosphorylation and inactivation of a protein factor required to initiate protein synthesis. Type II interferons, produced by T lymphocytes, inhibit the growth of cancerous cells in addition to having several immunoregulatory effects. As their names imply, the **tumor necrosis factors** (TNF) are toxic to tumor cells. Both TNF-α (produced by antigen-activated phagocytic white blood cells) and TNF-β (produced by activated T cells) suppress cell division. They may also have a role in regulating several developmental processes.

KEY CONCEPTS 16.5

Growth factors are a group of hormone-like polypeptides and proteins that promote cell growth, cell division, and differentiation.

Because hormones are so influential in the regulation of metabolic processes, it is not surprising that there are numerous hormone-related diseases. In general, such diseases are caused by either overproduction or underproduction of a specific hormone or the insensitivity of target cells.

Hormone Overproduction

The oversecretion of hormone molecules is most often caused by a tumor. Several types of pituitary tumor cause endocrine diseases. For example, one of the most common causes of Cushing's disease is an abnormal proliferation of ACTH-producing cells. *Cushing's disease* is characterized by obesity, hypertension, and elevated blood glucose levels. Patients with Cushing's disease develop a characteristic appearance: a puffy "moon face" and a "buffalo hump" caused by fat deposits between the shoulders. Occasionally, Cushing's disease is caused by adrenocortical tumors.

Depending on the patient's age, pituitary tumors that develop from somatroph cells (the cells that synthesize GH) cause either gigantism or acromegaly. *Gigantism*, marked by pronounced growth of the long bones, is caused by excessive secretion of growth hormone (GH) during childhood. In adulthood, excessive GH production causes *acromegaly*, in which connective tissue proliferation and bone thickening result in coarsened and exaggerated facial features and enlarged hands and feet.

Not all hypersecretion diseases are caused by tumors. For example, *Graves' disease*, the most common type of hyperthyroidism, is an autoimmune disease. For unknown reasons, autoantibodies are produced that bind to TSH receptors in the thyroid gland. (These antibodies are called *long-acting thyroid stimulators*, or LATS.) The resulting excessive production of thyroid hormone causes *thyrotoxicosis*, characterized by *goiter* (an enlarged thyroid gland) and *exophthalmos* (abnormal eyeball protrusion). Affected individuals become more sensitive to catecholamines. (Apparently, the number of catecholamine receptors is increased in the heart and the central nervous system.) During stressful situations (e.g., surgery or myocardial infarction) when large amounts of catecholamines are released, a life-threatening "thyroid storm" may occur. The consequences of a thyroid storm include agitation, delirium, coma, and heart failure.

Hormone Underproduction

Inadequate hormone production has a variety of causes. The most common are the autoimmune destruction of hormone-producing cells, genetic defects, and an inadequate supply of precursor molecules.

As described, in Addison's disease adrenal cortex function is inadequate. The most common cause of Addison's disease is the autoimmune destruction of the adrenal gland. (The major cause of adrenal insufficiency before the 1920s was tuberculosis of the adrenal gland.) Addison's disease also results from prolonged glucocorticoid therapy. (Recall that certain glucocorticoids are used as anti-inflammatory drugs.)

Hypothyroidism (thyroid hormone deficiency) may result from autoimmune disease (*Hashimoto's disease*) or from deficient synthesis of TSH or TRH (thyroid-stimulating hormone–releasing factor). Because adequate ingestion of iodine is a prerequisite for thyroid hormone synthesis, iodine deficiency also causes hypothyroidism. In children, thyroid hormone deficiency (called *cretinism*) causes depressed growth and mental retardation. Severe hypothyroidism in adults (*myxedema*) results in symptoms such as edema (abnormal fluid accumulation) and goiter. Hypothyroidism is usually treated with hormone replacement therapy.

Growth hormone deficiency may be hereditary or a consequence of a pituitary tumor or head trauma. Congenital GH deficiency results in shortened stature (*dwarfism*). Currently, affected children are treated with commercially produced GH, a recombinant DNA product. In *Laron's dwarfism*, exogenous GH has no effect on the patient's cells because of a defective GH response mechanism.

Diabetes insipidus is characterized by the passage of copious amounts of very dilute urine. It is caused by either an inadequate synthesis of vasopressin or the failure of the kidney to respond to vasopressin. The most common causes of diabetes insipidus are tumors and surgical procedures in which the neurohypophyseal nerve tracts are cut. In several forms of kidney disease, the organ's capacity to respond to vasopressin is compromised.

16.5 MECHANISMS OF HORMONE ACTION

Hormones typically initiate their actions within target cells by binding to a receptor. Water-soluble hormone molecules (e.g., polypeptides, proteins, and epinephrine) bind to receptor molecules on the outer surface of the target cell's plasma membrane. This binding process, which is reversible, triggers a mechanism that initiates a phosphorylation cascade either directly (e.g., insulin) or indirectly via second messenger molecules (e.g., glucagon). As a result, the activities of specific enzymes and/or membrane transport mechanisms are altered. Examples of well-researched second messengers include cyclic AMP (cAMP), cyclic GMP (cGMP), the phosphatidylinositol-4,5-bisphosphate derivatives diacylglycerol (DAG) and inositol trisphosphate (IP_3), and calcium ions. In contrast, the lipid-soluble hormones, such as the steroid and thyroid hormones, enter target cells, where they bind to specific receptor molecules. Each hormone-receptor complex then binds to specific regions of the target cell's DNA. Such binding modifies gene expression and leads to a change in the protein profile of the cell.

In Section 16.5, each type of hormone mechanism is briefly described. This is followed by a discussion of the insulin receptor, a critically important cellular component.

The Second Messengers

When a hormone molecule binds to a plasma membrane receptor, an intracellular signal called a second messenger is generated. The second messenger actually delivers the hormonal message. This process, referred to as **signal transduction**, also amplifies the original signal. In other words, a few hormone molecules initiate a mechanism that leads to the production of some multiple of second messenger molecules. The following second messengers are described: cAMP, cGMP, DAG, IP_3, and Ca^{2+}.

cAMP cAMP is generated from ATP by adenylate cyclase in response to a hormone-receptor interaction. The interaction between the receptor and adenylate cyclase is mediated by a G protein (Figure 16.11). (The receptor, G protein, and adenylate cyclase are all membrane-associated proteins.) **G proteins** are so named because they bind guanine nucleotides. A variety of G proteins have been characterized. The G protein that stimulates cAMP synthesis when hormones such as glucagon, TSH, and epinephrine bind is referred to as G_s. G_i inhibits adenylate cyclase and therefore decreases cAMP levels. In their unstimulated state, G proteins bind GDP. As a consequence of hormone binding and the resulting conformational change, the receptor interacts with a nearby G_s protein. As G_s binds to the receptor, GDP dissociates and is replaced by GTP. The activated G_s protein interacts with and stimulates adenylate cyclase. (G proteins usually contain three subunits: α, β, and γ. The α_s subunit binds guanine nucleotides, has GTPase activity, and activates adenylate cyclase when dissociated from the $\beta\gamma$-dimer.) The cAMP molecules formed by adenylate cyclase diffuse into the cytoplasm, where they bind to and activate cAMP-dependent protein kinase. Active protein kinase then phosphorylates and thereby alters the catalytic activity of key regulatory enzymes. Adenylate cyclase remains active only as long as it interacts with α_s-GTP. Once GTP is hydrolyzed to GDP, α_s-GDP dissociates from adenylate cyclase and reassociates with the $\beta\gamma$-dimer. The cAMP is quickly hydrolyzed by phosphodiesterase. This is a critical requirement of a second messenger; that is, once generated, the signal must be terminated rapidly.

The target proteins affected by cAMP depend on the cell type. In addition, several hormones may activate the same G protein. Therefore different hormones may elicit the same effect. For example, glycogen degradation in liver cells is initiated by both epinephrine and glucagon.

Some hormones inhibit adenylate cyclase activity. Such molecules depress cellular protein phosphorylation reactions because their receptors interact with G_i protein. When G_i is activated, its α_i subunit dissociates from the $\beta\gamma$-dimer and prevents the activation of adenylate cyclase. For example, because its receptors in adipocytes are associated with G_i, PGE_1 (prostaglandin E_1) depresses lipolysis. (Recall that lipolysis is stimulated by glucagon and epinephrine.)

cGMP Although cGMP is synthesized in almost all animal cells, its role in cellular metabolism is still relatively undefined. However, the following information is known. cGMP is synthesized from GTP by guanylate cyclase. Two types of guanylate cyclase are involved in signal transduction. In one type, which is membrane-bound, the extracellular domain of the enzyme is a hormone receptor. The other type is a cytoplasmic enzyme.

Two types of molecule are now known to activate membrane-bound guanylate cyclase: atrial natriuretic peptide and bacterial enterotoxin. *Atrial natriuretic factor* (ANF) is a peptide that is released from heart atrial cells in response to increased blood volume. The biological effects of ANF, that is, lowering of blood pressure

(continued on p. 552)

FIGURE 16.11

The Adenylate Cyclase Second Messenger System that Controls Glycogenolysis.

When the receptor is unoccupied, the G_s protein α_s subunit has GDP bound and is complexed with the β,γ-dimer. The binding of hormone (1) activates the receptor and leads to replacement of GDP with GTP (2). The activated subunit interacts with and activates adenylate cyclase. (3) The cAMP produced binds to cAMP-dependent protein kinase. Signal transduction ends when the ligand leaves the receptor, the bound GTP is hydrolyzed to GDP by the GTPase activity within the α_s subunit, and the α_s subunit dissociates from adenylate cyclase. Cyclic AMP is deactivated by hydrolysis to AMP, a reaction catalyzed by phosphodiesterase. (4) The α_s subunit then reassociates with the β,γ-dimer.

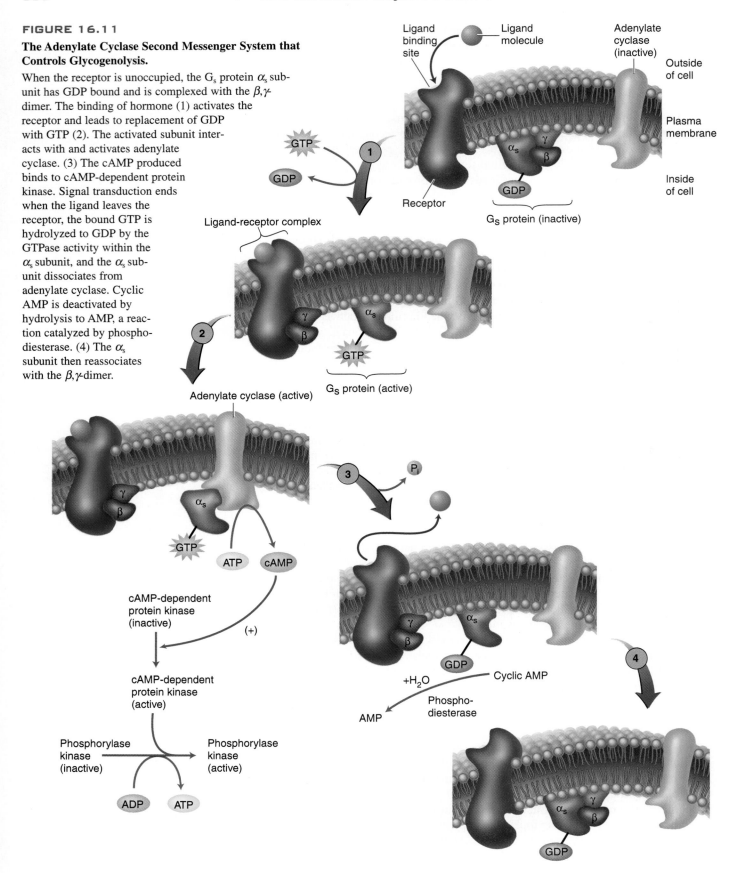

Diabetes mellitus is a group of devastating metabolic diseases caused by insufficient insulin synthesis, increased insulin destruction, or ineffective insulin action. All of its metabolic effects result when the body's cells fail to acquire glucose from the blood. The metabolic imbalances that occur have serious, if not life-threatening, consequences (Figure 16A). In insulin-dependent diabetes mellitus (IDDM), also called type I diabetes, inadequate amounts of insulin are secreted because the β-cells of the pancreas have been destroyed. Because IDDM usually occurs before the age of 20, it has (until recently) been referred to as juvenile-onset diabetes. Noninsulin-dependent diabetes mellitus (NIDDM), also called type II or adult-onset diabetes, is caused by the insensitivity of target tissues to insulin. Although these forms of diabetes share some features, they differ significantly in others. Once quite rare, diabetes is now the third leading cause of death in the United States, where it afflicts at least 5% of the population.

The most obvious symptom of diabetes is **hyperglycemia** (high blood glucose levels), caused by inadequate cellular uptake of glucose. Among diabetics the severity of hyperglycemia may vary considerably. Because the kidney's capacity to reabsorb glucose from the urinary filtrate is limited, glucose appears in the urine (**glucosuria**). (The kidneys filter blood and then reabsorb substances such as glucose, amino acids, and ions from the filtrate.) Glucosuria

x-ref

FIGURE 16A **The Metabolic Consequences of Insulin Deficiency or Resistance.**

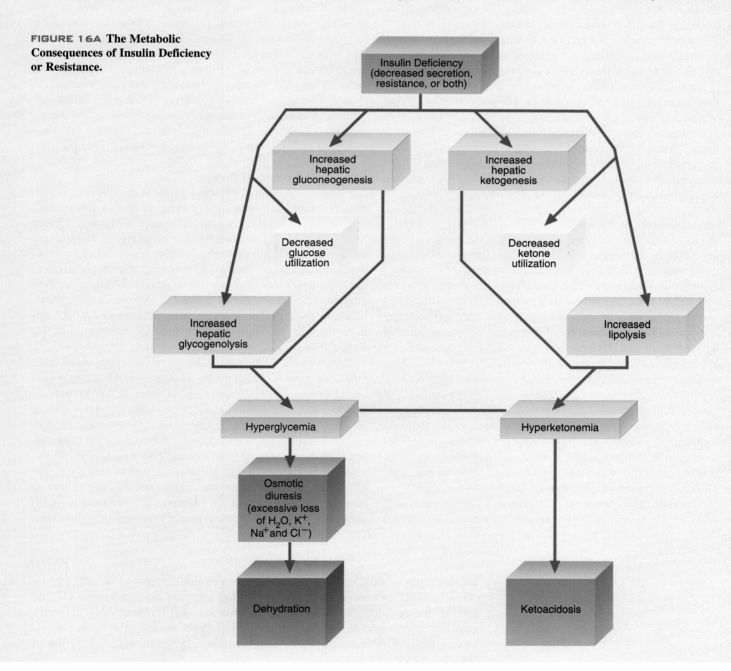

results in **osmotic diuresis**, a process in which an excessive loss of water and electrolytes (Na^+, K^+, and Cl^-) is caused by the presence of solute in the filtrate. In severe cases, despite the consumption of large volumes of fluid, patients may become dehydrated.

Without insulin to regulate fuel metabolism, its three principal target tissues (liver, adipose tissue, and muscle) fail to absorb nutrients appropriately. Instead, these tissues function as if the body were undergoing starvation. In the liver, gluconeogenesis accelerates because large amounts of amino acids are mobilized from muscle and synthesis of gluconeogenic enzymes (Section 8.2) increases. Glycogenolysis (ordinarily suppressed by insulin) produces additional glucose. The liver delivers these glucose molecules to an already hyperglycemic bloodstream. Increased lipolysis (Section 12.1) in adipose tissue (caused by the unopposed action of glucagon) releases large quantities of fatty acids into blood. In the liver, because these molecules are degraded by β-oxidation in combination with low concentrations of oxaloacetate (caused by excessive gluconeogenesis), large amounts of acetyl-CoA, the substrate for forming ketone bodies, are produced. Fatty acids not used to generate energy or ketone bodies are used in VLDL synthesis. This process causes hyperlipoproteinemia (high blood concentrations of lipoproteins) because lipoprotein lipase synthesis is depressed when insulin is lacking. In effect, many of the cells of diabetics "starve in the midst of plenty." In severe cases, despite increased appetite and food consumption, the failure to use glucose effectively eventually causes body weight changes.

Insulin–Dependent Diabetes

In most cases of insulin-dependent diabetes, the insulin-producing β-cells have been destroyed by the immune system. Although the symptoms of IDDM often manifest themselves abruptly, it now appears that β-cell destruction is caused by an inflammatory process over several years. The symptoms are not obvious until virtually all insulin-producing capacity is destroyed. As in other inflammatory and autoimmune processes, β-cell destruction is initiated when an antibody binds to a cell surface antigen. One of the most common autoantibodies found in type I diabetes is now believed to bind specifically to an antigen with glutamate decarboxylase activity. (Recall that glutamate decarboxylase catalyzes the synthesis of GABA from glutamate.) Autoantibodies to insulin and the tyrosine phosphatase IA-2 have also been detected. (Tyrosine phosphatase IA-2, one of several enzymes that remove phosphate groups from specific phosphotyrosine residues in key regulatory proteins, is found only in brain and insulin-producing cells in the pancreas.) The significance of this phenomenon is unknown.

The most serious acute symptom of type I diabetes is **ketoacidosis**. Elevated concentrations of ketones in the blood (**ketosis**) and low blood pH along with hyperglycemia cause excessive water losses. (The odor of acetone on a patient's breath is characteristic of ketoacidosis.) Ketoacidosis and dehydration, if left untreated, can lead to coma and death. IDDM patients are treated with injections of insulin obtained from animals or from recombinant DNA technology. Before Frederick Banting and Charles Best discovered insulin in 1922, most type I diabetics died within a year after being diagnosed. Although exogenous insulin prolongs life, it is not a cure. Most diabetics have a shortened life span because of the long-term complications of their disease.

Many researchers now believe that IDDM is caused by both genetic and environmental factors. Although the precise cause is still unknown, individuals who have inherited certain genetic markers are at high risk for developing the disease. Certain HLA antigens, that is, HLA-DR3 and HLA-DR4, are found in a large majority of type I diabetics. The *HLA* or *histocompatibility antigens*, found on the surface of most of the body's cells, play an important role in determining how the immune system reacts to foreign substances or cells. If a type I patient with both HLA-DR3 and HLA-DR4 antigens has an identical twin, the twin's risk for developing the disease is approximately 50%. That the risk is not 100% suggests that some environmental exposure is required for developing type I diabetes.

Non–Insulin–Dependent Diabetes

Non-insulin-dependent diabetes is a milder disease than the insulin-dependent form. Its onset is slow, often occurring after the age of 40. In contrast to type I patients, most individuals with type II diabetes have normal or often elevated blood levels of insulin. For a variety of reasons, type II patients are resistant to insulin. The most common cause of insulin resistance is the down-regulation of insulin receptors. (Defective insulin receptors or improper insulin receptor processing also causes type II diabetes in some patients.) Approximately 85% of type II diabetics are obese. Because obesity itself promotes tissue insensitivity to insulin, individuals who are prone to this form of diabetes are at risk for the disease when they gain weight.

Treatment of NIDDM usually consists of diet control and exercise. Often, obese patients become more sensitive to insulin (i.e., there is an up-regulation of insulin receptors) when they lose weight. Because sustained muscular activity increases the uptake of glucose without requiring insulin, exercise also decreases hyperglycemia. In some cases, oral hypoglycemic drugs are used. It is believed that these molecules promote additional insulin release from the pancreas.

When the failure of type II diabetic patients to control hyperglycemia is accompanied by other serious medical conditions (e.g., renal insufficiency, myocardial infarction, or infections), a serious metabolic state referred to as **hyperosmolar hyperglycemic non-**

via vasodilation and diuresis (increased urine excretion), appear to be mediated by cGMP. cGMP activates the phosphorylating enzyme protein kinase G. The role of this enzyme in mediating ANF effects is still unclear. ANF activates guanylate cyclase in several cell types. In one type, those in the kidney's collecting tubules, ANF-stimulated cGMP synthesis increases renal excretion of Na^+ and water.

ketosis (HHNK) can result. (Ketoacidosis is rare in type II diabetes.) Because of the additional metabolic stress, insulin resistance is exacerbated, and blood glucose levels rise. The patient can then become severely dehydrated. The resulting lower blood volume depresses renal function, which causes further increases in blood glucose concentrations. Eventually, the patient becomes comatose. Because the onset is slow, it may not be recognized until the patient is severely dehydrated. (This is especially true for elderly diabetics, who often have a depressed thirst mechanism.) For this reason, HHNK is often more life-threatening than ketoacidosis.

Long-Term Complications of Diabetes

Despite the efforts of physicians and patients to control the symptoms of diabetes, few diabetics avoid the long-term consequences of their disease. Diabetics are especially prone to develop kidney failure, myocardial infarction, stroke, blindness, and neuropathy. In *diabetic neuropathy*, nerve damage causes the loss of sensory and motor functions. In addition, circulatory problems often cause gangrene, which leads to tens of thousands of amputations annually.

Most diabetic complications stem from damage to the vascular system. For example, damaged capillaries in the eye and kidney lead to blindness and kidney damage, respectively. Similarly, the accelerated form of atherosclerosis found in diabetics leads to serious cases of myocardial infarction and stroke. It is now believed that most of this damage is initiated by hyperglycemia. High blood glucose levels promote the nonenzymatic glycosylation of protein molecules, a process that is referred to as the Maillard reaction. The **Maillard reaction** (named for the French chemist who discovered it in 1912) is initiated when a sugar aldehyde or ketone group condenses with a free amino group to form a Schiff base (Figure 16B). The Schiff base rearranges to form a stable ketoamine called the **Amadori product**. Amadori products eventually degrade into reactive carbonyl-containing products. These reactive products react with free amino groups and other amino acid side chains on nearby proteins to form a complex series of cross-linkages and adducts. (An **adduct** is the product of an addition reaction. In addition reactions, two molecules react to form a third molecule.) When levels of blood glucose are high, glycosylation end products accumulate and cause extensive damage throughout the cardiovascular system. When cells in the lining of blood vessels are damaged, a repair process involving macrophages and growth factors is initiated that inadvertently promotes atherosclerosis. The capacity of affected blood vessels to nourish nearby tissue is eventually compromised.

Diabetics are also damaged by another consequence of hyperglycemia: increased metabolism by the sorbitol pathway. In some cells that do not require insulin for glucose uptake (e.g., Schwann cells in peripheral nerve and ocular tissues such as lens epithelial cells), the hexose enters by facilitated transport down its concentration gradient. Glucose is then converted to sorbitol (p. 208) by the NADPH-requiring enzyme aldose dehydrogenase. The oxidation of some sorbitol molecules to form fructose by sorbitol dehydrogenase is coupled to the reduction of NAD^+. Increased concentrations of NADH promote the reduction of pyruvate to form lactate. The accumulation of sorbitol and the redox changes caused by sorbitol oxidation are associated with several pathological changes in diabetics, for example, nerve damage and cataract formation. The physiological significance of the sorbitol pathway in normal cells is unknown.

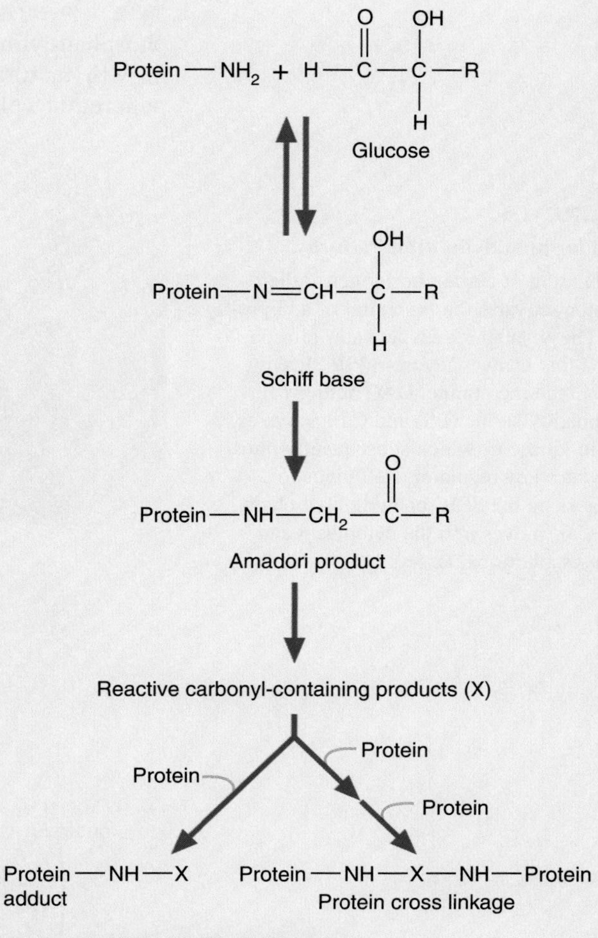

FIGURE 16B The Maillard Reaction.
Any amino group–containing molecule can undergo the Maillard reaction, so nucleotides and amines also react with glucose molecules. Proteins, however, are most susceptible to this process.

The term *diabetes,* derived from the Greek word *diabeinein* ("to go to excess"), was first used by Aretaeus (A.D. 81–138) to identify a group of symptoms that included intolerable thirst and "a liquefaction of the flesh and limbs into urine." After reviewing Special Interest Box 16.3, explain the physiological and biochemical basis for Aretaeus's findings.

QUESTION 16.8

The binding of *enterotoxin* (produced by several bacterial species) to another type of guanylate cyclase found in the plasma membrane of intestinal cells causes diarrhea. For example, one form of traveler's diarrhea is caused by a strain of *E. coli* that produces *heat stable enterotoxin*. The binding of this toxin to an enterocyte plasma membrane receptor linked to guanylate cyclase triggers excessive secretion of electrolytes and water into the lumen of the small intestine.

Cytoplasmic guanylate cyclase possesses a heme prosthetic group. The enzyme is activated by Ca^{2+}, so any rise in cytoplasmic Ca^{2+} causes cGMP synthesis. This guanylate cyclase activity is activated by NO (Special Interest Box 10.1). Some evidence suggests that binding NO to the heme group activates the enzyme. In several cell types (e.g., smooth muscle cells), cGMP stimulates the functioning of ion channels.

THE PHOSPHATIDYLINOSITOL CYCLE AND CALCIUM The phosphatidylinositol cycle (Figure 16.12) mediates the actions of hormones and growth factors. Examples include acetylcholine (e.g., insulin secretion in pancreatic cells), vasopressin, TRH, GRH, and epinephrine (α_i receptors).

FIGURE 16.12

The Phosphatidylinositol Pathway.

The binding of certain hormones to their receptor activates the α subunit of a G protein. The α subunit then activates phospholipase C that cleaves IP_3 from PIP_2, leaving DAG in the membrane. DAG acting with phosphatidylserine (PS) and Ca^{2+} activates protein kinase C, which subsequently phosphorylates key regulators. IP_3 binds to receptors on the SER, opening Ca^{2+} channels. Ca^{2+} moves into the cytoplasm and activates additional targets.

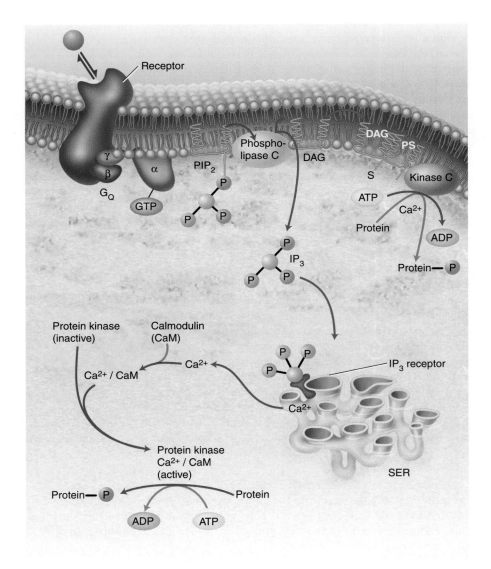

Phosphatidylinositol-4,5-bisphosphate (PIP_2) is cleaved by phospholipase C to form the second messengers DAG (diacylglycerol) and IP_3 (inositol-1,4, 5-triphosphate). Phospholipase C is activated by a hormone-receptor complex induced G protein activation. Several types of G protein may be involved in the phosphatidylinositol cycle. For example, G_Q (shown in Figure 16.12) mediates the actions of vasopressin.

The DAG product of the phospholipase C–catalyzed reaction activates protein kinase C. Several protein kinase C activities have been identified. Depending on the cell, activated protein kinase C phosphorylates specific regulatory enzymes, thereby activating or inactivating them.

Once it is generated, IP_3 diffuses to the calcisome (SER), where it binds to a receptor. The IP_3 receptor is a calcium channel. Cytoplasmic calcium levels then rise as calcium ions flow through the activated open channel. Recent evidence indicates that the IP_3-stimulated calcium signal is potentiated for a brief time by the release of another signal (as yet uncharacterized) that activates a plasma membrane calcium channel. Calcium ions are involved in the regulation of a large number of cellular processes including contributing to the activation of plasma membrane–associated protein kinase C. Because calcium levels are still relatively low even when the calcium release mechanism has been activated (approximately 10^{-6} M), the calcium-binding sites on calcium-regulated proteins must have a high affinity for the ion. Several calcium-binding proteins modulate the activity of other proteins in the presence of calcium. Calmodulin, a well-researched example of the calcium-binding proteins, mediates many calcium-regulated reactions. In fact, calmodulin is a regulatory subunit for some enzymes (e.g., phosphorylase kinase, which converts phosphorylase b to phosphorylase a in glycogen metabolism).

KEY CONCEPTS 16.6

When a water-soluble hormone binds to a plasma membrane receptor, an intracellular signal called a second messenger is generated.

QUESTION 16.9

The binding of a hormone molecule to its target cell receptor initiates a second messenger–dependent enzyme cascade that has five steps between the primary signal (hormone) and the activation of a certain regulatory enzyme (E_R). If each amplification step results in a tenfold increase in activated enzymes, how many molecules of E_R can be activated by a single hormone molecule?

QUESTION 16.10

Explain the sequence of events that occurs when epinephrine triggers the synthesis of cAMP. Once it is formed, cAMP breaks down rapidly. Why is this an important feature for a second messenger in a signal transduction process?

QUESTION 16.11

Cholera toxin causes a massive diarrhea that is life-threatening because it prevents the α subunit of G_s from converting GTP to GDP. Describe why this inhibition leads to the diarrhea. (*Hint*: Refer to Special Interest Box 5.1.)

QUESTION 16.12

Angina pectoris is a very painful symptom of coronary artery disease. In this condition, which is caused by the narrowing of the heart's coronary arteries, insufficient oxygen flow during exertion causes chest pain that radiates to the neck and left arm. Traditionally, angina pectoris has been treated with nitroglycerin. It is now believed that nitroglycerin (Figure 16.13) relieves the pain because it acts to promote vasodilation (increased blood vessel diameter) throughout the body. Vasodilation, which results from smooth muscle relaxation, reduces blood pressure, which in turn reduces the heart's work load. Suggest how nitroglycerin relieves angina pectoris. (*Hint*: Cytoplasmic calcium ions stimulate muscle contraction.)

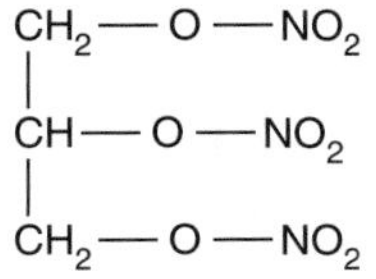

$$CH_2 - O - NO_2$$
$$CH - O - NO_2$$
$$CH_2 - O - NO_2$$
Nitroglycerin

FIGURE 16.13
Nitroglycerine.

QUESTION 16.13

Cancer often results from a multistage process involving an initiating event (mediated by a viral infection or a carcinogenic chemical) followed by exposure to tumor promotors. Tumor promotors, a group of molecules that stimulate cell proliferation, cannot induce tumor formation by themselves. The phorbol esters, found in croton oil (obtained from the seeds of the croton plant, *Croton tiglium*), are potent tumor promotors. (Other examples of tumor promotors include asbestos and several components of tobacco smoke.) In one of the tumor-promoting actions of the phorbol esters, these molecules mimic the actions of DAG. In contrast to DAG, the phorbol esters are not easily disposed of. Explain the possible biochemical consequences of phorbol esters in an "initiated" cell. What enzyme is activated by both DAG and phorbol esters?

Phorbol myrsitate acetate (PMA)
A phorbol ester

Steroid and Thyroid Hormone Mechanisms

The signal transduction mechanisms of hydrophobic hormone molecules such as the steroid and thyroid hormones result in changes in gene expression. In other words, the action of each type of hormone molecule causes the switching on or off of a specific set of genes, which in turn causes changes in the pattern of proteins that an affected cell produces. Steroid and thyroid hormones are lipid-soluble molecules that are transported in the blood to their target cells bound to several types of protein. Examples of steroid transport proteins include transcortin (also called corticosteroid-binding globulin), androgen-binding protein, sex hormone–binding protein, and albumin. In addition to albumin, the thyroid hormones are also transported by thyroid-binding globulin and thyroid-binding prealbumin.

Once they reach their target cells, hydrophobic hormone molecules dissociate from their transport proteins, diffuse through the plasma membrane, and bind to their intracellular receptors (Figure 16.14). These receptors are high affinity ligand-binding molecules that belong to a large family of structurally similar DNA-binding proteins. Depending on the type of hormone involved, initial binding to receptors may occur within the cytoplasm (e.g., glucocorticoid), or the nucleus (e.g., estrogen, androgens, and thyroid hormone). In the absence of hormone, several types of receptor have been observed to form complexes with other proteins. For example, unoccupied glucocorticoid receptors are found in the cytoplasm bound to hsp90. (Recall that hsp is the abbreviation for heat shock proteins. Refer to Special Interest Box 15.1 for a brief discussion of heat shock proteins.) The binding of hsp90 to the glucocorticoid receptor prevents the inappropriate binding of the unoccupied receptor to DNA. When corticosterone binds to the receptor, a change in the receptor's conformation makes it dissociate from hsp90. The two hormone-bound receptors associate to form a functional complex that then moves into the nucleus.

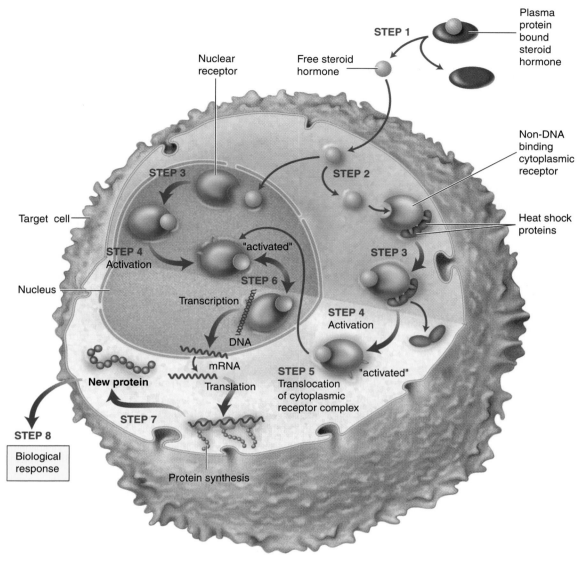

FIGURE 16.14

A Model of Steroid Hormone Action within a Target Cell.

Steroid hormones are transported in blood associated with plasma proteins. When they reach a cell and are released (1), the hormone molecules diffuse through the plasma membrane where they bind to receptor molecules in cytoplasm (2) or nucleus (3). After activation (4), a cytoplasmic hormone-receptor complex migrates to the nucleus (5). The binding of an activated hormone-receptor complex to HRE sequences within DNA (6) results in a change in the rate of transcription of specific genes, and therefore, in the pattern of proteins (7) that the cell produces. The net effect of the steroid hormone is a change in the metabolic functioning of the cell.

Within the nucleus, each hormone-receptor complex binds to specific DNA segments called **hormone response elements** (HRE). The binding of the hormone-receptor complex to the base sequence of an HRE via zinc finger domains (see p. 645) in the receptor either enhances or diminishes the transcription of a specific gene. (*Transcription,* the synthesis of an RNA copy of a gene, is the first major step in the expression of a gene. See Chapter 18.) Several HREs can bind to the same hormone-receptor complex so the expression of numerous genes is altered simultaneously. It has been estimated that each type of HRE can influence the transcription of as many as 50 to 100 genes. Consequently, the binding of a steroid hormone–receptor complex to its cognate HRE induces large-scale changes in cellular function.

KEY CONCEPTS 16.7

Hydrophobic hormones such as the steroids and thyroid hormones diffuse across cellular membranes and bind to intracellular receptors.

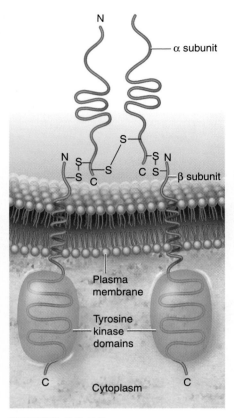

FIGURE 16.15

The Insulin Receptor.

The insulin receptor is a dimer composed of two pairs of α and β subunits. The subunits are connected to each other by disulfide bridges.

Once thyroid hormone enters a cell, it binds temporarily with a specific cytoplasmic protein. Thyroid hormone molecules migrate to the nucleus and mitochondria, where they bind to receptors. In the nucleus the binding of thyroid hormone initiates the transcription of genes that play crucial roles in a variety of cellular processes, such as those that code for growth hormone and Na^+-K^+ ATPase. In mitochondria, thyroid hormones promote oxygen consumption and increased fatty acid oxidation. (The mechanism by which this latter process occurs is not understood.)

The Insulin Receptor

The insulin receptor is a member of a family of cell surface receptors for various anabolic polypeptides, for example, EGF, PDGF, and IGF-I. Although there are several structural differences among this group, they do possess the following structural features in common: an external domain that binds specific extracellular ligands, a transmembrane segment, and a cytoplasmic catalytic domain with tyrosine kinase activity. When a ligand binds to the external domain, a conformational change in the receptor protein activates the tyrosine kinase domain. The tyrosine kinase activity initiates a phosphorylation cascade that begins with an autophosphorylation of the tyrosine kinase domain. Because most research efforts have been devoted to the insulin receptor, its structure and proposed functions are described next.

The insulin receptor (Figure 16.15) is a transmembrane glycoprotein composed of two types of subunits connected by disulfide bridges. Two large α subunits (130 kD) extend extracellularly, where they form the insulin-binding site. Each of the two β subunits (90 kD) contains a transmembrane segment and a tyrosine kinase domain.

Insulin binding activates receptor tyrosine kinase activity and causes a phosphorylation cascade that modulates various intracellular proteins. For example, insulin binding inhibits hormone-sensitive lipase in adipocytes. It apparently does so by activating a phosphatase that dephosphorylates the lipase. In addition, several models of insulin action suggest that several second messengers are employed, for example, inositol monophosphate or DAG, in activating protein kinase C.

Insulin binding appears to initiate a phosphorylation cascade that induces the transfer of several types of protein to the cell surface. Examples of these molecules include certain isoforms of the glucose transporter and the receptors for LDL and IGF-II. The movement of these molecules to the plasma membrane in the postabsorptive phase of the feeding-fasting cycle promotes the cell's acquisition of nutrients and growth-promoting signals.

The detection and measurement of hormones is useful in both clinical and research laboratories. Before the discovery of modern methods, hormones could be detected only indirectly. For example, in the old "rabbit test" for pregnancy, the patient's urine was injected into a female rabbit. The formation of corpora lutea within 24 hours in the animal's ovaries indicated that the patient was pregnant. (The corpus luteum is a structure formed from an ovarian follicle after ovulation.) The hormone that induces this transformation is now known to be human chorionic gonadotropin (HCG). (HCG has biological activity similar to that of luteinizing hormone (LH). It is produced only during pregnancy, by the placenta.) In addition to being awkward and time consuming, bioassays are usually insensitive and imprecise; for example, the rabbit test is useful only after several weeks of pregnancy and is either positive or negative, that is, this test can only demonstrate the presence or absence of a certain substance. It cannot be used to ascertain the concentration of substances. Eventually, a technique called the **radioimmunoassay** (RIA) was developed that could be used to detect and precisely measure vanishingly small amounts of antigen (any molecule for which an antibody can be obtained). In radioimmunoassays the concentration of a specific antigen is determined by measuring the competition of the unlabeled antigen with a known amount of the same antigen that is radiolabeled for binding to an antibody. (There must be too little antibody to bind all antigen molecules.) For many years, RIA was the method of choice in hormone-related research, as well as many other areas of biochemistry. Currently, a cheaper and safer technique called the enzyme-linked immunosorbent assay (ELISA) is preferentially used.

ELISA

Enzyme-linked immunosorbent assays are similar to RIA in that they use antibodies and are about as sensitive. For example, inexpensive ELISA-based pregnancy tests are now available that can measure HCG in urine accurately as early as 2 days after conception.

In general, ELISA (Figure 16C) involves the following steps:

1. An antibody specific for an antigen is attached to an inert surface, for example, the bottom of a well in a polystyrene lab dish.

FIGURE 16C Enzyme-Linked Immunosorbent Assay (ELISA).
Specific biomolecules such as protein hormones are easily identified and their concentrations measured with ELISA. In this technique an antibody that is specific for the biomolecule (called the antigen) is attached to a solid support such as a polystyrene lab dish. The specimen to be tested is added to the dish, which is subsequently rinsed gently to remove unattached material. A second antigen-binding antibody that is covalently linked to an enzyme is then added. The presence and concentration of the biomolecule of interest are determined by measuring the amount of substrate that is converted to product by the enzyme.

2. A small sample of the biological specimen (e.g., blood or urine) is added to the dish. If the sample contains the appropriate antigen, then antigen-antibody binding occurs.
3. A second antibody is added that also specifically binds to the antigen at a site different from that to which the immobilized antibody binds. This antibody is covalently linked to an assayable enzyme.
4. The dish is rinsed to remove any unbound antibody or enzyme-linked antibody molecules.
5. Enzyme assays are used to determine the amount of antigen present. The enzyme converts a colored reagent to a colorless one, or vice versa. The change in color is proportional to the antigen concentration in the sample.

In an alternate method, a specific antigen is attached to the lab dish. Any antibodies capable of binding the antigen present in the biological specimen bind to the immobilized antigen. After unbound antibody is rinsed off, another antibody linked to an assayable enzyme that is specific for binding to the first antibody is added to the dish.

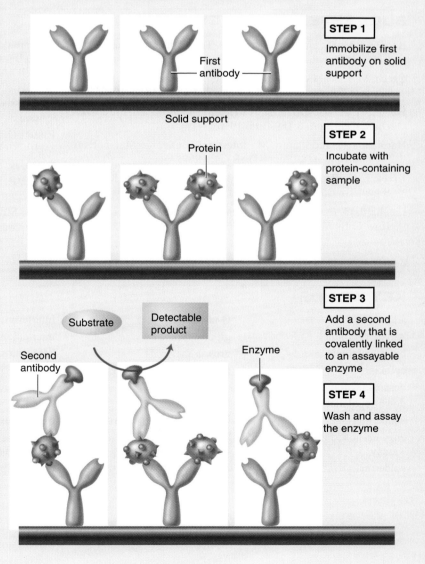

STEP 1
Immobilize first antibody on solid support

First antibody

Solid support

STEP 2
Incubate with protein-containing sample

Protein

STEP 3
Add a second antibody that is covalently linked to an assayable enzyme

Substrate

Detectable product

Second antibody

Enzyme

STEP 4
Wash and assay the enzyme

SUMMARY

1. Multicellular organisms require sophisticated regulatory mechanisms to ensure that all their cells, tissues, and organs cooperate. For example, the feeding-fasting cycle illustrates how a variety of organs contribute to the acquisition of food molecules and their use.

2. Hormones are molecules organisms use to convey information between cells. When target cells are distant from the hormone-producing cell, such molecules are called endocrine hormones. To ensure proper control of metabolism, the synthesis and secretion of many mammalian hormones are regulated by a complex cascade mechanism ultimately controlled by the central nervous system. In addition, a negative feedback mechanism precisely controls various hormone syntheses. A variety of diseases are caused by either overproduction or underproduction of a specific hormone or by the insensitivity of target cells.

3. Growth factors (or cytokines) are a group of polypeptides that regulate the growth, differentiation, or proliferation of various cells. They differ from hormones in that they are often produced by a variety of cell types rather than by specialized glandular cells.

4. Hormones and growth factors usually initiate their effects in a target cell by binding to a specific receptor. Water-soluble hormones typically bind to a receptor on the surface of the target cell. They alter the activities of several enzymes and/or transport mechanisms. Second messenger molecules such as cAMP, cGMP, IP_3, DAG, and Ca^{2+} often mediate a hormone's or growth factor's message. In general, the hydrophobic steroid and thyroid hormones bind to an intracellular receptors. The hormone-receptor complex subsequently binds to a DNA sequence referred to as a hormone response element (HRE). The binding of a hormone-receptor complex to an HRE enhances or diminishes the expression of specific genes.

SUGGESTED READINGS

Bootman, M. D., and Berridge, M. J., The Elemental Principles of Calcium Signaling, *Cell*, 83:675–678, 1995.

Czech, M. P., and Corvera, S., Signaling Mechanisms that Regulate Glucose Transport, *J. Biol. Chem.*, 274:1865–1868, 1999.

Hamm, H. E., and Gilchrist, A., Heterotrimeric G Proteins, *Curr. Opin. Cell Biol.*, 8:189–196, 1996.

Hartmann, F., and Plauth, M., Intestinal Glutamine Metabolism, *Metabolism*, 38(8)(Suppl 1):18–24, 1989.

Krebs, E. G., Role of the Cyclic AMP-Dependent Protein Kinase in Signal Transduction, *J. Am. Med. Assoc.*, 262:1815–1818, 1989.

Lienhard, G. E., Slot, J. W., James, D. E., and Mueckler, M. M., How Cells Absorb Glucose, *Sci. Amer.*, 266(1):86–91, 1992.

Maclaren, N. K., and Atkinson, M. A., Insulin-Dependent Diabetes Mellitus: The Hypothesis of Molecular Mimicry Between Islet Cell Antigens and Microorganisms, *Mol. Med. Today*, 3(2):76–83, 1997.

Rasmussen, H., The Cycling of Calcium as an Intracellular Messenger, *Sci. Amer.*, 261(4):66–73, 1989.

Rutter, G. A., Diabetes: The Importance of the Liver, *Curr. Biol.*, 10(20):R736–R738, 2000.

Smith, K. A., Interleukin-2, *Sci. Amer.*, 262(3):50–57, 1990.

Ulrich, P., and Cerami, A., Protein Glycation, Diabetes, and Aging, *Recent Prog. Hormone Res.*, 56:1–21, 2001.

Uvnas-Moberg, K., The Gastrointestinal Tract in Growth and Reproduction, *Sci. Amer.*, 261(1):78–83, 1989.

KEY WORDS

adduct, *553*

Amadori product, *553*

basal metabolic rate, *542*

cytokine, *547*

desensitization, *545*

down-regulation, *545*

endocrine, *541*

enzyme-linked immunosorbent assay, *559*

epidermal growth factor, *547*

G protein, *549*

glucosuria, *551*

growth factor, *547*

hormone response element, *557*

hyperglycemia, *551*

hyperosmolar hyperglycemic nonketosis, *552*

insulin resistance, *545*

insulinlike growth factor, *547*

interferon, *547*

interleukin-2, *547*

ketoacidosis, *552*

ketosis, *552*

Maillard reaction, *553*

mitogen, *547*

osmotic diuresis, *552*

platelet-derived growth factor, *547*

postabsorptive, *536*

postprandial, *536*

radioimmunoassay, *559*

second messenger, *533*

signal transduction, *549*

somatomedin, *547*

steady state, *531*

target cell, *532*

tumor necrosis factor, *547*

REVIEW QUESTIONS

1. Define the following terms:
   a. ketoacidosis
   b. hyperlipoproteinemia
   c. neurophysin
   d. growth factor
   e. chylomicron remnant
   f. down-regulation
   g. G protein

2. Which organ carries out each of the following activities?
   a. pH regulation
   b. gluconeogenesis
   c. absorption
   d. neural and endocrine integration
   e. lipogenesis
   f. urea synthesis

3. State the action of each of the following hormones:
   a. corticotropin
   b. insulin
   c. glucagon
   d. oxytocin
   e. LH
   f. GnRF
   g. somatostatin
   h. vasopressin
   i. FSH

4. What are the general functions of hormones in the body?

5. NADH is an important reducing agent in cellular catabolism, whereas NADPH is an important reducing agent in anabolism. Review previous chapters and show how the synthesis and degradation of these two molecules are interconnected.

6. Briefly discuss the major classes of second messenger that are now recognized.

7. Describe the general modes of hormone action.

8. What are phorbol esters and how do they promote tumors?

9. After about 6 weeks of fasting, the production of urea is decreased. Explain.

10. Extreme thirst is a characteristic symptom of diabetes. Explain.

11. During periods of prolonged exercise, muscles burn fat released from adipocytes in addition to glucose. Explain how the need for additional fatty acids by muscle is communicated to the adipocytes.

12. Bodybuilders often take anabolic steroids to increase their muscle mass. How do these steroids achieve this effect? (Common side effects of anabolic steroid abuse include heart failure, violent behavior, and liver cancer.)

13. What are four metabolic functions of the kidney? How are these affected by diabetes mellitus?

14. During periods of fasting, some muscle protein is depleted. How is this process initiated and what happens to the amino acids in these proteins?

15. The kidney has an unusually large demand for glutamine and glutamate. How does the metabolism of these compounds help maintain pH balance?

16. Hemoglobin molecules exposed to high levels of glucose are converted to glycosylated products. The most common, referred to as hemoglobin A_{1C} (HbA_{1C}) contains a β-chain glycosylated adduct. Because red blood cells last about 3 months, HbA_{1C} concentration is a useful measure of a patient's blood sugar control. In general terms, describe why and how HbA_{1C} forms.

THOUGHT QUESTIONS

1. Sustained metabolism of fats takes place only in the presence of carbohydrate. Explain.

2. The hypothalamus and the pituitary are two endocrine glands located near one another in the brain. The secretions of the hypothalamus exert a powerful effect on the pituitary, yet if the pituitary is surgically transported to a remote location such as the kidney, the secretions of the hypothalamus have no effect. Suggest a reason for this observation.

3. Explain how a second messenger works. Why use a second messenger rather than simply relying on the original hormone to produce the desired effect?

4. Dieters frequently fast in an attempt to reduce their weight. During these fasts, they often lose considerable muscle mass rather than fat. Why is this so?

5. You are being stalked by a large tiger. Explain how your metabolism responds to help you escape.

6. During fasting in humans, virtually all the glucose reserves are consumed in the first day. The brain requires glucose to function and adjusts only slowly to other energy sources. Explain how the body supplies the glucose required by the brain.

7. In uncontrolled diabetes, levels of hydrogen ions are elevated. Explain how these ions are generated.

8. Why is it important for hormones to act at low concentrations and be metabolized quickly?

9. Hormones are often synthesized and stored in an inactive form within secretory vesicles. Secretion usually occurs only when the hormone-producing cell is stimulated. Explain the advantages that this process has over making the hormone molecules as they are needed.

10. Steroid hormones are often present in cells in low concentrations. This makes them difficult to isolate and identify. It is sometimes easier to isolate the proteins to which they bind by using affinity chromatography. (Refer to Biochemical Methods 5.1.) Explain how you would use this technique to isolate a protein suspected of steroid hormone binding.

Nucleic Acids

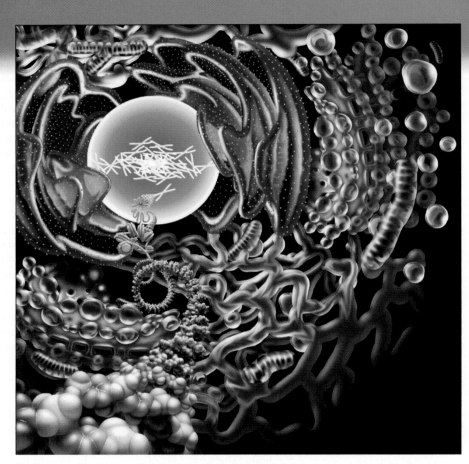

The Genome. All of the genetic information required to sustain living processes is encoded in the molecular structure of DNA. In eukaryotes most DNA is packed in complex chromosomes within the nucleus.

The determination of the structure of deoxyribonuleic acid by James Watson and Francis Crick in the early 1950s was the culmination of research that had begun almost a century before. The consequences of their seminal work are still unfolding as the genetic information of various organisms, including humans, is gradually being revealed. The structure and functioning of DNA and its companion molecule, RNA (ribonucleic acid), which have proven to be astonishingly complex, continue to fascinate researchers and students alike.

For countless centuries, humans have observed inheritance patterns without understanding the mechanisms that transmit physical traits and developmental processes from parent to offspring. Many human cultures have used such observations to improve their economic status, as in the breeding of domesticated animals or seed crops. It was not until the nineteenth century that the scientific investigation of inheritance, now referred to as **genetics**, began. By the beginning of the twentieth century, scientists generally recognized that physical traits are inherited as discrete units (later called **genes**) and that chromosomes within the nucleus are the repositories of genetic information. Eventually, the chemical composition of chromosomes was elucidated, and (after many decades of investigation) deoxyribonucleic acid (DNA) was identified as the genetic information. A complete set of this information in an organism, encoded in the nucleotide base sequence of its DNA, is now referred to as its **genome**.

The science devoted to elucidating the structure and function of genomes is called **molecular biology**. The discovery of the structure of DNA as a helical duplex of nucleotide polymers by James Watson and Francis Crick in 1953 has permitted life scientists to reexamine most biological phenomena using the research tools developed by molecular biologists and biochemists. In the past five decades, in one of the most fascinating and complex investigations of the twentieth century, molecular biologists formulated a general outline of biological inheritance and information transfer. This work revealed the following principles:

1. The information encoded within DNA, which directs the functioning of living cells and is transmitted to offspring, consists of a specific sequence of nitrogenous bases. DNA synthesis involves the complementary pairing of nucleotide bases on two strands of DNA. The physiological and genetic function of DNA requires the synthesis of relatively error-free copies.

2. The mechanism by which genetic information is decoded and used to direct cellular processes begins with the synthesis of another type of nucleic acid, ribonucleic acid (RNA). RNA synthesis occurs by complementary pairing of ribonucleotide bases with the bases in a DNA molecule.

3. Several types of RNA are involved in the synthesis of the enzymes, structural proteins, and other polypeptides required for the synthesis of all other biomolecules involved in organismal function.

   The flow of biological information is summarized by the following sequence:

This concept has been referred to as the "central dogma of molecular biology" because it describes the flow of genetic information from DNA through RNA and eventually to proteins.

Because of the informational nature of genetic processes, some descriptive terms have been borrowed from other information sciences. For example, DNA synthesis is often referred to as **replication** (to copy or duplicate). Similarly, the process in which DNA is used to synthesize RNA is called **transcription**. Each RNA molecule is called a **transcript**. Protein synthesis is referred to as **translation**. Several other terms (e.g., genetic code, codon, and anticodon), defined in Chapter 18, also come from the language of information transfer. As life scientists have used increasingly more powerful and sensitive analytical techniques in their investigations of genes and gene expression, new terms have been introduced. For example, the **transcriptome** and the **proteome** are the complete sets of RNA molecules and protein molecules, respectively, that are produced within a cell under specified circumstances (Figure 17.1). The term **metabolome** refers to the complete set of organic metabolites that are produced

DNA→RNA→Protein

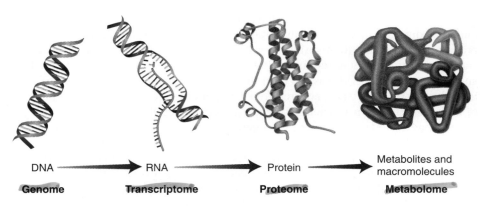

DNA	→	RNA	→	Protein	→	Metabolites and macromolecules
Genome		**Transcriptome**		**Proteome**		**Metabolome**

FIGURE 17.1

Biological Information.

Technological advances have made previously unimaginable access to the genomes of a growing number of organisms possible. The current challenge for biochemists and other life scientists is how to interpret not only the massive amounts of genetic information in living cells (the genome), but also how this information is expressed at the level of transcription (the transcriptome), protein synthesis (the proteome), and metabolism (the metabolome) so that specific biological and health problems can be solved.

within a cell under the direction of its genome. In addition to the macromolecules, these molecules include sugars, lipids, amino acids, and all of their derivatives.

Chapter 17, which focuses on the structure of the nucleic acids, begins with a description of DNA structure and the investigations that led to its discovery. This is followed by a discussion of current knowledge of genome and chromosome structure, as well as the structure and roles of the several forms of RNA. Chapter 17 ends with the description of viruses, macromolecular complexes composed of nucleic acid and proteins that are cellular parasites. In the following chapter (Chapter 18), several aspects of nucleic acid synthesis and function (i.e., DNA replication and transcription) are discussed. Protein synthesis (translation) is described in Chapter 19.

The strategies and techniques that are routinely used to isolate, purify, characterize, and manipulate nucleic acids are described in Biochemical Methods 17.1 and 18.1. The analysis of proteomes, referred to as **proteomics**, is described in Biochemical Methods 19.1.

QUESTION 17.1

What is the central dogma of molecular biology? How are the features of the central dogma related to terms such as genome, transcriptome, proteome, and metabolome?

17.1 DNA

DNA consists of two polynucleotide strands wound around each other to form a right-handed double helix (Figure 17.2). The structure of DNA is so distinctive that this molecule is often referred to as the *double helix*. As described (Sections 1.2 and 14.3), each nucleotide monomer in DNA is composed of a nitrogenous base (either a purine or a pyrimidine), a deoxyribose sugar, and phosphate. The mononucleotides are linked to each other by 3′,5′-phosphodiester bonds. These bonds join the 5′-hydroxyl group of the deoxyribose of one nucleotide to the 3′-hydroxyl group of the sugar unit of another nucleotide through a phosphate group (Figure 17.3). The antiparallel orientation of the two polynucleotide strands allows hydrogen bonds

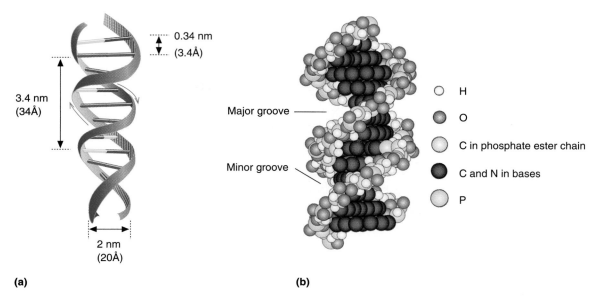

(a)

0.34 nm
(3.4Å)

3.4 nm
(34Å)

2 nm
(20Å)

(b)

Major groove

Minor groove

○ H

● O

○ C in phosphate ester chain

● C and N in bases

○ P

FIGURE 17.2

Two Models of DNA Structure.

(a) The DNA double helix is represented as a spiral ladder. The sides of the ladder represent the sugar-phosphate backbones. The rungs represent the pairs of bases. (b) In a space-filling model, the sugar-phosphate backbones are represented by strings of colored spheres. The base pairs consist of horizontal arrangements of dark blue spheres. Wide and narrow grooves are created by twisting the two strands around each other.

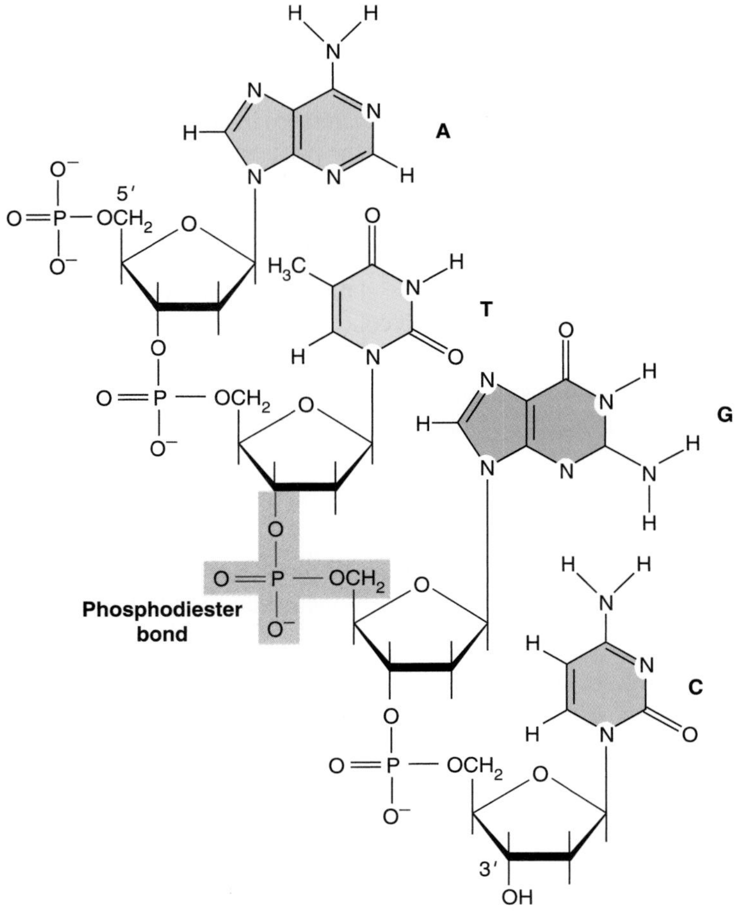

FIGURE 17.3

DNA Strand Structure.

In each DNA strand the deoxyribonucleotide residues are connected to each other by 3′-5′ phosphodiester linkages. The sequence of the strand section illustrated in this figure is 5′-ATGC-3′.

A

T

G

C

Phosphodiester bond

5′

3′

OH

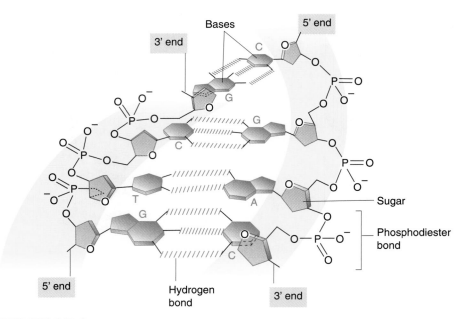

FIGURE 17.4

DNA Structure.

In this short segment of DNA, the bases are shown in orange and the sugars are blue. Each base pair is held together by either two or three hydrogen bonds. The two polynucleotide strands are antiparallel. Because of base pairing, the order of bases in one strand determines the order of bases along the other.

to form between the nitrogenous bases that are oriented toward the helix interior (Figure 17.4). There are two types of base pairs (bp) in DNA: (1) adenine (a purine) pairs with thymine (a pyrimidine), and (2) the purine guanine pairs with the pyrimidine cytosine. Because each base pair is oriented at an angle to the long axis of the helix, the overall structure of DNA resembles a twisted staircase. The dimensions of crystalline DNA have been precisely measured.

1. One turn of the double helix spans 3.4 nm and consists of approximately 10.4 base pairs. (Changes in pH and salt concentrations affect these values slightly.)

2. The diameter of the double helix is 2.4 nm. Note that the interior space of the double helix is only suitable for base pairing a purine and a pyrimidine. Pairing two pyrimidines would create a gap, and pairing purines would destabilize the helix.

3. The distance between adjacent base pairs is 0.34 nm. The relative dimensions of both types of base pairs are illustrated in Figure 17.5.

As befits its role in living processes, DNA is a relatively chemically inert molecule. Furthermore, several types of noncovalent bonding contribute to the stability of its helical structure:

1. Hydrophobic interactions. The base ring π cloud of electrons between stacked purine and pyrimidine bases is relatively nonpolar. The clustering of the base components of nucleotides within the double helix is a stabilizing factor in the three-dimensional macromolecule because it minimizes their interactions with water, thereby increasing entropy.

2. Hydrogen bonds. The base pairs, on close approach, form a preferred set of hydrogen bonds, three between GC pairs and two between AT pairs. The cumulative "zippering" effect of these hydrogen bonds keeps the strands in correct complementary orientation.

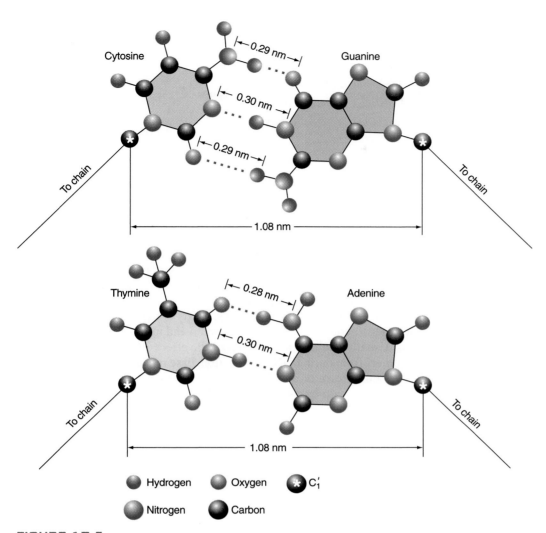

FIGURE 17.5

DNA Structure: AT and GC Base Pair Dimensions.

Two hydrogen bonds are formed in each AT base pair and three in each GC base pair. The equal overall dimensions of both types of base pairs allow the formation of identical helical conformations of the two polynucleotide strands.

3. Base stacking. Once the antiparallel polynucleotide strands have been brought together by base pairing, the parallel stacking of the nearly planar heterocyclic bases stabilizes the molecule because of the cumulative effect of weak van der Waals forces.

4. Electrostatic interactions. DNA's external surface, referred to as the *sugar-phosphate backbone*, possesses negatively charged phosphate groups. Repulsion between nearby phosphate groups, a potentially destabilizing force, is minimized by the shielding effects of divalent cations such as Mg^{2+} and polycationic molecules such as the polyamines and histones (see p. 580).

Describe how each type of noncovalent interaction contributes to DNA's stable helical structure.

QUESTION 17.2

When DNA is heated, it denatures, that is, the strands separate because hydrogen bonds are broken. The higher the temperature, the larger the number of hydrogen

QUESTION 17.3

bonds that are broken. After reviewing DNA structure, determine which of the following molecules will denature first as the temperature is raised.

   a. 5′-GCATTTCGGCGCGTTA-3′
      3′-CGTAAAGCCGCGCAAT-5′

   b. 5′-ATTGCGCTTATATGCT-3′
      3′-TAACGCGAATATACGA-5′

KEY CONCEPTS 17.1

DNA is a relatively stable molecule composed of two antiparallel polynucleotide strands wound around each other to form a right-handed double helix.

DNA Structure: The Nature of Mutation

DNA is eminently suited for information storage. However, despite its several stabilizing structural features, DNA is vulnerable to certain types of disruptive forces. Solvent collisions, thermal fluctuations, and other spontaneous disruptive processes can result in mutations, permanent changes in the base sequence of DNA molecules. In addition, a wide variety of xenobiotics, both natural and human-made, are known to alter DNA structure. Several examples of well-researched mutagenic factors are discussed.

As described (Figure 14.22), tautomeric shifts are spontaneous changes in nucleotide base structure that interconvert amino and imino groups and keto and enol groups. Usually tautomeric shifts have little effect on DNA structure. However, if tautomers form during DNA replication, base mispairings may result. For example, the imino form of adenine will not base pair with thymine. Instead, it forms a base pair with cytosine (Figure 17.6). If this pairing is not corrected immediately, a transition mutation results because cytosine has been incorporated during the replication process in a position that should carry thymine. In a **transition mutation** a pyrimidine base is substituted for another pyrimidine, or a purine is substituted for another purine. Transition mutations are **point mutations**, DNA base sequence changes that involve a single base pair. In the

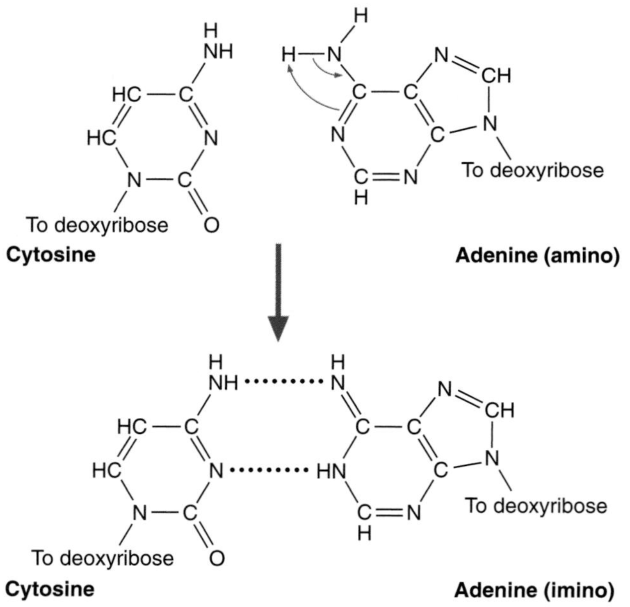

FIGURE 17.6

A Tautomeric Shift Causes a Transition Mutation.

As adenine undergoes a tautomeric shift, its imino form can base pair with cytosine. The transition shows up in the second generation of DNA replication when cytosine base pairs with guanine. In this manner an A-T base pair is replaced by a C-G base pair.

example described an AT base pair is replaced by a GC base pair in the second generation of DNA replication.

Several spontaneous hydrolytic reactions also cause DNA damage. For example, it has been estimated that several thousand purine bases are lost daily from the DNA in each human cell. In depurination reactions the N-glycosyl linkage between a purine base and deoxyribose is cleaved. The protonation of N-3 and N-7 of guanine promotes hydrolysis. If repair mechanisms do not replace the purine nucleotide, a point mutation will result in the next round of DNA replication. Similarly, bases can be spontaneously deaminated. For example, the deaminated product of cytosine converts to uracil via a tautomeric shift. Eventually, what should be a CG base pair is converted to an AT base pair. (Uracil is similar in structure to thymine.)

Some types of ionizing radiation (e.g., UV, X-rays, and γ-rays) can alter DNA structure. Low radiation levels may cause mutation; high levels can be lethal. Radiation-induced damage, caused by a free radical mechanism (either abstraction of hydrogen atoms or the creation of •OH, the hydroxyl radical), includes strand breaks, DNA-protein crosslinking, ring openings, and base modifications. The hydroxyl radical, formed by the radiolysis of water, as well as oxidative stress (Section 10.3), is known to cause some strand breakage and numerous base modifications (e.g., thymine glycol, 5-hydroxymethyl uracil, and 8-hydroxyguanine).

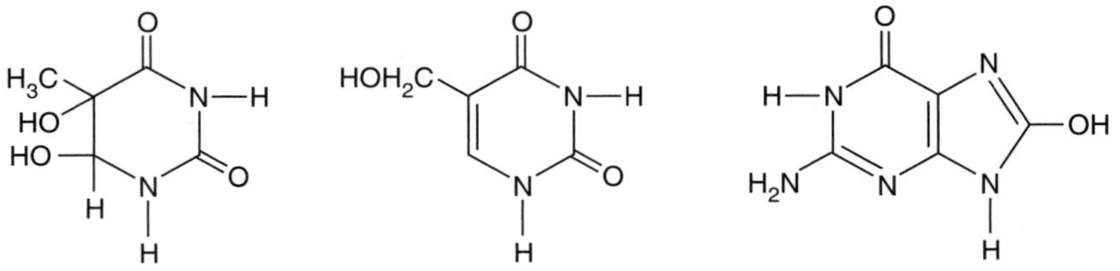

Thymine glycol   **5-Hydroxymethyl uracil**   **8-Hydroxyguanine**

The most common UV-induced product is pyrimidine dimers (Figure 17.7). The helix distortion that results from dimer formation stalls DNA synthesis.

A large number of xenobiotics can damage DNA. The most important of these molecules belong to the following classes:

1. Base analogues. Because their structures are similar to the normal nucleotide bases, **base analogues** can be inadvertently incorporated into DNA. For example, caffeine is a base analogue of thymine. Because it can base pair with guanine, caffeine incorporation can cause a transition mutation.

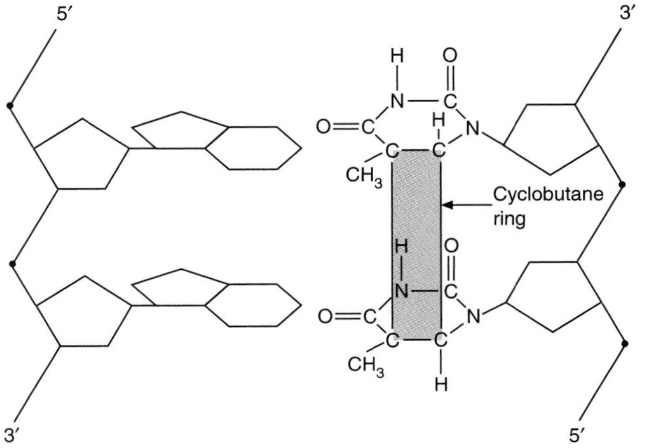

FIGURE 17.7

Thymine Dimer Structure.

Adjacent thymines form dimers with high efficiency after absorbing UV light.

2. Alkylating agents. Alkylation is a process in which electrophilic ("electron-loving") substances attack molecules that possess an unshared pair of electrons. When electrophiles react with such molecules, they usually add carbon-containing alkyl groups. Adenine and guanine are especially susceptible to alkylation, although thymine and cytosine can also be affected.

Alkylated bases often pair incorrectly (e.g., methylguanine pairs with thymine instead of cytosine) leading to possible transition mutations on subsequent rounds of replication. In the case of methylguanine, a GC pair becomes an AT pair. Transversion mutations may also occur when the alkylating group is bulky. (In a **transversion mutation**, another type of point mutation, a pyrimidine is substituted for a purine or vice versa.) Alkylations can also promote tautomer formation, which may result in transition mutations. Examples of alkylating agents include dimethylsulfate and dimethylnitrosamine. Mitomycin C is a bifunctional alkylating agent. It can prevent DNA synthesis by crosslinking guanine bases. DNA repair mechanisms (Section 18.1) usually remove the abnormal base ring prior to replication and reduce the mutation frequency.

3. Nonalkylating agents. A variety of chemicals other than the alkylating agents can modify DNA structure. Nitrous acid (HNO_2), derived from the nitrosamines and from sodium nitrite ($NaNO_2$), deaminates bases. HNO_2 converts adenine, guanine, and cytosine to hypoxanthine, xanthine, and uracil, respectively. The aromatic polycyclic hydrocarbons (e.g., benzo[a]pyrene) are also mutagenic. Once consumed, these molecules can be converted to highly reactive derivatives by biotransformation reactions such as those catalyzed by cytochrome P_{450} (Special Interest Box 10.1). The reactive derivatives can then form adducts of most bases. Damage occurs primarily because this chemical modification prevents base pairing.

4. Intercalating agents. Certain planar molecules can distort DNA because they insert themselves (intercalate) between the stacked base pairs of the double helix. Either adjacent base pairs are deleted or new base pairs are inserted. If not corrected, deletions or insertions cause so-called frame-shift mutations (described in Chapter 19). In addition, chromosomes may break. The acridine dyes are examples of intercalating agents. Quinacrine is an acridine dye used to treat malaria and intestinal tapeworms.

KEY CONCEPTS 17.2

DNA is vulnerable to certain types of disruptive forces that can result in mutations, permanent changes in its base sequence.

QUESTION 17.4

How will each of the following substances or conditions affect DNA structure?

a. ethanol b. heat c. dimethylsulfate d. nitrous acid e. quinacrine

QUESTION 17.5

Consider each of the following compounds. To what class of DNA-damaging xenobiotics does each belong?

Caffeine   **Benzo[a]pyrene**   **Ethyl chloride**

QUESTION 17.6

The accumulation of oxidative DNA damage now appears to be a major cause of aging in mammals. Animals that have high metabolic rates (i.e., use large amounts of oxygen) or excrete large amounts of modified bases in the urine typically have shorter life spans. The excretion of relatively large amounts of oxidized bases indicates a reduced capacity to prevent oxidative damage. Despite substantial evidence that oxygen radicals damage DNA, the actual radicals that cause the damage are still not clear. In addition to the hydroxyl radical, suggest other possible culprits. Some tissues sustain more oxidative damage than others. For example, the human brain is believed to sustain more oxidative damage than most other tissues during an average life span. Suggest two reasons for this phenomenon.

DNA Structure: From Mendel's Garden to Watson and Crick

To modern eyes, the structure of DNA is both elegant and obvious. DNA is now a cultural icon, synonymous with the concept of information storage and retrieval. As mentioned, the correct structure of DNA was proposed in 1953 by James Watson and Francis Crick (Figure 17.8). The investigation that led to this remarkable discovery is instructive for several reasons. First, as often happens in scientific research, the road to the elucidation of DNA structure was long, frustrating, and tortuous. Living organisms are so complex that discerning any aspect of their function is extraordinarily difficult. Adding to this obstacle is the propensity of scientists (and other humans) to reject or ignore new information that does not fit comfortably with currently popular ideologies. This latter problem is probably unavoidable, because the scientific method requires a certain degree of skepticism. (How does one differentiate, for example, between breakthrough concepts

FIGURE 17.8

The First Complete Structural Model of DNA.

When James Watson (left) and Francis Crick discovered the structure of DNA in 1953 using Rosalind Franklin's data, they were research students at the Henry Cavendish Laboratory of Cambridge University.

and erroneous ideas?) However, skepticism can often be confused with an unimaginative adherence to the status quo. Albert Szent-Gyorgyi (Nobel Prize in physiology or medicine, 1937), who identified ascorbic acid as vitamin C and made significant contributions to the elucidation of muscle contraction and the citric acid cycle, once remarked, "Discovery consists in seeing what everyone else has seen and thinking what nobody else has thought."

A second, more concrete reason for the length of the discovery process is that the development of new concepts often requires the integration of information from several scientific disciplines. For example, the DNA model was based on discoveries in descriptive and experimental biology, genetics, organic chemistry, and physics. Significant scientific advancements are usually made by imaginative and industrious individuals who have the good fortune to work when sufficient information and technology are available for solving the scientific problems that interest them. The most talented of these investigators often help create new technologies.

The scientific revolution that eventually led to the DNA model began quietly in the abbey garden of an obscure Austrian monk named Gregor Mendel. Mendel discovered the basic rules of inheritance by cultivating pea plants. In 1865, Mendel published the results of his breeding experiments in the *Journal of the Brunn Natural History Society.* Although he sent copies of this publication to eminent biologists throughout Europe, his work was ignored until 1900. In that year, several botanists independently rediscovered Mendel's paper and recognized its significance. This long delay was due largely to the descriptive nature of nineteenth century biology; few biologists were familiar with the mathematics that Mendel used to analyze his data. By 1900, many biologists were trained in mathematics. In addition, this latter group of scientists had a frame of reference for Mendel's principles, because many of the details of meiosis, mitosis, and fertilization had become common knowledge.

Amazingly, the substance that constitutes the inheritable units that Mendel referred to in his work was being investigated almost simultaneously. The discovery of "nuclein," later renamed nucleic acid, was reported in 1869 by Friedrich Miescher, a Swiss pathologist. Working with the nuclei of pus cells, Miescher extracted nuclein and discovered that it was acidic and contained a large amount of phosphate. (Although Joseph Lister published his findings on antiseptic surgery in 1867, hospitals continued to be a rich source of pus for many years to come.) Interestingly, Miescher came to believe (erroneously) that nuclein was a phosphate storage compound.

The chemical composition of DNA was determined largely by Albrecht Kossel between 1882 and 1897 and P. A. Levene in the 1920s as suitable analytical techniques were developed. Unfortunately, Levene mistakenly believed that DNA was a small and relatively simple molecule. His concept, referred to as the *tetranucleotide hypothesis,* significantly retarded further investigations of DNA. Instead, proteins (the other major component of nuclei) were viewed as the probable carrier of genetic information. (By the end of the nineteenth century it was commonly accepted that the nucleus contains the genetic information.)

While geneticists focused on investigating the mechanisms of heredity and chemists elucidated the structures of nucleic acid components, microbiologists developed methods of studying bacterial cultures. In 1928, while investigating a deadly epidemic of pneumonia in Britain, Fred Griffith performed a remarkable series of experiments with two strains of pneumococcus. One bacterial strain, called the smooth form (or type S) because it is covered with a polysaccharide capsule, is pathogenic. The rough form (or type R) lacks the capsule and is nonpathogenic. Griffith observed that mice inoculated with a mixture of live R and heat-killed S bacteria died. He was amazed when live S bacteria were isolated from the dead mice. Although this transformation of R bacteria into S bacteria was confirmed in other laboratories, Griffith's discovery was greeted with considerable skepticism. (The concept of transmission of genetic information between bacterial cells was not accepted until the 1950s.) In 1944, Oswald Avery and his colleagues Colin MacLeod and Maclyn McCarty reported their painstaking isolation and identification of the transforming agent in Griffith's experiment as DNA. Not everyone accepted this conclusion because their DNA sample had a trace of protein impurities. Avery and McCarty later demonstrated that the digestion of DNA by deoxyribonuclease (DNase) inactivated the transforming agent. (Eventually, it was determined that the DNA that transforms R pneumococcus into the S form codes for an enzyme required to synthesize the gelatinous polysaccharide capsule. This capsule protects the bacteria from the animal's immune system and increases adherence and colonization of host tissues.)

Another experiment that confirmed DNA as the genetic material was reported by Alfred Hershey and Martha Chase in 1952. Using T2 bacteriophage, Hershey and Chase demonstrated the separate functions of viral nucleic acid and protein. (*Bacteriophage*, sometimes called *phage*, is a type of virus that infects bacteria.) When T2 phage infects an *Escherichia coli* cell, the bacterium is directed to synthesize several hundred new viruses. Within 30 minutes after infection the cell dies as it bursts open, thus releasing the viral progeny. In the first phase of their experiment, Hershey and Chase incubated bacteria infected with T2 phage in a culture medium containing ^{35}S (to label protein) and ^{32}P (to label DNA). In the second phase the radioactively labeled virus was harvested and allowed to infect nonlabeled bacteria. Immediately afterwards, the infected bacterial culture was subjected to shearing stress in a Waring blender. This treatment removed the phage from its attachment sites on the external surface of the bacterial cell wall. After separation from empty viral particles by centrifugation, the bacteria were analyzed for radioactivity. The cells were found to contain ^{32}P (thus confirming the role of DNA in "transforming" the bacteria into virus producers), whereas most of the ^{35}S remained in the supernatant. In addition, samples of the labeled infected bacteria produced some ^{32}P-labeled viral progeny.

By the early 1950s it had become clear that DNA was the genetic material. Because researchers also recognized that genetic information was critical for all living processes, determining the structure of DNA became an obvious priority. Linus Pauling (California Institute of Technology), Maurice Wilkins and Rosalind Franklin (King's College, London), and Watson and Crick (Cambridge University) were all working toward this goal. The structure proposed by Watson and Crick in the April 25, 1953, issue of *Nature* was based on their scale model.

Considering how the scientific community responded to other concepts involving DNA as the genetic material, their acceptance of the Watson-Crick structure was unusually rapid. In 1962, Watson, Crick, and Wilkins were awarded the Nobel Prize in chemistry.

The information used to construct this model included the following:

1. The chemical structures and molecular dimensions of deoxyribose, the nitrogenous bases, and phosphate.

2. The 1:1 ratios of adenine:thymine and guanine:cytosine in the DNA isolated from a wide variety of species investigated by Erwin Chargaff between 1948 and 1952. (This relationship is sometimes referred to as **Chargaff's rules**.)

3. Superb X-ray diffraction studies performed by Rosalind Franklin (Figure 17.9) indicating that DNA is a symmetrical molecule and probably a helix.

4. The diameter and pitch of the helix estimated by Wilkins and his colleague Alex Stokes from other X-ray diffraction studies.

5. The recent demonstration by Linus Pauling that protein, another complex class of molecule, could exist in a helical conformation.

KEY CONCEPTS 17.3

The model of DNA structure proposed by James Watson and Francis Crick in 1953 was based on information derived from the efforts of many individuals.

FIGURE 17.9

**X-Ray Diffraction Study of DNA
by Rosalind Franklin and R. Gosling.**

Note the symmetry of the X-ray diffraction
pattern.

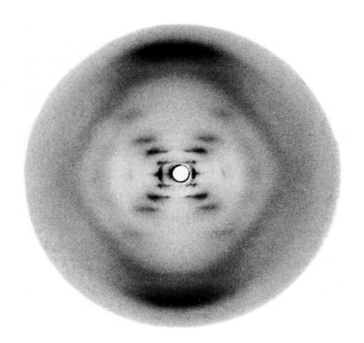

DNA Structure: Variations on a Theme

The structure discovered by Watson and Crick, referred to as **B-DNA**, represents the sodium salt of DNA under highly humid conditions. DNA can assume different conformations because deoxyribose is flexible and the C^1-N-glycosidic linkage rotates. (Recall that furanose rings have a puckered conformation.)

When DNA becomes partially dehydrated, it assumes the A form (Figure 17.10 and Table 17.1). In **A-DNA**, the base pairs are no longer at right angles to the helical axis. Instead, they tilt 20° away from the horizontal. In addition, the distance between adjacent base pairs is slightly reduced, with 11 bp per helical turn instead of the 10.4 bp found in the B form. Each turn of the double helix occurs in 2.5

FIGURE 17.10

A-DNA, B-DNA, and Z-DNA.

Because DNA is a flexible molecule, it can assume different conformation forms depending on its base pair sequence and/or isolation conditions. Each molecular form in the figure possesses the same number of base pairs.

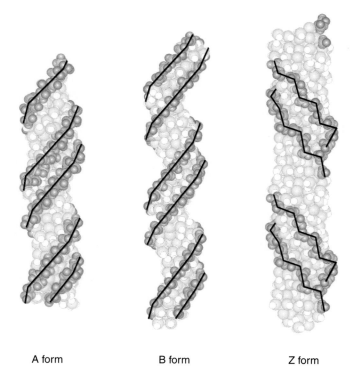

A form   B form   Z form

TABLE 17.1
Selected Structural Properties of B-, A-, and Z-DNA

	B-DNA (Watson-Crick Structure)	A-DNA	Z-DNA
Helix diameter	2.4 nm	2.6 nm	1.8 nm
bp per helical turn	10.4	11	12
Rotation per bp	3.4 nm	2.5 nm	4.5 nm
Helix rotation	Right-handed	Right-handed	Left-handed

nm, instead of 3.4 nm, and the molecule's diameter swells to approximately 2.6 nm from the 2.4 nm observed in B-DNA. The A form of DNA is observed when it is extracted with solvents such as ethanol. The significance of A-DNA under cellular conditions is that the structure of RNA duplexes and RNA/DNA duplexes formed during transcription resembles the A-DNA structure.

The Z form of DNA (named for its "zigzag" conformation) radically departs from the B form. **Z-DNA** (D = 1.8 nm), which is considerably slimmer than B-DNA (D = 2.4 nm), is twisted into a left-handed spiral with 12 bp per turn. Each turn of Z-DNA occurs in 4.5 nm, compared with 3.4 nm for B-DNA. DNA segments with alternating purine and pyrimidine bases (especially CGCGCG) are most likely to adopt a Z configuration. In Z-DNA, the bases stack in a left-handed staggered dimeric pattern, which gives the DNA a zigzag appearance and its flattened, nongrooved surface. Regions of DNA rich in GC repeats are often regulatory, binding specific proteins that initiate or block transcription. Although the physiological significance of Z-DNA is unclear, it is known that certain physiologically relevant processes such as methylation and negative supercoiling (discussed on p. 578) stabilize the Z form. In addition, short segments have been observed to form as the result of torsional strain during transcription.

Segments of DNA molecules have been observed to have several higher-order structures. Examples include cruciforms, triple helices, and supercoils. Each is briefly described.

As their name implies, **cruciforms** (Figure 17.11) are crosslike structures. They are likely to form when a DNA sequence contains a palindrome. (A **palindrome**

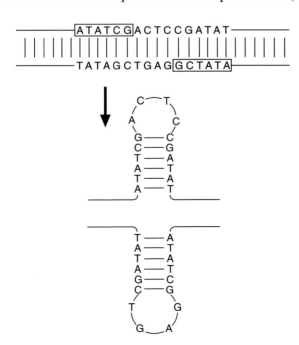

FIGURE 17.11
Cruciforms.

Cruciforms form because of palindrome sequences.

is defined as a sequence that provides the same information whether it is read forward or backward, e.g., "MADAM, I'M ADAM.") In contrast to language palindromes, the "letters" are read in one direction on one of the complementary strands of DNA and in the opposite direction on the other strand. One-half of the palindrome on each strand is complementary to the other half. The DNA sequences that form palindromes, which may consist of several bases or thousands of bases, are called *inverted repeats*. In one proposed mechanism, cruciform formation begins with a small bubble, or *protocruciform*, and progresses as intrastrand base pairing occurs. The mechanism by which bubble formation is initiated is unknown. The function of cruciforms is unclear but is believed to be associated with the binding of various proteins to DNA. DNA palindromes also play a role in the function of an important class of enzymes called the restriction enzymes (Biochemical Methods 18.1).

In certain circumstances (e.g., low pH) a DNA sequence containing a long segment consisting of a polypurine strand hydrogen-bonded to a polypyrimidine strand can form a triple helix (Figure 17.12). The formation of the *triple helix*, also referred to as **H-DNA**, depends on the formation of nonconventional base pairing (**Hoogsteen base pairing**), which occurs without disrupting the Watson-Crick base pairs. The significance of H-DNA is not understood. However, H-DNA may have a role in genetic recombination (Section 18.1).

Packaging large DNA molecules to fit into cells requires DNA supercoiling (discussed on p. 577). To undergo supercoiling, DNA molecules must be nicked

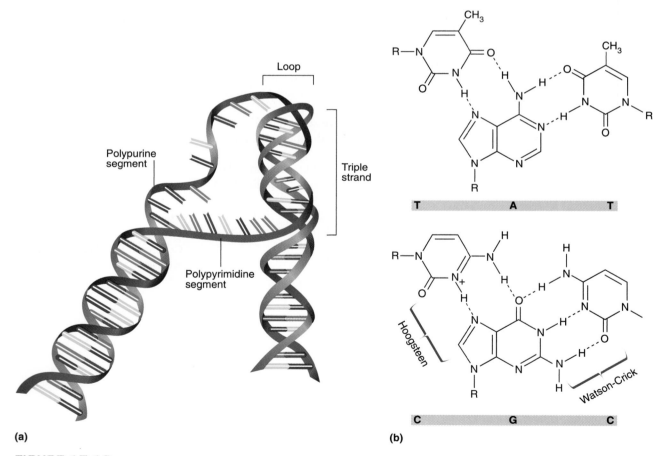

(a)   **(b)**

FIGURE 17.12

H-DNA

(a) DNA sequences with long segments such as (A-G)n bonded to (T-C)n can form H-DNA. (b) H-DNA formation depends on the formation of nonconventional (Hoogsteen) base pairing.

and then either overwound or underwound before resealing. Small changes in DNA shape depend on sequence. For example, four sequential AT pairs produce a bend in the molecule. Significant bending or wrapping around associated proteins, however, requires supercoiling.

DNA Supercoiling

DNA supercoiling, once considered an artifact of DNA extraction techniques, is now known to facilitate several biological processes. Examples include packaging DNA into a compact form, as well as replicating and transcribing DNA (Chapter 18). Because DNA supercoiling is a dynamic three-dimensional process, the information that two-dimensional illustrations can convey is limited. To understand supercoiling, therefore, consider the following thought experiment. A long, linear DNA molecule is laid on a flat surface. After the ends are brought together, they are sealed to form an unpuckered circle (Figure 17.13a). Because this molecule is sealed without under- or overwinding, the helix is said to be relaxed and remains flat on a surface. If the relaxed circular DNA molecule is held and twisted a few times, it takes the shape shown in Figure 17.13b. When this twisted molecule is laid back on the flat surface, it rotates to eliminate the twist. However, consider what happens if this molecule is cut before it is twisted. (The enzymes that perform this function in living cells, called topoisomerases, are discussed in Chapter 18.)

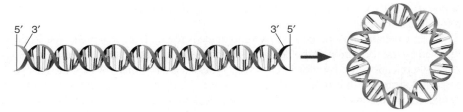

Linear double-stranded DNA molecule   Circular DNA molecule

(a)

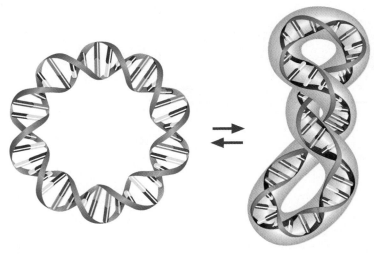

(b)

FIGURE 17.13

Linear and Circular DNA and DNA Winding.

(a) The formation of a relaxed circular DNA molecule. (b) When a relaxed molecule is twisted, it reverts to its flat structure once it is released.

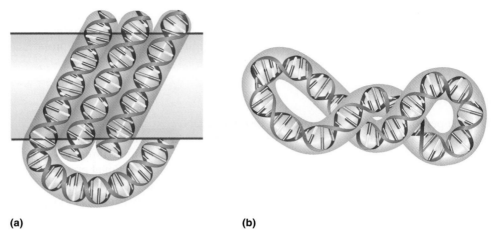

(a)   **(b)**

FIGURE 17.14

Supercoils.

Supercoils occur in two major forms: (a) toroidal and (b) interwound.

When DNA is underwound it twists to the right to relieve strain and *negative supercoiling* results. When DNA is overwound it twists to the left to relieve strain and *positive supercoiling* results. When DNA is negatively supercoiled, it usually winds around itself to form an interwound supercoil (Figure 17.14). Positively supercoiled DNA is usually found where DNA coils around a protein core to form a toroidal supercoil.

When a negatively supercoiled DNA molecule is forced to lie in a plane, the strain relieved by the formation of the negative supercoiling is reintroduced (Figure 17.15). During replication, topoisomerases (such as gyrase in *E. coli*) nick the DNA to relieve torsional strain so that replication can proceed. DNA negative supercoiling also explains the propensity of certain DNA sequences to form cruciforms and H-DNA.

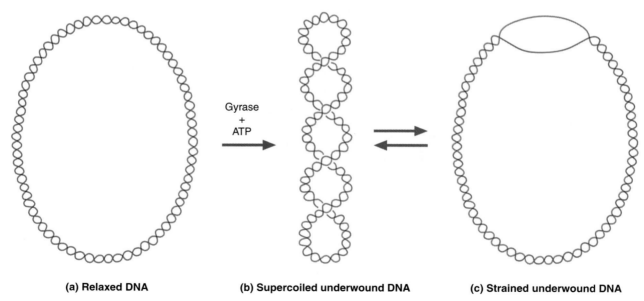

(a) Relaxed DNA   **(b) Supercoiled underwound DNA**   **(c) Strained underwound DNA**

FIGURE 17.15

The Effect of Strain on a Circular DNA Molecule.

Breakage and re-formation of phosphodiester linkages allow the conversion of a relaxed circular form (a) to the negatively supercoiled form (b). The strain relieved by the supercoiling process will be reintroduced when the underwound molecule is forced to lie in a plane (c).

The discussion of supercoiling also applies to the linear DNA molecules found in the nuclei of eukaryotic cells. Such molecules are constrained by their attachment to nuclear scaffolds, which are structural components of chromosomes.

Chromosomes and Chromatin

DNA, which contains the genes (the units of heredity), is packaged into structures called chromosomes. As it was originally defined, the term **chromosome** referred only to the dense, darkly staining structures visible within eukaryotic cells during meiosis or mitosis. However, this term is now also used to describe the DNA molecules that occur in prokaryotic cells. The physical structure and genetic organization of prokaryotic and eukaryotic chromosomes are significantly different.

PROKARYOTES In prokaryotes such as *E. coli*, a chromosome is a circular DNA molecule that is extensively looped and coiled so that it can be compressed into a relatively small space ($1 \ \mu m \times 2 \ \mu m$). Yet the information in this highly condensed molecule must be readily accessible. The *E. coli* chromosome (circumference = $1.6 \ \mu m$) consists of a supercoiled DNA that is complexed with a protein core (Figure 17.16).

In this structure, called the nucleoid, the chromosome is attached to the protein core in at least 40 places. This structural feature produces a series of loops that limit the unraveling of supercoiled DNA if a strand break is introduced. Compression is further enhanced by packaging with HU, a protein that binds prokaryotic DNA and facilitates bending and supercoiling. Approximately 60 bp are wound around each HU tetramer.

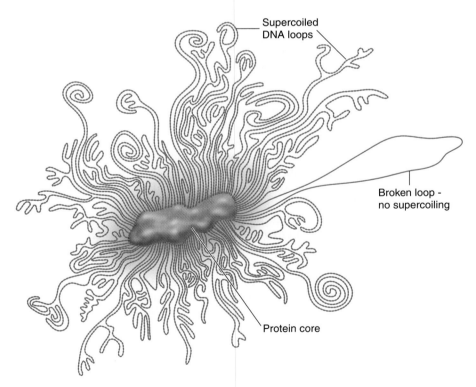

Supercoiled
DNA loops

Broken loop -
no supercoiling

Protein core

FIGURE 17.16

The *E. coli* Chromosome.

The circular *E. coli* chromosome is complexed with a protein core. Because the chromosome (3×10^6 bp) is highly supercoiled, the entire chromosome complex measures only $2 \ \mu m$ across. The attachment of each of the DNA loops to the protein core may prevent the unraveling of the entire supercoiled chromosome when a strand breaks.

In addition, the polyamines (polycationic molecules such as spermidine and spermine) may also assist in attaining the chromosome's highly compressed structure.

$$H_3\overset{+}{N}—CH_2—CH_2—CH_2—CH_2—\overset{+}{N}H_2—CH_2—CH_2—CH_2—\overset{+}{N}H_3$$
Spermidine

$$H_3\overset{+}{N}—CH_2—CH_2—CH_2—\overset{+}{N}H_2—CH_2—CH_2—CH_2—CH_2—\overset{+}{N}H_2—CH_2—CH_2—CH_2—\overset{+}{N}H_3$$
Spermine

(When the positively charged polyamines bind to the negatively charged DNA backbone they overcome the charge repulsion between adjacent DNA coils.)

EUKARYOTES In comparison to prokaryotes, the eukaryotes possess genomes that are extraordinarily large. Depending on species, the chromosomes of eukaryotes vary in both length and number. For example, humans possess 23 pairs of chromosomes with a total of approximately three billion bp. The fruit fly *Drosophila melanogaster* has four chromosome pairs with 180 million bp, and corn (*Zea mays*) has 10 chromosome pairs with a total of 6.6 billion bp.

Each eukaryotic chromosome consists of a single linear DNA molecule complexed with histones to form **nucleohistone**. (Small amounts of nonhistone proteins, RNA, and polyamines may also affect DNA packaging.) The histones are a group of small basic proteins found in all eukaryotes. The binding of histones to DNA results in the formation of **nucleosomes**, which are the structural units of eukaryotic chromosomes. Consisting of five major classes (H1, H2A, H2B, H3, and H4), the histones are amazingly similar in their primary structure among eukaryotic species. Histones from different species and different phases of the cell cycle do differ, however, in the degree to which they undergo various chemical modifications (e.g., phosphorylation, acetylation, methylation, ubiquination, and ADP-ribosylation). The significance of these modifications is currently under investigation. It is known, however, that a major role of histone modifications is regulating the accessibility of DNA to **transcription factors**, proteins that promote gene transcription when they bind to specific DNA sequences. Each eukaryotic chromosome also possesses two unique structural elements: centromeres and telomeres (see p. 585). A *centromere* is a specific AT-rich DNA sequence associated with nonhistone proteins to form the kinetochore, which interacts with the spindle fibers during cell division. *Telomeres* are CCCA repetitive regions of DNA at the ends of the chromosomes that postpone loss of coding sequences during DNA replication.

When eukaryotic cells are not undergoing cell division, the chromosomes partially decondense to form **chromatin**. In electron micrographs, chromatin has a beaded appearance. Each of these "beads" is a nucleosome, which is composed of a positively supercoiled segment of DNA forming a toroidal coil around an octameric histone core (two copies each of H2A, H2B, H3, and H4. See Figure 17.17b). The organization of the DNA around each histone core results from electrostatic interactions between the arginine residues of the core histones and the phosphodiester backbone of DNA. Approximately 140 bp are in contact with each histone octamer. An additional 60 bp of spacer (or linker) DNA connect adjacent nucleosomes (Figure 17.17a). One molecule of histone H1 also occurs in each nucleosome, although the exact location is still unclear (Figure 17.17c). H1 is believed to facilitate coiling of the beaded fiber into higher-order structures.

As chromatin is compacted to form chromosomes, the nucleosomes are coiled into a higher order of structure referred to as the *30-nm fiber* (Figure 17.18). The 30-nm fiber is further coiled to form *200-nm filaments*. The three-dimensional

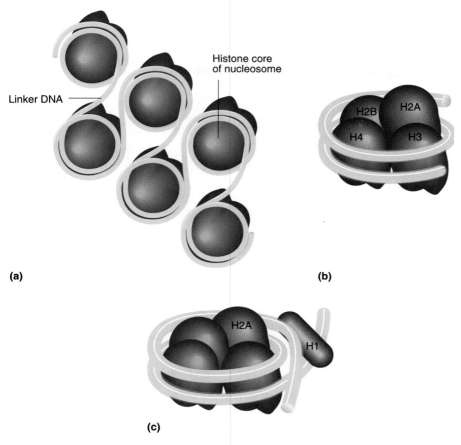

Linker DNA

Histone core
of nucleosome

H2B H2A
H4   H3

(a)   **(b)**

H2A
H1

(c)

FIGURE 17.17

Chromatin and Nucleosome.

(a) Nucleosomes are connected by linker DNA. (b) Each nucleosome core is composed of a histone octomer, around which is wrapped one and three-quarters turns of DNA. (c) A proposed structure of a nucleosome. The H1 histone aids in stabilizing the wrapping of DNA around the histone octomer. *Source:* (b) and (c) From Devlin, *Textbook of Biochemistry with Clinical Correlations,* 1992. Copyright © 1992, Wiley-Liss. Reprinted by permission of Wiley-Liss, Inc., a subsidiary of John Wiley & Sons, Inc.

KEY CONCEPTS 17.4

Each prokaryotic chromosome consists of a supercoiled circular DNA molecule complexed to a protein core. Each eukaryotic chromosome consists of a single linear DNA molecule that is complexed with histones to form nucleohistone.

structure of the 200-nm filaments is unclear but is believed to contain numerous supercoiled loops attached to a central protein complex referred to as a nuclear scaffold (Figure 17.19). Although the entire structural organization of eukaryotic chromosomes has not yet been elucidated, it is likely to consist of multiple levels of supercoiling.

ORGANELLE DNA Mitochondria and chloroplasts are semiautonomous organelles, that is, they possess DNA and their own version of protein-synthesizing machinery. These organelles, both of which reproduce by binary fission, also require a substantial contribution of proteins and other molecules that are coded for by the nuclear genome. For example, mitochondrial DNA (mtDNA) codes for several types of RNA and certain inner membrane proteins. The remainder of mitochondrial proteins are synthesized in the cytoplasm and transported into the mitochondria. Similarly, the chloroplast genome codes for several types of RNA and certain proteins, many of which are directly associated with photosynthesis. The activities of nuclear and organelle genomes are highly coordinated. Consequently, their individual contributions to organelle function

FIGURE 17.18

Chromatin.

Nuclear chromatin contains many levels of coiled structure.

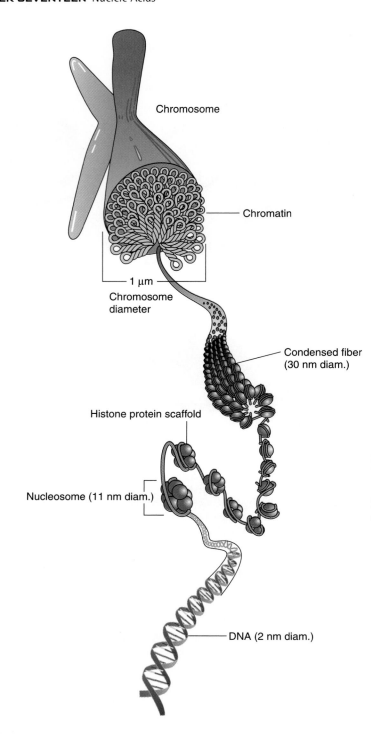

Chromosome

Chromatin

1 μm

Chromosome diameter

Condensed fiber (30 nm diam.)

Histone protein scaffold

Nucleosome (11 nm diam.)

DNA (2 nm diam.)

are often difficult to discern. Because mitochondria and chloroplasts are now believed to be the descendants of free-living bacteria, it is not surprising that they are susceptible to the actions of antibiotics (e.g., chloramphenicol and erythromycin) if their concentrations are sufficiently high. Many of these molecules are (or have been) used in clinical practice because they inhibit some aspect of bacterial genome function.

Compare the structural features that distinguish B-DNA from A-DNA, H-DNA, and Z-DNA. What is known about the functional properties of these variants of B-DNA, called the Watson-Crick structure?	QUESTION 17.7
What are the major protein components of prokaryotic and eukaryotic chromosome? What are their functions? What are the polyamines and what role do they play in DNA structure?	QUESTION 17.8
Explain the hierarchical relationships between the following: genomes, genes, nucleosomes, chromosomes, and chromatin.	QUESTION 17.9
Describe the evidence that James Watson and Francis Crick used to construct their model of DNA structure.	QUESTION 17.10

Genome Structure

The genome of each living organism is the full set of inherited instructions required to sustain all living processes, that is, the organism's operating system. Genomes differ in size, shape, and sequence complexity. Genome size, the number of base-paired nucleotides, varies over an enormous range from less than one million bp in some species of *Mycoplasma* (the smallest known bacteria) to greater than 10^{10} bp in certain plants. In general, prokaryotic genomes are smaller than those of eukaryotes. In contrast to prokaryotic genomes, which typically consist of single, circular DNA molecules, eukaryotic genomes are divided into two or more linear DNA molecules. The most significant difference between prokaryotic and eukaryotic genomes, however, is the vastly larger information-coding capacity and the presence of large quantities of noncoding DNA of eukaryotes. For this reason each type of genome will be considered separately. Short segments of the genomes of several eukaryotes are compared with that of *E. coli* in Figure 17.20.

PROKARYOTIC GENOMES Investigations of prokaryotic chromosomes, especially those of several strains of *E. coli*, have revealed the following:

1. Genome size. As described, most prokaryotic genomes are relatively small with considerably fewer genes than those of eukaryotes. The *E. coli* chromosome contains about 4.6 megabases (Mb) that code for about 4300 genes. (One Mb is 1×10^6 bases.)

2. Coding capacity. The genes of prokaryotes are compact and continuous; that is, they contain little, if any, noncoding DNA either between or within gene sequences. This is in sharp contrast to eukaryotic DNA, in which a significant percentage of DNA can be in noncoding form.

3. Gene expression. The regulation of many functionally related genes is enhanced by organizing them into operons. An **operon** is a set of linked genes that are regulated as a unit. About one-fourth of the genes of *E. coli* are organized into operons.

Recall that prokaryotes also often possess additional small pieces of DNA (see p. 37). These structures, called plasmids, are usually, but not always, circular. Plasmids typically have genes that are not present on the main chromosome. Although these genes are usually not essential for bacterial growth and survival, they may code for biomolecules that provide the cell with a growth

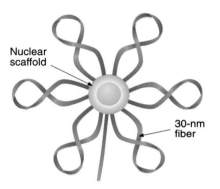

FIGURE 17.19

Chromatin.

In one proposal for the structure of 200-nm filaments, the 30-nm fiber is looped and attached to a nuclear scaffold composed of protein.

FIGURE 17.20

Comparison of 50-kb Segments of the Genomes of Selected Eukaryotes with the Prokaryote *E. coli* Genome.

As indicated, the genomes of organisms such as (a) humans, (b) *Saccharomyces cerevisiae*, (c) maize, and (d) *E. coli* can vary considerably in their complexity and gene density. Genes are indicated by letters and/or numbers. It should be noted that complex eukaryotes such as humans have genes that are interrupted with sequences such as introns and nonfunctional sequences called pseudogenes that resemble true genes. Also note that bacteria have few if any genome-wide repeats (repetitive, non-coding segments).

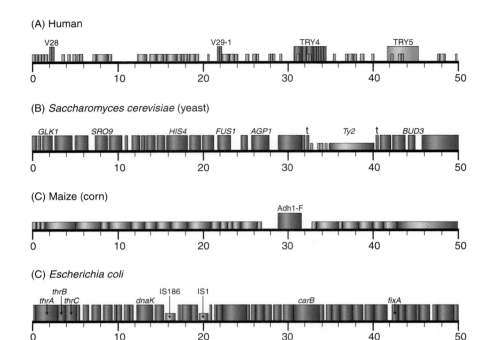

(A) Human

(B) *Saccharomyces cerevisiae* (yeast)

(C) Maize (corn)

(C) *Escherichia coli*

KEY

| ■ Gene | ■ Intron | ■ Human pseudogene | ■ Genomic-wide repeat | $\boxed{t}$ tRNA gene |

or survival advantage. Examples include antibiotic resistance, unique metabolic capacities (e.g., nitrogen fixation, and degradation of unique energy sources such as aromatic compounds) or virulence (e.g., toxins or other factors that undermine host defense mechanisms).

EUKARYOTIC GENOMES The organization of genetic information in eukaryotic chromosomes has proven to be substantially more complex than that observed in prokaryotes. Eukaryotic nuclear genomes possess the following unique features:

1. Genome size. Eukaryotic genomes do tend to be substantially larger than those of prokaryotes. However, in the higher eukaryotes genome size is not necessarily a measure of the complexity of the organism. For example, recall that the haploid genome of humans is 3000 Mb. The genomes of peas and the salamander are 5000 Mb and 90,000 Mb, respectively. For reasons that are still unclear, some species have accumulated vast amounts of noncoding DNA.

2. Coding capacity. Although there is enormous coding capacity, the majority of DNA sequences in eukaryotes do not appear to have coding functions, that is, they do not possess intact regulatory regions that initiate transcription (the production of RNA transcripts). The functions of these noncoding sequences are unknown; some of them probably have regulatory or structural roles. It has been estimated that no more than 1.5% of the human genome codes for proteins.

3. Coding continuity. Most eukaryotic genes investigated so far are discontinuous. Noncoding sequences (called **introns** or intervening sequences) are interspersed between sequences called **exons** (expressed sequences), which code for a gene product (any of various RNA molecules, some of which dictate the translation of proteins). Intron sequences are removed from hnRNA transcripts by a splicing mechanism (Section 18.2) to produce a functional RNA molecules.

By recent estimates, approximately 45% of the human genome is composed of repetitive sequences (repeated sequences of nucleotides). Although their significance is not understood, several types of repetitive sequences have been identified and investigated. There are two general classes: tandem repeats and interspersed genome-wide repeats. Each is briefly described.

Tandem repeats are DNA sequences in which multiple copies are arranged next to each other. These sequences were originally referred to as **satellite DNA** because they form a separate or "satellite" band when genomic DNA is broken into pieces and centrifuged to separate the fragments by density gradient centrifugation (Biochemical Methods 17.1). The lengths of the repeated sequences vary from less than 10 bp to over 2000 bp. Total lengths of the tandem repeats often vary between 10^5 and 10^7 bp. Certain types of tandem repeats apparently play structural roles in **centromeres** (the structures that attach chromosomes to the mitotic spindle during mitosis and meiosis) and **telomeres** (structures at the ends of chromosomes that buffer the loss of critical coding sequences after a round of DNA replication). Two relatively small types of repetitive sequences are referred to as minisatellites and microsatellites. **Minisatellites** have tandemly repeated sequences of about 25 bp with total lengths between 10^2 and 10^5 bp. In **microsatellites** there is a core sequence of 2 to 4 bp that is tandemly repeated from 10 to 20 times. The functions of these repetitive sequences are for the most part unknown. Because of their large number in genomes and their pleomorphic nature (i.e., they vary with each individual organism), minisatellites and microsatellites are used as markers in the diagnosis of genetic disease and in forensic investigations (Special Interest Box 17.1).

As their name implies, **interspersed genome-wide repeats** are repetitive sequences that are scattered around the genome. Most of these sequences are the result of **transposition** (Section 18.1), a mechanism whereby certain DNA sequences can be duplicated and move within the genome. **Transposable DNA elements**, referred to as **transposons**, excise themselves and then insert at another site. More commonly, however, transposition mechanisms involve an RNA transcript intermediate. These latter DNA elements are called **RNA transposons** or **retrotransposons**. The most abundant retrotransposon in humans is the *Alu sequence*. Alu sequences, whose lengths are about 280 bp, are present in about 500,000 copies. The function of Alu sequences and other retrotransposons is unknown. It is suspected that they are molecular parasites whose primary purpose is their own propagation.

KEY CONCEPTS 17.5

In each organisms' genome, the information required to direct living processes is organized so it can be efficiently stored and used. Genomes from different types of organisms differ in their sizes and levels of complexity.

QUESTION 17.11

Define the following terms:

a. tandem repeats

b. centromere

c. satellite DNA

d. introns

e. exons

f. microsatellites

g. transposition

QUESTION 17.12

Compare the sizes and coding capacity of prokaryotic genomes with those of eukaryotes. What other features distinguish them?

The techniques used in the isolation, purification, and characterization of biomolecules take advantage of their physical and chemical properties. Most of the techniques used in nucleic acid research are based on differences in molecular weight or shape, base sequences, or complementary base pairing. Techniques such as chromatography, electrophoresis, and ultracentrifugation, which have been used successfully in protein research, have also been adapted to use with nucleic acids. In addition, other techniques have been developed that exploit the unique properties of nucleic acids. For example, under certain conditions DNA duplexes reversibly melt (separate) and reanneal (base pair to form a duplex again). One of several techniques that exploit this phenomenon, called *Southern blotting*, is often used to locate specific (and often rare) nucleic acid sequences. After brief descriptions of several techniques used to purify and characterize nucleic acids, the common method for determining DNA sequences is outlined. More complex techniques are described in Biochemical Methods 18.1.

Once bacterial cells have been ruptured or eukaryotic nuclei have been isolated, their nucleic acids are extracted and deproteinized. This may be accomplished by several methods. Bacterial nucleic acid is often precipitated by treating cell preparations with alkali and lysozyme (an enzyme that degrades bacterial cell walls by breaking glycosidic bonds). Partially degraded protein is extracted by using certain solvent combinations (e.g., phenol and chloroform). Similarly, eukaryotic nuclei can be treated with detergents or solvents to release their nucleic acids. Depending on which type of nucleic acid is being isolated, specific enzymes are used to remove the other type. For example, RNase removes RNA from nucleic acid preparations, leaving DNA intact. DNA is further purified by centrifugation. All nucleic acid samples must be handled carefully. First, nucleic acids are susceptible to the actions of a group of enzymes called nucleases. In addition to a variety of such enzymes that are released during cell extraction, nucleases can also be introduced from the environment, for example, the experimenter's hands. Secondly, high molecular weight nucleic acids, primarily DNA, are sensitive to shearing stress. Purification procedures, therefore, must be gentle, applying as little mechanical stress as possible.

Techniques Adapted from Use with Other Biomolecules

Many of the techniques used in protein purification procedures have also been adapted for use with nucleic acids. For example, several types of chromatography (e.g., ion-exchange, gel filtration, and affinity) have been used in several stages of nucleic acid purification and in the isolation of individual nucleic acid sequences. Because of its speed, HPLC has replaced many slower chromatographic separation techniques when small samples are involved.

A type of column chromatography that uses a calcium phosphate gel called **hydroxyapatite** has been especially useful in nucleic acid research. Because hydroxyapatite binds to double-stranded nucleic acid molecules more tenaciously than to single-stranded molecules, double-stranded DNA (dsDNA) can be effectively separated from single-stranded DNA (ssDNA), RNA, and protein contaminants by eluting the column with increasing concentrations of phosphate buffer. The use of hydroxyapatite columns has recently been largely replaced by a form of affinity chromatography in which the column matrix molecules have been covalently bonded to avidin, a small protein that binds specifically to biotin. When an ssDNA binds to a biotinylated ssDNA, the resulting dsDNA binds to the column, while the rest of the sample passes through.

The movement of nucleic acid molecules in an electric field depends on both their molecular weight and their three-dimensional structure. However, because DNA molecules often have relatively high molecular weights, their capacity to penetrate some gel preparations (e.g., polyacrylamide) is limited. Although DNA sequences with less than 500 bp can be separated by polyacrylamide gels with especially large pore sizes, more porous gels must be used with larger DNA molecules. Agarose gels, which are composed of a crosslinked polysaccharide, are used to separate DNA molecules with lengths between 500 bp and approximately 150 kilobases (kb). Larger sequences are now isolated with a variation of agarose gel electrophoresis in which two electric fields (perpendicular to each other) are alternately turned on and off. DNA molecules reorient themselves each time the electric field alternates, resulting in a very efficient and precise separation of heterogeneous groups of DNA molecules.

Density gradient centrifugation (Biochemical Methods 2.1) with cesium chloride (CsCl) has been widely used in nucleic acid research. At high speeds, a linear gradient of CsCl is established. Mixtures of DNA, RNA, and protein migrating through this gradient separate into discrete bands at positions where their densities are equal to the density of the CsCl. DNA molecules with high guanine and cytosine content are more dense than those with a higher proportion of adenine and thymine. This difference helps separate heterogeneous mixtures of DNA fragments.

Techniques that Exploit the Unique Structural Features of the Nucleic Acids

Several unique properties of the nucleic acids (e.g., absorption of UV light at specific wavelengths and their tendency to reversibly form double-stranded complexes) are exploited in nucleic acid research. Several applications of these properties are briefly discussed.

Because of their aromatic structures, the purine and pyrimidine bases absorb UV light. At pH 7 this absorption is especially strong at 260 nm. However, when the nitrogenous bases are incorporated into polynucleotide sequences, various noncovalent forces promote close interactions between them. This decreases their absorption of UV light. This **hypochromic effect** is an invaluable aid in studies involving nucleic acid. For example, absorption changes are routinely used to detect the disruption of the double-stranded structure of DNA or the hydrolytic cleavage of polynucleotide strands by enzymes.

The binding forces that hold the complementary strands of DNA together can be disrupted. This process, referred to as **denaturation** (Figure 17A), is promoted by heat, low salt concentrations, and extremes in pH. (Because it is easily controlled, heating is the most common denaturing method in nucleic acid investi-

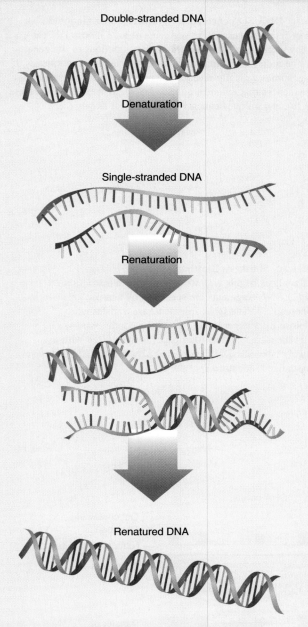

FIGURE 17A Denaturation and Renaturation of DNA.
Under appropriate conditions, DNA that has been denatured can
renature, that is, strands with complementary sequences re-form into
a double helix.

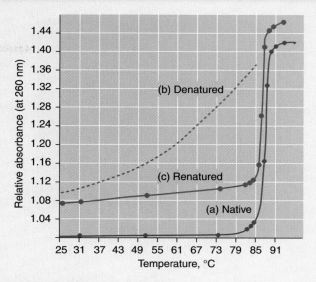

FIGURE 17B DNA Denaturation.
(a) When native DNA is heated, its absorbance does not change until
a specific temperature is reached. The "melting temperature" T_m
of a DNA molecule varies with its base composition. (b) When
the denatured DNA is chilled, its absorbance falls, but along a
different curve. It does not return to its original absorbance value.
(c) Reannealed (renatured) DNA can be prepared by maintaining
the temperature at 25°C below the denaturing temperature for an
extended period.

gations.) When a DNA solution is slowly heated, absorption at 260
nm remains constant until a threshold temperature is reached.
Then the sample's absorbance increases (Figure 17B). The
absorbance change is caused by the unstacking of bases and the
disruption of base pairing. The temperature at which one-half of
a DNA sample is denatured, referred to as the melting temperature (T_m), varies among DNA molecules according to their base
compositions. [Recall that DNA stability is affected by the number of hydrogen bonds between GC and AT pairs and base stacking interactions. (See p. 566). More energy is required, therefore,
to "melt" DNA molecules with high G and C content.] If the separated DNA strands are held at a temperature approximately 25°C
below the T_m for an extended time, renaturation is possible. Renaturation, or reannealing, requires some time because the strands
explore various configurations until they achieve the most stable
one (i.e., paired complementary regions).

DNA melting is extraordinarily useful in nucleic acid **hybridization**. Single-stranded DNA from different sources associates (or
"hybridizes") if there is a significant sequence homology (i.e., structural similarity). If a DNA sample is sheared into small uniform
pieces, the rate of reannealing depends on the concentration of DNA
strands and on the structural similarities between them. Reannealing rates have revealed valuable information about genome structure.
For example, organisms vary in the number and types of unique
sequences their genomes contain. (A unique DNA sequence occurs
only once per haploid genome.) The relative number of unique and
repeated sequences can be determined by constructing a C_0t curve.
(C_0t is a measure of renaturation in which C_0 is the concentration
of ssDNA in moles/liter and t is elapsed time in seconds.) C_0t curves
have demonstrated that the velocity of reannealing declines as
genomes become larger and more complex. The frequency of unique
and repeated DNA sequences determined by measuring reannealing rates for the mouse genome is shown in Figure 17C.

(continued on page 589)

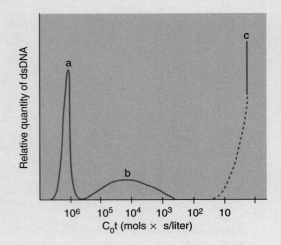

FIGURE 17C DNA Sequence Pattern of the Mouse Genome.
The degree of repetitiveness in the segments of total mouse DNA is determined by measuring C_0t values for several fractions of the genome. The dotted line indicates estimated values. The reassociation kinetics of eukaryotic genomes typically reveal three primary classes of sequences: repetitive sequences that reanneal quickly (a), sequences of intermediate complexity (b), and unique sequences that reanneal slowly (c).

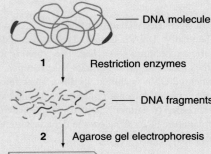

FIGURE 17D Southern Blotting.
(1) DNA analysis begins with its digestion by a restriction enzyme. (2) DNA fragments are separated by agarose gel electrophoresis. (3) The DNA fragments are transferred to nitrocellulose filter paper under denaturing conditions. (4) The ssDNA on the nitrocellulose filter paper is hybridized with a radioactively labeled ssDNA probe. (5) Any hybridized DNA can be visualized by autoradiography.

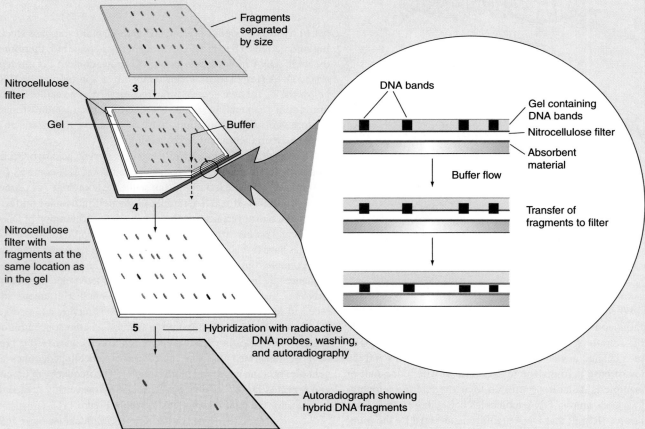

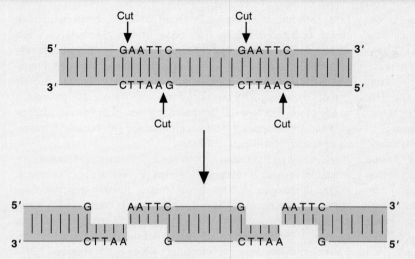

FIGURE 17E Restriction Enzymes.
Restriction endonucleases are enzymes isolated from bacteria that cut DNA at specific sequences. In this example the enzyme EcoRI (obtained from *E. coli*) makes staggered cuts that result in the formation of "sticky ends." Some restriction enzymes make "blunt cuts."

Hybridization can also be used to locate and/or identify specific genes or other DNA sequences. For example, ssDNA from two different sources (e.g., tumor cells and normal cells) can be screened for sequence differences. If one set of ssDNA is biotinylated, then the double-stranded hybrids bind to an avidin column. If any unhybridized sequence is present, it passes through the column. Then it can be isolated and identified. In **Southern blotting** (Figure 17D) radioactively labeled DNA or RNA probes (sequences with known identities) locate a complementary sequence in the midst of a DNA digest, which typically contains a large number of heterogeneous DNA fragments. A DNA digest is obtained by treating a DNA sample with restriction enzymes that cut at specific nucleotide sequences (Figure 17E). (Produced by bacterial cells, restriction enzymes protect bacteria against viral infection by cleaving viral DNA at specific sequences.) Once the DNA sample has been digested, the fragments are separated by agarose gel electrophoresis according to their sizes. After the gel is soaked in 0.5 M NaOH, a process that converts dsDNA to ssDNA, the DNA fragments are transferred to nitrocellulose filter paper by placing them on a wet sponge in a tray with a high salt buffer. (Nitrocellulose has the unique property of binding strongly to ssDNA.) Absorbent dry filter paper is placed in direct contact with the nitrocelluose filter/agarose gel sandwich. As buffer is drawn through the gel and filter paper by capillary action, the DNA is transferred and becomes permanently bound to the nitrocellulose filter. (The trans-

fer of DNA to the filter is the "'blotting" referred to in the name of this technique.) Subsequently, the nitrocellulose filter is exposed to the radioactively labeled probe, which binds to any ssDNA with a complementary sequence. For example, an mRNA that codes for β-globin binds specifically to the β-globin gene, even though β-globin mRNA lacks the introns present in the gene. Apparently, there is sufficient base pairing between the two single-stranded molecules that the gene can be located.

DNA Sequencing

The determination of DNA nucleotide sequences has provided valuable insights in such fields as biochemistry, medical science, and evolutionary biology. The analysis of long DNA sequences begins with the formation of smaller fragments using one type of restriction enzyme. Each fragment is then sequenced independently by the chain-terminating method. As with protein primary structure determinations, these steps are repeated with a different set of polynucleotide fragments (generated by another type of restriction enzyme) that overlap the first set. Sequence information from both sets of experiments then orders the fragments into a complete sequence.

DNA sequencing by the **chain-terminating method** (Figure 17F), developed by Frederick Sanger, uses restriction enzymes to cleave large DNA segments into smaller fragments. Each fragment *(continued on page 590)*

(continued on page 590)

FIGURE 17F The Sanger Chain Termination Method.
A specific primer is chosen so that DNA synthesis begins at the point of interest. DNA synthesis continues until a radioactive dideoxynucleotide is incorporated and the chain terminates. Afterwards the products of the reactions are separated by gel electrophoresis and analyzed by autoradiography. The fragments migrate according to size. The sequence is determined by "reading" the gel.

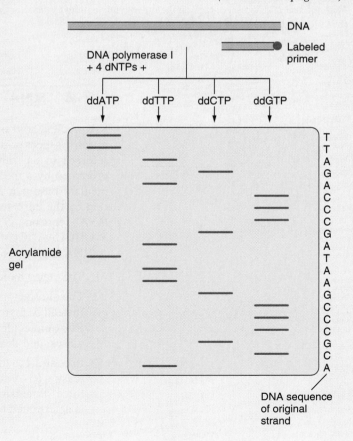

is separated into two strands, one of which is used as a template to produce a complementary copy. The sample is further divided into four test tubes. To each tube is added the substances required for DNA synthesis (e.g., the enzyme DNA polymerase and the four deoxyribonucleotide triphosphates). In addition, a ^{32}P-labeled primer (a short segment of a complementary DNA strand) is added to each tube. (By selecting the primer, the investigator can start DNA sequencing at specific sites.) Also present in each of the four tubes is a different 2′-3′ dideoxynucleotide derivative. (The dideoxy derivatives are synthetic nucleotide analogues in which the hydroxy groups on the 2′- and 3′-carbons have been replaced with hydrogens.) Dideoxynucleotides can be incorporated into a growing polynucleotide chain, but they cannot form a phosphodiester linkage with another nucleotide. Consequently when dideoxynucleotides are incorporated, they terminate the chain. Because small amounts of the dideoxynucleotides are used, they are randomly incorporated into growing polynucleotide strands. Each

tube, therefore, contains a mixture of DNA fragments containing strands of different lengths. Each newly synthesized strand ends in a dideoxynucleotide residue. The reaction products in each tube are separated by gel electrophoresis and analyzed together by autoradiography. Each band in the autoradiogram corresponds to a polynucleotide that differs in length by one nucleotide from the one that precedes it in any of the four lanes of the autoradiogram. Note that the smallest polynucleotide appears on the bottom of the gel, because it moves more quickly than larger molecules.

Recently, an automated version of the Sanger method has become available. Instead of using radiolabeled primers, it uses fluorescent tagged dideoxynucleotides. Because each dideoxy analogue fluoresces a different color, the entire procedure is carried out in a single test tube. Afterwards the reaction products are loaded and run on a single electrophoresis gel. After a detector scans the gel, a computer determines the sequence of the colored bands (Figure 17G).

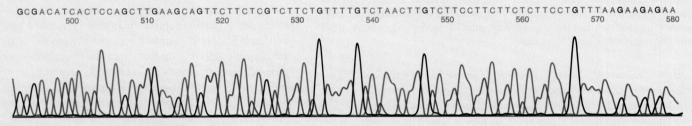

GCGACATCACTCCAGCTTGAAGCAGTTCTTCTCGTCTTCTGTTTTGTCTAACTTGTCTTCCTTCTTCTCTTCCTGTTTAAGAAGAGAA
500   510   520   530   540   550   560   570   580

FIGURE 17G Automated DNA Sequencing.
With the use of fluorescent tags on the dideoxynucleotides, a detector can scan a gel quickly and determine the sequence from the order of the colors in the bands.

17.2 RNA

Ribonucleic acid is a class of polynucleotides, nearly all of which are involved in some aspect of protein synthesis. RNA molecules are synthesized in a process referred to as transcription. During transcription, new RNA molecules are produced by a mechanism similar to DNA synthesis, that is, through complementary base pair formation. The sequence of bases in RNA is therefore specified by the base sequence in one of the two strands in DNA. For example, the DNA sequence 5′-CCGATTACG-3′ is transcribed into the RNA sequence 3′-GGCUAAUGC-5′. (Complementary DNA and RNA sequences are antiparallel.)

RNA molecules differ from DNA in the following ways:

1. The sugar moiety of RNA is ribose instead of deoxyribose in DNA.

2. The nitrogenous bases in RNA differ somewhat from those observed in DNA. Instead of thymine, RNA molecules use uracil. In addition, the bases in some RNA molecules are modified by a variety of enzymes (e.g., methylases, thiolases, and deaminases).

3. In contrast to the double helix of DNA, RNA exists as a single strand. For this reason, RNA can coil back on itself and form unique and often quite complex three-dimensional structures (Figure 17.21). The shape of these structures is determined by complementary base pairing by specific RNA sequences, as

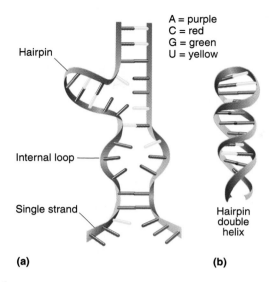

A = purple
C = red
G = green
U = yellow

Hairpin

Internal loop

Single strand

(a)

Hairpin
double
helix

(b)

FIGURE 17.21

Secondary Structure of RNA.

(a) Many different types of secondary structures occur in RNA molecules. (b) A hairpin structure.

well as by base stacking. In addition, the 2′-OH of ribose can form hydrogen bonds with nearby molecular groups. Because RNA is single stranded, Chargaff's rules do not apply. An RNA molecule's contents of A and U, as well as C and G, are usually not equal.

The most prominent types of RNA are transfer RNA, ribosomal RNA, and messenger RNA. The structure and function of each of these molecules is discussed. Examples of other less abundant (but no less important) types of RNA (heterogeneous RNA and small nuclear RNA) are also described.

Transfer RNA

Transfer RNA (tRNA) molecules transport amino acids to ribosomes for assembly into proteins. Comprising about 15% of cellular RNA the average length of a tRNA molecule is 75 nucleotides. Because each tRNA molecule becomes bound to a specific amino acid, cells possess at least one type of tRNA for each of the 20 amino acids commonly found in protein. The three-dimensional structure of tRNA molecules, which resembles a warped cloverleaf (Figure 17.22), results primarily from extensive intrachain base pairing. tRNA molecules contain a variety of modified bases. Examples include pseudouridine, 4-thiouridine, 1-methylguanosine, and dihydrouridine:

Ribose	Ribose	Ribose	Ribose
Pseudouridine	**4-Thiouridine**	**1-Methylguanosine**	**Dihydrouridine**

The structure of tRNA allows it to perform two critical functions involving the most important structural components: the 3′-terminus and the anticodon loop.

Ancient DNA

Evolutionary biology is essentially a historical science. Ever since the publication of *On the Origin of Species* by Charles Darwin in 1859, biologists have attempted to reconstruct the events and processes that gave rise to modern organisms by investigating fossils and the comparative anatomy of modern species. *Fossils*, the preserved part of ancient organisms, have been used to trace the lineage of modern organisms. For example, fossilized skeletal remains have allowed paleontologists to trace human lineage back about three million years. Most fossils were formed when recently deceased organisms were exposed to environmental conditions that slowed the process of decomposition. They were covered with sediment (fine soil particles suspended in water) or became embedded in bogs, tar pits, amber (a polymerized resin derived from plant essential oils), or ice. Arid desert conditions also promoted fossil formation.

Comparative anatomical studies have provided a wealth of information concerning the relationships of modern species. Consider for example, that the similar structures of the forelimbs of most vertebrates suggest a common history for these species. Species with greater structural similarities (e.g., humans and chimpanzees) are more closely related than those with obvious dissimilarities (e.g., whales and birds). As nucleic acid and protein sequencing techniques have become available, however, the structures of these molecules have provided more precise information concerning the relationships of modern species. For example, by using DNA and protein sequence studies, molecular evolutionary rates have been calculated by detecting changes in the DNA base sequences or the polypeptide amino acid sequences of different species. This information, in conjunction with fossil evidence, has been used to make estimates, referred to as an *evolutionary clock*, of the time required for evolutionary changes. In addition, DNA sequence information has provided an extraordinarily promising mechanism for comparing the genetic instructions of all existing species. Unfortunately, there are severe limitations on the conclusions that molecular paleontologists can infer from the study of modern DNA sequences, because they cannot check these sequences against a historical record. Or can they?

Although well-preserved nucleated cells have been observed in specimens since about 1912, DNA recovery became feasible only in the 1980s. The first successful extractions of ancient DNA (aDNA) took advantage of **cloning**, a recombinant DNA technique (Biochemical Methods 18.1), in which bacteria are used to generate a large number of copies of specific DNA sequences. These sequences are then investigated (via DNA sequencing and hybridization techniques) in terms of their relationship to comparable sequences of modern species. For example, in 1984, DNA was successfully cloned from a preserved quagga (an extinct animal that resembled both horses and zebras). Later investigations of quagga DNA confirmed its close similarity to both horse and zebra DNA.

DNA cloning was an important breakthrough, but its use is awkward when applied to fossilized specimens. The principal reason is that cloning requires larger amounts of DNA than are often present in fossils. (This is an important consideration because interesting genes, such as those that code for protein, generally appear in only two copies per cell.) Investigations of aDNA seemed as elusive as ever in the early 1980s. However, this situation changed dramatically in 1985, when a new technique, called **polymerase chain reaction (PCR)** became available. By using PCR (Biochemical Methods 18.1) as many as one billion copies of DNA sequences can be produced in a test tube. Because PCR is extraordinarily sensitive (a single DNA molecule can be amplified), it seemed made to order for studies of aDNA.

Since PCR has been applied to aDNA investigations, the molecular paleontologists have extracted DNA fragments from a wide variety of fossils, artifacts, and museum specimens. For example, DNA sequences have been isolated from such disparate sources as amber-embedded insects (over 100 million years old), fossil herbarium specimens (millions of years old), and Egyptian mummies (over 6000 years old). Comparisons of these and other DNA sequences to those of modern species have provided important information concerning how populations change over time and how much time has elapsed since species shared a common ancestor (i.e., the evolutionary clock).

And Not-So-Ancient DNA

The identification techniques used in aDNA investigations are similar to those used with DNA samples of more recent vintage. Such methods are especially valuable in diagnostic clinical medicine and forensic investigations.

In diagnostic clinical medicine, DNA identification has been applied to investigations of patient tissue specimens obtained during biopsies or autopsies. Because of imaginative adaptations of DNA extraction techniques, the DNA contained in slides and preserved tissue specimens can now be investigated for evidence of infectious or genetic diseases. For example, *in situ* DNA hybridization is an ultrasensitive technique in which specific DNA probes are applied directly to tissue embedded in paraffin. (Paraffin-embedded tissue specimens are used in microscopic studies. Such slides can be stored

The *3′-terminus* forms a covalent bond to a specific amino acid. The *anticodon loop* contains a three-base-pair sequence that is complementary to the DNA triplet code for the specific amino acid. The conformational relationship between the 3′-terminus and the anticodon loop allows the tRNA to align its attached amino acid properly during protein synthesis. (This process is discussed in Chapter 19.) tRNAs also possess three other prominent structural features, referred to

indefinitely.) This technique recently confirmed the presence of the bacterium *Helicobacter pylori* in preserved specimens taken from the stomachs of ulcer patients. (*H. pylori* is now believed to be a causative agent in most cases of gastric ulcer, as well as in stomach cancer.) *In situ* hybridization, as well as PCR combined with Southern blotting, has detected viral sequences (e.g., HIV) in preserved specimens. (These and other biochemical tests now indicate that some patients died of HIV in the United States at least as early as the 1950s, although AIDS was not recognized until 1981.)

DNA persists for many years in dried biological specimens (e.g., blood, saliva, hair, and semen) and in bone. Consequently, DNA can be used as evidence in any type of forensic investigation in which such specimens are available. DNA analysis techniques that are typically used to ascertain the identity of victims and/or perpetrators of violent crimes are referred to as **DNA typing**, or profiling. DNA typing involves the analysis of several highly variable sequences called markers. By using sets of variable sequences, investigators can provide unique, identifying genetic profiles for each individual human. In the past decade DNA typing has provided decisive information concerning guilt or innocence in numerous court cases. The techniques now available differ in their capacity to differentiate between individuals and in the speed with which results can be obtained. **DNA fingerprinting**, originally introduced in 1985 by the British geneticist Alec Jeffreys, is a variation of Southern blotting. In this technique, the banding characteristics of DNA minisatellites (see p. 585) from different individuals are compared, for example, crime scene specimen DNA with that of suspects (Figure 17H). When the quantity of DNA extracted from a crime scene sample is too minute to analyze, it is amplified by PCR. Consequently, the DNA from a single cell is now sufficient for DNA fingerprint analysis. The entire genome in each sample is isolated and treated with a restriction enzyme. Because of genetic variations, the DNA in minisatellite sequences from each individual fragments differently. (Such genetic differences are called **restriction fragment length polymorphisms**, or **RFLPs**.) After the restriction fragments are separated according to size on agarose gel electrophoresis and transferred to nitrocellulose filter paper, they are exposed to radiolabeled probes. Because the sizes of the fragments that bind to these probes differ from one individual to the next, banding patterns have been used successfully to either convict or acquit criminal suspects.

Although RFLP testing is an accurate method, it does have limitations. Among these are the substantial amounts of time (6 to 8 weeks), labor, and expertise required to obtain DNA profiles. A

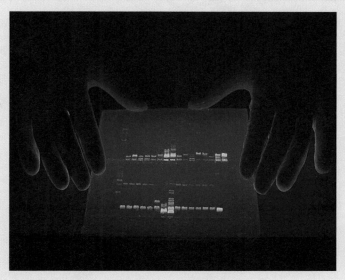

FIGURE 17H Forensic Use of DNA Fingerprinting. In many criminal cases RFLP (restriction fragment length polymorphism) analysis of biological evidence collected at a crime scene provides conclusive proof of guilt or innocence.

newer methodology that analyzes **short tandem repeats (STR)** (DNA sequences with between 2 and 4 bp repeats, called microsatellites, see p. 585) has significantly greater discriminating power than RFLP and is relatively rapid (several hours). After DNA is extracted from a specimen, several target STR sequences are amplified by PCR and linked to fluorescent dye molecules. A **DNA profile**, which results when the PCR products are separated in an electrophoretic gel, consists of the pattern and the number of repeats of each target sequence on the gel. Fluorescence detection increases the sensitivity of the technique. Unlike RFLP, STR-based DNA typing is easily automated. If the DNA profiles from individual samples are compared and determined to be identical, the samples are said to be a match. If compared profiles are not identical, they are said to have come from different sources. The results are reported in terms of the probabilities of a random match (the chance that a randomly selected person from the population will have an identical DNA profile to that of the specimen of interest such as that left at a crime scene). The use of multiple markers and the sensitivity of the methodology reduces the random match probability to at least 1 in several billion.

as the D loop, the TψC loop, and the variable loop. (ψ is an abbreviation for the modified base pseudouridine.) The function of these structures is unknown, but they are presumably related to the alignment of tRNA within the ribosome and/or the binding of a tRNA to the enzyme that catalyzes the attachment of the appropriate amino acid. The *D loop* is so named because it contains dihydrouridine. Similarly, the TψC loop contains the base sequence thymine,

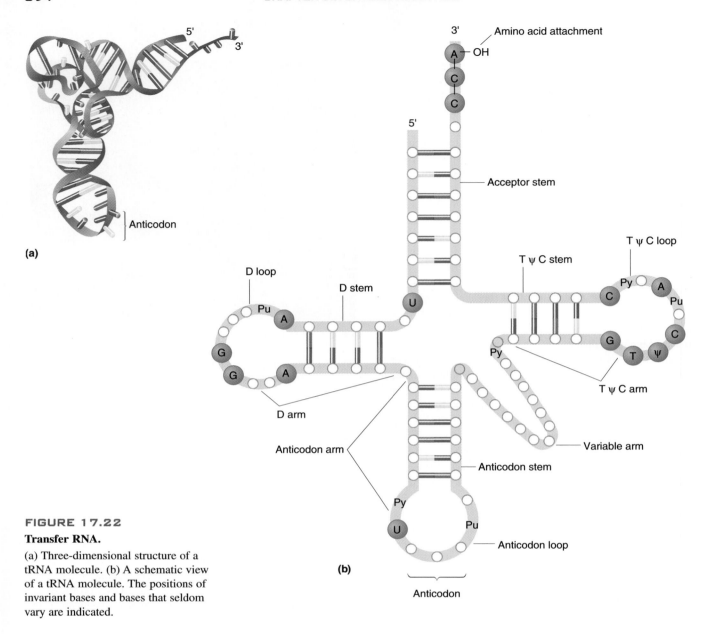

FIGURE 17.22

Transfer RNA.

(a) Three-dimensional structure of a tRNA molecule. (b) A schematic view of a tRNA molecule. The positions of invariant bases and bases that seldom vary are indicated.

pseudouridine, and cytosine. tRNAs can be classified on the basis of the length of their *variable loop*. The majority (approximately 80%) of tRNAs have variable loops with four to five nucleotides, whereas the others have variable loops with as many as 20 nucleotides.

Ribosomal RNA

Ribosomal RNA (rRNA) is the most abundant form of RNA in living cells. (In most cells, rRNA constitutes approximately 80% of the total RNA.) The secondary structure of rRNA is extraordinarily complex (Figure 17.23). Although there are species differences in the primary nucleotide sequences of rRNA, the overall three-dimensional structure of this class of molecules is conserved. As its name suggests, rRNA is a component of ribosomes.

As described, ribosomes are cytoplasmic structures that synthesize proteins. (Because they are composed of both protein and rRNA, the ribosomes are sometimes described as ribonucleoprotein bodies.) The ribosomes of prokaryotes and eukary-

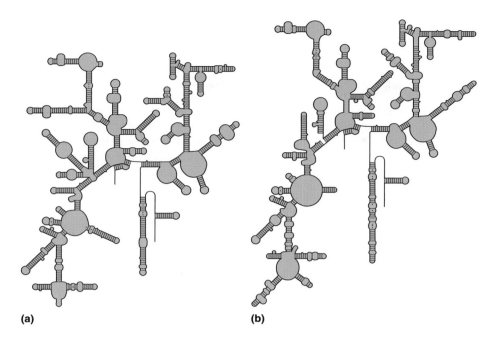

(a)   (b)

FIGURE 17.23
rRNA Structure.
Although their sequences differ, the three-dimensional structure of these 16-S-like rRNAs from (a) *E. coli* and (b) *Saccharomyces cervisiae* (yeast) appear remarkably similar.

otes are similar in shape and function, although they differ in size and their chemical composition. Both types of ribosome consist of two subunits of unequal size, which are usually referred to in terms of their S values. (S is an abbreviation for the Svedberg (or sedimentation) unit, which is a measure of sedimentation velocity in a centrifuge. Because sedimentation velocity depends on the molecular weight and the shape of a particle, S values are not necessarily additive.) Prokaryotic ribosomes (70 S) are composed of a 50 S subunit and a 30 S subunit, whereas ribosomes of eukaryotes (80 S) contain a 60 S subunit and a 40 S subunit.

Several different kinds of rRNA and protein are found in each type of ribosomal subunit. The large ribosomal subunit of *E. coli*, for example, contains 5 S and 23 S rRNAs and 34 polypeptides. The small ribosomal subunit of *E. coli* contains a 16 S rRNA and 21 polypeptides. A typical large eukaryotic ribosomal subunit contains three rRNAs (5 S, 5.8 S, and 28 S) and 49 polypeptides; the small subunit contains an 18 S rRNA and approximately 30 polypeptides. The functions of the rRNA and polypeptides in ribosomes are poorly understood and are being investigated.

Messenger RNA

As its name suggests, **messenger RNA** (mRNA) is the carrier of genetic information from DNA for the synthesis of protein. mRNA molecules, which typically constitute approximately 5% of cellular RNA, vary considerably in size. For example, mRNA from *E. coli* varies from 500 to 6000 nucleotides.

Prokaryotic mRNA and eukaryotic mRNA differ in several respects. First, many prokaryotic mRNAs are *polycistronic*, that is, they contain coding information for several polypeptide chains. In contrast, eukaryotic mRNA typically codes for a single polypeptide and is therefore referred to as *monocistronic*. (A **cistron** is a DNA sequence that contains the coding information for a polypeptide and several signals that are required for ribosome function.) Second, prokaryotic and eukaryotic mRNAs are processed differently. In contrast to prokaryotic mRNAs, which are translated into protein by ribosomes during or immediately after they are synthesized, eukaryotic mRNAs are modified extensively. These modifications include capping (linkage of 7-methylguanosine to the 5′-terminal residue), splicing (removal of introns), and the attachment of an adenylate polymer referred to as a poly A tail. (Each of these processes is described in Chapter 18.)

KEY CONCEPTS 17.6

RNA is a nucleic acid that is involved in various aspects of protein synthesis. The most abundant types of RNA are transfer RNA, ribosomal RNA, and messenger RNA.

Heterogeneous RNA and Small Nuclear RNA

Heterogeneous RNA and small nuclear RNA play complementary roles in eukaryotic cells. **Heterogeneous nuclear RNA** (hnRNA) molecules are the primary transcripts of DNA and are the precursors of mRNA. HnRNA is processed by splicing and modifications to form mRNA. **Splicing** is the enzymatic removal of the introns from the primary transcripts. A class of **small nuclear RNA** (snRNA) molecules (containing between 90 and 300 nucleotides), which are complexed with several proteins to form **small nuclear ribonucleoprotein particles** (snRNP or snurps), are involved in splicing activities and other forms of RNA processing.

QUESTION 17.13

The following terms apply to what types of RNA?

  a. poly A tail b anticodon c. codon d. splicing

QUESTION 17.14

List the four bases most commonly found in RNA.

QUESTION 17.15

Briefly describe the functions of each major type of RNA.

QUESTION 17.16

When a gene is transcribed, only one DNA strand acts as a template for the synthesis of the RNA molecule. This strand is referred to as a **antisense** (or noncoding) strand; the nontranscribed DNA strand is called the **sense** (or coding) strand. The base sequence of the sense strand is the DNA version of the mRNA used to synthesize the polypeptide product of the gene. (The strand of a DNA segment that acts as the sense strand differs from one gene to the next.) Investigations of this aspect of nucleic acid metabolism have revealed that bacteria and viruses use the synthesis of so-called antisense RNAs to control certain aspects of cell metabolism. When an antisense RNA is produced (by transcription from the antisense DNA strand), it binds specifically (through complementary base pairing) to the corresponding mRNA. This binding prevents polypeptide synthesis from the mRNA.

Because mRNA–antisense RNA binding is so specific, antisense molecules are considered promising research tools. Numerous investigators are using antisense RNA molecules to study eukaryotic function by selectively turning on and off the activities of specific genes. This so-called reverse genetics is also useful in medical research. Although serious problems have been encountered in antisense research (e.g., the inefficient insertion of oligonucleotides into living cells and high manufacturing costs), antisense technology has already provided valuable insight into the mechanisms of several diseases (e.g., cancer and viral infections). Consider the following sense DNA sequence:

5′-GCATTCGAATTGCAGACTCCTGCAATTCGGCAAT-3′

Determine the sequence of its complementary strand. Then determine the mRNA and antisense RNA sequences. (Recall that in RNA structure, U is substituted for T. So A in a DNA strand is paired with a U as RNA is synthesized.)

17.3 VIRUSES

Viruses lack most of the properties that distinguish life from nonlife. For example, viruses cannot carry on metabolic activities on their own. Yet under the appropriate conditions they can wreak havoc on living organisms. Often described as obligate, intracellular parasites, viruses can also be viewed as mobile genetic elements because of their structure, that is, each consists of a piece of nucleic acid enclosed within a protective coat. Once a virus has infected a host cell, its nucleic acid can hijack the cell's nucleic acid and protein-synthesizing machinery. As viral components accumulate, complete new viral particles are produced and then released from the host cell. In many circumstances, so many new viruses are produced that the host cell lyses (ruptures). Alternatively, the viral nucleic acid may insert itself into a host chromosome, resulting in transformation of the cell (Special Interest Box 17.3).

Viruses have fascinated biochemists ever since their existence was suspected near the end of the nineteenth century. Driven in large part because of the role of viruses in numerous diseases, viral research has benefited biochemistry enormously. Because a virus subverts normal cell function to produce new virus, a viral infection can provide unique insight into cellular metabolism. For example, the infection of animal cells has provided invaluable and relatively unambiguous information about the mechanisms that glycosylate newly synthesized proteins. In addition, several eukaryotic genetic mechanisms have been elucidated with the aid of viruses and/or viral enzymes. Viral research has also provided substantial information concerning genome structure and carcinogenesis (the mechanisms by which normal cells are transformed into cancerous cells). Finally, viruses have been invaluable in the development of recombinant DNA technology.

The Structure of Viruses

An enormous number of viruses have been identified since 1892, when the Russian researcher Dmitri Ivanovski first isolated the tobacco mosaic virus. Because their origins and evolutionary history are unclear, the scientific classification of viruses has been difficult. Often, viruses have been assigned to groups according to such properties as their microscopic appearance (e.g., rhabdoviruses have a bullet-shaped appearance), the anatomic structures where they were first isolated (e.g., adenoviruses were discovered in the adenoids, a type of lymphoid tissue), or the symptoms they produce in a host organism (e.g., the herpes viruses cause rashes that spread). In recent years, scientists have attempted to develop a systematic classification system based primarily on viral structure, although several other factors are also important (e.g., host and disease caused).

Viruses occur in a bewildering array of sizes and shapes. Virions (complete viral particles) range from 10 nm to approximately 400 nm in diameter. Although most viruses are too small to be seen with the light microscope, a few (e.g., the pox viruses) can be visualized because they are as large as the smallest bacteria.

Simple virions are composed of a *capsid* (a protein coat made of interlocking protein molecules called capsomeres), which encloses nucleic acid. (The term *nucleocapsid* is often used to describe the complex formed by the capsid and the nucleic acid.) Most capsids are either helical or icosahedral. (Icosahedral capsids are 20-sided structures composed of triangular capsomeres.) The nucleic acid component of virions is either DNA or RNA. Although most viruses possess double-stranded DNA (dsDNA) or single-stranded RNA (ssRNA), examples with single-stranded DNA (ssDNA) and double-stranded RNA (dsRNA) genomes have also been observed. There are two types of ssRNA genomes. A *positive-sense* RNA genome [(+)-ssRNA] acts as a giant mRNA, that is, it directs the synthesis of a long polypeptide that is cleaved and processed into smaller molecules. A *negative-sense* RNA genome [(−)-ssRNA] is complementary in base sequence to the mRNA that directs the synthesis of viral proteins. Viruses that employ (−)-ssRNA genomes must provide an enzyme, referred to as a reverse transcriptase, that synthesizes the mRNA.

In more complex viruses, the nucleocapsid is surrounded by a membrane envelope, which usually arises from the host cell nuclear or plasma membranes. Envelope proteins, coded for by the viral genome, are inserted into the envelope membrane during virion assembly. Proteins that protrude from the surface of the envelope, called spikes, are believed to mediate the attachment of the virus to the host cell. Representative viruses are illustrated in Figure 17.24.

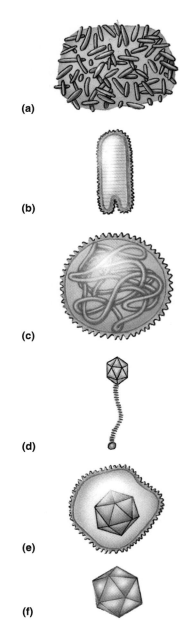

(a)

(b)

(c)

(d)

(e)

(f)

FIGURE 17.24

Representative Viruses.

(a) pox virus, (b) rhabdovirus, (c) mumps virus, (d) flexible-tailed bacteriophage, (e) herpes virus, (f) papilloma (wart) virus.

KEY CONCEPTS 17.7

Viruses are composed of nucleic acid enclosed in a protective coat. The nucleic acid may be a single- or double-stranded DNA or RNA. In simple viruses the protective coat, called a capsid, is composed of protein. In more complex viruses the nucleocapsid, composed of nucleic acid and protein, is surrounded by a membranous envelope derived from host cell membrane.

Evolution is the mechanism by which species diverge over time as random genetic changes and selection pressure promote adaptation to environmental conditions. Traditionally, evolutionary relationships have been traced by interspecies comparisons of the gross anatomical features of modern organisms. In recent decades these investigations have expanded to sequence comparisons of the primary sequences of homologous proteins and nucleic acids. Early efforts at establishing phylogenetic relationships by interpreting molecular sequences were based on the amino acid sequences of proteins such as the cytochromes. In recent years technological advances have made possible the more direct comparisons of nucleic acid sequences. As massive amounts of these data have been collected and analyzed, it has become apparent that evolution is both conservative and economical. In a sense each living organism is a palimpsest, that is, organismal processes are built step by step on systems already in place. (A *palimpsest* is a tablet or parchment on which information is inscribed, partially erased, and written again.) Consequently, organisms are in a very real sense historical documents. Taken together, this work has confirmed the assumptions of life scientists that there is a common ancestry of living species. This premise is based on the following observations:

1. The universal nature of the basic mechanisms by which genetic information is stored and transmitted.
2. The striking similarities among essential metabolic processes in currently existing organisms (e.g., in energy generation).
3. The near universal chirality of asymmetrical biomolecules (e.g., D-sugars and L-amino acids).

As evidence of common ancestry has accumulated, it has been used to construct the "tree of life" that illustrates phylogenetic relationships (those based on evolutionary history) between modern species. Originally conceived by Charles Darwin, the irregular branching of the tree of life is a metaphor for the history of modern species. At the base of the tree is the universal ancestor from which all later organisms are descended. The main branches of the tree diverge from the descendants of the universal ancestor (represented by the central trunk). The successive divergence of each branch signifies the evolutionary processes by which new species are created by the forces of natural selection. Each branch point rep-

resents the last common ancestor from which all of the species represented by the branch are descended. At the tip of each branch are currently existing species.

For many years the phylogenetic tree was believed to possess two main branches: the prokaryotes and the eukaryotes. Largely because of Carl Woese's pioneering work with rRNA sequences, it has become apparent that the relationships among currently existing life forms are more complex than the two-domain model implies. Small ribosomal subunit RNA molecules are especially useful for probing phylogenetic relationships because:

1. They are encoded by the genomes of all organisms.
2. Organisms in species already known to be closely related have similar rRNA sequences; those that are distantly related have dissimilar rRNA sequences.
3. Ribosomal RNA molecules are mosaics, that is, certain rRNA sequences are highly conserved among species, although others are variable.

For all of these reasons, rRNA sequences can be used to reveal subtle gradations in evolutionary relationships.

In 1977, based on his evaluation of rRNA sequences and other data, Woese proposed that the archaebacteria were entitled to a branch of their own on the tree of life, now referred to as the archaea (Figure 17I). Comparison of the genomes of archaeal species with

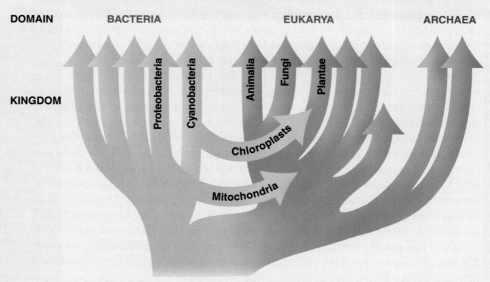

FIGURE 17I The Current Model of the Tree of Life.
This simplified diagram, which illustrates the phylogenetic relationships between modern organisms, is based on the degree of similarity of ribosomal RNA sequences. The endosymbiotic processes that have generated chloroplasts and mitochondria are indicated by the upper and lower diagonal arrows.

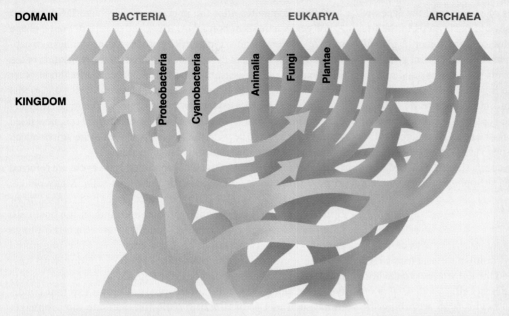

DOMAIN BACTERIA EUKARYA ARCHAEA

KINGDOM

Proteobacteria Cyanobacteria Animalia Fungi Plantae

FIGURE 17J A Proposed Modified Tree (or "Net") of Life.
This view of life's history reflects the analysis of a number of genes in addition to rRNA sequences. Several researchers suggest that lateral gene transfer (LGT) accounts for the evidence that most modern archaeal and bacterial genomes contain genes from multiple sources. Note that, with the exception of the endosymbiotic transfer of mitochondria and chloroplasts, LGT is not a factor in the recent history of the Eukarya.

genes within the three domains is the concept of lateral (or horizontal) gene transfer. Unlike the more familiar vertical gene transfer that occurs between parent and offspring, **lateral gene transfer (LGT)** involves the transfer of genes or gene fragments between unrelated organisms. In contrast to more common evolutionary mechanisms (e.g., gene duplication followed by random mutations) LGT can, under appropriate circumstances, rapidly improve an organism's capacity to exploit its environment. For example, the acquisition of a gene encoding an enzyme that degrades an abundant food molecule provides a substantial competitive advantage over those organisms that lack it. According to the standard model, endosymbiosis (see p. 54), the process by which α-protobacteria and cyanobacteria invaded or were consumed by larger host cells to later become the eukaryotic energy-transducing organelles (mitochondria and chloroplasts, respectively), is a major example of lateral gene transfer.

those of other organisms has revealed that in general they resemble eubacteria in some respects (e.g., metabolism and energy production) and eukaryotes in others (e.g., DNA replication, transcription, and translation).

In their attempts to use phylogenetic evidence to triangulate back from the genomes of organisms in the three-domain model to the genome of the universal ancestor, life scientists have encountered unexpected obstacles. Although the reliability of the rRNA data upon which the three-domain model is based has been backed up by lineages of certain ancient genes (e.g., specific nucleic acid–associated enzymes), the phylogenies of other genes indicate that there are inconsistencies. In other words, because individual genes evolve at different rates, phylogenetic trees based on the lineages of other genes do not entirely agree with the rRNA-derived tree. These differences are especially notable near the root of the tree, where the universal ancestor began to evolve into the ancestral organisms from which the modern species are descended. Instead of resolving the disparities, recent genomic data have appeared to increase the confusion.

According to one of the most prominent hypotheses put forward to explain the close similarities observed among certain

An alternate view that is currently being explored proposes that each of the three domains emerged independently from a community of primitive cells that frequently exchanged genes (Figure 17J). During the early stages of cellular evolution, genes, proteins, and metabolic processes were all relatively primitive. As evolution proceeded and the supply of preformed organic molecules was depleted (Special Interest Box 2.2), severe selection pressure caused by competition for limited resources promoted genetic mechanisms that ultimately created more complicated molecules and processes.

After the divergence of the three domains many organisms, especially the eukarya, were sufficiently complex that they became relatively resistant to LGT. Although it is less obvious in eukaryotes, LGT apparently still occurs. For example, susceptibility to certain cancers and autoimmune diseases in animals has been linked to viral causes either by the transfer of small bits of DNA with the virus or the disruption of host cell DNA by the integration of the viral genome. In prokaryotes, significant levels of LGT continue to occur, as evidenced by the spread of antibiotic resistance genes among modern species.

Despite the diversity in the structures of viruses and the types of host cell that are infected, there are several basic steps in the life cycle of all viruses: infection (penetration of the virion or its nucleic acid into the host cell), replication (expression of the viral genome), maturation (assembly of viral components into virions), and release (the emission of new virions from the host cell). Because viruses usually possess only enough genetic information to specify the synthesis of their own components, each type must exploit some of the normal metabolic reactions of its host cell to complete the life cycle. For this reason there are numerous variations on these basic steps. This point can be illustrated by comparing the life cycles of two well-researched viruses: the T4 bacteriophage and the human immunodeficiency virus (HIV).

Bacteriophage T4

The T4 bacteriophage (Figure 17K) is a large virus with an icosahedral head and a long, complex tail similar in structure to T2 (p. 573). The head contains dsDNA, and the tail attaches to the host cell and injects the viral DNA into the host cell.

The life cycle of T4 (Figure 17L) begins with adsorbing the virion to the surface of an *E. coli* cell. Because the bacterial cell wall is rigid, the entire virion cannot penetrate into the cell's interior. Instead, the DNA is injected by flexing and constricting the tail apparatus. Once the DNA has entered the cell, the infective process is complete, and the next phase (replication) begins.

Within 2 minutes after the injection of T4 phage DNA into an *E. coli* cell, synthesis of host DNA, RNA, and protein stops and phage mRNA synthesis begins. Phage mRNA codes for the synthesis of capsid proteins and some of the enzymes required for the replication of the viral genome and the assembly of virion components. In addition, other enzymes are synthesized that weaken the host cell's cell wall, so that new phage can be released for new rounds of infection. Approximately 22 minutes after viral DNA (vDNA) is injected, the host cell, now filled with several hundred new virions, lyses. Upon release, the virions attach to nearby bacteria, thus initiating new infections.

Bacteriophage that initiate this so-called **lytic cycle** are referred to as *virulent* because they destroy their host cells. Many phage, however, do not initially kill their hosts. So-called *temperate* or *lysogenic* phage integrate their genome into that of the host cell. (The term **lysogeny** describes a condition in which the phage genome is integrated into a host chromosome.) The integrated viral genome (called the **prophage**) is copied along with host DNA during cell division for an indefinite time. Occasionally, lysogenic phage can enter a *lytic* phase. Certain external conditions, such as UV or ionizing radiation, activate the prophage, which directs the synthesis of new virions. Sometimes, a lysing bacterial cell releases a few virions that contain some bacterial DNA along with the of phage DNA. When such a virion infects a new host cell, this DNA is introduced into the host genome. This process is referred to as **transduction**.

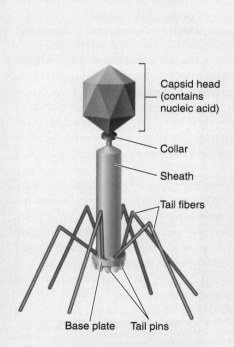

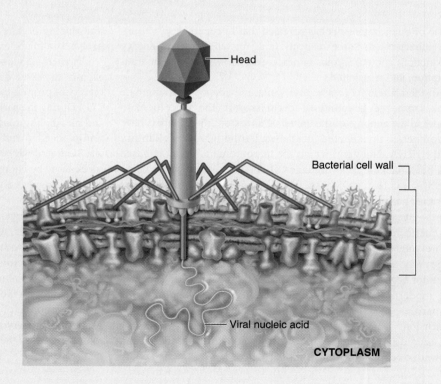

FIGURE 17K The T4 Bacteriophage.
(a) The DNA genome of the T4 bacteriophage induces the host cell to synthesize about 30 proteins. (b) Penetration of the cell wall of a host cell by a bacteriophage.

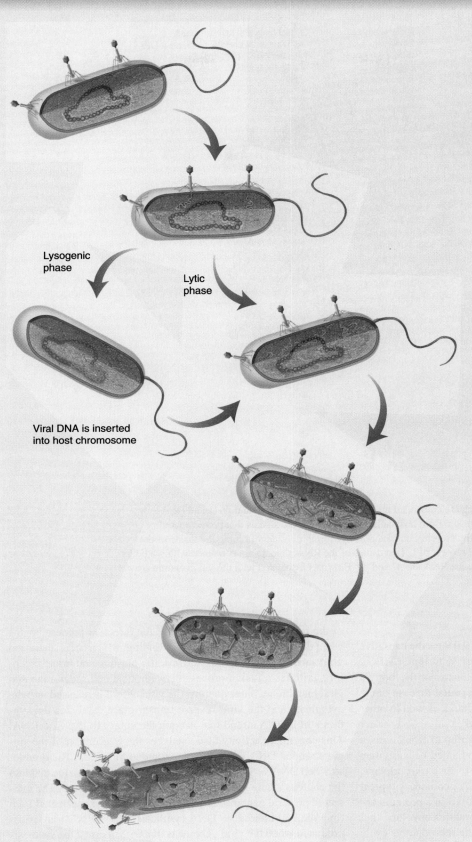

FIGURE 17L Life Cycle of the T4 Bacteriophage.
Viral particles adsorb onto the surface of the bacterial cell and the viral genome is injected (infection). If the virus enters the lytic phase, the machinery of the host cell is directed toward the production and release of new viral particles. If the virus enters the lysogenic phase, the viral genome is integrated into the host cell DNA and may enter the lytic phase at a later time.

Lysogenic
phase

Lytic
phase

Viral DNA is inserted
into host chromosome

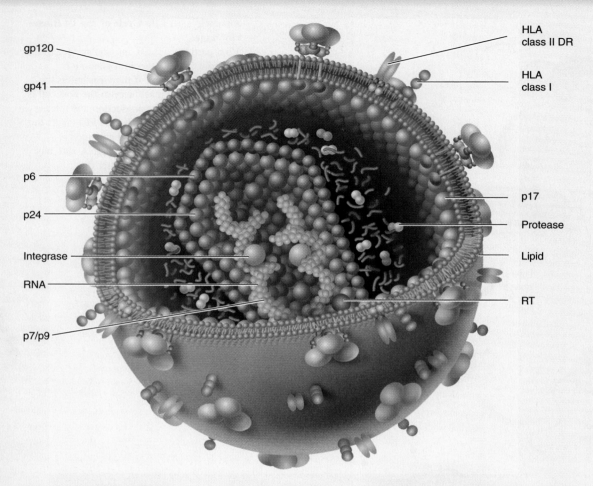

gp120

gp41

p6

p24

Integrase

RNA

p7/p9

HLA
class II DR

HLA
class I

p17

Protease

Lipid

RT

FIGURE 17M HIV Structure.
The surface of the virus is a lipid bilayer in which are embedded the viral glycoproteins gp120 and gp41 and HLA membrane proteins (human leukocyte antigens) taken from host cells. (HLA proteins are signals that protect that viral particle from the immune system, which ordinarily searches out and destroys foreign invaders.) Lining the inside of the envelope are hundreds of copies of the structural protein p17. Two copies of the RNA genome are contained within a bullet-shape capsid composed of p6 and p24. Bound to the RNA are p7 and p9. Enzymes associated with the viral genome are reverse transcriptase (RT), integrase, and protease.

HIV

The human immunodeficiency virus (HIV) is the causative agent of acquired immune deficiency syndrome (AIDS). AIDS is a lethal condition because HIV destroys the body's immune system, rendering it defenseless against disease-causing organisms (e.g., bacteria, protozoa, and fungi, as well as other viruses) in addition to some forms of cancer.

HIV (Figure 17M) belongs to a unique group of RNA viruses called the retroviruses. **Retroviruses** are so named because they contain an enzymatic activity referred to as reverse transcriptase, which synthesizes a DNA copy of an ssRNA genome. A typical retrovirus consists of an RNA genome enclosed in a protein capsid. Wrapped around the capsid is a membranous envelope that is formed from a host cell lipid bilayer. In the reproductive cycle

of the retrovirus HIV (Figure 17N), the infective process begins when the virus binds to a host cell. Binding, which occurs between viral surface glycoproteins and specific plasma membrane receptors, initiates a fusion process between host cell membrane and viral membrane. Subsequently, the viral capsid is released into the cytoplasm and the viral reverse transcriptase catalyzes the synthesis of a DNA strand that is complementary to the viral RNA. This enzymatic activity also catalyzes the conversion of the single-stranded DNA into a double-stranded molecule. The double-stranded DNA version of the viral genome is then translocated into the nucleus where it integrates into a host chromosome. The integrated proviral genome, acting like a prophage, is replicated each time the cell undergoes DNA synthesis. The mRNA transcripts produced when the viral genome is transcribed direct the synthesis of numerous copies of viral proteins. New virus, created as copies of the viral RNA genome are packaged with viral proteins, is released from the host cell by a "budding" process.

HIV contains a cylindrical core within its capsid. In addition, to two copies of its (+)-ssRNA genome, the core contains several enzymes: reverse transcriptase, ribonuclease, integrase, and protease. The RNA molecules are coated with multiple copies of two low-molecular-weight proteins, p7 and p9. (The numbers in these and other protein names indicate their kilodalton mass; for example, p7 is a protein with a 7-kD mass.) The bullet-shaped core itself is composed of hundreds of copies of p6, and copies of p17 form an inner lining of the viral envelope. The envelope of HIV contains two major viral proteins, gp120 and gp41, in addition to host proteins.

HIV infection occurs because of direct exposure of an individual's bloodstream to the body fluids of an infected person. Most HIV is transmitted through sexual contact, blood transfusions, and perinatal transmission from mother to child. Once HIV enters the body, it is believed to infect cells that bear the CD4 antigen on their plasma membranes. The principal group of cells that are attacked by HIV are the T-4 helper lymphocytes of the immune system. T-4 helper cells play a critical role in regulating the activities of other immune system cells. It is now known that T cell infection requires the interaction of the gp120-CD4 complex with a chemokine receptor. (Chemokines are immune system chemotactic agents. They stimulate T cells by binding to receptors on the T cell plasma membrane.) In the early stages of infection, the CCR5 receptor helps HIV enter T cells. Later the CXCR4 receptor is used. (Recent evidence suggests that humans with two copies of a defective CCR5 gene, a relatively small portion of the population, are resistant to HIV infection.) Other cells that are known to be infected by HIV include some intestinal and nervous system cells.

Once the gp120 envelope protein of HIV binds to the CD4 antigen and chemokine receptor on the T cell, the viral envelope fuses with the host cell's plasma membrane. The two RNA strands are released into the cytoplasm. Reverse transcriptase, a heterodimer with several enzymatic activities, then catalyzes the synthesis of an ssDNA using the vRNA as a template. The heterodimer's RNase activity then degrades the vRNA. The same protein produces a double-stranded vDNA by forming a complementary strand of the ssDNA. Viral integrase integrates the vDNA into a host cell chromosome. The proviral DNA remains latent until the specific infected T cell is activated by an immune response. The proviral DNA can then direct the cell to synthesize viral components. Newly synthesized viruses bud from the infected cell.

Recent research efforts using microarray DNA chips (Biochemical Methods 18.1) have provided some of the details of the molecular mechanisms by which HIV disrupts the functioning of T-4 lymphocytes. By monitoring the mRNA expression of over 6000 genes simultaneously, researchers tracked the consequences of HIV infection. They observed that within 30 minutes of infection, the expression of about 500 cellular genes had been suppressed and 200 had been activated. Within hours, host cell mRNA had largely been

replaced by viral mRNA. The virus had crippled the cell's capacity to generate energy and repair virally inflicted DNA damage.

Cell death is triggered by several mechanisms that include the following:

1. Viral activation of the genes that induce **apoptosis** (programmed cell death), a normal cell mechanism by which cells respond to external signals such as those that occur in developmental processes.
2. The simultaneous budding of numerous viral particles from the cell membrane may tear the membrane and cause massive leakages that cannot be repaired.
3. Massive release of new virus from a cell, directed by the provirus, may so deplete the cell that it disintegrates.
4. The binding of cell surface gp120 molecules to CD4 receptors on nearby healthy cells leads to the formation of large, nonfunctional multinucleated cell masses called *syncytia*.

HIV infection progresses through several stages, which may vary considerably in length among individuals. Initial symptoms, which usually occur soon after the initial exposure to the virus and last for several weeks, include fever, lethargy, headache and other neurological complaints, diarrhea, and lymph node enlargement. (Antibodies to HIV are detectable during this period.) Exaggerated versions of these symptoms, referred to as the AIDS-related complex (ARC), may often recur. Eventually, the immune system becomes so compromised that the individual becomes susceptible to serious opportunistic diseases and is said to have developed AIDS. The time required for the development of AIDS may vary from 2 years to 8 or 10 years. For reasons that are not understood, a few patients do not develop AIDS even after 15 years of HIV infection. (It has recently been suggested that some of these individuals are infected with attenuated HIV variants.) Some of the most common AIDS-related diseases include *Pneumocystis carinii* pneumonia, cryptococcal meningitis (inflammation of membranes that cover the brain and spinal cord), toxoplasmosis (brain lesions, heart and kidney damage, and fetal abnormalities), cytomegalovirus infections (pneumonia, kidney and liver damage, and blindness), and tuberculosis. HIV infection is also associated with several types of cancer, the most common of which is a rare skin cancer called Kaposi's sarcoma.

There is no cure for AIDS. Treatment seeks to suppress symptoms (e.g., antibiotics for the infections) and slow viral reproduction. Mortality rates have decreased since 1995 because of the introduction of a treatment protocol called highly active antiretroviral therapy (HAART) that consists of combinations of drugs from the following categories: (1) nucleoside reverse transcriptase inhibitors (NRTIs) (e.g., azidothymidine, also called zidovudine or AZT), (2) non-nucleoside reverse transcriptase inhibitors (NNRTIs) (e.g., efavirenz) and protease inhibitors (e.g., indinavir). Both NRTIs and NNRTIs inhibit vDNA synthesis catalyzed by reverse transcriptase. Protease inhibitors are a class of drugs that prevent the processing of viral protein that is required for the assembly of new virions.

Because the viral genome mutates frequently (i.e., its surface antigens become altered), developing an AIDS vaccine is problematic.

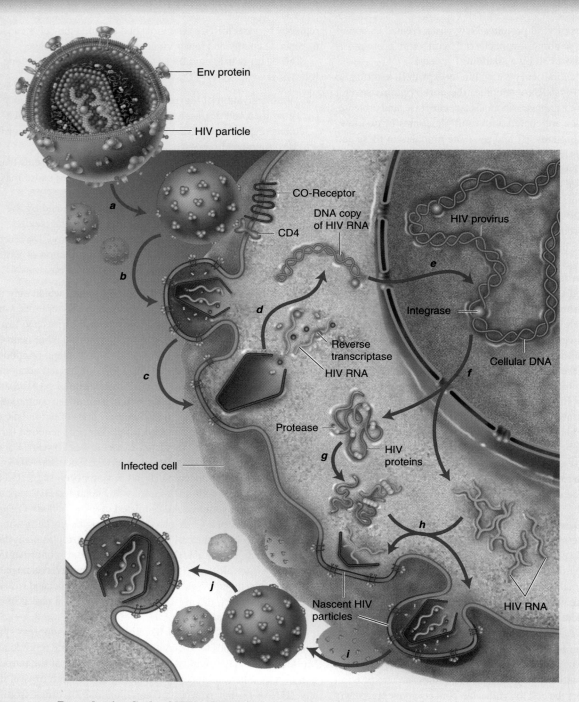

FIGURE 17N Reproductive Cycle of HIV, a Retrovirus.
After the viral particle binds to surface receptors on the host cell (a), its envelope fuses with the cell's plasma membrane (b), thus releasing the capsid and its contents [viral RNA (vRNA) and several viral enzymes] into the cytoplasm (c). The viral enzyme reverse transcriptase catalyzes the synthesis of a DNA strand complementary to the vRNA (d), and then proceeds to form a second DNA strand that is complementary to the first. Subsequently the double-stranded viral DNA (vDNA) transfers to the nucleus, where it integrates itself into a host chromosome with the aid of a viral integrase (e). The provirus (the integrated viral genome) is replicated every time the cell synthesizes new DNA. Transcription of viral DNA results in the formation of two types of RNA transcripts: RNA molecules that function as the viral genome (f) and molecules that code for the synthesis of viral protein (e.g., reverse transcriptase, capsid proteins, envelope proteins, and viral integrase) (g). The protein molecules are combined with the vRNA genome during the creation of new virus (h) that buds from the surface of the host cell (i) and then proceeds to infect other cells (j).

QUESTION 17.17

What is a capsid? Describe its components.

QUESTION 17.18

List all of the genome types found in viruses.

QUESTION 17.19

Name the substrate and product of the viral enzyme reverse transcriptase.

QUESTION 17.20

Describe the principal events that occur when an enveloped DNA virus infects a host cell and produces progeny.

QUESTION 17.21

Describe the events that occur when a retrovirus infects a host cell.

QUESTION 17.22

Recall that according to the central dogma, the flow of genetic information is from DNA to RNA and then to protein. Retroviruses are an exception to this rule. The alterations of the central dogma that are observed in retroviruses and other RNA viruses can be illustrated as follows:

DNA ⟷ RNA ⟶ Protein

Compare this illustration with that of the original central dogma (p. 563). Describe in your own words the implications of each component of these figures.

QUESTION 17.23

HIV screening tests detect the presence of HIV antibodies in blood serum. In the most common HIV test, an ELISA kit that contains HIV antigens detects these antibodies. Because of the life-altering significance of a positive HIV test, the presence of HIV infection is confirmed by an additional, more expensive test. In most laboratories this confirming test is a Western blot analysis. Western blot analysis is similar to Southern blotting (Biochemical Methods 17.1) with the following exceptions. Western blotting detects the presence of specific proteins instead of specific DNA sequences. In HIV analysis, these proteins are antibodies to gp41 and gp120. If there is antigen-antibody binding on a gel, it is detected with labeled "antihuman" antibodies. (Antihuman antibodies bind specifically to certain sites, other than the antigen-binding sites, that occur on all human antibodies.) After reviewing the discussion of ELISA (Biochemical Methods 16.1), describe how an HIV ELISA is conducted. Referring to the discussion of Southern blotting, describe in general terms how a Western blot analysis for HIV infection is performed. (*Hint*: An HIV Western blot begins by separating known HIV antigens on polyacrylamide gel electrophoresis.)

SUMMARY

1. The information required for directing all living processes is stored in the nucleotide sequences of DNA. DNA is composed of two antiparallel polynucleotide strands wound around each other to form a right-handed double helix. The deoxyribose-phosphodiester bonds form the backbones of the double helix, and the nucleotide bases project to its interior. The nucleotide base pairs form because of hydrogen bonding between certain bases: adenine with thymine, and cytosine with guanine. Mutations are changes in DNA structure, which may be caused by collisions with solvent molecules, thermal fluctuations, ROS, radiation, or xenobiotics.

2. DNA can have several conformations depending on the nucleotide sequence. In addition to the classical structure determined by Watson and Crick (B-DNA), A-DNA, H-DNA, and Z-DNA have also been observed. DNA supercoiling is now known to be a critical feature of several biological processes, such as DNA packaging, replication, and transcription.

3. Each eukaryotic chromosome is composed of nucleohistone, a complex formed by winding a single DNA molecule around a histone octomer to form a nucleosome. The DNA of mitochondria and chloroplasts is similar to the chromosomes found in prokaryotes.

4. The forms of RNA that occur in cells, that is, transfer, ribosomal, messenger, heterogeneous, and small nuclear RNAs, are all involved in the synthesis of proteins. RNA differs from DNA in that it contains ribose (instead of deoxyribose), has a somewhat different base composition, and is usually single-stranded. Transfer RNA molecules have specific amino acids attachd to them by specific enzymes and transport these to the ribosome, for incorporation into newly synthesized protein, where they are properly aligned during protein synthesis. The ribosomal RNAs are components of ribosomes. Messenger RNA contains within its nucleotide sequence the coding instructions for synthesizing a specific polypeptide. Heterogeneous nuclear RNA is the original transcript produced by complementary base pairing from a DNA template. It is then processed to form mRNA. The small nuclear RNAs are involved in splicing activities during mRNA synthesis.

5. Viruses are obligate intracellular parasites. Although they are acellular and cannot carry out metabolic activities on their own, viruses can wreak havoc on living organisms. Each type of virus infects a specific type of host (or small set of hosts). A virus does so because it can either inject its genome into the host cell or gain entrance for the entire viral particle. Each virus has the capacity to use the host cell's metabolic processes to manufacture new copies of itself, called virions. Viruses possess dsDNA, ssDNA, dsRNA, or ssRNA genomes.

6. HIV is a retrovirus that causes AIDS. Retroviruses are a class of RNA viruses that possess a reverse transcriptase activity that converts their RNA genome to a DNA molecule. This vDNA is then inserted into the host cell genome, causing a permanent infection. Eventually, HIV infection destroys the immune system of infected individuals.

SUGGESTED READINGS

Brown, T. A., *Genomes*, Wiley-Liss, New York, 1999.

Butler, J. M., *Forensic DNA Typing: Biology and Technology Behind STR Markers*, Academic Press, San Diego, 2001.

Calladine, C. R., and Drew, H. R., *Understanding DNA: The Molecule and How It Works*, Academic Press, San Diego, 1992.

Frank-Kamenetskii, M. D., and Mirkin, S. M., Triple Helix DNA Structures, *Ann. Rev. Biochem.*, 64:65–95, 1995.

Gallo, R. C., and Montagnier, L., The Chronology of AIDS Research, *Nature*, 326(6112):435–436, 1987.

Glick, B. R., and Pasternak, J. J., *Molecular Biotechnology: Principles and Applications of Recombinant DNA*, 2nd ed., ASM Press, Washington, D.C., 1998.

Herrmann, B., and Hummel, S. (Eds.), *Ancient DNA*, Springer-Verlag, New York, 1994.

Julian, M. M., Women in Crystallography, in Kass-Simon, G., and Farnes, P. (Eds.), *Women of Science: Righting the Record*, pp. 359–364, Indiana University Press, Bloomington and Indianapolis, 1990.

Kolata, G., *Flu: The Story of the Great Influenza Pandemic of 1918 and the Search for the Virus that Caused It*, Farrar, Straus, and Giroux, New York, 1999.

Lewin, B., *Genes VII*, Oxford University Press, New York, 2000.

Lykke-Anderson, J., Aagaard, C., Semionenkov, M., and Garrett, R. A., Archaeal Introns: Splicing, Intercellular Mobility and Evolution, *Trends Biochem. Sci.*, 22:326–331, 1997.

Maxwell, E. S., and Fournier, M. J., The Small Nuclear RNAs, *Ann. Rev. Biochem.*, 35:897–934, 1995.

Portugal, F. H., and Cohen, J. S., *A Century of DNA: A History of the Discovery of the Structure and Function of the Genetic Substance*, MIT Press, Cambridge, Mass., 1977.

Varmus, H., Retroviruses, *Science*, 240:1427–1435, 1988.

Watson, J. D., *The Double Helix*, Atheneum, New York, 1968.

Weintraub, H. M., Antisense RNA and DNA, *Sci. Amer.*, 262(1): 40–46, 1990.

Wolffe, A., *Chromatin: Structure and Function*, 3rd ed., Academic Press, San Diego, 1998.

KEY WORDS

A-DNA, *574*

alkylation, *570*

antisense strand, *596*

apoptosis, *604*

B-DNA, *574*

base analogues, *569*

centromere, *585*

chain-terminating method, *589*

Chargaff's rule, *573*

chromatin, *580*

chromosome, *579*

REVIEW QUESTIONS

Clearly define the following terms:
  - a. genetics
  - b. replication
  - c. transcription
  - d. sugar-phosphate backbone
  - e. bacteriophage
  - f. Chargaff's rule
  - g. palindrome
  - h. Hoogsteen base pairing
  - i. proteomics
  - j. Alu family
  - k. transcriptome
  - l. satellite DNA
  - m. transposon
  - n. hypochromic effect
  - o. DNA fingerprinting
  - p. STR DNA

2. There are several structural forms of DNA. List and describe each form.

3. Describe the higher-order structure of DNA referred to as supercoiling.

4. List three biological properties facilitated by supercoiling.

5. List three differences between eukaryotic and prokaryotic DNA.

6. Describe the structure of a nucleosome.

7. Describe the structural differences between RNA and DNA.

8. What are the three most common forms of RNA? What roles do they play in cell function?

9. Z-DNA derives its name from the zig-zagged conformation of phosphate groups. What features of the DNA molecule allow this distinctive structure to form?

10. There is one base pair for every 0.34 nm of DNA and the total contour length of all the DNA in a single human cell is 2 m. Calculate the number of base pairs in a single cell. Assuming that there are 10^{14} cells in the human body, calculate the total length of DNA. How does this estimate compare to the distance from the Earth to the Sun (1.5×10^8 km)?

11. A DNA sample contains 21% adenine. What is its complete percentage base composition?

12. The melting temperature of a DNA molecule increases as the G-C content increases. Explain.

13. What physical conditions result in DNA denaturation?

14. Organize the following into a hierarchy:
  - a. chromosome
  - b. gene
  - c. nucleosome
  - d. nucleotide base pair

15. Provide the complementary strand and the RNA transcription product for the following DNA segment:

5′-AGGGGCCGTTATCGTT-3′

16. Define the following terms:
  - a. telomere
  - b. minisatellite
  - c. DNA profile
  - d. intron
  - e. exon
  - f. retrotransposon
  - g. DNA denaturation
  - h. chain-terminating method
  - i. PCR
  - j. splicing

THOUGHT QUESTIONS

1. Under physiological conditions, DNA ordinarily forms B-DNA. However, RNA hairpins and DNA-RNA hybrids adopt the structure of A-DNA. Considering the structural differences between DNA and RNA, explain this phenomenon.

2. What structural features of DNA cause the major groove and the minor groove to form?

3. Explain in general terms how polyamines aid in achieving the highly compressed structure of DNA.

4. In contrast to the double helix of DNA, RNA exists as a single strand. What effects does this have on the structure of RNA?

5. Jerome Vinograd found that circular DNA from a polyoma virus separates into two distinct bands when it is centrifuged. One band consists of supercoiled DNA and the other relaxed DNA. Explain how you would identify each band.

6. 5-Bromouracil is an analogue of thymine that usually pairs with adenine. However, 5-bromouracil frequently pairs with guanine. Explain.

7. The flow of genetic information is from DNA to RNA to protein. In certain viruses, the flow of information is from RNA to DNA. Does it appear possible for that information flow to begin with proteins? Explain.

8. HIV is a retrovirus. Suggest reasons why the development of a vaccine to prevent HIV infection is such a difficult undertaking.

9. You wish to isolate mitochondrial DNA without contamination with nuclear DNA. Describe how you would accomplish this task.

10. Unlike linker DNA and deproteinized DNA, DNA segments wrapped around histone cores are relatively resistant to the hydrolytic actions of nucleases. Explain.

11. The set of mRNAs present within a cell changes over time. Explain.

12. DNA and RNA are information-rich molecules. Explain the significance and implications of this statement.

13. A 10-year-old murder case in a small town has finally been solved, thanks to the efforts of a detective who took advantage of a new statewide DNA database containing samples of DNA from convicted felons. Explain how this case was solved starting with the arrival of a crime scene unit at the murder scene. What technological advances made this case solvable?

Genetic Information

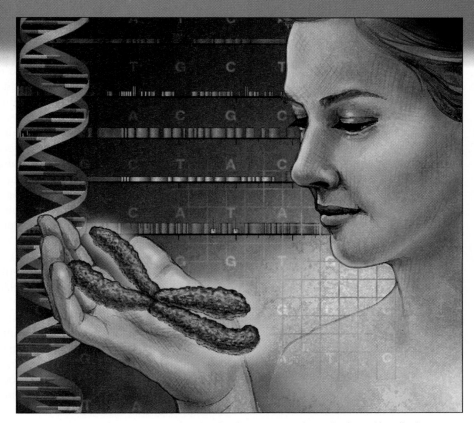

The Human Genome Project. The efforts of biochemists to uncover the mechanisms of heredity, begun over a century ago, have recently resulted in the sequencing of the human genome. Technology developed by biochemists and molecular biologists, such as electrophoresis and increasingly more sophisticated automated and computerized DNA sequence machines, have made this extraordinary achievement possible. The human genome project, performed by thousands of scientists in public and private laboratories around the world, took over 15 years to accomplish. Researchers are just beginning to interpret and utilize the tidal wave of information generated by this effort in their efforts to solve the medical and biological problems of humans.

Within the past several years, the investigation of genetic inheritance that began with Mendel has reached an explosive stage. Knowledge concerning how living cells store, use, and inherit genetic information constantly increases. Biochemists and geneticists are now painstakingly revealing the principles of nucleic acid function. The practical consequences of work done in the 1980s and 1990s have already been considerable. In addition to numerous medical applications (e.g., diagnostic tests and therapies), recombinant DNA technology has allowed investigators access to the inner workings of living organisms in ways never before possible. Not only are long-standing questions about living processes more accessible to investigation, but amazingly rapid progress is being made in understanding the molecular basis of a wide variety of diseases such as cancer and heart disease. Recent experience suggests it is impossible to predict the outcome of nucleic acid research. Whatever happens, it will certainly be exciting.

Any successful information-based system involves two aspects: conservation and transfer. The instructions required to produce a certain type of organization (e.g., the directions for building a house or for reproducing a living organism) must be stably stored so they are accurate and available for use by the system. To be useful, however, information must also be transferred. Consequently, the use of organic molecules by living organisms to store and transfer information presents an obvious paradox. Because information transfer involves changes in molecular configuration, if an organic molecule is sufficiently stable to conserve the information it contains, it is probably too rigid or unreactive for information transfer. Living organisms have solved this problem by partitioning these tasks between two molecular realms. DNA has structural features that maximize information storage and duplication; the proteins have diverse and flexible three-dimensional structures that maximize information transfer. DNA is actually a relatively inert molecule that is shielded somewhat from the frenetic pace of molecular collisions (i.e., within the nucleoid of prokaryotes and the nuclei of eukaryotes). All of the processes by which its information is utilized involve mechanical work performed by protein-based information processing devices. When activated, molecular machines composed largely of interlocking proteins and powered by cellular energy resources can bend, twist, unwind, and unzip DNA molecules during processes such as replication and transcription. The binding between specific DNA sequences and the appropriate molecular machine is made possible by protein components that "recognize" certain sequences. Recognition occurs because each protein possesses a unique set of contours that allow numerous atomic contacts with the DNA and/or protruding domains such as zinc fingers (Chapter 5, see p. 134) that plug directly into the double helix.

Chapter 18 provides an overview of the mechanisms that living organisms use to synthesize the nucleic acids DNA and RNA that direct cellular processes. The chapter begins with a discussion of several aspects of DNA **replication** (synthesis), repair, and **recombination** (the reassortment of DNA sequences). This is followed by descriptions of the synthesis and processing of RNA, the nucleic acid molecules involved in the conversion of gene sequences into polypeptides. Also included is an overview of several biotechnological tools that are used by biochemists to investigate living processes. Chapter 18 ends with a section devoted to gene expression, the mechanisms cells use to produce gene products in an orderly and timely manner.

18.1 GENETIC INFORMATION: REPLICATION, REPAIR, AND RECOMBINATION

Because of the strategic importance of DNA, all living organisms must possess the following features: (1) rapid and accurate DNA synthesis and (2) genetic stability provided by effective DNA repair mechanisms. Paradoxically, the long-term survival of species also depends on genetic variations that allow them to adapt to changing environments. In most species these variations arise predominantly from genetic recombination, although mutation also plays a role. In the following sections, the mechanisms that prokaryotes and eukaryotes use to achieve these goals are discussed. Prokaryotic genetic information processes are more completely understood than those of eukaryotes. Because of their minimal growth requirements, short generation times, and relatively simple genetic composition, prokaryotes (especially *Escherichia coli*) are excellent subjects for investigations of genetic mechanisms. In contrast, multicellular eukaryotes possess several properties that hinder genetic investigations. The most formidable are long generation times (often months or years) and extraordinary difficulties in identifying gene products (e.g., enzymes or structural components). A common tactic in genetic research is to induce mutations, then observe changes in or the absence of a specific gene product. Unfortunately, because of their considerable complexity, in higher organisms this method has identified very few gene products. Recombinant DNA techniques are being used to circumvent this obstacle.

DNA Replication

DNA replication must occur before every cell division. The mechanism by which DNA copies are produced is similar in all living organisms. After the two strands separate, each serves as a template for the synthesis of a complementary strand (Figure 18.1). (In other words, each of the two new DNA molecules contains one old strand and one new strand.) This process, referred to as **semiconservative replication**, was first demonstrated in an elegant experiment (Figure 18.2) reported in 1958 by Matthew Meselson and Franklin Stahl. In this classic work, Meselson and Stahl took advantage of the increase in density of DNA labeled with the heavy nitrogen isotope ^{15}N (the most abundant nitrogen isotope is ^{14}N). After *E. coli* cells were grown for 14 generations in growth media whose nitrogen source consisted only of $^{15}NH_4Cl$, the ^{15}N-containing cells were transferred to growth media containing the ^{14}N isotope. At the end of both one and two cell divisions, samples were removed. The DNA in each of these samples was isolated and analyzed by CsCl density gradient centrifugation. (Refer to Biochemical Methods 2.1 for a description of density gradient centrifugation.) Because pure ^{15}N-DNA and ^{14}N-DNA produce characteristic bands in centrifuged CsCl tubes, this analytical method discriminates between DNA molecules containing large amounts of the two nitrogen isotopes. When the DNA isolated from ^{15}N-containing cells grown in ^{14}N medium for precisely one generation was centrifuged, only one band was observed. Because this band occurred halfway between where ^{15}N-DNA and ^{14}N-DNA bands would normally appear, it seemed reasonable to assume that the new DNA was a hybrid molecule, that is, it contained one ^{15}N strand and one ^{14}N strand. (Any other means of replication would create more than one band.) After two cell divisions, extracted DNA was resolved into two discrete bands of equal intensity, one made up of $^{14}N,^{14}N$-DNA (light DNA) and one made up of hybrid molecules ($^{14}N,^{15}N$-DNA), a result that also supported the semiconservative model of DNA synthesis.

In the years since the Meselson and Stahl experiment, many of the details of DNA replication have been discovered. As mentioned, much of this work was accomplished by using *E. coli* and other prokaryotes.

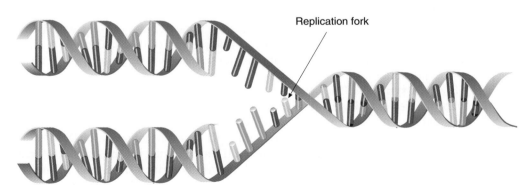

Replication fork

FIGURE 18.1

Semiconservative DNA Replication.

As the double helix unwinds at the replication fork, each old strand serves as a template for the synthesis of a new strand.

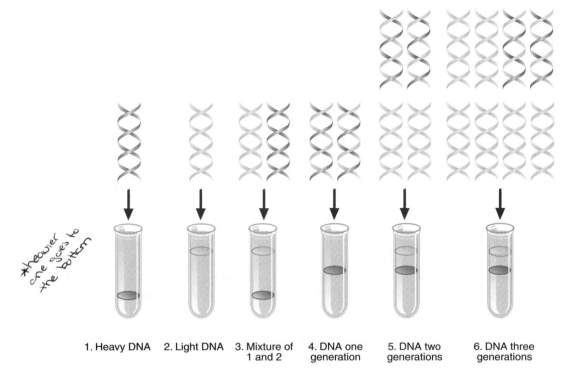

1. Heavy DNA   2. Light DNA   3. Mixture of 1 and 2   4. DNA one generation   5. DNA two generations   6. DNA three generations

*heavier one goes to the bottom

FIGURE 18.2

The Meselson-Stahl Experiment.

CsCl centrifugation of *E. coli* DNA can distinguish heavy DNA grown in ^{15}N media (1) from light DNA grown in ^{14}N media (2). A mixture is shown in (3). When *E. coli* cells enriched in ^{15}N are grown in ^{14}N for one generation, all genomic DNA is of intermediate density (4). After two generations, half of the DNA is light and half is of intermediate density (5). After three generations, 75% of the DNA is light and 25% of the DNA is of intermediate density (6).

DNA SYNTHESIS IN PROKARYOTES DNA replication in *E. coli* has proven to be a complex process that consists of several basic steps. Each step requires certain enzyme activities.

1. DNA unwinding. As their name implies, the **helicases** are ATP-requiring enzymes that catalyze the unwinding of duplex DNA. The principal helicase in *E. coli* is Dna B protein, a product of the dnaB gene.

2. Primer synthesis. The formation of short RNA segments called **primers**, required for the initiation of DNA replication, is catalyzed by **primase**, an RNA polymerase. Primase is a 60-kD polypeptide product of the dnaG gene. A multienzyme complex containing primase and several auxiliary proteins is called the **primosome**.

3. DNA synthesis. The synthesis of a complementary DNA strand by forming phosphodiester linkages between nucleotides base-paired to a template strand is catalyzed by large multienzyme complexes referred to as the DNA polymerases (Figure 18.3). DNA polymerase III (pol III) is the major DNA-synthesizing enzyme in prokaryotes. The pol III holoenzyme is composed of at least 10 subunits (Table 18.1). The core polymerase is formed from three subunits: α, ε, and θ The subunit τ allows two core enzyme complexes to form a dimer. The β-protein (also called the sliding clamp protein) is composed of two subunits (Figure 18.4). It forms a ring around the template DNA strand. The γ-complex, composed of five subunits, recognizes single DNA strands with primer and transfers the β-protein to the core polymerase. The γ-complex also promotes **processivity** of the polymerase complex; that is, it prevents frequent dissociation of

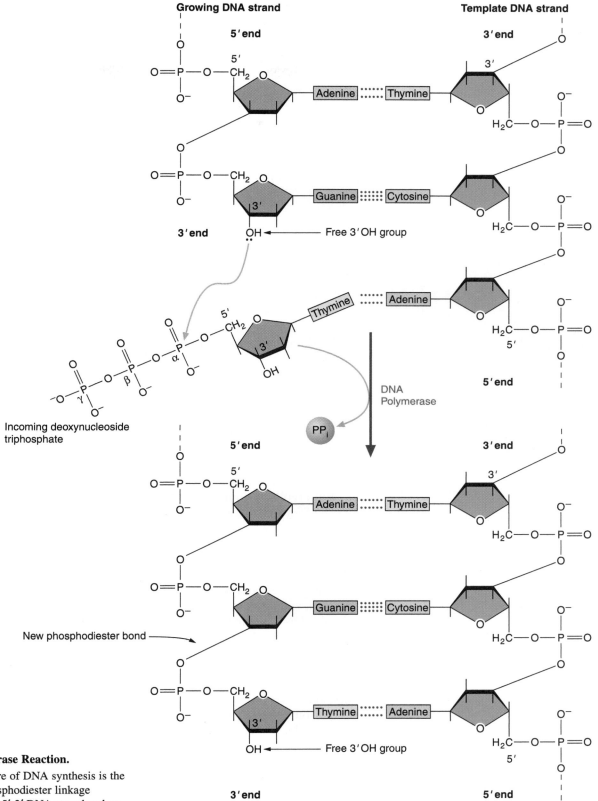

FIGURE 18.3

The DNA Polymerase Reaction.

The essential feature of DNA synthesis is the formation of a phosphodiester linkage between a growing 5′-3′ DNA strand and an incoming dNTP (deoxyribonucleoside triphosphate). The bond is created by a nucleophilic attack of the 3′-hydroxyl group of the terminal residue on the α-phosphate of the dNTP.

TABLE 18.1

Subunits of DNA Polymerase III

		Subunit	Function
Holoenzyme	Core	α	$5' \longrightarrow 3'$ polymerase
		ε	$3' \longrightarrow 5'$ exonuclease
		θ	Unknown
		τ	Dimerization of core, ATPase
		β	Sliding clamp
	γ-Complex	γ	
		δ	Clamp loading unit that
		δ'	hydrolyzes ATP to drive
		χ	binding of β-protein to DNA
		ψ	Also promotes processivity

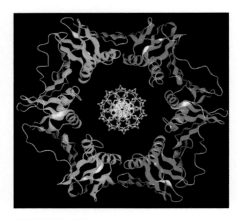

FIGURE 18.4

Cross Section of the β-Subunit of DNA Polymerase III.

The β-protein is a dimer (shown in red and orange) that encircles the DNA and acts like a clamp.

the polymerase from the DNA template. The tether created by the β-protein allows the holoenzyme to remain associated with the DNA template as replication proceeds. The DNA replicating machine, called the **replisome**, is composed of two copies of the pol III holoenzyme, the primosome, and DNA unwinding proteins. DNA polymerase I (pol I), also called the Kornberg enzyme after its discoverer Arthur Kornberg (Nobel Prize in physiology or medicine, 1959), is a DNA repair enzyme. Pol I also plays a role in the timely removal of RNA primer during replication. The function of DNA polymerase II (pol II) is not understood, although it appears to be similar to pol I. In addition to a $5' \longrightarrow 3'$ polymerizing activity, all three enzymes possess a $3' \longrightarrow 5'$ exonuclease activity (e.g., the ε subunit in pol III). (An **exonuclease** is an enzyme that removes nucleotides from an end of a polynucleotide strand.) Pol I also possesses a $5' \longrightarrow 3'$ exonuclease activity.

4. **Joining DNA fragments**. When replication is complete, an enzyme called **ligase** joins the ends of the newly synthesized DNA. This enzyme activity is higher on the discontinuous or lagging strand.

5. **Supercoiling control**. DNA *topoisomerases* prevent tangling of DNA strands. They function ahead of the replication machinery to relieve *torque* (rotary force) in the DNA that can slow down the replication process. The generation of torque is a very real possibility, because the double helix unwinds rapidly (as many as 50 revolutions per second during bacterial DNA replication). Topoisomerases are enzymes that change the supercoiling of the DNA (see p. 577) by breaking one or both strands, which is followed by passing the DNA through the break and rejoining the strands. The terms *topoisomerase* and *topoisomers* (circular DNA molecules that differ only in their degree of supercoiling) are derived from *topology*, a branch of mathematics that investigates the features of geometric structures that do not change with bending or stretching. When appropriately controlled, supercoiling can facilitate the unzipping of DNA molecules. Type I topoisomerases produce transient single-strand breaks in DNA; type II topoisomerases produce transient double-strand breaks. In prokaryotes, a type II topoisomerase called DNA gyrase helps to separate (decantanate) the replication products and to create the (−) supercoils required for genome packaging. (In contrast, eukaryotic type II topoisomerases catalyze only the removal of superhelical tension.)

The replication of the circular *E. coli* chromosome (Figure 18.5) begins at a precise initiation site referred to as *oriC* and proceeds in two directions. Helicases unwind the DNA duplex, two replisomes assemble, and replication proceeds outward in both directions. As the two sites of active DNA synthesis (referred to as **replication forks**) move farther away from each other, a "replication eye" forms. Because an *E. coli* chromosome contains one initiation site, it is considered a sin-

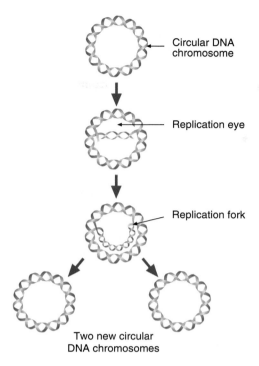

Circular DNA chromosome

Replication eye

Replication fork

Two new circular DNA chromosomes

FIGURE 18.5

Replication of Prokaryotic DNA.

As DNA replication of a circular chromosome proceeds, two replication forks can be observed by using autoradiography. The structure that forms is called a replication eye.

gle replication unit. A replication unit, or **replicon**, is a DNA molecule (or DNA segment) that contains an initiation site and appropriate regulatory sequences.

When DNA replication was first observed experimentally (using electron microscopy and autoradiography), investigators were confronted with a paradox. The bidirectional synthesis of DNA as it appeared in their research seemed to indicate that continuous synthesis occurs in the $5' \longrightarrow 3'$ direction on one strand and in the $3' \longrightarrow 5'$ direction on the other strand. (Recall that DNA double helix has an antiparallel configuration.) However, all the enzymes that catalyze DNA synthesis do so in the $5' \longrightarrow 3'$ direction only. It was later determined that only one strand, referred to as the *leading strand*, is continuously synthesized in the $5' \longrightarrow 3'$ direction. The other strand, referred to as the *lagging strand*, is also synthesized in the $5' \longrightarrow 3'$ direction but in small pieces (Figure 18.6). (Reiji

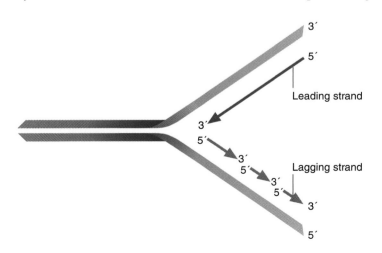

3′

5′

Leading strand

3′

5′

3′

5′

Lagging strand

3′

5′

3′

5′

FIGURE 18.6

DNA Replication at a Replication Fork.

The $5' \longrightarrow 3'$ synthesis of the leading strand is continuous. The lagging strand is also synthesized in the $5' \longrightarrow 3'$ direction but in small segments (now referred to as Okazaki fragments).

Okazaki and his colleagues provided the experimental evidence for discontinuous DNA synthesis.) Subsequently, these pieces (now called **Okazaki fragments**) are covalently linked together by DNA ligase. (In prokaryotes such as *E. coli*, Okazaki fragments possess from 1000 to 2000 nucleotides.)

The initiation of replication is a complex process involving several enzymes, as well as other proteins. Replication begins when *DnaA protein* (52 kD) binds to four 9-bp sites within the oriC sequence. When additional DnaA monomers (between 20 and 40 copies) bind to oriC, a nucleosome-like structure forms in a process that requires ATP and a histone-like protein (HU). As the DnaA-DNA complex forms, localized "melting" of the DNA duplex in a nearby region containing three 13-bp repeats causes a small segment of the double helix to open up. *DnaB* (a 300-kD helicase composed of six subunits), complexed with *DnaC* (29 kD), then enters the open oriC region. Then DnaC is released. The replication fork moves forward as DnaB unwinds the helix (Figure 18.7). Topoisomerases relieve torque ahead of the replication machinery. As DNA unwinding proceeds, DnaA is displaced. The single strands are kept apart by the binding of numerous copies of single-stranded DNA binding protein (SSB). SSB, a tetramer, may also protect vulnerable ssDNA segments from attack by nucleases.

A model of DNA synthesis at a replication fork is illustrated in Figure 18.8. For pol III to initiate DNA synthesis, an RNA primer must be synthesized. On the leading strand, where DNA synthesis is continuous, primer formation occurs only once per replication fork. In contrast, the discontinuous synthesis on the lagging strand requires primer synthesis for each of the Okazaki fragments. The primosome travels along the lagging strand and stops and reverses direction at intervals to synthesize a short RNA primer. Subsequently, pol III synthesizes DNA beginning at the 3′ end of the primer. As lagging strand synthesis continues,

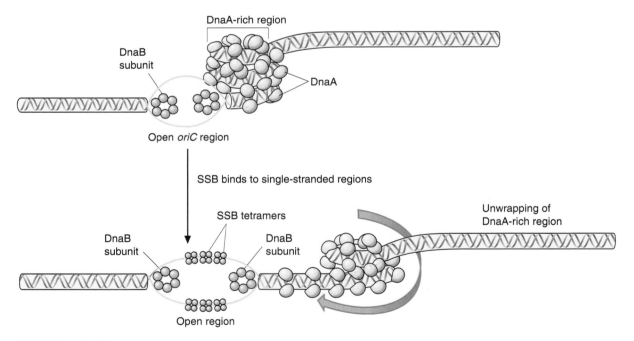

FIGURE 18.7

Replication Fork Formation.

After DnaA and DnaB binding, the DnaB helicase separates the duplex DNA strands at two replication forks. The binding of SSB to newly formed ssDNA prevents reassociation of the single strands. As the replication forks advance, the DnaA-DNA region unwraps. Subsequently DnaA is displaced.

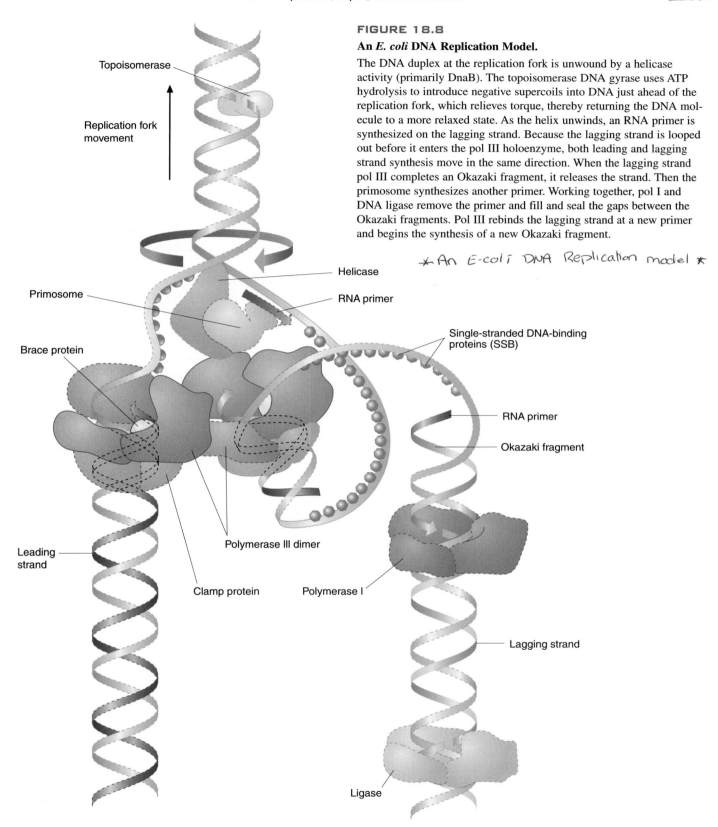

FIGURE 18.8

An *E. coli* DNA Replication Model.

The DNA duplex at the replication fork is unwound by a helicase activity (primarily DnaB). The topoisomerase DNA gyrase uses ATP hydrolysis to introduce negative supercoils into DNA just ahead of the replication fork, which relieves torque, thereby returning the DNA molecule to a more relaxed state. As the helix unwinds, an RNA primer is synthesized on the lagging strand. Because the lagging strand is looped out before it enters the pol III holoenzyme, both leading and lagging strand synthesis move in the same direction. When the lagging strand pol III completes an Okazaki fragment, it releases the strand. Then the primosome synthesizes another primer. Working together, pol I and DNA ligase remove the primer and fill and seal the gaps between the Okazaki fragments. Pol III rebinds the lagging strand at a new primer and begins the synthesis of a new Okazaki fragment.

★ An E-coli DNA Replication model ★

Topoisomerase

Replication fork movement

Helicase

Primosome

RNA primer

Single-stranded DNA-binding proteins (SSB)

Brace protein

RNA primer

Okazaki fragment

Leading strand

Polymerase III dimer

Clamp protein

Polymerase I

Lagging strand

Ligase

RNase removes the RNA primers and replaces them with DNA segments synthesized by pol I. DNA ligase joins the Okazaki fragments.

As illustrated in Figure 18.8, the synthesis of both the leading and lagging strands is coupled. The tandem operation of two pol III complexes requires that one strand (the lagging strand) be looped around the replisome. When the lagging strand pol III complex completes an Okazaki fragment, it releases the duplex DNA. Once it does so, the primosome moves in and synthesizes another RNA primer.

Despite the complexity of DNA replication in *E. coli*, as well as its rate (as high as 1000 base pairs per second per replication fork), this process is amazingly accurate (approximately one error per 10^9 to 10^{10} base pairs per generation). This low error rate is largely a consequence of the precise nature of the copying process itself (i.e., complementary base pairing). However, both pol III and pol I also proofread newly synthesized DNA. Most mispaired nucleotides are removed (by the $3' \longrightarrow 5'$ exonuclease activities of pol III and pol I) and then replaced. Several postreplication repair mechanisms also contribute to the low error rate in DNA replication.

Replication ends when the replication forks meet on the other side of the circular chromosome at the termination site, the *ter* (τ) region. The ter region is composed of a pair of 20-bp inverted repeat ter sequences separated by a 20-bp segment. Each ter sequence prevents further progression of one of the replication forks when a 36-kD *ter binding protein* (TBP) is bound. How the two daughter DNA molecules separate is not understood, although a type II topoisomerase is believed to be involved.

DNA SYNTHESIS IN EUKARYOTES Although the principles of DNA replication in prokaryotes and eukaryotes have a great deal in common (e.g., semiconservative replication and bidirectional replicons), they also have significant differences. Not surprisingly, these differences appear to be related to the size and complexity of eukaryotic genomes.

1. **Timing of replication**. In contrast to rapidly growing bacterial cells, in which replication occurs throughout most of the cell division cycle, eukaryotic replication is limited to a specific period referred to as the S phase (Figure 18.9). It is now known that eukaryotic cells produce certain proteins (Section 18.3) that regulate phase transitions within the cell cycle.

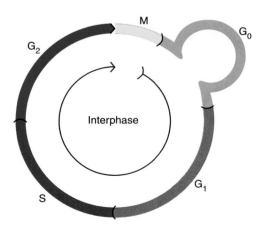

FIGURE 18.9

The Eukaryotic Cell Cycle.

Interphase (the period between mitotic divisions) is divided into several phases. DNA replication occurs during the synthesis or S phase. The G_1 (first gap) phase is the time between mitosis and the beginning of the S phase. During the G_2 phase, protein synthesis increases as the cell readies itself for mitosis (M phase). After mitosis, many cells enter a resting phase (G_0).

2. Replication rate. DNA replication is significantly slower in eukaryotes than in prokaryotes. The eukaryotic rate is approximately 50 nucleotides per second per replication fork. (Recall that the rate in prokaryotes is about 10 times higher.) This discrepancy is presumably due, in part, to the complex structure of chromatin.

3. Replicons. Despite the relative slowness of eukaryotic DNA synthesis, the replication process is relatively brief, considering the large sizes of eukaryotic genomes. For example, on the basis of the replication rate mentioned above, the replication of an average eukaryotic chromosome (approximately 150 million base pairs) should take over a month to complete. Instead, this process usually requires several hours. Eukaryotes use multiple replicons to compress the replication of their large genomes into short periods (Figure 18.10). Also in contrast to the prokaryotes, eukaryotes do not have replisomes. Instead DNA synthesis occurs within immobilized sites within the nucleus called replication factories. Each site contains a large number of replication complexes. As it is being synthesized, DNA is threaded through these complexes.

4. Okazaki fragments. From 100 to 200 nucleotides long, the Okazaki fragments of eukaryotes are significantly shorter than those in prokaryotes.

Although many eukaryotic replication enzymes are generally similar to their prokaryotic counterparts, they do possess their own distinctive properties. For example, eukaryotes have five DNA polymerases designated α, β, δ, ε, and γ. DNA polymerase α initiates the synthesis of both the leading and lagging strands. Further elongation of both strands is catalyzed by two DNA polymerase δ complexes in the same manner as that of pol III in *E. coli*, that is, one complex synthesizes

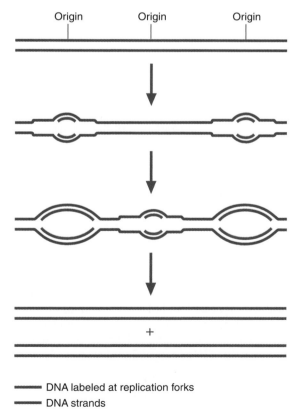

FIGURE 18.10

Multiple Replicon Model of Eukaryotic Chromosomal DNA Replication.

A short segment of a eukaryotic chromosome during replication.

KEY CONCEPTS 18.1

DNA is replicated by a semiconservative mechanism involving several enzymes. The leading strand is synthesized continuously in the 5′ ⟶ 3′ direction. The lagging strand is synthesized in pieces in the 5′ ⟶ 3′ direction; the pieces are then covalently linked.

the leading strand, while the other synthesizes the lagging strand. However, the two δ complexes do not form a dimer as in prokaryotic pol III. Replication protein A (RPA), the primary homologue to SSB in prokaryotes, prevents the separated DNA strands from re-forming the DNA helix during ongoing DNA synthesis. The removal of the RNA primer from each Okazaki fragment is catalyzed by FEN1 (also referred to as MF1), an activity associated with the δ complex. The Okazaki fragments are joined by a DNA ligase activity. DNA polymerase δ binds to *PCNA* (proliferating cell nuclear antigen), a sliding clamp protein similar in function to β-protein in *E. coli*. DNA polymerase β of eukaryotes is believed to be involved in DNA repair. DNA polymerase ε, which possesses a 3′ ⟶ 5′ exonuclease activity, plays an as yet poorly resolved role in both DNA replication and repair. DNA polymerase γ catalyzes the replication of the mitochondrial genome. The DNA polymerase activity of chloroplasts remains poorly characterized.

QUESTION 18.1

Explain the role of each of the following in DNA replication:

a. primase
b. primer
c. topoisomerase
d. polymerse
e. β-protein

QUESTION 18.2

Define the following terms:

a. replisome
b. primosome
c. replicon
d. leading strand
e. Okazaki fragment

QUESTION 18.3

Describe the basic features of the formation of the replication fork during DNA synthesis in *E. coli*.

QUESTION 18.4

Compare the replication processes of prokaryotes and eukaryotes.

DNA Repair

The natural (or background) rate of spontaneous mutation is remarkably constant among species at about 0.1 to 1 mutation per gene per million gametes in every generation. This estimate appears low, but most mutations are deleterious. Because of the complexity of living processes, most genetic changes can be expected to reduce the viability of the organism. For example, humans (an extraordinarily complex species) lose approximately 50% of conceptions because of genetic factors.

Mutations take many forms, from point mutations (single base changes) to gross chromosomal abnormalities (Section 17.1). As described, they are caused by factors that include the properties of the bases themselves and various chemical processes (e.g., depurinations and oxidative stress), as well as the effects of xenobiotics (foreign molecules). Not surprisingly, cells possess a wide range of DNA repair mechanisms.

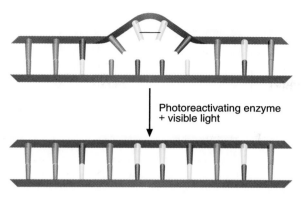

FIGURE 18.11

Photoreactivation Repair of Thymine Dimers.

Light provides the energy for converting the dimer to two thymine monomers. No nucleotides are removed in this repair mechanism. Many species, including humans, do not possess DNA photolyase enzyme activity.

The importance of maintaining the structural integrity of DNA is reflected in the variety of repair mechanisms employed by living cells. A few types of DNA damage can be repaired without the removal of nucleotides. For example, breaks in the phosphodiester linkages can be repaired by DNA ligase. In **photoreactivation repair**, or **light-induced repair**, pyrimidine dimers are restored to their original monomeric structures (Figure 18.11). In the presence of visible light, DNA photolyase cleaves the dimer, leaving the phosphodiester bonds intact. Light energy captured by the enzyme's flavin and pterin chromophores breaks the cyclobutane ring.

Most DNA repair mechanisms, however, involve the removal of nucleotides. In **excision repair**, mutations are excised by a series of enzymes that remove incorrect bases and replace them with the correct ones. Figure 18.12 illustrates the excision of a thymine dimer in bacteria. The process begins with the detection of a distorted DNA segment by a repair endonuclease, referred to as an excision nuclease or excinuclease. The excinuclease cuts the damaged DNA and removes a single-stranded sequence about 12 or 13 nucleotides long. (In eukaryotes, 27 to 29 nucleotide sequences are removed.) In *E. coli* the excinuclease is composed of three proteins: Uvr A, Uvr B, and Uvr C. Uvr A identifies the damaged site and associates with Uvr B to form a complex (A$_2$B). After A$_2$B bends the DNA, Uvr A dissociates from Uvr B–DNA. Uvr C then binds to Uvr B, cutting the damaged DNA strand 4 or 5 nucleotides to the 3′ side of the thymine dimer. Then Uvr C cuts the strand 8 nucleotides to the 5′ side. Uvr D, a helicase, releases Uvr C and the thymine dimer–containing oligonucleotide. The excision gap is repaired by pol I and DNA ligase. Although excision repair in humans is poorly understood, it is probably more complicated than the prokaryotic process because it involves at least 7 polypeptides. Several of these molecules are missing in xeroderma pigmentosum patients. (In xeroderma pigmentosum, skin lesions and skin cancer are caused by exposure to UV radiation.)

Recombinational repair can eliminate certain types of damaged DNA sequences that are not eliminated before replication. For some structural damage (e.g., pyrimidine dimers), replication is interrupted because the replication complex detaches, moves beyond the damage, and reinitiates synthesis. One of the daughter DNA molecules has a gap opposite the pyrimidine dimer. This gap can be repaired by exchanging the corresponding segment of the homologous DNA molecule. After the recombination process, the newly opened gap in the homologous molecule can easily be filled by DNA polymerase and DNA lig-

FIGURE 18.12

Excision Repair of a Thymine Dimer in *E. coli*.

Uvr A, a damage recognition protein, detects helical distortion caused by DNA adducts such as thymine dimers (a). It then associates with Uvr B to form the A_2B complex. After binding to the damaged segments, A_2B forces DNA to bend. Uvr A then dissociates (b). The binding of the nuclease Uvr C to Uvr B (c) and the action of the helicase Uvr D (d) results in the excision of a 12-nucleotide DNA strand (12-mer). After Uvr B is released (e) the excision gap is repaired by pol I (f).

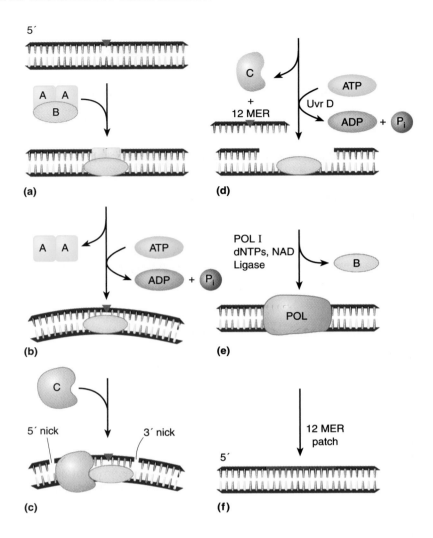

DNA is constantly exposed to chemical and physical processes that alter its structure. Each organism's survival depends on its capacity to repair this structural damage.

ase, because an intact template is present. The remaining damaged segment (the pyrimidine dimer) can then be eliminated by other repair mechanisms. Recombinational repair closely resembles genetic recombination, which is discussed next.

QUESTION 12.10   List three types of DNA repair. Explain the basic features of each.

DNA Recombination

Recombination, often referred to as genetic recombination, can be defined as the rearrangement of DNA sequences by exchanging segments from different molecules. The process of recombination, which produces new combinations of genes and gene fragments, is primarily responsible for the diversity among living organisms. More important, the large number of variations made possible by recombination can allow species opportunities to adapt to changing environments. In other words, genetic recombination is a principal source of the variations that make evolution possible.

There are two forms of recombination: general and site-specific. **General recombination**, which occurs between homologous DNA molecules, is most commonly observed during meiosis. (Recall that meiosis is the form of eukaryotic cell division in which haploid gametes are produced.) A similar process has been observed in some bacteria. In **site-specific recombination**, the exchange of sequences from different molecules requires only short regions of DNA homology. These regions are flanked by extensive nonhomologous sequences. Site-specific recombinations, which depend more on protein-DNA interactions than on sequence homology, occur throughout nature. For example, this mechanism is used by a bacteriophage to integrate its genome into the *E. coli* chromosome. In eukaryotes, site-specific recombination is responsible for a wide variety of developmentally controlled gene rearrangements. Gene rearrangements may be at least partially responsible for cell differentiation in complex multicellular organisms. One of the most interesting examples of gene rearrangement is the generation of antibody diversity in mammals. In a variation of site-specific recombination, sometimes referred to as **transposition**, certain sequences called **transposable elements** are moved from one chromosome or chromosomal region to another. Transposition differs from site-specific recombination in that a specific protein-DNA interaction occurs on only one of the two recombining sequences. The recombination of the second DNA sequence is nonspecific. The mechanism by which nonspecific recombination occurs is unresolved.

GENERAL RECOMBINATION General recombination requires the precise pairing of homologous DNA molecules. Figure 18.13 illustrates the currently accepted model for general recombination. This model, based on Robin Holliday's genetic investigations in fungi in 1964, involves the following essential steps:

1. Two homologous DNA molecules become paired.
2. Two of the DNA strands, one in each molecule, are cleaved.
3. The two nicked strand segments crossover, thus forming a Holliday intermediate.
4. DNA ligase seals the cut ends.
5. Branch migration caused by base pairing exchange leads to the transfer of a segment of DNA from one homologue to the other.
6. A second series of DNA strand cuts occurs.
7. DNA polymerase fills any gaps, and DNA ligase seals the cut strands.

As is often the case, most investigations of general recombination have been done with bacteria. Certain mutations in *E. coli* have proved to be useful in determining the molecular details of this process. Recombination in *E. coli* is initiated by RecBCD, an enzyme complex that possesses both exonuclease and helicase activities. After binding to a DNA molecule, RecBCD cleaves one of the strands and proceeds to unwind the double helix until it reaches 5′-GCTG-GTGG-3′ (the Chi site), a sequence that occurs frequently in *E. coli* DNA. Strand exchange is effected by monomers of RecA, an ATPase that coats one of the strands. Powered by ATP, the RecA-coated strand segment then interacts with a nearby double-helical DNA segment with a homologous sequence, thus forming a triple helix. Branch migration (Figure 18.14) is initiated as RuvA recognizes and binds to the Holliday junction. Two copies of RuvB, a hexamer with ATPase and helicase activities, then form a ring on either side of the junction. Branch migration is catalyzed by the RuvAB complex. This molecular machine separates, rotates, and pulls the strands in the two sets of helices, even after RecA dissociates. The migration ends when a specific sequence [5′-(A or T)TT (G or C)-3′] is reached. As RuvAB detaches, two RuvC proteins bind to the junction. Recombination ends as RuvC cleaves the crossover strands and the Holliday structure resolves itself to form two separate double-helical DNA molecules.

In bacteria, general recombination appears to be involved in several forms of intermicrobial DNA transfer:

1. Transformation. In **transformation**, naked DNA fragments enter a bacterial cell through a small opening in the cell wall and are introduced into the bacterial genome. (Recall Fred Griffith's experiment, see p. 572.)

FIGURE 18.13

General Recombination.

(a) One strand in each duplex is nicked, and each broken strand invades the other duplex. (b) Covalent bonds are formed (c), cross-linking the two duplexes. Branch migration then occurs (d). The bending of the chi structure in (e) and (f) makes later events easier to understand. In (g) the same strands are nicked as those in (a). The resulting heteroduplex (h) contains a patch. Nicking the opposite strands (i) results in the formation (j) of a spliced heteroduplex.

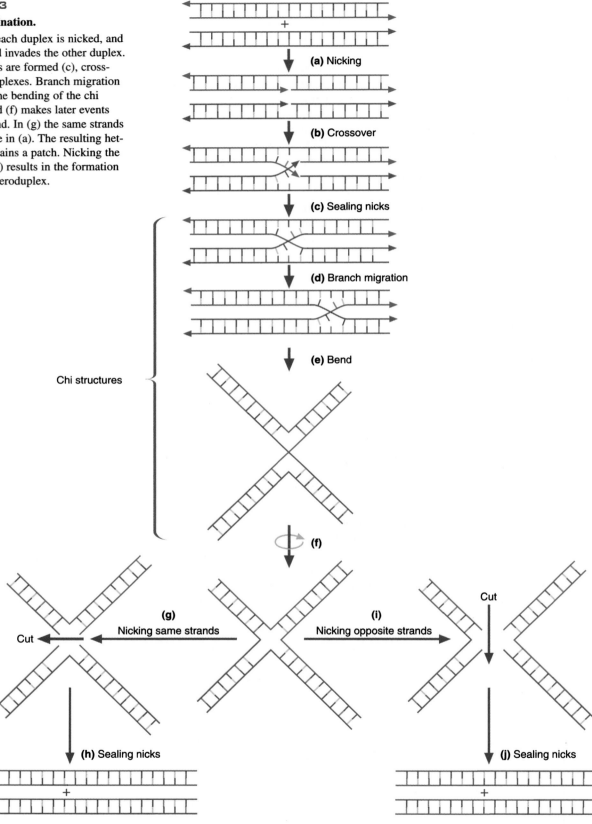

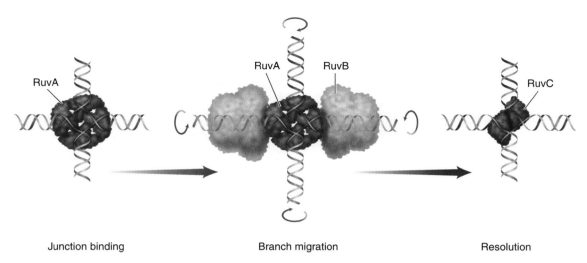

| Junction binding | Branch migration | Resolution |

FIGURE 18.14

Model of the Association of Ruv Proteins with a Holliday Junction.

RuvA, a tetramer, first binds to the Holliday junction point. Two hexameric RuvB rings then form on both sides of the DNA/RuvA complex with the DNA passing through the rings. Branch migration occurs as ATP hydrolysis drives the two RuvB rings to rotate the DNA helices in opposite directions. After branch migration, RuvA and RuvB detach and two RuvC proteins bind to the junction. RuvC, a nuclease, proceeds to cut the crossover strands thus resolving the Holliday structure.

   2. Transduction. Transduction occurs when bacteriophage inadvertently carry bacterial DNA to a recipient cell. After a suitable recombination, the cell uses the transduced DNA.

   3. Conjugation. Certain bacterial species are known to engage in **conjugation**, an unconventional sexual mating that involves a donor cell and a recipient cell. The donor cell possesses a specialized plasmid that allows it to synthesize a sex pilus, a filamentous appendage that functions in a DNA exchange process. (Recall that plasmids are small extrachromosomal circular DNA molecules that replicate independently of the cell's chromosome.) After the pilus attaches to the surface of the recipient cell, a fragment of the donor's genetic material is transferred. The transferred DNA segment can be integrated into the recipient's chromosome by recombination or it may exist outside it in plasmid form.

   General recombination in eukaryotes is believed to be similar to the process in prokaryotes. Several eukaryotic proteins have been discovered that closely resemble in both structure and function those observed in *E. coli*. For example, RAD51, found in yeast, performs the same functions as RecA, i.e., repairing double stranded breaks.

Bacterial conjugation has medical consequences. For example, certain plasmids contain genes that code for toxins. The causative agent of a new deadly form of food poisoning is a strain of *E. coli* (*E. coli* 0157) that synthesizes a toxin that causes massive bloody diarrhea and kidney failure. This toxin is now believed to have originated in *Shigella*, another bacterium that causes dysentery. Similarly, the growing problem of antibiotic resistance is the result, in part, of the spread of antibiotic-resistance genes among bacterial populations. Antibiotic resistance develops because antibiotics are overused in medical practice and in livestock feeds. Suggest a mechanism by which this extensive use promotes antibiotic resistance. (*Hint*: The high-level use of antibiotics acts as a selection pressure.)

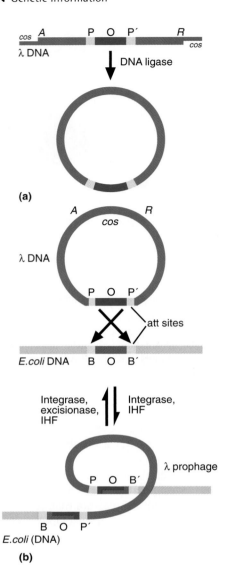

FIGURE 18.15

Insertion of the Bacteriophage λ Genome into the *E. coli* Chromosome.
(a) The λ DNA circularizes as the single-stranded cos sequences anneal. (b) Insertion occurs through site-specific recombination between short homologous phage and bacterial sequences called attachment (att) sites.

SITE-SPECIFIC RECOMBINATION As mentioned, site-specific recombination depends more on protein-DNA interactions than on DNA-DNA sequence homology. For example, the integration of bacteriophage λ into the *E. coli* chromosome (Figure 18.15) requires only short recognition sequences. A site-specific viral enzyme, called integrase, is largely responsible for promoting this recombinational event. The viral integrase, a topoisomerase, makes a staggered double strand cut in short similar sequences (called attachment or *att* sites) in both the viral and bacterial DNA. The integration process, which requires the bacterial protein IHF, involves the formation of a Holliday junction. The resolution of the junction results in the insertion of the λ genome into the bacterial chromosome.

TRANSPOSITION Barbara McClintock, a geneticist working with corn (maize), reported in the 1940s that certain genome segments can move from one place to another. Because chromosomes were believed to consist of genes in a fixed and unvarying order, it was not until 1967, when transposable elements (or **transposons**) were discovered in *E. coli*, that scientists began to comprehend that genomes are not as fixed as they had once thought. (In recognition of her monumental contribution to genetics, Dr. McClintock received the 1983 Nobel Prize in physiology or medicine.) Transposons (also referred to as "jumping genes") have now been observed in a wide variety of organisms in addition to bacteria, for example, various fungi, plants, and animals.

Several bacterial transposons, referred to as insertion elements (ISelements), consist only of a gene that codes for a transposition enzyme (i.e., transposase), flanked by short DNA segments called inverted repeats (Figure 18.16). (Inverted repeats are short palindromes.) More complicated bacterial transposable elements, called composite transposons, contain additional genes, several of which may code for antibiotic resistance. Because transposons can "jump" between bacterial chromosomes, plasmids, and viral genomes, transpositions are now believed to play an important role in the spread of antibiotic resistance among bacteria.

Two transposition mechanisms have been observed:

1. Replicative transposition. During transposition a replicated copy is inserted into the new location (target site), leaving the original transposon at its original site (Figure 18.16). In replicative transposition an intermediate referred to as a cointegrate forms consisting of the donor segment covalently linked to the DNA containing the target site. An additional enzyme, called resolvase, catalyzes a site-specific recombination that allows the resolution of the cointegrate into two separate molecules.

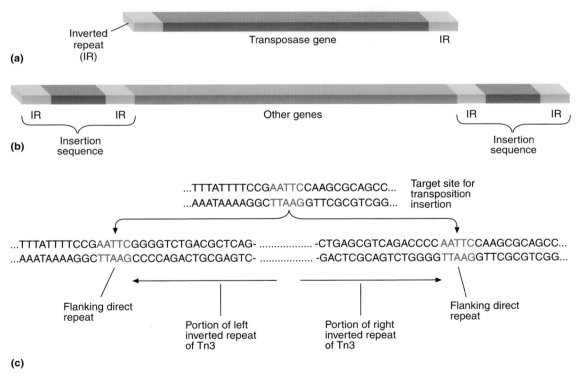

(c)

FIGURE 18.16

Bacterial Insertion Elements.

(a) An insertion sequence. (b) A composite transposon. (c) Insertion of a transposon (Tn3) into bacterial DNA. The insertion process involves the duplication of the target site. (Also refer to Figure 18.17.)

2. Nonreplicative transposition. The transposable element is spliced out of its original site (donor site) and inserted into the target site. Then the donor site must be repaired. If it cannot be repaired by the cell's DNA repair system, the consequences will be lethal.

Whether transposition is replicative or nonreplicative, short duplications of target site DNA segments are generated by the staggered cleavage catalyzed by transposase (Figure 18.17).

FIGURE 18.17

Formation of Duplicated Target Site Sequences During Transposition.

(a) Host DNA is cut (see arrows) in a staggered fashion (b). (c) The transposon (blue) is covalently attached at both ends to a strand of host DNA. (d) After the gaps are filled in by a DNA polymerase activity, there are nine base pair repeats of host DNA (red) flanking the transposon.

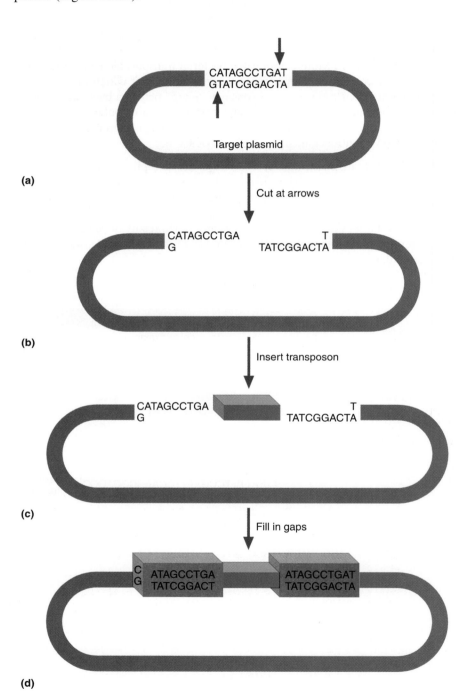

Some transposons found in eukaryotes resemble those found in bacteria. For example, the Ac element, the maize transposon that was first described by McClintock, is composed of a transposase gene flanked by short inverted repeats. (McClintock referred to the Ac transposon as a "controlling element" because it appeared to control the synthesis of the pigment anthocyanin in corn kernels.) Many other eukaryotic transposons, however, have somewhat different structures than those observed in bacteria. Instead of inverted repeats, many eukaryotic transposons, such as the Ty transposon in yeast, possess long terminal repeats (LTR), sometimes referred to as delta repeats. More important, the transposition mechanisms of Ty and many other eukaryotic transposons involve an RNA intermediate and bear a remarkable resemblance to the replicative phase of the retroviral life cycle. Ty contains genes for reverse transcriptase and RNase H, as well as several other genes required for its reverse transcription and integration into target DNA. Because these so-called retrotransposons lack the genes required to synthesize an envelope, they can move only within the genome, that is, they cannot exit as do the retroviruses.

The movement of transposons within a genome can cause several genetic changes. The insertion and excision of transposons often cause duplications, inversions, and deletions. Depending on the changes and their location, the effects of transposons can be viewed either as disruptive and damaging or as providing opportunities for genetic diversity. Some effects of transposition are observed as changes in gene expression, a topic that is discussed in Section 18.3.

KEY CONCEPTS 18.3

In general recombination, the exchange of DNA sequences takes place between homologous sequences. In site-specific recombination, DNA-protein interactions are principally responsible for the exchange of nonhomologous sequences.

Describe the differences between general recombination, site-specific recombination, and transposition.

QUESTION 18.7

Define the following terms:

a. transposition
b. conjugation
c. transduction
d. transformation
e. transposon

QUESTION 18.8

One of the fascinating aspects of complex organisms such as mammals is the existence of gene families (groups of genes that code for the synthesis of a series of closely related proteins). For example, several different types of collagen are required for the proper structure and function of connective tissues. Similarly, there are several types of globin gene. It is currently believed that gene families originate from a rare event in which a DNA sequence is duplicated. Some gene duplications provide a selective advantage by providing larger quantities of important gene products. In others, the two duplicate genes evolve independently. One copy continues to serve the same function, while the other eventually evolves to serve another function. Can you speculate about how gene duplications occur? Once a gene has been duplicated, what mechanisms introduce variations?

QUESTION 18.9

In the past few years, the life sciences have undergone both an intellectual and experimental sea change created by the sequencing of the genomes of dozens of species, including that of humans. This newfound wealth of information provided by **genomics** (the large-scale analyses of entire genomes) holds the promise not only for unprecedented insights into the inner workings of living organisms, but also for previously unimaginable capacities to accelerate research into all areas of interest to humans (e.g., health and disease, biodiversity, and agriculture). However, determinations of the structure and identity of genes in living organisms tell us nothing about how they operate. The goal of the newly emerging science of **functional genomics** is to understand how biomolecules work together within functioning organisms. Currently, attempts at closing this knowledge gap involve a variety of new "whole genome" technologies. For example, DNA chips (glass or plastic wafers holding a microarray of different DNA sequence probes) now make possible the simultaneous monitoring of the expression of thousands of genes in cultured cells. In the following discussion the basic tools used in genomic technology are described.

The isolation, characterization, and manipulation of DNA sequences, now considered to be commonplace, is made possible by a series of techniques referred to as **recombinant DNA technology**. The essential feature of this technology is that DNA molecules obtained from various sources can be cut and spliced together. These techniques have revolutionized biology and biochemistry because they have made investigations of genomes more accessible than ever before. For example, the large number of DNA copies required in DNA sequencing methods have been obtainable through molecular cloning and (more recently) the polymerase chain reaction. Commercial applications of recombinant DNA techniques have revolutionized medical practice. For example, human gene products such as insulin and growth hormone, as well as certain vaccines and diagnostic tests, are now produced in large quantities by bacterial cells into which recombinant genes have been inserted. Currently, several research groups are investigating the use of recombinant techniques in human gene therapy, a process in which (it is hoped) defective genes can be replaced by their normal counterparts.

Figure 18A illustrates the basic features of recombinant DNA construction. The process begins by using a restriction enzyme (Biochemical Methods 17.1) that generates sticky ends to cleave DNA from two different sources. The DNA fragments are then mixed under conditions that allow annealing (base pairing) between the sticky ends to occur. Once base pairing has occurred, the fragments are covalently bonded together by DNA ligase. After recombinant DNA molecules have been isolated and purified, it is usually necessary to reproduce them so that sufficient quantities are available for further investigation. Molecular cloning, a commonly used method for increasing the number of copies of DNA, is discussed next.

Molecular Cloning

In molecular cloning, a piece of DNA isolated from a donor cell (e.g., any animal or plant cell) is spliced into a vector. A **vector** is

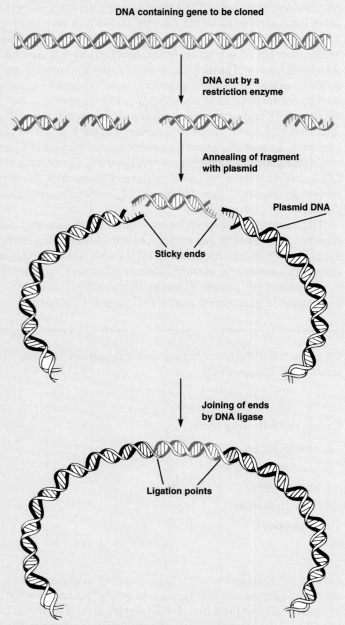

FIGURE 18A Recombinant DNA Construction.
Recombinant DNA molecules are created by treating DNA from two sources with restriction enzymes. Under hybridizing conditions, DNA fragments with sticky ends anneal together. Once base pairing has occurred, DNA ligase joins the fragments together.

a DNA molecule capable of replication that is used to transport a foreign DNA sequence, often a gene, into a host cell.

The choice of vector depends on the size of donor DNA. For example, bacterial plasmids are often used to clone small pieces (15 kb) of DNA. Somewhat larger pieces (24 kb) are incorporated into bacteriophage λ vectors, whereas cosmid vectors are used for DNA fragments as large as 50 kb. Bacteriophage λ can be used as a vector because a substantial portion of its genome does not code for phage production and can therefore be removed. The removed viral DNA can then be replaced by foreign DNA. **Cosmids** are cloning vehicles that contain λ bacteriophage cos sites incorporated into plasmid DNA sequences with one or more selectable markers. The *cos* sites allow delivery to the host cell in a phage head, the plasmid DNA facilitates independent replication of the recombined unit, and the selectable markers permit detection of successful recombinants. Still larger pieces can be inserted into bacterial artificial chromosomes and yeast artificial chromosomes. **Bacterial artificial chromosomes** (BAC), derived from a large *E. coli* plasmid called the F-factor, are used to clone DNA sequences as long as 300 kb. **Yeast artificial chromosomes** (YAC), which can accommodate up to 1000 kb, are constructed using yeast DNA sequences that are autonomously replicating (i.e., they contain a eukaryotic DNA replication origin).

As noted, forming recombinant DNA requires a restriction enzyme, which cuts the vector DNA (e.g., a plasmid) open. (Figure 18B)After the sticky ends of the plasmid have annealed with those of the donor DNA, a DNA ligase activity joins the two molecules covalently. Then the recombinant vector is inserted into host cells.

In some circumstances the introduction of a cloning vector into a host cell is a trivial process. For example, phage vectors are designed so they introduce recombinant DNA in an infective process called **transfection**, and some bacteria take up plasmids unaided. However, most host cells must be induced to take up foreign DNA. Several methods are used. In some prokaryotic and eukaryotic cells, the addition of Ca^{2+} to the medium promotes uptake. In others, a process called **electroporation**, in which cells are treated with an electric current, is used. One of the most effective methods for transforming animal and plant cells is the direct microinjection of genetic material. **Transgenic animals**, for example, are created by the microinjection of recombinant DNA into fertilized ova.

Once introduced, each type of cell replicates the recombinant DNA along with its own genome. Note that recombinant vectors must contain regulatory regions recognized by host cell enzymes.

As host cells that have been successfully transformed proliferate, they rapidly amplify the recombinant DNA. For example, under favorable conditions of nutrient availability and temperature, a single recombinant plasmid introduced into an *E. coli* cell can be replicated a billion times in about 11 hours. However, transformed and untransformed cells usually have identical appearances. (See Figure 18C for an example of an exception to this rule.) Consequently, researchers often design cloning protocols utilizing vectors with selectable **marker**

(continued on page 632)

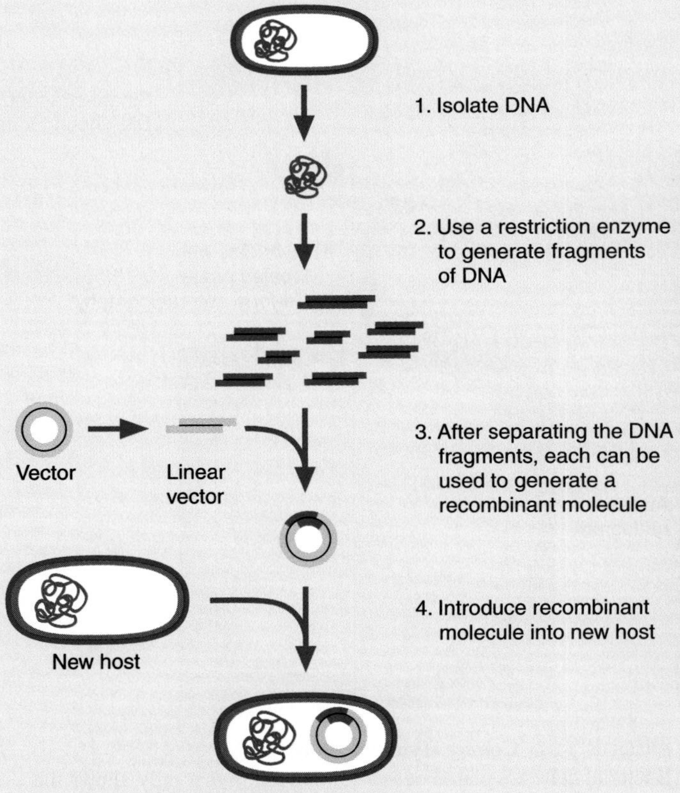

1. Isolate DNA

2. Use a restriction enzyme to generate fragments of DNA

Vector   Linear vector

3. After separating the DNA fragments, each can be used to generate a recombinant molecule

New host

4. Introduce recombinant molecule into new host

FIGURE 18B DNA Cloning.
In the cloning process, each clone is produced by introducing a recombinant molecule into a host cell, which then replicates the vector along with its own genome.

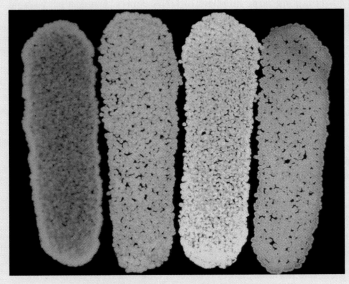

FIGURE 18C Four Recombinant *E. Coli* cells, each with a Different Variant of the Gene for Luciferase.
Luciferase is an enzyme found in organisms such as fireflies, mollusks, and several types of deep-sea fish. When in the presence of ATP and luciferin, luciferase catalyzes a light-emitting reaction. (The luciferins are a group of bioluminescent compounds that emit light when they are oxidized by O_2 in a reaction catalyzed by a luciferase.

genes (genes whose presence can be detected) so that they can easily identify transformed cells. Antibiotic resistance genes, for example, are usually incorporated into the plasmid vectors introduced into bacteria. When the treated bacteria are plated out on a medium containing the antibiotic, only the transformed cells will grow. With eukaryotic cells such as yeast, cells that lack an enzyme required to synthesize a nutrient may be used. For example, vectors containing the LEU2 gene are used to transform mutant yeast cells that lack a specific enzyme in the leucine biosynthetic pathway. Only those cells that have successfully transformed are able to grow in a leucine-deficient medium.

In another approach, the **colony hybridization technique** (Figure 18D, bacteria are screened by using a radioactively labeled nucleic acid probe, an RNA molecule or a single-stranded DNA molecule with a sequence complementary to that of a specific sequence within the recombinant DNA. Bacterial cells are plated out onto solid media in petri dishes and allowed to grow into colonies. Each plate is then blotted with a nitrocellulose filter. (Most of the original colonies remain on the petri dishes.) The cells on the nitrocellulose filter are lysed, and the released DNA is treated so that hybridization with the probe can occur. Once nonhybridized probe molecules have been washed away, autoradiography (Biochemical Methods 2.1) is used to identify the colonies on the master plate that possess the recombinant DNA.

Polymerase Chain Reaction

Although cloning has been immensely useful in molecular biology, the **polymerase chain reaction** (PCR) is a more convenient method for obtaining large numbers of DNA copies. Using a heat-stable DNA polymerase from *Thermus aquaticus* (Taq polymerase), PCR can amplify any DNA sequence, provided that the flanking sequences are known (Figure 18E). Flanking sequences must be known because PCR amplification requires primers. Priming sequences are produced by automated DNA synthesizing machines.

PCR begins by adding Taq polymerase, the primers, and the ingredients for DNA replication to a heated sample of the target DNA. (Recall that heating DNA separates its strands.) As the mixture cools, the primers attach to their complementary sequences on either side of the target sequence. Each strand then serves as a template for DNA replication. At the end of this process, referred to as a *cycle*, the copies of the target sequence have been doubled. The process can be repeated indefinitely, synthesizing an extraordinary number of copies. For example, by the end of 30 cycles a single DNA fragment has been amplified one billion times.

Genomic Libraries

Genomic libraries are collections of clones derived from fragments of entire chromosomes or genomes. They are used for a variety of purposes, the most important of which are the isolation of specific genes whose chromosomal location is unknown and in genome-wide sequencing efforts (gene mapping). Genomic libraries are produced in a process, referred to as **shotgun cloning**, in which a genome is randomly digested (Figure 18F). The range of fragment sizes, which is determined by the type of restriction enzyme and experimental conditions chosen, must be compatible with the vector. To ensure that all sequences of interest are represented in the library, DNA sam-

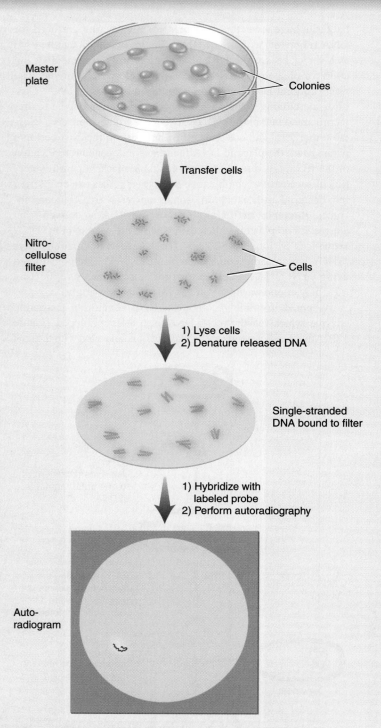

FIGURE 18D Colony Hybridization.
Bacterial cells are plated onto a solid medium that only allows the growth of transformed cells. Once the colonies become visible, the plate is blotted with a nitrocellulose filter. The cells clinging to the filter are lysed and the released DNA is denatured and deproteinized. After a labeled probe is added, unhybridized probe molecules are washed away. Cells that possess DNA sequences that hybridize with the probe are identified by comparing the autoradiogram of the filter with the master plate.

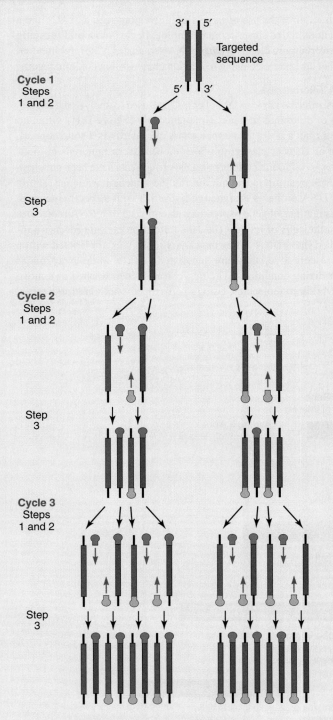

FIGURE 18E **Polymerase Chain Reaction.**
A single DNA molecule can be amplified millions of times by
replicating a three-step cycle. In the first step the dsDNA sample is
denatured by heating to 95°C. In step 2 the temperature is quickly
lowered to 50°C and an oligonucleotide primer is added. The primer
hybridizes to complementary sequences on the ends of the two strands.
During step 3, DNA synthesis occurs as the temperature is raised to
70°C, the optimal temperature of Taq polymerase. The cycle is then
repeated with both old and new strands serving as templates.

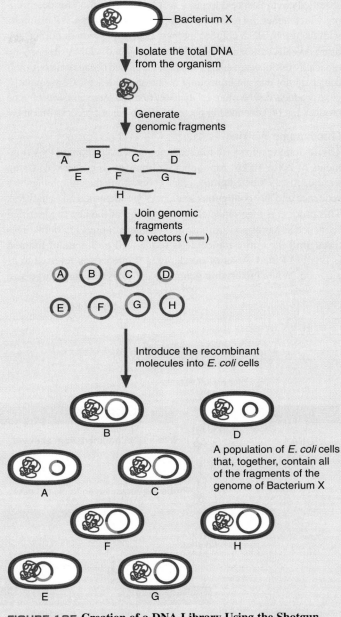

FIGURE 18F **Creation of a DNA Library Using the Shotgun
Method.**
After the organism's DNA is isolated from the organism and then
purified, it is cleaved into fragments with a restriction enzyme.
The fragments are then randomly incorporated into vectors and
the recombinant molecules are introduced into cells such as *E. coli*.
The collection of these cells is called a genome library.

ples are often only partially digested. The location of any gene can
be identified if an appropriate probe is available.

In a variation of genomic libraries, clone libraries of complementary DNA (cDNA) molecules, called **cDNA libraries**, are produced from mRNA molecules by reverse transcription. This technique
can be used to evaluate the transcriptome of certain cell types under

(continued on page 634)

specified circumstances. In other words, it is a method for determining which genes are expressed in a particular cell type. For example, with the use of DNA chip technology, gene expression in normal and diseased cells can be investigated and compared. cDNA libraries are especially useful when eukaryotic DNA is cloned because mRNA molecules do not contain noncoding or intron sequences. Consequently, gene products can be more easily identified and large amounts of gene product can be generated in bacteria, which cannot process introns.

Chromosome Walking

This technique allows for the sequencing of overlapping DNA fragments of 20 to 40 kb from genomic libraries (Figure 18G). Using a radioactively labeled probe, clones containing the complementary sequence and any contiguous sequences are identified and analyzed. This process is then repeated using the end of the newly identified sequence as a probe. Chromosome walking continues in both directions until the entire molecule is sequenced or a gene of interest is finally located. A set of overlapping sequences is referred to as a **contig**. When eukaryotic genomes are analyzed, their large size

often requires the use of large cloning vectors such as YACs and a technique called chromosomal jumping. In **chromosomal jumping** the overlapping clones contain DNA sequences of several hundred kb that are generated using restriction enzymes that cut infrequently.

DNA Microarrays

DNA microarrays, or DNA "chips," are used to analyze the expression of thousands of genes simultaneously (Figure 18H). Often no larger than a postage stamp, a DNA microarray is a solid support, such as glass or plastic, to which thousands, or hundreds of thousands, of oligonucleotides or ssDNA fragments have been attached. At each position in the microarray the attached sequence, acting as a DNA probe, is designed to hybridize with a specific gene. In investigations of gene expression, an entire set of mRNA molecules from the cells of interest (i.e., the transcriptome) are reverse-transcribed into cDNA. After the cDNA molecules are labeled with a fluorescent dye, they are incubated with a microarray under hybridizing conditions. The microarray is then washed to remove unhybridized molecules. Using microscopes, photomultiplier tubes,

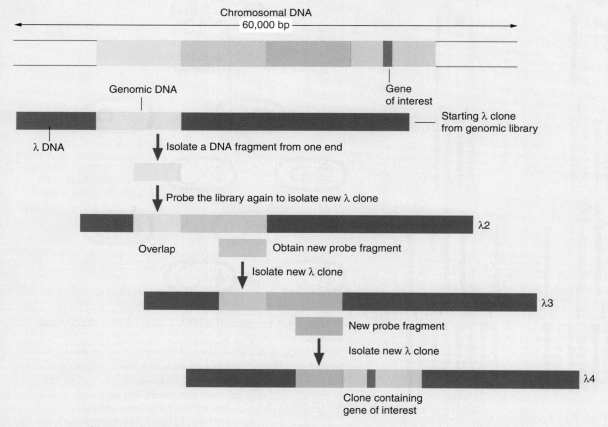

FIGURE 18G Chromosome Walking.
In chromosome walking DNA clones, which contain overlapping sequences, are systematically identified. They may then be mapped and sequenced. Unknown genes may also be searched for. The process begins when DNA is cleaved into pieces and cloned. (In this example, bacteriophage λ vectors are used.) One end of the starting clone is labeled and used as a probe to identify the clone in the λ library that contains both that sequence and an adjacent sequence. Repeating this step, the end of the second clone is labeled and used as a probe to identify yet another overlapping clone. The process continues until a collection of clones is obtained that together contain all of the sequences in the original DNA fragment.

and computer software, researchers can determine which genes are being expressed by identifying the positions on the microarray that are fluorescing. Using this technique scientists can observe changes in gene expression under a variety of circumstances. Examples include comparisons of normal and cancerous cells, and cells exposed to different nutrients or signal molecules.

Genome Projects

The Human Genome Project is an intensive international effort to determine the nucleotide sequence of the entire human genome. As this goal is being reached, the attention of researchers is shifting to **annotation** (i.e., functional identification) of the 30,000-odd human genes. Just as scientists have historically used structural and functional comparisons of other organisms in fields such as anatomy, biochemistry, physiology, and medicine to better understand human biology, the current effort to interpret human genome data is being aided immensely by comparisons with the information obtained in other genome projects. The genomes of well-researched organisms as diverse as bacteria (e.g., *E. coli*), yeast (e.g., *Saccharomyces cerevisiae*), the worm *Caenorhabditis elegans*, the fruit fly *Drosophila*, and various mammals (e.g., the mouse) have been used in genome structure analysis and in the assignment of recently discovered genes in other organisms.

When a genome is deciphered, no matter what type of organism it comes from, there are three basic steps in its analysis:

1. **Genetic linkage mapping**. Over most of the twentieth century geneticists constructed chromosome maps for selected organisms by analyzing recombination frequencies derived from genetic crosses. In humans the recombination frequency data used to determine the physical relationships and relative distances of genes on chromosomes were obtained by analyzing the pedigrees of large families. The low-resolution maps obtained from genetic mapping have been used as a framework for analyzing data obtained from more sophisticated methods.

2. **Physical mapping of genomes**. Physical maps are obtained by cleaving chromosomal DNA from the organism of interest with a restriction enzyme. Each DNA fragment is then cloned. The order of the fragments in the chromosome is determined by using probes made from known markers in the low-resolution genetic linkage map. The process is repeated with several other restriction enzymes so that ambiguities in the data can be resolved.

3. **DNA sequencing**. Individual cloned fragments from the entire chromosome or genome are sequenced. To make this feat feasible, DNA sequencing has been automated (Biochemical Methods 17.1). The nucleotide sequence of an entire genome is determined by linking the results of sequencing experiments from overlapping clones.

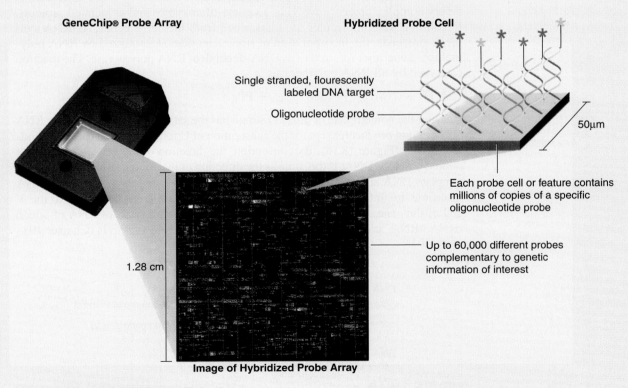

GeneChip® Probe Array

Hybridized Probe Cell

Single stranded, flourescently labeled DNA target

Oligonucleotide probe

50μm

Each probe cell or feature contains millions of copies of a specific oligonucleotide probe

Up to 60,000 different probes complementary to genetic information of interest

1.28 cm

Image of Hybridized Probe Array

FIGURE 18H DNA Microarray Technology.
DNA microarrays can be used to determine which genes are expressed in a specific cell type because each "chip" can accommodate from thousands to millions of DNA probes. (Oligonucleotide probes are synthesized on the chip surface using photolithographic techniques similar to those used in the manufacture of computer chips.) The microarray is incubated under hybridizing conditions with fluorescently labeled cDNA. The cDNA molecules are derived from mRNA extracted from the cells of interest.

QUESTION 18.10   Explain the use of marker genes in cloning.

QUESTION 18.11   Briefly outline the basic principles of PCR. Calculate the degree of amplification attained by 15 PCR cycles.

QUESTION 18.12   Describe how genomic libraries are generated. How do they differ from cDNA libraries?

QUESTION 18.13   What are DNA "chips"? Describe examples of the kinds of information their use can provide to researchers.

QUESTION 18.14   Define the following terms:

a. shotgun cloning
b. bacterial artificial chromosomes
c. electroporation
d. vector
e. recombinant DNA technology

18.2   TRANSCRIPTION

As with all aspects of nucleic acid function, the synthesis of RNA molecules is a very complex process involving a variety of enzymes and associated proteins. Recall that RNA molecules are transcribed from the cell's genes. As RNA synthesis proceeds, the incorporation of ribonucleotides is catalyzed by RNA polymerase, sometimes referred to as DNA-dependent RNA polymerase. The reaction catalyzed by all RNA polymerases is

$$NTP + (NMP)_n \longrightarrow (NMP)_{n+1} + PP_i$$

Because the nontemplate or plus (+) strand has the same base sequence as the RNA transcription product (except for the substitution of U for T), it is also called the **coding strand** (Figure 18.18). By convention, the direction of the gene, a segment of double-stranded DNA, is the same as the direction of the coding strand. Because the template DNA strand, also called the minus (−) strand, and the newly made RNA molecule are antiparallel, the polymerization proceeds from the 5′ end to the 3′ end of the gene. As noted, transcription generates several types of RNA of which rRNA, tRNA, and mRNA are directly involved in protein synthesis (Chapter 19).

```
DNA                                          (+)
5′— TTTGGACAACGTCCAGCGATC —3′   Nontemplate strand

3′— AAACCTGTTGCAGGTCGCTAG —5′   Template strand
                                             (−)
RNA
5′—UUUGGACAACGUCCAGCGAUC —3′
```

FIGURE 18.18

DNA Coding Strand.

One of the two complementary DNA strands, referred to as the template(−) strand, is transcribed. The RNA transcript is identical in sequence to the nontemplate (+) or coding strand, except for the substitution of U for T.

Transcription in Prokaryotes

The RNA polymerase in *E. coli* catalyzes the synthesis of all RNA classes. With a molecular weight of about 450 kD, RNA polymerase is a relatively large complex (Figure 18.19). It is composed of five types of polypeptide: α, β, β', ω, and σ (sigma). The core enzyme (α_2, β, β', and ω) catalyzes RNA synthesis. The transient binding of the σ factor to the core enzyme allows it to bind both the correct template strand and the proper site to initiate transcription. A variety of σ factors have been identified. For example, in *E. coli*, σ^{70} is involved in the transcription of most genes, whereas σ^{32} and σ^{28} promote the transcription of heat shock genes and the flagellin gene, respectively. (As its name suggests, flagellin is a protein component of bacterial flagella.) The superscript indicates the protein's molecular weight in kilodaltons.

The transcription of an *E. coli* gene is outlined in Figure 18.20. The process consists of three stages: initiation, elongation, and termination. Each is discussed briefly.

The initiation of transcription involves binding RNA polymerase to a specific DNA sequence called a **promoter**. Although prokaryotic promoters are variable in size (from 20 to 200 bp), two short sequences at positions about 10 and 35 bp away from the transcription initiation site are remarkably similar among various bacterial species. (In these sequences, called **consensus sequences**, specific nucleotides are usually found at each site. The sequences shown in figure 18.20 are named in relation to the transcription starting point, the −35 region and the −10 region. The −10 region is also called the *Pribnow box*, after its discoverer.) RNA polymerase slides along the DNA until it reaches a promoter sequence. Once the enzyme binds to the promoter region, a short DNA segment near the Pribnow box unwinds. Transcription begins with binding the first nucleoside triphosphate (usually ATP or GTP) to the RNA polymerase complex. A nucleophilic attack by the 3′-OH group of the first nucleoside triphosphate on the α-phosphate of a second nucleoside triphosphate (also positioned by base pairing in an adjacent site) causes the first phosphodiester bond to form. (Because the phosphate groups of the first molecule are not involved in this reaction, the 5′ end

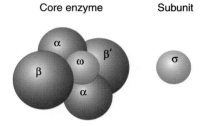

FIGURE 18.19

E. coli **RNA Polymerase.**

The *E. coli* RNA polymerase consists of two α subunits and one each of β, β', and ω subunits. The transient binding of a σ subunit allows binding of the core enzyme to appropriate DNA sequences.

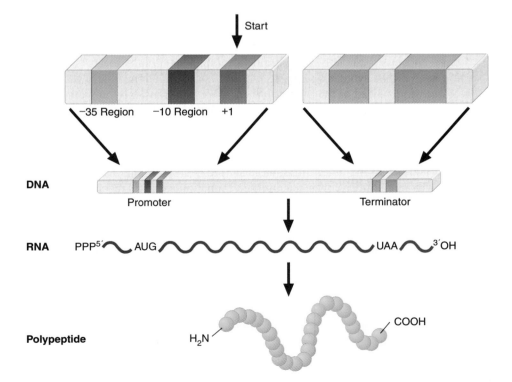

FIGURE 18.20

A Typical *E. coli* Transcription Unit.

If RNA polymerase can bind to the promoter, DNA transcription begins at +1, downstream from the promoter in bacteria. Translation of mRNA begins as soon as the ribosome binding site on the transcribing mRNA is available.

FIGURE 18.21

Transcription Initiation in *E. coli*.

(a) A transcription bubble forms as a short DNA segment unwinds. An RNA-DNA hybrid forms as transcription progresses. The bubble moves to keep up with transcription as DNA unwinds before it and rewinds behind it. (b) Transcription induces coiling. Positive supercoils form ahead of the bubble, while negative supercoils form behind it.

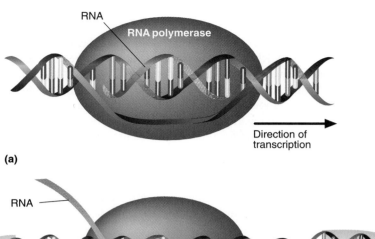

(a)

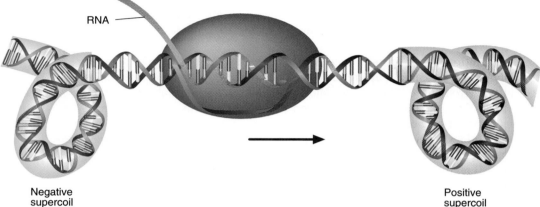

(b)

of prokaryotic transcripts possesses a triphosphate group.) If the transcribed sequence reaches a length of about 10 nucleotides, the conformation of the RNA polymerase complex changes; for example, the σ factor is released, and the initiation phase ends. As soon as an RNA polymerase has initiated transcription and moved beyond the promoter site, another RNA polymerase can move in, bind to the site, and start another round of RNA synthesis.

Once the σ factor detaches and the affinity of the RNA polymerase complex for the promoter site decreases, the elongation phase begins. The core RNA polymerase converts to an active transcription complex as it binds several accessory proteins. As RNA synthesis proceeds in the $5' \longrightarrow 3'$ direction (Figure 18.21), the DNA unwinds ahead of the *transcription bubble* (the transiently unwound DNA segment in which an RNA-DNA hybrid has formed). The unwinding action of RNA polymerase creates positive supercoils ahead of the transcription bubble and negative supercoils behind the bubble, which are resolved by topoisomerases. (As the bubble moves down the gene, it is said to move "downstream." Any structure or activity in the other direction is said to be "upstream.") The incorporation of ribonucleotides continues until a termination signal is reached.

Termination sequences contain palindromes. The RNA transcript of the DNA palindrome forms a stable hairpin turn. Apparently, this structure causes the RNA polymerase to slow or stop and partially disrupts the RNA-DNA hybrid structure. In some termination sequences, referred to as ρ (rho)-independent termination sites, several (about six) uridine residues follow the hairpin structure. Because U-A base pair interactions are weak, the short U-A sequence promotes the dissociation of the newly synthesized RNA from the DNA strand. In ρ-dependent termination the ρ factor (an enzyme that catalyzes the ATP-dependent unwinding of RNA-DNA helices) promotes the dissociation of the RNA polymerase complex from the RNA-DNA hybrid. ρ binds to RNA and not to RNA polymerase.

In prokaryotes, mRNA is used immediately in protein synthesis. In fact, protein synthesis begins while transcription is ongoing. However, mature rRNA and tRNA

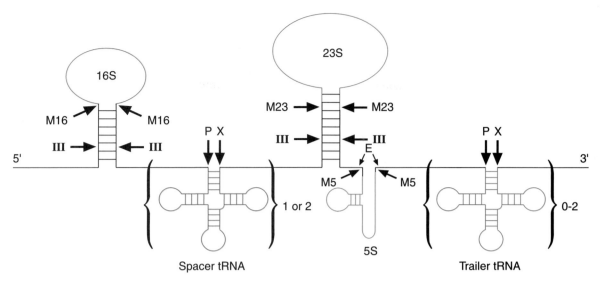

FIGURE 18.22

Ribosomal RNA Processing in *E. coli.*

Each rRNA operon encodes a primary transcript that contains one copy each of 16S, 23S, and 5S rRNAs. Each transcript also encodes one or two spacer tRNAs and as many as two trailer tRNAs. Posttranscriptional processing involves numerous cleavage reactions catalyzed by various RNases and splicing reactions. (Individual RNases are identified by letters and/or numbers, e.g., M5, X, and III.) RNase P is a ribozyme.

molecules are produced from larger transcripts by posttranscriptional processing. The RNA processing reactions for *E. coli* rRNA are outlined in Figure 18.22. The *E. coli* genome contains several sets of the rRNA genes 16S, 23S, and 5S. (Each set of genes is called an **operon**.) In the primary processing step, the polycistronic 30S transcript is methylated and then cleaved by several RNases into a number of smaller segments. Further cleavage by different RNases produces mature rRNAs. A few tRNAs are also produced. The other tRNAs are produced from primary transcripts in a series of processing reactions in which they are trimmed down by several RNases. In the last step of tRNA processing, a large number of bases are altered by several modification reactions (e.g., deamination, methylation, and reduction).

Transcription in Eukaryotes

Although DNA transcription in prokaryotes and in eukaryotes resembles each other, there are significant differences, apparently because of the greater structural complexity of eukaryotes. For example, most of the chromatin in eukaryotic cells is at least partially condensed at any time. Yet to be transcribed, DNA must be sufficiently exposed and accessible for RNA polymerase activity. Similarly, proper cell function depends on the timely transport of a wide variety of transcription products across the nuclear membrane into the cytoplasm. Eukaryotes appear to have solved these and other complex problems with equally complex solutions. For example, eukaryotic transcription is regulated by a vast number of transcription factors that must be precisely assembled before transcription can begin. Some of these factors influence transcription when bound to DNA sequences that are remote from the promoter region they influence. Transport problems appear to have been solved in part by certain processing reactions that allow each transcription product to be exported through a nuclear pore. (Nuclear pore complexes are complicated multisubunit structures. It is currently believed that transport through the pore complex occurs when RNA and protein molecules bind to specific receptors. Transcriptionally active nuclei may possess larger numbers of nuclear pores.)

KEY CONCEPTS 18.4

During transcription, an RNA molecule is synthesized from a DNA template. In prokaryotes this process involves a single RNA polymerase activity. Transcription is initiated when the RNA polymerase complex binds to a promoter sequence.

Principally because of the complexity of eukaryotic genomes, eukaryotic DNA transcription is not understood as completely as the prokaryotic process. However, the eukaryotic process is known to possess the following unique features:

1. **RNA polymerase activity**. Eukaryotes possess three nuclear RNA polymerases, each of which differs in the type of RNA synthesized, subunit structure, and relative amounts. RNA polymerase I, which is localized within the nucleolus, transcribes the large rRNAs. The precursors of mRNA and most snRNAs are transcribed by RNA polymerase II, and RNA polymerase III is responsible for transcribing the precursors of the tRNAs and 5S rRNA. Each polymerase possesses two large subunits and several (six to ten) smaller subunits. For example, the two large subunits of RNA polymerase II, the enzyme that transcribes the majority of eukaryotic genes, have molecular weights of 215 and 139 kD. The number of smaller subunits varies among species; for example, plants possess eight, whereas vertebrates have six. Some of the smaller subunits are also present in the other two RNA polymerases. In contrast to the prokaryotic RNA polymerase, the eukaryotic enzymes cannot initiate transcription themselves. Various transcription factors must be bound at the promoter before transcription can begin.

2. **Promoters**. The promoter sequences in eukaryotic DNA are larger, more complicated, and more variable than those of prokaryotes. Many promoters for RNA polymerase II contain consensus sequences, referred to as the TATA box, which occur about 25–30 bp upstream from the transcription initiation site. As illustrated in Figure 18.23, the binding of the transcription factor TFIID to the

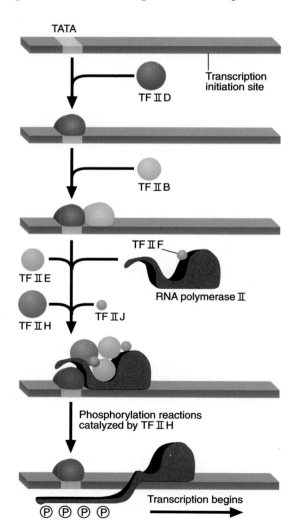

FIGURE 18.23

Transcription Initiation in Eukaryotes.

Before RNA polymerase II can begin transcribing a gene, transcription factors must assemble into a complex. The process begins when TFIID binds to a TATA sequence. TFIID, which consists of *TBP* (a TATA binding protein) and several associated proteins, binds to and unwinds the DNA duplex in the TATA sequence. Then TFIIB binds and, later, TFIIE, TFIIH, and TFIIJ. TFIIF binds directly to RNA polymerase II. Because of phosphorylation reactions catalyzed by TFIIH, the RNA polymerase II becomes active and begins transcription.

TATA box is the first step in the assembly of the RNA polymerase II transcription complex. The frequency of transcription initiation is often affected by binding certain transcription factors to upstream elements such as the *CAAT box* and the *GC box*. The activity of many promoters is affected by *enhancers*, regulatory sequences that may occur thousands of base pairs upstream or downstream of the gene they affect. (In yeast these sequences are called upstream activator sequences, or UAS.) The effects of enhancers can be complex. For example, a single gene may be controlled by the combined activities of several enhancers. Hormone response elements (Section 16.4) often act as enhancers.

3. Processing. Posttranscriptional processing occurs in both prokaryotes and eukaryotes. The most notable differences between the two types of organisms lie in the processing of mRNA. In contrast to prokaryotic mRNA, which usually requires little or no processing, eukaryotic mRNAs are the products of extensive editing. Throughout processing, pre-mRNA transcripts (hnRNA, see p. 596) are associated with about 20 different types of nuclear proteins in ribonucleoprotein particles (hnRNP). Shortly after the transcription of the primary transcript begins, a modification of the 5′ end called *capping* occurs. The cap structure (Figure 18.24), which consists of 7-methylguanosine linked to the mRNA through a triphosphate linkage, protects the 5′ end from exonucleases and promotes mRNA translation by ribosomes. For unknown reasons, RNA polymerase II transcribes well past the functional end of the primary transcript. After transcription has terminated, the transcript is cleaved at a specific site near the sequence AAUAAA. Immediately afterwards, from 100 to 250 adenylate residues are added by poly(A)polymerase to the 3′ end. This *poly A tail* is believed to have several functions, which include buffering the loss of critical sequences from the 3′-end of the mRNA through the action of 3′,5′-exonucleases and promoting the export of mRNAs into the

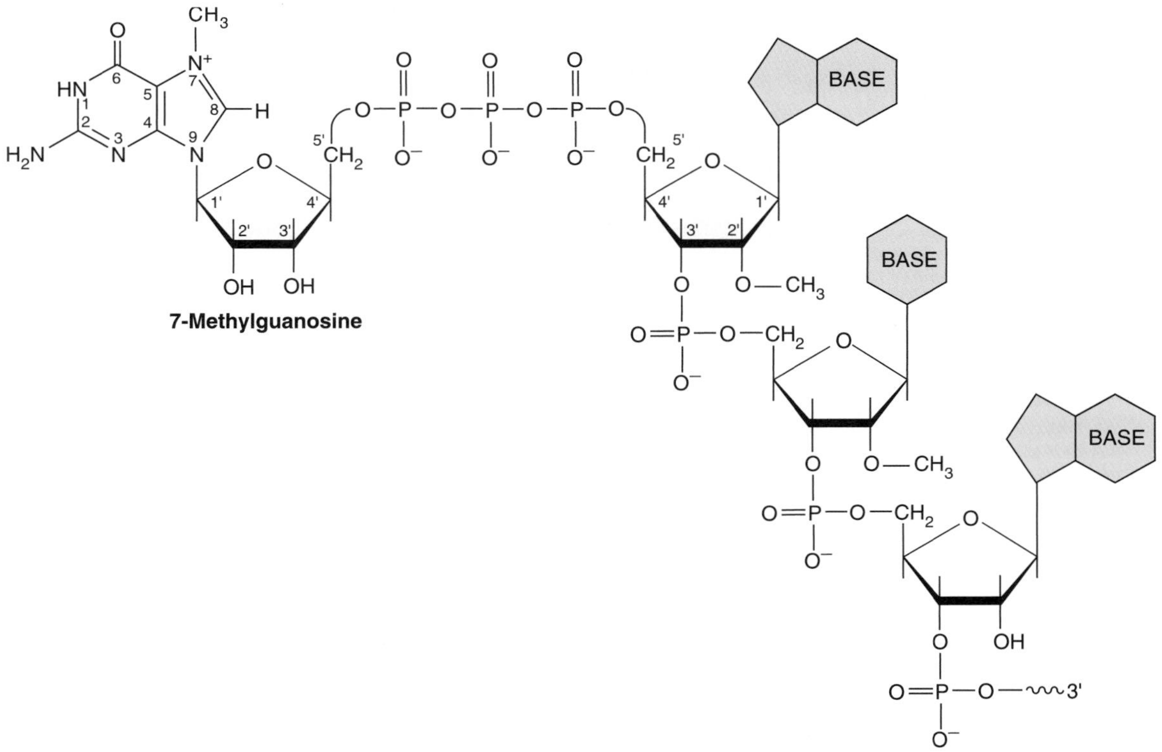

FIGURE 18.24

The Methylated Cap of Eukaryotic mRNA.

The cap structure consists of a 7-methylguanosine attached to the 5′ end of an RNA molecule through a unique 5′ ⟶ 5′ linkage. The 2′-OH of the first two nucleotides of the transcript are methylated.

cytoplasm and their translation by ribosomes. (Some mRNAs, such as histone mRNAs, do not contain poly A tails.) The most dramatic and complex processing reactions are those that remove introns. During this process (illustrated in Figure 18.25), which is referred to as **splicing**, each intron is excised in an unusual configuration called a *lariat*. Splicing takes place within a **spliceosome**, a multicomponent structure (40–60 S) containing several snRNAs, as well as several proteins. An unanticipated example of splicing, discovered in 1982 by Thomas Cech, is the self-splicing by pre-rRNA molecules in the protozoan *Tetrahymena*. These catalytic RNA molecules, now called **ribozymes**, have also been found in several other organisms.

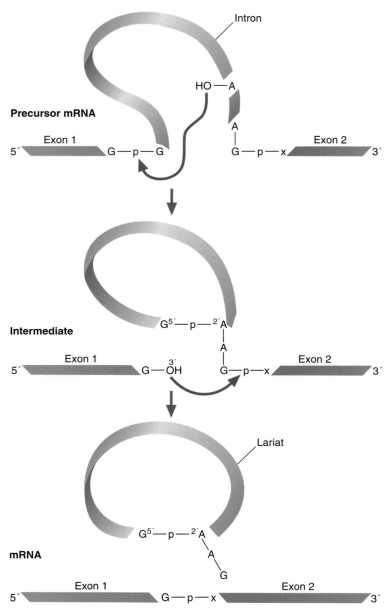

FIGURE 18.25

RNA Splicing.

mRNA splicing begins with the nucleophilic attack of the 2′-OH of a specific adenosine on a phosphate in the 5′ splice site. A lariat is formed by a 2′,5′ phosphodiester bond. In the next step, 3′-OH of exon 1 (acting as a nucleophile) attacks a phosphate adjacent to the lariat. This reaction releases the intron and ligates the two exons.

Define the following terms:

a. operon

b. promoter

c. spliceosome

d. coding strand

e. consensus sequences

What are the functions of RNA polymerases I, II, and III?

18.3 GENE EXPRESSION

Ultimately, the internal order most essential to living organisms requires the precise and timely regulation of gene expression. It is, after all, the capacity to switch genes on and off that enables cells to respond efficiently to a changing environment. In multicellular organisms, complex programmed patterns of gene expression are responsible for cell differentiation and intercellular cooperation.

The regulation of genes, as measured by their transcription rates, is the result of a complex hierarchy of control elements that coordinate the cell's metabolic activities. Some genes, referred to as **constitutive** or housekeeping **genes**, are routinely transcribed because they code for gene products (e.g., glucose-metabolizing enzymes, ribosomal proteins, and histones) required for cell function. In addition, in the differentiated cells of multicellular organisms, certain specialized proteins are produced that cannot be detected elsewhere (e.g., hemoglobin in red blood cells). Genes that are expressed only under certain circumstances are referred to as *inducible*. For example, the enzymes that are required for lactose metabolism in *E. coli* are synthesized only when lactose is actually present and glucose, the bacterium's preferred energy source, is absent.

Most of the mechanisms used by living cells to regulate gene expression involve DNA-protein interactions. At first glance, the seemingly repetitious and regular structure of B-DNA appears to make it an unlikely partner for the sophisticated binding with myriad different proteins that must occur in gene regulation. As noted in Chapter 17, however, DNA is somewhat deformable, and certain sequences can be curved or bent. In addition, it is now recognized that the edges of the base pairs within the major groove (and to a lesser extent the minor groove) of the double helix can participate in sequence-specific binding to proteins. Numerous contacts (often about 20 or so) involving hydrophobic interactions, hydrogen bonds, and ionic bonds between amino acids and nucleotide bases result in highly specific DNA-protein binding. Several examples of amino acid–nucleotide base interactions are illustrated in Figure 18.26.

The three-dimensional structures of most DNA regulatory proteins that have been analyzed have surprisingly similar features. In addition to usually possessing a twofold axes of symmetry, many of these molecules can be separated into families (Figure 18.27) on the basis of the following structures: (1) helix-turn-helix, (2) helix-loop-helix, (3) leucine zipper, and (4) zinc finger. DNA-binding proteins, many of which are transcription factors, often form dimers. For example, a variety of transcription factors with leucine zipper motifs form dimers as their leucine-containing a α-helices interdigitate (Figure 5.21c). Because each protein possesses its own unique binding specificity, and these and many other transcription factors can combine to form homodimers (two identical monomers) and heterodimers (two different monomers), a large number of unique gene regulatory agents are formed.

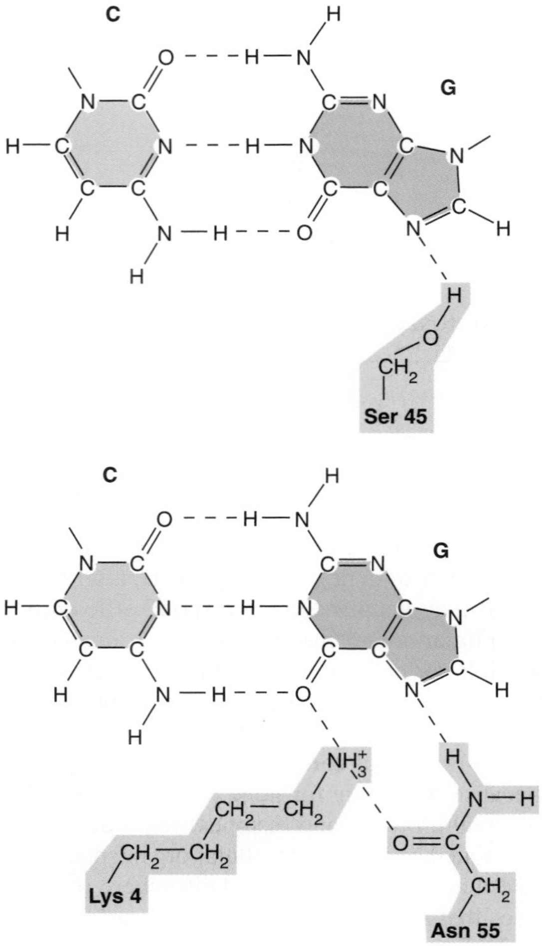

FIGURE 18.26

Examples of Specific Amino Acid–Nucleotide Base Interactions During Protein-DNA Binding.

These examples are taken from structure studies of the binding of λ repressors to DNA.

Considering the obvious complexity of function observed in living organisms, it is not surprising that the regulation of gene expression has proven to be both remarkably complex and difficult to investigate. For many of the reasons stated, knowledge about prokaryotic gene expression is significantly more advanced than that of eukaryotes. Prokaryotic gene expression was originally investigated, in part, as a model for the study of the more complicated gene function of mammals. Although it is now recognized that the two genome types are vastly different in many respects, the prokaryotic work has provided many valuable insights into the mechanisms of gene expression. In general, prokaryotic gene expression involves the interaction of specific proteins (sometimes referred to as regulators) with DNA in the immediate vicinity of a transcription start site. Such interactions may have either a positive effect (i.e., transcription is initiated or increased) or a negative effect (i.e., transcription is blocked). In an interesting variation, the inhibition of a negative regulator (called a *repressor*) activates affected genes. (The inhibition of a repressor gene is referred to as derepression.) Eukaryotic gene expression uses these mechanisms and several others, including gene rearrangement and amplification and various complex transcriptional, RNA processing, and translational controls. In addition, the spatial separation

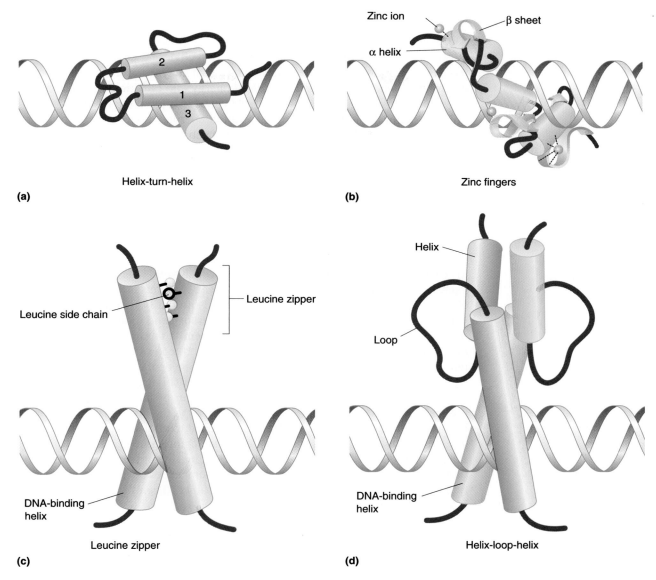

FIGURE 18.27

DNA-Protein Interactions.

Gene regulatory proteins contain specific structural motifs for interacting with DNA: (a) helix-turn-helix, (b) zinc fingers, (c) leucine zipper, and (d) helix-loop-helix.

of transcription and translation inherent in eukaryotic cells provides another opportunity for regulation: RNA transport control. Finally, eukaryotes (and prokaryotes) also regulate cell function through the modulation of proteins by covalent modifications.

In this section, several examples of control of gene expression are described. The discussion of prokaryotic gene expression focuses on the lac operon. The *lac operon* of *E. coli*, originally investigated by Francois Jacob and Jacques Monod in the 1950s, remains one of the best-understood models of gene regulation. Eukaryotic gene expression is less well understood. Despite our daunting ignorance, however, a significant number of the pieces in this marvelous puzzle have been revealed. The section ends with a brief discussion of recent discoveries concerning growth factor–triggered gene expression.

Gene Expression in Prokaryotes

As described, the highly regulated metabolism of prokaryotes such as *E. coli* allows these organisms to respond rapidly to a changing environment to promote growth and survival. The timely synthesis of enzymes and other gene products only when needed prevents wasting energy and nutritional resources. At the genetic level, the control of inducible genes is often effected by groups of linked structural and regulatory genes called operons. Investigations of operons, especially the lac operon, have provided substantial insight into how gene expression can be altered by environmental conditions.

The lac operon (Figure 18.28) consists of a control element and structural genes that code for the enzymes of lactose metabolism. The control element contains the promoter site, which overlaps the operator site. (In prokaryotes the *operator* is a DNA sequence involved in the regulation of adjacent genes that binds to a repressor protein.) The promoter site also contains the CAP site (described below). The structural genes Z, Y, and A specify the primary structure of β-galactosidase, lactose permease, and thiogalactoside transacetylase, respectively. β-Galactosidase catalyzes the hydrolysis of lactose, which yields the monosaccharides galactose and glucose, whereas lactose permease promotes lactose transport into the cell. Because lactose metabolism proceeds normally without thiogalactoside transacetylase, its role is unclear. A repressor gene i, directly adjacent to the lac operon, codes for the lac repressor protein, a tetramer that binds to the operator site with high affinity. (There are about 10 copies of lac repressor protein per cell.) The binding of the lac repressor to the operator prevents the functional binding of RNA polymerase to the promoter (Figure 18.29).

Without its inducer (allolactose, a β-1,6-isomer of lactose) the lac operon remains repressed because the lac repressor binds to the operator. When lactose becomes available, a few molecules are converted to allolactose by β-galactosidase. Allolactose then binds to the repressor, changing its conformation and promoting dissociation from the operator. Once the inactive repressor diffuses away from the operator, the transcription of the structural genes begins. The lac operon remains active until the lactose supply is consumed. Then the repressor reverts to its active form and rebinds to the operator.

Glucose is the preferred carbon and energy source for *E. coli*. If both glucose and lactose are available, the glucose is metabolized first. Synthesis of the lac operon enzymes is induced only after the glucose has been consumed. (This makes sense because glucose is more commonly available and has a central role in cellular metabolism. Why expend the energy to synthesize the enzymes required for the metabolism of other sugars if glucose is also available?) The delay in activating the lac operon is mediated by a catabolite gene activator protein (CAP). CAP is an allosteric homodimer that binds to the chromosome at a site

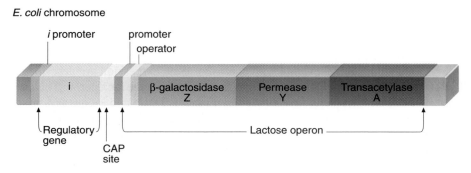

E. coli chromosome

FIGURE 18.28

The Lac Operon in *E. coli*.

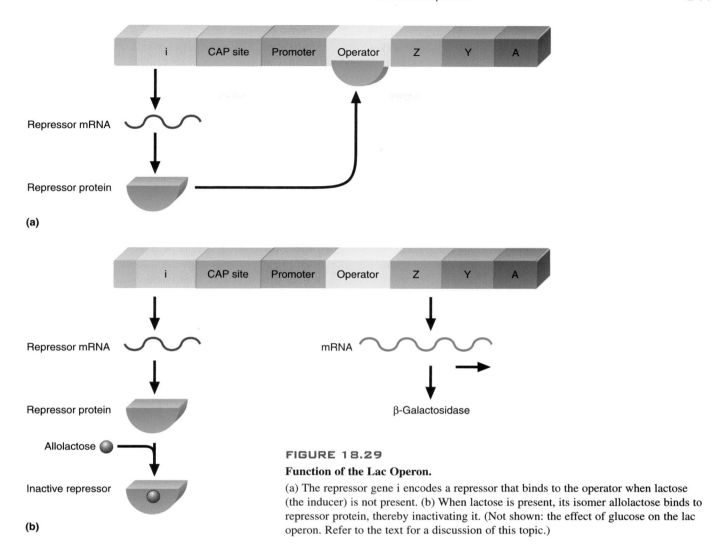

FIGURE 18.29

Function of the Lac Operon.

(a) The repressor gene i encodes a repressor that binds to the operator when lactose (the inducer) is not present. (b) When lactose is present, its isomer allolactose binds to repressor protein, thereby inactivating it. (Not shown: the effect of glucose on the lac operon. Refer to the text for a discussion of this topic.)

directly upstream of the lac promoter when glucose is absent. CAP is an indicator of glucose concentration because it binds to cAMP. The cell's cAMP concentration increases when the cell is in energy deficit, that is, when the primary carbon source (glucose) is absent. The binding of cAMP to CAP, which occurs only when glucose is absent and cAMP levels are high, causes a conformational change that allows the protein to bind to the lac promoter. CAP binding promotes transcription by increasing the affinity of RNA polymerase for the lac promoter. In other words, CAP exerts a positive or activating control on lactose metabolism.

KEY CONCEPTS 18.6

Constitutive genes are routinely transcribed, whereas inducible genes are transcribed only under appropriate circumstances. In prokaryotes, inducible genes and their regulatory sequences are grouped into operons.

Recently, several cases of infection caused by a rare, virulent strain of group A streptococcus have been reported. In approximately 25–50% of these cases (reported in Great Britain and the United States), infection resulted in necrotizing fasciitis, a rapidly spreading destruction of flesh, often accompanied by hypotension (low blood pressure), organ failure, and toxic shock. If antibiotic treatment is not initiated within 3 days of exposure to the bacterium, gangrene and death may result. Similar cases were reported in the 1920s. However, these earlier cases had a significantly lower fatality rate, although antibiotics were not then available. (Physicians reported treating affected areas by washing with acidic solutions.)

QUESTION 18.17

Group A streptococci are converted into the pathogenic form by becoming infected themselves with a certain virus. This virus's genome contains a gene that codes for a tissue-destroying toxin. Can you describe in general terms how a viral infection might cause a permanent change in the pathogenicity of a group A streptococcus bacterium? Considering the apparent difference in virulence between the bacterium in the 1920s and the present, is there any method for determining whether the same strain of group A streptococcus is responsible for both sets of cases? Preserved specimens of infected tissue from these early cases are available. (*Hint:* Refer to Biochemical Methods 18.1.)

Gene Expression in Eukaryotes

As noted, eukaryotic genomes are vastly larger and more complex than those of prokaryotes. Presumably, these differences can be accounted for, at least in part, by the obstacles that confront each type of organism. Eukaryotes typically lead more complicated lives than prokaryotes. This is especially true of multicellular eukaryotes. For example, among the higher animals and plants, numerous differentiated cells in each individual organism are derived from the genome in a fertilized egg. By its very nature, development requires an orderly and sequential expression of a vast number of genes. In addition, as described, the sustained life of multicellular organisms requires intercellular coordination, which involves changes in gene expression. In recent years, progress in the investigation of eukaryotic gene expression has been made largely because of molecular cloning and other recombinant DNA techniques (described in Biochemical Methods 18.1). Current evidence indicates that eukaryotic gene expression, as measured by changes in the amounts and activities of gene products produced, is regulated at the following levels: genomic control, transcriptional control (see pp. 639–642), RNA processing, RNA transport, and translational control. Each is discussed briefly.

GENOMIC CONTROL Although most cells in multicellular organisms possess the same set of genes, only a fraction is ever expressed in any one cell type. Expression is affected by several types of changes in the structural organization of the genome. Among the most commonly observed changes are DNA methylation and histone acetylation. The methylation of cytosines in certain 5′-CG-3′ sequences silences expression. For example, genes for transposases and sequences likely to undergo recombination are usually methylated. The addition of acetyl groups to side chain groups of lysine residues in histones H3 and H4 reduces their affinity for DNA. In general, histone acetylation promotes gene expression. A large number of proteins that affect gene expression have either acetylase or deacetylase activity.

A significant amount of gene regulation occurs through selective transcription. There appear to be two major influences on eukaryotic transcription initiation: chromatin structure and gene regulatory proteins. During interphase of the cell cycle, chromatin is observed in two forms. **Heterochromatin** is so highly condensed that it is transcriptionally inactive. A small portion of each cell's heterochromatin occurs in all of an individual organism's cells. Other portions of heterochromatin differ in a tissue-specific pattern. **Euchromatin**, a less condensed form of chromatin, has varying levels of transcriptional activity. Transcriptionally active euchromatin is the least condensed. Inactive euchromatin is somewhat more condensed (but less so than that observed in heterochromatin). The mechanism by which chromatin reversibly condenses is unknown. However, histone covalent modifications (e.g., acetylation as described, phosphorylation, and methylation to a lesser extent) are believed to be involved. As described, a wide variety of gene regulatory proteins affect transcription by either activating or repressing genes.

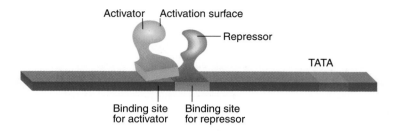

(a) Competitive DNA binding

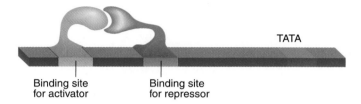

(b) Masking the activation surface

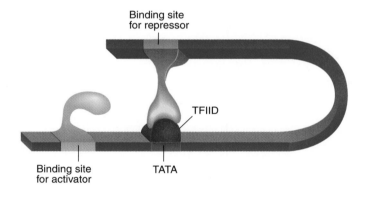

(c) Direct interaction with the general transcription factors

FIGURE 18.30

Proposed Mechanisms for Eukaryotic Gene Repression.

(a) Transcription factor proteins compete for binding to the same regulatory sequence.
(b) Both activating and repressing proteins bind to DNA but at different sites. The repressor blocks transcription by binding to and masking activating sites on the activator. (c) The repressor factor binds to a transcription factor bound to DNA, thus preventing a transcription complex from assembling.

Several proposed mechanisms for protein-mediated gene repression are illustrated in Figure 18.30. Transcription factors may also be involved in changes in transcription start sites.

Less common examples of genomic control include gene rearrangements and gene amplification. The differentiation of certain cells involves gene rearrangements, for example, the rearrangements of antibody genes in B lymphocytes (Figure 18.31). Transposition (see p. 626–29) is also believed to affect gene regulation. During certain stages in development, the requirement for specific gene products may be so great that the genes that code for their synthesis are selectively amplified. Amplification occurs via repeated rounds of replication within the amplified region. For example, the rRNA genes in various animals (most notably amphibians, insects, and fish) are amplified within immature egg cells (called oocytes).

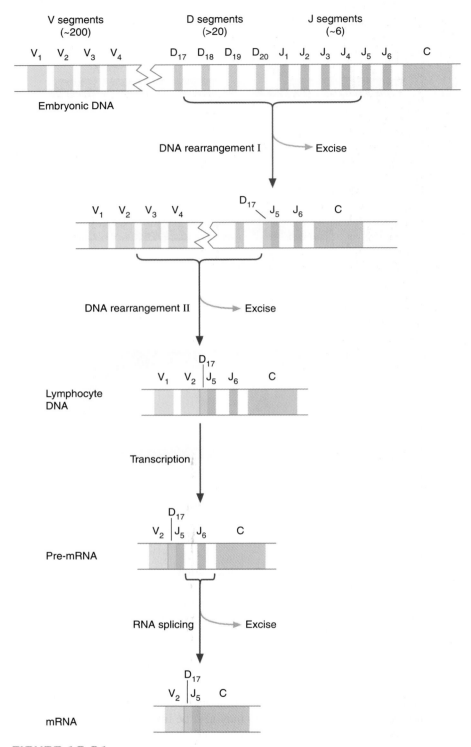

FIGURE 18.31

DNA Rearrangements.

Each of the heavy chains of antibodies contains protein sequences derived from one variant each of V (variable), D (diversity), and J (joining) gene segments. The DNA that codes for each type of heavy (H) chain within lymphocytes is generated by the rearrangements of several types of the gene segments. After several rearrangements and excisions, a specific D segment is directly adjacent to specific V and J segments. After the newly created gene is transcribed, several sequences separating the VDJ segment from the C (constant) segment are removed.

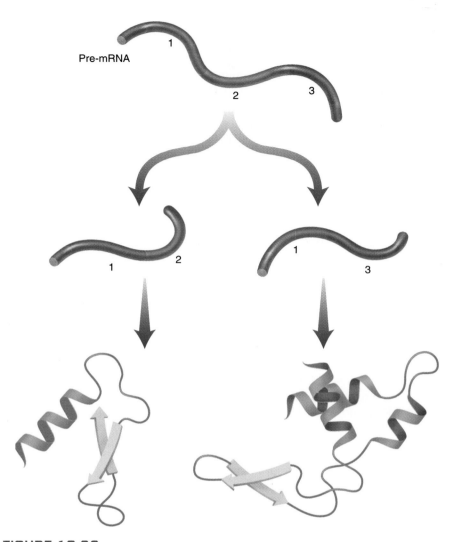

Pre-mRNA

FIGURE 18.32

RNA Processing.

The coding properties of an mRNA molecule depend on the types of processing events that its precursor undergoes. Different polypeptides can be synthesized from splicing different combinations of exons from the same pre-mRNA transcript.

rRNA amplification is apparently required because of the enormous requirement for protein synthesis during the early developmental stages of fertilized eggs.

RNA PROCESSING Cells often use alternative RNA processing to control gene expression. For example, alternative splicing—the joining of different combinations of exons—results in the formation of different mRNAs and, therefore, different proteins (Figure 18.32). For example, tissue-specific forms of α-tropomyosin, a cytoskeletal protein, are found in a wide variety of cells (e.g., skeletal, smooth and cardiac muscle, fibroblasts, and brain). The selection of alternative sites for polyadenylation also affects mRNA function. For example, such a change is involved in the switch, during the early phase of B lymphocyte differentiation, from producing membrane-bound antibody to that of secreted antibody. As noted, poly A tails have several roles in mRNA function (p. 641). In general, mRNAs with longer poly A tails are more stable, thereby increasing their opportunities for translation.

Some cells have been observed to change the coding properties of newly synthesized mRNA molecules. In this process, called **RNA editing**, certain bases are chemically modified, deleted, or added. For example, the mRNA for apolipoprotein B-100 in liver cells codes for a 4563 amino acid polypeptide which is a component of very low density lipoprotein (VLDL). Intestinal cells produce a shorter version of the molecule called apolipoprotein B-48 (2153 amino acid residues) that becomes incorporated into the chylomicron particles produced by these cells. The cytosine in a CAA codon that specifies glutamine is converted by a deamination reaction into a uracil. The new codon, UAA, is a stop signal in translation; hence a truncated polypeptide is produced during translation of the edited mRNA.

QUESTION 18.18

Explain the following terms:

a. euchromatin

b. heterochromatin

c. gene amplification

d. operator

e. repressor

RNA TRANSPORT As mentioned, eukaryotes regulate molecular traffic into and out of the nucleus. Nuclear export signals, for example, capping (see p. 641) and association with specific proteins, are believed to control the transport of processed RNA molecules through nuclear pore complexes (Figure 2.16). Recent evidence indicates that export requires the binding of the 5′ end of the mRNA within the hnRNP to cap-binding protein (CBP) and the presence of certain proteins that compose a nuclear export signal. As the CBP-bound domain of hnRNP is transported through the nuclear pore complex, certain hnRNP proteins are removed and retained within the nucleus. Once hnRNPs are in the cytoplasm, the remaining nuclear proteins are removed and exchanged for cytoplasmic RNP proteins. This latter process is driven to completion by GTP hydrolysis.

TRANSLATIONAL CONTROL Eukaryotic cells can respond to various stimuli (e.g., heat shock, viral infections, and cell cycle phase changes) by selectively altering protein synthesis. The covalent modification of several translation factors (nonribosomal proteins that assist in the translation process) has been observed to alter the overall protein synthesis rate and/or enhance the translation of specific mRNAs. For example, the phosphorylation of the protein eIF-2 affects the rate of hemoglobin synthesis in rabbit reticulocytes (immature red blood cells).

SIGNAL TRANSDUCTION AND GENE EXPRESSION All cells respond to signals from their environment by altering gene expression. In comparison to prokaryotes and even the unicellular eukaryotes, the informational processing capacity of multicellular eukaryotes is extraordinarily sophisticated. The regulation of gene expression within each of the millions or trillions of individual cells in a multicellular organism is now known to involve a vast array of signaling molecules. In most cases changes in gene expression are initiated by the binding of a ligand to either a cell surface receptor or an intracellular receptor. The mechanisms by which signal molecules switch certain genes on or off are intricate series of reactions that transmit information from the cell's environment to specific DNA sequences within the nucleus. Considering the enormous research efforts devoted to the investigation of cancer (Special Interest Box 18.1), the best understood examples of such signal transduction pathways are those that affect cell division.

In contrast to single-cell organisms in which cell growth and cell division are governed largely by nutrient availability, the proliferation of cells in multicellular organisms is regulated by an elaborate intercellular network of signal molecules. Complicating features of intracellular signal transduction mechanisms that have been revealed by research efforts in cell proliferation include the following:

1. Each type of signal may activate one or more pathways. The mechanisms by which signal molecules alter gene expression often involve the simultaneous activation of several different pathways. This accounts for the changes in cell metabolism and appearance that accompany alterations of gene expression during development.

2. Signal transduction pathways may converge or diverge. Depending on circumstances, the activation of several types of receptors may result in the same or overlapping responses. As mentioned above, signal molecules can also trigger several different pathways, any of which may also diverge.

In the eukaryotic cell cycle, cells repeatedly progress through each of the four phases (M, G_1, S, and G_2, refer to Figure 18.9). Investigations of mutant cell types reveal that checkpoints occur within G_1 (in yeast cells it is referred to as START), G_2, and M phases. The cell is prevented from entering the next phase until the conditions are optimal (e.g., sufficient cell growth in G_2 or alignment of chromosomes in M) and specific signals are received. The fixed, rhythmic activities observed in cell division are regulated so that each phase is completed before the next one starts. Progression is accomplished by a complex molecular machine, the principal mechanism of which is the alternating synthesis and degradation of a group of proteins called the cyclins. Cyclins, a group of regulatory proteins, bind to and activate the cyclin-dependent protein kinases (Cdks). The Cdks are a class of protein kinases that phosphorylate a variety of proteins, thus triggering the passage of the cell through a checkpoint to the next phase of the cell cycle. The regulation of cell division involves both positive and negative controls. Positive control is exerted largely by binding growth factors to specialized cell receptors. The initiation of cell division typically requires binding a variety of such factors. Cell proliferation is inhibited by *tumor suppressor genes*. Well-known examples of these genes include Rb (so named because of the role played by the loss of Rb gene function in retinoblastoma, a childhood eye cancer) and the p53 gene (a cyclin chaperone that arrests cell cycle progression). The arrest of the cell cycle is prolonged when a certain amount of DNA damage has occurred, as in overexposure to radiation. If DNA repair mechanisms are incomplete, a complex mechanism involving p53 leads to programmed cell death or **apoptosis**.

The positive effects exerted by growth factors are now believed to include gene expression that specifically overcomes the inhibitions at the cell cycle checkpoints, especially the G_1 checkpoint. The binding of growth factors to their cell surface receptors initiates a cascade of reactions that induces two classes of genes.

1. Early response genes. These genes are rapidly activated, usually within 15 minutes. Among the best-characterized early response genes are the jun, fos, and myc protooncogenes. **Protooncogenes** are normal genes that, if mutated, can promote carcinogenesis. (Refer to Special Interest Box 18.1.) Each of the jun and fos protooncogene families codes for a series of transcription factors containing leucine zipper domains. Both jun and fos proteins form dimers that can bind DNA. Among the best-characterized of these is a jun-fos heterodimer, referred to as AP-1, which forms through a leucine zipper interaction. Although myc gene expression is known to be critically important in normal cell function (i.e., the inappropriate expression of myc is found in several types of cancer), the biochemical function of this gene family remains unresolved. However, myc gene products probably act as transcription factors.

2. Delayed response genes. These genes are induced by the activities of the transcription factors and other proteins produced or activated during the early response phase. Among the products of the delayed response genes are the Cdks, the cyclins, and other components required for cell division.

As mentioned (see p. 555), many growth factors bind to tyrosine kinase receptors and some of these are linked via G protein-like mechanisms to DAG (diacylglycerol) and IP$_3$ (inositol triphosphate) generation (Figure 16.12). Epidermal growth factor (EGF) is one example of a growth factor of this type and its role in the activation of the transcription factor AP-1 is illustrated in Figure 18.33. The ras oncogene, first isolated from rat sarcomas, serves in its protooncogene form as the G-protein component of this system. Ras is activated when it binds to SOS/GRB2, a protein complex that has bound to the tyrosine kinase domain of the EGF receptor. SOS is a type of **guanine nucleotide–releasing protein (GNRP)** that causes ras to release GDP and bind a GTP when it is in turn bound to another protein called GRB2. Ras becomes inactive when GTP is hydrolyzed, a reaction catalyzed by **GTPase activating proteins**, or GAPs.

Phospholipase Cγ(PLCγ) is also activated when it binds to the EGF receptor. Like its counterpart in G protein linked receptors (Figure 16.12), active PLCγ hydrolyses PIP$_2$ to form DAG and IP$_3$. DAG activates protein kinase C (PKC), the enzyme that activates a variety of proteins involved in cell growth and proliferation.

Phosphorylation cascades are induced by both the activation of ras and increased levels of DAG and IP$_3$ in the cell following EGF binding. One of the key enzymes activated in the phosphorylation cascade is MAPKK (mitogen activated protein kinase kinase). (A **mitogen** is any molecule that stimulates cell division). Activated MAPKK then phosphorylates both a tyrosine and a tryptophan of MAP kinase. (This unusual reaction appears to ensure that MAP kinase is activated only by MAPKK.) Active MAP kinase then phosphorylates a variety of cellular proteins. Among those are jun, fos, and myc. Phosphorylated jun and fos proteins then combine to form the transcription factor AP-1. AP-1 subsequently promotes the transcription of several delayed response genes.

QUESTION 18.19

Identify which type of eukaryotic gene expression control is illustrated by the following examples:

   a. covalent modification of proteins

   b. histone acetylation

   c. joining of different sets of exons

   d. methylation of cytosine

QUESTION 18.20

The mechanism by which light influences plant gene expression is referred to as *photomorphogenesis*. Because of serious technical problems with plant cell culture, relatively little is known about plant gene expression. However, certain DNA sequences, referred to as *light-responsive elements* (LRE), have been identified. On the basis of the gene expression patterns observed in animals, can you suggest (in general terms) a mechanism whereby light induces gene expression? (*Hint*: Recall that phytochrome is an important component of light-induced gene expression.)

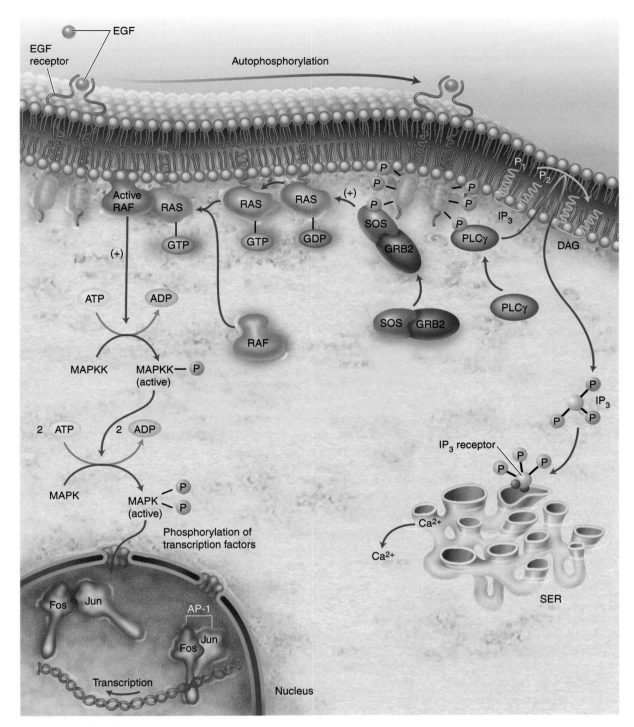

FIGURE 18.33

Eukaryotic Gene Expression Triggered by Growth Factor Binding.

The signal transduction pathway outlined in this figure illustrates selected events that are triggered when a growth factor (e.g., EGF) binds to its plasma membrane receptor. Subsequent changes in growth factor– or hormone-triggered gene expression that occur are typically mediated by several different mechanisms. In this example, only ras and PLCγ activation are illustrated. When EGF binds it promotes dimerization of the receptor and autophosphorylation of the tyrosine residues on its cytoplasmic domain. Once these residues are phosphorylated, the receptor binds a variety of cytoplasmic proteins. When GRB2 binds to the receptor, SOS (a GNRP) is activated and it in turn activates ras by promoting the exchange of GDP for GTP. Activated ras initiates a phosphorylation cascade by activating the protein kinase RAF, which in turn activates MAPKK. MAPKK activates MAPK, which in turn activates a number of transcription factors in the activation of a variety of transcription factors in the nucleus (e.g., fos and jun that form AP-1). When PLCγ binds to the EGF receptor it catalyzes the cleavage of PIP_2 into IP_3 and DAG. IP_3 stimulates the release of Ca^{2+} into the cytoplasm. As illustrated in Figure 16.12, in the presence of Ca^{2+} and DAG protein kinase C activates yet another series of protein kinases that in turn affect the function of several regulatory proteins.

Cancer is a group of diseases in which genetically damaged cells proliferate autonomously. Such cells cannot respond to normal regulatory mechanisms that ensure the intercellular cooperation required in multicellular organisms. Consequently, they continue to proliferate, thereby robbing nearby normal cells of nutrients and eventually crowding surrounding healthy tissue. Depending on the damage they have sustained, abnormal cells may form either benign or malignant tumors. Benign tumors, which are slow-growing and limited to a specific location, are not considered cancerous and rarely cause death. In contrast, *malignant* tumors are often fatal because they can undergo metastasis. (In *metastasis*, cancer cells migrate through blood or lymph vessels to distant locations throughout the body.) Wherever new malignant tumors arise, they interfere with normal functions. When life-sustaining processes fail, patients die.

Cancers are classified by the tissues affected. The vast majority of cancerous tumors are carcinomas (tumors derived from epithelial tissue cells such as skin, various glands, breasts, and most internal organs). In the leukemias, cancers of the bone marrow, excessive leukocytes are produced. Similarly, the lymphocytes produced in the lymph nodes and spleen proliferate uncontrollably in the lymphomas. Tumors arising in connective tissue are called sarcomas. Despite the differences among this diverse class of diseases, they also have several common characteristics, among which are the following:

1. **Cell culture properties**. When grown in culture, most tumor cells lack contact inhibition, that is, they grow to high density in highly disorganized masses. (Normal cells grow only in a single layer in culture and have defined borders in vivo.) In contrast to normal cells, cancer cell growth and division tend to be growth factor–independent, and these cells often do not require attachment to a solid surface. The hallmark of cancer cells, however, is their immortality. Normal cells undergo cell division only a finite number of times, whereas cancer cells can proliferate indefinitely.

2. **Origin**. Each tumor originates from a single damaged cell. In other words, a tumor is a clone derived from a cell in which heritable changes have occurred. The genetic damage consists of mutations (e.g., point mutations, deletions, and inversions) and chromosomal rearrangements or losses. Such changes result in the loss or altered function of molecules involved in cell growth or proliferation. Tumors typically develop over a long time and involve several independent types of genetic damage. (The risk of many types of cancer increases with age.)

The transformation process in which an apparently normal cell is converted or "transformed" into a malignant cell consists of three stages: initiation, promotion, and progression.

During the *initiation* phase of carcinogenesis, a permanent change in a cell's genome provides it with a growth advantage over its neighbors. Most initiating mutations affect protooncogenes or tumor suppressor genes. Protooncogenes code for a variety of growth factors, growth factor receptors, enzymes, or transcription factors that promote cell growth and/or cell division. Mutated versions of protooncogenes that promote abnormal cell proliferation

TABLE 1
Selected Oncogenes*

Oncogene	Function
sis	Platelet-derived growth factor
erbB	Epidermal growth factor receptor
src	Tyrosine-specific protein kinase
raf	Serine/threonine–specific protein kinase
ras	GTP-binding protein
jun	Transcription factor
fos	Transcription factor
myc	DNA-binding protein (transcription factor?)

*Abnormal versions of protooncogenes that mediate cancerous transformations.

are called **oncogenes** (Table 1). Because tumor suppressor genes suppress carcinogenesis, their loss also facilitates tumor development. Recall that Rb and p53 are tumor supressors. Other examples are FCC and DCC, which are both associated with susceptibility to colon cancer. The functions of most tumor suppressor genes are unknown. However, it now appears that p53 codes for a 21-kD protein that normally inhibits Cdk enzymes. Recent evidence indicates that other damaged or deleted tumor suppressor genes may code for enzymes involved in DNA repair mechanisms. The damage that alters function of protooncogenes and tumor suppressor genes is caused by the following:

1. **Carcinogenic chemicals**. Most cancer-causing chemicals are mutagenic, that is, they alter DNA structure. Some carcinogens (e.g., nitrogen mustard) are highly reactive electrophiles that attack electron-rich groups in DNA (as well as RNA and protein). Other carcinogens (e.g., benzo[a]pyrene) are actually procarcinogens, which are converted to active carcinogens by one or more enzyme-catalyzed reactions.

2. **Radiation**. Some radiation (UV, X-rays, and γ-rays) is carcinogenic. As noted, the damage inflicted on DNA includes single- and double-strand breaks, pyrimidine dimer formation, and the loss of both purine and pyrimidine bases. Radiation exposure also causes ROS to form. ROS may be responsible for most of radiation's carcinogenic effects.

3. **Viruses**. Viruses appear to contribute to the transformation process in several ways. Some introduce oncogenes into a host cell chromosome as they insert their genome. (Viral oncogenes are now recognized as sequences that are similar to normal cellular genes that have been picked up accidentally from a previous host cell. To distinguish viral oncogenes and their cellular counterparts, they are referred to as v-onc and c-onc, respectively.) Viruses can also affect the expression of cellular protooncogenes through insertional mutagenesis, a random process in which viral genome insertion inactivates a regulatory site or alters the protoonco-

gene's coding sequence. Most virus-associated cancers have been detected in animals. Only a few human cancers have been proven to be associated with viral infection.

Tumor development can also be promoted by chemicals that do not alter DNA structure. So-called **tumor promoters** contribute to carcinogenesis by two principal methods. By activating components of intracellular signaling pathways, some molecules (e.g., the phorbol esters) provide the cell a growth advantage over its neighbors. (Recall that phorbol esters activate PKC because they mimic the actions of DAG.) The effects of many other tumor promoters are unknown but may involve transient effects such as increasing cellular Ca^{2+} levels or increasing synthesis of the enzymes that convert procarcinogens into carcinogens. Unlike initiating agents, the effects of tumor promoters are reversible. They produce permanent damage only with prolonged exposure after an affected cell has undergone an initiating mutation.

Following initiation and promotion, cells go through a process referred to as progression. During *progression,* genetically vulnerable precancerous cells, which already possess significant growth advantages over normal cells, are further damaged. Eventually, the continued exposure to carcinogens and promoters makes further random mutations inevitable. If these mutations affect cellular proliferative or differentiating capacity, then an affected cell may become sufficiently malignant to produce a tumor. A proposed sequence of the events in the development of colorectal cancer is outlined in Figure 18I.

An Ounce of Prevention...

Because of the enormous cost and limited success of cancer therapy, it has become increasingly recognized that cancer prevention is cost-effective. Recent research indicates that the majority of cancer cases are preventable. For example, over one-third of cancer mortality is directly caused by tobacco use, and another one-third of cancer deaths have been linked to inadequate diets. Tobacco smoke, which contains thousands of chemicals, many of which are either carcinogens or tumor promoters, is responsible for most cases of lung cancer and contributes to cancers of the pancreas, bladder, and kidneys, among others. Diets that are high in fat and low in fiber content have been associated with increased incidence of cancers of the large bowel, breast, pancreas, and prostate. Other dietary risk factors include low consumption of fresh vegetables and fruit.

In addition to providing sufficient antioxidant vitamins, many vegetables (and fruits to a lesser extent) contain numerous nonnutritive components that actively inhibit carcinogenesis. Some carcinogenesis inhibitors (e.g., organosulfides), referred to as *blocking agents*, prevent carcinogens from reacting with DNA or inhibit the activity of tumor promoters. Other inhibitors, referred to as *suppressing agents* (e.g., inositol hexaphosphate), prevent the further development of neoplastic processes that are already in progress. Many nonnutritive food components (e.g., tannins and protease inhibitors) possess both blocking and suppressing effects. In general, these molecules very effectively protect against cancer because many of them inhibit the arachidonic acid cascade and oxidative damage. Apparently low-fat, high-fiber diets that are rich in raw

or fresh leafy green, cruciferous, and allium vegetables (e.g., spinach, broccoli, and onions), as well as fresh fruits, are a prudent choice for individuals seeking to reduce their risk of cancer.

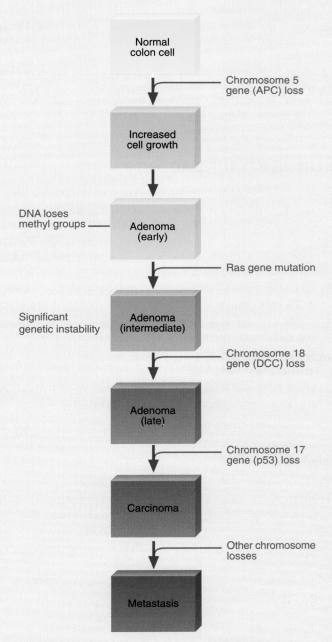

FIGURE 18I The Development of Colorectal Cancer. Colorectal cancer develops over a long time. Because somatic cells are diploid, the loss of the tumor suppressor genes APC, DCC, and p53 usually requires two mutations. Recent research suggests that mutations in genes involved in DNA repair processes may occur during the early stages of colon cancer. (An adenoma is a precancerous epithelial tumor.) Healthy diets that are high in folate, antioxidants, and certain polyunsaturated fatty acids found in fish provide significant protection against colorectal cancer.

SUMMARY

1. DNA structure and function are so important for living organisms that they must possess efficient mechanisms for the rapid and accurate synthesis of DNA. DNA synthesis, referred to as replication, occurs by a semiconservative mechanism, that is, each of the two parental strands serves as a template to synthesize a new strand.

2. There are several types of DNA repair mechanisms. These include excision repair, photoreactivation, and recombinational repair.

3. Genetic recombination, a process in which DNA sequences are exchanged between different DNA molecules, occurs in two forms. In general recombination, the exchange occurs between sequences in homologous chromosomes. In site-specific recombination, the exchange of sequences requires only short homologous sequences. DNA-protein interactions are principally responsible for the exchange of largely nonhomologous sequences.

4. The synthesis of RNA, sometimes referred to as DNA transcription, requires a variety of proteins. Transcription initiation involves binding an RNA polymerase to a specific DNA sequence called a promoter. Regulation of transcription differs significantly between prokaryotes and eukaryotes.

5. The control of transcription is still poorly understood. However, because of intensive research, many of the details in several examples of gene expression in both prokaryotes and eukaryotes are now known.

SUGGESTED READINGS

Cech, T. R., RNA as an Enzyme, *Sci. Amer.*, 255(5):64–75, 1986.

Davies, K., *Cracking the Genome: Inside the Race to Unlock Human DNA,* The Free Press, New York, 2001.

Glick, B. R., and Pasternak, J. J., *Molecular Biotechnology: Principles and Applications of Recombinant DNA,* 2nd ed., ASM Press, Washington, D.C., 1998.

Grunstein, M., Histones as Regulators of Genes, *Sci. Amer.,* 267(4): 68–74, 1992.

Hamadeh, H., and Afshari, C. A., Gene Chips and Functional Genomics, *Amer Sci.*, 88:508–515, 2000.

Kornberg, A., and Baker, T. A., *DNA Replication,* 2nd ed., W. H. Freeman, New York, 1992.

Lockhart, D. J., and Winzeler, E. A., Genomics, Gene Expression and DNA Arrays, *Nature* 405:827–836, 2000.

Loewenstein, W. R., *The Touchstone of Life: Molecular Information. Cell Communication, and the Foundations of Life,* Oxford University Press, New York, 1999.

McKnight, S. L., Molecular Zippers in Gene Regulation, *Sci. Amer.,* 264(4):54–64, 1991.

Moses, P. B., and Chua, N.-H., Light Switches for Plant Genes, *Sci. Amer,,* 258(4):88–93, 1988.

Ptashne, M., How Eukaryotic Transcriptional Activators Work, *Nature*, 335:683–689, 1988.

Vos, J.-M. H., *DNA Repair Mechanisms: Impact on Human Diseases and Cancer*, R. G. Landes, Austin, 1995.

Watson, J. D., Gilman, M., Witkowski, J., and Zoller, M., *Recombinant DNA,* 2nd ed., W. H. Freeman, New York, 1992.

Zweiger, G., *Transducing the Genome: Information, Anarchy, and Revolution in the Biomedical Sciences,* McGraw-Hill, New York, 2001.

KEY WORDS

annotation, *635*

apoptosis, *653*

bacterial artificial chromosome, *631*

cDNA library, *633*

chromosomal jumping, *634*

coding strand, *636*

colony hybridization technique, *632*

conjugation, *625*

consensus sequence, *637*

constitutive gene, *643*

contig, *634*

cosmid, *631*

DNA microarray, *634*

electroporation, *631*

euchromatin, *648*

excision repair, *621*

exonuclease, *614*

functional genomics, 630

general recombination, *622*

genomics, *630*

GTPase activating protein, *654*

guanine nucleotide–releasing protein, *654*

helicase, *612*

heterochromatin, *648*

ligase, *614*

light-induced repair, *621*

marker gene, *631*

mitogen, *654*

Okazaki fragment, *616*

oncogene, *656*

operon, *639*

photoreactivation repair, *621*

polymerase chain reaction, *632*

primase, *612*

primer, *612*

primosome, *612*

processivity, *612*

promoter, *637*

protooncogene, *653*

recombinant DNA technology, *630*

recombination, *610*

recombinational repair, *621*

replication, *610*

replication fork, *614*

replicon, *615*

replisome, *614*

ribozyme, *642*

RNA editing, *652*

semiconservative replication, *611*

shotgun cloning, *632*

site-specific recombination, *623*

spliceosome, *642*

splicing, *642*

transduction, *625*

transfection, *631*

transformation, *623*

transgenic animal, *631*

transposable element, *623*

transposition, *623*

transposon, *627*

tumor promoter, *657*

vector, *630*

yeast artificial chromosome, *631*

REVIEW QUESTIONS

1. Clearly define the following terms:
   a. chemoreceptors
   b. structural genes
   c. recombination
   d. semiconservative replication
   e. replisome
   f. oriC
   g. transcription
   h. protooncogene
   i. spliceosome
   j. oncogene

2. How does negative supercoiling promote the initiation of replication?

3. Explain how the Meselson-Stahl experiment supports the semi-conservative model of DNA replication.

4. List and describe the steps in prokaryotic DNA replication. How does this process appear to differ from eukaryotic DNA replication?

5. Indicate the stage of DNA replication when each of the following enzymes is active:
   a. helicase
   b. primase
   c. DNA polymerases
   d. ligase
   e. topoisomerase
   f. DNA gyrase

6. DNA is polymerized in the $5' \longrightarrow 3'$ direction. Demonstrate with the incorporation of three nucleotides into a single strand of DNA how the $5' \longrightarrow 3'$ directionality is derived.

7. Mutations are caused by chemical and physical phenomena. Indicate the type of mutation that each of the following reactions or molecules might cause:
   a. ROS
   b. caffeine
   c. a small alkylating agent
   d. a large alkylating agent
   e. nitrous acid
   f. intercalating agents

8. How can viruses cause mutations?

9. There are three principal mechanisms of DNA repair. What are they and how do they operate?

10. Describe two forms of genetic recombination. What functions do they fulfill?

11. Although genetic variation is required for species to adapt to changes in their environment, most genetic changes are detrimental. Explain why genetic mutations are rarely beneficial.

12. General recombination occurs in bacteria, where it is involved in several types of intermicrobial DNA transfer. What are these types of transfer and by what mechanisms do they occur?

13. What are the two types of transposition mechanisms and by what mechanisms do they occur?

14. Within cells, cytosine slowly converts to uracil. To what type of mutation would this lead in DNA molecules? Why is this not a problem in RNA?

15. A correlation has been found among species between life span and the efficiency of DNA repair systems. Suggest a reason why this is so.

16. What are the similarities and differences between cellular DNA replication and PCR?

17. Describe the purpose of marker genes in recombinant DNA technology.

18. Define and describe the roles of the following in replication:
   a. DNA ligase
   b. DNA polymerase III
   c. SSB proteins
   d. primase
   e. helicase
   f. RPA
   g. Okazaki fragments

19. Define and describe the roles of the following in transcription:
   a. transcription factors
   b. RNA polymerase
   c. promoter
   d. sigma factor
   e. enhancer
   f. TATA box

20. List the steps in the processing of a typical mRNA precursor that prepare for its functional role.

21. Describe the advantages and disadvantages for organisms that arrange genes in operons.

22. Determine the magnitude of amplification of a single DNA molecule that can be attained with PCR during five cycles.

THOUGHT QUESTIONS

1. In the Meselson-Stahl experiment, why was a nitrogen isotope chosen rather than a carbon isotope?

2. The bidirectional synthesis of DNA implies that one strand is synthesized in the $5' \longrightarrow 3'$ direction and the other in the $3' \longrightarrow 5'$ direction. However, all known enzymes that synthesize DNA do so in the $5' \longrightarrow 3'$ direction. How did Reiji Okazaki explain this paradox?

3. In eukaryotes the DNA replication rate is 50 nucleotides per second. How long does the replication of a chromosome of 150 million base pairs take? If eukaryotic chromosomes were replicated like those of prokaryotes, the replication of a genome would take months. Actually, eukaryotic replication takes only several hours. How do eukaryotes achieve this high rate?

4. There appears to be insufficient genetic material to direct all the activities of several types of eukaryotic cell. Explain how genetic recombination helps to solve this problem.

5. Mustard gas is an extremely toxic substance that severely damages lung tissue when it is inhaled in large amounts. In small amounts, mustard gas is a mutagen and carcinogen. Considering that mustard gas is a bifunctional alkylating agent, explain how it inhibits DNA replication.

6. Adjacent pyrimidine bases in DNA form dimers with high efficiency after exposure to UV light. If these dimers are not repaired, skin cancers can result. Melanin is a natural sunscreen produced by melanocytes, a type of skin cell, when the skin is exposed to sunlight. Individuals who spend long periods over many years developing a tan eventually acquire thick and highly wrinkled skin. Such individuals are also at high risk for skin cancer. Can you explain, in general terms, why these phenomena are related?

7. Phorbol esters have been observed to induce the transcription of AP-1–influenced genes. Explain how this process could occur. What are the consequences of AP-1 transcription? What role does intermittent exposure to phorbol esters have on an individual's health?

8. Because of overuse of antibiotics and/or weakened governmental surveillance of infectious disease, several diseases that had been thought to be no longer a threat to human health (e.g., pneumonia and tuberculosis) are rapidly becoming unmanageable. In several instances, so-called superbugs (microorganisms that are resistant to almost all known antibiotics) have been detected. How did this circumstance arise? What will happen in the future if this process continues?

9. Retinoblastoma is a rare cancer in which tumors develop in the retina of the eye. The tumors arise because of the loss of the Rb gene, which codes for a tumor suppressor. Hereditary retinoblastoma usually occurs during childhood. Such individuals inherit only one functional copy of Rb. Explain why nonhereditary retinoblastoma usually occurs later in life.

10. Explain the difference between the potential effects on an individual organism of errors made during replication and those made during transcription.

11. Explain how a reverse transcriptase activity within a cell can result in gene amplification.

12. Using the tools described in Chapter 18, describe in general terms how a researcher can map the genome of a newly discovered organism.

CHAPTER NINETEEN
Protein Synthesis

OUTLINE

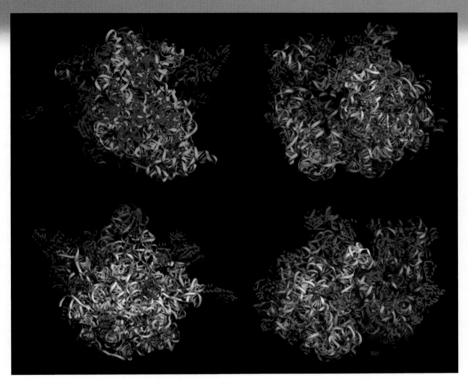

The Ribosome. Ribosomes are ribonucleoprotein molecular machines that synthesize proteins in all living cells. In this series of 90° rotations of the high resolution structure of the complete 70S ribosome of the thermophilic, bacterium *Thermus thermophilus*, proteins are illustrated in dark blue and magenta, and rRNA molecules are shown in cyan, gray, and light blue. The tRNAs are orange and red. Note that the ribosome is primarily composed of rRNA which perform most catalytic activities. Protein molecules largely serve supporting roles.

Proteins are the most dynamic and varied class of biomolecules. As described, in addition to providing structural components, proteins are largely responsible for promoting many of the most dynamic aspects of living processes. The roles that proteins play in living organisms are astounding. In addition to the unbelievably diverse catalytic proteins, protein receptors mediate the actions of untold numbers of signal molecules, many of which are also proteins. The uniqueness of each cell type is due almost entirely to the proteins it produces. It is not surprising, therefore, that the expression of most genes alters patterns of protein synthesis. Because of their strategic importance in the cellular economy, protein synthesis is a regulated process. Although control is also of major importance at the transcriptional level, control of the translation of genetic messages allows for additional opportunities for regulation. This is especially true in multicellular eukaryotes, whose complex lifestyles require amazingly diverse regulatory mechanisms.

Protein synthesis is an extraordinarily complex process in which genetic information encoded in the nucleic acids is translated into the 20 amino acid "alphabet'" of polypeptides. In addition to translation (the mechanism by which a nucleotide base sequence directs the polymerization of amino acids), protein synthesis can also be considered to include the processes of posttranslational modification and targeting. *Posttranslational modification* consists of chemical alterations cells use to prepare polypeptides for their functional roles. Several modifications assist in *targeting*, which directs newly synthesized molecules to a specific intracellular or extracellular location.

In all, at least 100 different molecules are involved in protein synthesis. Among the most important of these are the components of the ribosome, a supramolecular structure composed of RNA and protein that rapidly and precisely decodes genetic messages. Speed is required because organisms must respond expeditiously to ever-changing environmental conditions. In prokaryotes such as *E. coli*, for example, a polypeptide of 100 residues is synthesized in about 6 s. Precision in mRNA translation is critical because, as described previously, the accurate folding, and therefore the proper functioning, of each polypeptide is determined by the molecule's primary sequence.

Discovered in the 1950s, ribosomes have been the subject of biochemical and biophysical investigations ever since. An early revelation was that ribosome structure appears to be highly conserved. Although there are notable differences among the ribosomes of various species, similarities in the three-dimensional structures of rRNA and ribosomal proteins are even more striking. As noted (see p. 595), despite differences in rRNA base sequences, the secondary structure of these molecules is amazingly similar. Recent research has revealed an even more interesting facet of ribosomal structure and function. Because of the well-documented catalytic properties of protein, it was presumed from the beginning of ribosomal research that the principal aspect of ribosomal function, that is, peptide bond formation, is catalyzed by a ribosomal protein component. rRNA was believed to serve as a structural framework for the translation process. However, it is becoming increasingly apparent that rRNA, far from being an inert scaffold for protein synthesis, plays a central and very active role. For example, it has recently been discovered that peptidyl transferase, the enzymatic activity that catalyzes the formation of peptide bonds, resides in the 23S rRNA component of bacterial ribosomes. The removal of protein from bacterial ribosomes (while leaving remaining rRNA molecules relatively intact) does not substantially affect their capacity to catalyze protein synthesis. In addition, various evidence has linked specific rRNA molecules to roles in tRNA and mRNA binding, ribosomal subunit association, proofreading, and some regulatory aspects of translation (e.g., binding translation factors). Recent high-resolution structural studies of the *E. coli* ribosome have revealed that two-thirds of a ribosome's mass is composed of RNA.

As a result of these and many other conceptual breakthroughs, it is now apparent that the ribosome is perhaps the most sophisticated molecular machine in existence. Fueled by GTP and acting with a variety of translation factors, the ribosome is a dynamic ribonucleoprotein device that manufactures the most important functional biomolecules in living organisms.

Chapter 19 provides an overview of protein synthesis. The chapter begins with a discussion of the genetic code, the mechanism by which nucleic acid base sequences specify the amino acid sequences of polypeptides. This is followed by discussions of protein synthesis as it occurs in both prokaryotes and eukaryotes and a description of the mechanisms that convert polypeptides into their folded, biologically active conformations. Chapter 19 ends with an introduction to **proteomics**, a relatively new technology being developed to characterize the protein products of the genome.

KEY CONCEPTS 19.1

In modern organisms, proteins are synthesized by ribosomes, which translate an RNA base sequence into an amino acid sequence.

19.1 THE GENETIC CODE

It became apparent during early investigations of protein synthesis that translation is fundamentally different from the transcription process that precedes it. During transcription the language of DNA sequences is converted to the closely related dialect of RNA sequences. During protein synthesis, however, a nucleic acid base sequence is converted to a clearly different language (i.e., an amino acid sequence), hence the term *translation*. Because mRNA and amino acid molecules have little natural affinity for each other, researchers (e.g., Francis Crick) predicted that a series of adaptor molecules must mediate the translation process. This role was eventually assigned to tRNA molecules (Figure 17.22).

Before adaptor molecules could be identified, however, a more important problem had to be solved: deciphering the genetic code. The **genetic code** can be described as a coding dictionary that specifies a meaning for base sequence. Once the importance of the genetic code was recognized, investigators speculated about its dimensions. Because only four different bases (G, C, A, and U) occur in mRNA and 20 amino acids must be specified, it appeared that a combination of bases coded for each amino acid. A sequence of two bases would specify only a total of 16 amino acids (i.e., $4^2 = 16$). However, a three-base sequence provides more than sufficient base combinations for translation (i.e., $4^3 = 64$).

The first major breakthrough in assigning mRNA triplet base sequences (later referred to as **codons**) came in 1961, when Marshall Nirenberg and Heinrich Matthaei performed a series of experiments using an artificial test system containing an extract of *E. coli* fortified with nucleotides, amino acids, ATP, and GTP. They showed that poly U (a synthetic polynucleotide whose base components consist only of uracil) directed the synthesis of polyphenylalanine. Assuming that codons consist of a three-base sequence, Nirenberg and Matthaei surmised that UUU codes for the amino acid phenylalanine. Subsequently, they repeated their experiment using poly A and poly C. Because polylysine and polyproline products resulted from these tests, the codons AAA and CCC were assigned to lysine and proline, respectively.

Most of the remaining codon assignments were determined by using synthetic polynucleotides with repeating sequences. Such molecules were constructed by enzymatically amplifying short chemically synthesized sequences. The resulting polypeptides, which contained repeating peptide segments, were then analyzed. The information obtained from this technique, devised by Har Gobind Khorana, was later supplemented with a strategy used by Nirenberg. This latter technique measured the capacity of specific trinucleotides to promote tRNA binding to ribosomes.

The codon assignments for the 64 possible trinucleotide sequences are presented in Table 19.1. Of these, 61 code for amino acids. The remaining three codons (UAA, UAG, and UGA) are *stop* (polypeptide chain terminating) signals. AUG, the codon for methionine, also serves as a *start* signal (sometimes referred to as the *initiating codon*). The genetic code is now believed to possess the following properties:

1. Degenerate. Any coding system in which several signals have the same meaning is said to be degenerate. The genetic code is partially degenerate because most amino acids are coded for by several codons. For example, leucine is coded for by six different codons (UUA, UUG, CUU, CUC, CUA, and CUG). In fact, methionine (AUG) and tryptophan (UGG) are the only amino acids that are coded for by a single codon.

2. Specific. Each codon is a signal for a specific amino acid. The majority of codons that code for the same amino acid possess similar sequences. For example, in each of the four serine codons (UCU, UCC, UCA, and UCG) the first and second bases are identical. Consequently, a point mutation in the third base of a serine codon would not be deleterious.

TABLE 19.1

The Genetic Code

First position (5' end)		Second Position											Third position (3' end)
		U			**C**			**A**			**G**		
	U	UUU	Phe	UCU		UAU	Tyr	UGU		Cys		U	
		UUC		UCC	Ser	UAC		UGC				C	
		UUA	Leu	UCA		UAA	STOP	UGA		STOP		A	
		UUG		UCG		UAG		UGG		Trp		G	
	C	CUU	Leu	CCU	Pro	CAU	His	CGU				U	
		CUC		CCC		CAC		CGC		Arg		C	
		CUA		CCA		CAA	Gln	CGA				A	
		CUG		CCG		CAG		CGG				G	
	A	AUU	Ile	ACU	Thr	AAU	Asn	AGU		Ser		U	
		AUC		ACC		AAC		AGC				C	
		AUA		ACA		AAA	Lys	AGA		Arg		A	
		AUG	Met	ACG		AAG		AGG				G	
	G	GUU	Val	GCU	Ala	GAU	Asp	GGU				U	
		GUC		GCC		GAC		GGC		Gly		C	
		GUA		GCA		GAA	Glu	GGA				A	
		GUG		GCG		GAG		GGG				G	

3. **Nonoverlapping and without punctuation**. The mRNA coding sequence is "read" by a ribosome starting from the initiating codon (AUG) as a continuous sequence taken three bases at a time until a stop codon is reached. A set of contiguous triplet codons in an mRNA is called a **reading frame**. The term **open reading frame** describes a series of triplet base sequences in mRNA that do not contain a stop codon.

4. **Universal**. With a few minor exceptions the genetic code is universal. In other words, examinations of the translation process in the species that have been investigated have revealed that the coding signals for amino acids are always the same.

KEY CONCEPTS 19.2

The genetic code is a mechanism by which ribosomes translate nucleotide base sequences into the primary sequence of polypeptides.

QUESTION 19.1

The term *translation* refers to which of the following?

a. DNA ⟶ RNA

b. RNA ⟶ DNA

c. proteins ⟶ RNA

d. RNA ⟶ proteins

QUESTION 19.2

Explain the following terms:

a. codon

b. degenerate code

c. reading frame

d. open reading frame

e. universal code

As described, DNA damage can cause deletion or insertion of base pairs. If a nucleotide base sequence of a coding region changes by any number of bases other than three base pairs, or multiples of three, a frame shift mutation occurs. Depending on the location of the sequence change, such mutations can have serious effects. The following synthetic mRNA sequence codes for the beginning of a polypeptide:

5′-AUGUCUCCUACUGCUGACGAGGGAAGGAGGUGGCUUAUCAUGUUU-3′

First, determine the amino acid sequence of the polypeptide. Then determine the type of mutations that have occurred in the following altered mRNA segments. What effect do these mutations have on the polypeptide products?

a. 5′-AUGUCUCCUACUUGCUGACGAGGGAAGGAGGUGGCUUAUCAUGUUU-3′
b. 5′-AUGUCUCCUACUGCUGACGAGGGAGGAGGUGGCUUAUCAUGUUU-3′
c. 5′-AUGUCUCCUACUGCUGACGAGGGAAGGAGGUGGCCCUUAUCAUGUUU-3′
d. 5′-AUGUCUCCUACUGCUGACGGAAGGAGGUGGCUUAUCAUGUUU-3′

Codon–Anticodon Interactions

tRNA molecules are the "adaptors" that are required for the translation of the genetic message. Recall that each type of tRNA carries a specific amino acid (at the 3′ terminus) and possesses a three-base sequence called the **anticodon**. The base pairing between the anticodon of the tRNA and an mRNA codon is responsible for the actual translation of the genetic information of structural genes. Although codon-anticodon pairings are antiparallel, both sequences are given in the 5′ $\longrightarrow$ 3′ direction. For example, the codon UGC binds to the anticodon GCA (Figure 19.1).

Once the genetic code was determined, researchers anticipated the identification of 61 types of tRNAs in living cells. Instead, they discovered that cells often operate with substantially fewer tRNAs than expected. Most cells possess about 50 tRNAs, although lower numbers have been observed. Further investigation of tRNAs revealed that the anticodon in some molecules contains uncommon nucleotides, such as inosinate (I), which typically occur at the third anticodon position. (In eukaryotes, A in the third anticodon position is deaminated to form I.) As tRNAs were investigated, it became increasingly clear that some molecules recognize several codons. In 1966, after reviewing the evidence, Crick proposed a rational explanation, the **wobble hypothesis**.

The wobble hypothesis, which allows for multiple codon-anticodon interactions by individual tRNAs, is based principally on the following observations:

1. The first two base pairings in a codon-anticodon interaction confer most of the specificity required during translation. Recall that most redundant codons specifying a certain amino acid possess identical nucleotides in the first two positions. These interactions are standard (i.e., Watson-Crick) base pairings.

2. The interactions between the third codon and anticodon nucleotides are less stringent. In fact, nontraditional base pairs (i.e., non-Watson-Crick) often occur. For example, tRNAs containing G in the 5′ (or "wobble") position of the anticodon can pair with two different codons (i.e., G can interact with either C or U). The same is true for U, which can interact with A or G. When I is in the wobble position of an anticodon, a tRNA can base pair with three different codons, because I can interact with U or A or C.

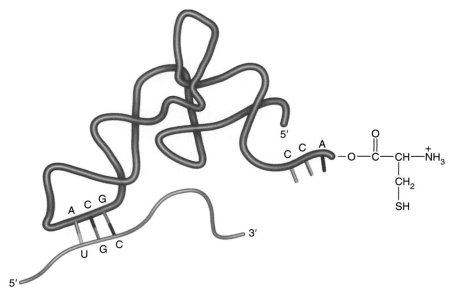

FIGURE 19.1

Codon-Anticodon Base Pairing of Cysteinyl-tRNA^cys.

The pairing of the codon UGC with the anticodon GCA ensures that the amino acid cysteine will be incorporated into a growing polypeptide chain.

KEY CONCEPTS 19.3

The genetic code is translated through base pairing interactions between mRNA codons and rRNA anticodons. The wobble hypothesis explains why cells usually have fewer tRNAs than expected.

A careful examination of the genetic code and the "wobble rules" indicates that a minimum of 31 tRNAs are required to translate all 61 codons. An additional tRNA for initiating protein synthesis brings the total to 32 tRNAs.

Chloroplast and mitochondrial genomes encode fewer tRNAs than nuclear genomes. Chloroplasts possess about 30 tRNAs, whereas mitochondria have only 24. It has been suggested that mitochondria can function with a reduced complement of tRNAs because of the smaller size of their genomes. Chloroplasts and mitochondria are also unique in another respect. Many of the known variations in the genetic code appear in these organelles. The most dramatic and best-documented of these variations are in animal mitochondria. In the mitochondria of humans and other vertebrates, for example, AUA and UGA code for methionine and tryptophan instead of isoleucine and a stop signal, respectively. Similarly, two codons (AGA and AGG), which ordinarily code for arginine, code instead for stop signals. The significance of these alterations is unclear at present.

The Aminoacyl–tRNA Synthetase Reaction: The Second Genetic Code

Although the accuracy of translation (approximately one error per 10^4 amino acids incorporated) is lower than those of DNA replication and transcription, it is remarkably higher than one would expect of such a complex process. The principal reasons for the accuracy with which amino acids are incorporated into polypeptides include codon-anticodon base pairing and the mechanism by which amino acids are attached to their cognate tRNAs. The attachment of amino acids to tRNAs, considered the first step in protein synthesis, is catalyzed by a group of enzymes called the aminoacyl-tRNA synthetases. The precision with which these enzymes esterify each specific amino acid to the correct tRNA is now believed to be so important for accurate translation that their functioning has been referred to collectively as the **second genetic code**.

In most organisms there is at least one aminoacyl-tRNA synthetase for each of the 20 amino acids. (Each enzyme links its specific amino acid to any appropri-

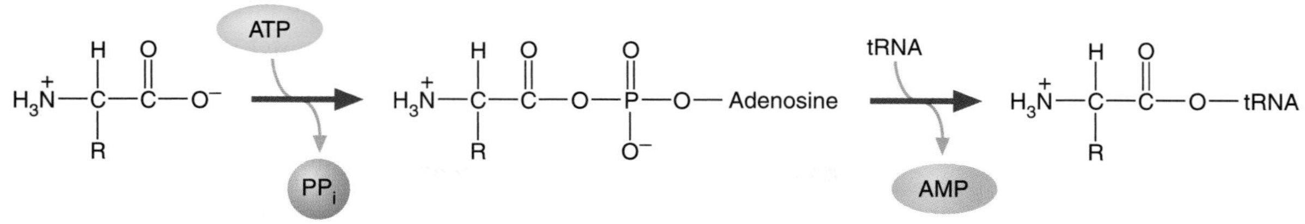

FIGURE 19.2

Formation of Aminoacyl-tRNA.

Each aminoacyl-tRNA synthetase catalyzes two sequential reactions in which an amino acid is linked to the 3′ terminal ribose residue of the tRNA molecule.

ate tRNA. This is important, because in most cells many amino acids have several cognate tRNAs each.) The process that links an amino acid to the 3′ terminus of the correct tRNA consists of two sequential reactions (Figure 19.2), both of which occur within the active site of the synthetase:

   1. **Activation**. The synthetase first catalyzes the formation of aminoacyl-AMP. This reaction, which activates the amino acid by forming a high-energy mixed anhydride bond, is driven to completion through the hydrolysis of its other product, pyrophosphate. (An **anhydride** is a molecule containing two carbonyl groups linked through an oxygen atom. The term **mixed anhydride** describes an anhydride from two different acids, for example, a carboxylic acid and phosphoric acid.)
   2. **tRNA linkage**. A specific tRNA, also bound in the active site of the synthetase, becomes attached to the aminoacyl group through an ester linkage. (Depending on the synthetase, the ester linkage may be through the 2′-OH or 3′-OH of the ribose moiety of the tRNA's 3′-terminal nucleotide. Subsequently, the aminoacyl group can migrate between the 2′-OH and 3′-OH groups. Only the 3′-aminoacyl esters are used during translation.) Although the aminoacyl ester linkage to the tRNA is lower in energy than the mixed anhydride of aminoacyl AMP, it still possesses sufficient energy to participate in acyl transfer reactions (peptide bond formation).

   The sum of the reactions catalyzed by the aminoacyl-tRNA synthetases is as follows:

Amino acid + ATP + tRNA $\longrightarrow$ aminoacyl-tRNA + AMP + PP_i

The product PP_i is immediately hydrolyzed with a large loss of free energy. Consequently, tRNA charging is an irreversible process. Because AMP is a product of this reaction, the metabolic price for the linkage of each amino acid to its tRNA is the equivalent of the hydrolysis of two molecules of ATP to ADP and P_i.

   The aminoacyl-tRNA synthetases are a diverse group of enzymes that vary in molecular weight, primary sequence, and number of subunits. Despite this diversity, each enzyme efficiently produces a specific aminoacyl-tRNA product relatively accurately. As was mentioned, the specificity with which each of the synthetases binds the correct amino acid and its cognate tRNA is crucial for the fidelity of the translation process. Some amino acids can easily be differentiated by their size (e.g., tryptophan versus glycine) or the presence of positive or negative charges in their side chains (e.g., lysine and aspartate). Other amino acids, however, are more difficult to discriminate because their structures are similar. For example, isoleucine and valine differ only by a methylene group. Despite this difficulty, isoleucyl-tRNA^ile synthetase usually synthesizes the correct product. However, this enzyme occasionally also produces valyl-tRNA^ile. Isoleucyl- tRNA^ile synthetase, as well as several other synthetases, can correct such a mistake because it possesses a separate *proofreading* site. Because of its size, this site binds valyl-tRNA^ile and excludes the larger isoleucyl-tRNA^ile. After its binding in the proofreading site, the ester bond of valyl-tRNA^ile is hydrolyzed.

Aminoacyl-tRNA synthetases must also recognize and bind the correct (cognate) tRNA molecules. For some enzymes (e.g., glutaminyl-tRNA synthetase), anticodon structure is an important feature of the recognition process. However, several enzymes appear to recognize other tRNA structural elements, for example, the acceptor stem, in addition to or instead of the anticodon.

QUESTION 19.4   The sequence of a DNA segment is GGTTTA. What is the sequence of the tRNA anticodons?

QUESTION 19.5   The amino acid sequence for a short peptide is Tyr-Leu-Thr-Ala. What are the possible base sequences of the mRNA and the transcribed DNA strand that code for it? What are the anticodons?

QUESTION 19.6   Explain the critically important roles of aminoacyl-tRNA synthetases in protein synthesis.

19.2 PROTEIN SYNTHESIS

An overview of protein synthesis is illustrated in Figure 19.3. Despite its complexity and the variations among species, the translation of a genetic message into the primary sequence of a polypeptide can be divided into three phases: initiation, elongation, and termination.

   1. Initiation. Translation begins with **initiation**, when the small ribosomal subunit binds an mRNA. The anticodon of a specific tRNA, referred to as an *initiator tRNA*, then base pairs with the initiation codon AUG. Initiation ends as the large ribosomal subunit combines with the small subunit. There are two sites on the complete ribosome for codon-anticodon interactions: the P (peptidyl) site (now occupied by the enitiator tRNA) and the A (aminoacyl) site. In both prokaryotes and eukaryotes, mRNAs are read simultaneously by numerous ribosomes. An mRNA with several ribosomes bound to it is referred to as a **polysome**. In actively growing prokaryotes, for example, the ribosomes attached to an mRNA molecule may be separated from each other by as few as 80 nucleotides.
   2. Elongation. During the **elongation** phase the polypeptide is actually synthesized according to the specifications of the genetic message. The message is read in the 5′ ⟶ 3′ direction, polypeptide synthesis proceeds from the N-terminal to the C-terminal. Elongation begins as a second aminoacyl-tRNA becomes bound to the ribosome in the A site because of codon-anticodon base pairing. Peptide bond formation is then catalyzed by peptidyl transferase. During this reaction (referred to as *transpeptidation*) the α-amino group of the A site amino acid (acting as a nucleophile) attacks the carbonyl group of the P site amino acid (Figure 19.4). Because of peptide bond formation, both amino acids are now attached to the A site tRNA. The now uncharged P site tRNA is released from the ribosome. (There is some evidence that a discharged tRNA lingers briefly in another site within the ribosome referred to as the E, or exit, site.) The next step in elongation involves **translocation**, whereby the ribosome is moved along the mRNA. As the mRNA moves, the next codon enters the A site, and the tRNA bearing the growing peptide chain moves into the P site. This series of steps, referred to as the *elongation cycle*, is repeated until a stop codon enters the A site.

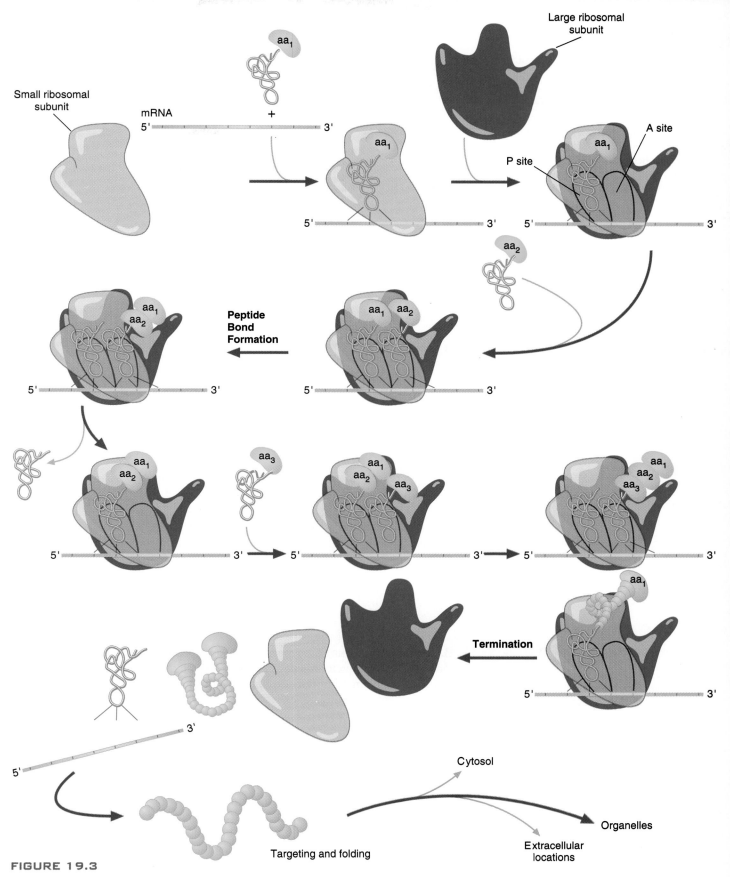

FIGURE 19.3

Protein Synthesis.

No matter what the organism, translation consists of three phases: initiation, elongation, and termination. The elongation reactions, which include peptide bond formation and translocation, are repeated many times until a stop codon is reached. Posttranslational reactions and targeting processes vary according to cell type.

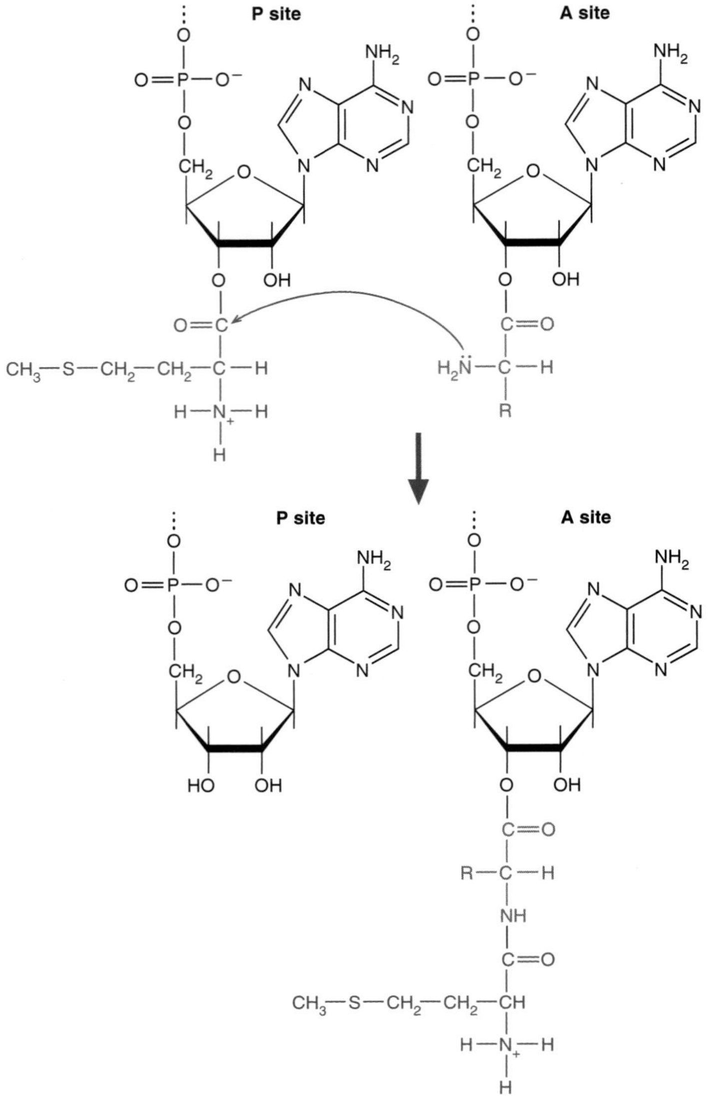

FIGURE 19.4

Peptide Bond Formation.

During the first elongation cycle peptide bonds form because of the nucleophilic attack of the A site amino acid's amino group on the carbonyl carbon of the methionine residue in the P site. Because a peptide bond has formed, both amino acids are now attached to the A site tRNA.

KEY CONCEPTS 19.4

Translation consists of three phases: initiation, elongation, and termination. After their synthesis, many proteins are chemically modified and targeted to specific cellular or extracellular locations.

3. Termination. During **termination** the polypeptide chain is released from the ribosome. Translation terminates because a stop codon cannot bind an aminoacyl-tRNA. Instead, a protein releasing factor binds to the A site. Subsequently, peptidyl transferase (acting as an esterase) hydrolyzes the bond connecting the now-completed polypeptide chain and the tRNA in the P site. Translation ends as the ribosome releases the mRNA and dissociates into the large and small subunits.

In addition to the ribosomal subunits, mRNA, and aminoacyl-tRNAs, translation requires an energy source (GTP) and a wide variety of protein factors. These factors perform several roles. Some have catalytic functions; others stabilize specific structures that form during translation. Translation factors are classified according to the phase of the translation process that they affect, that is, initiation, elongation, or termination. The major differences between prokaryotic and eukaryotic translation appear to be due largely to the identity and functioning of these protein factors.

Regardless of the species, immediately after translation, some polypeptides fold into their final form without further modifications. Frequently, however, newly synthesized polypeptides are modified. These alterations, referred to as **posttranslational modifications**, can be considered to be the fourth phase of

TABLE 19.2

Selected Antibiotic Inhibitors of Protein Synthesis

Antibiotic	Action
Chloramphenicol	Inhibits prokaryotic peptidyl transferase
Cycloheximide	Inhibits eukaryotic peptidyl transferase
Erythromycin	Inhibits prokaryotic peptide chain elongation
Streptomycin	Binding to 30S subunit causes mRNA misreading
Tetracycline	Binding to 30S subunit interferes with aminoacyl-tRNA binding

translation. They include removal of portions of the polypeptide by proteases, modification of the side chains of certain amino acid residues, and insertion of cofactors. Often, individual polypeptides then combine to form multisubunit proteins. Posttranslational modifications appear to serve two general purposes: (1) to prepare a polypeptide for its specific function and (2) to direct a polypeptide to a specific location, a process referred to as **targeting**. Targeting is an especially important process in eukaryotes because proteins must be directed to many destinations. In addition to cytoplasm and the plasma membrane (the principal destinations in prokaryotes), eukaryotic proteins may be sent to a variety of organelles (e.g., mitochondria, chloroplasts, lysosomes, or peroxisomes).

Although there are many similarities between prokaryotic and eukaryotic protein synthesis, there are also notable differences. In fact, these differences are the basis for the therapeutic and research uses of several antibiotics (Table 19.2). Consequently, the details of prokaryotic and eukaryotic processes are discussed separately. Each discussion is followed by a brief description of mechanisms that control translation.

Prokaryotic Protein Synthesis

Translation is relatively rapid in prokaryotes. For example, an *E. coli* ribosome can incorporate as many as 15 to 20 amino acids per second. (The eukaryotic rate, at about 50 residues per minute, is significantly slower.) Recall that prokaryotic ribosomes are composed of a 50S large subunit and a 30S small subunit. The large subunit contains the catalytic site for peptide bond formation. The small subunit serves as a guide for the translation factors required to regulate the process. Figure 19.5 provides a three-dimensional reconstruction of a functioning *E. coli* ribosome.

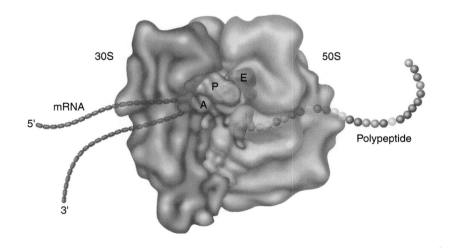

FIGURE 19.5

The Functional Ribosome.

In this three-dimensional reconstruction of an *E. coli* ribosome during protein synthesis, the large and small subunits are shown in blue and yellow, respectively. The relative positions of the mRNA, tRNAs, and the growing polypeptide chain are also illustrated. The tRNAs are identified as A and P to indicate their positions within the acyl and peptidyl sites where peptide bond formation occurs. The tRNA labeled E is in the exit position; that is, having discharged its amino acid during ongoing protein synthesis, it is in the process of leaving the ribosome.

INITIATION As described, translation begins with forming an initiation complex (Figure 19.6). In prokaryotes this process requires three initiation factors (IFs). *IF-3* and *IF-1* have previously bound to the 30S subunit. IF-3 prevents it from binding prematurely to the 50S subunit. IF-1 binds to the A site of the 30S subunit, thereby blocking it during initiation. As an mRNA binds to the 30S subunit, it is guided into a precise location (so that the initiation codon AUG is correctly positioned) by a purine-rich sequence referred to as the **Shine-Dalgarno sequence**. The Shine-Dalgarno sequence (named for its discoverers, John Shine and Lynn Dalgarno) occurs a short distance upstream from AUG. It binds to a complementary sequence contained in the 16S rRNA component of the 30S subunit. Base pairing between the Shine-Dalgarno sequence and the 30S subunit provides a mechanism for distinguishing a start codon from an internal methionine codon. Each gene on a polycistronic mRNA possesses its own Shine-Dalgarno sequence and an initiation codon. The translation of each gene appears to occur independently; that is, translation of the first gene in a polycistronic message may or may not be followed by the translation of subsequent genes.

In the next step in initiation, *IF-2* (a GTP-binding protein with a bound GTP) binds to the 30S subunit, where it promotes the binding of the initiating tRNA to the initiation codon of the mRNA. The initiating tRNA in prokaryotes is N-formylmethionine-tRNA (fmet-tRNA^{fmet}). (After a special initiator tRNA is charged with methionine, the amino acid residue is formylated in an N^{10}-THF-requiring reaction. The enzyme that catalyzes this reaction binds met-tRNA^{fmet} but not met-tRNA^{met}.)

The initiation phase ends as the GTP molecule bound to IF-2 is hydrolyzed to GDP and P_i. GTP hydrolysis presumably causes a conformational change that binds the 50S subunit to the 30S subunit. Simultaneously, IF-2 and IF-3 are released.

ELONGATION Elongation consists of three steps: (1) positioning an aminoacyl-tRNA in the A site, (2) peptide bond formation, and (3) translocation. As noted, these steps are referred to collectively as an elongation cycle.

The prokaryotic elongation process begins when an aminoacyl-tRNA, specified by the next codon, binds to the A site. Before it can be positioned in the A site, the aminoacyl-tRNA must first bind EF-Tu-GTP. The elongation factor *EF-Tu* is a GTP-binding protein (Special Interest Box 19.1) involved in positioning aminoacyl-tRNA molecules in the A site. After the aminoacyl-tRNA is positioned, the GTP bound to EF-Tu is hydrolyzed to GDP and P_i. GTP hydrolysis releases EF-Tu from the ribosome. Then a second elongation factor, referred to as *EF-Ts*, promotes EF-Tu regeneration by displacing its GDP moiety. EF-Ts is then itself displaced by an incoming GTP molecule (Figure 19.7).

After EF-Tu delivers an aminoacyl-tRNA to the A site, the formation of a peptide bond is catalyzed by peptidyl transferase. (Recall that the peptidyl transferase activity is now known to reside in the 23S rRNA component of the 50S subunit.) The energy required to drive this reaction is provided by the high-energy ester bond linking the P site amino acid to its tRNA. (During the first elongation cycle, this amino acid is formylmethionine.) As described, the now uncharged tRNA occupying the P site leaves the ribosome.

For translation to continue, the mRNA must move, or translocate, so that a new codon-anticodon interaction can occur. Translocation requires binding another GTP-binding protein, referred to as *EF-G*. GTP hydrolysis provides the energy required for the ribosomal conformation change that is apparently involved in moving the peptidyl-tRNA (the tRNA bearing the growing peptide chain) from the A site to the P site. The unoccupied A site then binds an appropriate aminoacyl-tRNA to the new A site codon. After EF-G is released, the ribosome is ready for the next elongation cycle. Elongation continues until a stop codon enters the A site.

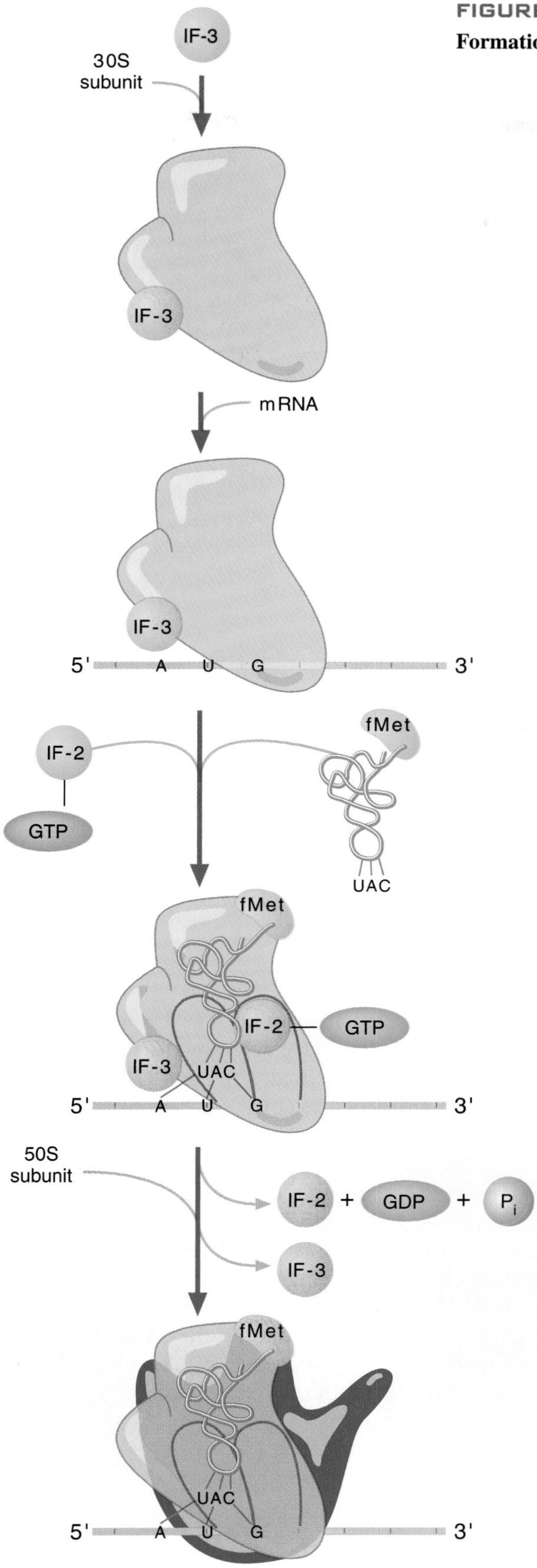

FIGURE 19.6

Formation of the Prokaryotic Initiation Complex.

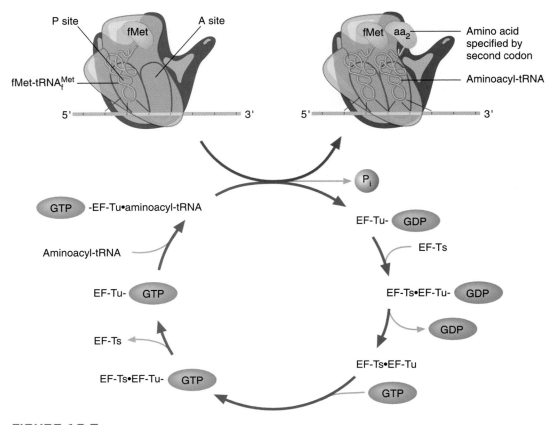

FIGURE 19.7

The EF-Tu-EF-Ts Cycle in *E. coli.*

Before EF-Tu can bind an aminoacyl-tRNA, its GDP moiety must be replaced by GTP. The binding of EF-Ts to EF-Tu (GDP) displaces GDP. EF-Ts is then itself displaced by an incoming GTP. EF-Tu (GTP) then associates with an aminoacyl-tRNA to form an EF-Tu (GTP) aminoacyl-tRNA complex, which proceeds to deliver the aminoacyl-tRNA to the A site of the ribosome.

TERMINATION The termination phase begins when a termination codon (UAA, UAG, or UGA) enters the A site. Three **releasing factors** (RF-1, RF-2, and RF-3) are involved in termination. The codons UAA and UAG are recognized by RF-l, whereas UAA and UGA are recognized by RF-2. (The role of RF-3 is unclear. It may promote RF-l and RF-2 binding.) This recognition process, which involves GTP hydrolysis, alters ribosome function. The peptidyl transferase, which is transiently transformed into an esterase, hydrolyzes the bond linking the completed polypeptide chain and the P site tRNA. Following the polypeptide's release from the ribosome, the mRNA and tRNA also dissociate. The termination phase ends when the ribosome dissociates into its constituent subunits.

POSTTRANSLATIONAL MODIFICATIONS As each **nascent** (newly synthesized) polypeptide emerges from the ribosome, it begins to fold into its final three-dimensional shape (see p. 692). As mentioned, most of these molecules also undergo a series of modifying reactions that prepare them for their functional role. Most of the information concerning posttranslational modifications has been obtained through research on eukaryotes. However, prokaryotic polypeptides are known to undergo several types of covalent alterations.

1. **Proteolytic processing**. Several cleavage reactions may occur. These include removing the formylmethionine residue and signal peptide sequences. (**Signal peptides**, or leader peptides, are short peptide sequences, typically near the amino

terminal, that determine a polypeptide's destination. In bacteria, for example, a signal peptide is required to insert a polypeptide into the plasma membrane.)

2. Conjugation. Most conjugated proteins in prokaryotes are lipoproteins. Blc, a recently discovered outer membrane lipoprotein in *E. coli*, is a type of lipocalin produced under stressful conditions that assists in membrane biogenesis and repair. (The *lipocalins* are a group of hydrophobic ligand-binding proteins previously only observed in eukaryotes.) Although once considered rare in prokaryotes, several glycosylated proteins have also been identified in archeans and certain gram-positive bacteria. The best-studied example is the cell surface glycoprotein of halophilic (salt loving) archean *Halobacterium salinarium*.

3. Methylation. A group of enzymes, referred to as the protein methyltransferases, use S-adenosylmethionine to methylate certain proteins. For example, one type of methyltransferase found in *E. coli* and related bacteria methylates glutamate residues in membrane-bound chemoreceptors. The methyltransferase and a methylesterase are components in a methylation/demethylation process, which plays a role in a signal transduction mechanism involved in chemotaxis. (Recall that the capacity of a living cell to respond to certain environmental cues by moving toward or away from specific molecules is referred to as chemotaxis.)

4. Phosphorylation. In recent years, protein phosphorylation/dephosphorylation catalyzed by protein kinases and phosphatases has been revealed to be widespread among prokaryotes. Many of the purposes of these reactions remain unclear. However, roles for transient phosphorylation have been identified in chemotaxis and nitrogen metabolism regulation.

TRANSLATIONAL CONTROL MECHANISMS Protein synthesis is an exceptionally expensive process. Costing four high-energy phosphate bonds per peptide bond (i.e., two bonds expended during tRNA charging and one each during A site–tRNA binding and translocation) it is perhaps not surprising that enormous quantities of energy are involved. For example, approximately 90% of *E. coli* energy production used in the synthesis of macromolecules may be devoted to the manufacture of proteins. Although the speed and accuracy of translation require a high energy input, the cost would be even higher without metabolic control mechanisms. These mechanisms allow prokaryotic cells to compete with each other for limited nutritional resources.

In prokaryotes such as *E. coli*, most of the control of protein synthesis occurs at the level of transcription. (Refer to Section 18.3 for a discussion of the principles of prokaryotic transcriptional control.) This circumstance makes sense for several reasons. First, transcription and translation are directly coupled; that is, translation is initiated shortly after transcription begins (Figure 19.8). Second, the lifetime of prokaryotic mRNA is usually relatively short. With half-lives of between 1 and 3 minutes, the types of mRNA produced in a cell can be quickly altered as environmental conditions change. Most mRNA molecules in *E. coli* are degraded by two exonucleases, referred to as RNase II and polynucleotide phosphorylase.

Despite the preeminence of transcriptional control mechanisms, the rates of prokaryotic mRNA translation also vary. A large portion of this variation is attributed to differences in Shine-Dalgarno sequences. Because Shine-Dalgarno sequences help select the initiation codon, sequence variations may affect the rate of translating genetic messages. For example, the gene products of the lac operon (β-galactosidase, galactose permease, and galactoside transacetylase) are not produced in equal quantities. Thiogalactoside transacetylase is produced at approximately one-fifth the rate of β-galactosidase. (Recall that the function of thiogalactoside transacetylase remains unknown and that lactose fermentation proceeds normally in mutant cells that are unable to produce this gene product.)

An interesting example of negative translational regulation in prokaryotes is provided by ribosomal protein synthesis. The approximately 55 proteins in prokaryotic ribosomes are coded for by genes located in 20 operons. Efficient bacterial growth

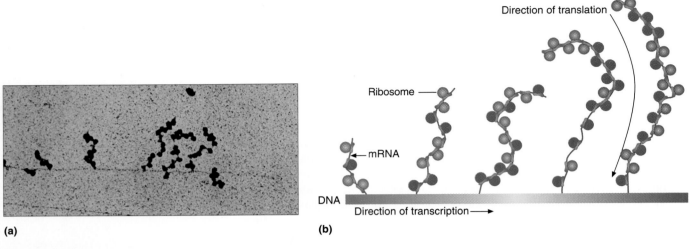

(a)   **(b)**

FIGURE 19.8

Transcription and Translation in *E. coli*.

(a) An electron micrograph of *E. coli* transcription and translation. In *E. coli*, as in other prokaryotes, transcription and translation are directly coupled. (b) Diagram of (a). Note polyribosomes.

KEY CONCEPTS 19.5

Prokaryotic protein synthesis is a rapid process involving several protein factors. Although most prokaryotic gene expression appears to be regulated at the transcriptional level, several types of translational regulation have been detected.

requires that their synthesis be coordinately regulated among the operons and with rRNA synthesis. For example, in the P_{L11} operon, which contains the genes for the ribosomal proteins L1 and L11, excessive amounts of L1 (i.e., more L1 molecules than can bind available 23S rRNA) inhibit P_{L11} mRNA translation (Figure 19.9). Apparently, L1 can bind to either 23S rRNA or P_{L11} mRNA. In the absence of 23S rRNA, L1 inhibits the translation of its own operon by binding to the 5′ end of P_{L11} mRNA.

Eukaryotic Protein Synthesis

Although the earliest work in protein synthesis (e.g., the discovery of aminoacyl-tRNA synthetases and the tRNAs) was done using mammalian cells, translation investigators directed their attention to bacteria in the 1960s. This change occurred for a variety of reasons, including the relative ease of culturing bacterial cells and the perception that bacterial gene expression is simpler and more accessible than that in the more complex eukaryotes. Only in the 1970s, when the principles of prokaryotic translation were understood, did the eukaryotic process again become a focus of attention. Not surprisingly, the large, complex genomes of eukaryotic cells (especially those in multicellular organisms) are now known to require sophisticated regulation of translation (Section 18.3). A large number of protein factors assist in translation. In addition, the posttranslational modifications of eukaryotic polypeptides appear significantly more complex than those observed in prokaryotes. Considering the structural complexity of eukaryotes, it is inevitable that polypeptide targeting mechanisms are also quite intricate.

In this section the features that distinguish the three phases of eukaryotic translation from its prokaryotic counterparts are described. This is followed by a discussion of several of the most prominent forms of eukaryotic posttranslational modifications and targeting mechanisms. The section ends with discussions of translational control mechanisms and protein folding.

INITIATION Most of the major differences between the prokaryotic and eukaryotic versions of protein synthesis occur during the initiation phase. Among the reasons for the additional complexity of eukaryotic initiation are the following:

1. mRNA secondary structure. Recall that eukaryotic mRNA is processed by the addition of a methylguanosine cap and a poly A tail and by the removal

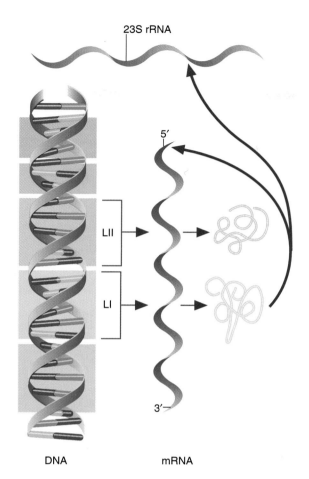

DNA   mRNA

FIGURE 19.9

A Negative Translational Control.

The P_{L11} operon of *E. coli* is controlled by the level of one of its gene products. The ribosomal protein L1 can bind to 23S rRNA or its own mRNA. When excessive amounts of L1 accumulate, then L1 binds to the 5′ end of its own mRNA, thereby inhibiting the translation of the operon.

of introns. In addition, eukaryotic mRNA does not associate with a ribosome until it leaves the nucleus and, as a result, is free to interact with a number of cellular proteins. (An mRNA that is complexed with these proteins is sometimes referred to as a ribonucleoprotein particle.)

2. mRNA scanning. In contrast to prokaryotic mRNA, eukaryotic molecules lack Shine-Dalgarno sequences, which allow for the identification of the initiating AUG sequence. Instead, eukaryotic ribosomes "scan" each mRNA. This scanning is a complex (and poorly understood) process in which ribosomes bind to the capped 5′ end of the molecule and migrate in a 5′ ⟶ 3′ direction searching for a translation start site.

Eukaryotes use a more complex spectrum of initiation factors than prokaryotes. There are at least nine eukaryotic initiating factors (eIFs), several of which possess numerous subunits. The functional roles of most of these factors are still under investigation.

Eukaryotic initiation (Figure 19.10) begins when the small 40S ribosomal subunit binds to a complex composed of *eIF-2* (a GTP-binding protein), GTP, and an initiating species of methionyl-tRNA^{met} (met-tRNA$_i$). (eIF-2-GTP, which mediates the binding of the initiating tRNA to the 40S subunit, is regenerated from inactive eIF-2-GDP by *eIF-2B*, a guanine nucleotide–releasing protein. After GDP is released from eIF-2, GTP binding occurs.) The small (40S) subunit is prevented from binding to the large (60S) subunit during this phase of initiation because it is associated with *eIF-3*, a multisubunit protein. (Subunit assembly is also prevented by the association of eIF-6 with the 60S subunit.) The complex consisting of the small subunit, eIF-2-GTP, eIF-3, and methionyl-tRNA^{met} is referred to as a *40S preinitiation complex*. Subsequently, mRNA binds to the 40S preini-

FIGURE 19.10

Formation of the Eukaryotic Initiation Complex.

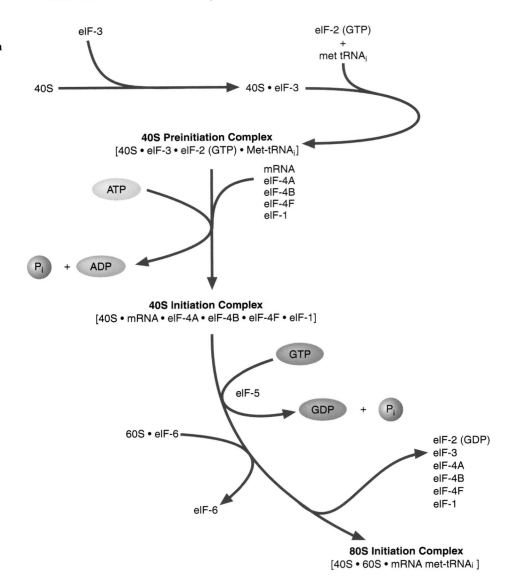

tiation complex to form a *40S initiation complex*. This is an ATP-requiring process that involves several additional initiation factors (e.g., eIF-4A, eIF-4B, eIF-1, eIF-4F). *eIF-4F* binds to the cap structure at the 5′ end of the mRNA, whereas the binding of *eIF-4A* (an ATPase) and *eIF-4B* (a helicase) is believed to reduce the secondary structure of the bound mRNA molecule. Identifying eukaryotic initiation factors has been confusing. For example, some factors have been revealed to be subunits of larger factors. *eIF-4E*, also referred to as cap-binding protein or *CBP* I, is one of several subunits of eIF-4F. eIF-4F is often referred to as *CBP II*.

Once the 40S initiation complex is formed, it scans the mRNA for a suitable initiation codon, which is usually an AUG near the 5′ end. The 40S complex then binds the 60S subunit (now dissociated from eIF-6) to form an *80S initiation complex*. The formation of the 80S complex involves the hydrolysis of the GTP associated with eIF-2, a process that requires *eIF-5*. The initiation phase ends as the initiation factors eIF-2, eIF-3, eIF-4A, eIF-4B, eIF-4F, and eIF-1 are released from the initiation complex.

ELONGATION Figure 19.11 illustrates the eukaryotic elongation cycle as it is currently understood. Several elongation factors (eEFs) are required during this phase

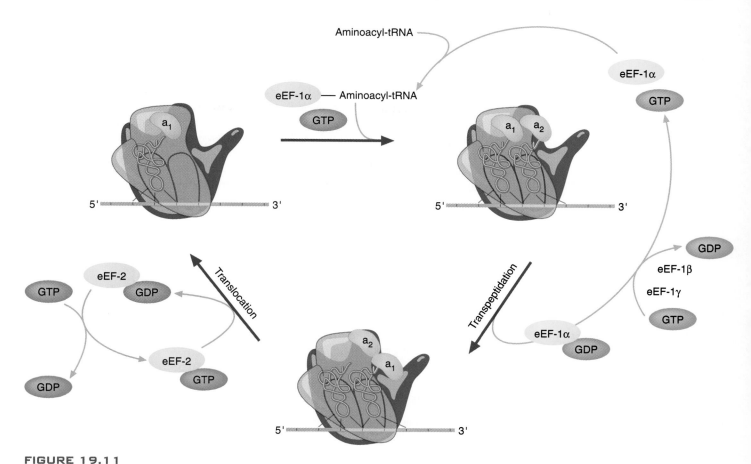

FIGURE 19.11

The Elongation Cycle in Eukaryotic Translation.

Elongation comprises three phases: (1) binding of an aminoacyl-tRNA to the A site, (2) transpeptidation, and (3) translocation.

of translation. *eEF-1α* is a 50-kD polypeptide that mediates the binding of aminoacyl-tRNAs to the A site. After a complex is formed between eEF-1α, GTP, and the entering aminoacyl-tRNA, codon-anticodon interactions are initiated. If correct pairing occurs, eEF-1α hydrolyzes its bound GTP and subsequently exits the ribosome, leaving its aminoacyl-tRNA behind. If correct pairing does not occur, the complex leaves the A site, thereby preventing incorrect amino acid residues from being incorporated. This process has been referred to as **kinetic proofreading**. In various fungi (e.g., yeast), another elongation factor, referred to as *eEF-3*, is also required in combination with eEF-1α for A site aminoacyl-tRNA binding.

During the next elongation step (i.e., peptide bond formation) the peptidyl transferase activity of the large ribosomal subunit catalyzes the nucleophilic attack of the A site α-amino group on the carboxyl carbon of the P site amino acid residue. Apparently eEF-1α dissociates from the ribosome immediately before transpeptidation. EEF-1β and eEF-1γ mediate the regeneration of eEF-1α by promoting an exchange of GDP for GTP. (Recall that a similar process involving EF-Tu and EF-Ts occurs in bacteria such as *E. coli.*)

Translocation in eukaryotes requires a 100-kD polypeptide referred to as *eEF-2*, which is also a GTP-binding protein. eEF-2-GTP binds to the ribosome at some as yet undetermined site during translocation. GTP is then hydrolyzed to GDP, and eEF-2-GDP is released. As noted, GTP hydrolysis provides the energy needed to physically move the ribosome along the mRNA. At the end of translocation a new codon is exposed in the A site.

TERMINATION In eukaryotic cells two releasing factors, *eRF-1* and *eRF-3* (a GTP-binding protein), mediate the termination process. When GTP binds to eRF-3, its GTPase activity is activated. eRF-1 and eRF-3-GTP form a complex that bind in the A site when UAG, UGA, or UAA enter. Then GTP hydrolysis promotes the dissociation of the releasing factors from the ribosome. This step is soon followed by the release of mRNA and the separation of the functional ribosome into its subunits. As described, the release of the newly synthesized polypeptide is catalyzed by peptidyl transferase.

QUESTION 19.7

Explain the following terms in reference to translation:

a. translocation
b. termination
c. elongation
d. polysome
e. releasing factors

QUESTION 19.8

Name and explain the roles of the protein factors that participate in the initiation phase of prokaryotic protein synthesis.

QUESTION 19.9

What types of chemical bonds are involved in the following molecules?

a. amino acids in polypeptides
b. nucleotides in a polynucleotide strand
c. codon-anticodon in translation

QUESTION 19.10

Explain the roles of the large and small subunits of ribosomes.

QUESTION 19.11

eEF-2 possesses a unique modification of a specific histidine residue called diphthamide. Mutant eukaryotic cells that cannot transform this histidine into diphthamide do not appear to be adversely affected. However, the ADP-ribosylation of the diphthamide residue of eEF-2 by toxins produced by *Corynebacterium diphtheriae* and *Pseudomonas aeruginosa* renders the factor inoperative.

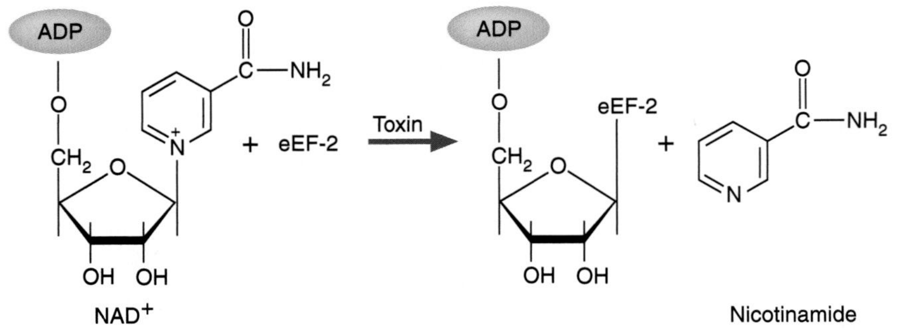

Cells die because they cannot synthesize proteins. The mechanism by which eEF-2 function is affected by ADP-ribosylation is unknown. Can you suggest any possibilities?

EF-Tu (Figure 19A), the protein factor that positions aminoacyl-tRNA complexes in the A site of prokaryotic ribosomes, is a well-researched example of a GTP-binding motor protein. Recall that *motor proteins* (Section 2.1) use nucleotide hydrolysis to drive changes in their own conformations that are often used to promote ordered conformational changes in adjacent molecules or subunits. In other words, motor proteins, often called NTPases, function as mechanochemical transducers. These NTP-hydrolysis driven conformational changes, which principally occur in localized structural units called switches, alter the affinity of the NTPase for other molecules.

EF-Tu possesses three domains. Domain 1 contains a GTP binding site and two switch regions. Domain 2 is connected to domain 1 through a pliable peptide segment. In its active GTP-bound form (EF-Tu-GTP), the elongation factor possesses a binding site for an aa-tRNA. After aa-tRNA binding, the entire structure is referred to as the ternary complex. All three domains of EF-Tu are involved in tRNA binding. For example, the TψC stem of tRNA molecules (Figure 17.22) interacts with several amino acid residues in domain 3. The binding of aa-tRNA projects the anticodon away from the ternary complex so it is free to interact with mRNA codons.

During protein synthesis, the interaction of EF-Tu-GDP (the inactive form) with EF-Ts releases GDP. The subsequent binding of GTP in the domain 1 nucleotide-binding site changes the conformation in the two switch regions. These changes bring domains 1 and 2 close together, forming a binding cleft. Once an aa-tRNA has been bound in the cleft, the ternary complex enters the ribosome where the aa-tRNA anticodon binds reversibly to an mRNA codon in the A site. When a ternary complex contains a cognate aa-tRNA, a conformation change in the ribosome triggers a conformation change in the EF-Tu nucleotide-binding site. The subsequent hydrolysis of GTP causes domains 1 and 2 to move apart, thus allowing the release of the aa-tRNA. Although the precise order of the domain 1 conformational changes remains unclear, the following details are known. Switch region 1, which is believed to contain a β-hairpin in the inactive EF-Tu-GDP, is converted to a helix (the "switch helix") in EF-Tu-GTP. In switch region 2 (near the GTP-binding site) the γ-phosphate of GTP causes a conserved glycine residue to flip 180°, moving it 4.6 Å. The dramatic movement of this residue forces the switch helix to migrate four residues along the polypeptide chain, a process that changes the axis of rotation of the helix by 45° and results in a 46 Å movement of the most distal portion of domain 1. This feature of EF-Tu function has been described as a timing mechanism. The timer is activated when the ternary complex binds to the ribosome. The relatively slow rate of GTP hydrolysis provides sufficient time for the dissociation of incorrect codon-anticodon pairings. In contrast, the binding of a cognate aa-tRNA is so tight that there is sufficient time for GTP to undergo hydrolysis. Thus, the functioning of the ternary complex provides another mechanism for proofreading during translation in addition to that described for the aa-tRNA synthetases.

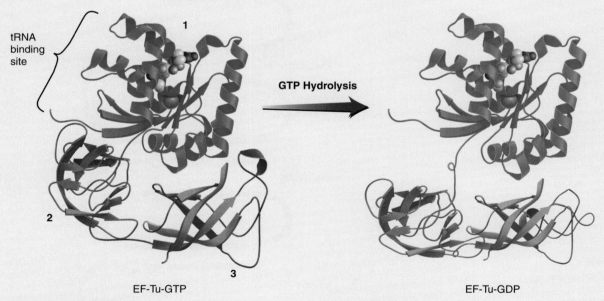

EF-Tu-GTP

GTP Hydrolysis

EF-Tu-GDP

tRNA binding site

FIGURE 19A EF-Tu.
EF-Tu is a GTPase that positions aminoacyl-tRNA complexes within the A site of prokaryotic ribosomes during protein synthesis. The binding of a GTP molecule (red) by EF-Tu causes domain conformation changes (not shown) that result in the creation of a binding cleft for an aminoacyl-tRNA complex. GTP hydrolysis causes domains 1 and 2 to move apart so that the aminoacyl-tRNA (not shown) is released. (A magnesium ion is illustrated with a yellow sphere.)

POSTTRANSLATIONAL MODIFICATIONS IN EUKARYOTES Most nascent polypeptides undergo one or more types of covalent modifications. These alterations, which may occur either during ongoing polypeptide synthesis or afterwards, consist of reactions that modify the side chains of specific amino acid residues or break specific bonds. In general, posttranslational modifications prepare each molecule for its functional role and/or for folding into its native (i.e., biologically active) conformation. Over 200 different types of posttranslational processing reactions have been identified. Most of them occur in one of the following classes:

1. **Proteolytic cleavage**. The proteolytic processing of proteins is a common regulatory mechanism in eukaryotic cells. Typical examples of proteolytic cleavage (the hydrolysis of specific peptide bonds in certain proteins by proteases) include removal of the N-terminal methionine and signal peptides (see p. 674). Proteolytic cleavage is also used to convert inactive precursor proteins, called **proproteins**, to their active forms. Recall, for example, that certain enzymes, referred to as proenzymes or zymogens, are transformed into their active forms by cleavage of specific peptide bonds. The proteolytic processing of insulin (Figure 19.12) provides a well-researched example of the conversion of a polypeptide hormone into its active form. The inactive insulin precursor produced by removing the signal peptide is referred to as proinsulin. Inactive precursor proteins with removable signal peptides are called **preproproteins**. The insulin precursor containing a signal peptide is referred to as pre-proinsulin.

2. **Glycosylation**. Although a wide variety of eukaryotic proteins are glycosylated, the functional purpose of the carbohydrate moieties is not always obvious (Section 7.4). In general, secreted proteins contain complex oligosaccharide

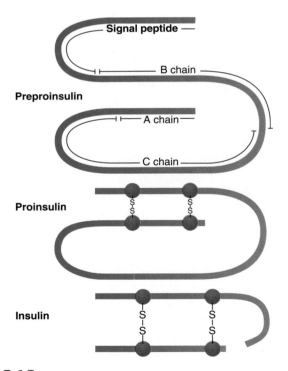

FIGURE 19.12

Proteolytic Processing of Insulin.

After the removal of the signal peptide, a peptide segment referred to as the C chain is removed by a specific proteolytic enzyme. Two disulfide bonds are also formed during insulin's posttranslational processing.

species, whereas ER membrane proteins possess high mannose species. The synthesis of the core N-linked oligosaccharide (common to all N-linked forms) is illustrated in Figure 19.13. The core oligosaccharide is assembled in association with phosphorylated dolichol. (Dolichol is a polyisoprenoid found within all cell membranes. Phosphorylated dolichol is found predominantly in ER membrane.)

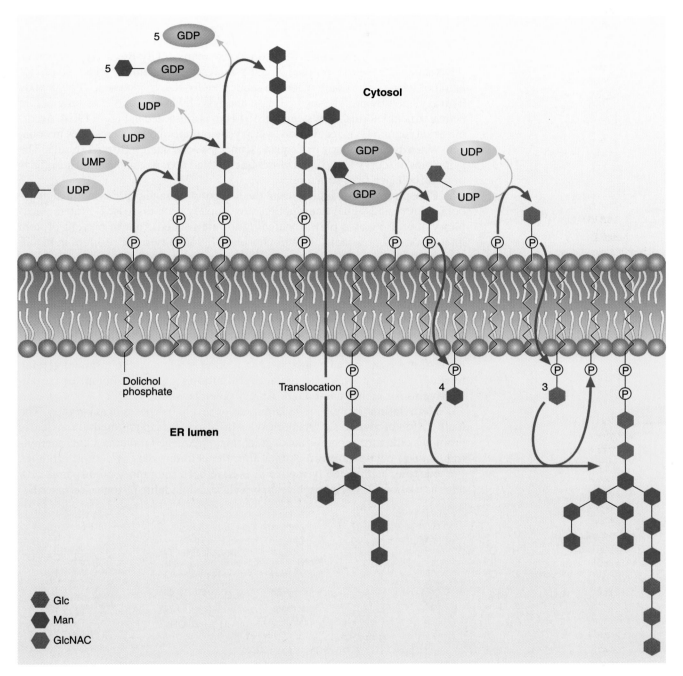

FIGURE 19.13

Synthesis of Dolichol-Linked Oligosaccharide.

In the first step, GlcNAc-1-P is transferred from UDP-GlcNAc to dolichol phosphate (Dol-P). The next GlcNAc and the following five mannose residues are then transferred from nucleotide-activated forms. After the entire structure flips to the lumenal side of the membrane, each of the remaining sugars (four mannoses and three glucoses) is transferred first to Dol-P and then to the growing oligosaccharide. Then N-glycosylation of protein takes place in the ER in a one-step reaction catalyzed by a membrane-bound enzyme called glycosyl transferase.

3. Hydroxylation. Hydroxylation of the amino acids proline and lysine is required for the structural integrity of the connective tissue proteins collagen (Section 5.3) and elastin. Additionally, 4-hydroxyproline is also found in acetylcholinesterase (the enzyme that degrades the neurotransmitter acetylcholine) and complement (a complex series of serum proteins involved in the immune response). Three mixed-function oxygenases (prolyl-4-hydroxylase, prolyl-3-hydroxylase, and lysyl hydroxylase) located in the RER are responsible for hydroxylating certain proline and lysine residues. Substrate requirements are highly specific. For example, prolyl-4-hydroxylase hydroxylates only proline residues in the Y position of peptides containing Gly-X-Y sequences, whereas prolyl-3-hydroxylase requires Gly-Pro-4-Hyp sequences (Hyp stands for hydroxyproline; X and Y represent other amino acids). Hydroxylation of lysine occurs only when the sequence Gly-X-Lys is present. (Polypeptide hydroxylation by prolyl-3-hydroxylase and lysyl hydroxylase occurs only before helical structure forms.) The synthesis of 4-Hyp is illustrated in Figure 19.14. Ascorbic acid (vitamin C) is required to hydroxylate proline and lysine residues in collagen. When dietary intake is inadequate, scurvy (Special Interest Box 7.1) results. The symptoms of scurvy (e.g., blood vessel fragility and poor wound healing) are effects of weak collagen fiber structure.

4. Phosphorylation. Examples of the roles of protein phosphorylation in metabolic control and signal transduction have already been discussed. Protein phosphorylation may also play a critical (and interrelated) role in protein-protein interactions. For example, the autophosphorylation of tyrosine residues in PDGF receptors precedes the binding of cytoplasmic target proteins.

5. Lipophilic modifications. The covalent attachment of lipid moieties to proteins improves membrane binding capacity and/or certain protein-protein interactions. Among the most common lipophilic modifications are acylation (the attachment of fatty acids) and prenylation (Section 11.1). Although the fatty acid myristate (14:0) is relatively rare in eukaryotic cells, myristoylation is one of the most common forms of acylation. N-myristoylation (the covalent attachment of myristate by an amide bond to a polypeptide's amino terminal glycine residue) has been shown to increase the affinity of the α subunit of certain G proteins for membrane-bound β and γ subunits.

6. Methylation. Protein methylation serves several purposes in eukaryotes. The methylation of altered aspartate residues by a specific type of methyltransferase promotes either the repair or the degradation of damaged proteins. Other methyltransferases catalyze reactions that alter the cellular roles of certain proteins. For example, methylated lysine residues have been found in such disparate proteins as ribulose-2,3-bisphosphate carboxylase, calmodulin, histones, certain ribo-

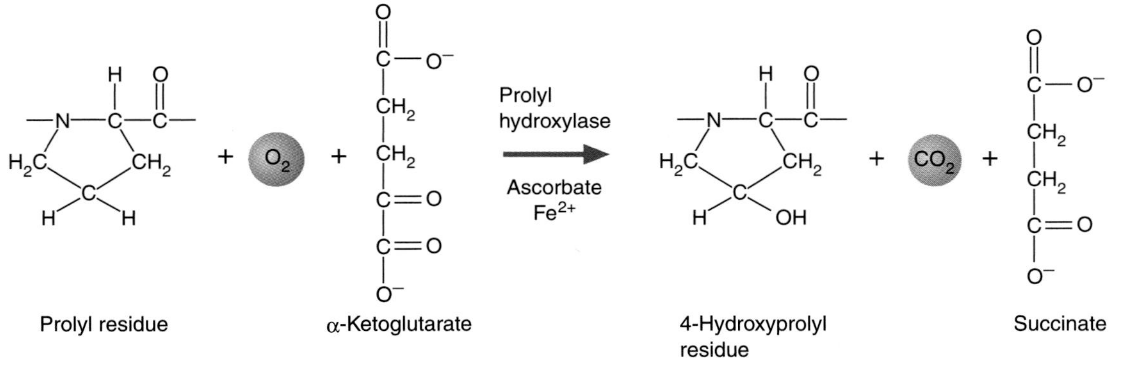

Prolyl residue α-Ketoglutarate   4-Hydroxyprolyl residue   Succinate

FIGURE 19.14

Hydroxylation of Proline.

Ascorbic acid and ferrous iron are cofactors of prolyl-4-hydroxylase, the enzyme that catalyzes the hydroxylation of the C-4 position of certain prolyl residues in nascent polypeptides. Ascorbic acid, acting as a reducing agent, prevents the oxidation of the iron atom cofactor of the enzyme.

somal proteins, and cytochrome c. Other amino acid residues that may be methylated include histidine (e.g., histones, rhodopsin, and eEF-2) and arginine (e.g., heat shock proteins and ribosomal proteins).

7. Disulfide bond formation. Disulfide bonds are generally found only in secretory proteins (e.g., insulin) and certain membrane proteins. (Recall that "disulfide bridges" are favored in the oxidizing environment outside the cell and they confer considerable structural stability on the molecules that contain them.) As described (Section 5.3), cytoplasmic proteins generally do not possess disulfide bonds because of the presence of various reducing agents in cytoplasm (e.g., glutathione and thioredoxin). Because the ER has a nonreducing environment, disulfide bonds form spontaneously in the RER as the nascent polypeptide emerges into the lumen. Although some proteins have disulfide bridges that form sequentially as the polypeptide enters the lumen (i.e., the first cysteine pairs with the second, the third residue pairs with the fourth, etc.), this is not true for many other molecules. Proper disulfide bond formation for these latter proteins is now presumed to be facilitated by **disulfide exchange**. During this process, disulfide bonds rapidly migrate from one position to another until the most stable structure is achieved. An ER enzymatic activity, referred to as protein disulfide isomerase, is now believed to catalyze this process.

8. Protein splicing. **Protein splicing** (Figure 19.15) is a posttranslational mechanism in which an intervening peptide sequence is precisely excised from a nascent polypeptide. In this self-catalyzed, intramolecular reaction, a peptide bond is formed between the flanking amino terminal and carboxy terminal amino acid

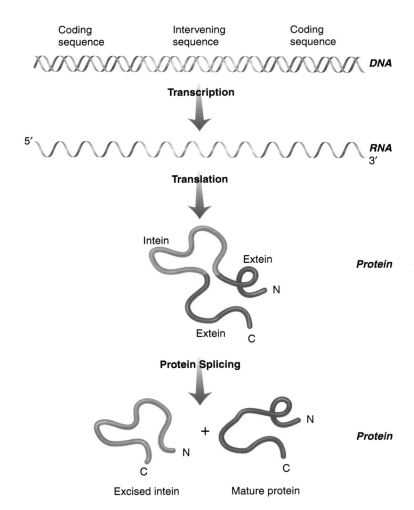

FIGURE 19.15

Protein Splicing.

Protein splicing is a posttranslational process in which an intein, a peptide sequence coded for by an intervening sequence in the gene, is precisely excised from the nascent polypeptide. In this autocatalytic reaction, a peptide bond is formed that links the N-terminal and C-terminal amino acid residues of the flanking peptide sequences, called exteins, to form the mature protein.

residues. The excised peptide segment is called an **intein**; the flanking segments that are spliced together to form the mature protein are called **exteins**. Protein splicing occurs without the aid of cofactors, metabolic energy sources, or auxiliary enzymes. First identified in 1990 during investigations of the catalytic subunit of an ATPase in the yeast *Saccharomyces cerevisiae*, protein splicing has been observed in certain unicellular organisms in all the domains of living organisms: archaea, bacteria, and eukarya. The mechanism of the splicing reaction and the overall significance of protein splicing in cellular processes remain unresolved.

TARGETING Despite the vast complexities of eukaryotic cell structure and function, each newly synthesized polypeptide is normally directed to its proper destination. Considering that translation takes place in the cytoplasm (except for certain molecules that are produced within mitochondria and plastids) and that a wide variety of polypeptides must be directed to their proper locations, it is not surprising that the mechanisms by which cellular proteins are targeted are complex. Although this process is not yet completely understood, there appear to be two principal mechanisms by which polypeptides are directed to their correct locations: transcript localization and signal peptides. Each is briefly discussed.

It is generally recognized that cells often have asymmetrical protein distributions within the cytoplasm. For example, mature *Drosophila* eggs contain a gradient of bicoid, a protein that plays a critical role in gene regulation during development. A high concentration of bicoid in the anterior portion of the egg is required for the normal development of anterior body parts (i.e., head segments), whereas the low bicoid concentration in the posterior portion of the egg cytoplasm promotes the development of posterior body parts. (If posterior cytoplasm is removed from one egg and substituted for anterior cytoplasm in a second egg, two sets of posterior body parts appear in the larva that develops from the recipient egg.) It is now believed that cytoplasmic protein gradients are created by **transcript localization**, that is, the binding of specific mRNA to receptors in certain cytoplasmic locations. It is known that bicoid mRNA is transported from nearby nurse cells into the developing oocyte (an immature egg cell). Once in the oocyte, bicoid mRNA binds via its 3′ end to certain components of the anterior cytoskeleton. After the mature egg is fertilized, translation of bicoid mRNA, coupled with protein diffusion, gives rise to the concentration gradient.

Polypeptides destined for secretion or for use in the plasma membrane or any of the membranous organelles must be specifically targeted to their proper location. Several types of these proteins possess sorting signals referred to as signal peptides. Each signal peptide sequence helps insert the polypeptide that contains it into an appropriate membrane. Signal peptides generally consist of a positively charged region followed by a central hydrophobic region and a more polar region. Although many signal peptides occur at the amino terminal, they may also occur elsewhere along the polypeptide.

The **signal hypothesis** was proposed by Gunter Blobel in 1975 to explain the translocation of polypeptides across RER membrane. Subsequent investigations revealed significant information concerning the insertion of polypeptides through the RER membrane. For this reason the discussion primarily focuses on this organelle. This is followed by a brief description of polypeptide uptake by other organelles.

As soon as about 70 amino acids have been incorporated into the polypeptide that emerges from a ribosome, a **signal recognition particle** (SRP) (a large complex consisting of six proteins and a small RNA molecule) binds to the ribosome (Figure 19.16). As a consequence of this binding, translation is temporarily arrested. The SRP then mediates binding of the ribosome to the RER via **docking protein**, a heterodimer also referred to as SRP receptor protein. Once binding to the RER has occurred, translation restarts, and the growing polypeptide inserts into the membrane. (The simultaneous translocation of a polypeptide

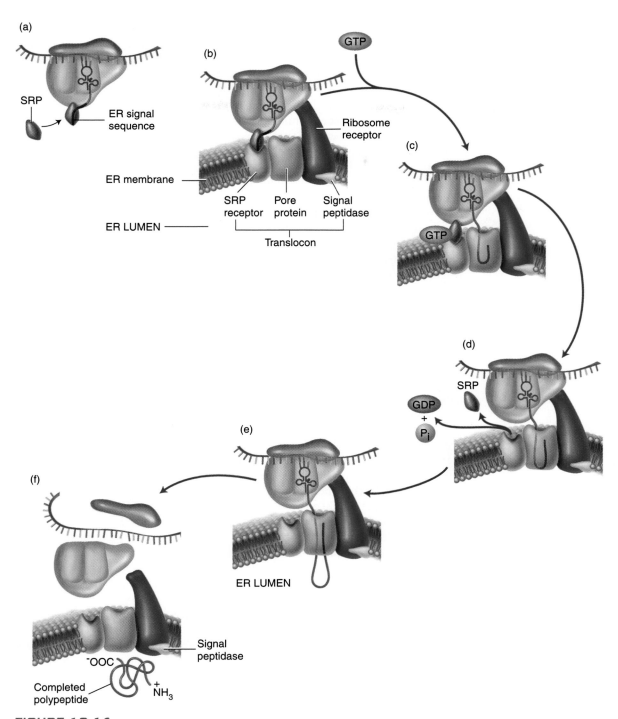

FIGURE 19.16

Cotranslational Transfer Across the RER Membrane.

(a) When the nascent polypeptide is long enough to protrude from the ribosome, the SRP binds to the signal sequence, causing a transient cessation of translation. (b) The subsequent binding of SRP to the SRP receptor results in the binding of the ribosome to the translocon complex in the RER membrane. (c) Polypeptide synthesis begins again as GTP binds to the SRP–SRP receptor complex. GTP hydrolysis accompanies the binding of the signal sequence to the translocon and (d) the dissociation of SRP from its receptor. (e) The polypeptide continues to elongate until (f) translation is terminated. The signal peptide is removed by signal peptidase in the RER lumen. The polypeptide is released into the lumen.

during ongoing protein synthesis is sometimes referred to as **cotranslational transfer**.) As translation begins, the SRP is released. An integral membrane protein complex, referred to as a **translocon**, is believed to mediate polypeptide translocation. The translocon consists of a hydrophilic transmembrane pore and several proteins that facilitate polypeptide translocation and processing. It is presumed that GTP hydrolysis provides the energy to push polypeptides across the RER membrane, because both the SRP and SRP receptor bind GTP. In **posttranslational translocation**, previously synthesized polypeptides are pulled across the RER membrane by an ATP-binding peripheral translocon-associated protein (hsp70).

The fate of a targeted polypeptide depends on the location of the signal peptide and any other signal sequences. As illustrated in Figure 19.16 soluble secretory protein transmembrane transfer is usually followed by removal of an N-terminal signal peptide by signal peptidase, a process that releases the protein into the ER lumen. Such molecules usually undergo further posttranslational processing. The initial phase of the translocation of transmembrane proteins is similar to that of secretory proteins. For these molecules, the amino terminal signal peptide serves as a *start signal* that remains bound in the membrane as the remaining polypeptide sequence is threaded through the membrane. So-called "single-pass" transmembrane proteins possess a *stop transfer signal* (or stop signal), which prevents further transfer across the membrane (Figure 19.17a). Membrane proteins with multiple membrane spanning segments (multi-pass) possess a series of alternating start and stop signals (Figure 19.17b).

Most proteins that are translocated into the RER are directed to other destinations. After they undergo initial posttranslational modifications, both soluble and membrane-bound proteins are transferred to the Golgi complex via transport vesicles that bud off from the ER and fuse with the *cis* face of the Golgi membrane (Figure 19.18). (Proteins that ultimately reside in the ER possess retention signals. In most vertebrate cells this signal consists of the carboxy terminal tetrapeptide Lys-Asp-Glu-Leu, often referred to with the one-letter abbreviations KDEL.) Within the Golgi complex, proteins undergo further modifications. For example, N-linked oligosaccharides are processed further, and O-linked glycosylation of certain serine and threonine residues occurs. Lysosomal proteins are targeted to the lysosomes by adding a mannose-6-phosphate residue. It is still unclear what signals direct secretory proteins to the cell surface (via exocytosis) or promote the delivery of plasma membrane proteins to their destination, although a "default mechanism" has been proposed. (In default mechanisms, the absence of a signal results in a specific sequence of events.) When protein modification is complete, transport vesicles exit from the *trans* face of the Golgi and move to their target locations.

As noted, although mitochondria and chloroplasts produce several of their own proteins, a variety of other proteins are produced on cytoplasmic ribosomes and subsequently imported. Again, specific signal sequences are required. The import of polypeptides into these organelles is complicated by the presence of several membranes. (Recall that both mitochondria and chloroplasts contain several compartments created by internal membranes.) Consequently, polypeptide transfer in these organelles often involves several signal sequences. An example of this import mechanism (the targeting of cytochrome c_1 to the inner membrane space of mitochondria) is illustrated in Figure 19.19.

QUESTION 19.12   List and describe the major classes of eukaryotic posttranslational modifications.

QUESTION 19.13   Explain the importance of the proper targeting of nascent polypeptides.

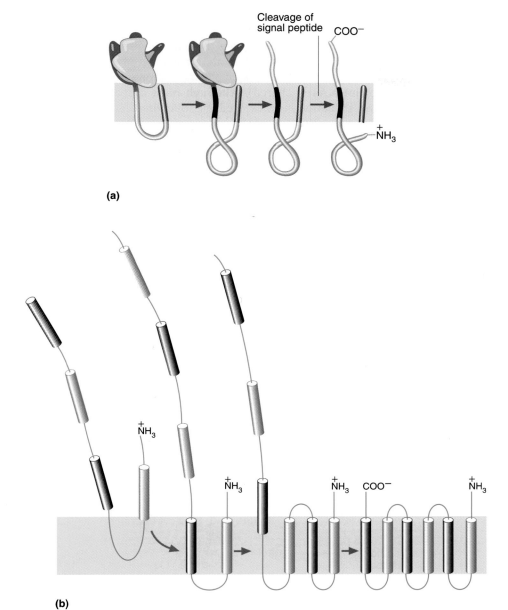

FIGURE 19.17

Cotranslational Transfer of Integral Membrane Proteins.

(a) Transfer of a single-pass transmembrane protein. (b) Transfer of a multi-pass membrane protein. For the sake of clarity the transfer apparatus has been omitted from the diagrams. In addition, the ribosome has been omitted from (b). The shaded segment is a signal peptide. The black segment is a stop transfer signal.

Explain the following terms:

a. signal peptide

b. docking protein

c. signal recognition particle

d. translocon

e. posttranslational translocation

QUESTION 19.14

TRANSLATION CONTROL MECHANISMS Eukaryotic translation control mechanisms are proving to be exceptionally complex, substantially more so than those observed in prokaryotes. In eukaryotes these mechanisms appear to occur on a continuum, from *global* controls (i.e., the translation of a wide variety of mRNAs is altered) to *specific* controls (i.e., the translation of a specific mRNA or small group of mRNAs is altered). Although most aspects of eukaryotic translational control are currently unresolved, the following features are believed to be important:

1. **mRNA export**. The spatial separation of transcription and translation afforded by the nuclear membrane appears to provide eukaryotes with significant opportunities for gene expression regulation. Only a small portion of RNA produced within the nucleus ever enters the cytoplasm. Most discarded RNA is probably nonfunctional introns. Export through the nuclear pore complex is known to be a carefully controlled, energy-driven process whose minimum requirements include the presence of a 5′-cap and a 3′ poly A tail.

2. **mRNA stability**. In general, the translation rate of any mRNA species is related to its abundance, which is in turn dependent on both its rates of synthesis and degradation. mRNA half-lives range from about 20 minutes to over 24 hours. Several features of mRNA structure are known to affect its stability, that is, its capacity to avoid degradation by various nucleases. The presence of certain sequences may confer resistance to nuclease action (e.g., palindromes that create hairpins) while other sequences may increase the likelihood of nuclease action particularly if present in multiple copies. The binding of specific proteins to certain sequences can also affect mRNA stability. Finally, reversible adenylation and deadenylation of the 3′ end of mRNA strongly influence both its stability and its translational activity. After processing within the nucleus, most mRNAs are transported into the cytoplasm possessing poly A tails containing between 100 and 200 nucleotides. As time passes, many poly A tails progres-

FIGURE 19.18

The ER, Golgi, and Plasma Membrane.

Transport vesicles transfer new membrane components (protein and lipids) and secretory products from the ER to the Golgi complex, from one Golgi cisterna to another, and from the trans-Golgi network to other organelles (e.g., lysosomes) or to the plasma membrane.

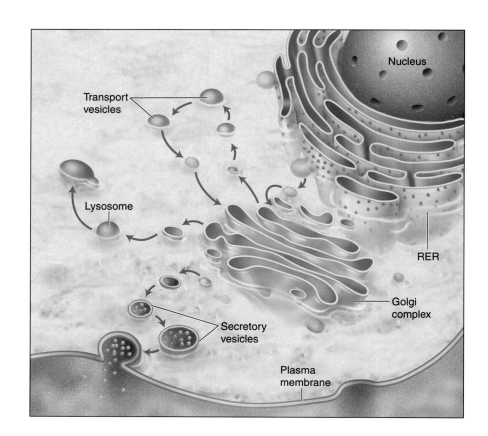

sively shorten to no fewer than 30 residues when the entire mRNA is degraded. In certain circumstances the poly A tail of some mRNAs is selectively elongated or shortened. For example, mRNAs in mature oocytes are "masked" by removal of most of their poly A tail nucleotides. After fertilization these mRNAs are reactivated by adding adenine nucleotides.

3. Negative translational control. The translation of certain specific mRNAs is known to be blocked by binding repressor proteins to sequences near their 5′ ends. A well-researched example is provided by ferritin synthesis control. Ferritin, an iron storage protein that is found predominantly in hepatocytes, is synthesized in response to high iron concentrations. Ferritin mRNA contains an iron response element (IRE) that binds an iron-binding repressor protein. When cellular iron concentrations are high, the large number of iron atoms binding to the repressor protein cause it to dissociate from the IRE. Then ferritin mRNA is translated.

4. Initiation factor phosphorylation. The phosphorylation of eIF-2 in response to certain circumstances (e.g., heat shock, viral infections, and growth factor deprivation) has been observed to decrease protein synthesis generally. However, the translation of certain mRNA increases. For example, hsp (heat shock protein) synthesis increases in response to heat shock and other stressful conditions. The specific mechanisms are unknown.

5. Translational frameshifting. Certain mRNAs appear to contain structural information that, if activated, results in a +1 or −1 change in reading frame. This **translational frameshifting**, which has been most often observed in cells infected by retroviruses, allows more than one polypeptide to be synthesized from a single mRNA.

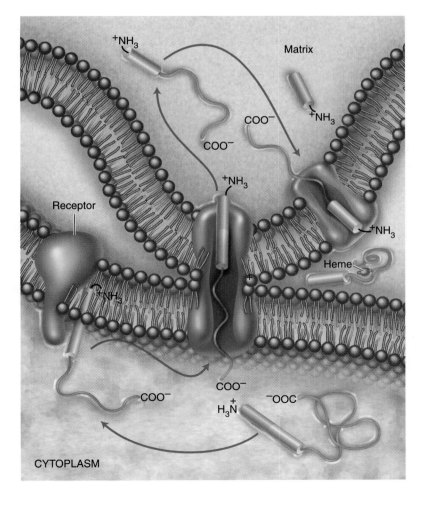

FIGURE 19.19

Posttranslational Transport of Cytochrome c_1 into a Mitochondrion.

After its synthesis in cytoplasm, cytochrome c_1 must be translocated into the mitochondrial inner membrane space. (Recall that cytochrome c_1 is a component of complex III of the ETC.) The targeting of cytochrome c_1 requires two sequences. The first targets the polypeptide to the matrix. After this sequence is removed by a protease, the second sequence targets the molecule to the inner membrane space. The second targeting sequence is then also removed. After folding and binding a heme, the molecule associates with complex III in the inner membrane.

Briefly outline the major mechanisms used by eukaryotes to control translation.

Explain the differences among preproproteins, proproteins, and proteins.

The mechanism involved in the posttranslational transport of proteins into chloroplasts has so far received only limited attention. However, the import of plastocyanin into the thylakoid lumen has been determined to require two import signals near the N-terminal of the newly synthesized protein. Assuming that chloroplast protein import resembles the import process for mitochondria, suggest a reasonable hypothesis to explain how plastocyanin (a lumen protein associated with the inner surface of the thylakoid membrane) is transported and processed. What enzymatic activities and transport structures do you expect are involved in this process?

The Folding Problem

The rapid and efficient folding of newly synthesized polypeptides into their highly specific native structures is an essential phase of information transfer in living organisms. The direct relationship between a protein's primary sequence and its final three-dimensional conformation, and by extension its biological activity, is among the most important assumptions of modern biochemistry. One of the principal underpinnings of this paradigm is a series of experiments reported by Christian Anfinsen in the late 1950s. Anfinsen (Nobel Prize in chemistry, 1972), working with bovine pancreatic RNase, demonstrated that under favorable conditions a denatured protein could refold into its native and biologically active state (Figure 5.24). This discovery suggested that the three-dimensional structure of any protein could be predicted if the physical and chemical properties of the amino acids and the forces that drive the folding process (e.g., bond rotations, free energy considerations, and the behavior of amino acids in aqueous environments) were understood. Unfortunately, several decades of painstaking research with the most sophisticated tools available (e.g., X-ray crystallography and NMR in combination with site-directed mutagenesis and computer-based mathematical modeling) resulted in only limited progress. (**Site-directed mutagenesis** is a recombinant DNA technique in which specific sequence changes can be introduced into a predetermined position in cloned genes.) Briefly, such work revealed that protein folding is a stepwise process in which secondary structure formation (i.e., α-helix and β-pleated sheet) is an early feature. Hydrophobic interactions appear to be an important force in folding. In addition, amino acid substitutions experimentally introduced into certain proteins reveal that changes in surface amino acids rarely affect the protein's structure. In contrast, substitutions of amino acids within the hydrophobic core often lead to serious structural changes in conformation.

The limitations of the traditional protein-folding model (i.e., interactions between amino acid side chains alone force the molecule to fold into its final shape) are highlighted by the following considerations:

1. Time constraints. The time to synthesize proteins routinely ranges from a few seconds to no more than a few minutes. According to one prominent protein-folding model, a newly made polypeptide tries out all possible conformations until the most stable one is achieved. Calculations of the time required for every bond in a small protein molecule to rotate until the final biologically active form is achieved indicate that an astronomical number of years would be required. Even when only a smaller number of possible bond rotations are considered, the time required for folding is still measured in years. Therefore most researchers have concluded that protein folding is not a random process based solely on primary sequence.

Protein misfolding and aggregation are now known to be an important feature of several human diseases. These maladies are referred to as **conformational diseases**, because they are believed to be caused, at least in part, by abnormal conformational changes in certain proteins. Prominent examples include Alzheimer's disease and Creutzfeld-Jacob disease. Each is discussed briefly.

Alzheimer's Disease

Alzheimer's disease (AD) is a progressive and ultimately fatal disease that is characterized by seriously impaired intellectual function. AD first manifests itself with short-term memory loss. Eventually, severe memory loss, disorientation, and agitation accompany a total loss of the patient's personality. Caused by neuronal death in brain regions related to memory and cognition, AD is diagnosed at autopsy by the presence of insoluble aggregates of extracellular proteinaceous debris called **amyloid deposits** (or senile plaques) along with other characteristic anatomical features. The core of amyloid deposits is primarily composed of a 30–42 residue peptide called amyloid β-protein (β-amyloid). β-Amyloid is generated by the proteolytic cleavage of amyloid precursor protein (APP), a transmembrane glycoprotein whose function is still unknown. Mutations in the APP gene (chromosome 21) are known to cause inherited cases of the disease. For reasons that are still unclear, β-amyloid, a soluble peptide produced by most cells types, aggregates to form the virtually insoluble fibrils that characterize the disease.

Creutzfeld–Jacob Disease

Creutzfeld-Jacob disease (CJD) is a rare neurodegenerative disease that has both inherited and infectious forms. CJD, which is characterized by dementia and impaired movement coordination, is one of several human diseases that were previously referred to as transmissible spongiform encephalopathies, but are now classified as prion diseases. The concept of the **prion** (*pro*teinaceous *in*fectious particle) was introduced by Stanley Prusiner to explain the mode of transmission of similar animal diseases such as scrapie in sheep and mad cow disease in cattle. Treatment of extracts of tissue from diseased animals known to contain the causative agent with conventional techniques that destroy nucleic acids failed to prevent transmission. Prusiner also observed that the brains of infected animals contained a protease-resistant protein that he later called a prion. By using gene cloning techniques, researchers eventually discovered that the prion protein is coded for by a normal gene on chromosome 20. The normal prion protein (PrP^C) is an α-helix-containing, protease-sensitive molecule that is soluble in certain nondenaturing solvents. Prion diseases are caused when the conformation of PrP^C is converted to PrP^{Sc}, a misfolded β-pleated-sheet-containing version that is protease resistant and insoluble. The mechanisms that give rise to this altered conformation are still unresolved. It is known that inherited CJD is caused by mutations in the prion gene. Infectious CJD occurs as a consequence of exposure to the infectious agent. For example, CJD has been linked to transplantation of infected corneas and the consumption of infected beef. Numerous investigations of infectious prion diseases indicate that a complex forms between the infectious agent PrP^{Sc} and the normal prion protein (PrP^C) once the latter has reached the cell surface. Acting as a template, PrP^{Sc} forces PrP^C to fold into the abnormal conformation. As PrP^{Sc} accumulates in the brain, holes appear in neural tissue that destroy function.

2. **Complexity**. The calculations required in the mathematical models of protein folding based on physical data (e.g., bond angles and degrees of rotation) are overwhelmingly complex. Therefore it appears unlikely at this time that they alone can resolve the principles of what appears in living organisms to be an astonishingly fast and elegant process.

In recent years important advances have been made by biochemists in protein-folding research by utilizing imaginative combinations of technologies such as site-directed mutagenesis, multidimensional NMR, and circular dichroism, to name a few. (**Circular dichroism** (CD) is a type of spectroscopy in which the relationship between molecular motion and structure is probed with electromagnetic radiation.) By utilizing these techniques in submillisecond timeframes, protein-folding researchers have determined that the process does not consist, as was originally thought, of a single pathway. Instead, there are numerous routes that a polypeptide can take to fold into its native state. As illustrated in Figure 19.20, an energy landscape with a funnel shape appears to best describe how an unfolded polypeptide with its own unique set of constraints (e.g., its amino acid sequence and post-translational modifications, and environmental features within the cell such as temperature, pH, and molecular crowding) negotiates its way to a low-energy folded state. Depending largely on its size, a polypeptide may or may not form intermediates (species

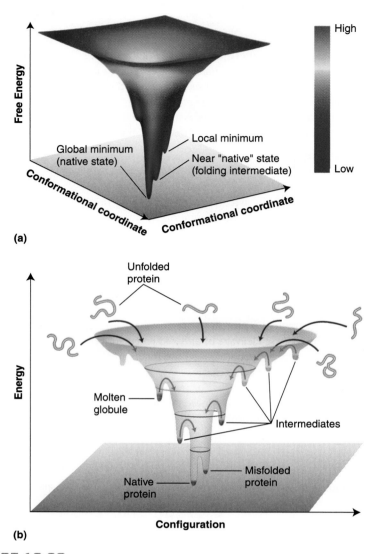

FIGURE 19.20

The Energy Landscape for Protein Folding.

(a) Color is used to indicate the entropy level of the folding polypeptide. As folding progresses the polypeptide moves from a disordered state (high entropy, red) toward a progressively more ordered conformation until its unique biologically active conformation is achieved (lower entropy, blue). In (b), a more realistic view of the energy landscape, polypeptides can fold into their native states by several different pathways. Many molecules form transient intermediates, whereas others may become trapped in a misfolded state.

existing long enough to be detected) that are momentarily trapped in local energy wells (Figure 19.20). Small molecules (less than 100 residues) often fold without intermediate formation (Figure 19.21a). As these molecules begin emerging from the ribosome, a rapid and cooperative folding process begins in which side chain interactions (solvent exclusion of hydrophobic regions, van der Waals forces, and hydrogen bonding) facilitate the formation and alignment of secondary structures. The folding of larger polypeptides typically involves the formation of several intermediates (Figure 19.21b, c). In many of these molecules or the domains within a molecule, the hydrophobically collapsed shape of the intermediate is referred to as a molten globule. The term **molten globule** refers to a partially organized globular state of a folding polypeptide that resembles

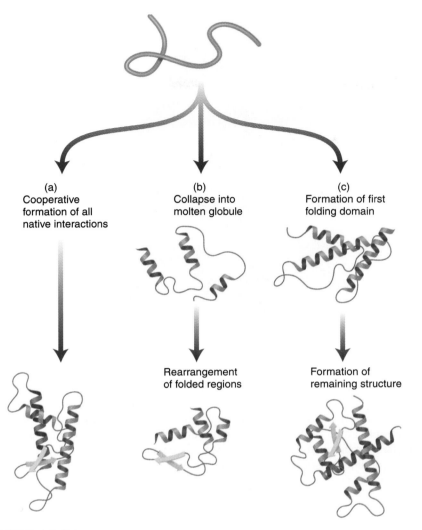

FIGURE 19.21

Protein Folding.

(a) In many small proteins, folding is cooperative with no intermediates formed. (b) In some larger proteins, folding involves the initial formation of a molten globule followed by rearrangement into the native conformation. (c) Large proteins with multiple domains follow a more complex pathway with each of the domains folding separately before the entire molecule progresses to its native conformation.

the molecule's native state. Within the interior of a molten globule tertiary interactions among amino acid side chains are fluctuating, that is, they have not yet stabilized.

It has also become increasingly clear that the folding and targeting of many proteins in living cells are aided by a group of molecules now referred to as the **molecular chaperones**. These molecules, most of which appear to be hsps, apparently occur in all organisms. Several classes of molecular chaperones have been found in organisms ranging from bacteria to the higher animals and plants. In addition, they are also found in several eukaryotic organelles, such as mitochondria, chloroplasts, and ER. There is a high degree of sequence homology among the molecular chaperones of all species so far investigated. The properties of several of these important molecules are described next.

FIGURE 19.22

Space-Filling Model of the *E. coli* Chaparonin called the GroES-GroEL Complex.

GroES (gold) is a seven-membered ring that sits on top of GroEL, which is composed of two stacked seven-member rings (green and red). Within GroEL is a cavity in which ATP-dependent protein folding occurs.

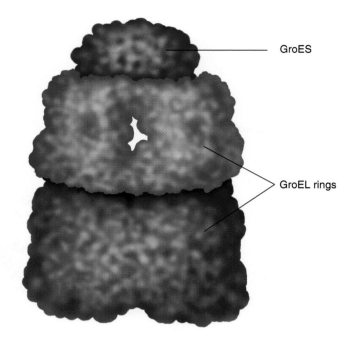

GroES

GroEL rings

MOLECULAR CHAPERONES Molecular chaperones apparently assist unfolded proteins in two ways. First, during a finite time between synthesis and folding, proteins must be protected from inappropriate protein-protein interactions. Some proteins must remain unfolded until they are inserted in an organelle membrane, for example, certain mitochondrial and chloroplast proteins. Second, proteins must fold rapidly and precisely into their correct conformations. Some must be assembled into multisubunit complexes. Investigations of protein folding in a variety of organisms reveal that two major molecular chaperone classes are involved in protein folding.

1. **Hsp70s**. The **hsp70s** are a family of molecular chaperones that bind to and stabilize proteins during the early stages of folding. Numerous hsp70 monomers bind to short hydrophobic segments in unfolded polypeptides, thereby preventing molten globule formation. Each type of hsp70 possesses two binding sites, one for an unfolded protein segment and another for ATP. Release of a polypeptide from an hsp70 involves ATP hydrolysis. Mitochondrial and ER-localized hsp70s are required for transmembrane translocation of some polypeptides.

2. **Hsp60s**. Once an unfolded polypeptide has been released by hsp70, it is passed on to a member of a family of molecular chaperones referred to as the **hsp60s** (also called the **chaparonins** or *Cpn 60s*), which mediate protein folding. The hsp60s form a large structure composed of two stacked seven-subunit rings that unfolded proteins enter (Figure 19.22). In an ATP-requiring process, hsp60 then facilitates the transformation of an unfolded molecule into a properly folded one. Many of the details of this process remain unresolved.

In addition to promoting the folding of nascent protein, molecular chaperones direct the refolding of protein partially unfolded as a consequence of stressful conditions. If refolding is not possible, molecular chaperones promote protein degradation. A diagrammatic view of protein folding is presented in Figure 19.23.

KEY CONCEPTS 19.7

All of the information required for each newly synthesized polypeptide to fold into its biologically active conformation is encoded in the molecule's primary sequence. Some relatively simple polypeptides fold spontaneously into their native conformations. Other larger molecules require the assistance of proteins called molecular chaperones to ensure correct folding.

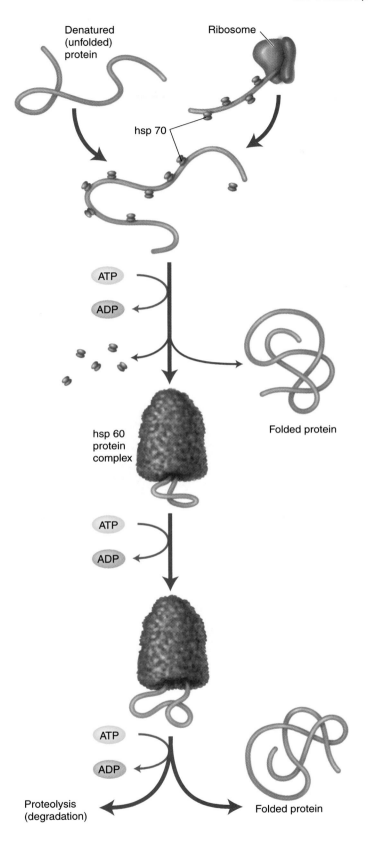

Denatured (unfolded) protein

Ribosome

hsp 70

ATP

ADP

Folded protein

hsp 60 protein complex

ATP

ADP

ATP

ADP

Proteolysis (degradation)

Folded protein

FIGURE 19.23

The Molecular Chaperones.

Molecular chaperones bind transiently to both nascent proteins and unfolded proteins (i.e., denatured by stressful conditions). The members of the hsp70 family stabilize nascent proteins and reactivate some denatured proteins. Many proteins also require hsp60 proteins to achieve their final conformations. If a protein cannot be salvaged, the molecular chaperones help destroy it.

Proteomics is a technology currently being developed to investigate the proteome, the functional output of the genome. The wealth of genetic data that are now available has provided primary sequence information for thousands of proteins. The identity and function of many of these molecules, however, are unknown. In addition, the structural alterations of many proteins cannot be predicted solely from nucleic acid sequences, either because of alternative mRNA splicing or the chemical modifications that occur after translation. As the sequencing efforts for many genome projects are coming to an end, the attention of biochemists and molecular biologists is shifting to the analysis of the protein products. The goals of proteomics are primarily twofold: to study the global changes in the expression of cellular proteins, and to determine the identity and functions of all the proteins in the proteomes of organisms. The potential applications for proteomic research are many and varied. In addition to providing opportunities to resolve basic biological problems (e.g., ascertaining the precise mechanisms by which cellular processes such as neuronal transport or mRNA splicing occur), proteomics-based technology also has obvious uses in biomedical research. Examples of the latter include investigations of the causes and diagnosis of genetic and infectious diseases, and the development of drugs.

Proteomic Tools

The investigation of proteomes is currently focused primarily on developing accurate, but relatively fast methods for identifying and characterizing proteins. Among the oldest technologies used in proteomics are two-dimensional gel electrophoresis and mass spectrometry. The use of each is discussed briefly.

Two-Dimensional Gel Electrophoresis. Protein expression is currently analyzed with two-dimensional (2-D) gels. Proteins separate according to charge (within a pH gradient) in the first dimension of the gel, and on the basis of molecular mass in the second (Figure 19B). As many as 3000 individual proteins can be visualized on a 2-D gel. Protein analysis of multiple gels (e.g., determinations of presence, absence, or relative concentration of specific proteins in healthy and diseased cells) is accomplished by comparisons of 2-D gel images with proteome databases with the aid of specialized computer software. Although 2-D gel technology has been improved in both speed and capacity, it also has limitations. In addition to being labor intensive, 2-D gels are not useful in the evaluation of certain types of proteins such as membrane proteins or proteins found in very low concentrations. Newer technologies are being developed to overcome these problems.

Mass Spectrometry. Mass spectrometry (MS) is a technique in which molecules are vaporized and then bombarded by a high-energy electron beam causing them to fragment as cations. As the ionized fragments enter the spectrometer they pass through a strong magnetic field that separates them according to their mass-to-charge (m/z) ratio. Each type of molecule is identified by the pattern of fragments that is generated, each pattern or "fingerprint" being unique. Because proteins do not vaporize, they are instead digested and then dissolved in a volatile solvent and sprayed into the vacuum chamber of the mass spectrometer. The electron beam ionizes these peptide fragments and the positively charged peptides are passed through the magnetic field. The peptide mass fingerprint that results is then compared to fragmentation information in protein databases. Although MS is highly accurate and automated, it is usually insufficient for identifying all the proteins in a sample. To improve protein identification, tandem MS (MS/MS), a method in which two mass spectrometers are linked in tandem, has been developed. In this technique the oligopeptide fragments produced in the first MS are then transferred to the second MS, where they are further fragmented and analyzed. MS/MS is used to rapidly sequence proteins.

Despite the recent advances in proteomic research techniques, several problems are a serious barrier to accomplishing the enormous task of characterizing entire proteomes. Among the most important of these is the continuing need for vastly improved efficiency and the lack of a technique equivalent to PCR for amplification of proteins found in very small amounts.

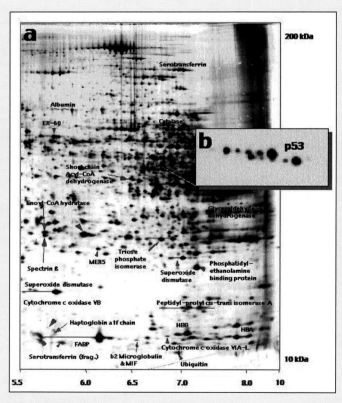

FIGURE 19B

Two-Dimensional Gel Pattern

After the extract of a liver sample was added to the gel, it was run first in a pH gradient (isoelectric points) and then in SDS-PAGE to separate the proteins according to molecular mass. The insert shows an enlargement of multiple versions of p53, a protooncogene product, that was revealed by treatment with antibodies.

SUMMARY

1. Protein synthesis is a complex process in which information encoded in nucleic acids is translated into the primary sequence of proteins. During the translation phase of protein synthesis, the incorporation of each amino acid is specified by one or more triplet nucleotide base sequences, referred to as codons. The genetic code consists of 64 codons: 61 codons that specify the amino acids and three stop codons. Translation also involves the tRNAs, a set of molecules that act as carriers of the amino acids. The base-pairing interactions between codons and the anticodon base sequence of tRNAs result in the ac- curate translation of genetic messages. Translation consists of three phases: initiation, elongation, and termination. Each phase requires several types of protein factors. Although prokaryotic and eukaryotic translational mechanisms bear a striking resemblance to each other, they differ in several respects. One of the most notable differences is the identity and function of the translation factors.

2. Protein synthesis also involves a set of posttranslational modifications that prepare the molecule for its functional role, assist in folding, or target it to a specific destination. These covalent alterations include proteolytic processing, modification of certain amino acid side chains, and insertion of cofactors.

3. Prokaryotes and eukaryotes differ in their usage of translational control mechanisms. In addition to variations in Shine-Dalgarno sequences, prokaryotes also use negative translational control, that is, the repression of the translation of a polycistronic mRNA by one of its products. In contrast, a wide variety of eukaryotic translational controls have been observed. These mechanisms range from global controls in which the translation rate of a large number of mRNAs is altered to specific controls in which the translation of a specific mRNA or small group of mRNAs is altered.

4. One of the most important aspects of protein synthesis is the folding of polypeptides into their biologically active conformations. Despite decades of investigation into the physical and chemical properties of polypeptide chains, the mechanism by which a primary sequence dictates the molecule's final conformation is unresolved: It has become increasingly clear that many proteins require molecular chaperones to fold into their final three-dimensional conformations. Protein misfolding is now known to be an important feature of several human diseases, including Alzheimer's disease and Creutzfeld-Jacob disease.

5. Proteomics is a technology that is used to investigate the proteome, the complete set of proteins produced from an organism's genome. The goals of proteomics are to study the global changes in the expression of cellular proteins over time and to determine the identity and the functions of all the proteins produced by organisms.

SUGGESTED READINGS

Arnez, J. G., and Moras, D., Structural and Functional Considerations of the Aminoacylation Reaction, *Trends. Biochem. Sci.*, 22(6):211–216, 1997.

Blackstock, W. P., and Weir, M. P., Proteomics: Quantitative and Physical Mapping of Cellular Proteins, *Trends Biotech.*, 17:121–127, 1999.

Chambers, G., Lawrie, L., Cash, P., and Murray, G. I., Proteomics: A New Approach to the Study of Disease, *J. Path.*, 192(3):280–288, 2000.

Craig, E. A., Chaperones: Helpers Along the Pathways to Protein Folding, *Science*, 260:1902–1903, 1993.

Ezzell, C., Proteins Rule, *Sci.Amer.* 286(2):40–47, 2002.

Ferreira, S. T., and De Felice, F. G., Protein Dynamics, Folding and Misfolding: From Basic Physical Chemistry to Human Conformational Diseases, *FEBS Lett.*, 498:129–134, 2001.

Noller, H. F., Hoffarth, V., and Zimniak, L., Unusual Resistance of Peptidyl Transferase to Protein Extraction Procedures, *Science*, 256:1416–1419, 1992.

Paulus, H., Protein Splicing and Related Forms of Protein Autoprocessing, *Annu. Rev. Biochem.*, 69:447–496, 2000.

Radford, S. E., Protein Folding: Progress Made and Promises Ahead, *Trends Biochem. Sci.*, 25:611–618, 2000.

Rothman, J. E., and Wieland, F. T., Protein Sorting by Transport Vesicles, *Science*, 272:227–234, 1996.

Schatz, G., and Dobberstein, B., Common Principles of Protein Translocation across Membranes, *Science*, 271:1519–1526, 1996.

Stansfield, I., Jones, K. M., and Tuite, M. F., The End in Site: Terminating Translation in Eukaryotes, *Trends Biochem. Sci.*, 20(12):489–491, 1995.

Weissman, J. S., All Roads Lead to Rome? The Multiple Pathways of Protein Folding, *Chem. Biol.*, 2:255–260, 1995.

Welch, W. J., How Cells Respond to Stress, *Sci. Amer.*, 268(5):56–64, 1993.

KEY WORDS

REVIEW QUESTIONS

1. List and describe four properties of the genetic code.

2. What two observations prompted the wobble hypothesis?

3. Describe the two sequential reactions that occur in the active site of aminoacyl-tRNA synthetases.

4. What are the major differences between eukaryotic and prokaryotic translation?

5. What are the major differences between eukaryotic and prokaryotic translation control mechanisms?

6. What are the three steps in the elongation cycle?

7. Describe how kinetic proofreading takes place.

8. Clearly define the following terms:
   a. targeting
   b. scanning
   c. codon
   d. reading frame
   e. molecular chaperones
   f. disulfide exchange
   g. proofreading site
   h. signal peptide
   i. glycosylation
   j. negative translational regulation

9. Describe the structure and function of the signal recognition particle.

10. Describe the function of the translocon in cotranslational transfer.

11. Describe how eukaryotic mRNA structure can affect translational control.

12. In general terms, describe the intracellular processing of a typical glycoprotein that is destined for secretion from a cell.

13. Describe the problems associated with determining a polypeptide's finale three-dimensional shape using its primary structure as a guide.

14. Describe the roles of the most prominent molecular chaperones in protein folding.

15. Define the following terms:
    a. proteomics
    b. preproprotein
    c. protein splicing
    d. site-directed mutagenesis
    e. prion
    f. nascent
    g. motor protein
    h. molten globule

16. Describe the process of protein splicing.

17. Why are tRNAs described as adaptor molecules?

18. What steps in the elongation cycle of protein synthesis require GTP hydrolysis? What role does it play in each step?

THOUGHT QUESTIONS

1. The three-dimensional structures of ribosomal RNA and ribosomal protein are remarkably similar among species. Suggest reasons for these similarities.

2. Explain the significance of the following statement: The functioning of the aminoacyl-tRNA synthetases is referred to as the second genetic code.

3. Although aminoacyl-tRNA synthetases make few errors, occasionally an error does occur. How can these errors be detected and corrected?

4. What are the three phases of protein synthesis? Describe the principal events in each phase. What specific roles do translation factors play in both prokaryotic and eukaryotic translation processes?

5. Determine the codon sequence for the peptide sequence glycylserylcysteinylarginylalanine. How many possibilities are there?

6. Indicate the phase of protein synthesis during which each of the following processes occurs:
   a. A ribosomal subunit binds to a messenger RNA.
   b. The polypeptide is actually synthesized.
   c. The ribosome moves along the codon sequence.
   d. The ribosome dissociates into its subunits.

7. Estimate the minimum number of ATP and GTP molecules required to polymerize 200 amino acids.

8. Discuss the role of GTP in the functioning of translation factors.

9. Posttranslational modifications serve several purposes. Discuss and give examples.

10. Describe how the base pairing between the Shine-Dalgarno sequence and the 30S subunit provides a mechanism for distinguishing a start codon from a methionine codon. What is the eukaryotic version of this mechanism?

11. Given an amino acid sequence for a polypeptide, can the base sequence for the mRNA that codes for it be predicted?

12. Because of the structural similarity between isoleucine and valine, the aminoacyl-tRNA synthetases that link them to their respective tRNAs possess proofreading sites. Examine the structures of the other α-amino acids and determine other sets of amino acids whose structural similarities might also require proofreading.

13. What advantages are there for synthesizing an inactive protein that must subsequently be activated by posttranslational modifications?

14. What factors ensure accuracy in protein synthesis? How does the level of accuracy usually attained in protein synthesis compare with that of replication or transcription?

15. Can you suggest a reason why ribosomes in all living organisms consist of two subunits and not one supramolecular complex?

16. Describe the probable mechanism by which sickle cell anemia originated. What is the specific failure in information transfer that occurred?

Solutions

CHAPTER 1

End-of-Chapter Questions

Review Questions

1. Among the many insights that have been discovered through biochemical research are that life is complex and dynamic, highly organized, self-sustaining, and information-based. Life adapts and evolves.

3. Eukaryotes are larger and considerably more complicated than prokaryotes. All multicellular organisms are eukaryotic.

5. Amino acids occur in peptides and proteins. Sugars occur in oligosaccharides and polysaccharides. Nucleotides are the components of the nucleic acids. Fatty acids are components of several types of lipid molecules, for example, triacyglycerols and phospholipids.

7. a. The functions of fatty acids include energy storage and membrane components.
   b. Sugars are energy sources and structural components.
   c. Nucleotides are involved in energy transformations. They are also components of DNA and RNA.

9. Cells use oxidation-reduction reactions to interconvert energy forms. Energy is captured as electrons are transferred from reduced molecules to more oxidized ones.

11. Saturated hydrocarbons contain only carbon-carbon single bonds, whereas unsaturated compounds contain carbon-carbon double or triple bonds.

13. Each molecule belongs to the following class:
    a. Amino acid
    b. Sugar
    c. Fatty acid
    d. Nucleotide

15. Organelles are specialized subcellular structures found in eukaryotes. They permit the concentration of reactants and products at sites where they can be efficiently used.

17. Examples of the following reactions include:
    a. Nucleophilic substitution—the reaction of glucose with ATP to produce glucose-6-phosphate and ADP
    b. Elimination—the dehydration of 2-phosphoglycerate to form phosphoenolpyruvate
    c. Oxidation-reduction—the conversion of ethyl alcohol to acetaldehyde
    d. Addition —the conversion of fumarate to malate

19. The common types of chemical reactions found in living cells are nucleophilic substitution, elimination reactions, addition reactions, isomerization reactions, and oxidation-reduction reactions.

21. In addition to being an important energy source, carbohydrates are important structural molecules in organisms and have a role in intracellular and intercellular communication.

23. Nucleotides participate in energy-forming and energy-generating reactions. Much of the energy available to drive biochemical reactions is stored in ATP molecules.

25. Examples of waste products produced by animals include carbon dioxide, ammonia, urea, and water .

27. The nucleotide base sequence of each type of mRNA molecule codes for the amino acid sequence of a specific polypeptide. Each tRNA molecule carries a specific amino acid which it subsequently delivers to the ribosome for incorporation into a polypeptide during protein synthesis. Ribosomal RNA molecules contribute to the structural and functional properties of ribosomes. Each polypeptide is manufactured as the base sequence information in the mRNA is translated by a ribosome. As base pairing occurs between the codon sequence of mRNA and the anticodon sequence of tRNA molecules, the amino acids are brought into close proximity and a peptide bond is formed.

Thought Questions

1. Although biochemical reactions and organic reactions must conform to the same physical laws, the precision with which biomolecules are transformed by living organisms far exceeds the capabilities of organic chemists. In addition, the integration of thousands of biochemical reactions within the living cell is extraordinarily complex.

3. Prokaryotes are single-celled organisms that are smaller and less complicated than the eukaryotes and have short life cycles. Biochemists make the useful assumption that the basic elements of living processes in ,the two types of organisms are similar. Finally, some prokaryotes are easier to obtain, manipulate, and investigate than are multicellular eukaryotes.

5. The C—H bonds of fatty acids are the most reduced form of carbon found in organic molecules. Oxidation of these molecules to form carbon dioxide—the most oxidized form of carbon—has the highest energy yield.

7. The new molecule forms three hydrogen bonds with guanine:

2-Amino-6-methoxypurine   **Guanine (G)**

9. Insulin produced by biotechnology is human insulin, in contrast to older forms that were isolated from the pancreases of cattle and pigs. Human insulin produces significantly fewer antigenic reactions and is, therefore, safer to use. In addition it is less expensive, after the original research and development costs are recouped, to use genetically altered microorganisms to produce the insulin.

CHAPTER 2

In-Chapter Questions

2.1 The volume of a prokaryotic cell is calculated as follows:

$$\pi r^2 h = 3.14 \times (0.5 \ \mu m)^2 \times 2 \ \mu m = 1.57 \ \mu m^3$$

The volume of a eukaryotic cell is calculated as follows:

$$4/3 \ \pi r^3 = 4 \times (3.14 \times 10^3)/3 = 4200 \ \mu m^3$$

By dividing the volume of the hepatocyte by the volume of the prokaryotic cell (4200 μm³/1.57 μm³) the number of prokaryotic cells that would fit within the heptocye is obtained: 2700

2.2 Without a means of disposal, the lipid molecules will accumulate in the cells. Cell function is eventually compromised and the cells die.

2.3 The cyanobacterium obtains a stable environment and a consistent supply of nutrients. The eukaryotic organism is assured a consistent supply of energy.

2.4 Refer to Figures 2.5 and 2.12

End-of-Chapter Questions
Review Questions

1. The cell is the basic unit of life that is separated from its environment by a plasma membrane.
3. Refer to Figure 2.5. The functions of the components of prokaryotic cells are:
   a. Nucleoid contains the bacterial chromosome.
   b. Plasmid is the site of extrachromosomal DNA.
   c. Cell wall provides protection and support.
   d. Pili allow attachment to other cells.
   e. Flagella allows locomotion.
5. a. Nucleus—eukaryotes
   b. Plasma membrane—eukaryotes and prokaryotes
   c. Endoplasmic reticulum—eukaryotes
   d. Mitochondria—eukaryotes
   e. Nucleolus—eukaryotes
7. Lysosomes digest all types of biomolecules. In addition to the normal processing of cellular molecules, lysosomes also destroy the components of foreign cells and other exogenous extracellular materials.
9. The evidence that supports the endosymbiotic hypothesis includes the following:
   a. Symbiosis has been observed between modern prokaryotes and eukaryotes.
   b. Mitochondria and chloroplasts are about the same size as prokaryotes.
   c. The ability of mitochondria and chloroplasts to synthesize DNA and proteins is similar to that of prokaryotes.
   d. Prokaryotes, mitochondria, and chloroplasts all reproduce by binary fission.
   e. The ribosomes of mitochondria and chloroplasts are similar in size and function to those of prokaryotes.
   f. Traces of RNA found in other eukaryotic cellular structures suggest that these also arose by symbiotic fusion.
11. In multicellular organisms such as animals, attachment of cells is impeded by a cell wall. For some eukaryotic cells, for example, macrophages, dramatic shape changes required for function would be impossible.
13. Among the roles of plasma membrane proteins are transport, response to stimuli, cell-cell contact, and catalytic functions.
15. The Golgi apparatus processes, sorts, and packages protein and lipid molecules for distribution to other regions of the cell or for export.

Thought Questions

1. The thick mucoid coat prevents antibodies from binding to surface cellular structures used by the immune system for recognition, thereby interfering with the immune response.

3. The DNA of eukaryotes is contained within the nucleus. The presence of DNA in mitochondria and chloroplasts argues strongly for their extracellular origin.
5. The presence of DNA or possibly RNA in the organelle would strongly suggest that it may once have been free living.
7. The volume of a ribosome is calculated as follows:

$$\pi r^2 h = 3.14 \times (0.007\ \mu m)^2 \times 0.02\ \mu m = 3.08 \times 10^{-6}\ \mu m^3$$

The volume of a bacterial cell (from question 2.1) is 1.57 μm³. The number of ribosomes that can fit in a bacterial cell is 1.57/3 × 10⁻⁶ = 5 × 10⁵ but because they occupy only 20% of the cell's volume divide by 5 to give 1 × 10⁵ ribosomes per bacterial cell.

CHAPTER 3

In-Chapter Questions

3.1 Ammonia "ice" would be expected to be less dense than liquid ammonia. Refer to Figure 3.8 for an analogous structure of water.
3.2 From left to right in the illustration, the noncovalent interactions are ionic, hydrogen bonding, and vander waals interactions.
3.3 The equation for osmotic pressure M is:
$\pi = iMRT$ where $\pi = 2.06 \times 10^{-3}$ atm
$$i = 1$$
$$R = 0.0821\ \text{L atm/mol K}$$
$$T = 298\ \text{K}$$
Substitute these values into the equation and solve for M.

2.06×10^{-3} atm =
$(1)M\,(0.0821\ \text{L atm/mol K})(298\ \text{K})$

$M = 2.06 \times 10^{-3}$atm/(0.0821 L atm/mol K)(298 K)
$= 2.06 \times 10^{-3}$ atm/24.4658 L atm/mol
$= 0.000084199$ mol/L

1.5 g/L $= 8.4199 \times 10^{-5}$ mol/L
1 mole $= 1.5$g/L/8.4199×10^{-5} mol/L $=$
$17816.84286 = 1.8 \times 10^4$ g/mol

3.4 The equilibrium shifts to the right to replace lost bicarbonate and the acid concentration increases. The resulting condition is called acidosis.

End-of-Chapter Questions
Review Questions

1. Both c and d are acid–conjugate base pairs.
3. The effective buffer range is between 7 and 8.
5. Molecules b and d can form hydrogen bonds with like molecules. Molecules a, b, and d can all form hydrogen bonds with water.
7. In a solution of 1 M sodium lactate, water flows into the dialysis bag. In solutions of 3 M or 4.5 M sodium lactate, water flows out of the dialysis bag.
9. $\pi = iMRT$ where $\pi = 0.01$ atm
$$i = 1$$
$$R = 0.0821\ \text{L atm/mol K}$$
$$T = 298\ \text{K}$$

Solving for M:
0.01 atm = (1)(0.0821 L atm/mol K)(298 K)(M)
M = 4.08×10^{-4} mol/L

Solving for the molecular weight of the protein:
0.056 g/0.030 L = 1.867 g/L

1.867 g = 4.08×10^{-4} mol
1 mol of the protein = 4575.98 g = 4600 g

11. a. Hydrogen bonds are electrostatic interactions between hydrogen covalently bonded to oxygen, nitrogen, or sulfur, and nearby oxygen, nitrogen, or sulfur atoms.
   b. pH = $-\log$ [H$^+$]
   c. A buffer is a mixture of a weak acid and its salt that resists changes in pH.
   d. Osmotic pressure is the pressure needed to stop the net flow of water across a membrane.
   e. Osmolytes are osmotically active substances that cells produce to restore osmotic balance.
   f. Isotonic refers to two solutions with the same osmotic pressure.
   g. Amphipathic molecules contain polar and nonpolar groups.
   h. Hydrophobic interactions are interactions between nonpolar groups.
   i. Dipoles have a net charge separation within a molecule.
   j. Temporary charge separation in a molecule that is produced by a nearby dipole is called an induced dipole.

13. Molecule d is capable of forming micelles because one end of the molecule is polar and the other end is nonpolar when very close together.

15. The buffering capacity of a system is increased by raising the concentrations of the buffer components but not changing their ratio.

17. A buffer is composed of a weak acid and its salt. Only c is a buffer.

19. The relationship between osmolarity and molarity is given by the equation $o = iM$ where o is osmolarity, i is the extent of ionization, and M is the molarity.

21. $K_a = 6.3 \times 10^{-8}$, therefore, $pK_a = 7.2$
   pH = pK_a + log [salt]/[acid]
   7.4 = 7.2 + log [salt]/[acid]
   log [salt]/[acid] = 7.4 − 7.2 = 0.2
   [salt]/[acid] = 1.58 : 1 or 1.6 : 1

23. The contribution from the ionization of water must be considered. The hydrogen ion concentration is 10^{-8} M from acid and 10^{-7} M from water for a total acid concentration of 1.1×10^{-7} M acid. The pH is therefore equal to $-\log 1.1 \times 10^{-7} = 6.96$.

Thought Questions

1. The highly concentrated sugar solution pulls water out of any bacterial cells present which kills them, thereby preserving the fruit.

3. The salts dissolved in the seawater pull water out of the plants. This is the reverse of the normal flow of water from the environment into the plant. Under these conditions the plants will die.

5. The pH scale is derived using the ionization constant of water. To establish the pH scale for another solvent, the ionization constant of that solvent would have to be used and the pH scale would be different from the pH scale for water.

7. The extreme electronegativity of the oxygen polarizes the O—H bond of water and makes the hydrogen electron deficient. Because the unshared pairs of electrons on the oxygen are available for bonding, an electrostatic interaction occurs.

9. The small water molecules can crowd closely around the ions and effectively disperse the charge thereby facilitating solution. The bulky R group of the alcohol prevents this close interac-

tion of solvent and solute. As a result the ionic compound does not dissolve as easily.

11. Hydration tends to make ionization easier. The hydrated acid group on the protein surface would have a higher K_a than one in the anhydrous interior of the protein.

CHAPTER 4

In-Chapter Questions

4.1 $\Delta G' = \Delta G^{o'} + RT$ in [ADP][P$_i$]/[ATP]
   where $R = 8.315 \times 10^{-3}$ kJ/mol $\cdot$ K
   $T = 310$ K
   [ADP] = 0.00135 M, [ATP] = 0.004 M,
   [P$_i$] = 0.00465 M
   $\Delta G^{o'} = -30.5$ kJ/mol

   $\Delta G' = -30.5$ kJ/mol + (8.315 J/mol $\cdot$ K)(310)
       In(0.00135 M)(0.00465 M)/(0.004 M)
   $\Delta G' = -30.5 + 2.577$ (ln 0.00157)
       $= -30.5 - 16.64$
       $= -47.14$ kJ/mol $= -47.1$ kJ/mol

4.2 Amount of ATP required to walk a mile
   = (100 kcal/mi)/7.3 kcal/mol
   = 13.7 mol/mi $\times$ 507 g/mol = 6945.2 g/mi = 6950 g/mi

   Amount of glucose required to produce 100 kcal through ATP
   = (100 kcal)/(.04)(686 kcal/mol)
   = 100 kcal/274.4 kcal/mol
   = 0.36 mol
   = 0.36 mol $\times$ 180 g/mole = 65.6 g of glucose

End-of-Chapter Questions

Review Questions

1. a. Thermodynamics is the study of the heat and energy transformations in a chemical reaction.
   b. Chemical reactions that absorb energy (i.e., have a negative free energy charge) are endergonic.
   c. Enthalpy is a measure of the heat evolved during a reaction.
   d. Free energy is a measure of the tendency of a reaction to occur.
   e. A high-energy bond is a bond that liberates large amounts of free energy when it is broken.
   f. Redox reactions are reactions that involve changes in the oxidation number of the reactants.
   g. A chemolithotroph is an organism that derives chemical energy from minerals.
   h. Phosphate group transfer potential is the tendency of a phosphate bond to undergo hydrolysis; the $\Delta G^{o'}$ of hydrolysis of phosphorylated compounds.

3. For a reaction to proceed to completion the total overall $\Delta G^{o'}$ must be negative and there must be a common intermediate, in this case P$_i$. This is true of a and b.

5. ATP + glutamate + NH$_3$ $\longrightarrow$ ADP + P$_i$ + glutamine
   ATP + H$_2$O $\longrightarrow$ ADP + P$_i$ $\Delta G^{o'} = -30.5$ kJ/mol
   Glutamine + H$_2$O $\longrightarrow$ glutamate + NH$_3$
       $\Delta G^{o'} = -14.2$ kJ/mol
   Reverse the second equation and add the $\Delta G^{o'}$ values.
   ATP + H$_2$O $\longrightarrow$ ADP + P$_i$ $\Delta G^{o'} = -30.5$ kJ/mol
   Glutamate + NH$_3$ $\longrightarrow$ glutamine + H$_2$O
       $\Delta G^{o'} = +14.2$ kJ/mol

ATP + glutamate + $NH_3 \longrightarrow$ ADP + P_i + glutamine
$$\Delta G^{o'} = -16.3 \text{ kJ/mol}$$

7. Under standard conditions the following statements are true: a, e, and f.

9. $\Delta G^{o'} = -RT \ln K_{eq}$
   $-7100 \text{ j/mol} = -(8.315 \text{ j/mol} \cdot \text{K})(298\text{K})(\ln K_{eq})$
   $\ln K_{eq} = 2.865$
   $K_{eq} = 17.56$

11. The following statements are true: a, b, c, and f.

Thought Questions

1. The energy liberated by the hydrolysis of 12.5 mol of ATP is

$$12.5 \text{ mol } (-30.5 \text{ kJ/mol}) = -381.3 \text{ kJ}$$

The energy required to produce 12.5 mol of ATP is 1142.2 kJ. The apparent efficiency of the process is

$$(381.3/1142.2) \times 100 = 33.4\%$$

3. Although only a few molecules are involved, the laws of thermodynamics are still obeyed.

5. ΔG is the most useful criterion of spontaneity because it reflects the change in entropy which must increase for a reaction to be spontaneous.

7. It is necessary to know the following information to determine the $\Delta G^{o'}$ of a reaction: temperature, concentrations of reactants and products, and $\Delta G^{o'}$.

9. Phosphoenolpyruvate has the greatest phosphate group transfer potential. On hydrolysis, the enol which has a resonance restricted structure, rapidly converts to the more stable, resonance hydridized keto form and drives the equilibrium to the right.

11. The ease of hydrolysis of ATP is partially explained by the electronic repulsions between the phosphate groups. The positive magnesium ion coordinates with these negatives charges, thus reducing their magnitude and therefore the repulsion between them.

CHAPTER 5

In-Chapter Questions

5.1 Amino acids a and b are neutral, nonpolar, c is basic, and d is an acidic amino acid.

5.2 Bacteria with surface polypeptides composed of D-amino acids are resistant to degradation because the proteases, the enzymes that immune system cells use to degrade protein in foreign cells, can only catalyze the hydrolysis of peptide bonds between L-amino acids. In other words, the active sites of proteases are stereospecific, i.e., they can only effectively bind peptides composed of L-amino acids.

5.3 The number of possible tetrapeptides is $20^4 = 160,000$.

5.4 The structure of the penicillamine-cysteine disulfide is

5.5 The complete structure of oxytocin is

At pH 4 the terminal amino group of the glycine would be protonated to give the molecule a +1 charge. The isoelectric point of oxytocin is 5.6. Therefore at pH9 the molecule will have a net negative charge.

5.6 The trait is recessive, and two copies of the aberrant gene are required for full expression of the disease. Primaquine induces; the production of excess amounts of the strong oxidizing agent hydrogen peroxide. In the absence of sufficient amounts of the reducing agent NADPH, the peroxide molecules cause extensive damage to the cell. No, a higher than normal peroxide level in blood cells is damaging to the malarial parasite and is selected for in geographical regions where malaria occurs.

5.7

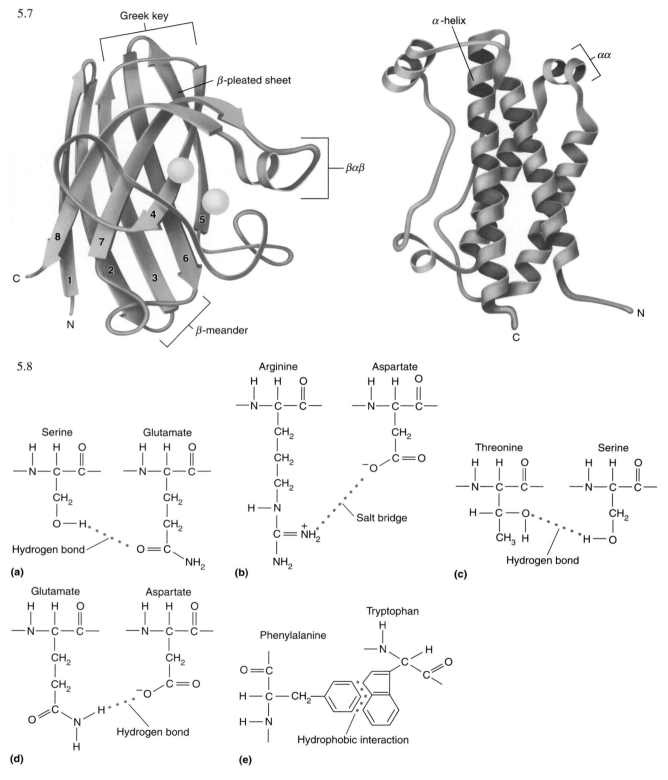

5.8

(a) Serine / Glutamate — Hydrogen bond

(b) Arginine / Aspartate — Salt bridge

(c) Threonine / Serine — Hydrogen bond

(d) Glutamate / Aspartate — Hydrogen bond

(e) Phenylalanine / Tryptophan — Hydrophobic interaction

5.9 Collagen is a major structural protein found in connective tissues. Consequently, the failure of collagen molecules to form properly weakens these tissues causing diverse symptoms. Examples include cataracts, easily deformed bones, torn tendons and ligaments, and ruptured blood vessels.

5.10 BPG stabilizes deoxyhemoglobin. In the absence of BPG, oxyhemoglobin forms more easily. Fetal hemoglobin binds BPG poorly and, therefore, has a greater affinity for oxygen.

5.11 Myoglobin, composed of a single polypeptide, binds oxygen in a simple pattern—it binds the molecule tightly and releases it only when the cells' oxygen concentration is very low. The binding of oxygen by hemoglobin, a tetramer, has a more complicated sigmoidal pattern that is made possible by the noncovalent interactions among its four subunits.

End-of-Chapter Questions
Review Questions

1. A polypeptide is a polymer containing more than 50 amino acid residues. A protein is composed of one or more polypeptide chains. A peptide is a polymer containing less than 50 amino acid residues.

3. The structure and net charge of arginine at various pH levels is as follows:

pH	Structure	Net Charge

1 $H_2N-C-N-CH_2-CH_2-CH_2-C-COOH$ + 2
   $\|$ H H
   $^+NH_2$ $^+NH_3$

4 $H_2N-C-N-CH_2CH_2CH_2-CHCOO^-$ + 1
   $^+NH_2$ H $^+NH_3$

7 $H_2N-C-NH-CH_2CH_2CH_2 CH-COO^-$ + 1
   $^+NH_2$ $^+NH_3$

10 $H_2N-C-NH-CH_2CH_2CH_2-CHCOO^-$ 0
   $^+NH_2$ NH_2

12 $H_2N-C-NHCH_2CH_2CH_2-CH-COO^-$ 0
   $^+NH_2$ NH_2

5. The name of the molecule is cysteinylglycyltyrosine. Its abbreviated structure is H_2N-Cys-Gly-Tyr-COOH.

7. Six examples of the major functions of protein in the body include catalysis, structure, movement, defense, regulation, and transport.

9. a. The carbon next to the carboxyl group in an amino acid is the α-carbon.
   b. The isoelectric point is the pH at which an amino acid is electrically neutral.
   c. A peptide bond is an amide bond between two amino acids.
   d. A hydrophobic amino acid is an amino acid with a nonpolar side group.

11. a. Polyproline—left-handed helix
    b. Polyglycine—β-pleated sheet
    c. Ala-Val-Ala-Val-Ala-Val—α-helix
    d. Gly-Ser-Gly-Ala-Gly-Ala—β-pleated sheet

13. a. Heat—hydrogen bonding (secondary and tertiary structure)
    b. Strong acid—hydrogen bonding (secondary and tertiary structure) and salt bridges (secondary and tertiary structure)
    c. Saturated salt solution—salt bridges (tertiary structure)
    d. Organic solvents—hydrophobic interactions (tertiary structure)

15. Refer to Biochemical Methods 5.1 (p.152) for protein purification techniques.

17. Refer to pp. 152–153 for a discussion of chromatographic separation techniques.

19. With overlapping fragments the segments can be fitted together because fragments that fit together have common sequences at their ends. If the segments are not overlapping, the order of the cannot be determined.

21. a. The isoelectric point is calculated by taking the average of the pK_a values for the amino group of glycine (9.6) and the carboxyl group of valine (2.32). The answer is pl = 5.96.
    b. At pH 1, the tripeptide is positively charged and will move to the negative electrode. At pH 5 the tripeptide has a net zero charge and will not migrate. At pH 10 and 12, the tripeptide has a −1 charge and moves toward the positive electrode.

Thought Questions

1. Hydrophobic amino acids such as valine, leucine, isoleucine, methionine and phenylalanine usually occur within the anhydrous core of proteins because of the hydrophobic effect. Hydrophilic amino acids such as arginine, lysine, aspartic acid, and glutamic acid occur most often on or near the surface of proteins where they interact with water molecules. Glycine and alanine are hydrophobic amino acids and so tend to occur in the interior of proteins. Glutamine has a polar side chain that can form hydrogen bonds and, therefore, is often on the surface of proteins.

3. The large size of enzymes is required to stabilize the shape and functional properties of the active site and shield it from extraneous molecules. In addition, structural features of the protein may function in recognition processes in signaling or binding to cellular structures.

5. The amide bond is stronger than the ester bond for two reasons. The N is closer to the size of C than the O is, which makes for greater covalency in the bond. Also, because the O and N differ in electronegativity, there is resonance hybridization in the amide bond. The amide bond, therefore, has partial double bond character.

7. The peptide bonds on the surface of the chymotrypsin are not readily cleaved by the enzyme.

CHAPTER 6

In-Chapter Questions

6.1 The amino acid residues forming the three-dimensional structure of the active site are chiral. As a result the active site is chiral. It can bind only one isomeric form of a hexose sugar, in this case the D-isomer.

6.2 a. Isomerase
    b. Transferase
    c. Lyase
    d. Oxidoreductase
    e. Ligase
    f. Hydrolase

6.3 The products of the degradation are the following compounds:

CH_3OH $H_2N-CH-C-OH$ $H_2N-C-C-OH$
           $\|$ $\|$
           O O
           CH_2 CH_2
                                 $C-OH$
                                 $\|$
                                 O

Methanol   **Phenylalanine**   **Aspartic acid**

Cleavage of the ester bond is catalyzed by an esterase; the amide bond is cleaved by a peptidase.

6.4

Enzyme — SH + I — CH$_2$ — $\overset{\overset{\displaystyle O}{\displaystyle \|}}{C}$ — NH$_2$ $\longrightarrow$ Enzyme — S — CH$_2$ — $\overset{\overset{\displaystyle O}{\displaystyle \|}}{C}$ — NH$_2$

6.5 Dialysis removes the formaldehyde, formic acid, and methanol that build up in the bloodstream. The bicarbonate neutralizes the acid produced and helps offset the resultant acidosis. The ethanol competitively binds with the alcohol dehydrogenase. This slows the dehydrogenation of the methanol and allows time for the kidneys to excrete it.

6.6 The acidic amino acids aspartate and glutamate and the basic amino acids lysine, arginine, and histidine can function as general acids or general bases, respectively. In addition, amino acids with OH or SH groups can function as weak acids. However, the only amino acid with a pK_a in the neutral range of physiological systems is histidine. In the presence of water, this is the only amino acid that can participate in proton transfer. In water deficient active sites and pockets of proteins, conditions may shift the pK_a of other amino acids (such as glutamate) closer to a pH at which proton transfer is possible.

6.7 a. Menkes' syndrome—injections of copper salts into the blood would avoid intestinal malabsorption and provide the copper necessary to form adequate levels of ceruloplasmin and offset the symptoms of the disease.

b. Wilson's disease—zinc induces the synthesis of metallothioein, which has a high affinity for copper. Some organ damage can be averted because metallothionein sequesters copper and prevents this toxic metal from binding to and inactivating susceptible proteins and enzymes. Penicillamine forms a complex with copper in the blood. This complex is transported to the kidneys where it is excreted.

6.8 a. Cofactor
b. Holoenzyme
c. Apoenzyme
d. Coenzyme
e. Coenzyme

6.9 The patient that failed to show improvement probably had a higher level of acetylating enzymes. The patient's dosage should be based on capacity to process the drug and not on body weight.

End-of-Chapter Questions

Review Questions

1. a. Activation energy is the minimum energy required to bring about a reaction.
b. A catalyst is a substance that alters the rate of a reaction but is not consumed by it.
c. The active site is that part of an enzyme directly responsible for catalysis.
d. A coenzyme is a small molecule needed to enable the enzyme to function.
e. Velocity of a chemical reaction is the change in concentration of a reactant with time.
f. Half-life is the time needed to consume half the reactant molecules.
g. Turnover number is the number of moles of substrate converted per second per mole of enzyme.
h. A katal is the amount of enzyme that transforms 1 mol of substrate per second. One katal is equal to 6×10^7 I.U.

i. A noncompetitive inhibitor is an inhibitor molecule that binds to an enzyme, but not at the active site.
j. Repression is the prevention of polypeptide synthesis.

3. Cells regulate enzymatic reactions by using genetic control (certain key enzymes are synthesized in response to changing metabolic needs), covalent modification (certain enzymes are regulated by the reversible interconversion between their active and inactive forms, a process involving covalent changes in structure), allosteric regulation (binding effector molecules to pacemaker enzymes alters catalytic activity), and compartmentation (preventing wasteful "futile cycles" by physical separation of opposing biochemical processes within cells).

5. Factors that contribute to enzyme catalysis include: proximity and strain effects, electrostatic effects, acid-base catalysis, and covalent catalysis. Refer to pages 177–180 for an explanation of each.

7. Negative feedback inhibition is a process in which the product of a pathway inhibits the activity of the pacemaker enzyme.

9. Refer to Table 6.3.

11. The activation energy for the reaction of glucose with molecular oxygen is quite high and, consequently, the reaction is relatively slow to occur .

13. At the start of a reaction, the concentrations of the reactants and products can be known precisely. Because equilibrium has not yet been established presumably only the forward reaction is taking place.

15. The amino acid residues that compose the active site are stereoisomers. Consequently, the active site is chiral and can bind only one form of an optically active compound.

Thought Questions

1. The data indicate that the reaction is first order in pyruvate and ADP and second order in P_i. The overall reaction is fourth order.

3.

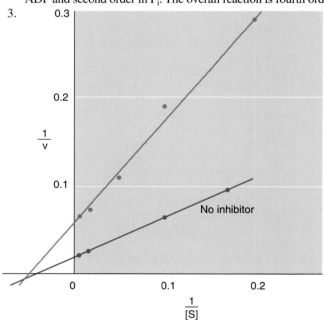

The intercept on the horizontal axis is $^{-1}/K_m$. The intercept on the vertical axis is $^1/V_{max}$. The slope is K_m/V_{max}. The type of inhibition being observed is noncompetitive.

5.

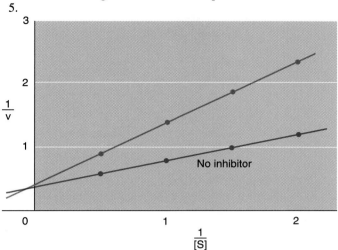

The inhibition is competitive.
7. Refer to Figure 6.19.
9. The inhibition is competitive.

CHAPTER 7

In–Chapter Questions

7.1 a. Aldotetrose, b. Ketopentose, c. Ketohexose
7.3 a.

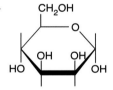

α-D-Mannose   β-D-Mannose

b.

α-D-Idouronic acid   β-D-Idouronic acid

c.

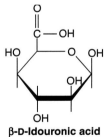

α-D-Fructofuranose   β-D-Fructofuranose

7.4

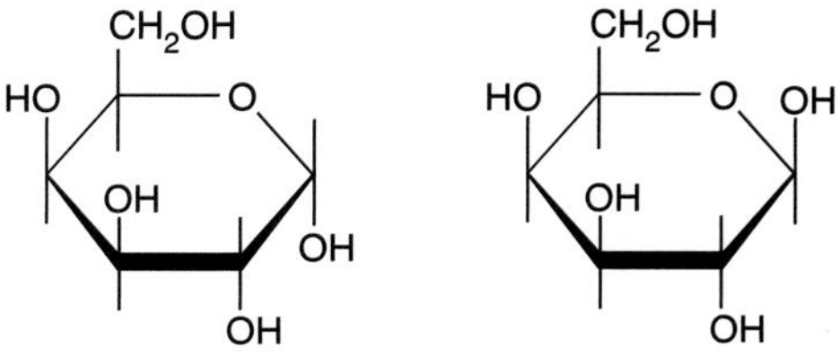

α-D-Galactose   β-D-Galactose

(a)

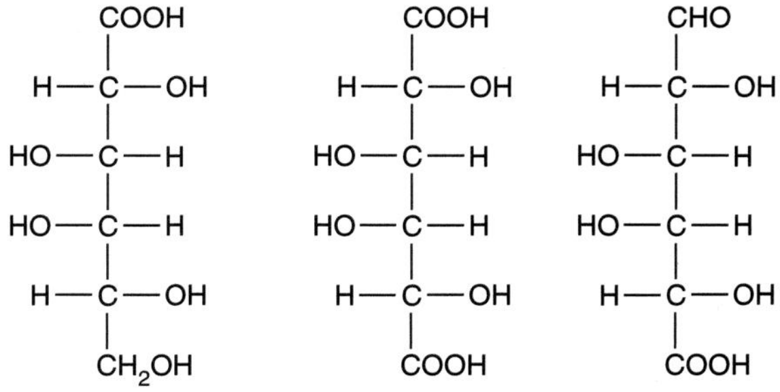

Aldonic acid   Aldaric acid   Uronic acid

(b)

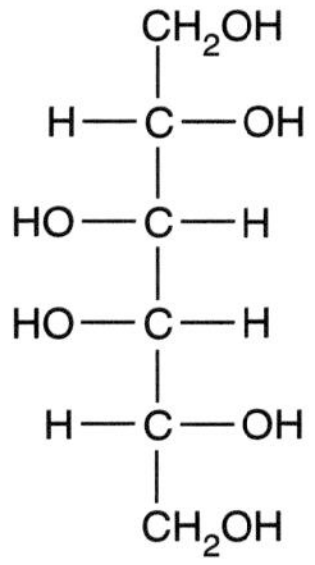

Galactitol   δ-Lactone of galactonic acid

(c)   (d)

7.5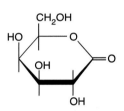

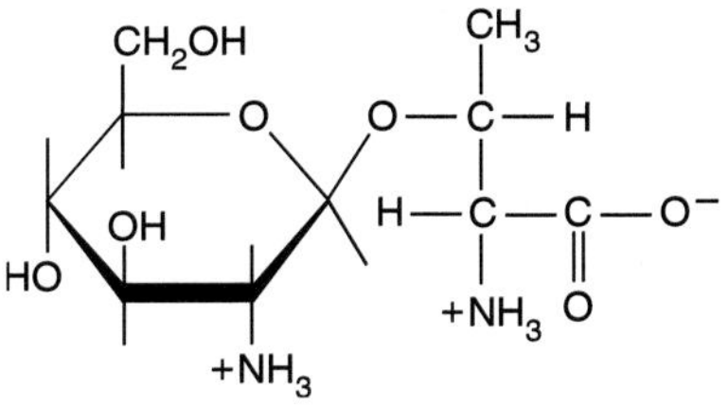

7.6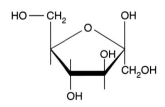

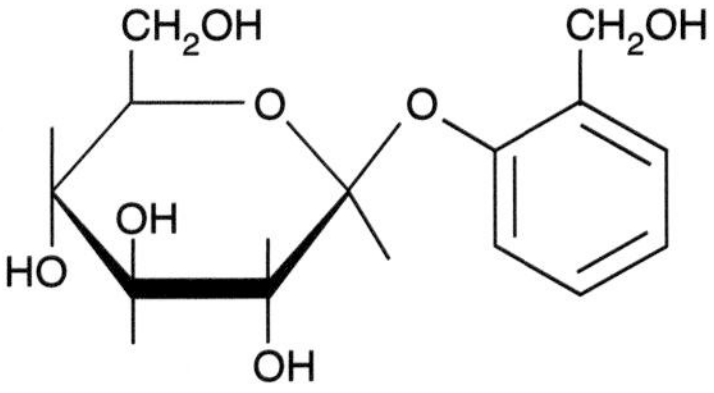

Carbohydrate   Aglycone

7.7 a. Glucose—reducing sugar
b. Fructose—reducing sugar
c. α-methyl-D-glucoside—nonreducing
d. Sucrose—nonreducing
Sugars a and b are capable of mutarotation.

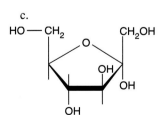

7.8 The larger, insoluble glycogen molecule makes a negligible contribution to the osmotic pressure of the cell. In contrast, each molecule of an equivalent number of glucose molecules contributes to osmotic pressure. If the glucose molecules were not linked to form glycogen, the cell would burst.

End-of-Chapter Questions
Review Questions

1. a.

α-D-Glucopyranose   **β-D-Glucofuranose**

b.

c.

d.

R =

e.

f.

3. In D-family sugars, the OH on the chiral carbon farthest from the carbonyl group is on the right side in a Fischer projection formula. So both (+)-glucose and (−)-fructose are D-sugars despite their rotation of plane-polarized light in opposite directions.

5. Heteropolysaccharides are made up of more than one type of monosaccharide residue but homopolysaccharides have only one. Examples of homopolysaccharides and heteropolysaccharides are starch and hyaluronic acid, respectively.

7. a. Nonreducing, b. Nonreducing, c. Reducing, d. Nonreducing, e. Reducing

9. Carbohydrate chains composed of sugars linked by $\alpha(1,4)$ glycosidic bonds coil into an α-helix.

11. a. Glucose, glucose, and fructose
    b. It is customary to move from left to right in the naming process. Therefore, the linkage between the two glucose molecules is α, and the linkage between the glucose and fructose is also α. This latter linkage is referred to as $\alpha 1\beta 2$
    c. Raffinose is a nonreducing sugar.
    d. Raffinose is not capable of mutarotation.

13. a. Carboxylic acid and hydroxyl groups bind large amounts of water.
    b. Hydrogen bonding is the primary type of bonding between water and glycosaminoglycans.

15. Chondroitin sulfate and proteoglycans are extensively negatively charged at physiological pH and as such are spread out, binding large amounts of water. The interwoven chains block the passage of large molecules. Smaller molecules can pass between the chains.

17. Proteoglycans are extremely large molecules that contain a large number of glycosaminoglycan chains linked to a core protein. They are found primarily in extracellular fluids where their high carbohydrate content allows them to bind large amounts of water. Glycoproteins are conjugated proteins in which the prosthetic groups are carbohydrate molecules. The carbohydrate groups stabilize the molecule through hydrogen bonding, protecting the molecule from denaturation, or shield the protein from hydrolysis. The carbohydrate groups on the glycoproteins on the surface of cells play an important role in a variety of recognition phenomena.

19. Both amylose, glycogen, and amylopectin possess polymers of glucose residues that are linked by α-1,4-glycosidic linkages. Glycogen and amylopectin also possess branches that are connected to the α-1,4-linked chain by α-1,6-

glycosidic linkages. Cellulose is an unbranched polymer of glucose residues linked by β-1,4-glycosidic linkages.

Thought Questions

1.

3. The thick proteoglycan coat acts to protect bacteria by preventing the binding of antibodies to their surface antigens.

5. a.

b. The polymer acts to immobilize water through extensive hydrogen bonding.

7.

CHAPTER 8

In-Chapter Questions

8.1 The large excess of NADH that is produced by these reactions drives the conversion of pyruvate to lactate.

8.2 Chromium is acting as a cofactor.

8.3 In the absence of O_2, energy is produced only through glycolysis, an anaerobic process. Glycolysis produces less energy per glucose molecule than does aerobic respiration.

Consequently, more glucose molecules must be metabolized to meet the energy needs of the cell. When O_2 is present, the flux of glucose through glycolysis is reduced.

8.4 At three strategic points, glycolytic and gluconeogenic reactions are catalyzed by different enzymes. For example, phosphofructokinase and fructose-1,6-diphosphatase catalyze opposing reactions. If both reactions occur simultaneously (i.e., in a futile cycle) to a significant extent, ATP hydrolysis in the reaction catalyzed by phosphofructokinase releases large amounts of heat. If the heat is not quickly dissipated, an affected individual could die of hyperthermia.

8.5 In gluconeogenesis pyruvate is converted to oxaloacetate. NADH and H^+ are required to reduce glycerate 1,3-bisphosphate to glyceraldehyde-3-phosphate. NAD^+ is the oxidized form of NADH also produced in this reaction. ATP is needed to provide the energy to carboxylate pyruvate to oxaloacetate and phosphorylate glyceraldehyde-3-phosphate to glycerate-1,3-bisphosphate. Both of these reactions also produce ADP and P_i. GTP converts oxaloacetate to phosphoenolpyruvate. This reaction is also the source of GDP and P_i. Water is involved in the hydrolysis reactions of ATP to ADP and P_i, the conversion of phosphoenolpyruvate to 2-phosphoglycerate, and the hydrolysis of glucose-6-phosphate to glucose. Six protons are formed when four molecules of ATP and two molecules of GTP are hydrolyzed.

8.6 Without glucose-6-phosphatase activity, the individual cannot release glucose into the blood. Blood glucose levels must be maintained by frequent consumption of carbohydrate. Excess glucose-6-phosphate is converted to pyruvate, which is then reduced by NADH to form lactate.

8.7 The enzyme deficiencies prevent the breakdown of glycogen. Because the synthetic enzymes are active, some glycogen continues to be produced and causes liver enlargement. Because of the liver's strategic role in maintaining blood glucose, defective debranching enzyme causes hypoglycemia (low blood sugar).

End-of-Chapter Questions

Review Questions

1. Phosphorylation of glucose upon its entry into the cells prevents leakage of the molecule out of the cell and facilitates its binding to the active sites of enzymes.

3. Ribose-5-phosphate is an aldopentose with a phosphate group at the fifth carbon. Ribulose-5-phosphate is a ketose with a carbonyl at carbon-2 and a phosphate group on the fifth carbon.

5. In glycolysis the entry level substrates are sugars and the product is pyruvate. The main purposes of the process are to provide the cell with energy and several metabolic intermediates. The substrates for gluconeogenesis are pyruvate, lactate, glycerol, and several amino or α-keto acids. Gluconeogenesis provides the body with glucose when blood glucose levels are low.

7. Such organisms utilize ethanol as a hydrogen acceptor, thus making possible the recycling of NAD^+.

9. Epinephrine promotes the conversion of glycogen to glucose by activating adenylate cyclase, an enzyme whose product, cAMP, initiates a reaction cascade that activates the glycogen degrading enzyme glycogen phosphorylase.

11. a. Lactate—stimulates gluconeogenesis
    b. ATP—stimulates gluconeogenesis
    c. Pyruvate—stimulates gluconeogenesis
    d. Glycerol—stimulates gluconeogenesis
    e. AMP—inhibits gluconeogenesis
    f. Acetyl-CoA—stimulates gluconeogenesis

Thought Questions

1. In such an individual, following a carbohydrate meal blood glucose levels would be higher than normal. Recall that the kinetic properties of hexokinase D allow the liver to remove excess glucose from blood. Skeletal muscle would accumulate some additional glycogen, but most excess glucose would be used to synthesize triacylglycerol in adipocytes, a process that is promoted by insulin. A significant amount of liver glycogen is synthesized from glucose produced by gluconeogenesis.

3. In liver, fructose is metabolized more rapidly than glucose because its metabolism bypasses two regulatory steps in the glycolytic pathway: the conversion of glucose to glucose-6-phosphate and fructose-6-phosphate to fructose-1,6-bisphosphate. Recall that fructose-1-phosphate is split into glyceraldehyde and DHAP, both of which are subsequently converted to glyceraldehyde-3-phosphate.

5. Two common oxidizing agents in anaerobic metabolism are NAD^+ and $NADP^+$.

7. Ethanol is the most reduced molecule and acetate is the most oxidized. The degree of oxidation of an organic molecule can be correlated with its oxygen content, that is, acetate has more oxygen than does ethanol.

CHAPTER 9

In-Chapter Questions

9.1 With a $\Delta E_o'$ value of -0.345 V, the oxidation of NO_2^- is spontaneous as written. The oxidation of ethanol is not spontaneous as written because its $\Delta E_o'$ value is positive ($+0.275$ V).

9.2 Reactions 3, 4, and 5 are redox reactions. In reaction 3, lactate is the reducing agent and NAD^+ is the oxidizing agent. In reaction 4, cyt b (Fe^{+2}) is the reducing agent and NO_2^- the oxidizing agent. In reaction 5, NADH is the reducing agent and CH_3CHO is the oxidizing agent.

9.3 The oxidation states of the functional group carbon (indicated in bold) are:

CH_3CH_2OH $\quad 0 - 1 - 1 + 1 = -1$
CH_3CHO $\quad 0 - 1 + 2 = +1$
CH_3COOH $\quad 0 + 1 + 2 = +3$

9.4 As it is incorporated into an organic molecule, the carbon atom in CO_2 is reduced.

9.5 Because of its symmetrical structure, a molecule of succinate derived from a ^{14}C-labeled acetyl-CoA is converted into two forms of oxaloacetate, one with a labeled methylene group and one with a labeled carbonyl group. ^{14}C-Iabeled CO_2 is not released until the third turn of the cycle when one-half of the original labeled carbon is lost (the carbonyl group derived from acetyl-CoA). The labeled carbon is further scrambled when succinyl-CoA is converted to succinate during the third and fourth turns of the cycle.

9.6 Pyruvate carboxylase converts pyruvate to oxaloacetate. If the enzyme is inactive, concentrations of pyruvate in the system rise and pyruvate is converted by NADH to lactate. Excess lactate is then excreted in the urine.

9.7 Fluoroacetate is converted to fluoroacetyl-CoA. This substance then reacts with oxaloacetate to produce fluorocitrate. Fluorocitrate is toxic because it inhibits aconitase, the enzyme that normally converts citrate to isocitrate, hence the buildup of citrate. In the plant fluoroacetate is stored , in vacuoles away from the mitochondria.

End-of-Chapter Questions

Review Questions

1. a. Aerotolerant anaerobes are organisms that grow without utilizing oxygen in energy generation.
   b. Anaplerotic reactions replenish substrates used in biosynthetic processes.
   c. Glyoxysomes are plant organelles that possess glyoxylate cycle enzymes.
   d. Reduction potential is the tendency for a specific substance to lose electrons.
   e. An electron donor and its acceptor are a conjugate redox pair.
   f. Coenzyme A is an acyl carrier molecule.
   g. Amphibolic refers to pathways that can function in both anabolic and catabolic processes.
   h. Lactic aciduria is a physiological condition in which lactic acid is found in urine.
   i. Carcinogenesis is the process whereby cells become genetically unstable and ultimately cancerous.
   j. Aerobic glycolysis is a process that occurs in tumors in which cells derive the energy required to drive their rapid cell divisions by mixed metabolism that involves a high rate of glycolysis and some level of oxidative phosphorylation.

3. a. Obligate anaerobes are organisms that not only do not use oxygen to generate energy, but live in oxygen-deprived, reduced environments.
   b. Aerotolerant anaerobes are organisms that generate energy via fermentation, but can exist in oxygen because they can protect themselves from oxygen's toxic effects.
   c. Facultative anaerobes are organisms that, depending on the availability of oxygen, can generate energy via fermentation or aerobic respiration.
   d. Obligate aerobes are organisms that require a continuous source of oxygen for energy generation.

5. The citric acid cycle is an important component of aerobic respiration. The NADH and $FADH_2$ produced during oxidation-reduction reactions of the cycle donate electrons to the mitochondrial ETC. Citric acid cycle intermediates are also used as biosynthetic precursors.

7. The glyoxylate cycle is a modified version of the citric acid cycle that allows certain organisms (e.g., plants or some microorganisms) to grow by utilizing two-carbon molecules such as acetyl-CoA, acetate, or ethanol. The glyoxylate cycle allows for the net synthesis of larger molecules from two-carbon molecules because the two decarboxylation reactions of the citric acid cycle are bypassed.

9. The top molecule is FMN; the lower molecule is $FMNH_2$.

Thought Questions

1. In substrate-level phosphorylation ADP is converted to ATP by the direct transfer of a phosphoryl group from a high energy compound. The only reaction in the citric acid cycle that involves this type of reaction is the cleavage of succinyl-CoA

to form succinate, CoASH, and GTP. Another example of a substrate level phosphorylation is the glycolytic reaction that converts phosphoenolpyruvate and ADP to pyruvate and ATP.

3. The biosynthesis of glutamate from pyruvate is shown below:

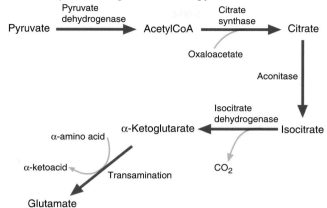

5. a. $NADH + H^+ + \frac{1}{2} O_2 \longrightarrow NAD^+ + H_2O \quad \Delta E_0' = +1.14 \text{ V}$

$$\Delta G^{0'} = -nF\Delta E_0'$$
$$= (-2)(96{,}485 \text{ J/V} \cdot \text{mol})(+1.14 \text{ V})$$
$$= -219{,}985.8 \text{ J/mol}$$
$$= -220 \text{ kJ/mol}$$

b. $\text{Cyt c } (Fe^{2+}) + \frac{1}{2} O_2 \longrightarrow \text{Cyt c } (Fe^{3+}) + H_2O \quad \Delta E_0' = 0.58 \text{ V}$

$$\Delta G^{0'} = (-2)(96{,}485 \text{ J/V} \cdot \text{mol})(+0.58 \text{ V})$$
$$= -112 \text{ kJ/mol}$$

CHAPTER 10

In-Chapter Questions

10.1 a. NADH
      b. $FADH_2$
      c. Cyt c (reduced)
      d. NADH
      e. NADH

10.2 DNP is a lipophilic molecule that binds reversibly with protons. It dissipates that proton gradient in mitochondria by transferring protons across the inner membrane. The uncoupling of electron transport from oxidative phosphorylation causes the energy from food to be dissipated as heat. DNP causes liver failure because of insufficient ATP synthesis in a metabolically demanding organ.

10.3 No, for ATP synthesis to occur, the proton concentration must be higher within the inside-out mitochondrial particles. ATP synthesis requires that protons move down a concentration gradient through the base of the ATP synthetase across the membrane.

10.4 Disregarding proton leakage and assuming that the glycerol phosphate shuttle is in operation, 38 ATP would be produced from the aerobic oxidation of a glucose molecule. If the malate shuttle is in operation, only 36 ATP would be produced.

10.5 Sucrose is a disaccharide composed of glucose and fructose. As described (p. 318) the oxidation of 1 mol of glucose yields a maximum of 31 mol of ATP. Fructose, which like glucose is also partially degraded by the glycolytic pathway, also yields a maximum of 31 mol ATP. The total maximum energy yield is 62 mol of ATP per mole of sucrose.

10.6 The larger selenium atom holds its electrons less tightly than sulfur. Selenium is more easily oxidized and therefore acts as a better scavenger for oxygen than does sulfur.

10.7 The SH groups reduce hydrogen peroxide or trap hydroxyl radicals to form water. An example of a nonsulfhydryl group–containing molecule that should be capable of this activity is vitamin C or any of a number of other antioxidants (carotenoids, flavonoids, tocopherols, etc.).

10.8 Low levels of G-6-PD in combination with a high level of oxidized GSH cause high oxidative stress. Without antioxidant protection, red cell membranes become fragile, a condition that eventually causes hemolytic anemia.

10.9 The phenolic groups of both molecules are responsible for their antioxidant activity because of the ease of formation of phenoxy radicals with subsequent neutralization of electron deficient ROS.

End-of-Chapter Questions
Review Questions

1. a. The chemical coupling hypothesis postulates that a high-energy intermediate generated by electron transport is used to drive ATP synthesis.
    b. Chemiosmotic coupling theory postulates that a proton gradient created by the mitochondrial electron transport system drives ATP synthesis.
    c. An ionophore is a hydrophobic molecule that inserts into membrane and dissipates osmotic gradient.
    d. Respiratory control is the regulation of aerobic respiration by ADP.
    e. Ischemia is inadequate blood flow.
    f. Aerobic respiration is the energy-generating process that uses O_2 as a terminal electron acceptor.
    g. A radical is a chemical species with an unpaired electron.
    h. Biotransformation is a series of enzyme-catalyzed processes in which hydrophobic, and usually toxic, molecules are converted into more soluble, and usually less toxic, products.
    i. An epoxide is a three-membered ring that contains an oxygen.
    j. Protonmotive force is the force arising from a proton gradient and a membrane potential.
    k. A proton gradient is a difference in the concentration of protons across a membrane.
    l. An uncoupler is a molecule that uncouples ATP synthesis from electron transport; it collapses proton gradients by transporting protons across a membrane.
    m. ROS are active oxygen species; reactive derivatives of molecular oxygen.

3. Processes believed to be driven by mitochondrial electron transport are ATP synthesis, the pumping of calcium ions into the mitochondrial matrix, and the generation of heat by brown fat.

5. According to the chemiosmotic theory, an intact inner mitochondrial membrane is required to maintain the proton gradient required for ATP synthesis.

7. The translocation of three protons is required to drive ATP synthesis. The fourth proton drives the transport of ADP and P_i.

9. Examples c, d, and f are all reactive oxygen species. Each of these species can act as a free radical, that is, it can attack various cell components producing such effects as enzyme inactivation, polysaccharide depolymerization, DNA breakage, and membrane destruction.

11. The major enzyme defenses against oxidative stress are provided by superoxide dismutase, catalase, and glutathione peroxidase.

Thought Questions

1. Once a labeled acetate molecule is converted to acetyl-CoA it is processed through the citric acid cycle (Figure 9.5). Because of the symmetrical structure of the intermediate succinate, $^{14}CO_2$ not released until two or more turns of the cycle. The number of moles of ATP produced from 1 mol of acetate is calculated as follows: Each turn of the cycle produces 10 ATP. Assuming that the release of $^{14}CO_2$ requires two turns of the cycle, a total of 20 ATP are generated. Because 2 ATP are required to convert acetate to acetyl-CoA, the net production of ATP is 18 mol.

3. Glutamine is readily converted to glutamate which then enters the citric acid cycle as α-ketoglutarate. The first turn of the cycle liberates one CO_2 and produces 2 NADH, 1 $FADH_2$, and 1 ATP. The four-carbon product is consumed in two further turns of the cycle, each of which generates 3 NADH, 1 $FADH_2$, and 1 ATP. The total yield is 8 NADH, 3 $FADH_2$ and 3 ATP. Assuming that each NADH yields 2.5 ATP and each $FADH_2$ yields 1.5, a grand total of 27.5 mol ATP are generated.

CHAPTER 11

In-Chapter Questions

11.1 The product of complete hydrogenation would be hard and therefore not useful as a margarine.

11.2 When soap and grease are mixed, the hydrophobic hydrocarbon tails of the soap insert (or dissolve) into the oil droplet. The oil droplet becomes coated with soap molecules. The hydrophilic portion of the soap molecules allows the soap-oil complex to be dispersed in water.

11.3 The phospholipid of the surfactant, which possesses a polar head group and two hydrophobic acyl groups, disrupts some of the intermolecular hydrogen bonds of the water thereby decreasing the surface tension.

11.4 Carvone and camphor are monoterpenes; abscisic acid is a sesquiterpene.

11.5 Bile salts are structurally similar to soap in that they contain a polar head group (e.g., the charged amino acid residue glycine) and a hydrophobic tail (the steroid ring system).

11.6 a. Simple diffusion
  b. Secondary active transport or facilitated diffusion
  c. Primary active transport or exchanger protein
  d. Primary active transport or gated channel
  e. Fat molecules (Triacylglycerals) are not directly transported across cell membranes. They must be hydrolyzed first.
  f. Simple diffusion

11.7 The main stabilizing feature of biological membranes is hydrophobic interactions among the molecules in the lipid bilayer. The phospholipids in the lipid bilayer orient themselves so that their polar head groups interact with water. Proteins in the lipid bilayer interact favorably in their hydrophobic milieu because they typically have hydrophobic amino acid residues on their outer surfaces.

11.8 The transport mechanisms discussed in the chapter fit into the following categories:
  sodium channel: uniporter
  glucose permease: passive uniporter
  Na^+-K^+-ATPase: antiporter

End-of-Chapter Questions
Review Questions

1. a. Lipids are naturally occurring substances that dissolve in hydrocarbon solvents.

  b. Autocrine regulators are hormones that act upon the synthesizing cell.
  c. Amphipathic molecules contain both hydrophobic and hydrophilic groups.
  d. A sesquiterpene is a terpene that contains three isoprene units.
  e. Lipid bilayer is the basic structural feature of biological membranes.
  f. Prenylation is the covalent attachment of isoprenoid groups to a protein.
  g. Fluidity is a measure of the resistance of membrane components to movement.
  h. Chylomicrons are large lipoprotein complexes with extremely low density.
  i. A membrane channel that is opened by changes in membrane voltage is a voltage-gated channel.
  j. Terpenes are a large group of molecules found primarily in the essential oils of plants; they are composed of isoprene units.
  k. CFTR is the abbreviation of cystic fibrosis transmembrane conductance regulator, the plasma membrane chloride channel in epithelial cells.
  l. Aquaporins are a group of water channel proteins in biological membranes.

3. a. Triacylglycerol
  b. Steroid
  c. Wax ester
  d. Unsaturated fatty acid
  e. Phosphatidyl choline
  f. Sphingolipid

5. Plasma lipoproteins improve the solubility of hydrophobic lipid molecules as they are transported in the bloodstream to the organs. The protein component serves to solubilize the lipoproteins in the blood. It also acts as a receptor that permits binding and uptake of lipoproteins by body cell.

7. Many transmembrane and peripheral proteins are attached to the cytoskeleton and therefore are not free to move in the phospholipid bilayer.

9. For a phospholipid to move from one side of the bilayer to the other, the polar head must move through the hydrophobic portion of the phospholipid membrane. This process requires a significant amount of energy and is therefore relatively slow.

11. Eicosanoids are derived from arachidonic acid. Medical conditions in which it is advantageous to suppress the synthesis of eicosanoids are anaphylaxis, allergies, pain, the inflammation caused by injury, and fever.

13. The indicated compounds are classified as follows:
  a. monoterpene
  b. monoterpene
  c. sesquiterpene
  d. polyterpene
  e. diterpene
  f. triterpene

15. Phospholipids are components of membranes.

17. Lipids are not directly involved in the active transport of ions. Lipids are directly involved in a, b, c, and d.

19. Each type of transport protein or carrier binds a specific molecule. As a result of this binding, a conformational change occurs in the carrier, which results in translocation of the ligand across the membrane. The rbc glucose transporter is an example of such a carrier.

21. a. Simple diffusion
  b. Diffusion through aquaporins

c. Facilitated diffusion

d. Facilitated diffusion

e. Active transport or gated channel

Thought Questions

1. The fluidity of the membrane allows for flexible movement. Any breaks that do occur expose the hydrophobic core of the membrane to an aqueous environment. Hydrophobic interactions spontaneously move the broken ends together and, in combination with certain other components of cell membrane resealing mechanisms (e.g., cytoskeleton and calcium ions), the membrane reseals.

3. Most of the cholesterol in plaque results from the ingestion of LDL by the foam cells that line the arteries. High blood plasma LDL therefore promotes atherosclerosis. Because the coronary arteries are narrow, they are especially prone to occlusion by atherosclerotic plaque.

5. For a phospholipid to move from one side of the bilayer to the other, the polar head must move through the hydrophobic portion of the phospholipid membrane. This process requires a significant amount of energy and is therefore relatively slow.

7. The hooves and lungs are subjected to much lower temperatures than the rest of the body. At these low temperatures, the membrane must be modified so that the membranes remain fluid. This can be done by increasing the unsaturation of the nonpolar tails of the membrane phospholipids.

CHAPTER 12

In-Chapter Questions

12.1 The triacylglycerols are emulsified in the small intestine by bile salts. They are then digested by lipases, the most important of which is pancreatic lipase. The products, fatty acids and monoacylglycerol, are transported into enterocytes and reconverted to triacylglycerol. Triacylglycerol is subsequently incorporated into chylomicrons, which are then transported into lymph via exocytosis and finally into the bloodstream for transport to the fat cells.

12.2 If there is a connection between female sex hormone levels and VLDL secretion, injection of estrogen into a male rat should have the following effects: There should be a timely and measurable increase in VLDL secretion. This process requires a concomitant increase in the synthesis of the components of VLDL, that is; apoproteins, triacylglycerols, phospholipid, and cholesterol. FABP synthesis should increase in response to the increased intracellular fatty acid concentrations.

12.3 a. Phospholipid

b. Acyl-CoA

c. Carnitine

12.4 Unlike the oxidation of glucose to form pyruvate, fatty acid oxidation, which involves the citric acid cycle and the electron transport system, cannot operate in the absence of O_2.

12.5 The yield from the oxidation of stearyl-CoA is calculated as follows:

8 $FADH_2$ × 1.5 ATP/$FADH_2$ =	12 ATP
8 NADH × 2.5 ATP/NADH =	20 ATP
9 Acetyl-CoA × 10 ATP/Acetyl-CoA =	90 ATP
	122 ATP

Two ATP are required to form stearyl-CoA from stearate to give a total of 120 ATP.

12.6 In cells without observable intact peroxisomes, the absence of ether-type lipids or the buildup of long-chain fatty acids suggests that the organelle is not present. In addition, his-tochemical tests can be employed to diagnose Zellweger syndrome. For example, radioactive isotope–labeled antibodies to peroxisomal marker enzymes can be used to determine if peroxisomal function is present in cells.

12.7 Propionyl-CoA can be reversibly converted to succinyl-CoA, an intermediate in the citric acid cycle. Oxaloacetate, a downstream intermediate of this cycle can be converted to PEP. PEP is then converted to glucose via gluconeogenesis. Adipic acid undergoes one round of β-oxidation to yield acetyl-CoA and succinyl-CoA. As just described, succinyl-CoA is sequentially converted to oxaloacetate, PEP, and then to glucose.

12.8 Because steroids inhibit the release of arachidonic acid, their use shuts down the synthesis of most if not all eicosanoid molecules, hence their reputation as potent anti-inflammatory agents. Aspirin inactivates cyclooxygenase and prevents the conversion of arachidonic acid to PGG_2, the precursor of prostaglandins and thromboxanes. Aspirin is not as effective an anti-inflammatory agent as the steroids because it shuts down only a portion of eicosanoid synthetic pathways.

12.9 Following the hydrolysis of sucrose, both monosaccharide products enter the bloodstream and travel to the liver, where fructose is converted to fructose-1-phosphate. Recall that the conversion of fructose-1-phosphate to glyceraldehyde-3-phosphate bypasses two regulatory steps. Consequently, more glycerol- phosphate and acetyl-CoA (the substrates for triacylglycerol synthesis) are produced. High blood glucose concentrations that result from this consumption of excessive amounts of sucrose trigger the release of larger than normal amounts of insulin. One of the functions of insulin is to promote fat synthesis.

12.10 a. β-hydroxybutyrate is a product of ketone body metabolism

b. Malonyl-CoA is the product of the reaction of acetyl-CoA and carboxybiotin that occurs during fatty acid synthesis.

c. Biotin is a carrier of CO_2 in fatty acid synthesis and several other reactions.

d. Acetyl ACP delivers acetate to the synthetic machinery of fatty acid synthesis.

12.11 The labeled carbon atoms in cholesterol are as follows:

Cholesterol

12.12 The structure of cortisol differs from that of cortisone in that the C-11 hydroxyl group is replaced by a carbonyl group. Glycyrrhizic acid inhibits 11-β-hydroxysteroid dehydrogenase, which prevents the deactivation of cortisol. Cortisone is administered to Addison's disease patients because it is converted to cortisol by 11-β-hydroxysteroid dehydrogenase a reversible enzyme.

12.13 The higher activity of HMG-CoA reductase in obese patients in combination with a high calorie diet increases the synthesis of cholesterol.

End-of-Chapter Questions

Review Questions

1.
  a. Biosynthesis from new materials is termed *de novo*.
  b. Oil bodies are structures in plants that store triacylglycerol.
  c. β-Oxidation is the main pathway for the degradation of fatty acids in which two-carbon fragments in the form of acetyl-CoA are removed from the carboxyl end of fatty acids.
  d. The rate at which molecules are degraded and replaced is called turnover.
  e. Thiolytic cleavage involves the cleavage by thiolase between the α and β carbons of a β ketoacid during β-oxidation to produce a molecule of acetyl-CoA and a new acyl-CoA.
  f. Autoantibodies are defense proteins that bind to surface antigens of the patient's own cells as if they were foreign.
  g. Acetyl-CoA molecules are condensed to produce acetone, β-hydroxybutyrate, and acetoacetate (i.e., the ketone bodies) in a process called ketogenesis.
  h. The enzyme-catalyzed process in which hydrophobic molecules are solubilized so they can be excreted is called biotransformation.
  i. A conjugation reaction may improve the water solubility of a molecule by converting it to a derivative that contains a water-soluble group.

3. Peroxisomes have β-oxidation enzymes that are specific for long-chain fatty acids, whereas mitochondria possess enzymes that are specific for short and moderate chain length fatty acids. In addition, the first reaction in the peroxisomal pathway is catalyzed by a different enzyme than the mitochondrial pathway. The $FADH_2$ produced in the first peroxisomal reaction donates its electrons to O_2 directly, (forming H_2O_2) instead of UQ as in mitochondria. The processes are similar in that acetyl-CoA is derived from the oxidation of fatty acids.

5. Fatty acid synthesis in plants differs from that in animals in the following ways: location (plant fatty acid synthesis occurs mainly in the chloroplasts, whereas in animals fatty acid biosynthesis occurs in the cytoplasm), metabolic control (in animals the rate-limiting step is catalyzed by acetyl-CoA carboxylase, whereas in plants, this does not appear to be the case), enzyme structure (the structures of plant acetyl-CoA carboxylase and fatty acid synthetase are more closely related to similar enzymes in *E. coli* than to those in animals).

7. Steroids are compounds containing the following ring system:

The term *steroid* is often used to designate all compounds containing this ring system. It is more accurately reserved for those derivatives that contain carbonyl groups. Sterols have a similar structure but also contain hydroxyl groups.

9. Insulin facilitates transport into adipocytes and stimulates fatty acid synthesis and triacylglycerol synthesis. It prevents lipolysis by inhibiting protein kinase.

11. The hydrophobic portions of the molecule are the long hydrocarbon tails of the molecule. The hydrophilic portion of the molecule is the phosphate ester functional group. The hydrocarbon tails are within the bilayer and the phosphate head group is on the surface of the molecule.

13. As described (see answer to question 12.5) each stearic acid molecule generates 120 ATP. Consequently, the three separate stearate products of the hydrolysis of tristearin yield 360 ATP. The glycerol product is transported to the liver where it is used in gluconeogenesis.

15. All the lipid molecules are originally synthesized from the isoprene units in isopentenyl pyrophosphate molecules. Steroid and terpene molecules are assembled by head-to-tail condensation of these groups.

Thought Questions

1. The reaction in which HMG-CoA reductase reduces HMG to form mevolonate is a rate-limiting step in cholesterol synthesis. Statins, inhibitors of this enzyme, act to lower blood levels of cholesterol in patients.

3. Carnitine is required for the transport of fatty acids into mitochondria where they are oxidized to generate energy. When carnitine levels are low, fat metabolism is impaired. Although glucose metabolism accelerates, an energy deficit occurs. In addition, accumulating acyl-CoA molecules become substrates for competing processes such as peroxisomal β-oxidation and triacylglycerol synthesis.

5. Severe dieting stimulates massive lipolysis. The large amounts of acetyl-CoA generated in this process trigger a vastly increased synthesis of the ketone bodies. When present in such large amounts, the ketone bodies overwhelm the buffering capacity of the blood and its pH falls.

7. Membrane phospholipids are synthesized on the cytoplasmic side of SER membrane. Because the polar head groups of phospholipid molecules make transport across the hydrophobic core of a membrane an unlikely event, a translocation mechanism is used to transfer phospholipids across the membrane to ensure balanced growth. Choline-containing phospholipids are found in high concentration on the lumenal side of ER membrane because a prominent phospholipid translocator protein called flippase preferentially transfers this class of molecule.

9. In the oxidation of butyric acid by the β-oxidation pathway, 1 mol each of $FADH_2$ and NADH and two mol of acetyl-CoA are produced.

CHAPTER 13

In-Chapter Questions

13.1
  a. LHCII is light harvesting complex II; consists of a transmembrane protein that binds numerous chlorophyll a and chlorophyll b molecules and carotenoids; a major component of thylakoid membrane.
  b. Lutein is a type of light-harvesting pigment found within the thylakoid membrane.
  c. PSII is photosystem II; a complex composed of proteins and pigment molecules located in the stacked regions of thylakoid membrane that oxidizes water molecules and donates energized electrons to electron carriers that eventually reduce photosystem I.
  d. MSP is manganese stabilizing protein; the oxygen-evolving component of PSII; located in the stacked regions of thylakoid membrane.
  e. CF_0CF_1 is the ATP synthase component of thylakoid membrane that penetrates into stroma.
  f. P700 is a special pair of chlorophyll molecules within photosystem I that absorb light at 700 nm.
  g. P680 is a special pair of chlorophyll molecules within photosystem II that absorb light at 680 nm.

h. O_2 generation occurs in the oxygen evolving complex which contains MSP and a critical tyrosine residue located in photosystem II. The electrons generated in the reduction of water are used to replace those transferred away from PSII when light energy is absorbed.

13.2 The energy of a photon is proportional to its frequency. Blue light has a higher frequency than green light and therefore has higher energy.

13.3 The presence of antenna pigments allows the light harvesting systems of chloroplasts to collect energy from a wider range of frequencies than those absorbed by the chlorophylls. Because their absorption spectra overlap, the energy absorbed by the antenna pigments is quickly transferred to the critical chlorophylls of PSI and PSII.

13.4 Excessive light promotes the formation of ROS, which damage proteins such as D_1. β-Carotene is an antioxidant that prevents some of this damage.

13.5 a. Plastocyanin is a component of the cytochrome b_6f complex; a copper-containing protein that accepts electrons from plastoquinone.

b. β-carotene is a carotenoid pigment that protects chlorophyll molecules from ROS.

c. Ferredoxin is a mobile, water-soluble protein that donates electrons to a flavoprotein called ferredoxin-NADP oxidoreductase.

d. Plastoquinone is a component of photosystem II that accepts electrons from pheophytin a to become plastoquinol.

e. Pheophytin a is a molecule similar in structure to chlorophyll that is a component of the electron transport pathway between PSII and PSI.

f. Lutein is a carotenoid that is a component of light harvesting complexes.

13.6 The process of phosphorylation changes the conformation and the binding characteristics of the complex and thereby alters its capacity to bind to PSII.

13.7 Uncoupler molecules stop ATP synthesis in both mitochondria and chloroplasts because they destroy proton gradients.

13.8 Soaking the chloroplasts in an acidic solution lowers their internal pH. Treatment with base establishes an artificial pH gradient across the thylakoid membrane. As protons flow down this gradient ATP is synthesized.

13.9 Of the herbicides discussed, paraquat and DCMU are most hazardous to humans. Paraquat generates free radicals that can attack cell components. DCMU poisons the electron transport complex.

13.10 The molecules of O_2 are generated from water .

13.11 Reactions of the Calvin cycle strongly resemble the pentose phosphate pathway.

13.12 The hydrolysis of glycerate-1,3-bisphosphate generates 1 mol of ATP. Recall that aerobic respiration is stimulated by relatively high ADP concentrations and inhibited by relatively high ATP concentrations. Any measurable increase in ATP concentration has the effect of depressing aerobic respiration. Also recall that ATP is an inhibitor of PFK-1 and pyruvate kinase, enzymes required to channel carbon skeletons into the citric acid cycle.

13.13 Photorespiration, a wasteful process that evolves CO_2, is favored by low concentrations of CO_2 and high concentrations of O_2 and high temperatures. In C4 plants, CO_2 is incorporated into oxaloacetate at night within specialized mesophyll cells, when the risk of water loss is lower. In the morning the CO_2 released within bundle sheath cells is incorporated into sugar molecules. Because the CO_2 concentration is high within these cells compared to O_2 concentration, photorespiration is avoided. In CAM plants, CO_2 is also incorporated into oxaloacetate and at night within mesophyll cells. Oxaloacetate is then reduced to form malate. In the morning, light stimulates the conversion of malate into pyruvate and CO_2. As in C4 plants, a high cellular concentration of CO_2 inhibits photorespiration.

End-of-Chapter Questions
Review Questions

1. a. A photosystem is a molecular complex that absorbs light and converts it to chemical energy.

b. A reaction center is a pigment-protein component of a photosystem that mediates the conversion of light energy into chemical energy.

c. Light reactions are a biochemical mechanism whereby the electrons energized by light are subsequently used in ATP and NADPH synthesis.

d. Dark reactions are light-independent reactions in which the ATP and NADPH generated during light reactions are utilized in the synthesis of carbohydrate.

e. Chloroplast photorespiration is a wasteful process in which O_2 is consumed and CO_2 is released.

3. The three primary photosynthetic pigments are (1) the chlorophylls, which absorb blue-violet and red wavelengths of light; (2) the carotenoids, which serve as antenna pigments and protect from ROS; and (3) the xanthophylls, which also serve as antenna pigments.

5. Excited molecules can return to the ground state by several means including: (1) fluorescence (light is absorbed at one wavelength and emitted at a longer wavelength), (2) resonance energy transfer (energy is transferred to neighboring chromophores with overlapping absorption spectra), (3) oxidation-reduction (an excited electron returns to its ground state by reducing another molecule), (4) radiationless decay (an excited molecule returns to the ground state and loses its excess energy as heat). Of these processes, oxidation-reduction and resonance energy transfer are important in photosynthesis. Oxidation-reduction is important in the transfer of electrons through the electron carriers such as the quinones and iron-sulfur complexes. Resonance energy transfer passes light energy to chlorophyll from accessory pigments.

7. During the light restrictions of photosynthesis, light-driven electron transport results in ATP and NADPH synthesis.

9. The oxygen evolving system is referred to as a clock because it involves five oxidation-reduction states that must be completed in order.

11. The optimal wavelengths for photosynthesis appear to be 400–500 nm and 600–700 nm.

13. Among the most notable metals present in photosynthesizing systems are magnesium, manganese, iron, and copper. Magnesium stabilizes the porphyrin ring in chlorophyll molecules without adding a redox element site during the reduction of water to oxygen. Manganese acts as a redox center in the transfer of electrons from the quinone electron carriers to the terminal electron acceptors in photosynthesis. Finally, copper acts as the terminal electron acceptor in PSII before the electrons are transferred to P700 in PSI.

15. The Z scheme is a mechanism whereby electrons are transferred from water to $NADP^+$. This process produces the reducing agent NADPH required for fixing carbon dioxide in the light-independent reactions of photosynthesis. Removal of the electrons from water also results in the production of oxygen. As electrons flow from PSII to PSI, protons are

pumped across the thylakoid membrane, a process that establishes the proton gradient that drives ATP synthesis.

17. The chloroplast contains the thyllakoid membranes in both appressed (stacked) and unappressed format. The ATPase is oriented in the membrane so that ATP synthesis is always exposed to the stromal compartment. LHCII and PSII are richly concentrated in the appressed regions to maximize light collection and electron transfer. PSI, which should not receive its excitation energy directly from PSII, is physically separated from it in the unappressed regions. Electron replacement of PSI and PSII is mediated by mobile carriers so physical separation is not a problem.

Thought Questions

1. The electrons energized by light absorption are used to generate NADPH. This NADPH is used in large quantity to fix CO_2 into carbohydrate molecules. If CO_2 is not available, NADPH accumulates and the chlorophyll molecules do not have electron transfer available as a way to return to the relaxed state. One avenue available is to release the energy as a photon, i.e to fluoresce.

3. Only light of a particular energy can be absorbed by photosynthetic pigments. Increasing the intensity of the light increases the number of photons present and hence can improve the rate of photosynthesis. Increasing the energy level of the light, that is, the energy of the photons, decreases the rate of photosynthesis by shifting the photons to energy levels that are not absorbed by the photosystems.

5. High oxygen concentrations promote photorespiration.

7. Chloroplasts possess DNA similar to that of modern cyanobacteria as well as prokaryotic-like protein synthesizing machinery. In addition, they multiply by binary fission as do bacteria.

9. Under conditions of high temperature, the carbon dioxide compensation point of C3 plants rises because the oxygenase activity of Rubisco increases more rapidly than the carboxylase activity.

11. The herbicides kill marine photosynthesizing organisms, thereby depressing worldwide O_2 production.

CHAPTER 14

In-Chapter Questions

14.1 a. CH_3NH_2
b. NH_3
c. CH_3CH_3

14.2 Refer to Figure 14.1.

14.3 Proteins are assembled on ribosomes in a linear sequence specified by the base sequence in an mRNA. If any of the amino acids is not present, the process stops when the base sequence that specifies its incorporation into the protein enters the ribosomal active site.

14.4 The transamination products are as follows:

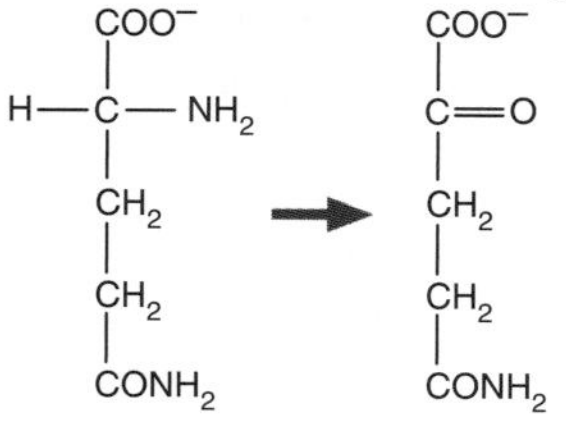

Glutamine

(a)

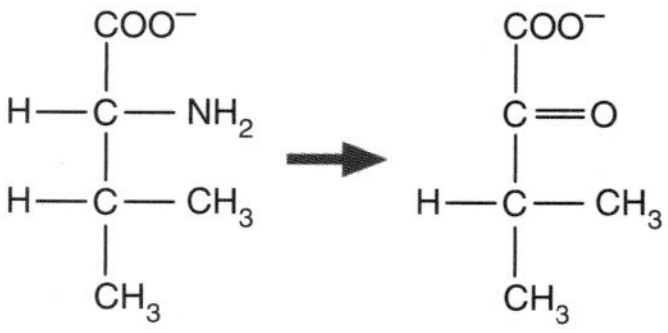

Isoleucine

(b)

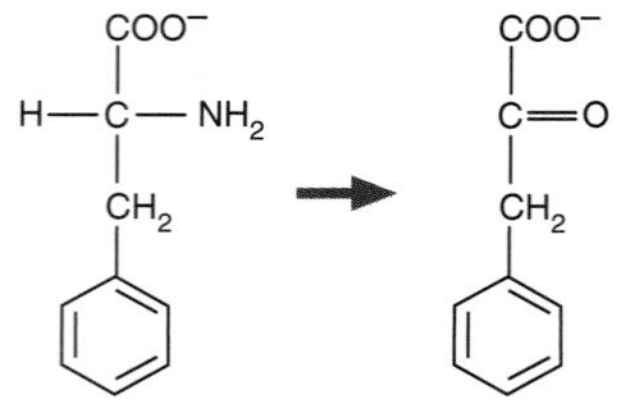

Phenylalanine

(c)

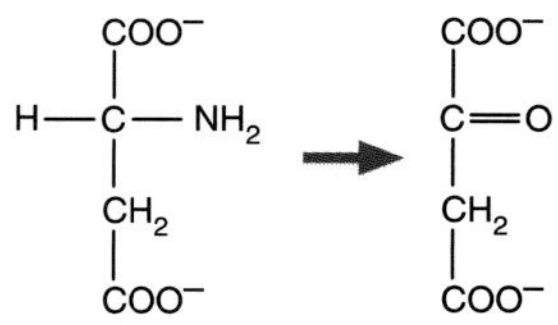

Aspartate

(d)

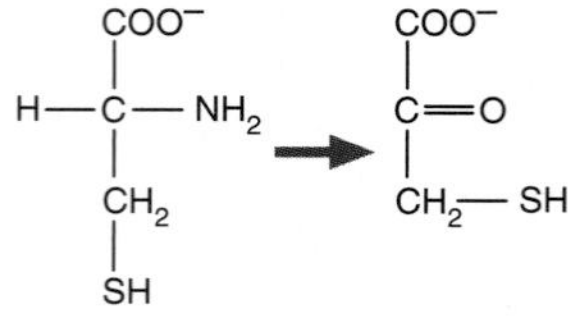

Aspartate

(e)

14.5 a. Glutamic acid
b. Tryptophan
c. Histidine
d. Tyrosine
e. Proline

14.6 Because of its close structural similarity to folic acid, methotrexate is a competitive inhibitor of the enzyme dihydrofolate reductase. (Recall that this enzyme converts folic acid to its biologically active form, THF.) Rapidly dividing cells require large amounts of folic acid. Methotrexate prevents the synthesis of THF, the one-carbon carrier required in nucleotide and amino acid synthesis. It is therefore toxic to rapidly dividing cells, especially those of certain tumors and normal cells that divide frequently such as hair and GI Tract cells.

14.7

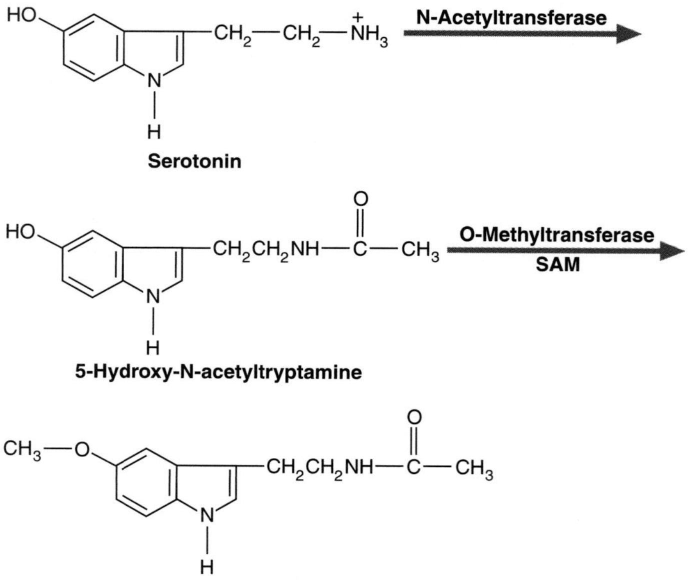

Serotonin

5-Hydroxy-N-acetyltryptamine

Melatonin

14.8 The reaction types involved in the synthesis of IAA from tryptophan are a deamination (removal of the α-amino group) and an oxidative decarboxylation.

14.9 The disruption of function that accompanies brain damage has numerous consequences. Among these are reduced ATP synthesis and the unbalanced intracellular ion concentrations that result from oxygen deprivation. Under these conditions, when an oxygen supply is reestablished, ROS production increases the extent of brain cell damage. Among the ROS formed are superoxide, hydroxyl radical singlet oxygen, and the peroxynitrite anion. The normal cellular defenses against ROS, such as NADPH, GSH, antioxidant vitamins such as vitamins C and E, and antioxidant enzyme systems, are quickly depleted because of disrupted synthesis and/or transport. When cells swell, they release the excitatory neurotransmitter glutamate, which stimulates some nearby cells to synthesize NO. Excessive and unregulated NO synthesis is extraordinarily dangerous to brain tissues under these circumstances, in part because the presence of elevated ROS concentrations causes the toxic peroxynitrite anion to be formed.

14.10 The source of the phosphoribosyl group of nucletotides is PRPP (5-phosphoribosyl-α-D-1-pyrophsophate). PRPP is generated from ribose-5-phosphate, a product of the pentose phosphate pathway.

14.11
a. Succinyl-CoA is condensed with glycine to produced δ-aminolevulinate (ALA), an intermediate in the heme biosynthetic pathway.
b. This molecule is the initial product of the condensation of succinyl-CoA and glycine to form ALA. ALA is formed when the carboxyl group is released.
c. Carbamoyl phosphate reacts with aspartate to form carbamoyl aspartate, a precursor of orotate, a precursor of orotate and subsequently UMP.
d. Phosphoribosylpyrophosphate reacts with orotate to give orotidine-5′-phosphate, a precursor of UMP.

End-of-Chapter Questions
Review Questions

1. a. Essential amino acids are molecules that cannot be synthesized by an animal and must be obtained from its diet.
   b. Nitrogen balance is a physiological condition in which the body's nitrogen intake is equal to the nitrogen loss.
   c. In *de novo* pathways, molecules such as amino acids or nucleotides are synthesized from simple precursors.
   d. Biogenic amines are amino acids or amino acid derivatives that act as neurotransmitters.
   e. An excitotoxin is a molecule that can sufficiently overstimulate a cell, causing its death.
   f. A retrograde neurotransmitter is a molecule that is released from a postsynaptic cell and diffuses back to and promotes an action in a presynaptic cell.
   g. An analogue is a compound that is similar in structure to a biomolecule such as a nucleotide.
   h. Auxins are a class of plant growth regulators.
   i. Pernicious anemia is a disease caused by a deficiency of vitamin B_{12}; symptoms include anemia.

3. While nitrogen incorporation into organic molecules is thermodynamically favored, the conditions in the biosphere (T, P, and pH) are such that transitions have low kinetic probability. Only a few organisms possess the machinery to surpass the energy barrier to nitrogen compound synthesis.

5. Glutamate is synthesized from α-ketoglutarate by two means: (1) transamination, catalyzed by the aminotransferases (pyridoxal phosphate is a required coenzyme) and (2) direct amination, catalyzed by glutamate dehydrogenase. NADPH provides the reducing power for this reaction.

7. The liver uses alanine, along with serine, to manufacture glucose. Phenylalanine, glycine and proline levels are not affected to any greater extent than that seen in other tissues. The concentrations of the branched chain amino acids isoleucine and valine in blood remain unchanged.

9. Glutamine is produced in the following reactions:
   a. α-ketoglutarate + NADH + NH_3 + H^+ ⟶ glutamate
      Glutamate + NH_3 + ATP ⟶ Glutamine + ADP + P_i
   b. Methionine is produced in the following series of reactions:
      L-aspartate + ATP ⟶ β-Aspartylphosphate + P_i + ADP
      β-Aspartylphosphate + NADPH + H^+ ⟶ Aspartate β-semialdehyde
      Aspartate β-semialdehyde + NADPH + H^+ ⟶ Homoserine
      Homoserine ⟶ Cysteine ⟶ Homocysteine ⟶ Methionine
   c. Homoserine produced as in part b above reacts as follows:
      Homoserine + ATP ⟶ Phosphohomoserine + ADP + H^+
      Phosphohomoserine + H_2O ⟶ Threonine + P_i
   d. Glycine is produced in the following series of reactions:
      3-phosphoglycerate + NAD^+ ⟶ 3-phosphohydroxypyruvate + NADH H^+
      3-Phosphohydroxypyruvate + Glutamate ⟶ 3-phosphoserine + α-ketoglutarate
      3-phosphoserine + H_2O ⟶ Serine + P_i
      Serine + THF ⟶ Glycine + N^5,N^{10}-methylene THF + H_2O
   e. Serine + L-homocysteine ⟶ Cystathione + H_2O
      Cystathione + H_2O ⟶ Cysteine + α-ketobutyrate + NH_4^+

11. The two most prominent one-carbon carriers are THF (the biologically active derivative of folic acid) and S-adenosymethionine (SAM). THF plays important roles in the synthesis of several amino acids and the nucleotides. SAM is a methyl donor in the synthesis of numerous biomolecules, for example, phosphatidylcholine, epinephrine, and carnitine.

13. The ten essential amino acids in humans are isoleucine, leucine, lysine, methionine, phenylalanine, threonine, tryptophan, and valine. In addition, histidine and arginine are essential for infants. These amino acids are essential because they cannot be synthesized in required amounts by humans and must be included in the diet.

15. No. The two fused rings of the purine ring system result in steric interaction with the pentose. Pyrimidine rings contain only one ring and steric interactions with the pentose are minimal.

17. The reactions involved in the synthesis of the purine are outlined in Figure 14.24 on p. 490.

19. In the γ-glutamyl cycle glutathione is excreted from the cell γ-Glutamyltranspeptidase converts GSH to a γ-glutamylamino acid and Cys-Gly. The γ-glutamylamino acid is transported

into the cell where it is converted to 5-oxoproline and the free amino acid. 5-oxoproline is eventually reconverted to GSH. The Cys-Gly is transported into the cell where it is hydrolyzed to cysteine and glycine. The location of the γ-glutamyltransferase on the plasma membrane facilitates the transport process because its function is linked to transport. The conversion of the transported glutamylaminoacid is imme-

diately converted the 5-oxoproline and the free amino acid, therefore driving the reaction in favor of transport (uptake of amino acids).

21. Depressed synthesis or the total absence of glycine causes the accumulation of 5-phospho-β-D-ribosylamine, the other substrate of the enzyme phosphoribosylglycinamide synthase.

23. Reaction of pyridoxyl phosphate with alanine:

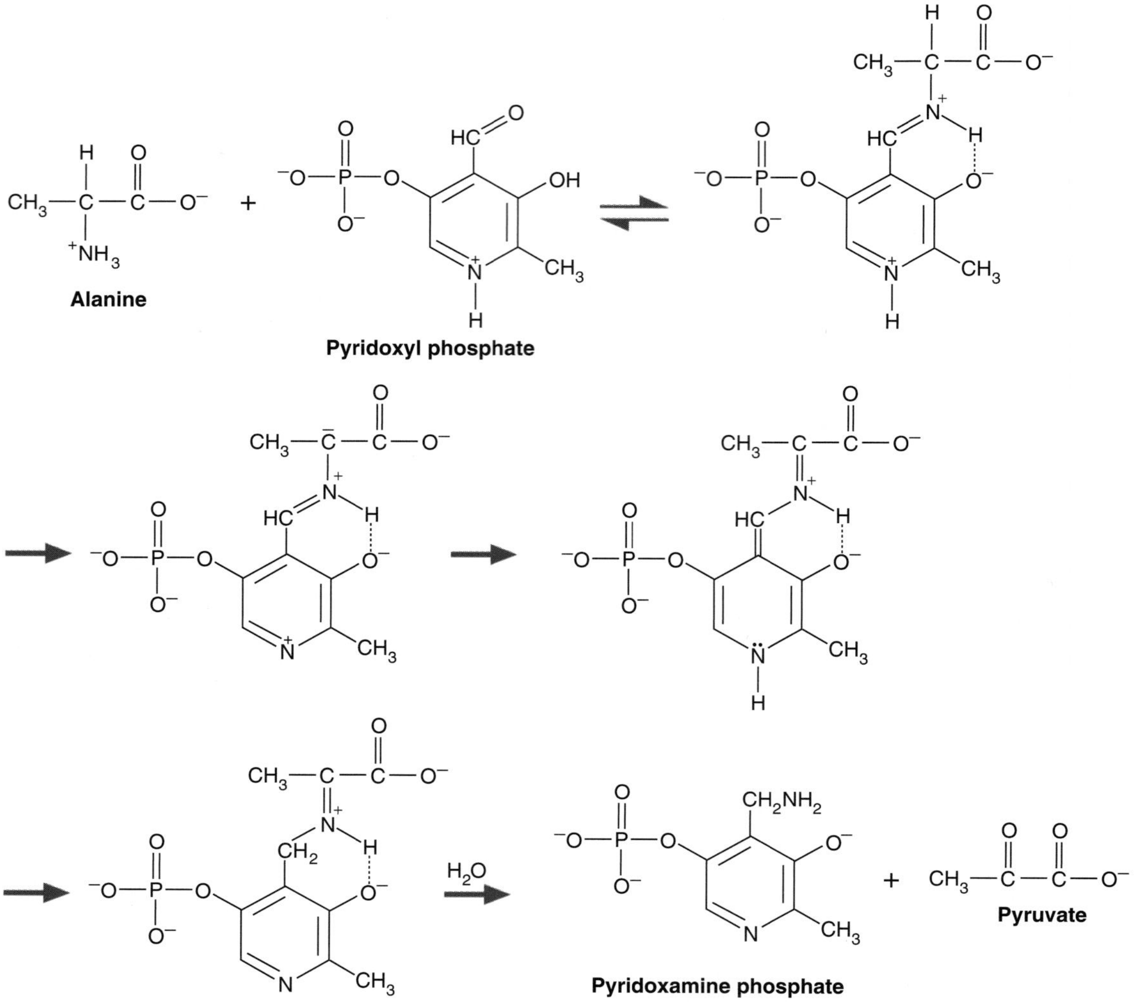

Alanine

Pyridoxyl phosphate

Pyruvate

Pyridoxamine phosphate

Reaction of pyridoxamine phosphate with α-ketoglutarate:

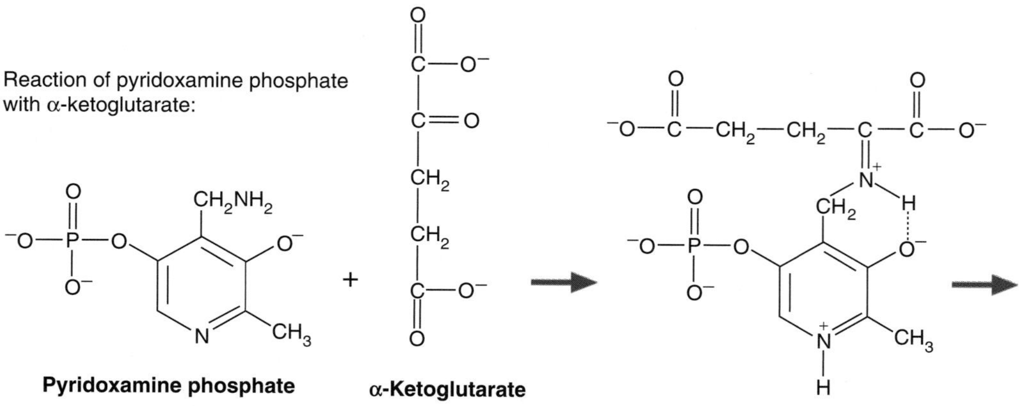

Pyridoxamine phosphate   **α-Ketoglutarate**

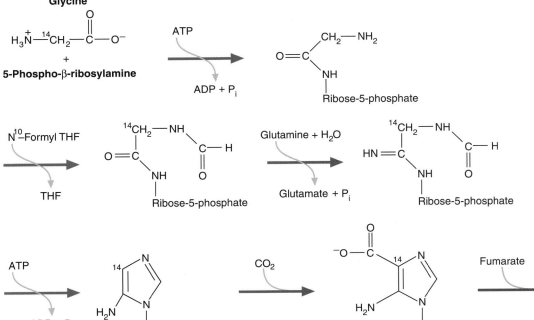

25. a. γ-Aminobutyric acid is an inhibitory neurotransmitter.
    b. Tetrahydrobiopterin is an essential cofactor in the hydroxylation of certain amino acids.
    c. Oxaloacetate is a citric acid cycle intermediate and the α-keto acid in transamination reactions involving aspartate.
    d. Phosphoribosylpyrophosphate is the source of the ribose moiety in nucleotide synthesis pathways.
    e. S-adenosylmethionine is a methyl transfer agent.

Thought Questions

1.

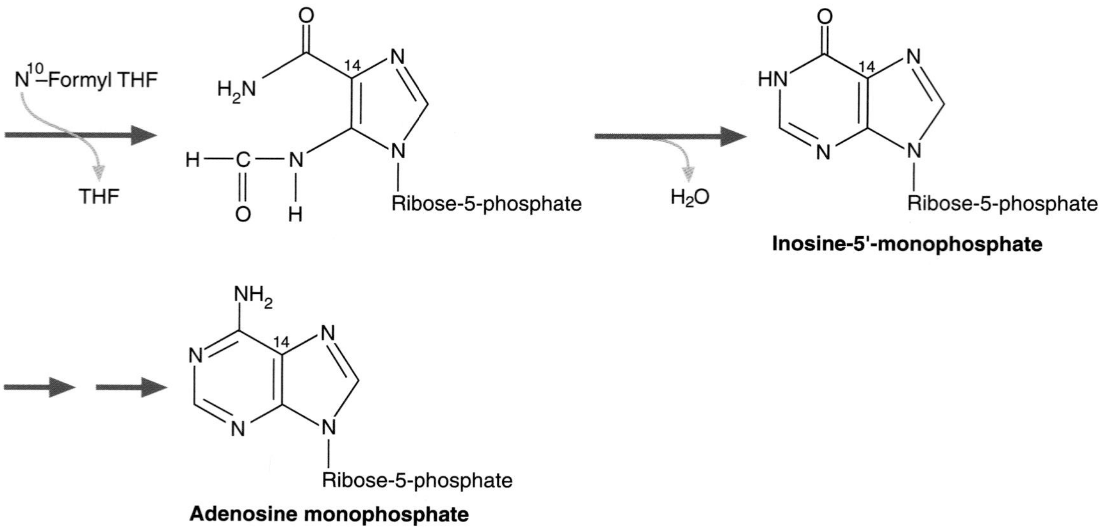

Inosine-5'-monophosphate

Adenosine monophosphate

3. Running is a physical activity that requires the rapid metabolism of both fatty acids and glucose. The most rapidly utilized source of glucose is blood glucose. Although certain amino acids can be absorbed easily in the intestine and used as substrates in gluconeogenesis in the liver, this process is slower than the immediate absorption of glucose into the blood.

5. $GSH + \cdot OH \longrightarrow GS\cdot + H_2O$

$2GS\cdot \longrightarrow GSSG$

7. The transamination of lysine:

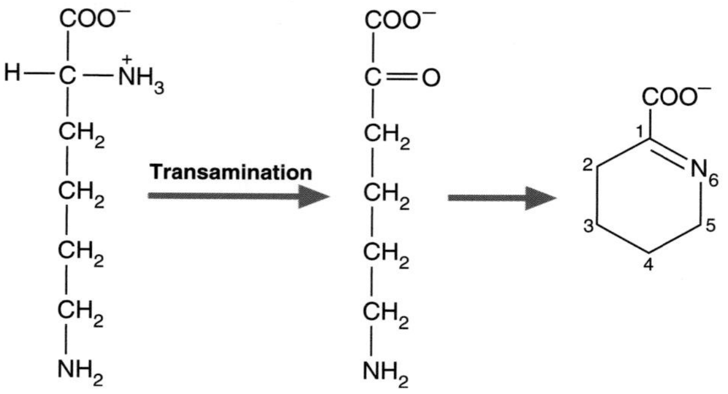

If lysine were transaminated, the ε amino group of the new keto acid (derived from lysine) would cyclize to form an intramolecular Schiff base. Consequently, the α-keto acid required to produce lysine by transamination cannot exist in appreciable quantities. Therefore, lysine cannot be produced by this reaction.

9. During the reaction catalyzed by serine hydroxymethyeltransferase the radiolabeled carbon atom of serine enters the THF pool as N^5,N^{10}-methylene THF. Because N^5,N^{10}-methylene THF is reversibly converted to N^5,N^{10}-methynyl THF and N^{10}-formyl THF (the coenzyme used in purine synthesis) some radiolabeled carbon atoms will enter the purine synthetic pathway. Because N^{10}-formyl THF is a required coenzyme in the reactions in which 5'-phosphoribosyl-N-formyl-glycinamide and 5-phosphoribosy-1-4-carboxamide-5- foramidoimidazole are synthesized ^{14}C will appear as C-2 and C-8 of the purine ring (shown below).

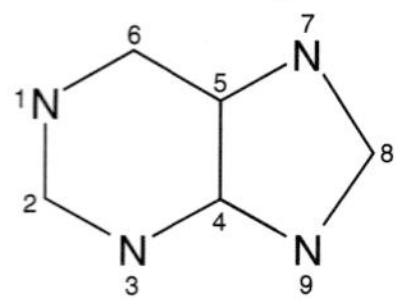

11. Breaking the bond perpendicular to the pyridinium ring generates a p orbital that can then interact with the π system of the pyridinium ring. The resulting delocalization of the charge stabilizes the carbanion.

CHAPTER 15

In-Chapter Questions

15.1 In newborn animals arginine will be an essential amino acid if the urea cycle is not yet fully functional.

15.2 Certain intestinal bacteria can release ammonia from urea molecules that diffuse across the membrane into the intestinal lumen. Treatment with antibiotics kills these organisms, thereby reducing blood ammonia concentration.

15.3

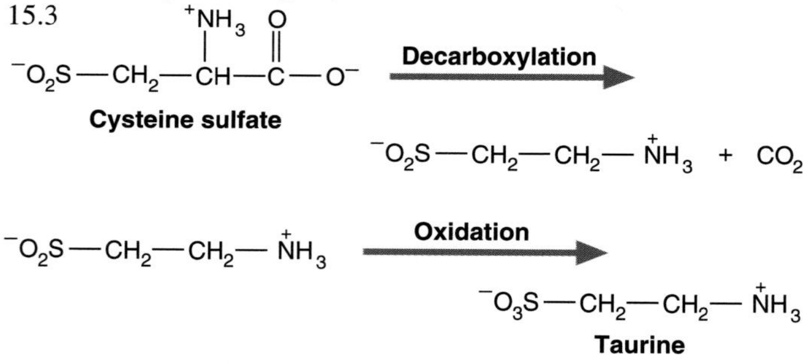

Cysteine sulfate

15.4 The following amino acids are both ketogenic and glucogenic: phenylalanine, isoleucine, lysine, tryptophan, and tyrosine.

15.5 a. Dopamine is inactivated in an oxidation reaction catalyzed by MAO to form 3,4-dihydroxyphenylacetaldehyde.

b. Serotonin is inactivated in a two-step pathway: After serotonin is oxidized by MAO to form 5-hydroxyindole-3-acetaldehyde, the product is further oxidized by MAO to form 5-hydroxyindole-3-acetate.

c. Epinephrine is inactivated by MAO to form 3,4-dihydroxyphenylglycolaldehyde.

d. Norepinephrine is oxidized by MAO to form 3,4-dihydroxyphenylglycolaldehyde.

15.6 Acetylcholine is normally degraded rapidly by cholinesterase. Drugs that block the action of cholinesterase prevent this hydrolysis. Consequently, acetylcholine molecules remain in the synaptic cleft for an extended time. There they can rapidly and reversibly bind and rebind to a reduced number of functional acetylcholine receptors. This process promotes the depolarization of the muscle cells.

15.7 Gout is caused by high levels of uric acid. Animals that do not suffer from gout possess the enzyme urate oxidase, which converts uric acid to allantoin. Unlike uric acid, which is relatively insoluble in blood, allantoin readily dissolves and is easily excreted.

15.8

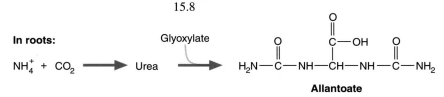

In roots:

$NH_4^+ + CO_2 \longrightarrow$ Urea $\xrightarrow{\text{Glyoxylate}}$ **Allantoate**

In leaves:

$\longrightarrow$ Urea + Glyoxylate

Urea $\longrightarrow$ $NH_4^+ + CO_2$

α-Ketoglutarate $\longrightarrow$ **Glutamate**

15.9 a. Urea is formed from ammonia, CO_2, and aspartate in the urea cycle.
  b. Uric acid is the oxidation product of purines.
  c. β-Alanine is produced in the degradative pathway of pyrimidines.

15.10 Suggested catabolic reactions of β-alanine and β-aminoisobutyrate.

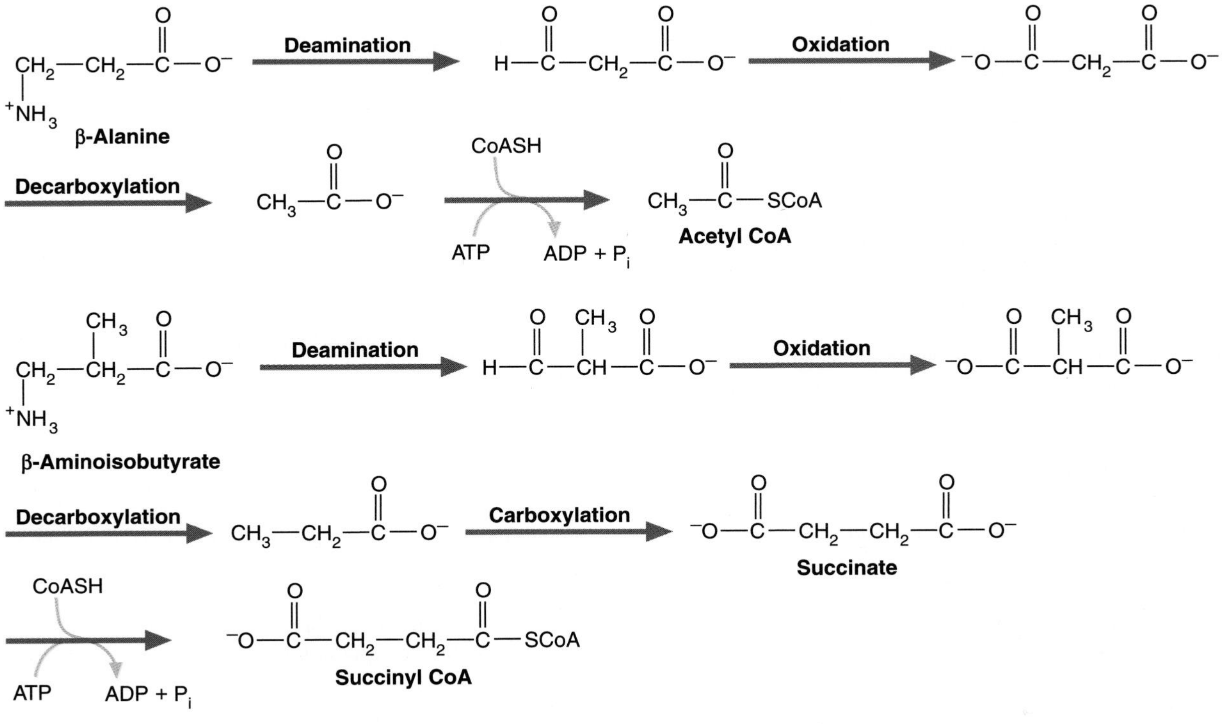

β-Alanine $\xrightarrow{\text{Deamination}}$ $\xrightarrow{\text{Oxidation}}$

$\xrightarrow{\text{Decarboxylation}}$ CH_3-C-O^- $\xrightarrow{\text{CoASH, ATP} \to \text{ADP} + P_i}$ **Acetyl CoA**

β-Aminoisobutyrate $\xrightarrow{\text{Deamination}}$ $\xrightarrow{\text{Oxidation}}$

$\xrightarrow{\text{Decarboxylation}}$ $CH_3-CH_2-C-O^-$ $\xrightarrow{\text{Carboxylation}}$ **Succinate**

$\xrightarrow{\text{CoASH, ATP} \to \text{ADP} + P_i}$ **Succinyl CoA**

End-of-Chapter Questions
Review Questions

1. a. In denitrification, nitrate is converted to atmospheric nitrogen.
    b. Ammonotelic animals excrete ammonia directly, that is, without conversion to less toxic molecules.
    c. The term *protein turnover* describes a process in which the proteins of living organisms are constantly synthesized and degraded.
    d. Ubiquination is a eukaryotic mechanism whereby proteins destined to be degraded are first covalently linked to the small protein ubiquitin.
    e. Ammonia intoxication is the potentially fatal condition caused by elevated blood ammonia.
    f. Humoral immune response is the production of antibodies by B lymphocytes.
    g. Hyperurecemia is high blood levels of uric acid.
3. The process of protein turnover promotes metabolic flexibility, protects a cell from the accumulation of abnormal proteins, and is a key feature of organismal developmental processes.
5. The metabolic products of amino acid degradation are acetyl-CoA, acetoacetyl-CoA, pyruvate, α-ketoglutarate, succinyl-CoA, fumarate, and oxaloacetate.
7. Refer to the following pages for a description of the degradation of each amino acid.
    a. lysine   p. 512
    b. glutamate   p. 514
    c. glycine   p. 511
    d. aspartate   p. 516
    e. tyrosine   p. 512
    f. alanine   p. 511
9. The first two reactions in the biochemical pathway that converts NH_4^+ to urea (i.e., the formation of carbamoyl phosphate and citrulline) occur in the mitochondrial matrix. Subsequent reactions that convert citrulline to ornithine and urea occur in the cytosol. Both citrulline and ornithine are transported across the inner membrane by specific carriers.
11. Individuals with PKU lack phenylalanine hydroxylase (phenylalanine-4-monooxygenase) activity so they cannot synthesize tyrosine from phenylalanine. Tyrosine is therefore an essential amino acid for these patients.
13. The term *Krebs bicycle* refers to two interlocking cyclic reaction pathways. The aspartate-arginosuccinate shunt of the citric acid cycle is responsible for regenerating the aspartate needed for the urea cycle from fumarate. The molecule that the two cycles have in common is arginosuccinate.
15. In ubiquination, the major mechanism of protein degradation, ubiquitin (a small hsp) is covalently attached to lysine residues of a protein with structural features such as oxidized amino acid residues that mark it for destruction. Once the protein is ubiquinated it is degraded by proteases in ATP-requiring reactions.
17. The following compounds yield uric acid when degraded: DNA, FAD, and NAD^+. All of these contain purines.

Thought Questions

1. Ammonia can be used to excrete waste nitrogen in certain aquatic species because they live in water. The toxic effects of ammonia are therefore immediately diminished after it is excreted into a large body of water. This strategy is not practical for mammals which are land animals who store liquid waste for intormittent excretion. Ammonia is very soluble in body fluids and would accumulate to toxic levels if not converted to a less toxic form. The toxic effects of ammonia, a relatively powerful base and a strong nucleophile, include depletion of glutamate and α-ketoglutarate, inhibition of amino acid transport and the Na^+-K^+ dependent ATPase, all of which contribute to brain damage.

3. If any of the urea cycle enzymes is missing, ammonia (the nitrogen-containing substrate of the cycle) cannot be metabolized. If any of the enzymes is defective (i.e., it does not catalyze its reaction at an appropriate rate), ammonia is metabolized slowly. Under both circumstances the body's ammonia concentration is excessively high.

5. 5,6,7.8-Tetrahydrobiopterin (BH_4) is required in the synthesis of neurotransmitters such as norepinephrine and serotonin, which are produced in brain. BH_4 is required in the reaction catalyzed by tyrosine hydroxylase in which tyrosine is hydroxylated to form L-dopa (a precursor of several neurotransmitters including dopamine and norepinephrine), and the reaction catalyzed by tryptophan hydroxylase in which tryptophan is hydroxylated to form 5-hydroxytryptophan. L-Dopa and 5-hydroxytryptophan must be supplied to individuals lacking the capacity to synthesize BH_4 because the latter molecule does not cross the blood-brain barrier.

7. a. Methylene ($-CH_2-$) groups
    b. Methyl groups
    c. Methyl groups
    d. Methyl groups

9. Caffeine (trimethylxanthine), which occurs in both beverages, is similar in structure to xanthine, the immediate precursor of uric acid. Some caffeine molecules are excreted as methyl uric acid, a molecule with solubility properties similar to those of uric acid. Recall that gout is caused by the precipitation of uric acid in the joints.

CHAPTER 16

In-Chapter Questions

16.1 Several series of signal transduction components are illustrated in Figure 8.17. The most prominent example is glucagon (signal), cAMP (messenger), and protein kinase (receiver).

16.2 a. The intestine engages in the digestion of food and transport of nutrients into blood.
     b. Liver has a key role in carbohydrate, lipid, and amino acid metabolism and monitors and regulates blood composition.
     c. During fasting or starvation, skeletal muscle provides amino acids to other organs. Cardiac muscle is the principal structural feature of the heart and is responsible for pumping blood carrying nutrients throughout the body.
     d. Adipose tissue functions as energy storage and in the generation of fatty acids for energy generation and glycerol, a substrate for gluconeogenesis in the liver.
     e. Kidney performs filtration of blood plasma and regulation of blood pH.
     f. Brain directs metabolic processes of the body and integrates sensory information required to obtain food.

16.3 Long-term fasting or low-calorie diets are interpreted by the brain as starvation. The brain responds by lowering the body's BMR. The majority of the energy is derived from fatty acid oxidation. The glucose needed for glucose-dependent tissues is generated via gluconeogenesis at the expense of muscle protein.

16.4 As blood glucose and insulin levels in blood drop back to normal, glucagon is released from the pancreas. Glucagon acts on the liver to prevent hypoglycemia by promoting glycogenolysis and gluconeogenesis. Glucagon stimulates glycogenolysis by triggering the synthesis of cAMP, which in turn initiates a cascade of reactions that lead to the activation of glycogen phosphorylase. Increased lipolysis, hydrolysis of fat molecules, provides glycerol molecules that are substrates for gluconeogenesis.

16.5 Cortisone, a steroid, is not degraded by the digestive system. In contrast, insulin, a protein, is inactivated because it is degraded to its amino acid components.

16.6 The types of noncovalent interactions that are involved in reversible binding of a signal molecule and its receptor include hydrogen bonding, hydrophobic interactions, and various types of electrostatic interactions, for example, salt bridges.

16.7 High levels of stress combined with lack of sleep or distorted wake/sleep cycles (e.g., shift work) leads to elevated levels of circulating cortisol. Cortisol depresses the synthesis and release of thymic hormones resulting in compromised development of T lymphocytes and a depressed immune response. Light inhibits the production of melatonin so its level falls when individuals have inadequate sleep. This reduces the melatonin-dependent inhibition of CRH release which amplifies the increase in circulating cortisol. The stress induced immunosuppression is therefore enhanced and prolonged.

16.8 The high blood glucose levels in untreated diabetics result in the loss of increasingly large amounts of glucose along with water in the urine, a condition that causes dehydration. In the absence of useable glucose, the body rapidly degrades fats and proteins to generate energy, Hence Aretaeus's observation that in this disease excessive weight loss and excessive urination are related.

16.9 Approximately 100,000 (or 10^5) molecules of target molecule (E_R) can be activated by a single molecule of hormone.

16.10 cAMP is generated from ATP by adenylate cyclase when a hormone molecule binds to its receptor. The interaction between the receptor and adenylate cyclase is mediated by a G protein G_S. As a consequence of hormone binding and the resulting conformational change, the receptor interacts with a nearby G_S protein. As G_S binds to the receptor, GDP dissociates. Then the binding of GTP to G_S allows one of its subunits to interact and stimulate adenylate cyclase, thus initiating cAMP synthesis. cAMP must break down quickly so the signaling mechanism can be precisely controlled.

16.11 The inhibition of GTP hydrolysis causes the subunit of G_S protein to continue activating adenylate cyclase. In intestinal cells, this enzyme activity opens chloride channels, causing loss of large amounts of chloride ions and water. The massive diarrhea caused by this process quickly leads to serious dehydration and electrolyte loss.

16.12 Nitroglycerin molecules are hydrolyzed in blood to yield NO. In turn NO relaxes smooth muscle cells in the walls of blood vessels. It is believed that NO activates guanylate cyclase, which subsequently promotes the intracellular sequestration of calcium ions that allows muscle cells to relax.

16.13 Both DAG and phorbol esters promote the activity of protein kinase C, which promotes cell growth and division. Phorbol esters provide initiated cells with a sustained growth advantage over normal cells. This condition is an early stage in carcinogenesis.

End-of-Chapter Questions
Review Questions

1. a. Ketoacidosis is a condition in which large amounts of ketone bodies occur in blood.
   b. In hyperlipoproteinemia, blood levels of lipoprotein are high.
   c. Neurophysins are proteins packaged with vasopressin and oxytocin in secretory granules that transport these hormones down axons from the hypothalamus to the posterior pituitary.
   d. Growth factors are a series of polypeptides and proteins that regulate the growth, differentiation, and proliferation of various cells.
   e. Chylomicron remnants are chylomicrons from which triacylglycerol molecules have been removed.
   f. In down-regulation, hormone receptor molecules are internalized by endocytosis.
   g. G proteins are multisubunit proteins that bind GTP. They mediate transmembrane signaling.

3. a. Corticotropin stimulates steroid synthesis in the adrenal cortex.
   b. Insulin promotes general anabolic effects, including glucose uptake by some cells and lipogenesis.
   c. Glucagon promotes glycogenolysis and lipolysis.
   d. Oxytocin stimulates uterine muscle contraction.
   e. LH stimulates the development of cells in the ovaries and testis and the synthesis of sex hormones.
   f. GnRF stimulates LH and FSH secretion.
   g. Somatostatin inhibits GH and TSH secretion, as well as the secretion of gastrin and glucagon.
   h. Vasopressin (or antidiuretic hormone) is involved in the regulation of blood osmolarity.
   i. FSH promotes ovulation and estrogen synthesis in ovaries and sperm development in testis.

5. NADPH, which is formed during the pentose phosphate pathway and reactions catalyzed by isocitrate dehydrogenase and malic enzyme, is used as a reducing agent in a wide variety of synthetic reactions (e.g. amino acids, fatty acids, sphingolipids and cholesterol). The degradation of some of these molecules (e.g., fatty acids and the carbon skeletons of the amino acids) results in the synthesis of NADH, a major source of cellular energy via the mitochondrial electron transport system.

7. The effects of hormone action include changes in gene expression (e.g., increased or decreased synthesis of specific proteins, including regulatory enzymes), and activation or inactivation of preexisting enzyme molecules.

9. For several weeks after the onset of fasting blood glucose levels are maintained via gluconeogenesis. During most of this period amino acids derived from the breakdown of muscle proteins are the major substrates for this process. Eventually, as muscle becomes depleted the brain switches to ketone bodies as an energy source. Consequently, the production of urea (the molecule used to dispose of the amino groups of the amino acids) declines.

11. One consequence of physical activity is the activation of the sympathetic nervous system, which in turn stimulates the adrenal gland to secrete epinephrine and norepinephrine. These hormones then activate the adipocyte enzyme hormone-sensitive lipase, which catalyzes the hydrolysis of triacylglycerol molecules to form glycerol and the fatty acids used to drive muscle contraction.

13. The functions of the kidney include (1) the excretion of water-soluble waste products, (2) the reabsorption of electrolytes, amino acids, and sugars from the urinary filtrate, (3) regulation

of pH, and (4) the regulation of the body's water content. In diabetes mellitus, the ability of the kidney to reabsorb glucose is overwhelmed and glucose spills over into the urine. The presence of glucose in the urine compromises the water recovery function of the kidney and dehydration results. This greatly affects the kidney's ability to maintain electrolyte and pH balance.

15. The metabolism of glutamine and glutamate generates ammonia which leaves the kidney in the urine, taking with it a proton. This process, along with the active transport of protons down a sodium gradient and into the kidney tubules, helps to maintain the blood pH at 7.4.

Thought Questions

1. The sustained generation of energy from fat requires ; large supply of citric acid cycle intermediates. Recall that oxaloacetate is derived from glucose via the enzymes of the glycolytic pathway and pyruvate carboxylase.

3. The second messenger is an effector molecule synthesized when a hormone (the first messenger) binds. It stimulates the cell to respond to the original signal. Second messengers also allow the signal to be amplified.

5. The recognition by the conscious centers in the brain that danger is imminent results in a discharge of epinephrine from the sympathetic nervous system and the adrenal medulla. The large amounts of glucose and fatty acids that flood into blood as a consequence of epinephrine's stimulation of glycogenolysis and lipolysis have several effects on the body. One effect, high blood glucose levels, provides the energy required for rapid decision-making processes in the brain. In addition, large quantities of glucose and fatty acids are required for strenuous physical activity if a decision is made to run away from the danger.

7. In uncontrolled diabetes mellitus massive breakdown of fat reserve results in the production of large quantities of ketone bodies. Two of the ketone bodies (acetoacetic acid and β-hydroxybutyric acid) are weak acids. The release of hydrogen ions from large numbers of these molecules overwhelms the body's buffering capacity.

9. The storage of preformed hormone molecules in secretory vesicles allows for a rapid response of the producing cells to metabolic signals. As soon as the appropriate signal is received the vesicles fuse with plasma membrane and (via exocytosis) release their contents into the bloodstream.

CHAPTER 17

In-Chapter Questions

17.1　According to the central dogma of molecular biology, the flow of information in living systems is summarized by the sequence

In this paradigm the information encoded into the base sequence of DNA perpetuates itself through replication, and is decoded in the form of RNA molecules. Information flow continues as RNA molecules become involved in the synthesis of the proteins, the molecules that constitute the structural and catalytic components responsible for the organism's functioning. Genome is the complete set of an organism's DNA. The transcriptome, the complete set of the RNA transcripts that are generated by a cell under specified conditions, can

be considered to be the cell's current operating instructions. The proteome is the complete set of the proteins that a cell makes and utilizes. Some of these proteins are structural. A subset of the proteome are the catalytic proteins, the enzymes, that catalyze the reactions that generate the complete set of organic biomolecules, referred to as the metabolome.

17.2　The types of noncovalent interactions that stabilize DNA structure are hydrophobic interactions, hydrogen bonds, base stacking, and electrostation interactions. The cumulative zippering effect of the hydrogen bonds between base pairs keeps the strands in the correct complementary orientation. The parallel stacking of the nearly planar bases is a stabilizing factor because of the cumulative effect of weak van der Waals forces.

In electrostatic interactions the potentially destabilizing force of the phosphate group charges (repulsion between nearby negative charges) is offset by the binding of magnesium ions and polycationic molecules such as the polyamines and histones.

17.3　The cytosine-guanine base pair with its three hydrogen bonds is more stable than the adenine-thymine base pair. The more CG bp there are, the more stable the DNA molecule. Structure b, with the least number of CG bp, will therefore denature first.

17.4　a. Ethanol will disrupt the hydrogen bonding in the base pairs and denature the ONA.
　　b. Heat, which easily disrupts hydrogen bonds, will cause DNA chains to separate and denature.
　　c. Dimethylsulfate is an alkylating agent that can cause transversion and transition mutations.
　　d. Nitrous acid deaminates bases.
　　e. Quinacrine is an intercalating agent that can cause frame shift mutations.

17.5　a. Caffeine is a base analogue.
　　b. Benzo[a]pyrene is a nonalkylating agent that is easily converted into a highly reactive form that forms adducts with bases.
　　c. Ethyl chloride is an alkylating agent.

17.6　The brain is especially sensitive to oxidative stress because it uses a greater proportion of oxygen than other tissues. Consequently, the chance of oxidative damage is also high. In addition, when most types of brain cells are irreversibly damaged by ROS, they cannot be replaced. In addition to hydroxyl radicals, other ROS that can contribute to oxidative stress in the brain include superoxide, hydrogen peroxide and singlet oxygen.

17.7　In A-DNA, the dehydrated form of DNA, the base pairs are no longer at right angles to the helical axis. Instead they tilt 20 degrees away from the horizontal as compared to B-DNA. The distance between adjacent base pairs is slightly reduced with 11 bp per helical turn instead of the 10.4 bp that occurs in the B-form. Each turn of the double helix of A-DNA occurs in 2.5 nm instead of the 3.4 nm of B-DNA. The diameters of A-DNA and B-DNA are 2.6 nm and 2.4 nm, respectively. The significance of A-DNA is unclear. It has been observed that its overall appearance resembles that of RNA duplexes and the RNA-DNA hybrids that form during transcription.

With a diameter of 1.8 nm, Z-DNA is considerably slimmer than B-DNA. It is twisted into a left-handed spiral with 12 bp per turn, each of which occurs in 4.5 nm instead of the 3.4 observed in B-DNA. Segments with alternating purine and pyrimidine bases are most likely to adopt the Z-DNA configuration. In Z-DNA the bases stack in a left-handed, staggered pattern that gives this form its flattened, non-grooved surface and its zig-zag appearance. The significance of Z-DNA is unresolved.

H-DNA (triple helix) segments can form when a poly-purine sequence is hydrogen bonded to a polypyrimidine sequence. H-DNA, which has been observed to form under low pH conditions, is made possible by nonconventional, Hoogsteen base pairing. H-DNA may play a role in recombination.

17.8 In the prokaryotic chromosome there is a protein core to which the circular DNA molecule is attached. In addition, HU protein binds to the DNA and facilitates its bending and supercoiling. In eukaryotic chromosomes, DNA forms complexes with the histones to form nucleosomes. The polyamines are polycationic molecules that bind to negatively charged DNA so the latter molecule can overcome charge repulsions between adjacent coils during the compression process.

17.9 The genome is the total set of DNA-encoded genetic information in an organism. A chromosome is a DNA molecule, usually complexed with certain proteins. Chromatin is the partially decondensed form of eukaryotic chromosomes. Nucleosomes are the repeating structural units of eukaryotic chromosomes formed by the interaction of DNA with the histones. A gene is a DNA sequence that codes for a polypeptide or an RNA molecule.

17.10 The information used to construct this model included the following:
1. The chemical structures and molecular dimensions of deoxyribose, the nitrogenous bases, and phosphate.
2. The 1 : 1 ratios of adenine : thymine and guanine : cytosine in the DNA isolated from a wide variety of species investigated by Erwin Chargaff (**Chargaff's rules**).
3. X-ray diffraction studies performed by Rosalind Franklin indicating that DNA is a symmetrical molecule and probably a helix.
4. The diameter and pitch of the helix estimated by Wilkins and his colleague Alex Stokes from other X-ray diffraction studies.
5. The recent demonstration by Linus Pauling that protein, another complex class of molecule, could exist in a helical conformation.

17.11 a. Tandem repeats are DNA sequences in which multiple copies are arranged next to each other; repeated sequence lengths vary from 10 bp to over 2000 bp.
   b. Centromeres are the structures that attach eukaryotic chromosomes to the mitotic spindle during mitosis and meiosis.
   c. Satellite DNA is DNA sequences arranged next to each other that form a distinct band when genomic DNA is digested and centrifuged; the original term for tandem repeats.
   d. Introns are noncoding intervening DNA sequences in a split or interrupted gene that are excised during mRNA formation.
   e. Exons are the coding regions in a split or interrupted eukaryotic gene.
   f. Microsatellites are core DNA sequences of 2 to 4 bp that are tandemly repeated 10 to 20 times.
   g. Transposition is the movement of a DNA segment from one site in the genome to another.

17.12 The genomes of prokaryotes are substantially smaller than those of eukaryotes. For example, the genome sizes of *E. coli* and humans are 4.6 Mb and 3000 Mb, respectively. Prokaryotic genomes are compact and continuous, that is, there are few, if any, noncoding DNA sequences. In contrast, eukaryotic DNA contains enormous amounts of noncoding sequences. Other distinguishing features of prokaryotic and eukaryotic DNA are the linkages of genes into operons in prokaryotes and intervening sequences in eukaryotic genes.

17.13 a. mRNA
   b. tRNA
   c. mRNA
   d. hnRNA

17.14 The most common bases found in RNA are adenine, uracil, cytosine, and guanine.

17.15 mRNA is the type of RNA that codes for the synthesis of polypeptides. rRNA along with ribosomal proteins are components of ribosomes, the molecular machines that synthesize proteins. Each type of tRNA molecule binds to a specific type of amino acid and transports it to the ribosome for assembly into proteins.

17.16 The antisence DNA sequence is 3′-CGTAAGCTTAACGTCT-GAGGACGTTAAGCCGTTA-5′; the mRNA sequence is 3′-CGUAAGCUUAACGUCUGAGGACGUUAAGCCGUUA-5′. The antisence RNA sequence is 3′-GCAUUCGAAU-UGCAGACUCCUGCAAUUCGGCAAU-5′.

17.17 The substrate of viral reverse transcriptase is an ssRNA genome. The product of this enzyme is a DNA strand that is complementary to the sequence of the ssRNA.

17.18 After a viral particle absorbs onto the surface of the bacterial cell, the viral genome is injected into the cell. If the virally infected cell enters the lytic phase, its molecular machinery begins synthesis of the components of new virus. If, instead, the viral genome is integrated into the host cell chromosome, the cell enters the lysogenic phase. At a later time the cell may enter the lytic phase.

17.19 After a retrovirus binds to the host cell, and its envelope fuses with the plasma membrane, the viral capsid is released into the cytoplasm. The viral reverse transcriptase synthesizes an ssDNA copy of the viral genome. This enzyme activity also converts the ssDNA into a double-stranded molecule. The dsDNA is then translocated into the nucleus where it integrates into a host cell chromosome. The integrated provirus is replicated each time the cell undergoes cell division. After the provirus is activated, new virus is created as newly synthesized vRNA and viral proteins are packaged with cell membrane and released from the host cell by a budding process.

17.20 The capsid is a protein coat made of interlocking protein molecules called capsomeres.

17.21 The genome types found in viruses are single- or double-stranded RNA or DNA.

17.22 In the original central dogma, the flow of genetic information is in one direction only, that is, from DNA to the RNA molecules, which then direct protein synthesis. The altered diagram indicates that the RNA genome of some viruses can replicate their RNA genomes (using a viral enzyme activity referred to as RNA directed–RNA polymerase) or undergo reverse transcription (i.e., synthesize DNA from an RNA sequence).

17.23 In an HIV ELISA assay, HIV antigens (e.g., gp41 or gp120) are attached to an inert support, such as a plastic lab dish. An aliquot of a patient's blood serum is added to the dish. After a short time the serum is removed and an antihuman antibody attached to an enzyme is added. When the enzyme's substrate is added to the dish, a color change indicates that the patient's blood contains antibodies to HIV. In the Western blot confirming test for HIV, the viral proteins are separated on polyacrylamide gel electrophoresis. The proteins are then transferred to nitrocellulose strips by a blotting procedure. Then the nitrocellulose strips are incubated with the patient's blood serum. Any HIV antibodies present in the serum bind to the viral proteins. The strips are then incubated with enzyme-linked antihuman antibodies. After the addition of the substrate, any color change is measured. The amount of the patient's HIV antibodies (if any) is then precisely determined.

End-of-Chapter Questions
Review Questions

1. a. Genetics is the study of inheritance.
   b. Replication is the process by which DNA strands are copied.
   c. Transcription is the synthesis of RNA from a DNA template.
   d. In polynucleotides the "backbone" is created by the 3′,5′-phosphodiester bonds that join the 5′-hydroxyl group of deoxyribose residues to the 3′-hydroxyl group of the sugar unit of another nucleotide.
   e. Bacteriophage are virus that attack bacteria.
   f. According to Chargaff's rules, there is a 1 : 1 ratio of adenine to thymine and guanine to cytosine, regardless of the source of DNA.
   g. A palindrome is a sequence that reads the same in both directions. In DNA a palindrome is an inverted repeat that has the potential to form a hairpin structure.
   h. Hoogsteen base pairing is a form of nonconventional base pairing that allows triplex H-DNA to form.
   i. Proteomics is the analysis of proteomes.
   j. The Alu family is a group of short DNA sequences that occur over 500,000 times in the human genome.
   k. Transcriptome is a complete set of RNA molecules that are produced within a cell under specified conditions.
   l. Satellite DNA is the original name for tandem repeats in which multiple copies of DNA sequences are arranged next to each other.
   m. A transposon is a DNA segment that carries the genes necessary for transposition; sometimes the term is reserved for transposable elements that can also contain genes unrelated to transposition.
   n. The hypochromic effect is a decrease in absorption intensity; used in the analysis of the nucleic acids.
   o. DNA fingerprinting is a variation of Southern blotting; the banding characteristics of minisatellites from different individuals are compared.
   p. STR DNA refers to short tandem repeats; repeated DNA sequences with 2 to 4 bp.
3. Supercoiling is a process in which DNA bends and twists to relieve torque, allowing DNA strands to be packaged in compact chromosomes.
5. Eukaryotic genomes are larger than those of prokaryotes. In contrast to prokaryotic genomes, which consist entirely of genes, the majority of eukaryotic DNA sequences do not appear to have coding functions. Most eukaryotic genes are not continuous (i.e., they usually contain introns) unlike those of prokaryotes.
7. RNA molecules differ from DNA in the following ways: (1) RNA contains ribose instead of deoxyribose, (2) the nitrogenous bases in RNA differ from those of DNA (e.g., uracil replaces thymine and several RNA bases are chemically modified), and (3) in contrast to the double helix of DNA, RNA is single-stranded.
9. The transition from B-DNA to Z-DNA can occur when the nucleotide base sequence is composed of alternating purine and pyrimidines (e.g., CGCGCG). Because alternate nucleotides can assume different conformations (syn or anti), these DNA segments form a left-handed helix. The phosphate groups in the backbone of this DNA conformation zig-zag hence the name Z-DNA.
11. According to Chargaff's rules, if a DNA sample contains 21% adenine then it also contains 21% thymine. If the A-T content is 42%, then the G-C content is 58%. Consequently the guanine and cytosine percentages are both 29% in the DNA sample.
13. The physical conditions that cause DNA denaturation include heat, low salt concentration, and extremes in pH.
15. The complementary DNA strand (written in the standard 5′ to 3′ direction is 5′-AACGATAACGGCCCCT-3′. The RNA strand is 5′- AACGAUAACGGCCCCU-3′.

Thought Questions

1. The principal structural difference between DNA and RNA is the 2′OH group of ribose in RNA molecules. In DNA, which lacks the 2′OH group in the deoxyribose sugar, hydrogen-bonded complementary strands can easily adopt the B-form double helix. In contrast, double-stranded regions of RNA molecules cannot adopt this conformation because of steric hindrance. Instead, they adopt the less compact A-helical form in which there are 11 bp per turn and the base pairs tilt 20° away from the horizontal.
3. The polyamines are positively charged at pH 7, which promotes binding to the negative charges of the DNA backbone. Polyamine binding overcomes the mutual repulsion of the adjacent DNA chains, packing the chains more closely.
5. Relaxed circular DNA with a nicked strand is less compact than supercoiled circular DNA and so has a lower effective density. As a result it will not migrate as far in the centrifuge tube as will the supercoiled DNA.
7. Because nucleotide base sequences and amino acid sequences are such completely different "languages," a complex mechanism is required for the "translation" of one type of information into another. In the absence of any evidence to the contrary it does not appear likely that information expressed in proteins can be utilized to direct the synthesis of nucleic acids.
9. Nuclei and mitochondria are separately obtained from source tissue by utilizing cell homogenization followed by density gradient centrifugation. The nucleic acids in each organellar fraction are then extracted with the aid of detergents, solvents, and proteases (to remove proteins). RNA is removed by treating each sample with RNase. The DNA from both types of organelles is then further purified by centrifugation.
11. Each cell is constantly receiving information from its environment. Cells adapt to changing conditions as information in the

form of nutrients, hormones, growth factors, and other types of molecules triggers changes in their molecular mechanisms that ultimately cause changes in gene expression. The transcriptome, the set of mRNA molecules produced under specified conditions, is a measure of the current status of gene expression.

13. At the crime scene, a forensic expert collects biological specimens such as blood, hair, and saliva. Once these specimens are delivered to the lab, they are analyzed and compared with the DNA of the victim. Any DNA not belonging to the victim is assumed to belong to a person, or persons, present during the time when the crime was committed. If a suspect is identified, his or her DNA profile (obtained from a swab of cheek cells or from a court-ordered blood sample) is compared with that obtained from crime scene specimens. If there is no obvious suspect, the crime scene specimens can be compared to the DNA profiles in the statewide database. This strategy has been remarkably successful in the identification of individuals later found guilty not only of recent murders, but also those from "cold cases" in which crime scene specimens had been preserved. The technology that makes this success possible includes PCR, RFLP, and STR-DNA analysis.

CHAPTER 18

In-Chapter Questions

18.1 a. Primase is an RNA polymerase that catalyzes the synthesis of short RNA segments called primers that are required to generate a starting point in DNA synthesis.
   b. A primer is a short oligonucleotide required as a starting point in the synthesis of a polynucleotide.
   c. Topoisomerases are enzymes that prevent tangling of DNA strands; they relieve torque that occurs during DNA synthesis ahead of replication machinery.
   d. Polymerase is a large multienzyme complex that forms phosphodiester bonds during polynucleotide synthesis.
   e. b-Protein is a component of DNA polymerase III; a "sliding clamp" protein that forms a ring around DNA during replication.

18.2 a. Replisome is a large protein complex composed of polIII, the primosome, and helicases that replicates DNA in *E. coli*.
   b. Primosome is a multienzyme complex involved in the synthesis of the RNA primers at various points along the DNA template strand during DNA replication.
   c. A replicon is a unit of a genome that contains a replication origin and regulatory regions required for replication.
   d. The DNA strand that is synthesized continuously in the 5′ to 3′ direction during DNA synthesis is the leading strand.
   e. An Okazaki fragment is one of several short oligodeoxynucleotide segments that are synthesized and then linked in the lagging strand simultaneously with the continuous synthesis of the leading strand.

18.3 Replication begins when DnaA monomers bind to oriC forming a nucleosome-like structure. This requires ATP and HU. Localized melting permits the binding of the DnaB/DnaC complex. Dissociation of DnaC allows the DnaB helicase to unwind the helix in preparation for DNA replication. The DnaA complex dissociates, SSB keeps the single strands apart and the replication fork is now ready for the assembly of the replisome.

18.4 Briefly, prokaryotic DNA replication consists of DNA unwinding, RNA primer formation, DNA synthesis catalyzed by DNA polymerase and the joining of Okazaki fragments by DNA ligase. Prokaryotic DNA replication differs from the eukaryotic process in that prokaryotic replication is faster, the Okazaki fragments are longer and there is usually only one origin of replication per chromosome (eukaryotes have many per chromosome).

18.5 In excision repair short damaged sequences (e.g., thymine dimers) are excised and replaced with correct sequences. After an endonuclease deletes the damaged single-stranded sequence, a DNA polymerase activity synthesizes a replacement sequence using the undamaged strand as a template. In photoreactivation repair a photoreactivating enzyme uses light energy to repair pyrimidine dimers. In recombinational repair damaged sequences are deleted. Repair involves an exchange of an appropriate segment of the homologous DNA molecule.

18.6 When antibiotics are used in large quantities, the bacterial cells that possess resistance genes (acquired through spontaneous mutations or through intermicrobial DNA transfer mechanisms such as conjugation, transduction, and transformation) survive and even flourish. Because of antibiotic use, which acts as a selection pressure, resistant organisms (once only a minor constituent of a microbial population) become the dominant cells in their ecological niche.

18.7 Genetic recombination promotes species diversity. General recombination, a process in which segments of homologous DNA molecules are exchanged, is most commonly observed during meiosis. In site-specific recombination, protein-DNA interactions promote the recombination of nonhomologous DNA. Transposition is an example of site-specific recombination in which genetic elements are moved from one site in the genome to another.

18.8 a. Transposition is the movement of a piece of DNA from one site in a genome to another.
   b. Conjugation is an unconventional sexual mating between bacterial cells; a donor cell transfers a DNA segment into a recipient cell through a specialized pilus.
   c. Transduction is the transfer of DNA between bacterial cells mediated by a bacteriophage.
   d. Transformation occurs when naked DNA fragments enter a bacterial cell and are introduced into a bacterial genome.
   e. A transposon is a DNA segment that contains the genes required for transposition.

18.9 Most gene duplications are apparently a consequence of accidents during genetic recombination. Examples of possible causes of gene duplication are unequal crossing-over during synapses and transposition. After a gene has been duplicated, random mutations and genetic recombination may introduce variations.

18.10 Marker genes are easily detectable genes that are incorporated into plasmid vectors. Detection of the marker gene in a cell in a mixed population of cells indicates that it has been transformed (i.e., it has incorporated the plasmid into its genome).

18.11 PCR begins by adding Taq polymerase, the primers, and the ingredients for DNA replication to a heated sample of the target DNA. As the mixture cools, the primers attach to their complementary sequences on either side of the target sequence. Each strand then serves as a template for DNA replication. At the end of this process, referred to as a *cycle*, the copies of the target sequence have been doubled. The process can be repeated indefinitely, synthesizing an extraordinary number of copies. After 15 replications, 32,768 (or 2^{15}) copies have been produced.

18.12 Genomic libraries are produced in a process, referred to as **shotgun cloning,** in which a genome is randomly digested. The range of fragment sizes, which is determined by the type of restriction enzyme and experimental conditions chosen, must be compatible with the vector. cDNA libraries are produced from mRNA molecules using reverse transcriptase. They are used to determine which genes are currently being expressed in a cell.

18.13 A DNA chip, or microarray, is a solid support such as glass or plastic to which thousands or hundreds of thousands of oligonucleotides or ssDNA fragments have been attached. Analysis of a microarray can provide information about gene expression.

18.14 a. Shotgun cloning is the fragmentation of genomic DNA by selected restriction endonucleases and the incorporation of the fragments into a chosen vector to create a library.
   b. A bacterial artificial chromosome is a derivative of a large *E. coli* plasmid used to clone DNA sequences as long as 300 kb.
   c. Electroporation is the treatment of cells with electric current that promotes uptake of foreign DNA.
   d. A vector is a DNA molecule capable of replication that is used to transfer foreign DNA sequences into a host cell.
   e. Recombinant DNA technology is a set of techniques used to cut and splice together DNA molecules from different sources.

18.15 a An operon is a set of linked genes that are regulated by the same promoter region.
   b. A promoter is a DNA sequence immediately before a gene that is recognized by RNA polymerase and signals the start point and direction of transcription.
   c. Spliceosome is a multicomponent complex containing RNA and protein that catalyzes RNA splicing.
   d. The DNA strand that has the same base sequence as the RNA transcript (with thymine instead of uracil) is the coding strand.
   e. Consensus sequences are the average of several similar sequences; they represent the most likely nucleotide that may occur in each position in the sequence; they are usually associated with a specific function.

18.16 RNA polymerase I, located in the nucleolus, transcribes the large rRNAs. The precursors of mRNAs and most snRNAs are transcribed by RNA polymerase II. Polymerase III is responsible for transcribing the precursors of the tRNAs and 5S rRNA.

18.17 Bacteria can permanently acquire the capacity to produce a toxin when the viral toxin gene becomes incorporated into the bacterial chromosome or a self-replicating plasmid. Comparison of modern group A streptococcus with the organism that caused a similar disease in the 1920s requires the production via PCR of a series of DNA probes from the modern organism. These probes are then used in an *in situ* hybridization

investigation of preserved specimens to determine any similarities or differences between the two organisms.

18.18 a. Euchromatin is the less condensed form of chromatin; it has varying levels of transcriptional activity.
   b. Heterochromatin is a highly condensed form of chromatin; it has no transcriptional activity.
   c. Gene amplification is a process in which multiple rounds of replication of a gene produce multiple copies.
   d. A DNA sequence within an operon involved in regulation to which a repressor protein can bind is an operator.
   e. A repressor is a protein that binds to the operator site, thus preventing transcription.

18.19 a. Translational control
   b. Genomic control
   c. RNA processing
   d. Genomic control

18.20 Because phytochrome has been demonstrated to mediate numerous light-induced plant processes, it appears reasonable to assume that it does so in part by interacting with light-response elements (LRE) in plant cell genomes. Presumably, phytochrome influences gene expression by binding, either alone or as part of a complex, to various LREs when its chromophore is activated by light.

End-of-Chapter Questions

Review Questions

1. a. Chemoreceptors are protein receptors on or near the external surface of a cell's plasma membrane that bind specific chemicals or nutrients, thus triggering chemotaxis.
   b. Structural genes are DNA segments that code for polypeptides.
   c. Recombination is the rearrangement of DNA sequences involving the exchange of segments from different molecules.
   d. In semiconservative replication, each new DNA molecule possesses one new strand and one old strand.
   e. A replisome is the protein complex that replicates DNA molecules.
   f. OriC is the replication initiation site on the *E. coli* chromosome.
   g. Transcription is the process in which an RNA molecule with a base sequence complementary to the template strand of DNA is synthesized.
   h. Protooncogenes are normal genes that code for molecules involved in cell cycle control.
   i. Spliceosomes are large complexes composed of proteins and snRNA in which exons are spliced together during RNA processing.
   j. Oncogenes are mutated protooncogenes that promote the formation of cancerous tumors.

3. The Meselson-Stahl experiment resulted in a single band of DNA on CsCl gradient centrifugation after one round of cell division. Its density was intermediate between heavy and light DNA. This could only occur if the newly synthesized dsDNA contained an old 'light' strand and a new 'heavy' strand.

5. a. Helicase is an enzymatic activity that relieves torque generated by supercoiling ahead of the replication machinery.
   b. Primase is an enzymatic activity that catalyzes the synthesis of RNA primers.

c. DNA polymerase is an enzymatic activity that catalyzes several reactions during DNA replication.

d. DNA ligase forms phosphodiester linkages between newly synthesized DNA fragments.

e. Topoisomerase is an enzymatic activity that prevents the tangling of DNA strands during DNA replication.

f. DNA gyrase facilitates the separation of DNA strands during prokaryotic replication.

7. a. ROS may cause single and double strand breaks, pyrimidine dimers, and the loss of purine and pyrimidine bases.

b. Because caffeine is a base analogue of thymine, it can cause transition mutations.

c. Small alkylating agents attach to the nitrogen atoms of the purines and pyrimidines, destabilizing glycosidic linkages (leading to depurination) and interfering with hydrogen bonding and promoting both transversion and transition mutations.

d. Large alkylating agents have the same effects as small alkylating agents but in addition they behave similarly to intercalating agents leading to frame shift mutations and breakage of the DNA chain.

e. Nitrous acid deaminates bases. For example, cytosine is converted to uracil.

f. Intercalating agents cause deletion or insertion mutations.

9. In excision repair, a series of enzymes remove damaged nucleotides and replace them with correct ones. Photoreactivation repair involves the use of light to repair thymine dimers. In recombinational repair, a mechanism that can eliminate certain types of damaged DNA sequences that have not been repaired before replication, the undamaged parental strands recombine into the gap left after the damaged sequence has been removed.

11. Most mutations are silent. Of those that do affect the functioning of an organism, most are deleterious because of the complex nature of living processes. Change in the properties of any of the thousands of different gene products is potentially disruptive. Only on rare occasions does a mutation improve the viability of an individual organism.

13. In replicative transposition, a replicated copy of a transposable element is inserted into a new chromosome location in a process that involves the formation of an intermediate called a cointegrate. In nonreplicative transposition, sequence replication does not occur, that is, the transposable element is spliced out of its donor site and inserted into the target site. The donor site must be repaired.

15. Because DNA is constantly exposed to disruptive processes, its structural integrity is highly dependent on efficient repair mechanisms, The life span of an organism is dependent on the health of its constituent cells, which is in turn dependent on the timely and accurate expression of genetic information. Consequently, the capacity of the organisms in a species to maintain the .integrity of DNA molecules is an important factor in determining life span.

17. Marker genes are useful in recombinant DNA technology because their function is known and their presence, which indicates that a successful recombinant event has occurred, is easily detected. For example, an antibiotic resistance gene, which codes for the synthesis of a substance that provides protection for a bacterium from the effects of an antibiotic, allows the growth of recombinant cells in a medium containing that antibiotic. Cells that do not contain the marker gene, that is, those in which the recombinant DNA is not present, do not survive.

19. a. Transcription factors are proteins that regulate or initiate RNA synthesis by binding to specific DNA sequences called response elements.

b. RNA polymerase is one of a group of enzymes that transcribe a DNA sequence into an RNA product.

c. A promoter is a DNA sequence immediately before a gene that is recognized by RNA polymerase and signals the start point and direction of transcription.

d. A sigma factor is a bacterial protein that facilitates the binding of the core enzyme of RNA polymerase to the initiation site during transcription.

e. An enhancer is a eukaryotic DNA sequence that can increase the expression of a gene.

f. TATA box is a consensus sequence in eukaryotic DNA that occurs within promoters for RNA polymerase II.

21. In relatively simple genomes, such as those in bacteria, operons provide a convenient mechanism for regulating genes. Proteins required in the same metabolic pathway or functional process are synthesized together because their genes are controlled by the same promoter.

Thought Questions

1. In the Meselson-Stahl experiment **all** of the nitrogen was N^{15}. To accomplish the same effect with a carbon isotope all of the carbon in the medium would have to be isotopically pure in both phases of the experiment. This would be prohibitively expensive.

3. DNA replication time is calculated as follows:

$$\frac{150,000,000 \text{ base pairs}}{50 \text{ bases/s}} = 3 \times 10^6 \text{ s} = 34.5 \text{ days}$$

Consequently, approximately one month is required for this DNA replication. Eukaryotic DNA synthesis is significantly faster than expected because each chromosome contains multiple replication units (replicons).

5. Mustard gas cross links the strands in DNA with permanent covalent bonds.

7. Recall that phorbol esters mimic the action of DAG, the normal cell metabolite that activates protein kinase C (PKC). PKC initiates a phosphorylation cascade that results in the activation of numerous molecules involved in cell growth and division, including jun and fos, which then combine to form AP-1. AP-1 is a transcription factor whose presence promotes cell division. Its formation causes an affected cell to have a growth advantage over nearby cells. Because phorbol esters are tumor promoters any exposure to them increases the risk that initiated cells may progress toward a cancerous state.

9. Because the Rb gene codes for a tumor suppressor, retinoblastoma only occurs when both copies have been damaged or deleted. Usually a long period of time is required for random mutations to cause this event. In hereditary retinoblastoma, in which an affected individual possesses only one functional Rb gene, the time necessary for a random mutation to inactivate the second Rb gene is significantly less than that required for the inactivation of both genes that cause the nonhereditary version of the disease.

11. Gene amplification, the selective duplication of certain genes, can occur via a reverse transcriptase–mediated event. The

creation of one or more cDNAs from an mRNA is followed by insertion of these sequences into the genome.

CHAPTER 19

In–Chapter Questions

19.1 Answer d is the process of translation.

19.2 a. A codon is a sequence of three nucleotides in mRNA that directs the incorporation of an amino acid during protein synthesis or acts as a start or stop signal.
   b. Degenerate code refers to two or more codons that code for the same amino acid.
   c. Reading frame is a set of contiguous triplet base sequences in an mRNA molecule.
   d. Open reading frame is a series of triplet base sequences in an mRNA molecule that do not contain a stop codon.
   e. Universal code means the coding signals for amino acids in protein synthesis are found throughout the living world.

19.3 The amino acid sequence of the beginning of the polypeptide is Met-Ser-Pro-Thr-Ala-Asp-Glu-Gly-Arg-Arg-Trp-Leu-Ile- Met-Phe. The mutation types in the altered mRNA sequences are (a) insertion of one base, (b) deletion of one base, (c) insertion of two bases, (d) deletion of three bases. The consequences of these mutations are altered amino acid sequences of the polypeptides produced from mRNA. In (a), (b), and (c) a frame shift occurs. Therefore the amino acid sequences past the mutation are different. In (d) no frame shift occurs because three bases are deleted. In this case, the only difference between the normal polypeptide and the mutated version is the deletion of a single amino acid.

19.4 Assuming that the DNA sequence given is the coding strand, the mRNA sequence is 5'-GGUUUA-3' and the anticodons are 5'-UAA-3'. If the DNA sequence is the template strand, the mRNA sequence is 5'-UAAACC-3' and the anticodons are 5'-GGU-3' and 5'-UUA-3'.

19.5 The possible choices for mRNA codon base sequences for the peptide are:

Tyr—Leu—Thr—Ala—

5'-UAU-3'	CUU	ACU	GCU
UAC	CUC	ACC	GCC
	CUA	ACA	GCA
	CUG	ACG	GCG
	UUA		
	UUG		

The possible choices for the DNA sequences that code for the peptide are:

Tyr—Leu—Thr—Ala—

3'-ATA-5'	GAA	TGA	CGA
ATG	GAG	TGG	CGG
	GAT	TGT	CGT
	GAC	TGC	CGC
	AAT		
	AAC		

The possible choices for the tRNA anticodons that code for the peptide are:

Tyr—Leu—Thr—Ala—

3'-AUA-5'	GAA	UGA	CGA
AUG	GAG	UGG	CGG
	GAU	UGU	CGU
	GAC	UGC	CGC
	AAU		
	AAG		

19.6 The aminoacyl-tRNA synthetases correctly attach each amino acid to its cognate tRNA and proofread the product.

19.7 a. Translocation is the movement of the ribosome along the mRNA during translation.
   b. Termination is the phase of translation in which the stop codon terminates translation and the newly synthesized polypeptide is released from the ribosome.
   c. Elongation is the phase during translation in which the polypeptide chain grows in length by one amino acid residue at a time.
   d. A polysome is an mRNA molecule with several ribosomes attached.
   e. Releasing factors are proteins involved in the termination of translation.

19.8 The proteins involved in the initiation of prokaryotic protein synthesis are: IF-1 (binds to the A site of the 30S subunit, blocking it during initiation), IF-2 (binds to the 30S subunit and promotes the binding of the initiating tRNA to the initiation codon of mRNA), and IF-3 (prevents the 30S subunit from binding prematurely to the 5OS subunit).

19.9 a. Amide linkage
   b. Phosphodiester bond
   c. Hydrogen bonds

19.10 The large subunit contains the catalytic site for peptide bond formation. The small subunit serves as a guide for the translation factors required to regulate protein synthesis. Together the two subunits come together and form a molecular machine that polymerizes amino acids in a sequence specified by the base sequence in the mRNA molecule.

19.11 The formation of an ADP-ribosylated derivative of eEF-2 affects the three-dimensional structure of this protein factor. Presumably protein synthesis is arrested because the ability of eEF-2 to interact with or bind to one or more ribosomal components is altered.

19.12 The major classes of eukaryotic post translational modifications are proteolytic cleavage (the hydrolysis of specific peptide bonds), glycosylation (attachment of sugar residues to specific amino acid residues in the protein), hydroxylation (the adding of OH groups to proline and lysine residues), phosphorylation (the addition of phosphate groups to specific amino acid residues on a protein), lipophilic modification (covalent attachment of lipid groups to a protein), methylation (attachment of methyl groups), disulfide bond formation (formation of -S-S- bonds between cysteine residues), and protein splicing (a specific segment of a polypeptide is removed and the remaining ends are joined covalently by an amide linkage).

19.13 To ensure that proteins end up in a location appropriate to their function in a timely and predictable way, it is necessary to have a targeting mechanism. The signaling process begins with specific signal sequences, which determine where translation will be completed. Specific localization sequences and/or posttranslational modification of the product protein then ensures delivery of the protein to its target location.

19.14 a. Signal peptide is a short sequence typically near the amino terminal of a nascent polypeptide that determines its cellular destination.
   b. Docking protein is a heterodimer that assists in the binding of ribosomes to the RER.
   c. Signal recognition particle is a large molecular complex consisting of six proteins and a small RNA molecule that binds to the ribosome and temporarily arrests translation.

d. Translocon is an integral membrane protein that mediates polypeptide translocation through or into a membrane.

e. Posttranslational translocation occurs when previously synthesized polypeptides are transported across an organelle membrane (mitochondrion, chloroplast, lysosome) with the aid of one or more signal sequences and accessory proteins.

19.15 The major mechanisms used by eukaryotes to control translation are mRNA export (transcription and translation are spatially separated; the export of processed mRNAs can be selectively blocked), mRNA stability (mRNAs have several destabilizing sequences that affect the molecule's longevity, i.e., its susceptibility to nucleases), negative translational control (the translation of some RNAs is specifically blocked by the binding of repression proteins near their 5′-ends), initiation factor phosphorylation (the phosphorylation of eIF-2 increases translation rate of certain mRNAs), and translocational frameshifting (certain RNAs have structural information that if activated results in a +1 or −1 change in reading frame; this allows for more than one polypeptide from a single mRNA).

19.16 A preprotein is the inactive precursor of a protein with a removable signal peptide. A proprotein in an inactive precursor protein. A protein is a fully functional product of translation.

19.17 After the synthesis of the plastocyanin precursor in cytoplasm, the first import signal mediates the transport of the protein into the chloroplast stroma. After this signal is removed by a protease. a second import signal mediates the transfer of the protein into the thylakoid lumen. Plastocyanin then binds a copper atom, folds into its final three-dimensional structure, and associates with the thylakoid membrane.

End–of–Chapter Questions
Review Questions

1. The genetic code is degenerate (several codons have the same meaning), specific (each codon specifies only one amino acid), and universal (with a few exceptions each codon always specifies the same amino acid). In addition, the genetic code is nonoverlapping and without punctuation (i.e., mRNA is read as a continuous coding sequence).

3. The sequential reactions that occur within the active site of aminoacyl-tRNA synthesis are: (1) the formation of aminoacyl-AMP , which contains a high-energy mixed anhydride bond, and (2) linkage of the aminoacyl group to its specific tRNA.

5. The major differences between prokaryotic and eukaryotic translation are speed (the prokaryotic process is significantly faster), location (the eukaryotic process is not directly coupled to transcription as prokaryotic translation is), complexity (because of their complex life styles, eukaryotes possess complex mechanisms for regulatory protein synthesis, e.g., eukaryotic translation involves a significantly larger number of protein factors than prokaryotic translation), and post translational modifications (eukaryotic reactions appear to be considerably more complex and varied than those observed in prokaryotes.)

7. Kinetic proofreading is a mechanism that ensures that the correct codon-anticodon pairing occurs in the A site of ribosomes. In eukaryotes eEF-1α mediates the binding of aminoacyl-tRNAs to the A site. When the correct pairing occurs eEF-1α hydrolyses its bound GTP and subsequently exits the ribosome.

If correct pairing does not occur the eEF-1α-GTP-aminoacyl complex leaves the A site, thereby preventing the incorporation of incorrect amino acids.

9. A signal recognition particle (SRP) is a large complex composed of protein and RNA that binds to a ribosome that has begun translating a polypeptide possessing a signal peptide component. Once the SRP has bound to the ribosome, translation is temporarily arrested. The SRP then mediates ribosomal binding to docking proteins on the surface of a membrane (e.g., RER membrane). Translation subsequently recommences and the growing polypeptide inserts into the membrane.

11. The major differences between prokaryotic and eukaryotic translation control mechanisms are related to the complexity of eukaryotic gene expression. Features that distinguish eukaryotic translation include mRNA export (spatial separation of transcription and translation), mRNA stability (the half-lives of mRNA can be modulated), negative translational control (the translation of certain mRNAs can be blocked by the binding of specific repressor proteins), initiation factor phosphorylation (mRNA translation rates are altered by certain circumstances when eIF-2 is phosphorylated), and translational frame-shifting (certain mRNAs can be frame-shifted so that a different polypeptide is synthesized).

13. One of the most significant problems associated with predicting the three-dimensional structure of a polypeptide based solely on its primary structure is that the calculations based on the forces that drive the folding process (e.g., bond rotations, free energy considerations, and the behavior of the amino acids in aqueous environments) are extraordinarily complex.

15. a. Targeting is a series of mechanisms that directs newly synthesized polypeptides to their correct cellular locations.

b. Scanning is a mechanism eukaryotic ribosomes use to locate a translation start site on an mRNA.

c. A codon is an mRNA triplet base sequence that specifies the incorporation of a specific amino acid into a growing polypeptide chain during translation or acts as a start or stop signal.

d. A reading frame is a set of contiguous triplet codons.

e. Molecular chaperones are molecules that assist in the folding and targeting of proteins.

f. Disulfide exchange is a mechanism that facilitates the formation of disulfide bridges in newly synthesized proteins.

g. Certain aminoacyl-tRNA synthetases possess a second active site, the proofreading site, which binds a specific tRNA if it is covalently bonded to an incorrect amino acid. After this binding, the tRNA–amino acid bond is hydrolyzed.

h. Signal peptides are short peptide sequences that determine a polypeptide's destination, for example, direct its insertion into a membrane.

i. Glycosylation is a posttranslational mechanism whereby carbohydrate groups are covalently attached to polypeptides.

j. In negative translational regulation, the translation of a specific mRNA can be blocked if a specific protein binds to a sequence near its 5′ end.

17. tRNA molecules are adaptor molecules because of their use to bind to specific amino acids and then to position those molecules in the ribosome according to the code on sequence of an mRNA. In other words, they bridge the gap between the base code of the nucleic acids and the amino acid sequence of polypeptides.

Thought Questions

1. Despite considerable species differences in the amino acid and nucleotide sequences of ribosomal proteins and RNA, respectively, the overall three-dimensional structures of these molecules are remarkably similar. This similarity is presumably due to high selection pressure. In other words, ribosomal function is such an important factor in species viability that evolution has conserved their tertiary structure.

3. When errors in ammo acid–tRNA binding do occur, they are usually the result of similarities in amino acid structure. Several aminoacyl-tRNA synthetases possess a separate proofreading site that binds incorrect aminoacyl-tRNA products and hydrolyzes them.

5. One possible codon sequence for the peptide sequence is GGUAGUUGUAGAGCU. The number of possible codons for the amino acids in this peptide sequence is as follows: glycine (4), serine (6), cysteine (2), arginine (6), and alanine (4). The total number of possible codon sequences for this peptide sequence is therefore 1152.

7. Five high-energy phosphate bonds are required to incorporate each amino acid into a polypeptide (i.e., 3 GTP and 2 ATP). The polymerization of 200 amino acids requires 600 GTP and 400 ATP.

9. Posttranslational modification reactions prepare polypeptides to serve their specific functions and direct them to specific cellular or extracellular locations. Examples of these modifications include proteolytic processing (e.g., removal of signal proteins), glycosylation, methylation, phosphorylation, hydroxylation, lipophilic modifications (e.g., N-myristoylation and prenylation), and disulfide bond formation.

11. While you can go directly and predictably from a nucleotide sequence to one and only one amino acid sequence, the reverse is not true because of the degeneracy of the genetic code.

13. Preproproteins contain signal sequences that direct them to the ER for translocation and Golgi for modification. The cleavage of an inactive proprotein and other posttranslational modification processes ensures that the protein is active only when it has been targeted to its site of function.

15. A two subunit ribosome is essential to ensure that all of the required elements are in place before the translational process begins. This is a physical ordering process much like an assembly line; the parts must be in place before the enzymatic activities are set in motion.

Glossary

abiogenesis the mechanism by which inanimate material was transformed on the early Earth into the first primitive living organisms

absorption spectrum a graph of a sample's absorption of electromagnetic radiation

acetal the family of organic compounds with the general formula $RCH(OR')_2$; formed from the reaction of a hemiacetal with an alcohol

acid a molecule that can donate hydrogen ions

acidosis a condition in which the pH of the blood is below 7.35 for a prolonged time

action spectrum measures the effect of wavelengths of light on the rate of photosynthesis

activation energy the threshold energy required to produce a chemical reaction

active site the cleft in the surface of an enzyme where a substrate binds

active transport the energy-requiring movement of molecules across a membrane against a concentration gradient

acyl group the functional group found in derivatives of the caboxylic acids

addition reaction a chemical reaction in which two molecules react to form a third molecule

adduct the product of an addition reaction

A-DNA DNA in which the base pairs are no longer at right angles to the helical axis; occurs when DNA becomes partially dehydrated

aerobic glycolysis the poorly regulated energy metabolism of tumor cells; involves a high rate of glycolysis and some level of oxidative phosphorylation

aerobic respiration the metabolic process in which oxygen is used to generate energy from food molecules

aerotolerant anaerobe organisms that depend on fermentation for their energy needs and possess detoxifying enzymes and antioxidant molecules that protect them from toxic oxygen metabolites

affinity chromatography a technique in which proteins are isolated based on their biological properties, that is, their capacity to bind to a special molecule (the ligand)

agonist a substance that boosts or amplifies the physiological effects of a neurotransmitter

aldaric acid the product formed when the aldehyde and CH_2OH groups of a monosaccharide are oxidized

alditol a sugar alcohol; the product when the aldehyde or ketone group of a monosaccharide is reduced

aldol addition a reaction between two aldehyde molecules (or two ketone molecules) in which a bond is formed between the α-carbon of one and the carbonyl carbon of the other

aldol cleavage the reverse of an aldol condensation

aldol condensation an aldol addition involving the elimination of a water molecule

aldonic acid the product when the aldehyde group of a monosaccharide is oxidized

aldose a monosaccharide with an aldehyde functional group

aliphatic hydrocarbon a nonaromatic hydrocarbon such as methane or cyclohexane

alkaloid a class of naturally occurring molecules that have one or more nitrogen-containing rings; many of the alkaloids have medicinal and other physiological effects

alkalosis a condition in which the blood pH is above 7.45 for a prolonged time

alkylation the introduction of an alkyl group into a molecule

alkyl group a simple hydrocarbon group formed when one hydrogen from the original hydrocarbon (e.g., methyl, CH_3—) is removed

allosteric enzyme an enzyme whose activity is affected by the binding of effector molecules

allosteric interaction a regulatory mechanism in which a small molecule, called an effector or modulator, noncovalently binds to a protein and alters its activity

allosteric transition a ligand-induced conformational change in a protein

Alzheimer's disease a progressive, fatal disease that is characterized by seriously impaired intellectual functions caused by neurological death

amethopterin a structural analogue of folate used to treat several types of cancer; also referred to as methotrexate

α-amino acid a molecule in which the amino group is attached to the carbon atom (the α-carbon) immediately adjacent to the carboxyl group

amino acid pool the amino acid molecules that are immediately available in an organism for use in metabolic processes

amino acid residue an amino acid that has been incorporated into a polypeptide molecule

ammonia intoxication elevated concentration of ammonia in the body that causes lethargy, tremors, slurred speech, protein-induced vomiting, and death

amphibolic pathway a metabolic pathway that functions in both anabolism and catabolism

amphipathic molecule a molecule containing both polar and nonpolar domains

amphoteric molecule a molecule that can react as both an acid and a base

amyloid deposit an insoluble aggregate of extracellular proteinaceous debris that occurs in the brains of Alzheimer's patients

amylopectin a type of plant starch; a branched polymer containing $\alpha(1,4)$ and $\alpha(1,6)$ glycosidic linkages

amylose a type of plant starch; an unbranched chain of D-glucose residues linked with $\alpha(1,4)$ glycosidic linkages

anabolic pathway a series of biochemical reactions in which large complex molecules are synthesized from smaller precursors

anabolism energy-requiring biosynthetic pathways

anaerobic occurring in the absence of molecular oxygen

analogue a substance similar in structure to a naturally occurring molecule

anaplerotic reaction a reaction that replenishes a substrate needed for a biochemical pathway

anhydride the product of a condensation reaction between two carboxyl groups or two phosphate groups in which a molecule of water is eliminated

annotation the functional identification of the genes in a genome

anomer an isomer of a cyclic sugar that differs from another in the arrangement of groups around an asymmetric carbon

antenna pigment a molecule that absorbs light energy and transfers it to a reaction center during photosynthesis

anticodon a sequence of three ribonucleotides on a tRNA molecule that is complementary to a codon on the mRNA molecule; codon-anticodon binding results in the delivery of the correct amino acid to the site of protein synthesis

antigen any substance able to stimulate the immune system; generally a protein or large carbohydrate

antioxidant a substance that prevents the oxidation of other molecules

antisense RNA an RNA molecule with a sequence complementary to that of an mRNA molecule

apoenzyme the protein portion of an enzyme that requires a cofactor to function in catalysis

apoprotein a protein without its prosthetic group

apoptosis programmed cell death

archaea one of three domains of living organisms; prokaryotic organisms that have the appearance of bacteria and many molecular properties that are similar to the eukaryotes

aromatic hydrocarbon a molecule that contains a benzene ring or has properties similar to those exhibited by benzene

atherosclerosis deposition of excess plasma cholesterol and other lipids and proteins on the walls of arteries, decreasing artery diameter

autocrine refers to hormonelike molecules that are active within the tissue or organ in which they are produced

autoimmune disease a condition in which an immune response is directed against an animal's own tissues

autotroph an organism that transforms light energy (from the sun) or various chemicals into chemical bond energy

bacteria one of the three domains of life; single-celled prokaryotes with diverse capacities to exploit their environments

bacterial artificial chromosome a derivative of a large *E. coli* plasmid used to clone DNA sequences as long as 300 kb

basal metabolic rate a measure of energy required to support essential life-sustaining metabolic activities

base a molecule that can accept hydrogen ions

base analogue a molecule that resembles normal DNA nucleotides and can substitute for them during DNA replication, leading to mutations

B cell a B lymphocyte; a white blood cell that produces and secretes antibodies that bind to foreign substances thereby initiating their destruction in the humoral immune response

bile salts amphipathic molecules with detergent properties that are important components of bile, a yellowish green liquid that aids the digestion of fat; conjugated derivatives of the bile acids cholic acid and deoxycholic acid

bioaccumulation the process that concentrates chemicals as they are processed through the food chain

bioenergetics the study of energy transformations in living organisms

biogenic amine an amino acid derivative that acts as a neurotransmitter (e.g., GABA and the catecholamines)

biomolecule the molecules that make up living organisms

bioremediation the use of biological processes to decontaminate toxic waste sites

biotransformation a series of enzyme-catalyzed processes in which toxic, and usually hydrophobic, molecules are converted into less toxic and more soluble metabolites

branched chain amino acid one of a group of essential amino acids with branched carbon skeletons (leucine, isoleucine, and valine)

buffer a solution that contains a weak acid or base and its salt and resists large pH changes when stronger acids or bases are added

calorie a unit of energy equal to the quantity of heat necessary to raise the temperature of 1 g of water by 1 degree C; equivalent to 4.184 J

Calvin cycle the major metabolic pathway by which CO_2 is incorporated into organic molecules

carcinogenesis the process whereby cells become genetically unstable and eventually cancerous

β-carotene a plant pigment molecule that acts as an absorber of light energy and as an antioxidant

carotenoid an isoprenoid molecule that either functions as a light-harvesting pigment or protects against ROS

catabolic pathway a series of biochemical reactions in which large complex molecules are degraded into smaller, simpler products

catabolism the degradation of fuel molecules and the production of energy for cellular functions

catecholamine one of a class of neurotransmitters derrived from tyrosine; includes dopamine, norepinephrine, and epinephrine

cDNA library a clone library of cDNA (complementary DNA) molecules produced from mRNA molecules by reverse transcription

cell fractionation a technique involving homogenization and centrifugation that allows the study of cell organelles

cellobiose a degradation product of cellulose; a disaccharide that contains two molecules of glucose linked by $\beta(1,4)$ glycosidic bond

cellular immunity immune system processes mediated by T cells, a type of lymphocyte

cellulose a polymer produced by plants that is composed of D-glycopyranose residues linked by $\beta(1,4)$ glycosidic bonds

chaparonin one of a family of molecular chaparones; also referred to as an hsp60

Chargaff's rules in DNA, the equality of the concentrations of adenine and thymine, and of cytosine and guanine

chemiosmotic coupling theory ATP synthesis is coupled to electron transport by an electrochemical proton gradient across a membrane

chemoheterotroph an organism that uses preformed food molecules as its sole source of energy

chemolithotroph an organism that uses specific inorganic reactions to generate energy

chiral molecule a molecule that has mirror-image forms

chitin an unbranched polymer in which N-acetyl glucosamine residues are linked by $\beta(1,4)$ glycosidic bonds; the principal structural component of the exoskeletons of arthropods

chlorophyll a green pigment molecule that resembles heme; a type of molecule that absorbs light energy

chloroplast a chlorophyll-containing plastid

chromatin the DNA-containing component of the eukaryotic nucleus; the DNA is almost always complexed with histones

chromophore a molecular component that absorbs light of a specific frequency

chromoplast a type of plastid in plants that accumulates the pigments that are responsible for the colors of leaves, flower petals, and fruits

chromosomal jumping a technique used to isolate clones that contain discontinuous sequences from the same chromosome

chromosome the physical structure, composed of DNA and some proteins, that contains the genes of an organism

chylomicron a large lipoprotein of extremely low density; transports dietary triacylglycerols and cholesteryl esters from the intestine to muscle and adipose tissue

circular dichroism a type of spectroscopy in which the relationship between molecular motion and structure is probed with electromagnetic radiation

cis **isomer** an isomer in which two substituents are on the same side of a double bond

cistron a DNA sequence that contains the coding information for a polypeptide and the signals required for ribosome function

citric acid cycle a biochemical pathway that degrades the acetyl group of acetyl-CoA to CO_2 and H_2O as three molecules of NAD^+ and one molecule of FAD are reduced

cloning a lab procedure that produces multiple copies of a gene

C4 metabolism a photosynthetic pathway that produces a four-carbon molecule and avoids photorespiration in eukaryotic photosynthesizing organisms

coding strand the DNA strand that has the same base sequence as the RNA transcript (with thymine instead of uracil)

codon a sequence of three nucleotides in mRNA that directs the incorporation of an amino acid during protein synthesis or acts as a start or stop signal

coenzyme a small organic molecule required in the catalytic mechanisms of certain enzymes

coenzyme A an acyl carrier molecule that consists of a 3′-phosphate derivative of ADP linked to pantothenic acid via a phosphate ester bond; pantothenic acid is linked to β-mercaptoethylamine by an amide bond

cofactor the nonprotein component of an enzyme (either an inorganic ion or a coenzyme) required for catalysis

colligative property a property of solutions that depends on only the number of dissolved particles in solution

colony hybridization technique a method used to identify bacterial colonies that possess a specific recombinant DNA sequence

conformational disease diseases caused by protein misfolding and aggregation

conjugate acid the cation (or molecule) that results when a base reacts with a proton

conjugate base the anion (or molecule) that results when an acid loses a proton

conjugated protein a protein that functions only when it carries other chemical groups attached by covalent linkages or by weak interactions

conjugate redox pair an electron donor and its electron acceptor form; for example, NADH and NAD^+

conjugation unconventional sexual mating between bacterial cells; donor cell transfers a DNA segment into a recipient cell through a specialized pilus

conjugation reaction a reaction that may improve the water solubility of a molecule by converting it to a derivative that contains a water-soluble group

consensus sequence the average of several similar sequences; for example, the consensus sequence of the −10 box of *E. coli* promoter is TATAAT

constitutive gene a routinely transcribed gene that codes for gene products required for cell function

contig one of a set of overlapping DNA sequences used to identify the base sequence of a region of DNA

cooperative binding a mechanism in which binding one ligand to a target molecule promotes the binding of other ligands

Cori cycle a metabolic process in which lactate, produced in tissues such as muscle, is transferred to liver where it becomes a substrate in gluconeogenesis

cotranslational transfer the insertion of a polypeptide across a membrane during ongoing protein synthesis

C3 plants plants that produce glycerate-3-phosphate, a three-carbon molecule, as the first stable product of photosynthesis

C4 plants plants that possess mechanisms that suppress photorespiration by producing oxaloacetate, a four-carbon molecule

Crassulacean acid metabolism a photosynthetic pathway that produces a four-carbon molecule (malate) and avoids photorespiration

Creutzfeld-Jacob disease a rare neurodegenerative disease characterized by dementia and impaired movement coordination; classified as a prion disease

cruciform a crosslike structure in DNA molecules likely to form when a DNA sequence contains a palindrome

cyclin one of a group of proteins that regulate the cell cycle

cyclin-dependent protein kinase one of a group of enzymes that activate the cyclins

cystic fibrosis an ultimately fatal autosomal recessive disease that is caused by a missing or defective chloride channel protein

cystic fibrosis transmembrane conductance regulator the plasma membrane glycoprotein that functions as a chloride channel in epithelial cells

cytokine a group of hormonelike polypeptides and proteins; also referred to as growth factors

cytoskeleton a set of protein filaments (microtubules, macrofilaments, and intermediate fibers) that maintain the cell's internal structure and allow organelles to move

decarboxylation reaction in which a carboxylic acid loses CO_2

denaturation the disruption of protein or nucleic acid structure caused by exposure to heat or chemicals leading to loss of biological structure

density gradient centrifugation a technique in which cell fractions are further purified by centrifugation in a density gradient

desensitization a process in which target cells adjust to changes in stimulation by decreasing the number of cell surface receptors or by inactivating those receptors

dialysis a laboratory technique in which a semipermeable membrane is used to separate small solutes from larger solutes

diastereomer a stereoisomer that is not an enantiomer (mirror image isomer)

dictyosome term often used for the Golgi complex in plants

differential centrifugation a cell fractionation technique in which homogenized cells are separated by centrifugal forces

dipole a difference in charge between atoms in a molecule resulting from the unsymmetrical orientation of polar bonds

disaccharide glycoside composed of two monosaccharide residues

disulfide bridge a covalent bond formed between the sulfhydryl groups in two cysteine residues

disulfide exchange an enzyme-catalyzed post-translational process in which correct disulfide bonds are formed, resulting in a biologically active protein

DNA fingerprinting a laboratory technique used to compare DNA banding patterns from different individuals

DNA microarray a DNA "chip" used to analyze the expression of thousands of genes simultaneously

DNA profile consists of the pattern and number of repeats each in STR sequences; used to identify individuals

DNA typing DNA analysis technique used to identify individuals; involves the analysis of several highly variable sequences called markers

docking protein also called signal recognition particle receptor; an RER transmembrane heterodimeric protein that binds an SRP bound to a ribosome thus triggering the resumption of protein synthesis

Donnan effect unequal distribution of ions across a membrane that results in the establishment of an electrical gradient; also known as a membrane potential

down regulation the reduction in cell surface receptors in response to stimulation by specific hormone molecules

effector a molecule whose binding to a protein alters the protein's activity

eicosanoid a hormonelike molecule that contains 20 carbons; most are derived from arachidonic acid; examples include prostaglandins, thromboxanes, and leukotrienes

electron spin resonance spectroscopy a technique that measures the differences in the energy levels of unpaired electrons that occur in a rapidly changing magnetic field

electron transport system a series of electron carrier molecules that bind reversibly to electrons at different energy levels

electrophile an electron-deficient species such as a hydrogen ion (H^+)

electrophoresis a class of techniques in which molecules are separated from each other because of differences in their net charge

electroporation a method of introducing a cloning vector into a host cell that involves treatment with an electrical current

electrostatic interaction noncovalent attraction between oppositely charged atoms or groups

elimination reaction a chemical reaction in which a double bond is formed when atoms in a molecule are removed

elongation the polypeptide chain growth phase during translation on ribosomes

Emerson enhancement effect an increased photosynthetic rate (measured by O_2 evolution per quantum of light) that occurs when blue wavelengths are used in addition to red wavelengths

enantiomer a stereoisomer that is a mirror image of another

endergonic reaction a reaction that does not spontaneously go to completion; the standard free energy change is positive and the equilibrium constant is less than 1

endocrine hormone a hormone secreted into the bloodstream that acts on distant target cells

endocytosis the process in which a cell takes up solutes or particles by enclosing them in vesicles pinched off from its plasma membrane

endoplasmic reticulum a series of membranous channels and sacs that provides a compartment separate from the cytoplasm for numerous chemical reactions

endothermic reaction a reaction that requires energy (as heat)

enediol the intermediate formed during the isomerization reactions of monosaccharides

energy the capacity to do work

enthalpy the heat content of a system; in a biological system it is essentially equivalent to the total energy of the system

entropy a measure of the randomness or disorder of a system; a measure of that part of the total energy in a system that is unavailable for useful work

enzyme a biomolecule that catalyzes a chemical reaction

enzyme induction a process in which a molecule stimulates increased synthesis of a specific enzyme

enzyme kinetics the study of the rates of enzyme-catalyzed reactions

enzyme-linked immunosorbent assay a technique involving antibodies that is used to detect and measure hormones and other molecules

epidermal growth factor a protein that stimulates epithelial cells to undergo cell division

epimer a molecule that differs from the configuration of another by one asymmetric carbon

epimerization the reversible interconversion of epimers

epoxide an ether in which the oxygen is incorporated into a three-membered ring

essential amino acid an amino acid that cannot be synthesized by the body and must therefore be supplied by the diet

essential fatty acid linoleic or linolenic acid which must be supplied in the diet because they cannot be synthesized by the body

euchromatin a less condensed form of chromatin that has varying levels of transcriptional activity

eukarya one of the three domains of life; contains single-celled and multicellular eukaryotic organisms

eukaryotic cell a living cell that possesses a true nucleus

excision repair a DNA repair mechanism that removes damaged nucleotides, then replaces them with normal ones

exergonic process a reaction that spontaneously goes to completion as written; the standard free energy change is negative, and the equilibrium constant is greater than 1

exocytosis the process in which an intracellular vesicle fuses with the plasma membrane, thereby releasing the vesicle contents into extracellular space

exon the region in a split or interrupted gene that codes for RNA and ends up in the final product (e.g., mRNA)

exonuclease an enzyme that removes nucleotides from the end of the polynucleotide strand

exothermic reaction a reaction that releases heat

extein peptide segments that are spliced together to form a mature protein during protein splicing

extracellular matrix a gelatinous material, containing proteins and carbohydrates, that binds cells and tissue together

extremophile an organism that lives under extreme conditions of temperature, pH, pressure, or ionic concentration that would easily kill most organisms

extremozyme an enzyme that functions under extreme conditions of temperature, pressure, pH, or ionic concentration

facilitated diffusion diffusion across a membrane that is aided by a carrier

facultative anaerobe organisms that possess the mechanisms needed for detoxifying oxygen metabolites; they can generate energy by using oxygen as an electron acceptor

fatty acid a long chain monocarboxylic acid that contains an even number of carbon atoms

fatty acid binding protein an intracellular water-soluble protein whose sole function is to bind and transport hydrophobic fatty acids

fermentation the anaerobic metabolism or degradation of sugars; an energy-yielding process in which organic molecules serve as both electron donors and acceptors

fibrous protein a protein composed of polypeptides arranged in long sheets or fibers

flavoprotein a conjugated protein in which the prosthetic group is either FMN or FAD

fluid mosaic model the currently accepted model of cell membranes in which the membrane is a lipid bilayer with integral proteins buried in the lipid, and peripheral proteins more loosely attached to the membrane surface

fluorescence a form of luminescence in which certain molecules can absorb light of one wavelength and emit light of another wavelength

fluorescence recovery after photobleaching a technique used to observe the lateral movement of molecules in cell membranes; after fluorescently labeled molecules are bleached, the movement of nearby unbleached molecules into the bleached area is tracked as a function of time

frame shift mutation deletion of one or more base pairs (but not multiples of three) from a DNA sequence

free energy the energy in a system available to do useful work

functional genomics the scientific discipline devoted to elucidating how biomolecules work together within functioning organisms

functional group a group of atoms that undergo characteristic reactions when attached to a carbon atom in an organic molecule or a biomolecule

futile cycle a set of opposing reactions that can be arranged in a cycle, but usually do not occur simultaneously; functioning of such reactions in both directions is avoided by metabolic control mechanisms to prevent energy waste

gel filtration chromatography a technique used to separate molecules according to their size and shape that employs a column packed with a gelatinous polymer

gene a DNA sequence that codes for a polypeptide, rRNA, or tRNA

gene expression the mechanisms by which living organisms regulate the flow of genetic information

general recombination recombination involving exchange of a pair of homologous DNA sequences; it can occur at any location on a chromosome

genetic code the set of nucleotide base triplets (codons) that code for the amino acids in protein as well as start and stop signals

genetics the scientific investigation of inheritance

genome the total genetic information possessed by an organism

genomics the large-scale analyses of entire genomes

globular protein a protein that adopts a rounded or globular shape

glucocorticoid a steroid hormone produced in the adrenal cortex that affects carbohydrate, protein, and lipid metabolism

glucogenic amino acid a molecule whose carbon skeleton is a substrate in gluconeogenesis

gluconeogenesis the synthesis of glucose from noncarbohydrate molecules

glucosuria the presence of glucose in urine; a symptom of diabetes mellitus

glycerol phosphate shuttle a metabolic process that uses glyceral-3-phosphate to transfer electrons from NADH in the cytosol to mitochondrial FAD

glycocalyx a carbohydrate-containing structure on the external surface of cells

glycoconjugate a molecule that possesses covalently bound carbohydrate components (e.g., glycoproteins and glycolipids)

glycogen a glucose storage molecule in vertebrates; a branched polymer containing $\alpha(1,4)$ and $\alpha(1,6)$ glycosidic linkages

glycogenesis the biochemical pathway that adds glucose to growing glycogen polymers when blood glucose levels are high

glycogenolysis the biochemical pathway that removes glucose molecules from glycogen polymers when blood glucose levels are low

glycolipid a glycosphingolipid; a molecule in which a monosaccharide, disaccharide, or oligosaccharide is attached to a ceramide through an O-glycosidic linkage

glycolysis the enzymatic pathway that converts a glucose molecule into two molecules of pyruvate; this anaerobic process generates energy in the form of two ATP molecules and two NADH molecules

glycoprotein a conjugated protein in which carbohydrate molecules are the prosthetic group

glycosaminoglycan a long unbranched heteropolysaccharide chain composed of disaccharide repeating units

glycoside the acetal of a sugar

glycosidic linkage an acetal linkage formed between two monosaccharides

glyoxylate cycle a modification of the citric acid cycle that occurs in plants, bacteria, and other eukaryotes; allows growth in these organisms from two-carbon substrates such as ethanol, acetate, and acetyl CoA

glyoxysome a type of peroxisome found in germinating seed in which lipid molecules are converted to carbohydrate

Golgi apparatus (Golgi complex) a series of curved membranous sacs involved in packaging and distributing cell products to internal and external compartments

G protein a protein that binds GTP which activates the protein to perform a function; the hydrolysis of GTP to form GDP inactivates the G protein

granum a folded portion of the thylakoid membrane

growth factor an extracellular polypeptide that stimulates cells to grow and/or undergo cell division

GTPase activating protein a protein molecule that hydrolyses GTP bound to a GTP binding protein

guanine nucleotide–releasing protein a protein that binds to a member of the Ras family of proteins and activates it by triggering the release of its bound GDP and subsequent binding of GTP

H-DNA a DNA sequence consisting of a polypurine segment hydrogen bonded to a polypyrimidine strand that forms a triple helix; involves the formation of nonconventional base pairing

heat shock protein a protein synthesized in response to stress (e.g., high temperature)

helicase ATP-requiring enzymes that catalyze the unwinding of duplex DNA

hemiacetal one of the family of organic molecules with the general formula RR′C(OR′)(OH) formed by the reaction of one molecule of alcohol with an aldehyde

hemiketal one of a family of organic molecules with the general formula RR′C(OR′)(OH) formed by the reaction of one molecule of alcohol with a ketone

hemoprotein a conjugated protein in which heme, an iron-containing organic group, is the prosthetic group

Henderson-Hasselbach equation kinetic rate expression that defines the relationship between pH, pK_a, and the concentrations of the acid and base components of a buffer solution

heterochromatin chromatin that is so highly condensed that it is transcriptionally inactive

heterogeneous nuclear RNA a primary transcript of DNA; precursor of an mRNA

heterokaryon a structure formed from the fusion of the membranes of two different cells; used to demonstrate membrane fluidity

heterotroph an organism that attains energy by degrading preformed food molecules obtained by consuming other organisms

high-density lipoprotein a type of lipoprotein with a high protein content that is believed to scavenge excess cholesterol from cell membranes and transport it to the liver

holoenzyme a complete enzyme consisting of the apoenzyme plus a cofactor

holoprotein an apoprotein combined with its prosthetic group

homeostasis the capacity of living organisms to regulate metabolic processes despite variability in their internal and external environments

homologous polypeptide a protein molecule whose amino acid sequences and functions are similar to those of another protein

Hoogsteen base pairing nonconventional base pairing that stabilizes H-DNA

hormone a molecule produced by specific cells that influences the function of distant target cells

hormone response element a specific DNA sequence that binds hormone-receptor complexes; the binding of a hormone-receptor complex either enhances or diminishes the transcription of a specific gene

hsp 60 one of a family of molecular chaperones that mediate protein folding by forming a large structure composed of two stacked seven-membered rings that facilitate the ATP-dependent folding of polypeptides; also called chaperonins or Cpn 60s

hsp 70 one of a family of molecular chaperones that bind to and stabilize proteins during the early stages of the folding process

humoral immune response the immunity that results from the presence of antibodies in blood and tissue fluid; also referred to as an antibody-mediated immunity

hydration a type of addition reaction in which water is added to a carbon-carbon double bond

hydrocarbon a molecule that contains only carbon and hydrogen

hydrogen bond the force of attraction between a hydrogen atom and a small, highly electronegative atom (e.g., O or N) on another molecule (or the same molecule)

hydrolase an enzyme that catalyzes reactions in which adding water cleaves bonds

hydrolysis a chemical reaction that involves the reaction of a molecule with water; the process by which molecules are broken into their constituents by adding water

hydrophilic molecules that possess positive or negative charges or contain relatively large numbers of electronegative oxygen or nitrogen atoms; dissolve easily in water

hydrophobic molecules that possess few if any electronegative atoms; do not dissolve in water

hydrophobic interaction the association of nonpolar molecules when they are placed in water

hydroxyapatite a calcium phosphate gel used in nucleic acid research; binds to double-stranded DNA more tenaciously than to single-stranded DNA

hyperammonemia a potentially fatal elevation of the concentration of ammonium ions in the blood

hyperglycemia blood glucose levels that are higher than normal

hyperosmolar possessing an osmotic pressure greater than that of normal blood plasma

hyperosmolar hyperglycemic nonketosis severe dehydration in noninsulin dependent diabetics; caused by persistent high blood glucose levels

hypertonic solution a concentrated solution with a high osmotic pressure

hyperuricemia abnormally high level of uric acid in blood

hypochromic effect the decrease in the absorption of UV light (260 nm) that occurs when purine and pyrimidine bases are incorporated into base pairs in polynucleotide sequences

hypoglycemia blood glucose levels that are lower than normal

hypotonic solution a dilute solution with a low osmotic pressure

inducible gene a gene expressed only under certain conditions

inhibitor a molecule that reduces an enzyme's activity

initiation the beginning phase of translation

inner membrane the innermost membrane of mitochondria

insulin-like growth factor a protein in humans that mediates the growth-promoting actions of growth hormone; has insulin-like properties (e.g., promotes glucose transport and fat synthesis)

intein a excised peptide segment generated during protein splicing

interferon one of a group of glycoproteins that have nonspecific antiviral activity (e.g., stimulation of cells to produce antiviral proteins) that inhibit the synthesis of viral RNA and proteins and regulate the growth and differentiation of immune system cells

intermediate fiber a component of the cytoskeleton containing a heterogeneous set of proteins

interspersed genome-wide repeats repetitive DNA sequences that are scattered around the genome

intron a noncoding intervening sequence in a split or interrupted gene missing in the final RNA product

ion-exchange chromatography a technique that separates molecules on the basis of their charge

ionophore a substance that transports cations across membranes

isoelectric point the pH at which a protein has no net charge

isomerase an enzyme that catalyzes the conversion of one isomer to another

isomerization the reversible interconversion of isomers

isoprenoid one of a class of biomolecules that contain repeating five-carbon structural units known as isoprene units; examples include terpenes and steroids

isothermic having a uniform temperature

isotonic solution solutions with exactly the same particle concentration; having identical osmotic pressure

isozyme one of two or more forms of the same enzyme activity with different amino acid sequences

katal measure of the rate of enzyme activity; 1 katal (kat) is equal to the conversion of one mole of substrate to product per second

ketal the family or organic compounds with the general formula $RR'C(OR')_2$; formed from reaction of a hemiketal with an alcohol

ketoacidosis acidosis caused by an excessive accumulation of ketone bodies

ketogenesis excess acetyl-CoA molecules are converted to acetoacetate, β-hydroxybutyrate, and acetone, known as the ketone bodies

ketogenic amino acid a molecule whose carbon skeleton is a substrate for synthesizing fatty acids and ketone bodies

ketone body acetone, acetoacetate, or β-hydroxybutyrate; produced in the liver from acetyl-CoA

ketosis accumulation of ketone bodies in blood and tissues

kinetic proofreading a mechanism suggested to account for the precision of codon-anticodon pairing during translation; correct base pairing allows sufficient time for hydrolysis of GTP bound to an elongation factor

kinetics the study of reaction rates

Krebs bicycle a biochemical pathway in which the aspartate required in the urea cycle is generated from oxaloacetate, an intermediate in the citric acid cycle

Krebs urea cycle the cyclic pathway that converts waste ammonia molecules along with CO_2 and aspartate into urea; named for its discoverer, Hans Krebs

lactone a cyclic ester

lactose a disaccharide found in milk; composed of one molecule of galactose linked in a $\beta(1,4)$ glycosidic bond to a molecule of glucose

lateral gene transfer the transfer of genes or gene fragments between unrelated organisms

leaving group the group displaced during a nucleophilic substitution reaction

Le Chatelier's principle a law that states that when a system in equilibrium is disturbed, the equilibrium shifts to oppose the disturbance

lectin a carbohydrate binding protein

leukotriene a linear derivative of arachidonic acid whose synthesis is initiated by a peroxidation reaction

ligand a molecule that binds to a specific site on a larger molecule

ligase an enzyme that catalyzes the joining of two molecules

light-independent reaction a photosynthetic reaction that can occur in the absence of light; also referred to as the Calvin cycle

light-induced repair DNA repair in which the damaged sequences are repaired utilizing light energy; also referred to as photoreactivation repair

light reaction a mechanism whereby electrons are energized and subsequently used in ATP and NADPH synthesis

limit of resolution the minimum distance between two separate points that allows for their discrimination

lipid any of a group of biomolecules that are soluble in nonpolar solvents and insoluble in water

lipid bilayer a biomolecular lipid layer that constitutes the structural framework of the cell membranes

lipogenesis the biosynthesis of body fat (triacylglycerol)

lipolysis the hydrolysis of fat molecules

lipoprotein a conjugated protein in which lipid molecules are the prosthetic groups; a protein-lipid complex that transports water-insoluble lipids in blood

lithotroph an organism that uses specific inorganic reactions to generate energy; also known as a chemolithotroph

London dispersion force a temporary dipole-dipole interaction

low-density lipoprotein a type of lipoprotein that transports cholesterol to the liver

lyase an enzyme that catalyzes the cleavage of C—O, C—C, or C—N bonds, thereby producing a product containing a double bond

lysogeny the integration of a viral genome into a host genome

lysosome a saclike organelle capable of degrading most biomolecules

lytic cycle a viral life cycle in which a virus destroys its host cell

macromolecule a biopolymer formed from the linkage of certain biomolecules via covalent bonds; examples include nucleic acids, proteins, and polysaccharides

Maillard reaction nonenzymatic glycosylation of molecules possessing free amino groups (e.g., proteins)

malate-aspartate shuttle a metabolic process in which the electrons form NADH in the cytosol are transferred to mitochondrial NAD$^+$

malate shuttle a metabolic process in which oxaloacetate is transferred by reversible conversion to malate from a mitochondrion to the cytoplasm

maltose a degradation product of starch hydrolysis; a disaccharide composed of two glucose molecules linked by an $\alpha(1,4)$ glycosidic bond

marker enzyme an enzyme known to be a reliable indicator of the presence of a specific organelle

mass spectroscopy a technique in which molecules are vaporized and then bombarded by a high-energy electron beam, causing them to fragment as cations

membrane potential potential difference across the membrane of living cells; usually measured in millivolts

messenger RNA an RNA species produced by transcription that specifies the amino acid sequence for a polypeptide

metabolism the total of all chemical reactions in an organism

metabolome the complete set of organic metabolites that are produced within a cell under the direction of the genome

metaloprotein a conjugated protein containing one or more metal ions

methotrexate a structural analogue of folate that is used in the treatment of several types of cancer; also called amethopterin

micelle an aggregation of molecules having a nonpolar and a polar component, leaving the polar domains facing the surrounding water

microfilament a component of the cytoskeleton composed of the protein actin

microsatellite sequences of 2 to 4 bp that are tandemly repeated from 10 to 20 times

microsome a membranous vesicle derived from fragments of endoplasmic reticulum obtained by differential centrifugation

microtubule a component of the cytoskeleton composed of the protein tubulin

mineralocorticoid a steroid hormone that regulates Na$^+$ and K$^+$ metabolism

minisatellite tandemly repeated sequences of about 25 bp with total lengths between 10^2 and 10^5 bp

mitochondrion an organelle possessing two membranes in which aerobic respiration occurs

mitogen a substance that stimulates cell division

mixed anhydride an acid anhydride with two different R groups

mixed terpenoid a biomolecule that is composed of nonterpene components attached to isoprenoid groups

mobile phase the moving phase in chromatographic methods

modulator a molecule whose binding to an allosteric site of an enzyme alters the enzyme's activity

molecular biology the science devoted to elucidating the structure and function of genomes

molecular chaperone a molecule that assists in protein folding; most are heat shock proteins

molecular disease a disease caused by a mutated gene

molten globule the partially globular site of a folding polypeptide that resembles the molecule's native state

monosaccharide a polyhydroxy aldehyde or ketone with the formula $(CH_2O)_n$ where n is at least 3

monounsaturated refers to a fatty acid with a single double bond

motor protein components of biological machines that bind nucleotides; nucleotide hydrolysis drives precise changes in the protein's shape

murein a complex polymer that contains two sugar derivatives: N-acetylglucosamine and N-acetylmuramic acid and several amino acids; also referred to as peptidoglycan

mutagen any chemical or physical agent that alters the nucleotide sequence of a gene

mutarotation a spontaneous process in which the α- and β- forms of monosaccharides are readily interconverted

mutation any change in the nucleotide sequence of a gene

nascent newly synthesized

negative cooperativity a mechanism in which the binding of one ligand to a target molecule decreases the likelihood of subsequent ligand binding

negative feedback a mechanism in which a biochemical pathway is regulated by binding a product molecule to a key enzyme in the pathway

nephrogenic diabetes insipidis an autosomal recessive disease in which the kidneys of affected individuals cannot produce concentrated urine

neurotransmitter a molecule released at a nerve terminal that binds to and influences the function of other nerve cells or muscle cells

neutral fat triacylglycerol molecules

nitrogen fixation conversion of molecular nitrogen (N_2) into a reduced biologically useful form (NH_3) by nitrogen-fixing microorganisms

nonessential amino acid an amino acid that can be synthesized by the body

nonessential fatty acid a fatty acid that can be synthesized by the body

nonpolar molecule a molecule that does not contain a dipole

nuclear envelope the double membrane that separates the nucleus from the cytoplasm

nuclear pore a channel through the nuclear envelope that allows molecules to pass between the cytoplasm and the nucleus

nuclease an enzyme that hydrolyzes nucleic acid molecules to form oligonucleotides

nucleic acid a macromolecule composed of nucleotides; DNA and RNA are nucleic acids

nucleohistone DNA complexed with histone proteins

nucleoid in prokaryotes, an irregularly shaped region that contains a long circular DNA molecule

nucleolus a structure found in the nucleus when the nucleus is stained with certain dyes; it plays a major role in the synthesis of ribosomal RNA

nucleophile an electron-rich atom or molecule

nucleophilic substitution a reaction in which a nucleophile substitutes for an atom or molecular group

nucleoplasm the material within the nucleus that consists of proteins called lamins that form a network of chromatin fibers

nucleoside a biomolecule composed of a pentose sugar (ribose or deoxyribose), and a nitrogenous base

nucleosome a repeating structural element in eukaryotic chromosomes, composed of a core of eight histone molecules with about 140 base pairs of DNA wrapped around the outside; an additional 60 base pairs connect adjacent nucleosomes

nucleotide a biomolecule composed of a pentose sugar (ribose or deoxyribose), at least one phosphate group, and a nitrogenous base

nucleus an organelle that contains the chromosomes

obligate aerobe an organism that is highly dependent on oxygen for energy production

obligate anaerobe an organism that grows only in the absence of oxygen

Okazaki fragment any of a series of deoxyribonucleotide segments that are formed during discontinuous replication of one DNA strand as the other strand is continuously replicated

oligomer a multisubunit protein in which some or all subunits are identical

oligonucleotide a short nucleic acid segment that contains fewer than 50 nucleotides

oligosaccharide an intermediate-sized carbohydrate composed of two to ten monosaccharides

oncogene a mutated version of a protooncogene that promotes abnormal cell proliferation

open reading frame a series of triplet base sequences in mRNA that do not contain a stop codon

operon a set of linked genes that are regulated as a unit

opioid peptide a molecule that relieves pain and produces pleasant sensations; produced in nervous tissue cells

optical isomer a stereoisomer that possesses one or more chiral centers

organelle a membrane-enclosed structure within a eukaryotic cell

osmolyte an osmotically active substance synthesized by cells to restore osmotic balance

osmosis the diffusion of a solvent through a semipermeable membrane

osmotic diuresis a process in which solute in the urinary filtrate causes excessive loss of water and electrolytes

osmotic pressure the pressure forcing the solvent, water, to flow across a membrane

outer membrane the porous external membrane of mitochondria

oxidation an increase in oxidation number caused by the loss of electron(s)

β-oxidation the catabolic pathway in which most fatty acids are degraded; acetyl-CoA is formed as the bond between the α and β carbon atoms is broken

oxidation-reduction (redox) reaction a reaction involving the transfer of one or more electrons from one reactant to another

oxidative phosphorylation the synthesis of ATP coupled to electron transport

oxidative stress excessive production of reactive oxygen species

oxidize the removal of electrons

oxidized molecule a molecule from which one or more electrons have been removed

oxidizing agent a substance that oxidizes (removes electrons from) another substance; the oxidizing agent is itself reduced in the process

oxidoreductase an enzyme that catalyzes an oxidation-reduction reaction

oxyanion a negatively charged oxygen atom

palindrome a sequence that provides the same information whether it is read forward or backward; DNA palindromes contain inverted repeat sequences

passive transport transport across membrane that requires no direct energy input

Pasteur effect the observation that glucose consumption is greater under anaerobic conditions than when O_2 is present

pentose phosphate pathway a biochemical pathway that produces NADPH, ribose, and several other sugars

peptide an amino acid polymer composed of fewer than 50 amino acid residues

peptide bond an amide linkage in an amino acid polymer

pernicious anemia an illness caused by a deficiency of vitamin B_{12}; symptoms include low red blood cell counts, weakness, and neurological disturbances

peroxisome an organelle that contains oxidative enzymes

pH optimum the pH at which an enzyme catalyzes a reaction at maximum efficiency

phosphate group transfer potential the tendency of a phosphorylated molecule to undergo hydrolysis

3′-phosphoadenosine-5′-phosphosulfate a high-energy sulfate donor molecule used in the biosynthesis of the sulfatides, a type of glycolipid

phosphoglyceride a type of lipid molecule found predominately in membrane composed of glycerol linked to two fatty acids, phosphate, and a polar group

phospholipid an amphipathic molecule that has a hydrophobic domain (hydrocarbon chains of fatty acids) and a hydrophilic (a polar head group) domain; an important structural component of membranes

phosphoprotein a conjugated protein in which phosphate is the prosthetic group

photoautotroph an organism that possesses a mechanism for transforming solar energy into other forms of energy

photochemistry the study of chemical reactions that are initiated by light absorption

photoheterotroph an organism that uses both light and organic biomolecules as energy sources

photophosphorylation the synthesis of ATP coupled to electron transport driven by light energy

photoreactivation repair a mechanism to repair thymine dimers using the energy of visible light

photorespiration a light-dependent process occurring in plant cells actively engaged in photosynthesis that consumes oxygen and liberates carbon dioxide

photosynthesis the trapping of light energy and its conversion to chemical energy, which then reduces carbon dioxide and incorporates it into organic molecules

photosystem a photosynthetic mechanism composed of light-absorbing pigments

pH scale a measure of hydrogen ion concentration; pH is the negative log of the hydrogen ion concentration in moles per liter

plasma membrane the membrane that surrounds a cell, separating it from its external environment

plasmid a circular, double-stranded DNA molecule that can exist and replicate independently of a bacterial chromosome; plasmids are stably inherited, but are not required for the host cell's growth and reproduction

plastid an organelle found in plants, algae, and some protists in which carbohydrate is stored or synthesized

platelet-derived growth factor a protein secreted by blood platelets during clotting; stimulates mitosis during wound healing

point mutation a change in a single nucleotide in DNA

polar head group a molecular group that contains phosphate or other charged or polar groups

polar molecule a molecule that has a permanent dipole resulting from an unsymmetrical electron distribution

polymerase chain reaction a laboratory technique used to synthesize large quantities of specific nucleotide sequences from small amounts of DNA using a heat-stable DNA polymerase

polypeptide an amino acid polymer with more than 50 amino acid residues

polysaccharide a linear or branched polymer of monosaccharides linked by glycosidic bonds

polysome an mRNA with several ribosomes bound to it

polyunsaturated refers to a fatty acid with two or more double bonds, usually separated by methylene groups

polyuria excessive urination; a symptom of diabetes insipidus and diabetes mellitus

postitive cooperativity a mechanism in which the binding of one ligand to a target molecule increases the likelihood of subsequent ligand binding

postabsorptive the phase in the feeding-fasting cycle in which nutrient levels in blood are low

postprandial the phase in the feeding-fasting cycle immediately after a meal; blood nutrient levels are relatively high

posttranslational modification a set of reactions that alter the structure of newly synthesized polypeptides

posttranslational translocation the transfer of previously synthesized polypeptide across the RER membrane

prenylation the covalent attachment of prenyl groups (e.g., fasnesyl and geranylgeranyl groups) to protein molecules

preproprotein an inactive precursor protein with removable signal peptide

primary structure the amino acid sequence of a polypeptide

primase an RNA polymerase that synthesizes short RNA segments, called primers, that are required in DNA synthesis

primer a short RNA segment required to initiate DNA synthesis

primosome a multienzyme complex involved in the synthesis of the RNA primers at various points along the DNA template strand during *E. coli* DNA replication

prion proteinaceous infectious particle; believed to be a causative agent of several acquired neurodegenerative diseases (e.g.,

"mad cow" disease and Creutzfeld-Jacob disease)

processivity the prevention of frequent dissociation of a polymerase from the DNA template

proenzyme an inactive precursor of an enzyme

prokaryotic cell a living cell that lacks a nucleus

promoter the sequence of nucleotides immediately before a gene that is recognized by RNA polymerase and signals the start point and direction of transcription

proprotein an inactive precursor protein

prostaglandin arachidonic acid derivative that contains a cyclopentane ring with hydroxyl groups at C-11 and C-15

prosthetic group the nonprotein portion of a conjugated protein that is essential to the biological activity of the protein; often a complex organic molecule

protein a macromolecule composed of one or more polypeptides

protein splicing a posttranslational mechanism in which an intervening peptide sequence is precisely excised from a nascent polypeptide

protein turnover the continuous degradation and resynthesis of proteins in an organism

proteoglycan a large molecule containing large numbers of glycosaminoglycan chains linked to a core protein molecule

proteome the complete set of proteins produced within a cell

proteomics the analysis of proteomes

proteosome a multienzyme complex that degrades proteins linked to ubiquitin

protomer a subunit of allosteric enzymes

protonmotive force the force arising from a gradient of protons and a membrane potential

protooncogene a normal gene that promotes carcinogenesis if mutated

purine a nitrogenous base with a two-ring structure; a component of nucleotides

pyrimidine a nitrogenous base with a single ring structure; a component of nucleotides

quantum theory a theory in physics that describes the behavior of particles (e.g., electrons) and their associated waves

quaternary structure association of two or more folded polypeptides to form a functional protein

racemization interconversion of enantiomers

radical an atom or molecule with an unpaired electron

reaction center the membrane-bound protein complex in a photosynthesizing cell that mediates the conversion of light energy into chemical energy

reactive oxygen species a reactive derivative of molecular oxygen, including superoxide radical, hydrogen peroxide, the hydroxyl radical, and singlet oxygen

reading frame a set of contiguous triplet codons in an mRNA molecule

receptor a protein on the cell surface that binds to a specific extracellular nutrient molecule and facilitates its entry into the cell; other receptors bind chemical signals and direct the cell to respond appropriately

recombination a process in which DNA molecules are broken and rejoined in new combinations

recombinational repair a repair mechanism that can eliminate certain types of damaged DNA sequences that are not eliminated before replication; the undamaged parental strands recombine into the gap left after the damaged sequence is removed

redox potential a measure of the tendency of an electron donor in a redox pair to lose an electron

reduce the transfer of electrons

reduced molecule a molecule that has gained one or more electrons

reducing agent a substance that reduces the oxidation number of another reactant; the reducing agent is itself oxidized in the process

reducing sugar a sugar that can be oxidized by weak oxidizing agents

reduction the lowering of oxidation number by the gain of electron(s)

reduction potential the tendency for a specific substance to lose or gain electrons

regulatory enzyme an enzyme that catalyzes a committed step in a biochemical pathway

releasing factor a protein involved in the termination phase of translation

replication the process in which an exact copy of parental DNA is synthesized using the polynucleotide strands of the parental DNA as templates

replication fork The Y-shaped region of a DNA molecule that is undergoing replication; results from separation of two DNA strands

replicon a unit of the genome that contains an origin for initiating replication

replisome the large complex of polypeptides, including the primosome, that replicates DNA in *E. coli*

resonance energy transfer the transfer of energy from an excited molecule to a nearby molecule, thereby exciting the second molecule

resonance hybrid a molecule with two or more alternative structures that differ only in the position of electrons

respiration a biochemical process whereby fuel molecules are oxidized and their electrons are used to generate ATP

respiratory burst an oxygen-consuming process in scavenger cells such as macrophages in which ROS are generated and used to kill foreign or damaged cells

respiratory control the control of aerobic respiration by ADP concentration

restriction fragment length polymorphism genetic variations that can be used to identify individuals

retrotransposon a transposition mechanism that involves an RNA transcript; an RNA transposon

retrovirus one of a group of viruses with RNA genomes that carry the enzyme reverse transcriptase and form a DNA copy of their genome during their reproductive cycle

ribosomal RNA the RNA present in ribosomes; ribosomes contain several types of single-stranded ribosomal RNA that contribute to ribosome structures and are also directly involved in protein synthesis

ribosome a protein-RNA complex where protein is synthesized

ribozyme self-splicing RNA found in several organisms

RNA editing the alteration of the base sequence in a newly synthesized mRNA molecule; bases may be chemically modified, deleted, or added

RNA transposon a transposition mechanism that involves an RNA transcript; also referred to as a retrotransposon

rough ER a type of endoplasmic reticulum involved in protein synthesis

salt bridge an electrostatic interaction in proteins between ionic groups of opposite charge

satellite DNA DNA sequences arranged next to each other; form a satellite band when genomic DNA is digested and centrifuged

saturated molecule a molecule that contains no carbon-carbon double or triple bonds

SDS polyacrylamide gel electrophoresis a method for separating proteins or determining their molecular weights that employs the negatively charged detergent sodium dodecyl sulfate

secondary metabolite a molecule derived from a primary metabolite; many serve protective functions

secondary structure folding of a polypeptide chain into local patterns such as α-helix and β-pleated sheet; secondary structure is maintained by hydrogen bonds between the amide hydrogen and the carbonyl oxygen of the peptide bond

second genetic code the precision with which amino acids are attached to their cognate tRNAs; catalyzed by the aminoacyl-tRNA synthetases; a principal reason for the accuracy of polypeptide synthesis

second messenger a molecule that mediates the action of some hormones

semiconservative replication DNA synthesis in which each polynucleotide strand serves as a template for the synthesis of a new strand

sense strand the DNA strand that RNA polymerase copies to produce mRNA, rRNA, or tRNA

serine protease one of a class of proteolytic enzymes that use the $-CH_2OH$ of a serine residue as a nucleophile to hydrolyze peptide bonds

Shine-Dalgarno sequence a purine-rich sequence that occurs on an mRNA close to AUG (the initiation codon) that binds to a complementary sequence on the 30S ribosomal subunit, thereby promoting the formation of the correct preinitiation complex

short-tandem repeats DNA sequences with between 2 and 4 bp repeats; can be used to generate DNA profiles that distinguish between individuals

shotgun cloning a cloning technique in which genomic libraries are created by the random digestion of a genome

signal hypothesis a mechanism that explains how secretory proteins are synthesized on ribosomes bound to the RER; a sequence of amino acid residues on the nascent polypeptide chain mediates the insertion of the molecule into the RER membrane

signal peptide a short sequence, typically near the amino terminal of a polypeptide, that determines its destination

signal recognition particle a large complex consisting of proteins on a small RNA molecule that mediates the binding of the ribosome to the RER during protein synthesis

signal transduction mechanisms by which extracellular signals are received, amplified, and converted to a cellular response

simple diffusion a process in which each type of solute, propelled by random molecu-

lar motion, moves down a concentration gradient

site-directed mutagenesis a technique that introduces specific sequence changes into cloned genes

site-specific recombination recombination of nonhomologous genetic material with a chromosome at a specific site

small nuclear ribonuclear particle a complex of proteins and small nuclear RNA molecule that promotes RNA processing

small nuclear RNA a small RNA molecule involved in removing introns from mRNA, rRNA, and tRNA

smooth ER a type of endoplasmic reticulum involved in lipid synthesis and biotransformation

solvation sphere a shell of water molecules that clusters around positive and negative ions

somatomedin a polypeptide that mediates the growth-promoting action of growth hormone

Southern blotting a technique in which radioactively labeled DNA or RNA profiles are used to locate a complementary sequence in a DNA digest

specific activity a measure of enzyme activity; the number of international units (I.U.) per milligram of protein (1 I.U. is the amount of enzyme that produces 1 μmole of product per minute)

sphingomyelin a type of phospholipid that contains sphingosine; the 1-hydroxyl group of ceramide (a fatty acid derivative of sphingosine) is esterified to the phosphate group of phosphorylcholine or phosphorylethanolamine

spliceosome a multicomponent complex containing protein and RNA that is involved in the splicing phase of mRNA processing

splicing the excision of introns during mRNA processing

spontaneous changes physical or chemical processes that occur with a release of energy

stationary phase the solid chromatography matrix

steady state a phase in an organism's life when the rate of anabolic processes is approximately equal to that of catabolic processes

stereoisomer a molecule that has the same structural formula and bonding patterns as another but has a different arrangement of atoms in space

steroid a derivative of triterpenes; contains four fused rings

sterol carrier protein a cytoplasmic protein carrier for certain intermediates during cholesterol biosynthesis

stroma a dense, enzyme-filled substance that surrounds the thylakoid membrane within the chloroplast

stromal lamella thylakoid membrane that interconnects two grana

structural gene a gene that codes for the synthesis of a polypeptide or a polynucleotide with a nonregulatory function (e.g., mRNA, rRNA, or tRNA)

substrate the reactant in a chemical reaction that binds to an enzyme active site and is converted to a product

substrate-level phosphorylation the synthesis of ATP from ADP by phosphorylation coupled with the exergonic breakdown of a high-energy organic substrate molecule

subunit a polypeptide component of an oligomeric protein

sucrose a disaccharide composed of α-glucose and β-fructose residues linked through a glycosidic bond between both anomeric carbons

sugar the basic unit of carbohydrates; a class of biomolecule containing hydroxyl groups and an aldehyde group or a ketone group

supersecondary structure a set of specific combinations of α-helix and β-pleated sheet structures in protein molecules

symbiosis the living together or close association of two dissimilar organisms

tandem repeats DNA sequences in which multiple copies are arranged next to each other; lengths of repeated sequences vary from 10 bp to over 2000 bp

target cell a cell that responds to the binding of a hormone or growth factor

targeting process that directs newly synthesized proteins to their correct destinations

tautomer an isomer that differs from another in the location of a hydrogen atom and a double bond (e.g., keto-enol tautomers)

tautomerization chemical reaction by which two isomers are interconverted by the movement of an atom or molecular group

T cell a T lymphocyte; white blood cell that bears antibody-like molecules on its surface; binds to and destroys foreign cells in cellular immunity

telomere structures at the ends of chromosomes that buffer the loss of critical coding sequences after a round of DNA replication

termination phase in translation in which newly synthesized polypeptides are released from the ribosome

terpene a member of a class of isoprenoids classified according to number of isoprene residues it contains

tertiary structure the globular, three-dimensional structure of a polypeptide that results from folding the regions of secondary structure; folding results from interactions of the side chains or R groups of the amino acid residues

thermodynamics the study of energy and its interconversion

thiolytic cleavage cleavage of a carbon-sulfur bond

thromboxane a derivative of arachidonic acid that contains a cyclic ester

thylakoid membrane an intricately folded internal membrane within the chloroplast

thylakoid space the internal compartment created by the formation of grana within chloroplasts; also called the thylakoid lumen

α-tocopherol a lipid-soluble molecule that acts as a radical scavenger; vitamin E

transamination a reaction in which an amino group is transferred from one molecule to another

transcription the process in which single-stranded RNA with a base sequence complementary to the template strand of DNA is synthesized

transcription factor proteins that regulate or initiate RNA synthesis by binding to specific DNA sequences called response elements

transcript localization the binding of mRNAs to certain cellular structures within cytoplasm so that protein gradients can be created within a cell

transcriptome the complete set of RNA molecules that are produced within a cell

transduction the transfer of genes between bacteria by bacteriophages

transfection a mechanism by which bacteriophage inadvertently transfers bacterial chromosome or plasmid sequences to a new host cell

transferase an enzyme that catalyzes the transfer of a functional group from one molecule to another

transfer RNA a small RNA that binds to an amino acid and delivers it to the ribosome for incorporation into a polypeptide chain during translation

transformation naked DNA fragments enter a bacterial cell and are introduced into the bacterial genome

transgenic animal an animal that results when recombinant DNA sequences are microinjected into a fertilized ovum

***trans* isomer** an isomer in which two substituents are on opposite sides of a double bond

transition mutation a mutation that involves the substitution of a different purine base for the purine present at the site of the mutation or the substitution of a different pyrimidine for the normal pyrimidine

transition state the unstable intermediate in catalysis in which the enzyme has altered the form of the substrate so that it now shares properties of both the substrate and the product

translation protein synthesis; the process by which the genetic message carried by mRNA directs the synthesis of polypeptides with the aid of ribosomes and other cell constituents

translational frameshifting a +1 or −1 change in reading frame allows more than one polypeptide to be synthesized from a single mRNA

translocation movement of the ribosome along the mRNA during translation

translocon an integral membrane protein that mediates translocation of a polypeptide

transposable DNA elements DNA sequences that excise themselves and then insert at another site

transposition the movement of a piece of DNA from one site in a genome to another

transposons (transposable elements) a DNA segment that carries the genes required for transposition and moves about the chromosome; sometimes the name is reserved for transposable elements that also contain genes unrelated to transposition

transsulfuration pathway a biochemical pathway that converts methionine to cysteine

transversion mutation a type of point mutation in which a pyrimidine is substituted for a purine or vice versa

triacylglycerol an ester formed between glyceral and three fatty acids

triose phosphate glyceraldehyde-3-phosphate and dihydroxyacetone molecules that are produced during photosynthesis

tumor necrosis factor a protein that suppresses cell division; toxic to tumor cells

tumor promoter a molecule that provides cells a growth advantage over nearby cells

turnover the rate at which all molecules in a structure are degraded and replaced with newly synthesized molecules

turnover number the number of molecules of substrate converted to product each second per mole of enzyme

ubiquination the covalent attachment of ubiquitin to proteins; it prepares proteins for degradation

ubiquitin a protein that is covalently attached by enzymes to proteins destined to be degraded

uncoupler a molecule that uncouples ATP synthesis from electron transport; it collapses a proton gradient by transporting protons across the membrane

uncoupling protein a molecule that dissipates the proton gradient in mitochondria by translocating protons; also called thermogenin

unsaturated molecule a molecule that contains one or more carbon-carbon double or triple bonds

urea cycle a cyclic pathway in which waste ammonia molecules, CO_2, and aspartate molecules are converted to urea

uronic acid the product formed when the terminal CH_2OH group of a monosaccharide is oxidized

van der Waals forces a class of relatively weak, transient electrostatic interactions between permanent and/or induced dipoles

vapor pressure the pressure exerted by a vapor in equilibrium with a liquid

vector a cloning vehicle into which a segment of foreign DNA can be spliced so it can be introduced and expressed in host cells

velocity the rate of a biochemical reaction; the change in the concentration of a reactant or product per unit time

very low density lipoprotein a type of lipoprotein with a very high relative concentration of lipids; transports lipids to tissues

vitamin an organic molecule required by organisms in minute quantities; some vitamins are coenzymes required for the function of cellular enzymes

vitamin B$_{12}$ a complex cobalt-containing molecule that is required for the N^5-methyl THF-dependent conversion of homocysteine to methionine

wax a complex mixture of nonpolar lipids including wax esters

weak acid an organic acid that does not completely dissociate in water

weak base an organic base that has a small but measurable capacity to combine with hydrogen ions

wobble hypothesis a hypothesis that explains why cells often have fewer tRNAs

than expected; freedom in the pairing of the third base of the codon to the first base of the anticodon allows some tRNAs to pair with several codons

work change in energy that produces a physical change

xenobiotics foreign and potentially toxic molecules

yeast artificial chromosome a cloning vector that can accommodate up to 100 kb; contains eukaryotic sequences that function as centromeres, telomeres, and a replication origin

Z-DNA a form of DNA that is twisted into a left-handed spiral; named for its zigzag conformation which is slimmer than B-DNA

Z-scheme a mechanism whereby electrons flow between PSII and PSI during photosynthesis

zwitterion neutral molecules that bear an equal number of positive and negative charges simultaneously

zymogen the inactive form of a proteolytic enzyme

Credits

Photos

CHAPTER 1

1.3: © Susan Detwiler; **1.21a-c:** © & Courtesy of David S. Goodsell, the Scripps Research Institute.

CHAPTER 2

2.10: © Charles C. Brinton, Jr. and Judith Carnahan; **2.14:** © Audrey M. Glauert & G. M. W. Cook; **2.15:** © P. Schulz/Biology Media/Photo Researchers, Inc.; **2.16:** © Don Fawcett/Photo Researchers, Inc.; **2.21:** © Gopal Murti/Phototake; **2.22b:** © M. M. Perry and A. B. Gilbert. *J. of Cell Science* 39:257-272, 1979. Company of Biologists Ltd.; **2.23:** © E. H. Newcombe and S. E. Frederick/Biological Photo Service; **2.24b:** © Don Fawcett/Photo Researchers, Inc.; **2.25a & b:** © C. R. Hackenbrook. Ultrastructural bases for metabolically linked mechanical activity in mitochondria. *J. of Cell Biology* 37: no. 2, 364-365 (1968). By permission American Society for Microbiology; © Rockefeller University Press; **2.26:** © Hermann Eisenbeiss/Photo Researchers, Inc.; **Figure 2A:** © Don Fawcett/Photo Researchers, Inc.; **2.27a:** © J. W. Shuler/Photo Researchers, Inc.; **2.27b:** © J. L. Carson/Custom Medical Stock Photo; **2.27c:** © Dr. Peter Dawson/Science Photo Library/Photo Researchers, Inc.

CHAPTER 3

Opener: NASA.

CHAPTER 4

Opener: © Tom Brakefield/Corbis.

CHAPTER 5

Opener: © & Courtesy of David S. Goodsell, the Scripps Research Institute; **5.16a:** From *Molecules of Life*, Purdue University; **5.26:** K. A. Piez in D. B. Wetlander, ed., 'The Protein Folding Problem, AAAS Selected Symposium 89, American Association for the Advancement of Science, Washington, D.C., 1984, pp. 47-61.' Courtesy of the Collagen Corporation. Reproduced by permission of American Association for the Advancement of Science; **Box 5B(a):** Courtesy of Dudley Library Archives; **Box 5B(b):** Courtesy of The Royal Society of Medicine.

CHAPTER 7

7.19a: © Leonard Lessin/Peter Arnold, Inc.; **7.20a, 7.21a:** © Leonard Lessin/FBPA; **Box 7B:** © Bellerophon Books.

CHAPTER 8

CO 8: © Mathew Klein/Photo Researchers, Inc.; **Box 8A:** The Egyptian Expedition of the Metropolitan Museum of Art, Rogers Fund, 1915. (15.5.19e) Photograph © 1989 The Metropolitan Museum of Art; **Box 8B(a & b):** © Phototake.

CHAPTER 10

10.14b: From D. F. Parsons, *Science* 1963. 140: page 985. © 2001 American Association for the Advancement of Science.

CHAPTER 11

11.2a: © Leonard Lessin/Peter Arnold, Inc.; **11.2b:** © Leonard Lessin/FBPA; **11.4:** © Leonard Lessin/Peter Arnold, Inc.; **11.8:** © Leonard Lessin/FBPA; **11.13a:** © Leonard Lessin/Peter Arnold, Inc.

CHAPTER 12

Opener: © corbisimages.com.

CHAPTER 13

Opener: © Govindjee.

CHAPTER 16

Opener: © corbisimages.com.

CHAPTER 17

Opener: From Elizabeth Pennisi, *The Human Genome*, Science Feb. 16 2001. 291: page 1177. © 2001 American Association for the Advancement of Science. Illustration by Cameron Slayden; **17.8:** From J. D. Watson, *The Double Helix*, p. 215, New York: Atheneum. © 1968 by J. D. Watson, A. C. Barrington photographer. Courtesy of Cold Spring Harbor Laboratory Archives; **17.9:** R. E. Franklin and R. Gosling. Molecular configuration in sodium thymonucleate. *Nature* 171: 740-741. © 1953 Macmillan Magazines Ltd. Reprinted with permission from *Nature*; **17H:** © Jean Claude Revy/Phototake.

CHAPTER 18

18.4: Reprinted from CELL, Vol. 79, 1994, pp 1233-1243, Krishna et al, "Crystal structure of..." © 2002, with permission from Elsevier Science; **18C:** © & courtesy of Dr. Keith V. Wood; **18H:** Courtesy of AFFYMETRIX, Inc.

CHAPTER 19

Opener: From Marat M. Yusupov, Gulnara Zh. Yusupova, Albion Baucom, Kate Lieberman, Thomas N. Earnest, J. H. D. Cate, and Harry F. Noller *Crystal Structure of the Ribosome at 5.5 Å Resolution*, Science May 4 2001. 292: page 885. © 2001 American Association for the Advancement of Science; **19B:** Banks R.E., Dunn M.J., Hochstrasser D.F., Sanchez J.-Ch., Blackstock W., Pappin D.J., Selby P.J. "Proteomics: new perspectives, new biomedical opportunities." *The Lancet*, Vol. 356, No 9243, p. 1749-1756, November 18th, 2000. Reprinted with permission from Elsevier Science.

Line Art

CHAPTER 1

1.14: From Geoffrey Cooper, The Cell: A Molecular Approach, 1997. Reproduced with permission of Sinauer Associates, Inc. **1.22:** Adapted from Genethics: Clash Between New Genetics and Human Values, by David Suzuki and Peter Knudtson, 1990, Harvard University Press.

CHAPTER 2

2.4: Copyright © 1998 From Essential Cell Biology by Bruce Alberts, et al. Reproduced by permission of Routledge, Inc., part of The Taylor & Francis Group. **2.5:** From David S. Goodsell, The Machinery of Life, 1998. Copyright © 1998 Springer-Verlag. Reprinted with permission of Springer-Verlag/Germany. **2.6:** From M. Hoppert and F. Mayer, "Prokaryotes," American Scientist, Vol. 87, November-December 1999. Copyright © David Goodsell. Reprinted with permission. **2.7:** From David S. Goodsell, The Machinery of Life, 1998. Copyright © 1998 Springer-Verlag. Reprinted with permission of Springer-Verlag/Germany. **2.8:** From Prescott, et al., Microbiology 4/e. Copyright © 1999 by The McGraw-Hill Companies. This material is reproduced with permission of The McGraw-Hill Companies. **2.9a & b:** From David S. Goodsell, The Machinery of Life, 1998. Copyright © 1998 Springer-Verlag. Reprinted with permission of Springer-Verlag/Germany. **2.11:** From Microbiology: An Introduction, by Gerard J. Tortora, Berdell R. Funke and Christine L. Case. Copyright © 1998 by The Benjamin/Cummings Publishing Company, Inc. Reprinted by permission of Addison Wesley Longman Publishers, Inc. **2.16:** From The World of the Cell, 4th ed. by Wayne M. Becker, Lewis J. Kleinsmith, and Jeff Hardin. Copyright © 2000 by Addison Wesley Longman, Inc. Reprinted by permission of Pearson Educa-

CHAPTER 16

16.1: Figure from J. Koolman and K.H. Rahn, Color Atlas of Biochemistry, 1999. Reprinted with permission of Thieme Medical Publishers. **16.11:** From The World of the Cell, 4th ed. by Wayne M. Becker, Lewis J. Kleinsmith, and Jeff Hardin. Copyright © 2000 by Addison Wesley Longman, Inc. Reprinted by permission of Pearson Education, Inc. **16.14:** From Thomas M. Devlin, Textbook of Biochemistry with Clinical Correlations, 1999. Copyright © 1999 John Wiley & Sons, Inc. This material is used by permission of John Wiley & Sons, Inc. **16.15:** From Geoffrey Cooper, The Cell: A Molecular Approach, 1997. Reproduced with permission of Sinauer Associates, Inc. **16B:** From Fundamentals of Biochemistry, by D. Voet, J. Voet, and C. Pratt. Copyright © 1999 John Wiley & Sons, Inc. This material is used by permission of John Wiley & Sons, Inc.

CHAPTER 17

17.1: From A.H. Fairlamb, "Brand New World of Post Genomes," Trends in Parasitology, Vol. 17, p. 255. Copyright © 2000 Elsevier Science. Reprinted with permission from Elsevier Science. **17.2b:** From Feughelman, et al., "Molecular Structure of DNA," Nature, Vol. 175, May 14, 1955,. Reprinted with permission from Nature. **17.4:** Copyright © 1999 From Molecular Biology of the Cell by Bruce Alberts, et al. Reproduced by permission of Routledge, Inc., part of The Taylor & Francis Group. **17.10:** From Principles of Biochemistry by A. Lehninger, D. Nelson and M. Cox. Copyright © 2000, 1993, 1982 by Worth Publishers. Used with permission. **17.12:** Data from H. Htun and J.E. Dahlberg, Science 43:1571, 1989. **17.15:** From Terence A. Brown, Genomes, 1999. Reprinted with permission of BIOS Scientific Publishers, Oxford, UK. **17.18:** From Biochemistry, 2nd ed. By Christopher K. Mathews and K.E. Van Holde. Copyright © 1996 by The Benjamin/Cummings Publishing Company, Inc. Reprinted by permission of Pearson Education, Inc. **17.20:** From Terence A. Brown, Genomes, 1999. Reprinted with permission of BIOS Scientific Publishers, Oxford, UK. **17A:** Copyright © Oxford University Press and Cell Press, 2000. Reprinted from Genes VII by Benjamin Lewin (2000), by permission of Oxford University Press. **17E:** From Prescott, et al., Microbiology 4/e. Copyright © 1999 by The McGraw-Hill Companies. This material is reproduced with permission of The McGraw-Hill Companies. **17F:** From Recombinant DNA by J.D. Watson, M. Gilman, J. Witkowski, M. Zoller © 1992, 1983 by J.D. Watson, M. Gilman, J. Witkowski, M. Zoller. Used with the permission of W.H. Freeman and Company. **17G:** From Prescott, et al., Microbiology 4/e. Copyright © 1999 by The McGraw-Hill Companies. This material is reproduced with permission of The McGraw-Hill Companies. **17H:** From The World of the Cell, 4th ed. by Wayne M. Becker, Lewis J. Kleinsmith, and Jeff Hardin. Copyright © 2000 by Addison Wesley Longman, Inc. Reprinted by permission of Pearson Education, Inc. **17I:** Reprinted with permission from "Phyloge-netic Classification and the Universal Tree," by W. Ford Doolittle, Science 284, 2124, 1999. Copyright © 1999 American Association for the Advancement of Science. **17J:** Reprinted with permission from "Phylogenetic Classification and the Universal Tree," by W. Ford Doolittle, Science 284, 2124, 1999. Copyright © 1999 American Association for the Advancement of Science. **17M:** From Scientific American, 279, Vol. 1, 1998. Reprinted with permission of L.E. Henderson and L.O. Arthur. **17N:** From "Improving HIV Therapy," Scientific American, Vol. 279. Reprinted with permission of the illustrator, Patricia J. Wynne.

CHAPTER 18

18.3: From The World of the Cell, 4th ed. by Wayne M. Becker, Lewis J. Kleinsmith, and Jeff Hardin. Copyright © 2000 by Addison Wesley Longman, Inc. Reprinted by permission of Pearson Education, Inc. **18.4:** Copyright © Oxford University Press and Cell Press, 2000. Reprinted from Genes VII by Benjamin Lewin (2000), by permission of Oxford University Press; Reprinted with permission from "Crystal Structure of DNA Recombination Protein RuvA and a Model for Its Binding to the Holliday Junction," by J.B. Rafferty, et al., Science 274, 415, 1996. Copyright © 1996 American Association for the Advancement of Science. **18.7:** Adapted from Biochemistry, by Reginal H. Garrett and Charles M. Grisham, 1996. Brooks/Cole Publishing. **18.8:** From Geoffrey Cooper, The Cell: A Molecular Approach, 1997. Reproduced with permission of Sinauer Associates, Inc. **18.15:** Adapted from A. Landy and R.A. Weisberg, in R.W. Hendrix, et al., Lambda II. Cold Spring Harbor Laboratory Press, 1983. **18.19:** From Principles of Biochemistry by Albert L. Lehninger, David L. Nelson, Michael M. Cox © 2000, 1993, 1982 by Worth Publishers. Used with permission. **18.21:** From Principles of Biochemistry by Albert L. Lehninger, David L. Nelson, Michael M. Cox © 2000, 1993, 1982 by Worth Publishers. Used with permission. **18.23:** Copyright © 1999 From Molecular Biology of the Cell by Bruce Alberts, et al. Reproduced by permission of Routledge, Inc., part of The Taylor & Francis Group. **18.26:** With permission from the Annual Review of Biochemistry, Volume 61 © 1992 by Annual Reviews www.AnnualReviews.org. **18.27:** From Geoffrey Cooper, The Cell: A Molecular Approach, 1997. Reproduced with permission of Sinauer Associates, Inc. **18.30:** Copyright © 1999 From Molecular Biology of the Cell by Bruce Alberts, et al. Reproduced by permission of Routledge, Inc., part of The Taylor & Francis Group. **18.31:** From The World of the Cell, 4th ed. by Wayne M. Becker, Lewis J. Kleinsmith, and Jeff Hardin. Copyright © 2000 by Addison Wesley Longman, Inc. Reprinted by permission of Pearson Education, Inc. **18.32:** From Terence A. Brown, Genomes, 1999. Reprinted with permission of BIOS Scientific Publishers, Oxford, UK. **18A:** From Prescott, et al., Microbiology 4/e. Copyright © 1999 by The McGraw-Hill Companies. This material is reproduced with permission of The McGraw-Hill Companies. **18B:** From E.W. Nester, Microbiology: A Human Perspective 3/e. Copyright © 2001 by The McGraw-Hill Companies. This material is reproduced with permission of The McGraw-Hill Companies. **18E:** From R.H. Tamarin, Principles of Genetics 6/e. Copyright © 1999 by The McGraw-Hill Companies. This material is reproduced with permission of The McGraw-Hill Companies. **18F:** From E.W. Nester, Microbiology: A Human Perspective 3/e. Copyright © 2001 by The McGraw-Hill Companies. This material is reproduced with permission of The McGraw-Hill Companies. **18G:** From Molecular Cell Biology by J. Darnell, H. Lodish, D. Baltimore, P. Matsudaira, S. Zipursky, and A. Berk. Copyright © 2000, 1995, 1990, 1986 by Scientific American Books. Used with permission of W.H. Freeman and Company. **18H:** From R.H. Tamarin, Principles of Genetics 6/e. Copyright © 1999 by The McGraw-Hill Companies. This material is reproduced with permission of The McGraw-Hill Companies.

CHAPTER 19

19.1: Adapted from Lodish, et al., Molecular cell Biology, 4th edition, 2000. **19.5:** From Molecular Cell Biology by J. Darnell, H. Lodish, D. Baltimore, P. Matsudaira, S. Zipursky, and A. Berk. Copyright © 2000, 1995, 1990, 1986 by Scientific American Books. Used with permission of W.H. Freeman and Company. **19.11:** Reprinted with permission from L.I. Slobin "Polypeptide Chain Elongation," in Translation in Eukaryotes, edited by H. Trachsel, 1991. Copyright CRC Press, Boca Raton, Florida. **19.15:** With permission from the Annual Review of Biochemistry, Volume 69 © 2000 by Annual Reviews www.AnnualReviews.org. **19.16:** From The World of the Cell, 4th ed. by Wayne M. Becker, Lewis J. Kleinsmith, and Jeff Hardin. Copyright © 2000 by Addison Wesley Longman, Inc. Reprinted by permission of Pearson Education, Inc. **19.20a:** From S.T. Ferreira and F.G. Felice, "Protein Dynamics, Folding and Misfolding," FEBS Letters 498:129-134, 2001. Reprinted with permission of Elsevier Science BV. **19.20b:** From S.E. Ranford, "Protein Folding," Trends in Biochemical Sciences, Vol. 25, pp. 611-618. Copyright © 2000 Elsevier Science. Reprinted with permission from Elsevier Science. **19.21:** From S.E. Ranford, "Protein Folding," Trends in Biochemical Sciences, Vol. 25, pp. 611-618. Copyright © 2000 Elsevier Science. Reprinted with permission from Elsevier Science. **19.22:** From Z. Xu, L. Horwich, and P.B. Stigler, "The Crystal Structure of the Asymmetric GroEl-Gros-Es-(ADP)7 Chaperonen Complex," Nature, Vol. 388, August 21, 1997, pp. 741.750. Reprinted with permission of Nature. **19A:** Adapted from K. Abel and F. Jurnak, Structure 4:229-239, 1996. **19B:** From R.E. Banks, et al., "Proteomics: New Perspectives and New Biomedical Opportunities," Lancet, 2000, Vol. 356, No. 9243. Reprinted with permission from Elsevier Science.

Index

Page numbers followed by f and t refer to figures and tables, respectively.

Names and Abbreviations of the Standard Amino Acids

Amino Acid	Three-Letter Abbreviations	One-Letter Abbreviations
Alanine	Ala	A
Arginine	Arg	R
Asparagine	Asn	N
Aspartic acid	Asp	D
Cysteine	Cys	C
Glutamic acid	Glu	E
Glutamine	Gln	Q
Glycine	Gly	G
Histidine	His	H
Isoleucine	Ile	I
Leucine	Leu	L
Lysine	Lys	K
Methionine	Met	M
Phenylalanine	Phe	F
Proline	Pro	P
Serine	Ser	S
Threonine	Thr	T
Tryptophan	Trp	W
Tyrosine	Tyr	Y
Valine	Val	V